INTRODUCTION TO
MODERN
INORGANIC
CHEMISTRY
5th edition

K M Mackay

R A Mackay

and

W Henderson

Department of Chemistry
University of Waikato
Hamilton
New Zealand

BLACKIE ACADEMIC & PROFESSIONAL
An Imprint of Chapman & Hall
London · Weinheim · New York · Tokyo · Melbourne · Madras

Published by Blackie Academic and Professional, an imprint of Chapman & Hall, 2–6 Boundary Row, London SE1 8HN, UK

Chapman & Hall, 2–6 Boundary Row, London SE1 8HN, UK

Chapman & Hall GmbH, Pappelallee 3, 69469 Weinheim, Germany

Chapman & Hall USA, 115 Fifth Avenue, New York, NY 10003, USA

Chapman & Hall Japan, ITP-Japan, Kyowa Building, 3F, 2-2-1 Hirakawacho, Chiyoda-ku, Tokyo 102, Japan

DA Book (Aust.) Pty Ltd, 648 Whitehorse Road, Mitcham 3132, Victoria, Australia

Chapman & Hall India, R. Seshadri, 32 Second Main Road, CIT East, Madras 600 035, India

First edition 1968
Reprinted 1969, 1970
Second edition 1972
Reprinted 1973, 1974, 1976, 1979
Third edition 1981
Reprinted 1983, 1984, 1986
Fourth edition 1989
Reprinted 1994, 1995
Fifth edition 1996

© 1996 K. M. Mackay,
   R. A. Mackay and
   W. Henderson

Typeset in 10/12 Baskerville by Thomson Press (India) Limited, New Delhi

Printed in Great Britain by Page Brothers, Norwich, Norfolk

ISBN 0 751403 733

A catalogue record for this book is available from the British Library

Library of Congress Catalog Card Number: 96–85910

♾ Printed on permanent acid-free text paper, manufactured in accordance with ANSI/NISO Z39.48-1992 and ANSI/NISO Z39.48-1984 (Permanence of Paper)

Contents

Preface to the Fifth Edition

For this edition, we acknowledge the ever-increasing interest in the natural chemistry of the elements by creating a new Chapter 20 on biological, medicinal and environmental inorganic chemistry. It is significant that the Periodic Table may still be used as an ordering principle, even in the complex chemistry of the ever-cycling processes of the cell. This chapter is partly new material, and partly a collection and ordering of chemistry which was included in the systematic chapters of earlier editions.

Innovation continues apace and a necessarily modest portion of the exciting new chemistry is added to the selected topics of Chapters 16, 18 and 19, including sections on the fullerenes and related species, coordinated dihydrogen, actinide organometallic chemistry, and expanded accounts of clusters. The survey of silicate structures has been moved from Chapter 5 into Chapter 18 and expanded with more emphasis on zeolites and on aluminium phosphate analogues. In these chapters, and throughout the systematic chemistry, the chemistry has been updated and many sections are expanded. A driving force for recent development has been the increasing power of instrumental techniques and Chapter 7 on experimental methods has been revised to reflect this.

The Mackays are delighted to welcome Bill Henderson to the team of authors. Cognoscenti of the earlier editions are challenged to trace his distinctive contributions. Finally we acknowledge gratefully the contribution of students and colleagues who have sent us comments and pointed out errors, and thank Dr T. Fujimoto of the Tsukuba Research Centre, Japan, for the mica image (Chapter 7) Professor D. Parker (University of Durham) and Professor S. Mann (University of Bath) for the MRI and magnetotactic bacterium images respectively (Chapter 20).

Note to the reader

Final decisions on the names of the elements from 104 onwards are still awaited. A review of alternative choices is included in section 16.12. As an interim measure the names recommended by IUPAC have been used in all other tables.

R. Ann Mackay
K. M. Mackay
B. Henderson

Preface to the
Fourth Edition

In the twenty years since this book was first published, the amount of known Inorganic Chemistry has doubled and contiguous areas such as organometallic chemistry have developed as disciplines in their own right. Such developments have been immensely exciting and have turned up fascinating ideas and materials.

It is well worth the effort required to enter the intriguing field of Modern Inorganic Chemistry, but an appropriate approach must be chosen to avoid being submerged by the volume of information and concepts. The total amount of information is rapidly increasing and although many of the advances lead to a simpler broader picture, any such saving is more than offset by the development of new areas of great interest. The modern reader cannot hope to become familiar with as high a proportion of current science as the students of twenty years ago covered. More and more, the modern student has to develop skills in finding and organizing material.

We suggest a twofold strategy for gaining an adequate grasp of Inorganic Chemistry without being overwhelmed by the amount of material. You need a broad and sufficient framework for the whole discipline, together with a knowledge in depth of selected areas within it. The framework alone is valuable, but insufficient to give confidence that you can attack a detailed field: in-depth assault on one topic is valuable but can be confusing, and difficult to generalize to the next. A student should also become thoroughly familiar with the sources of information available in a good chemistry library using Appendix A as a guide.

For Inorganic Chemistry, we are fortunate in having the Periodic Table as a ready-made framework. As our knowledge and experience of chemical relationships increases, the level of understanding of the Periodic Table also deepens, and this in turn improves and reinforces the base for attacking a new topic. Once you feel reasonably expert in one topic—however restricted it may be—you will have the confidence that you can gain similar mastery over the next, and the two will be found to mutually reinforce each other.

Although the Periodic Table is over a century old, and based on ideas formulated well before that, it is of immense current importance as one of the major unifying principles in science. A recent striking example of its use is shown by the search for high-temperature superconductors. As soon as the first material, an oxide of Cu, La and Ba, was announced, search for improved materials involved systematic permutations, based directly on the Periodic Table—for example, among Ca, Sr and Ba and among the lanthanide elements. These took the initial breakout of critical temperature to over 70 K. After a short pause, a second major step early in 1988 raised critical temperatures to 120 K by replacing the trivalent lanthanides by Tl and Bi chosen, again on the basis of Periodic Table relationships, as alternative trivalent elements of appropriate size and properties.

Such a pattern is seen in almost all developments: as soon as a new material or phenomenon is discovered, it is a major strategy to exchange elements according to the periodic relations to test the extent and variations of the effect.

In this textbook, we are aiming to bridge the gap between the level of presentation in Special Honours textbooks and the standard of understanding of Inorganic Chemistry reached in the schools. Some feeling for the advancing subject is appropriate at this level, and we have tried to provide this, both by an up-to-date account of the chemistry of the elements and by the use of three chapters which present specific topics of considerable current interest. We hope that this book will continue to be found useful for those who are studying chemistry as part of a more general course, for those who wish to proceed to a less sophisticated level than the Honours BSc, and for those who are looking for an introductory text which will later be followed by work in an advanced course.

The growth of Inorganic Chemistry has a twofold basis. Major developments in theory have provided an invaluable framework but the subject is too complex to evolve from theory alone. General experience of systematic chemistry—the reactions of the elements and their compounds—is vitally important to an understanding of the subject.

More than half of this textbook (Chapters 9 to 12, 14, 15, and 17) is a presentation of the basic systematic chemistry of the elements, following the Periodic Table arrangement and with emphasis on the compounds with oxygen and the halogens. A further fifth of the book complements this arrangement by periodic position with a discussion of selected topics in more depth. To the transition metal topics of Chapter 16, we have added in this edition new Chapters 18 and 19 which cover Main Group and general topics. Finally, the earlier chapters and Chapter 13 provide the introduction to theories, structures and methods which are needed to appreciate the rest.

The creation of Chapter 18 has modified Chapter 17, on systematic Main Group chemistry, but this is still the longest chapter in the book. We feel it is advantageous to present the chemistry of the elements Group by Group, to contrast with the school syllabus approach by period, and have therefore retained the format of a single chapter.

Other aspects include continued attention to molecules of biological importance, including metal-containing pharmaceutic and radiotherapeutic agents. Comments on industrial production and uses of inorganic materials have also been expanded from earlier editions, both to reflect many new applications and because this area is less prominent in school **syllabuses than it used to be, SI units are used, and we have comments on some points where usage is not yet settled.**

We are most grateful to many colleagues and readers who have commented on earlier editions, and have pointed out errors and infelicities. We are also pleased to acknowledge the inspiration of a number of new Honours textbooks on Inorganic Chemistry, as listed in Appendix A. The subject has benefited considerably from several quite different, and most exciting, expositions published in the last few years.

R Ann Mackay
K M Mackay

SI Units and Names

Many international scientific and engineering bodies have recommended a unified system of units and the SI (Système International d'Unités) has been generally adopted for use in scientific journals. The full scheme is not universally accepted, and some variant units are found. There are seven basic units:

Physical quantity	Symbol for quantity	Name of unit	Symbol for unit
length	l	metre	m
mass	m	kilogram	kg
time	t	second	s
electric current	I	ampere	A
temperature	T	kelvin	K
amount of substance	n	mole	mol
luminous intensity	I_v	candela	cd

Two supplementary units are radian (rad) for plane angles and steradian (sr) for solid angles. The SI system differs from the old metric system by replacing the centimetre and the gram by the metre and the kilogram.

Multiples of these units are normally restricted to steps of a thousand, and fractions by steps of a thousandth, i.e. multiples of $10^{\pm3}$, but with the multiples 10,100,1/10 and 1/100 retained.

Fraction	Prefix	Symbol	Multiple	Prefix	Symbol
10^{-1}	deci	d	10	deca	da
10^{-2}	centi	c	10^2	hecto	h
10^{-3}	milli	m	10^3	kilo	k
10^{-6}	micro	μ	10^6	mega	M
10^{-9}	nano	n	10^9	giga	G
10^{-12}	pico	p	10^{12}	tera	T
10^{-15}	femto	f	10^{15}	peta	P
10^{-18}	atto	a	10^{18}	exa	E

It is important to note that $1\,km^2$ implies $1\,(km)^2$, that is $10^6\,m^2$ and *not* $10^3\,m^2$.

A range of units derive from these basic units, or are supplementary to them. Those commonly used by the chemist are shown in Table 1. The unit of force, the newton, is independent of the Earth's gravitation and avoids the introduction of g (the gravitational acceleration) into equations. The unit of energy is the joule (newton × metre) and of power, the joule per second (watt).

The gram will be used until a new name is adopted for the kilogram as the basic unit of mass, both as an elementary unit to avoid 'millikilogram', and with prefixes, e.g. mg.

It is recommended that all units not compatible with SI should be abandoned progressively, and the majority of the incompatible units are steadily if slowly disappearing from the literature. There are, however, a few units with particular advantages whose use is likely to persist, and it is recognised that any system of units should combine logical construction and consistency with a reasonable degree of convenience. Among the traditional chemical units are:

Litre. If the proposal to restrict multiples to $10^{\pm3}$ is applied, then lengths would be confined to the metre and millimetre, giving volume units of m^3 and mm^3. It seems unlikely that a system restricted to values differing by 10^9 will find favour and the cm^3 and dm^3 are likely to be long with us. It is probable that the litre will survive as a convenient name for the cubic decimetre.

Ångstrom. The ångstrom unit $(10^{-8}\,cm = 10^{-10}\,m = 10^{-1}\,nm = 100\,pm)$ was originally introduced as a unit of length for use on the interatomic scale. It continues to be widely used by crystallographers, though some replacement by the picometre, pm, is occurring.

Atomic mass. This is expressed relative to $^{12}C = 12\cdot0000$ and is termed relative atomic mass, symbol A_r. *Relative molecular mass*, M_r, is the sum of the A_r values for each atom in the molecule.

Energy. As the joule is of the same order of magnitude, there are no reasons for retaining the calorie. Energies of chemical processes commonly fall into the convenient range of 10 to $10^3\,kJ\,mol^{-1}$, and $kcal\,mol^{-1}$ will disappear. Chemists also use two other units in energy measurement: the *wave number*,

TABLE 1. Units derived from the basic SI units, or supplementary to them.

Physical quantity	Name of unit	Symbol and definition
force	newton	$N = m\,kg\,s^{-2}$ or $J\,m^{-1}$
pressure	pascal	$Pa = m^{-1}\,kg\,s^{-2}$ or $N\,m^{-2}$
energy, work, heat	joule	$J = m^2\,kg\,s^{-2}$ or $N\,m$
power	watt	$W = m^2\,kg\,s^{-3}$ or $J\,s^{-1}$
electric charge	coulomb	$C = sA$
electric potential	volt	$V = m^2\,kg\,s^{-3}\,A^{-1}$ or $J\,s^{-1}\,A^{-1}$
resistance	ohm	$\Omega = m^2\,kg\,s^{-3}\,A^{-2}$ or $V\,A^{-1}$
capacitance	farad	$F = m^{-2}\,kg^{-1}\,s^4\,A^2$ or $s\,A\,V^{-1}$
frequency	hertz	$Hz = s^{-1}$
temperature, t	degree Celsius (centigrade)	$°C$ where $t/°C = T/K - 273.15$
area	square metre	m^2
volume	cubic metre	m^3
density	kilogram per cubic metre	$kg\,m^{-3}$
velocity	metre per second	$m\,s^{-1}$
angular velocity	radian per second	$rad\,s^{-1}$
acceleration	metre per (second)^2	$m\,s^{-2}$
magnetic flux density	tesla	$T = kg\,s^{-2}\,A^{-1}$ or $V\,s\,m^{-2}$
time	hour, year, etc. will continue to be used	

reciprocal centimetre or *Kayser*, written as cm^{-1}, and the *electron volt*, eV. The latter is strictly the energy acquired by one electron falling through a potential of one volt, but eV is commonly used for the molar unit found by multiplying the strict value by Avogadro's constant. The main advantages of the electron volt are its close relation to the methods used to measure certain parameters, such as ionisation potentials, and its larger size. At about $10^2\,kJ\,mol^{-1}$, it is convenient for the larger chemical energies and its use is likely to continue for many years. The cm^{-1} is not a unit of energy but is used in spectroscopy. The relations underlying this application are described in Chapter 7.

The SI system is for reporting precise measurements in a fundamentally self-consistent way, and does not require that other units should never be used. It is not intended to preclude the use of 'working units' and units used in a non-rigorous context. Laboratory workers continue to fractionally distil at pressures measured in millimetres of mercury, autoclave at hundreds of atmospheres and read temperatures in degrees Celsius.

In Table 1 are listed the units which are derived from the basic SI units or their supplements. In Table 2 we list those units which are contrary to SI and should be phased out. The first part includes those which differ only by powers of 10 and the rest have exactly defined conversion factors, except the electron volt and atomic mass unit which are given in terms of the best experimentally-determined conversion factors. Some of these units appear mainly in pre-1970 literature, but others continue to be used. The values of physical constants in SI units are listed in Table 3.

Units used in this edition

A textbook written entirely in SI units would have a number of disadvantages at the present time. All the details are not completely settled and the system has not yet been formally recognized by all scientific bodies in an agreed final form. Furthermore, many literature sources required by the student are not in SI units and scientists must be able to convert existing data into the system. With these considerations in mind, the following convention has been adopted in this edition.

Length. The SI system of m, mm, and smaller fractions is used. Notice that many interatomic distances (given originally in ångstroms $= 10^{-10}\,m$) were known to an accuracy expressed by two decimal places in ångstroms, e.g. 1·07 Å. It follows that the most convenient SI-allowed multiple is the picometre, pm, so that values become whole numbers of picometres, e.g. 107 pm, and more accurate values are distinguished by having figures after the decimal point as in 107·6 pm $= 1·076$ Å). This usage is more convenient than using fractions of the nanometre, as in 0·1076 nm.

Energy. Most values are given in $kJ\,mol^{-1}$ and kcal is not used. Where appropriate, cm^{-1} and eV are also used and conversion factors are included in tables in these units. Occasionally, particularly when dealing with magnetic resonance, frequencies in Hz are found in place of wave numbers in cm^{-1}. As 1 Hz is one wave per second and 1 cm^{-1} is one wave per cm, the two are connected by the speed of light. Table 4 shows the interconversion factors for all these units.

Temperatures. Temperatures are quoted in degrees Celsius

TABLE 2. Commonly occurring units which are contrary to SI

Unit	Quantity	Equivalent
(A) Units differing from SI units by powers of 10		
ångstrom (Å or A)	length	10^{-10} m $= 10^{-1}$ nm $= 10^2$ pm
litre (l or L)	volume	10^{-3} m^3 $=$ dm^3
dyne (dyn)	force	10^{-5} N
erg	energy	10^{-7} J
mho (siemens or reciprocal ohm)	conductance	Ω^{-1}
bar	pressure	10^5 Pa
poise	viscosity	$0{\cdot}1$ Pa s
tonne (t)	mass	10^3 kg
(B) Other units		
calorie (cal)	energy	I.T. cal $= 4{\cdot}186\,8$ J: 15° cal $= 4{\cdot}185\,5$ J: thermochemical cal $= 4{\cdot}184$ J
electron volt (eV) (electron volt per mole, also symbolized eV $= 96{\cdot}484$ kJ mol^{-1})	energy	$1{\cdot}602\,1 \times 10^{-19}$ J
atmosphere (atm)	pressure	$101{\cdot}325$ kN m^{-2}
millimetre of mercury (mmHg) or torr (Torr)	pressure	$133{\cdot}322$ N m^{-2}
atomic mass number (amu or u $= 1/12$ mass of ^{12}C)	mass	$1{\cdot}660\,41 \times 10^{-27}$ kg

TABLE 3. Values of physical constants.

Physical constant	Symbol	Recommended value
Speed of light in a vacuum	C_0	$299\,729\,458$ m s^{-1}
Atomic mass unit	u	10^{-3} kg mol^{-1} $1{\cdot}660\,565\,5 \times 10^{-27}$ kg
Mass of proton	m_p	$1{\cdot}672\,623\,1 \times 10^{-27}$ kg
Mass of neutron	m_n	$1{\cdot}674\,954\,3 \times 10^{-27}$ kg
Mass of electron	m_e	$9{\cdot}109\,389\,7 \times 10^{-31}$ kg
Charge on proton or electron$(-)$	e	$1{\cdot}602\,177\,33 \times 10^{-19}$ C
Boltzmann constant	k or k_B	$1{\cdot}380\,658 \times 10^{-23}$ J K^{-1}
Planck constant	h	$6{\cdot}626\,075\,5 \times 10^{-34}$ J s
Permeability of a vacuum	μ_0	$4\pi \times 10^{-7}$ J s^2 C^{-2} m^{-1}
Rydberg constant	$R_\infty = \dfrac{\mu_0^2 m_e e^4 c^3}{8h^3}$	$1{\cdot}097\,373\,177 \times 10^7$ m^{-1}
Bohr magneton	$\mu_B = \dfrac{eh}{4\pi m_e}$	$9{\cdot}274\,078 \times 10^{-24}$ J T^{-1}
Avogadro constant	N_A or L	$6{\cdot}022\,136\,7 \times 10^{23}$ mol^{-1}
Gas constant	R	$8{\cdot}314\,510$ J K^{-1} mol^{-1}
'Ice-point' temperature	T_{ice} or T_0	$273{\cdot}150$ K $(RT_0 = 2{\cdot}271\,081 \times 10^3$ J mol^{-1})
Permittivity of a vacuum	$\varepsilon_0 = (\mu_0 c^2)^{-1}$	$8{\cdot}854\,187\,82 \times 10^{-12}$ F m^{-1}
Faraday constant	$F = Le$	$9{\cdot}648\,456 \times 10^4$ C mol^{-1}
$RT \ln 10/F$ at 298 K		$5{\cdot}916 \times 10^{-2}$ V
Bohr radius	a_0	$5{\cdot}291\,770\,6 \times 10^{-11}$ m
Molar volume of ideal gas (273·15 K, 1 atm)	$V_0 = RT/P_0$	$2{\cdot}241\,038\,3 \times 10^{-2}$ m^3 mol^{-1}

TABLE 4. Conversion factors

	kJ mol^{-1}	cm^{-1}	eV	MHz	kcal mol^{-1}
kJ mol^{-1}	1	83·626	$1·036\ 4 \times 10^{-2}$	$2·506\ 2 \times 10^{6}$	0·239 4
cm^{-1}	$1·195\ 7 \times 10^{-2}$	1	$1·239\ 4 \times 10^{-4}$	$2·997\ 9 \times 10^{4}$	$2·858 \times 10^{-3}$
eV	96·484	8068·3	1	$2·418\ 8 \times 10^{8}$	23·063
MHz	$3·990\ 3 \times 10^{-7}$	$3·335\ 6 \times 10^{-5}$	$4·134\ 4 \times 10^{-9}$	1	$9·534\ 5 \times 10^{-8}$
kcal mol^{-1}	4·184	349·83	$4·335\ 9 \times 10^{-2}$	$1·048\ 7 \times 10^{7}$	1

(°C), or in Kelvin (K)—*note* no degree symbol, is used.

Electrical units. These do not occur widely in this text. Note that the interaction between charges is modified by the permittivity of a vacuum, ε_0. Thus the non-SI factors e^2/r, which occur in the wave equation, for example, now become $e^2/4\pi\varepsilon_0 r$.

Other parameters. SI units are used where exact values are stated. Note that the litre is used as the name of the strict SI dm^3. There is some discussion that the gram-molecule might be replaced by the kilogram-molecule but most chemists would find this unacceptable.

In the text, numerical values for physical properties, such as atomic radii or oxidation-reduction potentials, are chosen from consistent sets and are quoted mainly for purposes of comparison. Much more accurate values are often available for particular data, and any calculations of physical significance should use such values, which can be found in a number of critical complications of data of in the original literature.

Formulae are commonly used in place of chemical names where the former are clearer and less clumsy. Equations which are written unbalanced are used either to show the major product of a reaction, or to indicate the variety of products without being definite about their relative proportions.

In addition to the symbols for fundamental constants given in Table 3, a number of other symbols and abbreviations are to be found in the text. These are:

S	Entropy
H	Enthalpy (heat content)
K	Equilibrium constant
ccp	Cubic close packed
hcp	Hexagonal close packed
bcc	Body centred cube
m or m.p.	Melts or melting point
b or b.p.	Boils or boiling point
d	Decomposes
subl	Sublimes

The naming of geometrical shapes

Since the geometry of solids was first worked out by the ancient Greeks, we find that much of the terminology comes from Greek and occasionally causes misunderstanding. A regular solid is named for the number of faces, and the Greek root is *hedron*. Thus the tetrahedron is named for its four faces. The plural is formed by turning *-ron* into *-ra*, and similarly for the adjective. Thus 'tetrahedra' is the plural and 'tetrahedral' the adjective. Since we normally deal with atoms arranged in a geometrical form around a central atom, the chemist is usually more interested in the number of points or *vertices* (note the singular is *vertex*) in a figure, rather than the number of faces. While a tetrahedron has four points and four faces, the number of points and faces differs for all other solids. Thus an octahedron, familiar as the shape found for 6-coordination, has *eight* faces (hence octa-) and six vertices.

A further relation of interest is that of *capping* which means placing a further atom above a face so that it is at equal distances from each of the atoms which define that face. If we cap planar figures we get *pyramids*: capping a square gives a square pyramid and capping it again on the other side gives a square *bipyramid*. The capping atom is often termed the *apex* (plural *apices*). It may be placed at any distance from the face, but if the distance is equal to the length of the edge of the capped figure, then a special *regular* shape results. Thus a regularly capped triangle (or regular trigonal pyramid) is a tetrahedron and the octahedron is a regular square bipyramid. If regular figures are capped, further relationships emerge. Thus the eight atoms which regularly cap the faces of an octahedron, themselves form a cube, and if the six faces of a cube are regularly capped, the caps form an octahedron.

We also note some interrelationships involving the cube. If every second vertex of a cube is selected, these four points define a tetrahedron, so a cube may be seen as two identical interpenetrating tetrahedra (see figure 13.9). If these two tetrahedra are now altered regularly so that one is elongated and one is compressed (equivalent to pulling two pairs of atoms defining opposite face diagonals out of the face, forming triangles bent about the diagonal) the result is a *dodecahedron*. If a tetrahedron is regularly capped on all four faces, the eight atoms lie at the corners of a cube, the figure resulting corresponds to the formula A_4B_4 and is called a *cubane*.

Finally, one class of less regular figures also occurs among chemical structures—the *prisms*. A prism is formed by connecting two regular plane figures held in an eclipsed position. For example, a general pentagonal prism has two regular pentagonal faces and five rectangular ones. Again, special figures result if all the edges are equal—a cube is a special case of the square prism, for example. The *antiprism* is formed if the regular plane figures are placed in a staggered configuration, forming for example a square antiprism. The regular trigonal antiprism is the octahedron.

Examples of all these figures will be found in later chapters. It is extremely valuable to study solid models, or a good computer simulation program, as an aid to understanding the relationships between them.

Table 5 summarises the properties of the commoner shapes met in chemistry.

TABLE 5.   Some common shapes

Name	Significant geometrical properties	Examples (Figures)
Tetrahedron (a)	4 faces; 4 points; 6 edges	4·3a; 5·1c
(for relationship with a cube, see Figure 13.9)		
Trigonal pyramid	as above, 3 edges longer than others	17·65c
Trigonal bipyramid (b)	6 faces; 5 points; 9 edges	4·4a; 9·8a; 14·29
Square pyramid (c)	4 triangular + 1 square face; 5 points; 8 edges	13·11b; 14·13; 16·8; 17·65d
Octahedron (d)	8 faces; 6 points; 12 edges	4·5a; 5·1a; 15·11
face-bridged	8 bridging atoms	15·19
edge-bridged	12 bridging atoms	15·9
Trigonal prism (e)	2 triangular + 3 rectangular faces; 6 points; 9 edges	5·10c and plate II; 13·11a
monocapped (f)	cap on rectangular face	15·1
tricapped (g)	caps on all 3 rectangular faces	15·23; 17·40c
Pentagonal pyramid (h)	1 pentagonal + 5 triangular faces; 6 points; 10 edges	9·8b; 14·17a
Pentagonal bipyramid (i)	10 faces; 7 points; 15 edges	14·12; 14·17c
Cube (j)	6 faces; 8 points; 12 edges	5·1b; 5·12g; 15·16
cubane	A_4B_4 at alternate vertices	15·15
Square prism (k)	as cube, four edges different length	14·33
Square antiprism	2 square + 8 triangular faces; 8 points; 16 edges	12·5; 15·5b
monocapped (l)	cap on square face	18·21
Hexagonal bipyramid (m)	12 faces; 8 points; 18 edges	12·6
Pentagonal antiprism (n)	2 pentagonal + 10 triangular faces; 10 points; 20 edges	16·7a; 16·9
Hexagonal prism (o)	2 hexagonal + 6 rectangular faces; 12 points; 18 edges	5·12e (outline); 16·7b
Octagonal prism (p)	2 octagonal + 8 rectangular edges; 16 points; 24 edges	16·11a
Dodecahedron (q)	12 faces; 8 points; 16 edges	13·2; 15·2; 15·18
Icosahedron (r)	20 faces; 12 points; 30 edges	17·7d

Note: examples are chosen to give a variety of representations of the shapes.

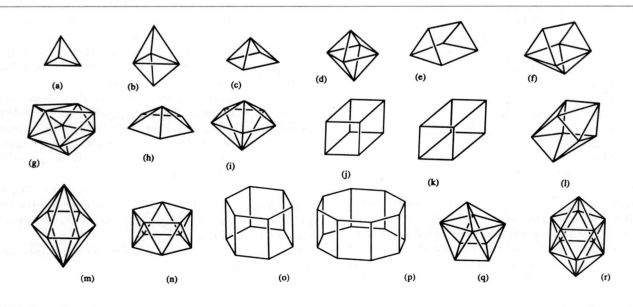

1 Introduction

1.1 Inorganic chemistry

Chemistry is one of the oldest and most wide-ranging branches of science, and hence of human knowledge and endeavour. For convenience, the subject is commonly divided into a number of fields, though the boundaries are broad and there is much overlap. Traditionally, there are three major divisions, *inorganic*, *organic*, and *physical* chemistry. Inorganic chemistry covers the properties and reactions of the chemical elements—now exceeding 110!

Since the chemistry of one element, carbon, is so enormously ramified—probably similar in extent to the total chemistry of the remaining 109 put together—it has traditionally made up the separate field of organic chemistry. The bridging discipline covering the organic chemistry of the inorganic elements, *organoelement* or *organometallic* chemistry, is an extensive field to which we shall make substantial reference. In addition, part of the chemistry of carbon, covering the element and simpler compounds like its oxides and oxyions or the carbides, is traditionally covered in inorganic chemistry. No attempt is made to draw rigid boundaries.

The detailed study of energy changes, reaction mechanisms, much of bonding theory, the chemistry of polymers, chemistry which occurs at surfaces and interfaces, the behaviour of metallic systems—all these fall into physical chemistry. Again, there are no rigid demarcations, and much of the most exciting work is done at the points of overlap. Inorganic chemistry also makes major contributions to interfaces with other sciences, in areas like geochemistry, inorganic biochemistry, materials science and metallurgy.

1.2 Origins and the discovery of the elements

The origins of inorganic chemistry are ancient. Observation followed by what we now describe as empirical experiments led to the slow development of new materials from the early stages in human history. Thus beads of glass and of ceramics are found in ancient Egyptian burials and pottery was made by the earliest civilizations. Considerable control was achieved: black or red pottery was made by reducing or increasing the proportion of air, and colours and glazes had developed to a high degree of sophistication by 500 BC.

As one informative indicator, we may review the discovery of the elements (Table 1.1).

If the discovery of the elements is plotted against time, a curve is obtained which mirrors the pattern of development in many sciences. A long slow period of completely empirical advance in the ancient world was followed by a phase mainly of preservation and rediscovery through the Arab alchemists and in India and China. For the century up to AD 1750, some of the basic ideas of what we now call chemistry were developed from more deliberate investigations. From AD 1750 up to the first half of this century, there was a sharply accelerating pattern of discovery as theory and technique advanced in parallel. Within this period we see individual spurts reflecting specific advances, like the 18th-century studies of gases, the early 19th-century use of electrolysis to isolate the very active metals, or the recognition of the Rare Gas Group which gave five new elements in five years. Eventually the pace slowed, in the decade to 1940, because there were 'no new worlds to conquer' and all the elements up to uranium had been identified. This was not the end of the story, as it turned out that further post-uranium elements could be synthesised. This phase is now slowing down, reflecting the decreasing intrinsic stability of the nuclei. Whether this is finally the end of the story of the elements is not yet clear (compare section 16.12). The overall pattern, found in many other developing fields, is of empirical discovery, acceleration fuelled by the interaction of greater understanding and improved methods, then maturity when the pace of change slows. Often, new accelerations start up from the mature phase, as unexpected observations or new ideas trigger off further developments.

It is worth emphasizing how much technical control was achieved by ancient craftsmen. Gold is usually found naturally as the free metal and, as shown by grave goods, has always been highly valued. Modern analysis of ancient artefacts shows that it was understood that the addition of small

TABLE 1.1 The discovery of the elements

Date range	Number of elements discovered	Comments
Prehistoric	3	C, S, Au which occur native, i.e. uncombined
ca. 3000 BC	5	Ag, Cu, Pb, Sn, Hg with readily processed ores
ca. 1000 BC	1	Fe requiring higher temperature reduction
ca. 500 BC	1	Zn ca 90% pure
Up to 1650	4	As, Sb, Bi: Zn rediscovered
1650–1700	1	First dated discovery: P in 1669
1700–50	3	Co, Ni, and native Pt
1750–75	7	First gases H, N, O, Cl and Ni, Mn, Bi*
1775–1800	5	Cr, Mo, W, Te, Ti(finally pure in 1910)
1800–25	18	Active metals, Li, Na, K, Mg to Ba: heavier metals Ce, Ir, Os, Pd, Rh, Zr: also B, Cd, I, Se
1825–50	9	Br, Si, Be, Al, V, La, Ru, Th, U
1850–75	5	Rb, Cs, Ga, Tl, Nb {Hc seen in solar spectrum}
1875–1900	approx. 11	5 inert gases: F, Ge: radioactive Po, Ra, Ac: some lanthanides
1900–25	approx. 10	Rn, Ta, In, Hf, Re, Pa, lanthanides
1925–50	11	2 lanthanides: radioactive Tc, Pm, Fr, At: man-made post-uranium Np, Pu, Am, Cm, Bk
1950–75	10	Last 2 purified lanthanides: 8 man-made
1975–today	ca 8	Man-made (a few atoms only)

Notes. 1. Compare Tables 2.5 and 2.6 for the names and periodic positions of the elements.
2. Many dates of discovery are approximate, as the existence of many elements was recognized anything from a few months up to a century before final purification. The very similar lanthanides present particular difficulties of definition.
3. Bismuth* known earlier but confused with lead.
4. See the reference by Ringnes (Appendix A) for a very readable account of the origins of the names of the chemical elements.

proportions of silver or copper gave a harder, more wear-resisting metal, and gave desirable variations in colour. The metal contents of gold coins held to highly consistent standards which correlated through many countries in a chain of related weights and gold contents—for example, from Macedonia to India in the 5th century BC. Copper has been known for about 7000 years and has been extracted from sulfide ores for around 5500 years. Small metal items like beads and bracelets are found in graves for several centuries before larger products like tools or weapons, suggesting a period where manufacture was difficult and not well understood. Analysis of ancient kilns shows that temperatures up to 1200 °C were achieved in Bronze age copper smelting. Once production methods had evolved to the larger scale, it was rapidly found that alloying copper with tin to give bronze, or with zinc to give brass, produced metals of superior properties. Tin, which is fairly inert to air, was probably prepared in a pure form many centuries before the more reactive zinc. Iron requires much more sophisticated treatment, and can be extracted only at the high temperatures achieved with air blown through a bed of charcoal. The carbon is also necessary to remove the oxygen from the ore. Even so, early temperatures did not reach the melting point, so casting was not possible and the metal was shaped by hammering. As iron weapons are much superior to bronze, it is thought that the initial discovery was kept secret for a millennium and knowledge of iron only became widespread after the destruction of the Hittite empire about 1200 BC. (recent archaeology suggests this picture is oversimplified). Compared with iron, isolation of the other elements found up to the 18th century is relatively easy, so iron-working represents a peak in technological achievement which lasted for something like 4000 years.

1.3 Development

While we have followed the tale of the elements, the growth of inorganic chemistry as a whole followed a similar pattern. Inorganic chemistry was the first of the chemical sciences to flower in the course of the Scientific Revolution, and most of the work leading to the formulation of the atomic theory was carried out on inorganic systems, especially the gases and simple compounds like the nitrates, carbonates or sulfates. A critical advance in technique was the development of ever more accurate measures of the quantity of material—both by weighing and by measurement of gases. Once it could be established that a particular substance had the same composition when prepared by different routes (for example, an oxide prepared from the metal and air, from heating the carbonate, from precipitating the hydroxide from solution and igniting) the way was open to following changes quantitatively, to formulating generalizations like 'the Law of Constant Composition', and ultimately to Dalton's atomic theory. It is worth remarking that even the most sophisticated modern experiment depends ultimately on accurate measurement of weight changes.

In the first half of the 19th century, not only had more than half of the elements been isolated but a great many of their simpler compounds had been studied. It is remarkable that explosive nitrogen trichloride or highly corrosive hydrogen fluoride were under study around 1800. By contrast, only a few simple organic compounds were known by 1820 and little progress was being made in organic chemistry as much of the effort was directed to extremely complex materials like milk or blood. By the middle of the 19th century came the period of spectacular advance in organic chemistry, followed around 1900 by a great upsurge of interest in physical chemistry. These advances meant nearly a century of comparative neglect of inorganic chemistry.

Of course, very important advances were made, including the formulation of the Periodic Table, the discovery and exploitation of radiochemistry, and the classical work on non-aqueous solvents and on the complex chemistry of the transition elements, but it was not until the nineteen-thirties that the modern upsurge of interest in inorganic chemistry got under way. Among the seeds of this renaissance were the work of Stock and his school on volatile hydrides of boron and silicon, of Werner and others on the chemistry of transition metal complexes, of Kraus and Walden on non-aqueous solvents, and the work of a number of groups on radioactive decay processes. At the same time, the theories which play an important part in modern inorganic chemistry were being formulated and applied to chemical problems. The discovery of the fundamental particles and the structure of the atom culminated in the development of wave mechanics, which is the basis of all modern approaches to valency and bonding. This theory is outlined in Chapter 2, and its application to molecular structure is given in Chapters 3 and 4. A little later, the effect of an atom's environment on the energy of its d electrons was brought into the treatment of transition metal compounds in the crystal field theory which is discussed in Chapter 13.

1.4 Recent advances

All these developments prepared the ground for the expansion of inorganic chemistry, starting in the 1950s, which was stimulated both by developments on the academic side in experiment and theory, and by the demand for new materials and for knowledge of many elements hitherto scarcely studied.

The advent of atomic energy focused attention on heavy transition elements and lanthanides (for example, the chemically very similar Zr and Hf have quite different neutron absorption properties). Similarly, the growth of electronics, followed by computers, led to growth in the chemistry of lesser-known Main Group elements involved in semiconductors, such as Ge, Ga, In and Se. The other significant change in the third quarter of this century has been the growth in the numbers of scientists and technologists. In the past, fields expanded only at the price of relative neglect in other areas. Recent developments have taken place simultaneously, adding a further factor of mutual reinforcement.

In the last quarter of the 20th century we are living through a continuing headlong expansion in inorganic chemistry, fuelled by new methods, new theories, new fields of interest like metals in biological systems, the search for new materials, new catalysts, more output for less pollution, and many other driving forces. Even unstable elements like technetium are finding uses in medicine and highly radioactive isotopes, such as americium-241, find a variety of applications, including household smoke detectors. Such interests have generated research in every corner of the Periodic Table, and there are now no elements, apart from the very unstable heavy ones, where there is not a very substantial body of knowledge available.

This rapid growth of inorganic chemistry continues to make it a very lively and exciting subject in which to work and teach but it does lead to problems from the student's point of view. Textbooks tend to be out of date by the time they are published and the treatment of each subject changes as new discoveries are made.

A striking example in the 1960s was the completely new chapter of chemistry which grew from nothing following the first announcement of a xenon compound in 1962. The pattern of development was typical. The tenet that the rare gases were inert was so well established that the announcement of the formation of a rare gas compound generated immense excitement and over 100 papers were published within a couple of years. After this initial surge of interest the production of new papers slowed but new discoveries continue to expand rare gas chemistry, including the recent evidence for an organoxenon compound.

A more recent example of this same pattern is provided by the announcement, late in 1986, of a superconductor whose critical temperature (the temperature below which the phenomenon of resistance-free conduction occurs) was around 40 K. This followed a long period where the highest critical temperature shown had risen only very slowly from about 5 K to around 23 K. The new superconductors were oxide phases involving copper and elements like the lanthanides and the alkaline earth metals. Excitement was enormous, as higher temperature superconductors have tremendous potential for all electrical devices. The claimed best critical temperature shot up to over 100 K, with preliminary manifestations even up to room temperature, all within the space of a few months. While the main interest has been in the physics, there was a very rapid exploration of the chemistry. A superconducting phase of major interest is $YBa_2Cu_3O_{7-x}$, where x is around 0.1, and the pattern of exploration is exactly what any inorganic chemist in the last hundred years would have followed—basically study of complex oxides of related elements, guided by the Periodic Table (see section 16.1 for a full review).

The recent discovery of polyhedral carbon species, commonly known as fullerenes and comprised solely of carbon atoms, has initiated a vast upsurge in the chemistry of these materials, with a wide variety of both inorganic and organic derivatives being prepared. In this regard, these materials can be viewed as yet another bridge linking inorganic with organic chemistry. These materials will be discussed in greater depth in section 19.3.

These three examples, one fairly academic and the other two based on expectations of enormous practical applications, are representative of many current advances. Some novel and unexpected discovery triggers a period of intense interest, where rapid and widespread exploration occurs, *heavily based on the pattern of previous knowledge*. To a substantial extent inorganic developments are rooted in the relationships of the Periodic Table and established systematic chemistry. The incidence and progress of the more novel new discoveries and growing points are unpredictable, and this should give pause to those who would attempt too closely to guide the development of science into areas deemed to be more 'relevant' to the problems of the day.

A textbook must attempt to reflect both the steadily-growing core of basic material, and the areas of current interest and excitement. An introductory text can only sample, while an advanced text will make a valiant attempt to cover all areas of current interest.

In the area of theory, the inorganic student is presented with a number of approaches, many partly overlapping, at different levels of sophistication. Because of the relative complexity of the material, chemistry is much less 'theory-led' than is physics. Although the power and sophistication of chemical computation is steadily increasing, it is still not possible to describe any but the most simple inorganic chemical observations exactly by theory. Thus we are faced by a range of theories of different power, and of different levels of approximation. While there have been arguments about which of several alternatives is the best, in most cases we are content to use overlapping and even apparently conflicting theories, depending on the specific application. For example, many species may be described in electrostatic terms—as charged ions, dipoles, etc.—and the energy changes calculated by electrostatics. The same species may equally well be described in terms of covalent bonding, with a theoretical approach which has its base in quantum

mechanics. Often, neither approach gives an exact answer because approximations are needed to bring the calculation within the compass of even the most powerful computer. In general, that approach is chosen which gives the most convenient answer to a specific problem, and different methods may be used to tackle different parts of the same problem. It follows that there is no one answer to a question like 'is this compound ionic or covalent?', but rather an understanding that either description is more or less useful, and more or less of an approximation.

A second type of case also causes difficulties at first acquaintance. This is the situation where a relatively simple approach allows the rationalization and systematization of a particular body of data. Because the approach is fairly simple, it is usually not complete—that is, there will be exceptions and anomalies. If the model is reasonably wide-ranging, it is worth retaining and using it even after cases appear which are not covered. On occasion, two different partial models will be used even though they overlap and are not fully compatible. Such situations are quite common, and usually do not greatly disturb the scientist working in the field. They can be confusing to the student on first acquaintance, as there is the feeling that only one can be 'right', and we tend to use fairly high-powered words like 'law', 'theory' or 'principle' to describe them. If they are seen as *partial* descriptions or models, to be used as convenient, many of these problems disappear for the chemist (however much they disturb the philosopher of science!).

In applying wave mechanics to chemistry, the two common approaches have been the *valence bond* and *molecular orbital* methods. Each is a different approximation to the wave equation for a system, and they converge to the same answer for very simple systems. For polyatomic molecules, approximation is essential and the theories are used side by side. Older preferences were for the valence bond approach which is closer to the classic picture of a molecule as linked together by discrete electron pair bonds. For example, the partial double-bonding in a species such as the nitrate ion was described in terms of 'resonance' between contributing forms, each of which was described in terms of single and double bonds:

$$^-O-N{\overset{O}{\underset{O}{<}}} \longleftrightarrow O{\leftarrow}N{\overset{O^-}{\underset{O}{<}}} \longleftrightarrow O{=}N{\overset{O}{\underset{O^-}{<}}}$$

This way of thinking is now favoured rather less than the molecular orbital approach, which discusses such species in terms of *delocalized* bonds extending over all the molecule (see section 4.4). In other areas, such as properties of excited states and of species with extended multiple bonding, the molecular orbital theory is more satisfactory. In this text, the structures of molecules and ions are described largely in terms of the molecular orbital theory.

Similarly, in transition metal chemistry, *ligand field theory*, which deals with compounds which have valency electrons in the d orbitals, subsumes the electrostatic *crystal field theory* and also wave-mechanical aspects giving a molecular orbital treatment of transition metal compounds with multi-centred bonds. Again the treatments are at a number of levels of generality and approximation, and are often used in tandem.

Transition element chemistry has become a major field of inorganic chemistry since the 1950s, largely as a result of the strong mutual stimuli of experimental and theoretical advances. This expansion has spread particularly widely in the last two decades into fields such as low oxidation state chemistry, organometallic compounds, multiple metal–metal bonding and metal cluster compounds.

The last decade has seen a similar expansion in Main Group element chemistry, building from the more traditional compounds into a substantial range of new species. In this case an important driving force has been the development of experimental methods for isolating and characterizing compounds which were unmanageable by older techniques. All these changes are reflected in the succeeding pages, but the reader is asked to realize that the rapid rate of growth means that any text is somewhat out of date by the time it is published and the review literature should be consulted for recent advances.

1.5 Inorganic nomenclature

The nomenclature of inorganic chemistry was put on a definitive basis by the publication of the IUPAC (International Union of Pure and Applied Chemistry) Rules in 1957, 1970 and 1990. These rules define a systematic method for naming all inorganic compounds, but they also allow the retention of a number of trivial names which are well established. (See section 2.12 and Table 2.7 for Periodic Table nomenclature.)

The principles of systematic naming are straightforward and are outlined below:

(i) The cation, or electropositive component, of a compound has its name unmodified.

(ii) If the anion, or electronegative constituent, is monatomic, its name is modified to end in -ide.

(iii) If the anion is polyatomic, its name is modified to end in -ate.

(iv) Where oxidation states are to be indicated, they are shown by means of Roman numerals following the name of the element.

(v) The stoichiometric proportions of constituents are denoted by Greek prefixes (mono, di, tri, tetra, penta, hexa, hepta, octa, nona, deca, undeca, and dodeca). Alternatively, numerals may be used, as in $B_{10}H_{14}$—decaborane-14. In addition, the multiplicative prefixes (bis, tris, tetrakis, etc.) may be used to indicate a multiplicity of complex groups, especially when these already contain a numeral, as in $Ni(PPh_3)_4$—tetrakis(triphenylphosphine) nickel.

(vi) In extended structures, a bridging group is indicated by the prefix μ, for example, $[(NH_3)_5Cr-OH-Cr(NH_3)_5]Cl_5$ which is μ-hydroxo-bis[penta-amminechromium(III)] chloride. A group bridging three atoms, e.g. the face-bridging Cl atoms in Figure 15.19, is labelled μ^3. A further term which is now commonly used is hapto, symbol η. Its use is best illustrated by considering cyclopentadiene (see 16.4). If this bonds equally through all five C atoms, as in ferrocene, Figure 16.7a, it is labelled penta-hapto, symbol η^5. A single C-element bond would be (mono)hapto, η^1, while if only one diene group bonded, involving two of the five carbons, the pentadiene group would be described as dihapto, η^2.

Notice that, according to rule (iii), all polyatomic ions have names ending with -ate. This must not be confused with the trivial naming of oxygen anions where the endings -ate, -ite, etc., are used to indicate the oxidation state (see below). In systematic naming, all such anions end with -ate and the oxidation state is shown by the stoichiometry or by Roman numerals. Thus, $SnCl_6^{2-}$ is hexachlorostannate(IV), $SnCl_3^-$ is trichlorostannate(II), SO_4^{2-} is tetraoxosulfate (or tetraoxosulfate(VI)) and SO_3^{2-} is trioxosulfate. Some examples are given in Table 1.2.

The system outlined above gives a means of providing an unambiguous systematic name for any inorganic compound. However, many of the systematic names of familiar compounds are clumsy, and a considerable number of common names are retained for general use. The following points may be particularly noted.

(i) The terminations -ous and -ic may be retained for cations of elements with only two oxidation states, as in ferrous and ferric or stannous and stannic.

(ii) In anions, the termination -ite to distinguish a lower oxidation state is retained in such cases as nitrite, sulfite, phosphite and chlorite. Similarly the hypo- . . . -ite method of showing an even lower oxidation state is retained for hyponitrite, hypophosphite and hypochlorite. Corresponding acid names end in -ous and hypo- . . . -ous.

(iii) The term thio- is used to denote the replacement of an oxygen atom by a sulfur one, as in $PSCl_3$—thiophosphoryl chloride. Similar use is made of seleno- and telluro-.

(iv) The terms ortho- and meta- are retained to indicate different 'water contents' of acids, as in orthophosphoric acid, H_3PO_4, and metaphosphoric acid, $(HPO_3)_n$, or H_5IO_6 orthoperiodic acid and HIO_4 periodic acid.

(v) As the last example shows, the prefix per- is used to indicate an oxidation state above the one indicated by the normal -ate or -ic termination of an anion or acid. (Per- should not be used for metals and cations.) This usage should be confined to the cases of perchlorate, perbromate, periodate, permanganate and per-rhenate. The prefix peroxo- should be used to denote the presence of the $-O-O-$ group derived from hydrogen peroxide, as in peroxodisulfuric acid (HO_3SOOSO_3H), although the old 'per' is still often found, especially in commercial use. The term 'perfluoro' is widely used in the organic literature to denote a complete substitution of H atoms by F atoms. Thus, the compound $C_{10}F_{18}$ is widely known as perfluorodecalin (decalin $= C_{10}H_{18}$) which is of interest as a blood substitute because of its high capacity for dissolving oxygen gas.

A further terminology will be found for Main Group compounds where organic and organometallic nomenclature overlaps. Hydrides are now given names ending in '-ane' and compounds are often named as hydride derivatives. Thus PH_3 is now *phosphane* instead of phosphine, AsH_3 is

TABLE 1.2 Examples of systematic inorganic nomenclature. Of the detailed rules for the order of citation of ligands in complexes we need only note that anionic ligands come before neutral and cationic ones and 'oxo' is often dropped out of the names of familiar oxyions.

NaCl	Sodium chloride	
SiC	Silicon carbide	Rules i, ii and v: note that mono-
As_4S_4	Tetra-arsenic tetrasulfide	is usually omitted as a prefix
Cl_2O	Dichlorine oxide	
OF_2	Oxygen difluoride	
$KICl_4$	Potassium tetrachloroiodate	
$FeCl_2$	Iron dichloride (or iron(II) chloride)	Rules i, iii, iv and v. Notice that
$Pb_2^{II}Pb^{IV}O_4$	Trilead tetroxide (or dilead(II) lead(IV) oxide)	the use of roman numerals in rule
$K_4[Fe(CN)_6]$	Potassium hexacyanoferrate(II)	iv, and the prefixes of rule v, often
$K_3[Fe(CN)_6]$	Potassium hexacyanoferrate(III)	provide alternative names; super-
$Na(SO_3F)$	Sodium trioxofluorosulfate	fluous information is avoided
NaH_2PO_4	Sodium dihydrogen tetraoxophosphate	
$Na(NH_4)HPO_4.4H_2O$	Sodium ammonium hydrogen phosphate tetrahydrate	Note that multiple groups are
BiOCl	Bismuth oxide chloride	written separately (often all run
$VOSO_4$	Vanadium (IV) oxide sulfate	together) in an order which is
$ZrOCl_2,8H_2O$	Zirconium oxide dichloride octahydrate	defined by the rules. Notice also
$Li(AlH_4)$	Lithium tetrahydroaluminate	that oxo- is often dropped out of
$NH_4[Cr(SCN)_4(NH_3)_2]$	Ammonium tetrathiocyanatodiamminechromate(III)	the names of familiar oxyanions
$[Co(CO_3)(NH_3)_4]Cl$	Carbonatotetra-amminecobalt(III) chloride	
$[Be_4O(CH_3COO)_6]$	(see Figure 10.9) μ_4-oxo-hexa-μ-acetatotetraberyllium	

TABLE 1.3 Some correlated study areas

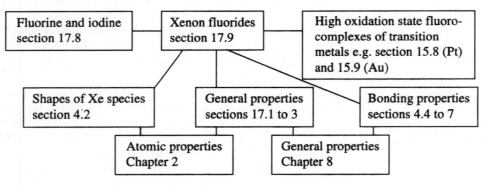

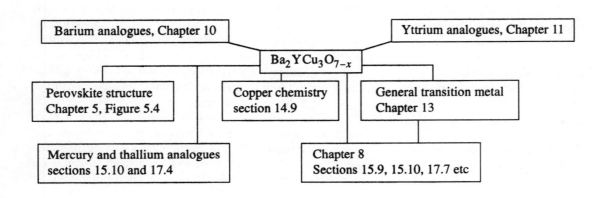

arsane, and we now often find the ligand Ph_3P called triphenylphosphane. While NH_3 and H_2O are systematically named nitrane or oxane respectively, it is unlikely that terms based on ammonia and water will disappear.

On the whole, the tendency is to use the simplest name available for well-known compounds, as long as this is accurate. New classes of compounds tend to be given their systematic name unless, as in the case of ferrocene for example, the discoverer happens to have hit on a suitable trivial name that has become widely accepted. Inorganic nomenclature need not cause any difficulty to the beginner who only has to translate the systematic name into a more familiar form, as long as the details above, especially about the various uses of the termination -ate, are familiar.

1.6 Approach to inorganic chemistry and further reading

The starting point for an understanding of inorganic chemistry lies in the electronic structure of the atom. The basis of this is provided by quantum theory and wave mechanics and is outlined in Chapter 2, using a relatively pictorial and qualitative approach. Combination of atoms into diatomic species is covered in Chapter 3, setting the basis for polyatomics discussed in Chapter 4. This Chapter also covers the shapes of covalent polyatomic molecules and ions. Chapter 5 deals with compounds in the solid state, both ionic and covalent materials, while Chapter 6 deals with solutions in water and in non-aqueous solvents. Chapter 7 gives a brief review of the experimental methods used in inorganic chemistry. The latter three-quarters of the book covers the chemistry of the elements. After placing this chemistry in the context of the Periodic Table in Chapter 8, and covering hydrogen as a forerunner in Chapter 9, Chapters 10, 11, 12, 14, 15, and 17 cover the different blocks of the Periodic Table in a Group-by-Group pattern. Chapter 13 provides a general context for the transition element chemistry of Chapters 14 to 16, while sections 17.1 to 3 serve a similar function for the elements of the Main Groups covered in the rest of 17 and 18. Finally, some themes of strong current interest are selected for more detailed discussion—for transition elements in Chapter 16, for Main Groups in Chapter 18, and for more general topics in Chapters 19 and 20.

Clearly, the arrangement sketched above is only one possible ordering, and does not bind the reader. You will find that any topic may be taken as starting point, and will eventually lead through most of the text. To illustrate, one possible network of relationships is sketched in Table 1.3 for two of the topics mentioned in section 1.4. Many others are possible, and you will start to feel some mastery of the subject once you have followed through a number of themes for yourself.

An introductory text can only provide a framework on which a more detailed knowledge may be built. We have tried to give a basic survey by concentrating on relatively well-known compounds, particularly the oxides, halides and hydrides. Added to these, we have tried to signal many other areas of great interest by relatively brief mention, and have extended the treatment of these in a limited number of cases.

Any textbook of inorganic chemistry nowadays covers only a few percent of established knowledge. Likewise, even the most dedicated student can only become familiar with a tiny part of the whole. It is therefore an important part of your training to become efficient in searching out material when you turn to a new area of the subject. What do you do when, with only a broad idea of copper chemistry and a shadowy recollection of the perovskite structure, your manager or pupils or others demand to know about these new superconductors which are in the news? What if chromium contamination suddenly becomes a problem, or your company decides to study indium telluride as a semiconductor when all you know about is silicon? Clearly, no-one expects to have an in-depth knowledge of the whole of inorganic chemistry, but you do need to know how to start finding out.

Since the selection and approach of any author is an individual one, your first step is to consult similar sources—find three or four other broad texts and build up a basic picture. Then, to get more depth, go to the more advanced texts, those written for British Honours students or for American graduate courses—a number of these are listed in Appendix A, and these will often give further references to reviews or papers. You should beware of becoming too entangled in specific detail early on in a survey, so you will normally focus on reviews. Inorganic chemistry is well-provided with review series (Appendix A) and most major topics will be found. Again, be aware that the depth of presentation may vary considerably, even within a single volume, and it is probably best to follow the references back to earlier treatments and start there. Among the more general presentations of inorganic topics are the Royal Society of Chemistry *Monographs for Teachers*, a number of articles in the journals, *Education in Chemistry* and *Journal of Chemical Education*, and articles in the 'news' publications *Chemistry in Britain* (Royal Society of Chemistry) and in *Chemical and Engineering News* (American Chemical Society). More advanced reviews are listed in the Appendix.

There have been a number of multivolume compendia of inorganic chemistry including *Mellor* and *Pascal* which are useful for older work but have not been updated. A third, *Gmelin*, is being systematically updated by a series of supplements and is reasonably current. All these are listed in Appendix A. Similar, shorter, and more recent multivolume works are *Comprehensive Inorganic Chemistry* *Comprehensive Heterocyclic Chemistry*, *Comprehensive Coordination Chemistry* and the first and second editions of *Comprehensive Organometallic Chemistry*, which should be consulted first. For information on specific compounds the *Dictionary of Inorganic Compounds*, *Dictionary of Organometallic Compounds*, and *Dictionary of Organic Compounds* may prove to be useful, especially if the compound in question is a common one.

PROBLEMS

1.1 Choose a simple broad topic, such as 'the chemistry of manganese'. Read the introduction to Appendix A.

(a) List all the pages in this book which give relevant information
(b) Similarly list the pages in any three other textbooks
(c) Find three reviews which cover some aspect of the topic (try to find an example of a broad survey and a detailed account of a specific area)
(d) Find three papers published in the last two years covering different parts of the topic.

1.2 Draw up the time plot suggested in section 1.2 for the discovery of the elements.

1.3 Plot the elements discovered at different times on the Periodic Table. Are there any patterns? Are there any other correlations, e.g. with reactivity as shown by ionization potentials, bond energies to oxygen, etc.?

2 The Electronic Structure and the Properties of Atoms

Introduction

2.1 Background

Although some 200 different sub-atomic particles have been discovered by the physicists, only three, the proton, the neutron, and the electron are of direct interest to the chemist. The masses and charges of these different particles are so minute that it is convenient to define much smaller units than the gramme and the coulomb which are used on the macroscopic scale. The unit of mass used on the atomic scale is approximately the mass of the proton or neutron (which are very nearly equal and about 10^{-24} g). The mass of the electron is very much less than that of the other two particles and is of the order of one two-thousandth a.m.u. The charges on the proton and electron are equal in size though opposite in sign, that on the electron being negative. (There is also a short-lived particle of the same mass as the electron, but with unit positive charge, called the positron or positive electron.) This electronic charge, which equals about $-1\cdot6 \times 10^{-19}$ C, is taken as the unit of charge on the atomic scale and is given the symbol $-e$ (but note that e is also used to represent the electron itself). The neutron has no charge. Thus, the proton has unit mass and unit positive charge, the neutron has unit mass and zero charge while the electron has negligible mass and unit negative charge. The exact values are given in Table 2.1, but this approximation suffices for most purposes.

TABLE 2.1 Properties of the fundamental particles

Particle*		Charge, coulomb	Relative mass
Proton	$^1_1$p	$+e = 1.602\,2 \times 10^{-19}$	$1\cdot007\,29$
Neutron	$^1_0$n	zero	$1.008\,660$
Electron	$^{\,0}_{-1}$e or β^-	$-e = -1.602\,2 \times 10^{-19}$	$0\cdot000\,548\,6$

*The symbols are explained on the next page. The positron has the symbol $^{\,0}_{+1}$e or β^+.

The foundation of the modern theory of atomic structure was laid by the work of Rutherford on the scattering of α-particles by very thin metal targets. He found that, when a beam of α-particles (mass = 4, charge = $+2e$) was directed at a target of thin metal foil, nearly all the particles passed through the target with scarcely any deflection, a few were deflected through large angles, and an even smaller proportion were reversed along their paths. These observations suggested a model of the atom as a small, dense, positively-charged core surrounded by a much larger and more tenuously-occupied region of electrons. The α-particles are so much more massive than the electrons that they would be relatively unaffected when they passed through the outer regions of the atoms and deflected only if they came close to the core. Since most of the α-particles passed straight through the target, which was about a thousand atoms thick, it followed that the cores must be very small. When the α-particle passed close to a core, it was strongly deflected by the charge repulsion, while the occasional particle which happened to be heading straight at the core (with a mass and charge about thirty times its own) was repelled back along its path. The effect is illustrated in a very diagrammatic form in Figure 2.1.

This work has been fully substantiated by later investigations. In the modern view, the atom consists of a tiny, dense, positively-charged nucleus containing protons and neutrons, surrounded by a much larger, more tenuous, cloud of electrons. Since the electron mass is so much smaller than the masses of the other particles, the relative atomic mass is approximately equal to the nuclear mass number, A, which equals the sum of the number of protons, Z, and the number of neutrons $(A - Z)$ or N. Since there are Z protons, the nucleus bears a charge of $+Ze$ and this is balanced by having Z electrons surrounding the nucleus so that the atom is electrically neutral. Z is termed the *atomic number* and is the most important single property of an atom.

The volume of an atom is essentially the space occupied by its electron cloud, the nucleus filling only a minute proportion of the whole (about 10^{-15} of the atomic volume). The

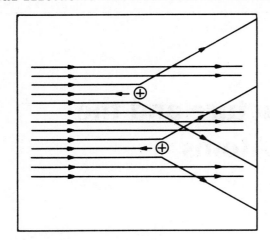

FIGURE 2.1 *Rutherford's experiment*
Diagrammatic representation of the deflections observed when a beam of alpha-particles hits a metal target.

radius of an atom is of the order of 100 pm, while the nuclear radius is about 10^{-3} pm. The radii of molecules extend to about 1000 pm, or even more for very complex or polymeric species. For comparison, the present-day electron microscope can resolve down to about 1000 pm and can thus be used to 'see' molecules of moderate size.

2.2 Isotopes

Chemical behaviour is determined by the interaction between the electron clouds of atoms whose character, in turn, depends on Z, the number of electrons present. That is, the atomic number determines the chemical properties of an element, and all the atoms of a particular element have Z protons in their nuclei and Z electrons. It is found, however, that these Z protons may be accompanied in the nucleus by varying numbers of neutrons so that atoms of the same atomic number may have different nuclear mass, A. For example, chlorine with $Z = 17$ has two forms, one with eighteen neutrons and nuclear mass $A = 35$, and the other with twenty neutrons and nuclear mass 37. Such atoms with the same atomic number Z, but with different nuclear mass A, are termed *isotopes*.* Since the chemical properties of an element depend only on Z, all the isotopes of an element undergo identical chemical reactions although the rates of reaction (and other effects which depend on mass) may show small differences. These mass effects are negligible except for the lightest elements such as hydrogen. However, the enrichment of uranium (for use in nuclear weapons and nuclear reactors) relies on the very small difference in the rates of diffusion between $^{235}UF_6$ and $^{238}UF_6$ molecules in the gas phase. The number of naturally-occurring isotopes of an element varies widely. Some elements such as $^{9}_{4}Be$, $^{31}_{15}P$, or $^{197}_{79}Au$ are found in only one isotopic form while others may form up to ten stable isotopes. For example:

* These nuclear properties are commonly shown by left sub- and super-scripts on the element symbol: $^{A}_{Z}X$. Thus the α-particle, which is the helium nucleus, is written $^{4}_{2}He^{2+}$. (The alternative form, $_{Z}X^{A}$ is also found, especially in American publications.)

Element	Z	A
Cadmium	48	106, 108, 110, 112, 113, 114, 116
Tin	50	112, 114, 115, 116, 117, 118, 119, 120, 122, <u>124</u>
Tellurium	52	120, 122, 123, <u>124</u>, 125, 126, 128, 130
Xenon	54	<u>124</u>, 126, 128, 130, 131, 132, 134, 136

This illustrates that, as well as isotopes with the same Z value and different A values, there also exist atoms with the same mass numbers but differing atomic numbers. Such atoms are called *isobars*. One example is provided by the isotopes of tin, tellurium, and xenon of mass 124 which are underlined in the table. Isobars are of less importance to the chemist than are isotopes.

The nuclear mass, as determined chemically, is the mean weight of the naturally-occurring mixture of isotopes. Thus chlorine, which consists of 75·4 per cent ^{35}Cl and 24·6 per cent ^{37}Cl, has a relative atomic mass of 35·453. Very extensive studies have shown that, where elements are not involved in natural radioactive decay processes, the isotopic composition and relative atomic mass of samples from widely varied sources is nearly always constant. The exceptions are those elements, e.g. copper and lead, with variable isotopic composition in naturally obtained samples, and also lithium, boron, and uranium in which an isotopic separation may result from methods used in commercial preparation of a sample. The relative ratios of $^{13}C:^{12}C$ and $^{18}O:^{16}O$ are widely used in geochemistry where they provide information on the conditions prevailing at the time of formation of the material.

2.3 Radioactive isotopes and tracer studies

Only certain combinations of protons and neutrons form stable isotopes. If the neutron/proton ratio becomes too large or too small, the nucleus is unstable and radioactive, and some nuclear process takes place to restore the balance. For example, the isotope of hydrogen of mass three called tritium, $^{3}_{1}H$, has too many neutrons to be stable. It spontaneously converts a neutron into a proton by emitting a negative electron to form a helium isotope, $^{3}_{2}He$. This type of process is written out as a nuclear equation; in this case

$$^{3}_{1}H \rightarrow \, ^{0}_{-1}e + \, ^{3}_{2}He$$

(In nuclear equations, both the charge on the nucleus and the mass are shown and each of these quantities must balance.)

The converse case is illustrated by magnesium-23 which has excess protons and converts a proton to a neutron by the emission of the positive electron (positron) to form a sodium isotope:

$$^{23}_{12}Mg \rightarrow \, ^{0}_{+1}e + \, ^{23}_{11}Na$$

Nuclei are not stabilized only by electron emission: other ways include the emission of an alpha-particle or the capture of an orbital electron (K capture). Sometimes a nucleus is so far from a stable ratio of protons to neutrons that the daughter nucleus produced by the first radioactive decay step is itself unstable and rearranges, and this process continues until a stable nucleus results.

All the atoms in a sample of an unstable material do not undergo a nuclear reaction simultaneously and instantaneously. It is found that a particular radioactive isotope is characterized by a half-life, given the symbol $t_{\frac{1}{2}}$, which is the time in which half the atoms present in a given sample have undergone transformation. Half-lives range from minute fractions of a second to millions of years. The half-life is a constant, characteristic of the isotope, and is not varied by any change in the environment of the atom. For tritium, $t_{\frac{1}{2}} = 12\cdot4$ years and for magnesium-23 it is $11\cdot6$ seconds. A very long half-life is found for uranium-238 ($t_{\frac{1}{2}} = 4\cdot5 \times 10^9$ years) and this forms the basis for dating geological materials millions of years old. By comparison, the much shorter half-life of carbon-14 ($t_{\frac{1}{2}} = 5730$ years) is utilized in the most important method for dating historical and prehistorical materials.

Since all the isotopes of an element have identical chemical properties, it is possible to study the course of many reactions by using an element in an isotopic form other than the natural one. To choose a very simple example, it could be shown that an acid ionizes in water by dissolving it in 'heavy water' formed from the isotope of hydrogen of mass two (heavy hydrogen or deuterium, ^2_1H, often given the special symbol D). When the acid is recovered from solution all the ionizable hydrogen atoms in the molecule will have been replaced by deuterium ions which are present in great excess in the solution. For example, acetic acid—CH_3COOH—is recovered from heavy water as CH_3COOD, showing that only the carboxylic hydrogen atom is acidic.

Radioactive atoms may be detected, by means of their characteristic radiations, in extremely small amounts and are therefore valuable for tracer studies such as these. One illustration is provided by a recent study of hypophosphorous acid, H_3PO_2. When this is dissolved in heavy water, only one of the three hydrogen atoms is replaced rapidly by a deuterium atom and this confirms that the other two hydrogen atoms are directly bonded to the phosphorus atom and do not ionize:

$$\begin{matrix} H & OH \\ & P \\ H & O \end{matrix} + D_2O \rightleftharpoons \begin{matrix} H & OD \\ & P \\ H & O \end{matrix} + HDO$$

However, when the radioactive form of hydrogen, tritium, was used to study the exchange, a further, very slow, reaction was discovered in which the hydrogen bonded to the phosphorus atom exchanged with the solvent by means of an isomerization reaction:

$$\begin{matrix} H & O^- \\ & P \\ H & O \end{matrix} \underset{slow}{\longleftrightarrow} H-P \begin{matrix} O^- \\ \\ OH \end{matrix} \xrightarrow[fast]{T_2O} H-P \begin{matrix} O^- \\ \\ OT \end{matrix} \underset{slow}{\longleftrightarrow} \begin{matrix} H & O^- \\ & P \\ T & O \end{matrix} \ etc.$$

Theory of the Electronic Structure of Hydrogen

2.4 Introduction
As the chemical properties of elements depend on the interaction of the electron clouds of their atoms, any fundamental

theory of chemistry must start by examining the electronic structure of atoms. The remainder of this chapter is devoted to this theme while the succeeding ones discuss the ways in which the electron clouds interact in the formation of molecules and ions. The modern theory of atomic structure—based on the work of Heisenberg and Schrödinger—attempts to describe the shape and arrangement of the electron clouds and to calculate the energy of any given configuration of electrons. If this aim could be carried out completely and accurately, the result would be a complete description of chemical phenomena derived purely from theory—and perhaps the end of chemistry as an experimental science! Thus, the course of a reaction or the shape of a molecule could be derived entirely from theory by calculating the energies of all the alternatives and then choosing the most favourable. The present state of theoretical chemistry is, perhaps fortunately, far from this idealized position. It is not possible to carry out calculations using theory alone for any systems other than the simplest. The difficulties are formidable as the energy of a reaction is a relatively small difference between two large quantities. For example, hydrogen atoms combine to form hydrogen molecules in an extremely vigorous and exothermic reaction:

$$2H \rightarrow H_2$$

The theory yields a value for the *total electronic energy* of the atoms or molecules. This is the gain in energy when, in this case, two protons and two electrons are brought from infinite separation to form the two hydrogen atoms on the left of the equation or else the hydrogen molecule on the right. The energy of formation of hydrogen molecules from hydrogen atoms is the difference between these total electronic energies: 2652 kJ mol^{-1} for 2H and 3109 kJ mol^{-1} for H_2, equal to $434\cdot3$ kJ mol^{-1} after allowing for minor effects. It can be seen that, even in this simple case, an error of one per cent in each of the total electronic energies can produce an error of about thirteen per cent in the resultant energy of formation.

In general, the total electronic energies of the reactants and products are much greater, and the energies of reaction are smaller, than in the case of hydrogen. Consider, for example, the question whether carbon and hydrogen combine as:

(a) $C_{(solid)} + H_{2(gas)} \rightarrow CH_{2(gas)}$

or (b) $C_{(solid)} + 2H_{2(gas)} \rightarrow CH_{4(gas)}$

The difference in the heats of the reactions (a) and (b) is under 42 kJ mol^{-1} while the total electronic energy of small, light molecules like CH_2 and CH_4 is of the order of 100 000 kJ mol^{-1}. It follows that if we try to predict the outcome of such reactions from absolute values of the total electronic energy, we must be able to calculate values which are accurate to 1 in 10^5. For larger molecules and those containing heavier elements, the demand is even higher. While calculations have improved rapidly in the last decade, such limits are still unobtainable and absolute calculations cannot yet be used in place of experiments. Fortunately, in comparisons—as between (a) and (b) above—many errors are common to both calculations and modern calculations do provide useful guidance to expected differences and comparisons.

The great value of the theory of the electronic structure of atoms and molecules in its present form lies, not in its use for absolute calculations, but in the correlation and rationalization it provides for a great mass of experimental results. In current theory, much use is made of experimental data (such as bond lengths and angles) to help to simplify and solve the equations and these solutions in turn help to suggest further experimental work. This provision of a wide, stimulating, and flexible theoretical framework and the strong interaction between theory and experiment has been one of the most exciting and vital aspects of chemistry in the last few decades. In the next few sections an attempt has been made to outline the basic steps in this theory.

2.5 The dual nature of the electron

Before examining the electronic structure of the atom, two important properties of the electron itself must be discussed. These are:

(i) the dual nature of the electron which partakes of the properties both of a particle and of a wave, and

(ii) the effect of Heisenberg's Uncertainty Principle when applied to the electron.

The classical picture of the electron is of a tiny particle whose position in space can be accurately defined by its co-ordinates x, y, and z. Its motion in an atom is then described by the variation of (x, y, z) with time. This was the way in which Rutherford and Bohr described the atom on a 'planetary' model with the massive central nucleus and the light electrons moving in 'orbits' around it. However, it was shown in the nineteen-twenties that moving particles should behave in some ways as waves and that this effect should be particularly marked for a particle as light as the electron. Experimental support for this prediction was soon found. It was shown, for example, that a beam of electrons could be diffracted by a suitable grating in exactly the same way as a beam of light. These wave properties were introduced into the theory of atomic structure by Schrödinger in his *wave mechanics* in which the electron in an atom is described by a wave equation.

The Uncertainty Principle places an absolute limit on the accuracy with which the position and motion of a particle may simultaneously be known. A formal statement is that the product of the uncertainty in the position Δx and of the uncertainty in the momentum Δp of a particle cannot be less than the modified Planck's constant, $h/2\pi$. (Planck's constant $h = 6.6261 \times 10^{-34}$ Js.) That is $\Delta x \Delta p \nless h/2\pi$. This limit is so small that the uncertainty is negligible for normal bodies but it is large for a particle as light as the electron. As a result, the idea that the electron has a definite position (for example, of its following a definite orbit) must be replaced by the concept of a probability distribution for the electron. In other words, the answer to the question, 'where is the electron in an atom?', becomes a statistical one.

When both these concepts—the wave nature of the electron and the Uncertainty Principle—were taken into account, it was found that a satisfactory description of an atom resulted from Schrödinger's Wave Equation whose solutions, named *wave functions*, were given the symbol ψ. The value

of the square of the wave function at any given point $P(x, y, z)$—written $\psi^2_{(x, y, z)}$—gave the probability of finding the electron at P. ψ^2 is sometimes termed the probability density of the electron. Put in rather crude terms, the old particle-in-an-orbit picture of classical atomic theory is replaced in wave mechanics by an electron 'smeared out' into a charge cloud, and the function ψ^2 describes how the density of the charge cloud is distributed in space.

2.6 The hydrogen atom

The theory of wave mechanics may now be examined in more detail for the simplest case, the hydrogen atom, where there is only one electron moving in the field of a singly-charged nucleus. Once the hydrogen atom is understood, the results for more complex systems may be derived similarly.

The time independent form of the Schrödinger Wave Equation for the hydrogen atom looks like this:*

$$\frac{-h^2}{8\pi^2 m}\nabla^2\psi - \frac{e^2}{4\pi\epsilon_0 r}\psi = E\psi \qquad (2.1)*$$

In this expression, e and m are the charge and mass, respectively, of the electron, h is Planck's constant, r is the distance from the nucleus which is taken as the origin of coordinates, ψ is the wave function, ϵ_0 is the permittivity of a vacuum and E is the total energy of the system. This equation looks rather formidable but it is just a statement in wave-mechanical terms of the principle of conservation of energy. The del-squared term corresponds to the kinetic energy, the term $-e^2\psi/r$ is the potential energy (the attraction between the charges on the electron $(-e)$ and on the nucleus $(+e)$ at a separation r), while the right hand side of the equation is the total energy.

A number of wave functions which satisfy equation (2.1) may be found. These are written ψ_1, ψ_2, ψ_3, etc. and are called *orbitals* by analogy with the orbits of the old planetary theory. To each solution or orbital there corresponds a certain value—E_1, E_2, E_3, etc.—of the total energy of the system. That orbital, say ψ_1, which corresponds to the lowest value of the total energy describes the electron distribution in the normal, most stable state of the hydrogen atom (called the *ground state*), and this is the orbital where the electron density is concentrated most closely to the nucleus. The expression for the ground state orbital is:

$$\psi_1 = \frac{1}{\sqrt{(\pi a_0^3)}}\exp{(-r/a_0)} \dots\dots\dots\dots (2.2)$$

where $a_0 = 0.529$ Å $= 52.9$ pm is the atomic length unit or Bohr radius. The corresponding energy E_1 is:

$$E_1 = -e^2/8\pi\epsilon_0 a_0 = -1310\,\text{kJ mol}^{-1} \qquad (2.3)$$
$$= -13.60\,\text{eV}$$

The other solutions to the wave equation describe states of hydrogen of higher energies than the ground state (called *excited states*) where the electron is in one of the orbitals concentrated further out from the nucleus. The hydrogen atom will be excited from the ground state to one of these

---

*∇^2 is an abbreviation for $\frac{\partial^2}{\partial x^2} + \frac{\partial^2}{\partial y^2} + \frac{\partial^2}{\partial z^2}$, where (x, y, z) are the space-coordinates of the electron. It is read 'del-squared'.

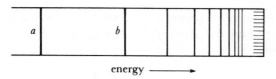

energy ⟶

FIGURE 2.2 *The spectrum of hydrogen*

This diagram shows the Balmer series in the visible spectrum. Transitions to upper states become increasingly close together until the continuum on the right of the diagram is reached. The Lyman, Paschen, Brackett, and Pfund series have similar appearances. The line labelled *a*, for example, results from the transition between E_2 and E_3 while that labelled *b* is the transition between E_2 and E_4.

higher-energy states if it absorbs energy, equal in amount to the difference between E_1 and the value of E of the higher orbital, which will promote the electron to this higher orbital. This energy absorption may be observed in the electronic spectrum of the hydrogen atom and the frequencies of electromagnetic radiation absorbed give experimental values for the various energy states E_1, E_2, E_3, etc., of the atom, which agree very well with the calculated E values. Thus Figure 2.2 shows that part of the hydrogen atom spectrum which involves transitions between E_2 and higher levels E_3, E_4, E_5, etc.

The wave equation (2.1) for hydrogen holds for all other 'hydrogen-like' species, that is for those with only one electron such as He^+ or Li^{2+}. The only modification to equation (2.1) required is to allow for the different nuclear charge Z in the potential energy term. Thus the general form of the wave equation for hydrogen and hydrogen-like atoms is:

$$\frac{-h^2}{8\pi^2 m}\nabla^2\psi - \frac{Ze^2}{4\pi\epsilon_0 r}\psi = E\psi \dots\dots\dots (2.4)$$

and the solutions are exactly the same as for hydrogen except that the value of Z will carry through the working.

Since the value of ψ^2 times unit volume at a given point P (x, y, z) is the probability of finding the electron in that volume at P, it follows that acceptable solutions of the wave equation must have certain properties. Suitable functions must, for example, be single-valued at all points P as there cannot be two or more answers to the question 'what is the probability of finding the electron at P?' Similarly, ψ must be continuous and finite. Further, the total value of ψ^2 summed over all the points in space (i.e. $\int_{-\infty}^{+\infty} \psi^2 \, dx \, dy \, dz$) must equal one, since the probability of finding the electron somewhere in space must be certainty (which is unity by definition). The result of these restrictions is to limit the acceptable solutions of the wave equation to those which can be determined by three quantum numbers n, l and m which may take only those values shown in Table 2.2

TABLE 2.2 The quantum numbers n, l, and m

Quantum number	Allowed values
n	1, 2, 3, 4, 5,....
l	$(n-1)$, $(n-2)$,....2, 1, 0
m	$+l$, $+(l-1)$, $+(l-2)$,....2, 1, 0, -1, -2, ...$-(l-2)$, $-(l-1)$, $-l$

A set of three quantum numbers is required to describe each orbital. Thus the ground state of the hydrogen atom (equation 2.2), has the electron in the orbital where $n = 1$, $l = 0$, $m = 0$. These quantum numbers arise naturally in the course of the mathematics because of the requirement that acceptable solutions are well-behaved functions. This is in contrast to the older theory where the quantum numbers had to be added, apparently arbitrarily, to the classical description. The detail of these calculations is too complicated to be shown here but the interested reader is referred to the sources cited at the end of the book.

When the complete set of allowed solutions of the wave equation is examined, it becomes apparent that the orbitals fall into families. The first type, of which ψ_1 is an example, is of the form $\psi = f(r)$. In other words, the value of the wave function, and hence of its square, the probability of finding the electron, depends only on the distance from the nucleus and is the same in all directions in space. These orbitals are spherical and they correspond to the cases where the quantum number l (and therefore m also) is zero. Orbitals are usually designated by a number equal to the value of n, and a letter corresponding to the value of l as follows:

$$l = 0, 1, 2, 3, 4, 5, \dots\dots$$
$$s \quad p \quad d \quad f \quad g \quad h \dots\dots$$

where the rather odd selection of letters at the beginning arises for historical reasons. Thus these orbitals which are functions of r only and have $l = 0$, are s orbitals and ψ_1 is the 1s orbital. There is an s orbital for each value of n and they increase in energy as n increases (see Table 2.3 and Figure 2.4).

A second type of solution to the wave equation has the form $\psi = f(r) f(x)$. Clearly these orbitals now have directional properties and will have different magnitudes in the $\pm x$ direction than in the rest of space. There are two other exactly similar types of orbitals, $\psi = f(r) f(y)$ and $\psi = f(r) f(z)$ which are concentrated in the y and z directions respectively. The most stable representatives of these orbital types for hydrogen are:

$$\psi_x = k. x \exp(-r/2a_0)$$
$$\psi_y = k. y \exp(-r/2a_0) \quad \dots\dots\dots (2.5)$$
$$\psi_z = k. z \exp(-r/2a_0)$$

where k is a constant compounded of a number of fundamental constants. These three orbitals are equal in energy and for each of them:

$$E = -e^2/32\pi\epsilon_0 a_0 \dots\dots\dots\dots (2.6)$$

This equality of energy for orbitals of this second type, in sets of three, is generally true. The three orbitals are entirely equivalent except for their direction in space.* They have $l = 1$ and are therefore p orbitals. The set of lowest energy in hydrogen, ψ_x, ψ_y and ψ_z, are 2p orbitals.† It follows from Table 2.2 that the p orbitals must occur in sets of three for each value of n as, when $l = 1$, m may take the three values $+1$, 0, -1.

The theory requires only that there should be three independent p orbitals and any set of three may be chosen. It

*Such a set of equal energy levels is termed *degenerate*.
†Since $l = n-1$, there are no p orbitals for $n = 1$.

is usually most convenient to choose the above set, ψ_x, ψ_y, and ψ_z, which coincide with the co-ordinate axes but different sets of three p orbitals may be useful on occasion. For example, a set making different angles with the axes may be chosen and these will be formed from appropriate combinations of ψ_x, ψ_y, and ψ_z. In particular, it should be noted that ψ_x, ψ_y, and ψ_z do not correspond directly to the m values, ± 1 and 0. On the normal convention of axial directions, the z-orbital is that for which $m = 0$ but the two orbitals for which $m = \pm 1$ have to be rearranged to give the two orbitals ψ_x and ψ_y.* There are further sets of higher energy p orbitals for the higher n values. These $3p$, $4p$, $5p$, etc., sets of orbitals again contain three independent orbitals of equal energy.

The next type of solution of the wave equation to be considered is the set of orbitals which are functions of two directions as well as of r, of the type $\psi = f(x)\,f(y)\,f(r)$. Such orbitals have l values equal to two and there are five independent orbitals of equal energy in a set, corresponding to the five m values, ± 2, ± 1, 0. These are d orbitals and, since five values of m can only arise when $n = 3$ or more (Table 2.2), the lowest energy d orbitals are the $3d$ set. The $4d$, $5d$, $6d$, etc., orbitals are of progressively higher energy and each set contains five orbitals of equal energy.

When $l = 3$, the 7 possible m values give rise to the 7 f orbitals and these arise when n is 4 or more. Likewise g, h, orbitals etc. are possible, but the known elements are accounted for without such orbitals being occupied. The primary interest in introductory chemistry lies in the s, p, and d orbitals. Table 2.3 lists all the orbitals with n values up to four.

TABLE 2.3 Atomic orbitals with n values up to four

n	l	m	Symbol	Number of levels with this n value
1	0	0	$1s$	1
2	0	0	$2s$	4
	1	$\pm 1, 0$	$2p$	
3	0	0	$3s$	9
	1	$\pm 1, 0$	$3p$	
	2	$\pm 2, \pm 1, 0$	$3d$	
4	0	0	$4s$	16
	1	$\pm 1, 0$	$4p$	
	2	$\pm 2, \pm 1, 0$	$4d$	
	3	$\pm 3, \pm 2, \pm 1, 0$	$4f$	

It is convenient to change from Cartesian co-ordinates (x, y, z) to polar coordinates (r, θ, ϕ). This does not alter the principles but simplifies the mathematics and makes it possible to separate the wave functions expressed in polar coordinates into a radial part involving only r, the distance from the nucleus, and an angular part $f(\theta, \phi)$ which expresses

*When the direction of a p orbital is to be distinguished a subscript is added to the symbol. For example, the orbitals of equation (2.5) are respectively, p_x, p_y and p_z.

the directional properties of the orbitals. Conventionally, θ is the angle from the z axis and ϕ is the angle from the x axis. The relations are

$$x = r \sin\theta \cos\phi$$
$$y = r \sin\theta \sin\phi$$
$$z = r \cos\theta$$
$$r^2 = x^2 + y^2 + z^2$$

Equations (2.5) then have the form

$$\psi_z = k'.r(\exp - r/2a_0).\cos\theta$$
$$\psi_x = k'.r(\exp - r/2a_0).\sin\theta\cos\phi \quad (2.7)$$
$$\psi_y = k'.r(\exp - r/2a_0).\sin\theta\sin\phi$$

The k' values are products of fundamental constants. If the nuclear charge is Z, but only one electron is present, that is, if the atom is hydrogen-like, then Z enter into the constants, and the exponential term becomes $\exp(-Zr/2a_0)$.

In hydrogen and hydrogen-like atoms, the energy of the electron depends only on the n value of the orbital in which it is found. That is, the order of energies is:

$$1s < 2s = 2p < 3s = 3p = 3d < 4s = 4p = 4d = 4f$$
$$< 5s = 5p = 5d = 5f < \ldots\ldots\ldots$$

The electron in the ground state is in the $1s$ orbital and is found in one of the higher energy orbitals only if the atom has been excited by the absorption of energy. The lowest excited state is where the electron is in an orbital of n value = 2, and this corresponds to the absorption of energy equal to $(e^2/8\pi\epsilon_0 a_0 - e^2/32\pi\epsilon_0 a_0)$, (see equations 2.3 and 2.6) which equals $10 \cdot 20$ eV ($984 \cdot 3$ kJ mol^{-1}) or absorption of radiation of wave number 82 273 cm^{-1}. The energies required for excitation to higher states can be calculated similarly.

These excitations can also be observed experimentally in the spectrum of hydrogen and the first major test of the theory to be applied was to see if the experimental and theoretical energies matched. The electronic spectrum of hydrogen was observed many years before the development of Schrödinger's theory, or before Bohr's earlier theory, and both theories do give energy levels which agree with the observed ones. Schrödinger's theory also gives close agreement for many-electron atoms where Bohr's theory is less successful.

If electromagnetic radiation consisting of a continuous range of frequencies—'white' radiation—is shone on to an absorbing system, then those frequencies which correspond to the difference, ΔE, in energy between two states of the system will be absorbed. (The relation between energy and frequency, ν, is $E = h\nu$.) Thus the absorption spectrum shows lines at frequencies which correspond to energy level differences in the irradiated species. Lines at the same frequencies may be *emitted* when the species return from the higher to the lower energy level and it is easier, for atoms, to observe this *emission spectrum*. When the hydrogen spectrum is examined several series of lines like that in Figure 2.2 are observed. Each series is named after its discoverer. They fall in the ultraviolet, visible, or infrared regions and each is similar in

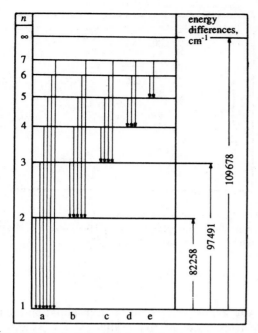

FIGURE 2.3 *Energy level diagram for hydrogen*
Energy level diagram showing the transitions corresponding to the various series of lines in the electronic spectrum of hydrogen. The transitions are lettered corresponding to the different series of lines in the hydrogen spectrum: (a) Lyman, (b) Balmer, (c) Paschen, (d) Brackett, (e) Pfund. Values shown are experimentally determined: compare the values calculated from the formula given in Table 2.4.

TABLE 2.4 The electronic spectrum of hydrogen

Series	m	m'	$\tilde{v}$	Corresponding orbital description
Lyman	1	2	82 303	$n = 1 \rightarrow n = 2$
		3	97 544	$n = 1 \rightarrow n = 3$
		4	102 879	$n = 1 \rightarrow n = 4$
		∞	109 737	Total electronic energy when electron is in the $1s$ orbital: dissociation energy for the ground state
Balmer	2	3	15 241	$n = 2 \rightarrow n = 3$
		4	20 576	$n = 2 \rightarrow n = 4$
		∞	27 434	Total electronic energy when electron is in $2s$ or $2p$ orbitals
Paschen	3	4	5 334	$n = 3 \rightarrow n = 4$
		5	7 803	$n = 3 \rightarrow n = 5$
		∞	12 193	Total electronic energy when electron is in $3s$, $3p$ or $3d$ orbitals
Brackett	4	5	2 469	$n = 4 \rightarrow n = 5$
		∞	6 859	Total electronic energy when electron is in any orbital with $n = 4$
Pfund	5	6	1 341	$n = 5 \rightarrow n = 6$
		∞	4 390	Total electronic energy of H atom when electron is in any orbital with $n = 5$

showing lines which become progressively less intense and closer together towards higher frequencies. They all show an upper frequency limit for line absorption, above which continuum is observed. In each series the wave numbers $\tilde{v}$ of the lines fit the general formula:

$$\tilde{v} = R(1/m^2 - 1/m'^2).$$

In this formula, m has an integral value which is different for each series of lines while $m' = (m+1), (m+2), (m+3) \ldots \ldots \infty$. R is a constant, the Rydberg Constant, which equals 109 740 cm^{-1} approximately. The values of m and examples of the transitions for the various series are shown in Table 2.4. These various series may be portrayed on an energy level diagram, as in Figure 2.3, where the zero value for the energy is taken as the start of the continuum corresponding to $m' = \infty$. The correlation between this diagram and the electronic levels of the atom, as derived from the wave equation, is obvious. The states whose line spectra correspond to $m = 1, 2, 3, \ldots$ are those whose n quantum number takes the values, 1, 2, 3, ... respectively. The transition in the Lyman series corresponding to

$$\tilde{v} = R(1/1^2 - 1/2^2)$$

(i.e. with $m = 1$, $m' = 2$) is the transition from the $1s$ level to the $2s$ (or $2p$) level. The Lyman series corresponds to transitions from the ground state, $1s$, to levels with higher n values; while the Balmer, Paschen, etc., series, for which $m = 2, 3$, etc., respectively, correspond to transitions when electrons which were already excited are further excited to higher

levels. Thus the frequencies of the Lyman series correspond to the energy gaps between the $1s$ level and the higher levels and, in particular, the start of the continuum corresponding to $m' = \infty$ corresponds to the complete removal of the $1s$ electron. Similar transitions from the $2s$ level are observed in the Balmer series, from the $3s$ level in the Paschen series, and so forth. As an illustration of the correlation between the calculated and observed energy differences, it is worth calculating the transitions in the Lyman series for $m' = 2$ (equal to the $1s-2s$ energy gap) and $m' = \infty$, (equal to the energy of the $1s$ orbital with respect to complete dissociation). For $1s$ to $2s$ the value is $\tilde{v} = R(1/1^2 - 1/2^2) = 3/4 \times 109\,740$ cm^{-1} $= 82\,303$ cm^{-1} (10·20 eV). The dissociation transition is $\tilde{v} = R(1/1^2 - 1/\infty^2) = R = 109\,737$ cm^{-1} (13·61 eV). The values calculated from the wave equation agree exactly.

The Bohr theory gives an equally good correlation with experiment for the hydrogen atom but wave mechanics is to be preferred for dealing with many-electron atoms.

The Lyman series occurs in the ultra-violet, the Balmer series in the visible, the Paschen series in the near infra-red, and the Brackett and Pfund series in the far infra-red regions of the spectrum.

Many Electron Atoms

2.7 The approach to the wave equation

When the wave equation for helium with two electrons and $Z = 2$ is considered, it is found to be more complicated than the wave equation for the hydrogen atom both in the kinetic energy term and in the potential energy term. As both electrons contribute to the kinetic energy there are now two del-squared terms, one applying to each electron. The potential energy term consists of three parts in place of the single term in equation (2.1). In helium there are two attractions—between each electron and the nucleus—and there is also a repulsion term between the two electrons.

The wave equation for helium is:

$$k(\nabla_1^2 + \nabla_2^2)\psi - \left(\frac{Ze^2}{4\pi\epsilon_0 r_1} + \frac{Ze^2}{4\pi\epsilon_0 r_2} - \frac{e^2}{4\pi\epsilon_0 r_{12}}\right)\psi = E\psi \ldots . (2.8)$$

where k is a product of universal constants, r_1 and r_2 are the distances of each electron from the nucleus and r_{12} is the interelectronic distance. This increase in complexity adds considerably to the difficulty of the calculation. Particular difficulty arises when dealing with the repulsion between the two electrons. These problems become even more complex as the analysis is extended to atoms with larger numbers of electrons and it is at present impossible to solve the wave equation of a many-electron atom directly. Methods of approximation have to be sought and many of these methods attempt to get round the problem of the interelectron terms by considering only one electron at a time. In other words, the many-electron atom is treated as a series of problems based on hydrogen-like situations and this, in turn, means that many of the results of the last section carry over for many-electron atoms with only minor modifications.

One method of approximation, for example, starts with the wave equation for only one of the electrons moving in a potential field determined by the nuclear charge and the averaged-out field of all the other electrons. This case is then like that of the hydrogen atom with a modified value for the nuclear charge and may be solved to give a 'first approximation' distribution function for this electron. Each electron is considered in turn and a set of first approximation distribution functions is obtained. These functions are then used to refine the value of the potential field of the nucleus plus electrons and then the analysis is repeated giving 'second approximation' functions. The whole process is continued until self-consistent results are obtained and a solution to the many-electron problem is found using the method for the hydrogen-like atom at each step in the calculation. The above method is called the *self-consistent field method*. Other, more accurate, approaches are available, some of which start off from the self-consistent field answers, but these cannot be followed out here. An introduction is given in Coulson's book listed in the references.

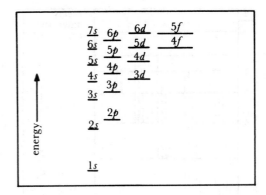

FIGURE 2.4   *Energy levels in many-electron atoms*
This diagram shows the relative energy levels of the orbitals at the Z values where they are about to be filled.

Since this and other systems of approximation use hydrogen-like fields, most of the results for the exactly-solved hydrogen atom apply in a qualitatively similar form. The electron distributions in many-electron atoms are described by wave functions ψ similar to those for hydrogen, and the square of the wave function again describes the probability distribution of the electron in one of these orbitals. These atomic orbitals are defined by the three quantum numbers n, l and m which are restricted to certain integral values by the same rules as for hydrogen, as in Table 2.2. One difference between many-electron atoms and the hydrogen atom is that the energy of an electron in an orbital depends on l as well as on n: that is, the order of energies now includes $s < p < d < f$ well as $n = 1 < 2 < 3 \ldots$. As a result, when there are d and f levels corresponding to an n value, these overlap in energy with the s and p levels of higher n values. The order of energies for a many-electron atom is approximately $1s < 2s < 2p < 3s < 3p < 4s = 3d < 4p < 5s = 4d < 5p < 6s = 5d = 4f < 6p < 7s = 6d = 5f$. This is illustrated in Figure 2.4. Two points are of importance in this figure: first, that for higher n values the ns level is approximately equal in energy to the $(n-1)d$ level and to the $(n-2)f$ level, and second, that the energy gaps between successive levels with the same l value become smaller as n values increase. This has important effects on the chemistry of the heavier elements.

Although the energy of an orbital in a many-electron atom depends on both the n and l values, there is no dependence on the value of m in the free atom, and the three p orbitals or the five d orbitals of a given n value are equal in energy as they are in the hydrogen atom. The total electronic energy of an atom is the sum of the energies of each electron in its orbital, with a correction for the interaction between the electrons. The spectra of many-electron atoms are much more complicated than that of hydrogen, as there are more energy levels and some overlap. However, most atomic spectra have now been successfully analysed and these give experimental values of the energy levels which agree with the calculated ones.

2.8 The electronic structures of atoms

The electronic structure of an atom may be built up by placing the electrons in the atomic orbitals. Clearly, the orbitals of lowest energy will be the first to be filled and the questions which have to be answered in order to carry out the building-up, or *aufbau*, process are:

(i) how many electrons in each orbital?

(ii) what happens when there are a number of orbitals of equal energy, such as the three $2p$ orbitals?

The answer to (i) depends on one other property of the electron, its *spin*. This property was discovered during a study of the spectra of the alkali metals, when it was found that the absorption lines had a fine structure which could be explained only if the electron was regarded as spinning on its axis and able to take up one of two orientations with respect to a given direction. This spin is included in the description of the state of an electron by introducing a fourth quantum number, m_s, which may take one of the two values $\pm\frac{1}{2}$. The spin is usually indicated in diagrams by using arrows, $\uparrow$ or $\downarrow$. The distribution of electrons in an atom is then determined by Pauli's Principle that no two electrons may have all four quantum numbers the same. Thus the answer to (i) is that each orbital, which is defined by the three quantum numbers n, l and m, can hold only two electrons and these will have $m_s = +\frac{1}{2}$ and $-\frac{1}{2}$. Such a pair of electrons of opposite spin in one orbital is termed 'spin-paired'.

The answer to (ii) is given by Hund's Rules which may be stated:

(a) electrons tend to avoid as far as possible being in the same orbital,

(b) electrons in different orbitals of the same energy have parallel spins.

The electronic structure of any atom may be worked out taking account of these factors as follows:

1. Orbitals are filled in order of increasing energy. This may be remembered from the diagram shown in Figure 2.5 where the orbitals corresponding to a given value of n are written out in horizontal rows and then filled in the order in which they are cut by the series of diagonal lines as shown. This device gives an order which is substantially correct, though there are a few slight anomalies for heavier atoms. In particular, one electron enters the $5d$ level before the $4f$ level is filled and some uncertainty exists about the distribution of electrons between the $6d$ and $5f$ levels in the heaviest atoms.

2. Each orbital (which is defined by specific values of n, l and m) may hold only two electrons with $m_s = \pm\frac{1}{2}$. In other words, each electron is described by the four quantum numbers, n, l, m and m_s, which may not all have the same values.

3. Where a number of orbitals of equal energy is available, the electrons fill each singly, keeping their spins parallel, before spin-pairing starts.

The use of these rules to build up the electronic structures of the elements may be illustrated by a few examples, but first two useful notations for describing these structures must be defined. A convenient way of showing the electronic structures on a diagram is to place cells on each level of the

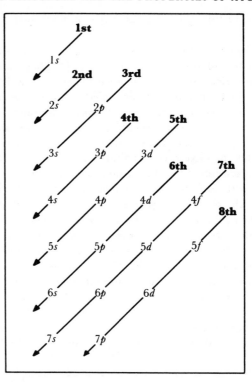

FIGURE 2.5 *The order of filling energy levels in an atom*
The levels are written out in a square array and filled in the order in which they are cut by a series of diagonals as shown.

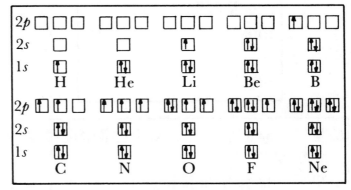

FIGURE 2.6 *The electronic structures of the lighter elements*
The electronic structures of the elements where the $1s$, $2s$, and $2p$ levels are being filled, illustrating the *aufbau* process.

appropriate part of Figure 2.4, corresponding to the number of orbitals in each level. The electrons are indicated by arrows and a cell with a pair of arrows shows a filled orbital. The diagrams for a number of the lighter elements are shown in Figure 2.6. A second notation is to write out the orbitals in order of increasing energy and to indicate the number of electrons by superscripts. To save this latter notation becoming too clumsy, it is convenient to show the configuration of inner, completed levels by the symbol of the appropriate rare gas element. Thus, helium is written, He $= (1s)^2$, lithium is Li $= (1s)^2(2s)^1$ or Li $=$ [He]$(2s)^1$ and sodium which is, in full, $(1s)^2(2s)^2(2p)^6(3s)^1$ is shortened to Na = $[Ne](3s)^1$.

The electron in the hydrogen atom is described by the four values of the quantum numbers $(1, 0, 0, \frac{1}{2})$. Then, in helium, the second electron enters the lowest energy orbital (rule 1) which is the $1s$ orbital singly-occupied in the hydrogen atom. This second electron must have its spin antiparallel with the first (rule 2) and corresponds to the quantum numbers $(1, 0, 0, -\frac{1}{2})$. In the next element, lithium, the third electron must enter the next lowest orbital, $2s$, as in Figure 2.6.

The operation of the third rule is illustrated by the case of carbon with $Z = 6$. Starting from the three electrons in lithium, the fourth completes the $2s$ orbital and the fifth enters the next lowest level, $2p$. The sixth electron then enters one of the two remaining empty $2p$ orbitals with its spin parallel to that of the previous electron. The third $2p$ level is occupied in the next element, nitrogen, and not until there is a fourth electron to be accommodated in the $2p$ level—at oxygen—does spin-pairing occur in that level. The orbitals with $n = 2$ are completely filled at neon, $Z = 10$. Sets of orbitals with the same value of the n quantum number are known as quantum *shells* and the electrons in that shell which

(a)

(b)

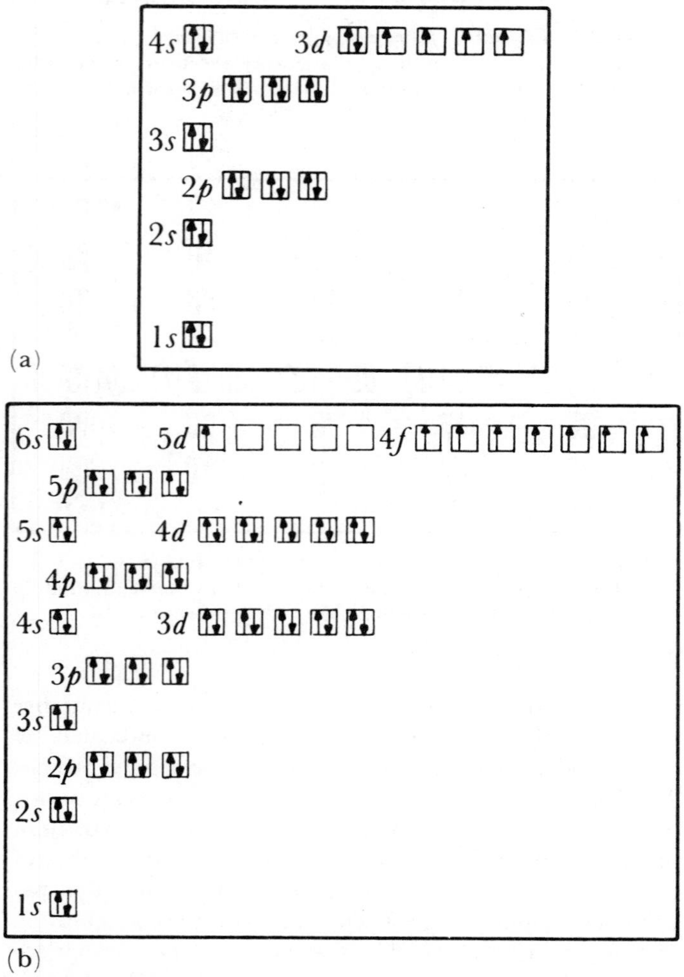

FIGURE 2.7   (a) *The electronic structure of iron*
(b) *The electronic structure of gadolinium*

is partly filled in a particular atom are termed the *valency electrons*.

The rules for deriving atomic structures are further illustrated below for some of the heavier elements.

Consider first iron, Fe with $Z = 26$, whose electronic structure is shown in Figure 2.7. The first ten electrons fill up to the neon structure while the next eight fill the $3s$ and $3p$ levels, repeating the pattern of the second shell, to form the argon core. This accounts for eighteen electrons. Figure 2.4 shows that the level which comes next in energy to the $3p$ level is $4s$ which is a little more stable than $3d$. The nineteenth and twentieth electrons fill the $4s$ level, leaving six electrons to be accommodated in the five $3d$ orbitals. The first five electrons enter these orbitals singly with all their spins parallel while the last electron pairs up with one of these. Thus the configuration is Fe = $[Ar](3d)^6(4s)^2$ with four unpaired spins. Note that $3d$ is now more stable than $4s$.

As a second example, take gadolinium, Gd with $Z = 64$, see Figure 2.7. Continuing from the configuration of iron, the next ten electrons fill the $3d$ and $4p$ levels to give the configuration of krypton, $Z = 36$. The next eighteen repeat this pattern in the $5s$, $4d$ and $5p$ levels giving the xenon ($Z = 54$) configuration. The next level in energy is $6s$ which is filled by the next two electrons and then come the $5d$ and $4f$ levels which are very close in energy, being almost identical in the range of Z values around $Z = 60$. In the event, the first electron enters the $5d$ level and the last seven electrons in gadolinium enter the $4f$ orbitals. As there are seven orbitals in an f level, each one is singly occupied and gadolinium has the configuration Gd = $[Xe](4f)^7(5d)^1(6s)^2$ with eight unpaired electrons (the single d electron and the seven parallel f electrons). As the $5d$ and $4f$ orbitals are so similar in energy, the elements in this region of the Periodic Table vary in the distribution of their electrons between the two. There is never more than one electron in the $5d$ level, but often there is none at all. Thus, europium with $Z = 63$, which precedes gadolinium, has the configuration Eu = $[Xe](4f)^7(5d)^0(6s)^2$.

The rest of the elements up to $Z = 86$ complete the $4f$, $5d$, and $6p$ levels to give the configuration of the heaviest rare gas, radon. The next two electrons fill the $7s$ level and then there is a close correspondence between the energies of the $5f$ and $6d$ levels, similar to that between the $4f$ and $5d$ levels. Here the energy gap is even smaller and there has been considerable difficulty and confusion about the levels being occupied in the heaviest elements. The present conclusion is that these elements are best regarded as paralleling the $4f$ ones in filling mainly the $5f$ level, but the first members of the set make more use of the d level than their lighter congeners. The full electronic structure of all the elements is given in Table 2.5.

Shapes of Atomic Orbitals

In the last section, the electronic structures of the elements were built up from a knowledge of the energy levels of the orbitals derived from the wave equation. Since the chemist is primarily concerned with the outermost electrons and these are found in s, p, or d orbitals, the shape and extension in

TABLE 2.5 The electronic configurations of the elements

Element	Symbol	Z	A_r	Inner shells	Electron configuration Valency shell		
						1s	
Hydrogen	H	1	1·007 94			1	
Helium	He	2	4·002 602			2	
						2s	2p
Lithium	Li	3	6·941	He		1	
Beryllium	Be	4	9·012 182	He		2	
Boron	B	5	10·811	He		2	1
Carbon	C	6	12·0107	He		2	2
Nitrogen	N	7	14·006 74	He		2	3
Oxygen	O	8	15·999 4	He		2	4
Fluorine	F	9	18·998 403 2	He		2	5
Neon	Ne	10	20·179 7	He		2	6
						3s	3p
Sodium	Na	11	22·989 768	Ne		1	
Magnesium	Mg	12	24·305 0	Ne		2	
Aluminium	Al	13	26·981 539	Ne		2	1
Silicon	Si	14	28·085 5	Ne		2	2
Phosphorus	P	15	30·973 762	Ne		2	3
Sulfur	S	16	32·066	Ne		2	4
Chlorine	Cl	17	35·452 7	Ne		2	5
Argon	Ar	18	39·948	Ne		2	6
					3d	4s	4p
Potassium	K	19	39·098 3	Ar		1	
Calcium	Ca	20	40·078	Ar		2	
Scandium	Sc	21	44·955 910	Ar	1	2	
Titanium	Ti	22	47·867	Ar	2	2	
Vanadium	V	23	50·941 5	Ar	3	2	
Chromium	Cr	24	51·996 1	Ar	5	1	
Manganese	Mn	25	54·938 05	Ar	5	2	
Iron	Fe	26	55·845	Ar	6	2	
Cobalt	Co	27	58·933 20	Ar	7	2	
Nickel	Ni	28	58·6934	Ar	8	2	
Copper	Cu	29	63·546	Ar	10	1	
Zinc	Zn	30	65·39	Ar	10	2	
Gallium	Ga	31	69·723	Ar	10	2	1
Germanium	Ge	32	72·61	Ar	10	2	2
Arsenic	As	33	74·921 59	Ar	10	2	3
Selenium	Se	34	78·96	Ar	10	2	4
Bromine	Br	35	79·904	Ar	10	2	5
Krypton	Kr	36	83·80	Ar	10	2	6
					4d	5s	5p
Rubidium	Rb	37	85·467 8	Kr		1	
Strontium	Sr	38	87·62	Kr		2	
Yttrium	Y	39	88·905 85	Kr	1	2	
Zirconium	Zr	40	91·224	Kr	2	2	
Niobium	Nb	41	92·906 38	Kr	4	1	
Molybdenum	Mo	42	95·94	Kr	5	1	
Technetium	Tc	43	98·906	Kr	6	1	
Ruthenium	Ru	44	101·07	Kr	7	1	
Rhodium	Rh	45	102·905 50	Kr	8	1	

(Contd.)

TABLE 2.5 (Contd.)

Element	Symbol	Z	A_r	Inner shells	Electron configuration — Valency shell		
					4d	5s	5p
Palladium	Pd	46	106·42	Kr	10	0	
Silver	Ag	47	107·868 2	Kr	10	1	
Cadmium	Cd	48	112·411	Kr	10	2	
Indium	In	49	114·82	Kr	10	2	1
Tin	Sn	50	118·710	Kr	10	2	2
Antimony	Sb	51	121.76	Kr	10	2	3
Tellurium	Te	52	127·60	Kr	10	2	4
Iodine	I	53	126·904 47	Kr	10	2	5
Xenon	Xe	54	131·29	Kr	10	2	6
					4f	5d	6s	6p
Cesium	Cs	55	132·905 43	Xe			1	
Barium	Ba	56	137·327	Xe			2	
Lanthanum	La	57	138·905 5	Xe		1	2	
Cerium	Ce	58	140·116	Xe	2		2	
Praseodymium	Pr	59	140·907 65	Xe	3		2	
Neodymium	Nd	60	144·24	Xe	4		2	
Promethium	Pm	61	146·92	Xe	5		2	
Samarium	Sm	62	150·36	Xe	6		2	
Europium	Eu	63	151·964	Xe	7		2	
Gadolinium	Gd	64	157·25	Xe	7	1	2	
Terbium	Tb	65	158·925 34	Xe	9		2	
Dysprosium	Dy	66	162·50	Xe	10		2	
Holmium	Ho	67	164·930 32	Xe	11		2	
Erbium	Er	68	167·26	Xe	12		2	
Thulium	Tm	69	168·934 21	Xe	13		2	
Ytterbium	Yb	70	173·04	Xe	14		2	
Lutetium	Lu	71	174·967	Xe	14	1	2	
Hafnium	Hf	72	178·49	Xe	14	2	2	
Tantalum	Ta	73	180·947 9	Xe	14	3	2	
Tungsten	W	74	183·84	Xe	14	4	2	
Rhenium	Re	75	186·207	Xe	14	5	2	
Osmium	Os	76	190·23	Xe	14	6	2	
Iridium	Ir	77	192·22	Xe	14	7	2	
Platinum	Pt	78	195·078	Xe	14	9	1	
Gold	Au	79	196·966 54	Xe	14	10	1	
Mercury	Hg	80	200·59	Xe	14	10	2	
Thallium	Tl	81	204·383 3	Xe	14	10	2	1
Lead	Pb	82	207·2	Xe	14	10	2	2
Bismuth	Bi	83	208·980 37	Xe	14	10	2	3
Polonium	Po	84	209·98	Xe	14	10	2	4
Astatine	At	85	209·987 1	Xe	14	10	2	5
Radon	Rn	86	222·017 6	Xe	14	10	2	6
					5f	6d	7s	
Francium	Fr	87	223·019 7	Rn			1	
Radium	Ra	88	226·025 4*	Rn			2	
Actinium	Ac	89	227·027 8	Rn		1	2	
Thorium	Th	90	232·038 1*	Rn		2	2	
Protactinium	Pa	91	231·035 88*	Rn	2	1	2	
Uranium	U	92	238·050 8	Rn	3	1	2	

(Contd.)

TABLE 2.5 (Contd.)

Element	Symbol	Z	A_r	Inner shells	Electron configuration Valency shell			
					5f	6d	7s	7p
Neptunium	Np	93	237·048 2*	Rn	5	:	2	
Plutonium	Pu	94	239·05*	Rn	6		2	
Americium	Am	95	241·06*	Rn	7		2	
Curium	Cm	96	(247)	Rn	7	1	2	
Berkelium	Bk	97	(247)	Rn	8	1	2	
Californium	Cf	98	(242)	Rn	10		2	
Einsteinium	Es	99	(252)	Rn	11		2	
Fermium	Fm	100	(257)	Rn	12		2	
Mendelevium	Md	101	(256)	Rn	13		2	
Nobelium	No	102	(259)	Rn	14		2	
Lawrencium	Lr	103	(260)	Rn	14	1?	2	
Dubnium	Db	104	(261)	Rn	14	1?	2	1?
Joliotium	Jl	105	(262)	Rn	14	3?	2?	
Rutherfordium	Rf	106	(260)					
Bohrium	Bh	107	(262)					
Hahnium	Hn	108	(265)					
Meitnerium	Mt	109	(266)					
(Ununnilium	Uun	110)						
(Unununium	Uuu	111)						
(Ununbium	Uub	112)						

(a) Values for relative atomic masses are based on 1993 revision. Recent studies of isotopic composition have shown variations from different sources (H, Li, B, C, O, Si, S, Ar, Cu, Pb) or variations introduced by commercial isolation (Li, B, U) which limits the accuracy of the atomic mass or requires that the isotopic composition of a particular sample should be determined. Elements involved in radioactive decay processes (Ar, Sr, Pb, Ra) may have different isotopic compositions in different geological specimens. The other changes are mainly increases in precision.
(b) Weights marked with an asterisk are those of the commonest long-lived isotope of a radioactive element: those in brackets indicate the most accessible isotope of the heavier elements.
(c) Relative atomic masses based on $^{12}C = 12.000\ 0$: in SI, $^{12}C = 12.000\ 0 \times 1.660\ 57 \times 10^{-27}$ kg.
(d) The names listed for elements from 102 onwards are those recommended by IUPAC in 1995 but they are not finally settled (see section 16.12 for fuller discussion.
(e) Elements 110, 111 and 112 are reported but have not so far been fully established.

space of these orbitals is of primary importance. Indeed, the general aim of understanding and predicting the shape and reactions of ions and molecules may be carried quite a long way by qualitative reasoning which is based largely on diagrams of the relevant atomic orbitals, as the next chapters show.

2.9 The s orbital
The solution of the wave equation for the 1s orbital of hydrogen is the wave function given in equation (2.2):

$$\psi_1 = \sqrt{(1/\pi a_0^3)} \exp(-r/a_0)$$

where r is the distance from the nucleus. The probability density of the electron distribution is the square of this wave function:

$$\psi_1^2 = 1/\pi a_0^3 \exp(-2r/a_0)$$

As these expressions are functions of r, the distance from the nucleus, only, they are spherically symmetrical around the nucleus. As they are exponential functions, it follows that the values of the wave function and of the electron density fall off smoothly and rapidly with increasing distance from the nucleus although they never quite reach zero (see Figure 2.8a). The contour diagram which results from joining points in space with the same value of ψ or ψ^2 has the general appearance of Figure 2.8b (where the values decrease outwards from the nucleus). As exponential functions never quite fall to zero, there is always a finite electron density outside any given one of these contour lines but it is possible to include the major part of the electron cloud within a boundary surface close to the nucleus, and it is common to represent the orbital by a single boundary contour (as in Figure 2.8c) enclosing an arbitrary fraction—say 90 per cent—of the electron density. Such a boundary diagram may be used to represent the electron density, ψ^2, or it may be the corresponding diagram of the orbital, ψ. Since both functions are exponential ones, the boundary diagrams for ψ and ψ^2 are similar in appearance, that for ψ being distinguished by showing the variation in the sign of the wave function in different regions of space (see Figures 2.10 and 2.11 for examples, the 1s orbital has the same sign throughout). As discussed in the next chapter, a bond is formed by the com-

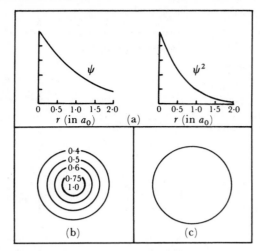

FIGURE 2.8   *The 1s orbital of hydrogen: (a) Plots of ψ_{1s} and ψ_{1s}^2 against r; (b) Contour representation of the 1s orbital (c) Boundary contour representation*

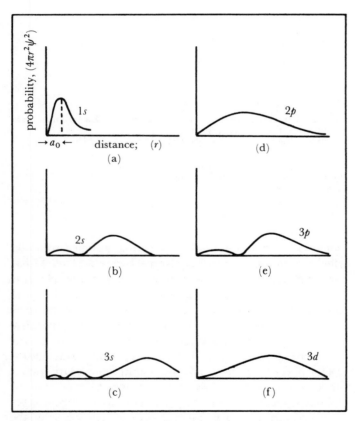

FIGURE 2.9   *Radial density plots of hydrogen orbitals: (a) 1s, (b) 2s, (c) 3s, (d) 2p, (e) 3p, (f) 3d*

bination of atomic orbitals, and the way in which they may combine depends on the signs of the various wave functions. For this reason, the chemist usually works initially with boundary diagrams of the wave functions, ψ. However, we must always remember that we are interested in the electron distribution around the nuclei—that is, having decided how

the wave functions, ψ, combine, we square the resultant to get ψ^2, the probability density of the electron distribution.

A further important function used in representing an orbital is the *radial density function*, $4\pi r^2 \psi^2(r)$. The volume of a spherical shell of thickness dr at a distance r from the nucleus is $4\pi r^2 dr$. Hence, $4\pi r^2 \psi^2(r) dr$ is the probability of the electron's being found at a distance between r and $(r+dr)$ from the nucleus. The plot of the variation of this function with distance from the nucleus for the hydrogen 1s orbital is shown in Figure 2.9a. The radial density in hydrogen is at a maximum at that distance, a_0, from the nucleus that Bohr calculated as the radius of the most stable orbit in the planetary theory. Note that the radial density function is the product of an r^2 function (increasing from zero at the nucleus) and the exponentially decreasing wave function, hence it starts at zero at the nucleus and passes through a maximum.

The s orbitals of higher n value resemble the 1s orbital, being spherically symmetrical and having the same sign for the wave function in all directions. They extend further into space and there are changes close in towards the nucleus where spherical nodes appear across which the wave function changes sign. As only the outer regions are of chemical interest, these nodes are rarely important. They appear as minima in the radial density plot; compare for example, the hydrogen 2s orbital in Figure 2.9b. The distance from the nucleus of the maximum probability increases rapidly to about $5.3a_0$ for 2s and nearly $14a_0$ for 3s so that these orbitals are much more diffuse (recall that the total area under the curve equals a probability of 1 for each orbital).

2.10 The p orbitals

From Table 2.3, the lowest-energy p orbitals are those where $n = 2$, given by equations (2.5) or (2.7) for hydrogen. These orbitals do have directional properties, in contrast to the s orbitals which are alike in every direction from the nucleus. Equations (2.7) allow us to see this directional property most readily, as the total wave function may be separated into two parts, one a function only of r (the radial part), and one a function of θ and ϕ (the angular part). The radial part falls off exponentially from the nucleus, just as for the s orbitals, and this mainly determines the energy of the electron in the p orbital. For our purposes, the more important component is the angular part. For p_z this is simply $\cos\theta$, the angle with the z axis, while p_x varies as $\sin\theta\cos\phi$ and p_y as $\sin\theta\sin\phi$, both involving the angle to the x axis as well. In Figure 2.10a are shown the boundary contour representations of these angular

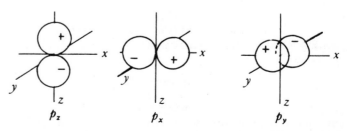

FIGURE 2.10a   *Boundary contour representations of the angular parts of the 2p orbitals of hydrogen.*

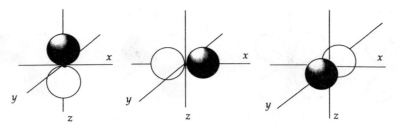

FIGURE 2.10b *Boundary contour representation of the 2p orbitals of hydrogen, using the convention in which the shaded and unshaded lobes represent opposite signs of the wave function.*

parts of the $2p$ orbitals. Only two regions of space are occupied, aligned along the plus and minus directions of one axis. The function changes sign across the nucleus, and there is a nodal plane through the nucleus where the value is zero. The three equations of (2.7) are chosen so that the three orbitals are identical apart from their orientation—along the z, x, and y axes respectively.

Several different conventions are in use to display atomic orbitals, and the use of shading to indicate the $+$ and $-$ parts of the wave function is becoming common. An alternative version of the p orbitals is shown in Figure 2.10b, using this shading convention.

Multiplying by the radial part of the wave function does not change the fundamental property that the p wave functions describe two lobes of opposite sign separated by a nodal plane, although the lobes are elongated into a fat tear-shape.

Thus when the wave function is squared to get the electron density, this is concentrated in two lobes (as sketched in Figure 2.11) and there is zero electron density in the nodal plane passing through the nucleus perpendicular to the axis of the lobes. Contours join points where the electron density decreases outwards by a factor of 4 from one to the next. For effective nuclear charges appropriate to B to F, the maximum electron density is found at distances ranging from about 0.3 to 1.3 a_0 from the nucleus.

Thus the p orbitals contrast with the s orbitals in being directional and having zero electron density at the nucleus. When the radial density function is plotted for the $2p$ orbital of

hydrogen, we find the maximum at $4a_0$ (i.e. $n^2 a_0$ as predicted also by the Bohr theory). For higher values of n, the p wave function is similarly oriented in two lobes with a nodal plane through the nucleus and there are also radial nodes of zero electron density, as shown by Figure 2.9e. As for the higher s orbitals, these radial nodes are too close in to the nucleus to be of significance in a qualitative treatment and we may disregard them.

Modern calculations are able to give detailed descriptions of the electron density in p orbitals for about two-thirds of the elements in the Periodic Table, though there are still difficulties in dealing with the largest numbers of electrons. Some consequences are discussed in section 18.9, here we need only the qualitative picture and concentrate on the sign changes of Figure 2.10.

2.11 The d orbitals

In a similar way, the angular parts of the d orbitals may be separated and are shown in Figure 2.12. Multiplying in the radial part does not change the basic general character of a four-lobed figure with alternating signs, and squaring the total wave function to get the electron density changes the shapes of the lobes but does not change the basic four-lobed character. The radial density function, for hydrogen shown in Figure 2.9f, has a maximum at $9a_0$. As for the p orbitals, this maximum contracts rapidly with nuclear charge and is found about one Bohr radius from the nucleus in the first row transition metals.

In a d orbital, there are two nodal planes at right angles through the nucleus and therefore zero electron density at the nucleus. The d_{z^2} orbital does not fit this description and it arises in the following way. The three orbitals d_{xy}, d_{yz}, and d_{zx} are identical except for their orientation in space, and the $d_{x^2-y^2}$ orbital is the same as the d_{xy} orbital but rotated through 45 degrees. There are two other possible orbitals corresponding to the $d_{x^2-y^2}$ orbital and directed along the other two pairs of axes (in an obvious notation, $d_{y^2-z^2}$ and $d_{z^2-x^2}$) but these would give six d orbitals in all, although there are only five independent solutions of the d type to the wave equation. It is easy to show that the three solutions of the d_{xy} type are independent while the three of the $d_{x^2-y^2}$ type are not. (If the orbital diagrams of this latter set of three are superimposed, taking account of the signs, the lobes all cancel out.) Any two of the three would be acceptable along with the three orbitals of the

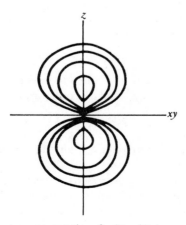

FIGURE 2.11 *Contour representation of a 2p orbital.*

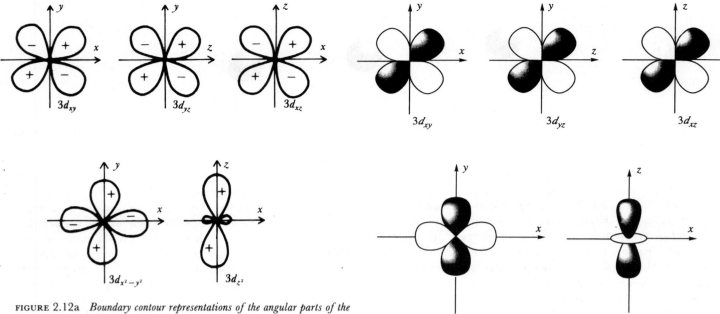

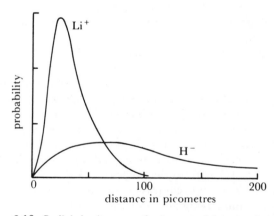

FIGURE 2.12a *Boundary contour representations of the angular parts of the 3d orbitals (note the changes of axes).*

FIGURE 2.12b *Representation in which the shading indicates opposite signs. Note: in subsequent diagrams the shading convention will be generally adopted.*

d_{xy} type, but it is convenient to use one of them, say $d_{x^2-y^2}$, together with a combination of the other two, d_{z^2}, where:

$$\psi_{d_{x^2}} = 1/\sqrt{3}(\psi_{d_{z^2-x^2}} - \psi_{d_{y^2-z^2}})$$

This gives a satisfactory set of five independent d orbitals and this is the set in common use. Naturally, the z-direction may be chosen to suit the situation. For example, if the atom is in an external magnetic field, it is useful to have the z-axis parallel to the field. The higher d orbitals are of the same basic shape as the $3d$ orbitals.

It might be noted that the five d orbitals described above do not all correspond directly to particular m values. While $m = 0$ gives d_{z^2}, d_{xz} and d_{yz} are formed from combinations of $m = \pm 1$, and d_{xy} and $d_{x^2-y^2}$ from combinations of $m = \pm 2$.

The f orbitals have shapes reminiscent of the d orbitals but with a further nodal plane through the nucleus. For example f_{xyz} is an 8-lobed figure with three planes at right angles. As these orbitals are little used in bonding they need not be discussed further.

The basic shapes of orbitals are unchanged by changes in the nuclear charge, although their extensions into space depend inversely on the value of Z. Figure 2.13 shows the radial density curve for the $1s$ orbital of helium-like species with different values of the nuclear charge. The radius of the electron cloud around an atom is the resultant of this contraction of the charge cloud with increasing Z, and the fact that orbitals further and further out from the nucleus are being occupied as the atomic number rises. There is a general increase in atomic size as Z increases, but this change occurs in a very irregular manner (compare the plot of atomic radius against atomic number in Figure 8.8). The greatest jumps in radius come when the outermost electron starts to fill the s level of a new quantum shell and there are less sharp

FIGURE 2.13 *Radial density curves for ions containing two electrons and with different nuclear charges*
This plot for isoelectronic ions illustrates the effect of increasing the nuclear charge from $Z = 1$ to $Z = 3$ on the spatial distribution of the electron probability density.

increases when any new level starts to be occupied and when spin-pairing occurs at p^4 and d^6 configurations.

2.12 The Periodic Table

As chemical behaviour depends on the interaction of the electron clouds of atoms, and especially on the interaction of the outermost parts of these clouds, atoms which have their outer electrons in the same type of orbital should have similar chemical behaviour. For example, a configuration such as s^2p^3 implies the same shape of electron cloud, whatever the n

values of the orbitals, although the extension of the electron cloud, and hence the atomic radius, clearly depends on the atomic number. If the elements are arranged so that those with the same outer electron configuration fall into Groups, the result is the Periodic Table of the elements. This is one of the most important of all scientific generalizations and grew from an increasing realization through the nineteenth century that elements fell into families whose properties followed a trend as their weight increased—for example, the 'triad' Cl, Br, I where the properties of Br are intermediate between Cl and I. As atomic weights became more accurate, more and more elements were fitted in and chemists like Newlands and Olding produced tabular arrays of elements which were close to the modern Table for the lighter elements. This development was brought to a climax some 150 years ago, simultaneously by Mendeléef and by Meyer, in the Periodic Law which defined the basic Periodic Table. This then served as the focus for the discovery and tabulation of new elements, and for much other chemical development, in the sixty years up to the development of quantum approaches. The electron configurations derived from the electronic theory of atoms then provided the theoretical rationalization and refinement of the Periodic Table.

The Periodic Table reflects the order of energy levels in the atoms as they are derived from the wave equation, and the form of the Periodic Table follows from the allowed values of the quantum numbers as given in Table 2.2. This is illustrated by the block form of the Periodic Table shown in Figure 2.14. The different 'blocks' hold sets of two, six, ten, and fourteen elements; these being, respectively, the elements where the s, p, d, and f levels are filling and these blocks thus follow from the number of orbitals of each type which are allowed by the quantum rules. The order of the blocks across the Table from left to right follows from the general order of energy levels: $ns < (n-1)d \leqslant (n-2)f < np$. The modern 'long' form of the Periodic Table is given as Table 2.6.

It is difficult to realize, nowadays when all this has become accepted, just how important a contribution a knowledge of electronic structure made to the chemist's understanding and use of the Periodic Table. In particular, the Table produced by Mendeléef, based on valency considerations, was in the 'short' form where the transition elements were introduced as sub-groups into the main groups of the s and p blocks. This practice was justifiable at the time as there are some resemblances, particularly when the transition elements are showing their maximum valencies, and there were many blanks in the knowledge of the chemistry of these elements. As such knowledge increased, it became apparent that more anomalies than analogies were introduced by the short form of presentation. The long form removes many of these difficulties, and the case for its adoption became overwhelming when the electronic structures of the elements were worked out. There are still a number of minor anomalies. In particular, no form of the Table completely reflects the differences in the ground state electron distribution between the s and d levels among the transition elements, nor those between the d and f levels among the inner transition elements (see Table 2.5), but these differences have no effect on the chemistry of these elements as the anomalies disappear in the valency states. In fact, all the problems and objections to the long form of the Periodic Table disappear when it is regarded as a very successful broad generalization about properties, based on the electronic structures of the elements but not reflecting them in every detail.

A number of special names are given to particular sections of the Periodic Table. There is, unfortunately, some confusion in nomenclature and Table 2.7 lists both the special names approved for general use and some of the cases where conflicting usages appear in text-books. In this book, Groups which do not have a trivial name listed in the table will be

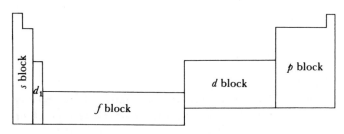

FIGURE 2.14 *Block diagram of the Periodic Table*

TABLE 2.6 Periodic Table of the elements

								Groups													Rare gas electrons
		1	2	3	4	5	6	7	8	9	10	11	12		13	14	15	16	17	18	
Period 1	1s	H																	H	He	2
2	2s	Li	Be											2p	B	C	N	O	F	Ne	2, 8
3	3s	Na	Mg											3p	Al	Si	P	S	Cl	Ar	2, 8, 8
4	4s	K	Ca	3d Sc	Ti	V	Cr	Mn	Fe	Co	Ni	Cu	Zn	4p	Ga	Ge	As	Se	Br	Kr	2, 8, 18, 8
5	5s	Rb	Sr	4d Y	Zr	Nb	Mo	Tc	Ru	Rh	Pd	Ag	Cd	5p	In	Sn	Sb	Te	I	Xe	2, 8, 18, 18, 8
6	6s	Cs	Ba	5d La*	Hf	Ta	W	Re	Os	Ir	Pt	Au	Hg	6p	Tl	Pb	Bi	Po	At	Rn	2, 8, 18, 32, 18, 8
7	7s	Fr	Ra	6d Ac†	Db	Jl	Rf	Bh	Hn	Mt	Uun	Uuu	Uub								
	*Lanthanides		4f Ce	Pr	Nd	Pm	Sm	Eu	Gd	Tb	Dy	Ho	Er		Tm	Yb	Lu				
	†Actinides		5f Th	Pa	U	Np	Pu	Am	Cm	Bk	Cf	Es	Fm		Md	No	Lr				

TABLE 2.7 Nomenclature of the elements

General sections of the Periodic Table

Main Group elements (or Typical elements or Representative elements)	Those elements with the outermost electrons in the s or p levels—the Groups headed by Li, Be, B, C, N, O, F, and He, plus H. Sometimes the two lighter elements in each group are excluded.
Transition elements	Those elements where the d or f levels are filling. On the fuller definition of Main Group elements, this class then includes the rest of the elements.
Inner Transition elements	Those elements where an f shell is filling. With this usage, 'transition elements' is then confined to elements of the d block.
A and B subgroups	This terminology is derived from the older short form. The Main and Transition elements were divided into subgroups, A and B. In Groups I and II the Main Group elements (the alkali and alkaline earth metals respectively) were termed A and the transition elements (copper and zinc Groups) were termed B. But in the other Groups conflicting usages exist; in one system, the Main Group was termed A throughout while in the other,* more common one, it was termed B. Thus Group IVB might be the carbon Group of Main elements or the titanium Group of Transition elements depending on the author and the reader has to check which convention was used.
M (Main) and T (Transition) Subgroups	This was an attempt to replace the A/B usages by an unambiguous nomenclature. It has not found official favour since the initials apply only to the English terms.
The 18 Group system*	Now recommended by IUPAC to finally eliminate the problems of ambiguity. At present subject to lively criticism as the simpler relations between Group number and maximum valency are lost. Use is indicated in Table 2.6.

In this text we stick by our original view that memorizing the group of elements that belong to a particular number is no easier than memorizing a group of elements! Thus we refer to the Periodic Groups either by the accepted trivial names listed below or by the name of the lightest element in the Group.

Trivial names for Groups of elements

Alkali metals*	Li, Na, K, Rb, Cs, Fr	
Alkaline earth metals*	Ca, Sr, Ba, Ra	
Chalcogens* or Calcogens	O, S, Se, Te, Po	
Halogens*	F, Cl, Br, I, At	
Inert gases or rare or noble* gases	He, Ne, Ar, Kr, Xe, Rn	
Rare earth elements*	Sc, Y, La, to Lu inclusive	These divisions are not usually strictly observed and these three names tend to be used interchangeably.
Lanthanum series* or Lanthanons	La to Lu inclusive	
Lanthanides or Lanthanoids*	Ce to Lu inclusive	
Actinium series*	Ac onwards	
Actinides or Actinoids*	Those elements where the $5f$ shell is being filled (see Chapter 11).	
Transuranium elements*	The elements following U	
Coinage metals	Cu, Ag, Au	
Platinum metals	Ru, Os, Rh, Ir, Pd, Pt	
Noble metals	An ill-defined term applied to the platinum metals, Au, and sometimes includes Ag, Re, and even Hg.	
Metal and non-metal	These two terms are widely used but it is not clear precisely where the boundary between them comes. The term metalloid or semi-metal is often applied to elements of intermediate properties such as B, Si, Ge, or As.	

*These usages are approved by the International Union of Pure and Applied Chemistry Rules for Nomenclature of Inorganic Chemistry.

TABLE 2.8 Ionization energies of the elements

Element	Ionization energies (electron volts, 1 eV $= 96 \cdot 48$ kJ mol^{-1})							
	1st	2nd	3rd	4th	5th	6th	7th	8th
H	13·60							
He	24·59	54·4						
Li	5·39	75·6						
Be	9·32	18·2	154					
B	8·30	25·2	37·9	259				
C	11·26	24·4	47·9	64·5	392			
N	14·53	29·6	47·5	77·5	97·9	552		
O	13·62	35·1	54·9	77·4	114	138	739	
F	17·42	35·0	62·7	87·1	114	157	185	954
Ne	21·56	41·0	63·5	97·1	126	158	207	239
Na	5·14	47·3						
Mg	7·65	15·0	80·1					
Al	5·99	18·8	28·5	120				
Si	8·15	16·3	33·5	45·1	167			
P	10·49	19·7	30·2	51·4	65·0	220		
S	10·36	23·3	34·8	47·3	72·7	88·0	281	
Cl	12·97	23·8	39·6	53·5	67·8	97·0	114	348
Ar	15·76	27·6	40·7	59·8	75·0	91·0	124	143
K	4·34	31·6						
Ca	6·11	11·9	50·9					
Sc	6·54	12·8	24·8	73·5				
Ti	6·82	13·6	27·5	43·3	99·2			
V	6·74	14·7	29·3	46·7	65·2	128		
Cr	6·77	16·5	31·0	49·1	69·3	90·6	161	
Mn	7·44	15·6	33·7	51·2	72·4	95	119	196
Fe	7·87	16·2	30·7	54·8	75·0	99	125	151
Co	7·86	17·1	33·5	51·3	79·5	102	129	157
Ni	7·64	18·2	35·2	54·9	75·5	108	133	162
Cu	7·73	20·3	36·8	55·2	79·9	103	139	166
Zn	9·39	18·0	39·7	59·4	82·6	108	134	174
Ga	6·00	20·5	30·7	63·6				
Ge	7·90	15·9	34·2	45·7	93·5			
As	9·81	18·6	28·4	50·1	62·6	128		
Se	9·75	21·2	30·8	42·9	68·3	81·7	155	
Br	11·81	21·8	36	47·3	59·7	88·6	103	193
Kr	14·00	24·5	37·0	52·5	64·7	78·5	111	126
Rb	4·18	27·3						
Sr	5·70	11·0	43·6					
Y	6·38	12·2	20·5	61·8				
Zr	6·84	13·1	23·0	34·3	80·4			
Nb	6·88	14·3	25·0	38·3	50·6	103		
Mo	7·10	16·2	27·2	46·4	61·2	68	127	
Tc	7·28	15·3	29·5	43	59	76	94	162
Ru	7·37	16·8	28·5	46·5	63	81	100	119
Rh	7·46	18·1	31·1	45·6	67	85	105	126
Pd	8·34	19·4	32·9	49·0	66	90	110	132
Ag	7·58	21·5	34·8	52	70	89	116	139
Cd	8·99	16·9	37·5					
In	5·79	18·9	28·0	54				
Sn	7·34	14·6	30·5	40·7	72·3			
Sb	8·64	16·5	25·3	44·2	55·5	108		

(Contd.)

TABLE 2.8 (Contd.)

Element	Ionization energies (*electron volts*, 1 eV = 96·48 kJ mol^{-1})							
	1st	*2nd*	*3rd*	*4th*	*5th*	*6th*	*7th*	*8th*
Te	9·01	18·6	28·0	37·4	58·8	70·7	137	
I	10·45	19·1	33·0					
Xe	12·13	21·2	32·1					
Cs	3·89	23·1						
Ba	5·21	10·0						
La	5·58	11·1	19·2					
Hf	6·65	14·9	23·3	33·3				
Ta	7·89	16·2	22·3	33·1	45			
W	7·98	17·7	24·1	35·4	48	61		
Re	7·88	16·6	26·0	37·7	51	64	79	
Os	8·7	16·9	25	40	54	68	83	99
Ir	9·1	16	27	39	57	72	88	104
Pt	9·0	18·6	28·5	41·1	55	75	92	109
Au	9·23	20·5	30·5	43·5	58	73	96	114
Hg	10·44	18·8	34·2					
Tl	6·11	20·4	29·8	50·5				
Pb	7·42	15·0	32·0	42·3	68·8			
Bi	7·29	16·7	25·6	45·3	56·0	88·3		
Po	8·42							
At	9·2							
Rn	10·75							
Fr	3·98							
Ra	5·28	10·2						
Ac	5·17	12·1						
Th	6·08	11·5	20·0	28·8				

named by the lightest element of the group—carbon Group, titanium Group, etc.

Further Properties of the Elements

In this section a number of important atomic properties are defined and discussed.

2.13 Ionization potential

If sufficient energy is available, it is possible to detach one or more electrons from an atom, molecule, or ion. The minimum amount of energy required to remove one electron from a gaseous atom, leaving both the electron and the resulting ion without any kinetic energy, is termed the *ionization potential*. Since energy has to be provided to remove the electron against the attraction of the nucleus, ionization potentials are always positive. The energies required to remove the first, second, third, etc., electrons from an atom are its first, second, third, etc., ionization potentials. It is clear that the successive ionization potentials will increase in size as it becomes increasingly difficult to remove further electrons from the positively charged ions. Ionization potentials of molecules and ions are defined in a similar way to those for atoms.

The ionization potentials of atoms reflect the binding energies of their outermost electrons and will be lowest for those elements where the valency electrons have just started to enter a new quantum level. If Table 2.8 of ionization potentials and Figures 8.4 and 8.5 are examined, it will be seen that the lowest first ionization potentials are shown by the alkali metals where the last electron has entered a new quantum shell and has its main probability density markedly further out from the nucleus than the preceding electrons. As a quantum shell fills, on going across a Period from the alkali metal, the outermost electrons become more and more tightly bound and the ionization potentials rise to a maximum at the rare gas element where that quantum shell is completed. The stability of the completed shell is also shown when the successive ionization potentials of any particular element are examined. As electrons are removed from a partly-filled shell the successive ionization potentials rise steadily, but there is a great leap in the energy required to remove an electron when all the valency electrons have been removed and the underlying complete quantum shell has to be broken. For example, it requires 518 kJ mol^{-1} to remove the single $2s$ electron from lithium but, when the closed $1s^2$ group has to be broken in order to remove a second electron, the second ionization potential shoots up to 7280 kJ mol^{-1}. Similarly, the successive ionization potentials of aluminium

TABLE 2.9 Electron affinities of the elements (after Zollweg) ($kJ\ mol^{-1}$)
(Exothermic changes are taken as negative)

H -74.5																	He +21.2
Li -59.8	Be -36.7											B -17.3	C -122.3	N +20.1	O -141.3	F -337.4	Ne +28.9
Na -52.2	Mg +21.2											Al -19.3	Si -131	P -68.5	S -196.8	Cl -349.2	Ar +35.7
K -45.4	Ca +186	Sc +70.5	Ti +1.93	V -60.8	Cr -93.5	Mn +93.5	Fe -44.5	Co -102	Ni -156	Cu -173	Zn -8.7	Ga -35.3	Ge -139	As -103	Se -203	Br -324.1	Kr +40.5
Rb (-37.6)	Sr +145	Y +38.6	Zr -43.5	Nb -109	Mo -114	Tc -95.5	Ru -145	Rh -162	Pd -98.5	Ag -193	Cd +26.1	In -19.3	Sn -99.5	Sb -90.5	Te -189	I -295.2	Xe +43.5
Cs (-36.7)	Ba +46.4	La -53.1	Hf +60.8	Ta -14.4	W -119	Re -36.7	Os -139	Ir -190	Pt -247	Au -270	Hg +18.6	Tl -30.4	Pb -99.5	Bi -91.5	Po -127	At -270	Rn

are:

1st $Al[Ne](3s)^2(3p)^1 \rightarrow Al^+[Ne](3s)^2$
needs $577\ kJ\ mol^{-1}$

2nd $Al^+[Ne](3s)^2 \rightarrow Al^{2+}[Ne](3s)^1$
needs $1815\ kJ\ mol^{-1}$

3rd $Al^{2+}[Ne](3s)^1 \rightarrow Al^{3+}[Ne]$
needs $2740\ kJ\ mol^{-1}$

but 4th $Al^{3+}[Ne] \rightarrow Al^{4+}[He](2s)^2(2p)^5$
needs $11\ 590\ kJ\ mol^{-1}$

Other examples are shown in the Table of ionization potentials. For all atoms the energies required to remove electrons from filled shells below the valence shell are so great that they cannot be provided in the course of a chemical reaction, and ions such as Li^{2+} or Al^{4+} are found only in high energy discharges. Thus the energy gap between the valence shell and the underlying filled shell, which is reflected in the successive ionization potentials, puts an effective upper limit on the valency of an element and this, of course, follows from the electronic structures of the elements as derived from the wave equation in the early part of this chapter.

2.14 Electron affinity

The electron affinity is the energy of the reverse process to ionization: the uniting of an electron with a gaseous atom or ion or molecule. The energy change in this process is the electron affinity of the species.* Electron affinities are difficult to measure experimentally and only a few have been directly determined with accuracy. Zollweg has compiled a tabulation of first electron affinities corresponding to

$$M_{(g)} + e^- = M^-_{(g)}$$

by examining measured and interpolated values normalised to the accurately known first electron affinities of elements like the halogens and oxygen. These first electron affinities are listed in Table 2.9.

The stability of the rare gas configuration is reflected in the high electron affinities of the halogens which are forming the anion with the rare gas electronic structure. Conversely,

the electron affinities for the rare gases, where the extra electron in the anion starts a new quantum shell, are all endothermic.

While most of the first electron affinities, shown in Table 2.9, are exothermic, the second electron affinities for elements forming doubly charged anions are always large and positive (i.e. energy has to be provided to add the second electron) with the result that the formation of doubly (or higher) charged anions requires the net addition of energy. For example:

$$O_{(gas)} + 2e^- = O^{2-}_{(gas)}$$

requires $703\ kJ\ mol^{-1}$. Such values are usually derived indirectly from the Born-Haber cycle (see section 5.4).

It should be noted that the largest exothermic electron affinity, that of chlorine, is smaller than the ionization potential of caesium which is the least endothermic of any atom. The result is that any electron transfer between a pair of atoms to form a pair of ions is an endothermic process:

$$Cl_{(gas)} + Cs_{(gas)} = Cs^+Cl^-_{(gas)} \quad \Delta H = +12.1\ kJ\ mol^{-1}$$

and usually a considerable amount of energy is required:

e.g. $I_{(gas)} + Na_{(gas)} = Na^+I^-_{(gas)} \quad \Delta H = +180\ kJ\ mol^{-1}$

It follows that the formation of an ionic compound from its component elements occurs only because of the additional energy provided by the electrostatic attractions between the ions in the solid. This is further discussed in Chapter 5.

2.15 Atomic and other radii

As the probability density distribution of an electron decreases exponentially from the nucleus, it never exactly equals zero. There is therefore no unambiguous definition of the radius of an isolated atom, though it may be taken as the radius of, say, the 95% contour. In a molecule, and even more so in a liquid or solid, the electrons are subject to the fields of all the neighbouring atoms and their distribution depends on the detailed chemical environment. The radius will differ from that of the single atom, and will also differ from compound to compound, though the latter variation is usually relatively small. Furthermore, only the distances between atoms (bond lengths) can be measured experimentally and the radii of the atoms forming the bond have to be deduced from these bond lengths.

*The convention used here is the accepted thermodynamic one of endothermic changes being positive and exothermic changes negative, the opposite convention may be found in older texts.

Although the atomic radius is not an exact concept, it is possible to compile sets of atomic radii which reproduce most of the observed interatomic distances to within ten per cent or so. These sets of atomic radii are valuable, as any marked discrepancy between the observed bond length and that calculated from the atomic radii suggests that there is some change in the type of bond or some other effect which should be investigated further. Moreover, when working with closely related compounds the bond lengths should agree much more closely than to ten per cent so that quite small discrepancies are meaningful and worth further study. A number of sets of values for radii are required depending on the nature of the bonding—covalent, ionic, or metallic.

2.15.1 Covalent species

A set of covalent radii may be derived by starting from the experimentally-measured bond lengths in the elements. If these bond lengths are divided by two they give reasonable values for the radii of the atoms, and then the atomic radii of elements which do not form single bonds in the elemental state may be deduced from the bond lengths in suitable compounds with elements of known radii. A few simple examples of the process of building up a set of atomic radii are shown below.

Element	Bond length, pm	Atomic radius, pm
F_2	142	F = 71
Cl_2	199	Cl = 99
Br_2	228	Br = 114
I_2	267	I = 134
C (diamond)	154	C = 77

Molecule		Found	Bond length, pm Calculated from above	Difference
CF_4	C—F	132	148	16
CCl_4	C—Cl	177	176	1
CBr_4	C—Br	191	191	0
CI_4	C—I	214	211	3

The agreement between experimental and calculated values is excellent for the heavier halogens but less good for fluorine. It is found from a wide number of fluorine compounds that better general agreement with experiment is found if the atomic radius of fluorine is taken as F = 64 pm. This is a purely empirical correction chosen to give the best fit with experimental data. In a similar way, the value used for the hydrogen radius is H = 29 pm although the bond length in the H_2 molecule is 74 pm. With these and similar empirical adjustments to the experimental values the Table of atomic radii shown in Table 2.10a was built up.

Most values calculated from Table 2.10a will agree to within 20–30 pm with the experimental bond lengths. The discrepancies are often wider when hydrogen or fluorine are involved as is illustrated by the measured values shown in Table 2.10b for the halides of the carbon Group elements.

TABLE 2.10a Atomic radii in covalent molecules, pm

Be	B	C	N	O	F	H
89	80	77	70	66	64	29
	Al	Si	P	S	Cl	
	126	117	110	104	99	
Zn	Ga	Ge	As	Se	Br	
131	126	122	121	117	114	
Cd	In	Sn	Sb	Te	I	
148	144	140	141	137	133	
Hg	Tl	Pb	Bi			
148	147	146	151			

	B	C	N	O
Double bond radii	71	67	62	62
Triple bond radii	64	60	55	

TABLE 2.10b Bond lengths of halides of the heavier elements of the carbon group, pm

	Silicon	Germanium	Tin
MF_4	154	167	?
MCl_4	201	208	231
MBr_4	215	231	244
MI_4	243	250	264

A self-consistent and semi-empirical set of values of this type is the best that can be done with a single set of figures for atomic radii. A number of suggestions have been made for modifying the calculated bond length to allow for environmental effects. One example is the Schomaker-Stevenson correction which allows for the polarity of the bond. This is:

$$r_{A-B} = r_A + r_B - 0.09|x_A - x_B|$$

where r_{A-B} is the bond length, r_A and r_B are the covalent radii, and x_A and x_B are the electronegativities (see section 2.16). This formula does improve the agreement between calculated and experimental values in many cases, especially for fluorides, but the discrepancies are still significant and it is probably better to accept the purely empirical nature of the atomic radius and seek for other evidence to establish the existence of special effects within the bond.

The discussion above applies only to single bonds. When double or triple bonds are present, the bond length is shortened and appropriate values of the atomic radius must be used. For example, in ethylene the C=C distance is 135 pm and in acetylene C≡C is 120 pm, corresponding to a double bond radius for carbon of 67 pm and a triple bond radius of 60 pm. Approximate values for other double and triple bond distances may be calculated by using the radii which are given in the last two lines of Table 2.10a. It will be noticed that the variations in bond lengths due to multiple bonding are larger than the uncertainties associated with the empirical nature of the set of atomic radii, but not by a very great margin. This means that attempts to deduce the bond order from variations in the bond length are legitimate but should be treated with some reserve unless closely similar compounds are being discussed.

In addition to the covalent radius just discussed, a further, much larger radius called the Van der Waals' radius is characteristic of atoms in covalent compounds. This radius represents the shortest distance to which atoms which are not chemically bound to each other will approach before repulsions between the electron clouds come into play. The Van der Waals' radius therefore governs steric effects between different parts of a molecule. Some values of Van der Waals' radii are shown in Table 2.11.

TABLE 2.11 Van der Waals' radii, pm

H		N	O	F
100		155	152	147
		P	S	Cl
	210	180	180	175
	Ge	As	Se	Br
	195	185	190	185
	Sn	Sb	Te	I
	210	205	206	198

2.15.2 Ionic species

For ionic radii, we look for a set of self-consistent values to reproduce the observed interionic distances in ionic solids, in the same way as covalent radii do in molecular compounds. It is more difficult, however, to devise a set of ionic radii as there is no obvious way of dividing the observed internuclear distances, r_{MX}, in ionic compounds into cationic (r_+) and anionic (r_-) radii. This is in contrast to the case of covalent or metallic radii where there are distances between like atoms which can be divided into two. In addition, ionic radii may be expected to vary with the environment. This is particularly the case with anions as the electrons are generally less tightly held. (Consider for example K^+ and Cl^-. Both have the same number of electrons but the positive charge on the potassium nucleus is two higher than that on the chlorine nucleus.) Careful studies and calculations suggest cations are relatively invariable in size, with less than a 3% contraction between the free ion and that ion in a symmetric crystal environment. By contrast anions vary quite substantially with, for example, the radius of Cl^- decreasing from 187 pm in Cs^+Cl^- to 125 pm in Cu^+Cl^-. There are grounds for questioning how good the ionic approximation is for compounds like CuCl (compare Chapter 5), but for compounds of the s element cations M^+ and M^{2+} (apart from Be^{2+}), a range of about 17 pm or around 10% is appropriate for halide ion radii. The H^- ion, with two electrons and a single nuclear charge, exhibits extreme variations and is discussed in Chapter 9. Overall, we can expect to produce a usable set of cation radii for species with rare gas configurations but not for those with d electron populations. To go with these, a set of anion radii may be chosen which can vary within 10%. As with covalent radii, these values are both empirical and experimental. They

are useful to (a) systematize a large number of experimental observations of inter-ionic distances, and (b) indicate anomalies which might suggest unusual bonding or other interesting phenomena worth further study.

There have been two quite different types of attack on the problem of dividing up the measured interionic distances into anion and cation radii. The older, classical, approach was to find some acceptable assumption to act as a starting point, and two sets of values produced respectively by Goldschmidt and by Pauling in the period 1926–28, are widely used.

Goldschmidt's method was to assume that in a compound with a large anion and a small cation, such as LiI, the anions would be in contact. Then half the I–I distance equals the radius of I^-. This value is then used, in compounds with larger cations such as NaI, KI, etc., to calculate cation radii for Na^+, K^+ etc., and these in turn allow the calculation of the radii of other anions such as Cl^- or O^{2-}. Finally, the radius of Li^+ is derived from some compound with a small anion such as LiCl or Li_2O. This method, with further refinements, was used to compile the set of empirical ionic radii shown in Table 2.12. The value of 145 pm for oxygen is quoted in this set although the lower values of 140 pm or 135 pm are usually more compatible with transition metal values.

An alternative approach, used by Pauling, was to assume that the radii of isoelectronic ions, such as Cl^- and K^+, varied inversely as their effective nuclear charge. This then gave a way of dividing the experimentally observed M–X distances and allowed a different set of internally consistent ionic radii to be built up. Either the Pauling or the Goldschmidt radii allow a reasonable prediction of interatomic distances in crystals but, of course, the two sets of values must not be mixed.

Recent evaluations of ionic radii use experimental evidence that was not available in the 1920s, particularly modern X-ray diffraction. As X-rays are scattered by electrons, the major contribution (see section 7.4) is from the highly concentrated inner electrons. Positions of the centres of these (and thus of the nuclei) can be observed accurately. The small number of outer electrons in the valence levels make only small contributions to the electron density map, and superimposed on these are a number of experimental uncertainties and errors, such as those arising from thermal motion.

In a number of cases, careful X-ray analysis has allowed these errors to be minimized. When the electron density from the inner electrons is subtracted out, positions of minimum electron density between the cation and anion may be determined and these are used to define cation and anion radii. For example, the minima suggest the following 'experimental' radii:

$$Li^+ = 92 \text{ pm}, \quad F^- = 109 \text{ pm in LiF}$$
$$Na^+ = 118 \text{ pm}, \quad Cl^- = 164 \text{ pm in NaCl}$$
$$Mg^{2+} = 102 \text{ pm}, \quad O^{2-} = 109 \text{ pm in MgO}$$
$$Ca^{2+} = 126 \text{ pm}, \quad F^- = 110 \text{ pm in CaF}_2.$$

TABLE 2.12 Ionic radii, pm

Ion	Symmetric Ion Radii		Crystal Radii (Shannon)			
	Goldschmidt	Johnson	C.N. 4	6	8	12
Li^+	68	92	73	90	106	
Na^+	98	118	113	116	132	153
K^+	133	145	151	152	165	178
Rb^+	148	156		166	175	186
Cs^+	167	168		181	188	202
Be^{2+}	30	69	41	59		
Mg^{2+}	65	102	71	86	103	
Ca^{2+}	94	126	114	126	148	
Sr^{2+}	110	138		132	140	158
Ba^{2+}	129	140		149	156	175
Al^{3+}		92	53	68		
Sc^{3+}		106		89	101	
Y^{3+}		114		104	116	
La^{3+}		120		117	130	150
F^-	133	112	117	119		
Cl^-	181	164		167		
Br^-	195	179		182		
I^-	216	202		206		
O^{2-}	145(135)	116	124	126	128	
S^{2-}	190	158		170		
Se^{2-}	202	174		184		
Te^{2-}	222	192		207		
Cu^+	96		74	91		
Ag^+	126		114	129	142	
Au^+	137			151		
Zn^{2+}	83		74	88	104	
Cd^{2+}	103		92	109	124	145
Hg^{2+}	112		110	116	128	
Tl^+	149			164	173	184
Pb^{2+}				133	143	163

It will be seen that these values are internally self-consistent, and they add up to give good agreement with other experimental interatomic distances (e.g. $Ca^{2+} + O^{2-} = 240$ pm: experimental value in CaO = 240 pm).

On such a basis, Ladd formulated a set of 'experimental' radii which Johnson has extended beyond directly measured cation radii by using the interatomic distance in metals, a, in the simple formula for cation radii r_+:

$$r_+ = 0.64 \times a/2$$

The 0.64 is an empirical constant (0.61 is better for first-row elements like lithium). Using the directly measured values for Li^+, Na^+, K^+, Mg^{2+} and Ca^{2+} together with the formula gives the Johnson values listed in Table 2.12 for 'spherical potential ions', broadly covering ions in symmetric environments. By difference, the corresponding anion radii were

derived but it should be noted that these are average values from a relatively wide range and apply only in a symmetric environment.

In what is essentially an update of the classical approach, Shannon has thoroughly considered all the factors which influence ionic radii including effects of coordination number, charge of oxidation state, covalent and metallic contributions and distortions, and crystal vacancies. He has defined a set of *crystal radii*, based on $O^{2-} = 126$ pm in six-coordination. These relate directly to more traditional radii based on $O^{2-} = 140$ pm by subtraction of 14 pm from the crystal radius. We tabulate crystal radii because, in Shannon's words, they 'correspond more closely to the physical size of ions in a crystal'.

Listed in Table 2.12 are Goldschmidt radii representing the early lists, Johnson values based on direct measurements of electron density minima and usable only for symmetric environments, and Shannon values for representative co-ordination numbers. In each case the values are optimized to give the best fit to observed data. Values from any one set may be used to predict new interionic distances, but values from different sets must not be mixed. See also Chapter 5 for a discussion of the ionic model of solids.

2.15.3 *Metals*

In metals, the environment of each atom is the same, so a set of metallic radii may be derived by halving the interatomic distances. The structures of metals (see section 5.6) are usually close-packed with a coordination number of twelve, and the metallic radii listed in Table 2.13 are for 12-coordination. Some metal structures involve 8-coordination, and Goldschmidt proposed to take 0.97 of the 12-coordinate radius as an estimate of the 8-coordinate one. Use of metallic radii should be confined to metals and alloys, and similar provisos apply to their use as for covalent or ionic radii.

TABLE 2.13 Metallic radii, pm

Li	Be								
157	112								
Na	Mg	Al							
191	160	143							
K	Ca	Ga	Ge						
235	197	153	139						
Rb	Sr	In	Sn	Sb					
250	215	167	158	161					
Cs	Ba	Tl	Pb	Bi					
272	224	171	175	182					
Sc	Ti	V	Cr	Mn	Fe	Co	Ni	Cu	Zn
164	147	135	129	137	126	125	125	128	137
Y	Zr	Nb	Mo	Tc	Ru	Rh	Pd	Ag	Cd
182	160	147	140	135	134	134	137	144	152
La	Hf	Ta	W	Re	Os	Ir	Pt	Au	Hg
188	159	147	141	137	135	136	139	144	155

When dealing with all these sets of radii it is essential to keep in mind that the experimentally determined quantities are the inter-atomic or inter-ionic distances which can be measured to high accuracy (usually to within a few tenths of a picometre). The values listed for the atomic or ionic radii are empirical and chosen to give the best fit over the widest range of experimental data. As a result, small deviations between calculated and measured values (of up to 5 pm or so) are significant only if very critically examined. Larger differences (of the order of 10 pm) suggest the presence of abnormal bonding, either multiple bonds or strong polarization effects in covalent compounds, or polarization and covalent contributions in ionic compounds.

2.16 Electronegativity

One parameter which is widely used in general discussion of the chemical character of an element is its electronegativity. This is defined as *the ability of an atom in a molecule to attract an electron to itself*. There is no direct way of measuring this ability though a number of indirect methods have been suggested, such as the proposal of Mulliken who defined the electronegativity of an atom as the average of its electron affinity and ionization potential (as the electron affinity is a measure of the tendency of the atom to gain an electron, and the ionization potential indicates its tendency to lose an electron). This is the most fundamental of a number of proposed definitions of electronegativity and

TABLE 2.14a Pauling's values of the electronegativity of elements

(H = 2.1)

El		El		El		El		El		El		El		El		El		El		El		El		El		El		El		El		El	
Li 1·0	Be 1·5															B 2·0	C 2·5	N 3·0	O 3·5	F 4·0
Na 0·9	Mg 1·2															Al 1·5	Si 1·8	P 2·1	S 2·5	Cl 3·0
K 0·8	Ca 1·0	Sc 1·3	Ti 1·5	V 1·6	Cr 1·6	Mn 1·5	Fe 1·8	Co 1·9	Ni 1·9	Cu 1·9	Zn 1·6	Ga 1·6	Ge 1·8	As 2·0	Se 2·4	Br 2·8
Rb 0·8	Sr 1·0	Y 1·2	Zr 1·4	Nb 1·6	Mo 1·8	Tc 1·9	Ru 2·2	Rh 2·2	Pd 2·2	Ag 1·9	Cd 1·7	In 1·7	Sn 1·8	Sb 1·9	Te 2·1	I 2·5
Cs 0·7	Ba 0·9	La 1·0	Hf 1·3	Ta 1·5	W 1·7	Re 1·9	Os 2·2	Ir 2·2	Pt 2·2	Au 2·4	Hg 1·9	Tl 1·8	Pb 1·9	Bi 1·9	Po 2·0	At 2·2
Fr 0·7	Ra 0·9	Ac 1·1														

Lanthanides range from 1·0 to 1·2
Actinides range from 1·3 to 1·4

TABLE 2.14b Electronegativity values after Zhang

H: **1** 2·25

Element	Values (oxidation states in **bold**)
Li	**1** 0·95
Be	**2** 1·45
B	**3** 1·95
C	**4** 2·55
N	3·05
O	3·65
F	4·2
Na	**1** 0·95
Mg	**2** 1·2
Al	**3** 1·5
Si	**4** 1·75; **3** 1·7
P	**5** 2·1; **4** 2·0
S	**6** 2·45; **5** 2·35
Cl	**7** 2·85
K	**1** 0·9
Ca	**2** 1·05
Sc	**3** 1·3
Ti	**4** 1·6; **3** 1·4; **2** 1·2
V	**5** 2·0; **4** 1·85; **3** 1·6; **2** 1·35
Cr	**6** 2·3; **4** 1·9; **3** 1·65; **2** 1·4
Mn	**7** 2·5; **6** 2·4; **4** 1·95; **2** 1·45
Fe	**6** 2·4; **3** 1·7; **2** 1·45
Co	**3** 1·75; **2** 1·45
Ni	**2** 1·5
Cu	**2** 1·5; **1** 1·25
Zn	**2** 1·45
Ga	**3** 1·55; **1** 1·1
Ge	**4** 1·8; **2** 1·4
As	**5** 2·05; **3** 1·6
Se	**6** 2·3; **4** 1·85
Br	**7** 2·55; **5** 2·1
Rb	**1** 0·9
Sr	**2** 1·0
Y	**3** 1·2
Zr	**4** 1·5; **3** 1·35; **2** 1·2
Nb	**5** 1·75; **4** 1·65; **3** 1·5
Mo	**6** 2·0; **4** 1·8; **2** 1·4
Tc	**7** 2·3; **4** 1·8; **2** 1·4
Ru	**4** 1·9; **3** 1·65; **2** 1·45
Rh	**4** 1·85; **3** 1·65; **2** 1·45
Pd	**4** 1·85; **2** 1·45
Ag	**1** 1·15
Cd	**2** 1·3
In	**3** 1·45; **1** 1·1
Sn	**4** 1·6; **2** 1·25
Sb	**5** 1·75; **3** 1·45
Te	**6** 1·95; **4** 1·6
I	**7** 2·15; **5** 1·8
Cs	**1** 0·9
Ba	**2** 1·0
La	**3** 1·2
Hf	**4** 1·55; **3** 1·45; **2** 1·3
Ta	**5** 1·9; **4** 1·75; **3** 1·55
W	**6** 2·15; **5** 2·0; **4** 1·85
Re	**7** 2·35; **6** 2·2; **4** 1·9
Os	**8** 2·6; **6** 2·3; **4** 1·95
Ir	**4** 1·9; **3** 1·7; **2** 1·5
Pt	**4** 1·9; **2** 1·5
Au	**3** 1·7; **1** 1·25
Hg	**2** 1·35; **1** 1·2
Tl	**3** 1·5; **1** 1·1
Pb	**4** 1·55; **2** 1·25
Bi	**5** 1·7; **3** 1·4
Po	**6** 1·9; **4** 1·6
At	**7** 2·05; **5** 1·75
Fr	**1** 0·9
Ra	**2** 0·95
Ac	**3** 1·25

Lanthanides **4** 1·4 to 1·5
3 1·2 to 1·35
2 1·05 to 1·2

Note. Values rounded to 0·05. Oxidation states in **boldface**.

it may be applied where electron affinity values are known. However, the values available (such as Table 2.9) are not all directly measured, and other measures of electronegativity are used. The classic one is the Pauling electronegativity, based on bond energies. Despite all attempts at improvement, the Pauling values are still the most generally used (Table 2.14a). A major difficulty is that the attraction for an electron is clearly not expected to be the same for different valencies of an element. Zhang has proposed a set of values, based on covalent radii and ionization potentials and geared to Pauling values, which are defined for each of the main oxidation states (see section 2.17) of the element. These values are one of the more general sets available, though they have some deficiencies for the Main Group elements. They are listed in Table 2.14b.

There has been a great deal of discussion, argument, and often confusion, about the significance of electronegativity values, largely because various authors have used the concept with different degrees of sophistication. The electronegativity is extremely valuable as a brief summary, within one parameter, of the general chemical behaviour of an atom but it must be used in a general way and little significance attaches to small differences in values between two atoms. The most electronegative elements occur in the top right-hand corner of the Periodic Table and electronegativity falls on going down a Group towards the heavier elements or on going to the left along a Period towards the alkali metals.

Electronegativities are most useful in the guidance they give to the electron distribution in a bond. In a bond $A-B$ between two atoms, the electron density in the bond may lie evenly between the two atoms or be concentrated more towards one atom, say B, than towards the other, when the bond is said to be *polarized*. In the limiting case, when the electron density of the bonding electrons is entirely on B, an electron has been fully transferred from A to B and an ionic compound, A^+B^-, forms. The electron density distribution in the bond may be predicted from the electronegativities of A and B. If A and B have the same electronegativities, it follows from the definition that A and B attract the electrons in the bond equally and no polarization results. If B is more electronegative than A, its attraction for the bond electrons is the stronger and polarization results, the degree of polarization being proportional to the difference in electronegativity. A large electronegativity difference favours the formation of ions and, as a rough guide, an ionic compound forms between A and B if they differ in electronegativity by more than two units. Thus elements with very high or very low electronegativities are more likely to form ionic compounds than those with intermediate values.

The electronegativity of an element depends on the other atoms attached to the one in question. Thus, carbon in H_3C-X is less electronegative than carbon in F_3C-X, as the highly electronegative fluorine atoms in the trifluoromethyl compound remove more electron density from the carbon in the $C-F$ bonds than do the hydrogen atoms in the $C-H$ bonds of the methyl compounds. As a result, the carbon atom in F_3C-X has more tendency to attract the electrons in the $C-X$ bond than has the carbon

atom in H_3C-X. It follows that the electronegativity values given in Table 2.14 represent the behaviour of the elements in an 'average' chemical environment and the effective electronegativity of an element in any particular compound depends in detail on its environment.

Another parameter related and complementary to electronegativity is *hardness*. Whereas the Mulliken electronegativity is defined as the average of the ionization potential (I) and electron affinity (A), hardness η can be approximated as half the difference between these two values:

$$\eta = (I - A)/2$$

The hardness of an atom is a parameter which attempts to quantify the ability of electrons to redistribute themselves within the atom and thus is a measure of the *polarizability* of the atom, as described earlier. Atoms with small ionization energies and small electron affinities, such as the heavy halogens and oxygen Group elements (i.e. those elements on the bottom right-hand side of the p-block), are termed 'soft'. Small atoms, such as sodium, oxygen and fluorine, are termed 'hard'. The hardness of the donor atoms of a ligand bonding to a metal atom is of great consequence in determining the strength of the bonding interaction and this topic is discussed in greater detail for transition metal complexes in section 13.7. The general rule is that 'like bonds to like', i.e. soft metal centres such as Hg(II) and Ag(I) have a strong preference for binding to soft donor atoms such as P, S, Se and I.

2.17 Coordination number, valency, and oxidation state

The three terms, coordination number, valency, and oxidation state, are used to describe the environment and chemical state of an atom in a compound. The three overlap somewhat in meaning and application, but the use of each has advantages in certain circumstances.

The simplest term to describe an atom in a compound is its *coordination number*, which is the number of nearest neighbours to the given atom, whatever the bonding between them. The coordination number is a purely empirical property of the element determined from the structure of the compound. This simplicity is the main advantage in the use of the term, as a compound may be described by the coordination numbers of its constituent atoms, however difficult it may be to determine the bonding between these atoms. The only difficulty in determining the coordination number comes when all the distances between like substituents and the central atom are not the same. In some cases, it may be difficult to decide whether some distance which is distinctly longer than the rest is part of the coordination number or not. However, few such cases cause any real problem, and causes of asymmetry in coordination are well understood.

When more information about the atom is required, the valency or the oxidation state must be determined. *Valency* is a familiar term and need not be described in detail. Basically, it describes the bonding of the atom and it is a theoretical term whose use demands more than the experimentally-determined properties of the compound in question. This problem is often disguised by familiarity, but it arises in an acute form in the many cases where a compound or class of compounds is discovered and the structures determined long before an adequate theoretical description of the bonding, and hence the valency, is available. One example is nickel carbonyl, $Ni(CO)_4$, which was known for many years before there was an adequate theory of its bonding. Its structure has been written at various times with the nickel-carbon monoxide bond as $Ni=C=O$, $Ni-C\equiv O$, $Ni\leftarrow C\equiv O$ and $Ni\rightleftharpoons C\equiv O$ implying that the nickel is, respectively, eight-, four-, zero- and zero-valent.

Apart from this type of problem, the valency nomenclature is sometimes clumsy (just because it gives a more complete picture of the molecule). For example, cobalt in the ion $[Co(NH_3)_6]^{3+}$ has to be described as having a covalency of six and an electrovalency of three. There are also occasions when the term valency conceals differences in properties. An example is given by ammonia, NH_3, and nitrite ion, NO_2^-. In both compounds the nitrogen atom is properly described as trivalent and yet it has to be oxidized to pass from one compound to the other, and it is more useful in some contexts to discuss ammonia and related compounds such as the amines, R_3N, separately from the nitrites and other trivalent oxy-compounds.

Considerations such as the above, led to the introduction of a narrower, more empirical term, *oxidation number* (or *oxidation state*). The oxidation number of an element in a compound may be simply determined from a number of empirical rules and it is quite independent of the nature of the bonding. Obviously, it gives less information about the chemical state of the element than does an accurate description in terms of valency but it is useful and convenient when that extra information is not required or available.

The oxidation number of an atom in a compound is defined by the following rules:
(i) The oxidation number of an atom in the element is zero.
(ii) The oxidation number of an atom in an ionic compound is equal to the charge on that atom (with the sign).
(iii) The oxidation number of an atom in a covalent compound is equal to the charge which it would have in the most probable ionic formulation of the compound.

The first two rules are perfectly clear but a little experience is required to find the artificial ionic form required by rule (iii). The electronegativities of the elements in the compound usually serve to make the most probable ionic formulation clear, as illustrated by the examples given below:

Compound	More electro-negative element	Ionic formulation	Oxidation numbers
BCl_3	Cl	$B^{3+}(Cl^-)_3$	B = III, Cl = $-$I
SO_2	O	$S^{4+}(O^{2-})_2$	S = IV, O = $-$II
NH_3	N	$N^{3-}(H^+)_3$	H = I, N = $-$III
NH_4^+	N	$[N^{3-}(H^+)_4]^+$	H = I, N = $-$III
NO_2^-	O	$[N^{3+}(O^{2-})_2]^-$	N = III, O = $-$II
CrO_4^{2-}	O	$[Cr^{6+}(O^{2-})_4]^{2-}$	Cr = VI, O = $-$II
$Cr_2O_7^{2-}$	O	$[(Cr^{6+})_2(O^{2-})_7]^{2-}$	Cr = VI, O = $-$II

Notice, in the last column, that the sum of the oxidation numbers of the atoms equals the overall charge on the species. Although atoms may be shown with large charges, e.g. Cr^{6+} or S^{4+}, this by no means implies the existence of such unlikely ions. To make this clear, it is usual to indicate the oxidation state by Roman numbers—Cr(VI) or S(IV).

In nearly all compounds, rules (i) to (iii) are equivalent to taking O = $-$II (except in peroxides, where O = $-$I, and in OF_2, where O = $+$II), H = $+$I (except in ionic hydrides) and halogens = $-$I (except in their oxygen compounds, not including OF_2, as mentioned above).

Oxidation and reduction are very simple to define in terms of oxidation numbers. Oxidation is any process which increases the oxidation number of an element while reduction corresponds to a decrease in the oxidation number. For example, the conversion of ammonia to nitrogen involves an increase from $-$III to zero in the oxidation number of the nitrogen and is an oxidation by three steps, the conversion of ammonia to nitrite involves an oxidation by six steps, while the conversion to nitrate involves a change of eight steps to nitrogen (V). On the other hand, the change from ammonia, NH_3, to ammonium ion, NH_4^+, involves no change in the oxidation numbers and is not an oxidation. The same applies to the change from chromate to dichromate which sometimes causes trouble in analytical calculations. Further examples are provided by the range of nitrogen compounds below:

Oxidation number of the nitrogen	Examples
$-$III	NH_3 or NH_4^+
$-$II	N_2H_4
$-$I	NH_2OH
0	N_2
I	N_2O or $N_2O_2^{2-}$
II	NO
III	N_2O_3 or NO_2^-
IV	N_2O_4
V	N_2O_5 or NO_3^-

In complex ions, if the ligand is a neutral molecule like ammonia in $[Co(NH_3)_6]^{3+}$ or water in $[Cu(H_2O)_4]^{2+}$, the metal has an oxidation number equal to the charge, Co(III) and Cu(II) respectively. Similarly, nickel in nickel carbonyl, $Ni(CO)_4$, has an oxidation number of zero. If the ligand is charged, then the oxidation number of the metal must balance with the total charge on the ion: Fe(II) in ferrocyanide, $[Fe(CN)_6]^{4-}$, Fe(III) in ferricyanide, $[Fe(CN)_6]^{3-}$, or Co(III) in $[Co(NH_3)_3Cl_3]$ and in $[CoF_6]^{3-}$.

The use of oxidation numbers simplifies the calculations involved in oxidation-reduction titrations. In the overall reaction, the change in oxidation state of the reductant must balance that of the oxidant. The reaction stoichiometry is thus readily worked out from the oxidation state changes of the reactants. A full account of the method is given in the standard analytical textbooks but the following examples illustrate the approach. Compare also section 6.3.

The oxidation of arsenite by permanganate in acid solution

$$MnO_4^- + AsO_3^{3-} \text{ to } Mn^{2+} + AsO_4^{3-}$$

The manganese change is from MnO_4^-, where the Mn = VII to Mn^{2+} with Mn = II; change in manganese oxidation state = -5.

The arsenic change is from AsO_3^{3-}, where the As = III to AsO_4^{3-} with As = V; change in arsenic oxidation state = $+2$. The reaction stoichiometry is therefore:

$$2MnO_4^- + 5AsO_3^{3-}$$

The equation may then be balanced by introducing hydrogen ions and water molecules in the usual way to give:

$$2MnO_4^- + 5AsO_3^{3-} + 6H^+ = 2Mn^{2+} + 5AsO_4^{3-} + 3H_2O$$

The reaction between iodate and iodide

$$IO_3^- + I^- \text{ to } I_2$$

In this case, the oxidant and the reductant end up in the same form. In iodate, the iodine is in the V oxidation state so that the change in going from iodate to iodine is by -5. The change in oxidation state from the $-I$ in iodide to the element is by $+1$ and the reaction stoichiometry is therefore:

$$IO_3^- + 5I^-$$

The balanced equation is:

$$IO_3^- + 5I^- + 6H^+ = 3I_2 + 3H_2O$$

If this reaction is carried out in concentrated hydrochloric acid, instead of in dilute acid as above, the final product is not iodine but iodine monochloride, ICl, in which the iodine has an oxidation state of $+I$. In this case, the change from iodate to ICl is -4 and the change from iodide is $+2$ so that the balanced equation becomes:

$$IO_3^- + 2I^- + 6H^+ = 3I^+ + 3H_2O$$

As far as most calculations are concerned, only the reaction stoichiometry has to be known and the use of oxidation numbers in the calculation gives this very rapidly and easily.

The oxidation state concept breaks down in those cases where an ionic formulation is ambiguous. One example is in the case of the metal nitrosyls which contain groups, $M-NO$, which could quite validly be formulated in three ways—as $(NO)^+$, $(NO)^-$, or with neutral NO groups. Similar difficulties are encountered in, for example, the hydrides of boron or phosphorus where the electronegativities (B = 2·0, P = 2·05, H = 2·1) are so close that doubts arise whether to write H^+ or H^-: in fact, the hydrogen is negatively polarized in most boron-hydrogen compounds and positively polarized in most phosphorus-hydrogen ones. In organic chemistry, also, the oxidation state concept is not very useful; it is more convenient to discuss reactions such as $CH_4 \rightarrow CH_3Cl$ in terms of substitution rather than in terms of a change in the carbon oxidation state.

The three concepts, coordination number, oxidation state, and valency, become less empirical and convey increasing amounts of information in that order.

PROBLEMS

Readers may best test and reinforce their understanding of this chapter by applying the various formulae, and manipulating numerical data. A number of cases should be worked out, and the questions given below are mostly illustrations of the type of example which you can make up.

2.1 How many protons, neutrons, and electrons are present in

^{24}Mg, $^{24}Na^+$, ^{99}Mo, ^{99}Tc, ^{129}Xe, $^{127}I^-$, $^{195}Pt^{2+}$, $^{197}Au^{3+}$

2.2 What is the minimum uncertainty in the position of

(a) a mass of 1 mg,   (b) a molecule of UF_6,
(c) a molecule of H_2,   (d) a neutron,
(e) an electron

where each is moving at half the velocity of light?

2.3 Show by substitution that ψ_1 (equation 2.2) satisfies the Schrödinger equation.

2.4 Calculate the value of the minimum energy, E_1, of He^+ (see equations 2.1 to 2.4).

2.5 Calculate and plot out the values of ψ_1 and ψ_1^2 for the hydrogen $1s$ orbital for distances from the nucleus of 0, 25, 50, 100 and 200 pm. Compare with Figure 2.8.

2.6 Choose various values of the atomic number, Z, and decide the electron configuration. Compare your answers with Table 2.5. Decide the number of unpaired electrons in each case.

2.7 Calculate the excitation and dissociation energies from various excited states of the H atom—e.g. the dissociation energy for $n = 4$ or the excitation energy from $n = 3$ to $n = 6$. Work out the values in kJ mol^{-1}, eV, and cm^{-1}.

2.8 Which of the following sets of quantum numbers represent permissible solutions of the Schrödinger Wave Equation?

n	l	m	m_s
3	1	0	$-\frac{1}{2}$
3	1	1	1
3	2	4	$\frac{1}{2}$
5	4	-3	$\frac{1}{2}$
2	-1	-1	$-\frac{1}{2}$

2.9 Write out all the permissible sets of quantum numbers for

(a) 4 electrons in $3p$ orbitals.
(b) 4 electrons in $5d$ orbitals.

2.10 For a p orbital, plot the general form of the radial part of the wave function, $r(\exp -r/2a_0)$. Multiply this by the angular part to generate the shape of the orbital. Confirm that the change of sign and the nodal plane of the angular part remain in the full wave function.

2.11 Find from the references the electron density contour plot of

(a) an electron in a $2p$ orbital: in (i) H and (ii) Be
(b) an electron in a $3p$ orbital in (i) H and (ii) Al
(c) an electron in two different $3d$ orbitals.

2.12 Plot electron affinity (Table 2.9) against Period position (parallel to Figures 8.4 to 8.7). Discuss any relationship between these curves and the ionization energy ones.

2.13 Calculate Mulliken electronegativities for a Group of elements and compare with the listed values. Which set correlate best with the chemistry (refer to Tables 2.8, 2.9, 2.14 and the appropriate section of systematic chemistry).

2.14 Look up the references listed on electronegativity. Use these to find earlier references: can you find six different sets of electronegativity values? Assess the value of (a) the general idea and (b) each specific set of values. How many significant figures do you think should be used in expressing electronegativities?

2.15 Draw to scale the covalent and ionic radii of the halogens. Draw scale diagrams of X_2, HX (X = halogen) including the Van der Waals radii. For ClF_3 (see Figure 4.4; bond lengths are 170 pm, axial, and 160 pm, equatorial) do the Van der Waals' radii of axial and equatorial F atoms overlap?

2.16(a) Determine the stoichiometry of the oxidation of NH_3 to *each* of the other nitrogen oxidation states with MnO_4^-

  (i) in acid, forming Mn^{2+}
  (ii) in alkali, forming MnO_2.
(b) Write balanced equations for NH_3 going
  (i) to N_2, and
  (ii) to NO_3^-

by reacting with MnO_4^- in acid.

2.17 De Broglie proposed that the electron wave may be described by the same equations that apply to a photon, that is, $hv' = E = mv^2$, where v is the velocity of the electron.

(a) From the values of the constants, and using the relationship of equation (7.2), calculate the wavelength of an electron moving at 1 %, 10 % and 90 % of the speed of light.

(b) What would be the velocity of an electron whose wavelength was
  (i) equal to
  (ii) one fifth of

the circumference corresponding to the Bohr radius of hydrogen?

3 Covalent Molecules: Diatomics

General background

3.1 Introduction

In the last chapter, a picture of the electron structure of atoms was built up from the theory of wave mechanics taken together with experimental data. This knowledge of atomic structure will now be used to examine the process of combining atoms to form molecules.

The ideas of valency and bonding grew up in the nineteenth century, following the acceptance of Dalton's atomic hypothesis and the establishment by experiment of regularity in the composition of materials expressed by the Laws of Constant Composition, Multiple Proportions and the like. This led to the idea of atoms forming a fixed number of links and hence combining to give molecules of fixed composition. As wider and wider ranges of compounds were studied, it was concluded that the number of links characteristically formed by a particular element was constant, or took only a limited number of values. This number is the *valency* (see 2.17) and, for example, the first great rationalization of organic chemistry came from the realization that carbon was tetravalent.

These ideas were developed to give results which are now part of general introductory chemistry. Thus, the link between atoms was clarified into the notion of a bond, and molecules were formulated so that the number of bonds matched the valencies of the atoms. The steps in formulating a new species were as shown in Figure 3.1.

Now, all this is elementary revision for the reader, but it summarizes an evolution of thinking about materials which took well over a century to refine from the first simple observations. These, in turn, stemmed from the major advance made by eighteenth-century chemists who started to measure the changes in weight of substances undergoing transformation. Since the processes exemplified in Figure 3.1 are part of the first steps in learning chemistry, we tend to overlook their significance as one of the major revolutions in thinking about the material world.

Given the idea of constant valency, the picture of bonds became elaborated. Thus, if carbon is tetravalent and oxygen is divalent, then a molecular formula CO_2 can be understood by writing two bonds between C and each O. Using the convention whereby a line between atoms signifies a bond, we write CO_2 as $O=C=O$, and similarly HCN satisfies known valencies when written $H-C\equiv N$ with a triple bond between C and N.

Part of the development of the ordering of elements into families with related properties, which culminated in the formulation of the Periodic Table, was the grouping together of elements with the same pattern of valencies. It is well-known that Mendeléef left gaps in his Table, to be filled by hitherto-undiscovered elements, and the valency of these new elements (e.g. in the form of formulae of oxides or halides) was one of their properties which he predicted.

As we saw in Chapter 2, once the electron configuration of the elements was understood, the Periodic Table was found to result directly. The question then promptly arose whether bonding and valency could also be explained in terms of the electron configurations of the atoms forming a molecule. At this point, we adopt the convenient subdivision of valency into two phenomena—*covalency*, where electrons are shared between bonded atoms, and *electrovalency*, where electrons are transferred from one atom to another to form *ions* which are then bonded by electrostatic forces. The formation of covalent bonds is discussed in this chapter and the next, while ionic compounds are the subject of Chapter 5.

The first widely-accepted electronic description of a covalent bond is that due to Lewis, of two electrons shared between two atoms and binding them together. Furthermore, for light elements at least, stable configurations were those where a total of eight electrons, either wholly-owned or shared, surrounded an atom. This was the 'octet rule', and gave a good guide to the probable formulae of stable species. Thus the idea of electron sharing and the octet rule went a long way to rationalize the observed valencies. In this way, carbon, with four outer electrons and fluorine with seven,

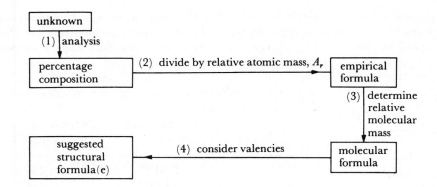

EXAMPLES

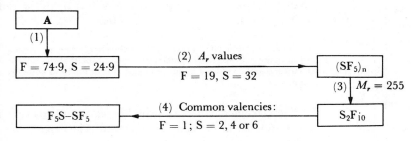

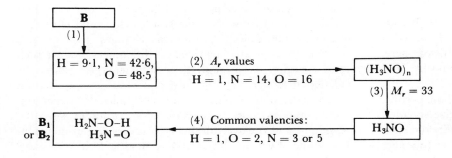

FIGURE 3.1 *Formulation of an unknown*
Stages shown by vertical steps involve experimental determinations, those shown by horizontal steps use known values. Note that in (2), as the atom ratios must be whole numbers, the calculations can be done with rounded-off values. While steps (1) to (3) are unambiguous, it may be possible to write more than one acceptable formula at step (4).

can achieve octets in CF_4

$$\cdot \ddot{C} \cdot + 4 \cdot \ddot{F} : \rightarrow : \ddot{F} : \overset{\overset{\displaystyle : \ddot{F} :}{\textstyle |}}{\underset{: \ddot{F} :}{C}} : \ddot{F} :$$

Hydrogen is stable with two electrons, hence

$$2H \cdot + \ddot{O} : \rightarrow H : \ddot{O} : H$$

Double bonds are readily explained by the sharing of four electrons

$$\cdot \ddot{C} \cdot + 2 \ddot{O} : \rightarrow \ddot{O} :: C :: \ddot{O}$$

and so on. Species may be charged, as in

$$4H \cdot + \cdot \ddot{N} : - 1e \rightarrow \left[H : \overset{\overset{\displaystyle H}{\textstyle |}}{\underset{H}{N}} : H \right]^+$$

Some species do not form an octet, but do share all the

electrons available to them, such as boron or beryllium in their halides

$$\cdot \dot{B} \cdot + 3 \cdot \ddot{Cl} : \rightarrow : \ddot{Cl} : \overset{: \ddot{Cl} :}{B} : \ddot{Cl} :$$

$$\cdot \dot{Be} \cdot + 2 \cdot \ddot{F} : \rightarrow : \ddot{F} : Be : \ddot{F} :$$

and these species, though stable, can complete their octets by co-ordinating with other molecules where unshared pairs are present as in

$$: \ddot{F} : B : \ddot{F} : + : \overset{H}{\underset{H}{N}} : H \rightarrow : \ddot{F} : \overset{: \ddot{F} : H}{\underset{: \ddot{F} : H}{B} : N : H}$$

Once again, these concepts are revision for the reader, but are important in representing the introduction of electron configuration into thinking about molecule formation. The octet rule is, of course, equivalent to emphasizing the stability of the rare gas configurations —those where the valence level *s* and all three p orbitals are filled with two electrons each. The octet rule is formally broken (a) if there are insufficient electrons to give an octet around each atom, as in BF_3 or (b) where there is a further energy level fairly close to the p level which can accept extra electrons. Thus we find that nitrogen forms only NF_3

$$: \overset{\displaystyle ..}{\underset{\displaystyle ..}{F}} :$$
$$: \overset{..}{F} : \overset{..}{N} : \overset{..}{F} :$$

which obeys the octet rule, as the next empty orbital on N is the $3s$ orbital which would require far too much excitation energy for occupation. However, if we move to the next element in the Group, phosphorus, we find PF_3 which obeys the octet rule and completely fills the P $3s$ and $3p$ orbitals, but also find that PF_5 is formed—with 10 electrons on P—as the $3d$ level can be used as well (compare Table 2.3 and Figure 2.4).

Thus in the examples in Figure 3.1, structure A contains S with 6 bonds and therefore 12 electrons around it. Similarly, while B_1 obeys the octet rule, B_2 needs 10 electrons on N which is not permissible.

3.2 Bond formation and orbitals

We must now examine the idea of a bond at a greater level of sophistication. One way of translating the Lewis idea into terms of wave mechanics will be discussed here, though a number of other approaches are possible and these are presented in the references.

Let us start with the simplest of all molecules, the positive ion of the hydrogen molecule, H_2^+. If we generalize equation (2.1) to cover one electron moving in the field of two nuclei, there is only one del-squared term in the kinetic energy part of the wave equation, while the potential energy part contains three terms—the attractions between each nucleus and the electron and the repulsion between the nuclei. The equation is similar to that of a hydrogen-like atom (see equation 2.4), but now the nuclear field is not spherically symmetrical. Although the problem is more difficult than that of the hydrogen atom, the wave equation for the hydrogen molecule ion can be solved exactly and the resulting energy levels and energy of formation agree with the experimental values.

Just as with atoms, major difficulties arise in the case of molecules as soon as more than one electron is involved. For example, the presence of the second electron in the hydrogen molecule, H_2, makes it a molecular analogue of the helium atom. In the wave equation for the hydrogen molecule (compare equation 2.8) there will be two del-squared terms in the kinetic part describing the two electrons and there are six components of the potential energy term. These are the four attractions between each electron and each nucleus, the repulsion between the nuclei, and finally the inter-electron repulsion term which provides the main source of difficulty in these calculations.

It is clear that such complexities increase rapidly as the number of electrons rises, and methods of approximation

have to be found. Before turning to these, it is of interest to record the results of the more rigorous calculations of electronic energies for some simple molecules. These are shown in Table 3.1.

TABLE 3.1 Calculations of total electronic energies

Molecule	Calculated value, kJ mol^{-1}	Experimental value, kJ mol^{-1}
H_2^-	2 887·2	2 893·8
H_2	3 081·2	3 081·1
CH_4	104 850	106 210
Other diatomic molecules like CO, N_2, or HF	Differences between calculated and experimental values lie in the range 0·5—1·5%	

The total electronic energy of a species is the energy evolved when all its constituent particles (nuclei and electrons) are brought from infinite separation and combined to form the molecule or ion in its equilibrium configuration.

The agreement is very close in these cases and gives grounds for confidence in the general approach. The calculations are very long and complex, however, and may require substantial computer resources. In addition, the quantity most readily derived from the calculation is the total electronic energy of the molecule, while the changes involved in a reaction are dependent on the small differences between such total energies. At present, these total energies cannot be calculated to the very high order of accuracy required for direct predictions of reaction paths or molecular structures. Simplifications must be introduced in order to solve the equations and the over-all theory becomes a semi-quantitative guide. Even in this relatively modest form, the wave mechanical approach to molecular structure has wrought an impressive change in the way in which chemists think about molecules.

In this text, the theory will be used as a general guide, and diagrams rather than calculations will be used to describe the processes of molecule formation. The reader is asked to keep in mind that these diagrams do mirror the calculations and that definite values may be found for the parameters—such as bond lengths and bond energies—which are qualitatively described here. Fuller accounts are given by Coulson for example (see references).

Diatomic Molecules

One well-known approximation method for the wave equation for molecules is the method of molecular orbitals. In this approach, the aim is to construct orbitals analogous to the atomic orbitals of Chapter 2 but centred on both nuclei. Then the electrons are fed in—two into each orbital in order of increasing energy—to build up the electronic structure of the molecule, just as the electronic structures of the atoms were built up.

The first problem is to find a way of constructing these

molecular orbitals. One starting point is to consider that, when the electron in a molecule is close to one nucleus, it is in almost the same environment as in the free atom. This suggests that the molecular orbitals may be derived from some combination of atomic orbitals. The simplest way of combining the orbitals is additively and a simple *linear combination of atomic orbitals* (LCAO) has been widely used.

3.3 The combination of *s* orbitals

In the case of H_2^+, the molecular orbitals are formed from the linear combinations (3.1) of the atomic orbitals ψ_X and ψ_Y on the two hydrogen atoms, X and Y:

$$\phi_B = \psi_X + \psi_Y$$
$$\phi_A = \psi_X - \psi_Y \quad \cdots\cdots\cdots\cdots (3.1)*$$

Consider the case where the wave functions ψ_X and ψ_Y represent $1s$ orbitals on the two hydrogen atoms. The result of combining these two atomic orbitals into the molecular orbital ϕ_B may be shown diagrammatically, using the curves for the $1s$ orbital which were given in Chapter 2 in Figure 2.8.

In Figure 3.2, the curves giving the complete cross-section

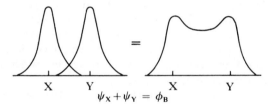

FIGURE 3.2 *The formation of the molecular orbital* ϕ_B (σ_s)
The nuclei X and Y are placed at the measured internuclear distance in the molecule and the wave function curves are superimposed to give the summation curve shown.

through each nucleus, X and Y, of their respective $1s$ orbitals are drawn out and the process of addition to give ϕ_B is indicated to the right. The two nuclei are brought together to a distance equal to the interatomic distance in the molecule, and the wave function curves are summed to give the wave function of the molecule. If we square, to get the electron probability density, the corresponding curve is a cross-section through the two nuclei of the electron distribution in the molecule.

---

*In these equations—and in all similar equations such as (3.2)—a fully rigorous discussion requires that the RHS is multiplied by a factor $\mathcal{N}$, called the *normalizing constant*. This is a numerical factor which adjusts the equation so that $\int \phi^2 d\tau = 1$; i.e. so that the probability of finding the electron described by the molecular orbital ϕ somewhere in space is unity (compare the requirement for atomic orbitals in Chapter 2). In this particular case, as the atomic orbitals ψ_X and ψ_Y are normalized (i.e. $\int \psi_X^2 d\tau = 1 = \int \psi_Y^2 d\tau$) $\mathcal{N} = 1/\sqrt{(2)}$ and the full version of equation (3.1) is

$$\phi_B = \frac{1}{\sqrt{(2)}} (\psi_X + \psi_Y).$$

However, as $\mathcal{N}$ is a numerical constant, its actual value rarely affects the qualitative discussion of molecular orbitals which we are giving here and we shall normally omit it.

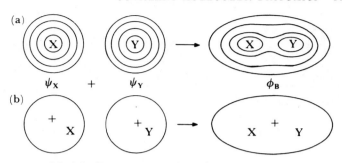

FIGURE 3.3 (a) *Contour representation of* ϕ_B (σ_s)
(b) *Boundary contour representation of* $\phi_B(\sigma_s)$

The contour line enclosing 90% of the electrons density is drawn. Such diagrams represent either the wave function or its square, the probability density function. The wave function diagram is the more useful and shows the sign of the wave function, as here.

The parts of ϕ_B outside the internuclear region follow an exponential curve and the electron density falls off rapidly on moving away from the nuclei. The orbital ϕ_B may be shown in projection as a contour map as in Figure 3.3a, or in the more convenient boundary diagram of Figure 3.3b where the contour line which encloses 90 or 95 per cent of the electron density is drawn (compare with atomic orbitals as in Figure 2.8c).

It is clear from these diagrams that, in the molecular orbital ϕ_B, there is an accumulation of electron density in the region between the two nuclei. This markedly decreases the repulsion between the two nuclear charges. At the same time, the electron has a high probability of being in the region where it experiences the attraction of both the nuclear charges, and is more strongly held than if it was in one of the contributing atomic orbitals and attracted by only one nuclear charge. The result is that the presence of electrons in this molecular orbital holds the two atoms together and a bond is formed. The orbital ϕ_B is termed a *bonding molecular orbital*.

The molecular orbital ϕ_A may be treated in the same way. The combination of atomic $1s$ orbitals ($\psi_X - \psi_Y$) is shown in Figure 3.4, and the contour representation and the boundary line diagram are given in Figure 3.5. In this molecular orbital, ϕ_A changes sign at the mid-point of XY and the electron probability density, ϕ_A^2, falls to zero here across a *nodal plane* perpendicular to the internuclear axis. Thus, if an electron is placed in the orbital ϕ_A, electron density is removed from the region between the nuclei and accumulated on the remote side of the atoms. The internuclear repulsion has full effect and no bond results. ϕ_A is termed an *antibonding molecular orbital*.

When the energies, E_B and E_A, of an electron in the molecular orbitals ϕ_B and ϕ_A are calculated, it is found that E_B is less than the energy of the electron in the constituent atomic $1s$ orbitals while E_A is greater than the atomic orbital energy by the same* amount, ΔE. That is, the molecular

---

*This is a first approximation. In a more quantitative treatment, the destabilization is somewhat greater than the stabilization so that complete occupation of the antibonding orbital (compare He_2 below) is *less* stable than reversion to filled atomic orbitals.

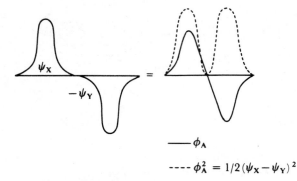

$$\phi_A$$

$$\cdots\cdots \phi_A^2 = 1/2\,(\psi_X - \psi_Y)^2$$

FIGURE 3.4 *The formation of the molecular orbital*
ϕ_A (σ_s^*)

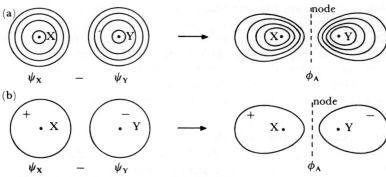

FIGURE 3.5 (a) *Contour representation of* ϕ_A (σ_s^*)
(b) *Boundary contour representation of* ϕ_A (σ_s^*)
(b) shows the molecular orbital wave function which changes sign across the nodal plane perpendicular to the internuclear axis.

orbital ϕ_B is stabilized, and the molecular orbital ϕ_A is destabilized, relative to χ_{1s}, by the energy ΔE. This may be shown on an energy level diagram as in Figure 3.6. The two atomic orbitals thus combine to form two molecular orbitals, one of which is more stable than the atomic orbitals while the other is less stable by the same amount of energy. The process of deriving the electronic structure of a molecule can then be formalized in steps similar to those used for atoms. The nuclei are first placed together at the appropriate distance, then molecular orbitals are constructed from the atomic orbitals, and finally the electrons are fed into the molecular orbitals in order of increasing energy. Just as with atomic orbitals, a molecular orbital holds no more than two electrons and, when there are a number of molecular orbitals of equal energy, the electrons enter them singly with parallel spins. The number of molecular orbitals formed must exactly equal the number of atomic orbitals used in their construction.

Although, in the above outline, it has been assumed that the inter-nuclear distance (the bond length) is a known factor—and it will usually be known experimentally—it

should be noted that in a full theoretical treatment it is possible to derive the optimum bond length by finding the value which gives the minimum total energy of the system. This has been done for a number of simpler cases with results that agree with the experimentally determined distances.

The formation of molecules can now be followed by using the energy level diagram of Figure 3.6. For the hydrogen molecule ion, H_2^+, the electronic structure is shown in Figure 3.7a. The orbital of lowest energy in the molecule is ϕ_B and the electron goes into this, gaining energy, ΔE, relative to its energy in the atom. The whole system, of two hydrogen nuclei and one electron, is thus more stable by ΔE as the molecule than as separate atoms.

In the hydrogen molecule, H_2, (Figure 3.7b) there are two electrons to be considered. These both enter ϕ_B and will have

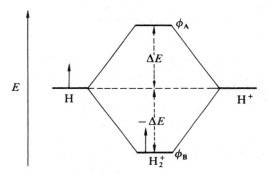

FIGURE 3.7a *The electronic structure of* H_2^+

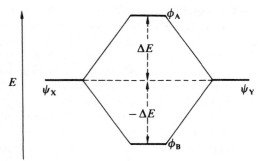

FIGURE 3.6 *Energy level diagram for* ϕ_B *and* ϕ_A (σ_s *and* σ_s^*)
This shows the energies of the molecular orbitals, ϕ_B and ϕ_A, relative to the energies of the constituent atomic orbitals. The atomic orbital energy levels are indicated by the horizontal lines to left and right while the molecular orbitals are shown in the centre of the diagram. Those atomic orbitals which contribute to a particular molecular orbital are connected to it by the finer sloping lines. This convention is followed in all diagrams of this type. The bonding orbital, ϕ_B, is more stable than the atomic orbitals by the same amount of energy as the antibonding orbital, ϕ_A, is less stable than the atomic orbitals.

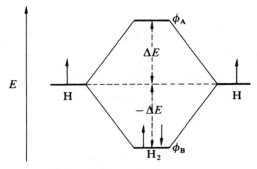

FIGURE 3.7b *The electronic structure of* H_2

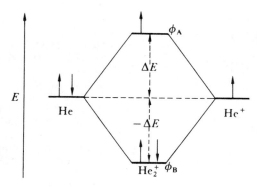

FIGURE 3.8 *The electronic structure of* He_2^+

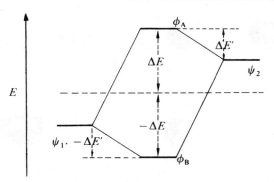

FIGURE 3.9 *Energy level diagram for the combination of s orbitals of unlike atoms*

Stabilization and destabilization is relative to the average energy level of the two atomic orbitals. Note that the differences $\Delta E'$ in energy from the atomic orbitals are also equal.

their spins paired. The gain in energy of the molecule over the two isolated atoms is $2\Delta E$ (less a relatively small term due to the interelectron repulsion) and this bond energy, in the hydrogen molecule, is regarded as that of a normal single bond. It follows that the bond in the ion, H_2^+, which has about half the energy of formation, may be regarded as a 'half-bond'. This accords with experiment. H_2^+ exists as a transient species in electric discharges and its bond energy, which may be determined from its spectrum, is about half the energy of H_2.

Next, consider a three-electron molecule such as the positive ion of diatomic helium, He_2^+, whose energy level diagram is shown in Figure 3.8. Since ϕ_B is filled by the first two electrons, the third must be placed in the antibonding molecular orbital ϕ_A. The energy of formation of He_2^+ is therefore $-2\Delta E + \Delta E$. The net gain in energy from rearranging two helium nuclei and three electrons as a molecule is thus ΔE, corresponding to a 'half-bond' again. (It should be remarked that the actual electron energies both in the atomic orbitals and in the molecular orbitals of helium are different from the energies of electrons in the corresponding orbitals in hydrogen, because of the difference in the nuclear charge. The difference between the atomic and molecular orbital energies, ΔE, will however, be of the same order of size.)

Finally, in a four-electron molecule such as He_2, two electrons would be placed in the bonding orbital ϕ_B and two in the antibonding orbital ϕ_A. The energy of formation is $-2\Delta E + 2\Delta E$ equal to zero, and no bond results. In fact, helium exists as a monatomic gas and the configuration $\phi_B^2 \phi_A^2$ exists only as $(1s)^2$ on each He atom.

An analysis such as this may be extended in an exactly similar way to all other s orbitals with higher n values, and also to the more general case of diatomic molecules where the two atoms are not the same. In this case, the two atomic orbitals which will combine to form the molecular orbital will be of different energies, and the most favourable combination will not be in the 1 : 1 ratio of equation 3.1, but some more general expression of the form:

$$\phi_B = \psi_1 + c\psi_2 \dots \dots \dots \dots (3.2)$$

will be needed. The value of the mixing coefficient c which will give the optimum energy has to be found in the course of

the calculation. The energy level diagram which corresponds to this more general case is shown in Figure 3.9. Here the bonding orbital is stabilized, and the antibonding orbital destabilized, by equal amounts of energy, ΔE, calculated from the mean energy of the contributing atomic orbitals ψ_1 and ψ_2. The gain in energy when an electron is taken from the more stable of the two atomic orbitals is the smaller amount labelled $-\Delta E'$. Clearly, this decreases as the energy difference between the atomic orbitals ψ_1 and ψ_2 increases. If $\Delta E'$ is too small no molecule results. It follows that useful molecular orbitals are formed only when the combining atomic orbitals are of similar energy. As a general rule, this limitation implies that only orbitals in the valency shells of atoms will combine to form molecular orbitals. Thus, in hydrogen chloride, the hydrogen $1s$ orbital is of too high an energy to combine with $1s$ or $2s$ orbitals on the chlorine atom (whose energy levels are greatly stabilized relative to those of hydrogen by the attraction of the nuclear charge of 17), and it is too stable to interact with the chlorine $4s$, or higher, orbitals. The hydrogen $1s$ orbital is comparable in energy with the chlorine $3s$ or $3p$ orbitals and could form molecular orbitals with these.

3.4 The combination of *p* orbitals

If the molecular axis in a diatomic molecule is taken as the z-axis*, then the p_z orbitals on the two atoms may combine to form molecular orbitals similar in type to those formed by s orbitals. Just as with the s orbitals, the p_z orbitals on the two atoms X and Y combine to form two molecular orbitals, ϕ_1 and ϕ_2, which are similar to those given by equations 3.1:

$$\phi_1 = \psi_X + \psi_Y$$
$$\phi_2 = \psi_X - \psi_Y \dots \dots \dots \dots (3.3)$$

where the atomic orbitals ψ_X and ψ_Y are here the p_z orbitals. The process of combination is illustrated using the boundary contour method in Figure 3.10. The electron density in the

---

*We are adopting here the usual convention that the *unique* direction in a molecule is taken as the z-axis. For example, the molecular axis in a linear molecule or the axis perpendicular to the molecule in a planar species would normally be labelled z.

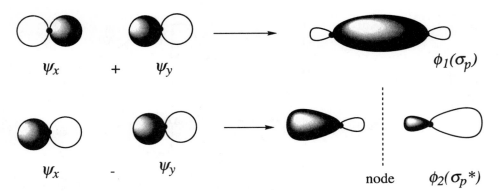

FIGURE 3.10 (left) *The combination of atomic p_z orbitals*

Two atomic p_z orbitals give rise to bonding and antibonding molecular orbitals which, respectively, concentrate or remove electron density between the nuclei. Note that a *plus* combination is defined here as the in-phase one, and *minus* as the out-of-phase one, even though this involves reversing the signs of some of the p orbitals.

first orbital, ϕ_1^2, which results from the addition of the two atomic p_z orbitals, is concentrated in the region between the two nuclei and this molecular orbital is bonding. In the second orbital, there is a nodal plane between the nuclei and the electron density, ϕ_2^2, is concentrated in the regions remote from the internuclear region, falling to zero between the nuclei. This orbital is therefore antibonding. Thus the combination of the two atomic p_z orbitals gives two molecular orbitals, one of which is bonding and one antibonding, just as in the case of the s orbitals. These molecular orbitals, ϕ_B, ϕ_A, ϕ_1, and ϕ_2, which are formed from s or p_z orbitals are symmetrical around the molecular axis and are termed sigma (σ) molecular orbitals by analogy with the symmetrical s atomic orbitals. To form a systematic nomenclature, subscripts are used to indicate the type of atomic orbital which goes to form the molecular orbital, and the antibonding molecular orbitals are starred. Thus the molecular orbitals discussed so far are named systematically as follows:

$$\phi_B = \sigma_s : \phi_A = \sigma_s^* : \phi_1 = \sigma_p : \phi_2 = \sigma_p^*$$

The formation of sigma molecular orbitals is not restricted to combinations of like atomic orbitals. The only necessary condition is that the two components are symmetrical in sign about the bond axis. Then, if they are of similar energy, they may combine to form a molecular orbital. For example, an s and a p_z orbital may combine to form a bonding sigma molecular orbital as shown in Figure 3.11. The corresponding sigma antibonding orbital also exists.

Orbitals with different symmetries about the bond axis cannot combine to form molecular orbitals. This is illustrated in Figure 3.12 for the cases of s with p_y and p_z with d_{yz}. It can be seen that the 'plus-plus' and 'plus-minus' areas of overlap cancel each other out.

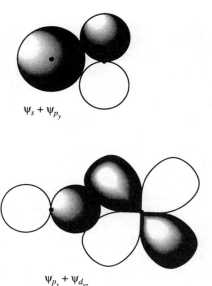

$$\psi_s + \psi_{p_y}$$

$$\psi_{p_z} + \psi_{d_{yz}}$$

FIGURE 3.12 *Combinations where no bond results*
Orbitals can only combine to give molecular orbitals if both are of the same symmetry with respect to the molecular axis; that is, if the signs of the wave functions change across the axis in the same way.

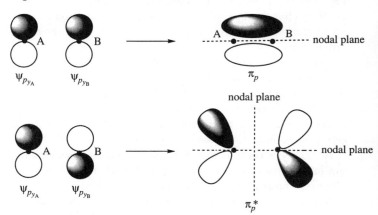

FIGURE 3.13 *The combination of atomic p_y orbitals*
When the z-axis is the molecular axis, the p_y orbitals are antisymmetrical with respect to this axis—that is, the wave function changes sign on crossing the axis. Such orbitals combine to give bonding, π, and antibonding, π^*, orbitals which are also antisymmetrical with respect to the molecular axis. The bonding orbital accumulates electron density between the nuclei, although less effectively than a σ orbital. The antibonding orbital has a nodal plane between the nuclei and perpendicular to the axis.

$$\psi_{p_z} + \psi_s \qquad \sigma_{s,p}$$

FIGURE 3.11 *The combination of an s and a p_z orbital*
Such an orbital is symmetrical around the internuclear axis and concentrates electron density between the nuclei. The corresponding antibonding orbital may be formed in a similar manner.

Although the p_y and p_x orbitals cannot enter into sigma bonding because they are antisymmetric (i.e. change in sign) across the bond axis, they can form a different type of bond by overlapping 'sideways-on' as shown in Figure 3.13. Such molecular orbitals, which have one nodal plane containing the bond axis, are called pi (π) molecular orbitals—again by analogy with atomic p orbitals. The first combination shown in Figure 3.13 accumulates charge in the region between the nuclei and is therefore bonding. The second mode of combining the atomic p orbitals has a nodal plane lying between the nuclei and perpendicular to that containing the bond axis. This second molecular orbital is antibonding. These are named, systematically, as π_p and π_p^* respectively.

The bonding π_p molecular orbital is of lower energy than the contributing atomic p orbitals, while the antibonding π_p^* molecular orbital is of higher energy by an equal amount. Because of the existence of the nodal plane containing the nuclei in the π orbitals, the π_p interaction has rather less effect on the electron density between the nuclei than has the σ_p interaction, and the energy gap between the bonding and the antibonding π molecular orbitals is less than that between the bonding and antibonding σ orbitals. The result of this is that the strength of a double bond, composed of a σ plus π contribution, is somewhat less than that of two σ bonds. This is of particular importance when accounting for the reduced stability of homonuclear double bonds between heavy p block elements (e.g. As = As, Sb = Sb, etc.) where the long bond distance and diffuse p orbitals result in poor overlap and a small π-bond energy, and thus σ-bonded compounds are more highly favoured.

Since the p_x and p_y atomic orbitals are both perpendicular to the molecular axis and identical with each other apart from their orientation, they both form π orbitals (with π_{p_x} perpendicular to π_{p_y}) in exactly the same way. The π_x and π_y levels are equal in energy as are the π_x^* and π_y^* levels. The combination of all the p atomic orbitals on two atoms to give the molecular orbitals of a diatomic molecule may therefore be shown on a composite energy level diagram, as in Figure 3.14.

This discussion may readily be extended to include d atomic orbitals which may combine with each other or with s or p orbitals to give σ or π molecular orbitals. It is also possible for two d orbitals to overlap each other with all four lobes to give a molecular orbital with two nodal planes containing the bond axis and mutually at right angles to each other. Such an orbital is termed a delta (δ) molecular orbital. These are found in metal–metal bonded dimeric complexes, typically of the second- and third-row transition elements. For example, molybdenum acetate, $[Mo_2(O_2CCH_3)_4]$, contains a Mo–Mo quadruple bond, formed from σ, two π, and a δ component. Complexes of this type are discussed in greater detail in section 16.7. Some examples of molecular orbitals involving atomic d orbitals are shown in Figure 3.15.

The discussion of covalent bonding between a pair of atoms given above may be summarized:

(a) Two atomic orbitals may combine to form two molecular orbitals centred on the nuclei, one of which concentrates electrons in the region between the nuclei and is bonding,

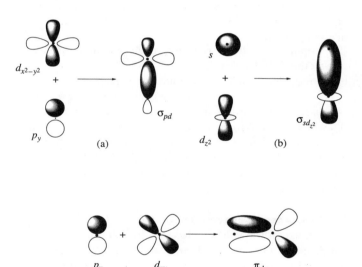

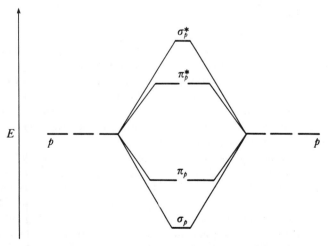

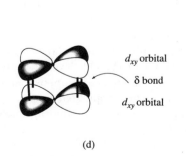

FIGURE 3.15 *Molecular orbitals involving d orbitals*
These diagrams indicate the formation of σ, π, and δ bonds, involving d orbitals. (a) sigma bond between p_y and $d_{x^2-y^2}$, (b) sigma bond between s and d_{z^2}, (c) pi bond between p_y and d_{xy}, (d) delta bond between two d_{xy} orbitals.

FIGURE 3.14 *Energy level diagram for the combination of p orbitals*
As the π interaction is weaker than the σ one, the π and π^* levels lie between the σ and σ^* ones.

while the other removes electron density from this region and is antibonding.

(b) Only those atomic orbitals which are of similar energy and of the same symmetry with respect to the inter-atomic axis may combine to form molecular orbitals.

(c) Any atomic orbitals which are symmetrical with respect to the bond axis may combine to form sigma molecular orbitals.

(d) Any atomic orbitals which are anti-symmetrical with respect to the bond axis may combine to form pi molecular orbitals.

Our next step is to try to describe the electronic structures of diatomic molecules in terms of these ideas. We must recognize that so far our discussion has been entirely qualitative. We have said nothing about the sizes of the ΔE terms, nor about the energy differences between, say, σ_p and π_p levels in Figure 3.14. Nor do we have any guidance about combining the energy levels of the orbitals of Figures 3.6 and 3.14 into a composite diagram. If full and accurate calculations could be carried out, all these questions would be answered from the wave equation. However, we have seen that this is not possible, so we have to turn to experimental evidence to help us to complete the picture.

There are three properties we are attempting to rationalize:

(1) Bond orders, reflected in measured bond lengths and heats of formation.

(2) Unpaired electrons, reflected in the magnetic properties (see section 7.10 for definitions).

(3) Energy levels, reflected in ionization energies.

It is convenient to break the discussion into two parts. First we discuss bond orders and magnetic properties in section 3.5 which we can do very successfully from a purely qualitative energy level diagram. Then we can get a more quantitative, accurate and sophisticated description by considering energy levels as given, especially, by photoelectron spectroscopy. This is taken up in section 3.6.

3.5 Bond orders of diatomic molecules

We can assign the valency electrons in a diatomic molecule, and hence predict the bond order and magnetic properties, if we simply combine Figures 3.6 and 3.14 by assuming the s diagram lies at lower energy than the p diagram, and there is no overlap or interaction. This gives us Figure 3.16.

We shall build up the electron configuration of the molecule by taking all the valency electrons from the atoms and feeding them into the molecular energy level diagram. We obviously fill the most stable level first, each orbital holds a maximum of two electrons with opposite spins, and orbitals of equal energy are first populated singly.

Consider first the case of the fluorine molecule, F_2. The fluorine atom has the electronic configuration $(1s)^2(2s)^2(2p)^5$, and the inner $(1s)^2$ shell is too tightly held to the nucleus to play any part in the bonding. We write the configuration [He] $(2s)^2(2p)^5$ to emphasize this (compare section 2.8). There are thus seven valency electrons from each atom to be fitted into the molecular orbitals of the fluorine molecule and, when these are filled in order of increasing

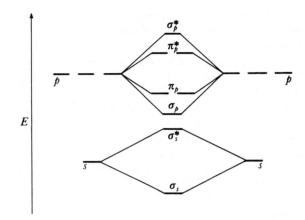

FIGURE 3.16 *Combined energy level diagram for combination of atomic s and p orbitals*

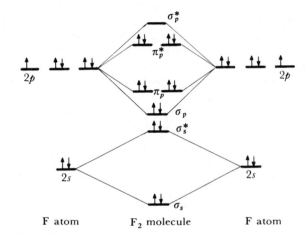

FIGURE 3.17 *Electronic structure of* F_2
An energy level diagram of the molecular orbitals is formed and all the valency electrons are placed in the molecular orbitals, according to Hund's Rules.

energy, the arrangement shown in Figure 3.17 results. This may be written as an equation:

$$2F[He]\,(2s)^2(2p)^5 \rightarrow F_2[He]\,[He]\,(\sigma_s)^2(\sigma_s^*)^2(\sigma_p)^2(\pi_p)^4(\pi_p^*)^4$$

To determine the bond order, we determine how many pairs of electrons in bonding orbitals there are in excess of the pairs in antibonding orbitals. In the fluorine molecule there are eight electrons in bonding orbitals (σ_s, σ_p, π_p) and six in antibonding sigma and pi orbitals. This gives an excess of two bonding over antibonding electrons and corresponds to the single bond in the fluorine molecule. All the electrons are paired, so fluorine is diamagnetic.

The molecular orbital energy level diagram for oxygen is shown in Figure 3.18. The outer electronic structure of the oxygen atom is $(2s)^2(2p)^4$, so that the oxygen molecule has two fewer electrons than the fluorine molecule and these will be missing from the highest level, the π^* one, leaving only two antibonding π electrons. As there are two electrons to enter the two π^* orbitals, which are of equal energy, they enter these orbitals singly, keeping their spins parallel in accordance with Hund's rules, giving the oxygen molecule

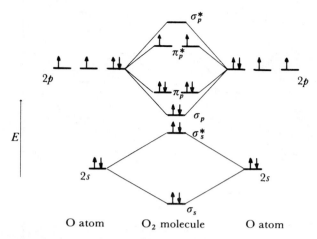

FIGURE 3.18 *Electronic structure of* O_2

two unpaired electrons. The structure is:

$$2O[He] (2s)^2(2p)^4 \rightarrow O_2[He][He] (\sigma_s)^2(\sigma_s^*)^2(\sigma_p)^2(\pi_p)^4(\pi_p^*)^2$$

There are eight electrons in bonding orbitals and four in the antibonding orbitals σ_s^* and π_p^*, giving a net count of four bonding electrons corresponding to the double bond. Since there are two unpaired electrons, O_2 is *paramagnetic*.

If this description of O_2 is compared with that derived from the Lewis electron pair theory, it will be seen that the molecular orbital treatment has the advantage. In order to complete the octet on oxygen, O_2 would be described as $\overset{..}{\underset{..}{O}} :: \overset{..}{\underset{..}{O}}$ thus predicting the double bond. However, the Lewis theory gives no reason for expecting unpaired electrons. This ready explanation of the paramagnetism of oxygen was one of the major early successes of the molecular orbital theory.

If diatomic neon, Ne_2, were to form, two further electrons would have to be added to the F_2 structure and they could go only into the highest energy antibonding orbital, σ_p^*. The number of antibonding electrons and the number of bonding electrons both equal eight, no bond results, and neon exists as a monatomic gas.

In a similar way, simple electron counting gives, for example

$$2N[He] (2s)^2(2p)^3 \rightarrow N_2[He][He] (\sigma_s)^2(\sigma_s^*)^2(\sigma_p)^2(\pi_p)^4$$
(diamagnetic, triple bond)

$$2C[He] (2s)^2(2p)^2 \rightarrow C_2[He][He] (\sigma_s)^2(\sigma_s^*)^2(\sigma_p)^4(\pi_p)^2$$
(paramagnetic, double bond)

As the example of C_2 shows, the mere fact that a diatomic species would have a high bond order does not prove that the element will exist in a stable form as a diatomic gas.

The electron counting procedure is simply extended to charged species by adding or subtracting electrons. For example, the superoxide ion O_2^- would be described

$$2O[He] (2s)^2(2p)^4 + e$$
$$\rightarrow O_2^-[He][He] (\sigma_s)^2(\sigma_s^*)^2(\sigma_p)^2(\pi_p)^4(\pi_p^*)^3$$
(paramagnetic, bond order $= 1\tfrac{1}{2}$)

Electron counting may also readily be applied to molecules where the atoms have higher n values than two for their valency levels. For example, the electronic structures of the other halogen molecules are exactly the same as for fluorine:

$$X_2(\sigma_s)^2(\sigma_s^*)^2(\sigma_p)^2(\pi_p)^4(\pi_p^*)^4$$

but here the atomic symbol X is to indicate that all the inner non-valency electron shells remain held by the nuclear attraction and play no part in the bonding. For example, for bromine

$$2Br[Ar] (4s)^2(4p)^5 \rightarrow Br_2[Ar][Ar] (\sigma_s)^2(\sigma_s^*)^2(\sigma_p)^2(\pi_p)^4(\pi_p^*)^4$$
(diamagnetic, single bond)

where the molecular orbitals are constructed from the $4s$ and $4p$ atomic orbitals and all the inner shells of the first, second, and third levels remain held by the atomic nuclei.

The analysis may be extended to heteronuclear molecules where the two atoms are in different periods. For example, in ClF, the larger nuclear charge of the chlorine atom lowers the energy of all its orbitals compared with the corresponding fluorine ones. As a result, the $3s$ and $3p$ levels of chlorine are approximately equal to the $2s$ and $2p$ levels of fluorine (and the Cl $2s$ and $2p$ levels are so tightly bound that they play no part in the bonding). The electronic diagram of ClF is then very similar to that of F_2 shown in Figure 3.17, except that the molecular orbitals are formed by overlap of F orbitals of the second shell with the corresponding Cl orbitals of the third shell:

$$Cl[Ne] (3s)^2(3p)^5 + F[He] (2s)^2(2p)^5$$
$$\rightarrow ClF[Ne][He] (\sigma_s)^2(\sigma_s^*)^2(\pi_p)^4(\pi_p^*)^4$$
(diamagnetic, single bond)

The structure of hydrogen halides follows similarly. For example, in HCl, the H $1s$ orbital is approximately equal to the Cl $3s$ and $3p$ orbitals in energy and is of correct symmetry to form a sigma bond either with the $3s$ or the $3p_z$ orbital (taking z as the H-Cl direction). The overlap is best with the $3p_z$ orbital, and a sigma bonding orbital is formed as in Figure 3.11, together with the corresponding antibonding orbital. The remaining valency orbitals of the chlorine are non-bonding and hold electron pairs. Thus HCl is described

$$H(1s)^1 + Cl[Ne] (3s)^2(3p)^5$$
$$\rightarrow HCl[Ne] (\sigma_{s,p})^2(\sigma_{s,p}^*)^0(3s)^2(3p_x)^2(3p_y)^2$$

If we now turn to consider heteronuclear species where the two atoms are in the same period, the only difference is in the relative levels of the atomic orbitals with those of the atom with highest nuclear charge lying lowest in energy. Consider nitric oxide and its cation, NO^+ (the nitrosonium ion). The energy level diagram for nitric oxide is given in Figure 3.19. The atomic orbital energy levels are no longer equal because of the difference in nuclear charges. The nitric oxide molecule has an odd number of electrons and the one of highest energy

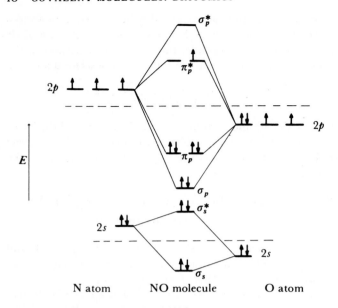

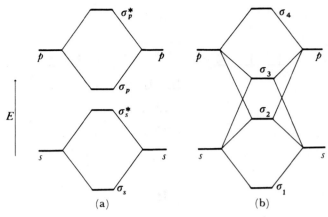

FIGURE 3.20

(a) *Sigma combinations without s-p mixing.*
(b) *Sigma combinations with s mixed with p.*

This shows the relation between the $\sigma_s \sigma_s^* \sigma_p$ and σ_p^* levels (a) and the levels σ_1 to σ_4 (b) which result when s and p contributions are allowed to mix. The main effect is to stabilize σ_2 relative to σ_s^* and to destabilize σ_3 relative to σ_p.

FIGURE 3.19 *Electronic structure of* NO
The energy level diagram differs from that of a homonuclear diatomic molecule, such as O_2, only in having the atomic levels at different energies. The splitting of the bonding and antibonding molecular orbitals of any particular type is symmetrical about the average energy of the constituent atomic orbitals.

occupies the antibonding π orbital.

$$N[He] (2s)^2(2p)^3 + O[He] (2s)^2(2p)^4$$
$$\rightarrow NO[He][He] (\sigma_s)^2(\sigma_s^*)^2(\sigma_p)^2(\pi_p)^4(\pi_p^*)^1$$

There are eight bonding electrons, in the sigma s and p levels and in the pi p level and three antibonding electrons, $(\sigma_s^*)^2$ plus $(\pi_p^*)^1$, leaving a net excess of five bonding electrons corresponding to a bond order of two and a half. If this molecule is ionized, the highest energy electron is the one removed and this is in the antibonding pi orbital. The excess of bonding over antibonding electrons goes up to six and the bond order rises from two and a half to three. (NO^+ is isoelectronic with N_2.) The cation is therefore more strongly bonded than the parent molecule. This can be seen experimentally in the infrared spectrum where the $N-O$ stretching frequency occurs at higher energy for NO^+ than for NO.

In all these cases, and many more, the observed magnetism matches with that predicted. Bond order is more difficult to quantify but all the observed changes are in the directions indicated by the predictions. For example, for oxygen species

	O_2^+	O_2	O_2^-	O_2^{2-}
Predicted bond order	$2\frac{1}{2}$	2	$1\frac{1}{2}$	1
Observed bond length (pm)	112.3	121.1	128	154.1

Thus the simple approach of feeding all the valency electrons into the levels of Figure 3.16, or related energy level diagrams like Figure 3.19, successfully rationalizes the observed bond orders and magnetism.

However, Figure 3.16 is not always an adequate description of the *energies* of the electrons in the molecular orbitals.

3.6 Energy levels in diatomic molecules

Closer consideration of Figure 3.16 raises two questions.

(1) Remembering that He_2, with a possible configuration $(\sigma)^2(\sigma^*)^2$, does in fact exist only as 2He atoms with $(1s)^2$, the general query arises whether distinct completely filled (bonding + antibonding) levels will persist, or will the better description be as non-bonding pairs on the atoms. That is, does the configuration $(\pi)^4(\pi^*)^4$ (implying *two* different orbital energies) describe the system better or worse than two filled p orbitals on each atom (implying only *one* orbital energy when the atoms are identical). Similarly, is the description $(\sigma_s)^2(\sigma_s^*)^2$ better or worse than $2 \times (s)^2$?

(2) We assumed that σ_s and σ_s^* were more stable than σ_p, and further that there was no interaction between them. However, since the bonding molecular orbitals σ_s and σ_p have the same symmetry, they may interact (or 'mix') with each other. In a similar manner, the antibonding molecular orbitals σ_s^* and σ_p^* have the same symmetry and also may interact with each other. It is important to note that only orbitals having the same symmetry can mix with each other in this way.

In fact, the separation of sigma molecular orbitals formed from atomic s orbitals and those formed from the p_z orbitals is a simplification: in principle, all orbitals of sigma symmetry on the atoms are expected to make some contribution to all the molecular sigma orbitals. That is, the two s and two p orbitals are, in general, combined in various proportions to give four molecular orbitals of sigma symmetry which we shall relabel σ_1, σ_2, σ_3, and σ_4, in order of increasing energy. The two extreme levels are similar to those in the simpler scheme with σ_1 close to σ_s and σ_4 close to σ_p^* in character. The major difference comes with σ_2 and σ_3, which both have substantial s and p contributions. The major effect of the mixing is to stabilize σ_2 below σ_s^* and to destabilize σ_3 relative to σ_p^*. The relation between the two schemes for the sigma molecular orbitals is indicated in Figure 3.20 parts (a) and (b).

The answers to queries (1) and (2) above come most readily from experimental data. While a number of techniques give information about energy levels, the most readily interpreted is photoelectron spectroscopy (see section 7.12).

To a good approximation, a particular photoelectron (PE) band may be regarded as giving the ionization energy of an electron in a particular orbital in the molecule. Thus we can use the ionization energies from the PE spectrum to put an energy scale on the molecular orbital energy level diagrams. Further, the fine structure of the PE band gives us the vibrational energy of the product ion A_2^+. If this is less than the vibrational energy of the parent molecule, A_2, it implies the ion has a lower bond order, and hence that an electron has been lost from a bonding molecular orbital in the ionization (cf. discussion of H_2 in section 7.12)

$$A_2 + h\nu \rightarrow A_2^+ + e$$

Conversely, equal or greater vibrational energies in A_2^+ implies the loss of a non-bonding or an antibonding electron.

As a simple example, let us consider the PE spectrum of HCl, shown schematically in Figure 3.21. Two bands are found using He(I) radiation. The one at 12·75 eV has the short vibrational series characteristic of the loss of a non-bonded electron while the long sequence of the 16·25 eV band is that expected where a bonding electron is lost. These assignments are supported by the vibrational separations

(Table 3.2). Thus the 12·75 eV ionization must be from the least strongly bound non-bonding level, the Cl(3p) level, while the 16·25 eV bond corresponds to ionization from the $\sigma_{s,p}$ orbital. The third occupied orbital, the Cl(3s) level, is too stable for excitation using He(I) radiation. We can present the results in a composite energy level/PE spectrum diagram as in Figure 3.22. Note that the levels of the empty orbitals, e.g. $\sigma_{s,p}^*$, cannot be located by photoelectron spectroscopy.

For a second example, consider O_2. Here an additional complication arises because of the unpaired electrons in the π^* orbital. If the ionization occurs from, say, the σ_p orbital, then the electron remaining in σ_p in the O_2^+ ion may have its spin parallel or antiparallel to the π^* electrons. Thus each energy state is doubled apart from that arising from π^* ionization. The analysis is indicated in Table 3.3. It will be seen that these observations could be fitted on to Figure 3.18.

However, if we turn to nitrogen, N_2, we find that we are forced to consider mixing among the sigma states, as in Figure 3.20b.

The He(I) photoelectron spectrum of N_2 is shown in Figure 3.23, and the data are listed in Table 3.4. It is immediately obvious that the 15·6 eV ionization is from an orbital that is only weakly bonding, and cannot therefore be

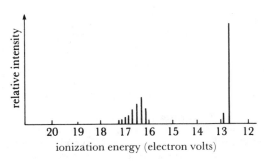

FIGURE 3.21 *The photoelectron spectrum of HCl*

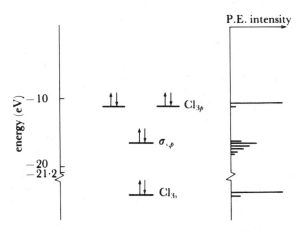

FIGURE 3.22 *Energy levels and photoelectron spectrum of HCl*

TABLE 3.2 Photoelectron bands of the hydrogen halides

	Molecular stretching frequency (cm^{-1})	Photoelectron energy (eV)	Vibrational features		Assignment
			Sequence	Separation (cm^{-1})	
HCl	2886	12·75	short	2660	non-bonding
		16·25	long	1610	bonding
HBr	2560	11·68	short	2420	non-bonding
		15·28	long	1290	bonding
HI	2230	10·2	short	2100	non-bonding
		13·8 approx	broad, bands overlap		bonding

Notes (a) p non-bonding band is actually doubled due to Jahn–Teller effect.
         (b) Excitation from 3s level is too energetic to be seen using He(I) radiation.

TABLE 3.3 Photo-ionization of oxygen, O_2
(Stretching frequency of neutral
$O_2 = 1555\,cm^{-1}$)

	Photoelectron energy (eV)	Vibrational energy (cm⁻¹)	Assigned to loss of electron which is	from
For O_2^+	12·07	1780	antibonding	π^*
	17·0	1010	bonding	π
	19·3	1110	bonding	σ_p
	26·9		antibonding	σ_s^*
	40·6		bonding	σ_s

Notes (a) Last two excitations used He(II) radiation.
(b) For comparison, loss of 1 s electrons needs > 540 eV.
(c) Levels doubled by spin effects have been averaged.

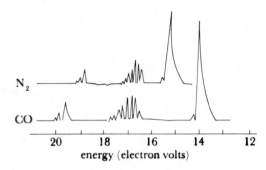

FIGURE 3.23 *Photoelectron spectra of* N_2 *and* CO

TABLE 3.4 Photoionization of N_2 and CO
(Stretching frequency of $N_2 = 2345\,cm^{-1}$,
CO $= 2157\,cm^{-1}$)

		Photoelectron energy (eV)	Vibrational separation (cm⁻¹)	Assignment of lost electron
For N_2^+	(1)	15·57	2150	weakly bonding
	(2)	16·69	1810	strongly bonding
	(3)	18·75	2390	weakly antibonding
	(4)	37·3		strongly bonding
	(5)	410		1s electron
For CO^+	(1)	14·01	2160	non-bonding
	(2)	16·53	1610	bonding
	(3)	19·68	1690	bonding
	(4)	38·3		strongly bonding
	(5)	293		C 1s electron
	(6)	542		O 1s electron

from the π level. Further, we expect the N_2 π level to be similar in energy to the O_2 π level which is 17·0 eV from Table 3.3. Thus the 16·69 eV level in N_2 has the expected attributes of the bonding π level. This leads to the conclusion that the weakly bonding level at 15·57 eV and the weakly antibonding level at 18·75 eV arise from the two higher sigma orbitals, and the least stable of these lies *above* the π level. This can only happen if σ_s and σ_s^*, and σ_p and σ_p^* interact with each other, as discussed earlier in this section. The net effect of this s–p mixing is to (a) raise σ_3 and lower

σ_2, and (b) average out their bonding and antibonding character to give relatively non-bonding combinations. It must be emphasized, however, that the π-levels are unaffected by this s–p mixing since they are orthogonal to the molecular axis. Thus the energy level diagram for N_2 has the π bonding level lying between σ_2 and σ_3 of Figure 3.20b. The electronic structure of N_2 is therefore given as in Figure 3.24a.

This interpretation is supported by the data for CO, Table 3.3. Carbon monoxide has the same number of valency electrons as N_2 (such species are termed *isoelectronic*) but its energy level diagram is a skew one, reflecting the unequal atomic levels, Figure 3.24b. We see that the π level of CO is of almost exactly the same stability as for N_2, but the electrons in σ_3 are less tightly bound (reflecting the greater contribution from $C2p_z$ in σ_3) while those in σ_2 are more tightly bound (O contribution greater) than in the respective orbitals of N_2.

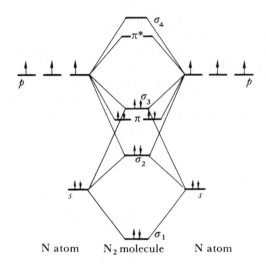

FIGURE 3.24 (a) *The electronic structure of* N_2
In the structure of N_2, the bonding π levels lie below σ_3. The bonding and antibonding contributions of σ_2 and σ_3 cancel, leaving σ, and the two π levels to make up a total bond order of three.

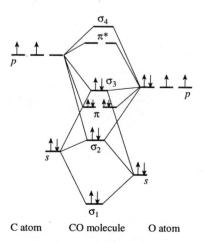

FIGURE 3.24 (b) *The electronic structure of* CO

In general, the importance of the mixing between s and p orbitals depends inversely on the difference in energy between them. With atoms to the right of the Periodic Table, like O or F, the separation into σ_s and σ_p orbitals is a good approximation, and the schemes for F_2 or O_2 shown in Figures 3.17 and 3.18 are satisfactory. Further to the left, the atomic s and p orbitals are closer in energy and the interaction between them is significant. Thus nitrogen is described as in Figure 3.24a:

$$2N[He](2s)^2(2p)^3 = N_2[He][He](\sigma_1)^2(\sigma_2)^2(\pi)^4(\sigma_3)^2$$

The bonding character added to σ_2 relative to σ_s^* exactly equals the loss of bonding character in σ_3 relative to σ_p and there is no nett bonding effect when σ_2 and σ_3 are both filled. Thus N_2 has six more bonding than antibonding electrons ($\sigma_1^2\pi^4$) making a triple bond.

In carbon monoxide and in the cyanide ion, the σ_3 level also lies above the bonding π orbitals. The species are triply bonded and the formation equations may be written:

$$C[He](2s)^2(2p)^2 + O[He](2s)^2(2p)^4$$
$$\rightarrow CO[He][He](\sigma_1)^2(\sigma_2)^2(\pi)^4(\sigma_3)^2$$

and $\quad C[He](2s)^2(2p)^2 + N[He](2s)^2(2p)^3 + e^-$
$$\rightarrow CN^-[He][He](\sigma_1)^2(\sigma_2)^2(\pi)^4(\sigma_3)^2$$

Finally, it is worth returning again for a somewhat closer look at the oxygen molecule, with an emphasis on how its electronic state influences its properties. As we discussed earlier in sections 3.5 and 3.6, the ground state of this molecule is that which places one unpaired electron in each of the π_p^* orbitals to give a total of two unpaired electrons, thereby accounting for the observed paramagnetism of the O_2 molecule. However, there are two other possibilities for assigning the two electrons to the π orbitals—the electrons can both be placed into one π orbital with opposite spin, or into the two π orbitals with opposite spins, respectively (a) and (b) below:

(a) $^1\Delta$ (b) $^1\Sigma$

These two other diamagnetic forms of oxygen, which are predicted by a simple extension of the rules for the occupation of molecular orbitals, have in fact been detected in the gas phase. The commonest form of O_2, with two parallel spin unpaired electrons, is termed 'triplet' oxygen, and given the spectroscopic symbol $^3\Sigma$, whereas the other two are termed 'singlet' oxygen $^1\Delta$ and $^1\Sigma$ (the terms singlet and triplet refer to the spin multiplicity, $2m_s + 1$). The $^1\Delta$ form of singlet oxygen is a chemically important species since it undergoes two-electron reactions, such as Diels–Alder type cyclizations,

similar to those undergone by the isoelectronic molecule ethylene. In contrast, the triplet form of oxygen undergoes radical reactions. Singlet oxygen is also responsible for the degradation of rubbers and other synthetic polymers by air, and perhaps the best known 'occurrence' of singlet oxygen is in the chemiluminescence demonstrations involving the compound luminol. The bond lengths of all three forms of the O_2 molecule have been determined and, as would be expected, all point to there being a nett double bond, but with the singlet excited states having slightly longer O–O bonds.

3.7 Summary

By combining atomic orbitals into molecular orbitals, we can describe the bonding in diatomic molecules in a way that gives greater depth to the early ideas of electron-pair bonds.

Using simple combinations of the s and p orbitals, as in Figure 3.16, we can feed in the valency electrons and arrive at correct indications of the bond orders and numbers of unpaired electrons for any diatomic species A_2 or AB, or any of their cations or anions. For O_2 or F_2, such an approach also gives the correct relative ordering of the energies of the molecular orbitals.

For the general case, particularly with molecules of atoms to the left of oxygen in the Periodic Table, a more accurate description of the relative energies of the molecular orbitals allows for mixing of s and p contributions to the sigma orbitals. This is often sufficiently large to alter the relative position of the π bonding orbital, as in Figure 3.24.

There are thus two stages in the description of a diatomic species:
(1) determine the bond order and magnetic properties by filling the levels of Figure 3.16;
(2) find the relative energy levels, and the best description of the orbitals, by reference to the photoelectron spectrum or other experimental data. In absence of such data, be prepared to use Figure 3.24 for molecules containing atoms to the left of oxygen.

The molecular energy level diagrams may also be used in the discussion of the excited states of the molecules. If the molecule absorbs energy equal to the difference between the energy of an occupied orbital and that of one of the higher, empty orbitals, an electron will undergo a transition into the upper orbital, e.g. $\pi \rightarrow \pi^*$ in N_2. The energy differences between molecular orbitals can thus be observed in the electronic spectrum of a diatomic molecule and such experimental values compared with the calculated energy levels, just as in the case of atomic orbitals. The electronic transitions for the light diatomic molecules such as nitrogen are of relatively high energy and occur in the far ultraviolet region of the spectrum.

Thus the photoelectron spectrum gives the absolute energies of the occupied orbitals, while the visible–UV spectrum can give the energy differences between occupied and empty orbitals. By combining both sets of data, the diatomic molecular energy level diagrams can be made quantitative.

PROBLEMS

In this chapter, we start to combine atoms into chemically much more significant entities—molecules. All the atomic properties discussed in Chapter 2 are significant in the resulting molecules as later chapters show. However, we emphasize here the atomic orbitals and show how we can think of them combining—considering them purely as waves with one crest reinforced by another or nullified by a trough. We turn the resulting wave into an electron density distribution and decide whether the attractions between the positive nuclear charges and this distribution will be greater or less than in the separate atoms. This brings us qualitatively to the idea of the energy of stabilization, and then to a quantitative picture using data from photoelectron spectra.

Question 1 revises basic ideas (see Figure 3.1).

Many variants of the other questions will suggest themselves. Thus the bond order of any diatomic A_2 or AB (real or hypothetical) or of any ions derived from these can be predicted. You can look for evidence of bond order changes in bond lengths, bond strengths (reflected by heats of formation or vibrational stretching frequencies) or from photoelectron spectra.

3.1 (a) The relative molecular mass of XCl_3 is 181. What is the relative atomic mass of X?

(b) A substance contains $16.1\%H$, $38.7\%C$ and $45.1\%N$. If the relative molecular mass is 31, decide its structural formula.

3.2 Using your values from question 2.5, draw out the plots corresponding to Figures 3.2 and 3.4 for two H atoms whose nuclei are separated by

(a) 50 pm and
(b) 200 pm.

3.3 Using z as the internuclear axis in AB, write down all the combinations of d orbitals on A with s, p or d orbitals on B which will give

(a) σ bonds,
(b) π bonds.

3.4 Work out the electron configurations, bond orders, and numbers of unpaired electrons for the species O_2^+, O_2, O_2^- and O_2^{2-}. (Compare your answers with section 3.5.)

3.5 Using Figure 3.16, determine the configurations of the diatomic molecules Li_2, Be_2, B_2, C_2, N_2, O_2 and F_2. Arrange in decreasing order of bond strength.

3.6 Repeat the exercise in 5 with all the possible mixed diatomics AB.

3.7 If Figure 3.18b applies, which of the unstable species in questions 5 and 6 might be stabilized?

3.8 Which of the species CN, CO, NO, NF and OF would be stabilized by

(a) loss of an electron,
(b) gain of an electron?

3.9 The (simplified) photoelectron spectrum of NO shows the following bands

Ionization energy (eV)	Vibrational separation (cm^{-1})
9.26	2260
16.4	1550
18.1	1180

Assign and compare with CO and O_2. (The stretching frequency of NO is $1890\,cm^{-1}$).

4 Polyatomic Covalent Molecules

4.1 Introduction

Once more than two atoms are present in a molecule, the problem of molecular shape arises. For example, an AB_4 molecule might be a tetrahedron, a square plane, or some less symmetrical shape. Since the molecule in its ground state adopts the shape which minimizes the total energy, a complete bonding theory would produce the shape of the molecule as one of its results. Unfortunately, such high-accuracy calculations are only possible for simple species with few electrons. Thus the shape, or at least the symmetry, has to be determined by experiment, and we then require of our bonding theory a *description* of the shape rather than a prediction.

While the shape may always be determined experimentally (see Chapter 7), it is convenient to be able to decide on the most likely shape of a polyatomic molecule or ion. It turns out that this can be done, quite simply, by considering the repulsive forces between electron pairs in the valency shell of the central atom (or atoms) of the species. This VSEPR theory (*v*alence *s*hell *e*lectron *p*air *r*epulsion) gives a good qualitative prediction of the shape of a molecule which turns out to be accurate for about 95 % of all main group compounds and is also basically correct for transition metal complexes.

We outline the VSEPR theory and show the shapes predicted by it in the next two sections before we turn to a discussion of bonding in polyatomics.

4.2 The shapes of molecules and ions containing sigma bonds only

The case of sigma-bonded species is the simplest and is discussed first.

4.2.1 *The arrangement of sigma bonds*

When two atoms are bound together, electron density is concentrated in the region of space between them (sections 3.3 and 3.4). If a central atom is bonded to a number of others, it is reasonable to expect the bonds from the central atom to be as far apart as possible in order to reduce the electrostatic repulsions between the electron-dense regions in the bonds. For a triatomic molecule, such as $BeCl_2$, a linear configuration $Cl - Be - Cl$ will minimize the repulsions between the electrons in the two $Be - Cl$ bonds. Similarly, when there are three attached atoms, as in BCl_3, an equilateral triangle with the $Cl - B - Cl$ angles all $120°$ is the expected form for the molecule. Table 4.1 shows the expected configurations for molecules of the types AB_n. In each case the configuration is the one of highest symmetry.

Coordination numbers greater than six are uncommon. Iodine is seven-coordinate in the pentagonal bipyramidal IF_7 and tellurium, for example, is seven- and eight-coordinate in complexes derived from the hexafluoride. There are many examples of the cases shown in Table 4.1, among ions as well as among molecules, and the tetrahedron and octahedron are especially common shapes.

If more than one type of atom is bonded to the central one, the configuration becomes less symmetrical although retaining the basic shape shown in the Table. (The tetrahedral configuration of all kinds of organic molecules is an obvious example.) In a molecule AB_rX_s, if the $A - X$ bond is shorter and stronger than the $A - B$ bond the electron density near A in the X directions will be greater than in the B directions. As a result, the BAB angles close up and the XAX angles open out relative to the values given in Table 4.1 for symmetrical AB_n cases. In addition, the AX bonds will tend to be as far apart in the molecule as possible; for example an octahedral AB_4X_2 species would be expected to be most stable in the *trans* configuration, compare Figure 13.15.

It is not always obvious, in a particular compound, in which direction away from the symmetrical configuration a distortion will occur. Bond lengths, bond polarities, and steric effects may all come into play and tend in opposite directions. However, most changes in bond angles due to unsymmetrical substitution are relatively small—c.f. the parameters of the silyl halides below—and the basic shape remains the one of Table 4.1.

TABLE 4.1 Shapes of AB_n molecules

n	Formula	Shape		Angles	Examples
2	AB_2	linear	○—●—○	B–A–B $= 180°$	$BeCl_2$, $HgCl_2$, XeF_2
2 .	AB_3	triangular		B–A–B $= 120°$	BF_3
4	AB_4	tetrahedral		B–A–B $= 109.5°$	$SiCl_4$, BH_4^-, CH_4
5*	AB_5 or AB_3B_2'	trigonal bipyramidal		B–A–B $= 120°$ B'–A–B' $= 180°$ B–A–B' $= 90°$	PCl_5, PCl_3F_2
6	AB_6	octahedral		B–A–B $= 90°$	SF_6, PCl_6^-

*All the B positions in all these configurations are equivalent, except in the trigonal bipyramid where the two apical positions (B') are not equivalent to the three equatorial (B) ones.

H – Si – H angle in silyl halides, SiH_3X:
    X = F, angle $= 109\frac{1}{2}°$: X = Cl, angle $= 110\frac{1}{2}°$:
        X = Br, angle $= 111\frac{1}{2}°$

4.2.2 The effect of lone pairs

The most important case of distortion by unsymmetrical substitution comes when an atom in the molecule is replaced by an unshared pair of electrons on the central atom. For example, ammonia NH_3, is not a plane triangular molecule as a casual glance at Table 4.1 would suggest, but is pyramidal with an H – N – H angle of $107°$. This angle is near the tetrahedral value and a count of the electrons on the nitrogen shows that there is a lone or unshared pair of electrons in addition to the three bond pairs. If the lone pair occupies a specific direction in space, the four electron pairs would still be expected to have a basically tetrahedral arrangement. Thus ammonia is an extreme case of an unsymmetrically substituted tetrahedron, AB_3X. Since a lone pair is subject to the attraction only of the central atom nucleus, rather than being shared between the central atom and a bonded atom, the electron density of a lone pair is concentrated close to the

central atom as shown schematically in Figure 4.1. Thus the lone pair electrons exert greater repulsion than the bond pairs, and the bond angles close up when a lone pair is present. For example, the X – M – X angles of all the nitrogen group

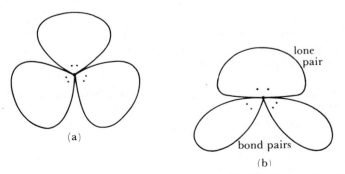

FIGURE 4.1 *Three electron pairs on a central atom:* (a) *three equal pairs,* (b) *two bond pairs and a lone pair*
As the lone pair is influenced only by the central nucleus, its density is concentrated much more closely to the central atom than the bond pairs and it thus dominates the stereochemistry.

trihalides such as NF_3 or SbI_3 lie within the range 97°–104° compared with the tetrahedral angle of $109\frac{1}{2}°$. Again, where geometrical isomerism is possible, lone pairs will tend to be as far apart as possible. Thus a central atom with two lone pairs and four bond pairs around it, is expected to form a square planar molecule, with the two lone pairs in the *trans* positions of Figure 4.5c.

When lone pairs are taken into account, the structure of any sigma-bonded species may be formulated in terms of two simple rules:

1. The basic shape of a molecule or ion depends on the number of electron pairs surrounding the central atom (lone pairs plus bond pairs) and is that which follows from Table 4.1, assuming that lone pairs occupy positions in space. It must be emphasized that the shape of the *molecule* is determined by the arrangement of the *atoms* themselves. Thus, the NH_3 molecule, while it has four electron pairs in a distorted tetrahedral arrangement, is properly described as trigonal pyramidal and *not* as tetrahedral.

2. Repulsions decrease in the order:— lone pair—lone pair > lone pair—bond pair > bond pair—bond pair, with the results that (a) lone pairs tend to be as far apart from each other as possible and (b) bond angles close up compared with those given in Table 4.1 for the regular structure of the same total number of electron pairs.

It is convenient to use the symbol E for a lone pair of electrons. The main types are listed in Table 4.2 and discussed below. In all these cases the ligands (a general term for any atom or group attached to the central atom) may be radicals like alkyl groups, or ions like cyanide, or neutral groups donating lone pairs like water or ammonia, as well as atoms as in the Table. Thus, $[Zn(H_2O)_4]^{2+}$ is a zinc ion (with no valency electrons of its own) surrounded by four electron pairs donated by the four co-ordinated water molecules and the shape is therefore tetrahedral.

4.2.3 *Three pairs*

Figure 4.2 shows the case where there are three electron pairs in the valence shell. When one of these pairs is un-

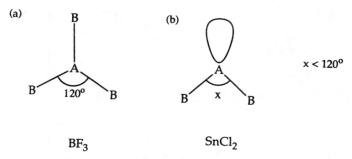

FIGURE 4.2 *Shapes of species with three electron pairs around the central atom:* (a) *three bonds,* (b) *two bonds and one lone pair*

shared, the V-shaped molecule of Figure 4.2b results. (It must be noted that lone pair, or other valency electron density is rarely found experimentally. The atom centres are normally determined and the electron positions deduced from the resulting structure. In this and the following Figures, the lone pairs are indicated schematically to show the relation to the basic structure.) Because of the greater repulsion of the lone pair, the bond angle is reduced from the value of 120°.

4.2.4 *Four pairs*

Shapes derived from the tetrahedron are shown in Figure 4.3 and illustrated by methane, ammonia, water, and their heavier analogues. CH_4, SiH_4 and GeH_4 are regular tetrahedra with bond angles of 109° 28′ (Figure 4.3a). In ammonia, one position is occupied by the lone pair of electrons, and the molecular shape is the trigonal pyramid of Figure 4.3b, with the $H-N-H$ angle reduced to 107° 18′ by the repulsion of the lone pair. Water has two unshared electron pairs on the oxygen atom, so the molecule is V-shaped as shown in Figure 4.3c and the increased repulsion reduces the $H-O-H$ angle to 104° 30′.

The bond angles in the heavy element analogues of ammonia and water all show considerable reduction from the tetrahedral angle, Table 4.3. The angles of the compounds of the oxygen Group are all smaller than those of the

TABLE 4.2 Shapes of species with lone pairs of electrons

Total number of electron pairs in the valence shell	Basic shape	Number of lone pairs	Formula	Shape	Examples
3	triangle	1	AB_2E	V-shape	$SnCl_2$ (gaseous)
4	tetrahedron	1	AB_3E	trigonal pyramid	NH_3, PF_3
		2	AB_2E_2	V-shape	H_2O, SCl_2
5	trigonal bipyramid	1	AB_4E	see Figure 4.4	$TeCl_4$
		2	AB_3E_2	T-shape	ClF_3
		3	AB_2E_3	linear	$(ICl_2)^-$
6	octahedron	1	AB_5E	square pyramid	IF_5
		2	AB_4E_2	square plane	$(ICl_4)^-$

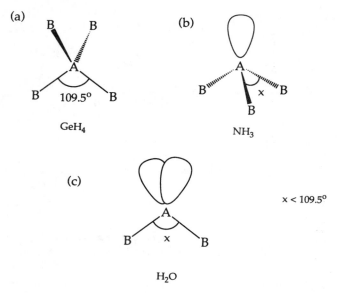

FIGURE 4.3 *Shapes of species with four electron pairs around the central atom:* (a) *four bonds,* (b) *three bonds and one lone pair,* (c) *two bonds and two lone pairs*

corresponding compounds of the nitrogen Group elements, reflecting the enhanced repulsion effect of the two lone pairs.

The variations in angle shown in Table 4.3 may be rationalized in terms of the electron densities in the bonds of these compounds: the higher the bond electron density at the central atom, the more resistance there is to the bond-closing effect of the lone pairs. Thus, comparing NH_3 and PH_3, the $P-H$ bond is longer than the $N-H$ bond and therefore the electron density is less. Furthermore, as phosphorus is less electronegative than nitrogen, the electron density is concentrated nearer the hydrogen in $P-H$ than in $N-H$. Both these factors mean that there is less electron density at the phosphorus atom in the $P-H$ bonds, than is the case at nitrogen in NH_3, and therefore the $\widehat{HPH}$ angle in PH_3 is much more drastically reduced by the lone pair repulsion. A similar effect occurs if the phosphorus, in turn, is replaced by the rather larger and less electronegative

TABLE 4.3 Bond angles in some AX_3E and AX_2E_2 compounds

		(XAX angles)			
NH_3	$107\cdot3°$	NF_3	$102\cdot5°$		
PH_3	$93\cdot8°$	PF_3	$97\cdot8°$	PCl_3	$100\cdot1°$
AsH_3	$91\cdot8°$	AsF_3	$96\cdot2°$	$AsCl_3$	$98\cdot7°$
SbH_3	$91\cdot3°$	SbF_3	$87\cdot3°$	$SbCl_3$	$99\cdot5°$
H_2O	$104\cdot5°$	OF_2	$103\cdot2°$	OCl_2	$111°$
H_2S	$92\cdot2°$	SF_2	$98°$	SCl_2	$100°$
H_2Se	$91\cdot0°$				
H_2Te	$89\cdot5°$				

All values measured on molecules in the gas phase. Most values are accurate to $\pm0\cdot5°$.

atoms, As or Sb, and the $\widehat{HAsH}$ and $\widehat{HSbH}$ angles show further contraction. The biggest relative change in size and electronegativity comes between nitrogen and phosphorus (compare Tables 2.10 and 2.14) and this is the point where the biggest change in bond angle occurs. Parallel changes are found for H_2O, H_2S, H_2Se and H_2Te.

If NH_3 and NF_3 are now compared, changes in bond length play only a minor role as fluorine is only a little bigger than hydrogen. However, the electronegativities fall in the order $F > N > H$ so that the electron density in the $N-F$ bond is concentrated towards the fluorine—the reverse of the polarization of the $N-H$ bond. Thus the lone pair repulsion closes up the $\widehat{FNF}$ angles more than the $\widehat{HNH}$ ones. Again, a parallel effect is observed in H_2O and OF_2.

The angles in the remaining halogen compounds of these elements are all around $100°$ and therefore larger than in the hydrides while electronegativity effects would lead us to predict angles smaller than in the hydrides. This may be a result of steric effects with the large halogen atoms coming into contact, or it may be an indication that π bonding is occurring between the halogen and the central atom (see section 4.3) which returns electron density to the central atom.

4.2.5 *Five pairs*

Five electron pairs around the central atom give the trigonal bipyramid as the basic shape, Figure 4.4. For this shape, the two apical (sometimes termed axial) positions (marked B') are not equivalent to the three equatorial (B) positions. When the ligands are different, for example in PF_3Cl_2, the more electronegative one occupies the axial position.

For trigonal bipyramids with one lone pair, this occupies the equatorial position giving the AB_2EB_2' configuration because this minimises the repulsions. In this configuration, the lone pair makes two angles of $90°$ to the $A-B'$ bond pairs and two angles of $120°$ to the $A-B$ bonds. If the lone pair was in the axial position, there would be three $90°$ angles, to the $A-B$ bonds, and one $180°$ angle to $A-B'$. As the electron density falls off exponentially, the repulsion effect is much greater at small angles than at large angles. The dominating factor is the additional $90°$ angle and the equatorial position of the lone pair is the only one found.

For two, or for three, lone pairs in the trigonal bipyramidal configuration, there are again a number of alternative structures possible. However, the only ones found are those where the lone pairs are equatorial. Thus, for two lone pairs, the structure is the T-shaped ABE_2B_2' one of Figure 4.4c, while three lone pairs give linear AE_3B_2'.

In AB_2EB_2' and ABE_2B_2' the repulsions from the lone pairs close up the bond angles, as shown for $TeCl_4$ and ClF_3 in Figures 4.4b and c. The repulsions of the three, symmetrically placed, lone pairs in AE_3B_2' balance out, so these species are strictly linear.

4.2.6 *Six pairs*

When six electron pairs are present around the central atom, the basic shape is the regular octahedron shown in

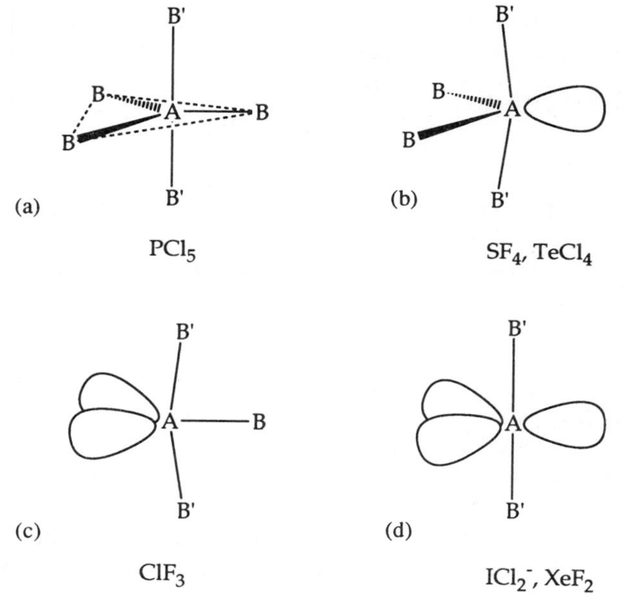

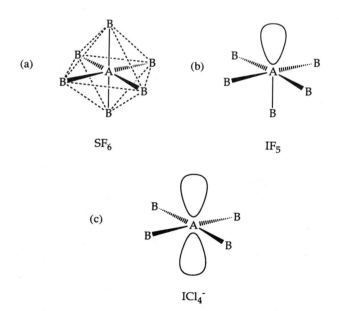

FIGURE 4.4 *Shapes of species with five electron pairs around the central atom:* (a) *five bonds,* (b) *four bonds and one lone pair,* (c) *three bonds and two lone pairs,* (d) *two bonds and three lone pairs*
The axial and equatorial positions are not equivalent in the trigonal bipyramid. All examples known so far with lone pairs have these lone pairs in equatorial positions.

Figure 4.5a. In this case, unlike the trigonal bipyramid, all the positions are equivalent and all the BAB angles are 90°. (In the conventional diagram of the octahedron the equatorial positions are commonly linked up to clarify the figure, and this tends to suggest that the apical and equatorial positions are non-equivalent, so care must be taken until such diagrams become familiar.)

FIGURE 4.5 *Shapes of species with six electron pairs around the central atom:* (a) *six bonds,* (b) *five bonds and one lone pair,* (c) *four bonds and two lone pairs*

If one unshared pair is present, the structure becomes the square pyramid of Figure 4.5b. When there are two lone pairs, the configuration which minimizes the lone pair—lone pair interaction is the square planar configuration in which the lone pairs are *trans* to each other at an angle of 180°, as in Figure 4.5c. There are no examples of more than two lone pairs in the octahedral case.

4.2.7 *Summary and examples*
With these ideas in mind, it is possible to put the process of determining the shape of any singly-bonded species on a formal basis in terms of simple rules.

1. Determine which atom is the central one. This is usually the least electronegative one, e.g. Cl in ClO_3^-, O in OF_2. If there are equally plausible choices for the central atom, then the shape should be worked out for each.

2. Determine the number of electron pairs on the central atom by adding up the number of valency electrons on the central atom plus one electron for each bond (this is the electron contributed by the second element to the bond) plus one electron for each negative charge, or minus one for each positive charge.

3. The shape then follows from the cases listed in Tables 4.2 and 4.3. Two other rules may be added to allow for special classes of compound:

4. When a co-ordinate bond is present, two electrons are added to the total around the central atom since the donor atom or group is providing both the electrons of the bond.

5. Transition elements follow the rules above but their valency shell *d* electrons are not always included when determining the number of electrons around the central atom. The shapes of species with partly-filled *d* shells is discussed in Chapter 13; here we need only note that the electron pair analysis applies rigorously to those cases, d^0, d^5 and d^{10}, where the *d* orbitals are symmetrically occupied and also, in practice, to all configurations except d^4, d^7 and d^9 (which are distorted) and d^8, which is commonly square planar.

  This approach is illustrated in the examples below.
(*a*) *The two coordinated species,* SCl_2, ICl_2^- *and* $Ag(CN)_2^-$.

(i) SCl_2: sulfur has the electron configuration $[Ne](3s)^2(3p)^4$. The electron count is

valency electrons from S = 6 electrons
add 1 bond electron per Cl = 2 electrons
total = 8 electrons = 4 pairs

SCl_2 has therefore four electron pairs and a V-shaped AB_2E_2 structure:

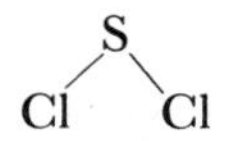

S
Cl   Cl

(ii) ICl_2^-: similarly

$$I = 7 \text{ electrons}$$
$$2Cl = 2 \text{ electrons}$$
$$\textit{add} \text{ for negative charge} = 1 \text{ electron}$$

ICl_2^- has thus five electron pairs and the structure is the linear $AB_2'E_3$ one: $(Cl-I-Cl)^-$.

(iii) $Ag(CN)_2^-$: silver has the electronic configuration $[Kr](4d)^{10}(5s)^1$ of which only the s electron is used. The number of electrons around the silver is thus:

$$Ag \; s \text{ electron} = 1$$
$$\textit{add} \text{ bond electron from each CN} = 2$$
$$\textit{add} \text{ negative charge electron} = 1$$
$$\text{total} = 4 \text{ electrons}$$

Thus there are two electron pairs on the Ag and the structure is linear AB_2: $(NC-Ag-CN)^-$.

(b) *The rare gas fluorides, XeF_2, XeF_4, and XeF_6.* The xenon atom is regarded as using all eight s^2p^6 electrons. The electron count then proceeds as follows:

XeF_2:

$$Xe = 8 \text{ electrons}$$
$$\text{bond electrons from } 2F = 2 \text{ electrons}$$

XeF_2 has thus five electron pairs around the xenon atom and the structure is the linear $AB_2'E_3$ type.

XeF_4:

$$Xe = \text{ electrons}$$
$$4F = 4 \text{ electrons}$$

XeF_4 is therefore the square planar AB_4E_2 structure derived from an octahedral arrangement of six electron pairs.

XeF_6:

$$Xe = 8 \text{ electrons}$$
$$6F = 6 \text{ electrons}$$

In XeF_6 there are thus seven electron pairs around the xenon. Infrared spectroscopy and electron-diffraction on XeF_6 show the molecule to be a distorted octahedron, with the distortion opening up one of the triangular XeF_3 faces in order to accommodate the lone pair of electrons on Xe. A number of other seven-electron pair structures are also known, such as IF_7, TeF_7^- and IOF_6^-, and all of these are pentagonal bipyramids.

(c) *PCl_5 and its dissociation ions, PCl_4^+ and PCl_6^-.* Phosphorus pentachloride exists as monomeric PCl_5 units in the gas phase but forms $PCl_4^+PCl_6^-$ in the solid by transfer of a chloride ion.

PCl_5:

$$\text{valency electrons P} = 5 \text{ electrons}$$
$$\text{bond electrons from } 5Cl = 5 \text{ electrons}$$

PCl_5 thus forms a trigonal bipyramid in the gas phase.

PCl_4^+:

$$P = 5 \text{ electrons}$$
$$4Cl = 4 \text{ electrons}$$
$$\text{Charge} = -1 \text{ electron for the plus charge}$$

Thus there are four electron pairs around the phosphorus in PCl_4^+ and the shape is tetrahedral.

PCl_6^-:

$$P = 5 \text{ electrons}$$
$$6Cl = 6 \text{ electrons}$$
$$\text{Charge} = 1 \text{ electron}$$

There are thus six electron pairs in PCl_6^- and the shape is a regular octahedron.

The ionization in the solid state probably derives from the impossibility of packing the five-coordinate structure regularly.

(d) *As an example of coordinate bonds, consider the adduct $BF_3.NH_3$.*

Boron has three valency electrons, therefore BF_3 has three electron pairs around the boron and is a planar molecule with FBF angles of 120°. Ammonia is a trigonal pyramid with a lone pair on the nitrogen as already seen. As the boron has four valency shell orbitals, one remains vacant in BF_3 and this can accept the electron pair of the ammonia to form the adduct.

$F_3B \leftarrow NH_3$: The electron count at the boron atom in the adduct is

$$\text{valency electrons B} = 3 \text{ electrons}$$
$$\text{one bond electron from } 3F = 3 \text{ electrons}$$
$$\text{two bond electrons from N} = 2 \text{ electrons (as the electron}$$
$$\text{pair is donated)}$$

The boron has therefore four electron pairs around it and the configuration (of the three F atoms and the N) is tetrahedral: the FBF angle is decreased towards the tetrahedral value of $109\frac{1}{2}°$.

The electron count at the nitrogen atom remains at four electron pairs as the two donated electrons remain in the nitrogen valence shell. The nitrogen is thus surrounded tetrahedrally by the three H atoms and the B atom.

(e) *Transition metal compounds.*
$TiCl_4$. Titanium has four valency electrons, d^2s^2. The electron count is

$$\text{valency electrons Ti} = 4 \text{ electrons}$$
$$\text{bond electrons from } 4Cl = 4 \text{ electrons}$$

Thus $TiCl_4$ is tetrahedral.

For the majority of transition metal compounds, this approach must be modified as the d electrons contribute less directly to the shape. To construct a general approach, ML_4 or ML_6 complexes of transition elements, M, and any ligand, L (such as halide, cyanide, water, ammonia) are regarded as formed from an ion M^{n+} and four, or six, ligands donating a pair of electrons each. Any d electrons on M^{n+} are not used and the shapes are ML_4 = tetrahedral and ML_6 = octahedral. (Notice that $TiCl_4$ can be treated in this way as Ti^{4+} and $4Cl^-$: this gives the correct shape and electron count, although it is misleading in implying ionic bonding.)

Compounds with other coordination numbers are treated similarly, leading to shape predictions following Table 4.1. This approach will be correct in about three-quarters of all transition metal compounds. However, the underlying d^n

configuration does affect the shape, especially when $n = 4, 7, 8$, or 9, and it is better to tackle transition metal compounds by the different approach of Chapter 13.

4.3 The shapes of species containing pi (π) bonds

The extension of the discussion of the preceding section to molecules and ions containing π bonds is basically simple. As the π electrons in a double bond follow the same direction in space as the σ electrons of that bond, the electrons used in pi bonding have no major effect on the shape of the molecule.

4.3.1 Electron counting procedure for pi electrons

The molecular shape is determined mainly by the sigma bonds and lone pairs and the presence of pi bonding causes only minor changes due to the additional electron density. Thus, to determine the shape of a molecule or ion which contains one or more π bonds, the electrons which the central atom uses to form the pi bonds must be subtracted from the total of electrons. Then the shape of the species is determined by the number of lone pairs and sigma bond pairs around the central atom. This is best understood in terms of an example. In phosphorus oxychloride, $POCl_3$, the central atom is phosphorus which forms a double bond to the oxygen, $Cl_3P = O$. The two electrons of this π bond come, one from the phosphorus and one from the oxygen. Thus, when counting the number of electrons around the phosphorus atom, this electron used in the π bond must be subtracted as it has no steric effect. The calculation proceeds

valency electrons at P = 5 electrons
sigma bond electrons from O + 3Cl = 4 electrons
less electron used by P in π bond = −1 electron

Therefore there are four electron pairs around the phosphorus atom which are shape determining. $POCl_3$ is therefore tetrahedral. (The effect on the bond angles of the π bond is discussed later.)

Next consider the carbonate ion, CO_3^{2-}. The most reasonable formulation for this ion in the normal valency form is

$$O = C \begin{array}{c} O^- \\ \\ O^- \end{array}$$

(note that in charged species involving terminal oxygen atoms it is usually more realistic to place the negative charge on the more electronegative ligand atom and not on the central atom). Then the electron count at the carbon atom is

valency electrons at C = 4 electrons
sigma bond electrons from 3O = 3 electrons
less electron used in 1 π bond = −1 electron

thus there are three electron pairs to determine the shape and the carbonate ion is planar.

As a final example, take the sulfite ion, SO_3^{2-}. This would

be formulated as $O = S \begin{array}{c} O^- \\ \\ O^- \end{array}$ and therefore:

valency electrons at S = 6 electrons
sigma bond electrons from 3O = 3 electrons
less electron used in 1 π bond = −1 electron

hence, four electron pairs and the sulfite ion has an unshared pair on the S atom and is a trigonal pyramid.

A list of cases involving π bonds is given in Table 4.4.

The main difficulty in dealing with π-bonded species lies in deciding how many electrons the central atom is using for π bonds. It will usually be found satisfactory to write down a formula with single and double bonds which satisfies normal ideas of valency, as in the examples above, and then count up the number of double bonds from the central atom.

In a few cases, notably ozone, O_3, and the nitrate ion, NO_3^-, it will be found that it is necessary to write some of the bonds as coordinate bonds to keep normal valencies or to avoid violating the octet rule. These two are best written as

$$O = O \rightarrow O \text{ and } O = N \begin{array}{c} O^- \\ \\ O \end{array}$$

. All the bonds are, despite this formalism, equivalent. The electron count then proceeds:

O_3	NO_3^-
O = 6 electrons (the central O)	N = 5 electrons
1O = 1 electron (the covalently bound O)	2O = 2 electrons (the two covalently bound Os)
1 π bond = −1 electron	1 π bond = −1 electron
total = 3 pairs	total = 3 pairs

Thus O_3 is a V-shaped molecule and NO_3^- is planar. Notice that, as the coordinate bonds originate from the central atom, the oxygen bound by the coordinate bond contributes no electrons to the count of those around the central atom.

Although a formula with localized double bonds is used to assist these calculations, it must not be thought that this type of formula gives a true description of the electron distribution in the species. If the ligand atoms are all equivalent, double bonds and charges are delocalized over the whole molecule or ion. For example, the three configurations:

are all equally likely for the carbonate ion, and it is found experimentally that all the C − O distances are identical and that there is a charge of $-\frac{2}{3}e$ on each oxygen

This delocalization is discussed in the next section.

Where π bonding is delocalized, as in this case, its only effect on the stereochemistry is to shorten all the bonds

TABLE 4.4 Structures of species with π-bonds

Number of valency electrons on the central atom	Number of σ-bonds	Number of π-bonds	Shape	Examples
Case 1: no lone pairs (Shape follows from column 2)				
4	2	2	linear	CO_2, HCN
	3	1	triangular	CO_3^{2-}
5	3	2	triangular	NO_3^-
	4	1	tetrahedral	POX_3, PO_4^{3-}, VO_4^{3-}
6	3	3	triangular	SO_3
	4	2	tetrahedral	CrO_4^{2-}, SO_4^{2-}
7	4	3	tetrahedral	IO_4^-, MnO_4^-
	6	1	octahedral	$IO(OH)_5$
Case 2: one lone pair (Shape is that due to the lone pair plus the number of bonds in column 2)				
5	2	1	V-shaped	NOCl, NO_2^-
6	3	1	trigonal pyramidal	$SOCl_2$, SO_3^{2-}
	2	2	V-shaped	SO_2
7	3	2	trigonal pyramidal	IO_3^-, XeO_3
	4	1	distorted (see Figure 4.6)	$IO_2F_2^-$
Case 3: two lone pairs (Shape is that due to the two lone pairs plus the bonds)				
7	2	1	V-shaped	ClO_2^-

compared with the single bond lengths. If, however, the π bond is localized, as in $Cl_3P{=}O$ or $Cl_2S{=}O$ for example, then it exerts a steric effect. As the double bond region contains two electron pairs and the bond is also shorter than a single bond, the electron density is high and so the repulsion between two double bonds, or between a double bond and a single bond is greater than that between two single bonds. The π bond thus has a steric effect similar to that of a lone pair and the bond angles between single-bonded ligands will be closed up because of the greater repulsion. For example, in $POCl_3$ the $Cl{-}P{-}Cl$ angle is reduced from the tetrahedral value to $103{\cdot}5°$. The similarity of the steric effects of a lone pair and of a double bond can be seen in the molecule $XeOF_4$ where all the bond angles are very close to $90°$.

4.3.2 Summary
The application of the VSEPR theory, as discussed in the last two sections, may be summarized:

(i) Empirical ideas of the repulsion of the electron-dense regions represented by chemical bonds leads to the principle that bonds around a central atom will be arranged in the most symmetrical manner possible for the coordination number. These are the configurations given in Table 4.1.

(ii) Lone or unshared pairs of electrons on the central atom must be taken into account when deciding the shapes. These lone pairs fill coordination positions in the basic configurations to give the shapes detailed in Table 4.2 and in the commentary thereon.

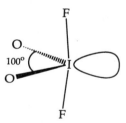

FIGURE 4.6 *The structure of* $IO_2F_2^-$

(iii) The direction of distortion from the basic shapes in unsymmetrically substituted species may be predicted by considering the electron density and polarization within the bonds. When lone pairs are present, they have a dominating effect and the order of repulsions is lone pair—lone pair > lone pair—bond pair > bond pair—bond pair.

(iv) The same analysis applies to species with π bonds. The electrons contributed by the central atom to the π bonds are subtracted and the configuration depends on the number of bonds plus lone pairs left on the central atom as in the cases in Table 4.4. Delocalized π bonds have little steric effect but localized ones are regions of high electron density and cause distortion.

Although it is at first necessary to work out each structure systematically using the rules above, it will be found that the process is readily short-circuited with practice. For example, a major question is always whether there are lone pairs present, and these are nearly always found in compounds where the central element is showing an oxidation state less than its Group maximum one. Similarly, structures of iso-electronic species are usually the same, e.g., CO_3^{2-} and NO_3^-, or SO_2 and O_3, and this often helps to link up with known compounds.

Finally, one warning: the methods above apply on the assumption that the molecular formula is known and does not allow for polymerization. Thus, beryllium dichloride exists as $BeCl_2$ only in the gas phase, in the solid it is the chloride-bridged polymer

Of course, given this structure, the geometry about the Be atoms can be predicted by VSEPR. There are now four electron pairs at each Be, made up from two donor pairs and the two pairs of each basic $BeCl_2$ unit, giving tetrahedral coordination.

Similarly, aluminium tribromide exists as the dimer Al_2Br_6, and SiO_2 is a three-dimensional polymer instead of a discrete molecule like CO_2. VSEPR gives no information about such alternative molecular forms. However, once the molecular formula is known, the methods given above lead to the structures of the large majority of simple molecules and ions, and also give the probable geometries in polymeric materials.

4.4 General approaches to bonding in polyatomic species

Once we know the shape of a polyatomic species, whether from experiment or by VSEPR theory, or other prediction, we wish to discuss its bonding. The molecular orbital approach may be extended to cover molecules of more than two atoms in two different ways. One way would be to take the n nuclei of an n-atomic molecule, together with their inner non-bonding electron shells, place them in their positions in the molecule, and then combine all the valency shell atomic orbitals on all the atoms into poly-centred molecular orbitals embracing all the nuclei. The total number of such *delocalized* molecular orbitals equals the total number of atomic orbitals used in their construction and each may hold up to two electrons. The valency electrons would then be fed into these n-centred molecular orbitals, in order of increasing energy, to form the molecule in a process which is exactly analogous to the building-up of atomic and diatomic molecular structures which has just been discussed. Thus we would place the atoms of water in their V-shape, and seek to form 3-centre molecular orbitals for H_2O from the hydrogen $1s$ and the oxygen $2s$, $2p_x$, $2p_y$ and $2p_z$ atomic orbitals.

An alternative approach is to think of each pair of bonded atoms separately as forming a two-centred bond just as in the diatomic molecules. The whole molecule is built up from such two-centred *localized* molecular orbitals. That is, we describe H_2O in terms of two $O-H$ bonds. This latter approach corresponds to the long-familiar ideas of electron pair bonds and it is the one which is mainly used to describe sigma bonding in polyatomic molecules. The picture of delocalized many-centred bonds is less familiar but it is very useful in the discussion of π-bonded species and must also be applied in the discussion of certain σ-bonded molecules, especially the 'electron-deficient' molecules typified by the boron hydrides.

In a complete treatment, both the method of localized bonds and the method of delocalized ones give exactly the same description of the electron density distribution in the molecule, and they are often used interchangeably in advanced work. We shall start by discussing sigma-bonded species in terms of localized, two-centre, orbitals. Later we indicate the multi-centred approach and discuss how experimental evidence from photoelectron spectroscopy allows us to choose which is the more satisfactory. Finally we describe π-bonded species.

4.5 Bonding in polyatomics: the two-centre bond approach

In tackling the shapes of sigma-bonded species, as described in Tables 4.1 to 4.3, we first note that there are already some steric properties in the atomic orbitals. For example, the three p orbitals are at $90°$ to each other. It is thus possible to describe bond angles of $90°$ in terms of overlap with p orbitals of the central atom and suitable orbitals on the ligands. Thus phosphane, PH_3, where the $H-P-H$ angle is just above $90°$, could be described as having each $P-H$ bond formed by overlap of a phosphorus $3p$ orbital with the $1s$ orbital of the hydrogen to give a situation very similar to that in a diatomic molecule. The phosphorus atom has the valency shell configuration $(3s)^2(3p_x)^1(3p_y)^1(3p_z)^1$. In the x direction, for example, the phosphorus $3p_x$ orbital (which will be written ψ_{p_x}) is of the correct symmetry to form a sigma overlap with the $1s$ orbital on that hydrogen atom lying in the x direction (written ψ_s). The resulting two-centred molecular orbitals are of the form given in equation 3.2:

$$\sigma_{s,\,p_x} = \psi_s + c\psi_{p_x}$$
$$\sigma^*_{s,\,p_x} = c\psi_s - \psi_{p_x}$$

where the value of c which gives the most favourable energy of combination would have to be determined in the course of the calculation (compare section 3.3). There are two electrons available, the hydrogen one and the one from the phosphorus p_x orbital, and these fill $\sigma_{s,\,p_x}$ leaving the anti-bonding orbital empty, and giving a single $P-H$ bond. This step is represented in Figure 4.7 and may be written as a formation equation:

$$P(3p_x)^1 + H(1s)^1 \rightarrow P-H(\sigma_{s,p_x})^2(\sigma^*_{s,p_x})^0$$

Thus the whole process of forming this $P-H$ bond is just the same as forming the bond in a heteronuclear diatomic molecule.

The $P-H$ bonds in the y and z directions are formed in the same way giving the phosphane molecule (Figure 4.8):

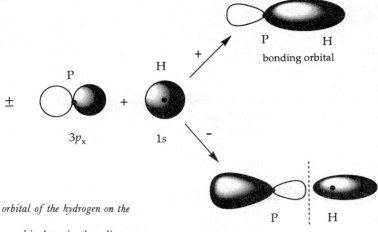

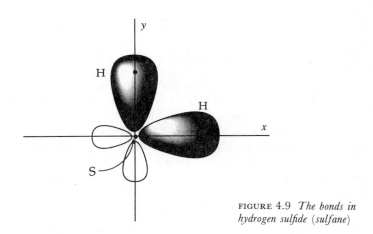

FIGURE 4.7 *The combination of the phosphorus $3p_x$ orbital with the $1s$ orbital of the hydrogen on the x-axis to give the molecular orbitals of a $P-H$ bond in PH_3*
The H atom in the *x* direction combines with a suitable phosphorus orbital to give bonding and antibonding two-centred molecular orbitals.

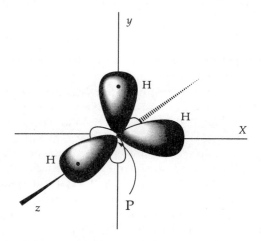

FIGURE 4.8 *The bonding orbitals in a phosphane molecule*
The overlap of the three phosphorus *p* orbitals with hydrogen *s* orbitals gives three P−H bonds at 90° to each other, reflecting the angle between the *p* orbitals.

FIGURE 4.9 *The bonds in hydrogen sulfide (sulfane)*

two bonds from $s + p$ overlap

$P(3s)^2(3p_x)^1(3p_y)^1(3p_z)^1 + 3H(1s)^1$
$\rightarrow PH_3(3s)^2(\sigma_{s,p_x})^2(\sigma_{s,p_y})^2(\sigma_{s,p_z})^2(\sigma^*_{s,p_x})^0(\sigma^*_{s,p_y})^0(\sigma^*_{s,p_z})^0$

The molecule may be built up in this way by considering each pair of bonded atoms in turn as if they were in a diatomic molecule and using the methods of section 3.5. The lone pair of electrons is accommodated in the *s* orbital (and they will tend to be concentrated in that spatial segment of the *s* orbital remote from the bond directions) and the H−P−H bond angle follows as a consequence of the 90° angle between the *p* orbitals.

A similar discussion applies to SbH_3, AsH_3, and to the hydrides of S, Se, and Te. In H_2S (Figure 4.9) the bonds are formed using two of the sulfur $3p$ orbitals, giving an H−S−H angle of 90°, and the third *p* orbital and the $3s$ orbital hold the two lone pairs. This analysis does not explain the small departures from the 90° angle in these molecules and cannot account for the much larger bond angles in water and ammonia.

6 Two-centred orbitals: hybridization

The use of pure atomic *s* and *p* orbitals fails to account for the shape of most of the molecules mentioned in the previous sections. Linear, triangular, or tetrahedral shapes, for example, cannot be explained using simple atomic orbitals and the concept of mixed or *hybrid* orbitals has to be introduced.

4.6.1 *Equivalent hybrids*

Consider first gas phase monomeric beryllium dichloride: the electronic configuration of the beryllium atom is Be = $[He](2s)^2(2p)^0$. The two electrons in the valency shell are paired in the *s* orbital, and the *p* orbitals are empty. In order to obtain divalency, the *s* orbital and a *p* orbital must be used and, to have two bonds at 180°, this $s + p$ configuration must be rearranged to form two equivalent orbitals. This is done by combining the two into two *sp* hybrids as shown in Figure 4.10. If the *s* orbital is superimposed on the *p* one as shown, the positive lobe of the *p* wave function is reinforced and the negative lobe is diminished. This description reflects diagrammatically the mathematical process:

$$sp = \frac{1}{\sqrt{(2)}}(s+p)$$

which gives the orbital shown in Figure 4.10a. The second hybrid, which is directed in the reverse sense, is $(1/\sqrt{2})(s-p)$ and is given in Figure 4.10b. The factors $1/\sqrt{2}$ arise as the total electron density in the two sp hybrids (found by squaring the two wave functions above and adding) must equal s^2+p^2, the electron density in the two constituent atomic orbitals. The total process is: $s+p = 2sp$ hybrids, two atomic orbitals giving two hybrid orbitals on the central atom. (Note that this process of mixing orbitals on the *same* atom, to give hybrid atomic orbitals, must be distinguished from the process of combining an s and a p orbital on different atoms to form σ molecular orbitals which was shown in Figure 3.11.) The two valency electrons are then placed one into each sp hybrid and each of these may overlap with a ligand orbital to form molecular orbitals.

The two sp hybrid orbitals are equivalent to each other and directed at 180°. They are used to form the bonds in a linear molecule. For example, the description of $BeCl_2$, given below, resembles the description of PH_3 apart from the use of hybrid orbitals.

BeCl₂. The chlorine atom has seven valency electrons and four valence shell orbitals. Six of the electrons are paired up in three of these orbitals, leaving the fourth singly-occupied. Let the $Cl-Be-Cl$ axis be taken as the z-direction. The singly-occupied chlorine orbital must be of sigma symmetry with respect to the $Be-Cl$ bond, and could be the chlorine s or p_z atomic orbitals or a hybrid such as sp pointing in the z direction. Molecular orbitals are formed by this chlorine σ orbital and the sp hybrid on the beryllium to give bonding and antibonding molecular orbitals, and the same happens for the second chlorine atom and the other sp beryllium hybrid. The complete process is first:

$$Be(s)^2(p)^0 \rightarrow Be(sp_A)^1\,(sp_B)^1 \text{ (designating the two chlorine atoms A and B)}$$

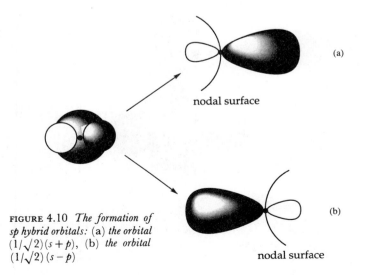

FIGURE 4.10 *The formation of sp hybrid orbitals:* (a) *the orbital* $(1/\sqrt{2})(s+p)$, (b) *the orbital* $(1/\sqrt{2})(s-p)$

nodal surface

(a)

nodal surface

(b)

(a)

Cl atom
3p orbital

Be atom
2 sp hybrid orbitals

Cl atom
3p orbital

(b)

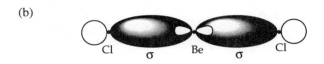

Cl σ Be σ Cl

BeCl₂ molecular orbitals

FIGURE 4.11 *Bonding in beryllium chloride:* (a) *the contributing atomic orbitals*, (b) *the bonding molecular orbitals*

that is, the beryllium valency electrons are placed one into each hybrid orbital.
Then $Be(sp_A)^1 + Cl_A(\sigma)^1 \rightarrow Be-Cl_A(\sigma_{Be-Cl})^2(\sigma^*_{Be-Cl})^0$ where σ_{Be-Cl} is of the form $\psi_{sp} + \psi_p$ and for the antibonding orbital σ^*_{Be-Cl} is $\psi_{sp} - \psi_p$.
Similarly for the bond to the second chlorine
$$Be(sp_B)^1 + Cl_B(\sigma)^1 \rightarrow Be-Cl_B(\sigma)^2(\sigma^*)^0$$

These are shown in Figure 4.11.

Each beryllium-chlorine bond is a normal single bond, just as in the diatomic molecules or in phosphane, except that it is formed using a hybrid atomic orbital on the beryllium instead of a simple one. The $Cl-Be-Cl$ angle of 180° follows from the angle between two sp hybrids.

As well as providing orbitals oriented at the correct angle, sp hybrids overlap more effectively than simple s or p orbitals, and therefore give stronger bonds. Pauling has proposed the S parameter, listed in Table 4.5, as a measure of this relative overlap for various orbitals.

A similar mixing process can be carried out with two p orbitals and the s orbital, to give three equivalent sp^2 hybrid orbitals directed at 120°. These are shown in Figure 4.12. If one of the three orbitals is directed along the x-axis, this has contributions from the s and p_x orbitals. The other two hybrids are composed of the s, p_x, and p_y orbitals. The expressions for the three hybrids are, respectively:

$$\sqrt{(\tfrac{1}{3})}s + \sqrt{(\tfrac{2}{3})}p_x$$
$$\sqrt{(\tfrac{1}{3})}s - \sqrt{(\tfrac{1}{6})}p_x + \sqrt{(\tfrac{1}{2})}p_y$$
$$\sqrt{(\tfrac{1}{3})}s - \sqrt{(\tfrac{1}{6})}p_x - \sqrt{(\tfrac{1}{2})}p_y$$

The coefficients are again chosen so that the total electron

TABLE 4.5 Values of Pauling's S parameter for various atomic orbitals

Orbital	s	p	sp	sp^2	sp^3
Relative strength	1·0	1·73	1·93	1·99	2·00

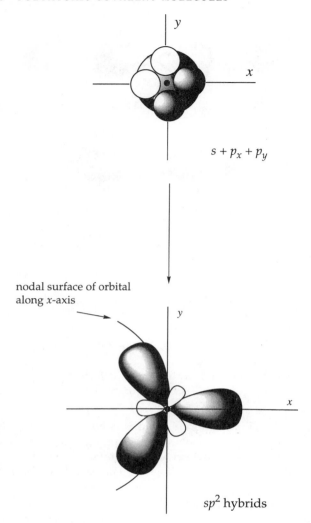

$s + p_x + p_y$

nodal surface of orbital along x-axis

sp^2 hybrids

FIGURE 4.12 *The formation of sp^2 hybrid orbitals*
One of the three nodal surfaces is shown. The small lobes are the negative part of these three orbitals.

density in the three hybrid orbitals adds up to $s^2 + p_x^2 + p_y^2$, the electron density in the three constituent orbitals.

The sp^2 hybrids are then used to form the bonds in a plane triangular molecule, such as BCl_3. The process may be described as:

(i) Formation of the sp^2 hybrids and placing one valency electron in each

$$B(2s^2)(2p)^1 \rightarrow B(sp^2)^1(sp^2)^1(sp^2)^1.$$

(ii) Each sp^2 hybrid is combined with a chlorine orbital of sigma symmetry to give one bonding and one antibonding two centre molecular orbital. These are of the form $sp_B^2 + c\sigma_{Cl}$ (bonding) and $sp_B^2 - c\sigma_{Cl}$ (antibonding). The electrons go into the bonding orbitals:

$$B3(sp^2)^3 + 3Cl(\sigma)^1 \rightarrow BCl_3 3(\sigma_{B-Cl})^6 3(\sigma_{B-Cl}^*)^0$$

The whole process is that three atomic orbitals on the boron, combined into three atomic hybrid orbitals, combine with three atomic orbitals on the three chlorines to give six molecular orbitals, three bonding and three antibonding.

The six valency electrons fill the bonding orbitals, giving the boron trichloride molecule as shown in Figure 4.13.

In tetrahedral configurations, all three p orbitals and the s orbital combine to form four sp^3 hybrids, which are shown in Figure 4.14. When five- or six-coordinated structures have

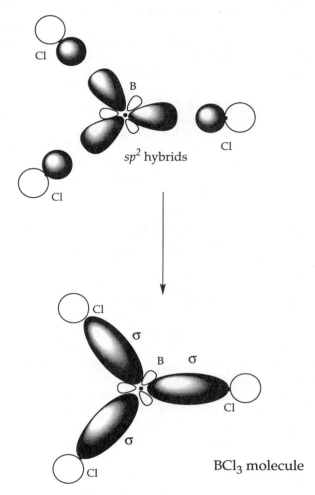

sp^2 hybrids

BCl_3 molecule

FIGURE 4.13 *The formation of boron trichloride*

sp^3

FIGURE 4.14 *The four sp^3 hybrid orbitals*
The four small negative lobes (compare Figure 4.12) have been omitted for clarity. The sp^3 hybrids are tetrahedrally directed.

to be formed, it becomes necessary to use *d* orbitals as well as *s* and *p* orbitals. Five-coordination arises from sp^3d hybridization and the appropriate *d* orbital is the d_{z^2} orbital. (This can be seen qualitatively; the three equatorial positions in the trigonal bipyramid correspond to three sp^2 hybrids and the two orbitals in the $\pm z$-direction arise by mixing the atomic orbitals lying on the *z*-axis, that is p_z and d_{z^2}.) In a similar manner, octahedral hybridization involves six sp^3d^2 orbitals and, here, the appropriate *d* orbitals are d_{z^2} and $d_{x^2-y^2}$. *d* orbitals may also contribute to other shapes. For example, sd^3 (using d_{xy}, d_{yz} and d_{zx}) is also tetrahedral. The above treatment of higher coordination numbers is useful as an initial approach. In more advanced treatments, the extent to which *d* orbitals are involved in five- and six-coordinated compounds is subject to discussion (compare section 18.9).

For linear, triangular, tetrahedral, and octahedral shapes, all the hybrid orbitals in a set are equivalent, have equal angles between them, and extend equally in space. For example, each sp^3 hybrid is equivalent to the other three and has one-quarter *s* character and three-quarters *p* character.

When hybrids are formed, energy is required to promote the electrons into the configuration required for the bonds, as in the s^2p^0 to $sp^1s p^1$ configuration discussed above for $BeCl_2$. If the energy of such a step can be calculated, and the energies of the remaining steps are known, it is possible to calculate the energy of formation of a particular molecule and perhaps predict the preferred product(s) of a reaction. Such an analysis is often not possible for lack of data, but the case of the reaction of carbon and hydrogen to form the CH_4 molecule has been worked out and it may be compared with the possible alternative reaction to form CH_2 (Figure 4.15). It will be noted that the difference in heats of formation favours CH_4 by a factor of 4/3 compared with CH_2 but the difference is only about one per cent of the total energies involved, which again highlights the difficulties of predicting chemical behaviour by direct calculation.

The carbon atom reacts in such a way that all its valency electrons and all its valency orbitals are involved in molecule formation, and this is generally the favoured mode of reaction, at least for lighter atoms. Where there are empty valence orbitals as in $BeCl_2$ or BCl_3, further reaction to use these orbitals is likely and such molecules act as electron pair acceptors: similarly, where there are one or more lone pairs, as in ammonia or water, these tend to form donor bonds to suitable acceptors. Many examples of such behaviour will be found in the later chapters.

4.6.2 *Non-equivalent hybrids*
In the above cases, a given number of simple atomic orbitals were combined to give the same number of hybrid orbitals which were all equivalent to each other. These modes of hybridization correspond to the basic shapes of Table 4.1. It is quite possible, however, to form non-equivalent hybrids. For example, an unsymmetrical beryllium compound, BeXY, would almost certainly have a more favourable bonding structure if non-equivalent '$s+p$' hybrids were used with, say,

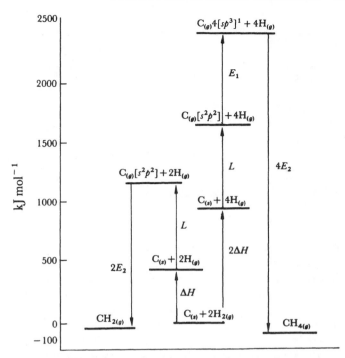

$E_1 \approx 782$ kJ mol^{-1} (promotion of electrons to four singly-occupied sp^3 hybrids)
ΔH = heat of dissociation of gaseous H_2 = 436 kJ mol^{-1}
L = heat of vaporization of $C_{(s)} \to C_{(g)}$ = 711 kJ mol^{-1}
E_2 = energy of C—H bond $\approx$ 607 kJ mol^{-1}

FIGURE 4.15 *Energy changes involved in the formation of CH_2 and CH_4*

the Be—X bond formed by a hybrid with rather more *s* character and the Be—Y bond from one with more *p* character.

In the same way, trigonal or tetrahedral hybrids may be non-equivalent. For example, in chloromethane, CH_3Cl, the C—Cl bond will be formed by an '$s+p^3$' hybrid which has a different amount of *s* character from the other three '$s+p^3$' hybrids which form the C—H bonds. The optimum amount of *s* character in such hybrids will vary from molecule to molecule, although the shape and basic hybrid structure is still tetrahedral.

Such non-equivalent hybrids become even more necessary when molecules containing lone pairs are under discussion. To get the bond angles of 107° in ammonia, the *p* character of the N—H bonds has to be significantly increased over the 3/4 *p* of the equivalent sp^3 hybrid, and the orbital holding the lone pair has proportionately more *s* character. Similar remarks apply to the case of water with a bond angle of 105°.

There is indeed a complete range of possible hybrids from the four equivalent sp^3 hybrids at 109½° in methane, through the non-equivalent '$s+p^3$' hybrids in CH_3Cl, NH_3 and H_2O, down to the hybrids in PH_3 and H_2S which correspond to bond angles of just over 90° and are mainly *p* orbitals with a little *s* character for the bonds and *s* plus a little *p* for the lone pair. Moving the other way, towards larger angles, there is the same continuous variation from equivalent sp^3 hybrids at 109½°, through non-equivalent arrangements, to

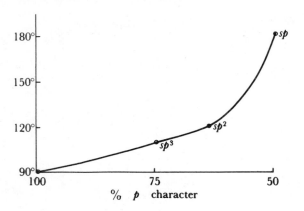

FIGURE 4.16   *Bond angle and p character*

the case of equivalent sp^2 hybrids at 120° plus an unused atomic p orbital.

There is thus a complete continuity of bond angle versus percentage p character in sp^n hybrids, as indicated in Figure 4.16.

In general terms, any of the structures, symmetrical or not, predicted by the methods of sections 4.3 and 4.4, may have their electronic structures explained in terms of molecular orbitals formed by appropriate equivalent or nonequivalent hybrids on the central atom together with appropriate ligand orbitals. Hybrids can be constructed to fit any set of bond angles. It is also possible, in a more detailed treatment, to derive theoretical heats of formation by calculations involving hybridization schemes, although the results are too inaccurate in general to give a direct guide to the course of a reaction.

It should be clear from the above discussion that the idea of hybridization is very valuable in providing a *description* of the electron density in a molecule but that it does not provide any *explanation* of the electron arrangements or of the shapes of molecules, in the present stage of development. It is common to see statements such as 'methane is tetrahedral because it is sp^3 hybridized' but this description is incorrect.

An interesting recent discussion by Smith (see references), has used Pauling's S parameter (a measure of hybrid orbital overlap and therefore bond strength) to predict structures and explains some observations which are exceptions to the VSEPR theory.

As with other wave-mechanical calculations, hybridization presents a potential method of predicting molecular shapes. It is possible, in principle, to find the optimum hybridization scheme for a molecule such as water which will give the most favourable energy of formation and hence the bond lengths and angles. However, just as in the cases discussed earlier, the necessity of introducing approximate methods to solve the wave equations adds so much uncertainty to the values which result that the predictions are rather crude in the present state of the art.

4.7 Delocalized, or multi-centred, sigma orbitals

Instead of constructing a set of 2-centred bonds we could describe a polyatomic species by constructing orbitals centred over all the atoms of the molecule. A full discussion is beyond our scope, but the general approach may be illustrated.

Consider first a very simple example, the BeH_2 monomer. The orbital combinations are shown in Figure 4.17. The beryllium $2s$ orbital combines, in phase, with the s orbitals on

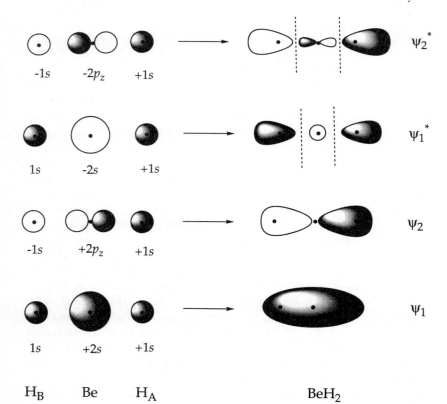

FIGURE 4.17   *Formation of molecular orbitals in BeH_2 (schematic)*

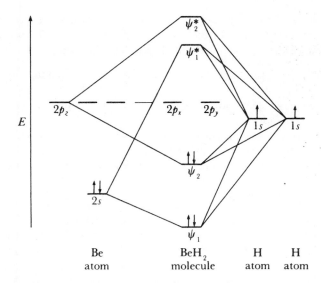

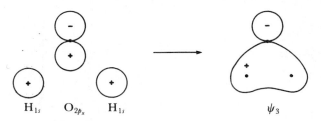

FIGURE 4.19 *Bonding orbital from in-phase overlap of $2p_x$ orbital on O with H $1s$ orbitals*

FIGURE 4.18 *Schematic energy level diagram for* BeH_2
Note that the z axis is the $H-Be-H$ axis (linear molecule). The p_x and p_y orbitals on Be are not involved in molecular orbital formation and are simply repeated in the BeH_2 column of the diagram.

hydrogens A and B to form the 3-centre orbital, ψ_1. The electron probability density in this orbital, ψ_1^2, will be concentrated in the regions of highest field between the nuclei, and will experience more attraction than in the isolated atoms. Thus ψ_1 is bonding. Similarly ψ_2, formed by the combination s (on A) $+ p_z - s$ (on B), is bonding. The corresponding out-of-phase combinations ψ_1^* and ψ_2^* are clearly anti-bonding. Since both the p_x and p_y orbitals on beryllium have nodes in the plane of the two hydrogens, no combinations between these and the hydrogen s orbitals are possible. Thus we can draw up a qualitative energy level diagram, Figure 4.18. Notice, first that six atomic orbitals (4 on Be, 1 on each H) give six BeH_2 orbitals (ψ_1, ψ_1^*, ψ_2, ψ_2^* together with p_x and p_y). Secondly, the four electrons are placed in the two bonding levels giving two bonds holding the three atoms together. This is equivalent to two $Be-H$ single bonds, but the bonding electrons are spread over all three atoms. Finally, the energy separation of ψ_1 and ψ_2 reflects approximately the separation of the beryllium $2s$ and $2p$ atomic levels.

For a second example, we can compare the water molecule, H_2O, with BeH_2. Because the nuclear charge on O is 8, compared with 4 for Be, and because the s orbital penetrates to the nucleus, the oxygen $2s$ orbital is markedly contracted and it will overlap poorly with the H s orbitals. Thus, in H_2O we will consider as a simplification that the oxygen $2s$ orbital has a very weak interaction with the H $1s$ orbitals, and the resulting orbital ψ_1 is effectively the oxygen $2s$ atomic orbital. The oxygen $2p$ orbitals are much more compatible in energy with the hydrogen $1s$ orbitals and make the main contributions to the bonding. Because the H_2O molecule is V-shaped, the overlap with $2p_z$ is reduced and the ψ_2 analogue will be less stable. On the other hand, the two hydrogen s orbitals may now overlap, in phase, with the second p orbital

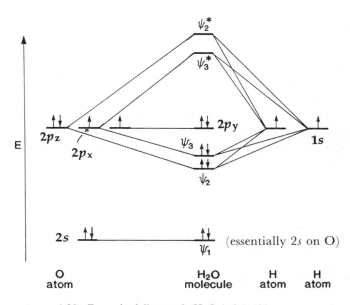

FIGURE 4.20 *Energy level diagram for* H_2O *(schematic)*
Compare with Figure 4.18 for BeH_2. The non-linear shape allows overlap with both the p orbitals in the plane of the molecule. As the oxygen s orbital is very tightly bound, ψ_1 is essentially non-bonding.

in the molecular plane, as shown in Figure 4.19. This orbital is clearly bonding and we shall label it ψ_3. We can approximate ψ_1 as non-bonding, leaving ψ_2 and ψ_3 as the main bonding orbitals. The out-of-phase combinations, ψ_2^* and ψ_3^*, are antibonding. The out-of-plane p_y orbital on O is non-bonding. Thus the energy level diagram is that of Figure 4.20. The eight valency electrons fill the two bonding orbitals, ψ_2 and ψ_3 and the two non-bonding orbitals, ψ_1 and p_y. Thus there are two bonds and two non-bonding pairs.

Let us compare this description with the photoelectron spectrum of H_2O, shown in Figure 4.21. First, we note that the vibrational structure is more complicated than in Figure 3.23, for example, since there are three modes of vibration for a triatomic molecule (cf Chapter 7). Band (1) at 12.6 eV has the sharp profile characteristic of the ionization of a non-bonded electron and the vibrational structure is analysed as the overlap of progressions of about 3200 cm^{-1} and 1400 cm^{-1} separations (compare frequencies in H_2O of 3650 cm^{-1} and 1595 cm^{-1}). Band (2), at 13·7 eV, shows a major drop in vibrational frequency, to 975 cm^{-1}, as expected for the loss of

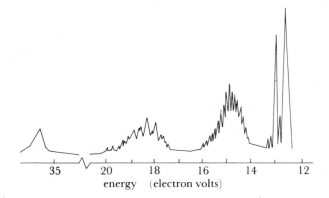

FIGURE 4.21 *The photoelectron spectrum of* H_2O

a bonding electron. Band (3) at 17·2 eV also corresponds to ionization of a bonding electron. There is then a large energy gap to band (4) at 36 eV.

All this is compatible with the H_2O energy level diagram of Figure 4.20. Band (4), corresponding to loss of the most tightly bound electron, is near the energy level of the 2s electron in oxygen and is assigned to ionization from ψ_1. [Note: ionization of the oxygen 1s electron needs 580 eV.] The gap from 36 to 17 eV corresponds to the $s - p$ energy gap. The two bonding levels come next, ionization from ψ_3 requiring 17·2 eV and from ψ_2 needing 13·7 eV. Finally the 12·6 eV ionization corresponds to the loss of a non-bonding electron from the oxygen $2p_y$ orbital (compare 12·1 eV for the loss of a π^* electron from O_2).

A rather similar picture emerges for ammonia, NH_3. The lowest ionization energy, at 10·2 eV, appears as a sharp band which shows a vibrational sequence under high resolution similar to the bending frequency of the parent molecule. The second band is of about twice the intensity, at 14·8 eV with a poorly-resolved vibrational spacing of about 1800 cm^{-1}, compared with the molecular stretching frequencies of 3340 cm^{-1} and 3440 cm^{-1}. The final band is weaker and very broad, centred about 27·5 eV. These can be assigned respectively to non-bonding (10·2 eV) and to doubly-degenerate bonding (14·8 eV) molecular orbitals involving the nitrogen p orbitals and to a bonding orbital (27·5 eV) involving the nitrogen s orbital.

Finally, consider the photoelectron spectrum of CH_4, shown in Figure 4.22. One band at 12.7 eV is very broad and

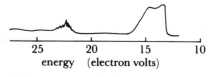

FIGURE 4.22 *The photoelectron spectrum of* CH_4 *(low resolution)*

would match ionization from a triply-degenerate bonding orbital using the three carbon p orbitals. The weaker band about 23 eV, would correspond to ionization of an electron from a bonding orbital from the carbon 2s orbital.

In these last two cases, we note that the observed spectrum is *not* that predicted from the two-centred bond model using hybridization, where we would expect the following:

for CH_4: four identical sp^3 hybrids on C forming four identical CH bonding orbitals. Thus we would expect only one ionization in the photoelectron spectrum.

for NH_3: non-equivalent sp^3 hybrids, one (with extra s character) holding the lone pair and the three others forming three identical NH bonds. Thus we would expect two ionizations in the photoelectron spectrum.

We can easily understand the observed spectra if we think of four-centred (for NH_3) or five-centred (for CH_4) molecular orbitals formed by the central atom s and p orbitals and reflecting the $s - p$ energy differences in N or O. The identity of the two bonding combinations involving the nitrogen p orbitals, or of the three bonding combinations involving the carbon p orbitals, follows from symmetry in a more detailed calculation.

Summary of sections 4.5 to 4.7

Sigma bonding in polyatomics may be described in terms of hybrid orbitals on the central atom and two-centred bonds, *or* in terms of multi-centred bonds. The two-centred description is easier to visualize, accords with classical ideas, and has some advantages in calculations. The multi-centred description appears to give the best explanation of the observed photoelectron spectra.

It should be emphasized that these are alternative models, each capable of considerable extension and refinement, and either may be used according to the problem being examined. Other models of bonding also exist and some are discussed in the references.

4.8 Pi bonding in polyatomic molecules

Polyatomic molecules which contain isolated localized π bonds, such as the $P = O$ bond in Cl_3PO or R_3PO or the $S = O$ bond in sulfoxides R_2SO, may be treated similarly to the π-bonded diatomic molecules of section 3.5. These π bonds which are localized between two atoms are formed by sideways overlap of atomic orbitals of suitable symmetry. In the case of phosphorus oxychloride or the phosphane oxides, we can form the σ bonds from sp^3 hybrids on the phosphorus atom (using all the $3p$ orbitals on the phosphorus), so that the π bond requires the use of a suitable phosphorus $3d$ orbital to overlap with a p orbital on the oxygen. This bond is shown in Figure 4.23. The sulfoxide is similar with a lone pair on the sulfur and a $d_\pi - p_\pi$ bond (Figure 4.24).

Similarly, the triple bond in hydrogen cyanide, $H - C \equiv N$, is localized between the carbon and nitrogen atoms and formed by the overlap of the p_x and p_y orbitals of the carbon and nitrogen atoms.

Such localized π bonds are obviously similar to those in diatomic molecules, but they are relatively uncommon in

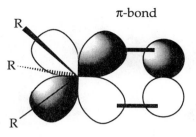

Phosphorus d-orbital   Oxygen p-orbital

$R_3P=O$

FIGURE 4.23 *Localized π bonds in a phosphane oxide*

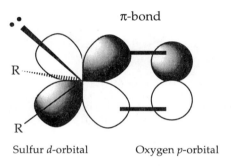

Sulfur d-orbital   Oxygen p-orbital

$R_2S=O$

FIGURE 4.24 *Localized π bonds in a sulfoxide*

inorganic chemistry. Most π bonds are *delocalized* over a number of atoms and are best discussed in terms of many-centred atomic orbitals.

For example, in the carbonate ion, CO_3^{2-}, all the experimental evidence shows that all three oxygen atoms are equivalent, and so a structure with a localized π bond,

such as $O = C \left\langle \begin{array}{c} O^- \\ O^- \end{array} \right.$, is incorrect and a description of the

ion must be found which keeps the equivalence of the oxygens. This is done by first deriving the shape of the molecule by the methods of section 3.7. We then describe the sigma bonding. This may be done in terms of hybrid orbitals and 2-centre bonds or in terms of delocalized sigma orbitals. It is easier to visualize (though not, perhaps, more logical!) if we use the 2-centred bond description for the sigma orbitals. Once we have accounted for the electrons and orbitals used in (a) sigma bonding and (b) to account for lone pairs, we then have available the remaining orbitals and electrons to form the π system. The carbonate ion is planar with no unshared pair on the carbon. Let the molecular plane be the xy plane. Then the carbon uses its $2s$, $2p_x$ and $2p_y$ orbitals to form the sigma bonds (using sp^2 hybrids), and the oxygen atoms

must also use their $2s$, $2p_x$ and $2p_y$ orbitals either in the sigma bonds or in accommodating non-bonding electrons (there is no chance of sideways overlap between oxygen orbitals in the plane of the molecule as the distances are too large). This accounts for all the s, p_x and p_y orbitals and for eighteen electrons, six in the three sigma bonds and four unshared electrons on each oxygen atom. The twelve s and p orbitals form the three bonding sigma orbitals and two orbitals on each oxygen to hold the nonbonding electrons, and all these are filled. The remaining three orbitals are the antibonding sigma orbitals and these remain empty and are of very high energy (see Figure 4.25). The ion has a total of twenty-four electrons (including two for the charge) so that six have still to be accommodated and the p_z orbitals on the four atoms have yet to be used. These four p_z orbitals are combined to form four, four-centred, π orbitals each holding two electrons. The pi orbital of lowest energy is of the form:

$$\psi_1 = (p_C + p_1 + p_2 + p_3)$$

where the constants have been omitted and the orbitals referred to are the carbon $p_z(p_C)$ and the three oxygen p_z orbitals respectively. This orbital is shown in Figure 4.26a. It has no nodes between C and O, concentrates electron density in the regions between the atomic nuclei, and is bonding. The orbital of highest energy, by contrast, has a node between the carbon atom and each oxygen and is antibonding. This orbital is of the form:

$$\psi_4 = (p_C - p_1 - p_2 - p_3)$$

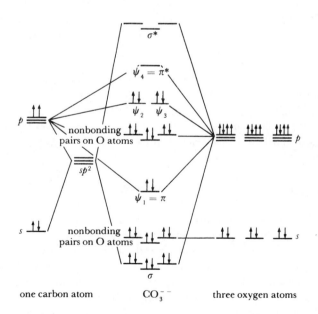

FIGURE 4.25 *Schematic energy level diagram for the carbonate ion*
For clarity in drawing, sets of three equal levels, such as σ, are staggered and hybridization of the oxygen orbitals has been neglected. Each oxygen atom has two pairs of non-bonded electrons—shown as the pair in the s orbitals and the pair of non-bonding p electrons. The delocalized π orbitals are ψ_1, ψ_2, ψ_3 and ψ_4. The bonding and antibonding sigma C – O orbitals are shown as σ and σ^*.

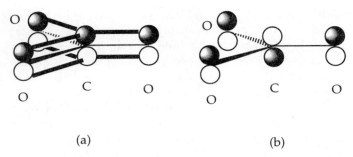

(a)   (b)

FIGURE 4.26 *Delocalized π orbitals in the carbonate ion:* (a) *the most strongly bonding orbital,* (b) *the most strongly antibonding orbital*

As there are four atomic p_z orbitals, there must be four molecular orbitals formed by them. The remaining two orbitals ψ_2 and ψ_3 are, in this case, of equal energy and lie between ψ_1 and ψ_4. They are nonbonding (see Figure 4.25).

The six remaining electrons enter ψ_1, ψ_2, and ψ_3, leaving the antibonding ψ_4 empty. The four electrons in ψ_2 and ψ_3 have no bonding effect, thus the carbonate ion is left with one effective π bond over the whole molecule with a resulting C–O bond order of $1\frac{1}{3}$ (one for the σ bond and $\frac{1}{3}$ for the π bond), which corresponds with that implied by the simple formula.

A more detailed discussion of π orbitals, especially of the orbitals of intermediate energy such as ψ_2 and ψ_3, is beyond the scope of this text, but the general properties of such π systems are readily recognized. Four generalizations about polycentred π orbitals are possible.

1. The number of many-centred molecular orbitals equals the number of component atomic orbitals (in the case of the carbonate ion this is four). Each molecular orbital may hold up to two electrons.

2. The molecular π orbital of lowest energy is that obtained by combining all the atomic orbitals with the same sign so that there are no nodes between pairs of atoms in the resulting π orbital. In the carbonate ion, this orbital is ψ_1.

3. The molecular π orbital of highest energy is generally one where there is a node between each pair of atoms, that is, where the sign of the wave function is reversed between each pair of atoms as in ψ_4 of the carbonate ion. Such an orbital is strongly antibonding and no π bonding can occur in a case where electrons have to be placed in this type of orbital.

4. The remaining molecular π orbitals are intermediate in energy and their energies fall symmetrically about the mean energy of the strongly bonding and the strongly antibonding orbitals. Thus, in the carbonate case, ψ_2 and ψ_3 are degenerate and nonbonding. In other cases there will be equal numbers of weakly bonding and weakly antibonding orbitals. Note that where there is **an odd number of contributing orbitals there must be at least one nonbonding level.**

The form of the intermediate molecular orbitals is not always clear from simple considerations, but these generalizations make it possible to work out whether there will be any π bonding in a molecule without knowing any more about the formation of the intermediate orbitals. The one necessary

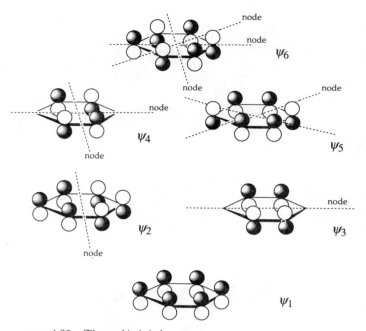

FIGURE 4.27.   *The π orbitals in benzene*
ψ_1 is bonding, ψ_2 and ψ_3 are degenerate and weakly bonding, ψ_4 and ψ_5 are degenerate and weakly antibonding and ψ_6 is strongly antibonding. The component p orbitals are shown as of uniform, small size for clarity.

condition is that it should be possible to leave the highest antibonding orbital empty and there will then be a net π bonding effect. If other, more weakly antibonding π orbitals also remain empty, the bonding effect is enhanced.

The classical example of a delocalized π system is, of course, the case of benzene. The π orbitals here are enumerated below and illustrated in Figure 4.27 and the reader can see how they fit the generalizations above.

1. There are six component atomic orbitals and six π orbitals result.

2. The orbital of lowest energy has the form:

$$\psi_1 = k(p_1 + p_2 + p_3 + p_4 + p_5 + p_6)$$

This has no nodes between atoms and is strongly bonding.

3. The orbital of highest energy has the form:

$$\psi_6 = k(p_1 - p_2 + p_3 - p_4 + p_5 - p_6)$$

Each pair of atoms is separated by a node and the orbital is strongly antibonding.

4. The other four molecular π orbitals fall into two sets of degenerate pairs. One pair has one node cutting the ring and is weakly bonding while the other pair of orbitals has two nodes cutting the ring and is weakly antibonding. The six π electrons fill the three bonding levels, leaving the three antibonding orbitals empty.

A few more inorganic examples will be discussed to illustrate how far these relatively simple principles will suffice to carry a discussion of π systems.

The nitrite ion, NO$_2^-$

In this ion, the two oxygen atoms are equivalent and the methods of section 4.3 show that the ion is bent with a lone

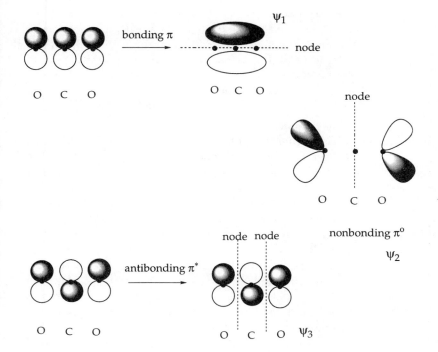

FIGURE 4.28 *The π orbitals in carbon dioxide*

pair on the nitrogen atom. Again, let the molecule lie in the xy plane, then only the p_z orbitals on each atom will be involved in π bonding. Of the eighteen electrons in the valency shells of the component atoms, four will form the sigma bonds, two give the nitrogen lone pair and there are four nonbonding electrons on each oxygen atom leaving four electrons to go into π bonds. There are three atomic p_z orbitals and thus there are three molecular, three-centre, π orbitals. The lowest energy orbital extends evenly over the molecule and is bonding, while the highest energy π orbital has a node between the nitrogen atom and each oxygen and is antibonding. The third π orbital is nonbonding and has its electron density largely on the oxygens. The four electrons enter the bonding and nonbonding π orbitals, giving a net effect of one π bond over the molecule or an $N-O$ bond order of $1\frac{1}{2}$.

Carbon dioxide, CO_2

This is a linear molecule with no lone pairs on the carbon, by the method of section 4.3. If the molecular axis is taken as the z-axis, the s and the p_z orbitals on the three atoms will form the sigma bonds or hold nonbonding electrons on the oxygen atoms. The molecule has a total of sixteen valency electrons, and four of these form the sigma bonds while there is one nonbonding pair on each oxygen atom leaving eight electrons for π bonding. The p_y and p_x orbitals on each atom are available to form π bonds. Consider first the p_y orbitals. There are three of these and therefore three delocalized molecular π orbitals will be formed. The lowest energy orbital will have no interatomic nodes and be of the form (omitting constants):

$$\psi_1 = (p_O + p_C + p_O)$$

while the highest energy orbital will have nodes between the carbon and each oxygen and have the form:

$$\psi_3 = (p_O - p_C + p_O)$$

The third molecular orbital, ψ_2, will be nonbonding. In the case of the p_x orbitals, exactly the same combinations will occur, as these orbitals are identical with the p_y ones apart from their direction in space. The resulting molecular orbitals are also identical; that is, the six atomic p_y and p_x orbitals combine to form six molecular orbitals, two of which are bonding and of the form of ψ_1, two nonbonding like ψ_2, and two antibonding like ψ_3. The eight valency electrons fill the bonding and nonbonding pairs giving a net effect of two π bonds over the molecule and a total $C-O$ bond order of two. The orbitals in carbon dioxide are shown in Figure 4.28 and the energy level diagram for the molecule is shown schematically in Figure 4.29.

Ozone, O_3

It is interesting to compare the cases of ozone and carbon dioxide. There are two more valency electrons in ozone and, if these are added to the energy level diagram of carbon dioxide in Figure 4.29, they would have to be placed in the antibonding π orbitals, reducing the net π bonding over the molecule from two to one and, if both electrons were placed in one of the π^* orbitals, say the one in the y direction, the whole π system in the y direction would have zero bonding effect. All this implies that the equivalence of the y and x directions in CO_2 will disappear in O_3, i.e., the molecule is no longer linear. This conclusion may also, of course, be derived by the methods of section 4.3. Ozone is isostructural and isoelectronic with the nitrite ion and has a bent structure,

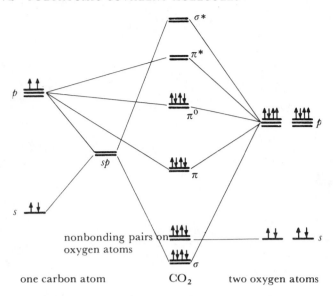

FIGURE 4.29 *Diagram of the energy levels in carbon dioxide*

with a nonbonding pair on the central oxygen atom and two nonbonding pairs on each of the two terminal oxygens. There are three, three-centre, π orbitals of which the bonding and nonbonding ones are occupied giving an $O-O$ bond order of $1\frac{1}{2}$. The schematic energy level diagram of ozone is shown in Figure 4.30 for comparison with carbon dioxide.

The nitrate ion, NO_3^-

This ion is planar and there are no unshared electrons on the nitrogen atom. The total number of electrons is twenty-four.

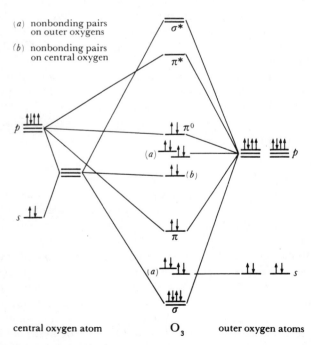

FIGURE 4.30 *Diagram of the energy levels in ozone*
(a) non-bonding pairs on outer oxygens, (b) non-bonding pairs on centre oxygen.

The ion is isoelectronic and isostructural with carbonate and has the same π electron configuration, with two electrons in a strongly-bonding π orbital delocalized over the whole molecule and four electrons in the degenerate pair of nonbonding orbitals. There is a net effect of one π bond in the ion and a $N-O$ bond order of $1\frac{1}{3}$.

Sulfur trioxide, SO_3

This is a plane triangular molecule with twenty-four valency electrons like the nitrate ion. It might therefore be described in the same way if the sulfur atom used only its $3s$ and $3p$ orbitals. There would then be six electrons in π orbitals but with a net bonding effect of only one π bond. If, however, the sulfur atom makes use of its $3d$ orbitals in addition, it is possible to construct π orbitals from six atomic orbitals (S $3p_z$, $3d_{xz}$, $3d_{yz}$, plus three O $2p_z$) instead of four. There would then be three bonding π orbitals and three anti-bonding ones, and the six π electrons can all be placed in bonding orbitals, giving three delocalized π bonding orbitals and a total $S-O$ bond order of about two (the approximation comes in as all three π orbitals are not equally stable).

In a similar way, SO_2 is isoelectronic and isostructural with O_3 but the possibility exists of the sulfur using d orbitals to increase the number of bonding π orbitals.

Whenever d orbitals of relatively low energy are available, there is the possibility that they will contribute to the bonding. This applies especially to molecules containing very electronegative elements bonded to elements in the third and higher periods and thus to all such oxygen compounds. The extent of d orbital participation will depend on the relative energies and the number of electrons to be accommodated and may be decided only by calculation (see section 18.9).

In particular, the use of d orbitals by elements of the $3p$ series has been a hotly debated area in recent years. A recent article has concluded that the bonding in SO_2 can indeed be best described by means of two $S=O$ double bonds and that resonance structures of the type

which are analogous and essential in the description of the bonding in O_3, are unimportant for SO_2.

Summary

Shapes of polyatomic molecules may be treated as follows:
1. Where only sigma bonds and unshared pairs of electrons are involved, the shape of simple molecules is determined by the number of electron pairs around the central atom. The bonding may be treated in terms of localized two-centred molecular orbitals formed between each pair of bonded atoms, or may, with more accuracy but less convenience, be

treated in terms of polycentred, delocalized orbitals. These treatments may often be extended quite considerably to more complex species (for example, those with more than one 'central atom') but the predictions become less secure as the complexity rises.

2. The basic shapes of simple species with pi bonds again depend only on the number of sigma and lone pairs on the central atom and this can be calculated by the methods of section 4.3. Localized pi bonds between pairs of atoms may be treated in terms of two-centred molecular orbitals just as in the case of diatomic molecules.

3. Where a number of equivalent atoms occur in a species with pi bonding, the pi electrons must be treated as de-localized over the whole molecule. Polycentred molecular pi orbitals are constructed from atomic orbitals of suitable symmetry, and these have the property that the molecular orbital of lowest energy and that of highest energy are readily distinguishable, and, as long as the latter remains unoccupied, a net pi bonding effect results. The treatment of delocalized pi bonding by relatively simple ideas is less complete that that of sigma bonding, and problems arise particularly in the cases where d orbital participation is possible. These more complicated cases can only be fully treated by detailed calculation which is beyond the scope of this text. However, the relatively simple ideas outlined in the earlier parts of this chapter allow a reasonably accurate description of the shapes of the large majority of simple polyatomic molecules and ions.

PROBLEMS

Many examples of molecular shapes can be found in Chapters 9 onwards. You should list a few examples and then work them out without immediate reference to the text. It is also important to comprehend molecular shapes in three dimensions. You should take every opportunity of looking at, or making, models of structures.

4.1 Determine the most probable shape of the fluorides of the Main Group Elements as listed in Table 17.2b, p. 322

4.2 Determine similarly the shapes of all the interhalogens and all the polyhalide ions given in Table 17.19, p. 371

4.3 Determine the most likely structures of all the xenon oxides, oxyfluorides and anions given in Table 17.22, p. 376

4.4 Discuss the following bond angles: $117°$ in O_3, $\sim 120°$ in SO_2, $101°$ in SiF_2, $180°$ in CO_2, $180°$ in N_3^-.

4.5 Arrange the following species in order of increasing OSO angle: SO_2, SO_3, SO_3^{2-}, SO_4^{2-}, H_2SO_3, H_2SO_4, SO_2Cl_2, SO_2F_2.

4.6 Discuss the ONO bond angles, $180°$ in NO_2^+, $134°$ in NO_2, $115°$ in NO_2^-.

Bonding in polyatomics should be worked out *both* on the 2-centre *and* on the poly-centre orbital basis. Again, the starting point is to consider the in-phase overlap of orbitals, thought of as waves.

The bonding material in this chapter should be correlated with the molecules described in the second half of the book. For example, you should go on to consider the bonding description of the compounds in questions 4.1 to 4.6 above. The following questions will suggest some approaches.

4.7 In a series of molecules PXH_2, the HPH angle was $90°$, $95°$, $100°$, $109\frac{1}{2}°$, $115°$, $120°$ and $135°$. What hybridization of phosphorus s and p orbitals is implied in each case? Use Table 4.5 to decide the relative strengths of these PH bonds.

4.8 Imagine a square-planar molecule MH_4. Assume M has valence shell s, p and d orbitals and take the molecular plane as xy. Write down all the combinations of M orbitals with hydrogen s orbitals which give delocalized sigma bonds. (Compare section 4.7.)

4.9 By analogy with Figure 4.26, draw out the molecular π orbitals formed by the out-of-plane p orbitals in a square planar AB_4 species.

4.10 In 9, if A can also use d orbitals, which of these π levels would be further stabilized? (Take the plane as xy, and put the B atoms on the x and y axes).

4.11 Assuming the species remain planar, decide the electron configurations of CO_3^{3+}, CO_3^{2+}, CO_3^+, CO_3, CO_3^- and CO_3^{3-} by comparison with CO_3^{2-} (Figure 4.25). Which of these would be:

(a) paramagnetic,   (b) more strongly bonded?

Which species would be stabilized by becoming non-planar, making ψ_2 more stable than ψ_3?

4.12 The photoelectron spectrum of formaldehyde H_2CO shows four bands above 21 eV:

	energy eV	vibrational sequence and separations (cm^{-1})		
1	10·8	short	2560	1590
2	14·1	long	2400	1210
3	15·9	long	—	1270
4	16·3	broad envelope		

The stretching frequencies for H_2CO are 2780 and 1740 cm^{-1}.

(a) Decide the probable shape of H_2CO.

(b) Discuss the sigma bonds, non-bonding electrons and pi bonds.

(c) Discuss the photo-electron spectrum and draw up an energy level diagram.

4.13 (a) The He(I) photoelectron spectrum of CO_2 shows the following bands: 13·8 eV (non-bonding): ca 17·6 eV (bonding), 18·1 eV (approx. non-bonding): 19·4 eV (approx. non-bonding).

(b) Other evidence indicates that the 13·8 and 17·6 eV ionizations are from π levels.

(c) It is calculated that there are two further ionizations at about 39 and 41 eV, both bonding.

Is this evidence compatible with Figure 4.29?
Do these results suggest modifications to the CO_2 discussion along the lines of section 4.7?

5 The Solid State

When we turn from single molecules to solids, the structures and energies depend on a wider range of forces.

The most direct continuation from our discussions of the small single molecule are those species where normal covalent forces are present throughout the crystal—the crystal is a giant molecule. The classical example is diamond where each C atom is bound by a normal C–C single bond to four neighbours arranged tetrahedrally and this structure extends indefinitely. Such crystals are typically hard and high-melting. An example of a compound of this type is SiO_2 in which each Si is bonded to $4O$ and each O to $2Si$ in an infinite array.

At the other extreme we find molecular solids where the individual compounds are strongly-bound covalent molecules, such as CH_4, but only very weak forces hold the molecules together in the solid. Such crystals are typically soft and low-melting.

We find a third class of solids where the components are ions and the major binding forces in the crystal are the electrostatic ones between the ions, as in NaCl. Such crystals are relatively hard and high-melting.

Finally we may pick out the class of metals which are related to the giant molecules but where the outer electrons are free to move throughout the crystal. Such crystals show a range of hardness and melting points, and are highly reflecting and conducting.

These four classes are idealized cases and a range of interactions apply in most solids (compare also the discussion based on Figure 5.11 in section 5.6). We start this chapter with ionic crystals, where the basic interactions are well-understood, and we go on to relate other types of solid to these.

Simple Ionic Crystals

5.1 The formation of ionic compounds

In the last two chapters, the Lewis electron pair theory was extended on the basis of wave mechanics to give a full description of the covalent bond. In this section, ionic bonding will be examined and the various factors which determine the formation and stability of ionic compounds will be discussed.

The basic process in the formation of an ionic compound is the transfer of one or more electrons from one type of atom to another: the resulting ions are then held together by electrostatic attraction.

Isolated ionic species, such as the two-atom entity M^+X^-, do not exist under ordinary conditions. In ionic compounds, we are dealing with an array of ions, which extends in three dimensions to the edges of the crystallite. If the ionic solid dissolves, the ions are indeed separated, but they are stabilized by interaction with the solvent molecules (see Chapter 6) and the free ion has only a transient existence.

The arrangement of the ions in the solid is the one which gives the highest electrostatic energy. To see what factors determine this arrangement, consider the process of bringing up successive anions around a given cation. If there are already n anions surrounding the cation, the addition of a further anion produces an extra attraction between its charge and the cation charge, and also produces a number of repulsions between its charge and the charges on the n anions already present. There are thus two opposing tendencies. One is to increase the attractive forces by making the coordination number of the cation as large as possible and this is balanced by the increase in the repulsive forces as more and more anions are added. When the two tendencies balance, the final structure results. An exactly similar argument holds, of course, for the number of cations to be found around an anion. The repulsions are at a minimum if the distribution of ions is as symmetrical as possible. Thus ions which are three-coordinated have their neighbours at $120°$ in a triangular arrangement, four-coordinated ions have a tetrahedral, six-coordinated ions an octahedral, and eight-coordinated ions a cubic arrangement. In addition, some coordination numbers, such as five, which do not pack regularly in a solid are not observed in ionic crystals.

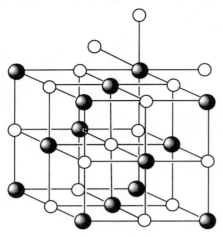

(a) sodium chloride

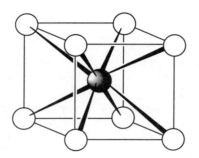

(b) cesium chloride

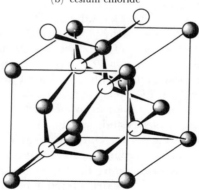

(c) zinc blende (ZnS)

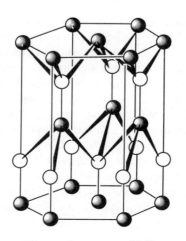

(d) wurtzite structure (ZnS)

The coordination numbers in a solid of given formula, such as AB, depend on the number of the larger ions which may be packed around the smaller one. The stoichiometry—in this example, 1:1—then determines the coordination number of the larger ion. As the formation of a cation involves the removal of electrons, cations are always smaller than the parent atoms (for example, the atomic radius of K is 203 pm, while the radius of K^+ is 133 pm). Conversely, the addition of electrons to atoms to form anions involves an increase in radius (for example, F = 71 pm but F^- = 133 pm in radius). As a result, anions are generally larger than cations and it is the number of anions which can pack around a cation which usually determines the coordination numbers and the structures. For example, sodium chloride crystallizes in a structure where the sodium ion is surrounded by six chloride ions (Figure 5.1a), and the chloride ion has, of course, six sodium ions around it. The larger cesium cation allows a coordination number of eight in the structure of cesium chloride (Figure 5.1b).

The number of anions which can pack around a given cation may be determined from the ratio of the radii of the cation and anion. This *radius ratio*, r_+/r_-, may thus be used to give an indication of the likely coordination number

FIGURE 5.1 *Structures of* AB *solids,* (a) *sodium chloride or rock salt,* (b) *cesium chloride,* (c) *zinc blende* (ZnS), (d) *wurtzite* (ZnS)

Note on crystal structure diagrams
It follows from the discussion in section 5.1 that ions, or atoms, in crystals are expected to be as close together as possible and that the available space will be filled as completely as possible. However, if such a 'spacefilling' situation is represented directly in a diagram it is very difficult to see the arrangement of the atoms. As a result, most diagrams give an exploded view of the crystal. This convention has been adopted here. In accurate diagrams, the centres of the atoms are positioned exactly but their diameters would be reduced. In most of the diagrams in this chapter, further slight distortions or alterations in perspective have been introduced to make the structural arrangement easier to interpret. In a number of cases, further conventions are used to assist understanding. Atoms or ions in the top layer may be distinguished from more distant layers by the thickness of the circles used to represent them or, as for example in Figure 5.3, by using darts to represent the bonds joining them with the broad end of the dart on the atom nearest the front of the unit cell. Thus in the rutile diagram, Figure 5.3a, the Ti atom at the body centre has three O atoms in front of it and three behind it.

In more elaborate structures, there is a conflict between showing clearly the coordination of each atom and avoiding an extremely elaborate diagram showing many repeat units. Compare the diagram of CdI_2 (Figure 5.10a), which shows the coordination but does not readily give the metal layers, and the photograph of a model of the same structure (Plate I) which gives a better impression of the metal occupying every second layer between I sheets by showing a larger portion of the structure. Any diagram is necessarily a compromise and a formalized representation of a three-dimensional, space-filling structure and the reader should search for as many representations as possible of difficult structures in the references given. The study of models in three dimensions greatly clarifies the more complex structures and every opportunity of examining such models should be taken (see references).

for a salt of given formula type. (Notice that the argument is exactly the same if the anion is smaller than the cation, except that it is the ratio of anion radius to cation radius, r_-/r_+, that is the important one.) If it is assumed (i) that ions are charged, incompressible spheres of definite radius (the validity of this assumption is discussed in section 5.5), (ii) that the central ion adopts the highest coordination number which allows it to remain in contact with each neighbour (this is a good approximation for the balance of attractive and repulsive forces referred to above), then the radius ratio limits corresponding to different coordination numbers may be calculated from purely geometrical considerations.

For example, in six-coordination, a cross-section through a site in the lattice appears as in Figure 5.2.

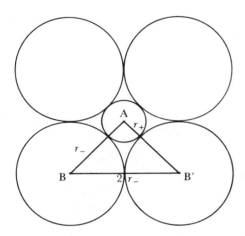

FIGURE 5.2 *Cross-section through an octahedral site*

When the anions just touch, BAB′ = 90°.

Then,

$$AB = AB' = (r_+ + r_-) \text{ and } BB' = 2r_-$$
$$\frac{AB}{BB'} = \frac{r_+ + r_-}{2r_-} = \cos 45° = 1/\sqrt{2}$$
$$\therefore r_+ = \sqrt{2}r_- - r_-$$
$$\text{or } r_+/r_- = \sqrt{2} - 1 = 0.41$$

Similar calculations may be carried out for all the coordination numbers. The results indicate the range of values for the radius ratio within which different coordination numbers should be stable. These are shown below.

r_+/r_- : 0.155 to 0.23 to 0.41 to 0.73 to higher values
C.N. :   3   4   6   8

The validity of this simple method of predicting the coordination number may be assessed by examining the structures of some AB and AB_2 compounds.

There are three structures of AB formula where the ions are all in sites of high symmetry and these are shown in Figure 5.1. By far the commonest is the sodium chloride (rock salt) structure where both ions are octahedrally coordinated. For larger ions, cubic coordination is found in the caesium chloride structure. Smaller ions show tetrahedral coordination in the ZnS structures. These each have both Zn and S atoms in tetrahedral coordination but the overall symmetry of the lattice is cubic in the zinc blende structure, 5.1(c), and hexagonal in the wurtzite structure, 5.1(d).

For AB_2 species, the standard structures (Figure 5.3) are:
(a) titanium dioxide (rutile) with coordinations of 6 (octahedral) and 3 (trigonal).
(b) calcium fluoride (fluorite) with coordinations of 8 (cubic) and 4 (tetrahedral).
(c) β-cristobalite (one form of SiO_2) with coordinations 4 (tetrahedral) and 2 (linear). This is the type compound for the most regular 4:2 structure, though SiO_2 is a giant covalent molecule, not an ionic species.

The structures and radius ratios of some ionic halides and oxides are listed in Table 5.1 calculated using the ionic radii of Table 2.12. Among the AB_2 structures, where the eight-coordinated fluorite structure is expected to be replaced by the six-coordinated rutile structure at a radius ratio of 0.73, the agreement with prediction is remarkably good. The lower limit for the stability of six-coordination comes at 0.41, and germanium dioxide, with a ratio of 0.38, occurs in two forms, one isomorphous with SiO_2 and the second with the rutile structure.

In the AB structures, the agreement is less good and the sodium chloride structure persists through a wider range of radius ratios than predicted, both at the upper end as shown in the Table and at the lower end of the range as shown by the lithium halides of radius ratios down to 0.28 for LiI, all of which have the rock salt structure.

The very simple model of ions as hard spheres may thus be used as a reasonable first approximation to suggest the structure expected for an ionic crystal. A clear indication of its limitations is given in the discussion of ionic radii in section 2.15, which should be re-examined.

A number of other structures where the coordination sites are of regular symmetry are shown in Figure 5.5. Of particular interest is the perovskite ($CaTiO_3$) structure, Figure 5.5(c), since many materials with very interesting magnetic, electrical, optical and catalytic properties adopt this structure. The reader is referred to section 16.1 for a discussion of the copper oxide superconductors, many of which have structures related to perovskite.

Less symmetrical configurations adopted by compounds in which the bonding is not purely ionic are discussed in section 5.5.

5.2 The Born-Haber cycle

What factors determine whether given elements combine to form an ionic solid? These may be found by considering the energy changes involved in the formation of an ionic solid from the elements and we shall use sodium chloride as

TABLE 5.1 Radius ratios and structures of some AB and AB₂ solids

Compound	r_+/r_-	AB Structure	Compound	r_+/r_-	AB₂ Structure
KF	1·00	sodium chloride	CaF₂	0·71	fluorite
KCl	0·73	sodium chloride	SrF₂	0·83	fluorite
KBr	0·68	sodium chloride	BaF₂	0·97	fluorite
KI	0·62	sodium chloride	MgF₂	0·49	rutile
RbF	0·92*	sodium chloride	TiO₂	0·48	rutile
RbCl	0·82	sodium chloride (and caesium chloride)	SnO₂	0·51	rutile
RbBr	0·76	sodium chloride	GeO₂	0·38	rutile and also a 4:2 form as in SiO₂
RbI	0·69	sodium chloride	CeO₂	0·72	fluorite
CsF	0·81*	sodium chloride			
CsCl	0·93	caesium chloride (and sodium chloride)			
CsBr	0·87	caesium chloride			
CsI	0·76	caesium chloride			

*These values are r_-/r_+ as the cations are larger than the anion.

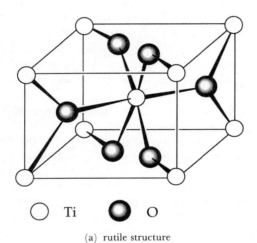

○ Ti   ● O

(a) rutile structure

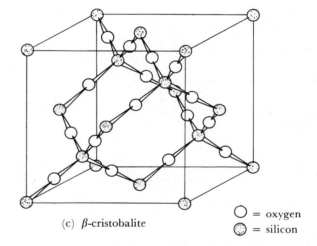

(c) β-cristobalite

○ = oxygen
◉ = silicon

FIGURE 5.3   *The structures of* AB₂ *solids,* (a) *rutile,* TiO₂, (b) *fluorite,* CaF₂, (c) β-*cristobalite* (SiO₂)

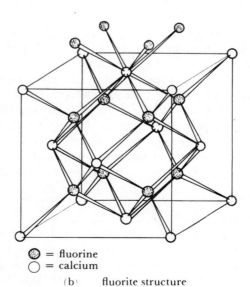

◉ = fluorine
○ = calcium

(b)   fluorite structure

an example. This has a measured heat of formation (H_f) of $-410\cdot9$ kJ mol^{-1}; i.e. for the reaction

$$Na_{(metal)} + \tfrac{1}{2}Cl_{2(gas)} \rightarrow Na^+Cl^-_{(solid)}, \ H_f = -410\cdot9 \text{ kJ mol}^{-1}$$

This reaction may be broken down into simpler steps whose energies are known, as in the diagram of Figure 5.4, by a method due to Born and Haber.

Known as the Born-Haber Cycle, this gives the net energy change calculated from the five simpler steps shown in the table below the diagram, and hence a calculated heat of formation.

In the case of sodium chloride, the values of all the quantities in the cycle are known so that a calculated heat of

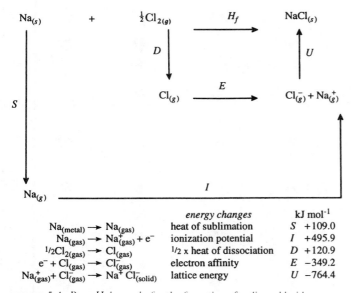

	energy changes		kJ mol⁻¹

$Na_{(metal)} \rightarrow Na_{(gas)}$ heat of sublimation S +109.0

$Na_{(gas)} \rightarrow Na^+_{(gas)} + e^-$ ionization potential I +495.9

$\frac{1}{2}Cl_{2(gas)} \rightarrow Cl_{(gas)}$ $\frac{1}{2}$ x heat of dissociation D +120.9

$e^- + Cl_{(gas)} \rightarrow Cl^-_{(gas)}$ electron affinity E −349.2

$Na^+_{(gas)} + Cl^-_{(gas)} \rightarrow Na^+ Cl^-_{(solid)}$ lattice energy U −764.4

FIGURE 5.4 *Born-Haber cycle for the formation of sodium chloride*

formation (equal to $-387 \cdot 8 \, \text{kJ mol}^{-1}$) may be found and compared with the experimental value. The close agreement confirms that the model of ions as incompressible spheres is a reasonable one in this case. Similar agreement between cycle values and experimental ones is found in many cases and this gives grounds for confidence in using the cycle in other ways. The most useful of these is the evaluation of one of the cycle quantities when the others are known. In particular, the electron affinity E which is difficult to measure is usually determined from the Born-Haber cycles of a series of appropriate salts. It can be seen from Figure 5.4 that:

$$H_f = S + I + D + E + U$$
$$\text{hence } E = H_f - U - S - I - D$$

The Born-Haber cycle may also be used to determine whether or not the bonding in a compound is purely ionic. If the calculated lattice energy and that derived experimentally do not agree, this is a strong indication that the assumption of ionic forces, which is made in order to do the calculation, is incorrect. This aspect is discussed further in section 5.5.

Our main reason for discussing the formation of an ionic solid in terms of the Born-Haber cycle is to allow a more detailed assessment of the contribution of each component of the energy cycle to the heat of formation of the ionic solid. The factors which determine whether a compound is ionic or covalent may thus be isolated and discussed. These factors will be examined in turn starting with the lattice energy U which is the most important exothermic term in the cycle. Then the endothermic terms—the ionization potential, the electron affinity, and the heats of atomization—will be examined.

5.3 The lattice energy

The energy change when the gaseous ions are brought together from infinite separation to their equilibrium distances in the solid is the lattice energy, U. That is, U is the heat of the reaction (for an AB solid)

$$A^{z+}_{(gas)} + B^{z-}_{(gas)} \rightarrow A^{z+}B^{z-}_{(solid)} \cdots \cdots \cdots U$$

This energy arises from electrostatic interactions between the ions. When the geometry of the solid array of ions is known, the lattice energy may be calculated. The method may be illustrated by considering the square two-dimensional array of ions shown in Figure 5.6. Any one cation in a square array of interatomic distance r has four anions at a distance r as nearest neighbours. The electrostatic potential energy between these ions is $4Z^+Z^-e^2/4\pi\epsilon_0 r$, where Z^+ and Z^- are the positive and negative charges on the ions (assuming an AB stoichiometry), e is the charge on the proton, and ϵ_0 is the permittivity of a vacuum. The next-nearest neighbours to the given cation are four cations at a distance $\sqrt{2}r$ and a repulsion exists between the given cation and these ions of $4(Z^+)^2e^2/4\pi\epsilon_0\sqrt{2}r$. Then come four more cations at $2r$, eight anions at $\sqrt{5}r$ and so forth. The total electrostatic energy of a cation in this square array is thus:

$$E = 4e^2(Z^+)(Z^-)/4\pi\epsilon_0 r + 4e^2(Z^+)^2/4\pi\epsilon_0\sqrt{2}r$$
$$+ 4e^2(Z^+)^2/8\pi\epsilon_0 r + 8e^2(Z^+)(Z^-)/4\pi\epsilon_0\sqrt{5}r + \dots$$
$$= -Z^2e^2/4\pi\epsilon_0 r(4 - 2\sqrt{2} - 2 + 8/\sqrt{5} \dots)$$

where $Z^+ = -Z^-$ has been put equal to Z for an AB crystal.

The convergent infinite series in the bracket may be evaluated from the geometrical properties of the array and its sum may be found. This sum is called the *Madelung constant*, A, after its first evaluator. A different geometrical array, for example a rectangular one, would have a different value of the Madelung constant and the whole analysis may be extended to three dimensions. For example, for a rock salt lattice the Madelung constant

$$A_{\text{NaCl}} = (6 - 12/\sqrt{2} + 8/\sqrt{3} - 6/2 + 24/\sqrt{5} \dots)$$
$$= 1 \cdot 748 \dots$$

The Madelung constant is the factor relating the electrostatic forces and the spatial arrangement of the ions in a crystal. It depends only on the geometry of the crystal and is independent of the nature or charge of the ions.

The electrostatic energy of a cation in an AB lattice may therefore be written:

$$E = \frac{-Z^2e^2A}{4\pi\epsilon_0 r}$$

The electrostatic energy of the anion in an AB crystal is the same as that for the cation. If a mole is considered, the electrostatic energy is E times Avogadro's constant, N. (In detail, the energy is $N \times (E_{\text{cation}} + E_{\text{anion}}) \div 2$: the division by two is to avoid counting each attraction twice over.) Thus, the electrostatic energy of a mole of an AB ionic compound is

$$NE = \frac{-Z^2e^2AN}{4\pi\epsilon_0 r}$$

So far, only the attractive electrostatic forces between the ions have been considered but Born introduced a second, repulsive,

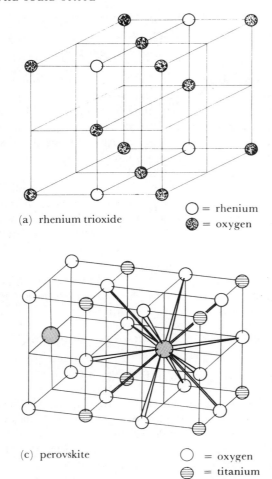

(a) rhenium trioxide

◯ = rhenium

◍ = oxygen

(c) perovskite

◯ = oxygen

⊜ = titanium

◎ = calcium

FIGURE 5.5 *Further examples of symmetrical structures*, (a) *rhenium trioxide*, (b) *corundum* (Al_2O_3), (c) *perovskite* ($CaTiO_3$), (d) *spinel* (Al_2MgO_4)

These structures all have the ions in positions of maximum symmetry with neighbours disposed regularly around them, with the exception of corundum and spinel where the oxygen positions are in a close-packed arrangement but the nearest neighbours are not of highest symmetry.

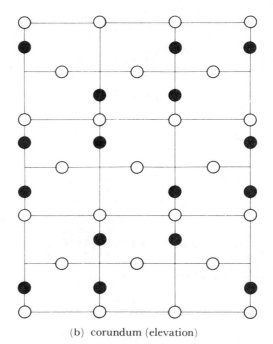

(b) corundum (elevation)

◯ = oxygen ions (close packed)

● = aluminium ions in $\frac{2}{3}$ of the octahedral sites

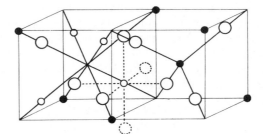

(d) spinel structure

◯ = oxide ion

○ = metal ion in octahedral site

● = metal ion in tetrahedral site

energy term into the equation to take account of the repulsion which arises at very short interionic distances when the ions start to interpenetrate (otherwise the expression for NE tends to infinity as r tends to zero). The repulsive force rises steeply as the interionic distance decreases and is represented by a term $E_{rep} = B/r^n$ where B is a constant similar to the Madelung constant and n is also a constant of the order of nine for sodium chloride. The total energy of the crystal, which is the lattice energy U, is then:

$$U = NE + NE_{rep} = \frac{-Z^2e^2AN}{4\pi\epsilon_0 r} + \frac{NB}{r^n}$$

The constant B can be eliminated, since the lattice energy is a minimum at the equilibrium value of $r = r_0$, so that by setting $dU/dr = 0$ for $r = r_0$, B can be expressed in terms of

the other constants. Then:

$$U = \frac{-Z^2e^2NA}{4\pi\epsilon_0 r_0} (1 - 1/n)$$

All the quantities on the right-hand side of this expression are known or may be found; r_0 comes from direct experimental evidence and n from calculation or experiment. The lattice energy U can thus be found by a combination of calculation and experimental determination of parameters.

It can be seen that the properties of an ionic solid which determine the lattice energy are the geometry, as reflected in the Madelung constant, the interionic distance, and the charges on the ions. The most important effect is that of the charge which appears as a squared term. The lattice energy

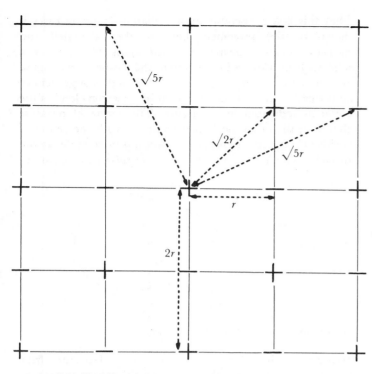

FIGURE 5.6 *Two-dimensional analogue of a crystal lattice*

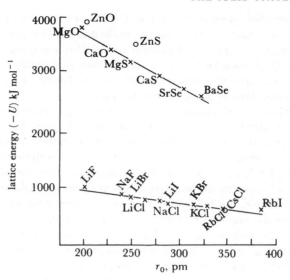

FIGURE 5.7 *Lattice energies* (plotted as $-U$)
Lattice energies are plotted as a function of r_0 for a number of cases of the sodium chloride structure and for CsCl, ZnO, and ZnS. The marked effect of the ionic charge is clear: it will also be noted that the variation in structure to 8:8 or 4:4 coordination has only a minor effect.

of an $A^{2+}B^{2-}$ solid is four times that of an $A^{+}B^{-}$ solid of the same geometry.

Of the geometrical factors, the Madelung constant has only a minor effect on the lattice energy as the values of the constant for different structures of the same stoichiometry are similar. For example, in AB structures the Madelung constants are as follows:

8-coordinated, CsCl structure, $A = 1\cdot763$
6-coordinated, NaCl structure, $A = 1\cdot748$
4-coordinated, wurtzite structure, $A = 1\cdot641$

When the stoichiometry changes, the Madelung constant changes to a greater extent, for example, $A = 5\cdot04$ for the fluorite structure; but changes in the stoichiometry imply changes in the charges and it is difficult to make comparisons between different formula types.

Variations in the second geometrical factor, the interionic distance, have a more important effect on the size of the lattice energy. The lattice energy depends inversely on r_0 so that, within a given structure type, the lattice energy will fall as the ionic sizes increase. It will be clear from section 5.1 that the ratio of anion to cation sizes should not vary too widely as r_0 varies or some other structure will become the more stable one. The effect of variations in charge and interionic distance is best seen by comparing the lattice energies of a series of compounds of the same structure formed by related elements, Figure 5.7. The marked effect of changes in charge is very obvious, as well as the linear fall in lattice energy as the ions (and therefore the interionic separation) increase in size. The values for AB compounds of different

structures (ZnO and ZnS having the wurtzite structure) are included in the figure to show the effect of the Madelung constant.

To summarize, the lattice energy, and therefore the probability that the formation of an ionic solid will be energetically favourable, is increased by (i) increasing the charges on the ions, (ii) decreasing the interionic separation, i.e. by having small ions, and (iii) by changes in the Madelung constant, though this effect is relatively small.

5.4 The endothermic terms in the formation of an ionic solid

In general, the largest of the energy contributions which have to be added to the system before an ionic solid is formed is the ionization potential of the cation, I. If the formation of an ionic solid is to be favoured, the ionization energy should be as small as possible. From Table 2.8, it will be seen that the ionization potentials of the elements have the following characteristics:

(i) The ionization potentials increase from left to right across a Period.

(ii) In any Group, the ionization potential decreases with increasing size down the Group.

(iii) For a given element, the first ionization potential is less than the second, which in turn is less than the third, etc.

It thus requires least energy to form cations of large atoms on the left of the Periodic Table, that is of the larger elements of the alkali and alkaline earth elements. The higher the charge on the ion, the greater is the energy required in its formation.

The energy of formation of the anion is the electron affinity. The electron affinities (compare section 2.14) may represent an endothermic or an exothermic contribution to the heat of formation of an ionic solid, but even the most exothermic electron affinity is less than the smallest ionization potential so that the formation of gaseous cation plus anion from the gaseous atoms is endothermic for any pair of elements. Only a few electron affinities are exothermic and these are all for the formation of singly-charged anions. The formation of doubly-charged anions is always a strongly endothermic process and this true *a fortiori* for anions of higher charge.

The heats of atomization of the elements in their standard states depend very much on the form in which the element exists. Little in the way of generalization can be said except that where elements in a Group occur in the same form—as for example, the halogens—there is a tendency for the heat of atomization to decrease with increasing atomic weight, but often with the lightest element anomalous. As the values for sodium chloride show, the heats of atomization commonly represent only a minor contribution to the energy balance.

The effect of the endothermic terms on the heat of formation of an ionic compound may be summarized by saying that least energy is required, and therefore formation of an ionic solid is most favourable, when the ions are (i) of low charge, (ii) large, so that the interionic distance in the solid is large, and (iii) formed from elements at the extremes of the Periodic Table.

If these factors are compared with those which lead to a high lattice energy, it is seen that the two main requirements for low endothermic energies—large ions of low charge—are exactly opposed to those which favour high lattice energies. The small ions of high charge which give the highest exothermic contribution are precisely those which require the highest endothermic energies of formation from the elements in their standard states. The formation of an ionic solid therefore depends on the detailed balance of all the energy contributions in each individual case. Once again, the direction of the chemical change depends on small differences between large values and is difficult to predict *a priori*. In general, ionic solids are formed by the *s* metals and the transition elements in the II or III oxidation states (though often as complexes) with anions from the halogen or chalcogen Groups. Simple ions with charges greater than two are less common, but the lanthanide elements form M^{3+} ions, and the existence of Th^{4+} is well established.

If the values of the ionization potentials are examined in detail, it will be seen that successive ionization potentials for an element rise in a fairly regular manner as long as only electrons in the valence shell are removed. Since lattice energies vary as the square of the charge on the ions, it might seem that, on balance, ion formation would be most favoured in cases where the charges are high. Unfortunately, a further complication appears which upsets this conclusion, in that highly-charged ions are those most likely to cause polarization and a departure from purely ionic bonding.

5.5 Bonding which is not purely ionic

In the discussion above, the assumption was made that ions were hard spheres and that the only forces (except at very short distances) were electrostatic ones between the charges.

That this is a reasonable approximation in many cases is shown by the agreement between the calculated and measured heats of formation of the solids. However, in a solid such as lithium iodide where the cation is very small and the anion is large (Figure 5.8) the high charge density on the cation distorts the rather diffuse electron cloud of the anion as indicated in the figure. The centre of negative charge in the anion no longer coincides with the centre of positive charge and an induced dipole is present in the anion. In such a case, the anion is termed *polarizable* and the cation,

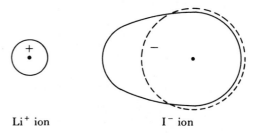

Li$^+$ ion   I$^-$ ion

FIGURE 5.8   *Diagrammatic representation of polarization in* LiI

polarizing. This polarization represents a departure from purely ionic bonding in the compound. Thus the elementary idea of ions as hard spheres has to be modified by allowing for distortions of the spherical electron clouds where ions have high charge densities.

In a similar way, the purely covalent bond discussed in the last chapter is uncommon. The pair of electrons which form a covalent bond are equally shared between the constituent atoms only if these are identical or, by coincidence, of the same electronegativity. If the two atoms differ in electronegativity, the bonding electrons are more strongly attracted by the more electronegative and a dipole is created in the bond (Figure 5.9). The pure covalent bond with the electrons equally shared between the two atoms and the pure ionic bond with the electron completely transferred from one atom to the other are the two extreme cases. There is a complete range of bond types lying between the two, the ionic bond distorting in the manner of Figure 5.8, and the covalent bond distorting in the manner of Figure 5.9, till the polarized covalent bond and the polarized ionic bond become indistinguishable.

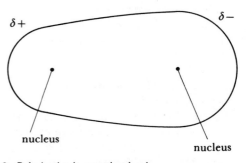

$\delta+$ $\delta-$

nucleus   nucleus

FIGURE 5.9   *Polarization in a covalent bond*

There is one reservation about this picture: the pure covalent bond does exist (between two atoms of the same element) but the pure ionic bond is an abstraction as there is bound to be some polarization between any pair of ions. However, since the environment of an anion in a crystal is symmetrical, the slight polarizations in a strongly ionic compound have only a minor effect on the bond energy and can be regarded as a normal attribute of ionic solids.

The polarizing power of the smallest ion (which is generally the cation) depends on the density of charge on it. The polarizing power is thus greatest for small, highly-charged ions like Mg^{2+} or Al^{3+}: so much so that the smaller congeners of these ions, Be^{2+} and B^{3+}, do not exist and beryllium and boron are covalently bound in their compounds. Polarizability is greatest for large ions like I^-, and especially for those with diffuse electron clouds like H^- or N^{3-}.

A marked degree of polarization shows up in the lattice energy values. When a dipole is induced in one ion, there is an ion-dipole attraction to be added to the ion-ion attraction which is used in the calculation of lattice energies on the purely ionic model. The actual lattice energy of a polarized solid should thus be higher than that calculated on a purely ionic model. The values in Table 5.2 illustrate this.

The differences for the alkali halides are less than one per cent and the direction of deviation is random. The alkali hydrides have experimental values which differ from the values calculated on an ionic model by up to eight per cent and the experimental values are all low. The difference is greatest for the lithium compound, showing the large polarizing effect of the small lithium ion. The sign of the deviation probably results from the unusual compressibility of the hydride ion (compare section 9.1).

The values of the silver and thallium(I) halides are given for comparison with those of the alkali metals of most similar radius, e.g. potassium and rubidium. The figures illustrate the effect of the filled d shell in such ions. The values for the silver and thallous halides (which have the sodium or cesium chloride structures) show marked discrepancies between the calculated and experimental lattice energies and these differences increase from the fluoride to the iodide. In these cases the experimental values are all higher than the calculated ones, showing the presence of the additional energy term due to polarization, which is largest for the iodides. It is interesting to note that gold(I) iodide, AuI, which has a markedly higher lattice energy (-1050 kJ mol^{-1}) than silver iodide, has had its structure determined. This consists of chains . . . $Au - I - Au - I - $ . . . and is clearly not ionic at all. The silver salt thus provides an intermediate case between the ionic alkali metal iodides and the covalent aurous iodide. This example illustrates that the structure of a solid may provide an additional criterion for a departure from ionic bonding.

It is characteristic of an ionic solid that the forces are equal in all directions so that a symmetrical arrangement of ions results. At the other extreme, in a molecular solid formed by a covalent compound, there are two quite different kinds of

TABLE 5.2 Calculated and experimental lattice energies, $-U$

| Compound | Lattice energy (kJ mol^{-1}) | | Difference |
	Experimental	Calculated	
LiF	1009·2	1019·2	−2·4
NaF	903·9	900·8	+3·1
LiH	905·4	979·1	−73·7
NaH	810·9	845·2	−34·3
KH	714·2	741·4	−27·2
AgF	954	920	+34
AgCl	904	833	+71
AgBr	895	816	+79
AgI	883	778	+105
KF	801·2	805·4	−4·2
KCl	697·9	702·5	−4·6
KBr	672·4	674·9	−2·5
KI	631·8	637·6	−5·8
TlCl	732	686	+46
TlBr	720	665	+55
TlI	695	636	+59
RbCl	677·8	677·8	0
RbBr	649·4	653·1	−3·7
RbI	613·0	619·2	−6·2

force: very strong interactions in the bonds between the atoms of the molecule, and very weak interactions between molecules. When the solid is intermediate between these extreme types, there is often stronger bonding in some directions than in others and this shows up in a lowered symmetry of the structure. Thus chains may be formed as in gold(I) iodide above, and another common deviation from ionic bonding is the formation of layer lattices as in CdI_2 (Figure 5.10a and Plate I). In general, departure from purely ionic bonding is shown by the adoption of lattices where ions do not have their neighbours in the most symmetrical possible environments. One example is nickel arsenide which is an AB structure with 6:6 coordination but, although the six neighbours of the nickel atom are disposed octahedrally, the six neighbours of the arsenic atom lie at the corners of a trigonal prism. Some of the commonly adopted structures of this less regular type are shown in Figure 5.10.

It must be noted that, although structures where the atoms are in environments of low symmetry indicate the presence of non-ionic contributions to the bonding, the converse is not true. Many compounds with symmetrical structures are not ionic. For example, many compounds of the transition metals with small atoms, the so-called 'interstitial' compounds, such as TiC or CrN, have the sodium chloride structure but there is no question of these com-

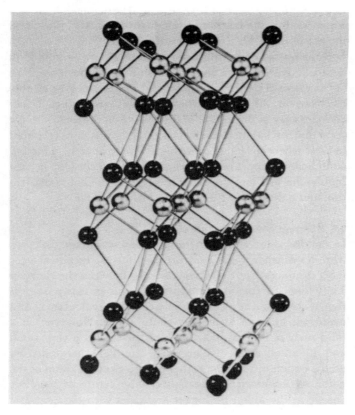

PLATE I *The CdI₂ structure*

Photograph of a model of the CdI_2 structure showing the layers of metal atoms (silver) sandwiched between layers of iodines (black). Compare with Figure 5.10a, which shows a smaller portion of the structure in a perspective that illustrates the hexagonal cell.

PLATE II *The nickel arsenide structure*

A model of the nickel arsenide structure showing the trigonal prism of Ni atoms around the black As atom. The octahedral coordination of the Ni is best seen for the atom at the centre of the model.

pounds containing C^{4-} or N^{3-} ions and their bonding has appreciable metallic character. Similarly, the silver halides have the sodium chloride structure although there is appreciable covalent character in the bonding as just discussed.

5.6 Metallic bonding

While elements to the right of the Periodic Table favour electron pair sharing and are covalently bound in their elemental form, those to the left, which readily lose their valency electrons, are metallic. A metal has been picturesquely described as 'an array of cations in a sea of electrons'. The cations usually assume one of three simple arrangements described below and the valency electrons become completely delocalized over the whole structure. The electrons are mobile, accounting for the typical metallic properties of high electrical and thermal conductivity, and there are no underlying directed bonds. From one point of view, metallic bonding is the limit of the process of delocalizing σ electrons. Consider, for example, a Li_{10} unit. From the $10\,s$ orbitals, we can form ten 10-centre orbitals by an extension of the steps outlined in section 4.7. Only five of the ten orbitals will be needed to hold the 10 valency electrons. Furthermore, the spread of energies between the most and least stable of the ten orbitals is limited, so that the energy gap between the highest filled orbital and the lowest empty one will be relatively small. If we now go to a Li_{1000} unit, and form orbitals delocalized over the whole 1000 atoms, the separation between successive orbitals becomes tiny. In the terminology used in metal theory, we are creating a *band* of orbitals, here an *s band*. There are sufficient electrons to half-fill this band, and the energy gap between the last (500th) filled orbital and the lowest empty one—number 501—will be so small that electrons are likely to spread out over the last few filled orbitals and the lowest few empty ones, due to their thermal energy. In the case of a crystal, delocalized orbitals with an infinite number of centres are constructed and then electrons are placed in these orbitals so that they are evenly shared by the infinite number of cations, the result is a wave-mechanical description of a metal where the electrons are standing waves over the whole crystal.

A band need not involve only *s* electrons. Thus, in calcium, for example, the *p* orbitals overlap to form a *p band*, whose range of energies overlaps with that of the *s* band. Thus, though there are 2 electrons per Ca atom, they do not simply fill the *s* band (which would give an insulator), but partly fill the overlapping *p* band.

Because the electrons at the top of any partly-filled band are free to move through the whole crystal, this accounts for the high electrical and thermal conductivity of metals. Similarly, since excitations of electrons into the upper part of the band can occur with a wide range of energies, electrons interact with all wavelengths of light, giving rise to metallic lustre. When transition elements are involved, *d bands* also occur.

Without going into further detail it can be summarized that there are three basic types of bonding: ionic, involving complete transfer of an electron from one atom to another;

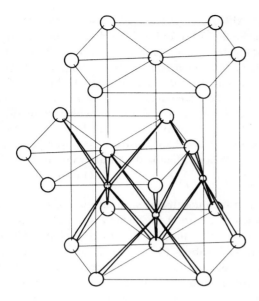

= iodine atom

= cadmium atom

(a)

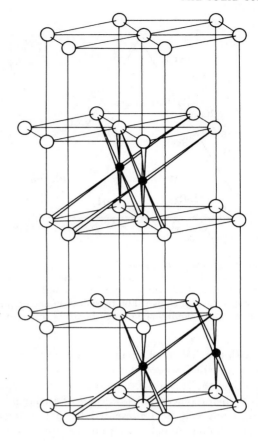

● = bismuth atom

= iodine atom

(b)

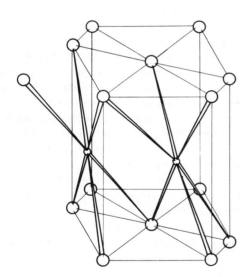

= As atom

= Ni atom

(c)

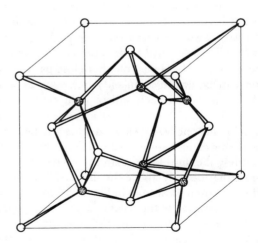

FIGURE 5.10. *Less regular structures:* (a) *cadmium iodide* (see also Plate I which shows three I−Cd−I layers), (b) *bismuth triiodide*, (c) *nickel arsenide* (see also Plate II which shows a larger portion of the structure), (d) Mn_2O_3

In these structures some or all the ions have their neighbours in coordination positions which are not of the highest possible symmetry and this reflects the directional, non-ionic character of part or all of the bonding forces.

= oxygen atom

= manganese atom

(d)

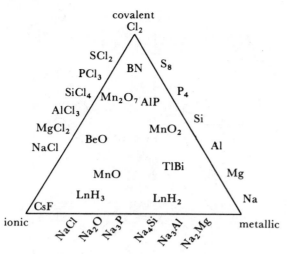

FIGURE 5.11 *Diagrammatic illustration of bond types*
Few bonds are purely ionic, covalent, or metallic and most have some characteristics of all three types and would lie within a triangular plot of the type shown. This presentation also emphasizes that there is no sharp boundary between bonds of different types.

covalent, with the sharing of a pair of electrons between two atoms; and metallic, where the electrons are completely delocalized over the crystal. If these three extreme types are thought of as being placed at the corners of a triangle, then some compounds will be represented by points near the vertices of this 'bond triangle', with bonding predominantly of one type. Some compounds will be represented by points along an edge of the triangle, with bonding intermediate between two types. Finally, the majority of compounds would be represented by points within the area of the triangle, showing that the bonding had some of the characteristics of all three types. This idea is illustrated schematically in Figure. 5.11.

There are three common metallic structures which are illustrated in Figure 5.12, and nearly all metals adopt one or other of these. Two are based on close-packing of spheres, i.e. the metal ions (assumed to be spherical) are arranged to fill the space as closely as possible. For close packing, a layer of spheres can be arranged in only one way, as shown in Figure 5.12a, in which each sphere is in contact with six neighbours. A second layer can be arranged on top of the first, again in only one way if close packing is to be preserved, and this is shown in Figure 5.12b. There are two possible ways of adding the third layer, both preserving close packing. These are either directly above the first layer (Figure 5.12c) or in a third position illustrated in Figure 5.12d. The first case then gives a repetition of the second layer for the fourth one, the first and third layers for the fifth one, and so on in an ABABAB . . . arrangement. This gives rise to *hexagonal close packing* (*hcp*) which is shown in Figure 5.12e and in Plate IIIa where the ABABA arrangement is more obvious. Plate IIIb shows another view of the hcp model illustrating the close-packed layer.

In the second case, the arrangement of the first three layers may be labelled ABC and the repeated pattern is then ABCABCABC . . ., to give *cubic close packing* (*ccp*). This is shown in Figure 5.12f and in Plate IV. In this structure, the close-packed layer is not parallel to the base of the cube, as it was in hcp, but lies along the body diagonal of the cubic array as indicated by the shaded plane in Figure 5.12f. Put another way, the close-packed layers of Figure 5.12d have to be turned through 45° to give the unit cube. Plate IVa shows a model of the ccp array while Plate IVb shows the model with spheres removed to illustrate the close-packed plane parallel to the body diagonal. An alternative name for ccp is *face-centred cube* (*fcc*) describing the orientation of Figure 5.12f.

The third common metal structure is the *body-centred cube* (*bcc*), shown in Figure 5.12g, which is not close-packed and in which the coordination number is eight. These high coordination numbers are typical of metal structures.

A number of common crystal structures are closely related to one or other of the close-packed structures above. In cubic close packing, there are two different kinds of interstitial sites, see Figure 5.12b. One of these is a tetrahedral site between four of the spheres, and the other is an octahedral site between six of the spheres. There are as many octahedral sites as there are spheres, and twice as many tetrahedral sites as spheres. Now suppose that a small atom was inserted into each octahedral site in a close-packed structure, leaving the large atoms in contact, the structure which results is of formula AB. The two kinds of atoms are six-coordinated and this is the same as the sodium chloride structure. Thus the sodium chloride structure may be described as derived from cubic close packing with all the octahedral sites occupied. Of course, in some compounds with the sodium chloride structure, the ions are of such relative sizes that the smaller ones force the larger ones out of contact with each other so that they are no longer close packed, but their relative positions remain the same and the structure is often described in terms of the close-packed form.

Many of the other common structures may be described in similar terms, with the larger ions—usually the anions—in cubic or hexagonal close packing, and the smaller ions occupying some fraction of the octahedral or tetrahedral sites in the close-packed structure. Table 5.3, at the end of the chapter, summarizes the common structures in terms both of the coordination numbers and of their relation to the close-packed structures.

Apart from metals and alloys, metallic bonding is found in a number of other types of compound. One class consists of the so-called 'interstitial' hydrides, carbides, nitrides and borides of the transition metals. These compounds are formed with a variety of compositions—W_2C, TiN, ZrH_2 etc.—which do not commonly fit any normal ideas of valency. The compounds have metallic properties, such as conductivity and magnetic ordering but have different structures from the parent metals. The bonding is still the subject of some controversy but most theories agree in leaving some valency electrons in conduction bands to give metallic properties. Thus these compounds may be pictured as, for example, ionic structures but with additional electrons

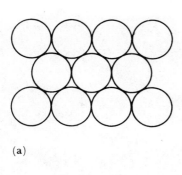

(a)

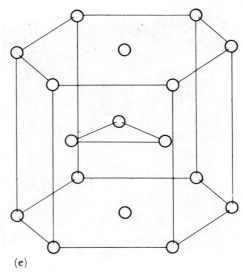

(e)

tetrahedral site
octahedral site

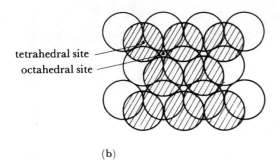

(b)

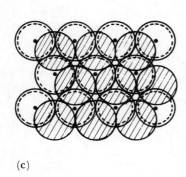

(c)

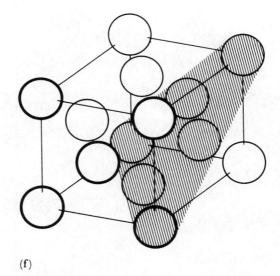

(f)

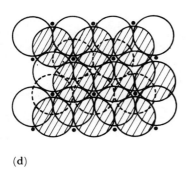

(d)

FIGURE 5.12 *Common metal structures and their construction:* (a) *a close-packed layer of spheres,* (b) *addition of a second layer to* (a) *in close-packing,* (c) *the addition of a third layer to* (a) *and* (b) *directly above* (a), (d) *alternative way of adding third layer,* (e) *hexagonal close-packing,* (f) *cubic close packing (fcc), close-packed layer shaded,* (g) *the body-centred cube*

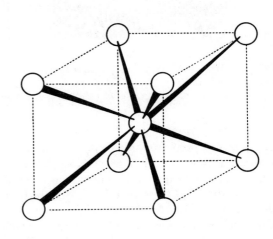

(g)

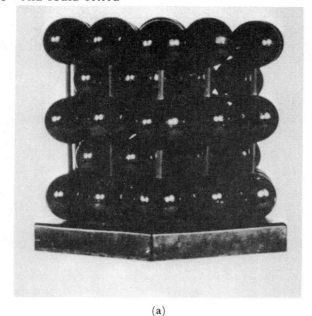

(a)

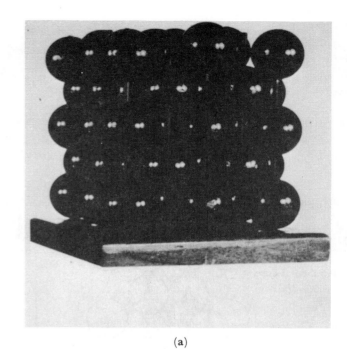

(a)

(b)

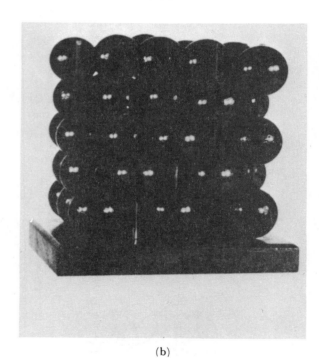

(b)

PLATE III (a) *Model illustrating hexagonal close packing. The ABABA arrangement is clearly seen.*
(b) *Another view of the model in* (a) *showing the close-packed layers.*

PLATE IV (a) *Model illustrating cubic close packing, in the face-centred cube orientation.*
(b) *The model of* (a) *with close-packed layer shown at 45° to fcc edges.*

providing metallic properties (see section 9.5 on the metallic hydrides). Many low oxidation state halides of the transition metals may belong to this class. These compounds, which exist only in the solid state, like ThI_2 or $NbCl_2$ may be regarded as containing M^{v+} cations (v = normal Group oxidation state of the metal) and v electrons which form anions and provide conduction electrons. Thus, ThI_2 would contain Th^{4+} ions, two I^- ions and two conduction electrons per formula unit.

There is a complete range of bond types involving metal atoms, from those involving only two metal centres, through polycentred clusters, giant clusters, microcrystallites, to metal crystals. There is much technological interest in the intermediate sizes—and an interesting question about what size is needed for characteristic metallic properties to appear (the current answer seems to be as low as a few hundred atoms). Bonds between two metal atoms are found in R_6M_2 for M = Pb, Sn, Ge, and R a variety of organic groups; $(CO)_5Mn-Mn(CO)_5$; a number of transition metals bonded to Hg, Ge, Sn, as in $Pt(SnCl_3)ClL_2$ (L = π-bonding ligand) and so on.

Examples of clusters start with triangles, as in $Re_3Cl_{12}^{3-}$ (Figure 15.25), include the $M_6X_8^{4+}$ ions of molybdenum and tungsten (section 15.4), the $M_6X_{12}^{2+}$ ions of niobium and tantalum (section 15.3), both of which contain octahedra of metal atoms, the Bi_9^{5+} cluster (section 17.6.4), and the clusters of sections 16.8 and 18.4.

The field of metal—metal bonding in clusters or in small molecules has attracted a lot of attention in the hope that these studies may throw light on the action of metallic catalysts.

5.7 Complex ions

Although the discussion so far has been concerned with simple ions, most of the points made apply to complex ions as well. Many of the structures of compounds containing complex ions are simply related to those discussed for simple ions. For example, the relationship between the calcium carbide structure of Figure 5.13, and the sodium chloride structure is obvious. Similarly, sodium nitrate, calcium carbonate, and potassium bromate all have structures in which the anion occupies the chloride ion position of the sodium chloride structure. Sodium iodate has a cesium chloride structure while potassium nitrate and lead car-

bonate have nickel arsenide structures. In most cases of complex ion salts, as in all the above examples, the actual symmetry of the lattice is lower than that of the sodium chloride lattice as the complex ions are not spherical. The calcium carbide lattice, for example, is elongated in the direction parallel to the axes of the carbide ions to give a tetragonal rather than a cubic lattice. However, there are a number of cases where complex ionic lattices are of high symmetry. This occurs if the complex ion is able to rotate in its lattice position at room temperature. The resulting, averaged-out, configuration is spherical as in the alkali metal borohydrides, MBH_4. These compounds have actual sodium or cesium chloride configurations at room temperatures as the BH_4^- ions are freely rotating. An interesting intermediate case is provided by the alkali metal hydroxides. These have lattices of low symmetry at room temperature, in which the OH^- groups are not rotating. When the temperature is raised, enough energy is present to allow the hydroxide ions to rotate and the alkali hydroxides undergo a transition to a high temperature form with the sodium chloride structure. Complex cations behave in exactly the same way; most ammonium salts, for example, have sodium chloride or cesium chloride structures as the NH_4^+ ion is freely rotating. Hydrated and other complex cations also typically form lattices of high symmetry with large anions. Thus $[Co(NH_3)_6][TlCl_6]$ has the sodium chloride structure while $[Ni(H_2O)_6][SnCl_6]$ crystallizes in the cesium chloride structure.

Compounds of more complicated formula type also mirror the structures of the simple ionic types. One example is provided by the many compounds which crystallize with the K_2PtCl_6 structure. This is related to the structure of calcium fluoride but with the cations in the fluoride positions and the large anion in the calcium site (called the *anti-fluorite* structure).

Compounds containing complex ions may also be found with the less symmetrical structures associated with significant non-ionic contribution to the bonding. Lead carbonate and one form of calcium carbonate have the ions in the same positions as in nickel arsenide while a number of complex fluorides, such as K_2GeF_6, have layer structures related to that of cadmium iodide.

When complex ions are present, a permanent dipole often exists, and the interaction with this dipole has to be added to the interactions between ions, and between induced dipoles and ions, which have already been discussed. One example is provided by the hydroxide ion where there is a permanent dipole $O^{\delta-}-H^{\delta+}$ in addition to the negative charge. In the high-temperature form where the hydroxyl group is freely rotating, the ion-dipole forces are equally directed, but when the hydroxyl groups become fixed in orientation, the existence of the dipole means that the forces between anion and cation differ in the direction of the dipole from those in directions perpendicular to the dipole. Salts containing small complex ions may thus be equivalent to those with simple ions when the complex ions can rotate freely, or the presence of permanent dipoles may introduce a directional

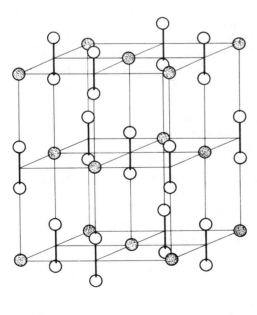

C_2^{2-} ions

Ca ions

FIGURE 5.13 *The calcium carbide structure*

element into the bonding in the crystal. Just as simple ions may occur in compounds which are ionic or which have non-ionic contributions to the forces in the crystal, so do small complex ions occur in symmetrical crystals which are ionic to a high degree of approximation and also in less symmetrical compounds with layer or other 'non-ionic' structures.

In addition to compounds with small, discrete, complex ions, there are very extensive series of compounds of large condensed ions, especially those containing condensed oxyanions such as polyphosphates or silicates (see section 18.6). Such compounds form a vast topic of their own and the structural problems involved have often been very difficult to study.

5.8 The crystal structures of covalent compounds

In the largely ionic or metallic compounds discussed above there is a reasonable uniformity of bond strength throughout the crystal. Either all the bonds are identical, as in salts of simple ions and in elemental metals, or, even when there are strongly bonded units within the solid, as in the salts of complex ions, these units are bonded together by strong ionic forces. Such relative uniformity of bond strength is reflected in the hardness and fairly high melting points which are common among the compounds discussed above. When compounds in which the bonding is largely covalent are examined, these properties are no longer found except in relatively few examples, like diamond or silica, which are very hard and high melting. The vast majority of covalent compounds form soft, low-melting solids of low crystal symmetry, and the actual structure of the solid form of these compounds is usually of relatively minor importance for an understanding of their chemistry.

The hard, high-melting covalent compounds are those where covalent, electron pair bonding extends throughout the crystal so that the whole crystal is a 'giant molecule'. An obvious example is diamond in which each carbon atom is bonded tetrahedrally to four others and this structure continues throughout. In order to melt or fracture such a crystal, strong covalent bonds must be broken and this requires considerable energy. Similar examples are silica, with four-coordinate silicon and two-coordinate oxygen throughout, silicon and silicon carbide and the tetrahedral form of boron nitride—all with the diamond structure—and oxides of other elements of moderate electronegativity, such as aluminium.

The structures of the elements in the carbon Group give a good illustration of this type of giant molecule and the effect of deviations from it. Carbon, silicon, germanium, and tin (in the grey form) all occur with the diamond structure but the valency electrons become increasingly mobile with increasing atomic weight so that silicon, germanium, and grey tin have conductivities increasing in that order. These conductivities are much higher than those of insulators, such as diamond, but many times less than the conductivities of true metals. Such compounds are called *semiconductors*. The valency electrons are largely fixed in the bonds but have a

small mobility, in other words, these elements are at the start of the trend from covalent to metallic properties.

White tin and lead have different structures from the diamond one (Figure 5.14) and tend much more towards metallic forms. White tin has the atoms in a distorted octahedral configuration with four nearest neighbours and two more a little further away, while lead has an approximately close-packed structure but with an abnormally large interatomic distance. These two elements represent further steps away from the covalent, low-coordinate structures towards the high-coordinate structures with close-packing which are typical of metals. Both have conductivities in the metallic range, and thus mobile electrons. However, even lead differs a little from the typical metal in its high interatomic spacing. These elements would all lie along the covalent-metal edge of the ionic-covalent-metallic triangle of bond types with diamond at the covalent apex, silicon, germanium and grey tin near the covalent end of the edge, and white tin and lead nearer the metallic end of the edge.

As well as diamond, carbon is found in two other forms—so far unique to it—graphite and fullerene. Graphite is a much softer solid than diamond with a pronounced horizontal cleavage so that it readily flakes. These properties reflect the bonding as the carbons are three-coordinated in a planar sheet, and the fourth electron and the fourth orbital form delocalized π bonds extending over the sheet. The sheets of carbon atoms are very strongly bound but the forces bonding the sheets together are relatively weak (shown, for example, by the C–C distances of 142 pm within the sheets and 335 pm between them), so the sheets readily slide over one another, giving graphite its lubricating properties. Thus the strength and external properties of the solid as a whole reflect the bonding, especially the weakest links in the solid.

Very recently, a new family of related structures has become established called fullerenes—these can be viewed as a new allotropic form of carbon. The simplest members are molecular structures, the type member being C_{60} which is a polyhedron consisting entirely of carbon atoms arranged in linked hexagons and pentagons. These molecules pack together efficiently and the molecule crystallizes in a face-centred cubic lattice. Such linked polyhedral structures were developed by the American architect Buckminster Fuller in the form of geodesic domes and many names have been coined such as buckminsterene, fullerene, 'buckyballs' and 'buckytubes'. The most-studied fullerene, C_{60}, has the shape of a soccer ball, see Figure 5.14(d), and contains twelve pentagons. This is the critical condition for forming closed polyhedra of this sort. Other members, such as C_{70}, C_{82}, etc., have the same 12 pentagons fused to an increasing number of hexagons. Larger extended clusters have also been identified, together with related polymeric forms including carbon 'nanotubes' made up of cylindrical graphite-like sheets with fullerene-like ends. These materials are currently the subject of intensive research around the world. In the few short years since their discovery, a whole new area of chemistry has opened up. A more detailed account is given in section 19.3.

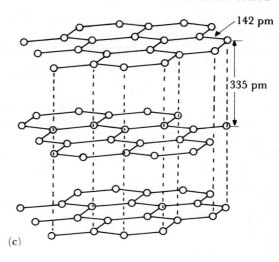

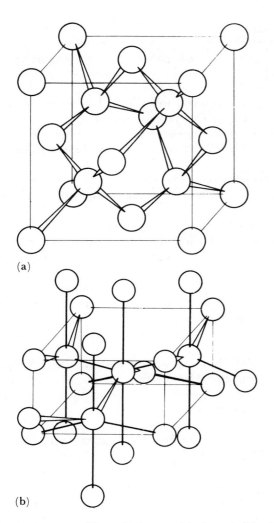

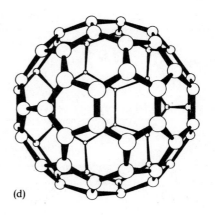

FIGURE 5.14 *Structures of Group 14 elements:* (a) *diamond,* (b) *white tin,* (c) *graphite,* (d) *the soccer ball-shaped fullerene,* C_{60}

When ordinary covalent compounds are considered, it is the weak intermolecular forces which are reflected in the crystal properties. In a simple molecule such as methane, the atoms are strongly linked together by directed bonds in the molecule, but the only forces between molecules are the very weak 'Van der Waals' interactions due to induced dipoles. Thus the compound melts readily and the solid is soft and readily fractured as only these weak interactions have to be overcome. Naturally, in such a case, properties of the solid give no information about the bonds in the compounds.

5.9 Defect structures and non-stoichiometric solids

Real crystals rarely show the ideal structures discussed in the earlier sections of this chapter. Some of the departures from ideality are trivial. For example, most solids are made up of a large number of small domains of ideal structure (called crystallites) which fit discontinuously at the edges and may have missing ions at these boundaries. Another case is found in many minerals where ions of similar size substitute one for another, for example Fe^{2+} for Mg^{2+}, so that there may be a continuum of compositions between the pure Fe and pure Mg species.

Within a crystallite, or single crystal, departures from the ideal arrangement may occur with maintenance of the overall stoichiometry. In the *Frenkel defect*, Figure 5.15a, an atom or ion is displaced from its regular position into a non-lattice site. This will usually be the cation, the smaller species. In the *Schottky defect*, Figure 5.15b, atoms or ions are missing. These, and other, rarer defects are often marked by colour and may markedly affect the conductivity of crystals. Solid state reactions, which involve atoms migrating through the lattice, will be markedly affected in rate. Clearly it requires much less energy for atoms or ions to move via vacant sites than by exchanging in a complete lattice. Since a defect corresponds to a loss of energy compared with the ideal arrangement, the number of defects will be temperature dependent. While defects can be 'frozen in', at equilibrium the proportion of defects will be low at temperatures well below the melting point, but can be high within a few tens of degrees of the m.pt., and the formula may depart markedly from stoichiometric if, say, cations are more readily removed than anions.

This leads us to compounds which are markedly non-stoichiometric. In the eighteenth century, Dalton and

TABLE 5.3 Summary of common structures and their relation to close-packing

Structure	Figure	Coordination	Description in terms of close-packing
—		FORMULA TYPE AB	
zinc blende ZnS	5.1c	4:4 both tetrahedral	S atoms ccp with Zn in half the tetrahedral sites (every alternate site occupied)
wurtzite ZnS	5.1d	4:4 both tetrahedral	S atoms hcp with Zn in half the tetrahedral sites
sodium chloride NaCl	5.1a	6:6 both octahedral	Cl atoms ccp with Na in all the octahedral sites
nickel arsenide NiAs	5.10c	6:6 Ni octahedral As trigonal prism	As atoms hcp with Ni in all the octahedral sites (note that the two types of position cannot be equivalent in hcp)
cesium chloride CsCl	5.1b	8:8 both cubic	Not close-packed. The AB_8 and A_8B arrangements are like bcc
		FORMULA TYPE AB_2	
β-cristobalite SiO$_2$	5.3c	4:2 tetrahedral and linear	The Si atoms occupy both the Zn and S positions in zinc blende (this is equivalent to two interpenetrating ccp lattices) and the O atoms are midway between pairs of Si
rutile TiO$_2$	5.3a	6:3 octahedral and triangular	Not close-packed. The Ti atoms lie in a considerably distorted bcc
fluorite, CaF$_2$ and anti-fluorite, Li$_2$O	5.3b	8:4 cubic and tetrahedral	Ca atoms ccp with F in all the tetrahedral sites O atoms ccp with Li in all the tetrahedral sites
cadmium iodide CdI$_2$	5.10a	6:3 layer lattice octahedral and the 3:coordination is irregular	I atoms are hcp and Cd atoms are in octahedral sites between every second layer The CdCl$_2$ structure is similar but the Cl atoms are ccp
		FORMULA TYPE AB_3	
rhenium trioxide ReO$_3$	5.4a	6:2 octahedral and linear	O atoms are in $\frac{3}{4}$ of the ccp sites and Re atoms are in $\frac{1}{4}$ of the octahedral sites
bismuth triiodide BiI$_3$	5.10b	6:2 layer lattice octahedral and the 2:coordination is non-linear	I atoms are hcp and Bi atoms occupy $\frac{2}{3}$ of the octahedral sites between every second layer The CrCl$_3$ structure is similar but the Cl atoms are ccp
		FORMULA TYPE M_2O_3	
corundum Al$_2$O$_3$	5.4b	6:4 octahedral, and at four of the six corners of a trigonal prism	O atoms are hcp with Al atoms in $\frac{2}{3}$ of the octahedral sites (cf. NiAs with $\frac{1}{3}$ Ni missing). Ilmenite (FeTiO$_3$) is the same structure with alternate layers of Fe and Ti atoms in the Al sites
manganese(III) oxide Mn$_2$O$_3$	5.10d	6:4 six of eight cube corners, and tetrahedral	Mn atoms ccp with O atoms in $\frac{3}{4}$ of the tetrahedral sites (cf. fluorite)

(Contd.)

TABLE 5.3 (Contd.)

Structure	Figure	Coordination	Description in terms of close-packing
		OTHER TYPES	
perovskite CaTiO$_3$	5.4c	Ca—12O (ccp) Ti—6O (octahedral) O—2Ti (linear) and 4Ca (square) giving distorted octahedron	Ca and O atoms together are ccp, with Ca in $\frac{1}{4}$ of the positions in a regular manner. The Ti atoms are in $\frac{1}{4}$ of the octahedral sites
spinel M$_2^{III}$M^{II}O$_4$	5.4d	M^{II}—tetrahedral M^{III}—octahedral	O atoms are ccp and $\frac{1}{8}$ of the tetrahedral sites and $\frac{1}{2}$ the octahedral sites are occupied by metal atoms. In a *normal* spinel, the M^{II} ions are tetrahedral and M^{III} octahedral. In *inverse* spinels, M^{II} ions are octahedral and half M^{III} are octahedral and half are tetrahedral

Notes: (i) Cubic close-packed = ccp: hexagonal close-packed = hcp: body-centred cube = bcc.
(ii) In both ccp and hcp, there are two tetrahedral sites and one octahedral site for each atom in the close-packed lattice.

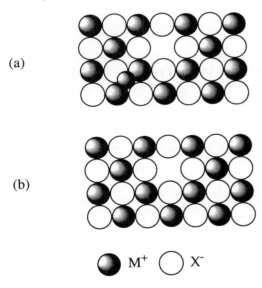

(a)

(b)

M$^+$ X$^-$

FIGURE 5.15(a) *A Frenkel defect and (b) a Schottky defect*

Berthollet argued whether materials had fixed or variable composition. Dalton's view of fixed composition carried the day, but we now know of many systems where variable composition is normal and the name *berthollide* is often used as a descriptive term for these.

A good example is provided by the transition metal hydrides (section 9.4). For example, titanium hydride can be regarded as forming from cubic close-packed Ti atoms with H atoms entering the tetrahedral sites. The hydrogen is taken up at elevated temperature and pressure. A whole range of uniform materials may be formed ranging from (say) TiH$_{0.1}$ to TiH$_{1.8}$ in overall composition (the limits depend on the temperature). A species of formula TiH has no special stability, but simply represents the stage where half the tetrahedral sites are occupied at random. At TiH$_{1.8}$, no metal phase remains, but 10% of the sites are empty. It requires prolonged treatment, or high H$_2$ pressures, to force the composition up to TiH$_2$. Since the defect structure has a higher entropy, if a lower enthalpy, the stable equilibrium composition may not be the stoichiometric one.

Similar non-stoichiometric phases are widely found among oxides, e.g. uranium and the actinides, ZnO, iron oxides (see section 14.6); among sulfides and among the lower-valent halides.

The structural, electrical, and optical properties of non-stoichiometric solids lead to a number of applications, but by far the most important is the controlled non-stoichiometry of semi-conductors. While silicon, or germanium, are intrinsically semi-conductors (see above, 5.8), their properties may be vastly modified and tailored by adding impurities. If, say, indium is added in small amounts to silicon or germanium, it will occupy a site in the lattice but, as it contains one electron less than the Group 14 atom, there will be an electron vacancy (or positive hole). If a potential is applied an electron will move into the vacancy, leaving a hole and so on—i.e. the effect is of a positive charge migrating across the crystal. Similarly, addition of a Group 15 atom such as arsenic or antimony gives an excess electron and this negative charge will move under a potential. These *p*-type and *n*-type semi-conductors are then united in various combinations to give all the components of modern electronics.

PROBLEMS

In this chapter, we are concerned with quite complex three-dimensional arrays of atoms. Read the *note on crystal structure diagrams* on p. 77, and also look at as many models and structural texts as you can.

You will need to use atomic parameters, especially from sections 2.13 to 2.16, and these should be revised.

5.1 (a) From the values in Table 2.12, and following the discussion on p. 32, calculate a set of ionic radii consistent with the 'experimental' values.

(b) Calculate radius ratios from the data derived in (a) for the solids listed in Table 5.1. Comment on the predictions which result. Treat Shannon radii similarly.

5.2 Calculate correct to 2 significant figures the Madelung constant of a rectangular array of spacings r and $3r$.

5.3 From Table 2.12 and section 5.3, calculate the approximate lattice energies of CaO, CaS and NaF based on the value in Figure 5.5 for NaCl.

5.4 Calculate lattice energies, as in question 5.3, for Na_2O, Na_2S, CaF_2, $CaCl_2$. (Assume the same Madelung constant as fluorite.)

5.5 The heats of dissociation of F_2 and O_2 are respectively 78·9 and 249·2 kJ mol^{-1}. The heat of formation of Ca atoms from the metal is 176 kJ mol^{-1}, and of S atoms from S_8 is 238 kJ mol^{-1}.

Work out the heats of formation of the solids in questions 5.3 and 5.4 (see Tables 2.8, 2.9). Discuss the order of stability of these solids. [See also question 9.5.]

5.6 Decide what approximate lattice energy is appropriate for the hypothetical ionic solids CaF and CaCl. Hence calculate their heats of formation.

Why do CaF and CaCl not exist as stable species?

5.7 The experimental heat of formation of CuCl is 136 kJ mol^{-1}, the calculated lattice energy is 880 kJ mol^{-1}, and the heat of formation of copper atoms is 337 kJ mol^{-1}. Discuss whether CuCl is likely to be ionic.

6 Solution Chemistry

Aqueous Solutions

Most work in inorganic chemistry is carried out in solution and the observed results depend to a large extent on the properties of the solvent. These will be examined in this chapter. Water is still by far the commonest solvent, but an ever-increasing number of reactions are carried out in other solvent media. These range from solvents like liquid ammonia, which have much in common with water, through more exotic media like anhydrous hydrogen fluoride or liquid bromine trifluoride, to molten salts and even molten metals. The major features of aqueous chemistry are discussed first and then follows the extension of these principles to nonaqueous solvents, with detailed discussion of the properties of a small number of representative solvent systems.

The discussion of solution behaviour is divided into three sections;

(i) solubility and solvolysis, which depend directly on the interaction between solvent molecules and the solute,
(ii) acid-base behaviour,
(iii) oxidation-reduction behaviour.

None of these types of behaviour is independent of the others but it is convenient to make these broad divisions. They will be discussed in turn, first of all as they apply to solutions in water.

6.1 Solubility

The energy changes which govern solubility may be discussed by first considering the process of dissolving an ionic solid in water. The principal enthalpy changes may be related by a simple cycle diagram such as Figure 6.1. The process of solution is treated as occurring in two steps: the ions in the solid are separated to infinity as gaseous ions, which requires the input of the lattice energy, U; the separated gaseous ions are hydrated by the water molecules with the evolution of the heats of hydration, H_{aq}, of the cation and anion. The heat of solution, H_s, is the difference between the lattice energy and the heats of hydration. This treatment should be compared with that used to analyse the formation of ionic solids (section 5.2).

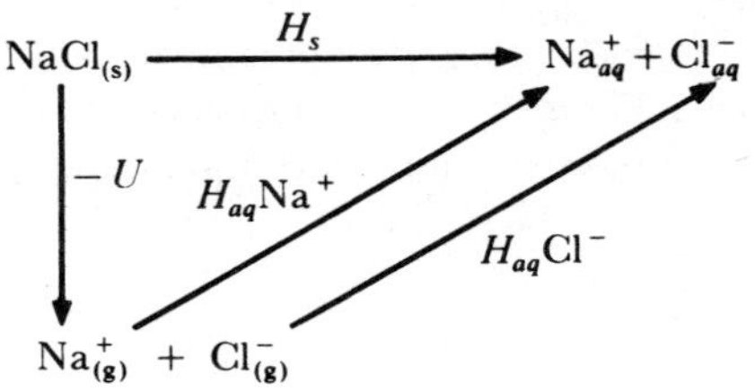

U = lattice energy of NaCl
H_s = heat of solution of NaCl
$H_{aq}Na^+$ = heat of hydration of Na^+
$H_{aq}Cl^-$ = heat of hydration of Cl^-

FIGURE 6.1 *Enthalpy changes in the solution of sodium chloride*
The process of solution may be split up into simple steps, whose heat changes are known or measurable, in the same way as the formation of an ionic solid was treated in the Born-Haber cycle.

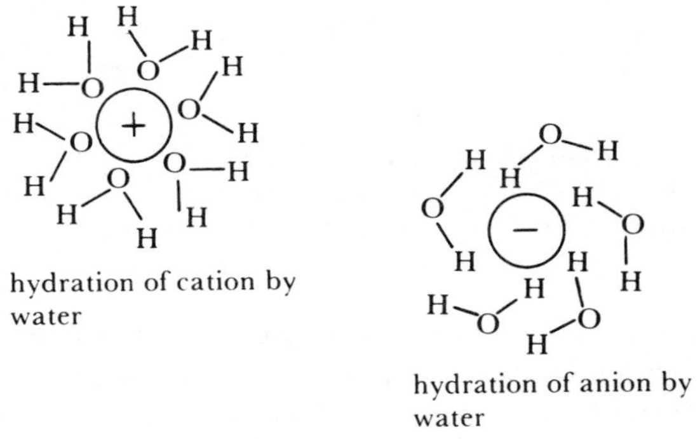

hydration of cation by water

hydration of anion by water

FIGURE 6.2 *Diagrammatic representation of the solvation of ions*

The factors affecting the lattice energy were discussed in section 5.3. It will be recalled that the lattice energy is greatest for small ions of high charge and is increased if polarization effects are present to add the attractions between the ionic charges and induced dipoles to those between the ions themselves.

The heats of hydration of the gaseous ions arise from the electrostatic attractions between the ionic charges (and dipoles if any) and the dipole of the water molecules. These interactions are shown schematically in Figure 6.2. In the water molecule, each $O-H$ bond is polarized in the sense $O^{\delta-}-H^{\delta+}$ by the uneven sharing of the bonding electrons between the very electronegative oxygen and the less electronegative hydrogen. In addition, the effective negative charge on the oxygen atom is increased by the two unshared pairs of electrons. In the case of an anion, the relatively positive hydrogen atoms interact with the negative charge to give the anionic heat of hydration. The cations interact with the relatively negative oxygen atoms of the water molecules and the major effect is probably that due to the lone pairs. The energy of hydration of the cation is usually the most important exothermic term as the cation-lone pair interaction is strong and is enhanced by the general small size and consequent high charge density of cations. The strength of the anion and cation interactions with the water molecules increases with the charge on the ions and is inversely proportional to their sizes. The effect of ion size on the hydration energies is seen in Figure 6.3 where the heats of hydration of the alkali halides are plotted against the anion radius.

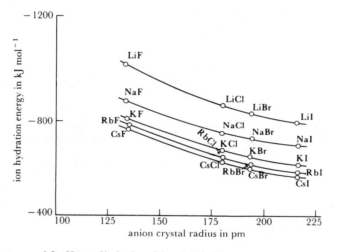

FIGURE 6.3 *Heats of hydration of the alkali halides*
The heats of hydration are plotted against the anion radius to illustrate the effect of ion size on the hydration energy.

It can be seen that the task of predicting the solubilities of ionic solids is very similar to that of predicting the formation of an ionic lattice which was discussed in the last chapter. The heat of solution, and therefore the probable solubility, is the result of the balance between the lattice energy of the solid and the heats of hydration of the gaseous ions. Both these factors are increased if the ions are small and of high charge, so that the difference between them may be expected to vary in a random manner and a general correlation of solubility with ion properties is impossible, each individual case having to be calculated separately. Solubility is yet another example of a phenomenon which depends on the small difference between large energies. There are, however, two factors which appear to have an over-riding effect on

the balance of energies involved in solution. One is the effect of the ionic charge. When this is increased, the ionic sizes and the crystal structure remaining constant, the lattice energy increases more than the heats of solvation of the ions. Thus, A^+B^- solids, for example, are much more soluble than $A^{2+}B^{2-}$ solids of the same structure. Highly-charged salts are generally of low solubility. A second effect which usually tends to decrease the solubility is the presence of polarization in the solid. A polarizing cation increases the lattice energy by the additional ion-dipole force which its presence introduces, and such a cation would also polarize the solvent molecules and increase the hydration energy in a similar manner. In this case, the effect of these forces appears to be largest in the solid and the presence of polarizing cations usually leads to low solubility, at least in cases like the silver and thallous salts which were discussed in Chapter 5.

The discussion above has been centred on heat changes whereas the true driving force is the free energy of solution which includes the entropy of solution as well as the heat. In general, entropy changes would be expected to favour the formation of solutions as the solute is much less ordered than the solid. (It will be recollected that a high degree of order corresponds to low entropy.) However, in a coordinating solvent such as water, the disorder of the solvent is reduced when ordered hydration sheaths are formed around the ions. If the ions are of high charge density, this increase in the ordering of the solvent may outweigh the decreased order of the solute. Thus, entropy changes should favour solution processes in general, but may oppose them for ions of high charge density which would be strongly solvated.

Another property of water, in addition to its power of hydrating ions, is important in determining its solvent power for ionic compounds. This is its very high dielectric constant, $D = 81 \cdot 1$. The attraction between opposite charges e_1 and e_2 at a distance r in a medium of dielectric constant D is $e_1 e_2/4\pi\epsilon_0 D r^2$. Thus the interposition of a medium of high dielectric constant reduces the forces between ions in solution, hindering their precipitation, and assists the passage into solution of ions in the solid. Water is among the best of all solvents for ions and this is a result of the strong solvating forces for ions combined with the high dielectric constant.

By contrast, water is a poor solvent for covalent compounds, especially if these have no dipole or only a relatively weak one. To return to the heat cycle of Figure 6.1, a covalent compound may go into solution as single molecules or as dimers or polymers, and the energy required to separate the appropriate species is analogous to the lattice energy U of an ionic compound. In the case of the covalent 'giant molecule' like diamond or silica, the heat required to separate out any particle from the solid is too large to be compensated by the solvation energy—because strong covalent bonds would have to be broken—and such compounds are completely insoluble in water. In the case of a molecular solid which will dissolve as molecules, the heat required is the very small amount of energy needed to overcome the weak van der Waals' forces and this is readily available. Such solids are not usually soluble in water although the heat of hydration, which would arise from the interaction of the water dipoles with the dipoles induced by them in the covalent molecules, should be greater than the heat required to break up the solid. This low solubility arises because there is a third energy term to be added to the cycle. This is the heat required to overcome the attractions between the water molecules themselves and allow the solute molecules to enter between them. This energy of the 'water structure', which arises from the attractions between the partial charges of the water dipoles, is negligible compared with the energies involved in the solution of an ionic solid and was neglected in the earlier discussion, but it becomes important when the much weaker interactions with covalent molecules are considered. Thus, before a covalent solid dissolves, enough energy must be available to separate the solute molecules and to separate the water molecules, and this can only be provided by the heat of hydration of the solute molecules. For non-polar covalent molecules, the energy of the water structure is the dominant term and such compounds are of low solubility. When polar molecules are involved, both the forces within the solid and the forces between the solute molecules and the water are increased, and the water structure energy no longer dominates the energy balance. Solubilities in such cases depend on the detailed balance of enthalpies but they tend to be higher than those of non-polar compounds. For example, the solubility of the non-polar molecule methane in water at room temperature is about 0·004 mole per litre, while the polar methyl iodide dissolves to the extent of 0·110 mole per litre.

This discussion may be summarized as follows:

(a) In the case of ionic solids strong forces are involved, those between ions in the solid and those between ions and dipoles in solution. Solubility depends on the detailed balance in each case, but salts with balanced numbers of ions with charges greater than one are of low solubility and compounds where there are appreciable polarization effects are also of low solubility.

(b) In the case of 'giant molecules' the strength of the forces binding the solid predominates and such compounds are insoluble in water.

(c) Covalent nonpolar solids are weakly bound and would be somewhat more strongly bonded to water molecules than to themselves, but there is insufficient energy to break down the 'water structure' and such compounds are insoluble in water.

(d) Polar covalent compounds present an intermediate range of interactions between those of (a) and (c). Solubilities vary but tend to be higher than those in class (c).

With some solutes, interaction with the water molecule goes further than simple coordination and new chemical species are formed. Such a reaction is termed *hydrolysis*. The distinction between hydration and hydrolysis is not completely clear-cut but hydrolysis implies a more extensive and less reversible interaction of the solute with the water molecule.

Some examples are given below:

$$SiCl_4 + 4H_2O = Si(OH)_4 + 4HCl*$$
$$O^{2-} + H_2O = 2OH^-$$
$$HCl + H_2O = H_3O^+ + Cl^-$$
$$BF_3 + 2H_2O = H_3O^+(HOBF_3)^-$$

6.2 Acids and bases

The qualitative properties of acids—sharp taste, solvent power, effect on the colours of dyes, etc.—and the corresponding properties of bases were first listed by Boyle and the succeeding centuries have seen a number of theories for acid-base properties developed and discarded. At present, three concepts are used to deal with acid-base phenomena in various solvents. These three theories overlap considerably although each has special uses and weaknesses. That of most value when dealing with aqueous solutions is the *protonic concept* of Brönsted and Lowry who characterized and related acids and bases by the equation:

$$A \text{ (acid)} \rightleftharpoons B \text{ (base)} + H^+ \text{ (proton)} \dots \dots (6.1)$$

An acid is a proton donor and a base is a proton acceptor. Since the proton in this theory is the hydrogen nucleus which has such a high charge density that it is never obtained free in the condensed state (see Chapter 9), it follows that this is a 'half-equation' and the acidic properties of a molecule are observed only when it is in contact with the basic form of a second species. That is, the observed equation is always:

$$A_1 + B_2 \rightleftharpoons A_2 + B_1 \dots \dots \dots \dots (6.2)$$

Such an equilibrium is displaced in the direction of the weaker acid and base, since the stronger acid is a stronger proton donor than the weaker acid and thus more of its molecules are in the basic form. The base corresponding to a given acid is called the conjugate base; clearly a strong acid has a weak conjugate base, and *vice versa*. The acid strength can be expressed in terms of the equilibrium constant of the above reaction:

$$K = \frac{[A_2][B_1]}{[A_1][B_2]} \dots \dots \dots \dots (6.3)$$

This gives a method of expressing relative acid strengths. If one acid-base pair is taken as the standard, then the strengths of all other pairs may be expressed in terms of the equilibrium constants involving the standard acid. It is most convenient to choose as the standard acid-base pair, the one involving the solvent. Thus in water, the acid strength is defined with respect to the pair $H_3O^+ - H_2O$ and the equilibrium used to define acid strengths is:

$$A + H_2O \rightleftharpoons B + H_3O^+ \text{ with } K' = \frac{[B][H_3O^+]}{[A][H_2O]} \dots (6.4a)$$

*These products undergo further reaction.

TABLE 6.1 Strengths of acids in water

Acid	Conjugate base	K	pK	
$HClO_4$	ClO_4^-	ca. 10^8	ca. -8	K very high
HCl	Cl^-	ca. 10^7	ca. -7	HBr and HI greater
HNO_3	NO_3^-	ca. 10^4	ca. -4	uncertain K but high
H_2SO_4	HSO_4^-	ca. 10^3	ca. -3	uncertain K but high
H_3O^+	H_2O	55·5	$-1·74$	
HSO_4^-	SO_4^{2-}	2×10^{-2}	1·70	compare with H_2SO_4
H_3PO_4	$H_2PO_4^-$	$7·5 \times 10^{-3}$	2·12	
HF	F^-	$7·2 \times 10^{-4}$	3·14	compare with HCl
$H_2PO_4^-$	HPO_4^{2-}	$5·9 \times 10^{-8}$	7·23	
NH_4^+	NH_3	$3·3 \times 10^{-10}$	9·24	typical weak base
HPO_4^{2-}	PO_4^{3-}	$3·6 \times 10^{-13}$	12·44	compare with H_3PO_4 and $H_2PO_4^-$
H_2O	OH^-	$1·07 \times 10^{-16}$	15·97	strong base
OH^-	O^{2-}	below 10^{-36}	above 36	

Providing dilute solutions are being used, the concentration of water $[H_2O]$ is essentially constant, so that the strength of an acid can be defined by:

$$K = [B][H_3O^+]/[A] \ldots\ldots\ldots (6.4b)$$

The strengths of bases are quite naturally expressed by means of the K values and there is no need for a separate scale of base strengths. Since strong acids have weak conjugate bases and *vice versa*, the order of base strengths is the inverse of the order of acid strengths. Some typical K values for acids and bases in water are shown in Table 6.1. Also included are pK values which are the negative logarithms, analogous to the well-known pH values.

The relative strengths of the acids and bases which lie between the two acid pairs of water

$$H_3O^+ \rightleftharpoons H_2O + H^+$$
$$\text{and} \quad H_2O \rightleftharpoons OH^- + H^+ \quad \ldots\ldots\ldots (6.5)$$

are well defined, but it is impossible to measure the acid strengths of acids stronger than H_3O^+ (or of bases stronger than OH^-) as they are completely converted to the water acid (or base), that is, the equilibrium in (6.4a) lies completely to the right. No acid stronger than the hydrated proton can exist in water. One way of determining the relative strengths of acids such as nitric acid or perchloric acid, which are completely dissociated in water, is to compare their catalytic powers in an acid-catalysed reaction. The catalysis of the inversion of sucrose gives an order of strengths as follows: $HClO_4 > HBr > HCl > HNO_3$. An alternative method is to measure the dissociation in a solvent which is more acidic than water.

An alternative definition of acids and bases was proposed by Cady and Elsey and is often termed the *solvent system* definition. They suggested that the acid and base in a particular solvent should be defined in terms of the ions formed in the self-dissociation of the solvent. For example, water dissociates slightly in the sense:

$$2H_2O \rightleftharpoons H_3O^+ + OH^- \ldots\ldots\ldots (6.6)$$
$$\text{acid} \quad \text{base}$$

and an acid in water is any substance which enhances the concentration of the solvent cation, H_3O^+, while a base is any substance which enhances the concentration of the solvent anion, OH^-.

This definition is closely related to that of Lowry and Brönsted, the dissociation of equation (6.6) corresponding to the two Brönsted acid pairs of water as given in equation (6.5). A similar correlation between the solvent system and Lowry-Brönsted definitions holds for any protonic solvent, but the solvent system definition has the advantage that it is readily extended to solvents which do not contain dissociable hydrogen atoms.

The third definition of acids and bases is that due to Lewis, who defines an acid as a lone pair acceptor and a base as a lone pair donor. This definition is of little advantage in discussing reactions in water so an account of it is deferred to section 6.5.

Strengths of oxyacids

Many compounds which show acidic properties fall into the class of oxy-acids, that is, they contain the Group $X - O - H$. Two general observations may be made about the strengths of these acids.

First, where there are a number of OH groups attached to the central atom, the pK values for the removal of the successive ionizable hydrogens increase by about five each time (compare the values for phosphoric acid and the two phosphate ions in Table 6.1).

Second, the strength of the acid depends on the difference $(x - y)$ between the number of oxygen atoms and the number of hydrogen atoms in the molecule H_yXO_x. When x is equal to y, pK is about $8·5 \pm 1$. If $(x - y) = 1$, pK is about $2·8 \pm 1$. If $(x - y)$ is two or more, the pK value is markedly less than zero. Examples of the last case are provided by the first three

TABLE 6.2 Strengths of oxyacids, H_yXO_x

$(x-y) = 1$	acid	HNO_2	H_2SO_3	H_3AsO_4	H_5IO_6
	pK	3·3	1·90	3·5	3·29
$(x-y) = 0$	acid	$HClO$	H_3BO_3	H_4GeO_4	H_6TeO_6
	pK	7·50	9·22	8·59	8·80

oxyacids in Table 6.1 while some examples of the first two cases are given in Table 6.2.

As oxygen is the second most electronegative element, when the central atom X has an oxygen atom attached to it as well as the OH group, that is in $O = X - O - H$, electron density will be withdrawn from X by the oxygen and this effect will be transmitted to the $O - H$ bond, making it easier for the hydrogen to dissociate as the positive ion. The acid strength should therefore rise with the number of $X = O$ groups, that is with $(x-y)$ as the Tables show. Two cases are known where simple acids do not fit this pattern. One is provided by the lower acids of phosphorus, phosphorous acid H_3PO_3 where $(x-y) = 0$ but pK = 1·8, and hypophosphorous acid H_3PO_2 with $(x-y)$ equal to −1 and pK = 2. In both these molecules, the value of $(x-y)$ does not correspond to the number of $X = O$ bonds as there are direct $P - H$ bonds present, one in phosphorous acid and two in hypophosphorous acid. Allowing for these, both acids have one $P = O$ bond, which should correspond to pK values in the range $2·8 \pm 1$, as is found.

The second exception is provided by carbonic acid, H_2CO_3, which is expected to have a pK value of about 2·8 and has an actual value of 6·4. In this case, as well as the acid dissociation equilibrium, there is a further non-protonic equilibrium in solution:

$$H_2CO_3 \rightleftharpoons H_2O + CO_{2(aq)}$$

When allowance is made for this dissolved carbon dioxide, the effective pK value of carbonic acid is about 3·6, which just falls within the expected range.

6.3 Oxidation and reduction

NOTE: In this section particularly, use is made without proof of various thermodynamic relationships. The reader will find these explained in any standard textbook of physical chemistry.

Oxidation, originally defined in terms of combination with oxygen and later generalized to include combination with other electronegative elements, is nowadays commonly defined as the *removal of electrons* from the element or compound which is oxidized. *Reduction*, similarly, has been defined in terms of removal of oxygen and electronegative elements, addition of hydrogen or electropositive elements and now in terms of *gain of electrons* by the element or compound in question. In cases involving ions, for example in:

$$Na + \tfrac{1}{2}Cl_2 = Na^+Cl^-,$$
$$\text{or} \quad 2Fe^{3+} + Sn = 2Fe^{2+} + Sn^{2+},$$

the direction of electron transfer is clear from the equations. When only covalent species are involved, it is usually clear that some electron rearrangement has taken place but there is often no obvious electron transfer, as in:

$$\tfrac{1}{2}H_2 + \tfrac{1}{2}Cl_2 = HCl$$

where the difference in electron density at the hydrogen atom, say, is the relatively small one due to the polarization of the $H - Cl$ bond compared with the unpolarized $H - H$ bond. It is in these cases that the definitions of oxidation and reduction in terms of oxidation numbers, using electronegativity values, discussed in Chapter 2 are so useful.

Some quantitative measure of oxidizing or reducing power is necessary and this is known as the *redox potential* of the reactant. The tendency to gain or lose an electron may be measured as an electrical potential under standard conditions and expressed relative to a suitable standard value (as only the relative values of potentials may be measured). The standard conditions used are a temperature of 25 °C, unit activity (which is usually taken as unit concentration) of the ions concerned and, for gases, one atmosphere pressure. The standard potential is provided by the hydrogen electrode (Figure 6.4), which consists of a platinum plate, coated with

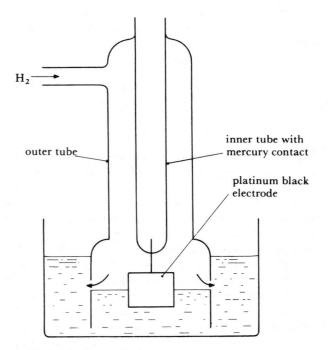

FIGURE 6.4 *The hydrogen electrode*

platinum black, partly dipping into a solution containing hydrogen ions at unit activity. Hydrogen gas at one atmosphere pressure is passed over the platinum plate and bubbled through the solution so that the platinum is in contact both with the gas and the hydrogen ions in solution and catalyses the attainment of equilibrium between them:

$$H^+ (aq, a = 1) + e^- \rightleftharpoons \tfrac{1}{2}H_2 (gas, p = 1 \text{ atm})$$
$$(a \text{ is activity})$$

The potential of this electrode is defined to be zero at 25 °C (298 K) and this is written $E°(298 K) = 0.000$ V.

The potential of any other oxidation-reduction or redox system is measured by immersing a platinum wire or other inert conductor in a solution which contains both the oxidized and the reduced form of the system at unit activity, and measuring the potential of this half-cell against the hydrogen electrode. For example, the standard potential of the ferric/ferrous system is measured from the cell:

$$Pt, H_2 (p = 1\ atm) \left| H^+ (a = 1) \right\| \begin{matrix} Fe^{3+} (a = 1) \\ Fe^{2+} (a = 1) \end{matrix} \right| Pt$$

(It is usually more convenient to use some other electrode of accurately known potential in place of the hydrogen electrode which is awkward to handle.) If the potential of a gaseous element is to be determined, an electrode similar to the hydrogen electrode is used, while the half-cell used to determine the potential of a metallic element consists simply of a rod of the metal (which is defined to have unit activity) immersed in a solution of the metal ions at unit activity. For example, the oxidation potential of iron going to ferrous ions is measured as the standard potential of the cell:

$$\text{hydrogen electrode} \| Fe^{2+} (a = 1) | Fe$$

In the half-equation describing the change in the redox system, the oxidized form is always written on the left hand side, that is the equations are in the form:

$$Ox + n\,e^- \rightleftharpoons Red \dots\dots\dots\dots (6.7)$$

where Ox stands for the oxidized form and Red for the reduced form,

$$\begin{array}{ll} e.g. & Fe^{3+} + e^- \rightleftharpoons Fe^{2+} \\ or & Fe^{2+} + 2e^- \rightleftharpoons Fe \end{array}$$

The short form for describing the electrode is written similarly: Ox|Red or, in the examples above, $Fe^{3+}|Fe^{2+}$ and $Fe^{2+}|Fe$.

The sign convention which applies for these potentials takes a negative sign for the potential Ox|Red to mean that the reduced form of the redox system is a better reducing agent than is hydrogen under standard conditions. This is tantamount to measuring the sign of the metal in the $M^{z+}|M$ electrode relative to the hydrogen electrode. Thus the standard potential for $Fe^{2+}|Fe$ is -0.41 V, showing that iron going to ferrous ions is a better reducing agent than hydrogen going to hydrogen ions. The potential for $Fe^{3+}|Fe^{2+}$ is $+0.77$ V so that ferrous ions going to ferric ions provide a much worse reducing agent than hydrogen. This is, of course, equivalent to saying that ferric going to ferrous is a useful oxidizing agent. The above sign convention is the one recommended by the International Union of Pure and Applied Chemistry (IUPAC) but the opposite convention (reversing the equation and the sign) is also found.*

*Equation (6.7) is a reduction from left to right, which is the normal way of reading, but it is an oxidation from right to left. It has been suggested that these standard electrode potentials may be called reduction potentials. The term redox is, however, recommended to emphasize the reversible nature of the reactions.

The full range of redox potentials extends over about six volts, from -3 V for the alkali metals which are the strongest reducing agents to $+3$ V for fluorine, the strongest oxidizing agent. The values at these extremes cannot be measured directly as these very reactive elements decompose water, but they can be derived by calculation. The range of useful redox reagents in an aqueous medium is about half of the total range, from about 1.7 V for strong oxidizing agents like ceric or permanganate to about -0.4 V for a strong reducing agent like chromous, or about -0.8 V for reduction by fairly active metals such as zinc. The values for many common oxidation-reduction couples are given in Table 6.3.

The potential of a redox system where the oxidized and reduced forms are not at unit activities is related to the standard potential, E^0, by the Nernst equation:

$$E = E^0 + \frac{RT}{nF} \ln \frac{[Ox]}{[Red]} \dots\dots\dots\dots (6.8)$$

where R is the gas constant, T the absolute temperature, F the Faraday constant, and n the number of electrons transferred in the oxidation process as in equation (6.7). Transferring to logarithms to the base ten and introducing the values of the constants

$$E = E^0 + \frac{0.059}{n} \log \frac{[Ox]}{[Red]} \text{ at } 25 °C \dots\dots (6.9)$$

This equation permits the extent of reaction between two redox systems to be calculated, and can be used, for example, to decide whether a particular reaction will go to completion or not. For the general reaction:

$$a\,Ox_1 + b\,Red_2 \rightleftharpoons b\,Ox_2 + a\,Red_1, \dots\dots (6.10)$$

$$E_1 = E_1^0 + \frac{RT}{nF} \ln \frac{[Ox_1]^a}{[Red_1]^a} \text{ and } E_2 = E_2^0 + \frac{RT}{nF} \ln \frac{[Ox_2]^b}{[Red_2]^b}.$$

At equilibrium, $E_1 = E_2$ so that:

$$\log \frac{[Ox_2]^b [Red_1]^a}{[Ox_1]^a [Red_2]^b} = \log K = \frac{n}{0.059} (E_1^0 - E_2^0) \dots (6.11)$$

where K is the equilibrium constant for the reaction. Consider as examples the oxidations of iodide ion, first by ceric ion and secondly by ferric ion. In the first case:

$$Ce^{4+} + I^- \rightarrow Ce^{3+} + \tfrac{1}{2}I_2$$

E^0 for I_2/I^- is 0.54 V and E^0 for Ce^{4+}/Ce^{3+} is 1.61 V hence:

$$\log K = \frac{1}{0.059} (1.61 - 0.54) = 18.48,$$

$$i.e.\ K \simeq 10^{18}$$

Thus iodide is completely oxidized to iodine by ceric ion.

In the second case, E^0 for Fe^{3+}/Fe^{2+} is 0.77 V so that for the reaction:

$$Fe^{3+} + I^- \rightarrow Fe^{2+} + \tfrac{1}{2}I_2,$$
$$\log K = 1/0.059 (0.76 - 0.52) = 4.068$$

Hence K is approximately 10^4 so that a small but significant

TABLE 6.3 Examples of standard redox potentials in acid solution

Couple	Reaction equation	E^0 (volts)
Li^+/Li	$Li^+ + e^- \rightleftharpoons Li$	$-3\cdot05$
M^+/M $(M = K, Rb, Cs)$	$M^+ + e^- \rightleftharpoons M$	$-2\cdot93$
Ba^{2+}/Ba	$Ba^{2+} + 2e^- \rightleftharpoons Ba$	$-2\cdot90$
Sr^{2+}/Sr	$Sr^{2+} + 2e^- \rightleftharpoons Sr$	$-2\cdot89$
Ca^{2+}/Ca	$Ca^{2+} + 2e^- \rightleftharpoons Ca$	$-2\cdot76$
Na^+/Na	$Na^+ + e^- \rightleftharpoons Na$	$-2\cdot71$
Mg^{2+}/Mg	$Mg^{2+} + 2e^- \rightleftharpoons Mg$	$-2\cdot38$
La^{3+}/La	$La^{3+} + 3e^- \rightleftharpoons La$	$-2\cdot37$
$\frac{1}{2}H_2/H^-$	$\frac{1}{2}H_2 + e^- \rightleftharpoons H^-$	$-2\cdot23$
Be^{2+}/Be	$Be^{2+} + 2e^- \rightleftharpoons Be$	$-1\cdot70$
Al^{3+}/Al	$Al^{3+} + 3e^- \rightleftharpoons Al$	$-1\cdot67$
Zn^{2+}/Zn	$Zn^{2+} + 2e^- \rightleftharpoons Zn$	$-0\cdot76$
Cr^{3+}/Cr	$Cr^{3+} + 3e^- \rightleftharpoons Cr$	$-0\cdot74$
S/S^{2-}	$S + 2e^- \rightleftharpoons S^{2-}$	$-0\cdot51$
H_3PO_3/H_3PO_2	$H_3PO_3 + 2H^+ + 2e^- \rightleftharpoons H_3PO_2 + H_2O$	$-0\cdot50$
$CO_2/H_2C_2O_4$	$2CO_2 + 2H^+ + 2e^- \rightleftharpoons H_2C_2O_4$	$-0\cdot49$
Fe^{2+}/Fe	$Fe^{2+} + 2e^- \rightleftharpoons Fe$	$-0\cdot41$
Cr^{3+}/Cr^{2+}	$Cr^{3+} + e^- \rightleftharpoons Cr^{2+}$	$-0\cdot41$
H_3PO_4/H_3PO_3	$H_3PO_4 + 2H^+ + 2e^- \rightleftharpoons H_3PO_3 + H_2O$	$-0\cdot28$
Sn^{2+}/Sn	$Sn^{2+} + 2e^- \rightleftharpoons Sn$	$-0\cdot14$
$H^+/\frac{1}{2}H_2$	$H^+ + e^- \rightleftharpoons \frac{1}{2}H_2$	$0\cdot00$
$S_4O_6^{2-}/S_2O_3^{2-}$	$S_4O_6^{2-} + 2e^- \rightleftharpoons 2S_2O_3^{2-}$	$0\cdot09$
Sn^{4+}/Sn^{2+}	$Sn^{4+} + 2e^- \rightleftharpoons Sn^{2+}$	$0\cdot15$
Cu^{2+}/Cu^+	$Cu^{2+} + e^- \rightleftharpoons Cu^+$	$0\cdot15$
Cu^+/Cu	$Cu^+ + e^- \rightleftharpoons Cu$	$0\cdot52$
$\frac{1}{2}I_2/I^-$	$\frac{1}{2}I_2 + e^- \rightleftharpoons I^-$	$0\cdot54$
H_3AsO_4/H_3AsO_3	$H_3AsO_4 + 2H^+ + 2e^- \rightleftharpoons H_3AsO_3 + H_2O$	$0\cdot56$
O_2/H_2O_2	$O_2 + 2H^+ + 2e^- \rightleftharpoons H_2O_2$	$0\cdot68$
Fe^{3+}/Fe^{2+}	$Fe^{3+} + e^- \rightleftharpoons Fe^{2+}$	$0\cdot77$
Hg_2^{2+}/Hg	$\frac{1}{2}Hg_2^{2+} + e^- \rightleftharpoons Hg$	$0\cdot80$
Hg^{2+}/Hg	$Hg^{2+} + 2e^- \rightleftharpoons Hg$	$0\cdot85$
$\frac{1}{2}Br_2/Br^-$	$\frac{1}{2}Br_2 + e^- \rightleftharpoons Br^-$	$1\cdot09$
IO_3^-/I^-	$IO_3^- + 6H^+ + 6e^- \rightleftharpoons I^- + 3H_2O$	$1\cdot09$
ClO_4^-/ClO_3^-	$ClO_4^- + 2H^+ + 2e^- \rightleftharpoons ClO_3^- + H_2O$	$1\cdot19$
$IO_3^-/\frac{1}{2}I_2$	$IO_3^- + 6H^+ + 5e^- \rightleftharpoons \frac{1}{2}I_2 + 3H_2O$	$1\cdot20$
$\frac{1}{2}Cr_2O_7^{2-}/Cr^{3+}$	$\frac{1}{2}Cr_2O_7^{2-} + 7H^+ + 3e^- \rightleftharpoons Cr^{3+} + 7/2H_2O$	$1\cdot33$
$\frac{1}{2}Cl_2/Cl^-$	$\frac{1}{2}Cl_2 + e^- \rightleftharpoons Cl^-$	$1\cdot36$
$HIO/\frac{1}{2}I$	$HIO + H^+ + e^- \rightleftharpoons \frac{1}{2}I_2 + H_2O$	$1\cdot45$
$BrO_3^-/\frac{1}{2}Br_2$	$BrO_3^- + 6H^+ + 5e^- \rightleftharpoons \frac{1}{2}Br_2 + 3H_2O$	$1\cdot52$
MnO_4^-/Mn^{2+}	$MnO_4^- + 8H^+ + 5e^- \rightleftharpoons Mn^{2+} + 4H_2O$	$1\cdot51$
$HBrO/Br_2$	$HBrO + H^+ + e^- \rightleftharpoons Br_2 + H_2O$	$1\cdot59$
H_5IO_6/IO_3^-	$H_5IO_6 + H^+ + 2e^- \rightleftharpoons IO_3^- + 3H_2O$	$1\cdot60$
Ce^{4+}/Ce^{3+}	$Ce^{4+} + e^- \rightleftharpoons Ce^{3+}$	$1\cdot61$
$HClO/\frac{1}{2}Cl_2$	$HClO + H^+ + e^- \rightleftharpoons \frac{1}{2}Cl_2 + H_2O$	$1\cdot64$
$HClO_2/HClO$	$HClO_2 + 2H^+ + 2e^- \rightleftharpoons HClO + H_2O$	$1\cdot64$
H_2O_2/H_2O	$H_2O_2 + 2H^+ + 2e^- \rightleftharpoons 2H_2O$	$1\cdot77$
$\frac{1}{2}S_2O_8^{2-}/SO_4^{2-}$	$\frac{1}{2}S_2O_8^{2-} + e^- \rightleftharpoons SO_4^{2-}$	$2\cdot01$
O_3/O_2	$O_3 + 2H^+ + 2e^- \rightleftharpoons O_2 + H_2O$	$2\cdot07$
$\frac{1}{2}F_2/F^-$	$\frac{1}{2}F_2 + e^- \rightleftharpoons F^-$	$2\cdot87$
$\frac{1}{2}F_2/HF$	$\frac{1}{2}F_2 + H^+ + e^- \rightleftharpoons HF$	$3\cdot06$

proportion of iodide would remain in equilibrium with the iodine. For a one electron change, an equilibrium constant of one million requires a difference in the standard potentials of the two redox systems of 0·354 V, so this is about the minimum difference necessary for a complete reaction.

The values of the standard potentials may also be used to determine the stability of different oxidation states in solution. If the potentials for the various oxidation states of iron in Table 6.3 are examined, it will be seen that iron going to ferrous ions is a much better reducing agent than ferrous ions going to ferric ions. Oxidation of metallic iron in aqueous solution therefore gives ferrous ions first and stronger oxidation is required to get to ferric ions. By contrast, the values for the different oxidation states of copper show that copper going to cuprous ions is a worse reducing agent than copper going to cupric ions (that is, an oxidizing agent strong enough to oxidize copper to cuprous, $E^0 = 0·52$ V, is more than strong enough to oxidize cuprous to cupric, $E^0 = 0·15$ V). The cuprous state is therefore avoided in aqueous solution. Put alternatively, cuprous ions in water disproportionate to cupric ions and metallic copper:

$$2\,Cu^+ \rightarrow Cu^{2+} + Cu$$

In this reaction, the Cu^+ ion and Cu may be taken as Ox_1 and Red_1 and then the Cu^{2+} ion and the Cu^+ ion become Ox_2 and Red_2: that is, the cuprous ion Cu^+ may be regarded as filling two roles, as both an oxidizing and a reducing agent, in the general equation (6.10). Substituting the standard potentials in equation (6.11) gives

$$K = \frac{[Cu^{2+}]\,[Cu]}{[Cu^+]^2} = \text{approx. } 10^6$$

so that only one part in a million of the original cuprous copper remains in solution as cuprous copper at equilibrium. This disproportionation of the copper oxidation states may be reversed by adding some reagent which forms a very stable complex with the cuprous ion, so stable that there is significantly less than one part per million of cuprous ion in equilibrium with the cuprous complex. The cuprous ion is thus removed from the cuprous-cupric equilibrium by complexing and the disproportionation reaction reverses. One example of such a complexing agent is cyanide ion which gives the very stable $Cu(CN)_2^-$ complex ion.

The above examples give some indication of the ways in which the redox potentials may be used. Let us examine the potential in more detail and analyze its components.

The type of cell used for determining these potentials, as in section 6.3

$$Pt,\ H_2\ |\ H^+\ ||\ M^+\ |\ M$$

implies the reaction:

$$\tfrac{1}{2}H_2 + M^+ \rightarrow H^+ + M$$

The electrode potentials of the left and right electrode are E_H^0 and E_M^0 respectively and the potential difference, the e.m.f., of the cell is $(E_M^0 - E_H^0)$. By convention we set $E_H^0 = 0$.

The electrode potential E^0 of the cell is related to the standard free energy change ΔG^0 of the reaction by the equation:

$$\Delta G^0 = -nFE^0$$

and ΔG^0 is related to the heat of reaction ΔH^0 and entropy of reaction ΔS^0 at temperature T by

$$\Delta G^0 = \Delta H^0 - T\,\Delta S^0$$

Let us first examine ΔH^0 in detail. This means the enthalpy of the products minus the enthalpy of the reactants which we may write

$$\Delta H^0 = H^0(M) + H^0(H^+) - H^0(M^+) - H^0(\tfrac{1}{2}H_2)$$

which can be regrouped as

$$\Delta H^0 = \{H^0(H^+) - H^0(\tfrac{1}{2}H_2)\} - \{H^0(M^+) - H^0(M)\}$$

and abbreviated as

$$\Delta H^0 = \Delta H^0(H) - \Delta H^0(M) \ \dots\dots\ (6.13)$$

The terms $\Delta H^0(H)$ and $\Delta H^0(M)$ may be analysed by means of a cycle, similar to those used to discuss lattice energies or solubilities (compare Figures 5.5 and 6.1). For the formation of a cation from the element in aqueous solution we may examine the contribution of various terms to $\Delta H^0(M)$.

where H_A is the heat of atomization of the element, I (see Table 2.8) is the ionization potential (or the sum of the first z ionization potentials if a Z^+ cation is formed) and H_{aq} is the heat of hydration of the gaseous ion. An exactly similar cycle may be constructed for the formation of an anion in solution except that the ionization potential must be replaced by the electron affinity, E (see Table 2.9), when the gaseous atom goes to the gaseous anion.

Thus, for a cation,

$$\Delta H^0(M) = H_A + I + H_{aq}$$

and for an anion,

$$\Delta H^0(X) = H_A + E + H_{aq}$$

The values of H_A, I (or E) and H_{aq} are known from experiment and, for hydrogen, equal respectively $+218$, $+1310$, and -1070 kJ mol^{-1}, giving $\Delta H^0(H) = 452$ kJ mol^{-1}.

A similar treatment can be carried through for the entropy terms ΔS^0.

However, we can simplify the discussion by recognizing:

(a) That when similar systems are being compared (say Fe^{2+}/Fe and Zn^{2+}/Zn) the entropy changes will be very similar in each, so that the *difference* in entropy contributions to the systems may be neglected to a first approximation. Note that this approximation is not valid when one component is a gas, as for $O^{2-}/\tfrac{1}{2}O_2$.

(b) That each system is being compared to the hydrogen electrode so that ΔS^0 for hydrogen subtracts out in a comparison between two different metals M.

Thus, for an approximate treatment, and especially for comparative purposes, the entropy changes may be neglected and the approximate relation

$$-nFE^0 \approx \Delta H^0 \ldots \ldots \ldots \ldots (6.13)$$

may be used.

We may now write from equations (6.12) and (6.13):

$$-nFE^0_M \approx 452 \text{ kJ mol}^{-1} - \Delta H^0(M)$$

and E^0_M will be negative if $\Delta H^0(M)$ is less than 452 kJ mol^{-1}.

The alkali metals have ΔH^0 values of about 188 kJ mol^{-1} reflecting their low ionization potentials and sublimation energies, so that their standard potentials are strongly negative. The fall in standard redox potentials from caesium or barium to sodium or magnesium shown in Table 6.3 reflects the regular decrease in ionization potential and sublimation energy with increasing size, while the anomalous position of lithium is mainly the result of the greater hydration energy of the small cation. In the case of the halogens, the sum of the atomization energy (half the bond energy of X_2) and the electron affinity is approximately constant, so that the fall of the redox potential from fluorine to iodine again reflects the fall in hydration energy as the size of the ion increases.

This treatment need not be confined to the case of elements, for example the Fe^{3+}/Fe^{2+} system depends on the energy changes:

$$
\begin{array}{ccc}
Fe^{3+} + e^- & \longrightarrow & Fe^{2+} \\
(\text{gas}) & & (\text{gas}) \\
\downarrow & & \downarrow \\
Fe^{3+} + e^- & \longleftarrow & Fe^{2+} \\
(\text{hydrated}) & & (\text{hydrated})
\end{array}
$$

where the overall change depends on the hydration energies of the Fe(II) and Fe(III) ions and on the third ionization potential of iron.

These cycles provide one method of calculating potentials which are unobtainable experimentally, either because the species are not stable in water or because the attainment of equilibrium is too slow.

It must be noted that all the above treatment of standard redox potentials gives no indication of the rates of reaction. Although it may seem from the values of the potentials that a certain molecule should be readily oxidized by another, the rate of reaction might be so slow that nothing is observed. For example, the values given in Table 6.3 show that iodine should oxidize thiosulfate to tetrathionate (as in the standard volumetric method) and that oxidation of thiosulfate to sulfate should also be significant. However, the reaction to tetrathionate is quantitative, showing that the oxidation to sulfate must be extremely slow. In other words, the quantitative nature of oxidation to tetrathionate depends on a kinetic factor and is not revealed by the thermodynamic approach involved in redox potentials.

The potentials which appear in Table 6.3 as involving simple ions do in fact apply to the hydrated forms of these ions, since the potentials are determined in aqueous solution. In some of these cases, the water molecules are strongly held and the heat of hydration is an important factor in the potential. If these coordinated water molecules are replaced by other ligands, the potential changes. For example, the potential of 0·77 V for Fe^{3+}/Fe^{2+} applies to the hydrated ions (approximately $Fe(H_2O)_6^{3+}/Fe(H_2O)_6^{2+}$). For the cyanide complexes, $Fe(CN)_6^{3-}/Fe(CN)_6^{4-}$ the potential drops to 0·36 V, and it rises as high as 1·2 V for the dipyridyl complexes, $Fe(dipy)_3^{3+}/Fe(dipy)_3^{2+}$. The potential also changes with the acidity when, for example, hydrated species in dilute acid change to hydroxy complexes in more alkaline media. The acidity is particularly important when oxyions are involved in the redox equation. Equations such as (6.8) or (6.9) include $[H^+]$ in equations from Table 6.3 which involve changing oxygen content by use of H^+. Thus the hydrogen ion concentration can enter in powers as high as 8 (for MnO_4^- going to Mn^{2+}). Standard conditions involve $[H^+] = 1$, i.e. pH = 0, and departures from this must be included in calculations of the redox potential.

If the redox potential in a solvent other than water is in question, it will clearly differ from that of the hydrated species in water. In the case of a simple ion, the difference lies in the heat of solvation compared with the heat of hydration, the contribution of the heat of atomization and the ionization potential (or electron affinity) remaining the same as in water. Although the solvation energy of any set of ions will neither be the same nor in the same order as in water, gross differences are less likely than small ones. In other words, ions which are strongly hydrated are likely to be strongly solvated by another ionizing solvent. This means that the general order of oxidizing powers should be similar in all solvents to that obtaining in water, although there may be marked differences in detail from one solvent to the next. Since, in general, thermodynamic measurements in non-aqueous solvents are much more sparse than those made in water, this generalization is usually the best that can be obtained. Similar remarks apply to oxidation-reduction reactions in systems which differ even more from that of aqueous solution, such as fused salts and other high temperature melts. In such systems, considerable changes take place in the relative stability and redox powers of chemical species, and, although the values derived from aqueous measurements may provide some guide, it is often better to go back to the fundamental parameters, such as ionization potentials, and base predictions on these.

For the use of redox potentials to display oxidation state stabilities, see the discussion of Ebsworth diagrams in section 8.6.

Non-Aqueous Solvents

Water is so familiar and accessible a solvent that its properties are usually taken for granted and underlie most general statements about inorganic compounds. Such general ideas as 'stability' usually imply 'stability in an environment containing air and water', and many inaccessible compounds or highly reactive oxidation states in this type of environment become stable and manageable if handled in a non-aqueous medium. In fact, one of the advantages of studying non-aqueous solvents has been the increased attention to, and wider insight into, chemistry in water that has resulted.

The practical reasons for using solvents other than water lie partly in the extended range of experimental conditions which are available and partly in the exclusion of water as a reactant. Many anhydrous compounds, for example, cannot be prepared from the hydrates and were unknown until the use of other solvents was introduced. Such compounds often have unusual and unexpected properties. The non-aqueous solvents range widely in intrinsic reactivity, from solvents such as sulfur dioxide which commonly act only as inert media for the reaction, to solvents such as anhydrous hydrogen fluoride which react with nearly all non-fluoride solutes. A number of selected examples of such solvent systems are discussed individually later but first the discussion of the early part of this chapter on solubility and solvent interaction and on acid-base behaviour will be extended to include non-aqueous solvents. It has already been indicated how the treatment of redox behaviour may be extended and further illustrations occur in the course of the discussion of specific solvents.

6.4 Solubility and solvent interaction in non-aqueous solvents

The extension of the discussion of solubility in section 6.1 to non-aqueous solvents is fairly easy. A very generalized energy diagram is shown in Figure 6.5. E_1 is the energy required to change the solute into the form in which it will exist in solution. In the case of an ionic solid, this means providing the lattice energy to separate the ions, while for a solute which dissolves as molecules it means providing the energy

required to separate the molecules or to form dimers or other species. E_2 is the energy required to separate the solvent molecules, breaking up any solvent 'structure' to allow the admission of the solute particles. E_3 is the energy given out when the separate solute particles associate with the solvent molecules. (It may be noted that the energy of hydration as commonly measured is not E_3 but the difference $[E_3 - E_2]$.) Two general cases exist. One, the generalization of the 'strong forces' case of section 6.1, is the case of ions or strongly polar molecules dissolving in a polar solvent. Here, the important balance of energies involves strong attractions between ions and permanent dipoles. The other extreme is the 'weak forces' case of weakly bound, nonpolar, covalent molecules dissolving in nonpolar solvents. Here, the balance of forces involves weak attractions between induced dipoles, and the order of magnitude of the effects is less than that in the strong forces case by a factor of about ten. The major conclusion is that 'mixed cases' will all represent situations where solubility is low. Thus, ions will not dissolve in nonpolar solvents because, although E_2 is low, E_3 is also low and the energetics are dominated by E_1. Conversely, nonpolar molecules are insoluble in polar solvents as E_1 and E_3 are both low and there is insufficient energy to supply E_2. This analysis is at the base of the very old generalization about solubilities that 'like dissolves like'.

Although a clear answer can be given in the mixed cases, it must be noted that the detailed variation of solubilities within the 'strong forces' or within the 'weak forces' class depends on the detailed balance of energies of similar magnitudes, so that a knowledge of solubilities in one solvent gives only a very general guide to the solubilities of similar compounds in similar solvents. This is illustrated by the solubilities given in Table 6.4.

TABLE 6.4 Solubilities in liquid ammonia and sulfur dioxide

| Compound | Solubility (millimole/litre at 0 °C) | | |
	Water	Ammonia	Sulfur dioxide
NaCl	6 100	2 200	insoluble
NaBr	7 710	6 210	1·4
NaI	10 720	8 800	1 000
KCl	3 760	18	5·5
KBr	4 490	2 260	40
KI	7 720	11 060	2 490
AgCl	0·005	20	0·05
AgBr	0·003	135	0·16
AgI	10^{-6}	4 000	0·68

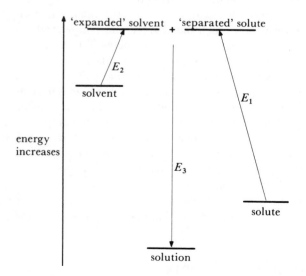

FIGURE 6.5 *Energy changes in the solution of a solid in a solvent*
This very generalized version of Figure 6.1 shows the energy changes for the solution of any solid in any solute, whether an ionic or a covalent system is involved.

For the alkali halides, the values in ammonia and sulfur dioxide follow the same order as in water, which is, in turn, the inverse order of the lattice energies, and most of the solubilities fall going from water to ammonia to sulfur dioxide

TABLE 6.5 Examples of solvolysis and other interactions

$SnCl_4 + 4H_2O$	$\rightleftharpoons Sn(OH)_4 + 4HCl$	hydrolysis*
$+ 8NH_3$	$\rightleftharpoons Sn(NH_2)_4 + 4NH_4Cl$	ammonolysis*
$+ 4HF$	$\rightleftharpoons SnF_4 + 4HCl$	solvolysis
$+ 2SeOCl_2$	$\rightleftharpoons SnCl_4.2SeOCl_2$ (or $2SeOCl^+ + SnCl_6^{2-}$)	solvation(?)
$LiBr + NH_3$	dissolves: $Li(NH_3)_4Br$ recovered	solvation
$+ SO_2$	$\rightleftharpoons Li_2SO_3 + SOBr_2$	solvolysis*
$KNO_3 + 4HF$	$\rightleftharpoons K^+ + H_2NO_3^+ + 2HF_2^-$	solvolysis

*(= further reaction of the product takes place)

but the relative sizes of the solubilities change enormously. The much greater solubilities of the iodides and the silver salts in ammonia and in sulfur dioxide show that polarization effects producing induced dipoles are more important in these two solvents than in water. It is worth noting that the solubility of the silver halides in liquid ammonia is in the reverse order to their solubility in aqueous ammonia.

The dielectric constants of non-aqueous solvents have an important effect on their power to dissolve ions. The general drop in solubilities of the alkali halides from H_2O to SO_2 reflects the drop in dielectric constant from 81 for water to 22 for ammonia and 12 for SO_2. Although important, the dielectric constant alone is a poor guide to solvent power for ions. The value for liquid hydrogen cyanide of 123 is one of the highest measured and yet this is a poor solvent for ions as the energy of solvation is low.

The general interaction between solvent molecules and ions in solution, which is analogous to hydration in water, is termed *solvation* and follows similar mechanisms. In liquid ammonia, for example, cations are ammonated by interaction with the lone pair of electrons on the nitrogen atom plus the interaction with the negative end of the $N^{\delta-} - H^{\delta+}$ dipoles. The anions interact with the relatively positive hydrogens.

In addition to solvation, the solute may react with the solvent molecules to break up the solvent molecule, the solute species, or both, in the reaction called *solvolysis*—analogous to hydrolysis in water. The distinction between solvation and solvolysis is not completely clear-cut but solvolysis implies a more extensive and less reversible interaction with the solvent molecule. Some examples are given in Table 6.5.

Supercritical fluids as solvents. A significant advance in solvent use has been the development of methods of using supercritical fluids. Up to a certain temperature a gas can be liquefied by applying pressure. The critical temperature is the highest one at which the liquid phase can exist, and increasing the pressure above this temperature fails to liquefy a material. The *critical point* is defined by the critical temperature and the critical pressure required to cause liquefaction. Above the critical point, the distinction between gas and liquid disappears, and we talk of supercritical fluids. Materials like the rare gases and carbon dioxide have valuable solvent properties in the supercritical state and are being explored as reaction media. Solubilities are often unusual and usually higher than for equivalent normal

solvents. Properties such as solubility can change rapidly with pressure. Thus preparations and separations can be much easier than in conventional solvents—once equipment for working at raised pressures is in place. For example, H_2 is 100 to 1000 times more soluble in supercritical CO_2 at 25°C and 300 atm than it is in normal organic solvents or in water, making this an attractive medium for hydrogenation reactions. In industrial applications supercritical fluids avoid the problems of removing and disposing of organic solvents. Even for conventional solvents, such as water, there can be considerable benefits in using them above the critical point.

6.5 Acid-base behaviour in non-aqueous solvents

Of the two theories of acids and bases discussed in section 6.2, the protonic theory of Lowry and Brönsted is restricted to solvents containing ionizable hydrogen atoms. The application of the theory is then exactly the same as for aqueous systems, except that the strengths of different acids are most conveniently measured relative to the acid-base pairs characteristic of the solvent. For example, K values in liquid ammonia may be measured relative to:

$$NH_4^+ \rightarrow NH_3 + H^+.$$

Such values would fall in a similar order to those obtained in water or in any other solvent, although there might be minor inversions.

The solvent system definition was developed for use in non-aqueous solvents and provides a definition of acid and base in non-protonic solvents. It is very convenient to use if work is being carried out on one particular solvent and it provides a guide to the properties to be expected for a new system, but it does suffer the disadvantage that the definition of acid or base changes from one solvent to another. A more important reservation which must be made is that it only suggests an acid-base system for a particular solvent and does not prove that the solvent ions do in fact behave as acid and base. This point must be proved experimentally. These remarks are extended and illustrated in the discussion of the sulfur dioxide solvent system (section 6.11).

As in the case of water (section 6.2), the solvent's acid-base properties place a limit on the range of the acid strengths of solutes in it. The strongest acid existing in a solvent is the solvent acid and all solutes which are more acidic are converted into the solvent acid—and similarly for bases—so that the number of solutes which are strong acids (i.e. those which are completely converted to the solvent acid) varies with the intrinsic acid strength of the solvent. In an acidic solvent such as glacial acetic acid, some solutes which are strong acids in water are not completely dissociated to the solvent acid, that is, they are weak acids in this medium. This is the case for HNO_3 or HCl in acetic acid. Indeed some molecules which are weakly acidic in water are basic in acetic acid. This latter effect is even more marked in strongly acidic solvents such as absolute sulfuric acid where even HNO_3 acts as a base and accepts a proton from the solvent. The relative strengths of those solutes which are strong acids in water may thus be differentiated by examining their properties in more acidic solvents. On the other hand, a basic solvent like liquid ammonia has a very weak solvent acid so that all solutes which are acids, strong or weak, in water are strong acids in ammonia. Base strengths in basic and acidic solvents are affected in an analogous manner.

An acid solute which *is* completely dissociated in an acidic solvent, such as perchloric acid in glacial acetic acid, is a much stronger acid than in water and finds valuable application in analysis for determining very weakly basic functions. In a similar way, the solution of a strong base in a basic solvent may be used for the estimation of very weakly acidic groups. Examples of these applications are given in sections 6.7 and 6.8.

Both the proton-transfer and solvent system definitions of acids and bases are in general use for non-aqueous solvents and they are practically equivalent when applied to a protonic solvent. The solvent system definition—for example:

	acid		base
$2 NH_3$	$\rightleftharpoons NH_4^+$	$+$	NH_2^-
$2 H_2SO_4$	$\rightleftharpoons H_3SO_4^+$	$+$	HSO_4^-
$2 HCN$	$\rightleftharpoons H_2CN^+$	$+$	CN^-

is equivalent in each case to a pair of Brönsted acids,

Acid	$\rightleftharpoons$ conjugate base $+ H^+$
NH_4^+	NH_3
NH_3	NH_2^-
$H_3SO_4^+$	H_2SO_4
H_2SO_4	HSO_4^-
H_2CN^+	HCN
HCN	CN^-

A third, much more extensive, theory of acids and bases was proposed by Lewis, who defined an acid as an electron pair acceptor and a base as an electron pair donor. This definition includes the Brönsted and solvent system definitions as special cases. For example, the proton is an electron pair acceptor and H_2O or NH_3 or H_2SO_4 or HCN, among our examples, act as donors. This definition emphasizes the similarity between all coordination reactions, for example:

Lewis acid	Lewis base	product
H^+	H_2O	H_3O^+
	OH^-	H_2O
	F^-	HF
BF_3	H_2O	$H_2O:BF_3$
	OH^-	$(HOBF_3)^-$
	F^-	BF_4^-

It includes solvation reactions:

Lewis acid	Lewis base	
M^{z+}	$x\, H_2O$	$[M(H_2O)_x]^{z+}$
	$x\, NH_3$	$[M(NH_3)_x]^{z+}$

and at least the initial steps of solvolysis reactions, for example:

$SnCl_4$	NH_3	$(SnCl_4:NH_3)$

but also includes many other types of reaction, including oxidation-reduction reactions. The Lewis theory is more extensively applied in America than in Europe but use of the general term 'Lewis acid' for acceptor molecules like BF_3 is very common, even by authors who make no other use of the theory.

The concept of *hard* and *soft* acids and bases (acceptors and donors) may be seen as an amplification of the Lewis approach. This idea is introduced in section 2.16 and its application to transition metal complexes is discussed in section 13.7.

6.6 General uses of non-aqueous solvents

It is now possible to outline some general reasons for choosing to work in a solvent other than water and to suggest points which will guide the choice of solvent. These are best seen by comparison with the properties of water itself.

(a) *Solvation and solvolysis*. The choice of a solvent is clearly also dependent on its solvent powers. Solvents of high polarity and coordinating power are good ionizing media, and the range runs from these down to solvents such as the hydrocarbons which take up only molecular solids. In preparation reactions, a solvent is often chosen because of the insolubility of a certain compound in it. For example, the reaction in water between KCN and $NiSO_4$ gives $Ni(CN)_2$ by direct precipitation but the product is very finely divided and difficult to filter and wash. The same reaction in liquid ammonia gives an easily handled precipitate of K_2SO_4, not $Ni(CN)_2$, and the potassium salt is readily removed and the nickel cyanide recovered by evaporation.

A solvent may also be chosen because of the way in which it reacts with solutes. Since water coordinates strongly, it is often impossible to remove it without decomposing the compound, and many anhydrous compounds are available only by syntheses which avoid water, or other strongly coordinating solvents. One example is anhydrous copper nitrate, $Cu(NO_3)_2$, which is a volatile solid prepared by reaction in N_2O_4. When solvolytic reactions are possible, the solvent may be chosen to avoid these, as in the use of liquid sulfur dioxide for reactions with non-metal halides. On the other hand, a solvent is often used because it does react with the solute. An important example is anhydrous hydrogen fluoride which solvolyses almost every compound placed in it and provides a standard method for preparing fluorine-containing compounds.

(b) *Acidity*. The effect of the acid-base properties has already been discussed. In water, the strongest acid is H_3O^+ and the strongest base is OH^-, all solutes of higher acidity or basicity being completely converted to these. If the effect of more acidic or more basic media is to be investigated, clearly a suitable non-aqueous solvent must be chosen. The most fully-investigated solvents from this point of view are glacial acetic acid, anhydrous hydrogen fluoride, and anhydrous sulfuric acid for acidic media, while ammonia and the amines have been the most-studied basic media.

One hundred percent sulfuric acid is so strong a proton donor that many substances which act as acids in water accept a proton (i.e. are bases) in sulfuric acid. The few proton donors include $H_2S_2O_7$ ('pyro' or di-sulfuric acid), fluorosulfuric acid FSO_3H and its trifluoromethyl analogue

CF_3SO_3H. The solutions of such species in sulfuric acid are extremely powerful protonating agents and allow the preparation of species such as the iodine cations I_n^+ (see Chapter 17), which are highly susceptible to attack by bases. An extreme case is the 'superacids' (section 6.9), and see also section 18.5.

At higher temperatures, fused salts provide suitable media. Acidic systems include ammonium salts of strong acids, acidic anion salts such as HSO_4^- or HF_2^-, and acidic oxides such as silica or phosphorus pentoxide. Basic systems include the oxides or hydroxides of the alkali metals.

(c) *Redox*. In a rather similar way, the oxidation-reduction properties of the solvent place limits on the reactions which may be carried out in it. Water is attacked rapidly by oxidizing agents with standard potentials greater than about $1·7\,V$, and by reducing agents with oxidation potentials below about $-0·4\,V$. Outside these limits, use of water as the solvent is undesirable. Non-aqueous solvents which are useful for carrying out oxidizing reactions include sulfuric acid and higher valency oxides such as dinitrogen tetroxide. The most important solvent for strong reductions is liquid ammonia, which dissolves the alkali and alkaline earth metals to give a very strongly reducing system.

A fairly wide variety of compounds have been used as non-aqueous solvents. Examples are listed in Table 6.6 with notes on their special uses. The general points made above are illustrated by more detailed discussion of glacial acetic acid and liquid ammonia, representing protonic acid and base, extremely acidic media, and two non-protonic solvents, bromine trifluoride and sulfur dioxide. The solvent system definitions of acid and base appear to hold in BrF_3, but they break down in SO_2. Table 6.7 shows the more important physical properties of these solvents.

6.7 Liquid ammonia

Liquid ammonia was one of the first non-aqueous ionizing solvents to be studied and is now the one which is most extensively used and best known. It is readily available in reasonable purity and is easily dried by the action of sodium. The liquid range is from $-77\,°C$ to $-33\,°C$, and this presents some handling problems, but these are partly compensated by the relatively high latent heat of vaporization. Liquid ammonia is normally handled in cooled vessels—and it is often used as its own coolant—or at room temperature under pressure. A wide range of special techniques have been developed for handling ammonia so that few problems are now met.

The dielectric constant and the self-ionization are both lower than for water, indicating that liquid ammonia will be a poorer ionizing solvent than water. Soluble salts include most ammonium salts, nitrates, thiocyanates and iodides. Fluorides and most oxy-salts are insoluble, and solubility among the halides increases $F^- < Cl^- < Br^- < I^-$. Calcium and zinc chlorides, which are extremely soluble in water, are quite insoluble in liquid ammonia, but they do take up a large amount of ammonia forming solid ammoniates with eight and ten molecules of ammonia respectively. Calcium chloride is thus a useful absorbent for traces of ammonia. Since ammonia is less highly associated than water, it is a better solvent for organic compounds, especially for those with fairly small carbon radicals. Unsaturated hydrocarbons, alcohols, esters, ammonium salts of acids, and most nitrogen compounds are soluble.

The self-ionization of ammonia is written:

$$2\,NH_3 \rightleftharpoons NH_4^+ + NH_2^-$$

and ammonium compounds behave as acids while amides are bases. Ammonium iodide, nitrate, or thiocyanate are very soluble and concentrated solutions will slowly dissolve metals with the evolution of hydrogen:

$$Mg + 2NH_4^+ \rightarrow Mg^{2+} + H_2 + 2NH_3$$

This reaction is rapid with the active alkali and alkaline earth metals but slow with less active metals such as magnesium or iron. As ammonia is more basic than water, these acids are weaker, so the solvent power for metals does not extend to the less active metals. The commonest base is potassium amide, KNH_2, which is much more soluble than sodium amide. Acid-base titrations between potassium amide and ammonium salts may be carried out and followed conductiometrically or by using phenolphthalein. The ammonia system brings out very weakly acidic functions in molecules. Thus, urea, $CO(NH_2)_2$, which is a weak base in water, acts as a weak acid in liquid ammonia and may be neutralized by amide.

This enhancement of weakly acidic functions by the basic solvent finds application in analysis. Ammonia itself is rarely used because of its low boiling point, but simple organic derivatives such as ethylamine, $CH_3CH_2NH_2$ (which is related to ammonia as ethanol, CH_3CH_2OH, is to water), or ethylenediamine, $H_2NCH_2CH_2NH_2$, are used as solvents

TABLE 6.6 Typical non-aqueous solvents

Solvent	Postulated self-ionization	Acids	Bases	Comments
NH_3	$NH_4^+ + NH_2^-$	ammonium salts	amides, e.g. KNH_2	See section 6.7 for discussion. A number of similar solvent systems have been studied including N_2H_4, NH_2OH, and organic amines such as $C_2H_4(NH_2)_2$ and CH_3NH_2.
$HCONH_2$ $HCON(CH_3)_2$ and CH_3CONH_2		protonic and Lewis acids	organic nitrogen bases, e.g. pyridine	H-bonding, high dielectric constant, make these good solvents for ionic compounds. Dimethylformamide is especially of current interest.
HF	$H_2F^+ + HF_2^-$	F^- acceptors, e.g. BF_3 $HF + BF_3 = H_2F^+BF_4^-$	alkali fluorides	Good ionizing solvent: many non-fluorine species react, e.g. $H_2SO_4 \rightarrow HSO_3F$. Used in the preparation of fluorine compounds, e.g. $AgNO_3 \rightarrow Ag^+ + H_2NO_3^+ + 2F^-$ $BF_3 + F^- \rightarrow BF_4^-$ then, $Ag^+ + BF_4^- \rightarrow AgBF_4$.
CH_3COOH	$CH_3COOH_2^+ + CH_3COO^-$	protonic acids	ionizable acetates	See section 6.8. Other carboxylic acids behave similarly, especially HCOOH.
H_2SO_4	$H_3SO_4^+ + HSO_4^-$	$H_2S_2O_7$ FSO_3H ($HClO_4$ very weak)	soluble bisulfates H_2O, HNO_3, H_3PO_4	Most common acids are bases in H_2SO_4
N_2O_4	$NO^+ + NO_3^-$	NOCl	alkali nitrates	Used to prepare anhydrous nitrates, nitrato- and nitro-complexes and nitrosyl compounds: commonly used in admixture with organic compounds such as ethyl acetate.
SO_2	Conflicting evidence, see section 6.11.			Useful as an inert reaction medium.
$SeOCl_2$	$SeOCl^+ + Cl^-$ (or $SeOCl_3^-$)	Cl^- acceptors e.g. $SnCl_4$	organic bases like pyridine ($\rightarrow C_5H_5NSeOCl^+Cl^-$)	Dissolves many elements with reaction and dissolves a variety of metal chlorides. Other salts are converted to the chloride.

Other oxyhalides, NOCl and $POCl_3$ for example, are similar. Some conductiometric and potentiometric titration data exist to support the self-ionization mode and acid-base behaviour which is postulated.

Solvent	Postulated self-ionization	Acids	Bases	Comments
BrF_3	$BrF_2^+ + BrF_4^-$	F^- acceptors	ionizable fluorides	Discussed in section 6.10. IF_5 behaves similarly.
$AsCl_3$	$AsCl_2^+ + AsCl_4^-$	Cl^- acceptors	Cl^-	$AsBr_3$, $SbCl_3$ and ICl are analogous

TABLE 6.7 Physical properties of non-aqueous solvents

	H_2O	NH_3	CH_3COOH	SO_2	BrF_3
m.p. (°C)	0	−77·7	16·6	−75·5	9·0
b.p. (°C)	100	−33·4	118·1	−10·2	126
dielectric constant	78·5 (18 °C)	23 (b.p.)	9·7 (18 °C)	17·3 (−16 °C)	—
specific conductivity ($\Omega\ m^{-1}$)	6×10^{-8} (25°C)	5×10^{-9} (b.p.)	$0·5$—$0·8 \times 10^{-8}$ (25°C)	4×10^{-8} (b.p.)	8×10^{-3} (m.p.)
heat of vaporization ($kJ\ mol^{-1}$)	40·7	23·6	24·3	25·0	42

for the determination of weak acids. Titrants used include methoxides of the alkali metals and soluble hydroxides which are easier to purify and standardize than amides. Applications include the determination of phenols and related compounds in ethylenediamine, and the determination of carbon dioxide which may be separated from other gases by solution in acetone and determined as a weak acid with sodium methoxide, CH_3ONa. Suitable indicators or potentiometric methods are used to determine the end points.

There are many examples of amphoteric behaviour in liquid ammonia. For example:

$$Zn^{2+} \underset{NH_4^+}{\overset{NH_2^-}{\rightleftharpoons}} Zn(NH_2)_2 \downarrow \underset{NH_4^+}{\overset{NH_2^-}{\rightleftharpoons}} Zn(NH_2)_4^{2-}$$

compare with

$$Zn^{2+} \underset{H_3O^+}{\overset{OH^-}{\rightleftharpoons}} Zn(OH)_2 \downarrow \underset{H_3O^+}{\overset{OH^-}{\rightleftharpoons}} Zn(OH)_4^{2-}$$

Amides and imides of the less-reactive metals dissolve in excess potassium amide:

e.g. $PbNH + NH_2^- + NH_3 = Pb(NH_2)_3^-$

and even sodium amide, which is insoluble in ammonia, dissolves in potassium amide to give the ammonosodiate:

$$NaNH_2 + 2\,NH_2^- = Na(NH_2)_3^{2-}.$$

Hydrolysis of heavy metal salts to basic salts (oxy- and hydroxy-compounds) is paralleled by reactions between ammonia and such salts. The analogues of the oxy- and hydroxy-compounds in the aqueous system are compounds containing the groups, amide $-NH_2$, imide $=NH$, and nitride $\equiv N$. For example, $MoCl_5$ dissolves in liquid ammonia and the compound $Mo(NH)_2NH_2$ may be isolated from the solution. If an ammonium salt is added, this compound dissolves, while the addition of potassium amide precipitates it.

The most striking property of liquid ammonia is its ability to dissolve the active metals to give fairly stable, blue solutions. As ammonia is more resistant to reduction than is water, the reaction:

$$M + NH_3 = MNH_2 + \tfrac{1}{2}H_2$$

is much slower than the reaction:

$$M + H_2O = MOH + \tfrac{1}{2}H_2$$

If the reagents are pure and dry, sodium solutions in liquid ammonia may be preserved for several weeks and even the much more reactive cesium gives solutions which may be kept overnight. These solutions are formed by all the alkali metals, by the alkaline earth metals, and by the reducible lanthanide elements such as samarium. In addition, very dilute solutions may be formed from less reactive elements, such as magnesium, beryllium, aluminium, and the other lanthanides, by electrolytic means. The alkali metals are also soluble in amines and dilute solutions are formed in certain ethers.

Dilute metal solutions are coloured a very deep blue and concentrated ones have a metallic, coppery appearance. The solutions are conducting and strongly reducing. In these solutions, the valency electrons are ionized and become solvated:

$$Na \rightarrow Na^+_{(ammoniated)} + e^-_{(ammoniated)}.$$

The unique properties of the solutions are associated with the presence of these readily available electrons.

The major application of metal solutions is in reduction reactions. In organic chemistry, the skeletal single bonds, $C-C$, $C-O$, $C-N$, are stable in these solutions as are isolated double bonds and single benzene rings, but nearly all other functional groups and most unsaturated compounds are reduced. In inorganic chemistry, the most interesting reductions have been the formation of polyanions, the reductions of hydrides, and the formation of transition metal complexes in unusually low oxidation states. Some examples are given in Table 6.8.

TABLE 6.8 Reduction reactions by metal solutions in liquid ammonia

Reactant	Reduction products
O_2	metal peroxides, O_2^{2-}, and superoxides, O_2^-
S, Se, Te, As, Sb, Bi, Sn, Pb and their oxides or halides	white binary compounds such as M_2S or M_4Pb and highly coloured polyanions such as M_2S_x ($x = 2$ to 7) (deep red), M_3Bi_3 (violet) or M_4Pb_9 (deep green) (cf. section 18.4.4)
metal oxides, halides, and dissociable complexes	metal
complexes which do not dissociate $Ni(CN)_4^{2-}$	$Ni_2(CN)_6^{4-}$; $Ni(CN)_4^{4-}$
e.g. $Co(CN)_6^{2-}$	$Co(CN)_4^{4-}$
$Ni(C \equiv CH)_4^{2-}$	$Ni(C \equiv CH)_4^{4-}$
hydrides of Groups IV and V	ions formed by removal of one or two hydrogens PH_2^- or SnH_2^{2-} and SnH_3^-
Ge_3H_8	$2GeH_3^- + GeH_2^{2-}$ (i.e. each $Ge-Ge$ bond is broken)

6.8 Anhydrous acetic acid

To some extent, glacial acetic acid is the converse solvent to liquid ammonia. It is readily available, easily purified, and differs from water in undergoing less self-ionization, and in having a markedly lower dielectric constant. It is a moderately strong acid and is used to investigate weakly basic functions, just as ammonia and the amines are used for weakly acid functions. The lower dielectric constant makes acetic acid a poorer solvent for ions, and most compounds which are insoluble in water are insoluble in acetic acid. Soluble compounds include most acetates, nitrates, halides, cyanides, and thiocyanates. The strong acids all dissolve in glacial acetic acid as do basic compounds like water and ammonia. A wide range of polar organic compounds is also soluble.

The self-ionization of acetic acid is:

$$2\,CH_3COOH \rightleftharpoons CH_3COOH_2^+ + CH_3COO^-$$

so that acetates of the reactive metals are bases. The normal protonic acids are acids in acetic acid by virtue of reactions such as:

$$HClO_4 + CH_3COOH \rightleftharpoons CH_3COOH_2^+ + ClO_4^-$$

(compare $HClO_4 + H_2O \rightleftharpoons H_3O^+ + ClO_4^-$), while ammonia, say, is a base as the acetate ion is produced by the reaction:

$$NH_3 + CH_3COOH \rightleftharpoons CH_3COO^- + NH_4^+$$

Of the mineral acids which are strong acids in water (that is are completely dissociated to the H_3O^+ ion), only perchloric acid is strongly dissociated in glacial acetic acid. The relative strengths of the common acids in acetic acid are $HNO_3 = 1 < HCl = 9 < H_2SO_4 = 30 < HBr = 160 < HClO_4 = 400$. The acetates which are strongest bases are those of the alkali metals and ammonium, while bismuth, lead, and mercuric acetates are ten to a hundred times weaker. Even the strongest bases and acids are poor electrolytes in acetic acid, largely as a result of the low dielectric constant of the solvent.

Acid-base titrations between the acetates and the mineral acids in acetic acid may be demonstrated by indicators or electrometrically. In addition, perchloric acid in glacial acetic acid is a widely used reagent for the estimation of the weakly basic functions of amines, amino-acids, metal salts of organic acids, and the like.

Amphoteric behaviour has also been demonstrated in acetic acid. For example, the addition of sodium acetate to a solution of a zinc salt precipitates zinc acetate, which redissolves in excess acetate:

$$Zn^{2+} \xrightleftharpoons[CH_3COOH_2^+]{CH_3COO^-} Zn(CH_3COO)_2 \downarrow$$

$$\xrightleftharpoons[CH_3COOH_2^+]{CH_3COO^-} Zn(CH_3COO)_4^{2-}$$

Copper and lead(II) acetates also show amphoteric behaviour, while lead(IV) tetra-acetate decreases in solubility

as sodium acetate is added to its solution, which corresponds to 'salting out' a poor electrolyte.

Acetic acid is a convenient reaction medium for the preparation of covalent hydrolysable compounds. For example, tin reacts smoothly and controllably with halogens in acetic acid:

$$Sn + 2X_2 = SnX_4$$

and the stannic halide is readily isolated by distillation or crystallization.

Other weak acids which have been studied as solvent systems include formic acid and hydrogen cyanide, but investigations in these systems have been largely confined to studies of solubility and acid-base relationships. Of the strong acids, attention has been concentrated on hydrogen fluoride and sulfuric acid, in which the principal type of reaction is solvolysis. This is also true of bromine trifluoride, 6.11.

6.9 'Superacid' media

We have mentioned above (section 6.6) the use of H_2SO_4 as a strongly-acidic medium. However, there are some systems which are even more powerfully proton-donating and these are of interest (a) in protonating species which are very weak H^+ acceptors and (b) in preparing species which are highly susceptible to base attack. (Clearly, the more strongly acidic the medium is, the more weakly basic it is.)

Probably the strongest proton donor of all simple substances is disulfuric acid, $H_2S_2O_7$, formed from SO_3 and H_2SO_4. However, this is a very difficult medium to handle as it is very viscous and the equilibria between various components are complex. The next strongest proton donor is fluorosulfuric acid, related to sulfuric acid by replacing an OH group by an F:

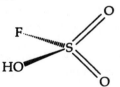

This change (a) lowers the viscosity by reducing the amount of H-bonding and (b) increases the acidity as F withdraws more charge than OH. (Compare section 6.2.) Handling properties are more convenient, with a m.pt. of $-89°$ and b.pt. of $163°C$. The low m.pt. allows ^1H nmr studies at low temperatures (see section 7.8) which have been extensively used to determine the site of protonation. At the higher m.pt. of H_2SO_4 ($10\cdot4°C$), exchange processes and rearrangements are already too rapid for the simple species to be observed. When free from HF, fluorosulfuric acid can readily be handled in ordinary glassware.

Using the Lowry-Brönsted approach, the self-ionization is

$$2HSO_3F \rightleftharpoons H_2SO_3F^+ + SO_3F^-$$

From the conductivity, the concentration of the solvent ions at $25°C$ is about 2×10^{-4} mol kg^{-1}, much lower than in H_2SO_4 but higher than in H_2O or CH_3CO_2H.

Acids in HSO_3F are species which increase the $H_2SO_3F^+$ concentration, either by proton donation

$$HA + HSO_3F = H_2SO_3F^+ + A^- \dots \dots (A1)$$

or by removing SO_3F^- ions as SbF_5 does

$$SbF_5 + 2HSO_3F = H_2SO_3F^+ + [SbF_5(SO_3F)]^- \quad (A2)$$

Bases are alkali metal fluorosulfates or proton acceptors

$$KSO_3F = K^+ + SO_3F^- \dots \dots (B1)$$

$$B + HSO_3F = BH^+ + SO_3F^- \dots \dots (B2)$$

Because the medium is so acidic, there are very few proton donors: for example, H_2SO_4 does not follow (A1) but acts as a weak base according to (B2) forming $H_3SO_4^+$. Similarly, although SO_3 dissolves to give HS_2O_6F, this analogue of $H_2S_2O_7$ does not ionize in HSO_3F, and so the 'fluoro-oleum' system contrasts with SO_3/H_2SO_4. Similarly HF, $HClO_4$ and other acids which are strong in water do not donate H^+ in HSO_3F: e.g.

$$KClO_4 + HSO_3F \rightarrow K^+ + SO_3F^- + HClO_4$$

Thus the only acids are the strong acceptors which follow equation (A2). Of these, two systems are important. SbF_5 behaves as a weak acid, i.e. ionization according to equation (A2) is not complete and a titration of SbF_5 with the base KSO_3F shows a conductivity minimum about $0.4:1$. However, if SO_3 is added, a range of species $SbF_{5-n}(SO_3F)_n$ is formed ($n = 1, 2, 3$) and such species act as strong acids. For example, $SbF_2(SO_3F)_3$ gives a $1:1$ titration with KSO_3F.

Having established the very high acidity of HSO_3F itself, it follows that these even more acidic systems HSO_3F/SbF_5 and $HSO_3F/SbF_5/SO_3$ have astonishingly high proton donor powers. They have been colourfully termed *superacids*, while the term 'magic-acid' has been applied to HSO_3F/SbF_5.

As examples, HSO_3F alone protonates organic acids to give $RC(OH)_2^+$, ketones to give $R_2C(OH)^+$, and halo-aromatics to give for example $FC_6H_6^+$ from FC_6H_5. The superacid systems will completely protonate trinitrobenzene and even alkanes to give intermediates such as CH_5^+. Such ions readily lose H_2 to give carbonium ions, e.g. CH_3^+.

In inorganic chemistry, fluorosulfuric acid is an excellent agent for preparing fluorides and fluorosulfates as in

$$As_2O_5 \rightarrow AsF_5 + AsF_2(SO_3F)_3$$
$$SiO_2 . xH_2O \rightarrow SiF_4$$
$$P_4O_{10} \rightarrow POF_3$$
$$BaTeO_4 \rightarrow TeF_5(SO_3F)$$
$$ClO_4^- \rightarrow ClO_3F$$

Also of importance is the use of this medium to prepare polyatomic ions using $S_2O_6F_2$ as oxidant, as in

$$Se + S_2O_6F_2 \rightarrow Se_4^{2+} + 2SO_3F^-$$
$$\text{or} \quad I_2 + \tfrac{1}{2}S_2O_6F_2 \rightarrow I_2^+ + SO_3F^-$$

Such polyatomic species are susceptible to base attack and need an acid medium to survive. Often they were first seen in the sulfuric acid system but only characterized in the more manageable fluorosulfuric acid one. For instance, I_2 was

reported to dissolve in oleum to give a blue solution twenty years before the blue species was identified by work in HSO_3F as I_2^+ and not I^+ as originally thought. Other examples are given in Chapter 17.

6.10 Bromide trifluoride

Bromine trifluoride is a much more restricted solvent than those discussed above but it is one to which the ideas of the solvent system theory appear to apply and it has been quite widely used as a preparative medium. It must be handled by special techniques which require experience but these are now highly developed and present few problems in a well-equipped laboratory. The main requirements are a rigorous exclusion of moisture and of materials such as tap greases which can be fluorinated or oxidized.

The alkali metal fluorides are soluble but most other ionic fluorides are rather insoluble. The more covalent fluorides of elements in higher oxidation states are also soluble. Most other compounds are either insoluble or converted to fluoro-compounds.

The self-ionization of bromine trifluoride is much more extensive than that of the other solvents in Table 6.6 and is presumed to follow the equation:

$$2 BrF_3 = BrF_2^+ + BrF_4^-$$

By the solvent system definition, solutes which increase the concentration of BrF_2^+ are acids and those which increase the concentration of BrF_4^- are bases. The bromofluorides of the alkali metals and a number of other elements, e.g. $AgBrF_4$, are known and act as bases, while the alkali metal fluorides add on a molecule of the solvent and may also be regarded as bases in the system. No solute is known which produces the solvent cation by dissociation (as the mineral acids produce H_3O^+ in water) that is, there are no *donor acids* known, but a number of solutes are known which react with the solvent molecule by removing a fluoride ion and these are termed *acceptor acids*. This is true of most covalent fluorides, which form fluoro-complexes, for example:

$$VF_5 + BrF_3 = BrF_2^+ + VF_6^-$$
$$SnF_4 + 2BrF_3 = 2BrF_2^+ + SnF_6^{2-}$$

Neutralization reactions are represented by the interaction of these acids and bases and may be followed conductio-metrically:

$$\underset{\text{acid}}{BrF_2^+VF_6^-} + \underset{\text{base}}{KBrF_4} = \underset{\text{salt}}{KVF_6} + \underset{\text{solvent}}{2BrF_3}$$

Solvolysis of some of these complex fluorides is also observed. For example, if hexafluorotitanate is treated with bromine trifluoride the reversible reaction

$$K_2TiF_6 + 4BrF_3 \rightleftharpoons (BrF_2)_2TiF_6 + 2KBrF_4$$

is observed leading to the formation of the solvent base in equilibrium.

The main use of bromine trifluoride is in the preparation of fluorides and fluorocomplexes of elements in the higher oxidation states. Among the transition metal compounds which have been prepared in bromine trifluoride are the hexafluoro-complexes of Ti(IV), V(V), Nb(V), Ta(V), Mn(IV), Ru(IV & V), Os(IV & V), Rh(IV), Ir(IV & V), Pd(IV), Pt(IV), and Au(III). The first preparation of gold trifluoride was also made in this solvent. Another interesting series of reactions involves the reactions of non-metal oxides with fluorocomplexes in bromine trifluoride to give fluorosulfonates, nitronium complexes, and nitrosonium complexes, such as:

$$SO_3 + KF \rightarrow KSO_3F$$
$$NO_2 + Au \rightarrow (NO_2)AuF_4$$
$$NO_2 + As_2O_3 \rightarrow (NO_2)AsF_6$$
$$NOCl + GeO_2 \rightarrow (NO)_2GeF_6$$

(Note that several of these reactions involve oxidations as well as fluorination.)

Other fluorine compounds, such as IF_5 or SeF_4, also allow synthesis of fluorides. For elements with a range of oxidation states, HF is the medium for preparing fluorides of the lowest oxidation states, IF_5 gives intermediate states while BrF_3 gives high oxidation fluorides. To force an element to form fluorides of very strongly oxidizing states it may be necessary to use the even more vigorous ClF_3, as a medium. Of course, most fluorides result from direct reaction with F_2, but these halogen fluoride solvents may be very useful when it is necessary to discriminate among a range of oxidation states.

6.11 Sulfur dioxide

The study of sulfur dioxide as a solvent extends back as far as that of liquid ammonia. Sulfur dioxide was found to be conducting and the postulated self-ionization was:

$$2SO_2 \rightleftharpoons SO^{2+} + SO_3^{2-}$$

According to this, ionizable sulfites, such as Cs_2SO_3, were bases, and acids were species which produced SO^{2+}. The thionyl halides were found to be ionized and the equilibrium:

$$SOX_2 \rightleftharpoons SO^{2+} + 2X^-$$

was postulated, so that these molecules were solvent acids. It was indeed found that thionyl chloride reacted with soluble sulfites in a 1:1 ratio and the titration curve was reasonably like the expected one. On the basis of these observations and others on solvate formation, solvolysis, and amphoteric behaviour, the above self-ionization was accepted as the basis for interpreting reactions in sulfur dioxide.

The existence of these sulfur dioxide results was an important reason for the general acceptance of the solvent system definition of acids and bases, since this was not confined to protonic solvents and appeared to fit for sulfur dioxide, the most extensively studied non-protonic system of the time. As other non-protonic solvents came to be studied, the self-ionization theory was applied and found to give a satisfactory basis for systematizing reactions and to be a useful guide in the study of the solvent, for the non-protonic solvents listed in Table 6.6.

However, the only self-ionization in that list which involves the separation of doubly-charged ions is that for sulfur dioxide. Table 6.7 shows that SO_2 has fairly low values for the dielectric constant and self-conductivity so the self-ionization theory may be on shaky ground in this case. This suspicion was confirmed by radio-isotope exchange experiments between SO_2 and thionyl compounds which showed negligible exchange of either ^{18}O or ^{35}S between $SOCl_2$ and SO_2. It follows that the concentration of any ion common to the two compounds is extremely low (compare the instantaneous exchange of hydrogen isotopes between protonic acids and water). Therefore ionization to SO^{2+} cannot be occurring and the 'neutralization' reactions between sulfite and thionyl compounds were probably not simple reactions between solvent ions but involved two-step reactions via SOX^+ intermediates.

These results for sulfur dioxide raise the question of the value of the solvent system concept of acids and bases. It is not needed for protonic solvents and the Lewis theory covers non-protonic ones. Whether it continues to be used will probably be decided in practice in terms of the perceived usefulness of the concept. As in other cases we have noted, it is not a matter of the absolute correctness of the approach but whether it is a useful means of rationalizing and predicting observations. It may be that the Lewis theory is just too general, as it covers all interactions of the solvent not just ionizing ones. The decision will be made gradually as practitioners change their usage. If the solvent system theory does continue to be used to interpret reactions, there will be marked reservations for non-protonic systems until the existence of self-ionization has been proved.

It has been shown that Brönsted acids, such as HCl, react with Brönsted bases in liquid sulfur dioxide. However, it is best to regard sulfur dioxide, not as a self-ionizing solvent which is the parent of an acid-base system, but as a relatively inert reaction medium. This has always been the main use of sulfur dioxide and it is extremely valuable as a solvent for reactions between weakly ionic or fairly polar molecules.

Among the compounds which are very soluble in sulfur dioxide are the alkali metal, or ammonium, iodides, thiocyanates, and carboxylic acid salts. Many classes of organic compounds are soluble, as are the more covalent halides and pseudohalides of the Main Group elements. Solvates of soluble salts are often recovered when the solvent is removed. Examples include $NaI.4SO_2$, $KSCN.SO_2$, or $AlCl_3.SO_2$. The sulfur dioxide is usually much less tightly bound than is water or ammonia in hydrates or ammoniates.

Sulfur dioxide finds its main use as a solvent for the reaction of readily hydrolysable halides and related compounds. Thus thionyl compounds may be prepared:

$$2SCN^- + SOCl_2 = SO(SCN)_2 + 2Cl^-$$

Chloro-complexes are readily prepared:

$$NOCl + SbCl_5 = (NO)^+(SbCl_6)^-,$$
$$3SOCl_2 + 2SbCl_3 = (SO)_3^{2+}(SbCl_6)_2^{3-} \text{ (in solution only)}$$

and a few solvolytic reactions are observed:

$$PCl_5 + SO_2 = POCl_3 + SOCl_2$$

A considerable range of organic reactions has also been conducted in sulfur dioxide.

PROBLEMS

This chapter is intended to extend your consideration of the conditions of chemical reactions from the familiar use of water as a medium to the use of other solvents which have many similarities to water.

Sections 6.1 to 6.3 are intended to bridge on to material normally covered in detail in physical chemistry courses. Only a few examples are given, but many others can be constructed, especially by applying the data of Table 6.3 to equation (6.11).

The material of sections 6.4 to 6.10 is best mastered by making your own correlations among the different solvents, and by relating to the chemistry discussion in later chapters. You should therefore look at the preparations of species of very low or high oxidation states.

6.1 Arrange the oxyacids of sulfur (Table 17.15) in their likely order of acid strengths from their formulae (p. 364).

Does knowledge of the structures alter the position of any of the compounds?

6.2 Find four couples from Table 6.3 which would be completely converted to the reduced form by Cr (but not by Cr^{2+}) going to Cr^{3+}.

6.3 Find three examples in the systematic chemistry chapters of preparations which had to be carried out in a non-aqueous solvent.

7 Experimental Methods

In the discussion of the chemistry of the elements which makes up the later part of this book, the structures of a number of compounds are described. It is the aim of this chapter to give a brief indication of the more important experimental methods of determining these structures, and a description of some of the more recently developed methods of separating compounds is also given. Fuller accounts of the individual methods are given in the references.

Separation Methods

7.1 Ion exchange

The use of clay minerals and zeolites for base exchange and water treatment has been established for many years, but a major advance was made when the use of synthetic resins containing specific functional groups was introduced. The materials in present use are cross-linked polystyrene resins and similar types containing active groups. These are:

sulfonic groups	$-SO_2OH$	giving strongly acid cation exchangers
carboxylic groups	$-COOH$	giving weakly acid cation exchangers
quaternary ammonium groups	$-NR_3^+$	giving strongly basic anion exchangers
amine groups	$-NR_2;$ $-NH_2;$ $-NHR$	giving weakly basic anion exchangers,

where the bond is to the polymer skeleton. The polymer skeleton of the resin acts as an inert, unreactive framework to support the functional groups and has a relatively porous structure with about ten per cent cross-linking.

The sulfonic acid resins* are strong acids and the hydrogen atoms ionize and are readily replaced by cations. If a solution of a salt, say NaX, is passed down a column containing the resin, the Na^+ is replaced:

$$c - ResinH + NaX \rightarrow c - ResinNa + HX$$

In a similar way, the strongly basic resins* containing quaternary ammonium groups have hydroxyl groups which ionize completely and are exchangeable with anions:

$$a - ResinOH + NaX \rightarrow a - ResinX + NaOH,$$
$$or \; a - ResinOH + HX \rightarrow a - ResinX + H_2O$$

This illustrates one important use of these resins, a strong acid cation exchanger used in series with a strong base anion exchanger will remove all dissolved salts from water and is widely used for water-treatment, especially in boilers to prevent scale formation.

Strongly acid cation exchangers in the sodium form will exchange the sodium ions for other cations:

$$c - ResinNa + M^{n+} = c - ResinM + nNa^+$$

These reactions are equilibrium reactions and the affinity of the resin for the cation in solution depends on the charge, the size of the ion, and the concentration. In 0·1 M solutions, the series $Th^{4+} > Fe^{3+} > Al^{3+} > Ba^{2+} > Pb^{2+} > Sr^{2+} > Ca^{2+} > Fe^{2+} > Co^{2+} > Mg^{2+} > Ag^+ > Cs^+ > Rb^+ > NH_4^+ = K^+ > Na^+ > H^+ > Li^+$ has been established, but in concentrated solutions the effect of valency is reversed and univalent ions are favoured over multi-valent ones. Thus, the sodium form of the resin may be used to remove, say, calcium ions from a dilute solution and the calcium can be recovered by treating the resin with a concentrated solution of a sodium salt. The use of a strongly basic anion exchanger for a similar purpose is fairly obvious.

The weakly acid cation exchangers behave as insoluble weak acids. They may be buffered so that exchange takes place at a controlled pH. In acid solution they are undissociated and have little exchange capacity, but in neutral or alkaline media they behave similarly to the strong acid resins as cation exchangers, being rather more selective for divalent cations. The hydrogen form is useful to produce a weak acid from one of its salts:

$$ResinH + CH_3COONa \rightarrow ResinNa + CH_3COOH$$

*Cation-exchange resins are abbreviated as c-resins and anion exchange resins as a-resins.

Similar remarks apply to the weakly basic anion exchange resins.

The uses of the ion-exchange materials will be fairly obvious from the above outline of their properties. In preparations or analysis, any specific ion may be replaced by another one or by a hydrogen or hydroxyl ion. In the latter cases the liberated acid or alkali may be titrated to determine the quantity of cation or anion respectively in the original solution. An example of a preparative application is the use of a strongly basic anion exchanger to prepare carbonate-free alkali metal hydroxides. A solution of, say, sodium hydroxide contaminated with carbonate is run down a column containing a strong base anion exchanger. This has a higher affinity for the doubly-charged carbonate ion which is then replaced by hydroxyl ion from the column. Pure sodium hydroxide solution is recovered.

An obvious extension of this technique is the removal of interfering ions in analysis, and the resins may also be used to concentrate trace constituents. The resins also find use as catalysts, especially as acid or base catalysts, and in the determination of dissociation constants and activity co-efficients. As the exchange process is an equilibrium, the resins may be used in the separation of isotopes. For example, a separation of ^{14}N and ^{15}N has been effected by the exchange between a cation exchange resin and ammonium hydroxide:

$$Resin^{14}NH_4^+ + {}^{15}NH_4OH = Resin^{15}NH_4^+ + {}^{14}NH_4OH$$

A band of NH_4^+ travelling down a long column of the resin by means of a large number of absorption-desorption steps gradually becomes enriched with $^{15}NH_4^+$ at the trailing edge and with $^{14}NH_4^+$ at the leading edge. By the time the band is extended to about forty times its original width, the tail fraction contains 99 per cent ^{15}N.

7.2 Chromatography

Chromatography is a process for separating a mixture which depends on the redistribution of the components between a stationary phase and a mobile one. The components may be adsorbed on the stationary phase or be held by more specific chemical bonds. The typical arrangement has the stationary phase in a column and the mixture is passed up or down the column in the gas or liquid phases. One form of chromatography involves the use of ion exchangers and follows on from the discussion in the previous section.

Suppose that a mixture of cations in solution have similar chemical properties, they will not be separated from each other by the simple ion exchange processes discussed above. If the cations are absorbed at the top of a cation exchange column and then treated with a weak complexing agent—one with an affinity for the cation similar to that between the cation and the column material—then the equilibrium,

cation on ion exchanger + complexing agent
$\rightleftharpoons$ cation in complex in solution,

will depend sensitively on the nature of the cation. If the band of mixed cations on the column is washed down (*eluted*) by

a solution of the complexing agent, those cations which form the strongest complexes will spend more time in solution than on the column, and will travel down the column faster than the cations whose complexes are weaker and which remain longer on the ion exchanger. As the cation band travels down the column, it starts to separate into its components and, if the column is long enough, the different components are recovered separately. The process is shown diagrammatically in Figure 7.1. As there are also random

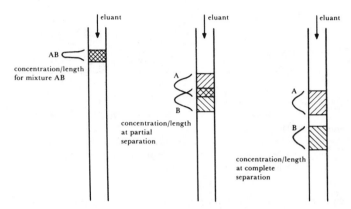

FIGURE 7.1 *Separation of a mixture by column chromatography*
The figure shows schematically the concentration gradients of the two components, expressed as Gaussian distributions, at different stages in the separation.

processes of diffusion affecting the ions, the distribution of cation concentration down the column is Gaussian. The classic example of this method of separation was the separation of the lanthanide elements, using citrate as the weak complexing agent. A similar use of anion exchange resins allows a separation of anions. If the ions are not too closely similar, the separation may be effected without the use of a complexing agent. For example, the halides may be separated on an anion exchange resin by eluting with a sodium nitrate solution, and an additional element of control is introduced by varying the concentration of the eluant.

Many other systems have been devised for chromatographic separation. Materials such as silica gel or cellulose which act by adsorption, or by a mixture of adsorption and chemical interaction with adsorbed water, are widely used, and one common application is in paper chromatography. A wide variety of ions and molecules may be separated by using suitable solvents moving over paper. For example, the alkali metals may be separated by using a mixture of alcohols as the solvent, and the various phosphate anions may be separated using a two-dimensional method with a basic solvent flowing in one direction followed by an acidic one in a direction at right angles, Figure 7.2. The analogous, but more controlled and sensitive, method of *thin layer chromatography* uses a uniform thin layer of silica spread on a glass plate as the separating medium.

Another modification of this method is in the separation of gas mixtures by passing them in a stream of nitrogen, or helium, through a column containing a high-boiling liquid

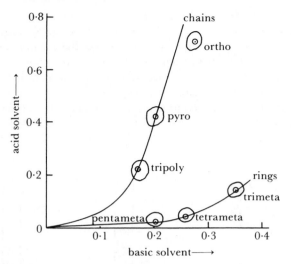

FIGURE 7.2 *Separation of phosphate anions by two-dimensional paper chromatography*
The ring anions fall on one curve and the chain ones on another. Higher members of each type would lie on the prolongations of these curves towards the origin.

supported on an inert material. For example, the silicon or germanium hydrides may be separated by this *vapour phase chromatography* by passing the hydride mixture through a column of silicone oil on powdered brick.

7.3 Solvent extraction

A mixture may be separated into its components by treating it with two or more immiscible liquids, most commonly with an aqueous phase and an organic phase. Ionic and strongly polar species remain in the aqueous phase while less polar species dissolve into the organic liquid. For example, germanium may be separated by forming the tetrachloride in contact with a solvent, such as carbon tetrachloride, in which it is soluble. As the hydrolysis of the tetrachloride is reversible in presence of hydrochloric acid, all the germanium ends up in the organic phase:

$$Ge(IV) \text{ in aqueous solution} + HCl \rightleftharpoons GeCl_4 + H_2O$$
(as oxy- or hydroxy-species)

$$\text{solution in } CCl_4$$

The nature of the solute may be altered by changing the pH, by adding complexing agents, or by adding common ions, while considerable variation in the organic phase is also possible. There is thus a fair chance that any given mixture may be separated by suitably varying the conditions.

Solvent extraction processes are very widely used in industry for the extraction of metals from ores or spent catalyst residues or the recovery of toxic metals from effluent streams. Perhaps the classic example of the application of this technique is the separation of uranium and plutonium from each other and from the fission products that accumulate in an atomic pile. One method involves the solution of the material in nitric acid, in which the uranium is in the VI state and is more covalent than the plutonium, which remains in the IV state. The uranium is extracted

into ether and then the plutonium and fission products are precipitated and strongly oxidized to convert the plutonium to Pu(VI). This can then be extracted from the fission products in turn. A variation, which particularly lends itself to a cyclic process of repeated extraction, is to separate the plutonium and uranium in the VI state into an organic solvent which is then carried into a second vessel where it is subject to mild reduction, in contact with water, which leaves the U(VI) unchanged but converts the plutonium to Pu(III), which is washed back into the aqueous phase by dilute nitric acid.

Three general classes of extractants are recognized, according to their structures and mechanism of extraction. Inert extractants are simple non-coordinating solvents, such as CCl_4 in the case of $GeCl_4$ extraction just described.

Solvating extractants are reagents which have one or more donor atoms available for metal complexation. These complexes allow extraction of the metal into the organic phase, from where it may be recovered. The principles of coordination chemistry are of great importance in designing a new extractant for a given purpose and extractants are therefore basically just ligands which have been specifically designed to make them highly soluble in organic solvents. Thus, extractants containing soft, polarizable donor atoms (see section 13.7 for a detailed discussion on hard and soft acid/base behaviour) will show a strong tendency to extract soft metal atoms. An example is phosphane sulfides, $R_3P{=}S$, used for the recovery of silver from photographic wastes. By contrast, extractants containing hard donor atoms (such as nitrogen or oxygen) typically extract hard metals, and a good example is tri-octyl phosphane oxide, $octyl_3P{=}O$, used in the separation of uranium from plutonium. Instead of solvating the metal itself, metals may often be extracted into organic solvents by solvating an associated proton, as in the case of the extraction of the $H^+FeCl_4^-$ ion pair into ether. Ionic extractants, such as organic soluble ammonium salts R_3NH^+ or R_4N^+, or sulfonium salts R_3S^+, work by a similar mechanism of forming an organic soluble ion-pair with the anionic metal complex, usually a halide, such as $FeCl_4^-$, or $CoCl_4^{2-}$.

The third major class of extractant is the acidic extractant, such as an organophosphorus acid or a carboxylic acid, which functions by replacing the acidic proton by a metal. These acids, designated H–A, typically exist as hydrogen-bonded dimers (see section 9.7 on the hydrogen bond), and in the presence of an excess of the extractant these dimers can persist in the extracted metal complex which can be, for example, of the type $M(HA_2)_n$ (n is typically 2 or 3). In addition, the neutral unionized acid dimer H_2A_2 can also act as a solvating ligand in these complexes. Organophosphorus acids are widely used in processes to separate the individual lanthanide elements, as described in section 11.3. Another type of extractant of this class are hydroxyoximes of the type shown above which are highly selective for the extraction of copper and used on a very large scale worldwide. The extractant functions by deprotonation of the phenolic –OH group; copper then coordinates to the resulting O^- group and the oxime nitrogen, forming a six-membered chelate ring.

Solvent extraction processes are not restricted to the extraction of metallic species and many other inorganic and organic species (such as phenolic compounds from effluents) can be recovered. Another of the classic examples of solvent extraction is the 'wet-acid' process for phosphoric acid purification. In this process crude H_3PO_4, produced by treating phosphate rock with sulfuric acid, is extracted into an organic solvent, such as methyl isobutyl ketone, from which the purified H_3PO_4 is recovered by extraction into water. Food-grade H_3PO_4 can be readily manufactured this way, resulting in considerable energy savings over the old 'thermal' process in which pure white phosphorus (P_4) is burned in air giving P_4O_{10} which is then dissolved in water giving H_3PO_4 (see section 17.6).

All these solvent extraction processes may be adapted for the separation of very similar solutes by using a continuous flow method, when fresh organic phase is brought into contact with the partially extracted aqueous phase and *vice versa*. This 'counter-current' method corresponds basically to a chromatographic separation involving two mobile liquid phases instead of one mobile phase and one held stationary on an absorbing phase. The close similarity of all these methods means that they may often be used interchangeably and the problem of separating chemically similar elements or compounds can now be treated by very powerful and versatile methods.

Structure Determination

7.4 Diffraction methods

The most accurate and powerful method of determining the structure of solids is by X-ray diffraction. Just as light can be diffracted by a grating of suitably spaced lines, so can the much shorter wavelength X-rays be diffracted. The regular array of atoms or ions in a crystalline solid provides a suitably spaced three-dimensional grating for the diffraction of a beam of X-rays, and the diffraction pattern can be detected by using a photographic film or other detector. When a beam of X-rays passes through a crystal it meets various sets of parallel planes of atoms. The diffracted beams from atoms in successive planes cancel unless they are in phase, and the condition for this is given by the Bragg relationship:

$$n\lambda = 2d\sin\theta$$

where λ is the wavelength of the X-rays, d is the distance between successive planes, and θ is the angle of incidence of the X-ray beam on the plane. From a set of distances d for different sets of planes in the crystal, the positions of the atoms can be derived. The angles θ can be measured and the values of d can be calculated if the wavelength of the X-rays is known. If the intensities of the diffracted beam in each direction can also be measured, complete structure determinations are possible. As each atom in the lattice acts as a scattering centre, the total intensity in a given direction of the diffracted beam depends on how far the contributions from individual atoms are in phase. The essential feature of complete X-ray structure determinations is a trial-and-error search for that arrangement of atoms which best accounts for the observed intensities of reflections. For large molecules, or structures of low symmetry, this process involves numerous calculations and is best carried out by computer.

There are a number of experimental methods for studying X-ray diffraction but two are widely used. One uses a single crystal of the compound which is mounted so that it can be rotated about a crystal axis. A monochromatic beam of X-rays (i.e. of a single wavelength λ) shines on the crystal. As it rotates, successive sets of planes are brought into reflecting positions and the reflected beams are recorded as a function of the crystal rotation. The positions and intensities of the reflected beams are measured and the complete structure can be derived from the resulting data.

A second method is used for the many cases when substances are available only as a crystalline powder and no single crystal is obtainable. The powder is packed in a thin capillary tube and illuminated by a monochromatic beam inside a circular camera. All possible crystal orientations occur at random, so that some crystallites will be correctly orientated to fulfil the Bragg condition for each value of d. The reflections are recorded as lines on a film placed round the inside of the camera and these positions give the d values. A complete analysis is possible only for crystals of high symmetry but the method is invaluable as a qualitative technique, e.g. in mineral analysis.

Although X-ray methods give the fullest and most reliable data, they are restricted to solids and are not very suitable for determining the positions of very light atoms. Two other diffraction methods are available, electron diffraction and neutron diffraction. By virtue of the wave properties of electrons, a beam of electrons may be diffracted. Electron diffraction is used to determine the structures of gaseous molecules and gives very accurate values for bond lengths and angles. The process of interpretation depends on matching the observed diffraction pattern with ones calculated for different model structures, so that it is not fully reliable as a method of structure determination but does give accurate parameters for molecules whose overall structure is known. Electron diffraction is a function of the atomic number and hydrogen atoms are, as a rule, not detectable.

A beam of neutrons behaves in a similar manner to a beam of electrons and gives information about the positions of light atoms and is thus complementary to the other two techniques. Neutron diffraction determines the positions of the nuclei, and the sensitivity does not vary much between elements. X-rays are diffracted by the electrons, and thus X-ray diffraction is dominated by the positions of largest electron density. In compounds which contain both light and heavy atoms, X-ray scattering from the low electron density around the light atom may not be detectable and only heavy atom positions are found. If a neutron diffraction is carried out as well, the heavy atom positions from the X-ray study allow a ready solution, and the light atom centres are easily determined. This is particularly so for hydrogen compounds, as H, and even more D, scatters neutrons efficiently. In addition, distinguishing between two adjacent atoms in the Periodic Table, such as iron ($Z = 26$) and cobalt ($Z = 27$), can be difficult by X-ray diffraction, but the different neutron scattering abilities of these two atoms makes distinguishing them by neutron diffraction a relatively straightforward matter. The disadvantages of the neutron diffraction method are the inaccessibility of neutron sources (commonly an atomic pile) and the requirement for large single crystals. Indeed, the difficulty of preparing crystals without twinning or other flaws, is the main limitation on both X-ray and neutron diffraction which are otherwise the most powerful structural tools.

Since the diffraction pattern has contributions from every atom in the unit cell, it becomes rapidly more difficult to solve the structure as the molecular size increases. While there have been tremendous successes in recent work on large biomolecules containing metal atoms, analysis is very slow and many systems are known only to relatively low spatial resolution, say at the 5 Å level.

In such cases, a further X-ray method is valuable. This is the extended X-ray absorption fine structure or EXAFS experiment. When an X-ray photon is absorbed by an atom (the plot of X-ray absorption rises steeply at this point giving the *X-ray absorption edge*) an electron is expelled. This may be back-scattered by neighbouring atoms, giving a fine structure on the absorption edge which gives information about the nearest neighbours of the absorbing atom. The technique is particularly valuable in probing large molecules or non-crystalline materials.

To illustrate, a recent study involved $\{[(RO)_2P(S)_2]_4Zn_2\}$ which has a known structure containing $Zn\{(S)_2P_2\}_2$ units. This compound is typical of antioxidants for engine oils. Their mode of action is difficult to investigate as the oxidized products are not crystalline. An EXAFS study showed the presence of O with four Zn nearest neighbours followed by six PS_2 groups and an indication of three Zn third neighbours. This suggests a central O bonded tetrahedrally to 4 Zn, bridged on each edge by a S—P—S unit (presumably retaining some of the P—OR ligands) and probably linked in turn to further OZn groups (compare Figure 10.9). Such information suggests how the antioxidant works and allows design of improvements.

7.5 Spectroscopic methods and the electromagnetic spectrum

If an atom, molecule, or ion, with two energy states differing in energy by ΔE is irradiated with continuous electromagnetic radiation, the radiation of frequency corresponding to ΔE will be absorbed and the species raised to the upper energy state. This absorption, or the consequent emission of radiation as the species returns to the ground state, may be detected and provides information about the energy states. The energy change is related to the frequency v' of the radiation by the relation

$$\Delta E = hv' \dots\dots\dots\dots\dots (7.1)$$

where h is Planck's constant equal to $6.625\,6 \times 10^{-34}$ J s. The frequency is related to the wavelength, λ, as their product equals the speed of light

$$v'\lambda = c = 2.998 \times 10^{10} \text{ cm s}^{-1} \dots\dots\dots (7.2)$$

Thus the energy change per mole obeys the relations

$$\Delta E = Nhv' = Nhc/\lambda = Nhc\tilde{v} \dots\dots\dots (7.3)$$

where $\tilde{v}$ is called the *wave number* or Kayser and is the number of cycles per centimetre, hence the unit is cm^{-1}. Equation (7.3) leads to the relation between different units given on pp. xiv and 119.

The electromagnetic spectrum spans a very wide range, from gamma and cosmic rays with energies in excess of 10^6 kJ mol^{-1} down to radiowaves corresponding to small fractions of a joule per mole. In different regions of the spectrum, the energy corresponds to differences in two states of a system spanning many kinds of transition. Thus a nuclear transformation involves very high energies, in the gamma ray region, while the reversal of an electron spin in a magnetic field of 0·1 tesla would involve a tiny energy corresponding to the shorter radiowaves. Nevertheless, both these processes, and many others of intermediate energy, may be used to yield information of value to the chemist. The major regions of the electromagnetic spectrum, and the transitions corresponding to the interaction with them, are listed in Table 7.1. There is no clear boundary between any two regions and the energy ranges and types of transition overlap: we give arbitrary round-figure ranges (which are thus slightly different for different units) whose boundaries are often set by arbitrary experimental factors. As there is such a wide range of energies, and so many different types of transition are involved, there have grown up a number of separate areas of study which have developed fairly independently and with their own conventions. In particular, this means that a variety of units have been used and some of the traditional ones in the important regions are indicated in Table 7.1. Notice that these include units of energy, wavelength, and frequency.

The gamma and X-ray regions of the spectrum give information about the nucleus and the inner, closed shell, electrons and are therefore of less direct interest to the chemist although the Mössbauer effect which involves gamma resonance absorption, and the derivation of atomic energy levels from X-ray spectra, as in Moseley's experiments, are important.

The average chemist is most likely to make direct measurements of *electronic spectra* in the ultraviolet and visible region, of *vibrational spectra* in the infrared region and in the Raman effect, and of *nuclear magnetic resonance (nmr) spectra* in the

TABLE 7.1 Principal regions of the electromagnetic spectrum

Region	Approximate range in			Transition excited
	Wavelength (m)	Energy		
		(SI)	(Commonly-used)	
gamma rays	$< 10^{-10}$	$> 10^6$ kJ mol^{-1}	$> 10^4$ eV	nuclear transformations
X-rays	10^{-8} to 10^{-10}	10^4 to 10^6 kJ mol^{-1}	100 to 10^4 eV	transitions of inner shell electrons
ultraviolet	4×10^{-7} to 10^{-8}	10^2 to 10^4 kJ mol^{-1}	1 to 100 eV or	transitions of valence shell
visible	8×10^{-7} to 4×10^{-7}		10^4 to 10^6 cm^{-1}	electrons including $d \rightarrow d$ and $f \rightarrow f$
infrared	10^{-4} to $2 \cdot 5 \times 10^{-6}$	1 to 50 kJ mol^{-1}	100 to 4000 cm^{-1}	molecular vibrations
microwave and far infrared	10^{-2} to 10^{-4}	10 to 1000 J mol^{-1}	1 to 100 cm^{-1}	molecular rotations
radio frequency	$\sim 10^{-2}$	~ 10 J mol^{-1}	3×10^4 MHz	electron spin reversal in magnetic field of 1 A m^{-1}
	~ 10	$\sim 0 \cdot 01$ J mol^{-1}	10 to 100 MHz	nuclear spin reversal in magnetic field of 1 A m^{-1}

1 eV = 96·49 kJ mol^{-1} = 23·06 kcal mol^{-1} and is equivalent to 8068 cm^{-1} and $2 \cdot 419 \times 10^8$ MHz: 1 A m^{-1} = 10^4 gauss
Ranges are rounded-off and are not converted exactly from one unit to the next.

radiofrequency region. These methods are therefore discussed in detail in the following sections. Pure rotational spectra, studied in the microwave region, give accurate values for the moments of inertia and thus provide values for bond lengths and angles for sufficiently small or symmetric molecules which must also have a permanent dipole moment to interact with the radiation. By using isotopic substitution, a number of independent parameters may be determined, but even so, only a limited number of molecules are simple enough to be analysed. Thus microwave spectroscopy gives information complementary to that derived from electron diffraction for small molecules in the gas phase. Electron spin resonance which is observed for species with unpaired electrons is also a tool available for only a restricted number of species. Rotational changes often accompany vibrational changes and are seen as a fine structure to the vibrational absorptions, and similarly, electronic transitions may have fine structures due to concomitant rotational and vibrational transitions.

Study of ultraviolet, visible and infrared spectra requires the same basic equipment. The sample must be placed in a beam of radiation which can be continuously varied in frequency and a detector is required to show absorption of energy. For example, visible spectra require a source of white light which is scanned in frequency by using a rotating prism, while the detector is a photocell whose output may be converted to a movement of a pen on a recorder. An infrared spectrum may be recorded using a heated element as source, an alkali halide prism or a grating to change the frequency, and a thermocouple as detector.

Vibrational transitions may also be detected in the Raman effect. In this, the sample is illuminated with strong monochromatic radiation, and some of the re-emitted quanta are found to have gained or lost a (smaller) quantum correspond-

ing to the energy of one of the fundamental vibrational modes. The spectrum of the scattered radiation thus consists of a very strong line corresponding to the incident radiation and a number of other lines whose energy differences from the primary line give the energies of vibrational transitions of the molecule. Raman spectroscopy requires an intense radiation source with a very narrow frequency spread. The laser is ideal and Raman spectroscopy has greatly expanded in scope since the advent of lasers.

7.6 Electronic spectra

Transitions of outer shell electrons fall in the general wave number range of 100 000 cm^{-1} to 10 000 cm^{-1}, that is in the ultraviolet, visible, and near infrared regions of the electromagnetic spectrum. The transitions involved are those between sigma, pi, or nonbonding, n, orbitals in the valence shell, such as those involved in Figures 4.25 or 4.29. Not all transitions are allowed, only

$$\sigma \rightarrow \sigma^*$$
$$n \rightarrow \sigma^*$$
$$\pi \rightarrow \pi^*$$
$$n \rightarrow \pi^*$$
$$\delta \rightarrow \delta^*$$

while sigma to pi or pi to sigma transitions are forbidden. It is clear from an energy level diagram like Figure 4.29 that this is the approximate order of decreasing energy. The largest energy is required for $\sigma \rightarrow \sigma^*$ transitions and these are found in the far ultraviolet regions in which the molecules of the atmosphere also absorb. Evacuated spectrometers have to be used and this region is less studied.

Most work is done in the 50 000 to 10 000 cm^{-1} region, 200 nm to 1 μm, where transitions involving nonbonding and pi electrons are involved. For lighter atoms, such transitions are usually in the ultraviolet, giving colourless compounds, but extended conjugation, giving rise to a large number of closely spaced pi orbitals, may lower the energy of the transition and give a coloured compound. In addition, heavy atoms have their outer orbitals closer together in energy so that the transitions again fall into the visible region. Such transitions account for the colours of many iodides, for example. In addition, transition elements with partly occupied d or f orbitals show bands, usually in the visible region, due to $d \rightarrow d$ or $f \rightarrow f$ transitions. Such transitions are formally forbidden and give rise to weak bands: they are discussed in more detail in Chapters 11 and 13.

While the position of an absorption band corresponds to the energy of the transition, its intensity depends on the nature and quantity of the absorbing material. The relation between the *absorbance*, A, the pathlength l and the concentration C is given by the Beer-Lambert law

$$A = kCl \qquad (7.4)$$

where k is a constant characteristic of the material. For a pathlength of 1 cm, and with the concentration expressed in moles per unit volume, the constant k becomes ϵ, the *molar extinction coefficient*. Allowed transitions generally have molar extinction coefficients in the range 10^3 to 10^5 l mol^{-1} cm^{-1} while d–d or f–f bands are much weaker with ϵ typically 10^{-1} to 10 l mol^{-1} cm^{-1}. The absorbance, formerly termed the optical density, is defined as

$$A = \log (I_0/I) \qquad (7.5)$$

where I_0 is the intensity of the incident light and I the intensity after the light has passed through the sample. Most modern spectrometers record the absorbance directly.

Electronic spectra are used primarily to give information about the energy levels of the valency orbitals, and for inorganic chemistry this is particularly widely studied for transition elements. Use may also be made, through Beer's law, of absorbance measurements in quantitative analysis. Here, measurements are preferably made on strong bands and at, or near, maxima in the absorption. Thus, for example, the strong allowed bands of CrO_4^{2-} would be used for estimating Cr, rather than the weak d–d transitions of Cr^{3+} ions. (Note that species like chromate or permanganate, which are in the Group oxidation state, have no d electrons not involved in bonding so that these colours are not due to d–d transitions.) In a more qualitative way, electronic spectra may also be used to indicate the presence of particular groupings of atoms. For example, π–π^* transitions in a benzene ring will be relatively constant in position and intensity as long as π bonding substituents are absent: thus the presence of a phenyl group would be indicated by such bands in the electronic spectrum. Such a use is more common in organic chemistry where such a chromophore is likely to occur in a fairly constant environment, but similar applications are of value particularly in organometallic chemistry.

7.7 Vibrational spectra

Vibrational modes of a molecule are excited by the absorption of quanta whose energies lie in the infrared region of the spectrum, from about 4000 cm^{-1} downwards. Vibrational transitions are also detected in Raman scattering. As the selection rules for infrared absorption differ from those governing Raman scattering, the two techniques are complementary and both infrared and Raman spectra need to be measured to obtain the maximum amount of information.

The information obtainable from vibrational spectroscopy depends on the size and symmetry of the molecule. For a diatomic molecule, assuming simple harmonic motion, the wavenumber is given by

$$\tilde{v} = \frac{1}{2\pi c} \sqrt{\frac{k}{u}} \qquad \ldots\ldots\ldots\ldots (7.6)$$

where k is the force constant (the proportionality between the extension of the bond and the restoring force) in N m^{-1} and u is the reduced mass ($1/u = 1/m_1 + 1/m_2$) of the two atoms. Thus the vibrational frequency is directly related to the force constant, which in turn is related to the bond strength. Absorptions of successive quanta of vibrational energy will continue until the molecule dissociates, and the frequency at which this occurs gives the bond energy. From the rotational fine structure, the moment of inertia may be derived and hence the bond length (if the atomic masses are known). The existence of isotopes of an element may be proved by observing different moments of inertia for the same compound. For example, hydrogen chloride is found to have a bond length of 128·1 pm and two moments of inertia corresponding to $H^{35}Cl$ ($I = 2\cdot649 \times 10^{-47}$ kg m^2) and $H^{37}Cl$ ($I = 2\cdot653 \times 10^{-47}$ kg m^2).

For polyatomic molecules, the position is more complicated. Vibrations involve not only bond stretching, but angle deformation and often twisting modes as well. An n-atom species has $3n - 6$ degrees of vibrational freedom ($3n - 5$ for a linear species) and a corresponding number of force constants are required to describe the vibrations. However, it is unlikely that all $3n - 6$ vibrations will be observable in the majority of cases. This is partly due to practical difficulties of detecting weak bands and resolving closely overlapping ones, and partly due to degeneracy. For example, a tetrahedral molecule like GeH_4 has $3 \times 5 - 6 = 9$ modes of vibration but the maximum number of bands observable in the infrared is two and in the Raman, four. This is because two of the modes are degenerate, giving only one fundamental, and two further groups of three are triply degenerate. Thus there are only four observable bands, two triply degenerate, one doubly degenerate and one non-degenerate and these are found in the Raman spectrum. Of these, only the triply degenerate modes involve a dipole change so that only these two are also observed in the infrared. Analysis in terms of force constants is thus difficult, though isotopic substitution may help.

For small and moderate-sized molecules, the number of bands expected in the vibrational spectrum is predictable from the symmetry of the molecule by the methods of *group theory* (this leads to the prediction above for GeH_4) and thus the spectrum may be used to determine which one out of a number of structures is the correct one. For example, a square planar AB_4 species has 3 infrared active bands (two of which are doubly-degenerate) and three Raman bands which do not coincide with the infrared modes. The ninth vibration is inactive in both the infrared and Raman. If this is compared with the prediction for a tetrahedral AB_4 species given above, it will be seen that these two shapes could readily be distinguished unless the intrinsic intensities or spacings were very unfavourable. In a similar way, the two possible structures for a species like perchloryl fluoride could be distinguished, as the $FClO_3$ structure, based on a tetrahedron, would have six bands active in both infrared and Raman (as three are doubly-degenerate) while the hypofluorite form, $FOClO_2$, has no degenerate bands and all nine vibrations would be seen in both infrared and Raman.

It should be noted that structural evidence of this sort may be used to support a structure but does not offer absolute proof. As some bands are inevitably difficult to observe, the assignment of six bands for perchloryl fluoride does not prove the first structure, as it is quite possible that the remaining three would be too weak or too close to others to be observed. The negative argument is much more definite —observation of seven or more fundamentals *would* disprove the first structure. Similarly, a species AB_4 showing three or more fundamentals in the infrared could not be a tetrahedron. These examples show the type of evidence that may be obtained by applying arguments based on the molecular symmetry, using group theory. The full discussion of these methods is beyond our scope, but the first step is to determine the molecular symmetry and this is discussed in Appendix C.

It will also be seen from these examples how valuable it is to be able to observe vibrational spectra both in the infra-red and in the Raman. Often different modes are active in different effects. The extreme example is provided by species which have a centre of symmetry (such as square planar AB_4) where no modes are both Raman and infrared active. Even where all the expected modes are active in both effects, it is likely that bands which are weak in the infrared will be strong in the Raman, and *vice versa*. For example, stretching modes involving similar heavy atoms are often weak in the infrared, as the dipole change is small, but are strong in the Raman because a fairly extended electron cloud is moving. Thus the $Si-Ge$ stretching mode in H_3SiGeH_3, which is allowed in both the infrared and the Raman, is too weak to be observed in the infrared but gives a strong Raman band at about $350 \, cm^{-1}$. A similar effect is found for many metal-metal stretching modes.

In more complicated species, this approach breaks down because the number of predicted bands becomes so large that detailed assignment is impossible. The vibrational spectrum is still useful, but in a more qualitative way. First, certain groups in a molecule may absorb in fairly constant regions of the spectrum and can thus be identified. For example, a CN group, with its triple bond, absorbs at about

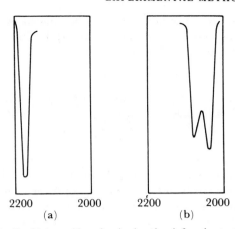

2200	2000	2200	2000
(a)		(b)	

FIGURE 7.3 $C-N$ *stretching bands in the infrared spectra of* (a) $Ni(CN)_4^{2-}$, *and* (b) $Ni_2(CN)_6^{4-}$ *ions*
The CN absorption occurs in a similar region for both compounds (a) at $2124 \, cm^{-1}$ and (b) at $2045 \, cm^{-1}$ and $2075 \, cm^{-1}$. The difference in symmetry between the two ions is reflected in the number of absorptions.

$2000 \, cm^{-1}$, a far higher wave number than modes involving single bonds and heavier atoms. Thus the presence of cyanide as a ligand may always be detected, and similarly of CO in metal carbonyls. Indeed, symmetry information may be derived from the number of bands in the $2000 \, cm^{-1}$ region, to distinguish *cis* and *trans* $ML_3(CN)_3$ for example, as these vibrations are little affected by the presence of other groups (compare also Figure 7.3). From the mass effect, modes involving hydrogen are also found at higher frequencies than those involving any other substituent and these also are readily distinguished. Such partial analyses are often valuable, and much information may be derived by comparing related species. Thus, the $Cl-F$ and $Cl-O-F$ alternatives for ClO_3F above, might be further distinguished by comparison with chlorine fluorides and other hypofluorite species to see if bands appear in regions characteristic of $Cl-F$ stretching or of $O-F$ stretching.

Finally, a purely qualitative approach may be made in which a complex species may be identified with a known compound if their spectra are identical. In this case, the more complex the spectrum, the more definite would be the identification.

7.8 Nuclear magnetic resonance (nmr)

When an atom, such as hydrogen, with a nuclear spin of $\frac{1}{2}$ is placed in a magnetic field, the spin may take up one of two orientations, either parallel or anti-parallel to the field vector. These two correspond to different energy states and, for hydrogen in a field of about one tesla (10^4 gauss) the energy difference is about $10^{-2} \, J \, mol^{-1}$. This corresponds to a frequency of about 40 MHz. In modern instruments, higher fields are common and proton resonance is commonly run at 60, 90 or 100 MHz with iron core magnets or at fields from 220 to 750 MHz using superconducting magnets. In the experimental arrangement commonly used, the sample is placed in a cylindrical tube in the field and irradiated with a fixed frequency. The field is slightly modulated by passing a current through coils and energy absorption is detected at resonance when

$$hv = \Delta E = g_N \beta_N B \ldots \ldots \ldots \ldots (7.7)$$

Here B is the magnetic flux density, β_N is the nuclear magneton $= eh/2\pi Mc = 5\cdot0505 \times 10^{-27}$ J T^{-1} and g_N is a constant, called the nuclear g factor, which is a characteristic of the element.

The great value of nmr to the chemist is that the magnetic field actually experienced by a particular nucleus is the sum of the applied field and fields induced in the electrons around the magnetic nucleus. Thus atoms of the same element which are in different chemical environments resonate at slightly different values of the external field and these differences in *chemical shift* may be detected and yield structural information. In other words, the position of the absorption depends on the atom to which the hydrogen is bonded and on the other bonds nearby. The classic case is that of ethanol, CH_3CH_2OH, where the absorption due to the hydrogen bonded to oxygen is in a different position to the absorptions of the carbon-bonded hydrogens. Furthermore, the hydrogens in the methyl group are in a different electronic environment to those in the methylene (CH_2) group, so the absorptions due to the two types of carbon-bonded hydrogens are separated. The three methyl protons are in identical environments, as are the two methylene ones, and the resulting spectrum consists of three peaks, in the ratio of $3:2:1$, for the three types of hydrogen atoms, and in positions typical of methyl, methylene, and hydroxyl hydrogens. An example of the ^{1}H nmr spectrum of an inorganic hydride, $SiH_3GeH_2GeH_3$, is shown in Figure 7.4. There are three main envelopes of absorption, each showing fine structure—see below—in intensity ratio $3:3:2$ which arise from the SiH_3, GeH_3 and GeH_2 protons respectively.

Such detailed information about the environment of an atom may be sufficient to determine the structure, or at least goes a long way towards this. For example, diborane B_2H_6 (cf. section 9.6), has had a number of structures proposed for it, including an ethane-like one (A) and a bridged structure (B).

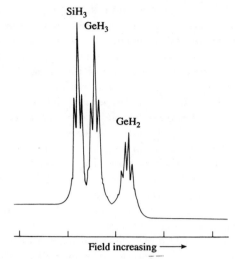

FIGURE 7.4 *The ^{1}H nuclear magnetic resonance spectrum of* $SiH_3GeH_2GeH_3$

The resonances are, in order of increasing field (1) the triplet due to the SiH_3 protons split by coupling to the two GeH_2 protons, (2) the triplet due to the GeH_3 protons, also split by the GeH_2 coupling, and (3) the multiplet due to the GeH_2 protons split by coupling both to the SiH_3 and to the GeH_3 protons. Notice that the coupling constant, $\mathcal{J}(SiH_3GeH_2)$ is almost identical to the constant $\mathcal{J}(GeH_3GeH_2)$, shown by the near-identity of the splittings in the two triplets.

with higher nuclear spins, although the spectra are more complicated in the latter cases. No resonance is possible when atoms have zero spin and this includes some common atoms such as ^{12}C and ^{16}O. Hydrogen is the easiest nucleus to observe but nmr studies have now been extended to practically all the suitable isotopes in the Periodic Table. Modern techniques which involve observation by pulsed irradiation followed by Fourier transformation have greatly enhanced sensitivities. Wide-ranging studies on lighter nuclei include ^{2}H (deuterium), ^{11}B, ^{13}C, ^{14}N, ^{19}F, ^{29}Si, and ^{31}P. Of these, carbon-13 is the most widely studied after hydrogen, and then phosphorus. One example of fluorine resonance is in the confirmation of the square pyramid structure of BrF_5 and IF_5 by the observation of two lines in the fluorine resonance, with intensities in the ratio $4:1$, corresponding to the basal and apical fluorines respectively.

Further information may be obtained from the fine structure of the nmr bands which arises from the effects of spin-spin coupling. If an atom with a nuclear spin is bonded to a second one which also has a nuclear spin, the local magnetic field will be affected by the orientation of the spin of the second nucleus. For example, in PH_3 the phosphorus-31 nucleus, which has a spin of $\frac{1}{2}$, may be aligned with or against the field, giving two different local resultant fields. As the energy difference between the two orientations is so small, there will be essentially equal numbers of molecules with each P spin alignment. Thus, in the proton resonance signal there will be two components corresponding to hydrogens bonded to P atoms with spins parallel or antiparallel to the external field. Thus the proton resonance

(A)   (B)

The hydrogen magnetic resonance spectrum of (A) would consist of only one line, as all the hydrogen atoms are equivalent, whereas the hydrogen spectrum of (B) would have two lines, in the ratio of $4:2$, corresponding to the terminal and bridge hydrogens respectively. The latter spectrum is observed, supporting the bridge structure.

Nuclear resonance may be observed for atoms other than hydrogen which have a spin of one half, and also for atoms

signal is a doublet with components of equal intensities. The phosphorus atom also shows a resonance signal, though at much lower frequency, and this is split by the spins of the three H atoms into four components. There are four possible arrangements of the proton spins—all parallel, which we can label $+\frac{1}{2}$, $+\frac{1}{2}$, $+\frac{1}{2}$, or with two, one or no spins parallel to the field and respectively one, two or three antiparallel spins, labelled $+\frac{1}{2}$, $+\frac{1}{2}$, $-\frac{1}{2}$: $+\frac{1}{2}$, $-\frac{1}{2}$, $-\frac{1}{2}$: $-\frac{1}{2}$; $-\frac{1}{2}$, $-\frac{1}{2}$. There are thus four different net fields and four components to the signal. In addition, while the $+\frac{1}{2}$, $+\frac{1}{2}$, $+\frac{1}{2}$ and $-\frac{1}{2}$, $-\frac{1}{2}$, $-\frac{1}{2}$ arrangements can only result in one way, any one of the three protons may be the antiparallel one in the other two combinations. Thus, a net spin of $+\frac{1}{2}$ results from any one of the three sets $+\frac{1}{2}$, $+\frac{1}{2}$, $-\frac{1}{2}$ or $+\frac{1}{2}$, $-\frac{1}{2}$, $+\frac{1}{2}$ or $-\frac{1}{2}$, $+\frac{1}{2}$, $+\frac{1}{2}$, and similarly for a net spin of $-\frac{1}{2}$. As all possible spin combinations are equally probable, there will be three times as many molecules in the sample with net spin $+\frac{1}{2}$ or $-\frac{1}{2}$ as with net spin 3/2 or $-3/2$. Thus the phosphorus signal becomes a quartet with relative intensities $1:3:3:1$. In a similar way, the signal of any atom bonded to n equivalent atoms of spin $\frac{1}{2}$ becomes an $(n+1)$ multiplet with intensities in the ratio of the binomial coefficients given by Pascal's triangle, Figure 7.5. An example of the application of this is shown in Figure 7.6 for the ^{31}P and ^{19}F nuclear magnetic resonance spectra of the octahedral PF_6^- ion. Thus, the ^{31}P nmr spectrum appears as a seven line multiplet, due to the phosphorus coupling to six equivalent fluorines, with the intensities of the lines $1:6:15:20:15:6:1$. The ^{19}F spectrum shows the expected doublet. It is worth noting that for resonances of this type having a large number of lines, the outer lines are relatively weak and could lead to incorrect assignment of the spectrum if the outer lines are not identified (for example, due to poor signal-to-noise in the spectrum).

Such spin-spin coupling is not limited to atoms which are directly bonded, as it is transmitted via the bonding electrons, and it may be observed for atoms separated by several bonds. Thus the CH_3 and CH_2 protons in ethanol couple to make the methyl signal a triplet (1:2:1 intensity ratio) and the

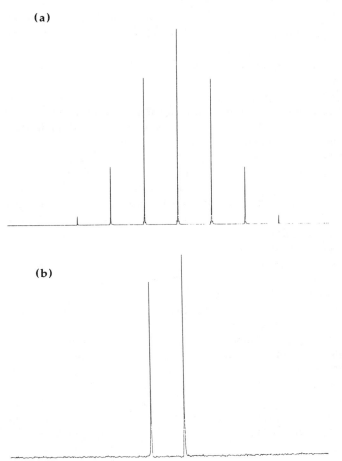

(a)

(b)

FIGURE 7.6   (a) ^{31}P and (b) ^{19}F nmr spectra of the octahedral PF_6^- ion The ^{31}P spectrum shows the expected 7 line multiple with peak intensities in the ratio $1:6:15:20:15:6:1$, see text. The splitting between the two lines in the ^{19}F spectrum is identical to the splitting between any adjacent pair of lines in the ^{31}P spectrum, this splitting being the one-bond $^{31}P-^{19}F$ coupling constant (708 Hz).

methylene signal a quartet. This latter may be further split by coupling to the OH proton. Similarly, the SiH_3 signal in $SiH_3GeH_2GeH_3$ is split into a 1:2:1 triplet by coupling to the GeH_2 protons and the GeH_3 signal is also a triplet for the same reason. The GeH_2 signal is a more complex multiplet as these protons are coupled both to the SiH_3 ones and to the GeH_3 ones. The resultant fine structure is seen in Figure 7.4.

The magnitude of the splitting of the lines caused by spin-spin coupling, i.e. the coupling constant $\mathscr{J}$, often holds a substantial amount of information for the chemist. A superscript is often used to denote the number of bonds over which the coupling is occurring, for example the one-bond phosphorus-31 to fluorine-19 coupling constant, $^1\mathscr{J}(^{31}P-^{19}F)$, is 708 Hz. Many factors can influence the magnitudes of coupling constants, such as the percentage of s-character in the bonding orbital, and three-bond couplings often obey a very strong torsion angle dependence. This is becoming of increasing importance in inorganic chemistry and while we do not have the space for a comprehensive discussion here, the

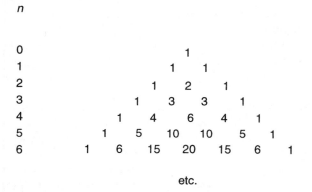

n

0						1						
1						1		1				
2					1		2		1			
3				1		3		3		1		
4			1		4		6		4		1	
5		1		5		10		10		5		1
6	1		6		15		20		15		6	1

etc.

FIGURE 7.5   *Pascal's triangle for predicting the intensities of lines in a multiplet formed by spin-spin coupling of a spin-1/2 nucleus with n identical spin-1/2 nuclei.*

reader is referred to the many nmr spectroscopy texts which cover this, in particular the book by Ebsworth, Rankin and Cradock (see Appendix A).

The magnitudes of spin-spin coupling constants prove to be very useful in the study of coordination complexes, and a good example comes from the study of square-planar platinum–phosphane complexes of the type $PtX_2(PR_3)_2$ (see section 15.8 for a discussion of platinum and palladium chemistry). First of all, however, we need to say something about the spectra of platinum-containing complexes, which are slightly more complicated since not all of the platinum nuclei are spin-active and the resulting observed spectrum is therefore composed of two sub-spectra. This is illustrated in Figure 7.7 which shows the typical ^{31}P nmr spectrum of the complexes $PtX_2(PR_3)_2$. There are several isotopes of platinum but the only one which is nmr-active is ^{195}Pt, with spin $= \frac{1}{2}$ and making up approximately 33% of platinum. When ^{31}P is bonded to ^{195}Pt the nmr spectrum shows a doublet due to spin-spin coupling. For the remaining 67% of ^{31}P nuclei which are bonded to non-magnetic Pt, there is no splitting and there is a singlet at the chemical shift position. Thus, as the chemical shift lies at the mid-point of the doublet, the spectrum shows three lines of intensities 33/2, 67, 33/2, i.e. a non-binomial triplet with relative intensities 1:4:1, easily confused at first sight with a 1:2:1 binomial triplet. The separation of the two **outer** lines is the one-bond ^{31}P–^{195}Pt coupling constant. Note that in the case of $SiH_3GeH_2GeH_3$ (Figure 7.4), it is the separation of the

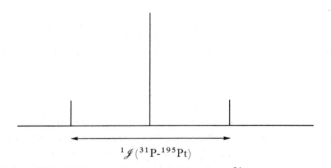

$$^1\mathcal{J}(^{31}P\text{-}^{195}Pt)$$

FIGURE 7.7 *A diagramatic representation of the ^{31}P nmr spectrum of square-planar platinum(II) complexes of the type $PtX_2(PR_3)_2$*
The spectrum is proton-decoupled, to simplify it, and is composed of two superimposed sub-spectra consisting of a doublet from molecules containing a ^{195}Pt atom and a singlet from molecules not containing ^{195}Pt. The separation of the two outer lines is the one-bond platinum–phosphorus coupling constant, $^1\mathcal{J}(PtP)$. The *trans-influence*, i.e. the ability of a ligand to weaken the Pt–ligand bond *trans* to it, plays a large part in determining the magnitude of $^1\mathcal{J}(PtP)$ in square-planar platinum complexes. Thus for a *cis*-$PtX_2(PR_3)_2$ complex, the PR_3 ligands are *trans* to X groups, and $^1\mathcal{J}(PtP)$ values vary from around 2000 Hz, when X is a high *trans*-influence alkyl group, to around 3600 Hz when X is a much lower *trans*-influence chloride ligand. For *trans*-complexes the value of $^1\mathcal{J}(PtP)$ is less informative since the PR_3 ligand is *trans* to another PR_3 ligand. While *trans*-influences in such complexes are relatively significant, the corresponding *cis*-influences are usually small. It is finally worth noting that the *trans*-influence is a thermodynamic phenomenon, and should not be confused with the *trans*-effect (section 13.9), which is a kinetic phenomenon.

successive pairs of lines of the SiH_3 or GeH_3 binomial triplets which gives the three-bond ^1H–^1H coupling constants.

The magnitude of the value of $^1\mathcal{J}(^{31}P$–$^{195}Pt)$ is highly dependent on the nature of the group *trans* to it, namely its *trans*-influence. The *trans*-influence of a ligand is its ability to weaken the bond *trans* to it. It is thermodynamic in origin and should not be confused with the *trans*-effect (see section 13.9) which is a kinetic effect. A phosphane ligand *trans* to a high *trans*-influence ligand (such as another phosphane, a hydride ligand, or an alkyl or aryl group) will have its bond to platinum weakened, and will therefore show a smaller value of $^1\mathcal{J}(^{31}P$–$^{195}Pt)$. Correspondingly, phosphanes *trans* to low *trans*-influence ligands, such as halides or carboxylates will show large values of $^1\mathcal{J}(^{31}P$–$^{195}Pt)$. It is therefore a relatively simple matter to follow substitution reactions at platinum centres using phosphorus nmr spectroscopy.

When coupled with other nmr techniques such studies allow the characterization of a range of coordination complexes containing nmr-active nuclei. The ^{103}Rh isotope is a useful one since it is spin $\frac{1}{2}$ and 100% abundant. The direct observation of ^{103}Rh nmr spectra is rather difficult, since the nucleus resonates at a very low frequency, but the additional splitting the ^{103}Rh nucleus provides in proton, carbon or phosphorus nmr spectra usually provides sufficient structural information. A good recent example of the utilization of ^{103}Rh coupling in nmr spectroscopy comes from the nmr identification of highly reactive rhodium–carbonyl intermediates in the rhodium-catalysed carbonylation of methanol. This process, developed by Monsanto, is used industrially to manufacture large quantities of acetic acid and is discussed further in section 15.7 on rhodium and iridium chemistry.

If the nucleus has a spin of more than $\frac{1}{2}$, the coupling splittings follow different rules. A nucleus of spin $n/2$ has $n+1$ orientations each equally likely, and this gives an $(n+1)$ multiplet of equal intensities. Thus the proton signal of the terminal hydrogens in diborane, above, when the coupling to ^{11}B with spin $= 3/2$ is taken into account, is a quartet with all components of equal intensity.

Thus the nmr investigation gives information about the relative numbers of magnetic nuclei of each type in a molecule, from the position and intensities of the signals, and shows something of the structure of the molecule from the spin-spin coupling.

Apart from its use in identifying compounds, information about reaction kinetics and exchange processes may be derived by studying the nmr signals over a range of temperatures.

While the nmr spectra of the majority of inorganic complexes, as well as organic compounds, such as ligands, are investigated in solution, there is a very large number of important materials which are highly insoluble. However, special methods are available to study nmr spectra of solid materials. In solution samples molecules are rapidly tumbling such that the nmr experiment observes their average values of chemical shift. Solids, however, are typically rigid and since most solids consist of microcrystals in all possible orientations, this typically results in very broad peaks in the solid-state nmr spectrum. This is overcome by spinning the sample at an angle (called the 'magic angle') to the applied magnetic field and sharp spectra can thus be obtained. Good examples come from the study of ^{29}Si and ^{27}Al nmr spectra of silicate and aluminosilicate minerals and glasses, and ^{13}C nmr spectra of a wide variety of materials such as coal and even freshly cut celery!

7.9 Further methods of molecular spectroscopy

While nmr is concerned with reversal of nuclear spin, *electron spin resonance* involves a very similar phenomenon, the reversal of the spin of an electron in a magnetic field. This involves a higher energy than the nuclear spin reversal, about 10 J mol^{-1} and thus a higher frequency of around 28 000 MHz. In an electron pair, the spins are already opposed and any reversal of one spin would be cancelled by that of the other (the energies involved are much lower than those required for excitation to the triplet state where the spins are parallel). Thus esr measurements can only be made on species with unpaired electrons like radicals and transition element compounds. For these, changes in the electron g factor (compare equation (7.7)), analogous to the nmr chemical shift, and spin-spin coupling to magnetic nuclei may be observed. One application is in the study of molecular orbitals. For example, if an electron is added to benzene to give the anion, $C_6H_6^-$, this electron will enter the lowest available orbital, which is one of the antibonding orbitals. The detailed structure of the electron resonance absorption then yields information about the interaction of the electron in this delocalized π-orbital with the atomic nuclei, and hence the distribution of this π-orbital over the atoms of the molecule may be determined and compared with the calculated one. In particular, an atom lying on a nodal plane should not interact with the electron, and this may be checked.

If an isotope, such as ^{119}Sn, has a metastable excited state which transforms to the normal ground state of the nucleus by emitting a gamma ray, the excited state can be used as a source and the ground state as target for the emission and resonant reabsorption of the gamma ray. This is the Mössbauer effect and is made use of in *Mössbauer spectroscopy*. The striking characteristic of the effect, which makes it so valuable, is the extremely well-defined energy of the transition. The line width is only about 10^{-12} of the energy, compared with about 1 cm^{-1} in 1000 cm^{-1} for a sharp infrared line for example. This means that the very small effect on the nuclear energy arising from the chemical environment may be detected in Mössbauer spectroscopy. Two interactions are important to us. First, as the nucleus has a slightly different radius in its ground and excited states, the resonance energy changes slightly with the electron density at the nucleus. This gives rise to a chemical shift whose magnitude reflects the s electron density as only electrons in s orbitals have any probability of being at the nucleus. A good example is provided by tin species where chemical shifts lie in the sequence Sn(IV) < Sn(0) < Sn(II).

In metallic tin (α-form) the element has the diamond structure with four more tin atoms surrounding each atom tetrahedrally. The configuration is thus s^1p^3 giving one s electron per tin atom. In tin (IV) compounds, all the valency electrons tend to be used in bonding so that the s electron population is less than one (and becomes 0 for the Sn^{4+} ion). On the other hand, in Sn(II) compounds, there is one unshared pair of electrons whose configuration approximates to s^2. Thus the order of chemical shifts in the Mössbauer follows from the configurations $s^0 < s^1p^3 < s^2$. Clearly, intermediate shifts give information about the s electron density, for example shifts between those for Sn^{4+} and for the metal show the direction of s electron drift in bonds to covalent tin(IV).

In the second phenomenon, the Mössbauer resonance may be split by interaction with the nuclear quadrupole moment. This arises where the nucleus has a spin of more than $\frac{1}{2}$, either in the ground state or in the metastable state. The nuclear quadrupole moment interacts with electric field gradients at the nucleus, and thus the quadrupole coupling indicates the degree of departure from spherical symmetry at the nucleus. That is, the quadrupole coupling gives information about the p and d electron densities. A further interaction which may be detected in the Mössbauer effect is the splitting of nuclear energy levels in a magnetic field. This may be imposed externally or arise internally from ferromagnetic or paramagnetic interactions and gives information about these.

The main limitation of Mössbauer spectroscopy is that only a limited number of elements have a suitable metastable nuclear state. All these have very short lifetimes and occur in the course of some decay sequence, so that the work requires an irradiating source and a suitable sequence of nuclear decay processes. The Mössbauer effect has been observed, or is predicted, for 49 elements but all with $Z \geqq 26$ (iron), except for potassium. It is experimentally easiest to study iron and tin but a fair amount of work has been done on others including Te, I, Xe, Au and several of the lanthanides. It has the advantage that the sample need only be a powder, so that it provides a method of studying insoluble, poorly crystalline materials of the heavier elements which are difficult or impossible to study in any other way.

One further spectroscopic technique which is making an impact on modern inorganic chemistry is *mass spectroscometry*. This is the refinement of Aston's method of determining isotope weights and has been extensively used by organic chemists in the last decade. More recently, extensive studies of metal carbonyls and organometallic compounds have appeared and other inorganic applications are becoming common.

In the experiment, a stream of the vapour of the substance to be studied is passed through a beam of electrons of energy usually in the region of 70 eV. These ionize the molecules, M,

$$M + e^- \rightarrow M^+ + 2e^-$$

and the resultant ions may fragment to daughter products,

radicals, and ions,

$$M^+ \rightarrow M' + M''^+$$

The ions are passed through a magnetic field, and in many cases an electric field as well, and resolved into species with the same m/e ratio. The resolved peaks are detected to give the mass spectrum. If the *parent ion* M^+ can be detected (in some cases it is not) its mass may be measured with high accuracy and yields the molecular weight of the molecule. The use of double focusing (magnetic and electrical fields) yields masses accurate to about 1 ppm and thus allows analysis. For example, CO may be distinguished from $^{14}N_2$ or from ^{28}Si. The fragmentation path may also give useful information, and related compounds often have similar fragmentations. For example, metal carbonyls $M(CO)_n$ lose CO groups stepwise so that ions $M(CO)_x^+$ are observed for all values of x from n to 0. By varying the energy of the electron beam, the minimum energy required to ionize the molecule, or one of its fragments, may be determined, leading to values of the ionization potentials and information on bond energies.

For traditional electron ionization mass spectroscopy involving organic or organometallic compounds, a vapour pressure of only a fraction of a millimetre of mercury is required to get the sample into the gaseous phase. The source can also be heated in order to increase the vapour pressure. However, of greater use to the inorganic chemist are the new softer ionization techniques, which can be used to study more thermally labile compounds and ones which are ordinarily considered to be involatile under normal mild conditions. As a lower energy is involved in the ionization process, much less fragmentation of the parent molecule is likely to occur, which can have significant advantages in many areas, including inorganic chemistry and biochemistry. The first of these ionization techniques utilizes Fast Atom Bombardment (FAB), in which ionization is effected by a beam of heavy atoms such as Xe or Cs^+, and this technique has now firmly found its place in the laboratory. The second, more recent, technique is that of electrospray, in which the sample is introduced as a solution into the spectrometer. This technique is particularly suited towards interfacing with other separation techniques, such as chromatography. Relatively few types of compounds have so far been studied using electrospray mass spectroscopy.

7.10 Fourier transform methods

All the spectroscopic methods discussed above were developed using a sequential, point-by-point scanning of the wavelength region of interest. Such methods are slow, needing anything from several minutes to many hours for accurate infrared or nmr studies, for example. When methods are of low intrinsic sensitivity, such as nmr, the slowness limits the chance of improving sensitivity by repeated scanning. A solution to this problem is to replace the point-by-point scan with the simultaneous observation of the whole spectral region. This excites all the transitions simultaneously, and the resulting signals are found as beating patterns which can be

analysed by the mathematical process of Fourier Transformation. With the development of computers, the Fourier calculations are rapidly performed, and the technique is now standard. As the observation time is short, this allows (a) much enhanced sensitivity (as the results of up to many thousand scans may be summed) and (b) experiments on a very short timescale.

The principal technique where the enhanced sensitivity was important is nmr, as indicated above. Here, because the energy of the transition is so low, the population of the excited state very easily becomes equal to the ground state and the signal disappears (called 'saturation'). The Fourier Transform method avoids this problem by using a short pulse of radiation, followed by a pause while the populations revert to thermal equilibrium. Measuring this relaxation time gives a further parameter not easily available from the older experiment. In modern studies every isotope with a nuclear spin is accessible, which means that practically every element in the Periodic Table yields at least some information. For the easier nuclei, the enhanced sensitivity allows work at lower concentrations, on metal carbonyls using ^{13}C at natural abundance, for example. Other major improvements arise because the computer may also be used to control and vary the experiment. For example, by irradiating with a sequence of pulses of different character, signals may be decoupled, enhanced in sensitivity, correlated, or separated from noise. Thus, in a complex molecule such as a polytungstate (section 15.4) two-dimensional nmr methods allow a structural determination in solution as it is possible to determine which atom bonds to which. In cases like this, where the species in solution may not be the same as the one isolated as a crystal, the method gives unique information. Other striking examples include studies on hydration and hydrolysis (a classic is of aluminium ions using ^{27}Al nmr), and the work on polyphosphorus compounds described in section 18.3.1.

The advantages of enhanced sensitivity or shorter timescale apply to all techniques, and are of particular advantage where the sensitivity of the method is intrinsically low, as in Raman or far-infrared spectroscopy. Of equal importance, even in methods of reasonable natural sensitivity, has been the reduction in the timescale of the experiment. This allows the study of short-lived species under actual conditions. Thus, infrared studies of the catalytic oxidation of hydrocarbons has allowed the detection of intermediates like CH_3O-M as transient intermediates on metal or metal oxide catalyst surfaces.

Finally, we note that, although Fourier Transform methods were a major motivation to add computers to spectrometers, computers have had other major impacts. Most intermediate and advanced level instruments are now operated more or less under computer control, and the data output is processed by computer. The possibility of accumulating a number of spectra to improve sensitivity is important even where point-by-point scanning is used, for example in Raman spectra. As well as allowing expansion, smoothing determination of maxima and so on, the computer is useful in matching with library spectra, subtracting background or solvent absorptions, and further calculation of quantities like concentrations or absorption coefficients.

7.11 Other methods

7.11.1 Magnetic measurements

If a sample of a compound is weighed in a magnetic field and then in absence of the field, a weight change will be observed. Most compounds are repelled by the field and show a decrease in weight; these are termed *diamagnetic*. The diamagnetism arises from the repulsion between the applied field and induced magnetic fields in the compound, and is a very small effect which occurs for all compounds. However, some compounds show a net attraction to a magnetic field and an increase in weight; these are termed *paramagnetic*. The paramagnetism arises where there are one or more unpaired electrons in the compound, and is a much larger effect than diamagnetism. An unpaired electron corresponds to an electric current, and hence to a magnetic field, by virtue of two effects, its spin, and its orbital motion. In most compounds, the effect of the orbital contribution is quenched out by the electric fields of surrounding atoms, and the spin-only magnetic moment is observed. This is given by:

$$\mu = 2\sqrt{[S(S+1)]}$$

where μ is the magnetic moment in units of Bohr magnetons, and $S = \frac{1}{2}n$ equals the number of unpaired spins multiplied by the spin quantum number. This formula holds, to within ten per cent, for most compounds, allowing a direct determination of the number of unpaired electrons. In some cases, particularly when the unpaired electrons are in an f orbital, the orbital contribution is not quenched out and a more complex formula:

$$\mu = \sqrt{[4S(S+1)+L(L+1)]}$$

which involves the orbital quantum number, L, holds (see Figure 11.8). In some cases, the determination of the number of unpaired spins gives direct structural information. For example, consider a nickel(II) compound $NiL_4.2S$, where L is any ligand and S is a molecule of solvent. If the solvent molecules are not coordinated to the nickel, the NiL_4 species could well be square planar and have no unpaired electrons, while coordinated solvent would mean an octahedral NiL_4S_2 species with two unpaired electrons (compare section 14.8.

It is possible to gain much more detailed information from magnetic measurements than is indicated above. Other effects such as ferromagnetism and anti-ferromagnetism are observed, and much valuable information results from studying the variation of magnetic moment with temperature, concentration, and field strength. However, the simple Gouy method of weighing the sample in a magnetic field is readily carried out and yields considerable information, especially in transition metal chemistry.

7.11.2 Dipole moments

The measurement of the dipole moment of a compound may also yield useful structural information. As any bond between atoms with different electronegativities is polarized, any molecule will have a dipole moment unless such bond dipoles are so arranged as to cancel out. As a simple example, if CO_2 is linear, the two $C-O$ dipoles oppose each other and no resultant moment is observed, while if the molecule is bent a resultant dipole is observed (Figure 7.8). The figure also illustrates, as a further example, how *cis* and *trans* isomers may be distinguished by dipole moment measurements. Care, however, must be exercised in interpreting dipole moments: thus, NF_3 has an almost zero dipole moment (whereas NH_3 has a marked moment), not because the molecule is planar as once thought, but because the bond and lone pair dipoles cancel. However, with care in interpretation, dipole moments have proved a very useful adjunct to structural determinations, and the method—like the magnetic measurements above—has the advantage that it does not destroy the sample, and adaptations are available which require only a small amount of material.

7.11.3 Scanning electron microscopy (SEM)

The basic difference between a conventional light microscope and an electron microscope is in the type of radiation used. The electron microscope uses a beam of electrons as the 'light' source (remember that electrons have wave as well as particle characteristics, see section 2.5) and the shorter wavelength therefore means that the resolution of an electron microscope is much greater, allowing smaller features to be resolved. While the applications of SEM to disciplines such as earth sciences and biology are immediately obvious, the technique is also becoming of increasing importance to chemists and especially materials scientists. For example, the degree of crystallinity of a solid sample, be it either a synthetic

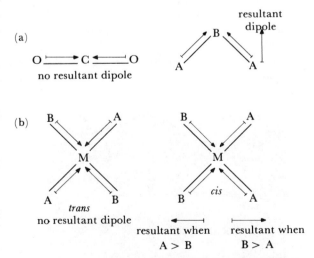

FIGURE 7.8 *The use of dipole moments to yield structural information*
Figure (a) shows how the two $C-O$ moments cancel in linear $O-C-O$ but would give rise to a resultant moment if the molecule was V-shaped. As CO_2 is observed to have no dipole moment, the linear structure is the correct one. Figure (b) shows how a *cis* square planar complex MA_2B_2 will have a resultant moment while the *trans* form will not.

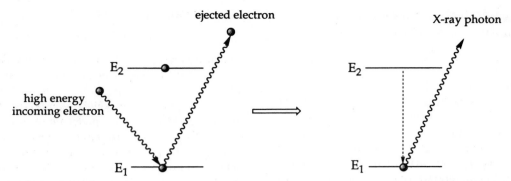

FIGURE 7.9 *The electronic process behind X-ray microanalysis*
A highly energetic electron (from the scanning electron microscope) ejects one of the core electrons of the atom. An electron from a higher

energy shell in the atom then fills the gap, emitting an X-ray photon in the process which is detected. The energy of the X-ray is equal to the separation of the two energy levels $E_2 - E_1$.

zeolite (see section 18.6), a copper oxide superconductor (see section 16.1), or some other material, can readily be determined.

In addition to providing information on the morphology of the sample, information may also be obtained on its elemental composition, and this is of direct benefit to the chemist. Irradiation of a sample with an electron beam results in electrons being ejected from the inner electron shells of atoms. When electrons from outer valence shells drop down to fill this newly-formed vacancy, they emit an X-ray photon, the energy of which is characteristic of the element concerned and the various energy levels involved. This is illustrated schematically in Figure 7.9. The analysis of these X-rays then provides information on the relative numbers of each type of atom in the sample, though it can only be applied with any accuracy for elements heavier than about fluorine. A typical X-ray spectrum is shown in Figure 7.10 for the potassium salt of a polyoxovanadate species (see

section 14.3.1). The spectrum consists of a rather broad background upon which are superimposed a number of sharp peaks, each being composed of a number of channels, or wavelengths, over which the X-rays are analysed. This analytical method, commonly called Energy-Dispersive X-Ray Microanalysis, is particularly useful in qualitative analysis for the number of elements present in a sample, though by the use of appropriate standards quantitative analyses may also be undertaken.

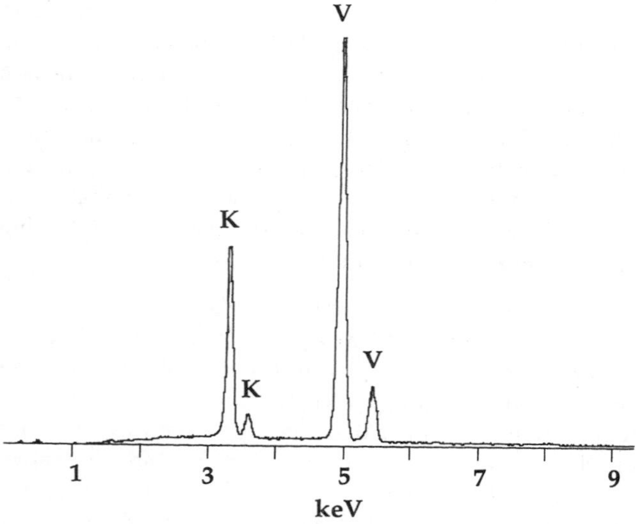

7.11.4 Scanning tunnelling microscopy (STM)

Scanning tunnelling microscopy is a very powerful and fairly recent (early 1980s) technique for studying surfaces of solids with atomic resolution. The principle behind this technique utilizes the fact that electron tunnelling occurs between a very sharp metal tip and a conducting or semiconducting surface—essentially the wave functions of the tip and the surface atoms overlap at very small distances. By scanning the metal tip over the surface of the solid, whilst keeping a constant flow of electrons (called the *tunnelling current*) between the tip and the solid, or *vice-versa*, a 'map' of the surface can be built up. The fact that atomic resolution can be attained makes this an exceedingly powerful technique. Figure 7.11 shows an image obtained from a mica sample (see section 18.6); the network of interlocking rings can be clearly distinguished.

FIGURE 7.10 *A typical X-ray spectrum of a compound containing vanadium and potassium*
Sharp, 'stepped' peaks due to X-rays emitted from vanadium and potassium atoms can be clearly seen superimposed on a broad baseline. The peaks due to oxygen atoms also present in the sample are very weak.

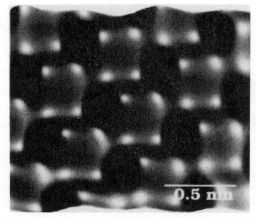

FIGURE 7.11 *Scanning Tunnelling Microscope image of a mica surface (see section 18.6).*

The early studies in this area tended to concentrate on materials such as metals and semiconductors, however, recently chemists have begun to add this to their armoury of techniques. Applications include the investigations of silicon single crystals (of importance to the semiconductor industry), copper oxide superconductors, and layered transition metal chalcogenides. This latter class of compounds includes molybdenum sulfide, MoS_2, a material which ordinarily has a layered structure but which has recently been found to exist in a range of polyhedral structures. Recent advances in this area are discussed in greater detail in section 19.3.3. These layered chalcogenides have been found to undergo a very subtle periodic distortion at low temperatures, called a charge–density wave. (These distortions arise due to the unequal electron occupation of degenerate valence bands of the solid which results in a distortion occurring to relieve the degeneracy, similar in many respects to the Jahn–Teller distortion in certain coordination complexes, see section 13.4.) STM provides a very powerful technique for the study of these charge–density wave distortions. Other areas of inorganic chemistry which have benefited from the STM technique include the study of zeolite morphology, metal colloids, and metal carbonyl clusters. A recent article, given in Appendix A, summarizes the application of STM to the above areas.

Determination of energy levels

7.12 Photoelectron spectroscopy

When a molecule, M, interacts with a quantum of ultraviolet radiation, it is possible for all the energy of the quantum to be used in expelling an electron from a valence level orbital

$$M + h\nu \rightarrow M^+ + e \dots\dots\dots\dots (7.8)$$

The quantum energy has first to provide the ionization energy, I, of the electron, and the remainder appears as the kinetic energy, E_{elect}, of the expelled electron. A small part of the energy, E_{vib}, may also be used in exciting vibrations of the molecular ion, M^+, and there are other small corrections which need not concern us. Thus the energy equation is

$$E = h\nu = I + E_{vib} + E_{elect} \dots\dots\dots (7.9)$$

If we use monochromatic radiation, $h\nu$ is fixed and is known. The energy of the electron can be measured by finding the size of repelling electric field which just stops the electron reaching a detector. Thus we can determine the ionization energy (equation 7.10).

$$I + E_{vib} = h\nu - E_{elect} \dots\dots\dots (7.10)$$

Since vibrational energy is quantized (and the quantum is much smaller than I) we observe a series of bands corresponding to $I + nh\nu_{vib}$ where n is the change in the vibrational quantum number and may take the values 0, 1, 2... etc. Thus the band of lowest energy of the series represents the value of I, and the separations give the vibrational energy of M^+. This is clearly seen in the photoelectron spectrum of H_2

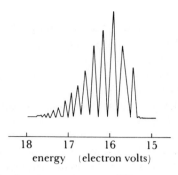

FIGURE 7.12 *The photoelectron spectrum of* H_2

(Figure 7.12) The lowest energy component about $15 \cdot 5\,eV$ gives the ionization energy and the separation, which corresponds to $2260\,cm^{-1}$ is the vibrational energy of H_2^+. Note that the vibrational spacing decreases, as more and more quanta are excited, as the ion moves towards its dissociation limit at about 18 eV.

If we compare with Figure 3.7b (and making certain assumptions which are discussed in the references) we can identify the ionization energy I, as the stabilization energy of an electron in the orbital ϕ_B. That is, I is the difference in energy between the ϕ_B level and the energy zero which (compare sections 2.6, 2.7) corresponds to the complete separation of electron and remaining ion. Consequently, Figures 3.7b and 7.12 may be combined as in Figure 7.13. In this way, we can use photoelectron spectra to determine by experiment the energies of occupied orbitals.

Note that photoelectron spectroscopy gives no direct information about the unoccupied level ϕ_A. However, ordinary electronic spectroscopy (section 7.6) does give the difference between ϕ_B and ϕ_A (the $\sigma \rightarrow \sigma^*$ transition, found in the ultraviolet for H_2). Thus these two techniques, in combination, can fix all the levels in molecular energy level diagrams.

In more complex molecules, where a number of valence level orbitals are occupied, the ionization from each is observed (in different molecules of the sample of course) and a spectrum with several bands is obtained as in Figures 3.23 or 4.21.

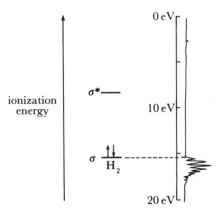

FIGURE 7.13 *Energy level diagram and photoelectron spectrum of* H_2

The energy source

Monochromatic radiation (i.e. all quanta of same energy) is needed, and the most convenient experimental source is the helium (I) emission, with a quantum energy of $21 \cdot 218$ eV ($2047 \cdot 4$ kJ mol^{-1}, wavelength $58 \cdot 43$ nm). This arises from the $1s \rightarrow 2p$ excitation. This energy is sufficient to ionize most of the valence level electrons. For more tightly bonded levels the helium (II) line ($40 \cdot 81$ eV or $30 \cdot 4$ nm) arising from He$^+$, is used but is experimentally more difficult. Thus ionization energies of more than 21 eV are generally known with rather lower accuracy and are not always determined.

If a quantum of much higher energy is used, in the X-ray region of the electromagnetic spectrum, then the most tightly bound electrons from the inner closed shells are ionized. This *X-ray photoelectron spectroscopy* (*XPS*) is of less direct interest to us than *ultraviolet photoelectron spectroscopy* (*UPS*), and it is the latter which is referred to when the general phrase *photoelectron spectroscopy* or *PES* is used. However XPS does give chemical information since, although the inner shell electrons, such as the $1s$ electrons of C, N or O, are not involved in bonding, there are small changes in their energies with changes in their chemical environment. For example, the oxygen $1s$ binding energy in RMn(CO)$_5$ changed from $538 \cdot 8$ to $540 \cdot 3$ eV as R varied from C_5H_5 to $SiCl_3$ and this change correlated with other measures of the degree of electron back-donation from Mn to CO in these molecules.

XPS and UPS are usually carried out with quite different instruments, and there have not been many studies on the same molecule. Jolly (see reference) has emphasized the advantages of using the two techniques in conjunction. Basically, XPS reflects the effects of the nuclear charge and the overall electron shells, while UPS depends on these core effects plus the valency level ones. XPS can thus be used to allow isolation of the valency effects in UPS, and hence greatly expand the approach to valency energy levels. UPS results are used in other chapters: X-ray PS methods are further described in the references.

Experimental

Ultraviolet photoelectron spectroscopy is carried out on a gas sample, at relatively low pressure, which is irradiated with a collimated, monochromatic beam. The expelled electrons are sorted according to their kinetic energies by passing them through a retarding field. The electron count is plotted against the ionization energy (which is the photon energy less the electron kinetic energy).

For X-ray photoelectron spectroscopy, solids may be used as well as gases. There is no convenient monochromatic source between the ultraviolet (40–50 eV) and the X-ray region (1000–1500 eV) and little work has been done on the intermediate region.

Vibrational structure

The spacing of the vibrational fine structure gives the vibrational frequency of the molecular ion, M$^+$, resulting from the ionization. If the electron was lost from a bonding orbital in M, then M$^+$ will be less tightly bound, it will re-

TABLE 7.2 Vibrational structure of photoelectron bands

Orbital from which electron is ionized	Vibrational frequency in M^+ compared with M	Number of vibrational components	Most intense vibrational components
bonding	markedly smaller	large	near centre
non-bonding	similar	small	first
anti-bonding	similar or larger	medium	near centre

Example NO: stretching frequency 1890 cm^{-1}

bonding π	1200	8	4th
non-bonding σ	1610	3	1st
antibonding $\pi*$	2260	5	2nd

quire less energy to stretch the bond, and thus the vibrational frequency in M$^+$ will be lower. For example, the frequency of 2260 cm^{-1} in H$_2^+$ compares with 4280 cm^{-1} for H$_2$ (compare section 3.3, especially Figures 3.7a and 3.7b).

Conversely, loss of an antibonding electron would lead to a higher stretching frequency for M$^+$. If a non-bonding electron is lost, there will be little change. Not all changes are as large as those for hydrogen, and the accuracy of measurement of the frequency from photoelectron spectra is much less than by other methods such as infrared spectroscopy. However, the vibrational frequencies are a very valuable guide to the assignment of ionizations to particular orbitals.

In addition to the frequencies, the number of vibrational components and their intensity pattern also vary with the bonding character of the lost electron. The detailed argument is given in the references, but their characteristics may be summarized as in Table 7.2.

Polyatomic molecules show much more complex vibrational structure, since a number of different vibrational energies may be involved. Often the individual components are not resolved but only the overall envelope is seen. Even in this case, non-bonding ionizations tend to give sharp asymmetric envelopes, steepened towards the leading edge while bonding and antibonding envelopes are rounder and more symmetric.

PROBLEMS

7.1 Consider how each method reviewed in this Chapter contributes to the basic scheme for identifying an unknown, given in Figure 3.1. Note that some methods, like chromatography, are precursors to the scheme in providing the unknown as a single substance, and other methods like photoelectron spectroscopy, extend the scheme to find bonding or other properties in terms of the structural formula.

7.2 If a compound occurs only (a) as a gas or (b) as a solid, discuss the limitations imposed on the experimental methods available for its study.

8 General Properties of the Elements in Relation to the Periodic Table

8.1 Variation in energies of atomic orbitals with atomic number

In Chapter 2 the derivation of the existence, shapes, and energies of the atomic orbitals from the wave equation was described and it was shown how the structure of the Periodic Table could be derived by filling the atomic orbitals in order of increasing energy. A more detailed discussion of the properties of the elements demands a closer look at the variation in energy of the atomic orbitals as the atomic number increases.

Consider first the s orbitals. The energy of an electron in the $1s$ orbital of a hydrogen-like atom of atomic number Z, is $-Ze^2/8\pi\epsilon_0 a_0$; so that the energy decreases as Z increases. As the nuclear charge and the number of electrons in the atom increase, account must be taken of the repulsive effect of the extra electrons, as well as that of the increased nuclear charge. This is done by replacing the actual nuclear charge Z, by the effective nuclear charge, Z^*, which is the resultant of the nuclear charge and the electron charges as experienced by an electron in a particular orbital. For example, the $2s$ electron in lithium experiences an effective charge which is the resultant of the nuclear charge of $+3$ and the charges of the two $1s$ electrons. The effect of the inner electrons in reducing the effective charge experienced by the outer one is termed *shielding*. If the shielding effect of the two $1s$ electrons in lithium were perfect, the outer electron would experience an effective nuclear charge, Z^*, of $1\cdot0$ $(3-2)$, but the $2s$ orbital has finite electron density at the nucleus (see Figure 2.9b) so that an electron in the $2s$ orbital penetrates the $1s$ shell and thus experiences a greater nuclear charge than that calculated from perfect shielding. The result of shielding by inner electron shells is that the effective nuclear charge experienced by the outer electrons in an atom is always markedly less than the actual nuclear charge, Z, but, as the shielding is not perfect, the effective nuclear charge increases as Z increases, but more slowly. A useful indication of the shielding effects is given by Slater's rules which are summarized in Table 8.1.

Other, more detailed, methods of allowing for shielding have been proposed, but the Slater scheme represents the principles and needs no information outside the quantum numbers.

Application of these rules shows that the effective nuclear charge experienced by one $1s$ electron in, say, carbon is $5\cdot7$ and in nitrogen, $6\cdot7$. Similarly, a $2s$ electron in carbon experiences an effective charge of $3\cdot25$, while a nitrogen $2s$ electron experiences a charge of $3\cdot9$. Such calculations, or

TABLE 8.1 Slater's rules for shielding contributions

The effective nuclear charge, Z^*, is given by $Z-\sigma$, where σ is the sum of the shielding contributions of all the other electrons in the atom, as follows

Principal quantum number, n, of shielding electrons	*Shielding contribution, σ*
n higher than principal quantum number of the electron under consideration	zero
n equal to principal quantum number of the electron under consideration	0·35 (for each electron) except that 0·30 is used for σ for a $1s$ electron acting on the second $1s$ electron
n is one less than the principal quantum number (a) for an s or p electron under consideration	0·85
(b) for a d or f electron under consideration	1·00
n is less by two, or more, than the principal quantum number of the electron under consideration	1·00

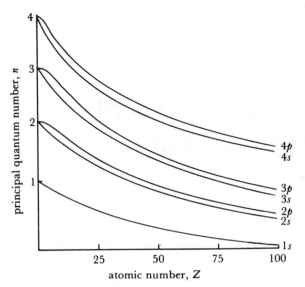

FIGURE 8.1 *Energies of the lower s and p orbitals as functions of Z*

direct experimental determination of energy levels by electron spectroscopy in one of its forms, allows the construction of diagrams showing the variation in energy levels with atomic number, such as Figure 8.1.

In this figure, the energies start at the calculated values for H, and fall off with Z (the scale is logarithmic) so that the electron is increasingly tightly held. One measure of the energy involved is given by the subshell binding energy measured by X-ray photoelectron spectroscopy. This energy is that required to remove one electron from a specific subshell in a free atom, say the energy to remove a $1s$ electron from a free C atom. For removal of the first electron, this is the same as the ionization potentials, but successive ionization potentials involve removal of electrons from ions (e.g. the 5th ionization potential of C is the energy to remove one $1s$ electron from a C^{4+} ion). Some binding energies are listed in Table 8.2.

TABLE 8.2 Free atom subshell binding energies (eV)

	H	He						
$1s$	13·6	24·6						
	Li	Be	B	C	N	O	F	Ne
$1s$	58	115	192	288	409	538	694	870
$2s$	5·4	9·3	12·9	16·6	20·3	28·5	37·9	48·5
$2p$			8·3	11·3	14·5	13·6	17·4	21·6
	Na	Mg	Al	Si	P	S	Cl	Ar
$1s$	1075	1308	1564	1844	2148	2476	2828	3206
$2s$	66	92	121	154	191	232	277	326
$2p$	34	54	77	103	134	168	207	249
$3s$	5·2	7·6	10·6	13·5	16·2	20·2	24·5	29·2
$3p$			6·0	8·2	10·5	10·4	13·0	15·8

Note: p values averaged over two multiplicities.

In hydrogen-like atoms, the p orbital has the same energy as the s orbital with the same value of n, but in all atoms with more than one electron, shielding effects come into play. The p

orbital is more shielded than the corresponding s orbital as it does not penetrate so far towards the nucleus. It accordingly experiences a smaller effective nuclear charge and is of higher energy. The curves of p orbital energies v. atomic number run roughly parallel with the curves of the corresponding s orbitals as Z increases (Figure 8.1). The gap in energy between the s and p orbital of a given n value is much smaller than that separating the p orbital and the s orbital with the next higher n value.

The case of the d orbitals is more complicated. Figure 8.2 shows the variations in energy with increasing Z of the $3d$ orbital with respect to the $3s$, $3p$, $4s$ and $4p$ orbitals. The $3d$ orbital has the same energy as the $3s$ and $3p$ orbitals in the hydrogen atom and, since it scarcely penetrates the first and second quantum shells at all, it is perfectly shielded from the increase of nuclear charge as these two atomic levels are filled. Thus the $3d$ level is subject to the same effective nuclear charge (about unity) for Z values up to $Z = 10$, and the plot of the $3d$ energy against Z remains level. On the other hand, the $4s$ and $4p$ orbitals—which are of considerably higher energy than the $3d$ orbital in the lightest elements—do penetrate the inner electron shells significantly, are less shielded, and drop steeply in energy as Z increases. When the $3s$ and $3p$ levels are filling, the $3d$ level still remains almost unaffected and the energy of the $4s$ level falls below it at about $Z = 15$. As a result of this, when the $3p$ shell is filled at argon, the next lowest energy level is $4s$ and not $3d$. The nineteenth and twentieth electrons therefore enter the $4s$ level, into which the $3d$ level does strongly penetrate. It experiences a marked increase in effective nuclear charge, and its energy falls from being nearly equal to that of the $4p$ level towards that of the $4s$ level. The twenty-first electron and the next nine enter the $3d$ level whose energy falls below that of the $4s$ level. As these two remain very close in energy, electrons readily switch between them; for example, copper, which might be $3d^9 4s^2$, is actually $3d^{10} 4s^1$, and gains the

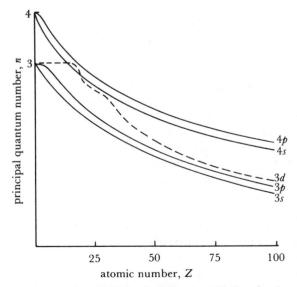

FIGURE 8.2 *Energies of the 3rd and neighbouring orbitals as functions of Z*

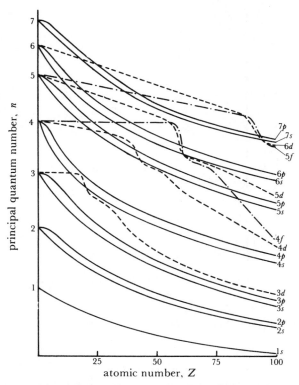

FIGURE 8.3 *The variation of the energies of the atomic orbitals with atomic number. Compare Figure 2.4 which shows relative energies at the Z values where the orbitals start filling.*

extra stability of the filled d shell by transferring an s electron. When the $3d$ shell is filled, the level next in energy is the $4p$ orbital. This filled d shell introduces an extra shielding effect on the higher orbitals, and the energy gap between the $4p$ and $5s$ levels at $Z = 36$, where the $4p$ level is filled, is larger than that expected by simple extrapolation from the $2p$–$3s$ and the $3p$–$4s$ differences.

A similar effect occurs for the $4d$ level relative to the $5s$ one, and the f levels show analogous relationships, Figure 8.3. As the f orbitals are even less penetrating than the d levels, an extra quantum shell is filled before they become involved. The $5s$, $5p$, $6s$, $6p$ and $5d$ levels all drop below the $4f$ level in energy, as Figure 8.3 shows. The filling order is $5s$, $4d$, $5p$, and then $6s$. Both the $4f$ and $5d$ levels are strongly affected by the filling of the $6s$ level and drop very steeply in energy below the $6p$ level, but remain almost equal in energy to each other. Lanthanum, which comes after the $6s$ level is filled, has its outer electron in the $5d$ shell, but the following element cerium has the outer configuration $4f^2 5d^0 6s^2$. The next electrons fill $4f$, with minor variations at the half-filled and filled levels (compare Table 11.1), then $5d$ and $6p$. For the heaviest elements, where $5f$ and $6d$ are even closer, the configurations follow the same general principles with variation in detail (Table 12.1).

8.2 Exchange energy

The energy of an electron in an orbital depends on other factors besides the attraction of the nuclear charge and the electrostatic interaction between the electrons. There is a second, quantum-mechanical, interaction between electrons which is known as the *exchange energy*. There is no classical analogue to this energy which derives from the indistinguishability of electrons and the arrangement of their spins. The exchange energy is a function of the number of pairs of electrons with parallel spins, i.e. $E_{ex} = K \times P$, where K is a constant and P is the number of pairs of parallel electrons. P is equal to the combination $_nC_2$ where n is the number of parallel spins, i.e. P has the following values:

n	1	2	3	4	5	6	7
$P = \dfrac{n(n-1)}{2}$	0	1	3	6	10	15	21

This energy is at the basis of Hund's Rule that electrons which enter orbitals of equal energy have parallel spins as far as possible. For example, the exchange energies for the various possible configurations of the electrons in the three p orbitals are shown below (note that the parallel and antiparallel electrons act as independent sets).

Number of electrons	Exchange energy if Hund's Rule is followed ($\times K$)		Exchange energy for maximum pairing ($\times K$)		Loss of energy in latter case ($\times K$)
1	↑	0	↑	0	0
2	↑ ↑	1	↑↓	0+0	1
3	↑ ↑ ↑	3	↑↓ ↑	1+0	2
4	↑↓ ↑ ↑	3+0	↑↓ ↑↓	1+1	1
5	↑↓ ↑↓ ↑	3+1	↑↓ ↑↓ ↑	3+1	0
6	↑↓ ↑↓ ↑↓	3+3	↑↓ ↑↓ ↑↓	3+3	0
	due to spins ↑ ↓		due to spins ↑ ↓		

Although the exchange energy is relatively small, it becomes significant when similar species are compared, as it changes with the number of electrons in a different way from the larger attractive and repulsive forces. An example of the stabilization due to exchange energy has already been noted in the case of the ground state of copper. The exchange energy of the actual configuration $d^{10}s^1$ is $20K$, from the two sets of five parallel electrons in the d shell. The energy of the alternative configuration d^9s^2 is $16K$ (from a set of five plus a set of four in the d shell; the two s electrons are antiparallel of course). The exchange energy gain thus favours the d^{10} configuration, but against this must be set the loss in orbital energy in moving the electron from the $4s$ orbital to the $3d$ one. In the case of copper, the gain in exchange energy more than balances this loss, while in the case of nickel, which has the configuration d^8s^2 but could be $d^{10}s^0$, the balance appears to lie the other way and the former configuration is ground state. That the balance of energies is very close is shown by the configurations in the nickel and copper Groups:

nickel $3d^8 4s^2$ palladium $4d^{10} 5s^0$ platinum $5d^9 6s^1$
copper $3d^{10} 4s^1$ silver $4d^{10} 5s^1$ gold $5d^{10} 6s^1$

The exchange energy contributes to the stability of filled shells, and also shows the greatest relative change at the half-filled shell electron count. Illustrations are to be found in the ground state configurations of transition and inner transition atoms where the d^5 and f^7 half-filled shells are favoured. Examples can be found in Table 2.5, including the configurations of chromium and gadolinium and their neighbours. This preference for half-filled and filled shell configurations is general in the Periodic Table, although the nice balance of energies means that configurations are not readily predictable (compare the ground state electronic configurations of the second transition series from ytterbium to cadmium in Table 2.5). The examples quoted so far have been confined to the ground states of atoms, but the stability of these special arrangements also shows up in the general chemistry of the elements. Manganese, for example, is particularly stable in the $+ \mathrm{II}$ state which is a d^5 configuration.

The interelectronic forces and the changes in nuclear charge play an important part in determining the stability and configurations of ions, but it must be noted that it is not possible to determine the detailed chemistry of an ion from the ground state configuration of its parent atom. For example, in most of the transition metals the $(n+1)\,s$ shell is filled while the nd level is only partly occupied. That is, in the atom the s shell is more stable than the d shell. However, when any of the transition elements form ions, it is always the s electrons which are lost first. Further, once one or more electrons are lost from an atom, the order of orbital stabilities is not necessarily the same as in the undisturbed atom. Thus, europium has the configuration, $4f^7 5d^0 6s^2$, while the next two elements have the configurations $f^7 d^1 s^2$ and $f^9 d^0 s^2$, yet all three lose three electrons to give a stable trivalent cation of valency configuration f^6, f^7 and f^8 respectively, just as if each element had had the outer configuration $f^n d^1 s^2$.

8.3 Stable configurations

With the reservations expressed above, it is possible to generalize about stable electronic configurations by considering the interplay of the nuclear attraction, the repulsion by electrons already present, particularly one in the same orbital, and other effects such as exchange energy. If a line is drawn in Figure 8.3 through the points at which the orbitals fill, it follows an irregular path. From H to He, it drops along the $1s$ curve so the energy is more negative and it takes nearly double the ionization potential to remove the first electron from He. The third electron leaps up to the $2s$ curve and the fifth to the $2p$ curve, and then the last electron becomes steadily more tightly bound until the second quantum shell is filled at Ne. Similar jumps occur as we continue through the Periodic Table, with the largest between the np and the $(n+1)s$ levels. If we add to this the more detailed effects of the electron configuration, we can understand the variation across a Period as shown in Figure 8.4. From the value of $24.6\,\mathrm{eV}$ to remove the first electron from He, the potential drops sharply to $5.4\,\mathrm{eV}$ for Li because the least tightly held electron is in a new quantum shell, further from the nucleus and quite effectively shielded by the inner $1s$ electrons. From Li to Ne, there is an overall strong increase in the ionization potential, as the effective nuclear attraction is increasing and

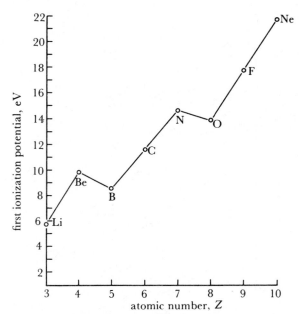

FIGURE 8.4 *Variations of first ionization potential across the First Short Period*

the electrons are being added in the same shell. The $1s$ orbital binding energies show the unshielded nuclear charge effect.

The finer detail in Figure 8.4 results from the differences between the s and p orbitals. As the electron in the p orbital has less probability of being close to the nucleus than an s electron, the first p electron, at B, is relatively less tightly bound than the Be s electron. The potential then rises for three successive elements as the nuclear charge increases and the electrons enter the three different p orbitals, maximizing exchange energy and minimizing interelectron repulsions. The fourth p electron has to be placed in an orbital already occupied so that there is no gain in exchange energy and a significantly increased repulsion. Thus it takes less energy to remove the first electron from O than from N. Alternatively expressed, the half-filled shell at N shows up relative to the configuration with one less electron, at C, or with one more electron, at O. Similar relative effects can be discerned at p^3 for other elements—e.g. compare the second and third potentials of F with its neighbours. Finally, filling the last three electrons into the $2p$ orbitals shows smoothly increasing ionization potentials.

The energy gaps between successive levels with the same l value decrease as the n values increase, so that all the atomic orbitals get closer in energy as the atomic number increases. This trend is not completely regular, and larger than average energy gaps occur between the $4p$ and $5s$ levels where the first set of d orbitals has been filled, and between the $6p$ and $7s$ levels where the first of the f levels comes. These energy jumps reflect the poorer-than-average shielding powers of d and f electrons.

Apart from the major discontinuities at the rare gases, there is also a gap in energy wherever the outermost electron enters a new atomic orbital. These gaps correspond to stabilization of the filled shell configurations, s^2, $d^{10}s^2$, and $f^{14}d^{10}s^2$,

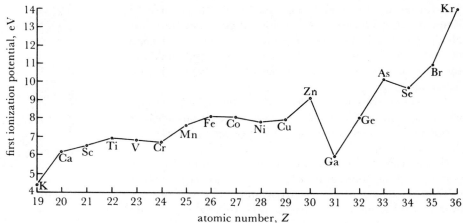

FIGURE 8.5 *Variation of the first ionization potential across the First Long Period*

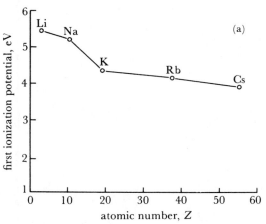

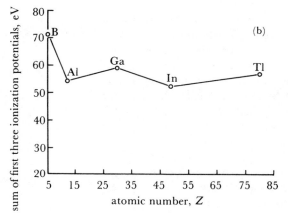

FIGURE 8.6 *Variation in ionization potentials in a Group (a) first potentials of the lithium Group; (b) sum of the first three potentials of the boron Group*

before the *p* orbitals are occupied, and also suggest the possibility of transfer of *s* electrons into the *d* shell to give the d^{10} configuration, or of *d* electrons into the *f* shell to give the f^{14} arrangement, which was discussed in the previous section.

As the ionization potentials measure the energy required to remove the least tightly bound electron from an atom or ion, values reflect the stability of the configuration from which the electron is being removed. Table 2.8 gives the ionization potentials of the elements.

The stability of the rare gas configurations can be seen, both from the high energies required to remove an electron from the rare gases themselves, and from the leap in the values of the potential when the rare gas configuration has to be broken (i.e. when the second electron is removed from an alkali metal, the third electron is removed from an

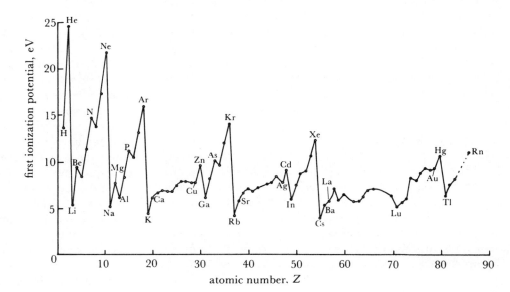

FIGURE 8.7 *Variation of the first ionization potential with Z*

alkaline earth, the fourth electron from a boron Group element, etc.). The very low first ionization potentials of the alkali metals, and, to a lesser extent, the low first and second potentials of the alkaline earths, show how loosely held are the first one or two electrons outside the rare gas configuration.

The relative stabilities of filled and half-filled shells show up when the variation of the first ionization potentials across a Period is plotted. Figures 8.4 and 8.5 show such plots across a short and a long Period respectively.

As the atomic size increases within a Group, where the outer electronic configuration is the same, it becomes easier to remove the outermost electrons and this is shown by a decrease in the ionization potentials down a Group. Figure 8.6 shows this for the first ionization potential of the alkali metals and for the sum of the first three potentials of the boron Group. In the latter, the effect of the insertion of the first d shell is shown by the potentials of gallium, and that of the presence of the f series by the value for thallium. These effects are most marked in the boron Group, which follows immediately after the transition series, but they continue to show in the rest of the p block of elements, both in the ionization potentials and in small discontinuities in the chemistry of gallium, germanium, arsenic, selenium, and bromine. The complete graph for all the elements is shown in Figure 8.7: ionization potentials increase from left to right across a Period and decrease with increasing atomic weight down a Group. The variation for elements filling an s or p level is much greater than that for the d or f elements.

Since the electron affinities measure the energy of the process converse to ionization, the energy of gaining an electron, the stable configurations are similarly reflected in the electron affinities, as the values in Table 2.9 show. Thus the addition of an electron to a halogen atom is exothermic, where the electron fills up to the rare gas configuration, while the addition of an extra electron to the rare gases is an endothermic process. Similarly, for the copper Group a large negative electron affinity marks the tendency to complete the $d^{10}s^2$ configuration. The values for the alkali metals, the chromium Group and the carbon Group show the tendency to attain, respectively, s^2, d^5 and p^3 configurations, while the values for the succeeding groups are markedly lower showing that the addition of a further electron to the filled or half-filled shell is a much less favoured process. However, less weight can be put on the trends in electron affinities as the values are more tentative than those for ionization potentials.

8.4 Atomic and ionic sizes

The definition and determination of the various sets of atomic and ionic radii is discussed in section 2.15. It is possible to derive an approximately self-consistent set of atomic radii which apply to all the elements and these are shown in Figure 8.8, plotted against the atomic number. A similar set for real or hypothetical cations and anions with the rare gas structures are shown for the Main Group elements in Figure 8.9. Figure 8.8 has many features in common with the ionization potential plot of Figure 8.7. The main discontinuities in size come between the rare gases and the alkali metals

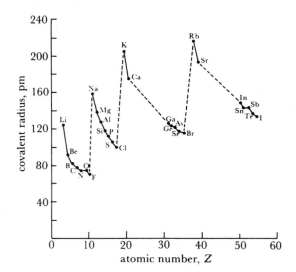

FIGURE 8.8 *Variation of 'covalent' atomic radii with* Z

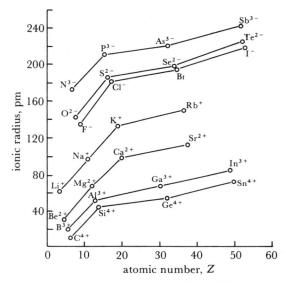

FIGURE 8.9 *Variation of ionic radii, corresponding to rare gas configurations, with* Z, *for Main Group elements*

where the outermost electron has to enter a completely new quantum shell. There is thus a marked increase in size and a marked decrease in ionization potential at this point. Due to the imperfect shielding of the valency shell electrons by each other, the effective nuclear charge increases, on average, across a Period. The outer electrons become more and more tightly bound and the atomic radius decreases while the ionization potential increases. The discontinuities at the filled shell configurations are particularly clear. The slow changes across the d and f series contrast markedly with the sharp changes in the s and p blocks, and the general decrease in size across the lanthanide series has a noticeable effect in reducing the sizes of the following elements.

The variation in atomic size may be generalized as a decrease on going from left to right across a Period and an increase in going down a Group. These changes are the exact reverse of the ionization potential changes, as would be expected. The changes in ionic radii generally reflect these changes in the atomic sizes.

The parallelism between the changes in radius and ionization potential is not, of course, accidental but follows from the existence and arrangement of the atomic orbitals. Both the size and the ionization potential would be expected to change as the number of electrons in the atom increases, and this change would be discontinuous whenever a new orbital was occupied. The inter-electronic forces, including both the electrostatic repulsions and the exchange forces, modify this pattern of change but leave the main outlines. The change embodied in the effective nuclear charge clearly affects both the extension of the electron cloud and the energy required to remove an outer electron. The property of electronegativity discussed in section 2.16 is a summarizing parameter which gives effect to the pattern of changes discussed above. It will be seen from the electronegativity values given in Table 2.14 that these increase towards the right of the Periods and decrease down the Groups. In addition, they reflect the other, smaller, variations which have been remarked; for example, the changes in the Main Groups are more pronounced than in the Transition Groups, and the discontinuity in properties of the elements from gallium to bromine when compared with the rest of their respective Groups is reflected in their electronegativity values.

8.5 Chemical behaviour and Periodic position

The detailed chemistry of the elements is discussed in the succeeding chapters. In this section, the skeleton of the Periodic properties is outlined to provide a framework for the more detailed account which follows.

Those elements where the outermost electrons are in a new quantum level, after a rare gas configuration, normally react by losing these loosely bound electrons and forming cations. This mode of behaviour is typical of the elements of the lithium, beryllium, and scandium Groups together with the lanthanide elements, which have the respective valency shell configurations, s^1, s^2, and d^1s^2. All these elements, with the exception of beryllium itself, lose these outer electrons completely with the formation of cations; M^+ in the lithium Group, M^{2+} in the beryllium Group, and M^{3+} for scandium, yttrium and the lanthanons.

The elements of the boron, carbon, nitrogen, oxygen, and fluorine Groups, where the outermost electrons are in p orbitals, show more complicated behaviour with varying oxidation states and in forming both covalent and ionic compounds. These p elements show a maximum oxidation state equal to the sum of the s and p electrons (the Group Oxidation State) and the other relatively stable oxidation states differ from the Group state by multiples of two. Thus the boron Group elements, with the configuration s^2p^1, show the Group oxidation state of III and the other stable state in this Group is I; the halogens, s^2p^5, show the Group state of VII and also the states V, III, I, and $-$I. The Group oxidation state is the most stable one for the lighter elements, especially of the earlier Groups, and a state two less than the Group state becomes the most stable for the heavier elements. Thus boron and carbon are stable in the III and IV states respectively, while their heaviest congeners are stable in the I and II states. In the Groups to the right of the Periodic Table,

where a larger variety of oxidation states is possible, the picture is more complex and a further trend becomes apparent. This is the tendency of the lighter elements and those in the halogen Group to form anions—especially N^{3-}, O^{2-}, and X^- in the halogens.

The elements where the d orbitals are filling have a Group oxidation state equal to the sum of the s and d electrons. This state is shown in the earlier Groups but involves too many electrons in the later Groups of the transition series where only low states occur. The highest oxidation state shown in the Periodic Table is VIII, by ruthenium, osmium and xenon. The common oxidation states of the d elements vary in single steps, instead of the double ones shown by the p elements. Thus in the Group with the same number of valency electrons as the halogens, the manganese Group d^5s^2, the oxidation states found are VII, VI, V, IV, III, II, I, and 0. Another distinction between the behaviour of a d Group and a p Group is that the heavier elements of a d Group are more stable in the higher oxidation states, in contrast to the trend in the p Group. Thus, the most stable state of manganese is II while its heavier congener, rhenium, is stable in the IV and VII states. The lower oxidation states of the transition metals, particularly the II state, often occur as cations, while the higher states are commonly bound to oxygen or to the halogens in covalent molecules or anions.

The f elements of the lanthanide series show only one stable oxidation state, the III state. This corresponds to the configuration $4f^1$ for cerium(III) and to $4f^n$ for the other elements up to $4f^{14}$ in lutetium(III). As already noted, three electrons are lost, as if the configuration was $4f^n5d^16s^2$, despite the fact that most of the elements do not have the d electron in the ground state. One or two of these elements do show fairly stable oxidation states other than the III state, and most of these correspond to the f^0, f^7 or f^{14} configurations. Thus, cerium shows a IV state corresponding to f^0, europium has a II state, and terbium a IV state, corresponding to f^7; while ytterbium has a II state corresponding to f^{14}

In the heaviest elements, where the $5f$ and $6d$ levels are filling, the pattern of oxidation states is less simple than with the lanthanides. As these two energy levels are very close, the earlier elements show a considerable variety of oxidation states with the maximum rising from III for actinium to VI at uranium and VII at neptunium and plutonium. The later actinide elements resemble the lanthanons more closely, and the III state becomes the most stable one at about curium (see Jørgensen in the references).

These patterns of behaviour lead to the division of the Periodic Table into four major blocks: the s elements, the p elements, the d elements, and the f elements, together with a number of Groups which serve to bridge these divisions. The s elements are those of the lithium and beryllium Groups; the boron, carbon, nitrogen, oxygen and fluorine Groups make up the p block; the lanthanides and actinides form the f block; and the remaining transition elements, the d block. As the typical behaviour of d elements depends on the presence of both d electrons and available d orbitals, the scandium Group (which always loses its solitary d electron and forms the M^{3+} ion) and the zinc Group (which always preserves the filled d^{10} configuration) are not properly *d* elements and are best regarded as bridging Groups. The scandium Group links the *s* elements and the *d* block, while the zinc Group links the *d* block to the *p* elements, and both Groups show the appropriate intermediate properties. The chemistry of the scandium Group links strongly, also, with that of the *f* elements. Indeed, the general chemistry of the lanthanides in the III state is almost identical with that of yttrium and lanthanum. There are bigger differences between the chemistry of actinium and the actinide elements. It is convenient to treat scandium, yttrium, lanthanum, actinium, and the lanthanides all together, and to treat the actinides independently. The remaining Group in the Periodic Table is the helium Group. This Group forms the division between the *p* block and the *s* block, and the recently-discovered chemistry of xenon shows strong links with that of iodine. Finally, there is the lightest element, hydrogen, which falls into no Group so far discussed and is best regarded as a unique introductory element to the Periodic Table. The arrangement of the following chapters reflects this division of the Periodic Table. The chemistry of hydrogen and its compounds is treated first, giving a microcosm of the properties of the elements. Then follow chapters on the *s* elements, the scandium Group and the lanthanides, the actinides, the *d* elements, and the *p* elements (Figures 2.14 and 8.10).

8.6 Methods of showing the stabilities of oxidation states

In many cases, as implied in the last section, elements show a number of oxidation states and some method of determining and portraying the relative stabilities of these states is necessary. Similarly, it is useful to be able to compare relative stabilities and show the trends in stabilities among related elements. Of course, in a full and complete description of the chemistry of an element, these stabilities are clear from the range of compounds of a given oxidation state and their ease of formation and decomposition. However, it is impossible to give a complete description of the known chemistry of any element within the space available in a general textbook so that methods of summarizing and illustrating the general behaviour are required. Stabilities vary a good deal with the chemical environment—the temperature, solid, liquid, or gaseous state of the compound, solvent, presence of air or moisture, and so forth. However, there are two

chemical states which are very common, as a solid and in solution in water.

The stability of an element in a particular oxidation state in the solid may be determined from the variety of ions or ligands with which it reacts to form solid compounds. Thus, a strongly oxidizing state will form compounds with non-oxidizable ligands only, and *vice versa* for a reducing state, while a stable state will give compounds with a wide variety of ligands. For example, consider the relative stabilities of the II and III states of iron, cobalt, and nickel as shown by the existence of the solid compounds of the II state with oxychloride anions.

Fe(II) $Fe(ClO_4)_2 : ClO_3^-$ and ClO_2^- oxidize to Fe(III)
Co(II) $Co(ClO_4)_2$ and $Co(ClO_3)_2 : ClO_2^-$ oxidizes to Co(III)
Ni(II) $Ni(ClO_4)_2$, $Ni(ClO_3)_2$, and $Ni(ClO_2)_2$

The oxidizing power of the oxychloride ions increases in the order, perchlorate < chlorate < chlorite, so the existence of the compounds shown above illustrate that the order of stabilities of the II state is Fe < Co < Ni.

A convenient set of ligands, for the purpose of demonstrating stabilities, is provided by the halides which form compounds with almost all oxidation states. An element in a stable oxidation state will form all four halides, a strongly-reducing state will tend not to have a fluoride, while a strongly-oxidizing state will tend not to have an iodide or bromide. Similarly, an oxidizing state will show an oxide but no sulfide, while a reducing state will form a sulfide but no oxide. Thus, the existence and stabilities of the oxides, sulfides, and halides of the elements in their different states provides a useful general guide to the stabilities, in the solid state, of the various oxidation states. In the following chapters, Tables 13.3, 13.4, 13.5, of transition element halides and oxides (sulfides are omitted here as their stoichiometry is often in doubt) and Table 17.2, of *p* block element oxides, sulfides, and halides, are used to provide a general, overall view of the chemistry of these elements.

Stabilities in aqueous solution are expected to be broadly similar to stabilities in the solid state, but to differ in detail due to differences between lattice energies and hydration energies (compare Chapters 5 and 6). The relative stabilities of the various oxidation states of an element in solution are given by the free energy changes of the set of half-reactions

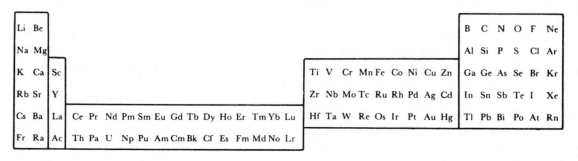

FIGURE 8.10 *Divisions of the Periodic Table*

connecting each pair of oxidation states. For example, the stabilities of the states of copper depend on the free energy changes of the half-reactions:

$$Cu^{2+} + e^- = Cu^+$$
$$Cu^+ + e^- = Cu$$
$$\text{and } Cu^{2+} + 2e^- = Cu$$

(These three free energies are not independent, of course, any one may be derived from the other two.)

Such free energies are related to the corresponding redox potentials, since $-\Delta G = nFE$, see p. 102. A full list of redox potentials for the half-reactions of all the elements is available, but a method is required for displaying these values to the best advantage. It has been suggested by Ebsworth (see references) that free energies may be usefully displayed graphically. In this, the oxidation states of the element are plotted against the free energy change, in one electron steps. The method is most readily discussed in terms of particular cases, for example uranium and americium whose potentials in acid solution have the values shown below:

	E^0 (volts)	
	M = U	M = Am
$MO_2^{2+} + e^- = MO_2^+$	0·05	1·64
$MO_2^{2+} + 4H^+ + 2e^- = M^{4+} + 2H_2O$	0·33	
$MO_2^+ + 4H^+ + e^- = M^{4+} + 2H_2O$	0·62	1·26
$M^{4+} + e^- = M^{3+}$	−0·61	2·18
$M^{3+} + 3e^- = M$	−1·80	−2·32
$MO_2^{2+} + 4H^+ + 3e^- = M^{3+} + 2H_2O$		1·69

The diagrams, Figure 8.11, are plotted by taking the value for the element itself as zero and plotting the free energy changes for the half-reactions against the oxidation states of the element. The free energies are given as $-\Delta G^0/F = nE^0$ where n is the number of electrons involved in the change.

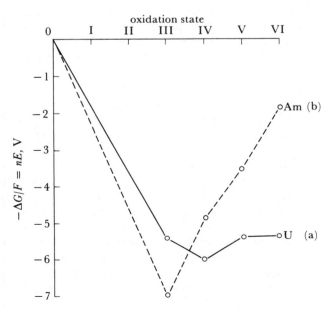

FIGURE 8.11 *Free energy versus oxidation state diagrams,* (a) *uranium,* (b) *americium*

Using americium as our example the diagram is constructed as follows (see Figure 8.11).
1. Take the value for the element as zero.
2. For the change

$$Am^{3+} + 3e^- = Am$$

E^0 is −2·32 V, therefore nE^0 is −6·96 V and is so plotted.
3. For the change

$$Am^{4+} + e^- = Am^{3+}$$

E^0 is 2·18 V, therefore nE^0 is 2·18 V and this is added to −6·96 V to give −4·78 V, which is plotted for Am(IV).
4. For the change

$$Am(V) \rightarrow Am(IV) \qquad (AmO_2^+ \rightarrow Am^{4+})$$

E^0 is 1·26 V and nE^0 is 1·26 V. Adding this to −4·78 V gives −3·52 V to be plotted for Am(V).
5. For Am(VI), continuing in the same way one obtains −1·88 V.

Plots like these provide a kind of cross-section of the chemistry of the elements. The lower a state lies on the diagram, the more stable it is. Also, since a change from one state to another which involves moving down a slope involves a negative change of free energy, it is thermodynamically favourable. Similarly, a change which involves an 'uphill' move is unfavourable. These points about changes apply to those involving any two states, not only states which are nearest neighbours. Thus the changes from U(0) to any of the states, U(III), U(IV), U(V), or U(VI) are favourable, and the changes from U(IV) to any of the states U(0), U(III), U(V), and U(VI) are unfavourable. Clearly, the greater the slope, the greater the driving force, so that Am(0) goes to Am(III) more readily than to Am(IV).

A particular oxidation state may be represented on such a diagram by a point which is (a) a minimum, (b) a maximum, (c) a concave point (i.e. one lying below the line joining the two neighbouring points), (d) a convex point (i.e. one lying above the line joining its neighbours), or (e) a 'linear' point (i.e. one lying on the line joining its neighbours). Table 8.3 lists the stability properties of the oxidation states represented by these different types.

The properties of the states represented by concave, convex, and linear points are not restricted to being with respect to the two nearest neighbour states but may hold with respect to any two states. For example, in Figure 8.11, Am(V) is linear, not only with respect to Am(IV) and Am(VI), but also with respect to Am(III) and Am(VI) and the disproportionation products will include Am(III). In a similar way, in the case of phosphorus (Figure 17.27) the points representing the 0, I, and III states are all convex with respect to the −III and V states. That is, the element (0 state), hypophosphorous acid (I state) and phosphorous acid (III state) all disproportionate to phosphine, PH_3, (−III) and phosphoric acid (V state).

Applying these results to uranium and americium, Figure 8.11 shows, at a glance, that uranium(V) is unstable in aqueous solution, uranium(IV) is stable, and uranium(III)

TABLE 8.3 Types of oxidation state in free energy diagrams

State represented by a point	Example from Figure 8.12	Further examples (Figure numbers in brackets)	Properties of the oxidation state
Minimum (a)	U(IV), Am(III)	Mn(II) (14.20) Cr(III) (14.15)	Stable relative to neighbouring states.
Maximum (b)	——	N(−I) (17.25)	Unstable relative to neighbouring states.
Concave (c)	U(III)	V(IV) (14.8) Re(IV) (15.19)	Relatively stable with respect to disproportionation into the neighbouring states.
Convex (d)	U(V)	Mn(III) (14.20) Re(III) (15.19)	Relatively unstable with respect to disproportionation.
Linear (e)	Am(V) and (IV)	All the intermediate states of Cl (17.51)	*Intermediate with respect to disproportionation.

*That is to say that, at equilibrium, there are approximately equal amounts of the original state and the states to which it has disproportionated.

is relatively stable. Uranium(VI) is also stable, with the U(VI) → U(IV) change being mildly oxidizing. By contrast, americium(VI) is unstable and the Am(VI) → Am(III) change is strongly oxidizing. The most stable state of americium is Am(III) and americium(IV) and (V) both disproportionate, with the V state a little more stable than the IV state.

The elements lying between uranium and americium—neptunium and plutonium—also show III, IV, V, and VI states and the VI state becomes more oxidizing and less stable in order from uranium, through neptunium and plutonium, to americium. This is clear from Figure 12.1, in which the curves for all four elements are plotted. The VI state of uranium lies lowest and the others in the order U < Np < Pu < Am. In a similar way, the relative stabilities of the II and III states of iron, cobalt and nickel, referred to above, are clear from Figure 14.1

These free energy diagrams will be extensively used in the following chapters but one or two reservations about them must be kept in mind. First, all the oxidation potential data are not of equal validity and some values may be in error so it is not wise to make too much of small effects—such as whether a point is slightly convex or slightly concave. It is unlikely, however, that there are any gross inaccuracies. The second point is that the properties listed in Table 8.3 are thermodynamic and there is no information about the rates of reactions. Thus a state may be thermodynamically unstable but persist in solution because its rate of reaction or disproportionation is slow. Similarly, the potential data apply only to systems in equilibrium, and the rate of attaining equilibrium may be slow. This applies particularly to systems involving solids—for example, an element. Many elements which are strongly reducing react only very slowly due to surface effects and the like. Throughout the later chapters, curves are plotted for potential data in acid solution, at a pH of 0. Similar data are available for alkaline solution, and a set of equivalent curves could be drawn for such media. These are less useful as many more states appear as solids (as hydroxides or hydrated oxides) in alkali.

The occurrence and stability of the solid halides, oxides, and sulfides of the elements, taken together with the free energy diagrams linking the different oxidation states, gives an adequate guide to the general chemistry of an element in its common compounds. In general, a stable state may be assumed to form compounds with all the common anions and ligands—all the oxyanions, pseudohalides, organic acid anions, hydride, nitride, carbide, amide, sulfur anions and so on. Unstable states will form a much more limited set of compounds, down to those states represented by only one or two examples. A state which is unstable and oxidizing will form no compounds with oxidizable ligands such as nitrite or organic groups, and similarly, a reducing state will form no compounds with oxidizing ligands, as with the chlorite and chlorate compounds of iron mentioned above. By avoiding cataloguing such compounds of the elements, space is preserved for the mention of the more unusual compounds formed and as many of these as possible have been discussed in the later chapters which may be regarded as an introduction and supplement to the systematic chemistry given in the major inorganic textbooks which are listed in the reading lists.

8.7 The abundance and occurrence of the elements

One of the most satisfying scientific constructs of the second half of the 20th century has been the picture of the genesis of the elements. Once 19th-century chemists had discovered that each element emitted a distinctive spectrum when heated, the way was open to identify elements in the stars by their spectral lines. In this way, helium (from the Greek *helios*, sun) was found in the solar spectrum about a decade before it was discovered on Earth. Such studies led to the evaluation of the abundance of the different elements in the Universe and a striking picture emerged. Hydrogen and helium together account for about 99% of the mass of the universe and over 99.9% of the atoms. At the other extreme, the heaviest elements are present to the extent of about 10^{-12} of that of hydrogen. The distribution of abundances has the following features:

(a) The abundance of the elements declines steeply and exponentially from H to a mass about 100, then more linearly to U

(b) Within this broad pattern, Li, Be and B are markedly scarce relative to their neighbours, by a factor of 10^{10} for Li and B, and 10^{12} for Be

(c) Less pronounced relative deficiencies, by factors of $10 - 10^3$, occur for elements around F, Sc, As, In and Ta

(d) There is a marked excess of around 10^3, over the trend line, for elements near Fe, and less marked excesses near Zr, Xe and Pt

(e) Finally there is a very pronounced alternation in abundances between successive elements

Table 8.4 gives some general values.

These abundance figures accumulated over more than a century of observation and analysis. They have been explained in considerable detail by a theory based on the genesis of the elements from hydrogen in the stars. This was refined by many workers over some 30 years up to the publication of the definitive review by Burbidge, Burbidge, Fowler, and Hoyle in 1957. More recent developments may be found in Fowler's Nobel lecture in 1984 (see references). The picture is extraordinarily satisfying as the model not only accounts for the general pattern of abundances, but explains most of the fine detail. The main features are:

(1) An initial primitive universe with about 75% H and 25% He.

(2) Condensation by gravity into early stars where H nuclei fused into He, and further fusion processes built up heavier elements. Eventually stellar condensation led to destruction of the star and scattering of the elements to be re-formed into further stars and the process continued.

(3) Eventually, the Sun and our planets condensed out of an accumulation of such cosmic dust. The Sun has proceeded along the H-fusion path while the cooling and evolution of the Earth has produced the crustal distribution of elements of the present day.

TABLE 8.4 Cosmic abundances of the elements

Element	Relative abundances			
	by number	% of total	by weight	% of total
hydrogen	4.0×10^{10}	92.8	4.0×10^{10}	75.5
helium	3.1×10^9	7.1	1.2×10^{10}	23.1
Li, Be, B	1.4×10^2	3.3×10^{-7}	1.3×10^3	2.4×10^{-6}
C, N, O, Ne	4.0×10^7	0.09	6.5×10^8	1.2
Na to Sc	2.7×10^6	0.006	7.3×10^7	0.13
Fe group (a)	6.4×10^5	0.0015	3.6×10^7	0.07
middle group (b)	1.1×10^3	2.6×10^{-6}	7.7×10^4	1.4×10^{-6}
heavy group (c)	28	6.5×10^{-8}	4.6×10^3	8.6×10^{-8}

(a) A_r from 50 to 62
(b) A_r from 63 to 100
(c) A_r over 100

It is the details of the fusion processes which account for the present abundances. Briefly, the first major process is the formation of He by a number of steps from 4 H nuclei. This evolves huge amounts of energy and is the major source of stellar radiation. Then follows further fusion of He nuclei (mass 4) into heavier elements. Such a process bypasses Li (isotopes of masses 6, 7) and B (10 and 11) but passes through the Be isotope of mass 8. This is extremely unstable (half-life about 10^{-16} sec) but survives long enough to fuse with a further He to give C (mass 12). Similar processes with alpha particles add further mass in steps of 4 to give O (16), Ne (20), Mg (24), Si (28), S (32), and so on up to Ti (48). The actual abundance of these elements represents the balance between their ease of formation, and their stability to further reaction. As the number of steps increases, the abundance falls, since each element depends on the previous formation of its precursor. Thus the general exponential fall-off in abundances is to be expected. All the elements of intermediate masses form from intermediate and usually less favoured processes, and the abundances are now understood in detail in terms of the energies of formation, and the liability to further reaction.

Different processes occur in stars of different masses and temperatures, at different times in the course of their evolution. Thus the sun is in the basic H-burning stage, as are the majority of stars in the galaxy.

Nuclear configurations around Fe are the most stable, with lighter elements combining exothermically towards mass 56, and heavy elements breaking down, again by exothermic processes, towards this mass range. Under conditions of very high temperature and pressure, an equilibration process takes place to give the relatively high ratios of the elements around this stable minimum.

Building nuclei of greater mass than Fe involves neutron capture, and two processes occur. In the slow process, which occurs over a period of years, there is sufficient time after the capture for the nucleus to rearrange, by electron emission, to give the most stable proton/neutron ratio. In the fast process, typically in a supernova explosion, several neutrons are captured in a very short period before nuclear reorganization occurs. These two processes thus give rise to the heavier isotopes, and with minor processes like proton capture, explain the distribution features in detail, as described in the review cited.

The above is only a brief outline of a very full theory. We have sketched it here since it is too little known that we can not only reach out into the universe and assess the abundance of the elements in distant stars, but we can also account for the observed abundances in terms of quantitatively known processes. While such cosmic, nuclear, processes are not themselves included in inorganic chemistry, they do create our starting materials, the elements.

The elements are now known to form simple combinations in interstellar space. In dark regions of the universe, matter is present which absorbs starlight, though the concentration is so minute—of the order of atoms per cubic metre—that it is of the order of a million million million times less than in an ultrahigh vacuum on Earth. Atoms do link into units which may be detected spectroscopically. Many of the species are radicals—it takes a long time for an OH unit to pick up another H atom, for example. Diatomic units involving the more abundant elements (Table 8.4) have all been identified, and multiatom units with as many as ten atoms are suggested. Entities involve H with C and O, and also the less abundant N and even S. We cannot further pursue this fascinating area of cosmochemistry here, but as more information is collected we can foresee the development of an interstellar inorganic chemistry!

The element abundances in the universe have a broad relation to the pattern of abundances in the Earth's crust, but further processes of loss (especially of H and He) and of concentration have taken place so that there are major differences in detail. The distribution of the elements on Earth is now reasonably well-evaluated (and we have preliminary figures for the Moon and Mars), though further refinement will occur, especially as we penetrate deeper into the crust. A good start has been made in understanding the geochemical processes which have collected and redistributed the elements into the rocks, minerals and ores which we find today. A start has even been made in reproducing on a laboratory scale the enormous pressures and temperatures of the mantle processes which form the igneous rocks. Table 8.5 shows the most abundant crustal elements; lesser abundances range all the way down to about three parts in 10^{-16} for actinium.

It is worth remarking on another interface zone—that between the atmosphere and space. Here the abundance of molecules is extremely low, the radiation level from the Sun is very high, and the component elements are those of the atmosphere. Concentrations of radicals like OH or OOH are relatively high, and species which are too reactive to accumulate near the Earth's surface, such as O_3 and H_2O_2, are important. The ozone problem is becoming of vital importance, as attack by Cl radicals derived from chlorofluorocarbons (CFCs) is reducing the ozone concentration and diminishing the vital shield against hard ultraviolet radiation. This and many other processes, like the greenhouse effect, have focused attention on the chemistry of the upper

TABLE 8.5 The most abundant crustal elements

	Mass, %		Mass, %
Oxygen	50	Sodium	2·6
Silicon	26	Potassium	2·4
Aluminium	7·5	Magnesium	1·9
Iron	4·7	Hydrogen	0·9
Calcium	3·4	Titanium	0·6

and lower atmosphere which is discussed in greater detail in Chapter 20.

The abundant elements are readily accessible, as are those elements which, though rare overall, occur in localized concentrated deposits. One example is boron, which occurs to the extent of only three parts per million in the crust but is found in concentrated deposits as borax. Also accessible are those elements which are found native or are readily recovered from their ores, for example the precious metals silver and gold. All the chemistry of common, or readily accessible, elements is well-explored.

A second group includes all those relatively rare elements which occur only in small proportions in the crust and are found only as trace constituents in the ores of more important minerals. We should also include the more expensive of the precious metals such as gold and platinum. Such elements form an intermediate group where intensive study has been more recent. Although all these elements are now well understood, interest often reflects their possible applications. For example, germanium, gallium, and indium have attracted attention for their semi-conductor applications, most of the heavier transition elements have been investigated for their organometallic compounds, catalytic properties and for their effects in atomic piles. Large-scale separation methods have been developed which have made pure samples of the individual lanthanide elements available.

The rarest elements are the artificial elements which have no naturally-occurring isotopes. These include all the post-uranium elements and a few lighter ones such as promethium and technetium. Supplies of many of these elements

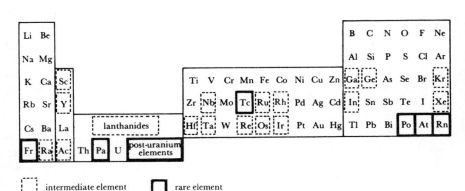

FIGURE 8.12 *Distribution of the elements*
The elements are divided into three classes, common, intermediate and rare, on the basis of their natural abundance combined with their accessibility. The boundaries between types are necessarily somewhat arbitrary.

FIGURE 8.13 *Methods used for the extraction of the elements*

(1) Reactive metals extracted electrolytically in a non-aqueous system. Main sources of alkaline earth minerals are insoluble sulfates and carbonates, while alkali metals often form deposits of soluble salts e.g. NaCl or KCl.

(2) Reactive metals of high charge with strong affinity for oxygen and occurring as oxyanions or double oxides. Separation is by electrolytic or chemical reduction, especially by active metal replacement.

(3) Elements occurring in sulfide ores or otherwise associated with sulfur. Extracted usually by roasting to oxide and then reducing or treating thermally.

(4) Elements occur native or in easily decomposed compounds yielding to thermal treatment.

(5) Non-metals which occur free, in the atmosphere, or as anions.

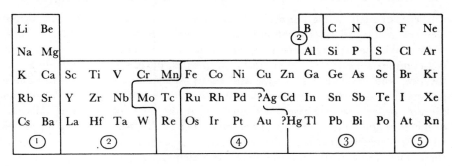

have become available recently from the fission products or synthesis products of nuclear reactors, while the attempt to synthesize ever heavier elements continues. Elements 111 and 112 were announced recently, with the identification based on only a few atoms. Figure 8.12 gives a broad picture of element accessibility.

8.8 The extraction of the elements

The extraction processes for producing elements from their ores may be divided into three classes in order of increasing power:

(i) mechanical separation and simple heat treatment,

(ii) separations involving chemical reduction,

(iii) separations by electrolytic reduction.

Although the days of the gold rush may be gone, mechanical separation on a huge scale is still the basic process of gold and diamond production; the final recovery of gold being by chemical reduction with zinc of a cyanide complex in solution. Also included in the first class of mild treatments are the recoveries, by distillation or thermal decomposition, of elements such as zinc or mercury. A further reaction of wide application in this class is the Van Arkel and De Boer process which is used to produce very pure metals on a small scale. In this process, elements which form volatile iodides are purified by a cyclic process in which the iodide is formed at a low temperature and decomposed on a heated wire, at a higher temperature, to the element and iodine. The iodine is recycled to form more iodide. For example, zirconium gives ZrI_4 when heated at 600 °C with iodine and this may be decomposed at 1800 °C, on a heated tungsten or zirconium filament, to zirconium and iodine.

Most commercial separations fall into the second class, commonly involving reduction by carbon, as in iron production. Reduction by other elements is also found on a small scale: examples include the preparation of pure molybdenum by hydrogen reduction and the reduction of titanium tetrachloride by magnesium in the Kroll process.

Electrolytic reduction represents the most powerful method available, but it is expensive compared to the chemical methods and is only used either for very reactive metals, such as magnesium or aluminium, or for the production of samples of high purity as in the electrolytic refining of copper (which has the additional advantage of allowing the recovery of valuable minor contaminants such as silver and gold). The main commercial application of electrolysis is, of course, in aluminium manufacture. Here, the ore bauxite, which is impure Al_2O_3, is purified by alkaline treatment, then dissolved in molten cryolite (Na_3AlF_6) and reduced electrolytically in this fused-salt system. The alkali metals and calcium and magnesium are also produced by electrolysis in

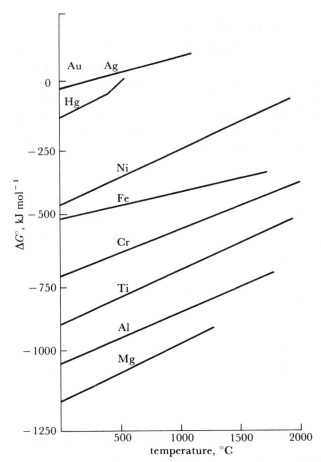

FIGURE 8.14 *The variation with temperature of the free energy of metal oxide formation according to equation (8.1)*

a fused salt melt, while copper and zinc are among those elements recovered by electrolysis in an aqueous medium.

Although the type of process chosen for any one element is a complex function of the chemical properties, nature of the ore, and relative economics, and many elements are processed in different ways in different parts of the world, the very general picture given in Figure 8.13 of the mode of occurrence and method of recovery of the elements is broadly true. The extraction of the elements reflects, in some measure, their general chemistry and thus elements which are treated similarly tend to lie together in the Periodic Table. (This is also true of their occurrence. Elements of similar chemistry would have behaved similarly when the rocks crystallized out of the primitive magma, and later processes, such as leaching out of soluble salts, would further tend to concentrate similar elements in similar forms.)

The process of extraction by carbon reduction illustrates several interesting general points in the chemistry of metal recovery and is worth discussing in fuller detail. The extent of the reduction of one element from its oxide by a second element depends on the difference in free energy of the two oxidation reactions of the type:

$$M + \frac{x}{2}O_2 \rightarrow MO_x \quad \ldots \ldots \ldots \ldots \quad (8.1)$$

It will be recalled that reactions which evolve free energy tend to occur spontaneously, so that the equilibrium between two elements and their oxides:

$$M'O + M'' \rightleftharpoons M''O + M' \quad \ldots \ldots \ldots \ldots (8.2)$$

(or the corresponding equations for different oxide stoichiometries), will favour that oxide whose free energy of formation (equation 8.1) is most negative. The free energy change, ΔG, is separable into two components, the heat change ΔH, and the energy involved in the change of entropy, $T\Delta S$, where T is the absolute temperature:

$$\Delta G = \Delta H - T\Delta S$$

In the formation of a metal oxide according to equation (8.1), the heat change is usually favourable but, as the reaction uses up a gaseous component (the oxygen) which has a relatively large entropy, the entropy term is unfavourable and this energy increases with increasing T. As a result, the free energy change for metal oxide formation in equation (8.1) falls off with rising temperature in a broadly similar way for any metal, as in the examples shown in Figure 8.14. It will be seen that the metal oxides can be divided into two classes: first those which are intrinsically unstable at normal temperatures, such as gold oxide, or at accessible temperatures such as silver oxide or mercury oxide, and the second class which contains the elements with a favourable free energy of oxide formation at any accessible temperature. Gold, silver and mercury, fall into the class (i) (p. 135) of elements which can be extracted by simple heat treatment alone, while the second class of elements contains those whose extraction process falls into class (ii) or (iii), requiring reduction. Any metal will reduce the oxide of any second metal which lies above it in Figure 8.14, according to equation (8.2), as the net change in the free energy will be negative (i.e. favourable) by an amount equal to the difference between the two curves at the appropriate temperature. For example, magnesium will reduce all the other oxides shown in Figure 8.14.

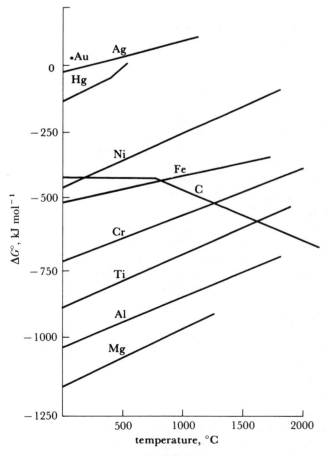

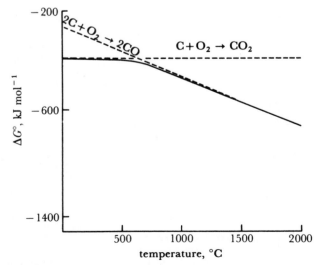

FIGURE 8.15 *The variation with temperature of the free energy of the reaction of carbon with oxygen*

FIGURE 8.16 *Combined diagram of the free energy changes of the reactions of metals and carbon with oxygen*

The formation of a metal oxide thus involves a free energy change which becomes less negative with increasing temperature, because the oxidation proceeds with the consumption of a gaseous component with a corresponding loss of entropy. The formation of carbon oxides is quite different. The main reaction between carbon and oxygen at higher temperatures is:

$$2C + O_2 \rightarrow 2CO$$

in which an excess of one mole of gas is produced for each mole of oxygen used. This reaction therefore involves a positive entropy change and its free energy becomes more negative as the temperature rises. At temperatures below 500 °C, the main reaction is:

$$C + O_2 \rightarrow CO_2$$

where there is no overall change in the amount of gaseous reactant. The entropy change is thus small and the free energy of this reaction is almost independent of temperature. Figure 8.15 shows the overall change in free energy as carbon is oxidized at increasing temperatures. The total free energy curve falls as T rises and will eventually cross every curve, of the type shown in Figure 8.14, for the free energy of metal oxide formation. It follows that, in principle, carbon may be used to reduce any metal oxide if a high enough temperature can be attained. In practice, of course, temperatures which would be high enough to allow carbon to reduce the more stable oxides such as TiO_2 or Al_2O_3 are not accessible economically on a large scale. The usefulness of carbon as a reducing agent may be seen in Figure 8.16, in which the curves of Figures 8.14 and 8.15 are superimposed. Of course, other chemical factors come into play in choosing reduction conditions, such as the formation by many elements of unwanted carbides, which limit the use of carbon in practice.

Once again there are important reservations to be kept in mind when applying such a thermodynamic analysis as that sketched above. Free energy changes are calculated on the assumption that the system is in equilibrium—which is far from the case in practice—and they give no indication of the kinetics of the reaction so that a particular reduction may have a favourable free energy change but be too slow. However, the thermodynamic analysis does distinguish reactions which will occur from those which will not, and it gives some indication of the conditions, for example of temperature, which are suitable. For a fuller account of this topic, and curves for sulfide and halide systems, the monograph by Ives (see references) should be consulted.

While thermal and electrolytic processes are still important for recovering the pure elements, hydrometallurgical processes using solvent extraction principles are becoming increasingly popular for the separation of aqueous solutions of metals into their components, from which the pure metals can be recovered if necessary. The basic principles behind such processes have already been briefly outlined in section 7.3 and space unfortunately does not permit a more detailed account. Specific examples used industrially include uranium extraction and nuclear fuel reprocessing (section 7.3),

separation of the various lanthanide elements (section 11.3), separation of platinum from palladium, and the recovery of copper from ores.

Another method used industrially for recovering metals from ores uses bacteria. This method is particularly suited to the recovery of metals from low-grade ores, and is therefore likely to grow in importance as reserves of the various metals become depleted. Acidophilic, or 'acid-loving', bacteria such as *Thiobacillus ferrooxidans* oxidize metal sulfide ores to the metal sulfate solution, obtaining energy by the oxidation of sulfide and Fe(II). The leach solution is then collected and can be subjected to the procedures described above for the recovery of its various metal components. This technique is well-placed for the recovery of gold from low-grade ores. A review of this inter-disciplinary extraction method is given in Appendix A for the interested reader.

PROBLEMS

This chapter should be related back to Chapter 2 and forward to Chapters 10 to 12 and 14 to 17. The emphasis is on *patterns* of behaviour in the Periodic Table which are used as a framework to correlate the detailed chemistry of the elements.

8.1 Replot diagrams such as 8.4 to 8.9 against the effective nuclear charge, Z^*, rather than against Z. Discuss the features which disappear and the ones which remain.

8.2 Plot the variation of the free atom binding energies from Table 8.2 against Z. Compare how each level varies, and compare also with the ionization potential plots for the same elements. How far do the changes reflect the changes in Z, Z^*, or the electron configurations?

8.3 Plot (a) the first ionization potential and (b) the $2s$ and $2p$ core binding energies for Na to Ar against the covalent radii. Discuss any relationships which emerge.

8.4 Choose a number of elements and look up their chemistry in the systematic chemistry chapters. Determine how valid are the generalizations of section 8.5, and where their properties fit into the patterns of this chapter.

8.5 The redox potentials of the heavier actinides are

M =	Cm	Bk	Cf	Es	Fm	Md	No
$M^{3+} + 3e$ $= M$ (eV)	−2·7	−2·4	−2·1	−2·0	−2·1	−2·2	−2·5

Use these together with the values in Table 12.2 to plot an Ebsworth diagram and discuss the relative stabilities of the oxidation states.

8.6 The redox data for xenon are reported as

in acid

$$H_4XeO_6 + 2H^+ + 2e = XeO_3 + 3H_2O \qquad 3\cdot0 \text{ volts}$$
$$XeO_3 + 6H^+ + 6e = Xe + 3H_2O \qquad 1\cdot8$$
$$XeO_3 + 2HF + 4H^+ + 4e = XeF_2 + 3H_2O \qquad 1\cdot6$$
$$XeF_2 + 2H^+ + 2e = Xe + 2HF \qquad 2\cdot2$$

in base

$$HXeO_6^{3-} + 2H_2O + 2e = HXeO_4^- + 4OH^- \qquad 0\cdot9$$
$$HXeO_4^- + 2H_2O + 4e = XeO + 5OH^- \qquad 0\cdot7$$
$$HXeO_4^- + 3H_2O + 6e = Xe + 7OH^- \qquad 0\cdot9$$
$$XeO + H_2O + 2e = Xe + 2OH^- \qquad 1\cdot3$$

Plot Ebsworth diagrams for species (a) in acid and (b) in base. Compare with the halogens. [Treat XeF_2 as you would Xe^{2+}].

8.7 From one of the major textbooks, find data on the abundance of the elements in the Earth's crust and in the ocean. Plot selected values, representing the full range of abundances, of the crustal figures against the ocean ones. Discuss how far they are in parallel, and discuss major differences. Carry out the same exercise versus the abundances in the cosmos.

8.8 Compare the isolation methods detailed in the systematic chemistry chapters with the discussion in section 8.8. How far do they fit, and how far are alternative methods used when very pure samples are needed?

8.9 Plot the pattern of the historical discovery of the elements (Chapter 1) on the Periodic Table. Discuss how far it reflects the distributions of Figures 8.12 and 8.13.

8.10 Extend the information of Figure 8.16 to other elements and take the two major prehistoric metals, copper and iron, as your markers. Which other elements were potentially accessible by the technology used for (a) copper and (b) iron? Which of these were actually known? Discuss reasons for selected omissions—why not Ni which is less demanding than Fe, for example. [If you become interested in primitive metallurgy, you may wish to read further and discuss whether the other ferrous metals were completely unknown.]

9 Hydrogen

9.1 General and physical properties of hydrogen

Hydrogen is the simplest of the elements with its one valency orbital and single electron. It can react only by gaining or sharing another electron and its behaviour is therefore fairly uncomplicated. Hydrogen combines with almost all the other elements, and, as its electronegativity value comes in the middle of the range, its bonds have a wide range of polarity from the strongly positive hydrogen of the hydrogen halides to the negative, anionic hydrogen in the active metal hydrides. The properties of the hydrides thus illustrate the variation in chemical properties with Periodic Table position. In addition, the small size of the hydrogen atom presents no steric barriers and the hydrides have the shapes expected from the electronic structure of the central atom. This small size does permit the close approach of other, non-bonded, atoms and special properties result, especially the *hydrogen bond* between atoms of high electronegativity values which is discussed later.

Hydrogen does not fit into any of the Groups of the Periodic Table and is best regarded as an introductory element to the Periodic classification. It does show significant analogies to three of the other Groups which are worth noting as a guide to its chemistry:

(1) in common with the halogens, it has a tendency to gain an electron and forms the hydride ion H^-.
(2) in common with the alkali metals, it may lose an electron to form a cation, though this statement must be treated with considerable reserve as will be seen.
(3) in common with the carbon Group elements, hydrogen has a half-filled valency shell and forms covalent bonds with a wide range of polarities.

This last comparison is especially apt, particularly if hydrogen is compared with carbon with one remaining free valency as in methyl, H_3C-. There is a very close relationship between the stability and properties of the hydrides and organometallic compounds of many elements, as the Table below illustrates, and it is often useful to regard the hydride as the parent compound of a homologous series of organometallic compounds, e.g. $HSnCl_3$, CH_3SnCl_3, $C_2H_5SnCl_3$, etc. As organometallics continue to form a field of rapid growth which cuts across the old boundaries of organic and

TABLE 9.1 Comparison of some metal-hydrogen and metal-alkyl compounds

Type	*R = H*	*R = Alkyl, e.g. methyl*
NaR	ionic, Na^+H^-	ionic, $Na^+CH_3^-$
AlR_3	polymeric solid linked by electron-deficient $Al\cdots H\cdots Al$ bridges	dimer linked by electron-deficient alkyl bridges, $Al\cdots CH_3\cdots Al$
SiR_4 to PbR_4	covalent gaseous molecules, $M-H$ decreases in stability $Si > Ge > Sn > Pb$	volatile covalent compounds, $M-R$ decreases in stability from Si to Pb though more stable than $M-H$
GeR_2	polymeric oxidizable solids of obscure structure	rings $(GeR_2)_n (n = 4, 5, 6)$ or polymeric solids
$RCo(CO)_4$	unstable complex hydride with σ $Co-H$ bond	unstable organometallic complexes with σ $Co-R$ bonds
$RPtX(PPh_3)_2$	hydride with $Pt-H$ bond stabilized by Pt–phosphane interaction	stable organometallic complex with σ $Pt-C$ bond stabilized by $Pt-PPh_3$ interaction

inorganic chemistry, this facet of hydrogen chemistry is of considerable current importance.

Hydrogen is the most abundant element in the universe and the other elements are built up from hydrogen by nuclear fusion processes in the stars. On earth, hydrogen occurs almost wholly in combination, especially as water and in organic compounds (Tables 8.2 and 8.5). Hydrogen is widely used, especially in the manufacture of ammonia and in hydrogenation reactions in petrochemical production. Most hydrogen is formed as part of the overall process, either from water, from oil fractions or from methane (natural gas or, on a smaller scale, from anaerobic fermentation).

Where electricity is cheap, electrolysis of water provides hydrogen—either as the direct product or as a byproduct in processes like chlorine manufacture by the electrolysis of brine. Although present costs usually make electrolysis a relatively expensive process, it has been suggested that in a future where accessible oil and coal stocks are running out, the conversion of electricity from solar energy or fusion power into hydrogen would give a fuel which is storable (unlike electricity) and which would easily replace hydrocarbons as a transport fuel, with the advantage of being clean-burning. These ideas have been ramified into a 'hydrogen economy' to replace the present 'oil economy'. The major drawback is that, although the energy derived from burning unit weight of hydrogen is very high, the energy per unit volume is low and any weight advantages would be lost by the weight of storage devices, such as high pressure cylinders. There has therefore been substantial interest in the storage of hydrogen as a metal hydride, see below.

An alternative production of H_2 from both water and hydrocarbons is by the steam-reforming process over a nickel catalyst at around 850 °C:

$$C_nH_{2n+2} + H_2O \rightarrow CO + H_2$$

Further reaction with steam can convert CO to CO_2 which is removed chemically. Alternatively, separation by molecular sieve is used. An alternative source of process hydrogen is from hydrocarbons by thermal cracking.

Pure hydrogen may be prepared by diffusion through palladium tubes (which pass only H_2) or, on a small scale, by hydrolysis of active metal hydrides like CaH_2.

There are three well established isotopes of hydrogen:

^1_1H normal or light hydrogen, mass = 1·008, natural abundance = 99·98%

^2_1H deuterium or heavy hydrogen, mass = 2·015, natural abundance = 0·02%

^3_1H tritium, mass = 3·017, natural abundance = 10^{-17}%, radioactive, with $t_{\frac{1}{2}}$ = 12·4 years, decay process $^3_1H \rightarrow ^3_2He + ^0_{-1}e$

Pure deuterium is normally separated from light hydrogen by the electrolysis of water. The lighter isotope is evolved preferentially and almost pure deuterium oxide remains by the time the bulk is reduced a millionfold. Tritium is most conveniently prepared by the irradiation of lithium with slow neutrons in a reactor:

$$^6_3Li + ^1_0n = ^3_1H + ^4_2He$$

and the tritium is separated by oxidation to T_2O. Both deuterium and tritium are most readily produced as oxide and most deuterated or tritiated compounds are made directly from this isotopically substituted water. For example, deuterated acids may be made simply by solution:

$$P_2O_5 + 3D_2O \rightarrow 2D_3PO_4$$
$$SO_3 + D_2O \rightarrow D_2SO_4$$

or tritiated ammonia by the reaction of a nitride:

$$Mg_3N_2 + 3T_2O \rightarrow 2NT_3 + 3MgO.$$

Since the chemical behaviour of all the isotopes of an element is identical (although rates of reaction may differ), deuterium or tritium substituted hydrides are widely used in studying the mechanisms of reactions involving hydrogen. (See section 2.3.)

Some of the important properties of hydrogen are listed in Table 9.2.

It has already been noted in section 2.15 that atomic and ionic radii show some variation with the chemical environment of the species. This variability is particularly marked in the ions and covalent molecules involving hydrogen, as Table 9.2 shows, since there is only a single nuclear charge on hydrogen and the 1s orbital is unshielded by any inner electron shells. The hydride ion, H^-, is especially sensitive to change of electric field intensity in its environment as it consists of two electrons in the field of the single nuclear charge and therefore has a very diffuse electron cloud. The free hydride ion has been calculated to have a radius of

TABLE 9.2 Properties of hydrogen

Heat of dissociation of H_2	$H_{2(gas)} = 2H_{(gas)}$, $\Delta H = 435·9$ kJ mol^{-1} (443·3 kJ mol^{-1} for D_2)
Ionization potential	$H_{(gas)} = H^+_{(gas)} + e^-$, $I = 13·595$ eV $= 1309$ kJ mol^{-1}
Electron affinity	$H_{(gas)} + e^- = H^-_{(gas)}$, $E = -68·99$ kJ mol^{-1} (exothermic)
Radius, anionic H^-	$= 112$ to 154 pm (measured in ionic hydrides)
	$= 208$ pm (calculated for free H^-)
cationic H^+	$= 10^{-3}$ pm
covalent H	$= 37·07$ pm (from H_2 bond length)
	28 pm (from bond lengths of hydrogen halides)
	32 pm (from MH_4 distances in carbon Group hydrides)

208 pm, twice as large as that of helium, which has two electrons in twice the nuclear field. The measured values of the hydride ion radius in ionic lattices are much smaller than this and show considerable variation, as the values in the alkali metal hydrides illustrate:

MH	Li	Na	K	Rb	Cs
H^- radius (pm)	126	146	152	153	154

(Compare these values with 133 pm for F^- and 145 pm for O^{2-}, where the larger numbers of electrons are offset by the increase in the nuclear charges and consequent greater density of the electronic cloud.) The covalent radius of hydrogen also shows considerable variability, although the effect is less pronounced than with the anionic radius, apart from the high value for the hydrogen molecule itself.

The most striking change in dimension would come if the hydrogen atom were to lose its electron to become the positive ion, H^+. This is the bare proton with a radius of about 10^{-3} pm and is a hundred thousand times smaller than any other ion (compare Li^+, radius 60 pm). The charge density on the proton is thus enormously higher than on any other chemical species and it would have a powerful polarizing effect on any other molecule in its neighbourhood. As a result, the free proton has no independent existence in any chemical environment and always occurs in association. Thus, if an acid dissociates in water, H_3O^+ and not (except as a shorthand) H^+ is formed, and this 'hydrogen ion' or 'hydroxonium' ion is then further solvated just as any other cation. A similar situation holds for any other protonic solvent, as discussed in Chapter 6 (although it must be noted that in the Brönsted definition of an acid it is the proton which is transferred). In water, the proton may well exist as the solvated species $H_9O_4^+$, shown in Figure 9.1, as it has been shown that this species can be extracted from an aqueous acid solution into an immiscible organic base. This is comparable with the isolation of, say, the nickel ion from aqueous solution as the green hexahydrate, $Ni(H_2O)_6^{2+}$. In a recent study, $H_5O_2^+$, the proton solvated by two water molecules, has also been identified. It has the structure shown in Figure 9.2.

One further physical modification of the hydrogen mole-

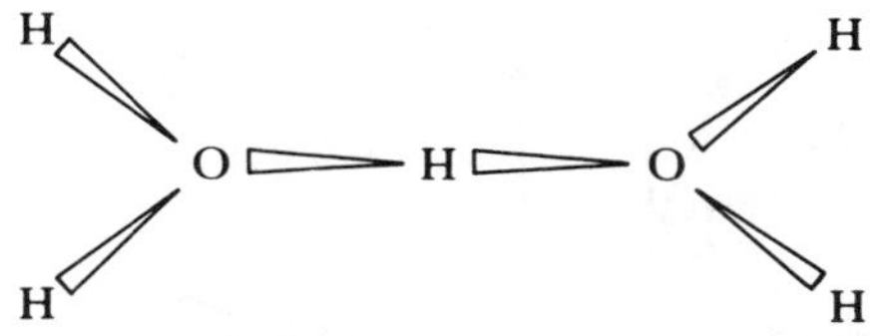

FIGURE 9.2 *The structure of* $H_5O_2^+$ *in* $[Co(en)_2Cl_2]Cl^-H_5O_2^+Cl^-$ The $(H_2O)-H-(OH_2)$ link is symmetrical and the hydrogens of the water molecule are *trans* to each other. The distance O–H–O is 243·1 pm, making the OH bridge length 121·6 pm. H–O in the H_2O molecule is 99·5 pm. Angles at oxygen are 109° to terminal H atoms and 114° for H (bridge)–O–H (terminal).

cule should be briefly mentioned: *ortho*- and *para*-hydrogen. These forms arise from the different ways in which the nuclear spins may be lined up. If the nuclear spins are parallel, the form is *ortho*-hydrogen, while if they are anti-parallel, the form is *para*-hydrogen. These two forms of molecular hydrogen have different physical properties, such as thermal conductivities, and the two co-exist at ordinary temperatures. (Other symmetrical molecules with nuclear spins, such as N_2 or Cl_2, have *ortho*- and *para*- forms but only H_2 and D_2 show significant differences in physical properties.) The more stable form at low temperatures is *para*-hydrogen and this makes up 100 per cent of hydrogen at absolute zero. At higher temperatures, the equilibrium proportion of *ortho*-hydrogen rises, and reaches its maximum of 75 per cent at room temperature. The equilibrium proportions at any given temperature may be calculated theoretically, and the interconversion may easily be followed experimentally. The interconversion is slow but subject to catalysis by a number of materials, especially by paramagnetic compounds, and this *ortho-para* conversion is frequently used in studies of catalysis.

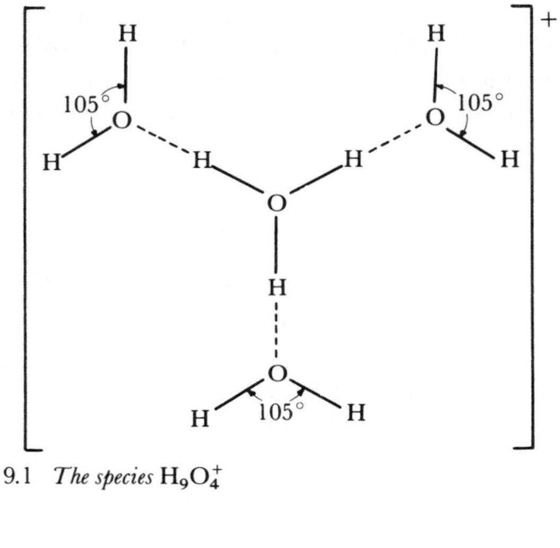

9.2 Chemical properties of hydrogen

As the hydrogen molecule bond energy, of 436 kJ mol^{-1}, is high, molecular hydrogen is fairly unreactive at ordinary temperatures. At higher temperatures it combines, directly or with aid of a catalyst, with most elements. Some of the more important reactions are shown in Figure 9.3.

Atomic hydrogen may be produced in high-intensity electric arcs and is very short-lived and reactive. It finds use in welding where the recombination of the atoms takes place on the metal surface, yielding up the heat of dissociation and at the same time providing a protective atmosphere against oxidation.

Hydrogen forms a molecular ion H_3^+ which has an equilateral triangular structure at equilibrium. This molecule, the simplest of all polyatomic molecules, has been of interest from a bonding point of view and it has also been proposed as a chain intitiator in reactions occurring in interstellar clouds. H_3^+ has also been recently detected for the first time outside of the laboratory—in the atmospheres of Jupiter, Uranus, and Supernova 1987A.

Binary compounds of hydrogen (i.e., those containing hydrogen and one other element) are termed hydrides whether they contain the hydride ion, H^-, or are covalent, and this term is commonly extended to less simple hydrogen

FIGURE 9.1 *The species* $H_9O_4^+$

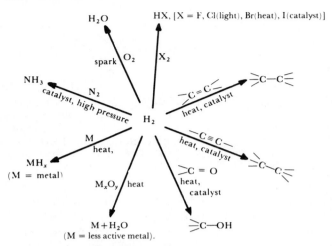

FIGURE 9.3 *Reactions of molecular hydrogen*

compounds, as in 'transition metal hydride complexes' and 'complex hydride ions'. It is convenient to discuss the chemistry of the hydrides in three groups, reflecting the three ways in which the hydrogen electron enters into bonding. These are (i) gain of an electron to form ionic compounds containing H^-, (ii) sharing the electron in covalent hydrides, (iii) forming metallic bonds with the electron delocalized in the so-called interstitial or metallic hydrides.

As the electron affinity of hydrogen is low compared with that of the halogens, the distribution of ionic hydrides in the Periodic Table is much more restricted than the distribution of ionic halides. Ionic hydrides are formed only by the elements of the alkali, alkaline earth and, possibly, the lanthanide Groups. Most other elements of the main Groups form covalent hydrides, while the transition elements give metallic hydrides. The transition metals also form hydrido-complexes such as $[ReH_9]^{2-}$ (Figure 15.23) or $RuH(CO)(PPh_3)_3$. In addition, the dihydrogen molecule can coordinate to metals. Metal hydride complexes are included in Chapter 15 and the $M—H_2$ compounds are discussed in section 16.11. Figure 9.4 shows the approximate distribution of the various types of binary hydride within the Periodic Table. The boundaries between classes are by no means sharp, and a number of intermediate types are observed. In addition, there is some doubt whether the hydrides of a number of elements, especially of the heaviest elements, exist at all.

In passing across a Period, the type of hydride changes from ionic compounds at one extreme to volatile covalent molecules at the other. In the middle of the short Periods, the transition between the two types is marked by solid hydrides, say of magnesium and aluminium, of polymeric structure with bonding of intermediate character. Among the transition elements in the long Periods, the type changes through interstitial hydrides to hydrides of dubious existence at the right of the transition block, before coming to the covalent hydrides of the *p* elements. In any Main Group of the Periodic Table, the stability of the hydrides tends to fall with increasing atomic weight.

9.3 Ionic hydrides

The gain of an electron by a hydrogen atom gives the helium configuration, $1s^2$ and is analogous to halide ion formation. However, the formation of the hydride ion is much less favourable than the formation of a halide ion, see Figure 9.5, as the electron affinity of hydrogen is much less exothermic and because more energy is needed to break the $H—H$ bond. As a result only the most active elements, whose ionization potentials are low, form ionic hydrides. The alkali metal hydrides and CaH_2, SrH_2 and BaH_2 are the only compounds which are clearly ionic. Magnesium hydride, MgH_2, is intermediate between the ionic hydrides and the solid covalent hydrides like AlH_3. The dihydrides of the lanthanide elements are metallic and only approach the ionic type as the hydrogen content rises towards the MH_3 stoichiometry. However, the two lanthanide elements which show a relatively stable II state in their general chemistry (compare Chapter 11)—europium and ytterbium—do form ionic hydrides, EuH_2 and YbH_2. These two compounds are isomorphous with CaH_2 and differ in structure from the other lanthanide dihydrides which have the fluorite structure.

The ionic hydrides are formed by direct reaction between hydrogen and the heated metal. They are all reactive and reactivity increases with atomic weight in a Group. The alkali metal hydrides are more reactive than those of the corresponding alkaline earth elements. The alkali metal hydrides have the sodium chloride structure (the radius of H^-, about 150 pm, is comparable with the halide radii, $F^- = 133$ pm, $Cl^- = 181$ pm). The alkaline earth hydrides, and EuH_2 and YbH_2, have the same structure as $CaCl_2$. The metal atoms are in approximately hexagonal close packing and each is surrounded by nine hydride ions in a slightly distorted lead dichloride structure. In the regular $PbCl_2$ structure, the metal atom is coordinated by six metal ions at the corners of a trigonal prism and three more beyond the

FIGURE 9.4 *Types of hydride in the Periodic Table*

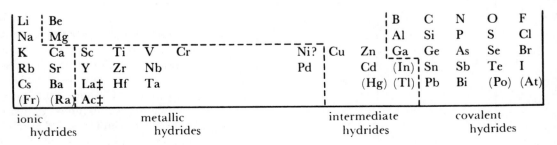

ionic hydrides		metallic hydrides						intermediate hydrides			covalent hydrides			
Li	Be									B	C	N	O	F
Na	Mg									Al	Si	P	S	Cl
K	Ca	Sc	Ti	V	Cr		Ni?	Cu	Zn	Ga	Ge	As	Se	Br
Rb	Sr	Y	Zr	Nb			Pd		Cd	(In)	Sn	Sb	Te	I
Cs	Ba	La‡	Hf	Ta					(Hg)	(Tl)	Pb	Bi	(Po)	(At)
(Fr)	(Ra)	Ac‡												

‡ lanthanide or actinide elements

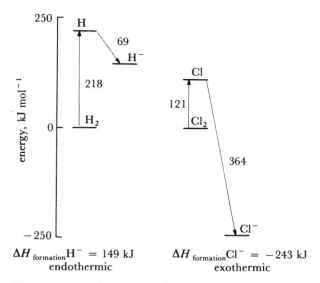

$\Delta H_{\text{formation}} H^- = 149 \text{ kJ}$
endothermic

$\Delta H_{\text{formation}} Cl^- = -243 \text{ kJ}$
exothermic

FIGURE 9.5 *Energies of formation of hydride and chloride ions*

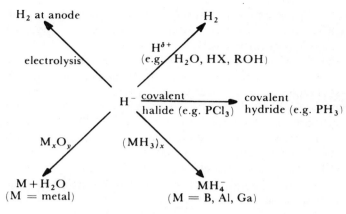

FIGURE 9.6 *Reactions of the hydride ion*

rectangular faces. In the dihydrides, the metal atom has seven hydride ions at equal distances and two more distant ions (e.g. CaH_2 has seven Ca—H distances of 232 pm and two of 285 pm). In all these ionic hydrides, the metal-metal distances are less than they are in the elements so that the hydrides are denser than the metals.

Evidence for the ionic nature of these hydrides is:

(i) Molten LiH shows ionic conductance and the melt gives hydrogen at the *anode* on electrolysis. The other hydrides decompose before melting, but they may be dissolved in alkali halide melts without decomposition and, on electrolysis, give hydrogen at the anode.

(ii) It has been possible, by a combination of X-ray and neutron diffraction, to construct an electron density map for LiH. This shows that 0·8 to 1·0 of an electron has been transferred to hydrogen from each lithium atom (i.e. to give Li^+H^-). Thus lithium hydride is almost completely ionic. As polarization effects are greatest in lithium hydride (see Chapter 4) it follows that the other alkali hydrides are also ionic with full transfer of an electron from each metal atom.

(iii) The crystal structures of the hydrides show no indication of directional bonding (chains, sheets or discrete molecules) and are reasonable for ionic compounds with the radius ratios of the hydrides.

(iv) Observed and calculated lattice energies are in good agreement (compare section 5.2).

The ionic hydrides react readily, and often violently, with water or any other source of acidic hydrogen and with oxidizing agents. The reaction with acidic hydrogen may be represented as:

$$X^{\delta-} - H^{\delta+} + H^- \rightarrow H_2 + X^-$$

Examples of this and other common reactions of the hydride ion are shown in Figure 9.6.

The ionic hydrides find use in the laboratory for drying solvents, and as reducing agents, though they have been largely superseded in the latter application by the advent of the complex hydrides. On the industrial scale, NaH and CaH_2 which are relatively inexpensive and easily handled, are commonly used as condensing agents in organic syntheses and as reducing agents, e.g.:

$$CaH_2 + MO \rightarrow CaO + M + M_2 \text{ (at } 500\text{–}1\,000\,°C)$$

$$CaH_2 + 2NaCl \rightarrow 2Na\uparrow + CaCl_2 + H_2$$

Complex hydride anions. One important reaction of the alkali metal hydrides is in the preparation of the complex hydride anions. If lithium hydride is reacted in ether with aluminium trichloride, the tetrahydroaluminate, $LiAlH_4$, results (this compound is usually called by its non-systematic name of lithium aluminium hydride):

$$4LiH + AlCl_3 \rightarrow LiAlH_4 + 3LiCl$$

Recent work has shown that the complex aluminium hydrides may also be made by direct reaction of the elements at high pressures, giving a much cheaper route as all the expensive lithium ends up in the product

$$Li + Al + 2H_2 \xrightarrow[\text{250 atm, 120–150°C}]{\text{ether solvent}} LiAlH_4$$

(also for Na or K).

$NaBH_4$, sodium tetrahydroborate (borohydride), Li_2BeH_4, $LiGaH_4$, and similar compounds may be prepared by one or other of these routes, as can the aluminium hexahydrides like Na_3AlH_6.

The alkali metal compounds MBH_4, $MAlH_4$, $MGaH_4$ and M_3AlH_6 contain EH_4^- (E = B, Al, Ga) anions, although there may be a weak interaction with the cation, especially in the case of lithium compounds, of the multicentred, electron-deficient type discussed in section 9.6. The EH_4^- ions are tetrahedral and AlH_6^{3-} is octahedral.

These complex hydrides, particularly $LiAlH_4$ and $NaBH_4$, are extremely useful reducing agents, especially the lithium compounds which are appreciably soluble in ether. Examples of their uses in the preparation of covalent hydrides are given in section 9.5 and organic applications include the reduction of aldehydes or ketones to alcohols and of nitriles to amines. The different complex hydrides vary in reactivity, thus BH_4^- is a milder reducing agent than AlH_4^-, and further modifications to the reactivity are possible by introducing organic substituents as in $Na^+[HB(OCH_3)_3]^-$. The introduction of such reagents in the last two decades has created a revolution in reductive preparations in both organic and inorganic chemistry. For example, yields in the preparation of the hydrides of the heavier elements of the carbon and nitrogen Groups have been raised to 80–90 per cent as compared with the 10–20 per cent which was common with the older methods using ionic hydrides.

Less electropositive elements also form borohydrides or aluminohydrides, but these are more closely related to the parent boron and aluminium hydrides, which are electron deficient covalent species and are discussed in section 9.6.

9.4 Metallic hydrides

Non-stoichiometric compounds between hydrogen and many transition metals have been known for a long time, but their nature and bonding have not been well understood. These compounds are brittle solids with a metallic appearance and with metallic conductivity and magnetic properties. Typical formulae are $TiH_{1.8}$, $NbH_{0.7}$ or $PdH_{0.6}$. These compounds were called 'interstitial' hydrides as it was originally conjectured that the hydrogen atoms were held in the interstices of the metal lattice. The problem was further complicated by the fact that most transition metals physically adsorb hydrogen so that many spurious compounds were reported.

It is now known that all the lanthanide and actinide elements, the elements of the titanium and vanadium Groups, chromium and palladium combine exothermically and reversibly with hydrogen to form metallic hydrides in which the metal atoms are in a different structure from those of the element. These metallic hydrides have the idealized formulae shown in Table 9.3, but, as usually prepared, are commonly non-stoichiometric and hydrogen-deficient.

There is thus a 'hydride gap' in the middle of the Periodic Table where binary hydrides are not formed. The exothermic heats of formation decrease from left to right towards this gap, and it has been calculated that hydrides of the ferrous metals would have heats of formation of around zero.

Many of these hydrides have structures where the hydrogens occupy the tetrahedral sites in cubic, close-packed metal lattices. With all these sites occupied, this corresponds to the formula MH_2 and the fluorite structure (compare Table 5.3). It is thought that VH, NbH and TaH have the hydrogens in tetrahedral sites and these compounds have slightly distorted body centred cubic metal lattices which are closely related to cubic, close-packed arrangements. In the lanthanide hydrides, and those of yttrium and the heavier actinides, the MH_2 phase can take up further hydrogen in octahedral sites. The lighter lanthanides form MH_3 phases in

TABLE 9.3 Metallic Hydrides (idealized formulae and structures)

ScH_2 (fluorite)	TiH_2 (fluorite)	VH (bcc) $VH_{1.6}$ (fluorite?)	CrH (anti-NiAs) CrH_2? (fluorite?)	—	—	—	(NiH)(c) (NaCl)	CuH (wurtzite)
YH_2 (fluorite) YH_3 (hexagonal)	ZrH_2 (fluorite)	NbH (bcc?) NbH_2 (fluorite?)	—		—	—	—	$PdH_{0.7}$ (NaCl)	—
LaH_2 (fluorite) LaH_3 (cubic)	HfH_2 (fluorite)	TaH (bcc)	—	—	—	—	—	—
	Lanthanide elements	MH_2 (fluorite) formed by Ce, Pr, Nd, Sm, (a), Gd, Tb, Dy, Ho, Er, Tm, (a), Lu MH_3 (cubic) formed by Ce, Pr, Nd, Yb MH_3 (hexagonal) formed by Sm, Gd, Tb, Dy, Ho, Er, Tm, Lu						
AcH_2 (fluorite)								
	Actinide elements	MH_2 (fluorite) formed by Th, (b), Np, Pu, Am MH_3 (hexagonal) formed by Np, Pu, Am MH_3 (cubic, complex str.) formed by Pa, U						

Notes: (a) EuH_2 and YbH_2 have orthorhombic, ionic hydrides discussed in section 9.3.
  (b) Thorium dihydride has a distorted fluorite structure: a second hydride, Th_4H_{15} also exists.
  (c) NiH exists only under hydrogen pressures of 10^4 atm.

this way which remain cubic, but the heavier elements undergo a structural change around the composition $MH_{2.5}$ to give a hexagonal lattice. UH_3, PaH_3 and Th_4H_{15} have more complex structures. In two of the hydrides, CrH and PdH, the hydrogens are in octahedral sites only. Palladium hydride (which has been prepared only up to the $PdH_{0.7}$ composition) forms a sodium chloride lattice but chromium hydride has a hexagonal, *anti*-nickel arsenide structure.

There has been much discussion of the bonding in these metallic hydrides, and in similar compounds such as some low-valency halides of the transition elements which also show metallic properties. One theory, which accounts for many of the facts, regards the metallic hydrides as modified metals. A metal with n valency electrons is regarded as forming M^{n+} cations and having the n electrons per metal ion in completely delocalized orbitals. Then, as hydrogen atoms enter the lattice, each acquires one of these delocalized electrons to form a hydride ion. Thus, a metal hydride, MH_x contains M^{n+} ions, xH^- ions and $(n-x)$ delocalized electrons. The relative numbers and sizes of the metal cation and hydrogen anions govern the structure of the hydride, while the remaining $(n-x)$ electrons give the hydride its metallic properties. For example, TiH_2 would be regarded as consisting of Ti^{4+}, two H^- and two conduction electrons per formula unit.

The transition metal hydrides are usually prepared by direct combination between the metal and hydrogen at moderate temperatures and, often, high pressures. They may be decomposed by raising the temperature. This reversible hydrogenation is made use of in two ways. One is to provide a convenient source of very pure hydrogen. The metal hydride is formed leaving any impurities in the hydrogen behind. Then, by heating the hydride to a higher temperature, pure hydrogen is evolved. The second use is to provide the metal in a finely divided and highly reactive form. Many of the transition metal hydrides differ sufficiently from the metal in lattice parameters so that when the hydride is formed from the bulk metal, it is produced as a fine powder. **The other hydrides are brittle and may be much more readily powdered than the metal. The powdered hydride is then heated to remove the hydrogen, leaving the metal in a suitable form and free of surface oxide for further reactions. In addition, the metal hydrides themselves often provide suitable starting materials for synthesizing other compounds of the metal.**

A number of the hydrides find industrial application, particularly in powder metallurgy where the hydrogen evolved during fabrication gives a protective atmosphere. Another application is as a moderating material in atomic piles. A metal hydride, such as zirconium hydride, provides a higher density of hydrogen than conventional moderators, such as water, and they may be used to higher temperatures.

A potential use of metal hydrides of high significance is as a portable, relatively safe, hydrogen store. Hydrogen is an excellent fuel for urban transport and may be produced directly from renewable energy sources. It is being examined, therefore, as a substitute for petrol. Work is proceeding on using metal hydrides as a hydrogen source in such applications, exploiting the readily reversible formation and decomposition of the metal hydrides.

Related to the simple hydrides are a wide range of mixed-metal hydrides. These often involve a more electropositive metal which, alone, forms an ionic hydride, but the mixed species retains metallic properties. Of particular interest are mixed metal hydrides formed by those metals which do not have a stable binary hydride phase such as Mg_2NiH_4. Similar mixed metal hydrides are known for most of the metals in the 'hydride gap'.

The Mg_2NiH_4/Mg_2Ni and also the simpler MgH_2/Mg systems have attracted interest as hydrogen and energy storage materials which are cheaper and offer a better H_2/weight ratio than single metals like Ti or U.

These compounds are intermediate between the binary metallic hydrides, and the hydrido-complexes of the transition metals which have covalent M—H bonds (compare the ReH_9^{2-} ion and related species, section 15.5.1). Mg_2FeH_6 also approaches the limit of an FeH_6^{4-} ion, but there is a significant H...Mg interaction remaining (section 14.6.3). One further, somewhat different, intermediate group is the metallic hydride-halides and similar species represented by ZrHCl. This shares the metallic properties of the binary hydride and of the 'subhalide' (section 15.2.3).

The hydrides of the remaining transition metals are quite different. Copper hydride, CuH, is formed endothermically by the reduction of copper salts with hypophosphorous acid and it decomposes irreversibly. In the zinc Group, ZnH_2, CdH_2 and HgH_2 are formed by the reaction of the halides with $LiAlH_4$ but stability decreases rapidly from Zn to Hg. A mixed hydride-halide, H_3Zn_2X, is formed similarly for X=Cl or Br; while cadmium gives CdHX. All these species probably resemble $(AlH_3)_x$ and belong to the electron-deficient class.

In 1989, the 'cold fusion' phenomenon burst onto the scene amidst a blaze of publicity: interest in palladium hydrides multiplied many times overnight. It was reported that electrolysis in D_2O using a palladium electrode produced excess heat which the investigators could not account for by any chemical reaction. They therefore ascribed it to a nuclear process: that D atoms were present at high local concentration in the Pd lattice and a few underwent nuclear fusion, liberating the extra energy. As the only established nuclear fusion is that at high temperatures in the H-bomb or the sun, the palladium process was termed 'cold fusion'. Such claims aroused enormous interest and soon enormous controversy. Many research groups, including the original authors, attempted to reproduce or extend the initial results and most failed. Many other, less innovative, explanations were put forward to account for the observations and the weight of opinion remains sceptical. Unexpected effects have been observed, but by the accepted standards of science the claimed phenomena have not yet been established. For such a striking departure from current understanding and expectation, it is essential that observations may be reproduced consistently by a number of independent investigators. This has not so far happened.

9.5 Covalent hydrides

The hydrogen atom may attain the inert gas structure by sharing an electron pair in a covalent bond. All the remaining binary hydrides fall into this group. MgH_2 has properties intermediate between ionic and covalent hydrides while CuH, ZnH_2 and CdH_2 are intermediate between metallic and covalent species. Covalent hydrides are formed by all the elements with an electronegativity down to about 1·5 (e.g. aluminium) which are just on the border for forming ionic hydrides. As hydrogen has an electronegativity of 2·1 it follows that bond polarities range from those, as in the hydrogen halides, where the hydrogen end is strongly positive, i.e. $H^{\delta+} - X^{\delta-}$, to those cases where the hydrogen end of the dipole is negative, as in $B^{\delta+} - H^{\delta-}$ or $Ga^{\delta+} - H^{\delta-}$. If the second element has an electronegativity less than about 1·2, the hydride becomes definitely ionic.

The covalent hydrides fall into two distinct classes. First there are the compounds of the carbon, nitrogen, oxygen and fluorine Groups which have normal electron pair bonds between the element and hydrogen. Secondly, there are compounds exemplified by the simplest boron hydride, B_2H_6, which do not have enough valency electrons to form electron pair bonds to all the hydrogens, and these are termed *electron-deficient*. Into this class fall the hydrides of beryllium, boron, aluminium and gallium. MgH_2, ZnH_2 and perhaps CdH_2 and CuH have some affinities with this class. Other members are the borohydrides and aluminohydrides of elements which are not sufficiently electropositive to give the complex hydride ions: examples are $Be(BH_4)_2$ and $Al(BH_4)_3$. The electron deficient hydrides are treated in section 9.6.

Covalent electron pair bonds are also formed between hydrogen and the transition elements, in compounds where the metal is also bonded to ligands capable of forming π-bonds, such as CO, phosphanes, arsanes, sulfanes, or NO. Such compounds as $(R_3P)_2PtH_2$—where R stands for a variety of aliphatic and aromatic substituents—or $(CO)_5MnH$, have metal to hydrogen covalent bonds of sufficient stability to allow for their isolation. Such covalent bonds to hydrogen seem to be most readily formed by those transition metals in the 'hydride gap' which do not form binary metallic hydrides. Some examples are given in Chapters 13 to 16.

Preparation

There are three general methods available for the preparation of covalent hydrides, although many others are available in specific cases. These general methods are:

(i) **simple direction combination, especially with the more reactive elements**:

$$H_2 + Cl_2 \xrightarrow{\text{light}} 2HCl$$

(ii) hydrolysis of a binary compound of the element with an active metal by any non-oxidizing dilute acid:

$$Mg_3B_2 \rightarrow B_2H_6$$
$$Al_4C_3 \rightarrow CH_4$$
$$Ca_3P_2 \rightarrow PH_3$$

(iii) reduction of a halide or oxide by an ionic hydride or by a complex hydride:

$$GeO_2 + BH_4^- \rightarrow GeH_4$$
$$SiCl_4 + LiH \rightarrow SiH_4$$
$$AsCl_3 + LiAlH_4 \rightarrow AsH_3$$
$$R_2SbBr + BH_4^- \rightarrow R_2SbH.$$

TABLE 9.4 Hydrides of the p elements

B_2H_6 (and many higher hydrides: see Table 9.5)	C_nH_{2n+2}, etc. (no limit to n is known)	NH_3 N_2H_4	H_2O H_2O_2	HF
$(AlH_3)_x$ (solid polymer)	Si_nH_{2n+2}[1],[2] (characterized up to $n = 8$; straight- and branched-chain isomers occur)	PH_3[2] $[P_xH_y]$[3]	H_2S H_2S_n (characterized up to $n = 6$)	HCl
$[(GaH_3)_x]$ (unstable oil)	Ge_nH_{2n+2}[1],[2] (characterized up to $n = 9$; straight- and branched-chain isomers occur)	AsH_3[2] $[As_xH_y]$[3]	H_2Se	HBr
	$[SnH_4]$ $[Sn_2H_6]$ $[PbH_4]$?	$[SbH_3]$ $[BiH_3]$?	$[H_2Te]$ $[H_2Po]$?	HI $[HAt]$?

[ ] = unstable at room temperature

[ ]? = existence unconfirmed or transient

1. Mixed hydrides $Si_xGe_yH_{2(x+y)+2}$ are also known.

2. Solids MH_x of uncertain composition are widely reported but few properties are established.

3. An extensive class of such hydrides is characterized: see 18.3

The reactions of all these hydrides may be carried out in ether solution and BH_4^- may also be used in an aqueous system. A fourth method of synthesis, in increasing use, involves the interconversion of hydrides in a suitable discharge. This is particularly valuable for forming longer chains from simple hydrides or for forming long chain halides which may then be reduced to the hydride. The discharge may be of radio or microwave frequency or may be of the ozonizer type.

(iv) $GeH_4 \rightarrow Ge_2H_6 + Ge_3H_8 + $ hydrides up to Ge_9H_{20}

$SiH_4 + GeH_4 \rightarrow GeH_3SiH_3 + Ge_2H_6 + Si_2H_6$

$$SiCl_4 \rightarrow Si_2Cl_6 \xrightarrow{LiAlH_4} Si_2H_6$$

$SiH_4 + PH_3 \rightarrow SiH_3PH_2 + (SiH_3)_2PH + Si_2H_5PH_2$, etc.

Of these four methods of preparation, (i) is of limited applicability but is being developed for the less reactive elements under high temperature and pressure, (ii) gives low yields but is the most direct way of obtaining the higher members of homologous series, (iii) is usually the best and most convenient method on a laboratory scale, while (iv) gives better yields than (ii) of the higher hydrides but requires a supply of the simple material. Higher hydrides, containing a chain of atoms of the central element, are known for many p block elements but, for most, chain lengths are short and stabilities are low. The exceptions are silicon and germanium, where hydrides M_nH_{2n+2} are characterized up to about $n = 10$. Both straight and branched chains are known, and hydrides containing mixed chains of silicon and germanium atoms are found. For example, pentagermane is found in all three isomeric forms $(GeH_3)_2GeHGeH_2GeH_3$, $GeH_3GeH_2GeH_2GeH_2GeH_3$ and $Ge(GeH_3)_4$, and Si_2GeH_8 occurs as $SiH_3SiH_2GeH_3$ and $SiH_3GeH_2SiH_3$. Mixed hydrides containing silicon or germanium and certain other atoms are also fairly stable. Examples include the silicon-phosphorus hydrides SiH_3PH_2, both isomers $(SiH_3)_2PH$ and $SiH_3SiH_2PH_2$, and $(SiH_3)_3P$.

The p element hydrides are listed in Table 9.4.

It is well-known that CH_4 is formed in the rumen and contributes to the greenhouse effect (compare Chapter 20). Interestingly, it has recently been found that PH_3 also occurs in the rumen and gut and even the less stable P_2H_4 has been detected in cultures of gut cells. It is thought the quantities are insufficient to account for dragon legends!

Properties

The thermal stabilities of the hydrides decrease in each Group as the atomic weight of the central element increases. Except in the case of carbon compounds (and, to some extent, boron ones), the thermal stability decreases fairly rapidly with increasing molecular weight for the members of a homologous series. The variation of stability across a Period is irregular, with the carbon Group and the halogens forming the most stable hydrides, e.g. the hydrides of Ga < Ge > As < Se < Br in stability.

The structures of the hydrides are as predicted from the number of electron pairs on the central element, cf. Chapter 4.

All the carbon Group hydrides are tetrahedral MH_4 molecules and the higher homologues are also based on tetrahedra. The nitrogen and oxygen Group molecules are based on tetrahedra with, respectively, one and two lone pairs. The bond angles decrease towards $90°$ with increasing atomic weight of the central element. The boron Group hydrides are discussed later. Hydrazine, N_2H_4, and hydrogen peroxide, H_2O_2, adopt the structures shown in Figure 9.7 which separate the lone pairs as much as possible.

The reactions of the hydrides are varied and many are familiar: for example, the reactions of hydrogen sulfide, hydrogen halides, water, and ammonia. In general terms, all the hydrides are reducing agents and react strongly with oxygen and halogens. Stability to oxygen varies from the relative stable germane, GeH_4, and hydrogen halides, to the cases such as silane, SiH_4, and phosphane, PH_3, which explode or inflame in air. With the halogens, reactions may also be violent, although iodine often reacts smoothly to cleave only one bond as in:

$$GeH_4 + I_2 \rightarrow GeH_3I + HI$$

The hydrogen halides also react with many hydrides to give

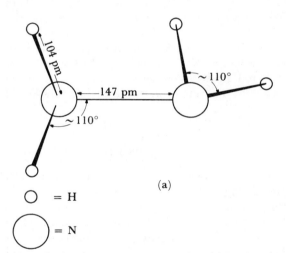

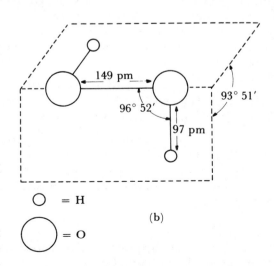

FIGURE 9.7 *The structures of* (a) *hydrazine,* (b) *hydrogen peroxide*

partial substitution:

$$SiH_4 + HX \xrightarrow{AlX_3} SiH_3X + H_2$$

The hydrides of most elements are strong reducing agents which reduce many heavy metal salts to the element:

$$PH_3 + AgNO_3 \rightarrow Ag$$
$$SnH_4 + CuSO_4 \rightarrow Cu, \text{ etc.}$$

Exceptions are provided by the hydrides of the first short Period which are relatively unreactive. One reason for this probably lies in the absence of an easy reaction path. Thus silane, SiH_4, reacts readily as the silicon $3d$ orbitals can provide a means of coordinating an attacking reagent in a reaction intermediate; in methane, CH_4, there is no such pathway. Reaction depends on the breaking of a $C-H$ bond which requires much more energy and is correspondingly slower.

One important reaction of many covalent hydrides is ionization to give a positive hydrogen species. The hydrides of the most electronegative elements are already polarized, with the hydrogen positive. Although the formation of the free proton is impossible, as shown above, the hydrogen bonded to an electronegative element can ionize to give a proton stabilized by bonding to one or more neutral molecules. In the hydrides of the most electronegative elements, F, O and N, this is accomplished by self-ionization in the liquid phase when the proton is associated with a molecule of the hydride (compare section 6.5 and Figures 9.1, 9.2).

$$2NH_3 = NH_4^+ + NH_2^-$$
$$2H_2O = H_3O^+ + OH^-$$
$$2HF = H_2F^+ + F^-.$$

The hydrides of less electronegative elements do not self-dissociate in the liquid phase (largely because the liquids are less efficient ion-supporters), but they do dissociate when dissolved in an ionizing solvent such as water. Such hydrides are those of the other halogens, and of the other chalcogens, which dissociate in water or ammonia. The energies involved in such ionizations may be illustrated by the case of hydrogen chloride. The formation of the free proton in the gas phase, $H \rightarrow H^+$ requires 1340 kJ mol^{-1}, and the hydration energy of the gaseous proton is about -1090 kJ mol^{-1}. For hydrogen chloride:

$$HCl_{(gas)} \rightarrow H^+_{(gas)} + Cl^-_{(gas)} \quad \Delta H = \quad 1380 \text{ kJ mol}^{-1}$$
$$H^+_{(gas)} + Cl^-_{(gas)} \rightarrow H^+_{(aq)} + Cl^-_{(aq)} \quad \Delta H = -1460 \text{ kJ mol}^{-1}$$

so ionization of hydrogen chloride in aqueous solution is exothermic by about 80 kJ mol^{-1}. The hydrogen halides are fully dissociated in water but most other hydrides which dissociate in water do so only very weakly. For instance, the dissociation constant for

$$H_2S + H_2O \rightleftharpoons H_3O^+ + HS^-$$

is only about 10^{-7}. The hydrides of phosphorus and the other members of the nitrogen Group do not dissociate measurably in water. This rapidly decreasing tendency to dissociate in solution clearly parallels the fall in the polarity of the element-hydrogen bond.

The formation of H_3O^+ or NH_4^+ may be considered as the result of the donation of an electron pair to the proton by the oxygen or nitrogen atoms, $H_3N: \rightarrow H^+$ or $H_2O: \rightarrow H^+$. This is one case of the general donor-acceptor (Lewis base-acid) behaviour of the hydrides. The hydrides with one or more lone pairs of electrons may donate them to suitable acceptor molecules to form coordination complexes. This is, of course, the reaction involved in the solvation of cations by water or ammonia, but it is a general reaction possible for all the hydrides of the nitrogen, oxygen, or halogen Groups. In general, donor power falls as the number of lone pairs on the central atom increases and as the size of the central atom increases. Thus the hydrogen halides are weak donors as there are three lone pairs on the halogen atoms, and the heavier analogues of ammonia and water, such as SbH_3 or H_2Se, are also weak. However, in super-acidic systems such as HF/SbF_5 mixtures, the sulfonium and selenonium-containing compounds $H_3S^+SbF_6^-$ and $H_3Se^+SbF_6^-$ can be isolated and characterized. It is also in reactions of this type that stable and well-established salts of the analogous, and very important, H_3O^+ ion were first prepared in 1975. Organic-substituted hydrides of the nitrogen and oxygen Group elements, especially the phosphanes, R_3P, and the sulfanes, R_2S (where R includes aliphatic and aromatic groups), do act as donors to a wide variety of species. Acceptor molecules with suitable empty orbitals available include the p acceptors of the boron Group (and beryllium), the nd acceptors such as the tetrafluorides of the carbon Group, and especially the $(n-1)d$ acceptors among the transition metals.

Acceptor molecules among the hydrides are largely confined to those of the boron Group, although there is some evidence of weak d orbital acceptor power in the carbon Group tetrahydrides. The complex hydride anions, BH_4^-, AlH_4^- and GaH_4^-, may be regarded as being formed by the acceptance of the electron pair on the hydride ion by the MH_3 species, $H^-: \rightarrow MH_3$. Donor-acceptor complexes are also formed between the boron Group hydrides and those of the nitrogen and oxygen Groups. Compounds such as H_3BNH_3 are formed at low temperatures but, on warming towards room temperatures, these compounds lose hydrogen and polymerize (in this case to the ring compound $B_3N_3H_6$ discussed in section 9.6). If organic derivatives are used, the complexes are more stable; both H_3BNMe_3 and Me_3BNH_3 are stable at room temperature. The much less-stable polymeric hydrides of aluminium and gallium also form such complexes and considerable stabilization of the $M-H$ bonds results. Complexes such as $R_3\overline{NAlH_3}$ and $R_3\overline{NGaH_3}$ are well characterized. These elements also make use of their d orbitals to accept more than one donor molecule. Thus, the compound $AlH_3.2NMe_3$ has the trigonal bipyramidal structure shown in Figure 9.8a. The interesting species $(C_5H_5)BeH$, Figure 9.8b, has a planar C_5H_5 ring π-bonded to Be (compare Figure 16.8) with Be-H on the five-fold axis.

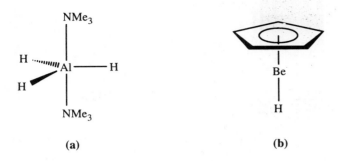

(a)   **(b)**

FIGURE 9.8(a) *The structure of* AlH$_3$.2NMe$_3$ (b) *The structure of* (C$_5$H$_5$)BeH

TABLE 9.5 Some boron hydrides

B$_2$H$_6$	B$_6$H$_{14}$	B$_{11}$H$_{15}$
B$_4$H$_{10}$	B$_7$H$_{13}$	B$_{12}$H$_{16}$
B$_5$H$_9$	B$_8$H$_{12}$	B$_{13}$H$_{19}$
B$_5$H$_{11}$	B$_8$H$_{14}$	B$_{14}$H$_{18}$
B$_6$H$_{10}$	B$_8$H$_{16}$	B$_{14}$H$_{20}$
B$_6$H$_{12}$	B$_8$H$_{18}$	B$_{14}$H$_{22}$
B$_{10}$H$_{14}$	*n*- and *iso*-B$_9$H$_{15}$	B$_{15}$H$_{23}$
	B$_{10}$H$_{16}$	B$_{16}$H$_{22}$
	B$_{10}$H$_{18}$	*syn*- and *anti*-B$_{18}$H$_{22}$
	B$_{10}$H$_{20}$	B$_{20}$H$_{16}$
		9 isomers of B$_{20}$H$_{26}$
		B$_{30}$H$_{38}$

Many of these ions and hydrides form isostructural sets, all based on four electron pairs:

four bond pairs	tetrahedron	BH$_4^-$, CH$_4$, NH$_4^+$
three bond pairs and one lone pair	pyramid	NH$_3$, H$_3$O$^+$
two bond pairs and two lone pairs	V-shape	H$_2$O, H$_2$F$^+$

9.6 Electron-deficient hydrides

9.6.1 Boron hydrides

Boron and related elements form a range of hydrogen compounds which cannot be accounted for by classical ideas of electron-pair bonds between two atoms. It is found that the simple monomeric species, such as BeH$_2$ or BH$_3$, do not occur but that more complicated compounds are formed. Thus the simplest boron hydride is B$_2$H$_6$ while beryllium and aluminium form high molecular weight polymers, (BeH$_2$)$_x$ and (AlH$_3$)$_x$. The most fully studied compounds of this class are the hydrides of boron and these are treated first.

The work of Stock, starting in 1909, on the boron hydrides, was one of the sources of the renaissance in inorganic chemistry in this century. He characterized the compounds listed in the first column of Table 9.5. It took 25 years, until 1958, before the next compound, B$_9$H$_{15}$, was added to Stock's list and only in the last decade has the chemistry of compounds larger than his B$_{10}$H$_{14}$ been much explored. The full extent of Stock's genius and experimental skill becomes apparent when it is realized that all these hydrides are inflammable, many are very unstable, and all the products identified by Stock and his team appeared as mixtures either from the hydrolysis of magnesium boride or from interconversions of these.

The modern development of this field started in the sixties and has continued rapidly ever since, expanding into further classes of compounds, like the carboranes which include C atoms in the skeleton. Table 9.5 lists the established binary boron hydrides. Clearly the field is extremely complex and can only be introduced here, and we shall focus on the simpler compounds from the first column.

The boron hydrides form part of an exceptional group of compounds. On simple valency grounds, the lowest hydride of boron is expected to be BH$_3$, but no such molecule has ever been discovered—despite considerable search. The simplest boron hydride is diborane, B$_2$H$_6$, which many studies have shown to have the bridge structure of Figure 9.9 (see section 7.8). There are only twelve valency electrons in B$_2$H$_6$ (3 per B and 1 per H), so that the molecule cannot be made up of electron pair bonds, which would require sixteen electrons for the structure of Figure 9.9. Such molecules, with insufficient electrons to form two-centre electron pair bonds between all the atoms, are termed *electron deficient*. The evidence of bond lengths and angles, and of the infrared stretching frequencies, suggests that the terminal B—H bonds are normal single bonds. This leaves the two bridging H atoms, together with four electrons, to be fitted into the picture. The most satisfactory description of the bridge was that proposed by Longuet-Higgins. He suggested that, in place of the normal electron pair bond centred on two nuclei, a three-centre bond should be considered, made up of the hydrogen

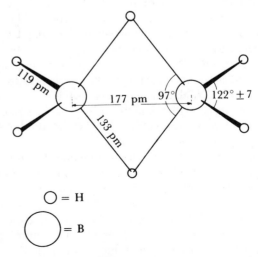

FIGURE 9.9 *The structure of diborane*

$1s$ orbital and appropriate hybrids on the two boron atoms. Such a bond may be indicated as B····H····B or as

If the boron atoms form sp^3 hybrid orbitals, two of which form the bonds to the terminal hydrogens, then the others may overlap with the hydrogen $1s$ orbitals as shown in Figure 9.10. Figure 9.11 gives the energy level diagram for such

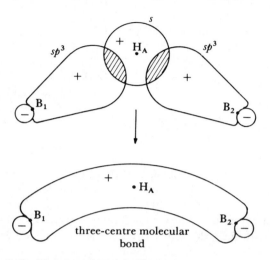

FIGURE 9.10 *Three-centred bonds in diborane*
A tetrahedral hybrid orbital on B_1, an s orbital on H_A, and a tetrahedral orbital on B_2 combine to form the bonding orbital shown, an antibonding orbital with nodes between each B and the H, and a non-bonding orbital which places the electron density mainly on the two B atoms. A similar set of three three-centre orbitals is formed by H_B with two other tetrahedral orbitals on B_1 and B_2.

bonds. The three atomic orbitals from B_1, H_A and B_2 combine to form a bonding, a nonbonding, and an antibonding, three-centred molecular orbital centred on these atoms. (Note that three atomic orbitals give three molecular orbitals.) The three atomic orbitals on B_1, H_B and B_2 give an identical set of three molecular orbitals. If two of the four electrons, left over after the terminal B—H bonds were formed, are placed in the $B_1H_AB_2$ bonding orbital and the other two in the $B_1H_BB_2$ bonding orbital, as in Figure 9.11b, the result is to use all the valency electrons and to fill only the three-centre molecular orbitals which are bonding. Diborane is thus described as having four two-centre B—H bonds and two three-centre B····H····B bonds, all sigma, all holding two electrons, and accounting for the twelve valency electrons.

The description of these three-centre bonds is one example of the polycentred sigma bonding discussed in Chapter 4. Here the three-centre description is by far the most satisfying of the alternatives possible.

Diborane serves as a model compound for the other boron hydrides and other hydrogen compounds which are formulated with similar polycentred bonding. For example, tetraborane, B_4H_{10}, is shown in Figure 9.12(a). This has six two-electron B—H bonds, one B—B bond, and four two-electron three-centre B····H····B bonds. The structures of higher boron hydrides may be built up similarly, but other types of polycentred bonds may also be present including

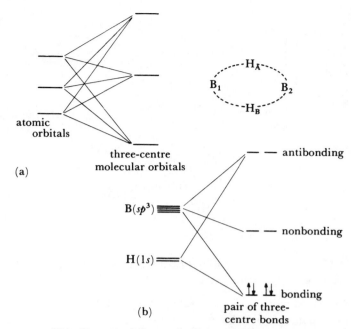

FIGURE 9.11 *Energy level diagram for three-centred bonds (diagrammatic),* (a) *general form,* (b) *electronic structure for the pair of three-centred bonds in diborane*
Figure (a) shows the generalized energy level diagram for three atomic orbitals combining to form three three-centred molecular orbitals. The energy level diagram (b) corresponds to the two sets of three-centred orbitals formed by the s orbitals on H_A and H_B and two suitably directed orbitals on each of B_1 and B_2 in the diborane bridge. The six atomic orbitals give six three-centred molecular orbitals and, in diborane, the two bonding orbitals are filled.

three-centred B····B····B bonds and even five-centred bonds involving five boron atoms, as found in B_5H_9, Figure 9.12.

In addition to the boron hydrides listed in Table 9.5, there is a wide range of boron hydride ions, of similar formulae and a variety of structures, and bonding in these species also involves multicentred bonding. Examples include $B_6H_6^{2-}$ and $B_{12}H_{12}^{2-}$ which have all the hydrogens present as terminal B—H bonds. The B_6 skeleton is a regular octahedron and the B_{12} one a regular icosahedron held together by multicentred B_n bonds. Some of these ions contain atoms other than boron. Thus there is an ion $B_9C_2H_{11}^{-}$ which can be regarded as derived from $B_{12}H_{12}^{2-}$ by removing one apex of the icosahedron (taking away one BH unit) and substituting two other B atoms by carbon atoms. The chemistry of such *carboranes* has expanded very rapidly in the last decade or so. Compounds range from relatively small molecules such as $C_2B_3H_7$ (isoelectronic with B_5H_9) and $C_2B_3H_8$ up to icosahedral species such as $C_2B_{10}H_{12}$. A very rich and varied chemistry has developed which is introduced in the references.

9.6.2 Wade's Rules

The wide variety of structures of the boron hydrides and their analogues have been rationalized by several different approaches. One of the simplest, which applies also to other clusters, is the Skeletal Electron Pair (SEP) approach which

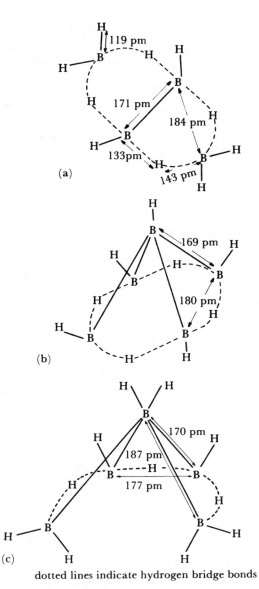

(a)

(b)

(c)

dotted lines indicate hydrogen bridge bonds

FIGURE 9.12 *Structures of* (a) *tetraborane,* (b) *pentaborane*-9, (c) *pentaborane*-11

(a) Tetraborane forms a boat-shaped molecule with four B····H····B three-centred bonds and a direct B—B single bond.

(b) Here the boron atoms form a square pyramid. The edges of the square base are bonded by three-centre B····H····B bonds while the apical B atom is bonded to the four base B atoms by five-centred bonds using boron orbitals directed into the body of the pyramid.

(c) Pentaborane-11 is an unsymmetrical square pyramid where, in addition to B····H····B bonds, there are three-centred bonds between three boron atoms in the triangular faces—the so-called 'central' B‿B bond.

was presented in a convenient form by Wade, and is thus often known as Wade's Rules. It is useful to introduce the ideas here in the context of the simpler boron hydrides—see section 18.4.2 for a fuller discussion.

This approach starts by building a boron hydride from n B atoms, each with one outward-pointing B—H bond external to the cluster. (Any further hydrogens are attached in other ways.) That is, any hydride B_nH_m is formulated as $(BH)_nH_{m-n}$. The external B—H is a normal two-electron, two-centre bond and thus uses one orbital (conveniently regarded as an sp_3 hybrid) from the B.

The skeletal electron pairs are then calculated by:

(a) finding the total number of valency electrons (i.e. 3 per B, 1 per H, plus or minus electrons for any charge)

(b) subtracting 2 electrons for each BH

(c) leaving the skeletal electron pairs, which bind the cluster.

Then the key predictions are:

(1) The structure observed for a molecule with $(n+1)$ skeletal electron pairs is the regular solid with n vertices, provided there are n B atoms. Such structures are closed clusters and given the descriptor *closo*. (Regular solids are those whose faces are all equilateral triangles.)

(2) If there are only $(n-1)$ B atoms in a cluster with $(n+1)$ SEP, the structure is that formed by removing one vertex from the *closo cluster* (called *nido*).

(3) Similarly, for $(n-2)$ B atoms, two vertices are removed (called *archno*). Any greater deficiency is rare.

(4) If there are more than n B atoms with $(n+1)$ SEP, the additional B go into capping positions over the triangular faces.

For mixed clusters, such as carboranes, the SEP approach also applies. Thus C is assumed to form an external C—H bond, and it contributes four valency electrons.

The theory has been considerably refined but this basic level suffices to account for simple species. Thus the two pentaboranes in Figure 9.12 are treated as follows:

$$\text{pentaborane-9}(B_5H_9) \text{ valency electrons} \quad \text{for } 5B = 15$$
$$\text{for } 9H = 9$$
$$\text{less a pair of electrons for } 5BH = -10$$

This leaves 14 electrons or 7 SEP and the structure is predicted to be related to the 6-vertex regular solid which is the octahedron. As there are only 5 B, one vertex is removed. Thus the predicted structure is the *nido*-octahedron or square pyramid as observed.

$$\text{pentaborane-11 } (B_5H_{11}) \qquad 5\,B = 15 \text{ electrons}$$
$$11\,H = 11 \text{ electrons}$$
$$\text{less} \quad 5\,BH = -10 \text{ electrons}$$

This gives 8 SEP, which is a structure based on the 7-vertex regular solid, the pentagonal bipyramid. If one vertex and one of the pentagonal plane atoms are removed, the result is the structure of Figure 9.12c.

In these two molecules the extra hydrogens bridge the edges in B_5H_9 and are also added as terminal bonds to the boron atoms in the open face for B_5H_{11}.

It is readily seen that the $[B_6H_6]^{2-}$ and $[B_{12}H_{12}]^{2-}$ ions count to 7 and 13 SEP respectively, giving the octahedron and the 12-vertex figure which is the icosahedron. B_4H_{10} counts to 7 SEP, so the structure should be the *arachno*-octahedron. Removing two adjacent vertices leaves two triangles joined by an edge as observed. A summary of the commoner shapes is given in Table 9.6.

TABLE 9.6 Shapes predicted by Wade's Rules for simple clusters

SEP $(n+1)$	Closo n vertices n atoms	nido $(n-1)$ atoms	arachno $(n-2)$ atoms
4	triangle		
5	tetrahedron	triangle	
6	trigonal bipyramid	tetrahedron	triangle
7	octahedron	square pyramid	square or linked triangles
8	pentagonal bipyramid	pentagonal pyramid	irregular (compare B_5H_{11})

It can be seen from Table 9.6 that while the simple theory rationalizes observed shapes, it is not predictive in detail. Thus the alternatives for 7 SEP and 4 atoms depend on whether the two removed vertices are adjacent or opposite, and the simple theory gives no guide. Likewise, placing two capping atoms often offers a choice. As the trigonal bipyramid is equivalent to the capped tetrahedron, we find the tetrahedron (and the triangle) occurring for different SEP counts.

Despite these ambiguities, the theory has the advantage of being extremely simple and of applying to the vast majority of the large number of boranes, carboranes, and related compounds. It accounts simply for the finding that some 6-atom molecules are octahedra and some are pentagonal pyramids, or that 5-atom molecules can be trigonal bipyramids or square pyramids, etc. It may be extended to transition metal clusters (see section 18.4.2).

9.6.3 Other electron-deficient hydrides

Electron-deficient bonding is by no means confined to boron compounds. Aluminium hydride is an insoluble polymer $(AlH_3)_x$, and a recent structural determination by X-ray and neutron diffraction shows that it contains Al atoms surrounded octahedrally by six H atoms. The structure is a giant molecule of AlH_2Al bridges and is analogous to the structure of AlF_3. This is an example of electron-deficient bonding analogous to that in diborane. The simple AlH_3 molecule has been prepared as a highly reactive molecular species in a noble gas matrix, by the reaction of Al atoms and hydrogen. Another example of $Al\cdots H\cdots Al$ bridging is found in the molecule $Al_2H_5NMe_2$ which has the structure

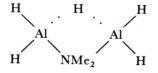

There is an exact boron analogue, $B_2H_5NMe_2$. In each molecule, the NMe_2 group may be regarded as replacing one of the hydrogen bridges in diborane or its aluminium analogue.

The somewhat elusive binary gallium hydride $[GaH_3]_n$ has recently been synthesized from dimeric $[H_2GaCl]_2$ and $LiGaH_4$ at low temperature. The principal product is the diborane-like dimer $[GeH_3]_2$ but aggregation occurs in the solid state with $n>2$ and possibly 4, though the material retains terminal Ga—H bonds. The mixed galloborane $H_2Ga(\mu\text{-}H)_2BH_2$ has been shown by electron diffraction to have the diborane structure.

The mixed gallium alkyl/halide/hydride compounds $Me_2Ga(\mu\text{-}H)_2GaMe_2$ and $Me_2Ga(\mu\text{-}Cl)_2GaMe_2$ have hydride or halide bridges in preference to alkyl ones.

Beryllium hydride, $(BeH_2)_x$, is also polymeric and insoluble. The structure is reasonably postulated to be a chain polymer with bridging hydrogen

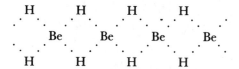

Hydrogen bridging to beryllium is illustrated in a number of other species, for example, the ion $Et_2BeH_2BeEt_2^{2-}$ with the structure

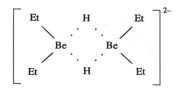

Electron deficient bonding is also found in some of the organic analogues of the hydrides. For example, trimethyl-aluminium exists as a dimer, $Al_2(CH_3)_6$, with a *methyl-bridged* structure very similar to that of diborane. This is based on three-centred bonds formed by an sp^3 hybrid on each aluminium together with an sp^3 hybrid on the carbon atom, as shown in Figure 9.13.

These electron-deficient bond systems are of reasonable strength. Thus, the boron hydrides resemble the silicon

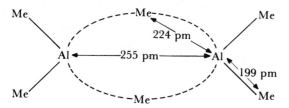

FIGURE 9.13 *The structure of* Al_2Me_6
This diagram gives the form of aluminium trimethyl dimer. The methyl group is bonded to two aluminium atoms in three-centred bonds formed by overlap of tetrahedral hybrid orbitals on each aluminium and on the carbon atom. This bonding resembles the $B\cdots H\cdots B$ bond in diborane, and also the central $B\diagdown B$ bonds found in pentaborane-11.

hydrides in thermal stability. The hydrides share the strong reducing properties of all hydrides and the boron hydrides react violently with oxygen. The electron-deficiency shows up mainly in the susceptibility of the hydrides to attack by electron pair donors, which provide the extra electrons to allow the formation of electron-pair two-centre bonds. Diborane, for example, readily reacts to give borine adducts, $H_3B \leftarrow D$, with suitable donors, D, such as the amines.

The boron Group of elements also makes use of the empty p orbital in the formation of the complex hydrides. The BH_4, AlH_4, and GaH_4 groups have been prepared and they decrease in stability in that order. Many elements form these complex hydrides and they range from the ionic compounds of the active metals, such as $Na^+BH_4^-$ or $Li^+AlH_4^-$, to the compounds of the less active metals which appear to be covalent and electron deficient. A structure with unsymmetric three centre $B\cdots H\cdots Be$ bonds has been reported for beryllium borohydride, $Be(BH_4)_2$.

For aluminium borohydride, $Al(BH_4)_3$, the Al atom is surrounded by six hydrogen atoms, forming three AlH_2B bridges while the low temperature form of $Zr(BH_4)_4$ probably contains zirconium surrounded tetrahedrally by four boron atoms linked by three-hydrogen bridges, of the form

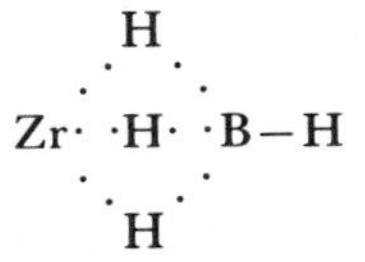

$$Zr \cdot \begin{matrix} \cdot H \cdot \\ \cdot H \cdot \\ \cdot H \cdot \end{matrix} \cdot B-H$$

Complex aluminium hydrides of the less electropositive metals probably have similar bridging structures. Even in $LiAlH_4$, the $Li-H$ distance is shorter than in LiH, suggesting some $Li\cdots H$ interaction which probably accounts for its solubility in ethers. The complex beryllium hydrides, Li_2BeH_4 and Na_2BeH_4 also behave more like electron deficient polymers than as species containing BeH_4^{2-} ions.

It has been seen that hydrogen-bridged, electron deficient species, where polycentred sigma bonding has to be postulated, occur quite widely. Although best known for boron, it is also established for compounds of beryllium, aluminium and zinc together with the $B\cdots H\cdots M$ links to many transition metals in their borohydrides. $M\cdots H\cdots M$ bridges are also established in metal-hydrido complexes such as $HCr_2(CO)_{10}^-$. It is also probable that a weaker interaction, of the form $Al-H\cdots Li$, occurs in complex beryllium, zinc, and aluminium hydrides. Thus electron deficient polycentred bonding is typical of hydrides of elements which have fewer valence shell electrons than valency orbitals. Its presence allows use of orbitals which would otherwise remain empty, and the interaction is sufficiently strong to give aluminium, and probably other elements, a coordination number greater than four. Similar conditions hold for the formation of methyl and similar bridges.

The boron group hydrides react readily with electron pair donors to revert to two-centre two electron bonding and adducts such as $H_3B.NMe_3$ are also readily formed (cf. also Figure 9.8a). If the donor group is itself a hydride, further

interaction may occur with elimination of H_2. For example, the product of the reaction between ammonia and diborane at low temperatures is the expected adduct, H_3BNH_3. On warming to room temperature, this compound loses hydrogen and gives a product of empirical formula, HBNH. The latter is actually the trimer, called borazole or borazine, whose structure is shown in Figure 9.14. The molecule, $B_3N_3H_6$ is isoelectronic with benzene, C_6H_6, and has a similar, planar structure with all the $B-N$ bonds the same and a delocalized π-bonding system. A wide variety of substituted borazines have been made, either by substitution reactions on borazine or, more commonly, by altering the composition of the starting adduct. Thus, $(CH_3)H_2BNH_3$ gives, on heating, $(CH_3)_3B_3N_3H_3$—called B-trimethylborazine (where the methyl groups are on the boron). The corresponding N-trimethylborazine is formed from the adduct $H_3BNH_2(CH_3)$. Many similar compounds can be made such as the chloro-boron derivative of Figure 9.14c. Later work showed the existence of an aluminium analogue $(MeAlNAr)_3$.

In contrast, if the ammonia is replaced by phosphane, PH_3, the adduct H_3BPH_3 loses less hydrogen to give $(H_2BPH_2)_3$ which has a different ring system. This is the analogue, not of benzene but of cyclohexane, C_6H_{12}, and the ring is no longer planar but chair-shaped. Similar compounds may be made by starting from substituted phosphanes.

9.7 The hydrogen bond
Compounds containing hydrogen bonded to a very electronegative element, especially F, O, or N, show properties

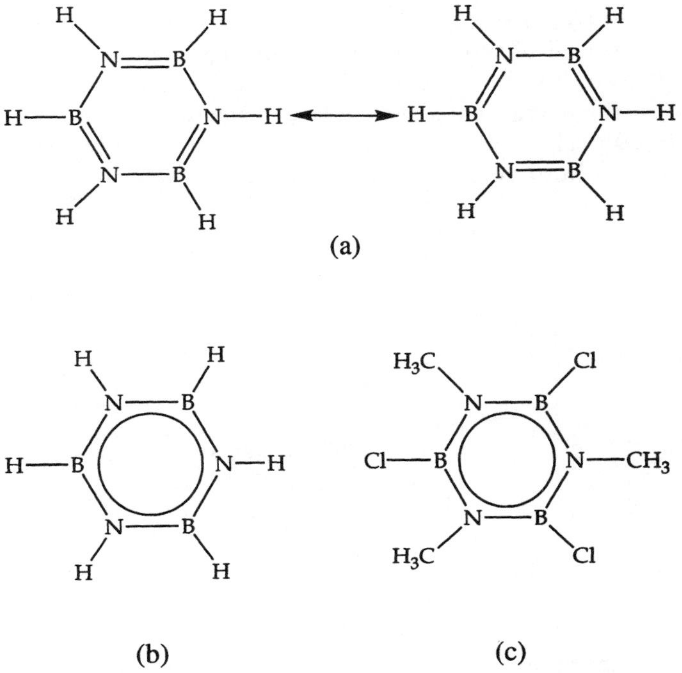

FIGURE 9.14 (a) *The structure of borazine* ($H_3B_3N_3H_3$), (b) N-*trialkyl borazine*, (c) B-*trichloro*-N-*trimethylborazine*

which are consistent with an interaction between the hydrogen bonded to one electronegative atom and a second atom. This secondary bond is relatively weak and is termed the hydrogen bond. It may be written $X-H\cdots Y$, where X is the atom to which the hydrogen is bonded by a normal bond (compare also Figures 9.1, 9.2). The importance of hydrogen bonding cannot be overstressed since substances such as water and DNA, both essential to life, are extensively hydrogen-bonded. While the double-helix structure of the DNA molecule is now well understood, the structure of liquid water still poses many challenges to the chemist, due to the formation of 'dynamic' hydrogen bonds. Protons are able to be transferred very rapidly from the oxygen atom of one water molecule to another, with a frequency of around $10^{12}\,s^{-1}$. The migration of a proton in water is accomplished *via* a cooperative cascade of proton transfers between adjacent water molecules, called the *Grotthus mechanism*. In contrast to liquid water, the various crystalline polymorphs of ice are better understood since experimental methods give much more direct information on the structure of the solid form.

Evidence for hydrogen bonding is widespread. Some of the important facts are:

(i) *Evidence of molecular association* from melting points, boiling points, heats of evaporation, and, in some cases, molecular weight determinations. For example, the boiling points of the hydrides of the carbon, nitrogen, oxygen, and fluorine Group elements are shown in Figure 9.15. For the carbon Group hydrides, the boiling point is a linear function of atomic weight. However, the positions of ammonia, water, and hydrogen fluoride are clearly anomalous. These unexpectedly high boiling points indicate an extra interaction between the molecules in the liquid, which is not found for the heavier hydrides. A similar anomaly appears in the latent heats of vaporization shown in Figure 9.16, and also in the melting points and the latent heats of fusion, showing that the interaction is also present in the solid. Figure 9.16 serves to give some idea of the size of the effect; the increases in the heats of evaporation, compared with those found by

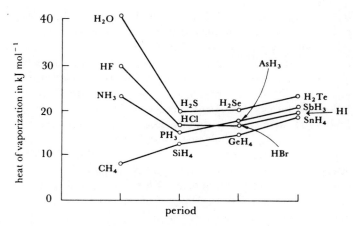

FIGURE 9.16 *Latent heats of vaporization of Main Group hydrides*

extrapolating from the heavier hydrides, is of the order of 20 ± 10 kJ mol^{-1}. In general, the hydrogen bond energies may be taken to lie in the range from 4 to 40 kJ mol^{-1}, about one tenth of normal bond energies but several times greater than normal weak interactions between molecules.

Molecular weight studies show clear evidence of association in many cases: perhaps the most striking case is the dimerization of the lower carboxylic acids. The heat of association of acetic acid:

$$CH_3COOH = \tfrac{1}{2}[CH_3COOH]_2$$

is found to be 28·9 kJ mol^{-1} (of monomer). Electron diffraction studies have shown that the structure of the acetic acid dimer is as given in Figure 9.17.

Other carboxylic and related oxy-acids behave similarly, for example dialkyl phosphoric acids, $(RO)_2P(O)OH$. Organic solvent-soluble long- and branched-chain analogues of these compounds are employed industrially in hydrometallurgical extraction and separation of various metals (e.g. the separation of cobalt from nickel).

(ii) *X-ray studies.* The evidence from boiling points, etc., above gives no indication of the atomic arrangements in the hydrogen bond and this information might be expected from X-ray studies. However, as hydrogen is so light compared with other nuclei, it is not an efficient scatterer of X-rays and it is not usually possible to determine the positions of hydrogen atoms in this way. One exception is boric acid, H_3BO_3 which contains only light nuclei and whose structure consists of loosely bound sheets with the configuration shown in Figure 9.18.

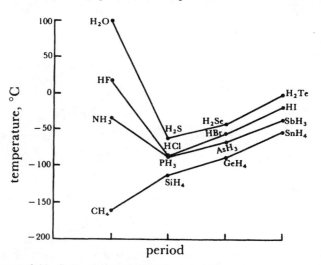

FIGURE 9.15 *Boiling points of Main Group hydrides*

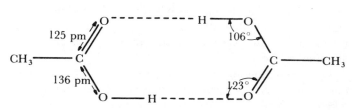

FIGURE 9.17 *The structure of the acetic acid dimer*

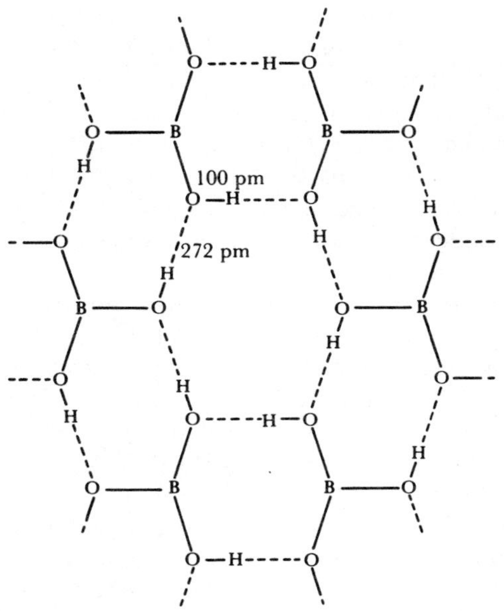

fraction, and refined X-ray methods, now available for crystal studies, have contributed to the structures in Figure 9.19.

(iv) *Infrared studies.* When an $X-H$ grouping is involved in hydrogen bonding, the characteristic infrared stretching of the $X-H$ bond is shifted and it is found that this is always to lower frequencies. There is a general correlation between the extent of this frequency shift and the hydrogen bond energy. Some examples are given in Table 9.7. As infrared spectra are readily measured, this frequency shift is usually the most convenient diagnostic test for hydrogen bonding.

(v) *Entropy data.* Evidence of hydrogen bonding may also be given by entropy data determined at low temperatures. As entropy is linked with disorder, a definition in terms of the number of possible arrangements of the system is possible. This is $S = -R\ln W$, where R is the gas constant and W is the probability of the state of the system. Now, a perfect crystal, when cooled to absolute zero (to eliminate thermal motions of the atoms), has only one possible arrangement of the atoms and the entropy at absolute zero will be $-R\ln 1 = 0$ (remember that certainty = a probability of one). However, if hydrogen bonding occurs in a crystal, more than one configuration may be possible at absolute zero. For example, in the acetic acid dimer, these are the two configurations of Figure 9.19a: similarly, ice has the oxygen atoms surrounded by four hydrogen atoms and a number of configurations are possible, such as those shown in Figure 9.19b. As a result, such hydrogen-bonded molecules have non-zero values of the entropy at 0 K. In the case of ice, it has been calculated that the entropy should be $3 \cdot 39$ J K^{-1} mol^{-1} and it has been measured as $3 \cdot 39 + 0 \cdot 21$ J K^{-1} mol^{-1}.

FIGURE 9.18 *Structure of boric acid,* H_3BO_3. *Hydrogen bonding holds the molecules in sheets*

In general, X-ray evidence gives only the $X-Y$ distance, and it is found that this is less than the sum of the van der Waals' radii, which would determine the distance in the absence of hydrogen bonding. See examples in Table 9.7.

(iii) *Neutron diffraction.* The scattering of neutrons does allow the location of hydrogen atom positions and a number of studies have appeared. Neutron diffraction, electron dif-

TABLE 9.7 Properties of hydrogen bonds

Bond	Compound	Bond length $X-Y$ distance, pm	Van der Waals' distance $X-Y$, pm	Depression of stretching frequency, cm^{-1}†	Bond energy, kJ mol^{-1}
$F-H-F$ (symmetrical)	$K^+HF_2^-$	226	270	2700	113
	$NH_4^+HF_2^-$	232	270		
$F-H\cdots F$ (unsymmetrical)	$(HF)_n$	255	270	700	28·0
$O-H-O$ (symmetrical, bent?)	$Ni(DMG)_2$*	240	280	1200	
$O-H\cdots O$ (unsymmetrical)	KH_2PO_4	248	280	900	
	$(HCOOH)_2$	267	280	600	29·7
	ice	276	280	400	18·8
$N-H\cdots N$	NH_4N_3	294 299	300		
	melamine	300	300	120	ca25
$N-H\cdots F$	NH_4F	263	285		
$N-H\cdots O$ (slightly bent)	$NH_4H_2PO_4$	291	290		
$N-H\cdots Cl$	$N_2H_4.HCl$	313	330	500	

*DMG = dimethylglyoxime
†The stretching frequency figures are rounded off to the nearest hundred wave numbers and represent averages of two frequencies in some cases.

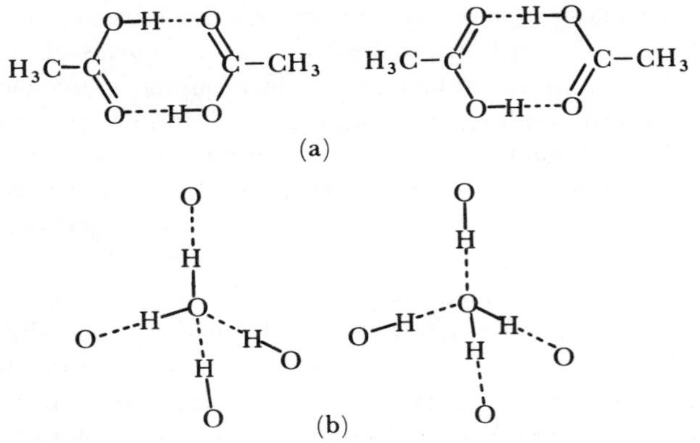

FIGURE 9.19 *Alternative configurations of hydrogen-bonded species leading to non-zero entropy at absolute zero:* (a) *acetic acid dimer,* (b) *ice*

The hydrogen bond can be shown to exist by the methods given above, and the next question is about its shape. The bond may be linear or bent and the hydrogen atom may lie symmetrically or unsymmetrically between X and Y. As far as present evidence goes, it appears that XHY are usually arranged linearly unless there is a good steric reason opposing this—as in the intramolecular hydrogen bonds in molecules like salicylic acid

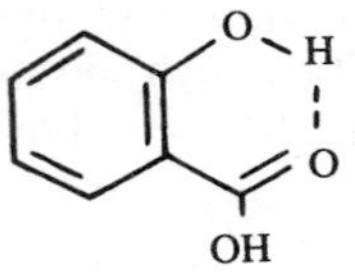

or nickel dimethylglyoxime (Figure 14.31).

The symmetry of the position of the H atom varies. In the bifluoride ion, FHF^-, the hydrogen is symmetrically placed between the two fluorine atoms, as is the hydrogen in the OHO links in the nickel dimethylglyoxime complex. In most other cases of known structure, however, the unsymmetrical position of the hydrogen is adopted, as in water or acetic

acid. Apart from structure determinations, it should be noted that the entropy determination will distinguish symmetrical structures as these will have no residual entropy. It is in fact found that there is no residual entropy for HF_2^- in potassium hydrogen fluoride. It appears that, in general, the stronger hydrogen bonds are the more symmetrical.

Table 9.7 summarizes the data available for some case of hydrogen bonds involving F, O, and N. One or two cases are worth further mention.

The solvated proton. As discussed above, the free H^+ ion cannot exist in a chemical environment, and there are a number of structural studies which demonstrate its occurrence as H_3O^+. Thus the solid monohydrates of perchloric, sulphuric, and hydrochloric acids have been shown to exist as $H_3O^+ ClO_4^-$, $H_3O^+ HSO_4^-$ and $H_3O^+ Cl^-$ respectively.

Recent studies of higher hydrates of protonic acids have shown that more complex forms of the hydrated proton exist, whose structures involve hydrogen bonding. For example, $HCl.2H_2O$ is best formulated as $H_5O_2^+ Cl^-$, while $HBr.4H_2O$ is $(H_7O_3)^+ (H_9O_4)^+ 2Br^-.H_2O$. The structures of these forms of the hydrated proton are summarized in Table 9.8. The $HAuCl_4$ hydrate was studied by neutron diffraction and the hydrogens were located.

The symmetric $H_5O_2^+$ ion has a short hydrogen bond length, similar to that in the nickel dimethylglyoxime molecule (Table 9.7). The unsymmetric $H_5O_2^+$ ion, $H_7O_3^+$ and $H_9O_4^+$ are structurally more like a H_3O^+ ion hydrogen bonded to one, two or three water molecules but with many of the hydrogen bond lengths distinctly short.

Hydrogen fluoride. This has been shown to have the zig-zag structure shown in Figure 9.20. There seems to be no

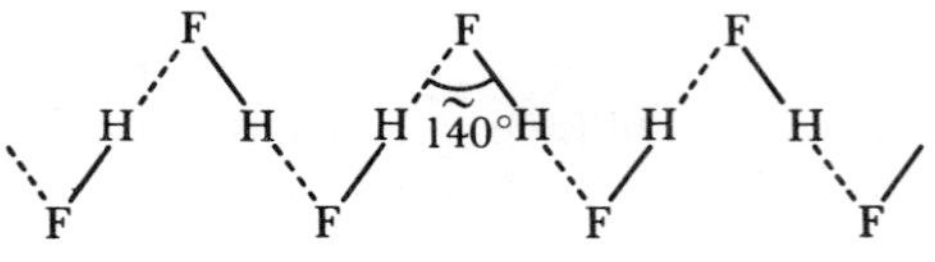

FIGURE 9.20 *The structure of hydrogen fluoride*

TABLE 9.8 Hydrogen bonds in hydrated protons

Ion	Structure	O····O length (pm)	Compound	Comments
$H_5O_2^+$	$(H_2O-H-OH_2)^+$ probably symmetrical cf. Figure 9.2	242·4	$HClO_4.2H_2O$	$HCl.xH_2O$ ($x = 2, 3$) contain similar ions with short hydrogen bonds (242–250 pm)
	$(H_2O-H····OH_2)^+$	257	$HAuCl_4.4H_2O$	Unsymmetric, $O-H = 99$ pm and $H····O = 148$ pm. $H_5O_2^+$ units are linked through H_2O by further hydrogen bonds of 274 pm
$H_7O_3^+$	$(H_2O····H-O-H····OH_2)^+$ | H	247 and 250	$HBr.4H_2O$	Central H_3O surrounded pyramidally by two outer H_2O and one Br
$H_9O_4^+$	similar to Figure 9.1	250 and two of 259	$HBr.4H_2O$	Central H_3O surrounded pyramidally by three outer H_2O

convincing explanation of the wide $H-F-H$ angle. The hydrogen bonding in hydrogen fluoride persists into the gas phase, where small polymers are found.

Ammonium fluoride. NH_4F shows a different structure from the other ammonium halides. The latter have the CsCl or NaCl structure (the transition to NaCl taking place below 200 °C), but the fluoride has the wurtzite structure (Figure 5.1d), in which each N atom forms four $N-H-F$ bonds of length 263 pm to its four neighbouring N atoms which are arranged tetrahedrally around it. The structure resembles that around the O atoms in ice (Figure 9.19).

Apart from F, O, and N, there are few atoms which form hydrogen bonds. Chlorine occasionally enters into hydrogen bonding and the HCl_2^- ion is formed in the presence of large cations such as Cs^+ or NR_4^+. Indeed, recent work has shown that HBr_2^- and HI_2^- also exist under such conditions. There is no evidence as to their degree of symmetry. One phase formed by trimethylamine and hydrogen chloride is a low-melting solid $Me_3N.5HCl$ which provides examples of two of the weaker hydrogen bond systems, $N-H \cdots Cl$ and $Cl-H \cdots Cl$, Figure 9.21. The compound is best formulated as $[Me_3NH]^+[Cl(HCl)_4]^-$ where Cl^- is at the centre of a trigonal bipyramid with two axial and two equatorial HCl groups and with $[Me_3NH]^+$ occupying the third equatorial position. The $Cl-H \cdots Cl$ bonds are nearly linear and vary in length from 343 to 353 pm, substantially longer than those in Table 9.6. The $N-H \cdots Cl$ bond is bent, with an angle of 141° at H, and the hydrogen bond length is 327 pm.

There is also some evidence for hydrogen bonding involving carbon, $C-H \cdots X$, if the carbon is bonded to electronegative groups as in HCN. One case has even been established of an $N-H \cdots C$ bond of length 258 pm. All these other hydrogen bonds are much weaker than those involving F, O, and N.

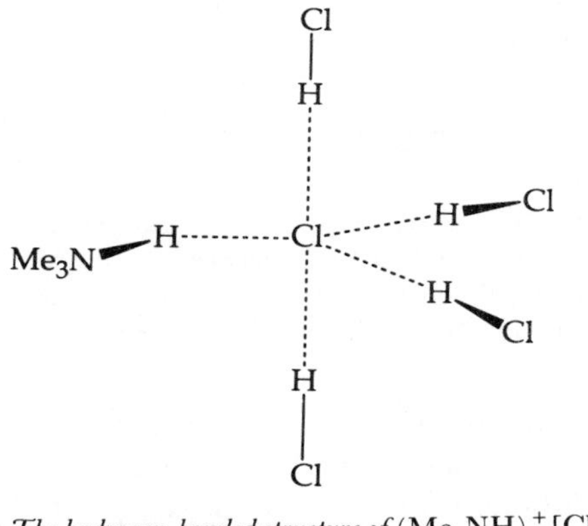

The fact that hydrogen bonds occur only between strongly electronegative elements, and that small elements form the strongest bonds, suggests an electrostatic origin for the interaction. The relatively positive hydrogen in $O-H$, $N-H$ or $F-H$ bonds interacts with the dipole in neighbouring atoms to form the bond. This happens only with hydrogen, probably because of its small size and lack of inner shielding electron shells.

PROBLEMS

It is important, in learning systematic chemistry, to correlate properties and highlight similarities and differences. The chemistry of the hydrides of the various elements should be compared with that of their other simple compounds.

9.1 Compare the shapes of the covalent hydrides with those predicted by the VSEPR method (Chapter 4).

9.2 (a) Assuming bond energies are additive, compare the calculated energy of formation from the elements of NH_3, PH_3, and AsH_3 with those of the analogous fluorides, using data in Table 17.3, p. 323. (The heat of atomization of $P = 334$ and $As = 289 \text{ kJ mol}^{-1}$.)

(b) Comment on the fact that the thermal stability of fluorides is generally higher than that of the corresponding hydrides.

9.3 Formulate the bonding in $Al(BH_4)_3$ (see p. 161). How does this compare with aluminium hydride?

9.4 Calculate the heats of formation of LiH, NaH and KH, using the experimental lattice energies in Table 5.2, p. 83.

9.5 The heat of formation, under standard conditions, of CaH_2 is -173 kJ mol^{-1}. Assuming the Madelung constant value is 5·5, calculate the lattice energy as in section 5.3 and compare with the Born-Haber cycle value (see also question 5.5).

9.6 You should also formulate answers to questions of the type:

Compare and contrast the chemistry of the hydrides with that of the fluorides/chlorides/iodides of the Main Group Elements/transition elements etc.

FIGURE 9.2   *The hydrogen-bonded structure of* $(Me_3NH)^+[Cl(HCl)_4]^-$

10 The 's' Elements

10.1 General and physical properties, occurrence and uses

Elements with their outermost electrons in an s level are those of the lithium Group (alkali metals) and of the beryllium Group (beryllium, magnesium and the alkaline earth metals). Many properties have been listed in the earlier chapters, including atomic masses and numbers (Table 2.5), ionic radii (Table 2.12), ionization potentials (Table 2.8), electronegativities (Table 2.14), and structural details of halides, etc. (Table 5.1). Structures of s element compounds include NaCl (Figure 5.1a), CsCl (Figure 5.1b), and CaC_2 (Figure 5.13).

The energetics of formation of alkali halides and alkaline earth chalcogenides are discussed in section 5.3.

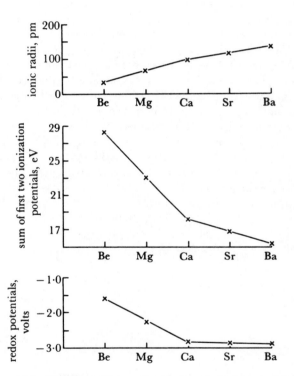

FIGURE 10.2 *The ionic radii, sum of the first two ionization potentials, and* M^{2+}/M *redox potentials of the beryllium Group of elements*
As in Figure 10.1 the major changes in properties come between the first and second members. The similarity between Ca, Sr, Ba is clear.

Other properties of the elements are given in Table 10.1 and important parameters in Figures 10.1 and 10.2.

Of the lithium Group elements, compounds of the first three are produced commercially. The common source is salt deposits from ancient seas, or brines from seawater evaporation. Lithium is also mined as *spodumene*, an aluminosilicate. The commonest form of production for all three is as the chloride, with Li and Na being also converted to the carbonate.

The cheaper metals, especially sodium, are utilized industrially in some processes requiring powerful reduction.

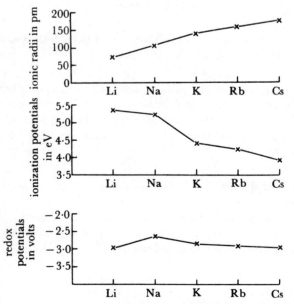

FIGURE 10.1 *The ionic radii, ionization potentials, and redox potentials of the alkali metals*
The relatively large changes between Li and Na, and—to a lesser extent—between Na and K are reflected in the general chemistry.

TABLE 10.1 The *s* elements

Element	Symbol	Electronic configuration	Abundance (ppm of the crust)	Accessibility*	Common coordination numbers
Lithium	Li	[He]$2s^1$	65	common	4, 6
Sodium	Na	[Ne]$3s^1$	28 300	common	6
Potassium	K	[Ar]$4s^1$	25 900	common	6
Rubidium	Rb	[Kr]$5s^1$	310	rare	6
Caesium[†]	Cs	[Xe]$6s^1$	7	rare	6, 8
Francium	Fr	[Rn]$7s^1$	—	very rare	?
Beryllium	Be	[He]$2s^2$	6	becoming common	2, 4
Magnesium	Mg	[Ne]$3s^2$	20 900	common	6
Caesium[†]	Ca	[Ar]$4s^2$	36 300	common	6
Strontium	Sr	[Kr]$5s^2$	300	common	6
Barium	Ba	[Xe]$6s^2$	250	common	6
Radium	Ra	[Rn]$7s^2$	trace	rare	

*See section 8.7 for a discussion of this point
[†]The name cesium is also widely found.

One example is the formation of titanium or zirconium from their tetrachlorides by reaction with sodium or calcium. Alloys of sodium and potassium (which are liquid at ambient temperatures for compositions containing 40 to 90% K) are used as the primary coolant in the experimental 'breeder' reactor.

Uses of the elements in the laboratory, and in fine chemical production, are based on their high reactivity. Substantial work has been done on Li/LiCl and Na/Na$_2$S fused salt batteries, which have advantages in weight and energy density. Problems of working with such media remain, but half of all lithium use by the year 2000 is expected to be in batteries and other electrochemical applications. Lithium compounds are also used in glasses, ceramics, and for specialized lubricants. Lithium therapy has made significant contributions to the treatment of mental illnesses. Sodium compounds in major use are the industrial alkalis, Na$_2$CO$_3$ and NaOH, and NaCl, especially for preparation of Cl$_2$ by electrolysis (and also further conversion to NaOCl or NaClO$_3$). The main use of potassium is as a primary fertilizer, where application is usually as KCl. The carbonates are the main vehicle for incorporating Na$_2$O and K$_2$O into glasses, and in cement manufacture.

The beryllium Group elements, with the exception of beryllium itself, occur mainly as carbonates or in brines and salt deposits. The principal source of beryllium is the aluminosilicate mineral beryl, Be$_3$Al$_2$Si$_6$O$_{18}$ (see section 18.6 for further discussion on silicate minerals). Beryl is found in a number of coloured gem forms including aquamarine (sea blue) and emerald (green), the most highly-prized. Beryllium is an element which presents an interesting dichotomy. On the one hand it is the most toxic non-radioactive element known (second in the Periodic Table to plutonium) and great care must be exercised in handling beryllium compounds. The reader may be interested to learn that more research papers have been published on theoretical aspects of beryllium chemistry than on practical aspects. On the other hand, however, the metal itself has a number of highly specialized uses. As a result of its transparency to X-rays, it is used in windows in X-ray apparatus, and its strength and lightweight nature have even led to its use as a component of the alloy from which golf clubs are made!

Calcium carbonate (lime) and magnesium carbonate are important in agriculture, as components of cements, and as the source of the oxides. MgO, prepared at high temperature, is inert and refractory and is extensively used as a furnace lining, especially in the steel industry. Barium carbonate finds a significant application in high-density concrete. Organomagnesium compounds are extensively used in laboratory syntheses and in the fine chemical industry.

The elements are all prepared by electrolysis of the fused halides. The relatively volatile rubidium and caesium are also conveniently prepared in the laboratory by heating the chlorides with calcium metal and distilling out the alkali metal. The alkali metals have melting points ranging from 180 °C–29 °C and boiling points in the range 1340 °C–670 °C, in both cases falling with increased atomic weight so that cesium has the second-lowest melting point of any metal. The beryllium Group elements are much less volatile, with melting points ranging from 1300 °C–700 °C and boiling points six or seven hundred degrees higher. Again there is a general tendency for volatility to increase with atomic weight.

Francium and radium both occur only as radioactive isotopes. Francium is inaccessible, with the longest-lived isotope, ^{223}Fr, having a half-life of only 21 minutes. Its chemistry is little known, but, in the few reactions which have been studied, it resembles the other heavy alkali metals. For example, it has an insoluble perchlorate and hexachloroplatinate, like rubidium and caesium. The longest-lived isotope of radium, ^{226}Ra, $t_{\frac{1}{2}} = 1600$ years, is much more stable and radium chemistry is well-established. Radium was used as a 'medicinal tonic' called Radithor in the late 1920s though, thankfully, the fad of consuming radioactive products is now long gone. The reader may also wish to refer here to section 20.2.3 which describes the application of radioactive technetium isotopes in diagnostic medicine; these techniques, however, expose the patient to little more radiation than does the average medical X-ray.

The chemistry of the elements of the s block is dominated by their tendency to lose the s electrons and attain a stable, rare gas configuration. (Beryllium is an exception to this, see section 10.7). This tendency is shown by the low ionization potentials and strongly negative redox potentials. The metals are among the most powerful of chemical reducing agents and combine directly, and usually violently, with most non-metals to yield ionic compounds. The cations formed by these elements (M^+ by the lithium Group, M^{2+} by Mg and the alkaline earths) are very stable and form salts with the most strongly oxidizing or reducing anions. Cesium is particularly useful to stabilize large anions, as in the formation of the bihalide ions, HX_2^-, mentioned in section 9.7.

The reactivity increases down each Group and the lithium Group elements are more reactive than the corresponding members of the beryllium Group. Although metallic lithium is the most reducing of the metals as judged by redox potentials, see Table 6.3, the heavier metals are considerably more reactive, cesium most of all. This poses much greater difficulties in handling, partly on account of its low melting point ($28.5\,^{\circ}\mathrm{C}$) when compared to lithium ($180\,^{\circ}\mathrm{C}$) or sodium ($98\,^{\circ}\mathrm{C}$).

The lightest elements are atypical, as their small size leads to a high charge density on the ions with resulting strong polarizing effects and high heats of solution. Lithium and beryllium show marked differences from their heavier congeners and sodium and magnesium also show distinct, though smaller, differences. There are many similarities between lithium and its diagonal neighbour in the beryllium Group, magnesium. Beryllium (section 10.7) differs markedly from all the other elements.

Salts of the s elements tend to be among the most soluble of their kind in water and other ionizing solvents, though solubilities vary widely with the anion. This is especially the case with the compounds of the alkaline earths where the double charge results in high lattice energies and high heats of solvation. AB_2 compounds tend to be soluble (e.g. the chlorides), while the insolubility of such AB compounds of the alkaline earths as the sulfates, oxalates, and carbonates, is well known in qualitative and quantitative analysis.

All the s elements dissolve in liquid ammonia to give intensely blue, conducting solutions which contain the metal ions and electrons which are free of the metal and appear to be associated with the solvent. (Beryllium and magnesium give only dilute solutions by electrolysis.) These 'solvated electrons' are very reactive and the metal-ammonia solutions are powerful reducing agents which can be used at low temperatures. (Compare section 6.7.) The heavier alkali metals may be recovered unchanged on evaporating off the ammonia, while lithium gives $Li(NH_3)_4$ and the alkaline earths yield hexammoniates, $M(NH_3)_6$. On standing, or in presence of catalysts such as transition metal oxides, the metals react with the ammonia to form the metal amide and

TABLE 10.2 Reactions of the s elements

Reaction	Notes
M is used for one mole of an alkali metal or a half-mole of a beryllium Group element.	
$2M + X_2 = 2MX$	X = halogen: alkali metals also form polyhalides, e.g. CsI_3, $KICl_4$ or KIF_6, see Chapter 15.
$2M + Y = M_2Y$	Y = O, S, Se, Te: higher oxides are also formed (section 10.2) and also polysulfides M_2S_{2-6}.
$3M + Z = M_3Z$	Z = N, P: Li reacts even at room temperature, Mg to Ra at red heat. The alkali metals also react with As, Sb, and Bi. Metal solutions in ammonia give polyanions such as $(Bi_9)^{4-}$ or $(Sb_5)^{3-}$.
$2M + 2C = M_2C_2$	Most rapidly with Li of the alkali metals. Ca, Sr, Ba and Ra at high temperatures. Be forms Be_2C.
$2M + H_2 = 2MH$	At high temperatures. Ionic hydrides. Not with Be, Mg.
$M + H_2O = MOH + \frac{1}{2}H_2$	By Be and Mg only slowly at room temperatures. $Be(OH)_2$ is amphoteric, all others are strong bases. $Mg(OH)_2$ is insoluble, others dissolve readily. All except beryllium hydroxide absorb CO_2 to give M_2CO_3, and alkali metals give $MHCO_3$ also.
$M + NH_3 = MNH_2 + \frac{1}{2}H_2$	With gaseous ammonia at high temperatures or liquid ammonia plus catalyst. Mg and Be only by reaction with amide of a more reactive metal, $Be + 2NaNH_2 \rightarrow Be(NH_2)_2 \rightarrow NaBe(NH_2)_3$.

hydrogen, e.g.:

$$M + NH_3 = MNH_2 + \tfrac{1}{2}H_2$$

(More accurately) $e^-_{(solvated)} + NH_3 = NH_2^- + \tfrac{1}{2}H_2$

Similar, but more dilute, solutions are formed in amines and methyl ethers. With aromatic hydrocarbons, such as benzophenone $PhC(O)Ph$ and naphthalene, alkali metals form highly coloured charge-transfer compounds, M^+Ar^-. These are used as 'soluble sodium' in reduction reactions and are also invaluable for rigorous drying of solvents like diethyl ether or tetrahydrofuran. The electron occupies the $\pi*$ orbital of the aromatic hydrocarbon and is highly reactive, though less so than in liquid ammonia.

The reactions of the elements with a number of typical reagents are given in Table 10.2.

10.2 Compounds with oxygen and ozone

The reaction between the s elements and oxygen may go further than to the simple oxide, and a number of higher oxides are formed when the metals are burned in air or are oxidized by O_2 in liquid ammonia. Peroxides, O_2^{2-}, are formed by all the elements except beryllium. In addition, sodium, potassium, rubidium, caesium, and calcium superoxides, O_2^-, have been prepared. The normal products of the combustion of the metals in an adequate supply of air are:

oxide	Li, Be, Mg, Ca, Sr
peroxide	Na, Ba (and Ra?)
superoxide	K, Rb, Cs.

The peroxides contain the ion $^-O-O^-$ and are salts of hydrogen peroxide (compare the relation of the hydroxides and oxides to water). The superoxides contain the ion, $O-O^-$. It will be recalled that oxygen, O_2, has its two outermost electrons unpaired in $\pi*$ orbitals. The superoxide ion and the peroxide ion have, respectively, one and two electrons more than in O_2. The superoxide ion has thus the configuration $(\pi*)^3$ and has the paramagnetism corresponding to one unpaired electron and a bond order of one and a half. The peroxide ion has the configuration $(\pi*)^4$, with no unpaired electrons and a bond order of one. It is isoelectronic with F_2 (see section 3.5 for further discussion and a list of the bond lengths of dioxygen ions). The MO_2 solids have the tetragonal lattice of calcium carbide. The increasing stability of the peroxides and superoxides with increasing cation size is noteworthy, and provides another example of the stabilization of large anions by large cations through lattice energy effects.

When the metals are treated with ozonized oxygen, or when ozone is passed into their solutions in liquid ammonia, the ozonides are formed. These are yellow or orange and contain the group $(O-O-O)^-$ which is paramagnetic with one unpaired electron. For example, the rubidium ozonide has an overall caesium chloride structure with a bent anion. The $O-O$ distance is 134 pm and the bond angle is $114°$. Red, crystalline complexes are formed between the alkali metal ozonides and crown ethers or cryptands (compare section 10.5) and in these adducts the geometry of the ozonide

anion is very similar to that in the parent MO_3 compounds.

The heavier elements, rubidium and cesium, also form oxides which are enriched in the metal (called suboxides). Established formulae include $Cs_{11}O_3$, Cs_7O, Rb_9O_2 and Rb_6O. The structures involve M_6O octahedra (which may be linked by shared faces) and may also include metallic regions. Thus $Cs_{11}O_3$ is three Cs_6O units sharing faces and Cs_7O is $[Cs_{11}O_3]Cs_{10}$. These compounds often form on surfaces and are usually metallic or highly coloured.

10.3 Carbon compounds

If acetylene is passed through a solution of an alkali metal in liquid ammonia, or is reacted with the heated metal, the following reactions take place:

$$M + C_2H_2 = MHC_2 + \tfrac{1}{2}H_2$$
$$MHC_2 + M = M_2C_2 + \tfrac{1}{2}H_2$$

The carbon compounds M_2C_2 and MHC_2 are termed acetylides and contain the discrete anions, $(C \equiv C)^{2-}$ and $(C \equiv CH)^-$, arising from the displacement of both or one of the relatively acidic hydrogens in the acetylene molecule. Acetylides also result from the direct reaction between carbon and heated lithium, sodium, magnesium, and alkaline earth metals. All form acetylene on hydrolysis. The structure of calcium acetylide (commonly called calcium carbide) has been determined (Figure 5.13) and is related to that of sodium chloride (Figure 5.1a).

These acetylides are the principal carbides (binary compounds of element and carbon) formed by the s elements, but two others of interest exist. Magnesium also forms a carbide of formula Mg_2C_3, which yields propyne, $HC\equiv C-CH_3$, as the main product on hydrolysis. This suggests the presence of a C_3 unit in this carbide. The product of direct combination between beryllium and carbon is the carbide Be_2C. This carbide probably contains single carbon atoms as the main hydrolysis product is methane. It has the anti-fluorite structure. All these carbides have many of the properties of ionic solids, with colourless crystals which are non-conducting at ordinary temperatures.

A second group of ionic compounds containing carbon is formed by the more reactive s elements. These are the metal alkyls and aryls such as ethyl sodium, $C_2H_5^-Na^+$, or phenyl-potassium, $C_6H_5^-K^+$. Such compounds are extremely reactive solids which inflame in air and react violently with almost all compounds apart from nitrogen and saturated hydrocarbons. They are involatile solids which decompose before melting and the evidence available indicates that they are ionized, R^-M^+. Crystal structures, supported by neutron diffraction, are reported for CD_3M species (D is preferred to H for neutron scattering). These show that methylpotassium is purely ionic, $K^+CD_3^-$, and the CD_3^- ion is pyramidal with $C-D = 109$ pm and $DCD = 106°$ (compare isoelectronic NH_3). The structure of methylsodium is more complex. Half the units exist as separate ions and half as $(CD_3Na)_4$ tetramers, like methyl-lithium (see below). The tetramers are arranged in chains in the crystal and these are linked together via $Na \cdots C$ contacts with the free ions.

Anions like $C_5H_5^-$ are more stable as the charge can delocalize in π orbitals, though they are still very air- and water-sensitive. They are widely used in the synthesis of sandwich compounds: thus $Na^+C_5R_5^-$ ($R = H$ or Me) is used to form ferrocene and all its many analogues (see section 16.4).

The corresponding lithium and magnesium compounds are covalent and much less reactive. The alkyl-lithiums, for example, are liquids or low-melting solids which are soluble in ethers or hydrocarbons. They are relatively involatile and exist as associated molecules with a highly polar $C-Li$ bond. Butyl-lithium, for example, is hexameric in hydrocarbon solvents and dimeric in ether. Methyl-lithium has been isolated as the tetramer, $(CH_3Li)_4$, which contains a tetrahedron of lithium atoms with a methyl group bridging each triangular face (Figure 10.3). The organolithium compounds find extensive uses in organic and organometallic syntheses. It is interesting that calculations first suggested that CLi_4, i.e. methane entirely substituted by Li, would be stable and even that CLi_6 species containing octahedral carbon should exist. These have both since been detected in the gas phase at 1000 K. It has even proved possible to synthesize such a startling compound on a bench scale by displacing HgX from $C(HgX)_4$ [X = halogen or alkyl] with an organolithium, RLi. The CLi_4 is a grey, extremely pyrophoric solid.

The principal class of organomagnesium compounds are the halides, or Grignard reagents, RMgX. These resemble the organolithiums in reactivity and these two types of reagent complement each other usefully. The more reactive ionic alkyls are much less useful as they are so difficult to handle, but they find some application where particularly vigorous conditions are required. Unsubstituted organomagnesiums, R_2Mg, are much less fully studied. They are bridged polymers with covalent properties. For example, $(C_5H_5)_2Mg$ is a low-melting, volatile solid.

10.4 Complexes of the heavier elements

The cations of the heavier *s* elements are very poor electron pair acceptors as their positive charge density is low. Solvates, such as hydrates or ammoniates, are not found for the heavier alkali metals although sodium gives a moderately stable tetrammoniate in the iodide, $Na(NH_3)_4I$. The alkaline earths are of higher charge density and more strongly hydrated. With large anions, octahedral $Mg(H_2O)_6^{2+}$ is found, while a 7-coordinated capped trigonal prism configuration is found around Ca in $2[Ca(H_2O)_7][Cd_6Cl_6(H_2O)_2]$. In the complex double azide, $CaCs(N_3)_3.H_2O$, the Ca is 7-coordinate to H_2O plus 6N from the azides, while the Cs is 9-coordinate.

In complexing agents for these elements, the best donor atom is oxygen sometimes together with nitrogen. Chelating agents with donor oxygen give a few complexes, of which the four-coordinated salicylaldehyde complex of the alkali metals shown in Figure 10.4 is typical. Potassium, rubidium, and cesium also give six-coordinated complexes $M(OC_6H_4CHO)(HOC_6H_4CHO)_2$.

Because of their higher charge, the alkaline earth elements form a rather wider variety of complexes with compounds containing donor oxygen or nitrogen atoms. One complexing agent of considerable value in the quantitative analysis of these elements is ethylenediamine-tetra-acetic acid (EDTA). This forms six-coordinated complexes of the type shown in Figure 10.5 and is especially useful in the determination of calcium and magnesium.

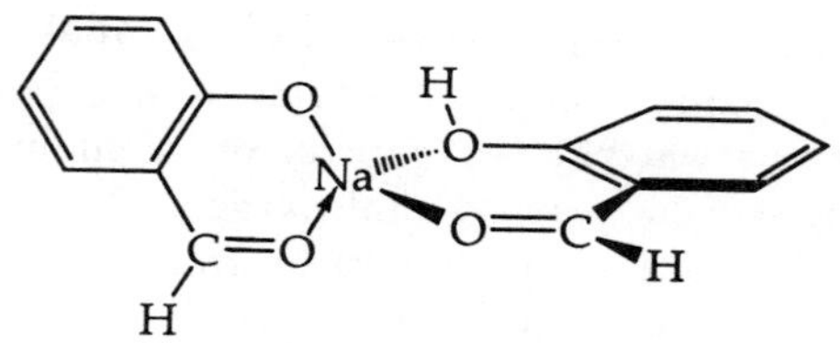

FIGURE 10.4 *The salicylaldehyde complex,* $Na(OC_6H_4CHO)(HOC_6H_4CHO)$

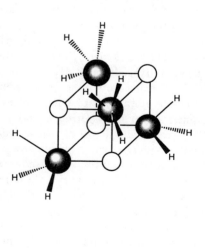

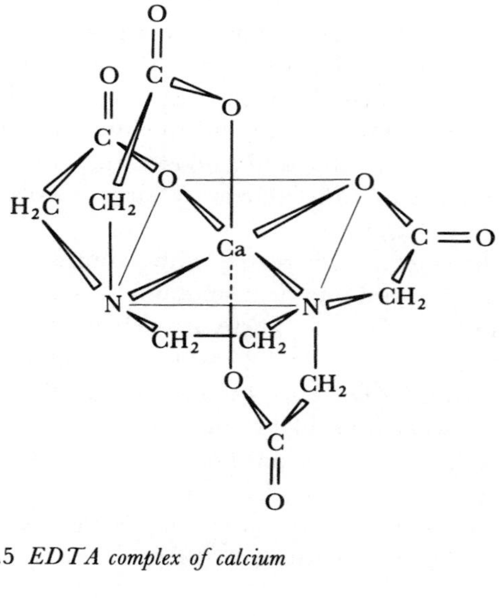

○ Li

● C

FIGURE 10.3 *The tetrameric structure adopted by methyl-lithium,* $(CH_3Li)_4$

FIGURE 10.5 *EDTA complex of calcium*

Calcium is also complexed by polyphosphates and this reaction is the basis of methods of removing hardness from water.

10.5 Crown ethers, cryptates, and alkali metal anions

An important class of strongly-complexing reagents has been developed since the mid-1960s. A typical example is the cyclic ether shown in Figure 10.6. This is called a *crown* ether, from the shape of the metal complex. Other similar reagents, called *cryptates* (because the metal is *hidden* in the structure) are also used. These poly-functional ligands provide a variety of O and/or N donor atoms which strongly complex the M^+ or M^{2+} ions of the s elements. They provide unusual solubilities in organic solvents, stabilize large anions, and give model systems for the movement of ions through biological membranes.

One example is the greatly enhanced solubility of alkali metals in ethers. In absence of cryptate ligands, sodium and potassium form extremely dilute blue solutions in ethers, analogous to their solutions in liquid ammonia (section 6.7) but at 10^{-4} of the concentration. When a crown ether or cryptate is added, the concentrations of alkali metal can be increased by a factor of 10^3 to 10^4. From such solutions have been isolated a golden compound, 2Na(Crypt 222) with interesting properties. (Crypt 222 is the cryptate of

(a)

(b)

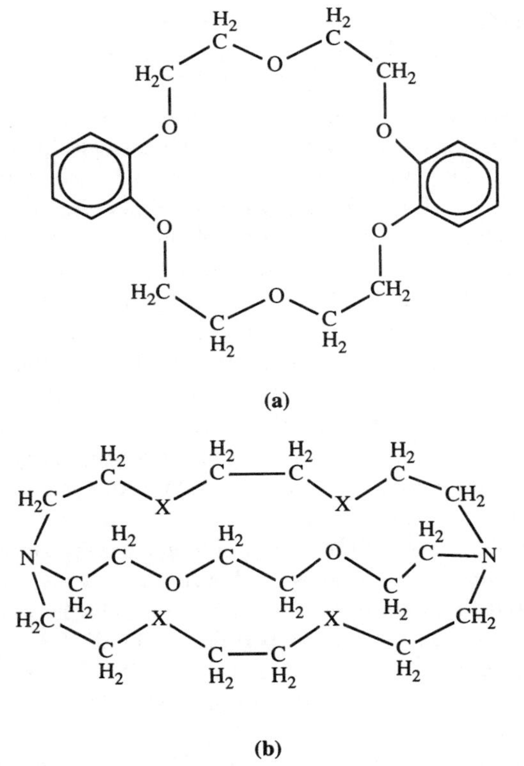

FIGURE 10.6 *Examples of macrocyclic polyfunctional ligands or cryptates* (a) a crown ether, (b) a mixed-donor cryptate: X can be O or S in various combinations. Such ligands complex strongly to the s elements through the oxygen atoms, forming complexes of the type (polyether)M^+ or (polyether)M^{2+} containing a large number of 5-membered rings. A variety of ring sizes, with various combinations of N and O donor atoms, is available.

Figure 10.6b with all X = O.) The ^{23}Na nmr spectrum showed two signals, one in the position of Na(crypt)$^+$ as seen in ordinary salts such as Na(crypt)$^+$Br$^-$. The other was in an unusual position 60 ppm upfield and this suggested the signal came from the Na$^-$ ion where the two s electrons would strongly shield the nucleus. Calculations on the gaseous Na$^-$ ion confirmed the position. The other alkali metal solutions, in the presence of cryptates or crown ethers, contained similar M$^-$ ions with upfield shifts as follows: K$^-$ (90 ppm), Rb$^-$ (190 ppm) and Cs$^-$ (250 ppm). The metal anions also show a distinctive optical spectrum and crystal structure studies have confirmed the formulation. Mixed cation–anion systems may be prepared, such as K(crypt)$^+$Na$^-$.

The stability of these anions is ascribed to the shielding effect of the cryptate on the cation. The large ligand completely isolates the cation so that electron transfer from Na$^-$ to Na$^+$ is prevented (otherwise sodium metal forms). The M$^-$ anions have the s^2 configuration, isoelectronic with the elements of the beryllium Group. A similar metallide anion is found for gold (the $d^{10}s^2$ auride anion, Au$^-$) which is discussed in section 15.9.2.

Perhaps the most spectacular species is illustrated by the dark blue Li(crypt). This is formulated as [Li(crypt)]$^+$e$^-$ and contains the stabilized electron acting as an anion. Such *electrides* are formed by all the alkali metals with suitable cryptates or crown ethers, for example, [Cs(15-crown-5)$_2$]$^+$e$^-$. (The ether, 15-crown-5 is analogous to Figure 10.6a but with a 15-membered ring with five O atoms in place of the 18-membered ring with six O.)

The large, well-shielded cation allows the isolation of other anions which would normally be unstable to reaction with the positive charge. One example is K(crypt)HgTe$_2$ containing the linear Te–Hg–Te$^-$ ion. The metal crypt solutions provide a major route to cluster anions of the p-block elements (compare section 18.4), where the large cation is an important contributor to the stabilization.

10.6 Special features in the chemistry of lithium and magnesium

Lithium differs from the other alkali metals in a number of ways. These variations stem from the small size of the lithium cation (radius 68 pm, c.f. Na$^+$ = 98 pm) and its resulting higher polarizing power. This has already been seen in the more covalent nature of its alkyls. Magnesium resembles lithium in much of its chemistry, as the higher charge is offset by the greater size, and these elements are one example of the diagonal similarity which holds in parts of the first two short Periods.

Lithium and magnesium resemble each other in the direct formation of the nitride and carbide (Table 10.2), in their combustion in air to the normal oxide, and in the properties of their organic compounds, as already remarked in the previous sections. They are also similar to each other, and different from the heavier elements, in the thermal stability of their oxysalts and in the mode of decomposition of these. For example, lithium and magnesium nitrates decompose on heating to give the oxide and dinitrogen tetroxide, e.g.:

$$2LiNO_3 = Li_2O + N_2O_4 + \tfrac{1}{2}O_2$$

while the other alkali metal nitrates give the nitrite on heating ($NaNO_2$ decomposing further to a mixture of oxides):

$$MNO_3 = MNO_2 + \tfrac{1}{2}O_2$$

and the alkaline earth nitrates give nitrite and oxide mixtures.

In the case of the carbonates, lithium and magnesium (and also calcium and strontium) give the oxide on heating:

$$MgCO_3 = MgO + CO_2$$

while the carbonates of the other alkali metals and of barium and radium are stable to heat.

Lithium and magnesium are more strongly hydrated than their heavier congeners and their salts are similar in solubility. Most salts are soluble but the carbonate and phosphate are insoluble in each case. The high solubility of lithium perchlorate in diethyl ether has led to its use in organic chemistry as a reagent for accelerating Diels–Alder cyclization reactions. Magnesium fluoride and magnesium hydroxide are rather insoluble while lithium fluoride and hydroxide are both markedly less soluble than the sodium salts. Lithium and magnesium halides are also similar in being soluble in organic solvents such as alcohol. With larger anions, solubilities tend to fall with cation size in both Groups. Thus the alkali metal perchlorates are relatively insoluble for K, Rb, Cs and Fr and the solubilities of carbonates and sulfates are in the order $Mg > Ca > Sr > Ba > Ra$. The fluorides and hydroxides present an exception to this order—for example, the solubilities of the fluorides are in the order $Mg < Ca < Sr < Ba$—and this must result from the effect of the relatively small OH^- and F^- anions on the lattice energies.

Lithium and magnesium are also more strongly complexed by nitrogen donors, in ammonia and amines, than are the heavier elements. The ammoniated salts $[Li(NH_3)_4]I$ and $[Mg(NH_3)_6]Cl_2$, for example, can be precipitated from liquid ammonia solution by double decomposition reactions.

The high hydration energy of lithium more than compensates for its relatively high ionization potential in the case of reduction reactions which are carried out in water. The result is that the redox potential in water of lithium is as strongly reducing as that of cesium, although the latter is much more reactive under anhydrous conditions.

Mg and Li form a wider range of complexes than the other s elements, and show a variety of coordination behaviour in addition to the monomers with 4-coordinate Li or 6-coordinate Mg. Thus, the dimer $[R_2NLi.OEt_2]_2$ contains the unit

where R is a bulky group. Similarly, $DMg(OR)Br$ (for D a donor such at Et_2O) is a dimer of the form

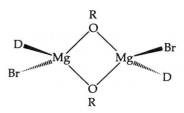

In the complex cation, $[Mg_2Cl_3.6THF]^+$ the Mg atoms are six-coordinated to three THF molecules and three bridging Cl:—$[(THF)_3Mg(Cl)_3Mg(THF)_3]^+$.

Magnesium and lithium can activate molecular N_2, which gives one route to transition metal complexes containing $M(N_2)$ units. One interesting example where the product contains magnesium is the species $[(Me_3P)_3Co(N_2)Mg(THF)_4]_2$. This has the central structural unit $Co-N\equiv N-Mg-N\equiv N-Co$ with an angle of $150°$ for $N-N-Mg$. The synthesis depends on the activation of N_2 by a fresh magnesium surface.

10.7 Beryllium chemistry

The size effects which produce the differences between the chemistry of lithium and the other alkali metals, have a much more marked effect on the chemistry of beryllium as compared with its heavier congeners. On passing from lithium to beryllium, the size of the atom grows less and the charge on the possible ion doubles. The result is to give such a high charge density on the hypothetical Be^{2+} ion that it is too polarizing to exist and all beryllium compounds are either covalent or contain solvated beryllium ions, such as $Be(H_2O)_4^{2+}$. Even the anhydrous halides are only feebly ionic; beryllium fluoride is one of the few metal fluorides which is not completely ionized in solution, while the conductivity of fused beryllium chloride is only one thousandth of that of sodium chloride under the same conditions.

The small size and strong hydration effects have a marked influence on the solubility of beryllium compounds as compared with those of the other elements of the Group. In water, the size effects increase the solvation energies more than the lattice energies, and the solubilities of compounds such as the sulfate, selenate, or oxalate are markedly greater than those of the corresponding calcium compounds. An extreme case is provided by beryllium fluoride which is a thousand times more soluble than magnesium fluoride.

In water, the hydrated beryllium ion differs from the other s element ions in being hydrolysed. Beryllium therefore more closely resembles aluminium, which also undergoes extensive hydrolysis in solution. $Be(H_2O)_4^{2+}$ exists in strongly acid solutions but in neutral or weakly acid solutions this is hydrolysed and polymerized, for example:

$$2Be(H_2O)_4^{2+} = (H_2O)_3Be-O-Be(H_2O)_3^{2+} + 2H_3O^+$$

and more complex species are formed. One stable form in dilute solutions is $[Be(OH)_3]_3^{3-}$ which is thought to have the ring structure of Figure 10.7. Beryllium hydroxide is eventu-

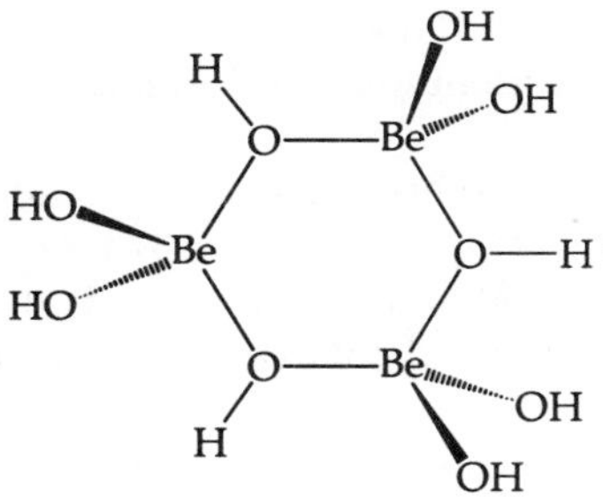

FIGURE 10.7   *Structure of the* $[Be(OH)_3]_3^{3-}$ *ion*

ally precipitated on going to more alkaline conditions, and this differs from the other s element hydroxides in being amphoteric and dissolving in excess base, probably with the formation of the species $Be(OH)_4^{2-}$.

As would be expected from the above, beryllium forms more, and more stable, complexes than the other s elements. There is a strong tendency to assume a coordination number of four—in which use is made of all the valency orbitals. Thus the halides are monomeric and linear in the gas phase (section 4.2), but they dissolve in ether and a dietherate, $BeX_2.2Et_2O$, is recovered. This contains beryllium in an sp^3 configuration with the ether oxygen atoms acting as lone pair donors (Figure 10.8a). A large number of similar four-coordinated complexes are formed with ligands such as ethers, aldehydes, and ketones. In the liquid and solid states, the beryllium halides polymerize in order to achieve four-coordination. In the solid, the unshared pair on a halogen atom of one molecule donates to the beryllium atom of the next and a long chain structure containing tetrahedral beryllium results (Figure 10.8b). Planar 3-coordinate Be is found together with tetrahedral Be in the dimer $[Be(OR)_2Cl]_2$, when R is the bulky tert-butyl group,

Figure 10.8c. Solid beryllium fluoride is a glassy material which also appears to contain chains but these are disordered and the material does not crystallize. The fluoride reacts with fluoride ion to form the tetrafluoroberyllate ion, BeF_4^{2-}, which is tetrahedral and usually isostructural with sulfate.

A more complex example, which illustrates the tendency of beryllium compounds to hydrolyse, and also to become four-coordinated, is provided by the compound called 'basic beryllium acetate'. This is a very stable compound formed by the partial hydrolysis of beryllium acetate:

$$4Be(OOCCH_3)_2 + H_2O$$
$$= Be_4O(OOCCH_3)_6 + 2CH_3COOH$$

It is a volatile solid which may be purified by sublimation. It is soluble in organic solvents such as chloroform, insoluble in water, and stable to heat and moderate oxidation. The structure is shown in Figure 10.9. The four beryllium atoms are placed tetrahedrally around the central oxygen and they are linked in pairs by the six acetate groups, each of which spans one edge of the tetrahedron. Similar compounds of other carboxylic acids have been prepared. In basic beryllium nitrate, $Be_4O(NO_3)_6$, the nitrate groups are linked to two beryllium atoms, $Be-O-N(O)-O-Be$, in the same way as the acetate groups in the basic acetate.

Complexes also exist in which nitrogen is the donor atom. For example, beryllium chloride readily takes up ammonia to give the ammine, $[Be(NH_3)_4]Cl_2$. This compound is very stable to thermal decomposition but the ammonia groups are readily displaced by water.

Beryllium also achieves four-coordination in the hydride and in the beryllium alkyls by polymerization through electron-deficient bridges. Thus beryllium dimethyl, $Be(CH_3)_2$, has a chain structure very similar to that of beryllium chloride, and the bridge bonding is the same as that in the aluminium trimethyl dimer of Figure 9.13.

Just as there is some resemblance in the properties of lithium and magnesium, so beryllium and its diagonal neighbour, aluminium, have similar properties. In this case, the resemblance in general chemistry is very close and beryllium is traditionally difficult to distinguish from aluminium in qualitative analysis. The basic resemblance is due to the high charge density on each element in the hypothetical

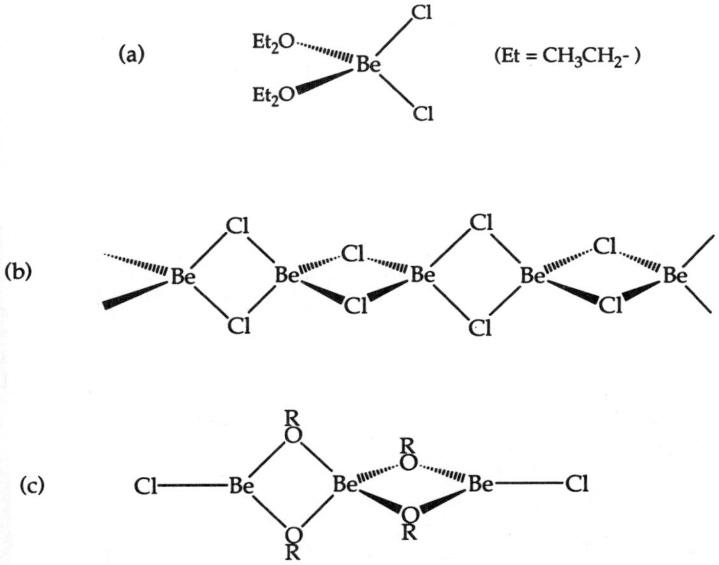

(Et = CH_3CH_2-)

FIGURE 10.8 *Beryllium chloride compounds:* (a) *etherate* $BeCl_2.Et_2O$, (b) $BeCl_2$ *polymer in the solid,* (c) $Be[Be(OR)_2Cl]_2$

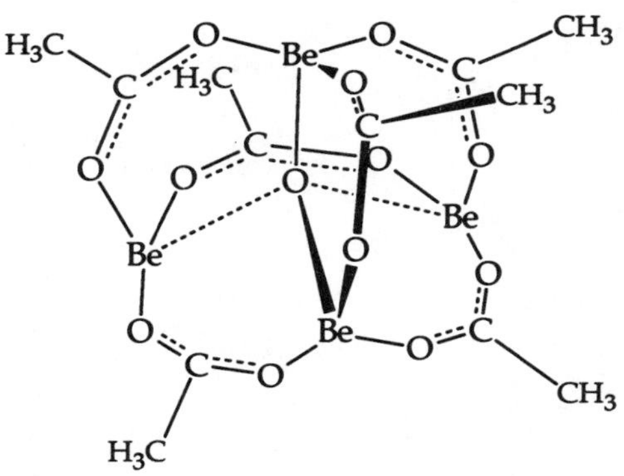

FIGURE 10.9 *Basic beryllium acetate,* $Be_4O(OOCCH_3)_6$

cations, and to the existence of an empty p orbital or orbitals giving the elements acceptor properties. Among the detailed resemblances there may be noted:

(i) Similar redox potentials, $Be^{2+}/Be = -1.70$ V, $Al^{3+}/Al = -1.67$ V.

(ii) Both metals dissolve in alkali with the evolution of hydrogen.

(iii) The hydroxides are amphoteric, and the salts are readily hydrolysed.

(iv) The halides form polymeric solids, beryllium halides forming chains and aluminium halides, dimers. The halides are also similar in their solubilities in organic solvents, their electron pair acceptor properties (e.g. $AlCl_3.OEt_2$), and their catalytic effects.

(v) Both metals form carbides by direct combination and these yield mainly methane on hydrolysis.

(vi) Both elements form hydrides and alkyls which polymerize through electron-deficient bridges.

PROBLEMS

10.1 The 'diagonal relationship' is a long-standing generalization about the similarity in chemical properties of elements with neighbours 'one down to the right' in the Periodic Table.

Discuss similarities and differences between the following sets:

(a) lithium and magnesium

(b) beryllium and aluminium

(c) potassium, strontium and lanthanum.

10.2 The melting points of the alkali metals are (°C): Li = 453; Na = 371; K = 337; Rb = 313; and Cs = 302. Discuss these values and find out whether (a) the boiling points and (b) the values for the Be Group, show similar trends.

10.3 The alkali metals form the commonest group of M^+ cations. Which other M^+ cations exist? Compare and contrast the properties of their compounds with those of the alkalis. Extend this discussion to NH_4^+ species.

10.4 'The chemistry of the s elements is dominated by their tendency to lose the s electrons.'

Discuss this statement, and comment on

(a) the parameters which reflect the tendency

(b) compounds formed only by the more reactive s elements

(c) 'exceptions that prove the rule'.

10.5 Compare and contrast the chemistry of the s^1 and s^2 elements.

10.6 Research the topic 'The biological activity of the alkali and alkaline earth elements'.

10.7 Magnesium and calcium are often found in different environments in silicate and aluminosilicate minerals. Find from the references three examples of this.

11 The Scandium Group and the Lanthanides

11.1 General and physical properties

The scandium Group of elements has the outer electronic configuration d^1s^2 and is formally part of the d block of the Periodic Table. However, as the chemistry is dominated by the III oxidation state, which involves the loss of the d electron, this Group is best regarded as forming a transitional region between the s elements and the main d block. Following lanthanum, the third member of the scandium Group,

the $4f$ shell fills for the next fourteen elements. As the general chemical behaviour in these *lanthanide elements* (or rare earths) is that of the III state and is very similar to that of the scandium Group proper, it is convenient to include these elements here. Properties are listed in Table 11.1, and earlier in Tables 2.5 (atomic masses and numbers), 2.8 (ionization potentials), 2.14 (electronegativities), and 6.3 (some redox potentials).

TABLE 11.1 Properties of the scandium elements and the lanthanides

Element	Symbol	Electronic configuration	Abundance ppm of crust	M^{2+} CN6	8	M^{3+} 6	8	M^{4+} 6	8	Oxidation states	Redox potentials $M^{3+} + 3e = M(V)$
Scandium	Sc	$[Ar]3d^14s^2$	5			88·5	101·0			III	1·88
Yttrium	Y	$[Kr]4d^15s^2$	28			104·0	115·9			III	2·37
Lanthanum	La	$[Xe]5d^16s^2$	18			117·2	130·0			III	2·522
Actinium	Ac	$[Rn]6d^17s^2$	trace			126				III	2·13
Cerium	Ce	$[Xe]4f^25d^06s^2$	46			115	128·3	101	111	III, IV	2·483
Praseodymium	Pr	$[Xe]4f^35d^06s^2$	5			113	126·6	99	110	III, (IV)	2·462
Neodymium	Nd	$[Xe]4f^45d^06s^2$	24		143	112·3	124·9			III, (IV?) (II?)	2·431
Promethium	Pm	$[Xe]4f^55d^06s^2$	unstable			111	123·3			III	2·423
Samarium	Sm	$[Xe]4f^65d^06s^2$	6	136	141	109·8	121·9			III, (II)	2·414
Europium	Eu	$[Xe]4f^75d^06s^2$	1	131	139	108·7	120·6			III, II	2·407
Gadolinium	Gd	$[Xe]4f^75d^16s^2$	6			107·8	119·3			III, (II?)	2·397
Terbium	Tb	$[Xe]4f^95d^06s^2$	1			106·3	118·0	90	102	III, (IV)	2·391
Dysprosium	Dy	$[Xe]4f^{10}5d^06s^2$	4	121	133	105·2	116·7			III, (IV?)	2·353
Holmium	Ho	$[Xe]4f^{11}5d^06s^2$	1			104·1	115·5			III	2·319
Erbium	Er	$[Xe]4f^{12}5d^06s^2$	2			103·0	114·4			III	2·296
Thulium	Tm	$[Xe]4f^{13}5d^06s^2$	0·2	117		102·0	113·4			III, (II?)	2·278
Ytterbium	Yb	$[Xe]4f^{14}5d^06s^2$	3	116	128	100·8	112·5			III, (II)	2·267
Lutetium	Lu	$[Xe]4f^{14}5d^16s^2$	0·8			100·1	111·7			III	2·255

*Bracketed states are unstable: states marked (?) are either unconfirmed or very unstable.

The organometallic compounds formed by the lanthanide elements are discussed in section 16.5.

When the 5*f* shell fills following actinium, which is the fourth member of the scandium group, the very small energy gap between the 5*f* level and the 6*d* level gives rise to a considerable variability in the chemistry of the actinide elements which are therefore treated separately in Chapter 12.

Of these elements, scandium and actinium are very rare and are not completely investigated. The lanthanide elements and lanthanum and yttrium all occur together, and there is a certain amount of segregation into larger and smaller ions. Thus the lighter (and larger) lanthanides make up ca. 90% of the two commercial ores *bastnasite* (a carbonate-fluoride), and *monazite* (a phosphate), with Ce ca.45%, La ca. 25%, and Nd ca. 15% as the major components. The best source of the smaller elements (yttrium and the heavier lanthanides) is the relatively rare phosphate ore *xenotime* (60% Y, 9% Dy, 6% Yb, 5% Er and other heavy elements 10%).

Separation of the lanthanides was difficult by classical methods, and chromatographic techniques are now used (see section 7.2). An interesting historical account of the discovery and isolation problems of the lanthanide elements is given in the Further Reading section (Appendix A). Cerium may be removed chemically by oxidation to the IV state, and promethium does not occur naturally. It is thus much easier to separate the lighter lanthanides with these two gaps in the series and this, coupled with the abundances, makes the earlier members much cheaper to obtain than the later ones.

The largest scale use of lanthanides is in the fabrication of special steels. For this, the elements are not separated, but the naturally occurring mix is converted to the chlorides and then reduced electrolytically to *mischmetal* which is added to steels, and also used as lighter flints. A second, rapidly gowing, use of mixed lanthanides is their addition to zeolites (see section 18.6) to increase their catalytic activity, especially in petroleum crackers. Gd, Sm, Eu, and Dy have large cross-sections for neutron capture and are used in control rods in nuclear reactors.

Because of their specific magnetic and optical properties, due to the *f* electrons (compare section 11.5), specific elements find magnetic, electronic, and optical applications. 'Didymium', the natural mixture of Pr and Nd, is used in protective glasses to shield from UV and high intensity light. Uses as activators in phosphors include Eu^{3+} for the red colour in TV, and with Tb^{3+} (green), in fluorescent tubes. Europium(III) complexes with certain cryptate, arylnitrogen or aminopolycarboxylate ligands show strong luminescence and high efficiency in converting light energy. These *light-conversion molecular devices* have interesting potential applications as luminescent materials. In medical X-ray intensifiers, Eu^{2+}, Tm^{3+} and Tb^{3+} act as activators and La or Gd oxide species as support media. Neodymium lasers are very important. The neodymium may be held in a glass or in yttrium aluminium garnet, and there is also a liquid laser using neodymium oxide dissolved in selenium oxychloride. A liquid system has many technical advantages,

but this was only the second liquid system to be announced and it was by far the most powerful. The key aspect seems to be the use of a solvent which does not contain light atoms, such as hydrogen, so that most of the input energy is emitted in the laser beam and not transferred to heat the solution.

Yttrium is used in the 90 K $YBa_2Cu_3O_{7-x}$ superconductor, and other lanthanides have been tried in similar formulations (see section 16.1). Yttrium and gadolinium in *garnet* oxides, $Ln_3M_5O_{12}$ (M = a trivalent element, especially Fe or Ga) are used in bubble devices for memory storage, microwave components, and other magnetic applications.

In the laboratory several lanthanide complexes are used as shift reagents for spreading out the signals in proton nmr. The advent of high-field nmr spectrometers (currently up to 750 MHz), together with powerful multi-dimensional nmr techniques, has resulted in a decline in the use of these reagents. However, lanthanide complexes, particularly those of gadolinium, are finding ever-increasing use as magnetic resonance imaging (MRI) agents. In this technique, gram quantities of gadolinium complexes are injected into the body and the proton nmr signals, largely of tissue water molecules, are spatially mapped; the role of the strongly paramagnetic (f^7) gadolinium ion is to locally change the relaxation times of the water molecules, thereby improving the contrast of the MRI technique. MRI is just one of a rapidly-growing array of inorganic-based diagnostic and therapeutic medical techniques, discussed in greater detail in Chapter 20.

The elements are electropositive and reactive, with the heavier elements resembling calcium in reactivity while scandium is similar to aluminium. Two of the elements, promethium and actinium, occur only as radioactive isotopes. Actinium is found associated with uranium and the most readily available isotope, ^{227}Ac, has a half-life of 22 years. It is, however, very difficult to handle as the decay products are intensely active and build up in the samples. Its chemistry fits in as that of the heaviest element of the scandium Group. The missing lanthanide, promethium, occurs only in radioactive forms with the longest-lived isotope, ^{145}Pm having $t_{\frac{1}{2}} = 30$ years. Its chemistry fits in with its place in the series. The analysis of samarium and neodymium isotope ratios provides a geochronological method for dating very old rocks.

Reactivity increases with increasing atomic weight in the scandium Group, just as in the *s* Groups. The elements are prepared by the reduction of the chlorides or fluorides with calcium metal. Some reactions of the metals are shown in Figure 11.1. These direct reactions with elements broadly parallel those of the *s* elements given in Table 10.2.

The hydrides, formed by direct combination, illustrate the transitional character of this Group (compare section 9.4). They form stable MH_2 and MH_3 phases, which usually occur in non-stoichiometric form, and the MH_3 formula for the most highly hydrogenated species is never fully attained. The hydrides have some salt-like properties and appear to contain the H^- ion. There are also available extra delocalized electrons giving metallic properties, so that the overall properties of the hydrides are a mixture of the ionic character

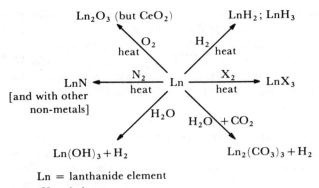

Ln = lanthanide element
X = halogen

FIGURE 11.1 *Reactions of the lanthanide elements*

shown by the *s* element hydrides and the metallic character of the hydrides of the *d* elements. The hydrides also resemble the ionic hydrides in their reactivity to oxygen and water. Mixed element hydrides involving lanthanides and transition elements show promise as hydrogen storage materials. The alloy $LnNi_5$ reacts reversibly with hydrogen gas to form $LnNi_5H_6$ at readily accessible temperatures and pressures, and the properties can be 'tailored' by varying Ln = La, Ce, Pr, or Nd or mixtures of these.

11.2 Chemistry of the trivalent state

As Table 11.1 shows, the oxidation state of +III is shown by all these elements and is the most stable state. Other states are found only where *f* electrons are present. For the group Sc, Y, La and Ac, the oxides, M_2O_3, and hydroxides, $M(OH)_3$, increase in basicity with increasing atomic mass. Scandium, because of its smaller size, is more easily hydrolysed in solution than the other ions, and its oxide is amphoteric. There does not appear to be a definite scandium hydroxide, although the species $ScO(OH)$ is well established [compare the existence of $AlO(OH)$]. The oxide, with water, forms the hydrous oxide, $Sc_2O_3 . nH_2O$, which dissolves in excess alkali to form anionic species such as $Sc(OH)_6^{3-}$. The other elements form oxides and hydroxides which are basic only. These hydroxides are precipitated from solution by the addition of dilute alkalis and do not dissolve in excess alkali. Yttrium oxide and hydroxide are strong enough bases to absorb carbon dioxide from the atmosphere, while lanthanum oxide 'slakes', with evolution of much heat like calcium oxide, and rapidly absorbs water and carbon dioxide. The lanthanide element basicities decrease towards lutetium, which is similar to yttrium in the properties of its oxide. Actinium compounds are more basic than the lanthanum ones.

Among the trihalides, the fluorides are insoluble and their precipitation, even from strongly acidic solution, is a characteristic test for these elements. Scandium fluoride dissolves in excess fluoride with the formation of the complex anion, ScF_6^{3-}. The fluorides of the heavier lanthanide elements are also slightly soluble in hydrogen fluoride, probably because of complex formation. All the elements form oxide-fluorides

MOF from the reaction of MF_3 with M_2O_3. More complex phases such as $Y_7O_6F_9$ are also reported.

The other trihalides are very deliquescent and soluble (compare $CaCl_2$). $ScCl_3$ is much more volatile than the other trichlorides. It resembles $AlCl_3$ in this, but it is monomeric in the vapour (aluminium trichloride is dimeric) and it has no activity as a Friedel-Craft catalyst. The chlorides are recovered from solution as the hydrated salts and these, on heating, give the oxychlorides, MOCl (with the exception of scandium which goes to the oxide). Actinium also forms oxyhalides but only by reaction with steam at 1000 °C, a treatment which produces oxide from the lower members of the Group. This is a further example of the increase in basicity from scandium to actinium. Bromides and iodides resemble the chlorides in general behaviour.

Among the oxysalts, most anions are to be found including strongly oxidizing ones. The carbonates, sulfates, nitrates, and perchlorates, for example, all resemble the calcium compounds. The carbonates, phosphates, and oxalates are insoluble while most of the others are rather more soluble than the calcium salts. Scandium carbonate differs from the others in dissolving in hot ammonium carbonate, with double salt formation, and this affords a method of separating scandium from yttrium and lanthanum. Double salts are very common and include double nitrates, $M(NO_3)_3.2NH_4NO_3.4H_2O$, and sulfates such as $M_2(SO_4)_3.3Na_2SO_4.12H_2O$. Such salts were used for separation of the lanthanides by fractional crystallization methods.

These salts are fully ionic and lanthanum is useful as one of the few available stable ions with a charge higher than $2+$. The scandium ion is more readily hydrolysed than the others, and polymeric species of the type $[Sc-(OH)_2-Sc-(OH)_2-]_n$ have been identified with the chain length increasing as the acidity falls. The other ions are only slightly hydrolysed in the sense:

$$M(H_2O)_6^{3+} + H_2O \rightleftharpoons M(H_2O)_5(OH)^{2+} + H_3O^+$$

with the tendency to hydrolyse increasing as the size decreases.

Although these ions have a charge of $+3$, the tendency to complex-formation is relatively slight. When compared with transition metal ions, such as Fe^{3+} or Cr^{3+}, which readily form complexes, this reluctance to complex may be ascribed to the greater size of the scandium Group ions, and to their low electronegativity which decreases any possible covalent contribution to the bonding. The best donor atom is oxygen, and insoluble complexes are formed by β-diketones such as acetylacetone (Figure 11.2). Of some importance are the water-soluble complexes formed by chelate ligands such as EDTA, and especially the complexes formed by hydroxy-carboxylic acids such as citric acid, $HOOC.CH_2.C(OH)(COOH).CH_2COOH$, which are used in the separation of these elements by ion exchange methods (see section 11.3). The lanthanide elements are often six-coordinated in complexes, but higher coordination numbers are known. Eight-coordination is found in $La(acac)_3(H_2O)_2$, where acac = acetylacetonato, and in $[Y(CF_3COCHCHCF_3)_4]^-$.

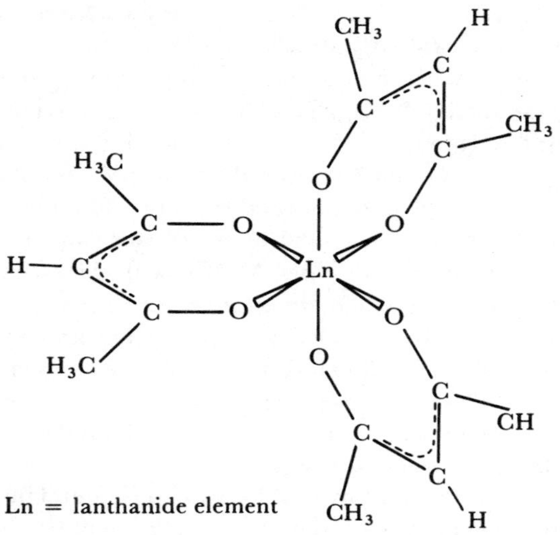

Ln = lanthanide element

FIGURE 11.2 *Acetylacetonato complex of the lanthanide elements*

Shapes include antiprismatic (compare Figure 15.5b) and dodecahedral (compare Figure 13.12). An interesting example is found in the $[HoAl_3Cl_{12}]_n$ polymer which contains square antiprismatic $HoCl_8$ units formed by four $Ho(\mu^2\text{-}Cl)_2Al$ bridges (compare Figure 17.12) and these four Al atoms bridge further in a polymeric structure.

$Nd(BrO_3)_3.9H_2O$ contains the 9-coordinate $[Nd(H_2O)_9]^{3+}$ ion and similar ions exist for other lanthanides. $Yb(NO_3)_3(H_2O)_3$ is also 9-coordinate with bidentate nitrates (as in the cerium compound below). The nine O atoms form a tricapped trigonal prism (compare Figure 15.23). EuN_9 coordination is also found in Eu_2L_3 (L = a pyridine/benzimidole long-chain ligand) which self-assembles into a triple helix. Ten-coordination is found in $La(H_2O)_4EDTA.3H_2O$. The EDTA occupies six positions and there are four water molecules attached to the lanthanum. Three water molecules lie on one side of the lanthanum, the two nitrogen atoms of the EDTA lie opposite them, while the four EDTA oxygens and the final water molecule lie in a rather distorted medial plane. The complex nitrate, $Ce(NO_3)_5^{2-}$, contains ten-coordinate Ce. The nitrates are present at the apices of a trigonal bipyramid and each nitrate is coordinated by two oxygens—

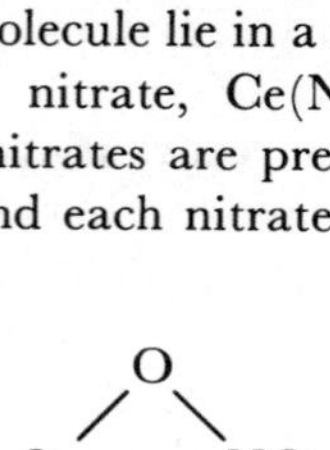

Yttrium, and the other lanthanides investigated, also form ten-coordinate $M(NO_3)_5^{2-}$ complexes but Sc in $Sc(NO_3)_5^{2-}$ is only nine-coordinate with one of the nitrate groups bonded through only one oxygen. In $Ce(NO_3)_6^{3-}$, the coordination number is twelve. In the ion $La(C_6H_9N_3)_4^{3+}$, the twelve nitrogen atoms form an almost regular icosahedron (compare Figure 17.7d) around the La atom.

Scandium, because of its small size, forms more stable complexes than the other elements. For example, the scandium acetylacetonate may be sublimed at about 200 °C,

while all the others decompose on heating. Continuing this trend to the heaviest element, actinium is less ready to form complexes than the others. Thus, the lanthanides can be extracted into an organic solvent by means of tributylphosphate, $OP(OC_4H_9)_3$, which forms a complex, but actinium extracts much less readily under these conditions.

11.3 The separation of the elements

One separation of scandium is mentioned above, and actinium occurs separately, but yttrium, lanthanum, and the lanthanides are commonly found together in minerals. From the radii given in Table 11.1 it will be seen that the lanthanides are very close in size, with a small but regular decrease from lanthanum to lutetium, and that yttrium is close to dysprosium and holmium. A similar gradation is found in the redox potentials of the elements, M_{aq}^{3+}/M, which range from -2.52 V for La to -2.25 V for Lu, in steps of about 0.01–0.02 V between each element and the next. Again yttrium, -2.37 V, fits in near dysprosium, -2.35 V. These resemblances in size and behaviour are much closer than those between elements in the same Group (compare Sr^{2+} and Ba^{2+} which differ by 19 pm) and mean that the chemistry of all the lanthanide elements is practically identical.

The slow decrease in size from La to Lu just about balances the normal increase in size between the elements in one period and the next. This is shown by the similarity between yttrium and the heavier lanthanides. The decrease is termed the *lanthanide contraction* and arises from the slow increase in effective nuclear charge as the *f* electrons are added. This accounts for the decrease in size and the increase in oxidation potential. As will be seen, the lanthanide contraction also affects the chemistry of the heavier transition metals.

As the elements and ions are so similar in size and properties, the separation of the individual lanthanide elements is extremely difficult. In the classical studies of the elements, fractional separations had to be adopted. These included fractional crystallization of double salts, such as the nitrates, fractional precipitation of the hydroxides and fractional decomposition of the oxalates. These processes were very slow, and as many as twenty thousand operations are reported in some cases before pure samples were obtained. The separation of the lighter elements, up to gadolinium, was relatively easy as cerium could be removed by oxidation to the relatively stable IV state, promethium was missing from the natural sources, and samarium and europium could be reduced to the II state. The heavier elements and yttrium were much more difficult to separate as such chemical aids were not available. Despite this, all the lanthanide elements had been separated and correctly characterized, before the advent of more powerful methods, in what was one of the most painstaking series of studies in chemistry.

The separation problem is greatly simplified by the use of ion exchange or solvent extraction techniques (see Chapter 7). One common ion exchange method uses the soluble citrate complexes. To a cation exchange resin, which will be written resin-H, a solution of the lanthanides is applied, and

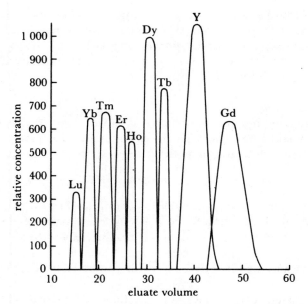

FIGURE 11.3 *Elution curve of the lanthanide elements from an ion exchange resin column*

the acid formed washed out:

$$3\text{Resin-H} + \text{Ln}^{3+} \rightarrow \text{Resin}_3\text{-Ln} + 3\text{H}_3\text{O}^+$$

Then citric acid, buffered with ammonia to constant pH, is added, and the equilibrium:

$$\text{Ln-Resin}_3 + 3\text{HCit} \rightleftharpoons 3\text{H-resin} + \text{LnCit}_3$$

is set up (Ln is used as a general symbol for any lanthanide element). As the buffered citrate flows down the column, the concentration of lanthanide ions changes and the equilibrium reverses many times. As the heavier ions are smaller, they will be more strongly complexed by the citrate and so will tend to spend more time in solution and less on the resin. As a result, the heavier lanthanide elements are washed down the column first and will eventually be eluted. If the conditions are correct, the different elements will be separated into pure components. Figure 11.3 gives an example of

such an elution curve. The whole process is analogous to the classical fractionations but with the numerous operations taking place *in situ* on the column. The process leads to considerable dilution: in one example 0·4 g of mixed oxides per litre was used and collection of about fifty litres of eluate gave each element in about 80 per cent purity. Each fraction would then be concentrated by precipitation of the oxalate and the exchange repeated to give pure samples.

Similar separations are possible using solvent extraction methods (also called hydrometallurgy), typically in a counter-current extraction technique. For example, by extraction of lanthanides from a strong nitric acid solution into tributylphosphate, 95% pure gadolinium has been prepared on a kilogram scale. Similar separation processes involving organophosphorus acids such as R(RO)P(O)OH, where R is, for example, 2-ethylhexyl, have also been developed. There is a gradual change in the pH at which the trivalent lanthanide extracts into the organic solvent, as shown in Figure 11.4, and this leads to a separation of the elements. Such processes are currently operated industrially. Solvent extraction processes are of considerable importance in the nuclear industry, as discussed in the next chapter, and have contributed to the characterization of the synthetic post-plutonium elements.

11.4 Oxidation states other than (III)

A number of the lanthanide elements exist in oxidation states other than III and the most stable of these are Ce(IV) and Eu(II). The cerium(IV) state corresponds to the loss of the four outer electrons to give an $f^0 d^0 s^0$ rare gas configuration, while the europium(II) state corresponds to the loss of only the two s electrons to retain the half-filled f^7 shell.

Cerium(IV)

Cerium is the only lanthanide which exists in the IV state in solution. Ceric oxide, CeO_2, which is colourless when pure, is the product resulting from heating the metal, or decom-

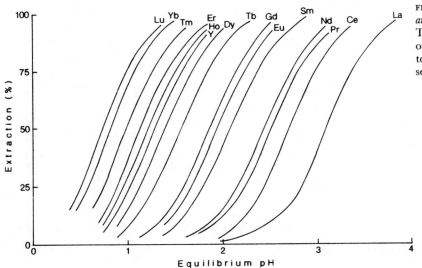

FIGURE 11.4 *Extraction curves for the trivalent lanthanides using an organophosphorus acid*
The separation of the curves is a measure of the selectivity of the extractant—two lanthanides with curves close together will require many repeat extraction stages to separate them completely.

posable cerium(III) oxysalts such as the oxalate, in air or oxygen. It is inert and insoluble in strong acids or alkalies. It does dissolve in acids in the presence of reducing agents to give cerium(III) solutions, and these, in turn, give cerium(IV) in solution on treatment with strong oxidizing agents such as persulfate. A yellow, hydrated, form of ceric oxide, $CeO_2 . nH_2O$, is precipitated from such cerium(IV) solutions by the action of bases. Another solid cerium(IV) compound known is the tetrafluoride, prepared by the action of fluorine on the trichloride or trifluoride. The aqueous chemistry of cerium(IV) resembles that of zirconium and hafnium, or of the four-valent actinides such as thorium.

The very high charge on the Ce^{4+} ion leads to its being strongly hydrated. The hydrated ion is acidic and hydrolyses to give polymeric species and hydrogen ions, except in strongly acidic solution. The solution of cerium(IV) in acid is widely used as an oxidizing agent, and its redox potential depends quite strongly on the acid used, ranging from 1·44 V in molar sulfuric acid to 1·70 V in perchloric acid. This variation probably arises from the formation of complex ions by association with the acid anion in nitric or sulfuric acids. As perchlorate shows no tendency to complex in this way, the potential in perchloric acid probably characterizes the plain hydrated ion, $Ce^{4+} . nH_2O$.

The high charge density means that cerium(IV) forms stronger complexes than the tripositive lanthanides. It is much more readily extracted by tributylphosphate, and the hexachloro-complex $CeCl_6^{2-}$ has been prepared as the pyridinium salt.

In more complex compounds, the oxidizing power of cerium(IV) is modified and a wider range of reactions can be studied. A range of water-soluble cerium(IV) salts has been prepared to act as models for the active photosynthetic centre (compare Chapter 20) in bacteria. In contrast to chlorophyll (see Figure 20.1), the bacterial centre has a double four-nitrogen ring. The cerium(IV) compounds are of the form $Ce(P)_2$ where P^{2-} stands for a porphin ring (see also Figure 20.6) with various substituents on the bridging

carbons. The structure has these rings staggered above and below the eight-coordinated central cerium atom so that the coordination is square antiprismatic CeN_8, as shown in Figure 11.5. Such complexes undergo a variety of redox and aggregation reactions.

Europium(II)

Europium has the most stable divalent state of all the lanthanides. Europium(II) chloride is prepared as a solid by the action of hydrogen on the trichloride:

$$EuCl_3 + H_2 \rightarrow EuCl_2 + HCl$$

Its dihydrate is very insoluble in concentrated hydrochloric acid (like $BaCl_2 . 2H_2O$) and this is used as a means of purification. Europium in solution is readily reduced to the II state, for example by magnesium, zinc, or alkali metal amalgams, and it resembles calcium in this state. Thus the sulfate or carbonate may be precipitated from solution. The oxide does not exist, but EuS, EuSe or EuTe can all be prepared. EuH_2 is ionic and isomorphous with CaH_2.

The redox potential, $Eu^{3+} + e^- = Eu^{2+}$, is -0.43 V, so that europium(II) is a reducing agent of similar power to Cr^{2+}. A careful magnetic investigation has shown that the magnetic properties of Eu(II) are identical with those of Gd(III), over a wide range of temperature, confirming the $f^7 d^0 s^0$ arrangement in the ion. The dichloride, dibromide, and diiodide all have moments of 7·9 Bohr magnetons corresponding to the seven unpaired electrons.

Europium(II), although the most stable of the lower oxidation states, is a strong reducing agent as the potential shows, and its solutions are readily oxidized in air. The solids are rather more stable. Europium, together with ytterbium which also has a II state, dissolves in liquid ammonia to give a concentrated blue solution. The other lanthanides are either insoluble or give only weak solutions on electrolysis.

Other IV states

Other elements which form the IV state are praseodymium, neodymium, terbium, and dysprosium. Of these, only Tb(IV) can be accounted for by the tendency to equally-occupied f orbitals, in this case f^7. All the states are very unstable and have only been prepared as solid compounds.

Ignition of praseodymium compounds in air gives a complex oxidation product of approximate composition Pr_6O_{11}. Heating finely divided Pr_2O_3 in oxygen at 500 °C and 100 atmospheres gives the stoichiometric oxide, PrO_2. No binary fluoride, PrF_4, has been prepared but solid solutions of PrF_3 in CeF_3, containing less than 90 per cent PrF_3, do react completely with fluorine to the composition PrF_4/CeF_4.

There is no firm evidence for the existence of Nd(IV) in oxide systems, but the fluorination of NdF_3 in presence of CsF gives compounds containing 10–20 per cent Nd(IV) in the form of a double salt.

A higher oxide of terbium, of approximate composition Tb_4O_7, has long been known as a product of ignition. A careful study has yielded three oxide phases, in the range

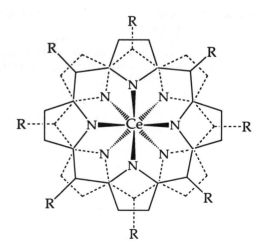

FIGURE 11.5   *The structure of an eight-coordinate cerium bisporphyrinate complex*

$TbO_{1.5 \text{ to } 1.8}$, in ignition products of oxalate or nitrate. The TbO_2 composition results from the reaction of atomic oxygen on Tb_2O_3. This, like PrO_2, has the fluorite structure. Structures containing linked $Tb^{IV}O_6$ octahedra are found in the oxyanion species M_2TbO_3, $M_{16}Tb_3O_{14}$, and $M_6Tb_2O_7$ where M is an alkali metal, especially Li. Terbium(IV) is also formed as fluoride by fluorination of the trifluoride. TbF_4 is isostructural with CeF_4. Terbium(IV) is probably the most stable of the IV states after cerium(IV), but it is an extremely powerful oxidizing agent and there is no question of its existence in an aqueous medium.

Dysprosium(IV) resembles neodymium(IV) in being found only in a fluorine system. Fluorination of DyF_3, in presence of CsF, gives materials containing up to 50 per cent Dy(IV).

Other II states

Elements found in the II state are neodymium, samarium, gadolinium, thulium, and ytterbium. Ytterbium(II) corresponds to the completed f^{14} level. All these elements are much less stable in the II state than is europium. Yb(II) and Sm(II) may be prepared in water but are oxidized by water on standing; the others are found only in the solid state. The order of stability is $Nd(II) \approx Gd(II) < Tm(II) < Sm(II) < Yb(II)$.

Divalent neodymium and gadolinium are prepared, as the dichloride or di-iodide, by the reaction of the metals with the fused trihalides. $NdCl_2$ is isostructural with $EuCl_2$. Controlled reduction of $NdCl_3$ gives the mixed (II)/(III) KNd_2Cl_5. Thulium di-iodide may be prepared in a similar manner by the action of Tm on TmI_3 at $600\,°C$. It is isostructural with YbI_2. These low valency halides tend to be non-stoichiometric and they have metallic conduction and other properties.

Samarium(II) occurs in a number of compounds, including the halides, sulfate, carbonate, phosphate, and hydroxide. It may be extracted from a lanthanide mixture, along with Eu(II), by reduction of the trichlorides with alcoholic sodium amalgam. The mixture of Eu(II) and Sm(II) chlorides is readily separated by controlled oxidation, which produces Sm(III) only. Samarium(II) iodide, largely as a result of its rather oxophilic nature, is beginning to find quite widespread application in the synthetic organic chemistry laboratory. This, together with the use of cerium(IV) salts as powerful oxidizing agents, are examples of the current upsurge in the use of lanthanide reagents in chemical syntheses.

A variety of ytterbium(II) compounds exist, including all those found for samarium, and also possibly the monoxide. The dihalides may be prepared by hydrogen reduction of the trihalides and, in the case of the di-iodide, by thermal decomposition. Yb(II) is more stable than Sm(II) and has been estimated to have an oxidation potential of $-1.15\,V$ with respect to the III state. YbH_2, like EuH_2, is ionic and isomorphous with CaH_2.

Low formal oxidation states are also found in a number of 'subhalides' with metallic properties. Among the best-established are the monochlorides LnCl formed by Sc, Y, and all the lanthanides. The structure consists of a four-layer repeat unit in the order Cl – Ln – Ln – Cl in cubic close packing, the same structure as ZrCl (section 15.2). Two electrons per metal atom are delocalized, giving metallic conductivity. Similar MBr phases are found for M = Y, La, and Pr. Small atoms such as C, O, or H may intercalate between the metal layers, and cations may be held between the Cl sheets of successive four-layer units. Thus phases such as $ScCCl_{0.56}$ or $K_{0.26}YClC_{0.4}$ are found.

Phases of composition approximating M_2Cl_3 are also known. Sc_7Cl_{10} consists of chains of Sc octahedra linked by shared Cl atoms. Similar phases, also with M–M bonding, are known for M = Y, La, and Gd.

Work on the complex chemistry of Ln(II) species is increasing. One interest in lanthanides is their possible incorporation into III–V or II–VI semiconductors. For Yb(II), this has led to the synthesis of various $Yb(ER)_2L_n$ ($n = 4$ for E = S, Se or 5 for E = Te) species where R is usually an aromatic group and L is a donor ligand such as pyridine. Such molecules are synthesized from the blue solution of Yb metal in liquid ammonia (see sections 6.7, 10.1) or from $YbCl_3$ with reduction. The structures (Figure 11.6a) have the two ER groups *trans* to each other and 4 L groups for E = S or Se giving an octahedral configuration. The larger Te–Yb distances allow 5 L groups to coordinate to Yb, forming a pentagonal bipyramid (Figure 11.6b). The Yb–E distances are 283 pm for E = S, 296 pm for E = Se and 328 pm for E = Te reflecting the changes in the ionic radii of E^{2-} (see Table 2.12).

Very gentle oxidation of lanthanide amalgams (solutions of the metal in mercury) in the presence of oxygen donors such as diglyme, DIME (see appendix B), gives complex ions, e.g. $[(DIME)_3Ln]^{2+}$ for Ln = Sm or Eu. There are three donor O in each DIME molecule so the coordination

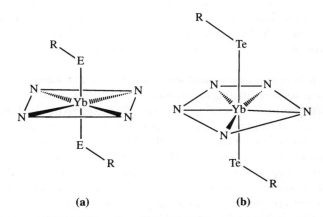

FIGURE 11.6 (a) *Structure of trans-*$Yb(EPh)_2(N)_4$, *where* E = S *or* Se *and* N = C_5H_5N (*pyridine*)
The octahedral Yb(II) ion is bound to *trans* phenylthiolato or phenylselenolato ligands.
(b) *Structure of* $Yb(TePh)_2(N)_5$
The 7-coordinate Yb(II) ion contains *trans* phenyltellurolato ligands separated by a ring of pyridine ligands in a pentagonal bipyramidal geometry.

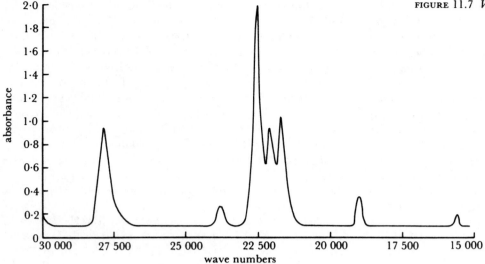

FIGURE 11.7 *Visible and near ultra-violet spectrum of holmium*

is nine-fold and approximately tricapped trigonal prismatic, as in Figure 15.23. Similar preparations using Yb in acetonitrile gave the complexes $[(DIME)_2Yb(CH_3CN)_2]^{2+}$ and $[(DIME)Yb(CH_3CN)_5]^{2+}$. In the first of these, three O from one DIME and the N from one CH_3CN form a square on each side of the Yb giving an overall square antiprismatic arrangement, while the second has a similar configuration except that one face is made up of four acetonitrile nitrogens.

It will be noted that, although there are a significant number of examples of other oxidation states, the III state is by far the most stable among the simple compounds of the lanthanide elements. The most stable of the other states, Ce(IV) and Eu(II), are very reactive and most other examples are found only as solid state species. The exceptions are complexes of large multidentate ligands which do modify the normal high reactivity of Ln(IV) or (II). This dominance of the III state in the lanthanides presents a marked contrast with the behaviour of the actinide elements and is probably a result of the larger energy gap between the $4f$ and $5d$ levels than that existing between the $5f$ and $6d$ levels.

11.5 Properties associated with the presence of *f* electrons

As is implied in the discussion in the earlier sections, the electrons in the $4f$ level are too strongly bound to be involved in the chemistry of the elements except under unusual conditions. In addition, the *f* orbitals appear to be too diffuse to enter into bonding generally, so that there are few chemical effects from the presence of *f* electrons or unfilled *f* orbitals. There are, however, electronic effects which show up in the spectra and magnetic properties of the lanthanides.

The lanthanide ions show absorptions in the visible or near ultraviolet regions of the spectrum, except La^{3+} with no *f* electrons and Lu^{3+} with no empty *f* orbitals. These colours are due to transitions between *f* levels, *f–f* transitions, and, as the *f* levels lie deep enough in the atom to be shielded from much perturbation by the environment, these transitions appear in the visible and near ultraviolet spectra as sharp

bands. This is in contrast to the $d-d$ transitions found for the transition elements, which usually appear as broad bands due to environmental effects. Figure 11.7 illustrates a typical lanthanide ion spectrum. As these bands are so sharp, they are very useful for characterizing the lanthanides and for quantitative estimations. The positions of the absorptions shift with the *f* configuration, giving rise to the visible colours of the different ions as shown in Table 11.2.

Solutions of samarium(II) are red, and of ytterbium(II) are green. Note that the much more intense colours of cerium(IV) are not due to *f–f* transitions, but to a different mechanism involving charge-transfer between ion and coordinated ligand.

All the *f* states, except f^0 and f^{14}, contain unpaired electrons and are therefore paramagnetic. These elements differ from the transition elements in that their magnetic moments do not obey the simple 'spin-only' formula (section 7.11). In the *f* elements, the magnetic effect arising from the motion of the electron in its orbital contributes to the paramagnetism, as well as that arising from the electron's spinning on its axis. (In the transition metals the orbital contribution is usually quenched out by interaction with electric fields of the environment—at least to a first approximation—but the *f* levels lie too deep in the atom for such quenching to occur.) When the moments are calculated on the basis of spin and orbital contributions, there is excellent

TABLE 11.2 Typical colours of lanthanide compounds

f^1 or f^{13}	Ce(III), Yb(III)	uv absorption
f^2 or f^{12}	Pr(III), Tm(III)	green
f^3	Nd(III)	blue-violet
f^4 or f^{10}	Pm(III), Ho(III)	pink or yellow
f^5 or f^9	Sm(III), Dy(III)	cream
f^6, f^7 or f^8	Eu(III), Eu(II), Gd(III), Tb(III)	uv absorption
f^{11}	Er(III)	pink

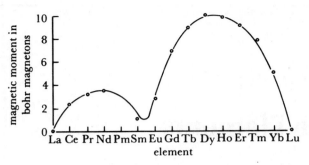

FIGURE 11.8 *Calculated* (—) *and experimental* (°) *values of the magnetic moments of the lanthanides*

agreement between experimental and calculated values, as Figure 11.8 shows.

The one case in which contributions to the bonding from the f orbitals is possible is in complexes of the heavier elements in which the coordination number is high. Use of the s orbital together with all the p and d orbitals of one valency shell permits a maximum coordination number of nine in a covalent species. Thus, higher coordination numbers imply either bond orders less than unity or else use of the f orbitals. In addition, certain shapes (such as a regular cube) of lower coordination number also demand use of f orbitals on symmetry grounds. These higher coordination numbers have only become clearly established recently, but their occurrence in lanthanide or actinide element complexes suggest the possibility of f orbital participation. Examples include the ten-coordinate complexes mentioned above, $LaEDTA(H_2O)_4$ and $Ce(NO_3)_5^{2-}$ or 10-coordinate $La_2(CO_3)_3.8H_2O$; 11-coordinate $Th(NO_3)_4.5H_2O$ (coordination by four bidentate nitrate

groups and three of the water molecules); and the 12-coordinate lanthanum atoms in $La_2(SO_4)_3.9H_2O$—with twelve sulfate O atoms around one type of La atom position.

PROBLEMS

11.1 Find out the details of one method which was used classically to separate the lanthanide elements and one method currently used, other than the examples given in section 11.3.

11.2 Investigate the names of the lanthanide elements: how far do these reflect their chemical similarities?

11.3 Compare and contrast the chemistry of lanthanum (a) with calcium (see also question 10.1), and (b) with thallium.

11.4 How far does the chemistry of their hydrides reflect the general characteristics of the lanthanide elements?

11.5 Find out from the literature

(a) further examples of coordination numbers of 8 or more.

(b) the use of lanthanide 'shift reagents' in nmr.

(c) examples of the application of the narrow-line $f-f$ spectrum (e.g. as wavelength standards, in estimation, in lasers).

11.6 To what extent does the chemistry of scandium parallel that of aluminium. Discuss why there should be similarities and differences.

12 The Actinide Elements

12.1 Sources and physical properties

All the elements lying beyond actinium in the Periodic Table are radioactive, and many of them do not occur naturally. Uranium and thorium are available as ores. Although actinium, protactinium, neptunium, and plutonium are available in small amounts in these ores their isolation is difficult and expensive and it is more convenient to isolate them from the fuel materials of nuclear reactors.

There are two main nuclear reactions which are used in the synthesis of elements beyond plutonium in the actinide series. One is the capture of neutrons, followed by *beta* emission, which increases the atomic number by one unit. The second method is by the capture of the nuclei of light elements, ranging from helium to neon, which increases Z by several units in one step.

Atomic piles provide intense neutron sources and samples can be inserted into piles for irradiation, so the first method is readily carried out, but it is a process of diminishing returns as each element has to be made from the one before. For example, in the irradiation of plutonium-239, less than one per cent of the original sample appears as californium-252, after capture of thirteen neutrons. The main stages are

$$^{239}Pu \rightarrow {}^{241}Pu \rightarrow {}^{245}Cm \rightarrow {}^{247}Cm \rightarrow {}^{251}Cf \rightarrow {}^{252}Cf$$

| % of original sample | 100 | 30 | 10 | 1·5 | 0·7 | 0·3 |

Neutron capture by a nucleus increases its neutron/proton ratio until this becomes too high for stability. Then a neutron is converted to a proton, with emission of a β-particle, and an increase of one in the atomic number, for example:

$$^{242}_{94}Pu + {}^{1}_{0}n \rightarrow {}^{243}_{94}Pu \rightarrow {}^{0}_{-1}e + {}^{243}_{95}Am$$

In the synthesis of heavier elements by successive neutron capture, not only does the yield of a heavier nucleus fall off sharply as the number of neutron addition steps from the starting material increases, but the process is made even less favourable by the general decrease in nuclear stability with increasing atomic mass. The heaviest elements may thus be obtained only by means which short-circuit the step-by-step addition of neutrons. One means is provided by an atomic explosion, where there is a vast flux of fast neutrons. A number of neutrons are added to the target nucleus simultaneously before the intermediate nuclei can decay. Thus, einsteinium and fermium were first discovered in the fall-out products of the first atomic bomb explosion.

A second, and more amenable, method for jumping a number of places in one operation is to bombard the starting material with species containing several nuclear particles. Bombardment by *alpha*-particles is the easiest way of achieving this, and many of the actinides, such as ^{248}Cm, ^{249}Bk, ^{249}Cf, and ^{256}Md, were first made in this way, e.g.:

$$^{253}_{99}Es + {}^{4}_{2}He \rightarrow {}^{256}_{101}Md + {}^{1}_{0}n$$

Alpha-particle bombardment requires the target to be the element with an atomic number of two less than the desired element, and such target elements will themselves be scarce and only obtained in small amounts where the desired element is one of very high atomic mass. To make the very heaviest elements, therefore, bombarding nuclei heavier than the α- particle are required. The last two actinide elements to be discovered were obtained by heavy nucleus bombardment in this way:

$$^{246}_{96}Cm + {}^{12}_{6}C \rightarrow {}^{254}_{102}No + 4{}^{1}_{0}n$$

$$^{252}_{98}Cf + {}^{11}_{5}B \rightarrow {}^{257}_{103}Lr + 6{}^{1}_{0}n$$

Table 12.1 indicates the current sources of each element.

The ease with which a radioactive element may be handled depends on the type and intensity of the radiation from the isotope and from its decay products which accumulate in the sample. This activity is indicated by the half-life, which is a measure of the rate of the decay process. The degree of activity of an isotope governs the extent to which its chemistry may be studied. In decay processes, the extremely energetic emitted particles break bonds and disrupt crystal structures. The energy given out appears largely as heat in the sample; for example, the heat produced in a millimolar solution of curium-242 salts in water would be sufficient to evaporate the solution to dryness in a short time. The breaking of bonds by emitted particles in a sample of a radioactive element is equivalent to a continuous process of self-reduction. As a result, it may be impossible to prove the existence of an oxidizing state for very reactive isotopes. For example, the evidence for the oxidizing IV states of a number of the heavy elements had to await the production of isotopes

TABLE 12.1 Properties of the actinide elements.

Z	Element	Symbol	Mass of most accessible isotope	Half-lives		Source	Electronic configuration Rn 5f 6d 7s		
89	Actinium	Ac	227		$21 \cdot 77$ y	natural	0	1	2
90	Thorium	Th	232·1 (natural mixture)	232	$1 \cdot 41 \times 10^{10}$ y	natural	0	2	2
91	Protactinium	Pa	231	231	$3 \cdot 28 \times 10^{4}$ y	natural and fuel elements	1 2 2 or 2 1 2		
92	Uranium	U	238·1 (natural mixture)	235 238	$7 \cdot 04 \times 10^{8}$ y $4 \cdot 47 \times 10^{9}$ y	natural	1 3 2 or 3 1 2		
93	Neptunium	Np	237	237	$2 \cdot 14 \times 10^{6}$ y	fuel elements	5	0	2
94	Plutonium(iii)	Pu	239	239 242 244	24360 y $3 \cdot 76 \times 10^{5}$ y $8 \cdot 3 \times 10^{7}$ y	fuel elements neutron bombardment	6	0	2
95	Americium	Am	241	241 243	433 y 7380 y	fuel elements neutron bombardment	7	0	2
96	Curium	Cm	242 244	242 244 247 248	163 days $18 \cdot 11$ y 8×10^{7} y $3 \cdot 4 \times 10^{5}$ y	neutron bombardment	7	1	2
97	Berkelium	Bk	249	247 249	1380 y 320 days	ion bombardment neutron bombardment	8 1 2 or 9 0 2		
98	Californium	Cf	252	249 252	351 y $2 \cdot 64$ y	neutron bombardment	10	0	2
99	Einsteinium	Es	254	253 254	$20 \cdot 5$ days 276 days	ion bombardment neutron bombardment	11	0	2
100	Fermium	Fm	253	253 257	$4 \cdot 5$ days 100 days	neutron bombardment	12	0	2
101	Mendelevium	Md	256	256 258	77 minutes 55 days	ion bombardment	13	0	2
102	Nobelium	No	254	254 255 259	3 s 3 minutes 1 h	ion bombardment	14	0	2
103	Lawrencium(iv)	Lr	257	256 260 261 262	35 s 3 minutes 39 minutes 214 minutes	ion bombardment	14	1	2

Notes: (i) The source given is the most recent for manageable quantities of the isotope in question, except that elements 100 onwards are not available in weighable amounts at present. The most accessible isotope of the heavier elements is usually the one produced by fewest steps and is not necessarily the longest-lived. Half-lives are given only for the most accessible and for the longest-lived isotopes. (ii) The isotopes Th-232, U-238, Np-237, Pu-239, and Am-241 are available in kilogram quantities; other isotopes from fuel elements in gram amounts, as are Am-243 and Cm-244; Ac-227, Bk-249, and Cf-249 are available at the 1–10 mg level and Es-253 somewhat less. From fermium onwards, quantities range from micrograms down to atoms. (iii) For plutonium, the abundance of isotopes in fuel elements is $239 > 240$ (half-life 7000 y) > 241 (14 y) $> 242 > 238$ (88 y). The choice of isotope for a particular study is a balance of abundance, half-life, and type of radiation. (iv) Alternative configuration $5f^{14}6d^{0}7s^{2}7p^{1}$.

which were more stable than the very short-lived ones originally available. The intense activity of the samples of these elements, and of their sources, demands special handling techniques which involve manipulating microgram amounts of the sample by remote control. Such work demands specialized facilities and training and is limited to a few laboratories in the world, the most famous being at the University of California where most of the transuranium elements were discovered.

In a number of cases, neutron bombardment, which is readily carried out, does not give rise to the isotope of longest half-life. An example is provided by berkelium where the most accessible isotope, Bk-249, has a much shorter half-life than Bk-247, which is only available in tiny amounts from ion bombardment.

The electronic configurations of these elements gave rise to considerable controversy in the early days of work in this field. The elements which were available before the advent of the atomic pile, especially thorium and uranium, strongly resemble the transition metals (hafnium and tungsten) in their chemistry, and the heaviest elements were accordingly placed in the d block. As new elements were studied, it became increasingly obvious that an f shell was being filled but it was not clear whether this started after actinium, paralleling the lanthanides, or later on in the middle of a d series. It became clear, however, that curium corresponded to gadolinium in properties and was thus the $f^7 d^1 s^2$ element, implying that the series are genuinely *actinides*. Later, it was possible to interpret the very complicated atomic spectra and determine the electronic configurations as given in the Table. Magnetic studies have also confirmed these. Element

103, lawrencium, completes the actinide series. Elements 104 and 105 are discussed in section 16.12.

12.2 General chemical behaviour of the actinides

Table 12.2 lists some important chemical properties of the actinides, with actinium included for comparison. The M^{4+}/M^{3+} redox potentials clearly show the increasing stability of the III state for the heaviest elements. Diagrams of the free energy changes per electron in oxidation-reduction reactions are shown in Figure 12.1, illustrating the stable oxidation states for the elements. The stabilization of the III state to the right of the actinide series is again shown here.

The most stable oxidation state of the elements up to uranium is the one involving all the valency electrons. Neptunium forms the VII state, using all its valency electrons, but this is oxidizing and the most stable state is Np (V). Plutonium also shows states up to VII and americium up to VI but the most stable states are Pu(IV) and Am(III). Later elements also tend to be most stable in the III state. This pattern of higher oxidation state stabilities has more in common with that of a d series where all the electrons are used in the Group oxidation state, until the middle of the series— Mn(VII) or Ru and Os(VIII)—and this state becomes more oxidizing across the series. Only for the later actinide elements does the III state become dominant and the resemblance to the lanthanides appears. Although the IV state of berkelium is strongly oxidizing, it is more stable than the IV states of curium and americium. In this, it is showing a parallel to the properties of terbium, where the IV state, corresponding to the f^7 configuration, has some stability. Americium does not form the II state in aqueous media, but it has been

TABLE 12.2 Chemical properties of the actinides and actinium

Element	Oxidation states coordination	Crystal radii (pm) M^{3+} 6	M^{4+} 6	8	Potential (V) $M^{4+} + e = M^{3+}$	$M^{3+} + 3e = M$	1st ionization energy (eV: 1 eV = 96.48 kJ mol^{-1})
Ac	III	126				− 2·13	5·17
Th	(III), IV		108	119		− 1·17	6·08
Pa	(III), IV, V	(118)	101	115		− 1·49	5·89
U	III, IV, V, VI	116·5	103	114	− 0·631	− 1·66	6·05
Np	III, IV, V, VI, VII	115	101	112	+ 0·155	− 1·79	6·19
Pu	III, IV, V, VI, VII	114	100	110	+ 0·982	− 2·00	6·06
Am	(II), III, IV, V, VI, (VII)*	111·5	99	109	+ 2·0	− 2·07	5·99
Cm	(II)*, III (IV)	111	99	109	+ 3·2	− 2·06	6·02
Bk	(II), III, IV	110	97	107	+ 1·6	− 1·97	6·23
Cf	(II)*, III (IV)	109	96	106		− 2·01	6·30
Es	(II), III	(108)				− 1·98	6·42
Fm	(II), III					− 1·95	6·50
Md	II, III					− 1·66	6·58
No	II, III					− 1·78	6·65
Lr	III					− 2·06	

*Transient only. Stable oxidation states are underlined: unstable states are in brackets.

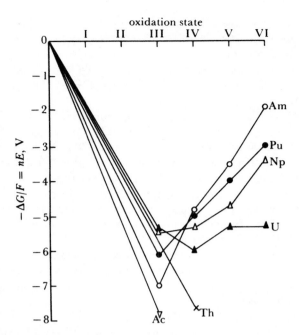

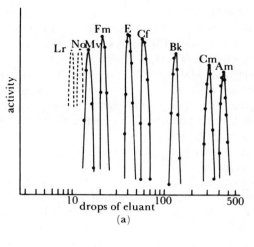

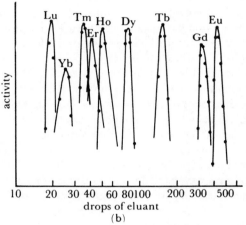

FIGURE 12.1 *Free energy changes per electron for actinide element oxidation-reduction couples in acid solution*
Here the free energy change (in volts) is plotted against the oxidation state for all the actinide elements, as far as data are available. The uranium and americium diagrams are discussed in section 8.6, Figure 8.11.

FIGURE 12.2 *Elution diagrams for* (a) *actinide and* (b) *lanthanide elements*

reported in a chloride melt, so that there is some slight resemblance to europium which attains the f^7 configuration in its moderately stable II state. An oxide phase, MO, exists for Pu and Am. The heavier actinides, though studied mainly by carrier methods, show some evidence for a II state in addition to the stable III state but attempts to oxidize Md^{3+}, No^{3+} or Lr^{3+} to the IV state were unsuccessful. These II states are distinctly more stable than for the corresponding lanthanides, and a significant difference between the two f series is evident. In particular, Md(II) is moderately stable and No(II) is markedly more stable than No(III), with No^{2+} requiring an oxidizing agent comparable with permanganate or ceric to form No^{3+}. The f^{14} configuration would probably be attained by No^{2+} but this state is relatively much more stable than the analogous Yb^{2+} in the lanthanides.

The regular trend in ionic radii resembles that shown by the lanthanide elements, and it is possible to talk of an *actinide contraction* similar to the lanthanide contraction and arising from a similar increase in effective nuclear charge due to poor screening by the f electrons. This actinide contraction means that the actinide elements should show similar ion-exchange behaviour to the lanthanides, and this has been made use of in a striking way in the identification of the newer heavy elements. The elements beyond curium have all very similar properties chemically, and the methods of synthesis mean that they are formed in only small amounts in the presence of excess target material. The identification of the heavier elements depends upon detecting their characteristic radiation (which can be predicted theoretically).

The method which was successfully adopted was to dissolve the irradiated targets and pass the actinides in solution through an ion-exchange column and count the radiation from each fraction. Due to the tiny scale of the experiments, the 'column' was a few beads of resin and the fractions were single drops. The order of elution, and the elution positions, of the tripositive actinide ions and the tripositive lanthanide ions are the same on the same resin, and this was used in the first identification of the elements from americium up to mendelevium on a weighable scale (i.e. apart from the use of carrier methods). A composite elution diagram of these elements is shown in Figure 12.2 along with a similar diagram for the heavier lanthanides. The one-to-one correspondence in positions is clear. The scale of the operations is made evident by the fact that the first identification of element 101, mendelevium, was based on the count of five decompositions, i.e. of the fission of five individual atoms.

The actinide metals resemble the lanthanides, are of low electronegativity, and are very reactive. The metals are produced by electrolytic reduction of fused salts or by treating the halides with calcium at high temperatures. They are all extremely dense, with densities ranging from 12 to 20 g/cm³. The direct reactions of the metals (for example, with oxygen, halogens and acids) are similar to those of the scandium Group elements. The metals also react directly with hydrogen with the formation of non-stoichiometric hydrides, such as Th_4H_{15}. Phases with idealized compositions MH_2 and MH_3 are most common. These hydrides are reactive and often form suitable starting materials for the preparation of other compounds. On heating, they decompose leaving the metal in a very finely divided and reactive form.

12.3 Thorium

Thorium has been known since 1828. Its principal source is the mineral, monazite, which is a complex phosphate of thorium, uranium, cerium, and lanthanides. Thorium is extracted by precipitation as the hydroxide, along with cerium and uranium, and then separated by extraction with tributyl phosphate from acid solution. The metal is made by calcium reduction of the oxide or fluoride, and pure samples can be prepared in the *Van Arkel* process by decomposing the iodide, ThI_4, on a hot filament.

The only stable oxidation state, is the IV state in which thorium resembles hafnium. This is very stable and, because of the large size, the Th^{4+} ion has a low enough charge density to be capable of existing without excessive polarization effects. This is the highest-charged monatomic ion known. The hydroxide is precipitated from thorium solutions and gives the oxide, ThO_2, on ignition. This is also formed directly from the metal and oxygen, and on ignition of oxy-salts. It is a stable and refractory material (m.p. 3050 °C) and is soluble only in hydrofluoric/nitric acid mixtures. The anhydrous halides, ThX_4, are prepared by dry reactions such as metal plus halogen or oxide plus hydrogen halide at 600 °C. The tetrafluoride is involatile but the others sublime in vacuum above 500 °C. Treatment of the halides with water vapour gives the oxyhalides, $ThOX_2$.

Dilute thorium solutions in strong aqueous acid contain the hydrated thorium ion, $Th^{4+} nH_2O$, but hydrolysis takes place on concentration or when the pH is raised. At a pH of about 6, the hydroxide, $Th(OH)_4$, is precipitated. This has a crystal structure containing chains of thorium atoms linked by oxy- and hydroxy-bridges and 8-coordinated Th.

The commonest salt is the nitrate, $Th(NO_3)_4.5H_2O$ which is very soluble in water, alcohols, and similar solvents. The fluoride, oxalate and phosphate are very insoluble and may be precipitated even in strong acid solution (compare hafnium and cerium(IV). Thorium also gives a borohydride, $Th(BH_4)_4$ which sublimes in vacuum at about 40 °C.

The coordination of thorium(IV) is variable and tends to be high. $ThCl_4$ and $ThBr_4$ have the distorted eight-coordination of UCl_4 while seven-, eight-, and nine-coordinate Th atoms are all found in $ThOCl_2$ (and in the isomorphous Pa, U, and Np analogues). Eight-coordination to sulfur is found in the complex $Th(S_2CNEt_2)_4$. Nine-coordination, by sharing fluorines, is found in $(NH_4)_4ThF_8$ and in Na_2ThF_6 and the ThF_9 arrangement is similar to that shown in Figure 15.23.

An alternative form of nine-coordination is found in the oxydiacetato(oda) complex $[Th(oda)SO_4(H_2O)_2]$. (See Appendix B for oda). Here the Th atom is coordinated by 9 O atoms in a capped square antiprism arrangement—compare Figure 12.5 with a ninth O atom above one of the square faces.

No state other than IV exists for thorium in solution, but there is evidence for the tri-iodide, ThI_3, formed by heating the metal with the stoichiometric amount of iodine in vacuum at 555 °C. Using a deficiency of iodine gives the di-iodide which can also be prepared electrochemically. It has been shown that ThI_3, and two different crystal modifications of ThI_2, can be also prepared by heating ThI_4 with thorium metal. The tri-iodide converts to the di-iodide on further heating. Both ThI_2 and ThI_3 react with water with the evolution of hydrogen and formation of thorium(IV). These compounds are metallic with bonding which can be explained on the same model as that used for the metallic hydrides, section 9.4. One non-metallic thorium(III) compound is the white ThOF, prepared by reducing a ThF_4/ThO_2 mixture with thorium metal at 1500 K. Thorium forms a range of alkoxide compounds $[Th(OR)_4]_n$, with the isopropoxide being dimeric in isopropanol and tetrameric in benzene solutions.

12.4 Protactinium

Protactinium was first identified in uranium in 1917. It is a product of uranium-235 decay and, in turn, gives actinium by alpha-particle emission:

$$^{235}_{92}U \rightarrow ^{4}_{2}He + ^{231}_{90}Th \rightarrow ^{0}_{-1}e + ^{231}_{91}Pa \rightarrow ^{4}_{2}He + ^{227}_{89}Ac$$

A further isotope occurs in the decay of neptunium 227, but both these naturally-occurring isotopes have low concentrations in equilibrium. The element is most readily obtained by synthesis:

$$^{232}_{90}Th + ^{1}_{0}n \rightarrow ^{233}_{90}Th \rightarrow ^{233}_{91}Pa + ^{0}_{-1}e$$

This isotope has a half-life of 27·4 days but is a beta-emitter and more readily handled than the alpha-emitting Pa 231. The latter is now available from the fuel elements of atomic piles and is commonly used. Because of its relative scarcity until recently, and because of the strong tendency of its compounds to hydrolyse and form polymeric colloid particles which are adsorbed on reaction vessels, protactinium chemistry is comparatively less well known.

The oxide system is complex and compounds range in composition from PaO_2 to Pa_2O_5. The pentoxide is obtained on igniting protactinium compounds in air and is a white solid with weakly acidic properties, being attacked by fused alkali. On reduction with hydrogen at 1500 °C, the black dioxide PaO_2 is formed.

Among the halides, two fluorides are known. The pentafluoride PaF_5 results from the reaction of bromine trifluoride on the pentoxide. It is a very reactive and volatile compound. The complex anion, PaF_7^{2-}, is known and was used in the classical isolation of the element. In this ion, the protactinium is nine-coordinate with Pa linked by two fluorine bridges to a neighbour on either side, giving a chain structure. The structure of the PaF_9 units is the same as that of the ReH_9^{2-} ion shown in Figure 15.23. In the complex Na_3PaF_8 the PaF_8^{3-} ion is a slightly distorted cube and the sodium ions are also eight-coordinated. The Na_3MF_8 compounds of uranium(V) and neptunium(V) are isostructural. The second fluoride, PaF_4, is a red, high-melting solid which results from the reaction of hydrogen and hydrogen fluoride on the oxide. The oxyfluoride, PaO_2F, and complexes $MPaF_5$ and M_4PaF_8 are known.

In recent studies, all the Pa^{IV} and Pa^V chlorides, bromides and iodides of the types PaX_4, PaX_5, $PaOX_2$, $PaOX_3$ and PaO_2X have been prepared, together with the complexes M_2PaX_6 and $MPaX_6$. $PaCl_5$ and PaF_5 are polymeric structures with pentagonal bipyramidal coordination, like β-UF_5. $PaBr_5$ consists of dimeric units with two bridging bromines giving six-coordinate Pa, while $PaBr_4$ is an infinite polymer with all the bromine atoms bridging pairs of Pa atoms giving $PaBr_8$ coordination. PaI_3 exists in the solid state. More complex oxyhalides of Pa(V) include Pa_2OX_8 (X = F, Cl) and Pa_3O_7F. There is only one reported alkoxide complex of protactinium, $[Pa(OEt)_5]_n$.

The solution chemistry is obscure because of the formation of colloids, but anionic complexes like $(PaOCl_6)^{3-}$ have been claimed. Lower oxidation states may be obtained in solution by reduction with zinc amalgam. The tetravalent state is stable in absence of air but evidence for the III state is slight and based on polarographic results. The absorption spectrum of $PaCl_4$ in water shows three maxima and is similar to that of Ce^{3+}, providing some evidence for the presence of a single f electron in Pa (IV).

12.5 Uranium

Uranium is the longest-known of the actinide elements, having been discovered in 1789, but it attracted little interest until the discovery of uranium fission in 1939. It is now of importance as a fuel, and its chemistry has been very fully explored in the course of the atomic energy investigations.

Natural uranium contains two main isotopes, ^{238}U 99.3 per cent and ^{235}U 0.7 per cent, and also traces of a third, ^{233}U. The vital isotope from the nuclear energy point of view is ^{235}U because this reacts with a neutron, not by building up heavier elements as in the examples discussed in section 12.1, but by fission to form lighter nuclei. This fission process releases considerable energy and more neutrons, which, in turn, fission uranium-235 nuclei and allow the building-up of a chain of fissions. The energy of such nuclear processes is about a million times the energy released in chemical reactions, such as the burning of a fuel or the detonation of a high explosive. This is the reason for the value of atomic fission as an energy source, and for the horror of fission as a source of explosive energy in a weapon. A typical fission process is:

$$^{235}_{92}U + ^1_0n \rightarrow ^{92}_{36}Kr + ^{140}_{56}Ba + 3^1_0n + \text{about } 8 \times 10^9 \text{ kJ mol}^{-1}$$

The nuclei formed in fission fall into two main groups, a lighter set with masses from about 70 to 110 and a heavier one with masses from 125 to 160. Splitting into approximately equal nuclei is about a thousand times rarer than splitting to an unequal pair such as that shown in the equation. The neutrons evolved in the fission are either used in other fissions, absorbed by non-fissionable nuclei such as uranium 238, or escape through the surface of the uranium mass. The essence of running an atomic pile is to ensure that one neutron per fission is available to cause another fission. More than one leads to a rapidly increasing chain reaction and explosion, while less than one means that the process dies out. The absorption of neutrons by uranium 238 leads to the formation of heavier elements, of which the most important is plutonium which is itself a nuclear fuel. In appropriate conditions, more plutonium can be produced from the uranium 238 than the amount of uranium 235 consumed, and such an arrangement 'breeds' nuclear fuel in the 'breeder reactor'.

The major problem of using nuclear power arises from the fission products, arising like ^{140}Ba above. These are neutron-rich isotopes which are themselves radioactive. The witches' brew which results is highly radioactive and contains isotopes with a wide range of chemical and radiochemical properties. At present, the fuel rods are allowed to stand in a strongly shielded store for a period to allow the short-lived component of the radioactivity to disappear. Then they are treated—all by remote control—to recover unchanged uranium, plutonium and other useful elements. Increasingly, uses are being found for some components, see the discussion of technetium (Chapter 20) for example. No method which is acceptable to general public opinion has yet been formulated to deal with the remaining mixture of unwanted radioactive material. Most of the spent fuel residues from current commercial power production remain in temporary storage awaiting a long-term solution to the storage problem. It is quite feasible to safely store the mixture until the short-lived isotopes have decayed to negligible proportions. Twenty half-lives reduce the amount of isotope by a factor of about one million, so storage for a few years deals with everything with half-lives of the order of days, or less. Likewise isotopes with extremely long half-lives have very low activities and thus do not create a major problem. The great difficulties arise from isotopes of intermediate half-lives, especially those of elements which are readily taken up by living organisms. Thus the major problems in Britain and Western Europe from the Chernobyl disaster were caused by ^{137}Cs which has a half-life of 30.2 years, and which mimics potassium in its metabolism. This continues to cause concern, particularly for sheep farming.

Most current work is directed towards immobilizing such residues in glasses or concrete, with a view to permanent storage deep underground in a geologically stable formation. This has often seemed a good idea—until a proposal arises to locate such a store near a particular community! There are general fears that radioactive material will eventually leach out and get into the environment. There are also justifiable doubts whether such stores could possibly be maintained in isolation for the thousands of years required. It may be that the ultimate solution will come from quite different directions—either an alternative source of power (possibly fusion power) which will remove the need for atomic reactors, or a radically different method of dealing with radioactive materials. It may be possible to convert the difficult isotopes by further irradiation, for example.

In the meantime, however, methods are still needed to 'manage' the intermediate half-life nuclides. A recent process using solvent extraction, the Truex or transuranium extraction process, has been developed in order to significantly reduce the volume of waste requiring deep burial underground. In this process the carbamoylmethylphosphine

oxide is used to extract, and thereby concentrate, the transuranic elements. Importantly, the process extracts all transuranic elements (in a range of oxidation states) including americium-241, which is responsible for much of the radiation in nuclear wastes.

In its chemistry, uranium resembles the three succeeding elements, neptunium, plutonium, and americium, in having four main oxidation states, III, IV, V, and VI. The most stable state drops from VI for uranium through V for neptunium, IV for plutonium, to III for americium. This is illustrated by the oxides and halides found for these elements, shown in Table 12.3, and also by the free energy diagrams for the changes in oxidation state in acid solution, Figure 12.1. In solution, the VI state is present as the uranyl ion, UO_2^{2+}, and the V state also occurs as an oxycation, UO_2^+. The IV and III states are present as simple cations, U^{4+} and U^{3+}. Since the change from U^{3+} to U^{4+}, and that from UO_2^+ to UO_2^{2+} (and the reverse changes) involve only transfer of an electron, these two pairs of redox reactions occur rapidly. On the other hand, oxidations such as U^{4+} to UO_2^{2+} involve oxygen transfer as well and are slow and often irreversible. In solution, the III, IV, and VI states all exist, but UO_2^+ has only a transitory existence. However, in a non-oxide medium, such as anhydrous hydrogen fluoride or a chloride melt, uranium(V), although still unstable, is well represented.

Uranium metal, produced by reduction of the tetra-fluoride with calcium or magnesium:

$$UF_4 + 2Ca \rightarrow U + 2CaF_2$$

is reactive and combines directly with most elements. It dissolves in acids but not in alkalis. Direct reaction with hydrogen at about 250 °C gives the hydride, UH_3, usually in a form with a small deficiency of hydrogen. This is a very reactive compound which is a useful starting material for the

preparation of uranium compounds of the III and IV states. The normal product is the IV compound, for example:

$$UH_3 + H_2Y \rightarrow UY_2 \ (Y = O, S \text{ at about } 400 °C)$$
$$UH_3 + HF \rightarrow UF_4$$
$$\text{but } UH_3 + HCl \rightarrow UCl_3 \ (Cl_2 \text{ gives } UCl_4)$$

The uranium-oxygen system is very complex as the oxidation states are of comparable stability and non-stoichiometric phases are common. The main uranium ore, *pitchblende*, is an oxide approximating to UO_2 in composition. The other stoichiometric oxides are U_3O_8, which is the ultimate product of ignition, and UO_3 which is obtained by the decomposition of uranyl nitrate, $UO_2(NO_3)_2$, at about 350 °C. The trioxide can be reduced to the dioxide by the action of carbon monoxide at 350 °C, and it goes to U_3O_8 on heating to 700 °C. All three oxides dissolve in nitric acid to give uranyl, UO_2^{2+}, salts. More recent studies have established U_3O_7 and U_4O_9 as definite compounds.

The known halides of uranium are listed in Table 12.3 and the interconversions of the fluorides and chlorides are shown in Figures 12.3 and 12.4, respectively. The hexahalides are octahedral. UF_6 is volatile, sublimes at 56 °C, and has been used in the separation of uranium isotopes by gaseous diffusion. The compound is a powerful fluorinating agent and is rapidly hydrolysed. Two complex fluoride ions, UF_7^- and UF_8^{2-}, contain uranium(VI). Uranium pentafluoride and pentachloride both readily disproportionate to give the hexahalide and the tetrahalide. Uranium(V) also occurs in fluoride complexes, UF_8^{3-} and UF_6^-. The latter is formed in HF solution, and may be the reason for the relative stability of the V state in this medium. UCl_6^- and UBr_6^- have also been prepared. The pentahalides all have polymeric structures. The pentafluoride has two forms. In α-UF_5, an octahedron of fluorines around the uranium is completed by the sharing of fluorine atoms and the formation of long chains of linked octahedra. In the β-form of UF_5, the uranium atom is seven-coordinated with three fluorines attached to only one uranium and the other four shared with four different uraniums in a polymeric structure. In U_2F_9, all the uranium atoms have nine fluorines at equal distances and this (black) compound

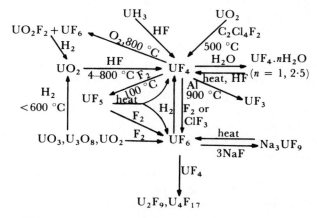

FIGURE 12.3 *Reactions and interconversions of uranium fluorides*

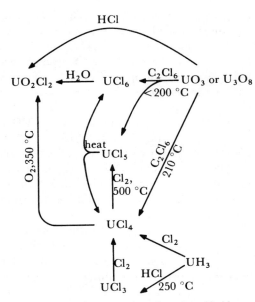

FIGURE 12.4 *Reactions and interconversions of uranium chlorides*

probably contains crystallographically equivalent uranium (IV) and uranium (V) atoms. UCl_5 is a dimer of two UCl_6 octahedra sharing an edge, with 2Cl bridges. This is the simplest form: polymeric forms also occur.

The most stable halides of uranium are the tetrahalides, to which the higher halides are readily reduced and the trihalides oxidized. The exception is UI_4 which slowly converts to UI_3 and iodine at room temperature. The tetrachloride has a distorted eight-coordinated structure. UCl_3 (like many other lanthanide and actinide trihalides), crystallizes with nine-coordinated U^{3+} ions, in a structure where the U atom has three chloride ions coplanar with it, three above this plane, and three below. UI_3, and the tribromides and triiodides of Np, Pu, and Am, have an eight coordinated layer lattice. All the halides form $U^{IV}X_6^{2-}$. The uranium III ion, UCl_4^-, is also known, though this is very readily oxidized.

The standard syntheses of UX_3 and other AnX_3 involve high temperature routes, producing unreactive polymeric solids which are not very useful starting materials for the preparation of other An(III) compounds. A much lower temperature route is provided by the direct reaction of a slight excess of the metal in THF at room temperature.

$$An + 1 \cdot 5X_2 \rightarrow AnX_3(THF)_4$$
$$(An = U, Np \text{ or } Pu; X = Br \text{ or } I)$$

The products are pentagonal bipyramidal in shape, with two X atoms at the apices and the third X and the four THF molecules in the equatorial positions.

One interesting product accessible through such species is the three-coordinate $An[N(SiMe_3)_2]_3$ which is soluble in hydrocarbons and sublimes at 60 °C. The low coordination is stabilized by the presence of the very bulky $N(SiMe_3)_2$ ligands.

$$AnI_3(THF)_4 + 3NaN(SiMe_3)_2 \rightarrow An[N(SiMe_3)_2]_3 + 3NaI$$
[An = U, red-purple; An = Np, dark blue; An = Pu, orange]

Oxyhalides of uranium(VI), UO_2X_2 and UOF_4 are also known. These are made from the oxides or halides by partial substitution:

$$UO_3 + 2HF \xrightarrow{400°C} UO_2F_2 + H_2O$$

Added to these are the more complex oxyfluoride phases, $U_2O_3F_6$, $U_2O_5F_2$ and $U_3O_5F_8$. Oxyhalides of uranium (V) are UO_2X (X = F, Cl, Br), UOX_3 (X = Cl, Br) and for uranium (IV), UOX_2 (X = F, Cl, Br).

A number of alkoxides of uranium have been described. The uranium(IV) alkoxides, $U(OR)_4$, are green sublimable compounds which, like their Th and Pa counterparts, are oligomeric in solution. Alkoxides in the V and VI states, such as $U(OR)_5$ and $UO_2(OR)_2$ are also known, as are the analogous amides, $U(NR_2)_4$.

In solution, hydrolysis occurs for all oxidation states, being least for the III state. Uranium(III) and (IV) exist as ions in strong acid, and uranium(IV) hydrolyses in more dilute solution in a similar manner to Th^{4+}. The U^{4+} ion also gives insoluble precipitates with similar reagents—F^-, PO_4^{3-}, etc. —as does Th^{4+}. Uranium(V) has a strong tendency to disproportionate to U^{4+} and UO_2^{2+}. It is most stable in fairly acid solution at a pH of about 3. Uranium(VI) also hydrolyses in solution, this time giving double hydroxy bridges so that polymers of the type . . . $UO_2(OH)_2UO_2(OH)_2$ . . . are formed. Uranium(VI), as uranyl, forms the only common uranium salts, and the most usual starting material is uranyl nitrate, $UO_2(NO_3)_2.nH_2O$, (n = 2, 3, or 6). This is soluble in water and in a variety of organic solvents.

The stereochemistry of uranium, and of the other actinides, shows a tendency to high coordination numbers, as illustrated by the halide structures above.

In $U(BH_4)_4$, 14-coordination of uranium occurs. The structure contains two of the four BH_4 groups bonded through three hydrogens to one U atom while the other two bridge two U atoms, sharing two hydrogens with each:

$$U \underset{H}{\overset{H}{<}} B \underset{H}{\overset{H}{>}} U$$

Thus the total structure is *cis* $(HBH_3)_2U(H_2BH_2)_{4/2}$ with 14 $U - H - B$ links. Borohydrides also form with U(III). In $U(BH_4)_3.3THF$, the BH_4 groups bonded through three bridging hydrogens, together with the three THF groups, make up 12-coordination around U.

A tendency to eight-coordination for M^{4+} actinide ions seems to be general. For example, uranium and thorium form an acetylacetonate, $M(C_5H_7O_2)_4$, which has the eight-coordinating oxygen atoms at the corners of a square antiprism, as shown in Figure 12.5. (The square antiprism is most readily visualized as a cube with the top face twisted 45° relative to the bottom one.) Another interesting 8-coordinate species is $M(NCS)_8^{4-}$. When the cation is Cs^+, the anion structure is a square antiprism (M = U, Pu). When a bulkier cation, in $(NEt_4)_4M(NCS)_8$, is used, the anion structure becomes a cube (M = Th, Pa, U, Np

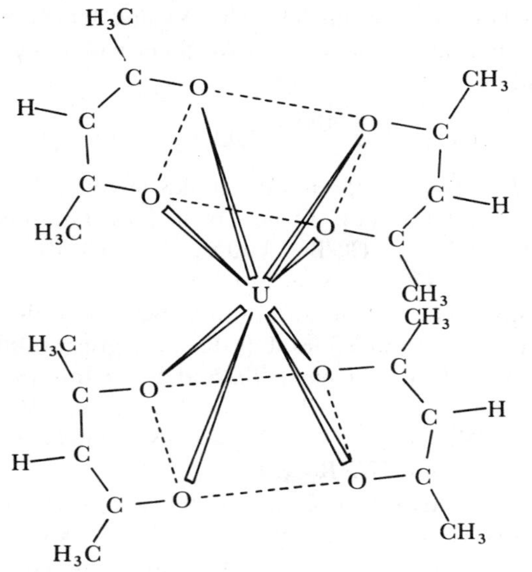

FIGURE 12.5 *The structure of uranium (or thorium) acetylacetonate,* $U(acac)_4$
The uranium (or thorium) atom is coordinated to eight oxygen atoms which are arranged in a square antiprism around it.

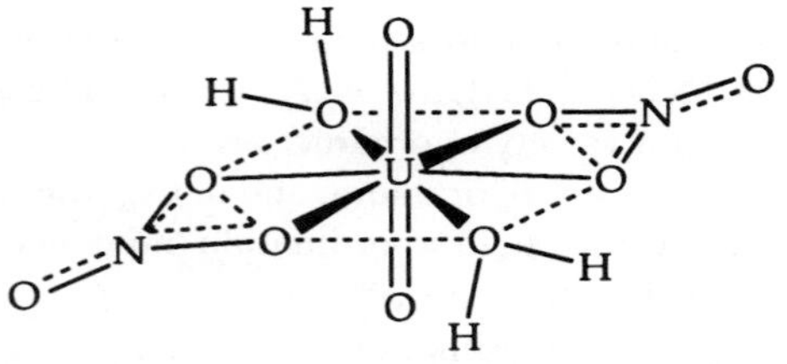

FIGURE 12.6 *The structure of uranyl nitrate hydrate,* $UO_2(NO_3)_2 . 2H_2O$
The UO_2 group is linear and two water molecules and the two nitrate groups are coordinated in the central plane to the uranium. As the nitrate groups here are bidentate, the uranium has a co-ordination number of eight.

and Pu). $(NEt_4)_4M(NCSe)_8$ (M = Pa, U) also contains a cubic anion.

In uranyl compounds, the UO_2 group is linear and complexes exist in which four, five, or six donor atoms lie coplanar with the uranium atom, giving six-, seven- or eight-coordination overall. An example is shown in Figure 12.6, where one form of hydrated uranyl nitrate has two water molecules and the two nitrate groups coordinated to uranium through both oxygens, all in the central plane, to give the eight-coordinated structure shown.

For a discussion of $M(C_8H_8)_2$ species, see section 16.6.

12.6 Neptunium, plutonium, and americium

The three elements, neptunium, plutonium, and americium, which follow uranium, also show the four oxidation states, III, IV, V, and VI. In addition, strong oxidation in alkaline media by reagents like peroxide, ozone, periodate, or XeO_3 produce Pu(VII) or Np(VII). Attempts to prepare

plutonium(VIII) have so far been unsuccessful. Compounds isolated include Li_5MO_6 and M'_3MO_5 (M = Np or Pu; M' = K, Rb or Cs) and also $Ba_3(NpO_5)_2$. On heating, these compounds lose O_2 and revert to M(VI) ternary oxides. The compounds are isostructural with their heptavalent iodine or rhenium analogues. The VII states also exist in strong alkaline solution but reduce rapidly to the VI state on acidification. The redox potential of the Np(VII)/Np(VI) couple at pH = 14 has been put at 0.61 V at 250 °C.

The III, IV, V, and VI states are present in solution as the M^{3+}, M^{4+}, MO_2^+ and MO_2^{2+} ions, as in the case of uranium, and they have a similar tendency to hydrolyse. The hydrolysis of Pu(IV) is of importance environmentally, since solid $Pu(OH)_4$ has an exceedingly low solubility product ($\log K_{sp} = -54$). This, together with the rather complex aqueous chemistry, can strongly limit the amount of plutonium in solution. The soluble PuO_2^+ ion is the principal form of dissolved plutonium in seawater, at a concentration of around 10^{-14} M.

The relative stability of the four oxidation states varies among the elements. The VI state becomes increasingly unstable and oxidizing from uranium to americium, and AmO_2^{2+} solutions are as strongly oxidizing as permanganate. The V state in solution, as MO_2^+, is the most stable state for neptunium, while PuO_2^+ and AmO_2^+, while less stable than NpO_2^+, are more stable than UO_2^+. Thus, UO_2^+ disproportionates in aqueous solution to uranium(VI) and uranium(IV) (see Figure 12.1) while NpO_2^+ is stable. In the compound, $Cs_7(NpO_2)(NpO_2)_2Cl_{12}$, the balance of charges shows that the Np(V) and Np(VI) oxycations coexist. When hydrofluoric acid is used instead of water as solvent, the relative stabilities of the V states of U and Np reverse and $CsUF_6$ dissolves unchanged while $CsNpF_6$ disproportionates to give $NpF_4 + NpF_6$. This example is a useful reminder that stabilities differ markedly in different media, and our generalizations apply only to stabilities in aqueous solutions in air. The IV state is the most stable one for plutonium, while americium resembles the following elements in being most stable in the III state.

This behaviour in solution is shown by the free energy diagrams, Figure 12.1, and is paralleled in the solid state, as Table 12.3 shows. The plutonium tetrahalides do not parallel the stability of plutonium(IV) in solution, with no bromide or iodide, and $PuCl_4$ existing only in a Cl_2 atmosphere. Thus the stability in solution probably reflects the high solvation energy of the Pu^{4+} ion. Similarly, although neptunium(V) is the most stable state in solution, the synthesis of NpF_5 caused great difficulty. It was finally achieved by using very reactive KrF_2 (compare section 17.9.5) to react with NpF_4 in anhydrous HF. Thus the general stability of Np(V) must depend on the formation of NpO_2^+.

As found for the transition elements (compare Chapters 14 and 15), oxyhalides are often found for oxidation states where the simple halides are missing, perhaps because of the lower coordination demand. We find in the VI state, MO_2F_2 for M = Np, Pu, Am and MOF_4 for M = Np and Pu: in the V

TABLE 12.3 Halides and oxides of uranium, neptunium, plutonium and americium

Uranium	UF$_6$		UF$_5$	U$_2$F$_9$	UF$_4$	UF$_3$
				U$_4$F$_{17}$		
	UCl$_6$		UCl$_5$		UCl$_4$	UCl$_3$
					UBr$_4$	UBr$_3$
					UI$_4$	UI$_3$
	UO$_3$	U$_3$O$_8$	U$_3$O$_7$	U$_4$O$_9$	UO$_2$	
Neptunium	NpF$_6$		NpF$_5$		NpF$_4$	NpF$_3$
					NpCl$_4$	NpCl$_3$
					NpBr$_4$	NpBr$_3$
						NpI$_3$
		Np$_3$O$_8$	Np$_2$O$_5$		NpO$_2$	
Plutonium	PuF$_6$			Pu$_4$F$_{17}$	PuF$_4$	PuF$_3$
					(PuCl$_4$)	PuCl$_3$
						PuBr$_3$
						PuI$_3$
					PuO$_2$	Pu$_2$O$_3$
Americium					AmF$_4$	AmF$_3$
						AmCl$_3$
						AmBr$_3$
						AmI$_3$
					AmO$_2$	Am$_2$O$_3$

state, NpO$_2$F occurs as well as MOF$_3$ for Np and probably Pu. Plutonium also forms PuOF.

The first preparations of all the hexafluorides were by direct reaction between metal and F$_2$. Later, it was found that O$_2$F$_2$ will react with any plutonium oxide, oxyfluoride, or lower fluoride to form PuF$_6$ at room temperature. Plutonium hexafluoride is a liquid at room temperature and is somewhat different in properties to uranium hexafluoride, the volatility of which permits separation of the U-235 and U-238 isotopes. PuF$_6$ is also a more powerful fluorinating agent than UF$_6$, and reacts with SF$_4$ producing SF$_6$, and also converts iodine to IF$_7$.

As elements, Np, Pu and Am closely resemble uranium. They are reactive metals which combine with most elements. Plutonium metal displays some rather interesting characteristics. At the melting point of liquid plutonium (640 °C), the solid is 2·4 per cent less dense than the liquid and therefore floats on the top; plutonium is one of only a handful of materials showing this property, the best known of which is water. The properties of metallic plutonium also differ somewhat from those of its neighbour, uranium. The difference in boiling points between Pu (3327 °C) and U (3818 °C) can be used as a means of efficiently separating these two metals. The metallic characteristics of plutonium (*viz.* metallic and electrical conductivities, and malleability) are also low, with the metallic conductivity of the pure metal being only one tenth that of silver. These properties have been explained by the way in which the 5*f* electrons are distributed. For the pre-plutonium metals, the electrons are delocalized, resulting in more metallic behaviour. whereas

the transplutonium metals show properties which are better explained by a localized, or non-metallic model. Plutonium is considered to show intermediate behaviour, which can be altered by changing the pressure.

All three elements, Np, Pu, and Am, form hydrides, but these resemble the lanthanide hydrides rather than uranium hydride. Stoichiometric hydrides, MH$_2$, are formed and also non-stoichiometric systems up to a composition, MH$_{2·7}$. The properties of the metals show increasing similarities to those of the lanthanide elements, the similarity is greatest for americium. Americium is the analogue of europium and americium(II) does exist, at least in a halide melt as AmX$_2$ (X = Cl, Br, I) and as AmO.

These three elements provide the most striking evidence, in their chemical properties, that the heaviest elements are, in fact, filling an *f* shell and are not a fourth *d* series. While thorium and uranium show similarities to hafnium and tungsten, neptunium and plutonium are clearly not the analogues of rhenium and osmium. Neptunium(VII) is strongly oxidizing while rhenium(VII) is stable and there is no indication as yet of an VIII state for plutonium analogous to osmium(VIII). Even the VI state of these two elements is relatively unstable and americium, with its stable III state, conforms to the expected pattern and links up with the chemistry of the succeeding elements.

Americium, as a result of its essentially monoenergetic radiation (it emits 5·4 and 5·5 MeV alpha and 0·6 MeV gamma radiations), has found a variety of industrial and even household uses. Many smoke detectors contain a small amount of Am-241 which serves to ionize the air in the detector. Industrial applications include the measurement of air humidity and the determination of the uniformity of thin films, by the passage of (poorly penetrating) alpha radiation through the material.

12.7 The heavier actinide elements

Although the first characterization was by tracer methods, the next four actinides are now available in sufficient quantity for their macroscopic chemistry to be studied. Curium is available in gram quantities, berkelium and californium in tens of milligrams, and einsteinium in tenths of a milligram. The oxygen and halogen compounds of these four elements are listed in Table 12.4.

The III state, the most stable for americium, is also the most stable state for these four actinides. Magnetic measurements have confirmed the f^7 configuration in curium (III) compounds. Curium (IV) does not exist in solution and, for example, attempts to oxidize Cm^{3+} with Na$_4$XeO$_6$ (which produces Pu(VII)) were unsuccessful. Thus curium, the f^7 actinide, shows a resemblance to gadolinium, the f^7 lanthanide, in its solution behaviour. Curium (IV) is found in the solid state in the compounds listed in Table 12.4, while a good range of curium (III) solid species are reported. For the next element, berkelium, the IV state, although strongly oxidizing, is rather more stable and is found in solution and in the solid state. In solution, green berkelium (III) species may be oxidized by bromate to yellow or orange berkelium (IV)

TABLE 12.4 Oxygen and halogen compounds of the heavier actinides

Curium	CmF_4	CmX_3 (X = F, Cl, Br, I)	
	$LiCmF_5$		
	$Na_7Cm_6F_{31}$		
	CmO_2	Cm_2O_3	
		$Cm(OH)_3$	
		$CmOX$ (X = F, Cl, Br, I)	
Berkelium	BkF_4	BkX_3 (X = F, Cl, Br, I)	
	$[BkF_6]^{2-}$		
	$Na_7Bk_6F_{31}$		
	$[BkCl_6]^{2-}$	$Cs_2NaBkCl_6$	
	BkO_2	Bk_2O_3	
		$Bk(OH)_3$	
		$BkOX$ (X = Cl, Br, I)	
Californium	CfF_4	CfX_3 (X = F, Cl, Br, I)	$[CfBr_2]$
	$Na_7Cf_6F_{31}$		CfI_2
	CfO_2	Cf_2O_3	
	[stable only	$Cf(OH)_3$	
	under high	$CfOX$ (X = F, Cl, Br, I)	
	O_2 pressure]		
Einsteinium	EsF_4(?)	EsX_3 (X = F, Cl, Br, I)	EsX_2
		Es_2O_3	(X = Cl, Br, I)
		$EsOX$ (X = Cl, Br, I)	

compounds whose oxidizing power is similar to that of cerium (IV). Californium (IV) is relatively unstable and the compounds listed in Table 12.4 have been prepared only as solids.

Both Bk(III) and Cf(III) are well established in solution and in solid compounds. Einsteinium does not form the IV state, the (III) state is stable, and the (II) state is well-established in the dihalides. Cf(II) compounds also exist but are less stable. These II states are strongly reducing (compare the potentials in Table 12.5).

As for the earlier actinides, coordination numbers above six are found. For example, the $[M_6F_{31}]^{7-}$ species have a complex structure of linked antiprisms.

^{249}Bk, while the half-life is only 320 days, is now readily available from spent fuels, allowing a more complete study of berkelium chemistry. It forms BkE (E = N, P, As, Sb), Bk_2E_3 (E = S, Se as well as O), both the cubic hydride BkH_{2+x} and the hexagonal BkH_3, and also the organometallic compounds $BkCp_2$ and $[BkCp_2Cl]_2$. In all this chemistry, there is a close parallel with the behaviour of terbium, the corresponding lanthanide.

The remaining four actinide elements have only been available in much smaller amounts and have been studied by

TABLE 12.5 Standard redox potentials for the later actinides (V)

	Es	Fm	Md	No	Lr
$M^{3+} + e = M^{2+}$	-1.55	-1.2	-0.2	$+1.4$	
$M^{2+} + 2e = M$	-2.20	-2.4	-2.4	-2.4	
$M^{3+} + 3e = M$	-1.98	-2.0	-1.7	-1.1	-2.1

carrier methods. In this way the behaviour of these elements has been elucidated by finding which one of a mixture of metal ions is accompanied by their characteristic radiation in the course of a chemical reaction. For example, Md accompanies Eu but not La when solutions of the M^{3+} ions are reacted with sodium amalgam and extracted. As this reaction forms Eu^{2+} but leaves La^{3+} unreduced, the formation of Md^{2+} is indicated. The scale of the experiments decreases very rapidly across these four elements. By the mid-eighties, it was possible to work on fermium with about 10^{11} atoms per experiment, for mendelevium using 10^6 atoms, for nobelium 10^3 and for lawrencium only about 10 atoms. Lawrencium chemistry, in particular, is so far based on only a handful of experiments each involving the detection (by their characteristic disintegrations) of a few atoms.

For Fm and Md, the (III) states are stable in solution and in solids, with a range of fermium solution chemistry already well explored. There is also a well-attested, strongly reducing Fm(II) state. The reduction potential ranges from about -1.6 V to -1.1 V depending on the medium. This Fm(II) state is more stable than Es(II) and this trend continues to Md(II) and No(II). Mendelevium(II) is mildly reducing while nobelium(II) is the stable state and No(III) is strongly oxidizing. Potentials are judged by reaction with known oxidizing and reducing agents, and also by polarography. The current best estimates are shown in Table 12.5. While the stability of No(II) could mark the completion of the f shell with 14 electrons in No^{2+}, these II states in the later actinides are markedly more stable than is found for the lanthanides. Thus the actinides differ from the lanthanides both in the range of higher oxidation states found at the beginning of the series and in the stability of lower states at the end. Evidence for Md^+ has been reported, but is subject to dispute.

Lawrencium behaves as expected for the $f^{14}d^1s^2$ configuration by forming only Lr(III)—$f^{14}d^0s^0$—and resisting oxidation or reduction. New lawrencium isotopes 260, 261, and 262, formed by bombarding ^{254}Es with neon ions, have half-lives up to 400 times longer than ^{256}Lr and ^{257}Lr which were the first ones prepared. This should allow Lr chemistry to be more fully established—particularly the intriguing possibility that the electron configuration is $5f^{14}7s^27p^1$ rather than $5f^{14}6d^17s^2$ because of relativistic effects (see section 16.13). However, recent studies using the ^{262}Lr isotope have suggested that any relativistic effects are insufficient to stabilize the Lr(I) state, since Lr(III) was resistant to reduction by the strong reducing agents Sm(II), Cr(II) and V(II). This places an upper limit of -1.56 V for the Lr(III)/Lr(I) couple in aqueous solution.

Recent work, particularly on the heavier elements, has thus filled in the picture of the actinide elements as a whole. The total view indicates that, while there are useful analogies with the chemistry of the lanthanides and of the 5d series transition metals, the chemistry of the actinides presents an individual pattern reflecting the relatively small difference between the 5f and 6d energies for all these elements.

Post-actinide elements are covered in section 16.12.

PROBLEMS

12.1 Reconsider question 8.5

12.2 Review the structures of compounds of the actinides where the coordination number is more than 6.

12.3 It used to be thought that the elements from actinium onwards formed the heaviest transition series. How far does the *chemical* evidence support this view

(a) in matching Th and U with Hf and W, or
(b) for the elements Th to Pu matching Hf to Os?

12.4 The other expected relationship is between lanthanides and actinides—which obviously fails for the early elements.

Discuss critically the parallels between Am to Lr and Eu to Lu including both chemical properties and numerical parameters.

12.5 As an alternative to 12.3 and 12.4, discuss the extent to which the chemistry of the elements actinium to lawrencium is unique in the Periodic Table. How far are parallels with the transition elements and lanthanide elements justified?

12.6 Give an account of the halides, oxyhalides and oxides of all elements in the Periodic Table with oxidation states of VI and above.

12.7 Outline the main oxidation states expected for elements 104 and 105, Rf and Ha. Speculate on the chemistry that might be found for element 126.

13 The Transition Metals: General Properties and Complexes

13.1 Introduction to the transition elements

The elements of the transition block are those with d electrons and incompletely filled d orbitals. The zinc Group, with a filled d^{10} configuration in all its compounds, is transitional between the d block and the p elements and is discussed later.

The Groups of the d block contain only three elements and correspond to the filling of the $3d$, $4d$ and $5d$ shells respectively. In between the $4d$ and $5d$ levels is interposed the first f level, the $4f$ shell, which fills after lanthanum. It has already been seen (Chapter 11) that the occupation of this level is accompanied by a gradual decrease in atomic and ionic radius from La to Lu and the total lanthanide contraction is approximately equal to the normal increase in size between one Period and the next. The result is that in the transition Groups there is the normal increase of about 20 pm in radius between the first and second members (filling the $3d$ and $4d$ shells), but the expected increase between the second and third members is just balanced by the lanthanide contraction so that these two elements are almost identical in size. This effect is illustrated by the radii given in Table 13.1, where the normal increases in the alkaline earths and in the scandium Group contrast sharply with the figures for the succeeding Groups.

As the pair of heavy elements have almost identical radii, and therefore very similar characteristics in other ways (e.g. ionization potentials, solvation energies, redox potentials, lattice energies), their chemistry is very similar. Thus each transition Group typically divides into two parts—the lightest element with its individual chemistry, and the pair of heavy elements with almost identical chemistries.

The three elements within each Group have a number of properties in common, of course. They show the same range of oxidation states in general, though these differ in relative stabilities. All the d and s electrons are involved in the chemistry of the earlier elements, so that the Group oxidation state is the maximum state shown. Once the d^5 configuration is exceeded, there is less tendency for all the d electrons to react, and the Group oxidation state is not shown by iron (though Os(VIII) and Ru(VIII) exist), nor by any elements of the cobalt, nickel, or copper Groups. Since, in the Group oxidation state, all the valency electrons are involved and since the properties of the elements then depend on valency and size only, there are similarities between the properties of Main Groups and Transition Groups of the same Group oxidation state. Thus sulfates and chromates, both MO_4^{2-}, are isostructural, while molybdenum and tungsten show higher coordination numbers with oxygen (especially six), just as does tellurium. The principal differences between the first and the heavier elements in a transition Group are those of size, and stability of oxidation states. The larger elements commonly show higher coordination numbers, and the higher oxidation states are more stable for the heaviest elements. Thus, chromium(VI) is strongly oxidizing while molybdenum and tungsten are stable in the VI state.

The effects of the lanthanide contraction die out towards the right of the d block. In the titanium and vanadium Groups, which immediately follow the lanthanides, the heavier elements are practically identical and their separation is more difficult than the separation of a pair of lanthanides. The next two Groups show clear differences between the two last elements, though these are still slight. In the platinum metals, the differences are increasing until, in the copper Group, there are few points of resemblance between silver and gold. Finally, in the zinc Group, the pattern approaches that in a p Group and zinc and cadmium resemble each other with mercury as the singular member.

Table 13.2 summarizes this discussion in terms of the oxidation states shown by the d elements, and the stabilities of these. The general behaviour is also illustrated by Tables 13.3 to 13.5 which give the oxides, fluorides, and other halides of the transition elements. A stable state will show all these compounds while a strongly oxidizing state will be more likely to have a fluoride than an iodide; similarly, a reducing state will be more likely to show a heavier halide than a fluoride. In fact, the highest oxidation state of a transition metal is typically obtained in an anionic complex, particularly oxides and fluorides. As a good illustration of this, the AgF_4^- and NiF_6^{2-} ions have been known for some

TABLE 13.1 Radii showing the effect of the lanthanide contraction (Pauling)

M^{2+}, pm	M^{3+}, pm	M^{4+} (calc), pm	atomic radii, pm		
Ca = 99	Sc = 70	Ti = 68	Ti = 132	V = 122	Cr = 117
Sr = 113	Y = 90	Zr = 74	Zr = 145	Nb = 134	Mo = 129
Ba = 135	La = 106	Hf = 75	Hf = 144	Ta = 134	W = 130

TABLE 13.2 Oxidation states of the transition elements

Group O.S.	Ti	V	Cr	Mn	Fe	Co	Ni	Cu	
	<u>IV</u>	V	VI (ox)	VII (ox)					d^0
	III (red)	<u>IV</u>	(V) (d)	(VI) (d)					d^1
	(II) (red)	III (red)	(IV) (d)	(V) (d)	(VI) (ox)				d^2
		(II) (red)	<u>III</u>	IV (ox)	(V) (ox)				d^3
	(0)	(I)	II (red)	(III) (ox)	(IV) (ox)				d^4
	(−I)	(0)	(I)	<u>II</u>	III	(IV) (ox)			d^5
		(−I)	0	(I)	<u>II</u>	<u>III</u>	(IV) (ox)		d^6
		(−II)	(−I)	0		<u>II</u>	(III) (ox)		d^7
			(−II)	(−I)	0	(I)	<u>II</u>	(III) (ox)	d^8
						0	(I)	<u>II</u>	d^9
							0	<u>I</u>	d^{10}

Group O.S.	Zr	Nb	Mo	Tc	Ru	Rh	Pd	Ag	
	<u>IV</u>	<u>V</u>	<u>VI</u>	<u>VII</u>	VIII (ox)				d^0
	(III) (red)	(IV) (d)	V	VI (d)	(VII) (d)				d^1
	(II) (red)	III (red)	IV	(V) (d)	VI				d^2
		(II) (red)	III	<u>IV</u>	(V)	(VI) (ox)			d^3
			(II)	(III)	IV				d^4
				?	<u>III</u>	IV			d^5
			0		II	<u>III</u>	IV		d^6
				0		II			d^7
					0	(I)	<u>II</u>	(III) (ox)	d^8
						0	(I?)	II (ox)	d^9
							0	<u>I</u>	d^{10}

Group O.S.	Hf	Ta	W	Re	Os	Ir	Pt	Au	
	<u>IV</u>	<u>V</u>	<u>VI</u>	<u>VII</u>	VIII (ox)				d^0
	(III) (red)	(IV) (d)	V	VI	(VII)				d^1
	(II) (red)	III (red)	<u>IV</u>	(V)	VI				d^2
		(II)	(III)	<u>IV</u>	(V)	(VI) (ox)			d^3
			(II)	III	IV	(V) (ox)	(VI) (ox)		d^4
				(II)	III	<u>IV</u>	(V) (ox)		d^5
			0	(I)	II	<u>III</u>	<u>IV</u>	V (ox)	d^6
				0	(I)	<u>(II)</u>	?		d^7
					0	(I)	<u>II</u>	<u>III</u>	d^8
						0	(I)		d^9
						(−I)	0	I	d^{10}

Notes: ox = oxidizing, red = reducing, unstable states bracketed, d = disproportionates, most stable state(s) for any given element underlined. State 0 usually in carbonyls and related complexes: the element itself is not counted as a 0 state here.

TABLE 13.3 Transition element oxides

			Oxidation state				Other compounds
+II	+III	+IV	+V	+VI	+VII	+VIII	
TiO	Ti$_2$O$_3$	$\underline{\text{TiO}_2}$ ZrO$_2$ HfO$_2$					
VO	V$_2$O$_3$	VO$_2$ NbO$_2$ (TaO$_2$?)	$\underline{\text{V}_2\text{O}_5}$ $\underline{\text{Nb}_2\text{O}_5}$ $\underline{\text{Ta}_2\text{O}_5}$				
CrO	Cr$_2$O$_3$	CrO$_2$ MoO$_2$ WO$_2$	Mo$_2$O$_5$ (W$_2$O$_5$?)	CrO$_3$ $\underline{\text{MoO}_3}$ $\underline{\text{WO}_3}$			
$\underline{\text{MnO}}$	Mn$_2$O$_3$ Re$_2$O$_3$*	MnO$_2$ TcO$_2$ ReO$_2$	(Re$_2$O$_5$)	TcO$_3$ ReO$_3$	Mn$_2$O$_7$ $\underline{\text{Tc}_2\text{O}_7}$ $\underline{\text{Re}_2\text{O}_7}$		Mn$_3$O$_4$ Also Tc$_2$S$_7$ Re$_2$S$_7$
FeO	Fe$_2$O$_3$ Ru$_2$O$_3$*	$\underline{\text{RuO}_2}$ $\underline{\text{OsO}_2}$		(RuO$_3$)* (OsO$_3$)*		RuO$_4$ $\underline{\text{OsO}_4}$	$\underline{\text{Fe}_3\text{O}_4}$
$\underline{\text{CoO}}$ RhO	(Co$_2$O$_3$)* $\underline{\text{Rh}_2\text{O}_3}$ Ir$_2$O$_3$	(CoO$_2$)* RhO$_2$ $\underline{\text{IrO}_2}$		(IrO$_3$)			Co$_3$O$_4$
$\underline{\text{NiO}}$ $\underline{\text{PdO}}$ (PtO)*	(Ni$_2$O$_3$)* (Pt$_2$O$_3$)*	(NiO$_2$)* (PdO$_2$)* $\underline{\text{PtO}_2}$		(PtO$_3$)*			Pt$_3$O$_4$
$\underline{\text{CuO}}$ AgO	(Ag$_2$O$_3$?) Au$_2$O$_3$						Cu$_2$O $\underline{\text{Ag}_2\text{O}}$ Au$_2$O

Most stable compounds underlined.
*Hydrous oxides of these states are reported.

TABLE 13.4 Transition element fluorides

			Oxidation state			Notes and other compounds
+II	+III	+IV	+V	+VI	+VII	
	TiF_3	$\underline{TiF_4}$				
	(ZrF_3)	$\underline{ZrF_4}$				
		$\underline{HfF_4}$				
VF_2	VF_3	$\underline{VF_4}$	VF_5			
	NbF_3	(NbF_4)	$\underline{NbF_5}$			Nb_6F_{15}
			$\underline{TaF_5}$			
CrF_2	$\underline{CrF_3}$	CrF_4	CrF_5	(CrF_6)		(CrF)
	MoF_3	MoF_4	MoF_5	$\underline{MoF_6}$		
		WF_4	$WF_5(d)$	$\underline{WF_6}$		
$\underline{MnF_2}$	MnF_3	MnF_4				
			TcF_5	$\underline{TcF_6}$		
		ReF_4	ReF_5	ReF_6	$\underline{ReF_7}$	
FeF_2	$\underline{FeF_3}$					
	$\underline{RuF_3}$	RuF_4	$\underline{RuF_5}$	RuF_6		
		$\underline{OsF_4}$	OsF_5	$\underline{OsF_6}$	(OsF_7)	*Note* No OsF_8
$\underline{CoF_2}$	CoF_3					
	RhF_3	RhF_4	(RhF_5)	RhF_6		
	IrF_3	$\underline{IrF_4}$	(IrF_5)	IrF_6		
$\underline{NiF_2}$	(NiF_3)	(NiF_4)				
$\underline{PdF_2}$	$[PdF_3]$	PdF_4				$PdF_3 = Pd^{2+}(PdF_6)^{2-}$
		$\underline{PtF_4}$	PtF_5	PtF_6		
$\underline{CuF_2}$						
AgF_2						Ag_2F, $\underline{AgF}$
	$\underline{AuF_3}.$		(AuF_5)			

Most stable compounds underlined. d = disproportionates. ( ) = compound well established but unstable at room temperature.

Structures: MF_2 rutile or distorted rutile
MF_3 ReO_3, i.e. octahedra linked through all corners; M = Au, linked planar AuF_4 units with longer bonds to next layers completing a distorted octahedron
MF_4 octahedra sharing edges; M = Zr, square antiprisms linked through all F
MF_5 octahedra sharing corners; chains or closed rings
MF_6 octahedron
MF_7 pentagonal bipyramid

TABLE 13.5 Transition element halides*

	Oxidation state				
+II	+III	+IV	+V	+VI	Notes
TiX_2	TiX_3	TiX_4			
(ZrX_2)	ZrX_3	$\underline{ZrX_4}$			ZrCl
$HfCl_2, Br_2$	$HfCl_3, Br_3$	$\underline{HfX_4}$			also HfCl?
VX_2	$\underline{VX_3}$	VCl_4, Br_4			VBr_4 very unstable
$(NbBr_2)$	NbX_3	NbX_4	$\underline{NbX_5}$		Nb_6X_{14}, Ta_6X_{14}
$TaCl_2$	$TaCl_3, Br_3$	TaX_4	$\underline{TaX_5}$		all $= (M_6X_{12})^{2+}(X^-)_2$(a)
					$Nb_3Br_8, Nb_3I_8, Nb_6I_{11}$
CrX_2	$\underline{CrX_3}$				
MoX_2	MoX_3	MoX_4	$MoCl_5$		MoX_2 and.WX_2
WX_2	WX_3	WX_4	WCl_5, Br_5	WCl_6, Br_6	$= (M_6X_8)^{4+}X_4^-$
$\underline{MnX_2}$					
		$TcCl_4$		$(TcCl_6)$?	
(ReX_2)	ReX_3	$\underline{ReX_4}$	$ReCl_5, Br_5$	$(ReCl_6)$?	$ReCl_3, ReCl_4$ are trimers
					ReX_2 in complexes only
$\underline{FeX_2}$	$FeCl_3, Br_3$				
	$\underline{RuX_3}$	$RuCl_4$			
		OsX_4	$OsCl_5$		$OsX_{3.5}$
$\underline{CoX_2}$					
	$\underline{RhX_3}$				
$(IrCl_2)$?	$\underline{IrX_3}$	$(IrCl_4)$			
$\underline{NiX_2}$					Platinum trihalides
PdX_2					may be mixtures of
$\underline{PtX_2}$	PtX_3?	$\underline{PtX_4}$			$Pt(II) + Pt(IV)$.
					$PtCl_2$ structure consists
					of Pt_6Cl_{12} units
$CuCl_2, Br_2$					Also $\underline{CuX}$
					$\underline{AgX}$
	$AuCl_3, Br_3$				AuCl, I

Most stable compounds underlined.

*The symbol X is used when the chloride, bromide and iodide all occur.

(a) All species $M_6X_{12}^{n+}(X^-)_n$ for $n = 2, 3$ and 4 occur for $M = Nb$, $X = Cl$; $M = Ta$, $X = Cl, Br$. $Nb_6I_{11} = (Nb_6I_8)^{3+}(I^-)_3$

time, but the parent binary fluorides AgF_3 and NiF_4 have only recently been synthesized as highly reactive, polymeric solids. A promising route for the synthesis of these high oxidation state compounds involves fluoride-ion removal from the anionic fluoro-complex by a powerful fluoride ion acceptor such as AsF_5.

This division between the lighter transition elements and the two heavier Periods is quite marked and is reinforced in practice by the relative inaccessibility of most of the heavier elements. The latter have therefore been less fully studied, especially the less available member of the pair (for example Hf, Nb, Tc). In addition, there are strong horizontal resemblances, especially among ions of the same charge, and horizontal trends in properties, with increasing number of d electrons in a given oxidation state, that make it convenient

to divide the discussion of the transition block into two sections, one on the first row elements (Chapter 14) and one (Chapter 15) on the heavier elements of the second and third rows. Selected transition metal topics of active current interest are reviewed in Chapter 16.

The pattern of oxidation state stabilities outlined above is complex and there are exceptions to most of the generalizations which can be made about it. The picture is complicated by the use of the term 'stability' in a number of different senses. In the most general sense, it is used to mean that a compound exists in air at around room temperatures: that is, that it is thermally stable at room temperature, that it is not oxidized by air, and that it is not hydrolysed, oxidized or reduced by water vapour. In turn, terms such as thermal stability, may cover a number of processes. Thus a higher oxide such as MO_2 might decompose thermally to $M + O_2$ or to $MO + \frac{1}{2}O_2$ and the free energy change of each process would have to be evaluated before conclusions could be drawn about the stability. Similarly, a compound may exist for a long time at room temperature, not because it is thermodynamically stable with respect to decomposition, but because the decomposition process occurred at a negligible rate. Thus, whether a compound can be kept 'in a bottle' depends on a wide variety of thermodynamic and kinetic factors.

Despite these difficulties, some attempts are being made to examine, predict, and rationalize stabilities, although most treatments to date are either limited in scope or are empirical. One example of a general approach which may be quoted is that of Sheldon who proposes that the preferred oxidation state of a transition element (defined as that which, in simple binary compounds such as the oxides or halides, is the most stable under normal laboratory conditions) is related to the quantity $rH/40$. Here, r is half the interatomic distance and H the heat of atomization of the metal—both well-known quantities. This expression leads to the following predictions for the most stable oxidation states of the transition elements:

VI   for W, Re, Os, Ir
V   for Nb, Mo, Tc, Ru, Ta
IV   for Ti, V, Zr, Rh, Hf, Pt
III for Cr, Fe, Co, Ni, Pd, Au
II   for Mn, Cu, Ag.

If these predictions are compared with Tables 13.2 to 13.5, it will be seen that they are surprisingly accurate for such a simple formula. The only really poor predictions are those for nickel, palladium and silver where the states predicted to be stable are non-existent or very unstable.

A much more fundamental and searching analysis is that, discussed in the next Chapter, on the stability of trihalides of the first row of transition elements, but this is limited to one particular decomposition reaction. Further work on rigorous thermodynamic analysis should lead to a greatly increased understanding of stabilities and Periodic trends.

13.2 The transition ion and its environment: ligand field theory

In the discussion of the energy levels of an atom given in the earlier chapters, the levels of a given p, d or f set were treated as of equal energy. This is true of isolated atoms, or of those in an electric field which is spherically symmetrical around the atom, but is not true when the atom lies in an unsymmetrical field. This may be readily seen by considering an atom which is strongly coordinated to two other groups in a linear configuration. If these groups lie in the $\pm z$ directions, the orbitals which point along the z axis will lie in the field of these ligands and be perturbed by them more than orbitals lying in other directions. As ligands are regions of negative charge (they coordinate through lone electron pairs and also have negative charges or the negative end of a dipole directed towards the central atom or ion) the z-directed orbitals on the central atom will be in a region of higher negative field than the non-z orbitals and electrons will avoid entering them as far as possible. This means that such orbitals as p_z, d_{z^2} and, to a lesser extent, d_{xz} and d_{yz}, are of higher energy than the remaining ones and the degeneracy of the p and d set is split in such a z-directed field in the manner shown in Figure 13.1. The size of such a splitting will depend on the size of the ligand field and this, in turn, depends on the distance to the ligand and thus on the intensity of the attraction between the central atom and the ligand. Such ligand fields occur in all chemical environments. Their effects are generally negligible, except for d orbitals, either because the fields are small (as in the case of f orbitals) or because the orbitals are equally populated as is usually the case for p orbitals. Thus our discussion of ligand fields is confined to d element chemistry.

The strength of the ligand field effect is marked in the case of the transition elements as their ions are small, and the M^{2+} and M^{3+} ions are thus centres of high charge density and are strongly coordinated by lone pair donors such as

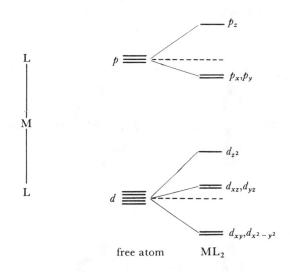

FIGURE 13.1 *Energy level diagram for a linear field*
The z axis is taken as the direction of the coordinated groups. The ligands are regions of high negative charge density, so orbitals on the central atom with components in the $\pm z$ direction are less stable than those with no such component. Among the d orbitals, the d_{z^2} orbital is most strongly destabilized as it has the greatest density in the z direction. The atom levels are those of an atom in a spherically symmetrical field of the average effect of the ligands.

water or ammonia. The case of the first row of the transition block is particularly interesting as the energy differences introduced by ligand field splittings are of the same size as the various energy losses involved in electron pairing. The effects of the fields of different ligands on ions of differing numbers of d electrons, and in different environments, show up in the numbers of unpaired d electrons. These are readily determined by magnetic measurements. Ligand field effects are also seen in a number of other properties such as ionic radii, lattice energies, reaction mechanisms, and electronic spectra, but it was the magnetic effects which first attracted attention and gave rise to the current interest in *ligand field theory*. The application of the theory may be examined in more detail in the case of octahedral complexes of the first row transition elements. This is the commonest geometry shown by these elements, in solution, in solvated or coordinated individual ions, and also in solids such as the oxides or fluorides. In later sections, the extension of the theory to other coordination numbers and to the heavier elements will be discussed.

13.3 Ligand field theory and octahedral complexes

Regular six-coordination is most readily pictured by placing the ligands at the plus and minus ends of the three coordinate axes. In the xy plane, the positions of such ligands relative to the d orbitals is shown in Figure 13.2a, while the corresponding diagrams for the xz and yz planes are shown in Figures 13.2b and c. In the xy plane, the orbital d_{xy} lies between the ligands while $d_{x^2-y^2}$ points directly at the ligands. An electron in the $d_{x^2-y^2}$ orbital is therefore most affected by the field of the ligands and is raised in energy relative to an electron in the d_{xy} orbital. Similarly, electrons in d_{z^2} are less stable than ones

in d_{xz} or d_{yz} (Figure 13.2b and c). If the alignments of the three orbitals, d_{xy}, d_{xz} and d_{yz}, relative to the ligands are compared, it will be seen that these are identical. It follows that in the full three-dimensional case, electrons in these three orbitals are identical in energy and are stabilized relative to the other two. Electrons in orbitals d_{z^2} and $d_{x^2-y^2}$ are also identical in energy and are destabilized. (It is easier to accept this if it is recalled that d_{z^2} is compounded of two orbitals similar to $d_{x^2-y^2}$.) The combined energy level diagram is therefore composed of two upper orbitals, of equal energy, and three lower orbitals, which are also degenerate (Figure 13.3). The energy zero is conveniently taken as the weighted mean of the energies of these two sets of orbitals; the lower trio are thus stabilized by $-2/5\Delta E$ while the upper pair are destabilized by $3/5\Delta E$, where ΔE is the total energy separation. The e_g and t_{2g} symbols are symmetry labels arising from Group Theory and are now the most commonly used symbols. It may help in remembering them that e signifies a doubly-degenerate state and t, a triply-degenerate one.

Consider the case of a d^1 system in an octahedral field, for example the hydrated titanium(III) ion, $Ti(H_2O)_6^{3+}$. The

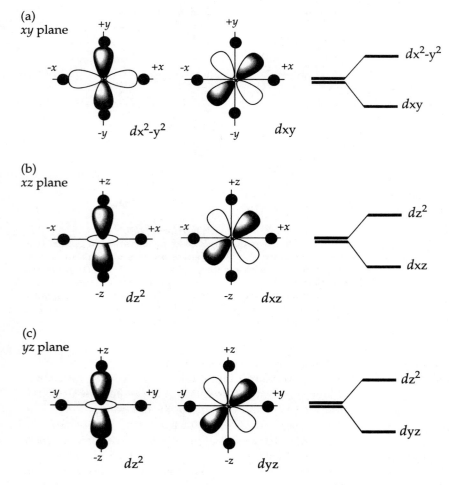

(a) xy plane

dx^2-y^2 dxy

(b) xz plane

dz^2 dxz

(c) yz plane

dz^2 dyz

● = ligand donor atom

FIGURE 13.2 *Positions of ligands and d orbitals in an octahedral complex:* (a) *the xy plane*, (b) *the xz plane*, (C) *the yz plane*

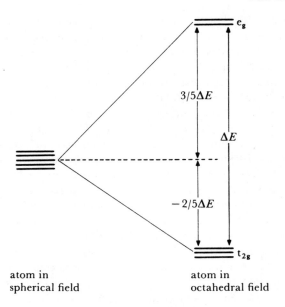

FIGURE 13.3 *Energy level diagram for the d orbitals in an octahedral field*
Note that electrons in the *d* levels of the free atom would be more stable than when the atom is in a spherical field. The energy gap ΔE is often labelled $10\,Dq$ or Δ_{oct}.

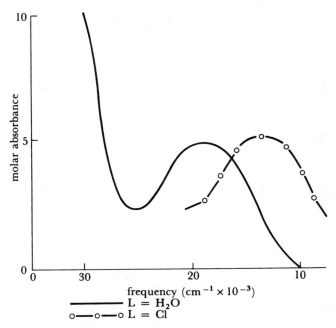

FIGURE 13.4 *Representation of the electronic spectra of* Ti(III) *complexes*
The full line (————) is a (slightly simplified) representation of the spectrum of $Ti(H_2O)_6^{3+}$ with a maximum at $20\,300\ cm^{-1}$. The broken line (–o–o–o–) represents the spectrum of $TiCl_6^{3-}$, with a maximum at $13\,000\ cm^{-1}$. The steeply-rising portion of the curve for the hydrate is the edge of the strong, allowed, charge-transfer transition in the ultraviolet.

orbitals are split as in Figure 13.3, and the single *d* electron naturally enters the lowest available one, here one of the t_{2g} set. In doing so it gains energy, equal to $-2/5\Delta E$, relative to the energy it would have had if the octahedral splitting had not occurred. In this case the energy gain is equal to about 90 kJ mol^{-1}. This energy gain, relative to the case of five equal *d* orbitals, is termed the *ligand field stabilization energy*, or, since it was first remarked in crystals, the crystal field stabilization energy or CFSE. This was first observed when it was found that calculations of the lattice energy of simple transition element oxides and fluorides, by the electrostatic method which was so successful for *s* element salts, gave answers which did not agree with the experimental values. Including the effect of the crystal field on the *d* orbital energies (section 13.7) led to full agreement. The CFSE is an additional energy increment to the system which has to be added to the other attractive and repulsive energies, both in solids and in calculation of solvation energies and the like.

The size of ΔE is most readily measured spectroscopically by observing the energy of the electronic transitions between the t_{2g} and e_g orbitals. The energy usually lies in the visible or near ultra-violet region of the spectrum and it is such $d-d$ transitions which are responsible for the colours of most transition metal compounds. The magnitude of ΔE depends on the ligand and on the nature of the transition metal ion. One of the simplest examples is that of titanium(III) complexes, where the configuration is d^1. The transition from the t_{2g} to e_g level of the single electron, gives rise to a single absorption band in the visible region (Figure 13.4). The position of this band gives the size of ΔE, compare the discussion in sections 7.5 and 7.6. When the ligand in Ti^{III}L$_6$

changes from L = H_2O to L = Cl$^-$, the position of the absorption band moves to lower energy, thus ΔE for Cl$^-$ is smaller than for H_2O. By examining a whole series of complexes with different ligands, L, the size of ΔE for each ligand may be determined, both for titanium(III) and for other metals in various oxidation states. It is found that the order of increasing ligand effects is approximately constant from one transition ion to the next and increases in the order:

$$I^- < Br^- < Cl^- < F^- < H_2O < NH_3 < en < NO_2^-$$
$$< CN^- \quad (en = ethylenediamine).$$

The ligand field increases by a factor of approximately two from halide to cyanide. A large number of other ligands are to be fitted into this series, of course, but these are representative ligands, and cyanide has the strongest ligand field of all common ligands. The series is termed the *spectrochemical series*. The main effects of the transition metal ion are those due to charge and to **Period**. The splitting increases by about 30 per cent between members of the same Group in successive Periods, and there is an increase of roughly 50 per cent on going from the divalent to the trivalent ion of any element. These trends are illustrated by the values shown in Table 13.6. The $d-d$ transitions for configurations other than d^1 are rather more complicated and are discussed in section 13.11.

In the case of titanium(III), there is no ambiguity as to the location of the *d* electron, and this is so for the d^2 and d^3 configurations also. The *d* electrons enter the t_{2g} set of

TABLE 13.6 Values of the ligand field splitting ΔE in octahedral complexes

Ion		$6Cl^-$	$6H_2O$	$6NH_3$	$6CN^-$
			(ΔE, kJ mol^{-1})		
Ti^{3+}	$3d^1$		243		
V^{3+}	$3d^2$		226		
Cr^{3+}	$3d^3$	163	213	259	314
Mn^{3+}	$3d^4$		(250)		
Fe^{3+}	$3d^5$		162		
Co^{3+}	$3d^6$		222	296	406
Mo(III)	$4d^3$	230			
Rh(III)	$4d^6$	243	322	406	
Ir(III)	$5d^6$	297			
Pt(IV)	$5d^6$	347			
V^{2+}	$3d^3$		151		
Cr^{2+}	$3d^4$		(170)		
Mn^{2+}	$3d^5$		92	approx.100	
Fe^{2+}	$3d^6$		126		393
Co^{2+}	$3d^7$		113	121	
Ni^{2+}	$3d^8$	88	100	130	
Cu^{2+}	$3d^9$		(150)	(180)	

Values for d^4 and d^9 configurations are approximate because of distortion in these octahedral complexes.

orbitals with parallel spins to give CFSE values of $-4/5\Delta E$ for d^2 and $-6/5\Delta E$ for d^3. However, in the case of d^4, two alternative configurations are possible. The first three electrons enter the three t_{2g} orbitals while the fourth may either remain parallel to the first three, thus producing maximum exchange energy, and enter the higher-energy e_g level, or it may pair up with one of the electrons already present in the t_{2g} level and produce maximum crystal field stabilization

energy. The first configuration is termed the *high-spin* or *weak field* configuration, while the arrangement with the paired electrons is the *low-spin* or *strong field* case. In the case of d^4, the CFSE for the $t_{2g}^3 e_g^1$ configuration is $-6/5\Delta E + 3/5\Delta E = -3/5\Delta E$, while the CFSE for the low-spin t_{2g}^4 configuration is $-8/5\Delta E$, so that the adoption of the low-spin configuration means the gain of ΔE in excess of the CFSE in the high-spin configuration. On the other hand, the exchange energy of four parallel electrons is $6K$ (see section 8.2) while that of the three parallel electrons in the low-spin configuration is only $3K$. Which configuration is actually adopted therefore depends on the relative sizes of ΔE and K. The K values are difficult to determine but remain approximately constant for atoms of the same quantum shell. In the case of the first transition series the loss of exchange energy usually lies within the range of values found for ΔE. Thus, for any configuration where alternative electronic arrangements are possible, there will be a particular value of ΔE where the change from high-spin to low-spin values takes place. A large value of ΔE obviously favours the low-spin arrangement, hence the alternative name of strong field configuration. Alternative electronic configurations are possible for d^4, d^5, d^6 and d^7 ions in octahedral complexes. Table 13.7 lists the values of the CFSE and exchange energy for all the d configurations, while Table 13.8 shows the differences for high- and low-spin configurations in the states d^4 to d^7. Notice that there are no examples known of intermediate configurations such as $t_{2g}^4 e_g^1$. In all cases the electrons are either paired as far as possible or parallel as far as possible. Table 13.9 gives the approximate magnetic moments, based on the 'spin-only' formula for each configuration. The 'spin-only' formula is usually a good approximation for transition metal ions although orbital coupling occurs to a small extent in most cases and is marked

TABLE 13.7 Crystal field stabilization and exchange energies in octahedral configuration

Number of d electrons	Electron configuration t_{2g}			e_g		CFSE ΔE	Exchange energy' K
1	↑					$-2/5$	0
2	↑	↑				$-4/5$	1
3	↑	↑	↑			$-6/5$	3
4 high-spin	↑	↑	↑	↑		$-6/5+3/5$	6
4 low-spin	↑↓	↑	↑			$-8/5$	3
5 high-spin	↑	↑	↑	↑	↑	$-6/5+6/5$	10
5 low-spin	↑↓	↑↓	↑			$-10/5$	3+1
6 high-spin	↑↓	↑	↑	↑	↑	$-8/5+6/5$	10
6 low-spin	↑↓	↑↓	↑↓			$-12/5$	3+3
7 high-spin	↑↓	↑↓	↑	↑	↑	$-10/5+6/5$	10+1
7 low-spin	↑↓	↑↓	↑↓	↑		$-12/5+3/5$	6+3
8	↑↓	↑↓	↑↓	↑	↑	$-12/5+6/5$	10+3
9	↑↓	↑↓	↑↓	↑↓	↑	$-12/5+9/5$	10+6
10	↑↓	↑↓	↑↓	↑↓	↑↓	$-12/5+12/5$	10+10

Exchange energies shown separately for the parallel and antiparallel sets.

TABLE 13.8 Balance of exchange and crystal fields energies for states of alternative configurations in the octahedral field

Number of d electrons	Gain in CFSE of low-spin relative to high-spin configuration	Loss in exchange energy of low-spin relative to high-spin configuration
4	ΔE	$3K$
5	$2\Delta E$	$6K$
6	$2\Delta E$	$4K$
7	ΔE	$2K$

TABLE 13.9 'Spin only' magnetic moments for octahedral arrangements

Number of d electrons	Magnetic moment, Bohr magnetons	
	High-spin	Low-spin
1	1·73	
2	2·83	
3	3·87	
4	4·90	2·83
5	5·92	1·73
6	4·90	0·00
7	3·87	1·73
8	2·83	
9	1·73	
10	0·00	

'Spin only' moment equals $2\sqrt{[S(S+1)]}$, where $S = \frac{1}{2}n$ = number of unpaired spins × spin quantum number.

in the cases of Co^{2+} and Co^{3+}. Apart from such exceptions, experimental magnetic moments usually agree with those calculated by the 'spin-only' formula (section 7.11) to within ten per cent, quite close enough to distinguish high-spin from low-spin configurations.

The crystal field stabilization energy and the possibility of alternative electronic configurations are the main phenomena which the theory of bonding in compounds of d elements has to treat. This theory may be formulated in two independent ways, as an electrostatic theory or as a molecular orbital theory. The results of these two approaches are very similar so that either may be used as seems most appropriate and the combined theory is termed the *ligand field theory*. The term crystal field theory is sometimes reserved for specific application to the electrostatic version, but usages vary among different authors.

These approaches will be examined briefly in turn. In the electrostatic approach, the bond energy is held to arise purely from the electrostatic attractions between the central ion and the charges, dipoles and induced dipoles on the ligands, with the repulsions between dipoles, induced dipoles, and charges, on different ligands taken into account.

To these forces is to be added the CFSE arising from the d electron arrangement in d orbitals split by the ligand field. The situation is similar to the electrostatic treatment of ionic crystals with the addition of the crystal field stabilization energy. As no covalent bonds enter into this treatment, the energy level diagram is simply that of the atomic energy levels in the transition element, of which the vital part is the d electron diagram as shown in Figure 13.3. This approach has the major advantage that the electrostatic calculations can actually be carried out, without drastic approximations, so that energies of formation and reaction mechanisms or stabilities can be predicted. The disadvantage of the theory is that it neglects the clear evidence that some covalent bonding does occur in transition metal compounds and it can throw no light on those cases, such as nickel carbonyl, where the central element is in a low, zero, or even negative oxidation state when the electrostatic forces would be weak or non-existent.

Covalent bonding is incorporated in the molecular orbital theory. This constructs seven-centred (in an octahedral complex) molecular orbitals from the six ligand orbitals holding the lone pairs which are to be donated to the central atom, together with six orbitals of suitable energy and symmetry, on the central atom. In a transition metal of the first series, these six atomic orbitals are the $3d_{x^2-y^2}$, $3d_{z^2}$, $4s$, and the three $4p$ orbitals. These are the metal orbitals directed towards the ligands and of the right energy. The bonding molecular orbitals from these combinations of atomic orbitals with ligand orbitals are shown in Figure 13.5; the numerical constants are chosen to weight the contributions of the ligand orbitals so that each adds up to unity. The antibonding orbitals corresponding to these six are those with the sign reversed between the central orbitals and the ligand combinations. Some of these are shown in Figure 13.6, which also illustrates that orbitals such as d_{xy} cannot form sigma bonds with any ligand combination. It will be noticed that these are delocalized polycentric sigma orbitals, similar to those discussed in section 4.7. The six atomic orbitals combine with the six ligand orbitals to form six bonding molecular orbitals and six antibonding molecular orbitals, while the remaining three $3d$ orbitals are non-bonding. The energy level diagram of the molecular orbitals in the octahedral complex is shown in Figure 13.7.

Let us place the 12 electrons from the ligand lone pairs in the six bonding orbitals. Then the nonbonding t_{2g} set of $3d$ orbitals and the antibonding e_g^* pair are the next five orbitals available to accommodate the d electrons on the metal. Thus the descriptions in the electrostatic and molecular orbital theories are very similar. Both theories produce five orbitals in a lower set of three and an upper set of two, separated by ΔE, to accommodate the d electrons. In the electrostatic theory these sets are, respectively, the atomic d_{xy}, d_{yz}, and d_{zx} orbitals and the d_{z^2} and $d_{x^2-y^2}$ atomic orbitals, while, in the molecular orbital theory, the lower set are the same atomic orbitals and the upper set are antibonding molecular orbitals composed of the d_{z^2} and $d_{x^2-y^2}$ atomic orbitals with ligand orbital contributions.

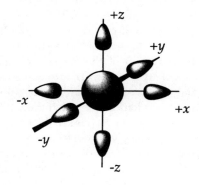

$$\psi_{a_{1g}} = \phi_s + 1/\sqrt{(6)}\,(l_x + l_{-x} + l_y + l_{-y} + l_z + l_{-z})$$

(a)

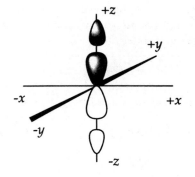

$$\psi_{t_{1u}} = \phi_{p_z} + 1/\sqrt{(2)}\,(l_z - l_{-z})$$

(d)

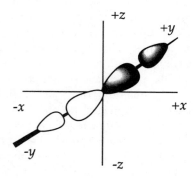

$$\psi_{t_{1u}} = \phi_{p_x} + 1/\sqrt{(2)}\,(l_x - l_{-x})$$

(b)

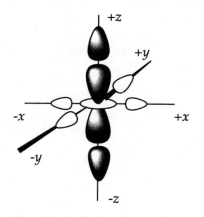

$$\psi_{e_g} = \phi_{d_{z^2}} + 1/(2\sqrt{3})\,(2l_z + 2l_{-z} - l_x - l_{-x} - l_y - l_{-y})$$

(e)

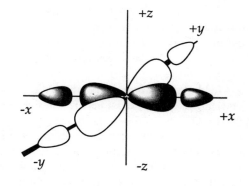

$$\psi_{t_{1u}} = \phi_{p_y} + 1/\sqrt{(2)}\,(l_y - l_{-y})$$

(c)

$$\psi_{e_g} = \phi_{d_{x^2-y^2}} + \tfrac{1}{2}(l_x + l_{-x} - l_y - l_{-y})$$

(f)

FIGURE 13.5 *The six bonding molecular orbitals formed by the six ligand orbitals and the s, p, and d_{z^2} and $d_{x^2-y^2}$ orbitals on the central atom in an octahedral complex*

The symbols a_{1g}, t_{1u} etc. are symmetry-indicating labels. Here they are used as a convenient way of distinguishing the orbitals. Note that there are three levels in sets labelled t and two in sets labelled e.

(a)

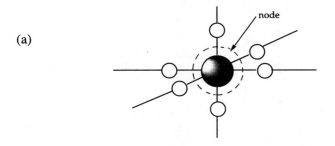

$$\psi^*_{a_{1g}} = \phi_s - 1/\sqrt{(6)}\,(l_x + l_{-x} + l_y + l_{-y} + l_z + l_{-z})$$

(b)

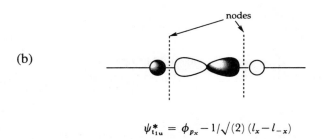

$$\psi^*_{t_{1u}} = \phi_{p_x} - 1/\sqrt{(2)}\,(l_x - l_{-x})$$

(c)

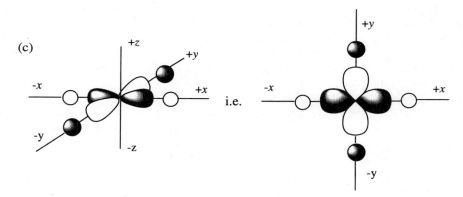

i.e.

$$\psi^*_{e_g} = \phi_{d_{x^2-y^2}} - \tfrac{1}{2}(l_x + l_{-x} - l_y - l_{-y})$$

(d)

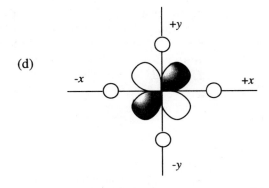

$$\phi_{d_{xy}} - \text{no } \sigma\text{-bonding combination possible}$$

FIGURE 13.6 *Some non-bonding and antibonding combinations in an octahedral complex*

Figures (a) to (c) are antibonding combinations of the *s*, one *p*, and one *d* orbital, with ligand orbitals, corresponding to some of the bonding orbitals in Figure 13.5. Figure (d) shows how the d_{xy} orbital is wrongly aligned to form any sigma bond; the other two t_{2g} orbitals are similarly non-bonding. The labels a^*_{1g} etc. are used to distinguish the antibonding orbitals corresponding to the bonding orbitals of Figure 13.5.

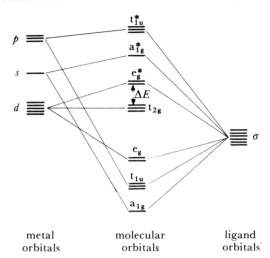

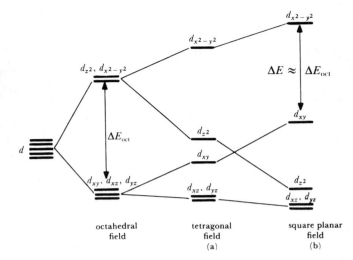

FIGURE 13.7 *Energy level diagram for the bonding, non-bonding, and anti-bonding molecular and atomic orbitals in a sigma-bonded octahedral complex* The levels are labelled as in Figure 13.5 and 13.6.

FIGURE 13.8 *Effect on orbital energies of removing ligands on the z axis, starting from an octahedron:* (a) *the tetragonal field with non-equivalent z ligands,* (b) *the square planar field with no z ligands*

Thus in their essential description of the varying magnetic properties and $d-d$ transitions, these two theories are identical. The molecular orbital theory has the advantage that it is easily extended to include π-bonding and it gives a prediction of the alteration of energy levels in such a case. It is also more useful for the interpretation of spectra as it provides information, not only about the d levels, but also about the higher energy antibonding orbitals to which excitations occur when higher energy quanta are absorbed. On the other hand, the molecular orbital theory suffers from the disadvantage of all wave mechanics, that it is impossible to calculate bond energies, heats of formation and the like directly.

In practice, these two theories may be used interchangeably, as most convenient. Both rely on experimental data to fix the energy levels; for example, ΔE is usually determined spectroscopically.

13.4 Coordination number four

Two configurations other than octahedral are common among the elements of the first transition series. Both involve four-coordination and are the tetrahedral and square planar configurations. The square planar configuration may be considered as derived from the octahedral one by removing the ligands on the z-axis. An elongation of the M—L distances on the z-axis leads to a decrease of the interaction between the ligand field and those metal orbitals with components in the z direction. The energy levels therefore split as indicated in Figure 13.8a with the d_{z^2} level and the d_{xz} and d_{yz} levels falling below the others. The $d_{x^2-y^2}$ and d_{xy} levels rise slightly as the metal-ligand distances in the xy plane shorten a little because of the decrease in repulsion from the z ligands. Such an intermediate case corresponds to elongation of the metal-ligand distances on the z-axis, to unsymmetric substitution on the z-axis in a complex such as MX_4Z_2, (both of which are tetragonal distortions) or to the

case of the five-coordinated square pyramidal configuration. If the z ligands are removed completely to give the square planar configuration, the energy level diagram of Figure 13.8b results. Here, the d_{z^2} and d_{xy} levels have crossed over. Notice that, as the configuration in the xy plane is similar to that in the octahedron, the energy separation between $d_{x^2-y^2}$ and d_{xy} remains practically the same as the octahedral ΔE.

In the tetrahedral arrangement, the d orbitals are split into a lower set of two (called e in this symmetry) and an upper set of three (t_2). No orbital points directly at a ligand in the tetrahedral case, but the d_{xy} type lies closer to the ligand than $d_{x^2-y^2}$ or d_{z^2}. Figure 13.9 shows that the distances are as half the side of a cube compared with half the face diagonal. This lack of direct interaction between orbitals and ligands in the tetrahedral configuration reduces the magnitude of the crystal field splitting by a geometrical factor equal to 2/3, and the fact that there are only four ligands instead of six reduces the ligand field by another 2/3. The total splitting in the tetrahedral case is thus approximately $2/3 \times 2/3 = 4/9$ of the octahedral splitting. In theory, alternative electron configurations are possible for the cases d^3, d^4, d^5 and d^6 in the tetrahedral field but the gain in

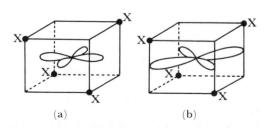

FIGURE 13.9 *Alignment of the d orbitals and the ligands in a tetrahedral complex:* (a) *the* $d_{x^2-y^2}$ *orbital,* (b) *the* d_{xy} *orbital*

TABLE 13.10 CFSE values in a tetrahedral field

Number of d electrons	Electronic configuration e_g		t_{2g}			CFSE (as ΔE_{tetr}) (assuming $\Delta E_{tetr} = 4/9\,\Delta E_{oct}$)	(as ΔE_{oct})
1	↑					$-3/5$	-0.27
2	↑	↑				$-6/5$	-0.53
3	↑	↑	↑			$-6/5+2/5$	-0.36
4	↑	↑	↑	↑		$-6/5+4/5$	-0.18
5	↑	↑	↑	↑	↑	$-6/5+6/5$	0.00
6	↑↓	↑	↑	↑	↑	$-9/5+6/5$	-0.27
7	↑↓	↑↓	↑	↑	↑	$-12/5+6/5$	-0.53
8	↑↓	↑↓	↑↓	↑	↑	$-12/5+8/5$	-0.36
9	↑↓	↑↓	↑↓	↑↓	↑	$-12/5+10/5$	-0.18
10	↑↓	↑↓	↑↓	↑↓	↑↓	$-12/5+12/5$	0.00

CFSE is reduced by the smaller size of the splitting, and the loss of exchange energy is never counterbalanced. Thus all tetrahedral complexes are high-spin. The CFSE values for these high-spin configurations are given in Table 13.10.

If Table 13.10 is compared with Table 13.7, it will be seen that the CFSE in an octahedral configuration is always greater than that in the corresponding tetrahedral configuration, except in the cases of d^0, d^5 and d^{10} where both values are zero. The configurations with the next smallest CFSE loss in the tetrahedral field compared to the octahedral field, are d^1 and d^6 if the octahedral state is high-spin. The adoption of four-coordination rather than six-coordination is expected, in general, to be accompanied by loss of energy of formation as only four interactions occur instead of six. In addition, there is commonly a loss of CFSE as well.

These considerations, and the data for the CFSE values, allow a prediction of the most probable cases in which four-coordination will be found for transition elements of the first series. The tetrahedral configuration is expected to be unfavourable compared with the octahedral one, except in the case of large ligands with low positions in the spectrochemical series, or in the ions with 0, (1), 5, (6), or 10 d electrons. The larger ligands will experience steric hindrance to formation of six-coordinated complexes and the interactions will also be reduced due to the increased metal-ligand distances. The low position in the spectrochemical series ensures that any loss of CFSE is not too serious due to the low intrinsic value of ΔE (similarly, a low charge on the transition metal ion affects the ΔE value so divalent ions should form tetrahedral complexes more readily than trivalent ones, and *a fortiori* for lower charges). Finally, the d configurations listed are those with lowest CFSE differences. In practice, tetrahedral complexes are typically formed by the halides (except fluoride) and related ligands.

The case of the square-planar configuration is rather different. Because the bond distances in the xy plane are essentially the same in octahedral and square planar con-

figurations, steric effects are negligible and the increase in attractions due to forming six bonds rather than four will normally overwhelmingly favour the regular octahedral complex. Changes in CFSE for low numbers of d electrons either favour the octahedral case or are small. Consider, however, the case of d^8. In an octahedral complex, there are two electrons in the e_g level, while the square planar configuration allows these to be paired in the d_{xy} orbital (Figure 13.10) with a gain in CFSE of about $2\Delta E_{oct}$. (This is only

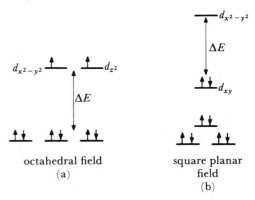

FIGURE 13.10 *The configuration of electrons in a d^8 complex:* (a) *octahedral, and* (b) *square planar*

approximate as the lower levels in the square planar case do not match the octahedral t_{2g} levels, but they are quite close.) This is offset by a reduction in exchange energy from the $13K$ of the octahedral case to $12K$ for two sets of four electrons. However, if ΔE is sufficiently large, it is possible for the gain in CFSE to overbalance the loss in bonding interactions and the small loss in exchange energy. It is found that square planar complexes are indeed formed by d^8 ions with ligands to the right of the spectrochemical series: for example Ni^{2+} forms a square planar cyanide complex, $Ni(CN)_4^{2-}$, while its hydrate or ammine are octahedral, e.g. $Ni(NH_3)_6^{2+}$. The larger ΔE values of heavier elements or of more highly-charged elements extend the scope of formation of square complexes. Thus all complexes, even halides, of platinum(II) and gold(III) (both d^8) are square planar.

This extra CFSE found in square planar complexes of d^8 elements also occurs for the d^7 and d^9 configurations. However, the CFSE gain is only ΔE. Table 13.11 lists the species which typically form square planar complexes. While such square complexes of the first row elements are comparatively rare, those formed by heavy transition elements are the majority of the representatives of these oxidation states, especially for the d^8 configurations.

TABLE 13.11 Species forming square planar complexes

d electron configuration	Species			Approx CFSE	Unpaired electrons
d^8	Ni(II)	Pd(II)	Pt(II)	$2\Delta E_{oct}$	0
		Rh(I)	Ir(I)		
			Au(III)		
d^9	Cu(II)	Ag(II)		ΔE_{oct}	1 (in $d_{x^2-y^2}$)
d^7	Co(II)			ΔE_{oct}	1 (in d_{xy})
d^6	Fe(II)			$\frac{1}{2}\Delta E_{oct}$	2
d^4	Cr(II)			$\frac{1}{2}\Delta E_{oct}$	4

The Cr(II) spin corresponds to one electron in each of the four stable orbitals, and the Fe(II) value indicates two filled and two half-filled orbitals. The CFSE are with respect to the octahedral configuration, those for d^4 and d^6 being for the weak field configuration.

In configurations other than d^8, the CFSE gain relative to octahedral is small, and most of these configurations are distorted in the octahedral case, so it is often difficult to decide what has happened. For example, copper(II) compounds often show four short bonds in a square plane with two longer ones, or even three sets of pairs of bonds with different lengths.

The distortion mentioned in the last paragraph arises whenever the d_{z^2} and $d_{x^2-y^2}$ orbitals are unequally occupied. If, for example, there is one electron in the d_{z^2} orbital, ligands on the z axis are more shielded from the nuclear field than are ligands on the x and y axes. The ligand-metal distances in the z direction are therefore shorter than those in the xy plane. If the electron is, instead, in the $d_{x^2-y^2}$ orbital the four distances in the xy plane are shorter. Such distortions, which are less simple than described here, are one manifestation of the Jahn-Teller Theorem, which states that if a system has unequally-occupied, degenerate energy levels it will so distort as to raise the degeneracy. Cases where distortions are expected are d^4 high-spin, d^7 low-spin, and d^9. Distortions involving t_{2g} levels are normally too small to be detected.

13.5 Stable configurations

With all the factors discussed in the last three sections in mind, it is possible to make some general remarks about the stabilities of various d configurations. These apply particularly to the first row elements which only occasionally show coordination numbers other than four or six (see section 13.6). The extent to which these apply in practice will become clear when the chemistry of the individual elements is discussed.

The traditionally stable configurations of the empty, half-filled, and filled shells should still be stable for d elements; the stability in the last two cases stemming from the high exchange energy as well as from the general symmetry of the electron clouds. In octahedral environments, d^5 will be unusually stable only in the high-spin arrangement. Hence, it should be more stable in the divalent ion Mn^{2+} than in the trivalent d^5 element, Fe^{3+}, and also more stable with ligands of relatively low field in each case. The configurations d^0 and d^{10} may well be found in stable tetrahedral complexes.

In the octahedral configuration, the states d^3 and low-spin d^6 are also expected to have special stability as they correspond to half-filled and filled t_{2g} orbitals. In this case, stability increases with increasing ΔE and should be most marked for trivalent ions, here Cr^{3+} (d^3) and Co^{3+} (d^6).

It will be noted that the d^4 configuration in an octahedral environment lies between two others that are especially stable, so that d^4 species are expected to be unstable, both to oxidation to d^3 and to reduction to d^5. In particular, the $M^{2+}d^4$ state should readily oxidize to the $M^{3+}d^3$ ion where the increase in charge increases ΔE and the CFSE, while the $M^{3+}d^4$ ion should be readily reduced to the $M^{2+}d^5$ ion which is favoured by the reduced ΔE. Such behaviour is indeed found, the ions in question being respectively Cr^{2+} and Mn^{3+}.

At the d^8 configuration, there is competition between the square planar and the octahedral arrangements. In nickel(II) the latter is more common but the former is dominant for Pd(II), Pt(II), and Au(III). In both configurations, d^8 tends to be more stable than either d^7 or d^9 which have neither the CFSE of square planar d^8 nor the equally-occupied e_g arrangement of octahedral d^8. There is some tendency for d^9 to go d^{10} but a number of factors come into play here as the oxidation states of the copper Group elements show. It might be noted that d^{10} species have some tendency to linear configurations as Cu(I) and Au(I), as well as Hg(II), show. This is in addition to their ready adoption of tetrahedral structures, and the two-coordinate species are found for relatively low charge densities on the ions, either in large ions or in M^+ species.

The discussion is summarized in Table 13.12.

13.6 Coordination numbers other than four or six

While the majority of compounds of the first row elements exhibit octahedral, tetrahedral or square planar coordination, there are now well-established examples of a range of other coordinations. For the heavier transition elements, octahedral and square planar configurations are common, but higher coordination numbers are well known. An alternative form of six-coordination, the trigonal prism of Figure 13.11 is also found but is rare. It is the configuration

TABLE 13.12 General stabilities of d configurations

Number of d electrons	Comments
0	Stable in earlier Groups as the Group oxidation state.
1 2	Tend to oxidize to d^0, e.g. Ti^{3+}, Ti^{2+}, V^{3+}
3	Stable, especially for trivalent ion, e.g. Cr^{3+}.
4	Unstable, e.g. Cr^{2+} oxidizes to Cr^{3+}; Mn^{3+} reduces to Mn^{2+}.
5	Stable, especially Mn^{2+}; Fe^{3+} is also relatively stable.
6	Stable in spin-paired state, e.g. Co^{3+} (except in hydrate); Fe^{2+} is also relatively stable. Pt(IV) very stable.
7	Relatively unstable, Co^{2+} oxidizes except as hydrate or halide complex; Ni^{3+} is scarcely known.
8	Stable, Ni^{2+} as octahedral and square planar complexes; Pd(II), Pt(II), Au(III) very stable square planar species.
9	Relatively unstable except in case of Cu(II).
10	Stable, e.g. Ag^+; zinc Group never show states above II.

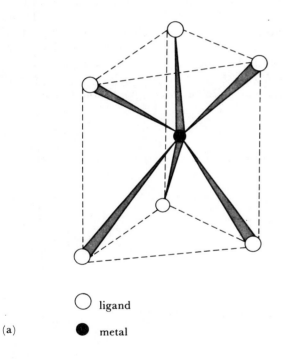

ligand
(a) metal

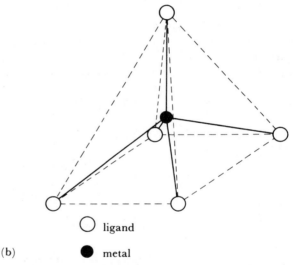

ligand
(b) metal

FIGURE 13.11 *Lower-symmetry coordinations* (a) *trigonal prism form of six-coordination* (b) *square pyramid form of five-coordination*
Trigonal prismatic coordination is often found when the donor atoms are sulfurs, as in MoS_2 or in the dithiolene complexes $M(S_2C_2R_2)_3$ where M = V, Mo, Re. In the square pyramid, note that the metal atom lies above the base plane making the base-L/metal/apex-L angle greater than 90°.

of arsenic in NiAs (Figure 5.10c) and is found for molybdenum in MoS_2.

Coordination numbers greater than six are most usually found among the compounds of the second or third row transition elements, and also for the lanthanides and actinides as illustrated in Chapters 11 and 12. For these higher coordination numbers, it is usual to find several alternative shapes, and the energy differences between them are commonly very small. There is at present no generally accepted, unambiguous, method available for predicting which shape will be adopted, either for the higher coordination numbers or for five-coordination.

Coordination Number Two. This is found in d^{10} configurations such as copper(I), silver(I), gold(I) and mercury(II). The arrangement is always linear, as in $Ag(NH_3)_2^+$ or $AuCl_2^-$, and is simply that derived from the electron pair repulsion theory.

Coordination Number Five. Five-coordination is known for most of the first row elements, but is uncommon for the heavier transition metals. As well as the trigonal bipyramid (Figure 4.4), square-pyramidal five-coordination is found for transition metal compounds. As the discussion of five electron pairs in Chapter 4 showed, there is little energy difference between the trigonal bipyramid and square pyramid, and the balance of stability is readily tipped one way or the other by additional stabilizations arising from the d electron configuration, the shape and properties of the ligands, or from interactions in the solids. The square pyramid configuration has the central metal atom raised *above* the plane of the four base atoms (Figure 13.11). Notice that this increases the bond angles at the central atom, and is the opposite distortion from that found in AB_5L species, like BrF_5, derived from the octahedron, where the repulsion of the lone pair decreases BAB angles so that the central atom lies *below* the base plane. (Compare BrF_5 with Figure 14.13.)

The small differences in energy between the two shapes is strikingly illustrated by the compound

$$(Cr(en)_3)[Ni(CN)_5)].\tfrac{3}{2}H_2O,$$

whose crystal structure contains both square pyramidal and distorted trigonal bipyramidal $Ni(CN)_5^{3-}$ ions. Table 13.13 gives some illustrative examples of five-coordinate complexes, chosen either for their relative simplicity or because a family of compounds of one type exist. In some of the structures, distortions from the ideal shapes occur, while many other cases are known where the structure is intermediate between the two geometries.

Coordination Number Seven. Seven-coordinate complexes are commonly found in one of three shapes. The most symmetric is the regular pentagonal bipyramid adopted by ReF_7, OsF_7, by the ions MF_7^{3-} (M = Zr, Hf) in their sodium salts, and also by the main group compound IF_7. Another example is the ytterbium complex $Yb(TePh)_2(pyridine)_5$ whose

structure is given in Figure 11.6b. As the in-plane angle of a regular pentagonal bipyramid is only $72°$, some distortion of the five equatorial ligands is likely to reduce steric interactions, either by buckling out of the plane or by variation of the M−F bond lengths in the equatorial plane. Some uranyl compounds, such as $UO_2F_5^{3-}$, also show this shape as the UO_2 unit has to be linear and the other five substituents lie in a plane at right angles to this, around the uranium atom (compare Figure 12.6 for the corresponding UO_2L_6 situation).

A second seven-coordinate arrangement found, is that formed by inserting an extra substituent into the triangular face of an octahedron, and spreading out the three ligands forming this face. Such a positioning is termed *capping* the face. One example is $NbOF_6^{3-}$ illustrated in Figure 15.5a (which emphasises that the seventh ligand is on the three-fold axis).

The third shape found for seven-coordination is that obtained by inserting a substituent above one of the rectangular faces of a trigonal prism—a capped triangular prism. This structure is shown in the ammonium salt of ZrF_7^{3-} (Figure 15.1) or by MF_7^{2-} (M = Nb, Ta). These structures are all very similar in energy and change from one to another is readily induced, for example by the change of counterion in the heptafluorozirconium complexes.

Coordination Number Eight. While the eight-coordinate shape of highest symmetry is the cube, this is common only in

TABLE 13.13 Examples of five coordination

	Trigonal bipyramid	*Square pyramid*
titanium(IV)	$TiCl_5^-$, $TiOCl_2(NMe_3)_2$	Y_2TiO_5
(III)	A	
vanadium(IV)	$VOCl_2(NMe_3)_2$	$VO(acac)_2$ (Figure 14.13)
(III)	A	
niobium(V)	$NbCl_5$ (in gas phase only)	
tantalum(V)	$TaCl_5$ (in gas phase only)	
chromium(III)	A	
(II)	B	
molybdenum(V)	$MoCl_5$ (in gas phase only)	$Mo_3O_{10}^{5-}$ includes Mo^VO_5 units
manganese(II)	B	C
rhenium(V)	$ReOX_4^-$ (X = Cl, Br, I)	
iron(0)	$Fe(CO)_5$	
(II)	B	C
(III)		$Fe(S_2CNEt_2)_2Cl$
cobalt(II)	B	C
nickel(II)	$Ni(CN)_5^{3-}$ (a), B	$Ni(CN)_5^{3-}$ (a)
platinum(II)	$Pt(SnCl_3)_5^{3-}$	
copper(II)	B, $Cu(bipy)_2X$	$[Cu(NO_3)_2(C_5H_5NO)_2]_2$, $Cu(acac)_2$ quinoline
zinc(II)	B	C, see also Figure 13.11

$A = M^{III}X_3L_2$ (X = halogen, L = ligand like NMe_3).
$B = [M^{II}(tetra\ N)Br]^+$ where tetra N = $(Me_2NCH_2CH_2)_3N$, compare Appendix B.
$C = (terpy)M^{II}Cl_2$ (terpy = terpyridyl, compare Appendix B).
(a) Pentacyanonickel(II) is found in both shapes in the $Cr(en)_3^{3+}$ compound.

extended structures like the ionic CsCl or CaF_2 structures or in metallic body-centred cube forms. A smaller repulsion between ligands results when one face of the cube is twisted by 45° relative to the opposite one to form the square antiprism. This structure is shown by TaF_8^{3-} (Figure 15.5b) and ReF_8^{2-}, or by $Zr(acac)_4$ and $U(acac)_4$ (Figure 12.5).

The other common eight-coordinate structure, the dodecahedron or bis-bisphenoid (bisdisphenoid), may also be derived from the cube. As every second corner of a cube defines a tetrahedron (as can be seen in Figure 13.9), the cube may be regarded as two interpenetrating tetrahedra. If one of these tetrahedra is flattened and the other elongated, the eight-coordinate structure of Figures 15.2 or 15.18 results. A tetrahedron distorted in this way is called a bisphenoid, hence the name bisbisphenoid. If the distortions are equal and produce equilateral triangular faces, the structure is a regular dodecahedron, shown in Figure 13.12 where the vertices labelled A define one bisphenoid, and those labelled B, the other. A number of structures approximate to a regular dodecahedron, for example the MF_8^{4-} species of Figure 15.2, but quite distorted forms also occur. These result particularly from the presence of bidentate ligands (see section 13.7) where the two donor atoms are close together. Thus $Co(NO_3)_4^{2-}$, with two oxygens bridging Co to each nitrate, forms a distorted dodecahedron with the $Co-O$ bond lengths equal to 204 pm in the elongated bisphenoid and averaging 255 pm in the flattened one. A similar distorted structure is found in the chromium(V) peroxy compound $K_3Cr(OO)_4$ where both oxygens of the $O-O$ group are bonded to Cr.

The eight-coordination of $UO_2(NO_3)_2.2H_2O$, Figure 12.6 shows a fourth shape—the hexagonal bipyramid. This is only found for such MO_2 species where the $O-M-O$ group has to be linear.

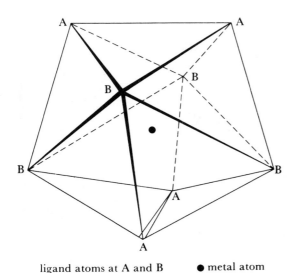

ligand atoms at A and B   ● metal atom

FIGURE 13.12 *The dodecahedron*
This eight-coordinate shape is related to that of the cube by elongating one of the interpenetrating tetrahedra (A vertices) and flattening the other (B vertices). For the regular figure, all the faces are equilateral triangles.

Coordination Number Nine. The one configuration found for nine-coordination is that formed by placing one ligand above each of the three rectangular faces of a trigonal prism. This structure is illustrated in Figure 15.23 for ReH_9^{2-} and is also shown by the lanthanide hydrates like $Nd(H_2O)_9^{3+}$ mentioned in Chapter 11.

Higher Coordination Numbers. As a transition element of the d block has only nine valence shell orbitals (five d plus one s plus three p) the maximum coordination number is limited to nine unless metallic bonding (compare close-packed structures with coordination number twelve, Chapter 5), purely electrostatic bonding, or electron deficient bonding (compare section 9.6) occurs. One example is $Zr(BH_4)_4$ which forms two hydrogen bridges to each BH_4 group at room temperature, but transforms to a three-bridge structure at $-160°C$, $Zr(H)_3BH$. Thus the Zr is twelve-coordinate in the low temperature form.

If f orbitals are available, this restriction no longer holds as illustrated by the ten-coordinate $La(H_2O)_4$ EDTA compound mentioned in Chapter 11.

13.7 Effect of ligand on stability of complexes

The type of ligand has a distinct influence on the stabilities of complexes and this must be discussed in more detail. The spectrochemical series gives one classification of the ligands and some points about this should be noticed. In an extended form of the series, it is seen that position depends largely on the donor atom in the order:

$$\text{Halogen or } S < O < N < P < C$$

Thus, nitrate which donates through O is much weaker than nitrite which donates through N. An even more striking example is the thiocyanate group, SCN, which is sometimes found coordinated through S, when it has a weak ligand field, and sometimes coordinates through N, when it shows a much stronger field. A similar example is shown by the isomeric form of nitrite, nitrito, which coordinates as ONO, donating through O, instead of the NO_2 form of nitrite which donates through N. The nitrito form has a weaker ligand field than the nitrite form. These features govern the effect of the ligands by way of their contribution to the crystal field stabilization energy, but this term may be only a minor contribution to the overall stabilization, and it is necessary to take care not to equate strong crystal field unthinkingly with stability.

A number of attempts have been made to find general relationships which indicate overall stability. One of the most extensive is Pearson's classification into *hard and soft* ligands and metal ions (or, using the Lewis nomenclature where a lone pair donor is a base and an acceptor is an acid, the metal ions are classified as hard or soft acids, and the ligands as hard or soft bases). In this, a ligand is a hard base if it is non-polarizable as for most ligands with a first row donor atom, and the ligand is a soft base if it is polarizable, as with sulfur or phosphorus ligands. Similarly, a metal ion is a soft acid if it has easily polarizable electrons, or is large, or has a low charge, while a hard acid is a metal ion

TABLE 13.14 Classification of ligands and ions under hard/
soft formalism

Ligands or bases

Hard	H_2O, ROH, R_2O, OH^-, OR^-, NO_3^-, RCO_2^-, SO_4^{2-}, CO_3^{2-}, $C_2O_4^{2-}$, PO_4^{3-} (denors through O) NH_3, NR_3, NHR_2, NH_2R, Cl^-
Soft	R_2S, RSH, RS^-, SCN^-, $S_2O_3^{2-}$ (S donors) R_3P, R_3As, I^-, CN^-, H^-, R^-
Borderline	py, Br^-, N_3^-, SO_3^{2-}, NO_2^-

Metal ions or acids (charges are formal only)

Hard	Mn^{2+}, Cr^{3+}, Fe^{3+}, Co^{3+}, Ti^{4+}, VO_2^+, VO^{2+}, Zr^{4+}, MoO^{3+} H^+, all s element ions, M^{3+} for M = Al, Ga, In, Sc, Y, Ln
Soft	Cu^+, Ag^+, Au^+, Hg_2^{2+}, Hg^{2+}, Pd^{2+}, Pt^{2+}, Pt^{4+} Tl^+, Tl^{3+}
Borderline	Fe^{2+}, Co^{2+}, Ni^{2+}, Cu^{2+}, Zn^{2+}, Ru^{2+}, Os^{2+}, Rh^{3+}, Ir^{3+} Sn^{2+}, Pb^{2+}, Sb^{3+}, Bi^{3+}

of high charge, small size and with valence electrons which are not polarizable. The classification extends beyond transition elements and some examples of hard and soft ligands and metal ions are given in Table 13.14.

The most important generalization about stabilities is then that soft ligands form stable complexes with soft metal ions, and hard ligand-hard ion complexes are also stable. Mixtures of hard ion-soft ligand or soft ion-hard ligand are less stable. A second general trend is, that the substitution of a ligand of a particular type tends to make the metal ion **behave** more as that type of acid. Thus a borderline ion like Fe^{2+}, becomes harder when a number of hard ligands are coordinated so that the intermediate $Fe(H_2O)_3^{2+}$ (say) is more likely to add further water molecules than would a ferrous ion coordinated by soft ligands.

This hard-soft classification has been greatly extended and ramified, and some features are still subject to controversy. However, the basic predictions derived from the simple version are most useful. One example is provided by thiocyanate, SCN^-, which can bond through S or through N (in isothiocyanates). The S-bonded form is a soft ligand, and is preferred in complexes of soft ions such as Hg^{2+}. The N-bonded form is harder and is most usually found in first row complexes. A neat example is $HgCo(NCS)_4$ which is an extended lattice formed of $Hg-SCN-Co$ links.

The next effect is the *chelate effect*. If a ligand contains more than one donor atom it may coordinate to more than one position on the cation giving ring formations. The existence of such chelation in a complex is accompanied by increased stability and this is shown by increased heats of formation and resistance to substitution, and also by the higher position in the spectrochemical series of the chelating ligand as compared to an analogous non-chelating ligand. An example is provided by the case of ethylenediamine,

$H_2NCH_2CH_2NH_2$, where both nitrogen atoms can donate, forming a five-membered ring. Such a reagent with two donor atoms is termed *bidentate*. It is observed that the treatment of ammonia complexes with ethylenediamine results in the displacement of the ammonia:

$$Co(NH_3)_6^{3+} + 3 \text{ en} = Co(en)_3^{3+} + 6NH_3$$
(en = ethylenediamine)

The driving force of such a replacement is probably the entropy change. The equation above has four particles on the left hand side and seven on the right, so that reaction proceeds with increase of entropy. For example, the change in free energy in the copper complexes is:

$$Cu(en)_2^{2+} - Cu(NH_3)_4^{2+}$$
difference in $\Delta G = 18\cdot0$ kJ mol^{-1}
difference in $\Delta H = 10\cdot9$ kJ mol^{-1}
difference in $T\Delta S = -7\cdot1$ kJ mol^{-1}
(Recall that $\Delta G = \Delta H - T\Delta S$)

This entropy term thus provides about 40 per cent of the free energy change, whereas the substitution of one monodentate ligand by another, for example the exchange of ammonia for water, usually has only a small entropy effect which is commonly neglected in calculations. The optimum ring size for elements of the transition series appears to be a five-membered ring. Thus the substitution of 1,3-propanediamine for ethylenediamine, increasing the ring size to six atoms (Figure 13.13b), results in a loss of free energy of

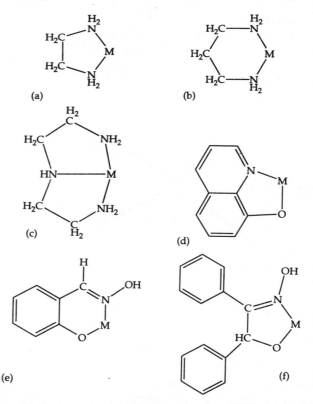

FIGURE 13.13 *Some chelating ligands:* (a) *ethylenediamine* (*en*), (b) *1,3-propanediamine*, (c) *diethylenetriamine* (*trien*), (d) *8-hydroxyquinoline* (*oxine*), (e) *salicylaldoxime*, (f) *α-benzoinoxime* (*cupron*)

formation of 16.3 kJ mol^{-1} in the case of the copper complex above and 35.6 kJ mol^{-1} in the case of the nickel complex Ni(diamine)$_3^{2+}$.

The chelating effect is increased still further if the chelating molecule has more donor atoms. The triamine (trien) $H_2NCH_2CH_2NHCH_2CH_2NH_2$ forms an even more stable complex than ethylenediamine. Chelating agents with up to six donor atoms are available, an example of the latter being ethylenediaminetetra-acetic acid, EDTA, shown in Figure 10.5. The common complexing and precipitating agents of analysis are nearly all chelating agents which form five- or six-membered rings with the metal atom which is being determined. Some examples are shown in Figure 13.13. Polydentate chelating agents may give rise to complexes of unusual coordination, for example, the less common five-coordination in the zinc complex shown in Figure 13.14.

Quantitative indication of the process of forming a complex comes from the evaluation of the stability constants which characterize the equilibria corresponding to the successive addition of ligands. That is, we can consider the steps

$$M + L \rightleftharpoons ML$$
$$ML + L \rightleftharpoons ML_2$$

and so on down to

$$ML_{n-1} + L \rightleftharpoons ML_n$$

These are characterized by equilibrium constants K_1, K_2 . . ., K_n such that

$$K_1 = [ML]/[M][L]$$
$$K_2 = [ML_2]/[ML][L]$$

and,

$$K_n = [ML_n]/[ML_{n-1}][L]$$

These constants K, are termed *stepwise formation constants*. An alternative formulation is to consider the overall formation reaction

$$M + nL \rightleftharpoons ML_n$$

characterized by the nth *overall formation constant* β_n.

$$\beta_n = [ML_n]/[M][L]^n = K_1K_2 \ldots K_n$$

In most preparations, the complex is formed in aqueous solution, and the stability constants refer to steps where L replaces coordinated water. The evaluation of formation constants usually calls for a good deal of experimental in-

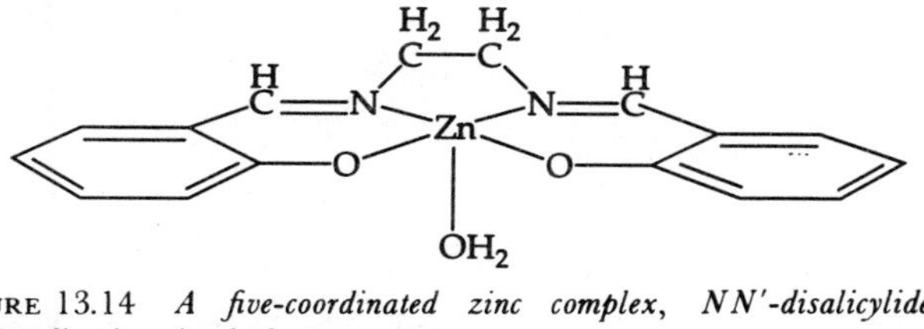

FIGURE 13.14 *A five-coordinated zinc complex, NN'-disalicylidene-ethylenediamine zinc hydrate*
The quadridentate ligand coordinates through its two nitrogen atoms and two oxygen atoms to the zinc, giving a structure of a shallow square pyramid, with the zinc atom 34 pm above the NNOO plane. The water molecule is coordinated on the opposite side of the zinc with the Zn – OH$_2$ bond in the direction of the axis of the pyramid.

genuity and will not be discussed here. We may simply note some values and their interpretation. For example, the logarithms of the successive formation constants of various nitrogen complexes formed in aqueous solution are shown in Table 13.15.

The values in Table 13.15 illustrate a number of features which give a quantitative indication of the stability properties discussed above. There is first a general tendency for K values to fall as the number of ligands increases. This is probably due, at least in part, to the statistical effect that the number of sites for substitutions is reduced as substitution proceeds.

The values for nickel show a steady progression to Ni(NH$_3$)$_6^{2+}$ and zinc, similarly, goes steadily to Zn(NH$_3$)$_4^{2+}$. The cobalt values show that the sixth ligand is unstable, probably reflecting the effect of the e_g electron, and this is even more marked for copper where the fifth NH$_3$ is only added in presence of a large excess of ammonia and the sixth is not taken up at all. This reflects the general tendency of the d^9 ions to form four strong bonds and two, much longer, weaker ones.

It will also be noted, comparing K_1 values for example, that the stability order for M^{2+} ions increases to copper. This is part of a more extensive sequence, the Irving-Williams order, which shows that stability constants vary

$$Mn(II) < Fe(II) < Co(II) < Ni(II) < Cu(II) > Zn(II)$$

towards a particular ligand.

The values also illustrate the effect of chelation on the stabilities. If we compare log β_2 for NH$_3$ with log K_1 for $H_2NCH_2CH_2NH_2$, we are comparing values where two N atoms are coordinated in each case. As $\beta_2 = K_1 . K_2$, log β_2 = log K_1 + log K_2. It will be seen that log K_1 for the ethylenediamine complexes is uniformly greater than log β_2 for ammonia, and likewise log K_2 and log K_3 for en are

TABLE 13.15 Stepwise formation constants of some nitrogen complexes

Ligand		Co^{2+}	Ni^{2+}	Cu^{2+}	Zn^{2+}
NH$_3$	log K_1	2·1	2·8	4·2	2·4
	log K_2	1·6	2·2	3·5	2·4
	log K_3	1·1	1·7	2·9	2·5
	log K_4	0·8	1·2	2·1	2·2
	log K_5	0·2	0·8	−0·5	
	log K_6	−0·6	0·03		
en	log K_1	6·0	7·5	10·6	5·7
	log K_2	4·8	6·3	9·1	4·7
	log K_3	3·1	4·3	−1·0	1·7
trien	log K_1	8·1	10·7	16·0	8·9
6-en	log K_1	15·8	19·3	22·4	16·2

en = ethylenediamine (Figure 13.13a): trien = diethylene-triamine (Figure 13.13c).

6-en = pentaethylenehexamine, the analogous 6-nitrogen molecule (compare Appendix B)

greater than the corresponding $\log \beta_4$ and $\log \beta_6$ values for NH$_3$. Similarly, the tridentate chelate, trien, gives a value of $\log K_1$ which exceeds $\log \beta_3$ for the analogous NH$_3$ species. Finally, the hexadentate ligand pentaethylenehexamine has a value of $\log K_1$ (coordination of all six nitrogens) which is greater than $\log \beta_3$ for the M(en)$_3^{2+}$ species and than $\log \beta_6$ for M(NH$_3$)$_6^{2+}$. It is also noteworthy that while copper and zinc fail to become six-coordinated to ammonia, they do form six-coordination to nitrogen when the ligand is a chelating amine.

The effect of chelate ring size is illustrated by the values for 1,3-propanediamine (Figure 13.13b) which has $\log K_1 = 10.0$ and $\log K_2 = 7.2$ for the copper compounds.

Anomalous changes in stability constants may often indicate electronic changes. One example is provided by the Cr(II) dipyridyl values where $\log K_1 = 4.5$, $\log K_2$ is 6.0, instead of the expected decrease, and $\log K_3$ is 3.5. In this case, the d^4 Cr^{2+} species is high-spin in its hydrate, and low-spin in Cr(dipy)$_3^{2+}$. The anomalous value for K_2 suggests that spin pairing occurs as the second dipyridyl is added.

A full set of stability constants is to be found in the Chemical Society special publication no. 25, of that title, which is given in the references.

The properties of the ligands affect the replacement reactions which they undergo. It has already been seen that a chelating ligand will replace a monodentate ligand with the same donor atom. Such chelating ligands also commonly come higher in the spectrochemical series than the corresponding simple one: en lies above ammonia for example. However, care must be taken, in general, to avoid equating high ligand field strength with high energy of formation or high stability of a complex. Obviously, in all configurations where there is some ligand field stabilization energy, the size of the field of a given ligand is an important factor in its reactions, but the CFSE varies with the metal atom, as has been seen. It is safe to expect a general correlation between stability and ligand field strength, for example, cyanide is expected to replace water in general, but the detailed behaviour in any one case is a balance of different effects. A most marked example of this is the case of the transition metal complexes containing metal-hydrogen bonds, such as (R$_3$P)$_2$PtH$_2$. Hydrogen lies high in the spectrochemical series with ΔE values approaching those of cyanide, yet many of its compounds with the transition metals are unstable and isolable compounds containing σ-M−H bonds (as opposed to metallic bonds in non-stoichiometric hydrides) are found only in the presence of a few stabilizing ligands. This instability may arise from the fact that, when metal-hydrogen bonds dissociate, hydrogen atoms are produced. These are very reactive and combine readily with each other. By contrast, when the normal metal-ligand bond dissociates, stable lone pair entities like water, ammonia, or halide ions are produced and the metal-ligand bond can reform in equilibrium with small quantities of ligand.

This is an extreme case, but it is common to find an unthinking equation of high ligand field with stability of complexes. This does not follow, particularly when com-

paring ligands of similar strength. Thus Co(NH$_3$)$_6^{3+}$ is prepared in water and is stable indefinitely to exchange of the coordinated ammonia for water, while Cr(NH$_3$)$_6^{3+}$ can only be prepared in liquid ammonia and rapidly exchanges the ammonia groups for water when dissolved in water.

13.8 Isomerism

The use of polydentate chelating agents aids the study of *stereoisomerism* in complexes. In octahedral complexes, both optical and geometrical isomerism is possible, as illustrated by the ethylenediamine complexes of cobalt in Figure 13.15. The existence of optical isomers may be shown by rotatory dispersion measurements and by the resolution of isomers in favourable cases. The scheme for the resolution of the tris-ethylenediamine complex of cobalt(III) is shown in Figure 13.16, making use of naturally occurring d-tartaric acid to separate the isomers. The existence of geometrical isomers is generally noticed when two compounds of different properties are found to have identical molecular formulae. In favourable cases the identity of the isomers may be determined by experiment, as in the separation of the optically active forms of the *cis*-Co(en)$_2$X$_2^+$ compound shown in Figure 13.15. It is also possible to distinguish geometrical isomers in some cases by substitution reactions with

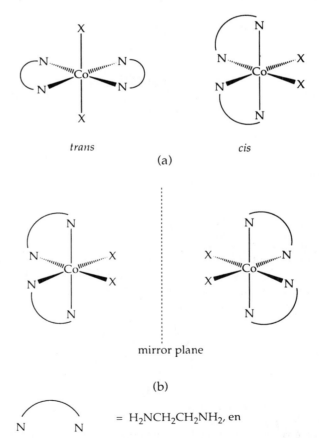

FIGURE 13.15 *Optical and geometrical isomerism in octahedral complexes: the example of* Co(en)$_2$X$_2^+$

Figure (a) shows the geometrical isomers, *cis* and *trans* forms, while Figure (b) shows the non-superimposable optical isomers of the *cis* form. The *trans* form has a plane of symmetry and is inactive.

$dl\text{-}[\text{Co(en)}_3]\text{Cl}_3$

| silver d-tartrate

$\text{AgCl} + [\text{Co(en)}_3]\text{Cl}$ tartrate
in solution

| evaporate to
crystallization

crystals   solution
d-form l-form

$+\text{KI}$ | recrystallize | evaporate
                                      further
                                      $+\text{KI}$

$d\text{-}[\text{Co(en)}_3]\text{I}_3$ $l\text{-}[\text{Co(en)}_3]\text{I}_3$

FIGURE 13.16 *Resolution into optical isomers of* $\text{Co(en)}_3\text{I}_3$

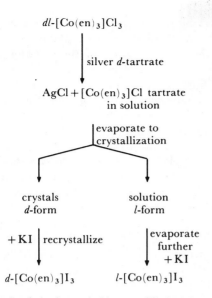

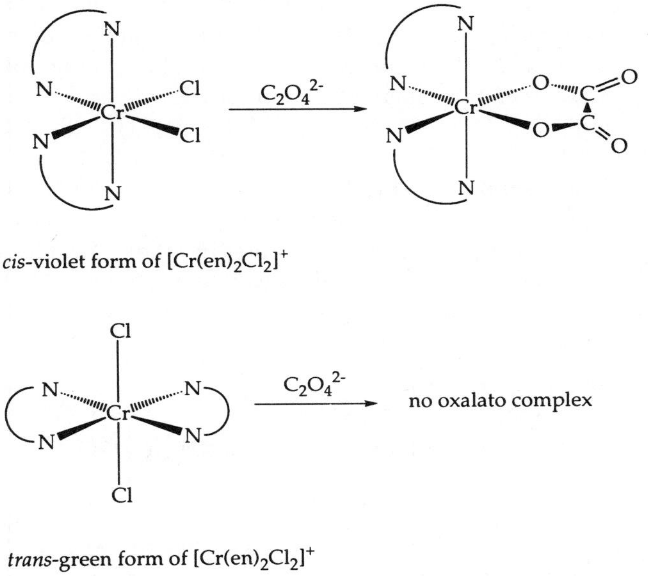

cis-violet form of $[\text{Cr(en)}_2\text{Cl}_2]^+$

trans-green form of $[\text{Cr(en)}_2\text{Cl}_2]^+$

FIGURE 13.17 *Differentiating between geometrical isomers by a replacement reaction with a bidentate ligand*

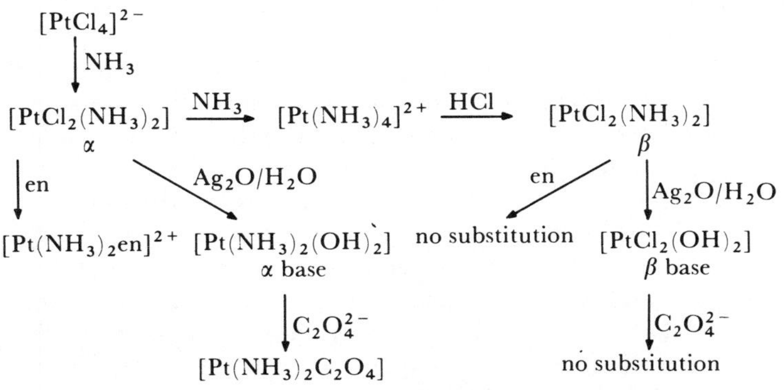

$[\text{PtCl}_4]^{2-}$

| NH_3

$[\text{PtCl}_2(\text{NH}_3)_2]$ $\xrightarrow{\text{NH}_3}$ $[\text{Pt(NH}_3)_4]^{2+}$ $\xrightarrow{\text{HCl}}$ $[\text{PtCl}_2(\text{NH}_3)_2]$
      α β

| en $\searrow$ $\text{Ag}_2\text{O}/\text{H}_2\text{O}$ en $\swarrow$ | $\text{Ag}_2\text{O}/\text{H}_2\text{O}$

$[\text{Pt(NH}_3)_2\text{en}]^{2+}$ $[\text{Pt(NH}_3)_2(\text{OH})_2]$ no substitution $[\text{PtCl}_2(\text{OH})_2]$
                                    α base β base

| $\text{C}_2\text{O}_4^{2-}$ | $\text{C}_2\text{O}_4^{2-}$

$[\text{Pt(NH}_3)_2\text{C}_2\text{O}_4]$ no substitution

Hence the α form of $[\text{PtCl}_2(\text{NH}_3)_2]$ is *cis* and the β form is *trans*.

FIGURE 13.18 *Stereoisomerism in square planar complexes of platinum*

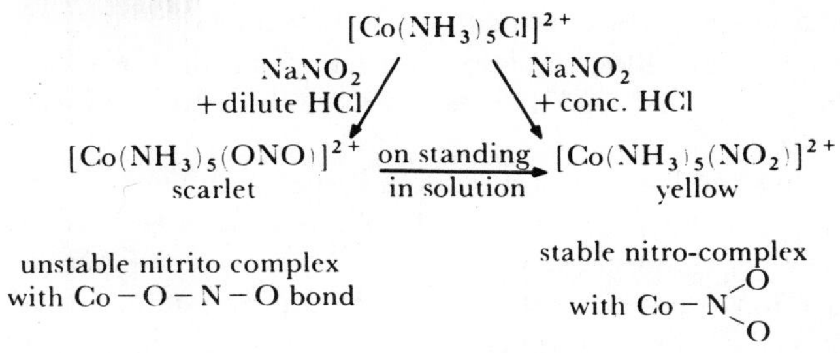

$[\text{Co(NH}_3)_5\text{Cl}]^{2+}$

NaNO_2 /   \ NaNO_2
+dilute HCl   +conc. HCl

$[\text{Co(NH}_3)_5(\text{ONO})]^{2+}$ $\xrightarrow[\text{in solution}]{\text{on standing}}$ $[\text{Co(NH}_3)_5(\text{NO}_2)]^{2+}$
scarlet   yellow

unstable nitrito complex   stable nitro-complex
with $\text{Co}-\text{O}-\text{N}-\text{O}$ bond   with $\text{Co}-\text{N}\begin{smallmatrix}\text{O}\\\text{O}\end{smallmatrix}$

FIGURE 13.19 *Linkage isomerism: nitro- and nitrito-pentammine-cobalt(III)*

a bidentate ligand. Such a ligand can replace two monodentate ligands lying *cis* but not two lying *trans*, as they are too far apart. Figure 13.17 shows an example of such a reaction scheme. It must be noted that such proofs of configuration by reaction are not fully definite, as change of configuration may occur during reaction. The final proofs depend on structural evidence, usually from crystallography, although other techniques such as nmr may be useful on occasion.

Stereoisomerism is not confined to octahedral coordination, of course. Figure 13.18 illustrates a classic case of isomerism among square planar platinum complexes.

Other types of isomerism occur among complexes. The case of linkage isomerism in which a ligand can coordinate through one or other of alternative atoms has been exemplified above by the cases of nitro-nitrito and thiocyanate-isothiocyanate. The nitro-nitrito isomerization occurs in the same compound, for example, both forms of the cobalt-pentammine are known, Figure 13.19. Another example is provided by the NCS group, which is found bonded through either N or S in certain compounds of palladium, platinum and copper. A good example is provided by the series LCu(NCS)_2 which is brown, LCu(NCS)(SCN) which is yellow-green, and LCu(SCN)_2 which is deep green. Here, L may be one of a number of bidentate nitrogen ligands.

A number of other types of isomer have been defined but the names are of little importance. Some examples are given in Table 13.16.

TABLE 13.16 Further types of isomerism in complexes

(a) $[\text{Co(NH}_3)_4\text{Cl}_2]\text{NO}_2$ and $[\text{Co(NH}_3)_4\text{Cl(NO}_2)]\text{Cl}$
      $[\text{Pt(NH}_3)_4\text{Cl}_2]\text{Br}_2$ and $[\text{Pt(NH}_3)_4\text{Br}_2]\text{Cl}_2$

(b) $[\text{Co(NH}_3)_6][\text{Cr(CN)}_6]$ and $[\text{Cr(NH}_3)_6][\text{Co(CN)}_6]$
      $[\text{Pt}^{\text{II}}(\text{NH}_3)_4][\text{Pt}^{\text{IV}}\text{Cl}_6]$ and $[\text{Pt}^{\text{IV}}(\text{NH}_3)_4\text{Cl}_2][\text{Pt}^{\text{II}}\text{Cl}_4]$

(c) $[\text{Pt(NH}_3)_2\text{Cl}_2]$ and $[\text{Pt(NH}_3)_4][\text{PtCl}_4]$

(a) is the case of isomers with one group either as a ligand or as a counter-ion; (b) and (c) show different cases of polymerization or coordination isomerism. Here, isomers of the same analytical formula contain species of different coordination, oxidation state, or degree of polymerization.

13.9 Mechanisms of transition metal reactions

A number of different classes of reaction of transition element complexes exists and we review some of the better-established cases.

The simplest type of reaction is that of oxidation-reduction by *electron-transfer*. For example, if a mixture of ferricyanide, $Fe(CN)_6^{3-}$, and ferrocyanide, $Fe(CN)_6^{4-}$, is made, and if an electron is lost from the $Fe(CN)_6^{4-}$ ion and gained by the $Fe(CN)_6^{3-}$ ion, an oxidation-reduction reaction has taken place, even though there is no change in the overall composition of the mixture. The presence of such reactions is most readily demonstrated by making one of the components with a radioactive isotope. If the radioactivity is found, at a later time, to be spread between both components the reaction is indicated (star indicates the radioactive isotope)

$$*Fe(CN)_6^{4-} + Fe(CN)_6^{3-} \rightleftharpoons *Fe(CN)_6^{3-} + Fe(CN)_6^{4-}$$

As there is no nett change in the reaction mixture, there is no heat change in such a reaction. There is, however, a requirement for activation energy. This is because the $Fe-CN$ bond length in ferricyanide is shorter than that in ferrocyanide. Thus the simple electron transfer between ions in their equilibrium configurations would produce ferrocyanide ions with compressed bonds, and ferricyanide ions with extended ones, that is the product ions would be vibrationally excited. The electron transfer takes place between matched ions, thus there is an activation energy required to stretch the bonds in ferricyanide and compress those in ferrocyanide to the intermediate matching configuration, by vibrational excitation. The matched anions must approach closely, but they do not have to be in actual contact for the electron transfer to occur.

Similar electron transfer reactions occur for a variety of matched ions, such as MnO_4^-/MnO_4^{2-} (permanganate/manganate), $IrCl_6^{2-}/IrCl_6^{3-}$ or $Mo(CN)_8^{3-}/Mo(CN)_8^{4-}$. The indications of the electron transfer mechanism are, first, that the reaction is second order (that is, the rate is a function of the concentration of each ion) and, second, the reactions are much faster than ones involving ligand exchange which are discussed below. Electron transfer reactions are favoured by the presence of ligands like cyanide or phenanthroline which allow electron delocalization from the metal through a conjugated system.

A slower group of electron transfer reactions is that exemplified by the $Co(H_2O)_6^{3+}/Co(H_2O)_6^{2+}$ system. Here, not only are the bond lengths different in the two oxidation states, but cobalt(III) is spin-paired with configuration t_{2g}^6 while cobalt(II) is high spin in the hydrate, configuration $t_{2g}^5 e_g^2$. Thus electron transfer between the ground state configurations would yield excited electronic states in the products as shown by the first row of configurations below the equation.

$$*Co(H_2O)_6^{3+} + Co(H_2O)_6^{2+} \rightleftharpoons *Co(H_2O)_6^{2+} + Co(H_2O)_6^{3+}$$

ground state initial configurations

$$t_{2g}^6 \qquad t_{2g}^5 e_g^2 \qquad t_{2g}^6 e_g^1 \qquad t_{2g}^5 e_g^1$$

pre-excitation needed

$$t_{2g}^5 e_g^1 \qquad t_{2g}^6 e_g^1 \qquad t_{2g}^5 e_g^2 \qquad t_{2g}^6$$

To overcome this, a further excitation energy is needed to produce an excited electronic configuration, such as that shown in the second line below the equation, in addition to the vibrational excitation needed to adjust the bond lengths. Thus, electron transfer reactions which involve changes in spin pairing need a higher excitation energy, and are slower than those which require only the matching of bond lengths.

A related class of reactions is that where oxidation-reduction takes place by *electron-transfer accompanied by transfer of a ligand*. In the transition state, the transferred ligand (sometimes in conjunction with other species) forms a link between the two metals, e.g. $L_5Cr-Cl-CoL_5'$. This type of mechanism is often termed *inner sphere* or *bridged*. Such reaction mechanisms are most readily established when the complex to which the ligand is transferred is *substitution-inert*. The rates of substitution processes vary enormously. For most d configurations substitutions take place very rapidly, but one or two configurations are substitution-inert, that is to say, reaction times are measurable in hours or days rather than in fractions of a second. Note that this inertness has nothing to do with the thermodynamic stability of the reactants or products but is a question of the reaction rate. The most common inert configurations are d^3 and d^6, for example Cr(III), Co(III), Mo(III), W(III), Re(IV), Rh(III), Ir(III), Ru(II), Os(II), Pd(IV), and Pt(IV). One striking example, used by Taube, is the study of the oxidation mechanism by using chromium(II) oxidized to chromium(III). As the latter is substitution-inert, any transfer of an atom or group during the oxidation process will be detected by its appearance in the chromium(III) complex. For example, in the reaction

$$Cr(H_2O)_6^{2+} + Co(NH_3)_5X^{2+}$$
$$Cr(II) \qquad \qquad Co(III)$$
$$= Cr(H_2O)_5X^{2+} + Co(NH_3)_5(H_2O)^{2+}$$
$$Cr(III) \qquad \qquad Co(II)$$

the transfer, in the oxidation process, of the group X has been demonstrated for $X^- = $ halide$^-$, NCS^-, N_3^-, SO_4^{2-} and PO_4^{3-}. Although the equation has been balanced by giving the cobalt(II) product as the pentammine hydrate, exchange in the labile cobalt(II) complex means that the hexahydrate would be recovered if the reaction was carried out in water.

The electron transfer reactions make up one group of transition metal reactions, the other major class are *ligand substitution reactions*. The mechanisms of ligand replacement reactions may be discussed in the light of ligand field theory. Two limiting modes of reaction for an octahedral complex are conceivable: either a ligand may be removed, leaving a five-coordinated intermediate which then picks up the substituting ligand, or else the incoming ligand may become coordinated to the original complex, giving a seven-coordinated intermediate, which subsequently expels one of the original ligands. That is, either:

$$MX_6 \xrightarrow{\text{slow}} MX_5 \xrightarrow{Y} MX_5Y \qquad (13.1)$$

or,

$$MX_6 + Y \rightarrow MX_6Y \rightarrow MX_5Y \qquad (13.2)$$

The first labelled S_N1 (*substitution, nucleophilic, first order*) and the second S_N2, as it is second order. That is, the rate of S_N1 reactions is governed by the first, dissociation, step which is far slower than the subsequent uptake of Y, so that the rate law is simply

$$\frac{d[MX_5Y]}{dt} = k[MX_6] \qquad (13.3)$$

It should be noted that reaction (13.1) is almost indistinguishable from that in which the solvent, S, enters the sixth position and is then displaced by the incoming ligand

$$MX_6 + S \xrightarrow{\text{slow}} MX_5S \xrightarrow{\text{Y, fast}} MX_5Y \qquad (13.4)$$

The rate law for (13.4) would be

$$\frac{d[MX_5Y]}{dt} = k'[MX_6][S] \qquad (13.5)$$

and, as the concentration of the solvent is effectively constant, $k'[S]$ may be replaced by a constant k'', giving equation (13.5) the same form as (13.3).

In the S_N2 reaction, again making the reasonable assumption that the first step is slow and the expulsion of a ligand from the seven-coordinate intermediate is fast, the rate law would be

$$\frac{d[MX_5Y]}{dt} = k[MX_6][Y] \qquad (13.6)$$

Whether a given Y will replace any particular ligand is governed by the usual balance of energies with the additional factor of the change in CFSE. If the latter is an important factor, it will allow not only a prediction of whether the substitution will occur but also a prediction of the path. The CFSE of the five- and seven-coordinated intermediates may be calculated and the most likely path determined. Consider the case of substitution in a low-spin cobalt(III) complex (which has six d electrons and a strong CFSE contribution). It may be shown that the possible intermediates have the CFSE values shown (all given in terms of the octahedral splitting ΔE_{oct}):

Shape:	octahedral	pentagonal bipyramid (7)	trigonal bipyramid (5)	square pyramid (5)
CFSE:	2·4	1·55	1·25	2·0

The square pyramid therefore corresponds to the lowest loss of CFSE of all the possible intermediates in this case, the loss being equal to about 92 kJ mol^{-1} if the ammine complex is the case in point. It can be calculated that the total activation energy for substitution in the cobalt(III) hexammine complex ion ranges from 521 kJ mol^{-1} for the trigonal bipyramidal intermediate, through 431 kJ mol^{-1} for the pentagonal bipyramid, to a range between 25 and 395 kJ mol^{-1} for the

square pyramid depending on whether the empty site in the last case is occupied by a solvent molecule or not. Thus the CFSE variation accounts for about 50 per cent of the total activation energy in each case. It would be predicted that the mechanism involving the five-coordinated square pyramidal configuration for the intermediate is the most likely and the activation energy found by experiment, about 140 kJ mol^{-1}, appears to bear this out.

It must be borne in mind that, while the kinetic study of reactions is a valuable guide to mechanisms, the order of the reaction is not necessarily the same as the molecularity of the critical step in the reaction path. One example is provided by systems where a rapid equilibrium is first set up

$$MX_6 + Y \rightleftharpoons MX_6.Y \xrightarrow{\text{slow}} MX_5Y + X \quad \ldots (13.7)$$

and the slow step is the elimination of X from the intermediate, associated, species $MX_6.Y$. The actual rate-determining step is the slow one, which is unimolecular but the rate law will involve the constants for the forward and reverse steps of the equilibrium and the concentration of both MX_6 and Y: that is, the rate law will have the overall form of a second order reaction. In this case, the $MX_6.Y$ does not involve coordination of Y to M but denotes some association with a definite lifetime, such as the formation of an ion pair. Such species become especially important when reactions are carried out in less polar solvents than water, such as acetone or methanol.

A further important class of reactions exists which are apparently S_N2, but where the basic step is unimolecular, and the overall kinetics result from the combination of this step with a pre-equilibrium. This is the class of reactions where the attacking ligand is OH$^-$ and there is a replaceable hydrogen present in the ligands on the metal. Thus the reaction

$$Co(NH_3)_5Cl^{2+} + OH^- \rightleftharpoons Co(NH_3)_5OH^{2+} + Cl^- \qquad (13.8)$$

obeys the rate law

$$\frac{d[Co(NH_3)_5Cl^{2+}]}{dt} = k[Co(NH_3)_5Cl^{2+}][OH^-] \quad (13.9)$$

(note that in the rate equation the square brackets indicate concentrations). Such reactions are about 10^6 more rapid than most ligand replacements. It is believed that this reaction is not S_N2, but instead involves the initial abstraction of a proton from one of the ammonia ligands in a fast pre-equilibrium,

$$Co(NH_3)_5Cl^{2+} + OH^- \rightleftharpoons Co(NH_3)_4(NH_2)Cl^+ + H_2O \qquad (13.10)$$

and this is followed by a slower, rate-determining step involving the expulsion of chloride from this amide complex

$$Co(NH_3)_4(NH_2)Cl^+ \xrightarrow{\text{slow}} Co(NH_3)_4(NH_2)^{2+} + Cl^- \qquad (13.11)$$

and the five-coordinate amido-intermediate rapidly reacts

with water to give the hydroxy complex

$$Co(NH_3)_5(NH_2)^{2+} + H_2O \xrightarrow{\text{fast}} Co(NH_3)_5(OH)^{2+} \quad (13.12)$$

Thus the rate-determining step is unimolecular, but the rate law is second order, with a constant which is a combination of the forward and backward constants of reaction (13.10) and the constant for the slow step (13.11). Thus, the overall reaction (13.8) is the sum of (13.10), (13.11) and (13.12) obeying the overall rate law (13.9) but the critical step is unimolecular. As the amido-complex $Co(NH_3)_4(NH_2)Cl^+$ is the conjugate base (see Chapter 6) of the starting complex $Co(NH_3)_5Cl^{2+}$, this mechanism is usually known as the $S_N1 \; CB$ (substitution, nucleophilic, unimolecular, conjugate base) mechanism. Such reactions are found whenever OH^- attacks a complex ion which contains ionizable hydrogen atoms.

These remarks about mechanisms apply, of course, to coordinations other than the octahedral one. In square planar complexes, an additional effect, the *trans effect*, becomes important. In reactions of square planar complexes of the general form:

$$(MLX_3) + Y = (MLX_2Y) + X$$

the group X which is displaced may be *cis* or *trans* to the ligand L and two isomeric products are possible, one or both of which may be observed. The common ligands may be arranged in an order of increasing ability to direct incoming substituents to a position *trans* to themselves:

$$H_2O < NH_3 < Cl^- < Br^- < NO_2^- < CN^-$$

The knowledge of *trans* directing powers opens up methods of synthesizing desired compounds. Consider the method of synthesizing the *cis* and *trans* isomers of $[Pt(NH_3)_2(Cl)_2]$. If the *cis* isomer is required, the starting material must be the tetrachloroplatinate ion:

The first ammonia enters to give $Pt(NH_3)Cl_3^-$ and then, as the *trans* effect of the chloride ion is greater than that of ammonia, the second ammonia molecule enters *trans* to chlorine giving the *cis* isomer. To prepare the *trans* isomer, the starting material must be the tetrammine. In the intermediate $Pt(NH_3)_3Cl^+$, the *trans* effect of the chloride is greater than that of the ammonia, and the second chloride enters *trans* to the first to give the *trans* isomer:

The examples given serve only to illustrate the type of work which is being carried out in this field and are by no means even a complete summary of the known results. The field is obviously of importance for the understanding of transition metal chemistry as all reactions of transition elements in solution are reactions between complexes. Thus, when simple reactions involving transition metal ions in solution are discussed, the species actually involved are the complexes formed between ion and solvent molecule.

For octahedral complexes, S_N1 mechanisms and electron transfers with or without ligand transfer are well established, together with those involving ion-pairing or conjugate base pre-equilibria which are unimolecular in their rate-determining step, although following a second order rate law, such as equation (13.6) for the overall sequence of reactions. There is, however, little firm evidence of true S_N2 mechanisms, going through a seven-coordinate intermediate, for ligand substitution in octahedral complexes. In square planar complexes, and also in five-coordinate species, there is little steric hindrance to an increase of coordination number, and both S_N1 and S_N2 mechanisms are available for these lower coordination numbers.

One mechanism which is related to this question is that of *oxidative addition* in which an increase in oxidation state is accompanied by an increase in coordination number. This occurs most readily where a low coordination number is stable in a low oxidation state of an element while a higher one is the preferred one for a higher oxidation state. One striking example is provided by platinum where the $+II$ state is d^8 and found in square planar configurations while the $+IV$ state is d^6 and most stable in the octahedral configuration (Table 13.12). Thus a change

$$Pt^{II}L_4 + X_2 \rightarrow Pt^{IV}L_4X_2$$

is particularly favoured, compare section 15.8. Such a reaction may lead to oxidations by relatively unreactive species under very mild conditions, as in the reaction at room temperature and normal pressure

$$Ir(CO)Cl(PPh_3)_2 + H_2 \rightarrow Ir(CO)(H)_2Cl(PPh_3)_2.$$

13.10 Structural aspects of ligand field effects

If the radii of divalent ions in a given Period are considered, it is found that the change of radius across a transition series follows a curve with two minima, such as that shown for the first transition series in Figure 13.20. A similar effect is observed for trivalent ions but the minima come one element later. A smooth curve may be drawn through the values for the radii of the d^0, d^5, and d^{10} ions representing the fall in radius with increasing number of d electrons as the increase in nuclear charge is imperfectly balanced by the shielding of a d electron (similar to the lanthanide contraction discussed in Chapter 11). The actual radii for the ions with empty, half-filled, and filled shells lie on this curve, as the d electron clouds are spherically symmetrical. The radii of ions corresponding to other d configurations lie below this curve, and the largest differences are found for those ions with the

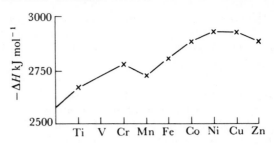

FIGURE 13.21 *The lattice energies of the difluorides of the elements of the First Transition Series*

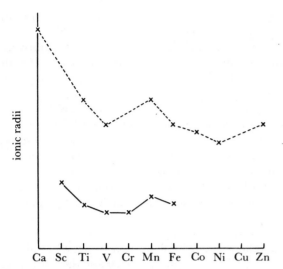

FIGURE 13.20 *Radii of the divalent (----) and trivalent (——) ions of the First Transition Series*

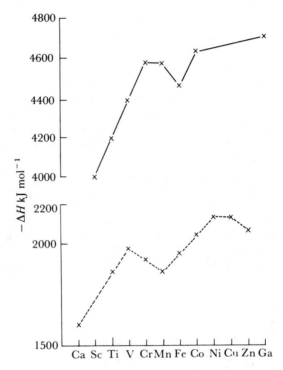

FIGURE 13.22 *The hydration heats of the divalent (----) and trivalent (——) ions of the First Transition Series*

largest excess of t_{2g} over e_g electrons. (The discussion and Figure refer here to high-spin configurations in the octahedral field—extensions to other cases are obvious, though less well documented.) Consider first the case of Ti^{2+} (divalent scandium is unknown). In this ion, there are two d electrons and these lie in the t_{2g} level, that is, in orbitals which are directed away from the ligands. Electrons in such orbitals have only a slight effect in screening the nuclear charge of the transition ion from the ligands—a much smaller screening effect than if they were disposed in a spherically symmetric shell. It follows that the metal-ligand distance is shorter than if these two electrons were spherically distributed, thus the octahedral radius is shorter, when measured, than that found by interpolation between the spherical d^0 and d^5 configurations. (It will be recalled from Chapter 2 that ionic radii are derived from measured ion-ion or ion-molecule distances.) The difference between measured and interpolated radii is even greater for V^{2+} where there are three electrons in the t_{2g} orbitals, and then the difference decreases for Cr^{2+} with one electron in the e_g set which is directed at the ligands. In the d^5 ion, Mn^{2+}, the configuration is $t_{2g}^3 e_g^2$ which is symmetrical, and the radius lies on the d^0—d^{10} curve. A similar pattern is repeated from d^6 to d^{10}.

Such a variation in radius will contribute a corresponding variation to the lattice energy of isostructural compounds of a given formula involving transition metal ions. Such a lattice energy curve is shown in Figure 13.21. The lattice energy rises as the ionic radii fall, as the lattice energy is inversely proportional to the interionic distance. Thus, the points for the d^0, d^5, and d^{10} ions lie on a smoothly rising curve, and those for intermediate configurations lie above this curve. If the crystal field stabilization energies are subtracted from the experimental values, the resulting points lie on the curve. Notice that the CFSE values and the effects on radii both result from the ligand field splitting and the resulting electron configuration. Whether the lattice energy values are re-

garded entirely as an effect of CFSE or the radial effects are included depends on whether calculated or experimental values are used (calculated values already take account of the variation in radius).

A similar curve is found if other thermodynamic values are considered. Figure 13.22 shows the hydration energies of the divalent and trivalent ions of the first row, that is the energy of:

$$M^{n+}_{(gas)} + 6H_2O_{(gas)} = M(H_2O)_6^{n+}{}_{(gas)}$$

In each case, a curve with two maxima and a minimum at the d^5 configuration is observed, and the values fall on a smooth curve rising from d^0 to d^{10} when the CFSE values are subtracted from the experimental values. Notice that such a curve may be used in reverse to find the crystal field stabilization energies for different configurations, and hence the value of ΔE. Such thermodynamically determined values of the ligand field splitting agree closely with those derived from spectroscopic data.

13.11 Spectra of transition element complexes

When only one d electron is present, a simple spectrum consisting of a single band is observed (Figure 13.4). The band is relatively broad both because the excited and ground states interact to some extent with their environment (usually with molecules of the solvent) and because the equilibrium bond lengths in the excited state would be greater than those in the ground state so that the upper state is produced with vibrational modes excited. These vibrational levels are too close to be resolved into individual bands but they give a general broadening of the $d-d$ absorption.

The effect of an octahedral ligand field on the d^1 ion may be shown as in Figure 13.23a. The t_{2g} and e_g levels are separated more widely as the ligand field increases. The symbol ^2D at the left is a symmetry label which describes the d^1 configuration in the absence of the ligand field splitting. We shall use such symbols here simply as labels and readers who study this subject further will find out how they are derived and their full significance.

Consider now the d^9 configuration in the octahedral field. We can describe this in a way which is related to the description of d^1 and use this as a method of determining the d^9 splitting diagram. In the ground state, d^9 can be regarded as derived from the filled d^{10} shell by forming a single hole in the e_g levels. Thus d^1 has a single electron in the t_{2g} level, outside a filled shell configuration, and d^9 has a single hole

in the e_g level. In d^9, when a t_{2g} electron is excited to the e_g level, the latter level becomes filled and the hole appears in the t_{2g} level. Thus, the excitation may be described as the transition of a hole, that is, an electron vacancy, from the e_g to the t_{2g} level, while the d^1 transition is described as the transition of an electron from t_{2g} to e_g. Thus the transition in d^9 is shown on an energy level diagram which is the *reciprocal* of that for d^1, as shown in Figure 13.23b.

Consider now a tetrahedral field. Here, the lowest level is the doublet e level and the upper state is the triplet t_2 level. Thus the energy level diagram for d^1 configuration in a tetrahedral field is the reverse of d^1 octahedral, and therefore qualitatively the same (though the ΔE value is only about 4/9) as that for d^9 octahedral. Similarly, d^9 tetrahedral is qualitatively the same as d^1 octahedral.

Yet another related configuration is that of d^6 high spin. Here, each d orbital is singly occupied and the sixth electron, in the octahedral field, is in one of the t_{2g} orbitals with an opposed spin. Now, transitions which require spin reversal are formally forbidden, and give rise to very weak bands if they can be observed at all, so that we may neglect any transition of one of the set of five electrons with parallel spins. Thus the only band we are likely to observe is that due to the excitation of the single antiparallel electron from the t_{2g} level to the e_g level. Thus the energy level diagram for d^6 high spin in the octahedral field is the same as that for d^1 in an octahedral field.

In a similar manner, the high spin d^4 configuration in the octahedral field may be described, by the hole formalism, in exactly the same way as d^9 octahedral. Furthermore, d^6 tetrahedral and d^4 tetrahedral are qualitatively described by the same diagrams as, respectively, d^4 octahedral and d^6 octahedral.

Thus, to summarize, by combining Figures 13.23a and 13.23b into one, as shown in Figure 13.24 we can describe all those configurations with one electron in excess of an empty or half-filled d shell and, by the hole formalism, all the configurations with one electron less than a half-filled or filled d shell. This description applies both to the octahedral field, and to the tetrahedral field which causes the reciprocal splitting. Thus Figure 13.24 describes qualitatively the effect

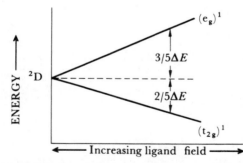

FIGURE 13.23a *Energy level diagram for the d^1 configuration in an octahedral field*

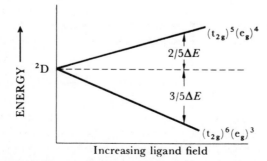

FIGURE 13.23b *Energy level diagram for the d^9 configuration in an octahedral field*

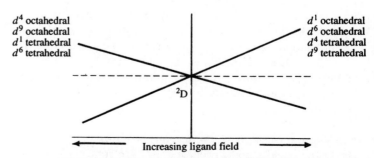

FIGURE 13.24 *Combined energy level diagram for single-electron or single-hole configurations in octahedral and tetrahedral fields.*
The zero crystal field level is shown at the origin and the left portion is the reciprocal of the right hand one. Such diagrams are termed *Orgel diagrams* after their originator.

of the ligand field on d^1 octahedral, d^6 octahedral, d^4 tetrahedral and d^9 tetrahedral, by the portion to the right of the origin, and on d^4 octahedral, d^9 octahedral, d^1 tetrahedral and d^6 tetrahedral by the portion to the left of the origin. For all these configurations, there is only one state above the ground state, separated by ΔE, so that there is only one transition and it occurs at a position in the spectrum corresponding to ΔE.

If the d^2 configuration is now considered, the energy level diagram is found to be more complicated. This is because, while the spin and orbital quantum numbers of a single d electron give rise to only one state, the ^2D state above, the combinations of these quantum numbers for two d electrons give rise to four different states. Some of these have the two electron spins antiparallel, so that transition to them would involve spin reversal. These states may therefore be neglected and we are concerned with two states for d^2, which are labelled ^3F (the ground state) and ^3P and are shown in the centre of Figure 13.25. In an octahedral field, the ^3F state splits up, in a similar manner to the ^2D state in Figure 13.23, but into three components while the ^3P state remains unsplit. Furthermore, interaction between the levels derived from ^3F and ^3P causes some of these levels to curve away from each other, to give the dependence of energy level on ligand field which is shown in Figure 13.25. This interaction depends on a single parameter, B—the Racah parameter—and this may be evaluated in the course of the assignment. Thus, the d^2 diagram, for an octahedral field, has three states above the ground state to which an electron may be excited without changing its spin. This means that the spectrum of a d^2 complex contains three bands, in place of the single band of d^1, and the positions of these three bands together give the

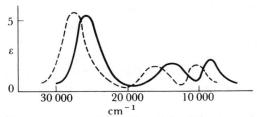

FIGURE 13.26 *The electronic spectra of* $Ni(H_2O)_6^{2+}$ *(———) and* $Ni(NH_3)_6^{2+}$ *(— — — —).*
Each shows the three bands expected for a d^8 system, corresponding to $\Delta E = 8500$ cm^{-1} for the hydrate and $\Delta E = 11\,300$ cm^{-1} for the ammine. The curve is idealized: the experimental spectrum shows a splitting for the middle band.

value of the ligand field for the complex. By the same arguments that applied to d^1, the energy level diagram is qualitatively the same for d^2 octahedral and for d^7 octahedral (high spin) and also for d^8 tetrahedral and d^3 tetrahedral, while the reciprocal diagram applies to d^8 and d^3 octahedral and also to d^2 and d^7 tetrahedral. All these levels are shown in Figure 13.25. A d^8 spectrum is illustrated in Figure 13.26.

Finally, the d^5 high spin configuration has all d orbitals half-filled with parallel spins in the ^6S state. Any electron transition must involve spin reversal so that the $d-d$ bands in a d^5 complex, such as a manganese(II) compound, are very weak. There are four excited states which involve only one spin reversal, and these split up into a total of ten states in the ligand field. Thus, a d^5 complex might show anything up to ten very weak absorptions in the visible and ultraviolet, but since these transitions are formally forbidden, their intensities are about 100 times weaker than normal $d-d$ transitions in an octahedral field. Hence most d^5 compounds look almost colourless, witness the very pale pink of the majority of manganese(II) species.

A further point about intensities is the difference between octahedral and tetrahedral complexes. A further selection rule forbids transitions between states with the same symmetry with respect to a centre of inversion (see Appendix C) and this applies to octahedral complexes but not to tetrahedral ones as the tetrahedron has no inversion centre. Thus tetrahedral complexes are commonly more strongly coloured than octahedral ones, as in the case of the cobalt(II) complexes where the blue of the $CoCl_4^{2-}$ species is much more intense than the pink of the octahedral aquo complex, $Co(H_2O)_6^{2+}$. All these intensities are much less than those of fully allowed transitions, see section 7.6, so that $d-d$ transitions are only observed where allowed bands of the ligands or of the total complex are sufficiently far into the ultraviolet that they do not swamp the weaker $d-d$ bands.

Low spin complexes may be analyzed in a similar way to the above, although the detailed arguments are more difficult and will not be treated here. Further effects found in the spectra must also be considered in a full discussion. For example, Jahn-Teller effects may cause absorption bands to split, as also will interaction between orbital and spin angular momentum. However, all these effects may be taken into

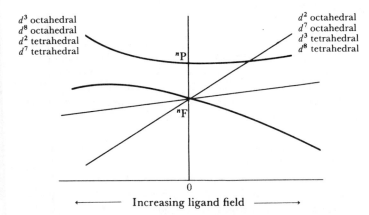

d^3 octahedral
d^8 octahedral
d^2 tetrahedral
d^7 tetrahedral

d^2 octahedral
d^7 octahedral
d^3 tetrahedral
d^8 tetrahedral

^nP

^nF

0

←———— Increasing ligand field ————→

FIGURE 13.25 *Combined Orgel diagram for two-electron and two-hole configurations in octahedral and tetrahedral fields.*
The multiplicities are, for d^2 and d^8, $n = 3$ and for d^3 and d^7, $n = 4$. States of lower multiplicity are omitted. On the d^3 octahedral side of the diagram, the upper state which is derived from the ^4P interacts with the highest level derived from the ^4F and these levels curve away from each other. The other two levels derived from ^4F do not interact with any other and vary linearly with the ligand field.

On the d^2 octahedral side of the diagram, the same two states also interact (derived now from ^3P and ^3F) but, as the ^3F state is now the lowest one, the interaction is less marked.

account and it is found that experimental spectra agree well with those predicted by application of ligand theory.

The major effects in the spectra of transition complexes may be summarized as follows. For configurations d^n where $n = 1, 4, 6$ and 9, one band is expected and its position gives ΔE directly. Configurations with $n = 2, 3, 7$ or 8 give rise to three bands whose positions depend on two parameters, ΔE and B, the latter measuring the interaction between energy levels derived from F and P states. As there are three bands depending on only two parameters, both ΔE and B may be evaluated unambiguously. Finally, when $n = 5$, a much larger number of transitions may be observed which may be analyzed to give the value of ΔE. Low spin configurations for $n = 4, 5, 6$ and 7 may be similarly analyzed to give ΔE values, though the treatment is more complicated. Apart from allowed transitions, the most intense absorptions are found in tetrahedral complexes, those in octahedral species are weak, while transitions in any d^5 high spin complex are extremely weak indeed.

13.12 π bonding between metal and ligands

The molecular orbital discussion of section 13.3 is readily extended to include π bonding between metal and ligands. Such π bonding helps to account for a number of the properties we have discussed, such as the positions of some of the members of the spectrochemical series and the existence of complexes of metals in low oxidation states where the charge would not provide the attractive forces required in the crystal field theory.

Let us consider an octahedral complex: similar considerations will apply for other shapes.

π bonds may be formed between ligand orbitals of suitable symmetry and metal d orbitals of the t_{2g} set or metal p orbitals. The most important interaction is with the t_{2g} set, and Figure 13.27 shows typical examples of the in-phase bonding components formed with ligand p or d orbitals. Clearly such orbitals will concentrate electron density between the metal and the ligands and form π bonding orbitals. The corresponding out-of-phase interactions will be antibonding.

Other π interactions are possible, such as those of t_{1u} symmetry involving the metal p orbitals (Figure 13.31), but these are less significant and can be neglected in a simple treatment. There are also combinations of the ligand π orbitals (labelled t_{1g} and t_{2u}) which are the wrong symmetry to combine with any metal orbitals. We neglect these also, and discuss only the three ligand combinations of the t_{2g} set.

There are two cases to consider (a) where the ligand π orbitals are empty and of relatively high energy and (b) where the ligand π orbitals are filled and relatively stable. Examples of (a) include phosphanes, PR_3, or sulfur ligands like SR_2, where the ligand π orbitals are the empty $3d$ orbitals on P or S. Examples of (b) are fluoride or oxide where π bonds can potentially be formed using the filled p orbitals not involved in sigma bonding.

(a) *Acceptor π ligands*   When the ligand π orbitals are empty, they will commonly be of higher energy than the metal d

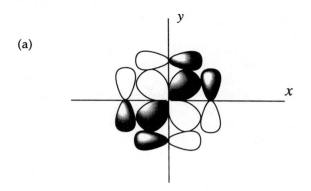

(a)

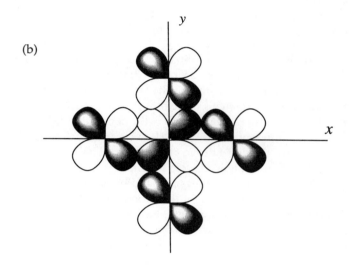

(b)

FIGURE 13.27   *π bonding involving t_{2g} orbitals*
(a) with ligand p orbitals, (b) with ligand d orbitals. Similar components occur in the xz and yz planes.

orbitals. The appropriate energy level diagram is shown in Figure 13.28. The effect of the π interaction is to produce the orbitals, labelled π, which are more stable than the originally non-bonding metal t_{2g} levels together with the corresponding π^* orbitals which lie among the other antibonding levels. The precise relationships among these orbitals will, of course, vary from one compound to the next.

Since the ligand π orbitals are empty, there are no more electrons to place in this scheme than there were in that of Figure 13.7, where only sigma bonds were formed.

Thus the effect of the additional π interaction in this case is to increase the separation, ΔE, between π and e_g^* compared with that between t_{2g} and e_g^* in the simple case. Thus ligands with empty orbitals capable of forming π bonds will have a large value of ΔE, and lie to the right of the spectrochemical series. This explains the positions of ligands such as the phosphines, or of CN^- which accepts into the empty π^* orbitals (Figure 13.29). Note in this latter case that although the π^* orbitals are antibonding as far as the interaction between C and N is concerned, the total $M-C-N$ π interaction is bonding overall—the increase in $M-C$ stabilization exceeds the loss of $C-N$ stabilization.

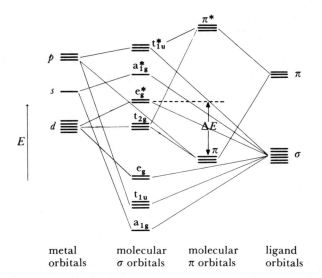

FIGURE 13.28 *Energy level diagram showing the interaction between the d orbitals in an octahedral complex with ligand orbitals of pi symmetry*
The significant effect of such pi-bonding is to create three more stable orbitals using the t_{2g} set and thus increase ΔE.
Note: the diagram indicates only the three ligand π orbitals most directly involved.

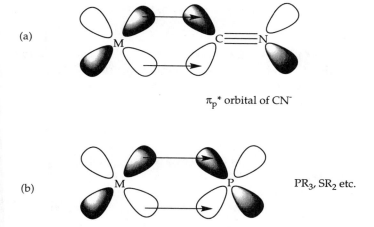

FIGURE 13.29 *π-bonding involving acceptor orbitals on the ligand*
(a) CN^-, (b) phosphane.

This type of interaction involves the transfer of charge from the metal to the ligands as electrons which would have been entirely on the metal in t_{2g} non-bonding orbitals are now shared in the π orbitals. Thus the metal can be regarded as a donor and the ligands as charge acceptors. This interaction clearly serves to offset the build up of charge on the metal which occurs in the sigma bonding. We come back to this in the discussion of carbonyl compounds in Chapter 16.
(b) *Donor π ligands* When the ligand π orbitals are filled, they will commonly be of very similar energy to the sigma levels, and will be more stable than the metal t_{2g} levels. The

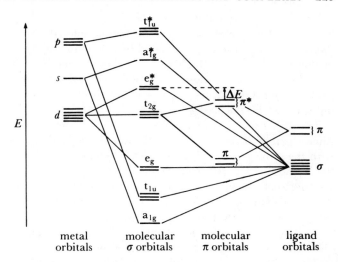

FIGURE 13.30 *Energy level diagram for molecular orbitals in an octahedral complex where pi-bonding occurs from donor ligand orbitals*
Here (contrast Figure 13.28) the pi interaction lowers the filled ligand pi levels to the molecular π orbitals, leaving the molecular π^* orbitals to hold electrons originally in the t_{2g} level. Thus ΔE is decreased.

appropriate change from Figure 13.7 is thus that shown in Figure 13.30. The ligand π orbitals interact with the t_{2g} level to form π orbitals in the complex which are more stable than the ligand level and π^* orbitals in the complex which are less stable than t_{2g}. The energy gap, ΔE, between π^* and e_g^* is thus less than that between t_{2g} and e_g^*.

Let us place electrons in this energy level diagram. The sigma bonding levels are filled as before (section 13.3). Since the ligand π levels are filled, we have these six electrons to accommodate and, if we regard them as entering the molecular π level, their stabilization compared with the ligand π levels will be the main driving force in forming the complex. However, the remaining non-bonding metal d electrons must enter π^* and e_g^*, corresponding to the situation of weak ligand field complexes.

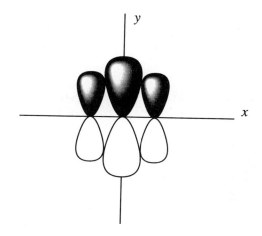

FIGURE 13.31 *Possible π bonding involving p orbitals*
One component of the t_{1u} set.

This type of reaction involves transfer of charge from the ligand to the metal both in the sigma and in the pi interactions. It is therefore favoured by a high formal charge on the metal, i.e. by a high metal oxidation state. It is not always clear whether π donor behaviour occurs, but it is more likely to contribute in, say, the nickel(IV) species NiF_6^{2-} than in the nickel(II) one, NiF_6^{4-}.

The discussion above applies to other shapes, such as square planar or tetrahedral, with differences in detail but with the same general effect on ΔE: (a) empty π orbitals on the ligand increase ΔE, correspond to ligands to the right of the spectrochemical series, and are commonly involved with the lower oxidation states from III downwards, (b) if full π orbitals on the ligand are delocalized onto the metal ΔE is decreased corresponding to ligands on the left of the spectrochemical series, and such π bonding commonly occurs for high oxidation states especially in oxyanions such as CrO_4^{2-} or MnO_4^-.

In all shapes, π bonding can also involve the metal p orbitals, or molecular orbitals derived from them like the t_{1u} levels of an octahedral complex. An example is shown in Figure 13.31. Such interactions are less significant as they involve less stable orbitals and so contribute less to the stabilization. They are included in more detailed treatments, particularly of excited states.

PROBLEMS

This chapter is mainly concerned with the properties of the d^n electrons of a transition element while Chapters 14 and 15 discuss their detailed chemistry. A great deal of reference back and forward between these chapters will be needed.

A first reading should concentrate on sections 13.1 to 13.5.

13.1 *General:* for each oxidation state of each element (as discussed in Chapters 14 and 15), decide the d configuration and hence use Chapter 13 to determine the expected configuration, stability, magnetism, reactivity and spectroscopic properties. Conversely, from Chapter 13 decide which configurations are expected to be stable or otherwise and determine from 14 and 15 how far these predictions are reasonable.

13.2 Draw a diagram analogous to Figure 13.1 for the case ML_3, where the ligands form an equilateral triangle around M (choose z as the axis perpendicular to the ML_3 plane).

13.3 Would Figure 13.3 be altered by choosing new x and y axes at $45°$ to the ones used in Figure 13.2?

13.4 Calculate CFSE values for $M(H_2O)_6^{3+}$ and $M(CN)_6^{3-}$ ions of the first series elements (compare Tables 13.6 and 13.7). Estimate the corresponding values for $M(H_2O)_4^{2+}$ (tetrahedral).

13.5 Calculate the change in CFSE on oxidizing M^{2+} to M^{3+} in water for V, Cr, Mn, Fe and Co.

13.6 The experimental magnetic moments (Bohr magnetons) of a number of complex ions are listed. Comment on (a) the d electron configuration and (b) the validity of the 'spin only' formula.

Moment	*Complex*
1·8	$Ti(H_2O)_6^{3+}$
1·8	$Co(NO_2)_6^{4-}$
2·2	$Mn(CN)_6^{4-}$
3·8	$Cr(H_2O)_6^{3+}$
4·8	$Cr(H_2O)_6^{2+}$
5·1	$Co(H_2O)_6^{2+}$
5·9	$Mn(H_2O)_6^{2+}$

13.7 Draw out each of the orbital combinations of a square planar complex ML_4, as was done for octahedral ML_6 in Figures 13.5 and 13.6. Thus draw up a full qualitative orbital diagram for ML_4 corresponding to Figure 13.7.

13.8 It is thought that the crossover point between high- and low-spin Co(III) comes just at $Co(H_2O)_6^{3+}$. Use this to evaluate K (Table 13.8). Assuming K is constant for the elements of the first transition series, decide which of the complexes in Table 13.6 are likely to be low spin.

13.9 It is valuable to see the relations across the transition series of the major energy parameters. For the elements Ca to Zn, plot 1st, 2nd, and 3rd ionization energies (from Table 2.8) against atomic number. Compare the curve for I_3 with figures 13.20 to 22.

13.10 The curves for I_1 and I_2 in question 13.9 are less regular than the discussion of this chapter would indicate. One reason arises from the variation in electron configuration between the $4s$ and $3d$ orbitals. While I_3 is for the loss of a d electron for each element, I_2 arises from $3d^n4s^1$ to $3d4s^0$ for Ca, Sc, Ti, Mn, Fe, and Zn but for the remaining elements, there is no electron in the $4s$ orbital in the M^+ ion and the ionization refers to the loss of a d electron. By plotting the $4s^1$ loss for the above six elements, and assuming a smooth curve, estimate the energy differences between the $3d^n4s^1$ and $3d^{n+1}4s^0$ configurations for the other elements.

Plot the sum of $I_1 + I_2$ against atomic number. Which elements are anomalous? What is your estimate of the relevant excitation energies?

13.11 Carry out the same exercise as in questions 13.9 and 13.10 for the second transition period, and discuss the results.

Note to questions 13.9 to 13.11. After you have completed these exercises, consult the valuable article by P. F. Lang and B. C. Smith in *Education in Chemistry*, March 1986, 50–53. (Note: these authors use the ionization *number* derived by dividing the ionization energy in wavenumbers by the Rydberg constant to get a dimensionless number, which is often convenient.)

13.12 Values for ionic radii (Shannon, compare Table 2.12) for six-coordination are (pm):

	Sc	Ti	V	Cr	Mn	Fe	Co	Ni	Cu	Zn
M^{2+}		100	93	94	97	92	79	83	87	88
M^{3+}	89	81	78	76	79	79	75	74		
M–F				179.5	181.1	176.9	175.4	172.9	171.3	174.5

(a) Use these values to plot out Figure 13.20 more precisely. (b) The bond lengths in certain difluorides in the gas phase, i.e. linear MF_2 molecules, are also shown above. Construct a similar figure to 13.20 and discuss differences from the curve for six-coordination in terms of the energy level diagram of Figure 13.1.

14 The Transition Elements of the First Series

14.1 General properties

The transition elements of the first series, dealt with here, are those elements, from titanium to copper, where the $3d$ level is filling. Table 13.2, in the last chapter, shows their oxidation states, while Tables 13.3–13.5 give their oxides, fluorides, and other halides. Other properties which are listed earlier include the electronic configurations, Atomic Weights and Numbers (Table 2.5) and the Ionization Potentials (Table 2.8). Some values of the Redox Potentials, especially of species used in quantitative analysis, are given in Table 6.3. The free energy diagrams of all the elements are shown on a small scale in Figure 14.1 so that a general comparison may be made. The individual diagrams for important systems are given on a larger scale in the later sections.

The lists of the oxides and halides and the free energy diagrams give largely complementary pictures of the stabilities of the various oxidation states, in the solid state and in aqueous solution respectively. At titanium, the Group oxidation state of IV is the most stable and the lower states become increasingly reducing. Moving along the series, the Group oxidation state becomes more unstable and more oxidizing so that, at manganese, only a few compounds of the VII state are known and all are strongly oxidizing. Beyond manganese, the Group state disappears and only a few, unstable, strongly oxidizing states greater than III exist. Among the lower states, either the II or the III state is the most stable state from chromium onwards, and the relative stability of these two states varies with the number of d electrons as was indicated in the last chapter, with the II state finally becoming the most stable at nickel and copper. This variation in stability is shown up in the free energy diagrams by the increasing height above zero of the Group state up to manganese, and by the increasing instability of the III state at the right of the Figure. (See also section 14.10.)

Oxidation states which lie between the Group state and the II or III states have a tendency to disproportionate. Some of them, such as Cr(IV) and (V), and Mn(V) and (VI), are very rare and poorly represented. The one state of moderate

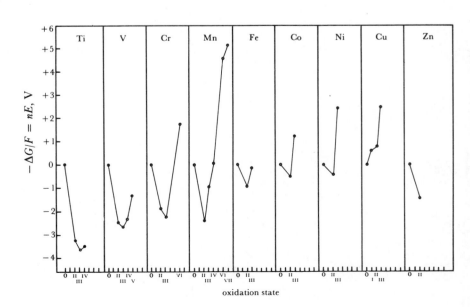

FIGURE 14.1 *Oxidation state free energy diagrams of the First Transition Series elements*
It is clear that the Group Oxidation state becomes increasingly unstable towards manganese. Thereafter, all states above III are unstable, and the variation in the relative stability of the II and III states (and I for copper) is of most importance. (Compare section 8.6.)

stability among these intermediate ones is manganese(IV), which owes part of its importance in chemistry to the insolubility of MnO_2 in neutral or basic solution.

All the elements of the first transition series are common and all are important commercially, titanium and copper in their own right, and the others largely as constituents or coatings of iron and steel. Most are produced in the blast furnace, and very pure nickel is produced in the Mond process (section 14.8). High grade copper for electrical applications is prepared by electrolysis.

The metals are reactive and electropositive with both properties decreasing towards the right of the series. However, a number form a thin layer of coherent and impermeable oxide on the surface and so resist attack. In particular, this is the action of chromium, hence its use in chromium plating to protect iron.

The $3d$ cations are relatively small, so that the charge density on an ion like Fe^{3+} is high. In water, the ion is strongly hydrated as $Fe(H_2O)_6^{3+}$, and the cation withdraws electron density from the water molecules to allow ionization:

$$[Fe(H_2O)_6]^{3+} + H_2O = [(H_2O)_5Fe(OH)]^{2+} + H_3O^+.$$

As a result, hydrates act as acids and their existence depends on the charge density. In particular, no transition element forms a simple M^{4+} ion and $[M(H_2O)_6]^{3+}$ species require strong acid media to prevent deprotonation. The divalent hydrates do exist in more neutral media. Thus, while it is often convenient to write M^{3+} or even M^{4+} species, this should be seen only as a shorthand for all the hydrated/hydroxy species in solution. When more than one OH group is present, a water molecule may be eliminated, forming an oxy species. For example, the di-hydroxy-vanadium (IV) ion, $[V(OH)_2(H_2O)_4]^{2+}$, goes to $[VO(H_2O)_5]^{2+}$, often written simply as VO^{2+} [called vanadyl(IV)], and the $+$ (V) species $[V(OH)_4(H_2O)_4]^+$ gives the vanadyl (V) species $[VO_2(H_2O)_4]^+$ or VO_2^+. The OH groups, or the O group formed by water elimination, may bridge between two metal atoms leading to progressively more complex and polymeric species until eventually the hydroxide, or hydrated oxide, precipitates. Thus in moderately basic media, chromium(III) exists as hydrated $[Cr(OH)]^{2+}$, $[Cr_2(OH)_2]^{4+}$, or $[Cr_3(OH)_4]^{5+}$.

The transition elements form non-stoichiometric hydrides, carbides, oxides, and similar compounds—often termed *interstitial*—where the small atoms are incorporated into the metallic lattice. It is now known that, although the metal atom arrangements in such compounds are commonly in one of the forms characteristic of metals (cubic or hexagonal close packed or body-centred cube), the arrangement of the metal atoms in the interstitial compound is rarely the same as that in the metal itself. The best theory of these compounds regards the included atoms as 'metallic'. They show larger coordination numbers than normal—e.g. six for carbon in TiC— and contribute valency electrons to the common pool of delocalized electrons in the metallic bonding (see also the discussion of metallic hydrides, section 9.4). Interstitial compounds show many metallic properties—hardness, met-

allic appearance, high conductivity with negative temperature coefficient, high melting point—which support this view. The interstitial compounds are among the hardest known and show extremely high melting points, e.g.:

	m.p. (°C)	hardness, Moh's scale
TiC	3140	8–9
HfC	4160	9
W_2C	3130	9–10

The binary compound $4TiC + ZrC$ has the highest recorded melting point of 4215 °C. (The hardnesses given above are on Moh's scale on which diamond $= 10$). The earlier metals, of the titanium, vanadium, and chromium Groups, form interstitial compounds with small metal-nonmetal ratios, such as MC, M_2C, MN, or MH_2 (all the formulae are idealized, nonstoichiometry marks most of these phases), and these have regular structures. Thus the MN and MC compounds are sodium chloride structure and the M_2C ones have defect NaCl structures. These compounds are very unreactive.

The elements of the later Groups form less well-defined compounds with complex metal-nonmetal ratios such as Fe_3C or Cr_7C_3. Such compounds are similar in general physical properties but are much more reactive chemically. They are attacked by water and dilute acids to give hydrogen and mixtures of hydrocarbons.

Borides, and also silicides, are similar to the carbides and nitrides in structures and properties. These provide a link between the interstitial compounds and metal alloys.

14.2 Titanium, $3d^2 4s^2$

The free energy diagram for titanium is shown in Figure 14.2. The titanium(IV) state is the most stable one and is well-characterized with a wide variety of compounds. The III state is reducing but reasonably stable; in water, it is a reducing agent somewhat stronger than tin(II). The II state is very strongly reducing. The only solid compounds are unstable polymerized solids and it rapidly decomposes in water.

Titanium metal is used in high-performance situations where a high strength/weight ratio is needed. It is also resistant to corrosion in harsh environments. Titanium is prepared by calcium or magnesium reduction of $TiCl_4$. Since Ti reacts readily with air to form the nitride, carbide, or hydride, which are all brittle compounds which ruin the metal quality, the final step in the commercial reduction is carried out in an atmosphere of argon. This was the first large-scale use of a rare gas in industrial metallurgy.

While titanium is inert at ordinary temperatures, it does combine with a variety of reagents on heating, see Figure 14.3. Elements such as carbon, silicon or iron increase the strength of titanium but also make it more brittle. They are often added to titanium in controlled amounts to modify its properties.

Apart from the use of the metal, the other major industrial production of titanium compounds is that of TiO_2 which is an

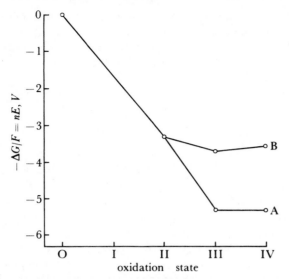

FIGURE 14.2 *The oxidation state free energy diagram of titanium*
The IV state is stable and the lower states reducing. Titanium redox potentials are uncertain: the Ti (IV) potential (A) is for Ti(OH)$^{3+}$ and the value (B) is for TiO$_2$.

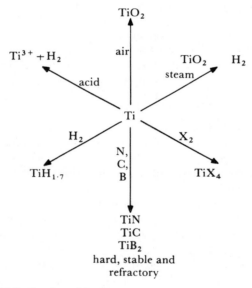

FIGURE 14.3 *Reactions of titanium*

excellent white pigment. TiO$_2$ is prepared by purifying the major ore, *rutile*, or by hydrolysis of TiCl$_4$. An alternative ore is *ilmenite*, an iron-titanium oxide.

Titanium compounds play an important role in Ziegler-Natta catalysis (see section 17.4.3). As well as the long-standing use of TiCl$_4$ in the original catalysts, 'second generation' catalysts based on TiCl$_3$, often supported by MgCl$_2$, give better molecular weight distributions and easier work-up, while homogeneous catalysts have been developed using (C$_5$H$_5$)$_2$TiR$_2$ (for R = Cl or alkyl) plus AlR$_3$. Such C$_5$H$_5$ compounds of titanium have also been tested as

anti-tumour agents. The bonding and chemistry of C$_5$H$_5$-metal compounds are discussed in section 16.4.

14.2.1 *Titanium(IV)*

Titanium dioxide is weakly acidic and weakly basic, dissolving in concentrated base or acid but easily hydrolysing on dilution:

$$\text{conc H}_2\text{SO}_4 \overset{\text{H}_2\text{O}}{\underset{}{\rightleftarrows}} \text{TiO}_2.n\text{H}_2\text{O} \overset{\text{H}_2\text{O}}{\underset{}{\rightleftarrows}} \text{conc KOH}$$

$$\text{(TiO)SO}_4 \qquad\qquad\qquad \text{K}_2\text{TiO}_3.n\text{H}_2\text{O}$$

titanyl sulphate   potassium titanate

No definite hydroxide, such as Ti(OH)$_4$, exists, the compound produced on hydrolysis being a hydrated form of the oxide. The titanyl group, TiO^{2+}, does not exist as the simple monomeric cation. It is present in acidic solutions, but is in equilibrium with oligomeric species such as hydrated Ti$_3$O$_4^{4+}$. The TiO^{2+} moiety may be distinguished only in solids, as in TiOCl$_4^{2-}$, and in the [TiO(ring)] species where ring = porphyrin or phthalocyanin (see Appendix B). In this behaviour, titanium(IV) contrasts with V(IV) and Cr(V), where the MO^{n+} unit does exist on its own. The structure of titanyl sulphate, for example, consists of (TiO)$_n^{2n+}$ chains held together in the crystal by the sulfate groups. The alkali metal titanates do not contain discrete anions but consist of TiO$_6$ octahedra linked into layers or three-dimensional arrays and holding the alkali metals in sites of high coordination (7 up to 10). Thus they are mixed oxides, not oxytitanium anions.

Titanium(IV) forms peroxy compounds which are the formula analogues of the oxy-compounds as shown in Figure 14.4. The yellow colour in acid is extremely intense and is the basis of the colorimetric determination of titanium. A more detailed study of the acid solutions shows that the principal species at pH < 1 is Ti(O—O) (OH)$^+$, and at pH = 1–2 are the ions

$$\left[\begin{array}{c} \text{O} \\[-2pt] \diagdown \\ \diagup \\ \text{O} \end{array} \text{Ti}-\text{O}-\text{Ti} \begin{array}{c} \text{O} \\[-2pt] \diagup \\ \diagdown \\ \text{O} \end{array} \right] (\text{OH})_x^{2-x} \; (x = 1 \text{ to } 6)$$

which then precipitate TiO$_3$$nH_2$O ($n$ = 1 or 2).

The titanium halides may all be made by direct reaction of the metal and halogen, and TiCl$_4$ may be converted to the others by the reaction of the hydrogen halide. All the tetra-halides are readily hydrolysed, although TiF$_4$ is more stable than the others, and the intermediate oxyhalides may be isolated under careful conditions:

$$\text{TiX}_4 \xrightarrow{\text{H}_2\text{O}} \text{TiOX}_2 \xrightarrow{\text{H}_2\text{O}} \text{TiO}_2.n\text{H}_2\text{O}$$

Titanium tetrachloride forms the hexachloride in concentrated hydrochloric acid, and salts, M$_2$TiCl$_6$, are known but are very unstable. The hexafluoride TiF$_6^{2-}$, is readily formed and is very stable. The shape is regular octahedral.

An interesting series of compounds, illustrating the three ways of linking octahedra by bridging groups, is provided by

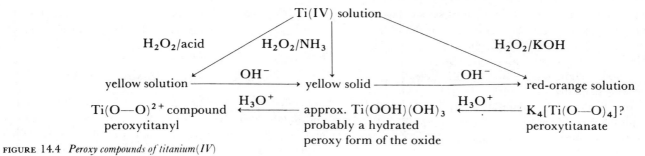

FIGURE 14.4 *Peroxy compounds of titanium(IV)*

the species formed by the action of TiF_4 on TiF_6^{2-} in liquid SO_2 as solvent. As the ratio varies from $1:3$ through $1:1$ to $3:1$, the ions $Ti_2F_{11}^{3-}$, $Ti_2F_{10}^{2-}$ and $Ti_2F_9^-$ are formed, with the structures shown in Figure 14.5 suggested by fluorine nmr.

The oxyfluoride may also accept F^- ions, giving complexes like $TiOF_3^-$, and, just as there are peroxy analogues of oxyanions, so peroxyfluoroanions such as $[Ti(O_2)_2F_2]^{2-}$ and $[Ti(O_2)F_5]^{3-}$ exist. If the O—O unit is seen as occupying an O position, then these are pseudo-tetrahedral and -octahedral respectively. Alternatively, since the O—O unit is bonded edge-on (compare, for example Figure 14.12), then these peroxyanions may be described respectively as 6- or 7-coordinated.

The tetrahalides act as acceptors to a wide variety of donor ligands such as R_3P, R_2O or py, to give complexes TiX_4L_2. These are usually in the *cis* configuration unless the ligands are bulky. $Ti_2Cl_{10}^{2-}$ has the same structure as the fluoride (Figure 14.5b). Among five-coordinate complexes are the unusual hydride derivative $TiCl_4.AsH_3$, where the arsane molecule is thought to occupy an equatorial position in a

trigonal bipyramid, and the compound $Et_4N^+(TiCl_5)^-$ where the $TiCl_5^-$ ion may be trigonal bipyramidal like the isoelectronic $SnCl_5^-$ ion. The oxychloride also forms a five-coordinate species $TiOCl_2.NMe_3$.

Two further simple titanium(IV) compounds are of interest. In the anhydrous nitrate, $Ti(NO_3)_4$, the nitrate groups form a tetrahedron around the titanium, but are each bonded through two oxygen atoms so that the titanium is eight-coordinate. The coordination is nearly regular dodecahedral. Although the perchlorate ion, ClO_4^-, does not usually coordinate, it is found that $TiCl_4$ reacts with anhydrous $HClO_4$ to form $Ti(ClO_4)_4$ which is a volatile solid. It has an 8-coordinate structure analogous to $Ti(NO_3)_4$. The only other volatile perchlorate is $Cu(ClO_4)_2$.

14.2.2 *Titanium (III) and lower oxidation states*

Titanium(III) compounds are readily formed by reduction and, as they contain a *d* electron, are coloured. Some preparations are shown in Figure 14.6. The existence of differently-coloured hydrates of titanium trihalides arises as both water and halide may be directly coordinated to the titanium ion. The violet trichloride is $[Ti(H_2O)_6]^{3+}Cl_3^-$ while the green form is $[Ti(H_2O)_5Cl]^{2+}Cl_2^-$.

Titanium(III) is much more basic than titanium(IV) and the purple, hydrated oxide, $Ti_2O_3.nH_2O$, which is precipitated from titanium(III) solutions by base, is insoluble in excess alkali. Anhydrous TiOCl is formed by heating TiO_2 with $TiCl_2$ at $700\,^\circ C$. The pure compound is stable in air.

The titanium(III) ion is a d^1 system with the electron in

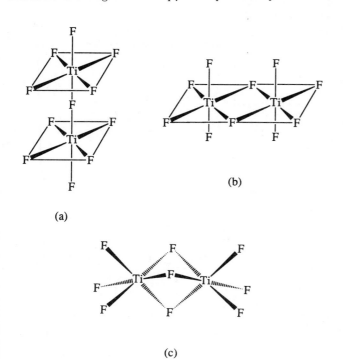

(b)

(a)

(c)

FIGURE 14.5 *The dinuclear fluorotitanium anions*
(a) $Ti_2F_{11}^{2-}$, (b) $Ti_2F_{10}^{2-}$, (c) $Ti_2F_9^{2-}$.

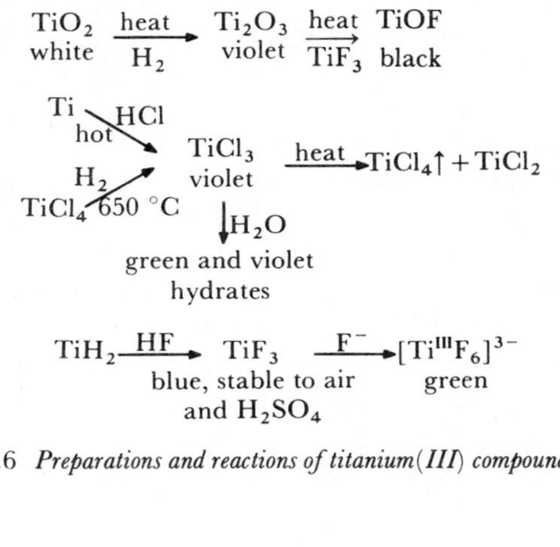

$$TiO_2 \xrightarrow[\text{H}_2]{\text{heat}} Ti_2O_3 \xrightarrow[\text{TiF}_3]{\text{heat}} TiOF$$
white   violet   black

$$Ti \xrightarrow[\text{hot}]{\text{HCl}}$$
$$H_2$$
$$TiCl_4 \; 650\,^\circ C$$
$$\searrow \; TiCl_3 \xrightarrow{\text{heat}} TiCl_4\uparrow + TiCl_2$$
violet
$$\downarrow H_2O$$
green and violet
hydrates

$$TiH_2 \xrightarrow{\text{HF}} TiF_3 \xrightarrow{F^-} [Ti^{III}F_6]^{3-}$$
blue, stable to air   green
and H_2SO_4

FIGURE 14.6 *Preparations and reactions of titanium(III) compounds*

a t_{2g} orbital. Only one $d-d$ transition is possible, Figure 13.23a, and a simple spectrum with one band in the visible region is observed, Figure 13.4. Magnetic measurements on titanium(III) compounds give values close to the spin-only value of 1·73 Bohr magnetons for one unpaired electron.

An interesting structure is found for $Ti(BH_4)_3$, which is volatile at room temperature. Each BH_4 group is linked to Ti by three bridging hydrogens, giving 9-coordinate Ti(III). Gas phase electron diffraction shows that the Ti and the 3 B atoms are coplanar.

Titanium(II) is a very unstable state represented only by solid compounds. The dihalides may be made by heating the trihalides, as indicated in Figure 14.6 above, but the dihalide disproportionates to metal plus tetrahalide at a temperature below that of its formation so that it is never obtained un-contaminated by metal. The oxide, TiO, is made by heating the dioxide with titanium.

The titanium-oxygen system is, in fact, an extremely complex one with at least twenty phases with different structures recognized. Some of the main ones are as follows:

Up to $TiO_{0.5}$	Ti atoms hcp with O in trigonal prismatic sites so $TiO_{0.33}$ is anti-BiI_3 and $TiO_{0.5}$ is anti-CdI_2 (layer structures with O in the Bi or Cd sites—see Table 5.3).
Around TiO	NaCl structures but defective with equal numbers of cation and anion sites empty (Schottky defects, see section 5.10). At $TiO_{1.0}$, at room temperature, one-sixth of the sites are vacant.
Ti_2O_3	corundum structure (Figure 5.4b)
Ti_3O_5, Ti_4O_7 etc to $Ti_{10}O_{19}$	TiO_6 octahedra sharing edges forming sections of rutile structure joined by shared faces.
TiO_2	Rutile (Figure 5.3b) together with two other forms.

Thus TiO_2 and Ti_2O_3 are stoichiometric phases but distinct species exist between these Ti(III) and Ti(IV) oxides. TiO is part of a range of structures while the 'lower' oxides—formally Ti in oxidation states of I or less—are better understood in terms of hcp Ti metal taking up oxygens and filling sites in a regular manner.

Titanium(II) is thought to be the active intermediate in one of the few reversible systems of *nitrogen fixation* so far announced. A number of halides (including $TiCl_4$) yield a species which fixes N_2 gas when reduced by organometallic compounds but these produce only NH_3 on hydrolysis with the concomitant destruction of the active intermediate (see under nitrogen, section 17.6). In a recent study, it was shown that nitrogen could be fixed and converted to ammonia by a cycle, involving titanium alkoxides, which also regenerated the starting reagent so that an overall catalytic system is possible. The main steps of the reaction are

$$Ti(OR)_4 + 2Na \rightarrow Ti(OR)_2 + 2RONa \quad (14.1)$$

$$Ti(OR)_2 + N_2 \rightarrow [Ti(OR)_2N_2]_n \quad (14.2)$$

$$[Ti(OR)_2N_2]_n + 4Na \rightarrow [\text{intermediate}] + 6ROH = 2NH_3 + Ti(OR)_4 + 4RONa \quad (14.3)$$

Thus, the overall reaction is

$$N_2 + 6e^- + 6ROH \rightarrow 2NH_3 + 6RO^-$$

The sodium source in equations (14.1) and (14.2) is the naphthalide complex, $Na^+C_{10}H_8^-$. In this, the valency electron of the sodium is transferred to the lowest antibonding π level of the aromatic hydrocarbon, where it is readily available for reductions but is in a more manageable form than if the metal was used alone (compare section 10.3). The critical step is the uptake of N_2, from the gas phase at normal temperature and pressure, by the titanium(II) alkoxide. The formulation of the reduced intermediate is still uncertain.

This cycle does not consume titanium and is thought to model the biological nitrogen-fixation process which also probably acts through the reversible formation of a low valency transition metal intermediate to which the nitrogen becomes coordinated. A model for the coordinated nitrogen is a sideways-bound N_2 (compare section 16.10), and such a unit is involved in nitrogen fixation by the organotitanium system $(C_5H_5)_2TiCl_2$-Mg which gives N^{3-} as the initial fixed form.

One compound of titanium in the oxidation state 0 has been reported. This is the compound $Ti(dipy)_3$, where dipy = 2,2'-dipyridyl (Figure 14.7). The oxidation states $-I$ and $-II$ are reported in the same system, in the compounds $Li(Ti\ dipy_3)3.5THF$ and $Li_2(Ti\ dipy_3).5THF$ respectively. These compounds result when $TiCl_4$ is reduced in the presence of dipyridyl by lithium in tetrahydrofuran (THF).

14.3 Vanadium, $3d^34s^2$

The free energy diagram for vanadium is shown in Figure 14.8. There are five valency electrons and oxidation states

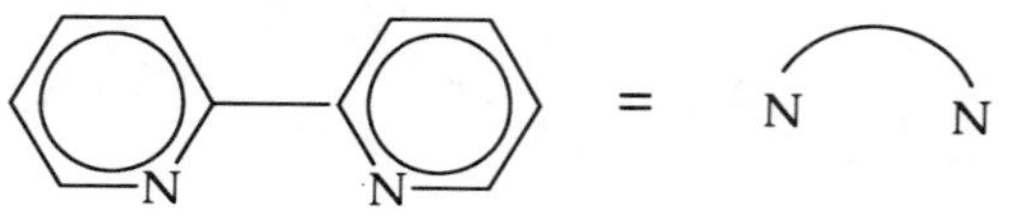

2,2' dipyridyl (dipy)

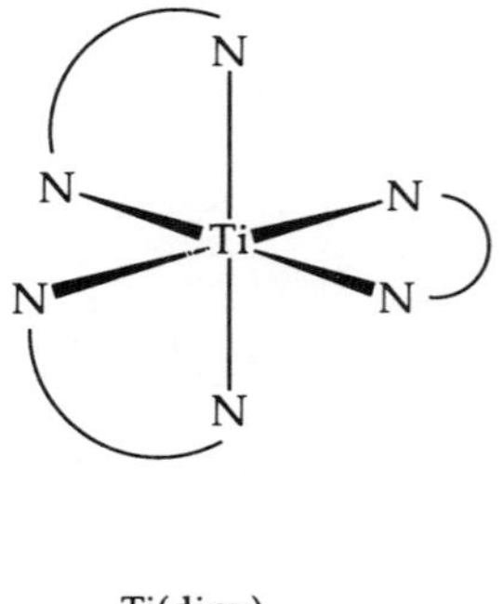

Ti(dipy)₃

FIGURE 14.7 *2,2'-dipyridyl, and its titanium*(0) *compound*, $Ti(dipy)_3$

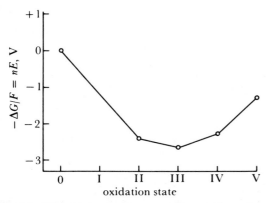

FIGURE 14.8 *The oxidation state free energy diagram of vanadium*
(Compare section 8.6)

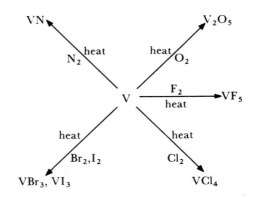

FIGURE 14.9 *Reactions of vanadium*

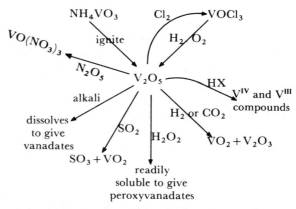

FIGURE 14.10 *Preparation and reactions of vanadium pentoxide*

from V to $-I$ are known, with the ones from II to V of importance. Vanadium(V) and (IV) are both stable, with the former mildly oxidising and represented mainly by oxy-species. Vanadium(IV) is stable, but shows disproportionation reactions in the solid when volatile species, like VF_5, can be driven off. Vanadium(III) is reducing but less so than Ti(III); it is stable to water and is only slowly oxidized in air. Vanadium(II) is strongly reducing; it attacks water and is rapidly oxidized in air. Again, in the solid state, its compounds tend to disproportionate, yielding vanadium(III) and the element. The oxidation state free energy diagram shows that none of the intermediate states disproportionates in solution.

The metal resembles titanium in readily forming carbides and nitrides, but as it is seldom used alone this is less of a disadvantage. Its main use is in steels and it is usually produced by reducing a mixture of V_2O_5 and Fe_2O_3 with aluminium to form ferrovanadium which is added directly in the steel manufacture. Vanadium metal is prepared by reduction with calcium of the pentoxide, V_2O_5. Pure samples of the metal are best prepared by the van Arkel process in which the iodide is decomposed on a hot filament under vacuum. The pure metal resembles titanium in being relatively unreactive at low temperatures. Although it is thermodynamically a strong reducing agent, vanadium easily becomes passive and reacts readily only with oxidizing agents such as nitric acid. When heated, vanadium does react, as shown in Figure 14.9. Note that only O_2 or F_2 give the (V) state.

Particular interest in aqueous vanadium chemistry since the early eighties has stemmed from the discovery that vanadium occurs in the nitrogen-fixing system of some bacteria. This is a second fixing system, independent of the better known one which involves molybdenum. Vanadium is also involved in peroxidases of some marine organisms and vanadate, in combination with hydrogen peroxide, mimics the effect of insulin (compare Chapter 20).

14.3.1 *Vanadium (V)*

Simple compounds found in the V state are V_2O_5, VF_5, VO_2X ($X = F$, Cl) and all the oxyhalides, VOX_3. The

pentoxide is prepared, and reacts, as shown in Figure 14.10. It is amphoteric, dissolving in acid and base to give a variety of species. In strong base, the mononuclear vanadate ion, VO_4^{3-}, is present and, as the pH is reduced, these units link up into binuclear species, and then into polynuclear ions until, at pH about 6, $V_2O_5.nH_2O$ is precipitated. In more acidic solutions, this dissolves to form cationic vanadyl species. The approximate species present at different pH values are shown in Table 14.1, together with the colour changes.

While the lower vanadium oxide systems are similar to the titanium species, being based on VO_6 units, vanadium(V) oxide and the vanadium(V) oxyanions are based on VO_4 tetrahedral units. An extensive series of polyvanadates is known, not all with determined structures. One species which has been characterized is $HV_4O_{12}^{3-}$. This is formed of four VO_4 tetrahedra linked into a ring by sharing oxygen. The $V-O-V$ bridges have relatively open angles at oxygen of $132–146°$. The H protonates one of the non-bridging oxygens. The parent tetravanadate $[V_4O_{12}]^{4-}$ has recently been found to have a very similar structure. Anhydrous MVO_3 species (M = alkali metal) have infinite chains of linked VO_4 tetrahedra while $KVO_3.H_2O$ is made up of edge-linked VO_5 units.

Complex oxyanions of vanadium(V) are becoming established whose building blocks are VO_6 units and which parallel the features of the polymolybdates and polytungstates

TABLE 14.1 Vanadium(V) species in basic and acidic solutions

pH	above 12	12−9	9−7	7−6·5
number of V atoms	1	2	3 → 4	5 → ∞
approximate formula	$(VO_4)^{3-}$	$[V_2O_6(OH)]^{3-}$	$(V_3O_9)^{3-}$	$V_5O_{14}^{3-}*$, $V_5O_{16}^{5-}*$
colour	←──────── colourless ────────→			red → red-brown

pH	6·5−2·2	below 2·2
number of V atoms	∞	10⇌1 (V_1 being favoured as the pH is lowered)
approximate formula	$V_2O_5.nH_2O*$	$V_{10}O_{28}^{6-}$ ⇌ VO_2^+ (or VO^{3+})
colour	brown	yellow

*These species are representative of solids precipitated at the appropriate pH; they do not necessarily correlate with the formulae of species in equilibrium with them in solution. The other formulae apply in solution; the V_{10} species is rather well established and, for example, V_9 or V_{11} give a much poorer fit to the data.

(section 15.4.1, especially Table 15.1). In particular, vanadium species $[XV_{12}O_{40}]^{n-}$, known as Keggin structures, are under study where an XO_4 tetrahedron is encapsulated in the structure of Figure 15.14a, formed by 12 linked MO_6 units sharing edges. In a recent determination the anion $[AsV_{12}O_{40}(VO)_2]^{9-}$ has been characterized which has the classical Keggin structure of a central AsO_4 tetrahedron within 12 VO_6 octahedra. A special feature is the presence of two VO^{3+} ions in cavities of the structure with a square pyramid $O=VO_4$ configuration. The presence of these additional cations avoids the very high charge which may well prevent the formation of the simple $[AsV_{12}O_{40}]^{15-}$ cluster.

Such compounds illustrate the strong tendency towards forming polymeric anions in this part of the Periodic Table. The most explored and striking exemplars are vanadium, molybdenum and tungsten, but their neighbours behave similarly though with a less extensive range of compounds.

In the early nineties, attention was focused on a further striking class of polyvanadates where neutral molecules and anions are enclosed within basket-like or shell structures. In the $[Ph_4P]^+$ salt, the anion $[V_{12}O_{32}]^{4-}$ may incorporate an acetonitrile molecule with the MeCN unit held 'head downwards' in an open basket structure (Figure 14.11a) and the interaction was shown to persist in solution by nmr spectroscopy. Even more surprising was the incorporation, despite the like charges, of a halide or carbonate anion within $[V_{15}O_{36}]^{5-}$, which could be described as a basket with a lid.

With the incorporation of a range of anions within polyvanadate structures (Figures 14.11b,c) it is becoming clear that the guest anion is acting as a template on which the polyvanadate builds up. An alternative view is that the polyvanadate unit is acting as a macro-ligand analogous to the cryptands and crown ethers (compare section 10.5).

Similar clusters may contain vanadium(IV) as well as (V), and there is even an example of one in which two ammonium ions and two chloride anions are encapsulated (Figure 14.11d). Related fully reduced species, all vanadium(IV), are well-established, with interesting magnetic and redox properties.

Vanadium forms a red peroxy cation in acid peroxide solution and this is probably the peroxy-analogue of the vanadyl cation, $V(O-O)^{3+}$. In alkaline solution, a yellow pervanadate is formed with two peroxy groups per vanadium; the simplest formulation is $[V(O-O)_2O_2]^{3-}$. The use of stronger base and 30% H_2O_2 gives the blue, fully-substituted $[V(O_2)_4]^{3-}$ and also blue $[V(O_2)_3]^-$.

Other ligands are found in peroxy complexes, as in $[VO(O_2)_2(C_2O_4)]^{3-}$ (Figure 14.12) and the related complex with bidentate CO_3^{2-} replacing the oxalate. In these complexes, the VOO triangles are unsymmetric with the shorter V−O bonds *trans* to the oxalate. Similar ions are known with F or Cl as ligands in $[VO(O_2)_2X]^{2-}$. Treatment of the pentoxide with H_2O_2 in presence of F^- in strong base gave deep blue $[V(O_2)_3F]^{2-}$, and the corresponding chloride is known. Thus from one to four O atoms in a vanadium-oxygen species may be replaced by an $O-O$ group.

The only pentahalide of vanadium is VF_5. This results from fluorine plus the metal at 300 °C or by disproportionation:

$$VF_4 \xrightarrow[N_2]{600\ °C} VF_5{\uparrow} + VF_3$$

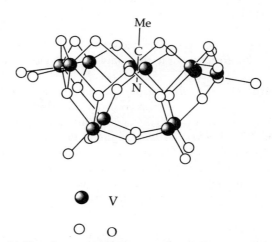

FIGURE 14.11a *Structure of the* $[V_{12}O_{32}]^{4-}$ *ion showing the incorporated acetonitrile* (CH_3CN) *molecule*

● V

○ O

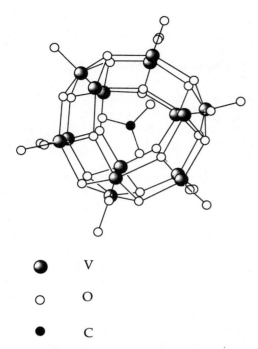

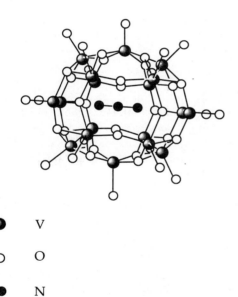

O   V

○   O

●   C

FIGURE 14.11b *Structure of the* $[V_{15}O_{36}]^{5-}$ *ion showing the included carbonate ion*

O   V

○   O

●   N

FIGURE 14.11c *Structure of the* $[H_2V_{18}O_{44}(N_3)]^{5-}$ *ion showing the encapsulated azide* (N_3^-) *ion*

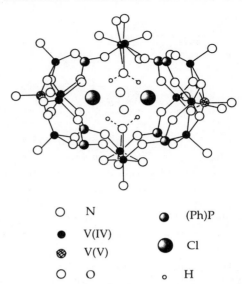

○ N		● (Ph)P
● V(IV)		
⊗ V(V)		● Cl
○ O		○ H

FIGURE 14.11d *A polyvanadate ion cluster with a central cavity which contains two* NH_4^+ *ions and two* Cl^- *ions*
The cluster can be viewed as two $[V_5O_9]^{3+}$ units which contain bottom. These are linked up to form the cavity by eight $PhPO_3$ right-hand sections, and two $V(IV)_2(\mu\text{-}O)_2$ units at the top and bottom. These are linked up to form the cavity by eight $PhPO_3$ units (two at each corner). This cavity contains the Cl^- to the left and right and the NH_4^+ in front and behind. The ammonium ions are linked by H-bonding, part of which is indicated. Note all the H were located by crystallography.
The overall formula is $[2NH_4^+, 2Cl^- \subset V_{14}O_{22}(OH)_4(H_2O)_2(PhPO_3)_8]^{6-}$ where $\subset$ indicates that the group is encapsulated within a cavity.

structure $VO_2F_2(F_{2/2})$. The O atoms are *trans* to the bridging fluorines. The hexafluoro- complex ion also exists and is made by reacting VCl_3 and an alkali halide in liquid BrF_3 which acts as a fluorinating and oxidizing agent:

$$Cl^- + VCl_3 + BrF_3 \rightarrow VF_6^- + ClF + BrF$$

Vanadium pentafluoride reacts with air to form the oxyfluoride, which also may be made by oxidizing or fluorinating

The pentafluoride is a volatile white solid melting at 20 °C. The crystal structure is composed of chains of VF_6 octahedra with two bridging fluorines in *cis* positions to each other. Such a structure is indicated by the useful nomenclature: $VF_4(F_{2/2})$ to show the four F atoms bonded only to one vanadium and two F atoms shared between two vanadiums, and also the overall six-coordination. A related structure is shown by the oxyfluoride ion, $VO_2F_3^{2-}$, which also contains six-coordinated vanadium with *cis* fluorine bridges in the

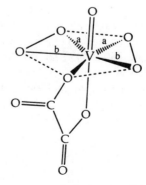

FIGURE 14.12 *The structure of the peroxy (oxalate) vanadate* (V), $K_3VO(O_2)_2(C_2O_4)$
Bond lengths to peroxide oxygens (a) are 186 pm, (b) are 192 pm. The four peroxy O atoms and one oxalate O lie approximately in a pentagon, with the second oxalate O and the V = O completing a pentagonal bipyramid.

suitable compounds:

$$VF_3 + O_2 \qquad VF_5 \qquad VOCl_3 + HF$$

$$\searrow \quad \downarrow air \quad \swarrow$$

$$VOF_3$$

m.p. 300 °C, very reactive

The oxychloride is made by the reaction of chlorine on any of the vanadium oxides, and the oxybromide similarly. The oxyhalides are relatively volatile. An electron diffraction study shows $VOCl_3$ is a trigonal pyramid in the gas phase, and the monomeric structure probably persists in the solid. A second oxyfluoride, VO_2F, is now established, together with the Cl analogue which adds Cl^- to give the tetrahedral ion $VO_2Cl_2^-$.

$VOCl_3$ is violently hydrolysed by water and is inert to metals, even to sodium. It shows no reaction with salts and dissolves many non-metals which suggests possible uses as a non-aqueous solvent.

The structures of two vanadium(V) sulfate species have been established. $[VO(SO_4)_2]^-$ consists of a $V=O$ unit linked by the O atoms of five different sulfate groups into a polymeric structure—that is, there are $O=VO_5$ octahedral units linked in chains via $\cdots-V-O-SO_2-O-V-\cdots$ linkages. In $V_2O_3(SO)_{42}$ similar VO_6 octahedra are present, this time linked into chains with successive pairs of V joined by triple bridges consisting of two $V-O-SO_2-O-V$ connections and one $V-O-V$ connection.

14.3.2 Vanadium (IV)

Vanadium(IV) is readily prepared by mild reduction of vanadium(V), for example by ferrous salts. The oxide, VO_2, is formed from V_2O_5 in this way and is dark blue. It reverts to the pentoxide on heating in air. It is also mildly amphoteric, but is much more basic than acidic. It dissolves in acid to give blue solutions of the vanadyl(IV) ion VO^{2+}, and a variety of salts containing this cation are known. It dissolves in alkali to form vanadites which are readily hydrolysed and relatively unstable. A number of vanadate(IV) anions are known, such as VO_3^{2-}, VO_4^{4-}, and $V_4O_9^{2-}$. These are generally prepared by reaction of VO_2 with alkali, or alkaline earth oxides in the fused state. The best known vanadyl(IV) salts are the sulfate, $VOSO_4$, and the halides, VOX_2. A number of complexes of vanadium(IV) are known, all derived from the vanadyl ion. Examples include the halides, VOX_4^{2-}, and corresponding oxalate and sulfate compounds. Neutral complexes are formed by the enol forms of β-diketones as

in vanadylacetylacetonate, $VO(acac)_2$, which has the square pyramidal coordination shown in Figure 14.13. Lone pair donors can coordinate weakly in the sixth position to complete the octahedron as in $VO(acac)_2C_5H_5N$.

The tetrachloride is formed by direct reaction and gives the tetrafluoride with HF. Alternatively, VF_4 may be prepared directly in crystalline form from vanadium powder and elemental fluorine in an autoclave. The structure of VF_4 contains layers of corner-sharing VF_6 octahedra, similar to the structure of SnF_4.

In an unusual reaction, a product with two different complex chloride ions of vanadium(IV) was formed when VCl_4 was treated with $Ph_4P^+Cl^-$ to give $(Ph_4P)_2^+(V_2Cl_9)^-$ $(VCl_5)^-$. The VCl_5^- ion is a trigonal bipyramid. The $(V_2Cl_9)^-$ ion has a structure often found for M_2R_9 species where three R groups bridge the two M, completing six-coordination at each (alternatively described as two octahedra sharing a triangular face—compare Figures 14.5(c) and 16.3(d)). In $(V_2Cl_9)^-$ there is also a long weak $V\cdots V$ interaction. Hydrolysis of the tetrahalides leads to VOX_2. The interrelationships are summarized in Figure 14.14. The tetrabromide is very unstable and VI_4 is known only in the vapour above VI_3. VCl_4 contains one d electron, which might be expected to lead to some distortion in the shape, but the molecule is tetrahedral in the gas. The liquid is dimeric. Notice, in Figure 14.14, the tendency for these halogen compounds of the IV state to disproportionate to give the III and V states.

14.3.3 Vanadium (III)

Vanadium(III) is produced by fairly strong reduction by hydrogen or red-hot carbon. It resembles other trivalent ions such as chromium or ferric in size and general properties, apart from reduction behaviour. The solid oxide and trihalides are known and the state exists in aqueous solution as the green $V(H_2O)_6^{3+}$ ion.

This state has no acidic properties: the action of alkali on the solution of V^{3+} precipitates hydrated $V(OH)_3$ which has

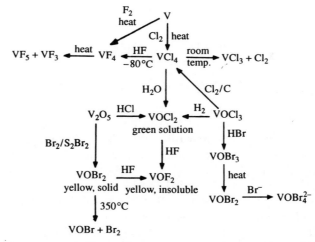

FIGURE 14.14 *Preparation and reactions of vanadium tetrachloride and oxychlorides*

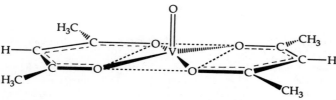

FIGURE 14.13 *Structure of vanadylacetylacetonate*, $VO(acac)_2$

no tendency to redissolve. The green hydroxide oxidizes rapidly in air, and the solutions are also oxidized in air.

All four trihalides exist. VBr_3 and VI_3 are the products of direct combination with the metal while VF_3 and VCl_3 result from the disproportionation of the tetrahalides, as shown in Figure 14.14. The trichloride itself disproportionates on heating and some of the vanadium chloride relationships are given in Figure 14.15. It will be seen that heating with removal of chlorine gives VCl_2, while heating in excess of chlorine gives the tetrachloride. VCl_3 forms a $2:1$ adduct with trimethylamine, $VCl_3.2NMe_3$, which is a trigonal bipyramid with axial NMe_3 groups. The triiodide also disproportionates

$$VI_3 \rightleftharpoons VI_2 + VI_4 \quad \text{(decomposes)}$$

The structure of VI_3 is a layer lattice like BiI_3 (Figure 5.10).

Vanadium(III) also forms a triacetate which exists as a dimer. Its structure is a sandwich one, with four bridging acetates and one terminal one on each vanadium, $AcOV(OAc)_4VOAc$. This is similar to the chromium(II) acetate (Figure 14.20) or copper acetate structure (Figure 14.33) with terminal acetates in place of the two water molecules.

Oxyhalides, VOX, are reported for $X = F$, Cl and Br. VOF has the rutile structure with both F and O in the oxygen positions of TiO_2. The corresponding $TiOF$ is isostructural.

The V^{3+} ion forms a variety of complexes, typical of transition ions. All the complexes are labile as expected of a d^2 ion where the empty t_{2g} orbital presents a low-energy path for attack. The commonest shape is octahedral and most complexes contain vanadium coordinated to oxygen, nitrogen or halogen. The hexafluoride and related complexes are stable, e.g. VF_6^{3-} and $VF_4(H_2O)_2^-$, but other halogen complexes are more readily oxidized. A cyanide, $V(CN)_6^{3-}$, can be made in alcohol but precipitates $V(CN)_3$ on addition of water. A second cyano-complex is $V(CN)_7^{4-}$, an example of the relatively rare coordination number seven. It has a pentagonal bipyramid structure with the five equatorial ligands close to a pentagon with bond angles averaging $72°$. The axial $C-V-C$ angle is slightly bent at $171°$. While the pentagonal bipyramid structure is found for molecular IF_7 (p. 370), 7-coordinate complexes commonly show one of the alternative structures of Figure 15.1 or Figure 15.4.

Among the complexes formed by the heavier vanadium(III) halides is $(V_2Cl_9)^{3-}$ and the corresponding bromide. This has the face-sharing octahedral structure, which we met above for the analogous vanadium(IV) complex $(V_2Cl_9)^-$, except that the structure is fully symmetrical with no $V\cdots V$ interaction. Analogous M(III) dimeric chloro-complexes are now known for Ti, Cr, and a number of heavy transition metals. The structure type recurs in a vanadium(II) species, formed by reducing VCl_3 with Zn in THF. This yields the $\lceil V_2Cl_3(THF)_6\rceil^-$ ion which has the structure $(THF)_3V(Cl)_3V(THF)_3$, again with three Cl bridges.

14.3.4 *Vanadium (II) and lower oxidation states*

The vanadium(II) state is strongly reducing and evolves hydrogen from water, although it is more stable in acid solution. The oxide, VO, is made by reaction of the pentoxide with vanadium and is commonly non-stoichiometric. All four halides, VX_2, exist. The chloride and iodide result from disproportionation of the trihalides, the dibromide from reduction of VBr_3 with hydrogen. VF_2 is made by the action of HF on VCl_2 or by the reduction of VF_3 by H_2/HF at $1200\,°C$. VF_2 forms blue crystals with the rutile structure while VI_2 exists in the CdI_2 layer lattice (see Figures 5.3 and 5.10). The dihalides dissolve in water to give violet solutions, from which $V(OH)_2$ is precipitated by alkali. The solutions soon turn green due to formation of $V(H_2O)_6^{3+}$. All these compounds are unstable and readily oxidized. The divalent ion in solution is probably octahedral and a few complexes are isolable, including the cyanide, $K_4V(CN)_6.7H_2O$. $K_4[V(CN)_7]$ is quantitatively reduced by potassium in liquid ammonia, first to $K_4[V(CN)_6]$ and then to the vanadium (I) species $K_5[V(CN)_6]$.

Low oxidation states are represented by a few compounds including dipyridyl complexes of vanadium I, 0, and $-$I, $V(dipy)_3^+$, $V(dipy)_3$, and $V(dipy)_3^-$, the latter being isolated as the etherate of the lithium salt, as in the case of titanium. The carbonyl, $V(CO)_6$, is known and differs from the sequence of first row carbonyls in being monomeric in the gas and therefore in not having 18 electrons on the vanadium. The anion, $V(CO)_6^-$, is also known and is a further case of $V(-I)$. In all the low oxidation state compounds, the geometry is octahedral.

Several vanadium compounds, including $VO(acac)_2$, $V(acac)_3$, and $(C_5H_5)_2VCl_2$, have been explored as homogeneous Ziegler–Natta catalysts (compare section 17.4.3).

14.4 Chromium, $3d^5 4s^1$

The oxidation state free energy diagram for chromium, Figure 14.16, illustrates the increased oxidizing power of the Group oxidation state of VI, as compared with the cases of titanium and vanadium, and the stability of the III state. This is further shown in the solid state by the existence of CrO_3 and the unstable CrF_6 only (Tables 13.3–5). The III state corresponds to the d^3 configuration and is very stable. Other states shown range to chromium ($-$II) in carbonyl anions but only the II state is well-represented, the IV and V states disproportionating very readily.

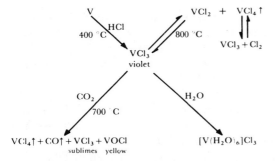

FIGURE 14.15 *The vanadium-chlorine system*

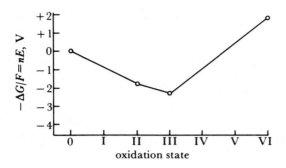

FIGURE 14.16 *The oxidation state free energy diagram for chromium*
It is clear that Cr(III) is stable, Cr(VI) is strongly oxidizing, and Cr(II) is reducing. The other states disproportionate, but data are not very reliable.

Chromium metal is formed by aluminium reduction of Cr_2O_3. Like titanium and vanadium it is unreactive at low temperatures, due to the formation of a passive surface coating of oxide. The main use of chromium is as a protective plating, and as an additive to steels. It dissolves in hot hydrochloric or sulfuric acids and reacts with oxygen or the halogens on heating to give chromium (III) compounds, and with hydrogen chloride to give $CrCl_2$ and hydrogen.

14.4.1 *Chromium (VI)*

The limited number of chromium(VI) compounds includes CrO_3, the chromium oxyions $Cr_nO_{3n+1}^{2-}$ ($n = 1, 2, 3, 4$), $CrOX_4$ (X = F, Cl), CrO_2X_2 and CrO_3X^- (X = F, Cl, Br, I) and peroxy compounds. CrO_3 reacts with F_2 to yield a volatile lemon-yellow solid when the reaction is carried out at 170 °C and 25 atmospheres. At normal pressures, the reaction gives CrO_2F_2 at 150 °C and $CrOF_4$ at 220 °C. The lemon solid was identified as CrF_6 but this has been a matter of controversy and the solid has also been interpreted as CrF_5, sometimes contaminated with $CrOF_4$. The second product, CrO_2F_2, is a monomer in the liquid and an F-bridged polymer, with 6-coordinate Cr, in the solid. In the monomer the OCrO angle is 108° and FCrF is 112°. Very similar values of 108.5° and 113° are found for CrO_2Cl_2. An electron diffraction study on the third product shows $CrOF_4$ has C_{4v} symmetry with a square plane of four fluorine atoms, the Cr=O group perpendicular to this and the Cr raised out of the F_4 plane.

CrO_3 is acidic and dissolves in alkali to give solutions of the chromate ion. On acidification, these give dichromate solution, which exists in strongly acid solutions down to a pH of 0. In very concentrated acid the trioxide is precipitated and there are no cationic forms of chromium (VI) in solution. This behaviour is similar to that of vanadium(V) but the chromium(VI) is more acidic and polymerization does not go so far. The equilibrium between chromate and dichromate is rapid and the two forms coexist over a wide range of pH. Many chromates of heavy metals, such as Pb^{2+}, Ag^+ or Ba^{2+}, are insoluble and these may be precipitated

from chromium(VI) solutions, even at a pH where the major part exists as dichromate, as the equilibrium is so rapidly established. The CrO_4^{2-} ion is tetrahedral while dichromate has two tetrahedral CrO_4 groups joined through an oxygen with the Cr−O−Cr angle about 115°. The polychromate series is much more limited than polyvanadates or polymolybdates, but the next two members, $Cr_3O_{10}^{2-}$ and $Cr_4O_{13}^{2-}$ have been characterized. They are chains of 3 and 4 CrO_4 tetrahedra linked by corners.

If a dichromate solution is heated in the presence of chloride ion and concentrated sulfuric acid, the red, hydrolysable chromyl chloride, CrO_2Cl_2, distils out (b.p. 117 °C). The chromyl bromide and iodide exist only in low-temperature matrices, and the fluoride is made only by the action of fluorine on chromyl chloride, so the latter compound distinguishes chlorine from the rest of the halogens and is used in this way in qualitative analysis. The chromyl halides are covalent compounds which hydrolyse in water to chromate, and there is no evidence for any CrO_2^{2+} cation. The intermediate halochromate ions, CrO_3X^-, are known for X = F, Cl, Br, with the chlorochromate the most stable. $CrOCl_4$ is also reported.

14.4.2 *Peroxy compounds of chromium*

A very complicated series of reactions occur between Cr(VI) species and hydrogen peroxide, which depend on pH and on Cr concentration. The ultimate product is Cr(III) and complete decomposition of the H_2O_2. When hydrogen peroxide is added to acidified chromate solution, a deep blue, transient colour appears. This is unstable in water but extracts into ether (the colour test for chromium) and a stable solid pyridine adduct, $CrO_5 \cdot C_5H_5N$, may be prepared. This is a peroxy-analogue of the trioxide. The structure of the pyridine species is a pentagonal pyramid, Figure 14.17a, with the two peroxy groups and the donor atom forming the pentagonal base. The dipyridyl analogue, CrO_5(dipy) has the related pentagonal bipyramid structure, Figure 14.17b.

In less acidic media, a violet species which is probably $CrO(O_2)_2(OH)^-$, i.e. the anion of $CrO_5 \cdot H_2O$, is found, and has been isolated as the violet, explosive, potassium salt $KHCrO_6$.

In alkaline solution, a red-brown species forms which is more stable, and well characterized as K_3CrO_8. This contains the tetraperoxy ion $Cr(O_2)_4^{3-}$ of chromium V. The paramagnetism corresponds to the d^1 configuration of Cr(V). The structure is dodecahedral, like $Mo(CN)_8^{4-}$ in Figure 15.18. Unlike the symmetrical CrO_2 triangle of the blue species, this has one Cr−O distance of 194 pm and one of 185 pm in each peroxy-chromium triangle. K_3CrO_8 is isostructural with K_3MO_8 (M = Nb, Ta), reinforcing the Cr(V) assignment.

In the presence of ammonia, a dark red-brown species $(NH_3)_3CrO_4$ may be isolated. This contains two peroxy groups and is a chromium(IV) compound, again with a pentagonal bipyramid structure as shown in Figure 14.17c. $CrO_4 \cdot 3KCN$ may be an analogue.

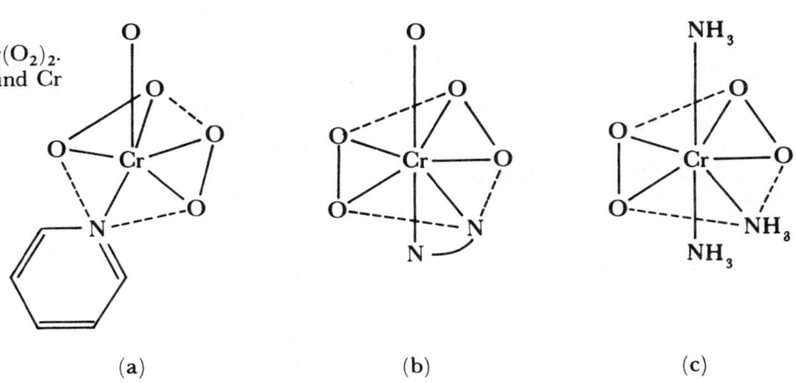

FIGURE 14.17 *Some chromium peroxy species*
(a) $(C_5H_5N)Cr(O_2)_2O$, (b) $(dipy)Cr(O_2)_2O$, (c) $(H_3N)_3Cr(O_2)_2$.
Note: $N \cup N$ indicates dipyridyl. The pentagonal plane around Cr is indicated in each figure.

(a) (b) (c)

14.4.3 Chromium(V) and chromium(IV)

There are few chromium (V) compounds and they readily convert to chromium (VI) or (III). They include the peroxide (see above), CrF_5, $CrOF_3$, some oxyhalide ions such as $CrOF_4^-$ and $CrOCl_5^{2-}$, and the oxyanion, CrO_4^{3-}, which results when potassium chromate is heated in molten KOH (compare $K_2Mn^{VI}O_4$). A recent structural study indicates that the CrO_4^{3-} ion is of distorted tetrahedral structure as might be expected for a d^1 species. Red CrF_5 is a *cis*-bridged polymer and is formed by heating CrO_3 with F_2 at 200° or by treating CrO_2F_2 with XeF_2. In purple $CrOF_3$, square pyramidal $CrOF_4$ units are linked into a three-dimensional array by sharing F corners. $CrOF_3$ disproportionates into Cr(VI) and Cr(III) in water, gives CrF_3 and O_2 on heating, and adds fluoride to form $CrOF_4^-$ which is probably also a fluorine bridged polymer.

An unusually stable class of Cr(V) complexes is the $CrOL_2^-$ group where $L = OCR_2COO$, formed by loss of two protons from α-hydroxybutyric acid and its relatives. The $Cr = O$ is axial and four O from the two ligands form the base of a square pyramid. This compound is stable as a solid in air, and stable in solution in presence of excess ligand. It finds important uses as a specific oxidizing agent.

Chromium(IV) compounds are also rare. CrO_2, CrF_4, and $CrOF_2$ exist and there are reports that $CrCl_4$ and $CrBr_4$ exist in the vapour phase at high temperatures in mixtures of the trihalides and halogen. The complex ion, CrF_6^{2-}, has been isolated and mixed oxides containing chromium(IV) are known. The structure of one of these, $Ba_2(CrO_4)$, has been partially determined and appears to contain discrete CrO_4^{4-} ions. Chromium (IV) is also found in some mixed oxidation state species. The sulfide Cr_5S_8 is formulated as $Cr_4^{III}Cr^{IV}S_8$, and the selenium analogue is similar. In a structural study on the oxygen species $M_2^ICr_3O_9$, linked chains of $Cr^{VI}O_6$ octahedra and $Cr^{IV}O_4$ tetrahedra were indicated.

14.4.4 Chromium (III)

By far the most stable and important oxidation state of chromium is chromium(III). It is the most stable of all the trivalent transition metal cations in water and a wide variety of compounds and complexes are known. The complexes are octahedral, inert to substitution, and have a half-filled t_{2g} set of orbitals.

The oxide is green Cr_2O_3, with the corundum structure, and is used as a pigment. It, and the hydrated form precipitated from Cr^{3+} solution by OH^- ions, dissolve readily in acid to give $Cr(H_2O)_6^{3+}$ ions and also in concentrated alkali to give the chromites. The species in the latter solutions are not identified but may be $Cr(OH)_6^{3-}$ and $Cr(OH)_5(H_2O)^{2-}$. They are readily hydrolysed and precipitate hydrated oxide on dilution. Cr_2S_3 is a black stable solid made by direct combination and is inert to non-oxidizing acids.

CrOF and all four anhydrous trihalides are known and may be prepared by the standard methods. The trichloride gives the dichloride and chlorine when heated to 600 °C, and sublimes in the presence of chlorine at this temperature. It is a red-violet solid which is rather insoluble in water, except in the presence of Cr^{2+} ions. It is thought that these assist the solution process by attaching through a Cl bridge to Cr^{3+} in the crystal—$Cr_{solid}^{3+} - Cl - Cr_{soln}^{2+}$—which then transfers an electron to give divalent chromium in the solid. This Cr^{2+} does not fit the crystal lattice and dissolves to repeat the process, $Cr_{solid}^{2+} - Cl - Cr_{soln}^{3+} \rightarrow Cr_{soln}^{2+}$. A similar effect is found in the case of chromium(III) complexes in solution. These are inert to substitution except in the presence of chromium(II) which may behave similarly in abstracting a ligand *via* a bridging and oxidation process.

The trihalides in aqueous solution give rise to violet $Cr(H_2O)_6^{3+}$ ions and to a number of aquo-halogeno ions, some of which are indicated in Figure 14.18.

A wide variety of chromium(III) complexes exist, all octahedral. The hexammine, $Cr(NH_3)_6^{3+}$, and similar complexes with variety of substituted amines and related molecules are found, and all possible aquo-ammine mixed complexes have been prepared. There are also a wide variety of aquo-X and ammino-X mixed complexes (X = acid radical like halide, thiocyanate, oxalate, etc.) and even aquo-ammino-X species. A commonly used example is Reinecke's salt, $NH_4[Cr(NH_3)_2(NCS)_4] \cdot H_2O$, where the large monovalent anion is widely used as a precipitant for large cations. Apart from the simple hydrated cation, hydrated chromium also occurs in an extensive series of alums, $M^ICr(SO_4)_2 \cdot 12H_2O$, where Cr^{3+} replaces Al^{3+} which is of similar size.

The unusual three-coordination appears to occur in the chromium(III) species $Cr(NR_2)_3$ which have been prepared for R = isopropyl and R = $SiMe_3$. The configuration may

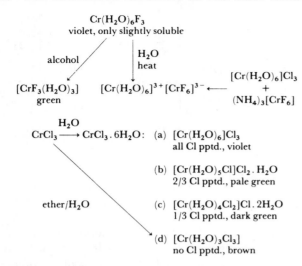

FIGURE 14.18 *Aquo-halogeno ions of chromium(III)*

be planar and its adoption is probably aided by the presence of the bulky ligands.

In anhydrous $KCrF_4$ the structure is based on distorted octahedral CrF_6 units. First, three of these are linked into a ring by sharing *cis* fluorine corners, and then these are linked further by the axial fluorines (perpendicular to the ring) into an infinite array.

The ion $[Cr\{Cr(OH)_2(NH_3)_4\}_3]^{6+}$ is linked through a central CrO_6 octahedron as shown in Figure 14.19.

An interesting problem is presented by the complex $[Cr(H_2O)_5NO]^{2+}$ which can be formulated either as a $Cr(III)$ complex with an NO^- ligand or as a $Cr(I)$ compound with NO^+. The NO stretching frequency in the chromium complex is $1747\ cm^{-1}$, compared with $1890\ cm^{-1}$ for NO, and this may be interpreted as evidence for a weaker N–O bond as expected for NO^- (compare section 3.5). However, other evidence led to the opposite view but a recent crystal structure shows bond distances appropriate to the $Cr(III)/NO^-$ formulation—in particular the $Cr–OH_2$ values are typical of $Cr(III)$. The Cr–NO value of 168 pm

suggests multiple bonding between Cr and N, and the $Cr–OH_2$ distance *trans* to NO is longer than the four *cis* values. All these observations are compatible with the $Cr(III)$ formulation. A similar problem arose with $[Fe(H_2O)_5NO]^{2+}$, the material formed in the 'brown ring' test for iron, which is also now accepted as the $Fe(III)/NO^-$ species.

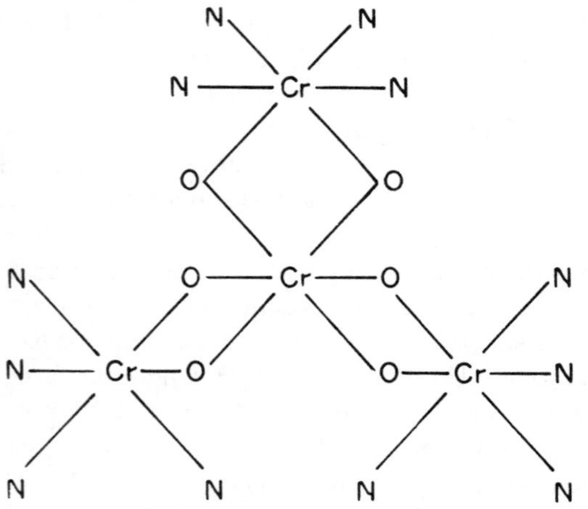

14.4.5 *Chromium (II) and lower oxidation states*

The mixed $Cr(II) – Cr(III)$ fluoride, Cr_2F_5, is found. The structure consists of chains of CrF_6 octahedra linked by apices with the $Cr(II)$ and $Cr(III)$ chains alternating. The $Cr(II)F_6$ unit is distorted with four short and two long CrF bonds while the $Cr(III)$ environment is regular. This is one example of a mixed valence compound where the two oxidation states are found in distinct environments (compare Ga_2Cl_4, p. 341 and Pb_3O_4, p. 339 and Figure 17.20).

A fair number of chromium(II) compounds are known, including all the dihalides and CrO. A considerable variety of complexes are also found, although all are unstable to oxidation. In water, the sky-blue $Cr(H_2O)_6^{2+}$ ion is formed. This is a strong reducing agent which is just too weak to reduce water. It has a number of uses as a reducing agent including the removal of traces of oxygen from nitrogen. The solutions are less stable when not neutral, and hydrogen is evolved. They react rapidly with the air. Chromium(II) complexes include the hexammine, $Cr(NH_3)_6^{2+}$, and related ions. The dipyridyl complex is found to disproportionate to give a chromium(I) species:

$$2Cr(dipy)_3^{2+} = Cr(dipy)_3^+ + Cr(dipy)_3^{3+}$$

The halides can be prepared by reaction of the appropriate HX on the metal at 600 °C or by the reduction of the trihalides with hydrogen at a similar temperature. The iodide, CrI_2, and CrS, may also be made by direct combination.

Among the chromium(II) salts, the hydrated acetate, $Cr_2(CH_3COO)_4.2H_2O$, is the commonest. It is readily pre-

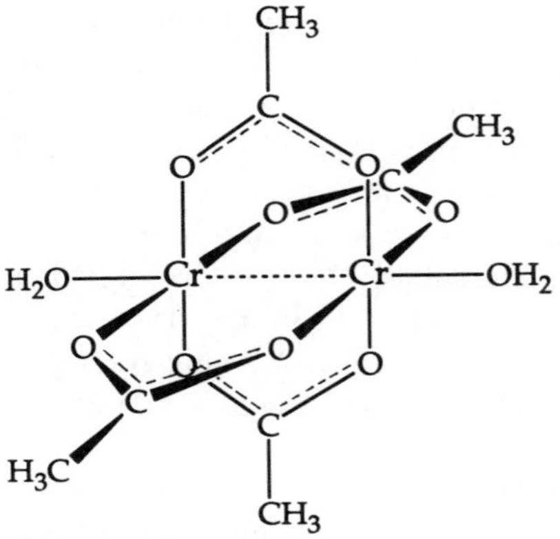

FIGURE 14.19 *The structure of the ion* $[Cr\{Cr(OH)_2NH_3)_4\}_3]_6^+$
A central octahedron, $Cr(OH)_6$ is linked to three outer $Cr(OH)_2(NH_3)_4$ octahedra by bridging OH groups. [N = NH₃; O = OH]

FIGURE 14.20 *Structure of chromium(II) acetate,*
$Cr_2(CH_3COO)_4.2H_2O$
The oxygen atoms of the acetate groups lie in a square plane around each Cr and the acetates link the edges together. The Cr–Cr distance is short, implying direct metal-metal bonding. Compare the copper acetate, Figure 14.33.

pared by adding chromium(II) solution to sodium acetate when it is precipitated as a red crystalline material. It has the unusual bridged structure shown in Figure 14.20, with a very short Cr—Cr distance suggesting metal-metal interaction, An interesting structure is found in $CrSO_4.5H_2O$. Four of the water molecules lie in a square plane around the Cr and the last two positions of the octahedral array are filled by O atoms from two different sulfate groups. These bridge two Cr atoms giving an infinite array of $\cdots O - Cr(H_2O)_4OSO_2O - Cr\cdots$ units with further hydrogen bonding linking SO_2 oxygens to coordinated water molecules on Cr atoms in parallel chains.

Lower oxidation states are represented by a few compounds including the chromium(I) dipyridyl complex above. The 0 state is shown in the stable, octahedral carbonyl, $Cr(CO)_6$, in carbonyl ions, $Cr(CO)_5X^-$, and in dibenzene chromium (Figure 16.7b). Carbonyl anions of chromium $(-I)$ and $(-II)$ are the compounds $Na_2[Cr_2(CO)_{10}]$ and $Na_2[Cr(CO)_6]$ respectively, prepared by sodium reduction of the hexacarbonyl. Chromium(I) is also found in the cyanide, $K_3Cr(CN)_4$, and the zero state in $K_6Cr(CN)_6$ which may be used to prepare $Cr(diphos)_3$ and $Cr(triphos)_2$ (for di- and tri-phosphine ligands see Appendix B).

14.5 Manganese, $3d^5 4s^2$

Manganese, with seven valency electrons, shows the widest variety of oxidation states in the first transition series. The oxidation state energy diagram, Figure 14.21 shows that Mn(II) is the most stable state, and comparison with the diagrams for previous elements shows that manganese(VII) is the most strongly oxidizing of all the high oxidation states known in aqueous solution. Many of the intermediate states are rare, with a strong tendency to disproportionate.

Manganese is abundant in the earth's crust and its principle ore, pyrolusite, is a crude form of the dioxide, MnO_2. The metal is obtained by reduction with aluminium, or in the blast furnace. The metal resembles iron in being moderately reactive and dissolving in cold, dilute non-oxidizing acids. It combines directly with most non-metals at higher temperatures, sometimes quite vigorously. Thus it burns in N_2 at 1200 °C to give Mn_3N_2 and in Cl_2 to give $MnCl_2$. The product of high temperature combination with oxygen is Mn_3O_4. The main use of manganese is in steel, for which ferromanganese is formed by reducing the mixed ores, as in the case of vanadium. It also finds limited uses in other alloys.

14.5.1 *The high oxidation states, manganese (VII), (VI), and (V)*

Manganese (VII) compounds are strongly oxidizing. Green Mn_2O_7 is produced by treating $KMnO_4$ with sulfuric acid. This heptoxide requires great care in handling, as it can decompose or react with explosive violence. Both in the gas phase and in the solid, the structure contains the linked tetrahedra of Figure 14.22. On treatment with chlorosulfonic acid, the explosively unstable oxychlorides result

$$Mn_2O_7 + ClSO_3H \rightarrow MnO_3Cl + MnO_2Cl_2 + MnOCl_3$$
$$ \text{black} \quad\quad \text{brown} \quad\quad \text{green}$$

corresponding to the VII, VI, and V states. They are more stable when handled in solution in carbon tetrachloride. The oxyfluoride, MnO_3F, has been reported, but the only stable representative of the VII state is the permanganate ion, MnO_4^-, which is most common as the potassium salt. This is a strong oxidizing agent in acid solution:

$$MnO_4^- + 8H_3O^+ + 5e^- \rightarrow Mn^{2+} + 12H_2O, \; 1\cdot 51 \, V$$

and is also strong in basic media:

$$MnO_4^- + 2H_2O + 3e^- \rightarrow MnO_{2(solid)} + 4OH^-, \; 1\cdot 23 \, V$$

when manganese dioxide is precipitated. However, in concentrated alkali the anion of the dibasic acid, manganate, is formed in preference, and this reverts to permanganate plus dioxide on acidification:

$$MnO_4^- + OH^- \rightarrow MnO_4^{2-} \underset{OH^-}{\overset{H^+}{\rightleftharpoons}} MnO_4^- + MnO_2$$

The permanganates are, of course, purple, while the manganate ion is an intense green, and is the only stable representative of the manganese(VI) state. Permanganate is about

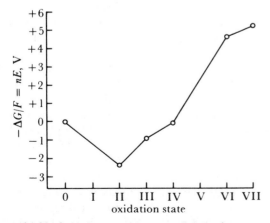

FIGURE 14.21 *Oxidation state free energy diagram for manganese*
The stability of Mn(II) is striking and it is clear that the III and VI states disproportionate. Compare Figure 14.1 and section 8.6.

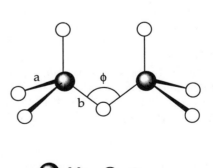

FIGURE 14.22 *The structure of dimanganese heptoxide.* a = 159 pm, b = 17 pm, ϕ = 121°.

the strongest accessible oxidizing agent which is compatible with water and even it undergoes slow decomposition in acid solution to give MnO_2 and oxygen.

In an interesting disproportionation of the manganese(VI) oxyion, $Mn_7O_{24}^{2-}$ has been isolated from the action of $BaMnO_4$ with sulfuric acid. The oxidation equivalent is 6·57 per manganese atom, and the compound is formulated as the mixed manganese(IV)–manganese(VII) species $[Mn(MnO_4)_6]^{2-}$. The structure shows a central Mn(IV) coordinated by six $Mn(VII)O_4$ tetrahedra each sharing one oxygen with the Mn(IV). The MnO_6 unit is octahedral and the distances are Mn(IV)$-O = 190$ pm, Mn(VII)$-O$ (bridge) $= 174$ pm and Mn(VII)$-O$ (terminal) $= 160$ pm.

The manganese(V) state is represented only by $MnOCl_3$ and the blue MnO_4^{3-} ion, formed by the action of alkaline formate on MnO_4^{2-}. Manganese(V) is relatively stable in alkaline melts, e.g. it is the major form of manganese in molten NaOH or KOH at about 400 °C. It has also been stabilized in the apatite-like structure of $Ba_5(MnO_4)_3Cl$ and similar blue or green species. Here the Mn(V) takes the place of P(V) in normal apatites. The MnO_4^{3-} ion is tetrahedral with Mn $- O = 170$ pm.

14.5.2 *Manganese (IV) and manganese (III)*

Manganese (IV) does not have a very extensive chemistry although black MnO_2 is well-known and is a common precipitate from manganese compounds in an oxidizing medium. It is very insoluble and usually only dissolves with reduction, as in its reaction with HCl:

$$MnO_2 + 4HCl \rightarrow MnCl_2 + Cl_2 + 2H_2O$$

A few complex salts are known including the hexachloride and fluoride, MnX_6^{2-} and also $Mn(CN)_6^{2-}$. The only simple halide reported is the fluoride, MnF_4.

The oxide Mn_5O_8 contains Mn(IV) and Mn(II). The structure is based on the CdI_2 layer type (Figure 5.10a) and contains Mn(IV) in a distorted octahedron of oxygens together with $Mn(II)O_6$ trigonal prisms.

The III state is represented by the oxide, Mn_2O_3, and the fluoride, MnF_3. The oxide stable at high temperatures, Mn_3O_4, is correctly formulated as a mixed oxide of the II and III states, $Mn^{II}Mn_2^{III}O_4$. The structure of MnF_3 is a polymeric one with all the F atoms bridging pairs of Mn atoms. There are two different MnF_6 sites, each a distorted octahedron. In one, there are three pairs of *trans* Mn–F bonds each of different length increasing from 183 pm, through 192 pm, to 210 pm. In the second type of site all six Mn–F distances differ and range from 180 to 208 pm. These distortions probably reflect the Jahn–Teller effect expected for the high spin d^4 configuration of Mn(III) (compare section 13.4).

No other simple trihalides exist but the red $MnCl_5^{2-}$ complex ion is known, as is the corresponding fluoride. $(NH_4)_2MnF_5$ contains chains of MnF_6 octahedra linked by *trans* bridging fluorides, in the structure $MnF_4(F_{2/2})$. Large cations stabilize MnF_6^{3-}. Complex ions of manganese(III)

include $MnCl_3L$ and $MnCl_3L_3$ types, where L is a nitrogen ligand, and the acetylacetonates $MnCl_2(acac)$, $MnCl(acac)_2$ and $Mn(acac)_3$ are also reported.

In solution manganese(III) is unstable both to reduction and to disproportionation:

$$2Mn^{3+} + 6H_2O \rightarrow Mn^{2+} + MnO_{2(solid)} + 4H_3O^+$$

If manganese(II) in sulfuric acid solution is oxidized with permanganate in the stoichiometric ratio, an intensely red solution results which contains all the manganese as manganese(III), presumably as a sulfate complex. This solution is as strongly oxidizing as permanganate and was once used as an alternative oxidizing agent for sulfate media. A rather similar oxidation of the acetate gives $Mn(OAc)_3.2H_2O$ as a solid. This manganese(III) acetate is readily prepared and used as a starting material for most manganese(III) studies. By heating the oxides at 500 °C for 14 days, red Na_5MnO_4 crystals were produced which contained compressed $Mn(III)O_4$ groups matching the MO_4 units found for M = Fe, Co, or Ni in the (III) state.

While the chemistry of simple Mn(IV) and Mn(III) compounds is limited, more complex species containing these oxidation states have been of increasing interest in biological chemistry. A number of metalloproteins containing manganese are known and their active sites have been increasingly identified. Parallel to this is the synthesis of new manganese compounds to act as models for the biological species. Compounds containing from one to four Mn are well-known and larger molecules with up to 12 Mn have been studied. The established oxidation states are Mn(II), (III), and (IV) with some indication of Mn(V). Many of the manganese–protein compounds are enzymes dealing with the control of dioxygen species in the organism—especially in protecting against damage by peroxide or superoxide radicals. Human superoxide dismutase has been shown to contain four separate Mn atoms, each in an identical protein sub-unit. These are linked in pairs through coordination to the Mn, and the pairs are further linked by four-helix interfaces. The characteristic Mn site in a superoxide dismutase is five-coordinated Mn(III) with a trigonal bipyramid formed from 3 N and an O from amino acids in the protein, with the fifth site occupied by OH^- or OH_2. Other enzymes contain two Mn(III) atoms in octahedral sites linked by three bridges which are different combinations of oxide and carboxylate bridges, Mn–O–Mn and Mn–O(CR)O–Mn. The structure of a similar Mn(IV) dimanganese compound, prepared as a model species, is shown in Figure 14.23a. Among the larger units is the active site in 'photosystem II' which is involved in the oxidation of H_2O to O_2 driven by visible light. One proposal for this site is the Mn_4O_4 cubane core, shown in Figure 14.23b, which may function by opening up and closing again in a four-step cycle, transferring an electron at each step. An example of a larger structure is given in Figure 14.23c, which is a simplified representation of $[Mn_4^{III}Mn_6^{IV}O_{14}\{N(CH_2CH_2NH_2)_3\}_6]^{8+}$. The structure can be seen as a series of Mn_3O_4 cubanes lacking a Mn corner and sharing Mn_2O_2 faces, terminated by N atoms

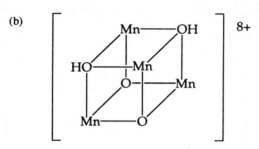

(a)

(b)

$$8+$$

● Mn

○ O

◗ N

FIGURE 14.23   (a) *Structure of the model compound* $[Mn_2O_2(O_2CCH_3)$ $(H_2O)_2(bipy)_2]^{3+}$. *(b) The* Mn_4O_4 *cubic unit of* $[Mn_4O_2(OH)_2]^{8+}$. *(c) The structure of the core portion of the polynuclear cation of* $[Mn_4^{III}Mn_6^{IV}O_{14}\{N(CH_2CH_2NH_2)_3\}_6]^{8+}$

from the ligand. The Mn atoms are all six-coordinated to O or to mixtures of O and N.

14.5.3   *Manganese (II) and lower oxidation states*

In contrast to all the above unstable or poorly represented

states, the manganese(II) state is very stable and widely represented. It is the d^5 state and all the compounds contain five unpaired electrons, except for the cyanide and related complexes, for example $Mn(CN)_6^{4-}$ and $Mn(CN)_5(NO)^{3-}$, which are low-spin with only one unpaired electron. All the high-spin manganese(II) compounds are very stable and resist attack by all but the most powerful oxidizing or reducing agents. They have all very pale colours, for example the hydrate $Mn(H_2O)_6^{2+}$ is pale pink, as the absorptions due to the d electron transitions are very weak. This is because a $d-d$ transition which involves the reversal of an electron spin (as it must be in a high-spin d^5 system) is an event of low probability, compared with one which is 'spin-allowed'. The absorptions in manganese(II) compounds are therefore about a hundred times weaker than the general run of transition compound absorptions.

By contrast to the high-spin compounds, the low-spin complexes are much more reactive and oxidize readily, for example:

$$Mn(CN)_6^{4-} \xrightarrow{\text{air}} Mn(CN)_6^{3-}$$

Manganese(II) is unstable in these low-spin compounds probably because the crystal field stabilization energy—though large enough to cause spin-pairing—is only a little greater than the loss of exchange energy due to spin pairing. In the trivalent state, the CFSE is increased because of the greater charge. A somewhat similar situation arises in the case of cobalt(III) in water.

The stability of the high-spin manganese(II) state is illustrated by the wide variety of stable compounds formed. Some of these are listed in Table 14.2.

A variety of complex ions exist, including $Mn(NH_3)_6^{2+}$ and octahedral complex ions with EDTA, oxalate, ethylenediamine, and thiocyanate but the hexahalide ions are unknown. The equilibrium constants for the formation of such ions in solution are low, as the Mn^{2+} ion is relatively large (Figure 13.20) and there is no CFSE. Thus one important source of the energy required to displace the coordinated water molecules is lacking.

A few examples exist of manganese in a square-planar environment. In $Mn(acac)_2.2H_2O$, the bidentate acetylacetonate groups form a square plane around the manganese and the water molecules complete a distorted octahedron: when the compound is dehydrated, it is probable that the $Mn(acac)_2$ remaining is truly square planar. The sulfate, $MnSO_4.5H_2O$, is isostructural with $CuSO_4.5H_2O$ and therefore contains square planar $Mn(H_2O)_4^{2+}$ units.

Some tetrahedral manganese units are found, especially the halogen anions, MnX_4^{2-}, which may be prepared as the salts of large cations. These tetrahedral complexes are unstable in water, or other donor solvents, and go to octahedral complexes of the solvent.

One or two examples of three-coordinate Mn(II) are known, where the low coordination is stabilized by bulky groups R. One example is the amide $(R_2N)_4Mn_2$, where $R = Me_3Si$. The structure is based on the unit

TABLE 14.2 Examples of high-spin manganese(II) compounds

MnX_2	X = F, Cl, Br, I	Isomorphous with the Mg halides.	Stable at red heat.
MnY	Y = O, S, Se, Te	Sodium chloride structure.	Very stable when dry but the hydrated forms slowly oxidize to MnO_2 in air.
$Mn(OH)_2$		This is a true hydroxide, not a hydrated oxide, isomorphous with $Mg(OH)_2$.	
$MnSO_4$		Very stable, even at red heat. The hydrate is isomorphous with copper sulphate.	
$Mn(ClO_4)_2$		Very soluble. Stable to 150 °C, then the perchlorate oxidizes the Mn^{2+} to the dioxide.	
$MnCO_3$		Insoluble. Very stable for a transition metal carbonate. It goes to MnO and CO_2 at about 100 °C.	
$Mn(OOCCH_3)_2$ and other organic acid salts		Stable: prepared by heating $Mn(NO_3)_2$ with the acid anhydride.	

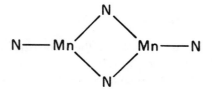

A similar species is found for Co(II), another stable + II state.

This very wide range of compounds and coordination behaviour underlines the stability of Mn(II) and its dominance in manganese chemistry.

In its low oxidation states, manganese forms the cyanide, $K_5Mn(CN)_6$, and the carbonyl halides, like $Mn(CO)_5Cl$, in the I state, the carbonyl, $Mn_2(CO)_{10}$, in the 0 state, and the anion, $Mn(CO)_5^-$, in the −I state. The latter is a trigonal bipyramid, as is the related $Mn(NO)(CO)_4$ with NO in an equatorial position. In its carbonyl chemistry, manganese forms an extensive range of polynuclear and substituted products.

14.6 Iron, $3d^6 4s^2$

When iron is reached, in the first transition series, the elements cease to use all the valency electrons in bonding, and the Group oxidation state of VIII is not found. The highest state of iron is VI and the main ones are II and III. The oxidation energy diagram, Figure 14.24, shows that the III state is only slightly oxidizing while the II state lies at a minimum and is stable in water. By comparison with the II and III states of other transition elements, the iron(II) and (III) states lie much closer together in stability, and this accords with the well-known properties of ferrous and ferric solutions which are readily interconverted by the use of only mild oxidizing or reducing agents.

Iron is the most abundant of the fairly heavy elements in the Earth's crust and is used on the largest scale of any metal. Its production in the blast furnace is well-known, and

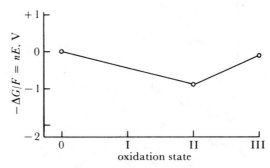

FIGURE 14.24 *Oxidation state free energy diagram for iron*
No potential values for higher states than III are known: the few Fe(IV) compounds known are strongly oxidizing. The II state is the more stable but, compare Figure 14.27, the III state is also relatively stable.

basically involves the reduction of the oxide by carbon. The resulting metal contains a small proportion of carbon, and the various types of iron and steel result from varying carbon contents, and from other metal additions. In particular, three forms of iron exist at different temperatures, the carbide phase is Fe_3C, called *cementite*. Thus, a steel cooled slowly from above 720 °C and containing about 1% C, consists of layers of alpha-iron interleaved with cementite to give the steel called *pearlite*, a soft malleable material. Alternatively rapid cooling of the same mixture prevents separation of the layers and *martensite* is formed which is hard but brittle. Technology has evolved to a point where a huge variety of special steels of varying properties is available. Iron is reactive in air and the process of rusting involves the formation of a coat of hydrated oxide on the surface in moist air. This is non-coherent and flakes away revealing fresh surfaces for attack. Iron combines at moderate temperatures with most non-metals and it readily dissolves in dilute acids to give iron(II) in solution, except with oxidizing acids which yield iron(III) solutions. Very strongly oxidizing agents, such as dichromate or concentrated nitric acid, produce a passive form of the metal, probably by forming a coherent surface film of oxide.

14.6.1 *The iron–oxygen system*

Iron forms three oxides, FeO, Fe_2O_3, and Fe_3O_4, which all commonly occur in non-stoichiometric forms. Indeed, FeO is thermodynamically unstable with respect to the structures with a deficiency of iron. The three oxides show a number of structures, some based on a cubic close packed array of oxide ions. Properties of the oxides are summarized in Table 14.3.

It will be seen that the structures of the cubic forms of all these oxides are related. If the cubic array of oxide ions is taken, then all the structures result from different dispositions of ferrous and ferric ions in the octahedral and tetrahedral sites. This explains the tendency to non-stoichiometry and the interconversion of oxides. If the oxide lattice has its octahedral sites filled up with Fe^{II} ions, the FeO structure builds up as the Fe/O ratio approaches one. If a small portion of the Fe^{II} is missing or replaced by Fe^{III} in the ratio of two Fe^{III} for every three Fe^{II}, defect FeO forms result, with a stability maximum at $Fe_{0.95}O$. If the process is continued until there are two Fe^{III} atoms for every Fe^{II} and if half these Fe^{III} ions enter tetrahedral sites, the structure becomes that of Fe_3O_4. Conversion of the remaining Fe^{II} to Fe^{III} gives the cubic form of Fe_2O_3. Compare section 5.9.

14.6.2 *The higher oxidation states of iron*

If hydrated ferric oxide in alkali is treated with Cl_2, a red-purple solution of iron(VI) is obtained containing the ferrate ion, FeO_4^{2-}. The sodium and potassium salts are very soluble but the barium compound may be precipitated. The ferrate ion is stable only in a strongly alkaline medium, in water or acid it evolves oxygen:

$$2FeO_4^{2-} + 10H_3O^+ \rightarrow 2Fe^{3+} + 3/2O_2 + 15H_2O$$

Ferrate is a stronger oxidizing agent than permanganate. It is a tetrahedral ion and the potassium salt is isomorphous with K_2CrO_4 and with K_2SO_4. The ferrate ion shows an unusual luminescence in the near-infrared region of the spectrum, suggesting a possible application in solid-state infrared laser materials.

A pentavalent oxyanion, FeO_4^{3-}, is also reported. Iron(V) also occurs six-coordinated to oxygen in the complex phase, $SrLa_3LiFeO_8$.

Iron(IV) is also rare. It is found in the perovskite $MFeO_3$ where M = Ca or Sr, but this only exists under pressure of O_2 and disproportionates to Fe(V) plus Fe(III). It also occurs in the complex $Fe(diars)_2Cl_2^{2+}$, where diars = the *ortho*-diarsane derivative of benzene, o-$C_6H_4(AsMe_2)_2$, which acts as a bidentate ligand through the lone pair on each arsenic atom. The complex is produced as the salt with a large anion, $FeCl_4^-$ or ReO_4^-, by oxidation of the iron(III) complex with 15M nitric acid. Analogous diphosphane and phosphane–arsane complexes of iron(IV) are reported, some with Br in place of the Cl.

14.6.3 *The stable states, iron (III) and iron (II)*

While a distinct chemistry of Fe (VI), (V), and (IV) has developed, it is a very minor part of iron chemistry and all the compounds are unstable outside a strongly oxidizing regime. The dominant oxidation states are Fe(II) and Fe(III).

Apart from the oxides, solid compounds of the II and III oxidation states are now represented by all of the halides, and salts of Fe(II) are known with nearly all stable anions. For a long time it was thought that the compound FeI_3 (ferric iodide) was a 'non-existent' compound, due to the fact that Fe(III) is mildly oxidizing, while iodide is a reducing anion. In fact, this is exactly the case in aqueous solution—addition of I^- to an Fe(III) solution immediately produces elemental iodine:

$$Fe^{3+} + I^- \rightarrow Fe^{2+} + 1/2I_2$$

TABLE 14.3 The iron oxides

FeO (black)	Prepared by thermal decomposition of ferrous oxalate at a high temperature, followed by rapid quenching to prevent disproportionation to $Fe + Fe_3O_4$.	Structure is sodium chloride—i.e. O^{2-} ions ccp and Fe^{2+} ions in all the octahedral sites.
Fe_2O_3 (brown)	Occurs naturally. Otherwise by ignition of hydrated ferric oxide, precipitated from a ferric solution by ammonia.	The structure has the oxide ions ccp with Fe^{3+} ions randomly distributed over the octahedral and tetrahedral sites. A second form is hexagonal.
Fe_3O_4 (black)	Occurs naturally as *magnetite*. It is the ultimate product of strong ignition in air of the other two oxides.	The structure is inverse spinel*: that is the O^{2-} ions are ccp, the Fe^{2+} ions are in octahedral sites, and the Fe^{3+} ions are half in octahedral sites and half in tetrahedral sites.

*It will be recalled from Chapter 5 that a normal spinel is an oxide AB_2O_4, with A a divalent metal ion and B a trivalent one. The oxide ions are cubic close packed and the A atoms are in tetrahedral sites while the B atoms are in octahedral sites. The inverse structure of magnetite may be ascribed to the much greater CFSE of d^6 (low spin) ferrous ions in octahedral sites instead of tetrahedral ones. The d^5 ferric ions have no CFSE in either type of coordination.

However, very recent studies have shown that by carrying out the synthesis of FeI_3 in non-aqueous media it can be isolated as an unstable black compound, of structure yet to be defined. The reaction involves oxidative decarbonylation of the iron(II) carbonyl iodide $(OC)_4FeI_2$ in hexane:

$$(OC)_4FeI_2 + 1/2 I_2 \rightarrow FeI_3 + 4CO$$

Several derivatives of the FeI_4^- ion, together with diphosphane and diarsane derivatives of FeI_3, have also been prepared and characterized. These results were all obtained in non-aqueous media and show that Fe(III) iodine compounds are not intrinsically unstable, although they are too reactive to prepare in water.

In solution, the relative stabilities of the III and II states vary widely with the nature of the ligand. As Fe^{II} is d^6 and Fe^{III} is d^5, changes in CFSE have an important effect on these relative stabilities. The effect is illustrated by the potentials below:

$$\begin{array}{ll} Fe(H_2O)_6^{3+} + e^- = Fe(H_2O)_6^{2+} & 0{\cdot}77\ V \\ Fe(CN)_6^{3-} + e^- = Fe(CN)_6^{4-} & 0{\cdot}36\ V \\ Fe(phen)_3^{3+} + e^- = Fe(phen)_3^{2+} & 1{\cdot}12\ V \end{array}$$

(phen = o-phenanthroline, $C_{12}H_8N_2$, a bidentate aromatic nitrogen ligand)

$$Fe(C_2O_4)_3^{3-} + e^- = Fe(C_2O_4)_2^{2-} + C_2O_4^{2-}\ 0{\cdot}02\ V$$

The cyanide and phenanthroline complexes are low-spin while the other two are high-spin, in all cases both in the II and III states. Since the ΔE value for the trivalent d^5 Fe^{III} ion is larger than that of the divalent d^5 ion, Mn^{II}, the cyanide of ferric iron is less unstable, from the CFSE versus exchange energy point of view, than the manganese(II) hexacyanide. There is a gain in CFSE on going from ferricyanide to the d^6 low-spin ferrocyanide, due to the additional t_{2g} electron, but this is relatively small. Ferricyanide acts as a mild oxidizing agent while ferrocyanide is stable. In addition, the d^5 ferricyanide is relatively labile to substitution and the cyanide may be replaced by water and other ligands, as in $Fe(CN)_5(H_2O)^{2-}$. Thus ferricyanide in solution is much more poisonous than is ferrocyanide.

In strong base, a number of Fe(III) oxyanions have been identified including FeO_4^{5-} (tetrahedron), $Fe_2O_6^{6-}$ (two tetrahedra sharing an edge), and $Fe_6O_{16}^{14-}$ which is a ring of six corner-linked tetrahedra.

In aqueous solution, ferric iron shows a strong tendency to hydrolyse. The hydrated ion, $Fe(H_2O)_6^{3+}$, which is pale purple, exists only in strongly acid solutions at a pH of about 0. In less acidic media, hydroxy complexes are formed:

$$Fe(H_2O)_6^{3+} + H_2O = Fe(H_2O)_5(OH)^{2+} + H_3O^+$$
$$Fe(H_2O)_5(OH)^{2+} + H_2O = Fe(H_2O)_4(OH)_2^+ + H_3O^+$$

These occur up to pH values of 2 to 3 and are yellow in colour, the typical colour of ferric salts in solution in acid. At lower acidities, above a pH of 3, bridged species are formed and the solutions soon form colloidal gels. As the pH is raised hydrated ferric oxide is precipitated as a reddish-brown gelatinous solid. This precipitate probably does not contain any of the hydroxide, $Fe(OH)_3$, and part of it is

probably in the form FeO(OH) and part as the hydrated oxide. The hydrated oxide readily dissolves in acid and is also slightly soluble in strong bases, so the ferric state is weakly acidic as well as moderately basic. The basic solutions in strong alkali probably contain the $Fe(OH)_6^{3-}$ ion which has been isolated as the strontium and barium salts.

Two interesting Fe(III)–sulfur species are the ion FeS_4^{5-} which contains discrete tetrahedral ions, and the complex ion $[(Et_3PFe)_6(\mu^3 - S)_8]^+$. The structure, Figure 14.25a, consists of an octahedral cluster of 6 Fe atoms, bridged on each of the triangular faces by the 8 S atoms, with a phosphane terminal on each Fe—compare the structure of Mo_6Cl_8 shown in Figure 15.19. A full series of oxidation/reduction steps have been demonstrated which vary the charge on the complex from 0 to 4^+.

(a)

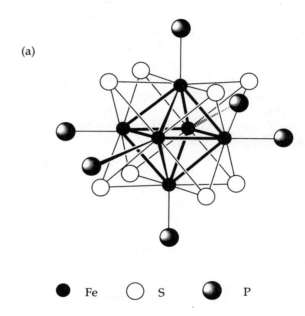

● Fe   ○ S   ◓ P

(b)

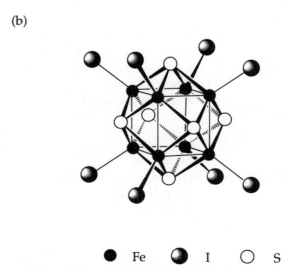

● Fe   ◓ I   ○ S

FIGURE 14.25 *Structures of some iron-sulfur clusters* (a) $(Et_3PFe)_6S_8]^+$ and (b) $[Fe_8S_6I_8]^{4-}$.

Ferric iron forms many complexes with ligands which coordinate through oxygen, especially phosphate anions and polyhydroxy-organic compounds such as sugars. It also forms intensely red thiocyanate complexes, used to detect and estimate trace quantities of iron. These colours are destroyed by fluoride due to the formation of FeF_6^{3-}. In contrast, Fe^{III} forms no ammonia complex (ammonia precipitates the oxide from aqueous solutions) and is only weakly coordinated by other amine ligands. If the ligand field is sufficiently strong to produce spin pairing, much more stable complexes are formed and this occurs with dipyridyl and phenanthroline.

Iron(III) forms the three-coordinate compound $Fe[N(SiMe_3)_2]_3$ like the chromium analogue. Structural studies show the presence of planar three-coordinate FeN_3 and $FeNSi_2$ groups indicating extended π bonding. The nitrate complex, $Fe(NO_3)_4^-$, is 8-coordinated in a dodecahedral configuration. The mixed valent $Fe(II)-Fe(III)$ fluoride, $Fe_2F_5 \cdot 7H_2O$, is known. On dehydration, it yields FeF_2 and FeF_3.

Ferrous iron also forms a variety of complexes. In aqueous solution it exists as the $Fe(H_2O)_6^{2+}$ ion, which is pale sea-green in colour. This is slowly oxidized by air in acid, and is very readily oxidized when the hydrated oxide is precipitated in alkali. The anhydrous halides combine with ammonia gas to give the hexammine, $Fe(NH_3)_6^{2+}$, but this is unstable and loses ammonia when brought into contact with water. Stable complexes are formed, however, by chelating amines such as ethylenediamine. All these examples are octahedral, and we note the most famous Fe(II) octahedral complex of all, haem, which exists in haemoglobin as discussed in Chapter 20.

Tetrahedral complexes are rare but the anions of the heavier halogens, $[FeX_4]^{2-}$, can be precipitated by large cations. The dimeric Fe(II) species $[Fe_2Cl_6]^{2-}$ has recently been prepared as a phosphonium salt and found to have a chloride-bridged structure very similar to that of Al_2Br_6 (Figure 17.12) with tetrahedral geometry about the Fe atoms.

Many Fe(II)–sulfur compounds have been synthesized as inorganic models for biological systems and these are discussed in Chapter 20. A more complex Fe(II)–sulfur species is the ion $[(IFe)_8(\mu^4\text{-}S)_6]^{4-}$ where a cube of 8 Fe atoms is bridged on each square face by a sulfur, and each Fe carries a terminal Fe–I bond (Figure 14.25b). The iodine can be replaced by other ligands such as phosphane. All the iron atoms are Fe(II), and the structure has been suggested as a further model for the active centre of iron–sulfur proteins. This structure is the 'inverse' of the $Fe_6^{III}S_8$ ion of Figure 14.25a (with Fe and S interchanged) since an A_6B_8 unit can be described either as a cube with six caps on the square faces or as an octahedron with eight caps on the triangular faces. The two descriptions are equivalent and whether the cube or the octahedron is seen as the basic structure depends on the point of view.

A well-known reaction among iron complexes is the formation of *Prussian blue* by the reaction of ferrocyanide with

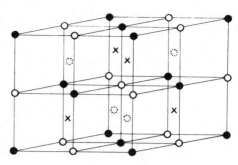

FIGURE 14.26 *The structure of Prussian blue and related compounds*
If none of the cube centre sites are occupied, the structure is that of ferric ferricyanide (both black and white Fe positions occupied by Fe^{III}): if every second cube centre site (marked with a dotted circle) is occupied by K^+, the structure is that of soluble Prussian blue (black = Fe^{II}, white = Fe^{III}): if all the centre sites are occupied by K^+ (crosses as well as dotted circles) the structure is that of dipotassium ferrousferrocyanide (black and white sites = Fe^{II}).

ferric solution, and of *Turnbull's blue* by the reaction of ferricyanide with ferrous. In addition, ferrous plus ferrocyanide gives a white potassium salt of ferrousferrocyanide, and ferric plus ferricyanide gives brown-green ferricferricyanide. It now appears that all these compounds are related structurally and that Prussian and Turnbull's blues are identical. The basic structure is that shown in Figure 14.26. Ferric ferricyanide, $Fe[Fe(CN)_6]$, contains a cubic array of iron(III) atoms of unit cell length equal to 510 pm. This is the structure in the Figure when the atoms are identical and the Figure contains eight unit cells. If every second atom becomes Fe^{II} and one potassium ion is placed in the centre of every alternate small cube, a structure of unit cell length equal to 1020 pm results. This corresponds to the so-called soluble Prussian blue, $K[Fe(Fe(CN)_6)]$ where one iron is Fe^{II} and one is Fe^{III}, the alternative cases being indistinguishable from the X-ray data. Insoluble, or true, Prussian Blue is $Fe_4^{III}[Fe^{II}(CN)_6]_3 \cdot 14H_2O$ and is identical to Turnbull's blue. The cyanide coordinates strongly to the ferrous iron, giving an $Fe^{II}C_6$ octahedral unit. Both the nitrogen of the cyanides and some of the water molecules interact with the ferric atoms which are found in $Fe^{III}N_6$ and $Fe^{III}N_4O_2$ environments. When Prussian Blue is heated in a vacuum at $400\,°C$, the water is lost and the compound isomerizes to ferrous ferricyanide which is stable when dry, but reverts to the ferric ferrocyanide when it takes up water. The isomerism appears to involve the reversal of the CN groups, and intermediate environments such as $Fe(CN)_3(NC)_3$ occur.

When all the iron is in the Fe^{II} state, the unit cell reverts to 510 pm in length. A potassium ion placed at the centre of each small cube gives the formula, $K_2FeFe(CN)_6$, of the ferrousferrocyanide ion. The insoluble blues correspond to the formation of alkali-free complexes by replacing the potassium by ferrous or ferric ions as follows. If the cyanides are regarded as lying in the cube edges and coordinated through carbon to one iron atom and through nitrogen to the next iron atom, the framework of Figure 14.26 corresponds to a superlattice of formula $Fe^{II}Fe^{III}(CN)_6^-$ in the case of the blue compounds. The compounds with ferric iron and ferrous iron then become $Fe^{2+}[FeFe(CN)_6]_2^-$ or $Fe^{3+}[FeFe(CN)_6]_3^-$. Other complex ferricyanides and ferrocyanides are related to these structures. Thus the cupriferricyanide ion, $CuFe(CN)_6^-$, is the same structure with the Fe^{II} ions replaced by Cu^{II} ions. Such structures probably hold for all ferrocyanide or ferricyanide complex ions of heavy metals apart from the alkali and alkaline earth metals.

In the II state, iron forms a polyhydride complex, $[FeH_6]^{4-}$, whose structure has been fully determined by X-ray crystallography supported by neutron diffraction to determine the position of the hydrogens (compare section 7.4). When $FeCl_3$ was treated with a Grignard reagent (RMgBr) and hydrogen in THF, a complex salt was isolated of formula $Mg_4(FeH_6)X_4 \cdot 8THF$ where X was $Br + Cl$ in ratio 7:1. The structure contains a regular octahedral $[FeH_6]^{4-}$ ion with an Fe–H distance of 160·9 pm, a reasonable value for a covalent bond. The 4 Mg atoms form a larger tetrahedron around the octahedron, lying above every alternate triangular face. There is also a Mg–H interaction shown by the distance of 204·5 pm which is appropriate for a Mg^{2+} – H^- ionic interaction. One other polyhydride complex, ReH_9^{2-} (see section 15.5), has been known for a long time, and a few others—$Li_4RhH_{4 \text{ and } 5}$, Sr_2RuH_6 (probably also octahedral), Mg_2NiH_4, and perhaps Cu and Ir species—have been studied to varying extents. The iron compound is interesting, both as adding a well-established member to this small class, and in showing the dual interaction of H with Fe and Mg.

Ferrocene, $(C_5H_5)_2Fe$, is discussed in section 16.4. It contains Fe(II) and is oxidizable to the ferricinium ion, $(C_5H_5)_2Fe^+$, which contains Fe(III).

14.6.4 Low oxidation states of iron

Iron (I) chemistry is limited to only a few compounds, many of them substituted carbonyls. In contrast, there is a well-developed chemistry of iron (0), mainly of the carbonyls and their derivatives and analogues which are discussed further in section 16.2. Three carbonyl compounds are known, $Fe_2(CO)_9$ and $Fe_3(CO)_{12}$ as well as the pentacarbonyl, and a fair number of compounds exist where the carbonyls are replaced by other π-bonding ligands, as in $(Ph_3P)_2Fe(CO)_3$. Trifluorophosphine behaves in a very similar manner to carbonyl and analogues of most carbonyl compounds exist. Thus, $Fe(PF_3)_5$ and $Fe_2(PF_3)_9$ are found, as are all the mixed carbonyl-phosphine analogues of the pentacarbonyl such as $(PF_3)Fe(CO)_4$ and $(PF_3)_3Fe(CO)_2$.

14.7 Cobalt, $3d^7 4s^2$

Figure 14.27 shows the oxidation state energy diagram for cobalt. Only the II and III states have any stability in water and the II state is far more stable than Co^{III}, except in the presence of complexing ligands. Hydrated Co^{3+} decomposes water. In the solid state, the only trihalide is CoF_3, which is a strong oxidizing and fluorinating agent.

Cobalt metal is usually found in association with nickel in

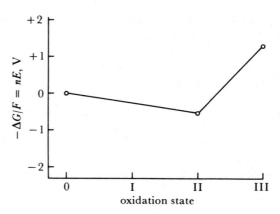

FIGURE 14.27 *Oxidation state free energy diagram for cobalt*
In aqueous acid conditions, cobalt(II) is much more stable than cobalt(III).

arsenical ores. It is relatively unreactive and dissolves only slowly in acids. It does not combine with hydrogen or nitrogen but it does react with carbon, oxygen, and steam at elevated temperatures, giving CoO in the latter cases. Cobalt is mainly used in high-performance alloys, in magnets, and in ceramics and paints where it provides a blue colour or, often, balances out yellow tinges arising from iron compounds.

14.7.1 Cobalt oxidation states greater than (III)

The highest oxidation state shown by cobalt is IV. An ill-defined, hydrated CoO_2 is formed when cobalt(II) solutions are oxidized in alkaline media. The species M_2CoO_3 (M = K, Rb, Cs), Li_8CoO_6, and Ba_2CoO_4 all contain cobalt(IV). The last has the same structure as K_2SO_4. Cobalt(IV) occurs in the well-defined dithiocarbamate complexes, $[Co(S_2NR_2)_3]^+$, formed by oxidation of the Co(III) species.

The green $(H_3N)_5Co-O-O-Co(NH_3)_5^{5+}$ was originally thought to be a peroxide (i.e. an O_2^{2-} derivative) with one Co(III) and one Co(IV) atom. It is now accepted as a superoxide (i.e. an O_2^- derivative) with two Co(III) atoms. It is prepared by oxidizing the brown cobalt(III) peroxide, $(H_3N)_5Co-O-O-Co(NH_3)_5^{4+}$.

14.7.2 Cobalt (III)

Cobalt(III) is strongly oxidizing in its simple compounds, and as the hydrated ion, but it forms a wide variety of stable octahedral complexes. It is the trivalent d^6 ion, therefore ΔE is large and spin-pairing is expected, to take advantage of the large CFSE of this configuration in the low-spin state, t_{2g}^6. Spin pairing appears to occur in most complexes, although CoF_6^{3-} is high-spin. The situation in the hydrate, $Co(H_2O)_6^{3+}$, is interesting. This is low-spin and diamagnetic but the CFSE gain appears to be only slightly greater than the loss of exchange energy, and the hydrated ion readily reduces to $Co(H_2O)_6^{2+}$, which is high-spin d^7. As the CFSE gain in d^6 is so high, it only takes a small increment in the ligand field to make the cobalt(III) state the more stable. Thus the hexammine, $Co(NH_3)_6^{3+}$, is prepared by air oxidation of cobalt(II) in aqueous ammonia. A still higher ligand field gives complexes in which the cobalt(II) state becomes strongly reducing. This is shown by the redox potentials for Co^{III}/Co^{II} complexes of the ligands shown below (all for octahedral complexes in both states):

ligand	H_2O	NH_3	CN^-
redox potential (V)	$+1\cdot84$	$+0\cdot1$	$-0\cdot8$

In the hydrates, cobalt(III) decomposes water by oxidizing it; in the ammines, cobalt(III) is stable to water; and in the cyanides, cobalt(II) decomposes water by reducing it. This range of activity resembles that already quoted for iron, but is more extreme because of the large ligand field effects.

Cobalt(III) complexes are extremely numerous and they undergo substitution only slowly so that a large variety of them are readily prepared and handled. It was the study of cobalt(III) and platinum(IV) complexes (both d^6), together with chromium(III) and square planar platinum(II) compounds which first led to the development of the ideas of complex chemistry at the start of this century by Werner and his school. By distinguishing different types of isomers and by proving the constancy through a series of chemical changes of certain groupings of atoms, Werner was able to formulate the idea of definite complex species and to determine the coordination numbers and shapes—all this before the development of any of the powerful modern techniques of structural determination. The variety and types of cobalt(III) complexes are best illustrated by some of these classical sequences of preparations and interactions in the field of cobalt-ammonia compounds, Figure 14.28. A number of other examples, including ethylenediamine complexes, are discussed in Chapter 13.

Although nitrogen is probably the commonest donor atom in cobalt complexes, a variety of oxygen complexes exist with ligands of the type of oxalate and acetylacetone,

$Co(C_2O_4)_3^{3-}$ or $Co(acac)_3$. Cobaltinitrite, $Co(NO_2)_6^{3-}$, is coordinated through nitrogen of course, but there is some evidence for the existence of the isomeric nitrito-form, $Co-O-N-O$, in solution in equilibrium with the nitro-compound. Essentially all cobalt(III) compounds are octahedral.

The cobalt(III) nitrosyl species of empirical formula $[Co(NH_3)_5NO]^{2+}$ has presented an interesting and long-standing problem. At the time of its discovery, in 1903, it was found to form two different isomers, one red and the other black. The crystal structure of the black form has now shown it to contain the mononuclear ion $[(H_3N)_5Co(NO)]^{2+}$ with a non-linear $Co-N-O$ group, angle about $120°$. The red isomer was recently shown to be a dimer, containing the hyponitrite group, N_2O_2, as a bridge. The structure is non-symmetric with both $Co-N$ and $Co-O$ bonds to the hyponitrite:

$$\left[(H_3N)_5Co-O \overset{\displaystyle N-N \overset{\textstyle Co(NH_3)_5}{\diagup}}{\underset{\displaystyle O}{\underset{\parallel}{}}} \right]^{4+}$$

Addition of CN^- to the black isomer gives two more hyponitrite species in the orange and yellow isomers of $(CN)_5Co-N_2O_2-Co(CN)_5^{6-}$. These have the bridges

$$Co \overset{O}{\underset{O}{\diagup}} N-N \overset{O}{\underset{Co}{\diagdown}} \quad and \quad Co \overset{O}{\underset{O}{\diagdown}} N-N \overset{O}{\underset{}{\diagup}} Co$$

Cobalt trifluoride is used as a fluorinating agent. CoF_2 reacts readily with F_2 to give CoF_3 and the latter is a strong fluorinating agent, though less reactive than fluorine. It provides a suitable way of moderating fluorination reactions. The compound to be fluorinated is streamed over CoF_3, giving the desired product and CoF_2. CoF_3 can then be regenerated by passing fluorine over the cobalt difluoride and the process may be continued in a cyclic manner.

One example of a simple cobalt(III) derivative is the anhydrous nitrate, $Co(NO_3)_3$. This is prepared in non-aqueous solvents and has bidentate nitrate groups giving octahedral CoO_6 coordination.

14.7.3 Cobalt (II)

Although cobalt(III) exists largely in complexes and has unstable simple compounds, cobalt(II) is just the reverse. It is perfectly stable in simple compounds and salts and forms a number of complexes with ligands of relatively weak ligand field. However, with ligands further along the spectrochemical series than water, the CFSE gain on achieving the low-spin d^6 configuration is sufficiently great to make cobalt (III) the preferred state.

Cobalt(II) oxide, halides, and sulfide are well-known and may be made by normal methods. Red or pink hydrated cobalt salts of all the common anions are known. On addition of base to Co^{II} solutions, the pink (occasionally blue) hydroxide is precipitated and this dissolves in concentrated

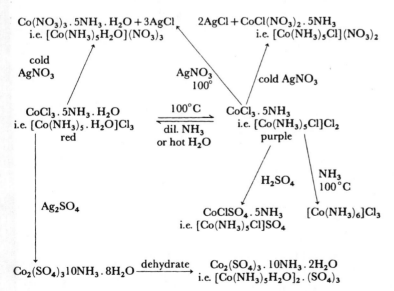

FIGURE 14.28 *Interconversions of cobalt(III) ammonia complexes*
This is an example of one of the series of interconversions which led Werner to the formulation of the concept of a complex.

alkali to give the deep blue $Co(OH)_4^{2-}$ anion. The latter may be precipitated as the sodium or barium salt.

Cobalt(II) oxide, halides, and sulfide are well-known The hydrate is octahedral, as is $Co(NH_3)_6^{2+}$. This, and a number of related complexes, may be prepared as long as an inert atmosphere is maintained to stop oxidation. $Co(CN)_5^{3-}$, and related species, are found in green and yellow forms, depending on the cation. In the green one, the coordination of the Co is square pyramidal with the sixth position occupied by a weakly coordinated cation or solvent molecule. The yellow form is a pure square pyramid. Similar species are found for Ni analogues, which also form distorted trigonal bipyramids in some cases.

The deep blue $CoCl_4^{2-}$, formed by addition of excess Cl^- to the pink hydrated solutions, is tetrahedral. The other halides form similar, blue anions as does thiocyanate, $Co(SCN)_4^{2-}$. These tetrahedral ions generally have to be precipitated from solution as salts of large cations. A related compound is the mercury complex, $CoHg(SCN)_4$, which is used as a calibrant in magnetic measurements. This contains cobalt(II) tetrahedrally coordinated by the nitrogen atoms of the thiocyanate groups while the sulfur atoms tetrahedrally coordinate mercury(II) ions to give a polymeric solid. Square planar cobalt(II) complexes are found for some chelating ligands, such as dimethylglyoxime and salicylaldehyde-ethylenedi-imine, which form stable planar complexes in general (compare nickel dimethylglyoxime).

The tetrahedral complexes of cobalt(II) have three un-paired electrons and the square planar ones have only one, both as expected for d^7. The octahedral complexes include both high-spin and low-spin cases, the former with three and the latter with one unpaired electron. The change-over appears to come between ammonia and nitro anion, NO_2^-, in the spectrochemical series.

An interesting cluster structure is found in $[(PhSCo)_8 (\mu^4\text{-}S)_6]$ which is the inverse of the $[(Et_3PFe)_6(\mu^3\text{-}S)_8]^+$ structure of the last section. In the cobalt(II) case, the 8 Co atoms form a cube with the six S atoms above the square faces—alternatively described as forming an octahedron enclosing the cube. The PhS-groups lie terminally and uniformly bent on each Co (compare Figure 14.25b).

14.7.4 Lower oxidation states of cobalt

Cobalt(I) is found in the hydride complex ion, CoH_5^{4-} which has a square pyramidal structure. Most compounds in the I, 0, and -I states form with π-bonding ligands. The commonest are the carbonyls (see section 16.2) and related species. The simplest carbonyl, $Co_2(CO)_8$, readily gives $Co(-I)$ in the anion $Co(CO)_4^-$ and the hydride $HCo(CO)_4$. It also reacts with organic isonitriles to give $Co(I)$ in the cation, $Co(CNR)_5^+$ which has a square pyramidal structure in the case $R = Ph$. The cation can also be prepared by reduction of the cobalt(II) compounds, $Co(CNR)_4X_2$.

Interesting cluster compounds are found among the more complex cobalt carbonyl anions. For example, in $Na_4Co_6(CO)_{14}$, the $Co_6(CO)_{14}^{4-}$ anion consists of an octa-

hedron of Co atoms, with each of the eight triangular faces bridged by a CO. The remaining six CO units are normal terminal groups, one on each cobalt. In other words, the structure is like that of $Mo_6Cl_8^{4+}$ shown in Figure 15.19, with Co in place of Mo, carbonyl in place of the chlorines together with six terminal CO groups.

Cobalt carbonyl hydride, $HCo(CO)_4$, has been shown to be the active species in the OXO process for the conversion of alkenes to alcohols in presence of a cobalt catalyst

$$RCH = CH_2 + CO + 2H_2 \rightarrow RCH_2CH_2CH_2OH$$

The reaction proceeds by initial insertion of H—Co into the double bond (an example of a *hydrometallation* reaction) followed by CO insertion into the reactive Co—C bond

$$RCH = CH_2 + HCo(CO)_4 \rightarrow$$
$$RCH_2CH_2Co(CO)_4 \xrightarrow{+CO} RCH_2CH_2COCo(CO)_4$$

It is thought that the acyl-cobalt compound then loses CO and the resulting tricarbonyl is cleaved by hydrogen and the product aldehyde is reduced to the alcohol.

$$H_2 + RCH_2CH_2COCo(CO)_n \rightarrow$$
$$(n = 3?)$$
$$RCH_2CH_2CHO + HCo(CO)_n$$
$$\searrow CO$$
$$\searrow HCo(CO)_4$$

Another interesting cobalt(I) species is the compound $CoH(N_2)(PPh_3)_3$, hydridodinitrogentris(triphenylphosphine)cobalt(I). This is a representative of a growing class of compounds which have N_2 as a ligand (compare section 16.10). The structure was one of the first N_2 complexes to be determined, and is shown in Figure 14.29, with trigonal bipyramidal coordination at the cobalt. The end-on, linear, Co—N—N bonding is analogous to the long-known Co–C–O unit, with which it is isoelectronic.

The ligand $L = CH_3C(CH_2PPh_2)_3$, with three donor

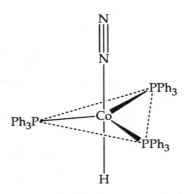

FIGURE 14.29 *The structure of the cobalt(I) nitrogen compound,* $HCo(N_2)(PPh_3)_3$.
The $Co—N_2$ unit is linear with bond length 180 ± 3 pm, the Co—H distance is 160 pm and the equatorial P atoms are bent slightly towards the hydrogen position as is usual in hydrido-complexes) with N—Co—P angles of 97 ± 2.

phosphorus atoms, forms halide complexes of cobalt I, CoLX, which are tetrahedral.

Cobalt(0) also occurs in $K_4Co(CN)_4$ which is the reported product of the reduction of the cobalt(III) hexacyanide with potassium in ammonia. Cobalt carbonyls are reviewed in section 16.2, and cobalt-carbon compounds in sections 16.3 and 16.4.

14.8 Nickel, $3d^8 4s^2$

The chemistry of nickel is much simpler than that of the other first row elements. The only oxidation state of importance is nickel(II) and these compounds are stable. Nickel(II) is the d^8 ion and is able to form stable square planar complexes as well as octahedral ones. Ligands with large crystal field favour square planar coordination, because of the more favourable CFSE. The heavier elements in the nickel Group are exclusively square planar in the II state.

The element occurs largely in sulfide or arsenic ores and is extracted by roasting to NiO and then reducing with carbon. Pure nickel is made by the Mond process in which CO reacts with impure Ni at 50 °C to give $Ni(CO)_4$. This is decomposed to Ni and CO at 200 °C to yield metal of 99·99 per cent purity. The alternative large-scale preparation involves reduction with carbon followed by electrolytic refining. Nickel is used in steels, especially stainless steel, and a number of alloys are widely familiar. Cupro-nickel, used in coins, is about 4Cu to 1Ni; addition of zinc gives nickel silver which is the base of EPNS tableware when silver-plated. Nickel is an important hydrogenation catalyst, and is used in the Ni – Fe battery. The metal resists attack by water

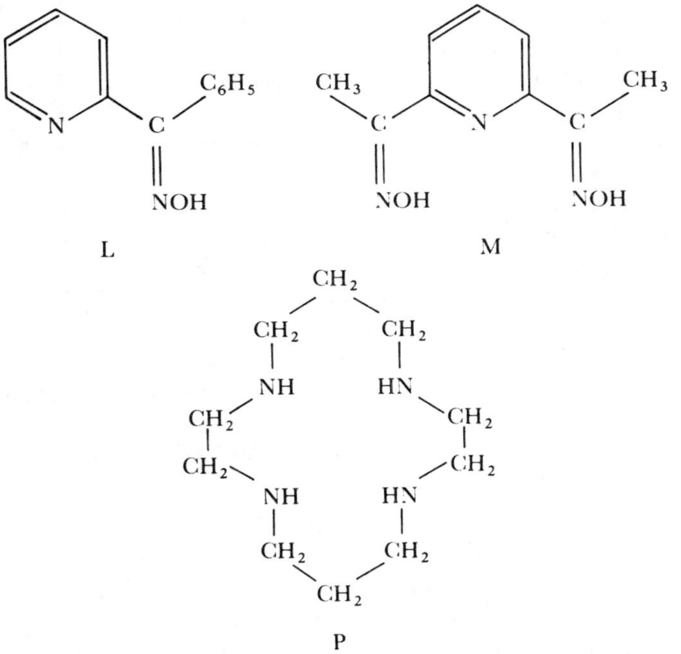

FIGURE 14.30 *Some ligands which stabilize* Ni(III) *or* Ni(IV) In each case, the hydrogen ionizes, giving L^-, M^{2-} or P^{4-}.

or air and is used as a protective coating. It has good electrical conductivity. The metal dissolves readily in dilute acids.

14.8.1 *Higher oxidation states of nickel*

Oxidation states above II are represented by obscure oxides and some complex compounds. Oxidation of $Ni(OH)_2$, suspended in alkali, by a moderately strong oxidizing agent like Br_2 gives a black solid which can be dried to the composition, $Ni_2O_3.2H_2O$. Attempts at further dehydration lead to decomposition to NiO. The action of powerful oxidizing agents in an alkaline medium give impure, hydrated products of approximate composition NiO_2. This is a powerful oxidant, forming permanganate from Mn^{2+} in acid, and decomposing water. If nickel is heated in oxygen in fused caustic soda, sodium nickelate(III), $NaNiO_2$, results.

The high oxidation states may be stabilized in complexes. Oxidation by chlorine in strong base allowed the formation of Ni(IV) which was isolated as the dimethylglyoxime complex $[(DMG)_3Ni]^{2-}$ ion (compare Figure 14.31 for the Ni(II)DMG complex). The NiF_6^{2-} ion is known in several salts such as K_2NiF_6, and with oxidizing anions, as in the iodine(VII) purple compound, $KNiIO_6.nH_2O$.

M_3NiF_6 (M = Na, K) is reported for Ni(III). Many other reported complexes have since been shown to contain nickel(II) and oxidized ligand, but trigonal pyramidal $Ni(PR_3)_2Br_3$ is quite well established.

By using ligands containing oxime groups (cf. Figure 14.31) and other donor atoms, relatively stable nickel(III) complexes, NiL_3, have been made. Here L contains one oxime and one donor N in a pyridine group, and the complex is neutral and six-coordinated. With a ligand M containing two oximes and a pyridine, NiM_2, containing six-coordinated Ni(IV) was prepared. Nickel(III) is also stabilized by large cyclic ligands (Figure 14.30), as in $[NiPX_2]^+$ or $NiL(SO_4)^+$. The products are formed by electrochemical oxidation of the Ni(II) analogue and are stable in dilute acid solutions: the oxidation potential is about 1 volt.

Nickel(III) is found in the green $(NiCl_2en_2)^+Cl^-$ complex which has a magnetic moment corresponding to one unpaired electron. Nickel(III) is also found in the phosphane complex $Ni(PEt_3)_2Br_3$ which is probably a trigonal bipyramid.

14.8.2 *Nickel (II)*

A wide variety of simple compounds of nickel(II) exist, including all the halides and all the oxygen Group compounds. Ni^{2+} forms salts with even strongly oxidizing anions such as chlorite, and with relatively unstable ions like carbonate. The relative stabilities of the II and III states of iron, cobalt, and nickel, as shown by the compounds formed with the oxychlorine anions are discussed in section 8.6. As chlorine in alkali oxidises nickel(II) to nickel(IV), it is to be assumed that the extremely powerfully oxidizing ClO^- anion will not form even a nickel(II) salt. The complex ions with ligands such as water and ammonia are octahedral, the green $Ni(H_2O)_6^{2+}$ ion being responsible for the typical colour of hydrated nickel salts. The hexammines and related complexes such as $Ni(en)_3^{2+}$ are generally blue. All possible mixed forms occur such as $Ni(H_2O)_2(NH_3)_4^{2+}$.

With ligands of high field strength, square planar, diamagnetic complexes are formed, such as the cyanide, $Ni(CN)_4^{2-}$ and the dimethylglyoxime complex used in quantitative analysis for nickel (Figure 14.31). In the latter, there may be interaction between nickel atoms in the crystal, where the flat molecules are stacked vertically above each other, so that the compound could be considered as a very distorted octahedron.

There are several examples of tetrahedral nickel(II) complexes. These involve halogen compounds, either as anions with large cations, as in $(Ph_4As)_2^+(NiCl_4)^{2-}$, or in

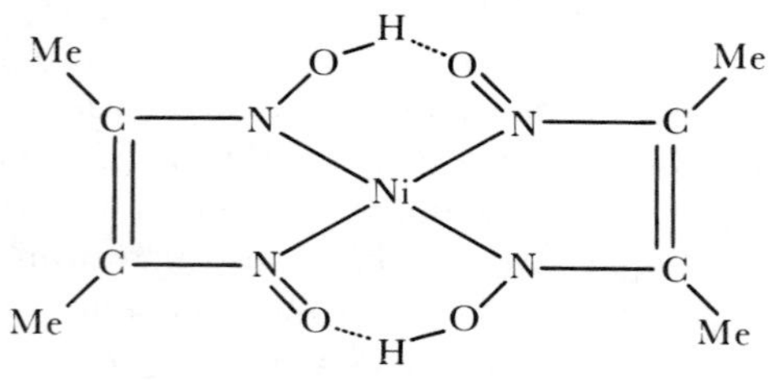

FIGURE 14.31 *Nickel dimethylglyoxime*
The dimethylglyoxime loses a proton and coordinates to Ni giving a five-membered ring. The complex is further stabilized by hydrogen bonding.

neutral complexes with phosphine or phosphine oxide and related ligands, as in $(PPh_3)_2NiI_2$ or $(Ph_3AsO)_2NiBr_2$. Such complexes are characteristically intensely blue due to a relatively intense absorption in the red end of the spectrum. These intense spectral lines distinguish tetrahedral nickel(II) from octahedral complexes, where the absorptions are relatively weak.

Since nickel(II) is found in octahedral, square planar and tetrahedral environments, and there is no large energy difference between these, there are many cases where more than one form of a complex occurs, differing in stereochemistry, or where an equilibrium exists between two forms in solution. The salicylaldimato complexes (compare Appendix B) of the type shown in Figure 14.32, provide examples of all three stereochemistries. When R = methyl or isobutyl, the complex of Figure 14.32 is square planar, but when R = isopropyl, the coordination about the nickel becomes tetrahedral. Further, if the diamagnetic square planar complex (e.g. R = Me) is dissolved in pyridine, it becomes paramagnetic with the moment expected for an octahedral

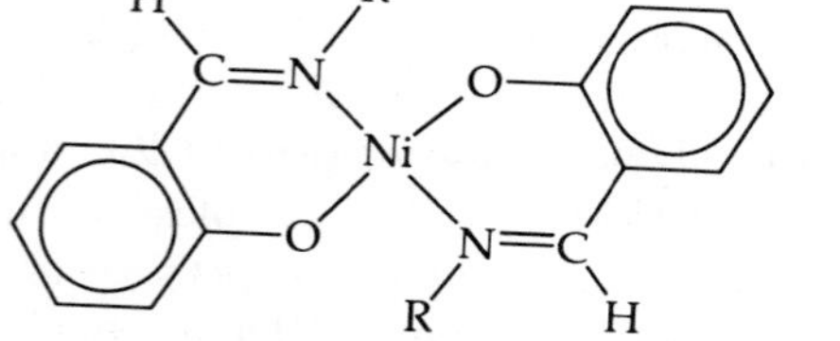

FIGURE 14.32 *Bis-*(N-(*alkylsalicylaldimato*)*nickel*(*II*) *complexes*

complex, and an octahedral dipyridine derivative may be isolated.

When a ligand atom is a moderately weak donor, it is possible to isolate complexes with the same formula but different coordinations. For example, NiL_2Cl_2 where

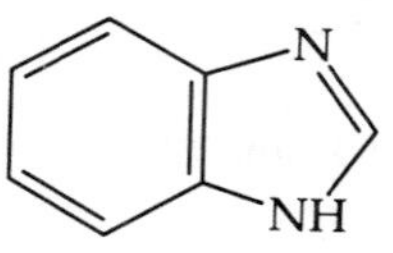

may be found in a blue tetrahedral form or in a yellow octahedral one. Similarly, $NiL_4'Br_2$, where L′ = benzimidazole, is found in an orange square planar form and a yellow octahedral one.

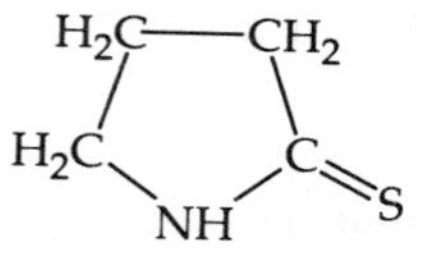

In the above cases, solids may be isolated and characterized, but often the two forms coexist in solution, even though only one is isolable. This is true of a wide range of complexes NiL_2X_2 which show tetrahedral-octahedral equilibrium (usually as blue-yellow changes) in solution, with the tetrahedral form most favoured when X = I. Such equilibria may be studied spectroscopically, and when L = phosphane with bulky groups, the exchange is slow enough to be followed by nmr at low temperatures.

One particularly striking example of the ease of interconversion of these stereochemistries is provided by the compound $Ni(PRR_2')_2Br_2$ where the substituents on phosphorus in the phosphane are R = benzyl ($C_6H_5CH_2$) and R′ = phenyl (C_6H_5). In the crystal of this compound, both square planar and tetrahedral nickel environments are found. Thus any small energy difference between the two coordinations is compensated for by the readier packing of the two different sorts of molecule.

Nickel(II) also forms a range of five-coordinate complexes (compare section 13.6), both in the square pyramidal configuration like NiX_2terpy (terpyridyl) and trigonal bipyramidal, as in many Schiff's base complexes (compare Appendix B). There is even the case, quoted in section 13.6 of the $Ni(CN)_5^{3-}$ ion adopting both coordinations in the same compound.

An interesting complex structure is found in $Ni_9Te_6(PR_3)_8$ which was isolated in the course of studies on metal–tellurium compounds as possible precursors to interesting electronic materials. The structure is related to that of Figure 14.25b, with an octahedron of Te atoms face-bridging a cube of Ni atoms, each carrying a terminal phosphane ligand. The ninth Ni atom lies at the centre of the whole structure, so there is a $Ni(Ni)_8$ cube having a common centre with a $Ni(Te)_6$ octahedron and the cube lies within the octahedron.

14.8.3 *Lower oxidation states of nickel*

Low oxidation states are represented by the cyanide complexes, $K_4Ni_2^I(CN)_6$ and $K_4Ni^0(CN)_4$, which result from the reduction of the $Ni^{II}(CN)_4^{2-}$ species with potassium in liquid ammonia. The nickel(I) compound may also be made by reduction with hydrazine in an aqueous medium. It is relatively stable but the nickel(0) complex is extremely reactive and not well characterized. Both are oxidized in air or water. Nickel(I) is also represented by the phosphane complexes $NiX(PPh_3)_3$ and NiLX, where $L = CH_3C(CH_2PPh_2)_3$). Nickel(0) is found in the carbonyl, $Ni(CO)_4$, in $Ni(PF_3)_4$, and in all the mixed trifluoro-phosphane-carbonyls. The carbonyl anion, $Ni_2(CO)_6^{2-}$, contains nickel($-$I). Although nickel carbonyl is one of the best-known carbonyls, the stability of such compounds is relatively low at this end of the d block and nickel has a much less rich carbonyl chemistry than the earlier elements.

A totally unexpected nickel(0) compound is formed by the very bulky phosphane with *tert*-butyl substituents, $Ni(PBu^t)_6$. The crystal structure shows that the Ni atom lies at the centre of a nearly planar ring of six P atoms with the *tert*-butyl groups enveloping the plane above and below. The structure is strongly determined by the steric bulk of the substituents and the P–P bonds are somewhat lengthened. The Ni–P distances are reasonable for zero-valent nickel.

14.9 Copper, $3d^{10}4s^1$

There are three possible oxidation states in the copper Group, the I state corresponding to d^{10}, the II state which is common to the whole transition series, and the III state corresponding to d^8 which would be stable in a square planar environment. The elements in this Group show these states, with II the stable state of copper, I the stable state of silver, and III the stable state of gold. Copper also exists in the I state, which is quite stable in solids, and one or two copper(III) compounds are reported. This wide variation in stabilities in this Group is probably a resultant of size, exchange energy, and ligand field effects, and is discussed in section 14.10.

Copper is found in sulfide ores and as the carbonate, arsenide or chloride. Extraction involves roasting in air to the oxide, reduction, and purification by electrolysis. In a clever application of biotechnology, processes have been developed to recover copper from low-grade ores and mine tailings by using sulfide bacteria which obtain their energy from the oxidation of sulfide to sulfate, which goes into solution for further recovery. Solvent extraction processes employing hydroxyoximes are also used industrially to recover and purify copper (see section 7.3).

Pure copper has an electrical conductivity second only to that of silver and its major application is in the electrical industry. The element is inert to non-oxidizing acids but reacts with oxidizing agents. With oxygen it combines on heating to give CuO at red heat and Cu_2O at higher temperatures. It also reacts with halogens and dissolves in hot nitric or sulfuric acid.

14.9.1 *Copper(IV) and copper(III)*

One example of copper(IV) is established. When the copper(II) complex, $CsCuCl_3$ is heated with F_2 at 250 °C, the product is $Cs_2Cu^{IV}F_6$. The ion is a slightly distorted octahedron with two opposite Cu–F distances of 177 pm and the other four Cu–F a little shorter at 175 pm. Such a structure is appropriate for a d^7 ion, compare cobalt(II).

The (III) state of copper may be obtained by oxidation in

alkali, yielding $MCuO_3$ for M = alkali metal. The structure is linked square planar CuO_4 units. Reaction of the Cu(II) complex $CsCuCl_3$ with F_2 gives $CsCuF_4$ containing the square planar $Cu^{III}F_4^-$ ion, while similar treatment of mixed MCl plus $CuCl_2$ gives alkali metal salts of octahedral CuF_6^{3-}, which have a very short Cu – F distance of 183 pm. A further example of copper(III) is the periodate, $K_7Cu(IO_6)_2.7H_2O$. It will be noticed, if the higher oxidation states of iron, cobalt, nickel, and copper are compared, that the methods of preparing such compounds tend to be similar, for example, by oxidation in alkaline media and the use of oxidizing anions. Cu(III) is also important in the new ceramic superconductors of formula $YBa_2Cu_3O_{7-x}$ (where x is about 0.1), which are discussed in detail in section 16.1.

14.9.2 *Copper(II)*

Copper(II) is the main state in aqueous solution and the compounds are paramagnetic with one unpaired electron. The hydrated ion may be written $Cu(H_2O)_6^{2+}$ although the structure is not regular. There are four near ligands in a square plane and the other two are further away as a result of the unequal occupation of the two e_g orbitals. This distorted shape is common for copper(II) compounds as the examples of bond lengths below show:

Compound	Distances, pm	
	shorter	longer
CuF_2	4 of 193	2 of 227
$CuCl_2$	4 of 230	2 of 295
$CsCuCl_3$	4 of 230	2 of 265
$CuCl_2.2H_2O$	2 of 231 (Cl)	2 of 298
	2 of 201 (O)	
K_2CuF_4	4 of 192	2 of 222

The $[CuCl_6]^{4-}$ ion was originally thought, from an X-ray structure, to show the opposite distortion with only two short bonds. It is now clear that the ions are disordered in the crystal, so that the four 'longer' bonds are actually an average of two long and two short bonds. Several experimental techniques now agree that the configuration is actually the normal one with four Cu–Cl distances in the range 228–238 pm and two at 283 pm, comparable with the values tabulated above.

In contrast, $KAlCuF_6$ does contain $[CuF_6]^{4-}$ ions with the unusual compressed geometry, unlike the K_2CuF_4 compound listed above. In $KAlCuF_6$, there are four F ions in the equatorial plane at a distance of 212 pm and two much shorter bonds of 188 pm. The four equatorial F have Al^{3+} ions as neighbours and the plane is quite distorted with the Cu slightly above the plane, one FCuF angle of 120° and three of 80°.

Mixed complexes of water and ammonia are found up to $Cu(H_2O)_2(NH_3)_4^{2+}$, but replacement of the last two water molecules is impossible in aqueous solution and $Cu(NH_3)_6^{2+}$ can only be prepared in liquid ammonia. It is similarly possible to form $Cu(H_2O)_4en^{2+}$ and $Cu(H_2O)_2en_2^{2+}$ but formation of $Cuen_3^{2+}$ is difficult. These, and similar amine complexes, are all a much deeper blue than the hydrated ion.

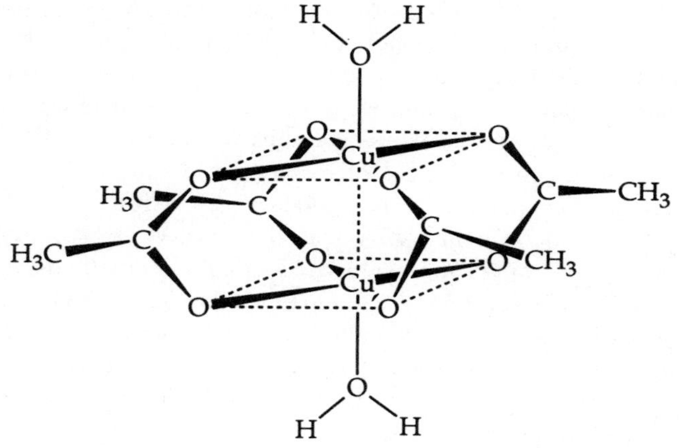

FIGURE 14.33 *Structure of hydrated copper(II) acetate*
The two square planar CuO_4 units are linked by the acetate residues and also by Cu–Cu bonding.

Halide complexes are also distorted but are tetrahedral with a flattened structure, for example, $CuCl_4^{2-}$, which can be precipitated from a chloride medium by large cations. Such tetrahedral complexes are generally green or brown.

Among the salts, mention must be made of the unusual structures of copper(II) acetate and of anhydrous copper nitrate. The acetate is dimeric and hydrated, $Cu_2(CH_3COO)_4.2H_2O$, and has the structure of Figure 14.33. The copper atoms are surrounded by a square plane of oxygen atoms and the acetate groups bridge the planes together. Similar structures are found for copper derivatives of other carboxylic acids.

Anhydrous copper nitrate cannot be made by dehydrating the hydrated salt as this decomposes to the oxide. However, copper metal dissolves in liquid N_2O_4/ethyl acetate mixture to give $Cu(NO_3)_2.N_2O_4$, and $Cu(NO_3)_2$ results when the solvating molecule is pumped off. The structure in the solid is shown in Figure 14.34. The structure consists of chains of copper atoms bridged by NO_3 groups and cross-linked by other nitrate groups. Anhydrous copper nitrate is slightly volatile as single molecules.

There are some apparently three-coordinated copper compounds but all are of more complex structure. For example, $KCuCl_3$ contains dimeric planar $Cu_2Cl_6^{2-}$ ions. In the lithium salt, these dimeric ions are joined into longer chains by long chloride bridges. A bromide complex contains the $Cu_3Br_8^{2-}$ ion which is completely planar with three square planar units linked through their edges. A very distorted octahedron around the Cu is completed by weak interactions with two Br from parallel ions above and below in an extended stack.

As with nickel, copper(II) appears in a variety of geometries and examples exist of the same compound occurring in two different structural modifications. One example is provided by complexes $CuCl_2L_2$, where L is an N-oxide such as pyridine-N-oxide, C_5H_5NO. These occur in a green form, which is thought to be *trans*-square planar and also in a yellow form in which the coordination is tetrahedral.

Copper(II) is found in five-coordination which is usually square pyramidal although $CuCl_5^{3-}$, $CuBr_5^{3-}$ and $CuCl_2Br_3^{3-}$ are trigonal bipyramids (with axial Cl in the latter).

As many of these structures are distorted, and often show a variety of bond lengths to the same ligand atom, copper stereochemistry presents an extremely difficult field. Normal spectroscopic methods are inadequate to distinguish distorted shapes from those of lower coordination number, and even X-ray crystallography may give inconclusive results because of the problems of correlating several different interatomic distances with short, long, or non-existent bonds. Three-coordinate copper(II) is found in the halides CuX_3^- with large cations.

14.9.3 *Copper(I)*

A mixed Cu(I)–Cu(II) species is found in the intensely blue $Cu_2Cl_4^-$ ion, which consists of an infinite chain of distorted tetrahedral units sharing edges. One type of Cu has Cu–Cl = 234·2 pm and two ClCuCl angles of 92·8° and four of 118°, values appropriate to Cu(I). Its neighbour has Cu–Cl = 225·5 pm, and angles in pairs of 98°, 100°, and 134°, all features indicating Cu(II). The blue colour should not be confused with that of Cu(II) in solution which is due to the hydrated ion. Cu(II) chloride species are generally yellow or red. The intense colour is due to electron transfer between the two oxidation states, and intense colours are characteristic of such systems (compare the discussion on Prussian Blue, section 14.6).

The I state of copper is represented largely by solid com-

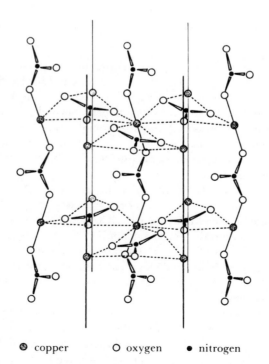

⊕ copper   ○ oxygen   ● nitrogen

FIGURE 14.34 *Solid anhydrous copper nitrate*, $Cu(NO_3)_2$
This structure consists of cross-linked chains of nitrate groups bonded to two copper atoms. A second crystal form of copper nitrate has been discovered recently.

pounds which are insoluble in water. For example, if iodide is added to a copper(II) solution, the cupric iodide initially formed rapidly decomposes to give a precipitate of CuI. Similarly, one method of determining copper quantitatively is by precipitation of CuSCN. The stability of copper(I) in solution is low and the redox potentials

$$Cu^+ + e^- = Cu \quad 0.52 \text{ V}$$
$$Cu^{2+} + e^- = Cu^+ \quad 0.15 \text{ V}$$

show that Cu^+ is unstable to disproportionation:

$$2Cu^+ = Cu + Cu^{2+}, E = 0.37 \text{ V}, K = \frac{[Cu^{2+}]}{[Cu^+]^2} = 10^6$$

The reason for the marked instability of Cu^+ in water is not completely clear, as the d^{10} configuration, with its very high exchange energy, would be expected to be reasonably favoured. One possible explanation lies in the low hydration energy which is likely for the Cu^+ hydrated ion, as compared with that for the Cu^{2+} ion. The cuprous ion is larger and has only half the charge so that its charge density is markedly lower, and hence the energy of interaction with the water dipole is less. In addition, in known complexes of Cu^+, especially the ammine, Cu^+ is only two-coordinated as in $Cu(NH_3)_2^+$. If the hydrate is reasonably supposed to have the same formula, then there would be only two interactions instead of the four strong and two weaker interactions in the Cu^{2+} hydrate. Thus Cu^+ has about half the interaction energy with half the number of water molecules compared with Cu^{2+}. When this case is compared with that of Ag^+, where the behaviour is quite the reverse and Ag^{2+} is quite rare and unstable, it will be seen that the greater size of silver tends to decrease these differences between the two oxidation states and the exchange energy probably then becomes the dominant term. Passing to gold, this trend again alters with the size, and the large CFSE term for a trivalent species of the third transition series helps to make square planar Au(III), with a d^8 system, the preferred state, although Au(I) also occurs. It is probable that the marked variation in chemistry in this Group, where the three elements differ markedly more than in any other Group, is a function of the increase in size coupled with the existence of stable electronic configurations on either side of the divalent d^9 state.

In the complex copper(I) ion, $Cu_4I_6^{2-}$, there is no Cu—Cu bonding. Cu lies almost symmetrically in the triangular faces of the I_6 octahedron.

Apart from the insoluble CuCl, CuBr, CuI, and the cyanide and thiocyanate, copper(I) gives soluble complexes with these groups as ligands. Two cyanide complexes exist, soluble $Cu(CN)_4^{3-}$, and the compound $KCu(CN)_2$ which has a chain structure containing three-coordinate Cu^I:

$$----Cu-C-N-Cu(CN)-C-N-Cu----$$

The red cuprous oxide, Cu_2O is well-known. This is the compound precipitated in Fehling's test for sugars; it is produced by the reduction of the blue cupric tartrate complex by glucose or related molecules.

It is still uncertain whether copper can exist in the 0 state as a number of reported compounds have been reformulated. One possible example is the diamagnetic species

$$Cu(NC-(CH_2)_4-CN)_2.$$

For zinc, $3d^{10}4s^2$, see section 15.10, The zinc Group.

14.10 The relative stabilities of the dihalides and trihalides of the elements of the first transition series

Although, at present, the factors affecting the stabilities of oxidation states are incompletely understood, answers can be obtained in a rigorous manner from the thermodynamic parameters involved, providing that the problem is precisely formulated. One case discussed is the thermal stabilities of the trihalides (excluding fluorides) of the first row transition elements to decomposition according to the equation

$$MX_{3(solid)} \rightarrow MX_{2(solid)} + \tfrac{1}{2}X_2$$

at 25 °C. This problem is only a small part of the question of the relative stabilities of the II and III states of the first row elements but its solution points the way for further work. Other compounds, other conditions, and even other decomposition routes of the trihalides, would have to be considered to extend our understanding of the general problem.

The observed stabilities show a fairly regular trend from scandium to zinc, with the trihalide stable relative to the dihalide at the left hand end of the transition series, and unstable with respect to the dihalide for the elements to the right, especially nickel, copper and zinc. There is an anomalous order of stability in the middle of the series with Mn < Fe > Co in the stability of the trihalide relative to the dihalide. This order holds for X = Cl, Br, and I. In the detailed analysis, it turns out that the change of entropy in the decomposition reaction is small and nearly the same for all the elements. Thus the free energy change in the reaction depends on the changes in enthalpy. These are best analyzed by breaking the reaction down into a number of simpler steps in a Haber cycle:

(recall that exothermic processes are negative in sign.)

Thus $\Delta G = -U_3 - I + U_2 + E - T\Delta S$, where ΔG is the free energy of the decomposition reaction, U_2 and U_3 are the lattice energies of the di- and tri-halide, $T\Delta S$ is the entropy energy, and E and I are the appropriate electron affinity and the third ionization potential of the metal. As the changes in the chlorides, bromides and iodides are parallel, attention may be focused on the chlorides.

As we are comparing the chlorides of one element with the next, the value of E is a constant for the series, and as $-T\Delta S$ is found to vary only slightly, we can take the sum of E and $-T\Delta S$ as a constant along the series. Thus the *variation* in ΔG along the series is a consequence of the variations in the terms $-I$, $-U_3$ and U_2. The values of the ionization energies, I, are well established (Table 2.8) and the lattice energies, U, may be evaluated by Born-Haber cycles, as discussed in Chapter 5, in terms of atomic properties.

The variations across the series of $-I$, U_2, and $-U_3$ are plotted in Figure 14.35, together with the resultant value of ΔG which is the combination of the constant terms E and $-T\Delta S$ with $(U_2 - U_3 - I)$. The lattice energy curves, U_2 and U_3, are similar to those shown in Figures 13.21 and 13.22 and clearly reflect ligand field stabilization energy changes. In particular, the d^5 species—$MnCl_2$ and $FeCl_3$—are less stable than their neighbours. (Note that U_2 and U_3 are plotted with opposite signs.) However, for the difference $(U_2 - U_3)$, these variations partly cancel (as shown in the top curve of Figure 14.35) as they occur one element later for the trihalides than for the dihalides. It is clear that the most important single factor is the third ionization potential of the metal, which is exothermic for the cycle step shown

$$M^{3+}(g) + e^- \rightarrow M^{2+}(g).$$

The free energy curve ΔG is clearly seen in the Figure to be dominated by the value of $-I$, moderated to some extent by the variation in $(U_2 - U_3)$. The change in free energy of the reaction

$$MCl_{3(solid)} \rightarrow MCl_{2(solid)} + \tfrac{1}{2}Cl_{2(gas)}$$

is positive for the elements at the left of the transition series and varies through near-zero values to negative at the right hand end. That is, the trichlorides (and similarly for the tribromides and triiodides) are stable with respect to decomposition at the left of the series while the decomposition to the dihalide is favoured towards the right. The variation of I, in turn, may be analyzed as a general increase across the transition series as the nuclear charge increases (that is, $-I$ becomes more negative) as expected from the incomplete shielding effect of the d electrons in the same shell. To this is added the major break between $Mn(d^5 \rightarrow d^4)$ and $Fe(d^6 \rightarrow d^5)$ which again reflects the exchange energy effect in the d^5 configuration.

Thus, although this decomposition reaction depends on the total effect of a large number of factors, it is seen that the pattern is set by exchange energy effects in the d^5 state as reflected in the value of I moderated by CFSE effects on the lattice energies of the halides.

PROBLEMS

The best way to build up a clear picture of systematic chemistry is to correlate one body of facts in as many ways as possible. Look at the chemistry of neighbouring elements, of other oxidation states of the same element, of the same d^n configuration, of ions of the same charge etc. The problems below are a guide to many similar ones which you can devise.

14.1 From Figure 14.1 (compare section 8.6) determine the most stable oxidation state in solution for each of the transition elements of the first series. Compare Tables 13.2 to 13.5 and decide how far stability in the solid state matches stability in solution. Discuss any difference.

14.2 For each element of the first transition series, examine how far the detailed chemistry (as given in the appropriate sections 14.2 to 14.9) matches the general deductions drawn in question 14.1 (see also Chapter 13, question 1).

14.3 The oxidation states of the first transition series are demonstrated by the oxides and fluorides in Tables 13.3 and 13.4. Collect together descriptions of mixed species (the oxyfluorides) from this chapter and compare the picture of oxidation states which results with that given by the tables. Do the same exercise for the heavier halides. Such an approach may be extended to the complex ions: compare the fluoro-complexes with the oxyfluoro- ones.

14.4 Compare and contrast the chemistry of every third element of the first transition series with that of its neighbour on either side.

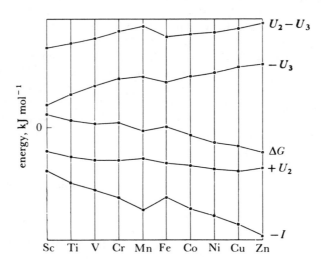

FIGURE 14.35 *The variation in the principle energy terms involved in the process* $MX_3 = MX_2 + \tfrac{1}{2}X_2$

The resultant ΔG follows the $-I$ curve, moderated by the lattice energy difference $U_2 - U_3$. For clarity, the zeros of the different curves have been shifted: the $-I$ curve ranges from -2430 to -3850 kJ mol^{-1}, the U_2 curve from -2340 to -2720 kJ mol^{-1}, the ΔG curve from $+375$ to -630 kJ mol^{-1}, and the $-U_3$ curve from $+5270$ to $+5860$ kJ mol^{-1}. (Recall that lattice energies are exothermic as defined so that $+U_2$ is exothermic and therefore has negative values while $-U_3$, the energy for the reverse process to forming the trihalide lattice, is endothermic and has positive values.)

14.5 (a) A potential of -1.9 volts has been measured for $Fe^{3+} + 4H_2O = FeO_4^{2-} + 8H^+ + 3e$. Extend the Ebsworth diagram for Fe and discuss the properties of Fe(VI) compared with Mn(VI) and Cr(VI).

(b) Survey all oxidation states of VI or above throughout the Periodic Table. Discuss their occurrence, stability and properties.

14.6 Survey a general topic throughout the transition elements. Examples of such topics include

(a) octahedral complexes (compare particular Groups of elements)

(b) complexes of ligands with alternative donor atoms (e.g. $-NCS$ and $-SCN$)

(c) coordination numbers greater than six

(d) shapes adopted for five coordination

(e) the relative stabilities of octahedral, tetrahedral and square planar configurations

(f) particular classes of compound such as 'bridged oxygen species', or 'complexes with metal–sulfur bonds', or 'peroxides'.

(g) complexes of bidentate oxyanions (e.g. acetates, nitrates)

(h) low oxidation states

14.7 Look up the paper 'Thermochemistry of the Potassium Hexafluorometallates(III) of the Elements from Scandium to Gallium' by P. G. Nelson and R. V. Rearse, in *J. Chem. Soc., Dalton Transactions*, 1983, pages 1977–1982. Compare the results with Figure 14.35. Discuss how far the interpretations are compatible and whether there is an effect of the halogen.

Many such questions can be devised. Surveys should start with the first transition series in Chapter 14 and include material from Chapters 9 and 13. Many should be broadened to the heavy transition elements in Chapter 15, and often to the rest of the Periodic Table (see Chapters 10 to 12 and 17). The special topics of Chapters 16 and 20 will also often be relevant.

15 The Elements of the Second and Third Transition Series

See section 16.12 for elements of the fourth transition series (the 'superheavy' elements).

15.1 General properties

This Chapter deals with the remaining elements of the transition block, those in which the $4d$ or $5d$ level is filling. The effect of the lanthanide contraction is to make the chemistry of the heavier pair of elements in the same Group very similar. A summary of the oxidation states of these elements, in relation to those of the first series, is given in Table 13.2, while Tables 13.3, 4, and 5 give the known oxides and halides. Other properties already listed include electronic configurations, atomic weights and numbers (Table 2.5), and Ionization Potentials (Table 2.8).

In these elements, higher oxidation states are more stable, in general, than in the first series. This is shown both by the relatively non-oxidizing behaviour of states like $Re(VII)$ and $Mo(VI)$ or $W(VI)$, and by the existence of high oxidation states to the right of the series, which are not found for the first row elements: examples include $Os(VIII)$ and $Ir(VI)$. Complementary to this increased stability of high oxidation states is a decrease in the stability of many lower oxidation states, as shown by such examples as the strong reducing properties of $Zr(III)$ or $Nb(IV)$ and the virtual non-existence of compounds of $W(II)$. Some similarities do, of course, exist between the lighter and heavier elements, especially in the properties of the higher oxidation states to the left of the d block and in those of the lower oxidation states at the right hand end. The maximum range of oxidation states comes in the middle of the block, rising to the VIII state of osmium and ruthenium.

Some of these elements are rather rare or inaccessible, especially those in the Groups immediately following the lanthanides, which are difficult to separate. Both hafnium and niobium, which occur along with their more abundant congeners, zirconium and tantalum, are not well studied and there are gaps in their chemistries where they are presumed to be similar to their congeners but without full proof. Other rare elements are technetium, which has only unstable isotopes and has to be made artificially, and the platinum metals and gold which, though accessible, are expensive to work with. The elements become less reactive towards the right of the transition block, and the tendency to reduction to the metal is an important characteristic especially of the platinum metals and gold, and also, to a lesser extent, of rhenium and silver. Few of these elements find large scale applications, though molybdenum and tungsten are components of highly resistant steels. The precious and semi-precious metals, apart from their uses in coinage and jewellery, find some application in precision instruments, electrical apparatus, and in surgery. Platinum, palladium and rhodium are used as catalysts in a number of industrial processes. Zirconium and niobium, the former especially, find use as 'canning' materials for the fuel in nuclear reactors. For this purpose, they must be separated from their congeners, hafnium and tantalum, which 'poison' the reactors by capturing neutrons. Most elements are extracted by carbon or metal reduction of the oxides or chlorides.

Many of these elements form non-stoichiometric carbides, nitrides, and hydrides, similar to those already discussed for the first series elements.

As these elements are larger than the first row members, higher coordination numbers are found. Although six-coordination to singly-bonded ligands is still common, seven- and eight-coordination are found, as in ZrF_7^{3-} or $Mo(CN)_8^{2-}$. To pi-bonding ligands such as oxygen, six becomes a common coordination number as well as four; compare the polymeric oxyanions of Mo and W, which are based on MO_6 groups, with the vanadates and chromates, which contain MO_4 units.

The increase in ligand field splittings which is shown in these larger atoms means the CFSE values increase markedly in all configurations, and the normal electronic configurations are the low-spin ones. Evidence for high-spin complexes is very rare.

15.2 Zirconium, $4d^2 5s^2$, and hafnium, $5d^2 6s^2$

Only the IV oxidation state is stable for these elements and potential data are not available for the lower states. The M/M^{IV} values for the two elements are similar and differ from that of titanium by a factor of about two; the elements being better reducing agents than titanium. $Ti^{IV}/Ti = -0.89$ V, $Zr^{IV}/Zr = -1.56$ V, and $Hf^{IV}/Hf = -1.70$ V. The IV state occurs in solution and in a variety of solid compounds, but the III and II states decompose water and are only found in solid products.

Zirconium ores occur, including the oxide and zircon, $ZrSiO_4$, but there are no discrete sources of hafnium which is always found as a minor component in zirconium minerals. Zirconium was known for nearly a hundred and fifty years before hafnium was discovered in its ores and compounds. The elements are extracted as tetrahalides and reduced with magnesium in a process similar to that for titanium. Here also, argon has to be used to provide an inert atmosphere, as the metals combine with nitrogen.

The covalent radii, 145 pm and 144 pm respectively for Zr and Hf, and also the radii of the hypothetical ions Zr^{4+} (74 pm) and Hf^{4+} (75 pm), are so close that the chemistry of this pair of elements is virtually identical. Separation is even more difficult than for the lanthanides but ion exchange and solvent extraction procedures are now available. One example of these is the separation of the tetrachlorides dissolved in methanol on a silica gel column. Elution is by an anhydrous HCl/methanol solution, with the zirconium coming off the column first.

In the IV state, these elements show a general resemblance to titanium(IV) but differ in the acidity of the dioxides and the solvolytic behaviour of the oxycation. The dioxides are soluble in acid solution and addition of base precipitates gelatinous hydrated oxide, $ZrO_2 \cdot nH_2O$. No true hydroxide exists. The hydrated oxide gives ZrO_2 (or HfO_2) on heating and these are hard, white, insoluble, unreactive materials with very high melting points (above 2500 °C). The more abundant zirconium dioxide is used for high temperature equipment, such as crucibles, because of these properties. ZrO_2, doped with Y_2O_3, finds important application as a solid state electrolyte in oxygen sensors. These are used, for example, in motor vehicle exhausts, to measure the oxygen content of the exhaust gases in a feedback loop which allows optimized combustion and minimum emissions. The hydrated oxide is quite insoluble in alkali and the dioxides therefore have no acidic properties. The solids M_2ZrO_3 contain ZrO_5 square pyramids linked into chains by sharing edges.

In solution, zirconium(IV) and hafnium(IV) hydrolyse less than titanium(IV). The main species in solution is frequently written as the oxyion ZrO^{2+} (zirconyl) or HfO^{2+} (hafnyl) but it is doubtful whether these simple species exist. The species in solution are probably $M(OH)_n^{(4-n)-}$ and trimeric or tetrameric hydroxy species. Tracer experiments indicate the presence of Zr^{4+} ions in very dilute solution in perchloric acid. Compounds containing the MO group are common, for example $ZrO(NO_3)_2 \cdot 2H_2O$ or $HfO(OOCCH_3)_2$, but their

structures may not be simple. Thus, $ZrOCl_2 \cdot 8H_2O$ contains the cation $[Zr_4(OH)_8(H_2O)_{16}]^{8+}$ which has dodecahedral coordination around the zirconium. There is also evidence for the possibly polymeric, ion $Zr_2O_3^{2+}$ in solution. The complexity of these hydrolysis products is typical of the solution chemistry of this part of the Periodic Table, although the larger atoms mean that there is less extensive hydrolysis than for titanium(IV). However, the elements are not large enough to allow the formation of the M^{4+} ions. Simple zirconium or hafnium compounds, such as the tetrahalides or the tetra-acetate, $Zr(OOCCH_3)_4$, are covalent compounds, not salts.

The halides may be made by standard methods and show the expected reactions:

$$MO_2 + C + Cl_2 \longrightarrow MCl_4 + CO_2$$

$$M + X_2 \longrightarrow MX_4$$

M heat MX_6^{2-} $\xrightarrow{H_2O}$ MOX_2

(also MF_7^{3-} and   (stable)

$MX_3 + MX_2$ $M_2'^{II}MF_8$ in the case $X = F$)

ZrF_4 forms a solid with 8:2 coordination where each Zr atom is surrounded by 8F atoms in a square antiprism configuration (compare Figure 15.5b). The other tetrahalides are volatile solids. In the vapour phase, the structure is monomeric and tetrahedral, while in the solid the Zr or Hf atoms are in octahedral coordination. The structure of one form of $ZrCl_4$ is a chain of linked octahedra ($ZrCl_2Cl_{4/2}$) like

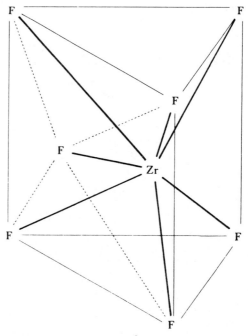

FIGURE 15.1 *The structure of the ZrF_7^{3-} ion in $(NH_4)_3ZrF_7$*
This species is basically a trigonal prism with an extra ligand attached to one face. The NbF_7^{2-} and TaF_7^{2-} ions adopt the same structure.

the NbI_4 structure shown in Figure 15.8. The halides hydrolyse vigorously to the oxyhalide, which is stable to further hydrolysis. The most stable complex halides are the fluorides, and the hexa-, hepta-, and octa-fluorides are known. The octahedral ZrF_6^{2-} ion is found for example in Li_2ZrF_6. In the formally similar K_2ZrF_6, the structure involves bridging fluorines and ZrF_8 coordination. In the mixed potassium-cupric hexafluorozirconate, the structure is even more complex and is formulated as $K_2Cu(H_2O)_6[Zr_2F_{12}]$. The $(Zr_2F_{12})^{4-}$ anion has the unusual pentagonal bipyramidal coordination around each zirconium, and the two bipyramids share an edge—$F_5ZrF_{2/2}ZrF_5$. The pentagonal bipyramid is also found in the sodium salts Na_3ZrF_7 and Na_3HfF_7. These contain pentagonal bipyramid MF_7 groups, as in IF_7, but the ammonium salt of the heptafluorides has the same structure as the isoelectric niobium and tantalum heptafluorides. This is shown in Figure 15.1 and may be described as a trigonal prism with the seventh fluoride added beyond the centre of one of the rectangular faces. In the MF_8 groups, the eight fluorines adopt the bisdisphenoid configuration shown in Figure 15.2, and described in section 13.6. If the cation is $Cu(H_2O)_6^{2+}$, in place of the alkali metals, the antiprismatic form is found for ZrF_8^{4-}, like the octafluorotantalate (Figure 15.5b).

The chloro complex ion $[Hf_2Cl_{10}]^{2-}$ contains two octahedral $HfCl_6$ units, linked in a dimer by sharing an edge $[Cl_4Hf-(\mu-Cl)_2-HfCl_4]^{2-}$ (compare Figure 14.5b).

Zr and Hf, as well as U, form volatile borohydrides $M(BH_4)_4$, with 12-coordination where each BH_4 group is bonded to M through three shared H atoms. In contrast, the larger uranium allows even higher coordination (see p. 180). The ion $M(BH_4)_5^-$ may be stabilized by large

cations and appears to be 10-coordinate with $M \overset{H}{\underset{H}{\diamondsuit}} B$ bridges. Similar bridges and 10-coordination are found in the hydride dimer $[Zr_2H_4(BH_4)_4(PMe_3)_4]$ which is formed by treating $Zr(BH_4)_4$ with excess PMe_3 at 195 K. Each Zr atom is bonded to two PMe_3 molecules and also to two BH_4 units via two shared hydrogens (as above). The two Zr are linked by four bridging hydrogens making the Zr 10-coordinate and giving overall $[\{H_2B(\mu-H)_2\}_2Zr(PMe_3)_2(\mu-H)_4Zr(PMe_3)_2\{(\mu-H)_2(BH_2)\}_2]$. A quadruple hydrogen bridge of this kind is very unusual, though triple and double hydrogen bridges are well-known for zirconium, e.g. in the compounds $[Zr_2H_3(BH_4)_5(PMe_3)_2]$ and $[Cp_2ZrH_2]_2$, respectively. The $Zr\cdots Zr$ distance shortens progressively from 346 pm to 312 pm to 298 pm as the number of bridging hydrogens increases from 2 to 3 to 4, respectively.

The lower oxidation states of zirconium and hafnium are very strongly reducing. Mixtures of di- and tri-halides result from reduction by the element, or by reduction with H_2 at 400–500 °C. It may be noted that treatment of a mixture of $ZrCl_4$ and $HfCl_4$ with zirconium metal gives $ZrCl_3$, which is involatile, and leaves $HfCl_4$ unreacted. The latter may then be sublimed out of the reaction mixture, providing a method of separating the elements. The III and II states do not exist in solution. One striking example of the reducing power of the III state is provided by the production of the blue potassium solution in liquid ammonia, in the following reaction:

$$ZrCl_3 + 4KNH_2 \xrightarrow{NH_3} Zr(NH_2)_4 + K + 3KCl$$

$HfCl_4$ gives $HfCl_3$ when heated with hafnium metal, and the trichloride is stable to 350 °C in the presence of $HfCl_4$. Hafnium dichloride is said to disproportionate to HfCl and $HfCl_4$ when heated to 627 °C. In contrast there is no lower hafnium iodide with iodine content less than $HfI_{3.2}$.

The MX_3 species have either the BiI_3 layer structure (Figure 5.10b) or the ZrI_3 structure which consists of a chain of octahedra sharing faces.

Zirconium also forms halogen-bridged clusters containing 6 Zr atoms which are of intermediate formal oxidation state. These are analogous to the long-established niobium and tantalum clusters, compare section 15.3.3. If we start from the $[Nb_6Cl_{12}]^{2+}$ cluster of Figure 15.9a and replace each Nb by Zr, we remove six electrons. In the known Zr_6 clusters, these missing electrons are partly offset by an encapsulated atom lying at the centre of the Zr_6 octahedron. Thus we find complexes like $[Zr_6Cl_{12}N]^{3+}(Cl^-)_3$, $[Zr_6Br_{12}C]^{2+}(Br^-)_2$ or $Cs^+[Zr_6I_4^{4}B]^-$. Where charges are balanced by anions, these coordinate in external apical positions to the Zr, so the fully symmetric complex would have 18 halogens, 12 bridging the edges and one coordinated to each Zr apex (Figure 15.3, compare Figure 15.9b). The encapsulated or interstitial atom is either a light element like H, Be, B, C, or N, or a transition metal such as Mn, Fe or Co.

It was initially thought that the encapsulated atom was

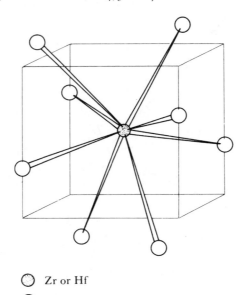

○ Zr or Hf

○ F

FIGURE 15.2 *The structure of ZrF_8^{4-} and HfF_8^{4-}*
This form of eight-coordination is similar to that found in the octacyanomolybdate ion (Figure 15.18). Contrast this with the structure of the octafluoride of tantalum (Figure 15.5b).

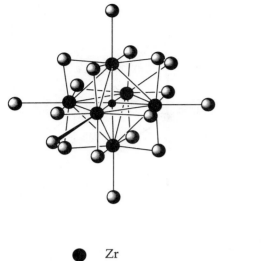

●   Zr

◗   Cl, Br, I

●   interstitial atom Z e.g. Be, B, C (see text)

FIGURE 15.3   *Structures of the interstitial clusters* $[Zr_6ZCl_{12}]^{n+}$

essential to the stability of the Zr_6 clusters but more recent work has demonstrated the existence of 'empty' clusters such as $[(R_3PZr)_6Cl_{12}]$ with an apical phosphane on each Zr. Further examples have various combinations of coordinated phosphane and halogens.

Treatment of Zr with $ZrCl_4$ in an inert container yields a new phase, ZrCl, which has metallic properties. The structure is a novel layer form, of nearly cubic symmetry, which can be regarded as the $CdCl_2$ structure (see Table 5.3) with an extra layer of metal atoms inserted giving a four-layer $Cl-Zr-Zr-Cl$ arrangement. A Zr atom is approximately ccp with 6 Zr neighbours in the same layer, three Zr in the neighbouring layer on one side and 3 Cl atoms as neighbours on the other side (compare section 5.6). ZrBr is similar.

The zero-valent state is rare but the violet dipyridyl, $[Zr(dipy)_3]$, is quite well-established. This is formed, like the corresponding Ti compound, by reduction of $ZrCl_4$ with Li in ether in the presence of dipyridyl. A further possibility is a cyanocomplex of Zr(0), formed by reduction in liquid ammonia (compare section 6.7), which is formulated as $[Zr(CN)_5]^{5-}$. However, this may be the dimer.

Compounds such as $Zr(benzyl)_4$ and $Zr(C_5H_5)_4$ show promise as homogeneous catalysts in the Ziegler–Natta process (see section 17.4.3).

15.3 Niobium, $4d^45s^1$, and tantalum, $5d^36s^2$

These two elements resemble zirconium and hafnium in the very close similarity of their chemistries, although here it is the lighter niobium which is the rarer element. The different ground state electronic configurations appear to have no effect on the chemistry in the valency states. Rather more is

known of the lower oxidation states of these elements than was the case with zirconium and hafnium, but the V state is by far the most stable and well-known. The element/M^{5+} potentials show the same trend as in the preceding Group:

$$V^V/V = -0.25 \text{ V}, \ Nb^V/Nb = -0.65 \text{ V}, \ Ta^V/Ta = -0.85 \text{ V}$$

A potential of about -1.1 V has been estimated for Nb^{III}/Nb and one of about -0.1 V for Nb^V/Nb^{III}, both in sulphuric acid solution where complex species are probably formed in both states.

Niobium and tantalum generally occur together and are separated by fractional crystallization of fluoro-complexes:

$$\text{ore} \rightarrow M_2O_5 + KHF_2/HF \rightarrow \underset{\text{less soluble}}{K_2TaF_7} \xrightarrow{\text{electrolysis}} Ta$$

$$\underset{\text{more soluble}}{K_2NbOF_5} \xrightarrow{\text{Al}} Nb$$

The metals are very resistant to acid attack but will react slowly with fused alkalis and with a variety of non-metals at high temperatures. They have very high melting points (Ta above 3000 °C) and find some use in high temperature chemistry. Tantalum is also used in surgery as it can be inserted in the body, as in fracture repair, without causing a 'foreign body' reaction.

15.3.1   *The (V) state*

In the V state, the oxides, Nb_2O_5 and Ta_2O_5, may be prepared by igniting the metals, their carbides, sulfides, or nitrides, or any compound with a decomposable anion. The oxides are inert substances which are generally brought into solution by alkali fusion, or treatment with concentrated HF. The oxides are therefore amphoteric but the acidity is very slight and the niobates are decomposed, even by as weak an acid as CO_2. The product of alkali fusion contains one metal atom and is written as NbO_4^{3-} (or TaO_4^{3-}) and termed orthoniobate (or orthotantalate). A monatomic metaniobate, $NaNbO_3$, is also known which has the perovskite structure. A number of more complex species are also known which contain 2, 5, or 6 metal atoms, for example, $M_4^INb_2O_7$ or $M_8^ITa_6O_{19}$. The latter $Ta_6O_{19}^{8-}$ ion has also been shown to be present in solution. The structure of none of these species is known, nor are the formulae unambiguous but may include water molecules and hydroxyl ions. It does seem clear, though, that niobium and tantalum share with vanadium the tendency to form polymeric oxyanions. The neutralization of these niobate or tantalate solutions with acid leads to the precipitation of white gelatinous precipitates of the hydrated pentoxides. These dissolve in hydrofluoric acid, probably as fluoro-complexes, but there is no evidence of cationic forms of niobium or tantalum in solution analogous to the vanadium oxycations.

The pentafluorides may be prepared by the reaction of fluorine on the metal, pentoxide, or pentachloride, or by HF on the pentachloride. Both are volatile white solids melting below 100 °C and boiling near 230 °C. The structures are

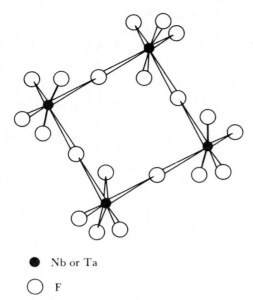

● Nb or Ta

○ F

FIGURE 15.4 *The structure of* NbF_5 *or* TaF_5
Most other pentafluorides of the transition metal adopt similar, tetrameric, structures.

tetrameric in the solid with MF_6 units and one bridging fluoride between each pair of metal atoms (Figure 15.4), isostructural with MoF_5. In the liquid phase, there is evidence for similar *cis*-bridging, but forming polymers rather than tetramers. In the vapour above NbF_5, 98 % is the trimer Nb_3F_{15} and about 2 % is monomeric NbF_5.

The pentoxide dissolves in HF with formation of fluoro-complexes and these are also formed by the pentafluorides and F^-. Crystallization from solutions of moderate F^- concentration gives the salts $M^INb(or\ Ta)F_6$ containing octahedral MF_6^- ions. In the presence of excess F^-, TaF_7^{2-} and $NbOF_5^{2-}$ are formed. A larger excess of F^- leads to the formation of NbF_7^{2-} and, at very high F^- concentrations, $NbOF_6^{3-}$ and TaF_8^{3-} are formed. The octafluoroniobate has not been reported. The structures of the MF_7^{2-} ions are those based on the trigonal prism shown in Figure 15.1. $NbOF_5^{2-}$ has an octahedral structure, while $NbOF_6^{3-}$ illustrates another form of seven-coordination. This, Figure 15.5a, is based on the octahedron with the seventh ligand placed at the centre of one triangular face. The TaF_8^{3-} ion is the square antiprism (Figure 15.5b) which reduces inter-actions between ligands to a minimum in eight-coordination. This corner of the *d* block gives a variety of seven- and eight-coordinated structures which is not found elsewhere in the Periodic Table.

The other pentahalides all exist and can be prepared by standard methods. All six compounds are volatile covalent solids with boiling points below 300 °C. The vapours are all monomeric and electron diffraction results indicate the structures are probably the expected trigonal bipyramids. In the solid, an X-ray study shows that $NbCl_5$ is a dimer,

Cl_4Nb $\overset{Cl}{\underset{Cl}{\diamondsuit}}$ $NbCl_4$, with two bridging chlorides giving an

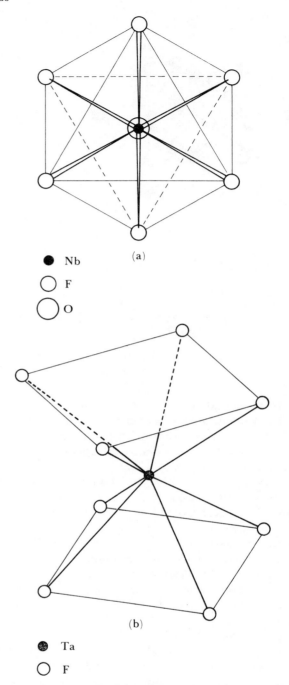

● Nb

○ F

○ O

(a)

● Ta

○ F

FIGURE 15.5 *Higher coordination numbers* (a) $NbOF_6^{3-}$ *and* (b) TaF_8^{3-}

octahedral configuration around each niobium. In solution in non-donor solvents, such as CCl_4, the dimeric form is retained. Tantalum pentachloride and both bromides have the same solid structure but the pentiodides are of a different, and unknown, structure. Much less is known about complexes of these heavier halides but there is good evidence for the existence of MCl_6^- species.

The pseudohalogen analogues, $M(NCS)_5$ and $M(NCS)_6^-$ are also known both for M = Nb and M = Ta, the penta-thiocyanates are dimers like the pentachlorides.

The halides all hydrolyse readily to the hydrated pent-oxides and MOX_3 for X = Cl, Br, I have been isolated as intermediate products for both metals. A better preparation is by reaction between the pentahalide and oxygen. Treatment of M_2O_5 with aqueous HF gives MO_2F which yields progressively on heating MOF_3 then M_3O_7F. The corresponding heavy halide compounds were also established so that the complete series MOX_3, MO_2X, and M_3O_7X is known for all the halides and both metals.

The mononuclear oxyhalides are covalent but less volatile than the pentahalides. The vapours appear to be mono-meric and tetrahedral but the solids are polymeric. The structure of $NbOCl_3$ is shown in Figure 15.6, with the pentahalide for comparison. The oxyhalide contains planar,

binuclear Cl_2Nb $\diagdown$ $\diagup$ $\overset{Cl}{\diagup}$ $\underset{Cl}{\diagdown}$ $NbCl_2$ groups which are linked

into long chains by oxygen bridges between the niobium atoms.

Apart from the oxides and halides, there are few important compounds of these elements in the V state. One interesting one is the cyclopentadienyl hydride, $(\pi\text{-}C_5H_5)_2TaH_3$, which provides another example of the stabilization of transition metal-hydrogen bonds by the presence of a pi-bonding ligand. The structure, shown in Figure 15.7, is derived from the ferrocene one (Figure 16.7a) by bending the rings towards each other and inserting the three hydrogens in the central plane pointing away from the rings. The ring — M — ring angle is 139° (Ta) and 142° (Nb) while the H_BMH_B angles are 126° in each case. The bond to the central hydrogen H_A bisects this angle and all three M–H distances are equal. Nmr studies confirm two different types of H atom environment occupied in the ratio 2:1.

A similar compound with angled cyclopentadiene rings is $[Nb(Cl)(O)(C_5Me_5)_2]$ where the Cl and the O occupy the H_B positions of Figure 15.7, implying an Nb=O interaction.

The gas phase structures of two volatile five-coordinate tantalum comounds, $TaMe_5$ and $Ta(NMe_2)_5$, have been determined by electron diffraction and show, unexpectedly, a square pyramidal geometry. With the reservation that electron diffraction is not an absolute method (compare section 7.4) and the amide particularly is very demanding for the method, the similar conclusions are mutually reinforcing. In the study on the amide the expected trigonal bipyramid structure was also tested and showed a poorer fit. Further, $SbMe_5$ was also measured for comparison with the pentamethyl and did show the expected trigonal bipyramid. Thus the TaR_5 square pyramid structures seem firmly based. In each case the Ta is above the plane of the four base atoms and the Ta–apex bond is shorter than the Ta–base one (Ta–Me values of 211 pm and 218 pm; Ta–N distances of 194 pm and 204 pm]. It is suggested that, because the metal d orbitals are involved in the bonding, the simple VSEPR analysis fails. In particular, the least stable bonding orbital in the trigonal bipyramid form (which involves largely metal

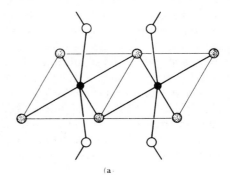

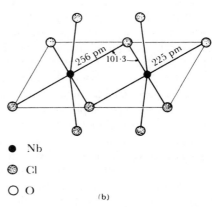

● Nb

◉ Cl

○ O   (b)

FIGURE 15.6   *The structures of* (a) $NbOCl_3$ *and* (b) $NbCl_5$

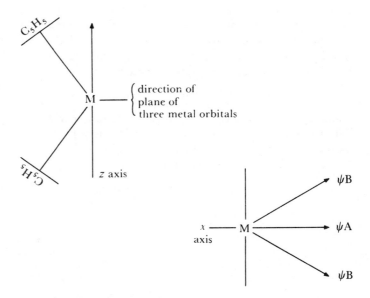

FIGURE 15.7 *Bonding in dicyclopentadienyltantalumtrihydride,* $\pi\text{-}(C_5H_5)_2TaH_3$

This is a diagrammatic representation of the three orbitals which become available on the central atom when the C_5H_5 – Ta – C_5H_5 angle is reduced from the linear arrangement in ferrocene. The two H atoms bonded to the B orbitals are identical and differ from the central A hydrogen.

p_z overlap with the apical ligand orbitals) is significantly stabilized by the change to square pyramid without offsetting losses, if the ligands are of relatively low electronegativity and there is no significant π bonding.

The electron diffraction structure shows that Me_3TaF_2 is a trigonal bipyramid with the three methyl groups equatorial (Ta–C = 213 pm; Ta–F = 186 pm). While TaF_5 is mainly tetramer in the vapour, it has been possible to determine the monomer which is also trigonal bipyramid. Clearly, the balance between the two geometries is a very delicate one.

15.3.2 *The (IV) state*

Compounds in the (IV) state include NbO_2, all the tetrahalides, MX_4, and some oxyhalides, MOX_2, for both metals. In addition there are a number of niobium oxide phases in between Nb_2O_5 and NbO_2, such as Nb_8O_{19} and $Nb_{11}O_{27}$. There is more uncertainty but TaO_2 probably does exist. The state is unstable, reducing, and shows some tendency to disproportionate. The dioxides are prepared by high temperature reduction of the pentoxides. They dissolve only in hot alkali, with reduction of the solvent.

All the tetrahalides exist except the tetrafluorides: the iodides are best known. NbI_4 results from prolonged heating of NbI_5 at 270 °C, when iodine sublimes off leaving the tetraiodide. TaI_4 is most easily made by heating TaI_5 with the metal. These iodides are diamagnetic solids, volatile at 300 °C. The diamagnetism arises because the metal atoms are linked by long bonds in dimers, as shown in Figure 15.8, thus pairing the single electron on each. The structure consists of chains of MI_6 octahedra, each joined by the edges to its neighbours, and with the metal atoms placed unsymmetrically in the octahedra and linked in pairs. $NbCl_4$ has a similar structure with a short Nb–Nb distance of 303 pm and the long (non-bonded) one equal to 379 pm. The tetrachlorides are made by reduction of the pentachlorides and both $NbCl_4$ and $TaCl_4$ disproportionate,

$$\text{e.g.}\quad TaCl_4 \xrightarrow{400\,°C} TaCl_3 + TaCl_5\uparrow$$

Similarly, on hydrolysis, $TaCl_4$, gives a precipitate of tantalum(V) oxide and a green solution of the trichloride, which is fairly stable unless heated. The IV state is also reported from the electrolytic reduction of niobium(V) in 13M hydrochloric acid. An orange Nb^{IV} solution results which probably contains the oxyhalide ion, $NbOCl_4^{2-}$. This solution disproportionates to niobium (III) + (V). MOX_2 oxyhalides are known for X = Cl, Br, and I. A significant complex is the chloro-species, $(Me_3P)_4M_2(Cl)_4(\mu-Cl)_4$. Here each M atom is bonded to two phosphanes and two terminal Cl atoms, and the two are linked together by four Cl bridges. The two M atoms are close enough to be bonded, thus the complex shows 9-coordination and parallels the NbI_4 interaction above.

15.3.3 *Lower oxidation states of niobium and tantalum*

In the (III) state, MX_3 are known for X = F, Cl, and Br. Heating NbI_5 at 430 °C forms the mixed oxidation state, Nb_3I_8, (and further heating gives Nb_2I_{11}). TaI_3 is not reported. The other niobium trihalides are formed by reducing the pentahalide with hydrogen at about 500 °C. Tantalum trichloride is obtained from the tetrachloride as above, while the tribromide is made by hydrogen reduction. Most of the trihalides are brown or black and strongly reducing, although the reactivity depends greatly on the thermal history of the sample. In strong hydrochloric acid, electrolytic reduction of niobium(V) to niobium(III) is reported. The III state solutions are yellow or blue, depending on the conditions used.

Nb(IV) and (III) are found in the octacyano complexes $Nb(CN)_8^{n-}$, where $n = 4$ or 5. Like the octacyanides of Mo and W (section 15.4) and Figure 15.18), both dodecahedral and square antiprismatic structures are found.

$TaCl_2$ is the only known tantalum dihalide. It is a non-stoichiometric, green-black solid resulting from the disproportionation of $TaCl_3$ at 600 °C. It is much more reactive than the trichloride, and attacks water under all conditions. Small quantities of $NbBr_2$ have been produced from the reaction of the pentabromide with hydrogen in an electric discharge. Nothing is known of its properties. Electrolytic reduction of niobium in 10 M HCl gives a violet solution colour attributed to niobium(II) but this state is not otherwise reported in solution.

Apart from the tetraiodides, nothing is known of the structures or degrees of polymerization of these other halides. They are all relatively involatile and the volatility drops from the tetrahalide to the trihalide to the dihalide; the

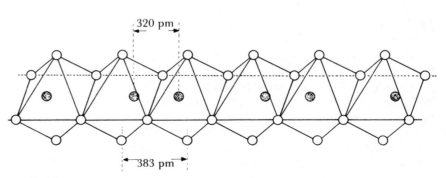

Nb or Ta
○ I

FIGURE 15.8 *The structure of niobium or tantalum tetraiodide*

structures are thus probably polymeric, although the ability of $TaCl_3$ to dissolve in water must be recalled.

In addition to these halides of simple stoichiometry, niobium and tantalum form compounds of formula M_6X_{14} for $X = Cl$, Br, and I. Related compounds Nb_6F_{15}, Ta_6Cl_{15}, Ta_6Br_{15} and Ta_6Br_{17} are also found. These result from sodium amalgam reduction of the pentahalides (a better preparation is by reduction with cadmium metal at red heat followed by precipitation of CdS). The compounds are soluble in water and alcohol and the action of Ag^+ on the compounds M_6X_{14} precipitates only one-seventh of the halide as AgX. The compounds are therefore salts of the complex cation $(M_6X_{12})^{2+}$. The structure of this ion has been determined by X-rays. The metal atoms (Figure 15.9a) form an octahedron and are bridged in pairs along the octahedron edges by halogen atoms. (Compare with the structure of the dihalides of molybdenum and tungsten.) The octahedron of metal atoms is an example of the *metal cluster* compounds which are attracting current attention. It is held together by poly-centred metal−metal bonding as well as by the bridging halogen atoms.

A variety of similar clusters is reported which are related to the $[M_6X_{12}]^{2+}$ structure, either by having charges of $+3$ or $+4$, or by containing terminal M−X or M−L groups in addition. The nicely symmetric structure of $[Nb_6Cl_{18}]^{4-}$, Figure 15.9b, is the result of complete substitution. Such species are also formed by Ta, may be partly substituted, or have different halogens in the bridging and terminal positions. In the hydrate $Ta_6Cl_{14}.4H_2O$, the $[Ta_6Cl_{12}]^{2+}$ ion is elongated along one axis, apparently to optimize hydrogen bonding. More complex species have been characterized, such as clusters linked together through halogen bridges.

A related structure is seen in $K_4Al_2[Nb_{11}O_{20}F]$ which has two Nb_6 octahedra sharing a common Nb apex. There are 12 edge-bridging O atoms and 8 terminal Nb−O bonds on the equatorial Nb atoms. The remaining two apices, which are *trans* to the common Nb, form Nb−F−Nb bonds which then link up the paired octahedra into polymeric units.

The oxide of formula NbO is gray and has metallic properties. The structure is cubic with 8 Nb vacancies at the corners and an O vacancy at the centre of the NaCl structure. This gives square planar NbO_4 coordination.

The lowest oxidation states are limited to a few carbonyl and organometallic species. For example, the $(-I)$ oxidation state is found in the carbonyl anion $[M(CO)_6]^-$ which is formed by Nb and Ta (also V). These elements do not, however, form a neutral carbonyl like vanadium.

Multiple M−M bonded compounds are discussed in section 16.7.

It is thus seen that this Group is quite similar to the titanium one. The heavier elements are broadly similar to the lighter ones in the Group oxidation states of IV and V respectively, but the oxides are less hydrolysed in solution and less amphoteric. The heavier elements form complexes with fluorine of high coordination numbers. The vanadium Group has a wider range of oxidation states and the lower ones are not quite so unstable for niobium and tantalum as they are for zirconium and hafnium, but they are much less stable than the Group state and are strongly reducing.

15.4 Molybdenum, $4d^55s^1$, and tungsten, $5d^46s^2$

These two elements, although there is still a close resemblance, show much more distinct differences in their chemistries than the two earlier pairs in these series. This is illustrated by the ready separation of the two elements in qualitative analysis, where tungsten appears in Group I of the conventional scheme and molybdenum in Group II.

This is because tungsten(VI) which is soluble in neutral and alkaline solution precipitates an insoluble hydrated oxide, $WO_3.nH_2O$, in acid solution. Molybdenum(VI) oxide, on the other hand, dissolves again due to formation of chloro-complexes and molybdenum is first reduced by H_2S and then precipitates as MoS_2.

Examples of all oxidation states from VI to −II are found for these elements. The Group state of VI is the most stable, but the V and IV states are well-represented and occur in aqueous solution. Strong reducing properties are not shown until the III and II states are met. This behaviour is illus-

(a)   (b)

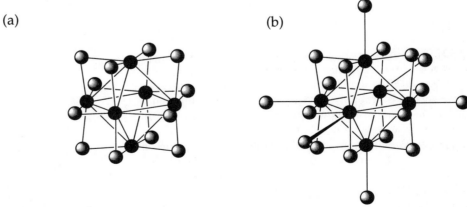

FIGURE 15.9   *The structures of* (a) $[Nb_6Cl_{12}]^{2+}$ *and* (b) $[Nb_6Cl_{18}]^{4-}$ The octahedra of niobium atoms are bonded, not only by the bridging chlorines but by metal−metal bonding within the cluster.

● Nb   ◗ Cl

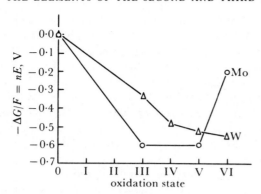

FIGURE 15.10 *Oxidation state free energy diagrams for molybdenum and tungsten*
This illustrates the greater relative stability of the lower oxidation states of molybdenum compared with those of tungsten.

trated by the occurrence of the various halides and oxides (Tables 13.3 to 5) and by the free energy changes in solution shown in Figure 15.10. This diagram shows the differences in behaviour of the two elements, in particular the greater relative stability of Mo^V and the similar effect at W^{IV}, both of which lie at shallow minima. These two elements show little resemblance to chromium: the stability of the VI state contrasts with the strong oxidizing nature of chromium(VI) while the III state, which is so stable for chromium, is very much less stable for molybdenum and tungsten.

The elements occur as oxyanions and molybdenum is also found as the sulfide, MoS_2, which has found recent popularity as a solid lubricant, as it has a layer structure which allows the planes of atoms to slide over each other easily, see also section 19.3.3. The ores are converted to the trioxides *via* the oxyanions and these are reduced to the metals by reduction with hydrogen (carbon cannot be used as the very stable carbides would result). Both metals are relatively inert, with very high melting points and with fairly high electrical conductivity. The metals are most readily attacked by alkaline peroxide or other fused, oxidizing alkaline medium. Attack by aqueous alkali and by acids is only slight.

Molybdenum and tungsten find extensive uses. Molybdenum is used in stainless steels, as a catalyst, and as an electrode material. One of the most important catalytic applications is in dehydrosulfurization of petroleum products. A Mo catalyst usually with added Co, supported on Al_2O_3, is used to treat a stream of crude petroleum with hydrogen to remove sulfur-containing organics by converting them to H_2S, and to hydrogenate olefinic bonds. The active species is probably MoS_2, formed in situ. Tungsten is used as the filament in light bulbs, in alloys, and, as WC, in applications where its hardness and wear-resistance are important. Molybdenum is found as MoS_2 which is converted to the oxide and used in the blast furnace or reduced with iron oxide by Al to give ferromolybdenum. Pure molybdenum is prepared by hydrogen reduction of ammonium molybdate (either the $Mo_2O_7^{2-}$ or $Mo_7O_{24}^{6-}$ species, see below) which is crystallized from a solution of the oxide in ammonia. Tungsten is recovered by dissolving the oxide in fused NaOH

and extracting the tungstate. Acidification yields 'tungstic acid'—a hydrated oxide which gives the metal on hydrogen reduction.

15.4.1 *Molybdenum (VI) and tungsten (VI)*

In the VI state, the oxides, the hexafluorides, and the hexachlorides all occur and tungsten also forms WBr_6. Uranium is the only other element to form a hexachloride, and hexabromides are unknown apart from the tungsten compound. There is also an extensive aqueous chemistry of Mo^{VI} and W^{VI} and a vast collection of polymeric oxyanions.

The oxides, MoO_3 and WO_3, are the final products of igniting molybdenum or tungsten compounds. Both oxides are insoluble in water but dissolve in alkali to give oxyanions. MoO_3 also dissolves in acids to give oxy-cations or related hydrolysed species but WO_3 is insoluble in acid. The simplest product of the solution in alkali is the MoO_4^{2-} ion (molybdate or tungstate) which is tetrahedral. The imido analogue $[W(NBu^t)_4]^{2-}$ has been prepared. This ion is isoelectronic with $[WO_4]^{2-}$ and has a similar structure. The molybdates and tungstates of most metals except the alkalis, NH_4^+, Mg^{2+}, and Tl^+, are insoluble. The neutralization of the alkaline solutions leads to the precipitation of the hydrated oxides, $MoO_3.2H_2O$ which is yellow, and $WO_3.2H_2O$ which is nearly white. These are definitely hydrates, as written above, not the hydrated acids, $H_2MoO_4.H_2O$. As with the anhydrous oxides, which are derived from the hydrates by ignition, the hydrated oxides differ in their behaviour to acid, the

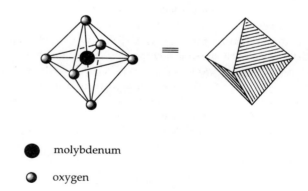

● molybdenum

○ oxygen

FIGURE 15.11 *The MO_6 octahedron*
The unit from which the polymolybdates and tungstates are built up.

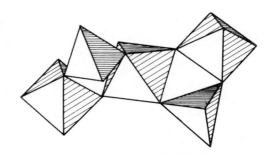

FIGURE 15.12 *Structure of the $[Mo_2O_7]^{2-}$ polymeric anion*

TABLE 15.1 Iso- and hetero-poly molybdates and tungstates

Iso-compounds

(a) Molybdenum in solution.

$$MoO_4^{2-} \xrightarrow{pH = 6} Mo_7O_{24}^{6-} \xrightarrow{\text{lower pH}} Mo_8O_{26}^{6-} \xrightarrow{\text{strong acid}} Mo_n \text{ species?}$$

(7− and 8− ions may be protonated and hydrated.)

(b) Solid molybdenum compounds (all except the 1− and 2− species are heavily hydrated).

MoO_4^{2-}	Tetrahedron.
$(Mo_2O_7^{2-})_n$	Chain of MoO_6 octahedra sharing opposite corners and linked in pairs through adjacent corners by MoO_4 tetrahedra (Figure 15.12). The NH_4^+ salt has two octahedra sharing an edge and successive pairs are linked by two tetrahedra sharing corners. $Ag_2M_2O_7$ is formed of octahedra only.
$Mo_7O_{24}^{6-}$	Linked octahedra (Figure 15.13).
$Mo_8O_{26}^{6-}$	Linked octahedra.
$Mo_{10}O_{34}^{8-}$	The ammonium salt is a Mo_8O_{28} unit built up of linked octahedra with two MoO_4 tetrahedra linked to opposite corners.

(c) Tungsten in solution.

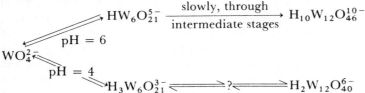

(d) Solid tungsten compounds (all are heavily hydrated as salts except the 1− and 2− species).

WO_4^{2-}	Tetrahedron.
$(W_2O_7^{2-})_n$	Linked tetrahedra and octahedra: see Mo compound.
W_6 species	Structures not known.
$W_{12}O_{46}^{20-}$	See Figure 15.14a, linked octahedra.
$W_{12}O_{40}^{8-}$	Linked octahedra; isomorphous with the 12-hetero-acids (q.v.).

Hetero-compounds

(a) MO_6 (M = Mo, W) octahedra surrounding a central (hetero) $M'O_6$ octahedron (all compounds are heavily hydrated).

$M'M_6O_{24}^{(12-x)-}$ (x = formal positive charge on hetero-atom M')	M′ = I^{7+}, Te^{6+}, or a number of trivalent ions such as Co^{3+}, Al^{3+} or Rh^{3+}. Structure is a ring of six linked MO_6 octahedra with a central octahedral site for M′ (Figure 15.14b).
$M'M_9O_{32}^{(10-x)-}$	M′ = Mn^{4+}, Ni^{4+}. Structure consists of three sets of clusters of three MO_6 groups giving a central octahedral hole for M′.

(b) MO_6 octahedra surrounding a central (hetero) $M'O_4$ tetrahedron.

$M'M_{12}O_{40}^{(8-x)-}$	M′ = P^{5+}, As^{5+}, Si^{4+}, Ge^{4+}, Ti^{4+}, Zr^{4+}, Sn^{4+}. Structure contains four sets of three MO_6 octahedra joined by edges and defining a central tetrahedral site. The structure of $W_{12}O_{40}^{8-}$ is the same as these with the central site empty.

Other heteroacids with tetrahedral $M'O_4$ groups occur with M′/M ratios of 1/11, 1/10, 2/18 and 2/17 and containing mainly P^{5+} or As^{5+}.   For example:

$P_2Mo_{18}O_{62}^{6-}$	This is the 12-acid structure as above with the three MO_6 octahedra at the base removed to give the 'half-unit' PM_9O_{34} and two of these are linked, sharing six oxygens, to give the P_2M_{18} unit.

(c) More complex hetero-compounds.

Much greater complication is possible as illustrated by the phosphorus-molybdenum compounds:

$(MoP_2O_{11})_n$	Chains of MoO_6 octahedra formed by linking corners. The chains are cross-connected by P_2O_7 groups into a three-dimensional structure.
$(MoP_2O_8)_n$	Layers formed by MoO_6 octahedra sharing oxygens and these layers linked up by $(PO_3)_n$ chains which run perpendicular to the planes of the layers.

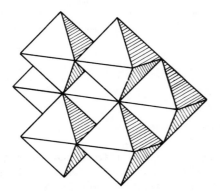

FIGURE 15.13   *The $Mo_7O_{24}^{6-}$ structure*

This structure is more compact than that found in the similar heteropolyion (Figure 15.14b).

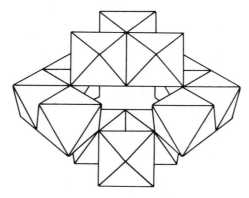

FIGURE 15.14a   *The $W_{12}O_{46}^{20-}$ structure*

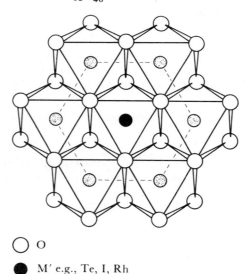

○   O

●   M' e.g., Te, I, Rh

◉   Mo or W

FIGURE 15.14b   *The $M'M_6O_{24}^{(12-x)-}$ structure*

This consists of a ring of six linked MO_6 octahedra forming a central octahedral site for the M' metal ion.

molybdenum compound dissolving and the tungsten compound being relatively insoluble.

These simple, mononuclear oxyanions are found only in strong alkali. In less basic solutions, condensation occurs, as

we have seen for other elements (compare Table 14.1). These condensed polyoxo-anions formed at lower pH values are called *isopoly-molybdates* or *tungstates* if only Mo or W is present, and *heteropoly* oxyanions when they contain other oxyanions as well, such as silicate or phosphate. This gives a rich and varied field of condensed structures, probably only exceeded in complexity by the silicates. It is not possible to deal with this group of compounds in anything like full detail but Table 15.1 lists some typical formulae. The polyacids are built up from MO_6 octahedra, Figure 15.11, and most of the structures so far determined are rings, double-rings, or clusters. On the basis of older experimental results, it was thought that Mo and W behaved similarly and passed, on increasing condensation, through stages which were termed ortho-, para-, and meta-acids (ortho being the normal MO_4^{2-}). It is now known that some of these stages correspond to a number of different compounds which are not represented by the same formulae for Mo as for W. The ions or other species are most conveniently named on the basis of the number of Mo or W atoms. The structures of a number of ions are known, and ions with the same number of metal atoms appear to exist in solution, but there is no final proof that the solution equilibria do correspond to the solids found, and there is certainly evidence for a number of intermediate species in solution. Problems arise as it is difficult to obtain unambiguous analyses for alkali metals and Mo or W in the heavier species, and because mixed crystals are readily formed.

The polyacids, and especially the heteropolyacids, form a variety of unusual environments and are at present attracting attention as a means of studying unusual coordinations or unusual oxidation states such as Ni^{IV}.

A recently-reported structure, of $CeMo_{12}O_{42}^{8-}$, contains twelve-coordinated cerium(IV) and the novel structure of MoO_6 octahedra linked in pairs by sharing a face, that is, with $Mo(O)_3Mo$ bridges. Th(IV) probably behaves similarly. One peroxide related to the polymolybdates has yielded a structure. In $[Mo_5O_{10}(O-O)_8]^{6-}$, a central MoO_4 tetrahedron shares each O in a $Mo-O-Mo$ link to four outer Mo atoms each bonded to two edge-on peroxy groups and one O.

Recently these polyoxoanions have been studied as 'ligands' towards organometallic species and a number of hybrid complexes have been prepared. The polyoxoanions in these cases are considered as small and, most importantly, *soluble* pieces of metal oxide, which allow the nature of the bonding interactions between the metal oxide species and the organometallic fragment to be thoroughly studied by such techniques as X-ray crystallography and solution nmr spectroscopy. These complexes are models for the interactions between organometallic complexes supported on a metal oxide support, such materials being important catalysts. Polyoxoanions themselves are also used as industrial catalysts, for example in the oxidation of methacrolein to methacrylic acid.

Apart from these oxygen compounds, the VI state is found in halides, oxyhalides, and complexes related to these. Both molybdenum and tungsten form MF_6, MOF_4, and MO_2F_2

compounds but molybdenum has a much lower affinity for the heavier halogens than has tungsten, as Table 15.2 shows.

TABLE 15.2 Halides and oxyhalides of molybdenum(VI) and tungsten(VI)

MX_6		MOX_4		MO_2X_2	
MoF_6	WF_6	$MoOF_4$	WOF_4	MoO_2F_2	WO_2F_2
$MoCl_6$	WCl_6	$MoOCl_4$	$WOCl_4$	MoO_2Cl_2	WO_2Cl_2
	WBr_6		$WOBr_4$	MoO_2Br_2	WO_2Br_2
					WO_2I_2

The hexahalides are made by direct reaction. They are volatile and unstable to oxygen and moisture, with which they react readily to give oxyhalides. Tungsten hexabromide is thermally unstable and decomposes on gentle warming. The oxyhalides are also volatile, covalent compounds which hydrolyse to the trioxides, the MOX_4 type more rapidly than the MO_2X_2 compounds. Tungsten forms analogous compounds with the heavier elements of the oxygen group; WSF_4, $WSCl_4$, and $WSeCl_4$ are all relatively volatile solids with a square pyramidal structure in the gas phase, where the W lies above the X_4 plane, just as in the oxygen compounds. WO_2Cl_2, which is yellow, disproportionates above $200\,°C$ to WO_3 and red $WOCl_4$. Molybdenum also forms a compound of formula $MoO_2Cl_2.H_2O$, which may well be the hydroxy compound, $MoO(OH)_2Cl_2$.

The hexahalides are octahedral, and WOF_4 has the tetrameric structure of NbF_5 with F atoms in the bridging positions (Figure 15.4). In contrast, $MoOF_4$ is formed into chains of octahedral units, linked by shared fluorines in both structures. In the gas phase, both MOF_4 molecules are square pyramidal monomers.

Molybdenum and tungsten form a variety of complex halides in the VI state, of which the fluorine compounds are the most widely represented. Examples include $M^I_2WF_8$ and M^IMF_7 (for both M = Mo and W). The latter provide further examples of seven-coordination and the structures of a number of salts have been determined. Because all three established seven-coordination shapes (compare Figures 15.1, 15.5a and 15.22) are very similar in energy, different counterions may give different structures (compare the discussion on tetrahedral or square planar Ni, section 14.8.2). By varying the cation $(M^I)^+$ it was expected that the intrinsic shape could be decided. For several $(M^I)^+$, including Cs^+, NO_2^+ and Me_4N^+, the shapes of $[MoF_7]^-$ and $[WF_7]^-$ were all found to be capped octahedral, as found for $[NbOF_6]^{3-}$ (see Figure 15.5a). This contrasts with the seven-coordinate Main Group structures so far determined which are all pentagonal bipyramids.

Mononuclear oxyhalides of both elements are also known. Of these $[MO_2F_4]^{2-}$, $[MO_2Cl_4]^{2-}$ and $[MO_3F_3]^{3-}$, and probably $[MOF_5]^-$, are octahedral monomers. More complex oxyfluoro compounds are built up of six-coordinate units linked together. A dinuclear example is the anion $[O_2F_3W–F–WF_3O_2]^{3-}$ with one fluorine bridge. $[WO_2F_3]^-$ is a chain polymer with two terminal *cis* F, two similar O,

and two *trans* O atoms bridging to the next W atoms while the Mo analogue is similar but with F bridges. The chlorine analogue, $[WO_2Cl_3]^-$, is a dimer linked through two bridging oxygens. One well-established tetrahedral species is MoO_3Cl^-. Other oxy-species are known, including molybdenyl sulfate, MoO_2SO_4, formed from molybdenum trioxide and sulfuric acid. Such compounds are probably molecular, rather than salts of the oxycation.

Molybdenum and tungsten are sufficiently stable in the VI state to form sulfides, MS_3. These are precipitated as hydrated compounds when H_2S is passed through slightly acid M^{VI} solutions. In stronger acid, the H_2S reduces the VI state to the IV state, giving the disulfides MS_2. Sulfur-containing polyanions are found including $W_3S_9^{2-}$ which has two tetrahedral WS_4 units sharing edges with a central, WS_5 unit which is a distorted square pyramid. The $W_3OS_8^{2-}$ analogue has the O at the apex position of the square pyramid, and Mo analogues occur. In the interesting mixed oxidation state sulfur anion, $W_3S_8^{2-}$, the structure shows a central planar WS_4 unit linked by sharing edges to two outer WS_4 tetrahedra. These outer units are W(VI), and the simple WS_4^{2-} ion is now well-established, while the unique, central, square planar unit is W(II).

Considerable interest in molybdenum–sulfur compounds derives from the work on nitrogen-fixing organisms (see section 16.10 and section 20.1.4), where the active site contains Mo and S. This interest has led to the intense study of molybdenum and tungsten sulfur chemistry. Starting from the simple thiomolybdate and thiotungstate ions, S_4^{2-}, an extensive array of more complex molecules has been built up, with emphasis on those containing the cubane M_4S_4 unit (Figure 15.15a), or a fragment of this such as that of Figure 15.15b.

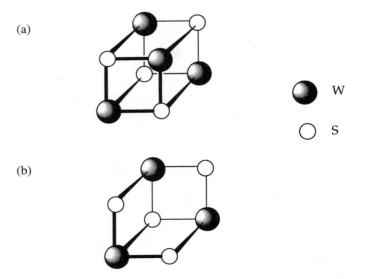

(a)

(b)

W

S

FIGURE 15.15 *Major structural units in complex tungsten sulfur clusters:* (a) *cubane* (b) *cubane missing one corner*
(b) may alternatively be described as W_3S trigonal pyramid with each W–W edge bridged by S.

The VI state is also sufficiently stable to allow the preparation of a polyhydrido-complex, $WH_6(PR_3)_3$. This shows nine-coordination and multiple substitution by hydrogen as in the similar rhenium hydrides discussed in section 15.5. The structure is based on the tricapped trigonal prism of Figure 15.23, with two of the phosphane groups replacing two H in one of the long edges, and the third phosphane taking the place of the opposite face-bridging H.

Two interesting types of compound are found which fall between the VI and V states. Mild reduction of the trioxides gives intensely blue oxides whose composition is intermediate between M_2O_5 and MO_3. These blue oxides appear to contain both M^V and M^{VI} in an oxide lattice, and the intense colour arises from the existence of two oxidation states in the same compound. Similarly intense colours are observed in other cases like this, for example in magnetite, Fe_3O_4, and in Prussian blue. The second compound is the product of reduction of sodium tungstate, of formula Na_nWO_3 (n lying between 0 and 1), called tungsten bronze. The colour varies from yellow to blue-violet as n varies from about 0·9 to 0·3. The structure of these bronzes is based on the ReO_3 structure shown in Figure 15.16. Here, the metal atoms lie at the corners of a cube and the oxygens are at the mid-points of the edges. If a M^I ion lies at the centre of each cube, the structure is that of perovskite, $M'MO_3$. The sodium bronzes are compounds where sodium ions appear at random in the cube centres. The metallic appearance and conductivity of the bronzes arises as the sodium valency electron is, delocalized over the structure. It may be noted that the structure of WO_3 itself is a distorted form of the ReO_3 structure.

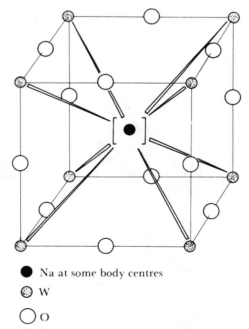

● Na at some body centres

◉ W

◯ O

FIGURE 15.16 *Tungsten bronze: the relation between the structures of rhenium trioxide, tungsten bronze, and perovskite*
The figure shows the ReO_3 structure, with the black circle indicating the site occupied, at random, by sodium ions in the bronze. If all the black circles are occupied, the structure is that of perovskite.

15.4.2 The V state

The V state is prepared by mild reduction of the VI compounds and is coloured, usually green or red. The oxides are not acidic. Although W^V does exist, there are fewer tungsten than molybdenum compounds in this state.

Mo_2O_5 may be made by reacting MoO_3 with Mo at 750 °C. It is violet in colour and insoluble in water and dilute acids. It now seems likely that W_2O_5 does not exist and reports of its preparation apply to oxygen-deficient WO_3. Mo_2S_5 is also known from the reaction of H_2S on Mo^V solutions.

Among the halides, MF_5, MCl_5 (M = Mo, W), and WBr_5 are found. The existence of WBr_5 parallels the behaviour in the VI state. WF_5 is unstable and disproportionates to $WF_6 + WF_4$ above 50 °C. Molybdenum pentafluoride is prepared by reduction with the carbonyl:

$$5MoF_6 + Mo(CO)_6 \rightarrow 6MoF_5 + 6CO$$

The pentachloride is the product of direct reaction between Mo and Cl. The tungsten pentahalides are formed by mild reduction of the hexahalides. All the pentahalides are covalent, relatively volatile solids. Solid MoF_5 has a tetrameric structure like NbF_5 (Figure 15.4) while $MoCl_5$ and WCl_5, in the solid, are dimeric like $NbCl_5$. Among the oxyhalides of the V state are $MoOF_3$, $MOCl_3$(M = Mo, W), $WOBr_3$ and WO_2I.

Most complexes of the V state are halogen or oxy-halogen compounds or contain pseudohalogen groups like CN or SCN. Reaction of $M(CO)_6$ in liquid IF_5 in the presence of alkali halides gives the two formula types $M^IMo(W)F_6$ and $M^I_3Mo(W)F_8$. The former are precipitated as salts of large cations and contain MF_6^- units. It is not known if the latter contain MF_8 units. Oxyhalide complexes result from the reduction of solutions of M^{VI}. The commonest types are MOX_5^{2-} and MOX_4^-, where X includes Cl, Br, SCN, and the Mo compounds are found in greater variety than the W ones. An interesting analogue is the $MoNCl_4^-$ ion which has a square pyramidal structure and a molybdenum-nitrogen triple bond.

The M^V octacyanides, $Mo(CN)_8^{3-}$ and $W(CN)_8^{3-}$ are obtained from strong oxidation of the M^{IV} octacyanides and have similar structures. These are discussed below.

15.4.3 The IV state

The two elements are quite similar in the IV state. This is usually formed by stronger reduction from the VI state than is used to reach the V state. The two dioxides, the disulfides and all the tetrahalides are known. Complexes are not very numerous and are similar in type to the complexes of the V state.

The dioxides, MO_2, are prepared by reduction of the trioxides by hydrogen or by careful oxidation of the metals. They are readily oxidized by halogens or oxygen and are reduced to the metal in hydrogen at temperatures above 500 °C. The dioxides are insoluble in nonoxidizing acids, but dissolve in nitric acid with oxidation to M^{VI}.

Both elements form the disulfide, MS_2. Molybdenum

disulfide is an important naturally-occurring molybdenum source. It has a layer lattice in which the Mo atoms are surrounded by six S atoms at the corners of a trigonal prism, see section 19.3. The outer S planes of neighbouring layers are only weakly cross-linked and the material is a solid lubricant. Among the sulfur complex ions of the (IV) state are the trinuclear species $[M_3S_4]^{4+}$ together with oxygen analogues. Indeed, for Mo the whole series $Mo_3S_nO^{4+}_{(4-n)}$ is known for $n = 0, 1, 2, 3, 4$. The structures are the cornerless cubane of Figure 15.15b. In $[W_3S_4(NCS)_9]^{5-}$, each W is approximately octahedrally coordinated to 3 NCS groups, two μ-S, and the central μ^3-S.

All the tetrahalides are known, although the bromide and iodide are poorly characterized. The IV halides tend to disproportionate, thus:

$$MoO_2 + Cl_2 \rightarrow MoCl_4 \rightarrow MoCl_3 + MoCl_5$$

$$WCl_6 + H_2 \rightarrow WCl_4 \xrightarrow{\text{heat}} W_6Cl_{12} + WCl_5$$

The WX_4 compounds are slightly more stable than the MoX_4 ones. All are coloured, involatile solids which are readily oxidized. The tetrachlorides have polymeric structures in which the metal coordination is octahedral and these are linked into chains by sharing Cl atoms. The M–M distances indicate some metal-metal bonding. $MoCl_4$ exists in a second form where six octahedra link into a hexameric ring by sharing opposite edges. The bridges show variable Mo–Cl distances between 243 and 251 ppm, the terminal distance is 220 ppm, while the Mo–Mo distance of 367 pm is too long for significant metal-metal bonding. The chlorocomplex, $Mo_2Cl_{10}^{2-}$, also has two octahedra sharing an edge.

Tungsten(IV) can be produced in solution by reduction of tungstate by tin and HCl. From the dark green solution the salt $K_2[W(OH)Cl_5]$ can be crystallized. This was formulated more recently as a dimer with bridging oxygen: $K_4[Cl_5W(O)WCl_5]$ and presumably hydrated. Some octahedral MX_6^{2-} complexes are also found, including the fluorides, chlorides, and thiocyanates. Both elements form seven-coordinate complexes $MCl_4(PR_3)_3$ with the face-capped octahedral structure of Figure 15.5a. Three Cl form one face, three P the opposite face and this is capped by the last Cl.

The complex ion $[W_4F_{18}]^{2-}$ has a closed cluster structure based on WF_6 octahedra. Each is linked to the other three by W–F–W bridges. The structure can alternatively be described as a tetrahedron of WF_3 units linked along each of the six edges by a W–F–W bridge.

Also in the IV state is the hydride-cyclopentadienyl class of compounds, $(C_5H_5)_2Mo(W)H_2$ and $(C_5H_5)_2WH_3^+$. The latter is isoelectronic with the tantalum trihydride and appears to have the same kind of structure. The structure of $(C_5H_5)_2MH_2$ shows similar ring–M–ring angles of 146°, but the HMH angle is only 76°. It is suggested that this, and related $(C_5H_5)_2ML_2$ species, better fit an alternative combination of orbitals than that of Figure 15.7, namely that of Figure 15.17. Contrast with Figure 15.7. $\psi_{A'}$ now lies along the y axis and is filled with a non-bonding electron pair. As

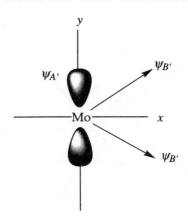

FIGURE 15.17 *Central plane orbitals in* $(C_5H_5)_2MoH_2$
Contrast with Figure 15.7. $\psi_{A'}$ now lies along the y axis and is filled with a non-bonding electron pair. As a result, the two $\psi_{B'}$ orbitals make a much smaller angle than the two ψ_B orbitals of Figure 15.7.

a result, the two $\psi_{B'}$ orbitals make a much smaller angle than the two ψ_B orbitals of Figure 15.7.

The best-known complex ion of the IV state is the octa-cyanomolybdate or tungstate, $M(CN)_8^{4-}$. This is an extremely stable grouping and is attacked only by permanganate or ceric which oxidize it only as far as the corresponding V octacyanide ion. The structure in the solid is dodecahedral and is shown in Figure 15.18. A similar arrangement is shown in Figure 15.2. This arrangement of ligands stabilizes the d_{xy} orbital relative to all the rest, as the energy level diagram (15.18b) shows. The configuration is thus suitable for d^1 and d^2 arrangements with large ligand fields and the CFSE must be a major factor in the stability of these octacyanides.

The $M(CN)_8^{3-}$ ion, which results from oxidation of $M(CN)_8^{4-}$, is also found in the dodecahedral configuration. However, as we have seen in other cases of high coordination, the energy difference between alternative structures is small, and the octacyanides are also found in square antiprismatic coordination (as in Figure 15.5b) when the cation is large. Examples include $Mo(CN)_8^{4-}$ in the $Cd(N_2H_4)_2^{2+}$ salt, and both $M(CN)_8^{3-}$ ions when the cation is $Co(NH_3)_6^{3+}$.

In the tungsten IV ion $(W_3O_4F_9)^{5-}$ the structure is a tetrahedron formed of the three W atoms and one O. Each W–W edge is bridged by an oxygen and there are three terminal F atoms per W.

15.4.4 *The lower oxidation states*

The lower oxidation states are unstable and strongly reducing. There is no evidence for simple cations in the III or II state. Tungsten(III) is not stable and compounds are limited to WCl_3 and some chloro-complexes. WCl_3 consists of Cl^- ions and the cluster $W_6Cl_{12}^{6+}$ which is isostructural with $Nb_6Cl_{12}^{2+}$ (Figure 15.9). The complex ion $W_2Cl_9^{3-}$ has two octahedral sharing a face, together with a W–W bond. WBr_3 does not contain W(III) but is a polybromide $(Br_4)^{2-}$ containing the $W_6Br_8^{6+}$ cluster related to the $Mo_6Cl_8^{4+}$ structure (Figure 15.19). By contrast, Mo^{III} is relatively stable and well-represented. The oxide Mo_2O_3 is not known, but

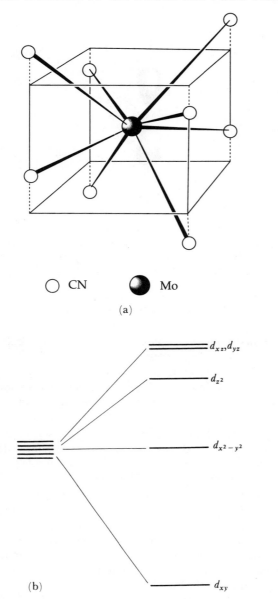

CN   Mo

(a)

d_{xz}, d_{yz}

d_{z^2}

$d_{x^2-y^2}$

d_{xy}

(b)

FIGURE 15.18 (a) *the structure of the octacyanomolybdate(IV) ion,*
$Mo(CN)_8^{4-}$, (b) *the energy level diagram for this structure*
(b) shows how this structure stabilizes one *d* orbital relative to the rest.

the sulfide exists. All the trihalides are reported, being derived from the higher oxidation states by strong reduction, or disproportionation. $MoCl_3$ is fairly stable, being only slowly oxidized in air and slowly hydrolysed by water. In solution, in presence of excess chloride, the complex anion $MoCl_6^{3-}$ may be prepared. A considerable number of other representatives of the type MoX_6^{3-} are known including the fluoride and thiocyanate. These are all octahedral complexes with three unpaired electrons, as expected. Some complex cations of Mo^{III} are known, including $Mo(dipy)_3^{3+}$ and $Mo(phen)_3^{3+}$, and neutral compounds like $Mo(acac)_3$ also occur. All these compounds are probably octahedral.

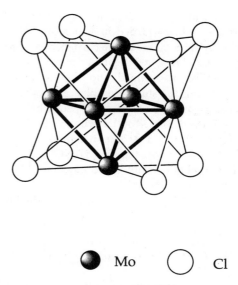

Mo   Cl

FIGURE 15.19 *The structure of the ion,* $Mo_6Cl_8^{4+}$
This structure also contains metal-metal bonding in the Mo_6 cluster.

The II state is represented by the complex halides $M_6X_{12}(M = Mo, W: X = Cl, Br, I)$ and by a few complexes with π-bonding ligands. The dihalides are relatively inert and insoluble materials. It was soon found that only a third of the halogen could be replaced, for example with $AgNO_3$ or by OH^-, and structural evidence has led to the formulation of the dihalides as hexamers:

$$'MX_2' = M_6X_{12} = (M_6X_8)^{4+}X_4^- \xrightarrow{Y^-} M_6X_8Y_4$$

The structure of the $Mo_6Cl_8^{4+}$ group is shown in Figure 15.19; the other compounds are isomorphous. The 6 Mo form an octahedron whose faces are bridged by the 8 Cl atoms (which themselves thus lie approximately in a cube circumscribing the octahedron). Compare this structure with that of Figure 15.9 where the octahedron of metal atoms is bridged along each edge, giving the $M_6X_{12}^{n+}$ cluster unit. In both these types of compound, there is polycentred metal-metal bonding as well as the bonding involving the bridging chlorines.

$(W_6Cl_8)^{4+}$ oxidizes to $(W_6Cl_{12})^{6+}$, the W(III) cluster, but oxidation of the molybdenum(II) chloride stops at the species $[Mo_6Cl_{12}]^{3+}$, corresponding to a formal oxidation state of $2\frac{1}{2}$. These species are reminiscent of the $(Ta_6Cl_{12})^{n+}$ clusters where a range of charges is found. Addition of 6 Cl atoms in terminal positions on the metals gives the $M_6Cl_{14}^{2-}$ cluster. Removal of one MoCl unit from this gave the $Mo_5Cl_{13}^-$ species where the Mo atoms now form a square pyramid with the open square face edge bridged by Cl. These examples illustrate the extensive chemistry which is developing for such clusters. Linked clusters, compounds containing atoms in the octahedral site in the centre, and other variations are being explored.

Most other compounds of the divalent elements are carbonyl complexes such as $(C_5H_5)W(CO)_3Cl$ and

$Mo(diars)_2(CO)_2X^+$. There is also an involatile acetate $Mo(CH_3COO)_2$, formed from $Mo(CO)_6$ and glacial acetic acid. This is a dimer and, like the red Mo(II) chloro-complex $K_4Mo_2Cl_8.2H_2O$, involves multiple Mo—Mo bonding. This is discussed in section 16.7. Reduction of Mo(IV) cyano-complexes yields $Mo(CN)_7^{5-}$ which has a regular pentagonal bipyramidal structure (compare $V(CN)_7^{4-}$, p. 237).

The I state occurs in π-bond compounds such as $Mo(C_6H_6)_2^+$ (compare the analogous Cr compound) or $[C_5H_5Mo(CO)_3]_2$.

The 0 state is represented by the octahedral and rather stable carbonyls, $Mo(CO)_6$ and $W(CO)_6$, which are white solids. Related compounds are also quite common, for example, $Mo(CO)_5I^-$, $W(PF_3)_6$, or $py_3Mo(CO)_3$. The $-$II state is shown by the carbonyl anion $M(CO)_5^{2-}$.

15.5 Technetium, $4d^65s^1$, and rhenium, $5d^56s^2$

The comparative chemistry of these two elements has substantially developed since the 70's when technetium found extensive use in medicine. Since all Tc isotopes are radioactive, thorough study is quite recent. Tc and Re, in sharp contrast to Mn, are particularly stable in the (VII) state which has only weak oxidizing properties. The lower states are also quite stable, especially Tc(IV) and Re(III) and (IV). The V and VI states are relatively unknown and tend to disproportionate. Low oxidation states are strongly reducing and not well-known, especially the II state which was so stable for Mn. These characteristics are illustrated by the oxides and halides in Tables 13.3–13.5 and by the oxidation state free energy diagrams in Figure 15.20.

The longest-lived isotopes of technetium, ^{97}Tc and ^{98}Tc, have half-lives of the order of two million years. These are prepared by neutron bombardment of molybdenum isotopes. However, the most accessible source is nowadays the isotope ^{99}Tc which is one of the fission products of uranium and thus occurs to a considerable extent in spent fuel elements. This has a half-life of 212 000 years and is a weak β-emitter. It is therefore only mildly radioactive and relatively easy to handle. The metastable isotope, technetium-99 m has a very useful application in medical imaging, discussed in Chapter 20.

Work on the usual isotopes of technetium involves their separation from the uranium and fission products in the fuel rods by oxidation, followed by distilling out the volatile Tc_2O_7. This may be separated from Re_2O_7, which is also volatile, by fractional precipitation of the sulfides. At an acid concentration above 8M HCl, Re_2S_7 precipitates but Tc_2S_7 does not.

Rhenium is a very rare element in the earth's crust and does not occur in quantity in any ore. However, it is found in molybdenum ores and can be quite readily recovered from these so that it is not too inaccessible and its chemistry is quite well known. It is left in oxidized solution as the perrhenate ion, ReO_4^-, whence it may be precipitated as the insoluble potassium salt. The metal is obtained by hydrogen reduction.

15.5.1 The (VII) oxidation state

The Group oxidation state is represented by the heptoxides, the acid and salts of the MO_4^- ion, by the heptasulfides, by oxyhalides and by the MH_9^{2-} complexes discussed below. Rhenium alone forms the heptafluoride, ReF_7. The structure of this compound has been the subject of much recent discussion as it is one of only two stable binary compounds of AB_7 stoichiometry (the other is IF_7). A recent low-temperature neutron diffraction study confirms that the configuration is a distorted pentagonal bipyramid (see Table 13.4).

The heptoxides result from heating the metals in oxygen or air. Tc_2O_7 is a yellow solid melting at 120 °C and boiling at 310 °C. It is stable up to the boiling point. Re_2O_7 is also yellow and the solid sublimes; the calculated boiling point is 360 °C. In the vapour, the structure is $O_3M-O-MO_3$, with tetrahedral coordination. The crystal structures of Tc_2O_7 and Re_2O_7 differ, with the technetium oxide having the same structure as the vapour, with a linear Tc—O—Tc bridge. The Re_2O_7 solid contains regular ReO_4 tetrahedra and ReO_6 octahedra with three short and three long bonds; the overall arrangement indicates that $ReO_3^+ReO_4^-$ (compare N_2O_5 in the solid) may be a reasonable formulation. Technetium heptoxide is somewhat more oxidizing than the rhenium compound and these two heptoxides differ strikingly in stability from Mn_2O_7 which is also volatile but which rapidly decomposes at room temperature.

The heptoxides dissolve in water to give colourless solutions of the acids. Pertechnic acid, $HTcO_4$, is produced as dark red crystals on evaporation, but perrhenic acid, $HReO_4$, cannot be isolated although the colour of the solution changes to yellow-green on concentration and lines due to the acid appear in the Raman spectrum of the concentrated solution. These colours are due to the lowering of the symmetry on passing from the tetrahedral anion ReO_4^- to the acid $(HO)ReO_3$ on concentration. Perrhenic acid resembles periodic acid in having a second form H_3ReO_5 (compare HIO_4 and H_5IO_6), and salts derived from this are readily prepared. The normal perrhenates are formed in dilute solution while salts of the tribasic form of the acid are formed in media of higher basicity:

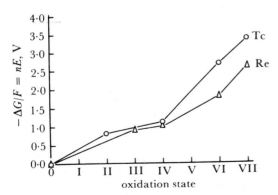

FIGURE 15.20 *The oxidation state free energy diagram for technetium and rhenium*

$$ReO_4^- + K^+ \rightarrow KReO_4 \text{ (yellow)}$$

$$ReO_4^- + K^+ + OH^- \rightarrow K_3ReO_5 \text{ (red)}$$

or $$Ba(ReO_4)_2 \xrightarrow[H_2O]{Ba(OH)_2} Ba_3(ReO_5)_2$$

The peracids are strong acids, with perrhenic acid lying between perchloric acid and periodic acid in strength. Permanganic acid is stronger than perrhenic acid, so it is likely that pertechnic acid is also stronger. The structure of Na_3ReO_5 shows a square pyramid ReO_5 unit with the axial Re–O at 184 pm a little longer than the four Re–O base values of 179 pm. A third anion is found in $Ba_5(ReO_6)_2$ which probably contains isolated ReO_6^{5-} ions. The technetium analogue of the tribasic anion has not been isolated but TcO_5^{3-} may exist in fused sodium hydroxide. Some reactions of perrhenic acid are shown in Figure 15.21. Reactions of $HTcO_4$ are similar, as far as is known, apart from the lack of H_3TcO_5 and some differences in solubilities. Thus $KTcO_4$ is twice as soluble as $KReO_4$ and does not precipitate so readily. Both these potassium salts are very stable and can be distilled at temperatures of over $1000\,^{\circ}C$ without decomposition. $KMnO_4$, on the other hand, loses oxygen above $200\,^{\circ}C$. The perrhenates, even of organic bases such as strychnine, may be isolated while permanganate readily oxidizes such compounds.

The imido analogue of the perrhenate ion, $[Re(NBu^t)_4]^-$, has been synthesized as the $Li(tmen)^+$ salt. It is an air- and moisture-sensitive solid, reflecting the greater reactivity resulting from the less electronegative nitrogen substituents compared with ReO_4^-.

A further interesting oxygen derivative is $MeReO_3$ made by reacting $SnMe_4$ with Re_2O_7 —another indication of the very low oxidizing power of Re(VII). The methyltrioxyrhenium catalyses many reactions including olefin oxidations and metatheses.

The VII state is found in halides and oxyhalides as shown in Table 15.3.

TABLE 15.3 Halides and oxyhalides of the VII state

Re	Tc	Mn
ReF_7		
$ReOF_5$		
ReO_2F_3	TcO_2F_3	
ReO_3F	TcO_3F	MnO_3F
ReO_3Cl	TcO_3Cl	MnO_3Cl ?
ReO_3Br		

Technetium heptafluoride is not formed by direct combination at 400 °C, the reaction which gives ReF_7, but TcF_6 is found instead. The halides and oxyhalides are all colourless or pale yellow compounds which are either liquids or low-melting solids. The oxyhalides result from halogenation of the oxide or oxyion, or from the action of oxygen or water on the fluoride in the case of the rhenium oxyfluorides. The structure of TcO_2F_3 shows a chain of octahedral units linked by

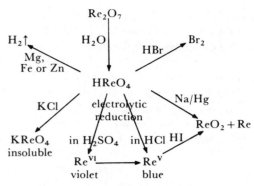

FIGURE 15.21 *Reactions of perrhenic acid*

Tc–F–Tc bridges. The two O atoms on each Tc are *cis* and the bridging fluorines are those *trans* to the O.

The structure of the $[ReOF_6]^-$ ion in both the Cs^+ and the NO_2^+ salts is a pentagonal bipyramid, Figure 15.22, with the Re–O bond axial and shorter, at 167 pm, than the axial Re–F (192 pm). The equatorial Re–F bonds vary slightly from 187–191 pm and the central pentagon is slightly puckered. One sulfur analogue of the oxyhalides is reported. Treating ReF_7 with Sb_2S_3 gives maroon $ReSF_5$ which is very sensitive to water.

Few complexes of the VII state are known, but oxidation of the octacyanide of Re(VI) gives salts of what appears to be the $Re(CN)_8(OH)_2^{3-}$ ion, which may contain Re(VII) in ten-coordination.

A particularly striking example of the stability of the (VII) state is provided by the hydrogen complexes, K_2ReH_9 and K_2TcH_9. It has long been known that treatment of perrhenate solutions with potassium/ethylenediamine/water gave a water-soluble reactive rhenium species. This was originally identified as a 'rhenide' anion, analogous to the halide ions in the corresponding Main Group but it has now been shown to contain Re(VII) coordinated to nine hydrogen atoms. The compound is colourless and diamagnetic, which accords with Re(VII). It was soon afterwards shown that a similar Tc compound existed.

The structure of the ion is shown in Figure 15.23. The positions of the metal atoms were determined by X-rays and then the hydrogen positions were found from neutron diffraction. The Re — Re distances rule out the possibility of any metal-metal bonding. The Re atoms are at the centre of

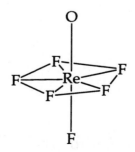

FIGURE 15.22 *The structure of the $[ReOF_6]^-$ ion*
The structure is based on a pentagonal bipyramid with a slightly puckered central pentagon.

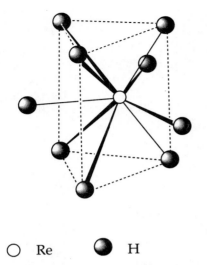

○ Re   ● H

FIGURE 15.23 *The structure of the enneahydridorhenium (VII) ion,* ReH_9^{2-}

trigonal prisms of six H atoms with the other three H atoms beyond the rectangular faces. This lies inside a similar prism of K atoms. The Tc compound is isostructural. The existence of such compounds, like the $WH_6(PR_3)_3$ species indicated in the last section, highlights the lack of oxidizing tendency in the high oxidation states of these heavy elements.

There has been a recent surge in interest in the chemistry of organometallics in high oxidation states, and that of technetium and rhenium in the VII oxidation state serves to illustrate this very well. High oxidation state organometallic compounds were once thought to be rare, but are actually turning out to be quite common. These compounds are of interest as catalysts, particularly in alkene epoxidation reactions which are of industrial importance. The reaction

of the heptoxides M_2O_7 (M = Tc or Re) with $SnMe_4$ gives the very interesting tetrahedral M(VII) compounds $MeMO_3$; the rhenium compound is air-stable and water-soluble, whilst the technetium analogue is moisture-sensitive. The pentamethyl cyclopentadienyl analogue of the rhenium complex $(\pi\text{-}C_5Me_5)ReO_3$ can also be prepared by oxidation of $(p\text{-}C_5Me_5)Re(CO)_3$. A wide range of Re(VII) derivatives containing different alkyl groups, oxo ligands and halides have also been prepared. The compound Me_3ReO_2 is another example of a rhenium(VII) oxo alkyl compound (compare Figure 15.24).

15.5.2 *The (VI) state*

The VI state is represented by a number of compounds but shows a marked tendency to disproportionate, especially in aqueous solution. In contrast, in concentrated sulfuric acid, Re(VII) may be reduced to pink Re(VI) and then to blue Re(V) in distinguishable steps. ReO_3 is well-known and there is one report of an oxide of composition $TcO_{3.05}$. The rhenium trioxide is made by reaction of rhenium on the heptoxide and is red. In vacuum at 300 °C it disproportionates to ReO_2 plus Re_2O_7. It is inert to acids and bases in a non-oxidizing medium, and rhenates, ReO_4^{2-}, have to be made by fusing mixtures of perrhenate and rhenium dioxide. Technates exist in alkaline solution and pink $BaTcO_4$ may be precipitated, indicating less tendency to disproportionate than for rhenium(VI). Solutions of TcO_4^{2-} (violet) and ReO_4^{2-} (olive-green) result from controlled cathodic reduction of the MO_4^- ions. The analogous, highly reactive, tetrathiorhenate(VI) anion ReS_4^{2-} has been investigated by spectroelectrochemistry.

The halides, oxyhalides, and related complexes, of the VI state are given in Table 15.4. The VI state of Re is well-represented although $ReCl_6$ (and $TcCl_6$) readily lose Cl_2.

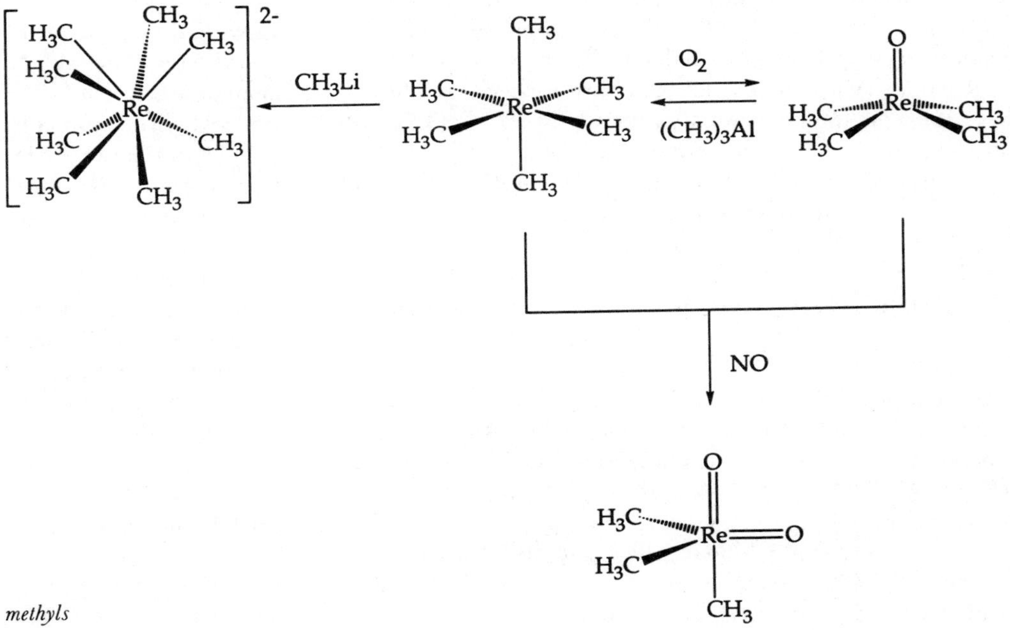

FIGURE 15.24 *Some rhenium VI and VII methyls*

TABLE 15.4 Halogen compounds of the lower oxidation states of technetium and rhenium

VI		V		IV		III	
ReF_6	TcF_6	ReF_5	TcF_5	ReF_4			$[ReF_3]$
$(ReCl_6)$	$(TcCl_6)$	$ReCl_5$		$ReCl_4$	$TcCl_4$		Re_3Cl_9
		$ReBr_5$		$ReBr_4$	$(TcBr_4)$		Re_3Br_9
				ReI_4			Re_3I_9
Also ReX_2 in complexes, and possible compound ReI							
$ReOF_4$	$TcOF_4$	$ReOF_3$					
$ReOCl_4$	$(TcOCl_4)$		$TcOCl_3$				
$(ReOBr_4)$			$TcOBr_3$				
ReO_2Br_2?							
ReF_8^{2-}	TcF_8^{2-}	ReF_6^-	TcF_6^-	ReF_6^{2-}	TcF_6^{2-}	$ReCl_6^{3-}$	
	TcF_7^-			$ReCl_6^{2-}$	$TcCl_6^{2-}$	$Re_3X_{12}^{3-}$	
				$ReBr_6^{2-}$	$TcBr_6^{2-}$	$Re_3X_{11}^{2-}$	
				ReI_6^{2-}	TcI_6^{2-}	$Re_3X_{10}^-$	
				$Re_2X_9^-$		$Re_2X_8^{2-}$	$Tc_2Cl_8^{2-}$
				$(X = Cl, Br)$		$(X = Cl, Br)$	
$ReO_2F_4^{2-}$?							
$ReOCl_6^{2-}$		$ReOCl_5^{2-}$	$TcOCl_5^{2-}$	$Re(OH)Cl_5^{2-}$	$Tc(OH)Cl_5^{2-}$		
$ReOCl_5^-$		$ReOBr_5^{2-}$	$TcOBr_5^{2-}$	$Re_2OCl_{10}^{4-}$			
		$ReOX_4^-$	$TcOX_4^-$				
		$(X = Cl, Br, I)$	$(X = Cl, Br)$				

Tc(VI) is much less-represented with only the oxide and oxyfluoride of reasonable stability. The fluorides are prepared by direct combination: ReF_6 may be purified from ReF_7 by heating with Re metal. $ReOF_4$ results from the reaction of ReF_6 with rhenium carbonyl. $TcOF_4$ exists in two forms: the blue variety contains infinite chains of octahedra while the green form contains trimers of octahedra. In both cases the octahedra are formed by sharing fluorines, $(TcOF_3F_{2/2})_n$. The other oxyhalides result from the reaction of the halides with air. $ReOCl_4$ is a square pyramidal monomer in the solid, as is the isoelectronic $TcNCl_4^-$, formed from the reaction of azide on TcO_4^-. The sulfur analogue, $ReSF_4$, is also known.

Rhenium(VI) methyl compounds are also well established, and some of the inter-relationships between Re(VI) and Re(VII) methyls are summarized in Figure 15.24. $ReMe_6$ is a distorted octahedron, as expected for a d^1 species, and $ReMe_8^{2-}$ is a square antiprism.

15.5.3 The (V) state

The V state is also relatively unstable and disproportionates:

$$3M^V = 2M^{IV} + M^{VII}$$

For example, $ReCl_5$ reacts with HCl to give ReO_2, perrhenic acid, and $ReCl_6^{2-}$. $ReCl_5$ exists as a dimer with octahedral rhenium, like other pentahalides such as $NbCl_5$ (Figure 15.6). Where known, the MOX_4^- complexes are square pyramidal.

The complex of the V state which is of most interest is the cyanide, $Re(CN)_8^{3-}$. This is the d^2 octacyanide, isoelectronic with the Mo^{IV} compound, and it appears to have the same dodecahedral structure. On oxidation it gives the d^1 compound of Re^{VI}, $Re(CN)_8^{2-}$. Related compounds are produced from the dioxides in alkaline cyanide solutions. The complex ions $Tc(OH)_3(CN)_4^{3-}$ and $Re(OH)_4(CN)_4^{3-}$ are reported to be formed under similar conditions. It is interesting that Re is oxidized in this system while Tc remains in the IV state.

The diarsane ligand already encountered (p. 245) appears to stabilize the relatively unstable II, III, and V states of Tc and Re. Both metals form the $M(diars)_2Cl_4^+$ ion in which the element is in the V oxidation state and eight-coordinate. The $M^{II}(diars)_2Cl_2$ and $M^{III}(diars)_2Cl_2^+$ compounds are also known. In $ReH_5(PR_3)_3$ the skeleton forms a dodecahedron (compare Figure 15.18a) with two of the PR_3 groups in the same edge.

The related $ReH_7(PR_3)_2$ has been the subject of recent investigation. With the discovery of coordinated dihydrogen compounds (see section 16.11) the question arises whether a polyhydride might contain $M–H_2$ units in place of 2 MH bonds. Initially it was thought that the properties of $ReH_7(PR_3)_2$ indicated formulation as a $Re(V)–H_2$ species, but further work, including a neutron diffraction structure, has confirmed the original formulation as a Re(VII) heptahydride.

15.5.4 The (IV) state

The IV state is the second most stable oxidation state for both elements. The dioxides, MO_2, disulfides, MS_2, and most of the halides are known. A number of complexes exist including the very important class of hexahalides, MX_6^{2-}.

These are formed by dissolving the dioxides in the hydrohalic acid and are a most useful starting point for preparations. The halide complexes $Re_2X_9^-$ are ReX_6 octahedra linked through a shared face: $X_3Re(\mu-X)_3ReX_3^-$.

The dioxides are formed by reducing the heptoxides, or from the metal by controlled oxidation. Apart from dissolving in the hydrogen halides, ReO_2 dissolves in fused alkali to give the oxyanion, rhenite, ReO_3^{2-}. This precipitates the dioxide when treated with water. It does not appear that the technetium dioxide is soluble in alkali.

A further example of the smaller tendency of Tc to enter the V state is provided by the reaction of the MI_6^{2-} complexes with KCN. TcI_6^{2-} reacts in methanol with KCN to give the Tc^{IV} cyanide complex, $Tc(CN)_6^{2-}$, but ReI_6^{2-} undergoes oxidation in the course of the reaction to the Re^V octacyanide $Re(CN)_8^{3-}$. It is easy to see that, if the octacyanide is to be formed, the M^V (d^2) compound will be more stable than the M^{IV} (d^3) compound, as the third electron is in an unstable orbital in the dodecahedral field (Figure 15.18b). It is therefore likely that this difference in behaviour is to be ascribed to a steric effect favouring eight-coordination for Re rather more than for Tc.

A dimeric hydride with a Re-Re bond of 253 pm is formed in the IV state: $Re_2H_8(PR_3)_4$ has coplanar Re and P atoms and results from the reduction of $ReCl_5$ with $LiAlH_4$ in presence of the phosphane.

15.5.5 *The lower oxidation states*

In the III state, technetium is quite unstable but rhenium forms the oxide, $Re_2O_3.nH_2O$, and the heavier halides. The oxide is formed by hydrolysis of the rhenium(III) chloride.

Rhenium trichloride and tribromide, formed by heating the pentahalides in an inert gas, are dark red solids which have been shown to be trimers, Re_3X_9. The structure is similar to that of the $Re_3Cl_{12}^{3-}$ ion shown in Figure 15.25. In these species, a triangle of Re atoms is bridged along each edge by a halogen atom, and there are two more halogen atoms on each rhenium situated above and below the Re_3 plane giving a basic Re_3Cl_9 unit. In the ion $Re_3Cl_{12}^{3-}$, the structure is completed by adding a terminal halogen, in the plane of the triangle, to each rhenium. The related ions (see Table 15.4) $Re_3Cl_{11}^{2-}$ and $Re_3Cl_{10}^-$ lack one and two, respectively, of these in-plane terminal chlorines. In $[Re_3Cl_{10}(H_2O)_2]^-$ two of the in-plane terminal positions are occupied by water molecules and other donors will behave similarly. The corresponding bromides have similar structures. In the parent trihalide, the basic Re_3Cl_9 units are joined together by forming $Re(Cl)_2Re$ bridges, using the in-plane positions and one of the Cl atoms which are perpendicular to the plane from each Re. A range of related complexes are also found, of type $Re_3X_9L_3$ where neutral ligands occupy the in-plane terminal positions. Earlier reports of dimeric forms of rhenium trihalides have been shown to be unfounded. ReI_3 is a black solid, which results from heating the tetraiodide, and further heating leads to another black compound of formula ReI. The structure is also trimeric and similar to that of the trichloride. The overall linking of Re_3 units gives long chains in the iodide, while the chloride and bromide have layer structures.

A further type of complex halide ion of the III state of rhenium is formed by reduction of perrhenate in acid solution and contains dimeric units, $(Re_2Cl_8)^{2-}$. Similar species are $Mo_2Cl_8^{4-}$, $Re_2Br_8^{2-}$, a number of derivatives such as $Re_2Cl_6[PEt_3]_2$, and dinuclear carboxylic acid derivatives. Technetium forms $Tc_2Cl_8^{n-}$ for $n = 2$ or 3. Such compounds contain $M-M$ bonds of high order which are discussed in more detail in section 16.4. Other complexes of the trichlorides with donor ligands are found, for example, $(Ph_3PO)_2ReCl_3$. Rhenium(III) also gives hydrides of formula $(C_5H_5)_2ReH$ and $(C_5H_5)_2ReH_2^+$.

Representatives of lower oxidation states include complexes of MX_2 with donor ligands such as py_2ReI_2 and $(diars)_2TcCl_2$. Rhenium(I) is found in the stable dinitrogen complexes, $ReCl(N_2)L_4$, where the ligands L are various phosphines, CO or PF_3. Oxidation with Ag^+ or Fe^{3+}, gives the rhenium(II) species, $ReCl(N_2)L_4^+$.

Both elements form a cyanide in the I state corresponding to the manganese compound, $K_5M(CN)_6$. The I state is also found in carbonyl halides and similar species, for example, in $Re(PF_3)_4Cl$.

The 0 state appears in the carbonyls, $Tc_2(CO)_{10}$ and $Re_2(CO)_{10}$, which have the same structure as the manganese carbonyl. The anion, $Re(CO)_5^-$, is formed with the alkali metals and contains $Re(-I)$.

15.6 Ruthenium, $4d^75s^1$, and osmium, $5d^66s^2$

These are the first two of the six platinum metals, i.e. the six heavier members of the iron, cobalt, and nickel Groups. These elements, together with rhenium and gold, are broadly similar in that the element is fairly unreactive—'noble'—and decomposition of compounds to the element is fairly ready. The platinum metals, gold, and silver are commonly found together and a number of schemes are in current use for their separation. One method involves extracting the mixed metals with aqua regia and then treating the soluble and insoluble portions as in Figure 15.26. Osmium may occur in

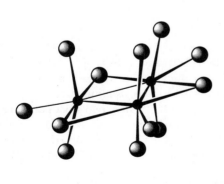

● Re ⬤ Cl

FIGURE 15.25 *The structure of the* $[Re_3Cl_{12}]^{3-}$ *ion*

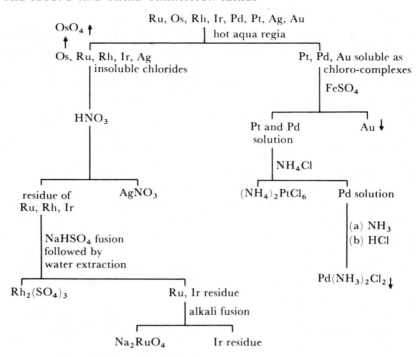

FIGURE 15.26 *A reaction scheme for the separation of the platinum metals, gold, and silver*

either fraction and is removed as the volatile tetroxide, while ruthenium ends up in the VI state in fused alkali.

These two elements share with xenon the highest observed oxidation state of VIII, and their oxidation states range downwards to $-$ II. The VI and IV states are stable while the VII and V states are poorly represented and tend to disproportionate. Osmium is most stable in the IV state and ruthenium in the III state.

Some inter-relations among the oxides and oxyions of the different oxidation states are shown in the reaction diagrams of Table 15.5, and for halogen compounds in Table 15.6.

15.6.1 *The higher oxidation states*

Note from the tables that osmium is more stable in the VIII state than ruthenium, the metal being oxidized directly to the tetroxide. The tetroxides are volatile (b.p. about 100 °C), tetrahedral, covalent molecules which are strongly oxidizing. Osmium(VIII) also occurs bonded to N in $Os(NBu^t)_4$, with the bulky *tert*-butyl substituents contributing to the stability. The structure is a distorted tetrahedron with OsNC angles of 156°, all in accord with an Os=N formulation. The mixed oxygen–nitrogen species $Os(NBu^t)_2O_2$ and $Os(NBu^t)O_3$ have similar tetrahedral structures. Further osmium(VIII)–oxygen species include the hydroxide complexes $OsO_4(OH)_2^{2-}$, which is a *cis* octahedron, and $OsO_4(OH)^-$, which contains a distorted trigonal bipyramidal group with OsO_4 units bridged axially by hydroxides.

Osmium (VIII) and (VII) are also represented among halogen compounds although OsF_8 does not exist and OsF_7 survives only under a high pressure of F_2. The VIII state is found in the oxyfluorides OsO_2F_4 and OsO_3F_2. OsO_2F_4

(previously identified incorrectly as $OsOF_6$) is formed by the action of KrF_2 on OsO_4 and has a *cis* octahedral structure. KrF_2 is a powerful oxidizing and fluorinating agent, and oxygen in this reaction is oxidized from $O(-II)$ to $O(0)$.

$$OsO_4 + KrF_2 \rightarrow OsO_2F_4 + 2Kr + O_2$$

OsO_3F_2 is formed by the reaction of OsO_4 with liquid ClF_3 and has a structure of linked octahedra. The Os bears three O and one terminal F and is linked through Os—F—Os bridges to the octahedra on either side, with the two bridging F occupying *cis* positions.

This compound undergoes an interesting reaction where it is *reduced* by reaction with fluorine to Os(VII) and Os(VI) complexes, and oxygen is again oxidized, this time from $O(-II)$ to $O(II)$.

$$OsO_3F_2 + F_2 \rightarrow OsOF_5 + OsF_6 + OF_2$$

Halo-complexes of Os(VIII) include $[OsO_4F_2]^{2-}$ which is a *cis* octahedron and $[OsO_4Cl]^-$ which is a trigonal bipyramid with the Cl axial with a very long bond.

As well as the MO_4^- ions, osmium forms a second oxyanion in the VII state, OsO_6^{5-}. These two oxyanions compare with those found for rhenium. A second oxyfluoride of osmium(VII), OsO_2F_3, forms from OsO_3F_2 and $OsOF_4$, and disproportionates back to these (VIII) plus (VI) species at 60 °C.

In the VI state, $OsOX_4$ (X = F, Cl or CH_3), are square pyramids in the gas phase with the oxygen in the axial position, like the rhenium and tungsten analogues. In contrast, the MOX_4 species formed by the Main Group elements are trigonal bypyramids with the oxygen in an

TABLE 15.5

Oxidation state

VIII $\quad$ Os $\xrightarrow{\text{burn in air}}$ MO_4 $\xleftarrow[\text{oxidation}]{\text{strong}}$ Ru(VI) in acid solution

VII $\quad$ Os $\xrightarrow[\text{fusion}]{Na_2O_2}$ OsO_4^- $\searrow^{OH^-}$ RuO_4^- (unstable)

$\qquad\qquad\qquad$ mild $\downarrow$ $\qquad\qquad$ $\downarrow OH^-$

$\qquad\qquad\qquad$ reduction

VI $\qquad\qquad\qquad$ $OsO_2(OH)_4^{2-}$ or RuO_4^{2-} $\xrightarrow[\substack{\text{(various} \\ \text{conditions)}}]{Cl^-}$ $\begin{array}{l} MO_2Cl_2 \\ OsOCl_4 \\ OsOCl_6^{2-} \text{ and } OsO_2Cl_4^{2-} \end{array}$

$\qquad$ Also M (VI) solution $\xrightarrow{OH^-}$ $MO_3.nH_2O$
$\qquad$ (anhydrous MO_3 occurs only in presence of O_2 at high pressure)

V $\qquad$ No oxygen compound of the V state exists

IV $\quad$ OsO_4 $\xrightarrow[\text{reduction}]{\text{mild}}$ MO_2 $\xleftarrow[\text{air}]{\text{burn in}}$ Ru $\quad$ (Also MS_2, MSe_2 and MTe_2)

$\qquad\qquad\qquad$ $\swarrow^{HX}$ $\qquad$ $\nwarrow$ air

$\qquad\qquad$ OsX_4 (X = Cl, Br, I)

III $\qquad\qquad\qquad\qquad$ $Ru_2O_3.nH_2O$ $\xleftarrow{OH^-}$ Ru(III) in solution

TABLE 15.6

Oxidation state

VI $\quad$ $M + F_2$ $\longrightarrow$ MF_6 (readily with Os, only under careful conditions for Ru)

$\qquad\qquad\qquad\qquad$ $\downarrow\substack{\text{mild} \\ \text{reduction}}$

V $\quad$ $Ru + F_2$ $\xrightarrow[\text{product}]{\text{normal}}$ MF_5 $\xrightarrow{F^-}$ MF_6^-

$\qquad\qquad\qquad\qquad\qquad\qquad\quad$ $\downarrow OH^-$

$\qquad\qquad\qquad\qquad\qquad\qquad$ $MF_6^{2-} + O_2\uparrow$

IV $\quad$ $\left.\begin{array}{l} M + X_2 \text{ or} \\ MO_2 + HX \end{array}\right\}$ $\rightarrow OsX_4$, RuF_4, $RuCl_4$

$\qquad\qquad\qquad$ Os—stable $\qquad\qquad\qquad X^- \searrow$

$\qquad\qquad\qquad$ Ru—less stable $\qquad\qquad\qquad$ all MX_6^{2-} (except RuI_6^{2-})

$\qquad$ The OsX_4 give stable solutions in cold water,
$\qquad$ although HX is evolved on warming

III $\quad$ $M + X_2 \rightarrow OsCl_{3.5(?)}$, all four $\quad RuX_3$ $\xrightarrow{X^-}$ $\begin{cases} RuX_6^{3-} \\ OsCl_6^{3-} \text{ only} \end{cases}$
$\qquad$ (Normal reaction for $\qquad\qquad$ stable
$\qquad$ Ru, only with halogen deficit for Os)

equatorial position. RuF_6 is more reactive than OsF_6. It is thermally stable at room temperature but reacts vigorously with most materials. The structure is a fully symmetric octahedron in a low-temperature matrix. The osmium(VI) polyhydride, $OsH_6(PR_3)$, can be made by the reaction of osmium(IV) chloro-complexes with $LiAlH_4$. The thiosulfato complex $[OsO_2(S_2O_3)_2]^{2-}$, prepared by reduction of OsO_4 with aqueous sodium thiosulfate, has a distorted tetrahedral geometry about osmium, with the two thiosulfate ligands bonded through S.

In the unstable V state both metals form the pentafluoride and its anion. Both MF_5 compounds are tetramers with non-linear M–F–M bridges, similar to NbF_5 (Figure 15.4). This tetrameric form is adopted, with minor variations, by most of the pentafluorides of the heavier transition metals. In the gas phase the major form is a similar fluorine-bridged trimer, again with four terminal F and two M–F–M bridges formed by *cis* fluorines. Dimers form the minor gas phase species. $OsCl_5$ is isomorphous with rhenium pentachloride, having a dimeric structure with two chloride bridges. The complex ions $OsCl_6^-$ and $OsBr_6^-$ have slightly distorted octahedral structures.

The bulky aromatic group mesityl (which is 2,4,6-trimethylbenzene) helps to stabilize less-accessible species and allows the formation of Ru–C bonds in both the V and IV states in the compounds $[Ru(mes)_4]^+(PF_6)^-$ and $Ru(mes)_4$ respectively. The structures are tetrahedral.

15.6.2 *The lower oxidation states*

The lower oxidation states are very accessible. Only mild reduction of osmium tetroxide is required to give the dioxide, and ruthenium burns in air to give RuO_2 directly. The dioxides are stable compounds with the rutile structure. The stability of the IV state is also illustrated by the existence of the disulfides and of the heavier chalcogenides.

In Table 15.6, the stable states of Ru(III) and Os(IV) show clearly. One dimeric anion in the IV state is $[Os_2Br_{10}]^{2-}$ which has a structure with two octahedra sharing an edge. It reacts with a range of ligands to yield Os(III) compounds. A fuller study has thrown doubt on the compounds originally reported as osmium trihalides. These have now been characterized as the oxyhalides, Os_2OX_6, which are Os(IV) compounds. The Os(III) halide complexes OsX_6^{3-} X = Cl, Br, I can be prepared in presence of the large cation $[Co(en)_3]^{3+}$.

Deep pink RuF_4 has a structure built up of apex-sharing octahedra forming a puckered sheet array with the bond length to the bridging F equal to 200 pm. The non-bridging F(Ru–F = 182 pm) are *trans* to each other, lying above and below the sheet. This fluorine bridging is very similar to that found in $(RuF_5)_4$ and also in RuF_3 where all the F bridge, giving a three-dimensional array.

Both elements form the mixed oxyhalide, $M_2OX_{10}^{4-}$ (M = Ru, Os(IV) : X = Cl, Br). The oxygen bridges the M atoms, completing the octahedra, e.g. $[Cl_5Ru-O-RuCl_5]^{4-}$.

Apart from those given above, these elements form a variety of complexes in the IV, III, and II states. In the IV state, osmium complexes are more extensive and stable than the ruthenium ones. The commonest are the halide, hydroxy-halide, amine, and diarsine complexes. In the III state, the situation is reversed and there are more ruthenium species, mainly octahedral. Both elements give hexammines, $M(NH_3)_6^{3+}$, and ruthenium gives the whole range of mixed halogen-ammonia complexes down to $Ru(NH_3)_3X_3$. A variety of other ruthenium complexes occurs, including those with substituted amines, and complex chlorides with 4, 5, 6, and 7 chlorine atoms in the anion. A very interesting series is provided by the complexes $M(NCS)_6^{3-}$ for M = Ru or Os. The thiocyanate group can bond M–NCS as a 'hard' ligand or M–SCN as a 'soft' ligand (section 13.7) and these metals are clearly on the hard-soft borderline in the +III state since a number of members of the series $M(NCS)_n(SCN)_{6-n}^{3-}$ are formed. For Ru, $n = 1$ to 4 are indicated while Os gives $n = 2$ to 4.

Although there are few simple compounds of the II state, there are a variety of complexes of both ruthenium and osmium. All are formed by reduction of metal solutions, in the IV or III states, in presence of the ligands. Examples include $M(dipy)_3^{2+}$, the very stable $M(CN)_6^{4-}$, and a variety of ammine and arsane complexes. As the II state is the d^6 configuration, there will be a large CFSE contribution to octahedral complexes of these heavy elements. An important group of compounds are those containing NO bonded to ruthenium(II), such as $[Ru(NO)X_5]^{2-}$, where X = halogen, OH, CN and many more. Another interesting group of Os(II) compounds is derived from the carbonyls, see below and section 16.2. Oxidative fluorination gives a range of species in solution identified spectroscopically as *cis*-$[Os(CO)_4F_2]$, $[Os(CO)_5F]^+$ and $[Os_2(CO)_8F_3]^+$—the latter being two octahedra linked by Os–F–Os. In the solid CO is lost, giving the tetramer $[Os(CO)_3F_2]_4$.

In lower oxidation states, there are a number of compounds including $Os(NH_3)_6Br$ and $Os(NH_3)_6$—Os(I) and Os(0) respectively—formed by reduction with potassium in liquid ammonia. Other representatives of the 0 state include the carbonyls, $M(CO)_5$ and $M_3(CO)_{12}$. Among the more complex carbonyls, there is the very interesting derivative $Ru_6C(CO)_{17}$. This contains an isolated carbon atom which is situated at the centre of a distorted octahedron of Ru atoms giving a CRu_6 environment which is also found in most interstitial metallic carbides (section 17.5). Four of the Ru atoms carry three terminal CO groups and the other two have two terminal carbonyls and are bridged by the seventeenth one. The Ru_6C cluster involves delocalised poly-centred bonding. There is a reported anion, $Ru(CO)_4^{2-}$, which would contain ruthenium(−II) and be isoelectronic with nickel carbonyl.

Nitrogen complexes (see also section 16.10). A little after the first reports of nitrogen fixation in the systems metal halide/organometallic reducing agent which were discussed under *Titanium* (section 14.2), there was reported the first compound in which it was clearly shown that nitrogen was coordinated as a ligand. This was the ruthenium(II) complex

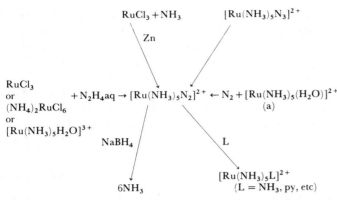

(a) The ruthenium(II) complex is made *in situ* by zinc amalgam reduction of ruthenium(III) pentammine chloro complex.

FIGURE 15.27 *Preparations and some reactions of* $[Ru(NH_3)_5N_2]^{2+}$

$[Ru(NH_3)_5N_2]^{2+}$. This compound was first made by the action of hydrazine (which was the source of the coordinated N_2) on the trichloride, but a number of other routes have since been reported including the reaction of a ruthenium(II) complex with gaseous nitrogen. Some of these preparations and some reactions are summarized in Figure 15.27.

This series of reactions shows that the ligand nitrogen may be derived from a number of sources, including the direct fixation of atmospheric nitrogen. The coordinated nitrogen is replaceable by a range of ligands, presenting a useful route to otherwise unobtainable ruthenium(II) complexes. In these cases, the nitrogen comes off as N_2 gas, so that the cycle of reactions does not lead to fixation of N_2 as a compound. The only indication of the latter process is in the reaction with $NaBH_4$ which yields six volumes of ammonia, that is, one N atom out of the two in the coordinated group is converted to NH_3.

A number of other metals in this part of the periodic table have been shown to form nitrogen complexes (Table 16.6). Osmium forms the analogue of the ruthenium complex, $[Os(NH_3)_5(N_2)]^{2+}$, by similar routes. The osmium complex is much more stable and resists replacement of the nitrogen. In $(NH_3)_5Ru(N_2)Ru(NH_3)_5^{4+}$, the NN unit bridges two metal atoms. Structural studies on these compounds indicate a linear $M-N-N-M$ bridge, in contrast to bridging carbonyl which bonds to both metals through carbon in a non-linear unit.

$$M \diagdown \quad \diagup M$$
$$C$$
$$|$$
$$O$$

15.7 Rhodium, $4d^85s^1$, and iridium, $5d^96s^0$

In this Group, the Group oxidation state of IX is not shown and a strong tendency for lower oxidation states to become stable is seen, as compared with the previous Groups of heavy transition elements. The highest state found is VI, while the most stable states are iridium(IV) and (III) and

rhodium(III). In the separation from the rest of the platinum metals, Figure 15.26, both elements are found in the fraction insoluble in aqua regia. Iridium remains as the insoluble residue, after alkaline fusion, while rhodium is extracted in aqueous solution as sulfate after fusion with sodium bisulphate.

The highest oxidation state is represented by the hexafluorides, MF_6, formed by direct reaction. As in the previous Group, the heavier element is more stable in the high oxidation states and IrF_6 is much more stable than the rhodium fluoride. However, the highly reactive RhF_6, formed by high temperature and pressure fluorination of RhF_5, is thermally stable at room temperature in passivized nickel vessels. No oxygen species exist in the VI state: the reported IrO_3 is possibly a peroxide rather than an iridium(VI) compound.

The V state is also unstable and is found only in the IrF_6^- ion, whose salts are prepared by fluorinating mixtures of iridium trihalides and alkali halides in bromine trifluoride solution. Mixed fluoride-chloride Ir(V) complexes are made similarly with up to 3Cl, all *cis*. The first Ir(V) oxygen species was found in $KIrO_3$ and related salts. In strong base, the higher coordinated IrO_6^{7-} ion is formed. As with Re, Os, and other elements, we see the formation of oxyanions in the same oxidation state with two different oxygen coordination numbers. As for ruthenium, stable organometallic compounds of iridium are available using mesityl as ligand. The iridium(V) species is the tetrahedral anion $[Ir(mes)_4]^+$, while the neutral species $[Ir(mes)_4]$ and $[Ir(mes)_3]$ form in the IV and III states. The latter, like $[Rh(mes)_3]$, is pyramidal with CMC angles around 108°. The hydride $IrH_5(PR_3)_2$ is formally Ir(V) and extends the heavy metal polyhydride series. It readily transfers H, e.g. to $Pt_2Cl_4L_2$.

The IV state is fairly stable for iridium but represented by only a few rhodium compounds. The products formed on igniting the metals in air are IrO_2 and Rh_2O_3, respectively, and RhO_2 can be made only by strong oxidation of rhodium(III) solutions. It can be dehydrated without reverting to rhodium(III) only by heating under a high pressure of oxygen. IrO_2 with alkali metal bases gives black Cs_4IrO_4 or red Na_4IrO_4. The crystal structures show the presence of square planar IrO_4^{4-} ions. Both elements form the tetrafluoride, MF_4, and $IrCl_4$ is also reported. Rhodium gives only the halogen complexes, RhF_6^{2-} and $RhCl_6^{2-}$, but a wider variety of iridium(IV) complexes occur. Fairly stable hexahalides, IrX_6^{2-}, are formed for X = F, Cl, and Br, and other complexes include the oxalate, $Ir(C_2O_4)_3^{2-}$, which can be resolved into optical isomers.

In the III state both elements form all four trihalides and the oxides, M_2O_3. Ir_2O_3 can only be formed, by addition of alkali to iridium(III) solutions, as a hydrated precipitate in an inert atmosphere. Attempts to dehydrate it lead to decomposition, and it oxidizes in air to IrO_2. Rhodium trifluoride results, along with some tetrafluoride, from fluorination of $RhCl_3$, but IrF_3 can only be prepared by reduction of IrF_6 with Ir, as fluorination of iridium(III) compounds gives the tetra- or hexa-fluoride. The other trihalides are made by direct reaction with halogen. They are insoluble in water and probably polymeric.

A considerable number of rhodium(III) complexes are known, and these are generally octahedral. The hydrate, $Rh(H_2O)_6^{3+}$, the ammine, $Rh(NH_3)_6^{3+}$, and a large variety of partially substituted hydrates and ammines are found. For example, hydrates containing 1, 2, 3, 4, and 5 chloride ions in place of water molecules are formed, as well as $RhCl_6^{3-}$ and other hexahalides. Other halide complexes include RhX_5^{2-}, and RhX_7^{4-}, especially for the bromides. The dinuclear complex halides, $Rh_2X_9^{3-}$, have the face-sharing octahedral structure $[Cl_3Rh(\mu - Cl)_3RhCl_3]^{3-}$.

A similar structure is found for the iridium chloro- and bromo-analogues. An interesting example of a mixed metal-metal bond is provided by $(R_3As)_3Rh^{III}(HgCl)^+Cl^-$, which contains the Rh – Hg bond. A similar compound is the dimeric iridium (III) species $[Ir_2Cl_6(SnCl_3)_4]^{4-}$ which contains Ir – Sn bonds. Rhodium also forms a compound with $SnCl_3^-$ as a ligand, this time in the I state but again a dimer, $[Rh_2Cl_2(SnCl_3)_4]^{4-}$. In such compounds, the $SnCl_3^-$ species behaves rather like a halide ion. Such compounds with bonds from transition to non-transition metals serve to link the long-established Main Group metal-metal bonded compounds with the newer field of interest in bonds between transition metal atoms.

The complexes of rhodium(III) are relatively inert to substitution so that isomers may be isolated, as in the resolution into optical isomers of the cis-$Rh(en)_2Cl_2^+$ ion. There is also an extensive series of iridium complexes which are generally similar to the rhodium compounds. Iridium also gives the hydrides, $(R_3P)_3Ir(H)_n(Cl)_{3-n}$, for $n = 1, 2$, and 3. Rhodium forms the analogues with $n = 1$ and 2. The related carbonyls, $(R_3P)_2Rh(H)(Cl)_2(CO)$ and $(R_3P)_2Ir(H)_3(CO)$ are known.

These elements form only a few compounds in the II state. Simple compounds are restricted to polymeric $IrCl_2$ and the reported oxide, RhO, which is not well-established. There is a slowly growing number of simple complexes such as $[Ir(CN)_6]^{4-}$ and $Ir(NH_3)_4Cl_2$. An important binuclear type is the rhodium(II) carboxylate $[Rh_2(OOCR)_4]$ which acts as a catalyst in a number of applications. The molecules have the same structure as the Cr or Cu analogues, see Figures 14.20 or 14.33, but without the extra water molecules. As well as the acetate (R = Me), the fluoroacetate (R = CF_3) is particularly stable. Compounds where the carboxylate is optically active are particularly interesting as these act as enantioselective catalysts.

Low oxidation state compounds include the carbonyl monohalides of the I state, and the carbonyl anion, $Rh(CO)_4^-$, in the – I state. These are paralleled by trifluorophosphine analogues like $KIr(PF_3)_4$ which is oxidized from the – I state to the + I state by iodine to yield $Ir(PF_3)_4I$. Compounds in the 0 state include $Ir(NH_3)_5$ and $Ir(en)_3$ which are formed by reduction in liquid ammonia, like the similar osmium compounds. The 0 state is also represented in the carbonyls which include $M_2(CO)_8$ and the interesting polymeric molecule $Rh_6(CO)_{16}$. This has a structure with an octahedron of rhodium atoms, each with two terminal carbonyl groups, and the other four carbonyl groups are found in the middle of opposite faces. The structure is held together by delocalized metal–metal bonding in the Rh_6 cluster.

Among the most fully explored reactions are those of the square planar compounds of the I state, which add neutral molecules to give octahedral compounds of the III state in the process which has been called $oxidative\ addition$. For example,

$$IrCl(CO)(PPh_3)_2 + XY \rightarrow XIrY(Cl)(CO)(PPh_3)_2$$

where XY = HCl, CH_3I, Cl – HgCl, etc. Molecules such as H_2, O_2 or SO_2 are also taken up and may be lost again in a reversible reaction. While H_2 gives two M – H bonds, the O_2 molecule is not split and the two O atoms are cis to each other at a distance O – O = 130 pm, suggesting coordination as superoxide, O_2^-.

A very important application of oxidative addition is in catalysis. The Rh(I) complex $[Rh(CO)_2I_2]^-$ catalyses the carbonylation of CH_3OH to acetic acid in the Monsanto process, shown in Figure 15.28, via an oxidative addition (step 1) of CH_3I, forming $[Rh(CO)_2(CH_3)I_3]^-$ as an intermediate. One of the CO groups then inserts into the Rh–CH_3 bond in step 2, to give an Rh–C(O)CH_3 (acetyl) group. After the addition of another CO ligand in step 3 to regenerate the six-coordinate complex $[Rh(COCH_3)(CO)_2I_3]^-$, this then undergoes a reductive elimination of $CH_3C(O)I$ in step 4, regenerating the original $[Rh(CO)_2I_2]^-$ catalyst. The cycle is completed by the $CH_3C(O)I$ reacting with water in step 5, producing the product CH_3COOH, together with HI. Reaction of the HI with methanol feedstock produces the methyl iodide required for the oxidative addition reaction in step 1, thus continuing the catalytic cycle. Development of this work has led to similar Rh catalysts for other reactions and modification of the catalyst, for example by bonding to an insoluble polymer backbone, to attempt to improve the control and activity. Square planar Rh(I) species with at least one CO group, usually a halide, and other ligands like phosphanes, are the compounds most studied. Recent work using low-temperature nmr spectroscopy has been successful in detecting the intermediate complexes, such as $[Rh(CO)_2(CH_3)I_3]^-$, which are produced transiently in very small amounts in the reaction mixture.

In a different type of reaction, of potential catalytic interest, hydride complexes of Ir and Rh, of the type $(Me_5C_5)ML_2$. (where L = CO, 2H, phosphane or similar ligands), activate the usually inert C – H bond. An organic group R – H adds to the metal forming R – M – H with elimination of L. This is seen as the first step in a potentially very important controlled conversion of hydrocarbons to functional organic compounds, and it is particularly important that such activation has been established for CH_4, as natural gas is a premium starting material for chemical synthesis.

The I state is also found in five-coordinate compounds, usually involving phosphane ligands. The structure of

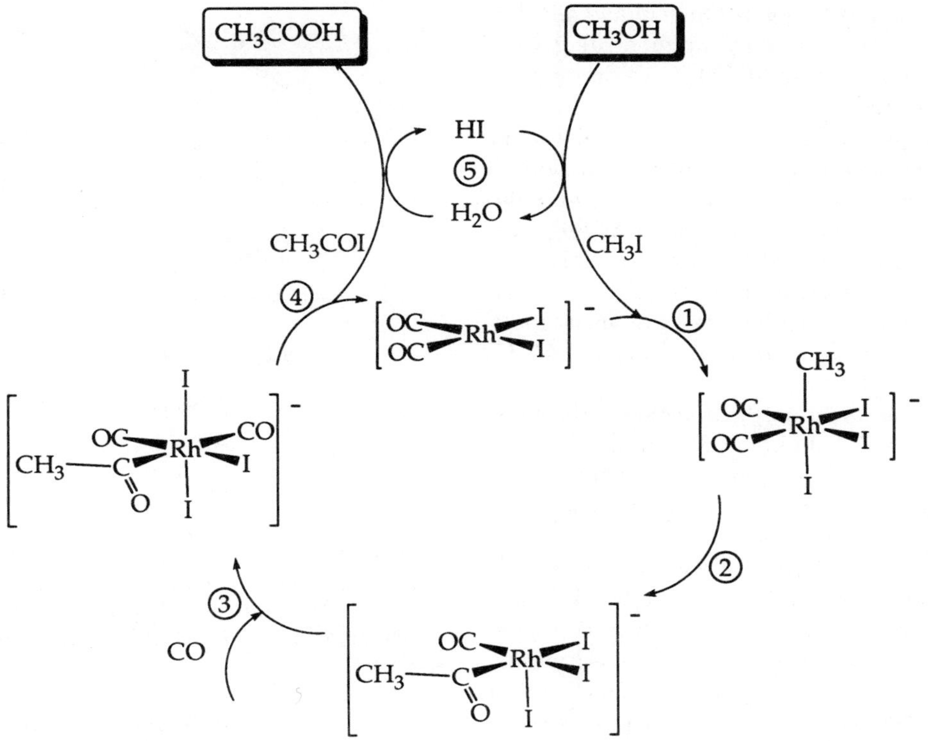

FIGURE 15.28 *The Monsanto process for the manufacture of acetic acid by carbonylation of methanol*

HRhCO(PPh$_3$)$_3$ is trigonal bipyramidal, with the P atoms in the equatorial plane.

15.8 Palladium, 4d^{10}5s^0, and platinum, 5$d^9$6s^1

This pair of elements continues the trends already observed. The higher oxidation states are unstable and the VI and V states are represented only by a few platinum compounds. The IV state is stable for platinum and well-represented for palladium. This corresponds to d^6 and many octahedral Pt^{IV} complexes occur with high CFSE. The II state is the other common oxidation state. In this palladium and platinum are almost invariably four-coordinated and square planar. Palladium continues the trend, which probably starts at technetium or molybdenum and is clearly seen for ruthenium and rhodium, in being less stable than the heaviest element of the Group in the higher oxidation states and in having a well-developed lower oxidation state.

Platinum metal is particularly important as a catalyst, both in the laboratory and in general. One of the best-known uses is that of Pt in car exhaust converters to reduce the proportion of pollutants (such as nitrogen oxides) emitted. Major applications in industry include the hydrogenation of benzene to cyclohexane, the dehydrogenation of petroleum hydrocarbons to aromatics and, in conjunction with rhodium, the conversion of ammonia to nitric acid. This last is one case of increasing use of 'bimetallic' catalysts which are usually superior to the individual metal. Usually the metal is dispersed as fine particles on a support such as a alumina with

a high surface area. Another important application of platinum complexes is in cancer chemotherapy where the complex *cis*-PtCl$_2$(NH$_3$)$_2$ is very widely used for treating certain types of cancers and this is discussed in greater detail in Chapter 20.

15.8.1 *The* (*VI*), (*V*), *and* (*IV*) *oxidation states*

Only platinum forms a hexafluoride, PtF$_6$, and attempts to isolate the palladium compound have failed. The VI state of platinum may also occur in the reported oxide, PtO$_3$. There is also a compound of unknown structure which may contain Pt(VIII). This is PtF$_8$(CO)$_2$ which is reported to be formed by the reaction of CO under pressure on PtF$_4$. It is difficult to see why such a system should be oxidizing, but spectroscopic evidence shows no bridging carbonyl groups and no adduct molecules such as F$_2$CO.

The V state is found in PtF$_5$ and in the PtF$_6^-$ ion. The latter was first found as a product of the reaction of O$_2$ and PtF$_6$ which yielded the unexpected oxygen cation in O$_2^+$PtF$_6^-$. The now famous first report of a rare gas compound was of the Xe$^+$PtF$_6^-$ complex and a number of other compounds of the anion have since been made.

The highest state for palladium, and the first stable state for platinum, is the IV state. This is represented by all four PtX$_4$ halides and by PtO$_2$. Palladium forms PdF$_4$ and PdO$_2$. The latter is found as the poorly characterized hydrated oxide but PtO$_2$ is the most stable oxide of platinum. It is obtained as a hydrated precipitate from the action of carbonate on Pt^{IV} solution and it is soluble in acid and alkali in this condition. It can be dehydrated by careful heating when it becomes insoluble. On heating to 200 °C, it decomposes giving platinum metal and O_2.

PtF_4 is the major product of fluorination of platinum, although PtF_5 and PtF_6 also result from the direct reaction. PdF_4 is also formed by direct fluorination though here the main product is PdF_3. The other platinum tetrahalides are formed by direct halogenation and $PtCl_4$ also results when H_2PtCl_6, the product from the aqua regia solution of the metal, is heated. These heavier halides of platinum are quite stable, even PtI_4 does not decompose until about 180 °C, when it goes to PtI_2 and iodine.

Palladium(IV) complexes are only a little more stable than the simple Pd^{IV} compounds. The common examples are the halides. PdX_6^{2-}, all of which are known except the iodide, and the tetrahalide amine, $Pd(amm)_2X_4$, where amm = ammonia, pyridine or related ligands.

By contrast, platinum(IV) forms a large number of complexes which are always octahedral, and are stable and inert in substitution reactions. As platinum(II) also gives a wide range of complexes, platinum is probably the most prolific complex-forming element of all. All the common types of complex are found, for example all members of the set between $Pt(NH_3)_6^{4+}$ and PtX_6^{2-} are known for a variety of amines as well as ammonia, and for X = halogen, OH, NCS, NO_2, etc. However, fluoride is found only in PtF_6^{2-}.

One very interesting Pt(IV) complex is the sulfide $Pt(S_5)_3^{2-}$ which contains three S_5^{2-} units forming PtS_5 six-membered rings by linking into *cis* positions in the octahedron. As this species with three bidentate ligands has non-superimposable mirror images, the optical enantiomers may be resolved (compare $Co(en)_3$ in section 13.8: also section 19.2 for metal–sulfur ring compounds).

The Pt^{IV} octahedral complexes may readily be obtained from the Pt^{II} square planar complexes if the attacking ligand is also an oxidizing agent. Thus $Pt^{II}L_4 + Br_2 \rightarrow$ *trans*-$Br_2Pt^{IV}L_4$ for a variety of ligands (L). The halogen atoms simply add on opposite sides of the square plane.

An interesting class of complexes is the deeply coloured, usually green, type of compound which apparently contains trivalent platinum, such as $Pt(en)Br_3$ or $Pt(NH_3)_2Br_3$. These are not Pt^{III} compounds at all but chains made up of alternate $Pt^{II}(NH_3)_2Br_2$ and $Pt^{IV}(NH_3)_2Br_4$ units as shown in Figure 15.29. The Pt–Br distances on the vertical axis are 250 pm for Pt^{IV} – Br and 310 pm for the weak Pt^{II} – Br inter-action.

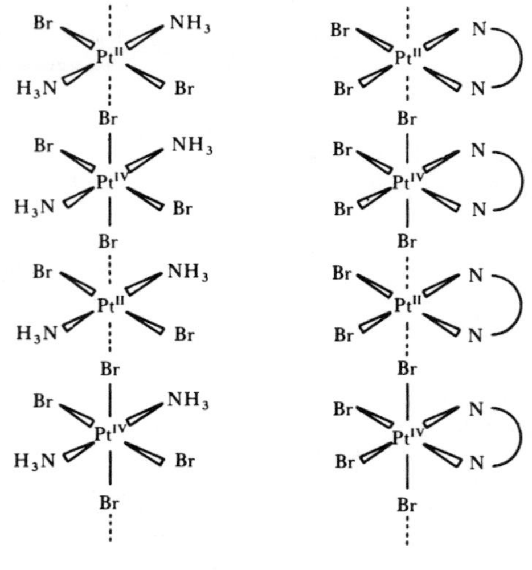

15.8.2 The (III), (II), and lower oxidation states

As in the example above, many compounds which appear to be in the (III) state are instead mixed valency ones. Thus PdF_3 is $Pd^{2+}PdF_6^{2-}$ and similarly for PtX_3. The interesting Pt_3I_8 consists of two octahedral $Pt(IV)I_6$ units linked to a central square planar $Pt(II)I_4$ by shared edges. Hydrated oxides, M_2O_3, are reported but their identity is not proven.

However, Pd(III) was confirmed in the complex ions

FIGURE 15.29 *Structures of* (a) $Pt(NH_3)_2Br_3$, (b) $Pt(en)Br_3$
These structures consist of chains of alternate square planar platinum(II) units, PtN_2Br_2, and octahedral platinum(IV) units, PtN_2Br_4. (N = NH_3 or half an **ethylenediamine molecule**).

$Na^+[PdF_4]^-$, and Ag_3PdF_6. In NaK_2PdF_6, a structural study shows that the PdF_6^{3-} ion has four shorter Pd – F bonds of 195 pm, and two longer ones of 214 pm, just as expected for a low spin d^7 state. Pt(III) has been observed by electron spin resonance after X-irradiation of a Pt(II) complex, $[Pt(en)_2]^{2+}$, in the presence of a large anion.

In a second class of formally + III compounds, a Pd – Pd or Pt – Pt bond exists in a dimer in compounds of the type X-$Pt(bident)_n$ Pt – X. Here bident = a bidentate ligand which bridges the two Pt atoms, holding them close enough together for a Pt – Pt bond to form. The M(III) compounds are made when the bridged Pt(II) complexes are treated with X_2 (compare the analogous gold compounds, Figure 15.32). There may be between n = 2 and n = 4 bidentate groups present. Such Pt(III) compounds are well-established and a few Pd analogues are known.

Like nickel, these elements have a very stable II state, but occur only in the square-planar configuration in this state. All the dihalides except PtF_2 are found. Palladium forms a stable PdO while platinum gives a hydrated PtO which readily oxidizes to PtO_2. Palladium dichloride has a chain structure of linked square planar $PdCl_4$ units:

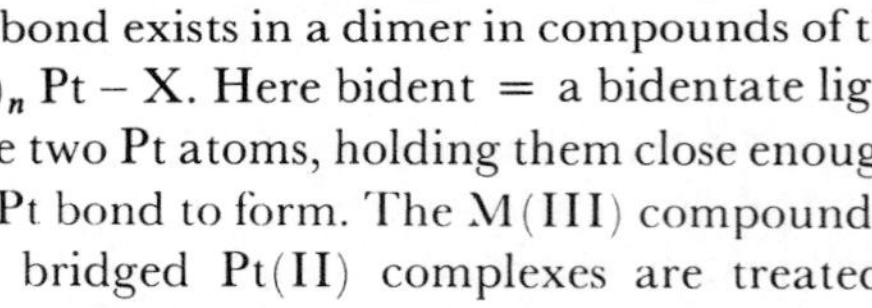

A second form of palladium dichloride, and the only form of the platinum compound, consists of M_6Cl_{12} units with the chlorine atoms placed above the edges of an octahedron of platinum atoms. This structure is reminiscent of the tantalum(II) and molybdenum(II) halide complex ions and is a further example of a metal cluster compound.

The complexes of palladium(II) and platinum(II) are

abundant and include all common ligands. Some examples have already been given in Chapter 13 (see Figure 13.18). Palladium(II) complexes are a little weaker in bonding and react rather more rapidly than the platinum(II) complexes, but are otherwise very similar. The commonest donor atoms are nitrogen (in amines, NO_2), cyanide, the heavier halogens and phosphorus, arsenic, and sulfur. The affinity for F and O donors is much lower, but the stability of Pt(II) is underlined by the isolation from nitric acid of $Pt(NO_3)_4^{2-}$ where the nitrate groups are bonded through a single O forming a square planar PtO_4 coordination shell. Another example is provided by the planar PtH_4^{2-} ion, a red-violet compound synthesized by the action of NaH on Pt under hydrogen. This is a further case of the polyhydride ions remarked in earlier sections.

While $Pt(NH_3)_4^{2+}$ ions are colourless in solution and $PtCl_4^{2-}$ is red, the salt $[Pt(NH_3)_4][PtCl_4]$—named *Magnus's green salt*—is strongly coloured. The square-planar units stack, alternate cation and anion, directly above each other in the crystal with a Pt—Pt distance of 325 pm. Similar species with small ligands which allow short Pt—Pt distances are also abnormally coloured, betokening a metal-metal interaction in the crystal. Such species have aroused interest as possible 'one-dimensional conductors'. If some electrons are removed, as in the partly oxidized compound $K_2Pt(CN)_4 \cdot 0 \cdot 3Br$, the crystals become strongly conducting but only in the direction parallel to the Pt chains.

A further interesting square-planar platinum II compound is Zeise's salt, one of the first organometallic compounds (see Figure 16.5a). Similar ethylene complexes are formed with a variety of other ligands and for Pd as well. The dimeric species $[C_2H_4PdCl_2]_2$ rapidly gives CH_3CHO and Pd in water. When linked with $CuCl_2$ to reoxidize the Pd, this becomes the basis of the *Wacker process* for converting ethylene to acetylene:

$$C_2H_4 + PdCl_2 + 3H_2O \rightarrow Pd + CH_3CHO + 2H_3O^+ + 2Cl^-$$
$$Pd + 2CuCl_2 \rightarrow PdCl_2 + 2CuCl$$
$$2CuCl + 2H_3O^+ + 2Cl^- + \tfrac{1}{2}O_2 \rightarrow 2CuCl_2 + 3H_2O$$

which adds up to $C_2H_4 + \tfrac{1}{2}O_2 \rightarrow CH_3CHO$. The process is viable as the reaction between Pd and $CuCl_2$ is quantitative and efficient, so only low Pd concentrations are required.

One or two examples do exist of coordination numbers other than four in the II complexes: $Pd(diars)_2Cl^+$ for example is a trigonal pyramid and $Pt(NO)Cl_5^{2-}$ is octahedral (though the latter might be formulated as an NO^- compound of Pt^{IV} rather than as the NO^+ compound of Pt^{II}). The $SnCl_3^-$ complexes also provide examples. Platinum(II) forms a square-planar ion $PtCl_2(SnCl_3)_2^{2-}$ (as does ruthenium) but it also gives the ion $Pt(SnCl_3)_5^{3-}$ which has a trigonal bipyramidal $PtSn_5$ skeleton. This species reacts with hydrogen under pressure to form the hydride ion $HPt(SnCl_3)_4^{3-}$. Palladium forms the ion $PdCl(SnCl_3)_2^-$, but it is not yet known if this is a monomer or forms a chloride-bridged dimer which would be square-planar around the palladium. Such compounds are commonly precipitated as salts of very large cations (like Ph_4As^+) from acid chloride

solutions containing palladium, or platinum, and tin. Compounds with analogous M—Ge bonds, such as $(R_3P)_2M(GePh_3)_2$, are also well-known.

The I state is uncommon and the established examples are M—M bonded dimers. This probably occurs in the reported $[Pd_2(CN)_6]^{4-}$ and has been found crystallographically in $[(Me_3P)_3Pd-Pd(PMe_3)_3]^{2+}$. The structure shows a square planar arrangement around each Pd, with the two planes mutually perpendicular. On heating to $100\,°C$, a Me group cleaves from a phosphane and the Pd(II) complex $[(Me_3P)_3PdMe]^+$ is formed. Platinum(I) is found in the Pt—Pt bonded carbonyl anion dimer, $[PtCl_2(CO)]_2^{2-}$, again with square planar coordination.

In the 0 state, there are no stable carbonyls analogous to $Ni(CO)_4$, though $M(CO)_n$ species are formed in low-temperature matrix studies. The isoelectronic $M(CN)_4^{4-}$ ions, and $M(PF_3)_4$ species, do exist and CO-containing molecules occur such as $Pt(CO)_n(PR_3)_{4-n}$ $(n = 1, 2)$. There are also reports of $Pt(NH_3)_5$ and $Pt(en)_2$ analogous to the iridium compounds. It has also been shown that the phosphane complexes, $(Ph_3P)_4$Pt and $(Ph_3P)_3$Pt, are derivatives of Pt^0 rather than hydrides of higher states as originally reported. The structure of $Pt(PPh_3)_3$ has been shown to contain the planar PtP_3 group.

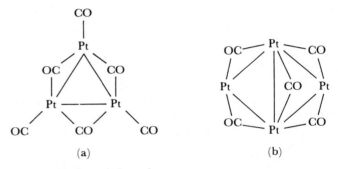

FIGURE 15.30 *Some platinum clusters*
(a) The basic building unit of $[Pt_3(CO)_6]_n$ (known for n up to 12),
(b) The $Pt_4(CO)_5$ skeleton of $[(R_3P)Pt]_4(CO)_5$.

Platinum forms a number of irregular clusters containing carbonyls. The family of polymetallic 2-carbonyl anions of the general type, $[Pt_3(CO)_6]_n^{2-}$, are built up of the basic unit shown in Figure 15.30a. The Pt_3 triangles stack up above each other in twisted trigonal prisms, and species with n up to 12 have been characterized. The cluster of Figure 15.30b is found in $Pt_4(PR_3)_4(CO)_5$. The four Pt atoms, each with a terminal PR_3, form a tetrahedron which has five out of the six edges bridged by CO groups. See also section 16.8.

15.9 Silver, $4d^{10}5s^1$, and gold, $5d^{10}6s^1$

The major part of silver chemistry is that of the d^{10} state, Ag(I). Gold is most often found in Au(III) square planar complexes, but there are also a variety of Au(I) compounds, mostly complexes. Simple compounds of gold readily give the element, and silver is also reduced to the element fairly readily. These elements differ quite widely in their chemistries, and also differ from copper which occurs in the II state. Reasons for this were discussed in the copper section (14.9).

The elements are found uncombined and in sulfide and arsenide ores. They may be recovered as cyanide complexes which are reduced to the metal, in aqueous solution, by the use of zinc. Gold is inert to oxygen and most reagents but dissolves in HCl/HNO_3 mixture (aqua regia) and reacts with halogens. Silver is less reactive than copper and is similar to gold, except that it is also attacked by sulfur and hydrogen sulfide.

It is easiest to treat the two elements separately as there is little resemblance in their chemistries.

15.9.1 Silver

There are a limited number of examples of Ag^{III} compounds. Oxidation in basic solutions gives $[Ag(IO_6)_2]^{7-}$ or $[Ag(TeO_6)_2]^{9-}$, and copper(III) compounds are obtained similarly. Fluorination of a mixture of alkali and silver halides gives $M^I AgF_4$ and red diamagnetic AgF_3 may be precipitated by addition of a fluoro Lewis base to $[AgF_4]^-$ solutions in anhydrous HF:

$$AgF_4^- + L \rightarrow AgF_3 + LF^-$$

(examples of L include BF_3, PF_5 or AsF_5)

AgF_4^- also reacts with the silver(II) ion, AgF^+, to give the maroon mixed-oxidation state compounds $Ag^{II}Ag^{III}F_5$ which, with AgF_3, gives $Ag^{II}Ag_2^{III}F_8$. This latter compound is also formed by the thermal decomposition of AgF_3.

Electrolytic oxidation gives Ag_2O_3 which has the same structure as Au_2O_3, i.e. square planar MO_4 units linked by shared corners into a network. A similar oxidation in base yielded square planar $[Ag(OH)_4]^-$. The mixed Ag(II)–Ag(III) oxide, Ag_3O_4, has also been isolated from anodic oxidation work. All silver(III) compounds are strongly oxidising.

When Ag_2O is oxidized by persulfate, black AgO is obtained. This is a well-defined compound which is strongly oxidizing. It is diamagnetic, which excludes its being an Ag^{II} compound which would contain the paramagnetic d^9 configuration. A neutron diffraction study has shown the existence of two units in the lattice, linear $O-Ag-O$ and square planar AgO_4 groups. This strongly suggests a formulation as $Ag^I Ag^{III}O_2$. The square planar configuration is expected for the d^8 Ag^{III} while the linear one is common for d^{10} as in Ag^I (compare Ag^I with Cu^I or Au^I which both show linear configurations and compare the III state with Au^{III}). AgO is stable to heat up to 100 °C and it dissolves in acids giving a mixture of Ag^+ and Ag^{2+} in solution and evolving oxygen. Reaction of Cl_2 with $CsAgCl_2$ gives the analogous mixed-oxidation chloride $Cs_2Ag^I Ag^{III}Cl_6$. The structure shows square planar Ag^{III} (Ag–Cl = 221 pm) and linear Ag^I(Ag–Cl = 227 pm).

The only simple compound of silver(II) is the fluoride, AgF_2. This is formed from the action of fluorine on silver or AgF. It is a strong oxidizing or fluorinating agent and can be used in reversible fluorinating systems in a similar way to CoF_3.

Silver ions of the II state can exist in solution but only transiently. They are produced by ozone on Ag^+ in perchloric acid. The potential for $Ag^{2+} + e^- = Ag^+$ has been measured as 2·00 V in 4M perchloric acid. This makes Ag^{2+} a much more powerful oxidizing agent than permanganate or ceric. Silver also gives a number of complexes in the II state. These are usually square planar and paramagnetic corresponding to the d^9 state, and known structures, for example, of $Agpy_4^{2+}$, are isomorphous with the Cu^{II} analogues. Other examples include the cations $Ag(dipy)_2^{2+}$ and $Ag(phen)_2^{2+}$. These cations are stable in the presence of non-reducing anions such as nitrate, perchlorate or persulfate. The interesting species $Ag(MF_6)_2$ (M = Nb or Ta) consists of a central AgF_6 octahedron sharing three cis F with MF_6 octahedra. The AgF_6 octahedron has two long and four short Ag – F bonds consistent with d^9 Ag(II).

The normal oxidation state of silver is Ag^I and the chemistry of this state is already familiar, for example, from the precipitation reactions of Ag^+ in qualitative analysis. Salts are colourless (except for anion effects) and generally insoluble apart from the nitrate, perchlorate, and fluoride. Structures are usually similar to those of the alkali metal equivalents, for example, the silver halides have the sodium chloride structure. As discussed in Chapter 5, their lattice energies are higher because the Ag^+ ion is more polarizable, hence the lower solubilities.

One very striking structure containing Ag is provided by the central core of the $[Ag_4(S-R-S)_3]^{2-}$ ion (R = o-xylyl) which contains an octahedron of the 6 S atoms with the 4Ag atoms in the centres of alternate faces, so that the Ag is planar and 3-coordinated.

A highly important application of Ag(I) salts is in photography. The basic process is the activation by light of Ag centres in the AgBr film, and these nucleate the formation of Ag metal particles when reduction (development) occurs. The unreacted salt is removed (fixing) by forming soluble Ag^+ complexes, typically sulfur species. Colour photography is based on the same general process with the interposition of dyes to filter out the three primary colours.

Silver (I) forms a wide variety of complexes. With ligands which do not π-bond, the most common coordination is linear, two-coordination as in $Ag(NH_3)_2^+$, while π-bonding ligands give both 2- and 4-coordinate complexes and 3-coordination is found for some strongly π-bonding ligands such as phosphanes.

A number of very interesting complexes of silver(I) (and other late transition metals) with alkyl halides have been described recently, as detailed in the review given in Appendix A. Solvents such as dichloromethane and 1,2-dichloroethane are often used in synthetic inorganic chemistry because of their non-coordinating nature. However, the recent isolation of complexes of the type $[Ag(CH_2Cl_2)(OTeF_5)]$ and $[AgNO_3(CH_2I_2)]$, containing coordinated alkyl halides, clearly points to the term 'non-coordinating' now being of limited relevance. The isolation of such complexes is even more remarkable given that the silver(I) ion is a very powerful halide-abstracting group and the isolated halocarbon complexes may therefore be thought of, in some respects, as 'frozen intermediates' in the abstraction process.

Silver also has the distinction of being one of the few transition metals which, until recently, did not form an isolable carbonyl complex, although $[Ag(CO)_n]^+$ species can be detected in zeolite hosts or in strongly acidic solutions. In 1991 the complex $[Ag(CO)\{B(OTeF_5)_4\}]$ was reported and its carbonyl stretching frequency of $2204\ cm^{-1}$ is one of the highest for any known metal carbonyl. This indicates that the degree of backbonding from filled silver d-orbitals to CO π^* orbitals is very small (see section 16.2.2 and Figure 16.2 for a discussion of the bonding in metal carbonyls). The $[Ag(CO)_2]^+$ and $[Ag(CO)_3]^+$ ions have also been recently reported (the latter under a high pressure of CO) with similar carbonyl stretching frequencies of 2198 and $2192\ cm^{-1}$, respectively.

It is noteworthy that ions containing the poorly coordinating $OTeF_5$ group have been successfully employed in the isolation of all of these unusual halocarbon and carbonyl silver species, and this suggests that a range of other novel complexes should be accessible using this type of group. The reader is referred to sections 17.5.5 and 17.7.3 for further chemistry involving the $OTeF_5$ group.

A formal 1/2 oxidation state is found in Ag_2F. This has a layer lattice of the anti-CdI_2 type and the $Ag-Ag$ distance is similar to that in silver metal, which accounts for the low formal oxidation state. The properties are metallic.

15.9.2 Gold

The highest oxidation state found for gold is the (V) state, which has the d^6 configuration. Thus Au(V) corresponds to Pt(IV). Powerful oxidation of AuF_3 with F_2 and XeF_2 gives AuF_6^- as the $Xe_2F_{11}^+$ salt (compare section 17.9). The AuF_6^- ion is a slightly distorted octahedron linked by long weak $Au-F\cdots Xe$ bonds to the two Xe atoms of the $Xe_2F_{11}^+$ ion. Other cations, including the larger alkali metals and O_2^+, stabilise the AuF_6^- ion. The KrF_2AuF_5 species has a stronger bridge and is best seen as $FKr{-}FAuF_5$ with a more covalent interaction. AuF_5 is formed by gentle heating of the krypton compound. The pentafluoride is extremely reactive and a violent fluorinating agent. It decomposes to AuF_3 and F_2 at $200\,°C$ and forms $HAuF_6$ which melts at $88\,°C$. Among unusual adducts are $BrF_5.AuF_5$ which is oxidized by KrF_2 to $BrF_7.AuF_5$ although free BrF_7 does not exist (section 17.8).

Gold(III) is the most common oxidation state of gold in compounds. The interrelations are illustrated in Figure 15.31 starting off from the solution in aqua regia. The simple compounds readily revert to the element but the complexes are more stable. Square planar four-coordination is the most common shape. $AuCl_3$ and $AuBr_3$ form planar bridged dimers (as R_2AuX below with $R = X$) while AuF_3 forms a polymer made up of *cis*-bridged AuF_4 units—i.e. $(AuF_2F_{2/2})_n$. There is evidence for a few six-coordinate complexes such as $AuBr_6^{3-}$. Examples of the tendency to four-coordination are provided by the alkyls such as R_2AuX. These contain sigma $Au-C$ bonds which are among the

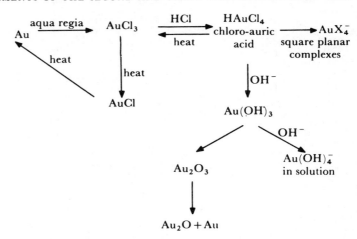

FIGURE 15.31 *Reactions of gold(III)*

most stable metal–carbon bonds in the transition block. The halides have dimeric structures:

$$R_2AuX_2AuR_2 \text{ (bridged dimer)}$$

while, when $X = CN$, tetrameric structures are found:

$$
\begin{array}{c}
R_2Au-CN-AuR_2 \\
| \qquad\qquad | \\
\left(\begin{array}{c}C\\N\end{array}\right) \qquad \left(\begin{array}{c}C\\N\end{array}\right) \\
| \qquad\qquad | \\
R_2Au-CN-AuR_2
\end{array}
$$

Formal Au(II) results from $Au-Au$ bonding in the same way as Pt(III) in the last section (see Figure 15.32). Ligands with two donor atoms close together, such as $Ph_2PCH_2PPh_2$ bridge two Au atoms holding them close together. Then stepwise oxidation by successive additions of halogen converts the Au from (I) through (II) with the Au–Au bond to (III). However, relatively few examples of gold(II) complexes have been prepared, and good sigma donors and pi acceptors (such as dithiolate ligands) are required to stabilize the mononuclear, paramagnetic (d^9) gold(II) centre. The Au^{2+} ion has recently been generated in a solid matrix by heating gold(III) fluorosulfate, $Au(OSO_2F)_3$, or in solution by the reduction of the same compound with gold metal in $HOSO_2F$ solution.

Gold(I) is less stable but forms the simple halides and Au_2O. AuI and AuCl have similar chain structures

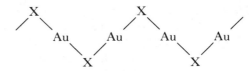

Complexes are usually linear, e.g. $[AuX_2]^-$ (X = halogen, CN etc.) or R_3PAuX. X includes CH_3 as well as halides etc.

The R_3PAu^- group itself forms a range of complexes, often

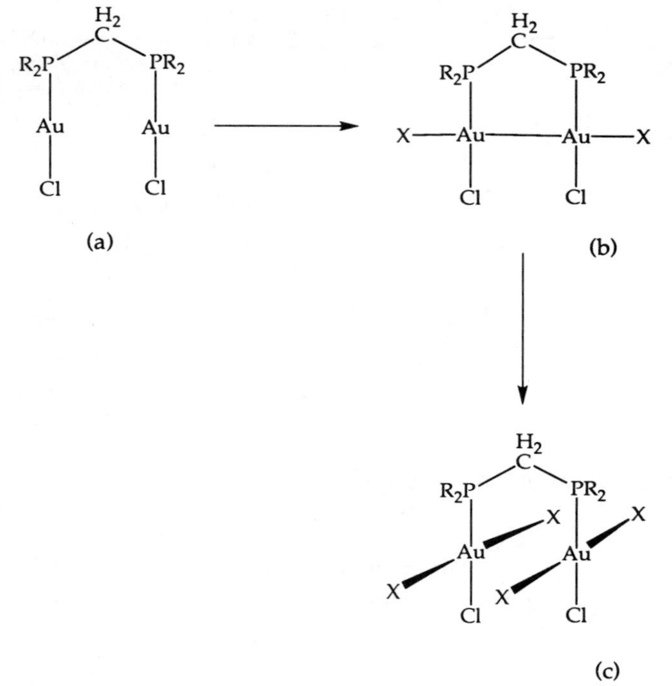

FIGURE 15.32   *Transformations of the* $R_2P-CH_2-PR_2$ *bridged digold compound*

The short 'bite' of the bidentate phosphane holds the two Au atoms close together in a rigid conformation.

(a) Two gold(I) centres linked through the diphosphane. (b) Oxidation forms the Au–Au bond giving gold(II). (c) Further oxidation gives two gold(III) centres linked through the diphosphane.

of high coordination number, for example $R_3PAuV(CO)_6$ or $(R_3PAu)_3Re(CO)_4$. The long-known carbonyl Au(CO)Cl can now be readily made and is a valuable source of Au(I) complexes since the CO is very weakly bonded. Planar three-coordinate Au(I) is found in $(R_3P)_2AuCl$ and $[(R_3P)_3Au]^+$.

In an unusual complex containing a linear chain of 5 Au atoms, a central Au(I) is bonded to two Au–Au units, similar to those of Figure 15.29b, which are formally Au(II). The Au(II)–Au(II) distance is 276 pm while the Au(II)–Au(I) value is 264 pm.

Gold(I)–sulfur compounds are of interest as photosensitizers of silver halide layers in photography. The hydrogensulfido complex $[Au(SH)_2]^-$ has recently been prepared. It is suggested that this species is responsible for the transport of considerable quantities of gold in hydrothermal ore solutions.

Au has a relatively high electron affinity (compare Table 2.9) and Au(-I) has the $d^{10}s^2$ configuration which is fairly stable as found in Hg(0) (compare relativistic effects, section 16.13). A very striking consequence is to have evidence for the Au$^-$ ion. The first indications appeared half a century ago, particularly the observation that CsAu has relatively little metallic character and is best formulated as a salt, Cs$^+$Au$^-$, containing the *auride* Au$^-$ anion, rather than as an alloy. In line with this the compound crystallizes in the cesium chloride structure, rather than the more densely

close-packed structures which are usually adopted by metals. However, there has been recent discussion as to the relative contributions of ionic and covalent character in the bonding in CsAu and an equal contribution from both of these bonding types is currently the most favoured opinion. A range of other derivatives containing anionic gold has been prepared in liquid ammonia or ethylenediamine (compare section 6.7), and isolated in complexes with various large cations, especially (crypt)M$^+$. The solid-state mixed-metal oxide derivative Cs_3AuO has been prepared by reaction of Cs_2O with CsAu and this compound also contains anionic gold. A related nitride auride, Ca_3AuN, has also been described.

15.10 The zinc Group

The Zinc Group does not fit the general picture of the transition Groups as developed in the last two chapters. It shares with beryllium the property of belonging to one block of the Periodic Table and having many of the properties characteristic of another. In this case, the three elements of this Group resemble the three heavy elements of the Boron Group.

The elements, zinc, cadmium, and mercury, in this Group have the outer electronic configuration $d^{10}s^2$ and have the common oxidation state of II, corresponding to the loss of the two s electrons. In addition, mercury shows a well-established I state and cadmium and zinc form analogous but very unstable I compounds. Thus the heaviest element is more stable than the lighter ones in a low oxidation state, a characteristic of the Main Groups. However, a recent theoretical study has suggested that HgF_4 should be thermodynamically stable or, at worst, only slightly endothermic with respect to gas-phase HgF_2 and fluorine, whereas CdF_4 and particularly ZnF_4 are much less likely to be stable entities. It was suggested that a relativistic destabilization of the Hg(II)–F bonds, rather than a stabilization of HgF_4, is the reason for this and that the use of powerful fluorinating agents, such as KrF_2, might be suitable for the oxidative fluorination of HgF_2 to HgF_4.

The radii also reflect this transitional character. Shannon crystal radii for M(II) are (pm)

Coordination no.	4	6	8
Zinc	74	88	104
Cadmium	92	109	124
Mercury	110	116	128

The significant increases between successive Group members for the 4-coordinate radii is like that found for a Main Group while the pattern of changes in 8-coordination is that typical of a Transition Group. (See also relativistic effects, section 16.13).

The M^{II}/M redox potentials show a large difference between Cd and Hg which can be ascribed in part to the higher solvation energy of Cd^{2+}, which is an effect of smaller size. The potential values for $M^{2+} + 2e^- = M$ are:

$$Zn = -0.762 \text{ V}, \quad Cd = -0.402 \text{ V}, \quad Hg = 0.854 \text{ V}$$

These values show the relatively high electronegativity of

zinc and cadmium and reflect the reducing power of these elements. By contrast, mercury is unreactive and 'noble'.

The elements are all readily accessible as they occur in concentrated ores and are easily extracted. Zinc and cadmium are formed by heating the oxides with carbon and distilling out the metal (boiling points are 907 °C for zinc and 767 °C for cadmium). Mercury(II) oxide is decomposed by heating alone, without any reducing agent, at about 500 °C, and the mercury distils out (b.p. = 357 °C). Mercury is the lowest melting metal with a melting point of −39 °C, while zinc and cadmium melt at about 420 °C and 320 °C respectively. Mercury is monatomic in the vapour, like the rare gases. The element, and many of its compounds, are very poisonous and, as mercury has a relatively high vapour pressure at room remperature, mercury surfaces should always be kept covered to avoid vaporization.

15.10.1 The (I) state and subvalent compounds

In the I state, mercury exists as the dimeric ion, Hg_2^{2+}. This has been demonstrated by a number of independent lines of evidence:

(i) The Raman spectrum of aqueous mercurous nitrate shows an absorption attributed to the $Hg-Hg$ stretching vibration.
(ii) The crystal structures of mercury(I) salts show the existence of discrete Hg_2 units: see Figure 15.33 for the case of Hg_2Cl_2.

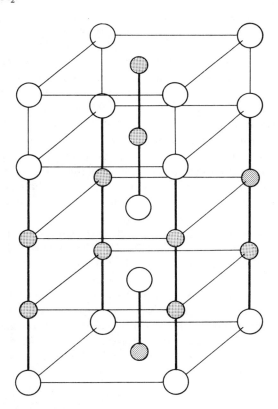

tetragonal unit cell of Hg_2Cl_2

○ Cl   ● Hg

FIGURE 15.33 *The crystal structure of mercurous chloride,* Hg_2Cl_2

(iii) All mercury(I) compounds are diamagnetic, whereas an Hg^+ ion would have one unpaired electron.
(iv) E.m.f. measurements on concentration cells of mercurous salts show that two electrons are associated with the mercurous ion. For example, the e.m.f. of the cell:

| Hg | mercury(I) nitrate in (0·05M) 0·1M HNO_3 ‖ mercury(I) nitrate in (0·005M) 0·1M HNO_3 | Hg |

was found to be 0·029 V at 25 °C. Now $E = \dfrac{RT}{nF}\ln(c_1)/(c_2)$, where the concentrations are c_1 and c_2, n is the number of electrons, and the other constants have their usual significance. If values are put in for the constants and conversion to logarithms to the base ten is carried out, the equation becomes $E = (0·059/n)\log(0·05/0·005)$: hence $n = 2$.
(v) Conductivities and equilibrium constants also fit for a dimeric ion of double charge and not for Hg^+.

The redox potentials involving mercury(I) are

$$Hg_2^{2+} + 2e^- = 2Hg_{(liquid)} \qquad E = 0·789 \text{ V}$$
$$2Hg^{2+} + 2e^- = Hg_2^{2+} \qquad E = 0·920 \text{ V}$$

It follows that, for the disproportionation reaction:

$$Hg + Hg^{2+} = Hg_2^{2+},$$

the potential is $E = 0·131$ V. Then, as $E = (RT/nF)\ln K$, where K is the equilibrium constant for the reaction:

$$K = [(Hg_2^{2+})]/[(Hg^{2+})] = \text{about } 170$$

In other words, in a solution of a mercury(I) compound there is rather more than $\frac{1}{2}$ per cent mercury(II) in equilibrium. Thus, if another reactant is present which either forms an insoluble mercury(II) salt or a stable complex, the mercury(II) ions in equilibrium are removed and the disproportionation goes to completion. If OH^- is added to a solution of Hg_2^{2+}, a grey precipitate of HgO mixed with Hg is formed. Similarly, sulfide precipitates HgS and CN^- ions give a precipitate of mercury and the undissociated cyanide of mercury(II), $Hg(CN)_2$.

The existence of this disproportionation reaction means that a number of mercury(I) compounds, such as the sulfide, cyanide or oxide, do not exist. The main compounds of mercury(I) are the halides and a number of salts. Hg_2F_2 is unstable to water giving HF and HgO plus Hg. The other halides are very insoluble. Mercurous nitrate and perchlorate are soluble and give the insoluble halides, sulfate, acetate and other salts by double decomposition reactions. The nitrate and perchlorate are isolated as hydrates which contain the hydrated ion $[H_2O-Hg-Hg-H_2O]^{2+}$.

A three-atom chain exists in $Hg_3(AlCl_4)_2$ where the basic structure is a Z-shaped molecule with Cl bridging between approximately tetrahedral $AlCl_4$ units and approximately linear $Hg-Hg-Hg$ chains.

In the reaction with fluoro-arsenic or antimony species, a number of poly-mercury ions result. For example,

$$3Hg + 3AsF_5 \rightarrow Hg_3(AsF_6)_2 + AsF_3$$
$$4Hg + 3AsF_5 \rightarrow Hg_4(AsF_6)_2 + AsF_3$$
$$6Hg + 3AsF_5 \rightarrow [Hg_{2\cdot85}AsF_6]_n + AsF_3$$

(the last reaction in SO_2).

The crystal structures show a linear symmetric Hg_3^{2+} ion with $Hg-Hg = 255\cdot2$ pm. In Hg_4^{2+}, the structure is centrosymmetric and nearly linear with angle $176°$.

$$Hg \xrightarrow{257} Hg \xrightarrow{270} Hg \text{——} Hg$$

Infinite chains are found in $[Hg_{2\cdot85}AsF_6]_n$ which contains chains of Hg atoms passing in two directions at right angles through a lattice of octahedral AsF_6^- ions. The average $Hg-Hg$ distance is 264 pm. Similar mercury chains are found where the As is replaced by Nb or Hf, but such compounds change on standing to give golden crystals which contain planes of close-packed Hg atoms separated by layers of MF_6^- ions.

Alkali metals dissolve in mercury to form amalgams and clusters of mercury atoms have been identified in the solids isolated from such amalgams. All these clusters involve formal oxidation states for mercury of around I. CsHg, KHg and Na_3Hg_2 phases have all been shown to contain rectangular Hg_4 clusters. Even more striking is $Rb_{15}Hg_{16}$ which contains slightly distorted Hg_4 squares and Hg_8 cubes (angles, 86–94°; Hg–Hg distances, 294–298 pm, except for two edges of the square at 304 pm). A parallel is to be drawn with the gold clusters discussed in section 16.8.

Mercury(II) and (I) carbonyl cations have been isolated in liquid antimony fluoride from Hg^{2+} and CO at $100°C$. The $[Hg(CO)_2]^{2+}$ and $[Hg_2(CO)_2]^{2+}$ ions were isolated as the $[Sb_2F_{11}]^-$ salts, following a similar preparation of the analogous $[Au(CO)_2]^+$ cation. While neutral binary carbonyls are not known for the heavy elements to the right of the d block (compare section 16.2), reasonably stable M–CO species have now been isolated as ions, or in presence of additional ligands, for most of these elements.

Though the existence of the dimeric form of the mercury(I) state is without doubt, there is now also some evidence for the monomeric ion, Hg^+; both in solution and in solids.

The existence of a cadmium(I) ion, Cd_2^{2+}, has been conclusively proved in a molten halide system. $CdAlCl_4$ is obtained as a yellow solid and shown to be $Cd_2^{2+}(AlCl_4)_2^-$. This reacts violently with water to give cadmium metal and Cd^{2+} in solution. Cd^I may also exist in the deep red melts of Cd in cadmium(II) halides. The Cd_2^{2+} ion is quite unstable to water. A recent report indicates that zinc(I), Zn_2^{2+}, exists under similar conditions.

15.10.2 The (II) state

The most stable state for all three elements is the II state. In this, zinc and cadmium resemble magnesium and many of the compounds are isomorphous. Mercury(II) compounds are less ionic and its complexes are markedly more stable than those of zinc and cadmium. All three elements resemble the transition elements more than the Main Group elements in forming a large variety of complexes.

The halides of all three elements are known. All the fluorides, MF_2, are ionic with melting points above $640°C$. HgF_2 crystallizes in the fluorite lattice and is decomposed in contact with water. The structures of ZnF_2 and CdF_2 are unknown: these compounds are stable to water and are poorly soluble, due both to the high lattice energy and to the small tendency of the fluorides to form complex ions in solution. In this the zinc and cadmium fluorides resemble the alkaline earth fluorides. The chlorides, bromides, and iodides of zinc and cadmium are also ionic, although polarization effects are apparent and they crystallize in layer lattices. The structures are approximately close-packed arrays of the anions with Zn^{2+} in tetrahedral sites in the zinc halides while, in the cadmium halides, Cd^{2+} ions occupy octahedral sites. The zinc and cadmium halides have lower melting points than the fluorides and are ten to thirty times more soluble in water. This is due, not only to lower lattice energies, but also to the ready formation of complex ions in solution. A variety of species result, especially in the case of cadmium halides. Thus a 0·5M solution of $CdBr_2$ contains Cd^{2+} and Br^- ions and, in addition, $CdBr^+$, $CdBr_2$, $CdBr_3^-$ and $CdBr_4^{2-}$ species, the most abundant being $CdBr^+$, $CdBr_2$ and Br^- (these species are probably hydrated). Hydrolysis also occurs and species such as $Cd(OH)X$ are observed. The tetrahalides, ZnX_4^{2-} and CdX_4^{2-} may be precipitated from solutions of the halides in excess halide by large cations. These are tetrahedral ions, as are all four-coordinated species in this group.

By contrast, $HgCl_2$, $HgBr_2$, and HgI_2 are covalent solids melting and boiling in the range $250°C–350°C$. $HgCl_2$ is a molecular solid with two $Hg-Cl$ bonds of 225 pm and the next shortest $Hg-Cl$ distance equal to 334 pm, so that there is little interaction between the mercury atom and these external chlorines. In the bromide and iodide, layer lattices are formed, but in the bromide, two $Hg-Br$ distances are much shorter (248 pm) than the rest (323 pm) so that this is a distorted molecular lattice. HgI_2 forms a layer lattice in which there are HgI_4 tetrahedra with the $Hg-I$ distance equal to 278 pm. In the gas phase, the mercury halogen distances in the isolated HgX_2 molecules are, for $X = Cl$, 228 pm; $X = Br$, 240 pm; $X = I$, 275 pm. Thus the $Hg-Cl$ distances are the same in the solid and gas, underlining the molecular form of the solid. The $Hg-Br$ distance is a little longer in the solid while the $Hg-I$ distance is markedly longer in the solid, showing the increasing departure from a purely molecular solid on passing from the chloride to the iodide. Mercury also forms halogen complexes, and the same species are found for mercury as for cadmium. The stability constants for the mercury complexes are much higher than those for the zinc and cadmium species. Halo-mercury ions are found, e.g. $(HgI)^+$ which exists as infinite chains, with IHgI angles near linear and HgIHg angles around 90°. In presence of a large cation, the $HgCl_5^{3-}$ ion may be isolated, whose shape is a trigonal bipyramid with short axial $Hg-Cl$ distances of 233 pm and long $Hg-Cl$ equatorial distances of 303 pm. This can be seen as a linear $Cl-Hg-Cl$ unit weakly coordinated by three further Cl^- ions.

The oxides are formed by direct combination. ZnO is white and turns yellow on heating. CdO is variable in colour from yellow to black. The colours in both cases are due to the formation of defect lattices, where ions are displaced from their equilibrium positions in the crystal lattices to leave vacancies. These may trap electrons whose transitions give rise to colours in the visible region. HgO is red or yellow, depending on the particle size. Zinc oxide and the hydroxide are amphoteric. $Zn(OH)_2$ is precipitated by the addition of OH^- to zinc solutions and dissolves in excess alkali. $Cd(OH)_2$ is precipitated similarly but is not amphoteric and remains insoluble in alkali. Mercury(II) hydroxide does not exist; the addition of alkali to mercuric solutions gives a precipitate of the yellow form of HgO. The elements all form insoluble sulfides and these are well-known in qualitative analysis. ZnS is somewhat more soluble than CdS and HgS, and has to be precipitated in alkali rather than the acid conditions under which yellow CdS and black HgS precipitate.

Most of the oxygen Group compounds of Zn, Cd, and Hg have the metal in tetrahedral coordination in the zinc blende structure (see Figure 5.1c) or in the related wurtzite structure (Figure 5.4a). Both these are found for ZnS, with the wurtzite form the stable one at high temperatures. ZnO, ZnS, ZnTe, CdS, and CdSe all occur in both the wurtzite and the zinc blende forms. ZnSe, CdTe, HgO, HgSe and HgTe are found in the zinc blende form only. HgS occurs in two forms, one is zinc blende and the other is a distorted NaCl lattice. The only other example of six-coordination is CdO which forms a sodium chloride lattice. These structures again illustrate the strong tendency for these elements to form tetrahedral coordination. A growing range of sulfides and selenides of cadmium and zinc have been prepared which are polynuclear species containing segments of the tetrahedral ZnS structure. A small unit is seen in the Cd_4Se_{10} skeleton of $[(RSe)_6(CdX)_4]^{2-}$. Figure 15.34 shows the structure for X = SeR (note the SeR groups occupy both terminal and bridging positions). Other terminal groups, such as X =

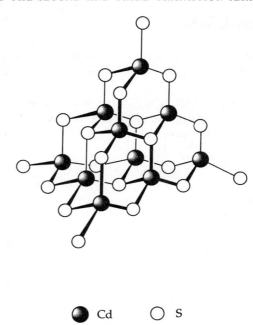

FIGURE 15.35 *The $Cd_{10}S_{20}$ skeleton of $[S_4Cd_{10}(SR)_{16}]^{4-}$*

halogen, are found, SR may replace SeR, and Zn may replace Cd. The basic skeleton is the same structure as P_4S_{10} (Figure 17.32b).

In $[S_4Cd_{10}(SR)_{16}]^{4-}$, four of these units are fused together, see Figure 15.35. Here, four of the S atoms are bonded only to Cd, six edges are bridged by SR, and the remaining SR are terminal on Cd. Again, Zn and Se analogues are known. Even larger polytetrahedral clusters are known, and the ultimate product of successive fusion is the sulfide. Such compounds may reflect the coordination of Cd in biomolecules.

This formation of four and six-coordinated species is also found in the complexes of these elements. Zinc occurs largely in four-coordination in complexes like $Zn(CN)_4^{2-}$ or $Zn(NH_3)_2Cl_2$ and is also found in rather unstable six-coordination, as in the hexahydrate and the hexammine, $Zn(NH_3)_6^{2+}$. Cadmium forms similar four-coordinated complexes but is rather more stable than zinc in six-coordination, due to the larger size. Mercury is commonly found in four-coordination, though a few octahedral complexes such as $Hg(en)_3^{2+}$ are also found.

All three elements are also found in linear two-coordination especially in their organo-metallic compounds, and in the halides and similar compounds. The organic compounds R_2M (M = Zn, Cd or Hg) are well-known and mercury also forms $RHgX$ compounds with halides. The so-called RZnX and RCdX species are, like the Grignard reagents, RMgX, more complex and their structures are not fully understood. They are polymeric with some evidence for MX_2 and MR_2 groups and are usually coordinated by the ether used in their preparation.

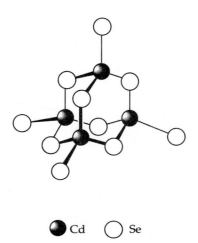

FIGURE 15.34 *The structure of the Cd_4Se_{10} skeleton of the $[Cd_4(SeR)_{10}]^{2-}$ ion*

PROBLEMS

15.1 Carry out, for the transition elements of the second and third series, the exercises given in questions 14.1 and 14.2.

15.2 For each element of the second transition series in turn, compare and contrast the chemistry with that of the other two elements in the same Group.

15.3 Discuss the lanthanide contraction and its effect on transition metal chemistry.

15.4 Continue general surveys along the lines suggested for Chapter 14 Problems, e.g. question 14.5.

16 Transition Metals: Selected Topics

The preceding three chapters have covered the basic themes of transition metal chemistry. In this chapter we select a few topics—each where work is proceeding actively—to give a further indication of the transition chemist's current interest. This selection is necessarily arbitrary and many areas of interest have been omitted. The treatment of each is limited to an introductory review.

Readers may wish to omit this chapter at first reading.

16.1 'Warm' superconductors

Superconductivity was first discovered in pure metals in 1911. When a metal such as Hg or Pb was cooled nearly to absolute zero, it was found to offer no resistance to electric current. This happens at a *critical temperature*, T_c, and work for the next 50 years led to a slow rise in the highest critical temperature found, up to around 20 K for alloys. By 1973, the T_c of Nb_3Ge at 23·7 K was the best that had been achieved, and similar but cheaper niobium-tin alloys, T_c about 22 K, were being used in magnet windings. The great advantage of superconductors is that, once a current has been created in a coil, it continues with no loss. Thus the first major application was in producing high-field electromagnets, used in nmr, in particle accelerators, and similar devices. Unfortunately, the low temperatures needed to achieve criticality needed expensive liquid helium as a coolant. Large-scale commercial applications, such as power transmission, were out of the question.

Work in raising T_c had thus stalled, with no significant rise for years despite the study of a wide range of metals and alloys (but also including mixed oxide phases), until Bednorz and Mueller, working in Switzerland in 1986, announced a mixed metal oxide, $La_{5-x}Ba_xCu_5O_{5(3-y)}$ with a T_c of 35 K. The species was a member of the perovskites, and turned attention away from metals and alloys in the quest for superconductors. The discovery created enormous interest, and reports of new records in T_c poured out from the USA, Japan, China, and all round the world. Extra sessions on superconductivity were slipped into conferences, and short-notice seminars were filled to overflowing and continued into the night. The huge interest cumulated in the award of the 1987 Nobel Prize in Physics to Bednorz and Mueller—the most rapid award ever.

Of course, in all the excitement and pressure, confusions arose. The preparation of the materials demands a specific regime to control the oxygen content, and lots of private 'recipes' flew about. Many samples were mixtures where only one phase was superconducting. In such systems, the resistance drops over a temperature range before true superconductivity sets in, and many of these high temperatures appeared as new 'records' in the popular press.

Judged by two reasonably strict criteria—that the resistance should drop to zero sharply over a range of only a degree or so, and that the sample should show the Meissner effect in which it expels magnetic field lines(most spectacularly, a small sample floats in a magnetic field)—there are two well-established classes of superconductors. One is based on the original species, reformulated as $La_{2-x}Ba_xCuO_{4-y}$, where x is 0·15 to 0·2, and y is undefined but small. These are cases of the established La_2CuO_4 layer perovskite with the K_2NiF_4 structure where up to 10% of the La^{3+} is replaced by Ba^{2+}, and there is a small oxygen deficiency. Of course, the changes were immediately rung on the constituents, especially replacing Ba by Ca or Sr, or substituting lanthanides, and T_c was raised to around 50 K very quickly.

The second system was first announced by Wu, Chu, and colleagues early in 1987, and is of general formula $YBa_2Cu_3O_{7-x}$, called the 1-2-3 type from the numbers of metal atoms. This compound also has a perovskite structure. It is superconducting when x is around 0·1, but semiconducting when x is 0.5 or more, and the value of T_c is 93 K. Again, analogues have included substitution of Y by Ho, the lanthanide with the same radius but paramagnetic. Despite theories to the contrary, the presence of paramagnetic ions did not destroy the superconductivity and the holmium species has T_c of 91 K. Sm, Eu, Nd, Dy, and Yb have also been substituted so the M^{3+} radius is not critical. Values of T_c

around 90 K are well-established for this series, and the Meissner effect is exhibited. The great advantage is that T_c is above the temperature of liquid nitrogen which boils at 77 K. Thus superconductivity is accessible much more cheaply and widely. Stop press news early in 1988 was the discovery of two new systems, based on Bi or Tl in the place of the lanthanide elements, which bring T_c up to around 120 K. The structure is a perovskite, related to those of Figure 16.1, but with CuO and BiO or TlO layers as the major feature. As an example, the thallium compound $Tl(Ba, Ca)_2Ca_3Cu_4O_{10.5+\delta}$ has four consecutive copper layers and adding more layers suggests that this will increase T_c further. By growing an oriented thin film of a cuprate compound belonging to the BiSrCaCuO family onto a $SrTiO_3$ single crystal, a material having eight adjacent cuprate layers has been produced and studies suggest this material may have a T_c as high as 250 K. If such materials can be prepared in bulk form, then very interesting applications may arise from them.

The 90 K superconductors are perovskite phases which are brittle solids. Most have been made only as powders, and much effort is going into the problems of finding forms which can be fabricated on a large scale. Because the rewards are large, the problems are likely to be solved faster than the pessimists' estimate of twenty years.

Out of all the syntheses, the main route has been to grind together the metal oxides or carbonates in stoichiometric amounts and then heat in air, cool, regrind, heat in oxygen, anneal in oxygen and finally cool. In a modification which allows easier control of composition, the carbonates or citrates are precipitated from a stoichiometric mixture of soluble salts such as the nitrates. One preparation then continues: 'dry at 140 °C overnight, heat in air at 950 °C for 24 h, cool in air, grind and form into pellets and reheat in a flow of oxygen at 950 °C for 16 h, cool in oxygen to 200 °C over 1 h and finally cool in air'. These details show how critical the oxygen content is. Another synthetic route which is attracting an increased amount of interest uses molecular metal–organic precursors containing a precise ratio of the different desired metals which can then be converted to the desired mixed-metal oxide. Metal alkoxide compounds are of particular interest in this area.

The key to the superconducting properties is thought to lie in the copper chemistry, and to understand this we have to look at the structures in more detail. If we start with the perovskite structure of Figure 5.5c, we need to insert a number of different metal atoms in the twelve-coordinate site. Thus the unit cell will be stretched from the cubic one of the simple perovskite along one axis to accommodate several units with different metal atoms. By crystallographic convention, this long axis is the c axis, parallel to the Z coordinate, so we turn the Figure 5.5c structure through 90° and extend it by a number of units.

The formula La_2CuO_4, on which the first class of superconductors is based, is created by adding an extra layer of O atoms to the perovskite formula, $CaTiO_3$ (recall that the atoms on the edges of the unit cell are shared between four cells, so each adds 1/4 to the contents of the cell). Some 10% of

the La ions are replaced with Ba ions, and to compensate for this loss of charge, a corresponding number of Cu atoms change from Cu(II) to Cu(III). It is thought that the superconductivity is due to electron transfer between the two copper states, mediated by the oxygens, perhaps also involving copper(I), and assisted by the small oxygen deficiency. From general copper chemistry, we recall that Cu(II) has a tetragonally-distorted coordination sphere.

For the 1-2-3 structure, we need a tripled perovskite structure. To arrive at a formula $YBa_2Cu_3O_7$ we insert a Y^{3+} ion in the centre cell, with Ba^{2+} ions in the two outer cells, and also remove the oxygen atoms in the plane of the yttrium (which reduces the composition to $YBa_2Cu_3O_8$ from the $(ABO_3)_3$ formula of the tripled perovskite). If we further remove two of the four oxygens in the top and bottom faces of the unit cell, we arrive at the stoichiometry $YBa_2Cu_3O_7$ which is that shown in Figure 16.1(a). The final superconductor stoichiometry $YBa_2Cu_3O_{7-x}$ results from removing between 10% and

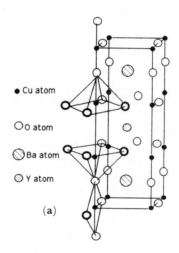

• Cu atom

◯◯ O atom

▨ Ba atom

▨ Y atom

(a)

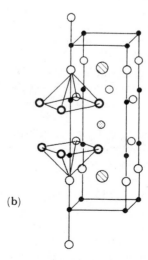

(b)

FIGURE 16.1 *Barium-yttrium-copper-oxide perovskite phases:* (a) *the structure of the superconductor* $YBa_2Cu_3O_{7-x}$, (b) *the structure of the semiconductor phase* $YBa_2Cu_3O_6$

50% of the remaining oxygens in the end faces, with 50% at the limit of the superconducting composition range. If we remove all the oxygens in these two faces, the structure of Figure 16.1b results, for formula $YBa_2Cu_3O_6$, and this phase is only a semiconductor.

In the structure of $YBa_2Cu_3O_{7-x}$, there are two different copper positions. In those copper layers which lie between the Ba and the Y, the removal of the oxygens from the Y layer leaves the Cu 5-coordinated in a square pyramid. (Remember that the Ti atoms, which are at this position in the basic perovskite structure, are octahedrally coordinated to O). The second copper site is in the end layers (i.e. between two layers of Ba in successive unit cells) and these Cu atoms are in square planar 4-coordination in $YBa_2Cu_3O_7$, linear 2-coordination in $YBa_2Cu_3O_6$, and somewhere between these two in $YBa_2Cu_3O_{7-x}$.

From the stoichiometry, as neither Ba nor Y are likely to have variable charges, seven charges are shared among three Cu atoms in $YBa_2Cu_3O_7$. The square planar positions are allocated to Cu(III), reasonable for the d^8 configuration, while the Cu(II) are allocated to the square pyramidal positions, again reasonable in view of the usual distortions found for Cu(II). As x increases and O is removed, some of the square planar Cu(II) sites are converted into trigonal, and eventually linear Cu(I) configurations (completely at the $YBa_2Cu_3O_6$ stoichiometry). Again, the superconductivity is thought to result from the ready transfer of electrons between the different copper oxidation states, mediated by the oxygens.

Since the current theory of superconductivity in the classical metal and alloy case involves electrons moving as pairs, the coexistence of Cu(I) and Cu(III) is thought to be important. However, there are many current suggestions and no generally accepted theory for the high-temperature superconductors at present.

These observations suggest that this type of superconductivity is particularly associated with copper. The combination of three accessible oxidation states, with different preferred coordination in each, and perhaps the tetragonal distortion associated with Cu(II) are important. If the rest of the transition elements are reviewed, this combination is unique. For example, while d^4 configurations are also distorted, choices such as Mn(III) or Cr(II) may be less good because the neighbouring oxidation states vary substantially in stability and do not show the same variation in preferred coordination. On the other hand, variations in the M^{2+} and M^{3+} species have minor effects, presumably as long as they are reasonably big. No doubt many other changes will be tried, S for O, doping of some M(I) sites with Ag or M(III) sites with Au.

Since the early discoveries of superconducting metal oxides, a steady increase in T_c has been observed in different systems. The maximum value now observed has exceeded 150 K. One interesting observation made recently is that an increase in T_c can be effected with an increase in pressure. When the compound $HgBa_2Ca_2Cu_3O_{8+\delta}$ is placed under a pressure of 150 000 atmospheres it shows a T_c of 153 K. It has been predicted that even higher pressures might increase T_c further. One way of mimicking the effects of increased pressure might be to replace the large barium atom by a somewhat smaller strontium atom. Of course, pressures such as these would rule out any direct applications within the forseeable future but, nevertheless, these studies serve to indicate that higher superconducting temperatures should be possible, given the right system. It is possible that the high toxicity of thallium and mercury compounds might ultimately inhibit their applications and the search for even higher T_c superconductors is beginning to look at a number of other systems.

16.2 Carbonyl compounds of the transition elements

A brief indication of the carbonyls of the individual elements is given in Chapters 14 and 15 in the discussions of the low oxidation states.

16.2.1 Formulae

Over a century ago, Mond found that nickel reacted with carbon monoxide to form an unusual volatile compound, $Ni(CO)_4$, which we call nickel carbonyl. $Ni(CO)_4$ is readily decomposed to $Ni+4CO$ by gentle heating. The formation and decomposition of $Ni(CO)_4$ became the *Mond process* for nickel refining which was used for many years (compare section 14.8).

It has since been found that most transition metals form one or more carbonyl compounds. Although a number of specific syntheses are available, the preparation from metal or readily reducible compound and CO applies generally

$$M + nCO \rightarrow M(CO)_n \ldots \ldots \qquad (16.1)$$

though usually requiring raised temperatures and pressures. Table 16.1 lists the simpler carbonyls formed by each element. A very large number of more complicated species are reported, particularly for Re, Ru, Os, Co, Rh, and Pt. The first M — CO compound to be reported was the mixed ligand $Pt(CO)_2Cl_2$.

The carbonyls, on the basis of equation (16.1), are best regarded as forming from neutral CO and the metal in an oxidation state of zero. A convenient guide to the formulae is provided by the *18-electron rule*. Each CO provides 2 electrons and these, with the metal valence shell electrons, add up to the rare gas configuration of eighteen electrons in nearly every case. Thus nickel(0) has 10 valency electrons $[d^8s^2]$ and the 8 from the four CO groups give the rare gas configuration. Similarly, iron with 8 electrons and chromium with 6 require respectively 5 and 6 CO groups to make up the 18. For manganese, with 7 electrons of the d^5s^2 configuration, we see that $Mn(CO)_5$ would have only 17 electrons while $Mn(CO)_6$ would have 19 electrons. In this case we find that two $Mn(CO)_5$ units join together forming a Mn — Mn bond and the two shared electrons of the bond count for each Mn. Thus there are 10 CO electrons, 7 Mn electrons plus 1 shared from the second Mn making 18. Similarly cobalt forms $Co_2(CO)_8$ where the count at each Co atom is 8 from 4 CO groups, 9 valency

TABLE 16.1 Properties of the simpler carbonyls

$V(CO)_6$ dark green d. 70° octahedron paramagnetic	$Cr(CO)_6$ colourless d. 130°, s octahedron	$Mn_2(CO)_{10}$ yellow m 154°, s Fig. 16.3a	$Fe(CO)_5$ yellow m −20°, b 103° trigonal bipyramid	$Co_2(CO)_8$ orange m 51°, s Fig. 16.3b, c	$Ni(CO)_4$ colourless m −19.3°, b 42° tetrahedron
			$Fe_2(CO)_9$ orange d. 100° Fig. 16.3d	$Co_4(CO)_{12}$ black d. 60°	
			$Fe_3(CO)_{12}$ green d. 140°, s Fig. 16.4a		
	$Mo(CO)_6$ colourless d. 180°, s octahedron	$Tc_2(CO)_{10}$ colourless m 160°, s as Fig. 16.3a	$Ru(CO)_5$ colourless m −16° trig. bipyramid	$Rh_4(CO)_{12}$ red d. 150°	
			$Ru_3(CO)_{12}$ orange m. 150° Fig. 16.4b		
	$W(CO)_6$ colourless d. 180°, s octahedron	$Re_2(CO)_{10}$ colourless m 177°, s as Fig. 16.3a	$Os(CO)_5$ colourless m 2° trig. bipyramid	$Ir_4(CO)_{12}$ yellow d. 210° tetrahedron of $Ir(CO)_3$ units	$Pt(CO)_4$ transient existence
			$Os_3(CO)_{12}$ yellow m 224° Fig. 16.4b		

electrons and one from the Co—Co bond. The only exception to the 18-electron rule is provided by vanadium which forms the $V(CO)_6$ monomer with only 17 electrons. Possibly this is due to the steric problems of 7-coordination in a dimer.

The carbonyls are readily reduced, by such reagents as sodium amalgam or hydrides, to form anions. These all obey the 18-electron rule and we find species such as $Co(CO)_4^-$ or $Fe(CO)_4^{2-}$. Vanadium carbonyl very readily forms $V(CO)_6^-$ where the extra electron completes the eighteen. Addition of acid to the anions forms the carbonyl hydrides, such as $HMn(CO)_5$ or $H_2Fe(CO)_4$. These also all obey the 18-electron rule. The M – H bond is relatively weak and the simple hydrides readily lose H_2 forming the parent carbonyl.

Cationic species are rarer, but the large majority obey the 18-electron rule, for example $[Mn(CO)_6]^+$. The recently-reported $[Pt(CO)_4]^{2+}$ is square-planar d^8 and a rare exception to the 18-electron rule (as is often found for this configuration).

As the known carbonyls, carbonyl ions, and also a range of other derivatives such as hydrides and halides, almost all follow the 18-electron rule, we can turn the argument round and use the 18-electron rule as a guide to suggest the formulation of more complex species. Thus, iron forms two more carbonyls in addition to $Fe(CO)_5$. In $Fe_2(CO)_9$, there are 18 electrons from the CO groups, 16 from the two Fe atoms, and so we need one Fe—Fe bond to complete 18 electrons at each iron. Similarly, $Fe_3(CO)_{12}$ with 24 CO electrons and 24 Fe electrons needs 3 Fe—Fe bonds (i.e. 2 per Fe atom) to complete 18 electrons per iron.

16.2.2 Bonding

Since the metal is zero-valent, and the CO group is a very poor lone pair donor, it is clear that a simple electrostatic approach cannot account for the carbonyls. We find that the carbonyls are readily accommodated by the molecular orbital approach with π bonding between metal and ligand, as in case (a) of section 13.12, where the ligand has empty orbitals of π symmetry which accept electron density from the metal.

The basic $M-CO$ bond is illustrated schematically in Figure 16.2. The CO molecule has three occupied σ orbitals (see Table 3.4 and Figure 3.24b) of which σ_1 is strongly bonding between C and O. σ_2 and σ_3 are weakly bonding or non-bonding between C and O, and are directed outwards on O and C respectively. Thus σ_3 is suitable for donation to the metal as in Figure 16.2a. As the M accepts electrons from σ interactions with a number of CO groups, the electron density on M builds up. However, the π^* orbitals on CO are empty and suitably aligned to form π bonds as in Figure 16.2b. This removes electron density from the metal (making it a better acceptor) and increases electron density on the CO (making it a better donor) so that the σ and π processes reinforce each other. The term *synergic* has been applied to describe this mutual reinforcement. Notice that this process weakens the CO π bonds, as it populates the CO π^* orbital, but this is more than compensated by the enhanced $M-C$ bonding so the overall process is favourable.

This CO weakening is reflected by a drop in the CO stretching frequency (compare section 7.7). As the CO frequencies are usually very strong bands in the infrared spectrum, this gives a conveniently-observed way of studying carbonyl bonding. In this way we find the infrared absorption due to the carbonyl stretching at $2037 \, cm^{-1}$ in $Ni(CO)_4$, distinctly less than the frequency of $2157 \, cm^{-1}$ in free CO. If we increase the negative charge on the metal, we expect more π donation from M to π^* (as in Figure 16.2b), a weaker CO bond and a reduced carbonyl frequency. Thus for $Co(CO)_4^-$ and $Fe(CO)_4^{2-}$ (which are isostructural with $Ni(CO)_4$) the frequencies drop to $1918 \, cm^{-1}$ and $1788 \, cm^{-1}$ respectively, reflecting the increased negative charge. The relatively high carbonyl stretching frequencies in the recently-prepared silver carbonyls like $[Ag(CO)_2]^+$ (section 15.9.1) indicate little back-bonding between Ag and carbonyls.

The measurements of other parameters, such as $M-C$ bond lengths, all support this picture. Therefore the bonding model (a) of section 13.12, as applied in Figure 16.2, is an adequate way of describing the metal carbonyls. More detailed treatments, taking account of all the CO groups and the symmetry of the complex and extending to other substituents, are very successful, but all are based on the core concepts described here.

We note that the model helps to account for the distribution of the simple carbonyls: to the left of the d block the metals have insufficient electrons to form the π bonds while to the right the elements lack empty orbitals to accept the sigma electrons.

16.2.3 Structures

The simple carbonyls have the symmetric shapes expected (Table 16.1). In $M_2(CO)_{10}$ (M = Mn, Tc, Re) the second metal atom completes the octahedron, Figure 16.3a.

The other dimeric carbonyls of the first transition series, $Co_2(CO)_8$ and $Fe_2(CO)_9$, introduce a new type of carbonyl bonding, the bridging carbonyl group. The structure of $Co_2(CO)_8$, in the solid state, is shown in Figure 16.3b. Each Co is bonded to three ordinary terminal CO groups and to the second Co by a Co—Co bond. The two remaining CO groups bond to both Co atoms, bridging the Co—Co bond. Although sometimes written in a keto form,

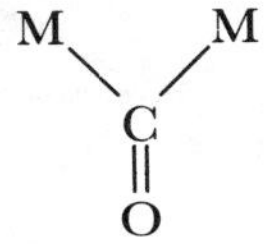

this bridging group has no ketonic properties. The bonding is best regarded as a three-centre overlap, similar to that of the B_3 faces in B_5H_{11} (Figure 9.12c).

In terms of electron counting, the bridging group contributes one electron to each metal, so two groups bridging two metal atoms have the same effect as if each was terminal on one metal. Thus the 18-electron rule does *not* allow us to distinguish bridging and non-bridging structures. This is well illustrated by the existence of the second isomer of cobalt carbonyl, Figure 16.3c, which is in equilibrium with the bridged form in solution.

$Fe_2(CO)_9$ has the structure shown in Figure 16.3d, with three bridging CO groups placed symmetrically around the Fe—Fe bond. The $Co_2(CO)_8$ form of 16.3b is similar to this

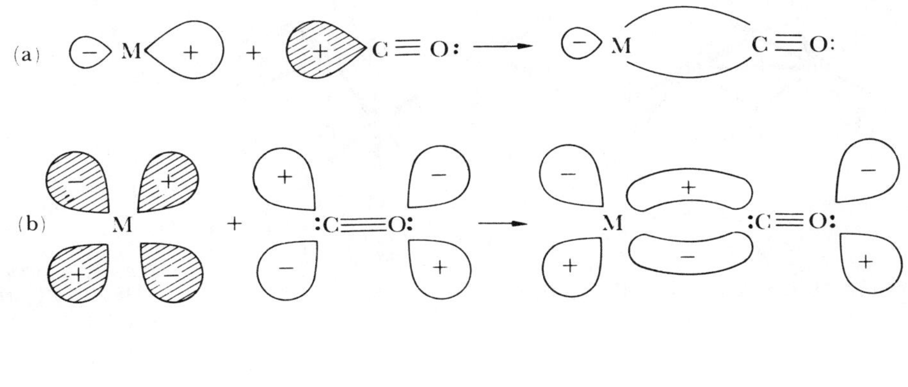

FIGURE 16.2 *The sigma and pi-bonds between a metal and CO*

(a) the sigma bond formed when the lone pair on the CO donates into a suitable metal orbital; (b) the pi-bond formed when electrons from a filled metal orbital donate back into the empty π^* antibonding orbital on the CO. Cross-hatching indicates the orbitals which originally held the donated electrons.

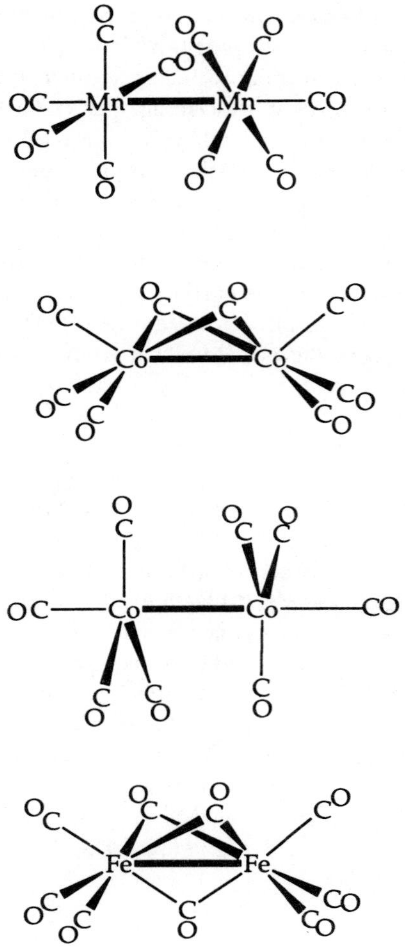

FIGURE 16.3 *Binuclear carbonyls*
(a) $Mn_2(CO)_{10}$, (b) $Co_2(CO)_8$ in solid, (c) $Co_2(CO)_8$ in solution, (d) $Fe_2(CO)_9$.

with one bridging CO removed. For $Fe_2(CO)_9$, the electron count is 8 valency electrons plus 6 from three terminal CO groups plus 3 from three bridging CO groups plus 1 from the Fe − Fe bond making 18 at each iron atom.

The presence of bridging carbonyls is clear from the infrared spectrum where the stretching frequencies of bridging groups are found at lower energies. Thus the terminal CO groups of $Co_2(CO)_8$ give vibrations between 2028 and

2104 cm^{-1} while the bridging modes are at 1898 and 1867 cm^{-1}.

The structures of the more complex carbonyls are built up in similar ways. For example, the $M_3(CO)_{12}$ species of the iron group have the structures of Figure 16.4. $Fe_3(CO)_{12}$ involves bridging carbonyls and can be envisaged as $Fe_2(CO)_9$ with one bridging CO replaced by a $Fe(CO)_4$ group. The Fe$_3$ skeleton is an isosceles triangle with the three Fe−Fe bonds indicated by the 18-electron rule. $Ru_3(CO)_{12}$ and $Os_3(CO)_{12}$ form equilateral triangles with no bridging carbonyls. A similar pattern is seen for the $M_4(CO)_{12}$ species of the cobalt group, where the metal atoms form a tetrahedral cluster. For M = Co and Rh, the base triangle face consists of terminal $(CO)_2M$ units with a bridging CO along each edge and the apex is occupied by an $M(CO)_3$ group. For M = Ir, the structure is symmetric with four $Ir(CO)_3$ terminal units in a regular tetrahedron. It is found generally that bridging CO groups are commoner for the lighter elements.

Hydrides with more complex structures are well known. Figure 16.5 shows some of the simpler examples. The edge-bridging hydrogens (as in Figure 16.5c) are analogous to the B − H − B bridges in the boron hydrides (Figure 9.10). In $HFe_3(CO)_{11}^-$, the H replaces one of the bridging CO groups of Figure 16.4a.

The simplest stable carbonyl of iridium is the tetranuclear species $Ir_4(CO)_{12}$ which has all the CO in terminal positions and the Ir$_4$ core bonded together into a regular tetrahedron. The CO are oriented equally about the three-fold axes so the overall symmetry is $\mathbf{T_d}$. The rhodium analogue has an alternative configuration where one $Rh(CO)_3$ unit is bonded to a triangle of three $Rh(CO)_2$ units with the last three CO edge-bridging this triangle (overall $\mathbf{C_{3v}}$). $Co_4(CO)_{12}$ has the same structure as the Rh carbonyl.

Larger carbonyl clusters are a well-established and rapidly growing field which is included in section 16.8.

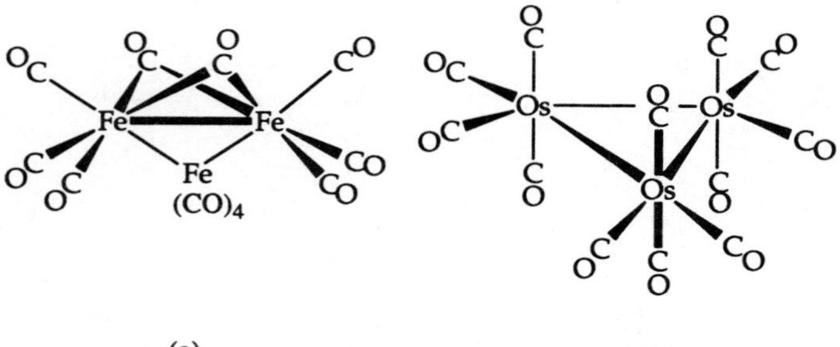

FIGURE 16.4 *Structures of* (a) $Fe_3(CO)_{12}$, (b) $Os_3(CO)_{12}$.

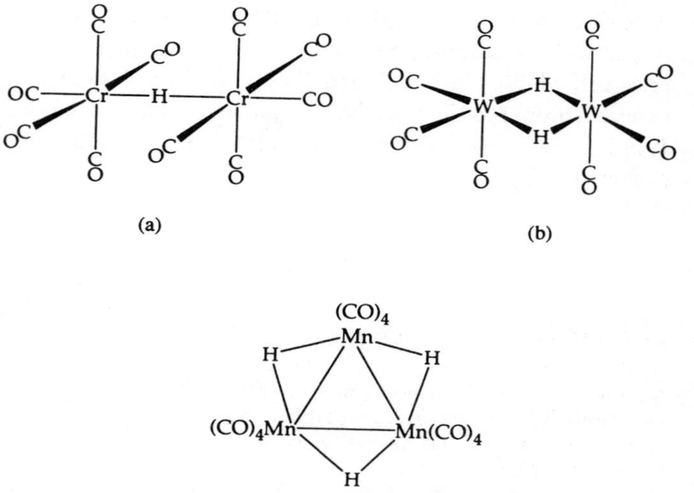

FIGURE 16.5 *Some carbonyl hydride species*
(a) $HCr(CO)_{10}^-$, (b) $H_2W_2(CO)_8^{2-}$, (c) $H_3Mn_3(CO)_{12}$.

16.2.4 Related species

There are three other groups of compound which are closely related to the carbonyls. First the nitrosyls, where the NO group behaves similarly to CO but acts as a 3-electron donor. Thus $Cr(NO)_4$ is an 18-electron species and we can construct the isoelectronic series from this through $Mn(CO)_3(NO)$, $Fe(CO)_2(NO)_2$, and $Co(CO)_3(NO)$ to $Ni(CO)_4$, all of which have now been reported.

Another ligand closely related to CO is PF_3. Unlike organic phosphines, such as $P(CH_3)_3$, PF_3 is a poor σ donor, as the electron density of the phosphorus lone pair is attracted by the electronegative fluorines. However, this same attraction makes the P a good π acceptor, using its empty $3d$ orbitals. As a result of these effects, PF_3 turns out to be a ligand very similar to CO, and PF_3 can replace terminal CO in most formulae, so we find $Ni(PF_3)_4$, $Fe(PF_3)_5$ or $Cr(PF_3)_6$ for example. The properties of such species are very similar to those of the carbonyl analogues. It has been very striking to find that $V(PF_3)_6$ can be made and parallels the carbonyl. It is a volatile paramagnetic molecule, corresponding to the same 17-electron count as the exceptional $V(CO)_6$, and it very readily forms the $[V(PF_3)_6]^-$ anion, completing the 18 electrons.

There is also a huge range of mixed-ligand compounds containing the carbonyl group. For example, we find the reaction

$$(CH_3)_3SnW(CO)_5^- + NO^+ \rightarrow (CH_3)_3SnW(CO)_4NO$$

or the reaction of $[(C_5H_5)Fe(CO)(PR_3)I]$ to form other $[(C_5H_5)Fe(CO)(PR_3)X]$ species with halogens.

Another class of related ligand, isoelectronic with carbon monoxide, is the isocyanides (also known as isonitriles), RNC, which are conveniently discussed here. One advantage of isocyanides over carbon monoxide as a ligand is that the steric and electronic properties of the former may be readily varied by changing the R group. Nevertheless, the number of isocyanide compounds is substantially less than that of carbonyls. Analogues of carbonyl species such as $Ni(CNR)_4$ and $Cr(CNR)_6$ have been prepared, and in dinuclear and cluster compounds bridging isocyanides akin to bridging carbonyl ligands are also known. The isocyanide group is in general a better sigma donor than CO, and stabilizes complexes in higher oxidation states than does CO, for example the complex $[Mn(CNR)_6]^{2+}$, which has no known carbonyl analogue. The isoelectronic cyanide ligand CN^-, while invariably a carbon donor ligand, is a much poorer π-acceptor than CO or RNC and is best considered as a pseudohalide ligand.

16.3 Metal-organic compounds

Many of the early attempts to make metal-organic compounds were unsuccessful, and molecules which were isolated, such as $CH_3CH_2Mn(CO)_5$, decomposed easily (in this case, at $-30°C$). The discovery in 1951 of the very stable organometallic compound, ferrocene $(C_5H_5)_2Fe$, was thus of considerable moment. This was quickly found to have an unusual structure and the interest created led to a very rapid expansion of the organometallic field. The two chemists most involved in the organometallic revival, Fischer and Wilkinson, were jointly awarded the 1973 Nobel prize for this work.

16.3.1 Metal-carbon sigma bonding

The simplest system to consider is that of a metal bonded to a methyl group. Here, the only bonding interaction of any significance is a sigma bond using appropriate orbitals on M and CH_3. Such bonds are very similar to the transition metal, $M-H$, bond and also to the relatively stable bonds to main group metals as in $Sn(CH_3)_4$. On the whole, sigma bonded metal-organics are unstable. A number of simple compounds exist, notably $Ti(CH_3)_4$, $Nb(CH_3)_5$ and $W(CH_3)_6$ with a few related species, which are well-characterized but which decompose at or below normal temperatures, often violently. In the presence of other ligands, stability increases. Thus $Ti(CH_3)_4$ starts to decompose above $-78°C$ but $(CH_3)_2TiCl_2$ can be prepared at $-20°C$ and diamine complexes are stable at $0°C$. The higher alkyls are generally less stable, but aromatic derivatives are more stable, than the methyls.

It is instructive to examine the factors affecting the metal-carbon bond stability. A number are of importance:

(1) *the metal-carbon bond is probably relatively weak*
There is little quantitative data but we note the thermal instability and the fairly low force constants for the metal-carbon stretching vibration.

(2) *organic products of M-C cleavage are highly reactive*
Whether $M-CH_3$ cleaves to give methyl radicals or organic ions, these will react rapidly with the solvent or with each other. Contrast this with the cleavage of most metal-ligand bonds, e.g. $M-OH_2$ or $M-Cl$, which give unreactive species such as H_2O or Cl^-.

(3) *bonds formed by the organic cleavage products are relatively strong*
We expect to find bonds such as $C-C$, $C-O$, $C-N$, $C-halogen$, depending on the system.

Thus we expect that, in a model reaction such as

$$2M-CH_3 \rightarrow H_3C-CH_3 + M \text{ products} \ldots \quad (16.2)$$

the forward reaction will be exothermic and thermodynamically preferred. Further, any equilibrium involving the first cleavage step, e.g.

$$M-CH_3 \rightarrow M\cdot + \cdot CH_3 \ldots \ldots \quad (16.3)$$

will be driven to the right as the $\cdot CH_3$ will rapidly be removed by further reaction.

However, even if compounds are thermodynamically unstable with respect to their decomposition products, their lifetime depends on the rate of the reaction—i.e. on their *kinetic stability*. Lifetimes range from the extremely short to the indefinitely long. The main factor affecting kinetic stability is the size of the energy input—the activation energy—required to get the molecule into a state where reactions such as 16·2 or 16·3 occur. Such an input may involve bond weakening or breaking, formation of an intermediate, population of antibonding orbitals and so forth. If the required activation energy is large, the compound may be stable indefinitely, even if the overall decomposition reaction has a strongly favourable free energy change. Similar comments apply to stability to other reactions such as oxidation or hydrolysis.

With these comments in mind we can list some factors which will generally be expected to lead to relatively stable metal-carbon sigma bonds.

(1)   If we make the reasonable assumption that population of the antibonding orbitals leads to bond breaking, any factor that increases the energy gap between the highest filled orbitals and the antibonding ones, will improve stability. Such factors are:

(a) a d^0 configuration—no electrons in the relatively high nonbonding orbitals such as t_{2g} in an octahedral complex.

(b) in any other configuration, all factors increasing the ligand field ΔE (compare sections 13.3 to 13.5, and 13.12).

(2)   All factors which decrease the kinetic contribution such as:

(a) steric hindrance from large ligands in higher coordination numbers.

(b) substitution-inert configurations such as d^6 octahedral (see p. 210).

(c) absence of low-energy pathways such as β-elimination (in the interaction with β-hydrogen below).

(3)   Factors which directly increase the metal-carbon bond strength. These include π contributions to the M − C bond as discussed below.

Of course, such contributions to stability do not necessarily have only one effect. Thus, π bonding ligands L in $L_nM − (CH_3)_x$ will increase ΔE (see 1b) but will probably increase the metal-carbon bond energy as well. This energy, in turn, affects the stability through both its thermodynamic effect (decreasing the free energy change for the forward reaction of equation 16.1), and through its kinetic effect increasing the activation energy input for M − C bond cleavages as in equation 16.2).

Table 16.2 lists some examples of metal-carbon sigma bonds, and it will be seen that the factors outlined above apply in most cases. Note that there are many further examples with π-bonding ligands such as C_5H_5, CO, phosphines etc.

The stability order $C_6H_5 > CH_3 > C_2H_5$ is accounted for by two features:

(a) aromatic ligands (and unsaturated ones generally) allow for additional M − R π interactions between metal d orbitals and ligand π^* ones. Note, for example, that a metal-acetylide, M − C ≡ CH, is isoelectronic with the carbonyls, M − C ≡ O, discussed in the last section.

Further, since the charge distribution is usually $M^{\delta+} − C^{\delta-}$, any organic ligand able to delocalize the negative charge should help stability. This applies to aromatic ligands, halogen-substituted ones etc., culminating in the generally high stability of M − CF$_3$ groups.

(b) It is commonly found that systems which do not contain H on the carbon β to the metal atom are distinctly the more stable. Thus, in equivalent species, M − CH$_3$, M − CH$_2$ − C(R)$_3$, M − CH$_2$ − Ar, M − CH$_2$ − SiR$_3$ or M − CH$_2$ − NR$_2$ are more stable than M − CH$_2$CH$_3$ or M − CH$_2$ − CHR$_2$ in general. This effect arises because the metal may interact with a β − H atom

$$M \cdots\!\!\!\!\!{\Large\diagdown} \begin{matrix} {}_{\overset{H}{\vdots}} \\ C \\ C \end{matrix} \rightleftharpoons M - H + \begin{matrix} | & | \\ C = C \\ | & | \end{matrix}$$

This provides an additional pathway for reaction (and the reverse reaction provides a synthesis).

16.3.2 *Metal-carbon multiple bonding*

An interesting development has been the evidence that metal-carbon double and triple bonds may occur in the *carbenes* M = CR$_2$, and *carbynes* M ≡ CR respectively. One of the first examples of a carbene was $(CO)_5WC(R)OMe$ whose crystal structure shows the CR(OMe) group completing an octahedron at W with the structure

TABLE 16.2 Some examples of metal-carbon sigma bonds

(CH$_3$)$_4$Ti	(CH$_3$)$_5$M	(CH$_3$)$_6$W	RM(CO)$_5$
CH$_3$TiCl$_3$	M = Nb, Ta	(CH$_3$)$_8$W^{2-}	M = Mn, Tc, Re
diamine Ti(CH$_3$)$_4$	(C$_5$H$_5$)$_2$VR	(C$_5$H$_5$)M(CO)$_3$R	
	V(C$_6$H$_5$)$_6^{4-}$	M = Cr, Mo, W	
		RCr(H$_2$O)$_5^{2+}$	
(C$_5$H$_5$)M(CO)$_2$R	RCo(CN)$_5^{3-}$	(PR$_3$)$_2$MR$_2$	(PR$_3$)AuR
M = Fe, Ru, Os	RM(X)$_2$(CO)(PR$_3$)$_2$	(M = Ni, Pd, Pt)	(PR$_3$)AuR$_3$
(diphos)Ru(R)(H)	(M = Rh, Ir)	[(CH$_3$)$_3$PtX]$_4$	

Notes (a) R = alkyl or aryl: stability usually decreases
R = aryl > CH$_3$ > C$_2$H$_5$

(b) Stability increases down Group e.g. M = Ni < Pd < Pt

(c) Note the common configurations are d^0 or d^{10} or the substitution-inert ones.

$(CO)_5M - C$
R
O
CH₃

$(R = CH_3, C_6H_5)$
$M = Cr, C - Cr = 204 \text{ pm}$
$W, C - W = 205 \text{ pm.}$

The $W - C$ bond length of 205 pm is greater than $W - CO$ (190 pm). The bonding is formulated as (a) a $W - C$ sigma bond plus (b) a π bond formed by a carbon p orbital overlapping with, say, d_{xz} on W (z is $W - CR_2$ axis). As the d_{yz} overlap would be identical, there is cylindrical symmetry around the z axis and the CR_2 group is free to rotate. A range of metal-carbenes is now known.

More recently a further family of compounds has been reported, the carbynes, with an $M - CR$ unit. An example is $I(CO)_4WC(C_6H_5)$. For this and related compounds, the structure is octahedral with the halogen *trans* to the CR and with the dimensions

For $R = CH_3$ or C_6H_5 and $X = Br$ or I

$X(CO)_4M - CR$
$Cr - C = 169 \text{ pm}$
$W - C = 190 \text{ pm.}$

The $M - C - R$ angle is near 180° for $R = CH_3$ but about 170° for $R = C_6H_5$.

The bonding is expressed as an overlap of p_x and p_y orbitals on C with metal d_{xz} and d_{yz} orbitals, in addition to the sigma component. Metal p_x and p_y may also contribute.

Thus we have three classes of metal-carbon compound with the bonding along the $M - C$ axis. These are analogues of $C - C$, $C = C$, and $C \equiv C$ bonding respectively although the $M - C$, $M = C$ and $M \equiv C$ systems are much weaker. A further, very widespread, class involving metal-carbon π bonding is that with unsaturated organic ligands bonding 'sideways-on' to the metal, and this is discussed in section 16.4.

The carbene or carbyne metal complexes essentially consist of a highly reactive organic fragment stabilized by coordination to the transition metal centre. They are examples of a general effect which is now very important in organometallic chemistry—that highly reactive species can be stabilized on coordination. A wide and diverse range of such stabilized reactive entities is now established, including highly strained cyclic alkynes, such as cyclobutyne and cyclohexyne, cyclobutadiene, benzyne, and trimethylene-methane $[C(CH_2)_3]$. Inorganic entities are also stabilized, such as silylenes $[R_2Si]$ and the corresponding germylenes and stannylenes.

A major reason for interest in metal–carbon bonded species comes from this stabilization, which applies to intermediates in many metal-catalysed organic reactions. Examples are the oxo-process (section 14.7) the Monsanto acetic acid process (section 15.7) or the Wacker process (section 15.9).

16.4 π-bonded cyclopentadienyls and related species

A further large class of pi-bonded transition metal complexes, which we can refer to only briefly here, is the group con-

taining unsaturated organic molecules as ligands. The two classical examples are *Zeise's salt*, $K[PtCl_3(C_2H_4)] \cdot H_2O$, and *ferrocene*, $Fe(C_5H_5)_2$.

In Zeise's salt, the three Cl atoms and the mid-point of the $C = C$ bond form a square plane around the Pt atom, and the ethylene molecule lies perpendicular to this plane (Figure 16.6a). The bonding is illustrated in Figure 16.6b. As in the carbonyls, sigma donation is postulated from the ligand, this time from the filled $C - C$ π orbital, and π donation from the metal into the ligand π^* orbital.

In the ferrocene molecule the metal atom is sandwiched between the organic parts, $C_5H_5 - Fe - C_5H_5$, and the planes of the organic rings are parallel (Figure 16.7a). A very similar situation is found for other aromatic systems, as in dibenzene chromium (Figure 16.7b). An interesting extension of this class is the triple sandwich (Figure 16.7c). Many more layers in such sandwich complexes have been constructed of late and even a hexadecker sandwich has been described. This consists of five cobalt atoms and six cyclopentadienyl or boron-substituted cyclopentadienyl ligands.

Cyclopentadienyl compounds are formed quite generally. Most transition metals form $(C_5H_5)_2M$ species though some, such as $[(C_5H_5)_2Ti]_2$, are less simple than was first thought. Related cyclopentadienyls are also widely used as π ligands, especially C_5Me_5 [often labelled cp*] where steric effects might be significant. One of the most substantial is C_5Ph_5, 'supercp', in which the steric demand often forces alternative geometries. The first perfluorinated analogue, $(C_5F_5)Rucp^*$, was recently discovered.

The related ions, such as $(C_5H_5)_2Fe^+$, are also readily formed and redox relationships between ions and neutral species reflect the 18-electron rule. One interesting case is cp_2^*Fe where the neutral compound is staggered, like the H analogue, but the cation is in the eclipsed configuration.

Molecules are found with one cp substituent together with other ligands, such as $C_5H_5Mo(CO)_3Cl$, and the reader is referred to the review in Appendix A for a detailed account of compounds of this type.

Dibenzene compounds are less widespread, but are typified by $(C_6H_6)_2Cr$, Figure 16.7b, which has the Cr sandwiched between two parallel benzene rings. Other aromatic systems appear in similar compounds including the cation of cycloheptatriene $(C_7H_7)^+$ and the dianion

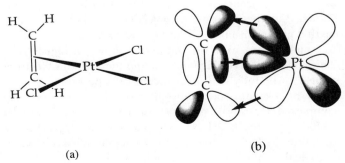

(a)   (b)

FIGURE 16.6 *Zeise's salt anion*, $PtCl_3(C_2H_4)^-$
(a) structure, (b) Pt–ethylene bonding.

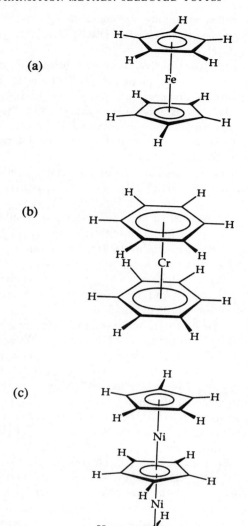

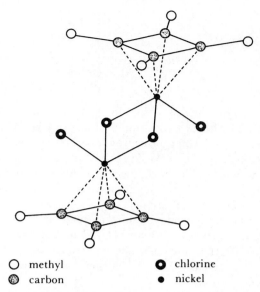

○ methyl   ◉ chlorine
⊛ carbon   ● nickel

FIGURE 16.8 *Di(tetramethylcyclobutadienylnickeldichloride),*
$(C_4Me_4NiCl_2)_2$

(a)

(b)

(c)

FIGURE 16.7 *The structure of metal sandwich compounds*
(a) ferrocene, $(C_5H_5)_2Fe$, (b) dibenzene chromium, $(C_6H_6)_2Cr$,
(c) tris(cyclopentadienyl)dinickel cation $(C_5H_5)Ni(C_5H_5)Ni$
$(C_5H_5)^+$.

of cyclobutadiene $(C_4H_4)^{2-}$. The latter system has long
appeared as a hypothetical aromatic system with no evidence
of its actual occurrence. The parent hydride is still unknown, but the substituted molecule with four phenyl or
methyl groups in place of the hydrogens has been attached
to a nickel atom, as in the compound shown in Figure 16.8.
The four-membered ring is planar and the aromatic electrons
appear to be fully delocalized. A very wide range of other
aromatic systems containing heteroatoms has also been
complexed to transition metals in a π-manner. Examples
include pyridine (C_5H_5N), phosphabenzene (C_5H_5P) and
the cyclic P_5 ligand analogous to cyclopentadienyl which is
described in more detail in section 18.3.1.

The bonding in these aromatic sandwich compounds
may be described briefly. The basic bond in ferrocene

is a single bond of π symmetry between the iron atom
and each ring. This bond is formed by overlap of the d_{xz}
and d_{yz} orbitals on the iron (the z axis is the molecular axis,
and these two d orbitals are of equal energy) with that
aromatic orbital on each ring which has one node passing
across the ring. These two metal d orbitals and the two ring
orbitals combine to give two bonding π orbitals, of equal
energy, and two antibonding orbitals. Each of these orbitals
is 'three-centred' on each ring and the iron atom. There are
four electrons in ferrocene which fill the two bonding orbitals.
In the corresponding cobalt and nickel compounds, the extra
electrons enter the antibonding orbitals, making these compounds less stable and giving cobaltocene one unpaired
electron and nickelocene two unpaired electrons (as there are
two degenerate π* orbitals). One component of the π bond is
shown in Figure 16.9. Other overlaps add smaller contri-

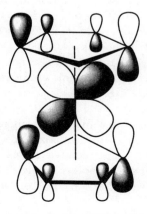

FIGURE 16.9 *One component of the principal ring-metal-ring bond in*
ferrocene
This figure shows the interaction between the ring orbitals and the
d_{xz} orbital of the iron (where the z axis is the ring-metal-ring axis).
The ring-d_{yz}-ring interaction is similar.

butions to the bonding, but this is the main interaction. The basic feature is that there is a single metal-ring bond and the rings are aromatic, undergoing aromatic substitution and similar reactions. Analogous orbitals can be constructed for the species with other C_nH_n rings.

Many compounds are known which contain both carbonyl and pi-bonded organic ligands. The wide variety of types is illustrated by the species $\pi\text{-}C_5H_5M(CO)_x$ (M = V, Nb, Ta, $x = 4$; M = Mn, Tc, Re, $x = 3$; M = Co, Rh, Ir, $x = 2$). The dimers $[\pi\text{-}C_5H_5M(CO)_x]_2$ are found with M = Cr, Mo, W and $x = 3$ or with M = Fe, Ru, Os and $x = 2$, or with M = Ni, Pt and $x = 1$.

Mixed metal species include

$$\pi\text{-}C_5H_5(CO)_2FeMo(CO)_3(\pi\text{-}C_5H_5) \text{ and}$$
$$\pi\text{-}C_5H_5(CO)_3MoMn(CO)_5,$$

and other pi ligands include, in addition to cyclopentadienyl, C_5H_5, the groups C_4H_4, C_6H_6 or C_7H_7 or derivatives of these. Just as C_5H_5 may be regarded as contributing five electrons to the central atom, these latter groups contribute respectively four, six or seven electrons. Also well-known are allyl derivatives like $\pi\text{-}C_3H_5Mn(CO)_4$ where the C_3H_5 group contributes three electrons and diene derivatives where the substituent contributes four electrons, as in $\pi\text{-}C_6H_8Re(CO)_3H$.

Boron hydride analogues of ferrocene exist. For example, the $B_9C_2H_{11}^{2-}$ ion, which presents an open face consisting of a pentagon of three boron and two carbon atoms, can replace the $C_5H_5^-$ ions. Compounds $(B_9C_2H_{11})Fe(C_5H_5)^-$ and $(B_9C_2H_{11})_2Fe^{2-}$ are formed which are ferrocene analogues, and these are oxidizable to Fe(III) compounds, such as $(B_9C_2H_{11})_2Fe^-$, which are analogues of the ferricinium ion. This and many other *carborane* ions have been shown to replace the cyclopentadienyl ion in a variety of other compounds such as the cobalticinium ion and $(C_5H_5)Mn(CO)_3$.

16.5 The organometallic chemistry of the lanthanides

An organic chemistry of the lanthanides has developed steadily since the establishment in the 1950s of the cyclopentadienyl $(C_5H_5)^-$ compounds formed by the reaction

$$LnCl_3 + 3NaC_5H_5 \rightarrow Ln(C_5H_5)_3 + 3NaCl.$$

Such species are known for all lanthanides, though the route to $Eu(C_5H_5)_3$ had to be indirect to avoid reduction to the europium(II) species $Eu(C_5H_5)_2$. The compounds are air and moisture sensitive and all except $Sc(C_5H_5)_3$ readily form $Ln(C_5H_5)_3D$ species with a range of donor molecules including $D = NH_3$, THF, or R_3P. The properties are basically ionic. The commonest structures have the C_5H_5 rings and are planar with the Ln^{3+} ion lying above the centre with all five $Ln - C$ distances equal. Further weaker interactions occur leading to polymeric units. Thus in $Sc(C_5H_5)_3$, one ring bridges two Sc atoms (in the 1, 3 positions) giving a structure with long chains $\{-(C_5H_5)_2Sc-C_5H_5-Sc(C_5H_5)_2-C_5H_5-\}_\infty$. In a similar way, the related $Nd(C_5H_4Me)_3$ is a tetramer in the solid.

Closely related halides are also well-known, e.g.

$$LnX_3 + 2NaC_5H_5 \longrightarrow Ln(C_5H_5)_2X$$

or

$$LnX_3 + Ln(C_5H_5)_3 \longrightarrow Ln(C_5H_5)_2X.$$

The structures are dimers with halogen bridges

and similar bridges form to other elements as:

These halides may be converted to other organometallics such as $[(C_5H_5)_2LnR]_x$ where R = H or an organic group. These compounds are usually dimers with electron deficient bridges:

(compare 9.6 and figures 9.9, 9.13). If R is a bulky group such as $C(CH_3)_3$ or $CH(SiMe_3)_2$, monomers are formed and these are also stabilized by donor molecules in $(C_5H_5)_2$ LnR.D species. One interesting reaction is the abstraction of H from a $C(CH_3)$ substituent:

$$3(C_5H_5)_2Ln[C(CH_3)_3].THF + LiCl$$
$$\rightarrow [Li(THF)_4][\{(C_5H_5)_2Ln\}_3(H)_3Cl]$$

where the anion has a core structure containing two types of bridging H

Bulky groups allow the isolation of species with direct Ln—C bonds of formulae LnR_3 or $[LnR_4]^-$. The latter is stabilized by bulky cations such as $(Li(THF)_4)^+$. Such compounds are found for ligands R = $C_6H_3Me_2$, $C(CH_3)_3$, CH_2SiMe_3 etc.

These organolanthanide compounds are broadly parallel to those formed by the s elements, but they do allow a unique opportunity to assess steric effects. The Ln^{3+} ions are large and compounds are clearly most stable when the ligands are very bulky. As the size of Ln^{3+} varies in small steps from La^{3+} to Lu^{3+}, effects sensitive to size can be seen. Thus $[Ln(C_6H_3Me_2)_4][Li(THF)_4]$ could be isolated only for Ln = Yb or Lu. Similarly, $[Ln\{C(CH_3)_3\}_4][Li(THF)_4]$ was formed for Ln = Sm, Er, Y, Yb, and La.

Two interesting types of compound resulted when the ligands were made more bulky. First, the cyclooctatetraene

($C_8H_8^{2-}$) sandwich compounds first found for the actinides (see $U(C_8H_8)_2$, Figure 16.11a) were paralleled in the species $K(diglyme)^+Ln(C_8H_8)_2^-$. The crystal structure of the cerium compound shows Ce in the centre of a sandwich formed by parallel planar C_8H_8 rings. Divalent lanthanide species $M(C_8H_8)$ and $M(C_8H_8)_2^{2-}$ are known for M = Yb, Eu and Sm.

A second bulky ligand of interest is the $C_5Me_5^-$ ring. This is too bulky to allow formation of $Ln(C_5Me_5)_3$ but it does stabilise Ln–R in $(C_5Me_5)_2LnR$ species. In addition, it has given the most manageable compounds known to date of divalent lanthanides as in

$$SmI_2 \cdot THF + C_5Me_5K \rightarrow [(C_5Me_5)(THF)_2SmI]$$
$$\downarrow C_5Me_5K$$
$$(C_5Me_5)_2Sm(THF)_2$$
$$\downarrow 75\,C$$
$$(C_5Me_5)_2Sm.$$

The iodide is again a dimer with

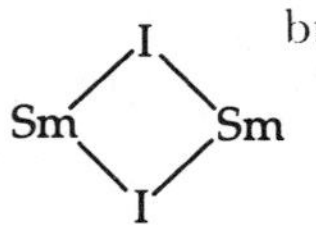

Sm〈I〉Sm

bridges.

In $(C_5Me_5)_2Sm(THF)_2$, the coordination around Sm is roughly tetrahedral. When the THF molecules are removed, the structure remains bent with the ring-Sm-ring angle = 140°. Similarly, in $(C_5Me_5)_2Yb$, the angle = 158°. $(C_5H_5)_2Sm$ was prepared more recently.

Whereas the predominant oxidation state for organic derivatives of the lanthanides is the trivalent state (with the exceptions caused by the stability of the f^0, f^7 or f^{14}

configurations, typified by Eu(II), Ce(IV), etc.), there is a developing chemistry of zerovalent organometallic compounds of the lanthanides. In an attempt to prepare the first example of a zerovalent sandwich complex $Sm(p\text{-}C_6Me_6)_2$, analogous to dibenzene chromium (Figure 16.7b), the reaction of $SmCl_3$, Al, $AlCl_3$ and C_6Me_6 was found to give the Sm(III) complex $(\pi\text{-}C_6Me_6)Sm(AlCl_4)_3$, Figure 16.10a. The synthesis of zerovalent π-arene complexes was eventually achieved by employing the technique of metal vapour synthesis in which metal atoms (from a heated filament) are co-condensed with the organic molecule of interest at low temperature. By this method a number of zerovalent π-sandwich complexes have been synthesised from 1,3,5-tri-t-butyl benzene, and 2,4,6-tri-t-butyl pyridine, such as the complexes shown in Figure 16.10b.

Organolanthanide chemistry presents interesting parallels with the chemistry of the s elements and B or Al species, a way of undertaking detailed study of steric effects, and a number of unusual species not found elsewhere in the Periodic Table.

16.6 Actinide organometallic chemistry

Interest in the organometallic chemistry of the actinides is relatively recent. The larger size gives a preference for higher coordination and the main examples are formed by the IV and III states.

The cp_4E molecules are formed by Th, Pa, U and Np and the centroids of the four rings are arranged tetrahedrally around E. As well as cp and cp*, a variety of cyclopentadienyls form similar compounds. Th and U also form cp_3ER and cp_3EX compounds in the IV state. In these molecules electron transfer from E into the ring π orbitals is relatively unimportant in the bonding and the main interaction is donation from ring π into the $6d$ and $5f$ orbitals. Both are significant, but the f contribution is weaker as the orbitals are large and dispersed. There are similar cp_2EX_2 and cp_2ER_2 species but these are less common.

In oxidation state III all the actinides from Th to Cf have been shown to form cp_3E, again with a variety of cp analogues. The centroids are arranged in a plane triangle around around E. Cp_3Cm and cp_3Cf are the only well-characterized organometallic compounds of these elements.

In a benzene π-complex, $(C_6Me_6)U(BH_4)_3$, the ring is planar and the centre, together with the three borohydrides, gives a tetrahedral array at U. The U–B distance indicates $U(\mu\text{-}H)_3BH$ coordination through three bridging hydrogens.

While carbonyl chemistry of the actinides is not well developed, we should note the uranium carbonyl, $cp_3^{**}UCO$ (where cp** is the relatively bulky trimethylsilyl substituted ring, $Me_3SiC_5H_4$). The CO and the three ring centres lie tetrahedrally around U. The U–CO bonding (compare section 16.2.2) consists of σ donation by CO into the $6d_{z^2}$ orbital on the U and back-donation from U $5f$ orbitals into the CO π^* orbitals.

A further effect of the large size is the ability to stabilize a larger ring system. There was great interest in the discovery,

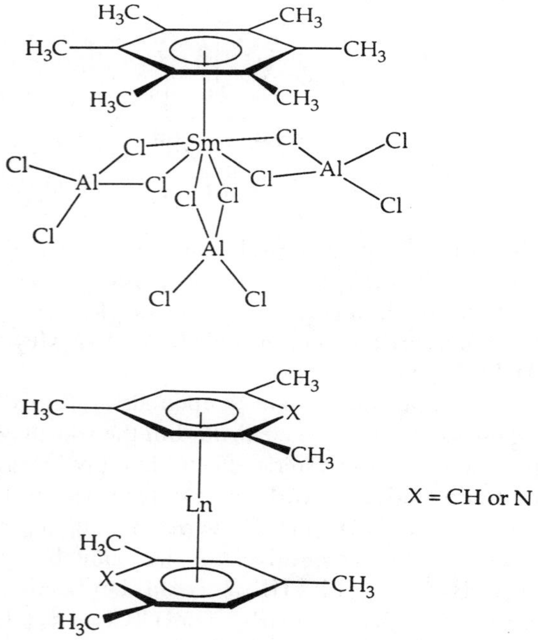

FIGURE 16.10 *π-arene complexes of the lanthanide elements*
(a) $(\pi\text{-}C_6Me_6)Sm(AlCl_4)_3$. (b) $(\pi\text{-arene})_2Ln$ (arene = tri-t-butyl benzene or tri-t-butyl pyridine)

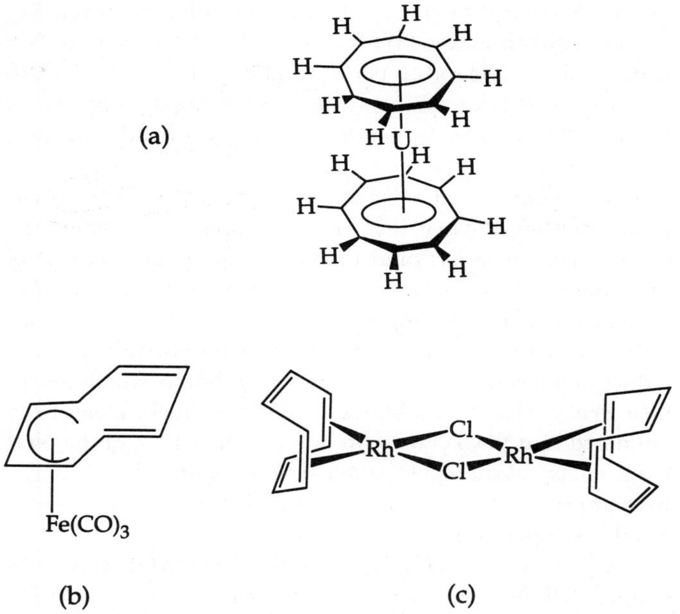

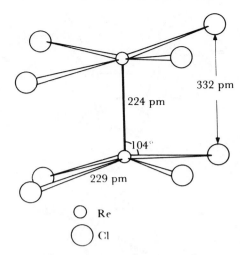

(a)

Fe(CO)₃

(b)   **(c)**

FIGURE 16.11 *The coordination modes of cyclooctatetraene*
(a) as a planar dianion in uranocene, $U(C_8H_8)_2$ (i.e. octahapto),
(b) as a delocalized diene (tetrahapto) in $(C_8H_8)Fe(CO)_3$, and (c)
as four alkenes (dihapto) in the $[(C_8H_8)RhCl]_2$ dimeric complex.

in 1968, of *uranocene*, $U(C_8H_8)_2$, which has two planar 8-membered rings forming a sandwich structure analogous to ferrocene (Figure 16.11a). Similar $E(C_8H_8)_2$ compounds were soon reported for Sc and other *f* elements including Th, Pa, Np and Pu. In contrast, the cyclooctatetraene dianion prefers alternative coordination modes to transition metals, as illustrated in Figures 16.11b and c.

The $E(C_8H_8)_2$ species are very sensitive to oxygen and water but have now been made with a wide range of substituted rings. Half-sandwiches, such as $Th(C_8H_8)Cl_2$ $(THF)_2$, are also found.

Simple alkyls and aryls are rare though the anion $[ThMe_7]^{3-}$ has been stabilized with a complex cation and $Th(CH_2Ph)_4$ is also relatively stable. E-alkyl bonds are more stable in the presence of π ligands as noted above.

16.7 Multiple metal-metal bonds

The discovery in 1964 of the ion $Re_2Cl_8^{2-}$ brought into the focus of chemists' attention the existence of metal-metal bonds of high bond order. The $Re_2Cl_8^{2-}$ ion has the eclipsed structure shown in Figure 16.12. There are two extraordinary features of this structure: the extremely short $Re-Re$ distance of 224 pm (compare 274 pm in Re metal) and the eclipsed configuration which makes the $Cl-Cl$ distance between opposite halves of the molecule markedly less than twice the Cl van der Waals' radius (compare p. 31).

It was soon demonstrated that a wide range of similar rhenium(III) species existed, of general type $Re_2X_8^{2-}$ where X was a univalent anion, and with $Re-Re$ distances in a very narrow range around 222 pm. Such species often occurred with further ligands, such as H_2O, on the Re atoms in the axial position *trans* to the $Re-Re$ bond but this sub-

332 pm

224 pm

104°

229 pm

○ Re

◯ Cl

FIGURE 16.12 *Structure of the* $Re_2Cl_8^{2-}$ *ion*

stitution had only a minor effect on the $Re-Re$ distance. A further series of compounds with very short $M-M$ bond lengths were those with carboxylic acid groups bridging the two metals, of the type $Re_2(O_2CR)_4^{2+}$. Mixed species like $Re_2Cl_4(O_2CCH_3)_2$ are also found. The dinuclear species bridged by four acetate groups, $M_2(O_2CCH_3)_4$, was already well-known e.g. for $M=Cr$ (Figure 14.20) or $M=Cu$ (Figure 14.33). The molybdenum acetate, $Mo_2(O_2CCH_3)_4$ has a short $Mo-Mo$ distance of 211 pm, thus extending the class of multiple metal-metal bonds to Mo(II). The chromium compound has $Cr-Cr$ of 238 pm (compare the metal, 258 pm) so it also belongs to this class. (Note that not all dimeric acetates show multiple bonds—the copper compound has a long $Cu-Cu$ distance of 265 pm compared with 256 pm in the metal).

The Cr_2 molecule has been identified in the gas phase from photolysis of $Cr(CO)_6$ and has an extremely short $Cr-Cr$ bond length of 168 pm, which was discussed as a sextuple bond.

A further group of short metal-metal bonded species was found for neutral compounds M_2X_6 where $M=Mo(III)$ or $W(III)$ and $X=$ halide, NR_2^-, OR^- etc. Other formulae include the rather rare $M-C$ sigma bond in $W_2(CH_3)_8^{4-}$, bridging carbonate or sulfate analogous to the acetates e.g. $Re_2(SO_4)_4^-$, and the cyclooctatetraene derivatives $M_2(C_8H_8)_3$ $(M=Cr$ or $W)$ with one bridging C_8H_8 group (Figure 16.13a).

The first stable technetium species was $Tc_2(O_2CCMe_3)_4Cl_2$ (Figure 16.13b) which shows axial ligands and seems to require them for stability as $Tc_2Cl_8^{4-}$ is unstable.

Bonding

The unusual properties of $Re_2Cl_8^{2-}$ were explained by Cotton by postulating a quadruple $Re-Re$ bond, Figure 16.14. If we consider the $Re-Re$ axis as z, and x and y to lie along $Re-Cl$ directions, then the square planar $ReCl_4$ unit will be bonded using the Re s, p_x, p_y and $d_{x^2-y^2}$ orbitals. The $Re-Re$ sigma bond is then formed using the d_{z^2} orbital, on each Re. (If we

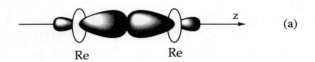

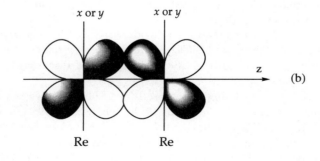

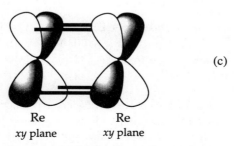

FIGURE 16.13   *Unusual metal-metal bonding*
(a) $W_2(C_8H_8)_3$, (b) $Tc_2(O_2CCMe_3)_4Cl_2$

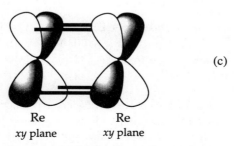

(a)

(b)

(c)

FIGURE 16.14   *The quadruple bond in* $Re_2Cl_8^{2-}$
(a) sigma component, (b) pi (one of the two components), (c) delta component.

include p_z contributions as well, a second orbital on each Re, pointing outwards along the z axis, is suitable for bonding the additional axial ligands.) The d_{xz} and d_{yz} orbitals on each Re may then overlap to give two Re—Re π bonds. This leaves the d_{xy} orbital on each rhenium to form a δ bond (compare p. 45).

Thus a bond of order 4 may be formulated with the occupancy $\sigma^2\pi^4\delta^2$. We see that the δ component explains the eclipsed configuration (and indeed was proposed because of this configuration) whereas a *pair* of π bonds allows any configuration. The δ component makes a relatively weak contribution to the total bond strength since the two nodal planes reduce the electron density (compare π bonds which are in turn weaker than sigma bonds—see Figure 3.13). It has been estimated that the δ contribution is less than 15 % of the total bond energy. Calculated stabilization energies in $Mo_2Cl_8^{4-}$ were approximately $-7\cdot3$, $-6\cdot1$ and $-4\cdot9$ eV for σ, π and δ electrons respectively.

If we take the $Re(III)_2X_8^{2-}$ and the carboxylate species as bases, it will be clear that the corresponding Cr(II), Mo(II) or W(II) species $M_2X_8^{4-}$ and $M_2(O_2CR)_4$ are isoelectronic and will also be formulated with a quadruple bond. If we then move to Mo(III) or W(III), and at the same time take two X^- ligands away, we leave enough electrons to form $\sigma^2\pi^4$.

Thus we obtain the well-represented M_2X_6 class. As the triple bond does not impose barriers to rotation, these species are staggered in configuration like ethane. We can add a donor group (retaining the triple bond), as in $Mo_2(OR)_6\cdot2NHMe_2$, and find that the M_2X_8 configuration is adopted with approximately planar $M(OR)_3(NHMe_2)$ units, but the two ends of the molecule are again staggered.

A triple bond also results if 2 electrons are added to the quadruple bond since the lowest empty orbital is δ^*. Thus we get the $\sigma^2\pi^4\delta^2\delta^{*2}$ configuration for $Re_2Cl_4(PEt_3)_4$ (regard the 4P lone pairs as replacing $2Cl^-$ lone pairs and $2Cl\cdot$ single electrons—hence the two additional electrons).

Finally, if we remove one electron from the quadruple bond we get $\sigma^2\pi^2\delta^1$, corresponding to a bond order of 3·5. Two pairs related in this way are $Mo_2(SO_4)_4^{4-}$ with $Mo_2(SO_4)_4^{3-}$ and the unstable $Tc_2Cl_8^{2-}$ with $Tc_2Cl_8^{3-}$. In most cases, oxidation causes structure change as in the reaction of $Re_2X_8^{2-}$ with halogens to form the face-sharing bis-octahedral structure of $Re_2X_9^-$.

Quite often, reactions of these multiply-bonded species occur with little change to the M_2 unit. A striking case is the addition of axial donor groups: for example the Mo—Mo distance in $Mo_2[O_2CCF_3]_4$ is 208 pm and in $(C_6H_5NMo)_2[O_2CCF_3]_4$ it is 213 pm. Some M_2X_6 species undergo insertion reactions

$$W_2(NMe_2)_6 + 6CO_2 \rightarrow W_2[O_2CNMe_2]_6$$
$$Mo_2(NMe_2)_6 + 4CO_2 \rightarrow Mo_2[(NMe_2)_2(O_2CNMe_2)_4].$$

The substituted species have very similar M—M bond lengths but the M coordination number increases. For example, $W_2(O_2CNMe_2)_6$ has two bridging W—OC(NMe₂)—OW

TABLE 16.3 Species with multiple metal-metal bonds

	$M-M$ distance (pm)	Bond order	Comments
$V_2[C_6H_3(OMe)_2]_4$	220	3	Few examples
$Cr_2[O_2CR]_4$	185–254	4	Many examples: very variable bond length
$Cr_2[C_8H_8]_3$			
$Cr_2[C_5Me_5(CO)_2]_2$	228	3	
$Mo_2[O_2CR]_4$ $Mo_2X_8^{4-}$	209–214	4	Many examples including mixed species and extra D
$Mo_2[SO_4]_4^{3-}$	216	3·5	Compare 211 pm in $Mo_2[SO_4]_4^{4-}$
Mo_2X_6	220–224	3	Many examples including extra D
$W_2[Cl_3Me_5]^{4-}$	226	4	Also $W_2Me_8^{4-}$
$W_2[C_8H_8]_3$	238	4	
W_2X_6	227–230	3	Many examples including extra D
$Tc_2[O_2CCMe_3]_4Cl_2$	219	4	$Tc_2Cl_8^{4-}$ very unstable
$Tc_2Cl_8^{3-}$	212	3·5	
$Re_2X_8^{2-}$	220–224	4	Many examples including mixed species and extra D
$Re_2[O_2CR]_4^{2+}$	223	3	
$Re_2O_8^{8-}$	226	3	in $La_4Re_2O_{10}$
$Ru_2(O_2CR)_4Cl$	228	4 (3 unpaired electrons)	Chain structure linked through anion Cl^-
$Rh_2(O_2CCH_3)_4$	239	3?	Compare 246–255 for Rh=Rh and 280 ± 15 for Rh–Rh

Notes 1. Where a range is given, this shows the commonest bond lengths. Unusual species greatly widen the range—e.g. 218 pm in $Re_2Me_8^{2-}$ and 226 pm in $Re_2Cl_5(diphos)_2$.

2. X = halide, OR^-, NR_2^- and other monovalent monodentate anions. O_2CR = carboxylic anion or other bridging bidentate species. D = neutral donor such as H_2O, R_3N, R_3P.

3. Multiple bonds are found in rather different types of compound,
    (a) For Fe in $[(C_4R_4)_2Fe](\mu-CO)_3$ and $[(R_3P)_3Fe]_2(\mu-H)_3$ which have Fe–Fe = 218 pm and 234 pm respectively (bond order 3 postulated).
    (b) For Ir, in $[(R_3P)_2(H)Ir]_2(\mu-H)_3^+$ a triple bond is also postulated.
    (c) Long triple bonds are found for Cr, Mo, W in species of the type $(C_5H_5)_2M_2(CO)_4$.

units together, one bidentate and one monodentate ligand on each W, making W five-coordinate to oxygens.

These metal-metal bonds of high order complete a natural progression which starts from very weak interactions—e.g. Cr–Cr of 391 pm in $Cr_2Cl_9^{3-}$ and the $Cu_2[O_2CCH_3]_4$ case mentioned above—through single bonds (compare Figures 16.3, 16.4), and double bonds to these cases of order 3 to 4. Double bonds are fairly widespread and we quote just one type

$$(C_5H_5)M \overset{X}{\underset{Y}{=}} M(C_5H_5)$$

found for a number of metals M (e.g. Fe, Rh) and a range of bridging groups X and Y including CO, NO, NR_2 and organic groups. The steps in bond order are not nearly as sharp as they are in main group chemistry (e.g. between

C–N, C=N, and C≡N) but the whole range of species represents a continuum of metal-metal interactions.

16.8 Transition metal clusters

Many areas of interesting chemistry involve compounds where a number of metal atoms are bonded together directly and the M–M bonds are often supported by bridging ligands. If we use a broad definition of a cluster as any species with more than two M atoms linked together, then we would include the compounds of Figures 15.25, 15.30, 16.4 and 16.5 as clusters. Such compounds represent only the tip of the iceberg of a large field of substantial current interest. Clusters are known under ambient conditions ranging in size from three to over a hundred metal atoms and an even larger variety has been detected by spectroscopic and other means. The development has stemmed partly from improved synthesis and characterization—in particular in the greatly increased accessibility and improved performance of X-ray crystallography. In addition, the work has been driven by the possibility of using soluble metal clusters in place of heterogeneous catalysis on metal surfaces. With the smaller clusters this idea largely failed, mainly because a group of a few metal atoms does not reproduce the properties of bulk metal or even of metal surfaces. However, the larger clusters which have become available more recently do begin to act as bulk metal and interesting developments can be expected. Thus, in a Pt_{309} cluster, it was found that 147 core atoms do behave like bulk metal whereas the 13 core atoms in Au_{55} still differ in charge density from the bulk. Work with clusters of this size is right at the limit of current techniques, falling within the limits in which X-ray crystallography and scanning electron microscopy are most effective (section 7.11). Much work has been reported on smaller clusters, with intensive study of species containing from 4 to about 20 metal atoms.

We will look briefly at three classes of these metal clusters.

16.8.1 *Halide clusters*

We have already met a few examples of transition metal clusters in the M_6X_n species formed by Zr (Figure 15.3), Nb and Ta (Figure 15.9), or Mo and W (Figure 15.19). Further halogen-bridged clusters are reported, including $[Ti_6C](\mu\text{-}Cl_{12})Cl_6$ which resembles Figure 15.9b but also contains a central C atom in an octahedral site.

Extended halogen-bridged structures are known, including chains and sheets of linked octahedra, and this area extends into the layer and three-dimensional structures of the lower halides and oxides. An exotic example is $[Y_{16}Ru_4I_{20}]_n$ where a central tetrahedron of Ru atoms is sheathed by 16 Y which form fused octahedra containing Ru at the centre. The whole metal cluster is sheathed by Y–I bonds and the I atoms bridge to other clusters giving a three-dimensional network.

However, a major focus in cluster work has been on species like the carbonyls which have no or few bridging groups and are clearly held together by M–M cluster bonding.

16.8.2 *Carbonyl and related clusters*

To sample this extensive field of chemistry we look first at some moderate-sized clusters and then at a small selection of larger species.

In Table 16.1 we have noted the formation of $M_4(CO)_{12}$ carbonyls by Co, Rh and Ir and there are a variety of related compounds such as $[H_2Ir_4(CO)_{10}]^{2-}$, $Ru_2Co_2(CO)_{13}$ or $[Fe_4(CO)_{13}]^{2-}$ which all contain a tetrahedron of metal atoms with various combinations of terminal and bridging CO. In contrast, $[HFe_4C(CO)_{12}]^-$ and $[Os_4N(CO)_{12}]^-$ have a butterfly structure with two triangles sharing an edge (Figure 16.15a). Similarly, $Os_5(CO)_{16}$ contains a trigonal bipyramidal Os_5 skeleton (Figure 16.15b) while $Os_5C(CO)_{15}$ has a square pyramid of Os atoms (Figure 16.15c). Many examples of octahedral clusters of metal atoms are known including $M_6(CO)_{16}$[M = Co, Rh, Ir (Figure 16.15d), M = $M(CO)_2$, remaining CO bridging] and $[M_6(CO)_{18}]^{2-}$ or $M_6C(CO)_{17}$ [M = Fe, Ru, Os (Figure 16.15d), M = $M(CO)_3$, and C at the centre of the octahedron]. Two

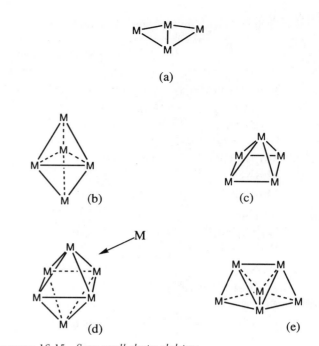

FIGURE 16.15 *Some small cluster skeletons*
(a) M_4 butterfly, (b) M_5 trigonal bipyramid, (c) M_5 square pyramid, (d) M_6 octahedron, also indicating a capping site for M_7, (e) M_6 capped trigonal bipyramid.

octahedra joined by a Rh–Rh$(\mu^2\text{-}CO)_2$ bridge occurs in $[Rh_{12}(CO)_{30}]^{2-}$. Interstitial atoms are now well-established with B, C and N found in smaller clusters while larger units, such as C_2 or even C_3, and heavier atoms like Si, Ge, Sn, P, As, or Sb, are found interstitially in larger clusters. Even interstitial H is now well-established, usually in an octahedral site as in $[HRu_6(CO)_{18}]^-$.

Rationalization of such variations has been approached by use of the 18-electron rule (see section 16.2.1) or by an extension of SEP theory (compare section 9.6). You can determine that the 18-electron rule works for the four- and five-metal examples above but fails for the six-metal ones.

For example, the two Fe_4 species give electron counts of 60 for $[Fe_4(CO)_{13}]^{2-}$ and 62 for $[HFe_4C(CO)_{12}]^-$. To achieve $4 \times 18 = 72$ electrons thus requires six Fe–Fe bonds in the former case and five in the latter. The tetrahedron allows 6 M–M bonds along the six edges, whereas the pair of triangles sharing one edge need only 5 M–M bonds. The cluster $[Fe_6(CO)_{18}]^{2-}$, for example, has 86 electrons, compared with 108 required by the 18-electron rule which would require 11 M–M bonds. However, the octahedron has 12 edges (and the structure is fully symmetrical) so the idea of each metal being joined by a normal bond breaks down.

To apply Wade's rules we note that the modification required is to count 12 electrons per metal atom for external bonding, in place of the 2 electrons per B atom in the boranes. The calculation of the number of skeletal electron pairs for $[Fe_6(CO)_{18}]^{2-}$ for example, then proceeds: from the 86 electrons subtract $6 \times 12 = 72$ electrons for external bonding, leaving 7 SEP. Thus (compare Table 9.6) the expected structue is an octahedron. Note that in electron counting of these species atoms like C or N which are encapsulated are taken to add all their valency electrons to the count and to be incorporated in the skeletal bonding without requiring any specific allocation of electrons.

A further example is the two Os_5 cases quoted above: $Os_5(CO)_{16}$

5 Os valency electrons = 5 × 8		= 40
16 CO contributing 2 electrons each		= 32
less 12 electrons per Os		= −60
Giving 12 electrons or 6 SEP		

Thus the structure is the 5-vertex cluster, a trigonal bipyramid (Figure 16.15b: all $M = Os(CO)_3$ apart from one equatorial $Os(CO)_4$ unit).

$Os_5C(CO)_{15}$

5 Os	40 electrons
C	4 electrons
15 CO	30 electrons
less 12 per Os	−60 electrons
Giving 7 SEP	

Thus the structure is an octahedron less one vertex, a square pyramid (Figure 16.15d: all $M = Os(CO)_3$ and the C lying just below the base).

The series $[Os_6(CO)_{18}]^{2-}$, $Os_7(CO)_{21}$ and $[Os_8(CO)_{22}]^{2-}$ is interesting as the counts are each 7 SEP. Thus the expected structures are octahedra and the extra metal atoms in the two latter cases are in capping positions (compare Figure 16.15d).

A significant test case is $Os_6(CO)_{18}$. While there are a large number of octahedral M_6 complexes, this molecule is one of a few examples of the capped trigonal bipyramid (bicapped tetrahedron) shown in Figure 16.15e with all $M = Os(CO)_3$. The electron count (48 + 36 less 6 × 12) gives only 6 SEP thus predicting a trigonal bipyramid with the extra atom taking up a capping position.

Out of the large number of carbonyl clusters with 10 or more metal atoms we can look briefly at only one or two structural themes. Figure 16.16 shows the skeleton of the $[H_5Os_{10}(CO)_{24}]^-$ ion and related clusters. The electron

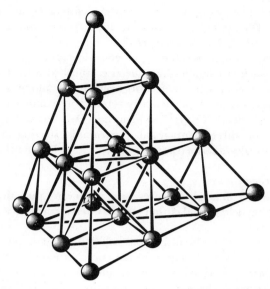

FIGURE 16.17 *Structure of the metal core of osmium atoms in the giant cluster* $[Os_{20}(CO)_{40}]^{2-}$

count gives 7 SEP so this is an octahedron with four caps. These are the outer apices in the figure and each is an $Os(CO)_3$ unit. All the rest are $Os(CO)_2$ units defining the octahedron. An alternative way of looking at this structure is as a large tetrahedron with atoms placed at the mid-points of each edge or, alternatively, built up of layers containing 1,3 and 6 atoms. A big brother is known, $[Os_{20}(CO)_{40}]^{2-}$ (Figure 16.17), which is a giant tetrahedron containing layers of, successively, 1,3,6 and 10 atoms. In a rather similar way $[Ni_{38}Pt_6(CO)_{48}]^{6-}$ is close to a giant octahedron with 10 atoms in each face (or layers of 1,4,9,16,9,4 and 1 atoms).

Such layers are in the positions corresponding to close-packing. A further family of structures can also be regarded as being built up from close-packed layers. The simplest example is $[Rh_{13}(CO)_{24}]^{5-}$, Figure 16.18, which has a central Rh inside a regular cluster—alternatively seen as close-packed layers of 3,7 and 3 atoms. Larger clusters of

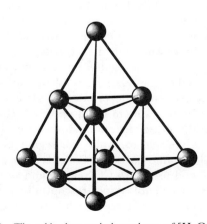

FIGURE 16.16 *The cubic close-packed metal core of* $[H_4Os_{10}(CO)_{24}]^-$

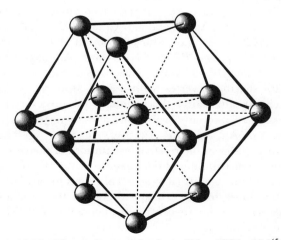

FIGURE 16.18 *The metal core of the cluster* $[Rh_{13}(CO)_{24}H_x]^{(5-x)-}$

this type are for example, a Rh_{22} species with layers arranged 3,6,7,3, $[Pt_{26}(CO)_{32}]^{2-}$ with 7, 12, 7 layers and a Pt_{38} species with 7, 12, 12, 7 layers.

A further theme is stacked polygons. Compounds with up to five stacked triangles have been established and stacked squares or pentagons have also been found, along with a hexagonal prism. A particularly symmetric member of this family is $[Pt_{19}(CO)_{22}]^{4-}$ which has $\mathbf{D_{5h}}$ symmetry and consists of three pentagonal bipyramids sharing apices (alternatively described as a 1, 5, 1, 5, 1, 5, 1 system).

16.8.3 Gold clusters

Different structural features are seen in the rapidly developing field of gold clusters which often incorporate silver atoms also.

Gold(I)–sulfur chemistry has yielded the complex ion $[Au_{12}S_8]^{4-}$ which has an unusual cubic structure in which the S atoms are at all the corners and the 12 Au atoms are in the centres of the edges. The structure is slightly distorted with angles at S in the range 87–93° and the Au atoms in essentially linear configuration, as preferred. The Au$\cdots$ Au distances are 318–335 pm which suggests a weak interaction (see section 16.13 for discussions of Au$\cdots$Au interactions).

In lower formal oxidation states gold forms an extensive range of cluster compounds. In $[Au_4(PR_3)_4]^{2+}$ gold atoms, each with one terminal phosphane, form a regular tetrahedron with very short Au–Au edges. Short distances are also found, for example in $[Au_6(PR_3)_6]^{2+}$ which has formal oxidation state 0.33. The gold atoms, again each with one terminal phosphane, form an axially compressed octahedron. This form is paralleled in $[Au_7(PPh_3)_7]^+$ in which the 7 Au form a pentagonal bipyramid with a short axial Au–Au distance.

Further examples of the poly-gold clusters include $[Au_9L_8]^{3+}$, $[Au_{11}L_7]^{3+}$, $[Au_{13}L_{12}]^{3+}$ and related species where L includes phosphanes and halides. The Au_{13} species has one Au atom at the centre of an Au_{12} icosahedron. The other two species also have a centred Au with the remaining atoms forming an incomplete icosahedron. Evidence has been presented for the much larger $[Au_{55}(PPh_3)_{12}Cl_6]$ cluster containing an inner Au_{13} icosahedron surrounded by an outer shell of Au, Au–P, and Au–Cl units. Very large mixed silver–gold clusters have also been found. A closely related pair are $[Au_{13}Ag_{12}(PR_3)_{12}Cl_7]^{2+}$ and $[Au_{13}Ag_{12}(PR_3)_{10}Br_8]^+$ (Figure 16.19) whose structures consist of two icosahedra sharing a vertex with a gold atom at the centre of each. An alternative description of this structure is as a succession of alternate pentagons and single atoms in an array: Ag, 5Au, Au, 5Ag, Au, 5Ag, Au, 5Au, Ag. The two structures differ in the relative orientations of the pentagons and in the distribution of the ligands. In the chloride ten of the phosphine ligands (R = para-Tol) bond to the Au atoms and two more to the two apical Ag. In the bromide the halogens bridge the two central Ag_5 pentagons and also bond to the terminal Ag.

A number of larger clusters belong to this family, including

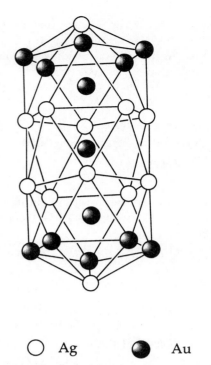

○ Ag   ● Au

FIGURE 16.19   *The metal core of the cluster* $[(R_3P)_{10}Au_{13}Ag_{12}Br_8]^+$ *(R = p-tolyl)*

$[Au_{18}Ag_{17}(PPh_3)_{12}Br_{11}]^{2-}$ and $[Au_{22}Ag_{24}(PPh_3)_{12}Cl_{12}]$ whose structures are also based on icosahedra sharing vertices. These large and magnificent structures show how giant molecules can be assembled using clusters (in these cases the centred M_{13} icosahedron) as the building blocks to form clusters of clusters.

The L–Au(I) unit readily attaches to other metal clusters so that gold is incorporated in clusters with many metals. Examples are VAu_3L_9, containing a VAu_3 trigonal pyramid, and $Fe_4Au_2(C)L_{14}$ in which the 2 Au atoms occupy neighbouring positions in an octahedron enclosing the C. In such compounds L is generally CO or a phosphane, or similar ligands in combination.

16.8.4 Very large clusters

Both the metal carbonyl and the gold cluster work builds up to cluster sizes where metal surface and even bulk metallic properties start to appear. One approach was that of Schmid (see Appendix A) who argued that, while clusters probably needed a surface layer of ligands for stability, the larger the cluster the smaller the relative size of the surface. Thus by progressively reducing the proportion of ligands, such as phosphanes, he was able to build up larger clusters. Then, working in the other direction, downwards in size from colloids, it is possible to probe the region between large clusters and metallic particles. Colloids such as palladium have been found to absorb CO readily and surface Pd–CO groups have been identified spectroscopically by infrared and ^{13}C nmr. One driving force for this work is the study of catalysis.

16.9 Metal-dioxygen species

Several areas of work have converged to create the current interest in species containing O_2 groups bonded to metals. Compounds resulting from the action of hydrogen peroxide have long been of interest, e.g. in analysis—see section 14.4 for the Ti, V and Cr species as examples. Secondly, the reversible uptake of oxygen gas by haemoglobin (see Figure 20.4) and other oxygen-carrying proteins has been shown to depend on dioxygen-metal coordination. Much work has gone into the study of such systems and into the syntheses of simpler model compounds which might aid understanding. A third area of interest lies in catalysis of oxidation by O_2: here the basic species is again likely to be an O_2-M unit formed on the metal surface in the case of a heterogeneous catalyst or formed by homogeneous catalysts such as metallo-enzymes.

There has been considerable debate about the formulation of metal-dioxygen species. First we recognize that it may

be difficult to distinguish, let us say, M(I) plus neutral O_2 from M(II) plus O_2^- (superoxide) from M(III) plus O_2^{2-} (peroxide), since it may not be possible to determine the degree of electron transfer from M to the O_2 species. An example is given by the cobalt species on p. 248, where green $[(NH_3)_5Co-O-O-Co(NH_3)_5]^{5+}$ was originally formulated as a superoxo species containing one Co(IV) atom. Secondly, the method of synthesis does not, as was thought, give any guidance. For example, direct addition of molecular O_2 is found (16.4)

$$[IrCl(CO)(PR_3)_2] + O_2 \rightarrow [(O_2)IrCl(CO)(PR_3)_2] \quad (16.4)$$
$$\text{A} \qquad\qquad\qquad\qquad \text{B}$$

but A is iridium (I) and B is best formulated as a peroxo complex of iridium (III). Indeed, it has now been demonstrated that the same species results either by the action of O_2

TABLE 16.4 $M-O$ and $O-O$ bond lengths (pm) in peroxo MO_2 units

Ti	V	Cr	Mn	Fe	Co	Ni
145–46	144–47	140–46			142–45	obs.
185–89	187–88	181–92			187–90	
Zr	**Nb**	**Mo**	**Tc**	**Ru**	**Rh**	**Pd**
obs.	148–51	138–55	obs.	obs.	142–47	obs.
	197–204	191–97			202–03	
Hf	**Ta**	**W**	**Re**	**Os**	**Ir**	**Pt**
obs.	obs.	150	obs.	obs.	130–52	145–51
		193			200–07	201

(a) Top value is range of $O-O$ distances in the reported compounds and the lower value is the $M-O$ range.
(b) Obs. = peroxo species observed but no structural data.

TABLE 16.5 Parameters in metal-dioxygen species

Bond Type	Metal (a)	Average $O-O$ distance (pm)	Average $O-O$ frequency (cm^{-1})	Number of examples (b)
End-on (1)	Fe(III), Co(III), Rh(III)	125	1134	9
Side-on (2)	Table 16.4	145	881 (3d elements)	33
			872 (4d elements)	66
			850 (5d elements)	75
Planar bridge (3a)	Co(III)	131	1110	5
Twisted bridge (3b)	Mn(III), Fe(III), Co(III), Rh(III), Mo(VI), Co(II)?	144	807	7
Double bridge (4)	V(IV)	149		1

(a) Metal species forming this structure type.
(b) Number of examples giving the stretching frequencies.
(c) Stretching frequencies (cm^{-1}) of $O_2 = 1556\,cm^{-1}$, $O_2^- = 1145\,cm^{-1}$, $O_2^{2-} = 770\,cm^{-1}$.

or by the traditional reaction of hydrogen peroxide in a number of cases such as (16.5)

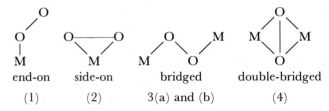

$$[Co^{III}(H_2O)_2(diars)_2]^{3+}$$
$$+ H_2O_2$$
$$[(O_2)Co^{III}(diars)_2]^+ \dots (16.5)$$
$$[Co^I(diars)_2]^+ + O_2$$

(diars = bidentate arsane ligand, see Appendix B).

However, there is now sufficient evidence from molecular structure determinations to show that $M - O_2$ units exist as one or other of the forms (1) or (2).

end-on	side-on	bridged	double-bridged
(1)	(2)	3(a) and (b)	(4)

There are also two bridged forms: 3(a) where the $M - O - O - M$ unit is all in one plane and 3(b) which is non-planar with a twist angle generally about $145°$ between the two MOO planes. A final, double-bridge species has recently been reported.

The *side-on* form (2) is by far the commonest, and the structures of about fifty species have been determined. The two $M - O$ distances are usually equal within experimental limits, and $M - O$ increases with metal size while $O - O$ distances are relatively constant. Table 16.4 lists the ranges found. The $O - O$ distances fall in the range 130–155 pm and 90% are between 140 and 152 pm, while the overall average is 146 pm (compare 149 pm in O_2^{2-} and 147 pm in H_2O_2).

The number of established end-on structures of type (1) is much smaller but these are characterized by much shorter $O - O$ bond lengths (compare $O_2^- = 128$ pm). This bond length evidence is supplemented by the $O - O$ stretching frequencies, which have been assigned for a much larger number of complexes than have had crystal structures determined. Furthermore, when the $M - O - O - M$ bridged structures are examined it is found that the values for the planar species (3a) fall into the same range as the end-on MO_2 species (1), while those for the non-planar species (3b) agree with the side-on MO_2 species (2). The values are summarized in Table 16.5.

Thus, apart from the double-bridge example, there are only two basic types of metal-dioxygen species:
(A) those with a shorter $O - O$ distance and stretching frequency above 1075 cm^{-1}.
(B) those with a longer $O - O$ distance and stretching frequency below 950 cm^{-1}. Type A shows the end-on structure or the planar bridge while type B shows the side-on structure or twisted bridge. While there are a few marginal cases, most known species fall firmly into one or other class.

The species of type A are formulated as *superoxo* compounds, formally derived from the superoxide ion O_2^-. The species of type B are *peroxo-*, based on the peroxide ion O_2^{2-}. Bonding to the metal mainly involves the filled π^* orbitals (compare section 3.4) of the dioxygen unit. These electrons are markedly stabilized by their delocalization into metal orbitals and the reduced π^* density in the O_2 unit leads also to some stabilization of the $O - O$ bond (seen especially in the slight shortening of the $O - O$ bond and increase in stretching frequency for the peroxo-compounds). These descriptions for A and B are supported by a number of other physical properties and by theoretical analysis (compare references).

16.10 Compounds containing $M - N_2$ units and their relationship to nitrogen fixation

There is considerable chemical interest in the activation of a molecule as stable and strongly-bonded as N_2, and in comparisons between complexes of N_2 and the isoelectronic species, especially CO and C_2^{2-}. By studying metal–dinitrogen complexes we can begin to obtain an understanding of the fundamental chemical steps which occur during nitrogen fixation by the so-called *nitrogenase* enzymes. This is of great technological importance—much energy is utilized annually in sysnthesising ammonia fertilizer from dinitrogen at high temperatures and pressures. If we can begin to understand the fundamental chemistry occurring in these enzymes we can design catalysts to synthesize ammonia under very mild (and hence inexpensive) conditions.

A discussion on the nature of the structures of the active sites of the nitrogenases is given in the section on bioinorganic chemistry in Chapter 20, together with some of the attempts at directly mimicking the complexes responsible for the nitrogen fixation process. In this section we look more closely at the interaction of the dinitrogen molecule with transition metal centres and what reactions the coordinated N_2 molecule undergoes, which is of relevance to the problem of nitrogen fixation. Of particular importance is the reduction of a coordinated N_2 molecule to reduced, but still metal-bonded, nitrogen-containing species and ultimately to the desired, fully reduced product—ammonia.

16.10.1 *Metal compounds with coordinated dinitrogen species*
There are two distinct classes of reaction which fix dinitrogen into other chemical compounds:

(A) Under non-aqueous conditions and with powerful reducing agents, N_2 can be fixed in the form of air-sensitive and often poorly-characterized materials which react with H_2O or other proton sources to yield NH_3, or partly-reduced species such as N_2H_4. Of course, the simplest species of this type, such as Li which forms Li_3N with N_2, have long been known (compare section 10.5). More complex systems are illustrated by the titanium species discussed in section 14.2. A number of similar systems are well established but all depend on active organometallics as the reducing agents.

Thus these systems do not appear to provide a suitable route for industrial N_2 fixation, but they model the take-up of N_2 under very mild conditions and allow study of the steps of the subsequent conversion. Thus $TiCl_3$ and Mg react with N_2 at $25\,^\circ C$ forming a species postulated as $TiNMg_2Cl_2$, via Ti^{II} and subsequent reaction with Mg. Under even milder conditions, at $-78\,^\circ C$, $(Me_5C_5)_2Ti$ gives $(Me_5C_5)TiN_2$ which changes structure at -62° and evolves N_2 quantitatively at $20\,^\circ C$. The two structures suggested are $(Me_5C_5)_2Ti-N\equiv N$ and $(Me_5C_5)_2Ti-\overset{N}{\underset{N}{|||}}$. [$Me_5C_5$ is the fully-substituted analogue of C_5H_5: the structures are probably 'bent sandwiches' of the type found in $(C_5H_5)_2TaH_3$, see Figure 15.7). Other model reactions include the formation of N_2H_4 from N_2 via $Ti(OR)_2$ polymer [$R = (CH_3)_2CH$], and the ready formation of a substituted diimine complex $(C_5H_5)_2Ti(N_2Ph_2)$ perhaps with the sideways link

The compound $[Me_5C_5ZrN_2]_2N_2$ contains both linear $Zr-N\equiv N$ units ($N\equiv N$ of 112 pm) and bridging $Zr-N\equiv N-Zr$ ($N\equiv N$ of 118 pm). On addition of HCl, both N_2 and N_2H_4 are evolved and tracer studies showed half the hydrazine nitrogen was from the terminal N_2 and half from the bridge. The intermediate $(Me_5C_5)_2Zr(N_2H)_2$ was postulated.

We may add to this class the work in aqueous systems which are a long way away from biological conditions of moderate pH and solution reactions. An example is provided by the observation that a coprecipitate in base of Mo(III–V) with 10% Ti(III) converts N_2 via N_2H_2 to hydrazine, with the active species probably Mo(IV). This type of system may become important in larger-scale synthesis, as it is cheap and accessible, even though it does not model the natural fixation route. We note again the formation of hydrogenated N–N compounds.

(*B*) *In aqueous media*, relatively stable complexes are formed, mainly by elements in the middle of the transition series especially in d^6 states. These contain N_2 as a ligand but the complexes are unreactive or release unchanged N_2 when attempts are made to reduce them under mild or moderate conditions. Thus these compounds allowed a detailed study of the modes of $M-(N_2)$ bonding.

The first of these compounds to be made was $[Ru(NH_3)_5N_2]^{2+}$ (see Figure 15.27). Many others are now known (see Table 16.6 for some examples). Common preparations involve strong reduction in presence of N_2 and other ligands, as in

$$WCl_4(PR_3)_2 + Na/Hg + N_2 \rightarrow W(N_2)_2(PR_3)_4$$

or replacement of labile ligands, which often occurs reversibly under mild conditions.

$$Ru(NH_3)_5(H_2O)^{2+} + N_2 \rightleftharpoons Ru(NH_3)_5(N_2)^{2+} + H_2O$$
$$CoH_3(PR_3)_3 + N_2 \rightleftharpoons CoH(N_2)(PR_3)_3 + H_2$$

A wide range of such coordinated dinitrogen compounds has been made, with various types of $NN-M$ bonding, and differing stabilities. The next step in modelling fixation is to consider that it is unlikely that the naturally-occurring reduction

$$N_2 \rightarrow 2NH_3$$

is a direct one-step process. Instead, it seems probable that a number of intermediates follow the formation of the first complex, $S-N_2$ (S = substrate: here the metal atom site of nitrogenase).

A scheme which is consistent with most of the currently accepted observations is shown in Figure 16.20. Here each addition step involves the supply of both H^+ and an electron, and the paths branch at some points. Thus, in addition to complexes containing bound N_2 in all its modes, work has also been carried out on diazenido (NNH), diazene or diimine (NHNH), hydrazido (NHNH$_2$), and hydrazine complexes, as well as single-nitrogen species containing $M-N$, $M-NH$, and $M-NH_2$ groups.

To illustrate this, we cite one series of experiments.

The addition of H to N_2 bound to W has been established with reactions such as

trans $[W(N_2)_2(diphos)_2] + 2HCl$
$$\rightarrow [W(N_2H_2)Cl_2(diphos)_2] + N_2 \quad (A)$$

The Mo complexes behave similarly, as does HBr: diphos in a bidentate ligand such as $R_2PCH_2CH_2PR_2$. The complex (A) is 7-coordinate but will lose a halide ion.

$$(A) + BF_4^- \rightarrow [W(N_2H_2)Cl(diphos)_2]^+ [BF_4^-] + Cl^- \quad (B)$$

Structural studies show that the N_2H_2 group differs between (A) and (B). In (A), it behaves as a 2-electron donor—the

TABLE 16.6 Some representative stable N_2 complexes

Terminal N_2	Bridging N_2	
	$[(C_5H_5)_2Ti]_2N_2$	
$(C_6H_6)Cr(CO)_2N_2$		
$P_4M(N_2)_2$	$[(C_6H_6)PM]_2N_2$	M = Mo, W
$(C_5H_5)M(CO)_2N_2$		M = Mn, Re
$(C_5H_5)Re(CO)(N_2)_2$		
$P_4ReCl(N_2)$		
$P_3M(H)_2N_2$		M = Fe, Ru, Os
$[(NH_3)_5MN_2]^{2+}$	$[(NH_3)_5M]_2N_2$	M = Ru, Os
$P_3(H)CoN_2$	$(P_3Co)_2N_2$	
$P_2MCl(N_2)$		M = Rh, Ir
$P_3NiH(N_2)$	$[P_2Ni]_2N_2$	

Notes (1) PR_3 (phosphane) or $\frac{1}{2} \times$ (diphosphane) = P.
(2) All are d^6 except the compounds of Ti(d^3), Co(d^8 and d^9), Rh and Ir(d^8) and Ni(d^9 and d^{10}).

unsymmetrically coordinated diimine HN = NH in the form

However, the loss of the halide ligand in (B) would give the metal a 16 electron configuration if the N_2H_2 group remained a 2-electron donor. Instead it forms the isomeric di-anion of hydrazine, H_2N-N^{2-} and acts as a 4-electron donor in the form

$$M = N - NH_2$$

In both (A) and (B), the N_2H_2 group is stable to further reduction. However, if the diphosphine is replaced by two mono-phosphines, such as PMe_2Ph or $PMePh_2$, then such Mo or W dinitrogen complexes are reduced completely to NH_3 plus some N_2H_4 as in

$$trans-[(N_2)_2W(PMePh_2)_4]+H_2SO_4$$
$$\rightarrow 1\cdot 9NH_3 + \text{trace } N_2H_4.$$

The Mo analogue yields $0\cdot 7NH_3$ per mole of complex. The overall scheme postulated, via successive additions of H^+, is

$$M-N\equiv N \rightarrow M-N=NH \rightarrow M=N-NH_2$$
$$\rightarrow M-NHNH_2 \rightarrow M=NH+NH_3:$$
$$M=NH \rightarrow M-NH_2 \rightarrow M(VI)+NH_3.$$

The metal goes from (0) to (VI), and a number of the intermediates have been isolated.

Experiments such as this have been carried out to clarify all the steps in the general scheme of Figure 16.20, and several other minor paths have been found. The main thrust of the work, in addition to synthesis, has been the study of the protonation-deprotonation reactions of these nitrogen complexes leading to overall nitrogen fixation reactions such as

$$L_xM + N_2 \rightarrow L_xMN_2 + 6H^+ \rightarrow L_xM + 2NH_3$$

There has also been considerable exploration of parallels between these nitrogen complexes and C_2H_4 complexes or the various dioxygen species (compare previous section).

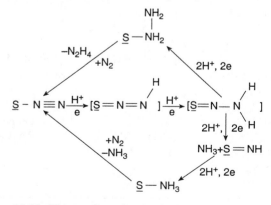

FIGURE 16.20 *The transformations involved in the conversion of substrate (S) bound dinitrogen into NH_3 or N_2H_4. Each step involves transfer of a proton and the corresponding electron.*

The work on model fixation reactions may be summarized by saying that separate laboratory reactions are now known which reproduce many of the observed or postulated steps in the reaction of N_2 with *nitrogenase*, such as those in Figure 16.20. These have not been combined into one catalytic cycle, different steps have been demonstrated with different nitrogen complexes, and S represents a variety of transition metal + ligand combinations, but not yet the natural substrate in *nitrogenase*.

The whole fixation problem has turned out to be a very complex one. Three decades of work have led to an enormous increase in understanding and development of quite new classes of compound.

16.10.2 *Bonding between N_2 and M*

Although many fewer N_2 complexes are found, their general properties are similar to the carbonyls (section 16.1) and a similar linear bond

$$M-N\equiv N \quad \text{cf.} \quad M-C\equiv O$$

is expected. This is indeed found in all structural studies of stable terminal nitrogen complexes (compare Figure 14.29). The $M-N-N$ angle is usually very close to 180° and the NN bond, in accurate structures, is in the range 112 ± 2 pm compared with 109·8 pm in the N_2 molecule. Again, as with carbonyls, the stretching frequency is found in the triple bond region, usually in the range 1990–2160 cm^{-1} compared with 2345 cm^{-1} in N_2.

This all points to a bonding model for $M-N\equiv N$ very similar to $M-C\equiv O$ (see Figure 16.2). However, there are significant differences. First (see Table 3.4), the highest filled orbital in N_2 is at $-15\cdot 57$ eV compared with $-14\cdot 01$ eV in CO. This is the sigma orbital involved in the sigma $M \leftarrow NN$ (or $M \leftarrow CO$) bond. That is, the electrons donated *to* the metal are more tightly held in N_2 than in CO, and the sigma bond to the metal provides less energy in the nitrogen complexes. This is partly compensated by the lower energy of the empty π^* orbital (-7 eV for N_2, -6 eV for CO) involved in the π donation *from* the metal $M \rightarrow NN$ or $M \rightarrow CO$, but the difference of 0·6 eV (ca. 58 kJ mol^{-1}) is an indication of the expected increased stability of the carbonyl over the dinitrogen system. This is reflected by the much smaller number of N_2 complexes known and by the fact that most contain only one N_2 group. In the $M-N-N$ unit, the sigma donation is more important for the $M-N$ bond formation than is the pi back donation, and the end-on coordination should be appreciably stronger than the side-on configuration. The main effect of the pi back donation in both configurations is to weaken the $N-N$ bond, providing the required activation. In the side-on configuration, the N_2 unit donates both sigma and pi electrons to the metal, considerably weakening the $N-N$ bond, and it is suggested that this may be the active intermediate in the reduction reaction.

Although we have not discussed them in detail, the terminal metal acetylides, $M-C\equiv CH$, are also very similar to the nitrogen complexes.

When we come to bridging groups, we find a linear $M-N\equiv N-M$ system quite different from bridging carbonyls (compare Figure 16.3b and d) but found for acetylides $M-C\equiv C-M$. Here each $M-N$ bond is similar to that in the terminal complexes, and a fully-linear system is expected. As charge transfers are relatively small, and the total effect of the upper energy levels in $M-N\equiv N-M$ is approximately non-bonding for NN, we find the bond-lengths and stretching frequencies in the bridged nitrogen complexes are generally similar to those in the terminal species, again comparable with the acetylides and in contrast to the bridging carbonyls. The carbonyl type of bridging *is* found when the $N-N$ bond order is reduced as in the imine complex $[Pt(PR_3)_2N_2H]_2^{2+}$ which contains the bridges

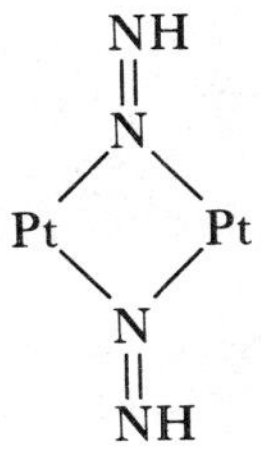

$$
\begin{array}{c}
NH \\
\| \\
N \\
Pt \diagup \quad \diagdown Pt \\
N \\
\| \\
NH
\end{array}
$$

Although 'sideways on' nitrogen, $M\text{---}\begin{smallmatrix}N\\ \||\\ N\end{smallmatrix}$, was postu-

lated in reactions, stable complexes with this bonding are rare. One report is of *trans* $(R_3P)_2Rh(Cl)N_2$ and the analogues $(R_3P)_2Rh(Cl)O_2$ and $(R_3P)_2RhCl(C_2H_4)$ $[R = CH(CH_3)_2]$. Each contains a $Rh\text{---}\begin{smallmatrix}X\\ |\\ X\end{smallmatrix}$ unit $(X = N, O$ or $CH_2)$ with the X_2 perpendicular in the $RhClP_2$ plane and its mid-point completing the square-planar configuration expected for $Rh(I)$. The reaction with Rh is relatively weak.

16.11 Metal–dihydrogen complexes

16.11.1 Discovery

In section 15.5.1 we discuss the formation of the remarkable $[ReH_9]^{2-}$ complex which was unambiguously proved by X-ray and neutron diffraction (compare section 7.4) to contain 9 $M-H$ bonds. Following this and other determinations of complexes L_xMH, it became accepted that H could act as a ligand even in complexes of metals in states which are relatively strongly oxidizing. Polyhydrides of the more stable high oxidation states such as $Ta(V)$, $W(VI)$ or $Os(VI)$ are noted in Chapter 15. It was also widely accepted that metal–dihydrogen units, $M-H_2$, were intermediates in the activation of H_2 by metal catalysts, and low-temperature matrix studies gave evidence for species such as $Pd(H_2)$ and $Cr(CO)_5(H_2)$.

In 1983 startling evidence appeared that the H_2 molecule could coordinate as an entity in complexes which were stable at room temperature. During a study of SO_2 complexes a purple compound turned yellow under N_2 and yielded the dinitrogen complex $M(CO)_3L_2(N_2)$, from which the N_2 could be displaced to form the desired SO_2 complex. The

purple compund was later identified as the 16-electron intermediate $M(CO)_3L_2$ $[M = Mo, W: L = a$ bulky phosphane such as $P(iso\text{-}Pr)_3]$. The significant discovery was that H_2 behaved in exactly the same way as N_2 to give a yellow complex containing 2 H on addition to the purple intermediate. Furthermore, addition of N_2 resulted in replacement of H_2 by nitrogen to give the dinitrogen complex. Two interpretations were possible: (i) that two normal $M-H$ bonds were formed giving $M(CO)_3L_2(H)_2$ (Figure 16.21b) and involving oxidation to $M(II)$ which happens to be yellow, or (ii) that a new type of molecule is formed, $M(CO)_3L_2(H_2)$ (Figure 16.21a), in which dihydrogen is bound as a molecule as N_2 is in dinitrogen complexes.

The experimental resolution of this question is interesting as it highlights the need that often arises to use a combination of techniques. First, the infrared spectrum showed no band in the regions $2300-1700\ cm^{-1}$ and $900-700\ cm^{-1}$ where $M-H$ stretching or bending occur, but there were bands at 1570, 960, and $465\ cm^{-1}$ which shifted in the D_2 analogue, as expected from the mass effect (see section 7.7) for vibrations involving hydrogen. An X-ray crystal structure study can often not detect H in presence of heavy atoms, and there were additional experimental problems involving disorder. The incomplete structure that emerged did show the three CO and the two phosphane ligands in octahedral positions, suggesting that the hydrogen occupied the sixth site. The hydrogen was located using neutron diffraction on the W complex, which showed an $H-H$ unit coordinated sideways on to the metal, parallel to the $P-W-P$ axis (Figure 16.21a). However, the disorder left some remaining doubt and since this totally new mode of bonding by H_2 required further proof, nmr studies were undertaken.

Metal hydrides have a characteristic resonance position and the H signal is split by coupling to the P atoms and to the magnetic ^{183}W isotope (section 7.8). Any coupling between two $M-H$ bonds would be fairly small and give a doublet. For W complexes sharp lines would be expected. The yellow hydrogen complex showed a resonance close to the $M-H$ range but the signal was unexpectedly

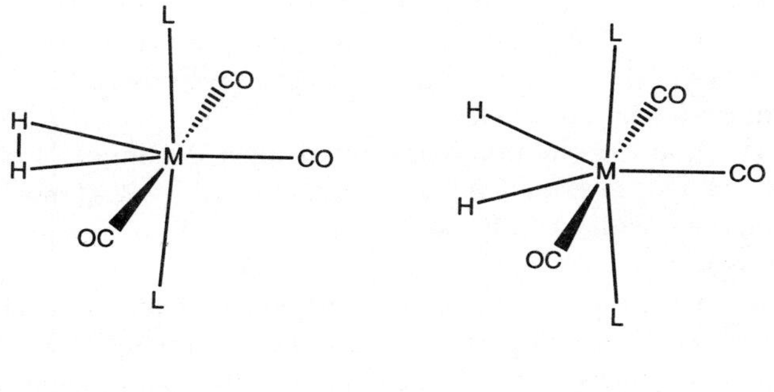

(a) (b)

FIGURE 16.21 *Complexes formed between a metal entity and* H_2
(a) coordinated dihydrogen, (b) metal dihydride with $M-H$ bonds.

broad. The HD analogue was therefore made for nmr study. D has a spin $I = 1$ and therefore gives a 1:1:1 triplet on coupling. This was indeed found with $\mathcal{J}(H-D) =$ 33.5 Hz whereas for coupling between separate M–H and M–D bonds, $\mathcal{J}$ is 1–2 Hz. In addition, the HD complex now showed a band in the infrared at 2360 cm^{-1}, which was assigned as the HD stretch [the H–H stretch was later detected in the region 2700–3000 cm^{-1} in $M(H_2)$ complexes]. This adds up to strong supporting evidence for the complex and also shows that the $W-H_2$ structure persists in solution. Additional nmr effects, especially the relaxation times, were also found to be characteristic of dihydrogen coordination.

Other dihydrogen complexes were quickly identified including hydrogen–dihydrogen species such as $[MH(H_2)L_4]^+$ (for M = Fe, Ru, Os) and even $[IrH_2(H_2)_2L_2]^+$. In most examples L_x = phosphane or a combination of phosphane with ligands like CO or cp. The series $M(CO)_3L_2(H)_2$ was completed by the synthesis of the M = Cr species, which was stable only under an atmosphere of H_2.

Re-examination of some of the established polyhydrides led to their re-formulation as dihydrogen complexes. Nmr and neutron diffraction established the structure $MH_2(H_2)$ $(PR_3)_3$ (M = Fe, Ru)—a much more comfortable status than the original formulation as a 7-coordinate Fe(IV) tetrahydride. However, many polyhydrides remain established as solely M–H compounds. While there was some question about Re(VII) species, the current conclusion is that the ReH_7L_2 analogues of $[ReH_9]^{2-}$ are indeed Re(VII) heptahydrides and not dihydrogen complexes of Re(V).

16.11.2 *Properties and bonding*

The H–H bond distance is best determined by neutron diffraction or nmr methods. The observed values range from 85 pm to 102 pm (compare $H_2 = 74$ pm, Chapter 9). The values of the HD coupling in the nmr vary from 35 to 22 Hz, and the coupling decreases as the bonds get longer—as expected since the coupling is mediated by the bonded electrons [$\mathcal{J}(HD) = 43.2$ Hz for the free molecule]. The H–H stretching frequency ranges from 3080 to 2690 cm^{-1}, with the corresponding H–D values around 2300 cm^{-1} (compare free H_2 at 4390 cm^{-1} and HD at 3820 cm^{-1}). The other infrared bands are the $M(H)_2$ stretches in the 1500 cm^{-1} and 900 cm^{-1} regions and lower frequency bending modes.

The heat of formation of the dihydrogen W complex from the purple intermediate is measured as -42 kJ mol^{-1} compared with -57 kJ mol^{-1} for the corresponding N_2 compound.

All these properties indicate an interaction which weakens the H–H bond compared with the H_2 molecule. As H_2 is extremely simple, with no accessible orbitals other than σ and σ^* (Figure 3.7b), the interaction with the metal must involve only these. The proposed model is shown in Figure 16.22 and involves a side-on orientation with

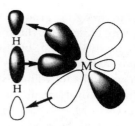

FIGURE 16.22   *Sigma donation and π back-donation proposed for the* $M-H_2$ *interaction*

donation from the H_2 sigma bonding orbital into an empty metal d orbital, supported by back-donation from a filled metal orbital into σ^*. The model is closely analogous to the long-standing one for ethylene–M coordination (Figure 16.6b), except that there the π orbitals were involved. The effect of such an electron transfer is to weaken the H–H interaction, especially by back-donation, in order to gain the M–H interaction. The system is very delicately balanced but, as the H–H bond is very strong, some weakening is tolerable before dissociation occurs. The back-donation is thought to be important for the formation of $M(H_2)$ complexes since acceptors which lack filled d orbitals do not form them. More detailed calculations show that the sigma interaction is the major one, and give values for the parameters which are a good match to the observed ones.

For many complexes there is nmr evidence in solution for dissociation of the dihydrogen complex into 2 MH bonds, i.e. an equilibrium between forms corresponding to Figure 16.21a and b. The dihydrogen complexes thus capture the initial stage of hydrogen activation and support the model proposed for the mechanism of metal catalysis. It is reasonable to apply a similar model to other cases of activation of strong single bonds, such as C–H. It has been observed that C–H bonds in ligands such as phosphanes sometimes lie in an abnormal position close to the transition metal in what is termed an *agostic* interaction, suggesting bond formation of a type very similar to that proposed for dihydrogen. One example is found in the 16-electron purple intermediate above. This clearly models the first stage of C–H activation. Similar activation is seen in Wilkinson's compound, $RhCl(PPh_3)_3$, where an H of one of the phenyl groups of the phosphane ligand coordinates agostically to Rh, explaining to the good hydrogenation catalyst properties of the complex.

16.12 Post-actinide 'superheavy' elements

In the elements up to about $Z = 100$, the mode of decay is principally by alpha- or beta-particle emission and it is the likelihood of these events which governs the half-lives. It will be seen from Table 12.1 that the half-lives tend to decrease as Z increases. However, at about the atomic number of the heaviest elements, another mode of decomposition, spontaneous fission, becomes the dominant one. In this mode half-lives again decrease with increasing atomic number, and it was estimated that, by about element 110, the half-life would have dropped to about 10^{-4} second. When Lr-262 was prepared, it was found to have a half-life of 216 minutes, which is about 500 times longer that expected, suggesting that the older calculations of expected stability were over-pessimistic. However, the expectation of ever-decreasing half-life with increasing Z still remains valid, and the other experimental difficulties of synthesizing new nuclei remain enormous.

At present, there are major groups of experimenters in the United States and in the USSR who have worked for many years on the synthesis of new heavy elements, and who were later joined by a group in Germany. Because of the experimental problems, there have been conflicting claims to be the first to synthesize new elements. The Russian team made first claims for elements 104 and 105 but, because their approach, by heavy ion bombardment, gave isotopes which decayed by spontaneous fission, it was less certain that they had actually made a new element compared to the American experiments which gave isotopes decaying by alpha-particle emission into directly-identifiable daughter products. The very substantial experimental problems, coupled with Cold War suspicions, left a very confused situation.

By international agreement, the final decision on names is made by the International Union of Pure and Applied Chemistry (IUPAC) but the discoverer of an element has the right to recommend the name. Thus various names have been applied and become fairly established over the years, such as rutherfordium or kurchatovium for element 104.

Until priorities were finally decided, and to allow new elements to be named systematically in the interim, IUPAC proposed the following terminology to be used. The ten numbers are shown by

0	nil	1	un	2	bi	3	tri	4	quad
5	pent	6	hex	7	sept	8	oct		
9	en								

Names are formed from these number-elements in order, with rules about elision, and the symbol is formed by the three initials. Thus element 104 is unnilquadium and the symbol is Unq. Needless to say, these names did not find widespread approval in the chemical community, but it is unfortunate that they were not adopted as much confusion would have been avoided.

Recently further experiments settled some of the uncertainties in earlier work. Before recommendations could be considered, the priority of discovery had to be determined. A joint commission of IUPAC and IUPAP (Physics) reviewed all the work and reported where credit should lie for all the elements beyond fermium. In some cases, such as 103, 104 and 105, credit was shared.

The appropriate IUPAC Commission then considered the three sets of proposals for names made by the American, Russian and German groups, together with the credits for discovery.

In balancing three sets of recommendations, the Commission would obviously not please everyone and a particularly controversial factor was their decision (by 16 to 4) to continue the convention that elements should not be named after living persons although the American team wished to honour Glenn T. Seaborg in this way.

The outcome is still rather unsatisfactory for chemists who wish for an unambiguous answer remote from these areas of high politics! The Commission has recommended names for elements 104 to 109, as listed in Table 2.5, and these have been sent out for further comment. Meanwhile, up to four names are in circulation for any one element, and some equivalences are given in Table 16.7. It will probably be 1997 before the final recommendations are made and become official.

The actinide series is complete at lawrencium, $Z = 103$. The same techniques of bombarding actinides with the accelerated nuclei of light atoms produced the next two elements, 104 and 105. Thus, bombarding californium with ^{15}N nuclei gave an isotope of 105, Unp-260, which has a half-life of 1·6 s and decays to Lr-256 with emission of an alpha-particle.

The routes to the heavier elements have used preferred targets whose nuclei are close to the 'magic numbers', corresponding to high stability. Thus elements 106 to 109 have been approached through lead or bismuth targets corresponding to $Z = 82$ or $N = 126$, and the energy of the bombarding particle is kept relatively low in the 'cold fusion' approach. Thus an announcement of the synthesis of element 109 from Darmstadt involved bombarding Bi-209 with accelerated Fe-58 nuclei. Only one in 10^{14} interactions gave element 109, which was formed at the rate of one atom in a

TABLE 16.7 Some alternative names for the transactinide elements*.

Element	Proposed names	Proposed symbols	Proposer
104	Dubnium	Db	IUPAC
	Rutherfordium	Rf	USA
	Kurchatovium	Ku	Russia
	Unnilquadium	Unq	
105	Joliotium	Jl	IUPAC
	Neilsbohrium		Russia
	Hahnium	Ha	USA
	Unnilpentium	Unp	
106	Rutherfordium	Rf	IUPAC
	Seaborgium	Sg	USA
	Unnilhexium	Unh	
107	Bohrium	Bh	IUPAC
	Neilsbohrium	Ns	Germany
	Unnilseptium	Uns	
108	Hahnium	Hn	IUPAC
	Hassium	Hs	Germany
	Unniloctium	Uno	
109	Meitnerium	Mt	IUPAC, Germany
	Unnilennium	Une	
110	Ununnilium	Uun	
111	Unununium	Uuu	
112	Ununbium	Uub	

*Synthetic work on superheavy elements continues apace and initial reports of the preparation of a few atoms of elements up to 112 (the heavy partner of Hg) have appeared (using the heavy atom bombardment method).

week. Identification was purely by use of nuclear properties (using a new velocity filter technique for separation), and four independent properties matched calculation. Decay was by successive alpha emissions to Uns-262, then to Unp-258, and then electron capture gave Unq-258. Similarly, three atoms of element 111 were formed at the end of 1994 by fusing nickel and bismuth. The element undergoes four stages of α decay down to ^{253}Lw. Early in 1996, one atom of element 112 was observed by the German group.

Table 16.8 summarizes the properties of the isotopes of those elements reported by 1996.

TABLE 16.8 Superheavy element isotopes

Element	Isotope	half-life	Other isotopes
104	257	0.8 s	253, 254(?), 255, 256, 258, 260,
	259	1.7 s	262
	261	65 s	
105	258	4 s	255, 257, 260, 261
	262	34 s	
	263	27 s	
106	265	2–39 s	259, 260, 261, 263
	266	20–30 s	
107	261	12 ms	262
	264	0.45 s	
108	265	1.8 ms	264, 269
	267	74 ms	
109	266	3.4 ms	
	268	72 ms	
110	269	0.17 ms	273, also 267?
	271	1.1 ms	
111	272	1.5 ms	271
112	277	0.28 ms	

Clearly these observations on only a few atoms cannot give chemical information, but past experience suggests that quantities will increase, or longer-lived isotopes will be made, and a chemistry will develop.

For element 104, tracer studies are possible, and it does appear to behave as the $6d$ congener of hafnium. Thus it forms the chloride $UnqCl_4$ comparable to $HfCl_4$, and unlike the involatile lanthanide and actinide trichlorides. Element 104 was manipulated in solution as the hydroxybutyrate anion complex, and again its tracer behaviour followed hafnium. Element 105 formed a chloride which was more volatile than $HfCl_4$ but less volatile than $NbCl_5$ and the bromide showed similar parallels, with an estimated boiling point of 430 °C. The last observation involved 'eighteen fission events', that is, only 18 atoms were being observed. Tracer experiments have shown that element 105 adsorbs to glass in a similar fashion to the other group members niobium and tantalum, whereas the Group 4 elements zirconium and hafnium do not show this property. However, in solvent extraction experiments element 105 remained with niobium in the aqueous phase while tantalum extracted into the organic phase. This suggested that the properties of the superheavy elements cannot always be derived by simple extrapolation of the $4d$ and $5d$ element congeners.

The recent discovery of two isotopes of element 106, which may have half-lives as long as 30 s, suggests that chemical studies of this element should be possible in the near future.

With all the reservations appropriate to the tiny scale of work, these observations are in accord with the view that the post-actinide elements are starting to fill the $6d$ shell. In this case, 109 will match iridium, the $6d$ shell would be full at element 112, and then the $7p$ shell should be occupied until element 118 which would be a new rare gas. If it should be possible to go further, there is the interesting possibility that a g shell, the $5g$ level, would start to be occupied somewhere about element 123.

It has been proposed that nuclei with Z values close to a 'magic number', stable proton configurations should be significantly more stable than their neighbours. The work has given much more information about nuclear stabilities, and has been basic to the syntheses up to element 109. Element 112 brings us very close to this region. The nearest nuclear configurations which might provide the so-called 'islands of stability' are around elements 114 and 126. Element 114 would probably be 'ekalead', the highest member of the carbon Group and would be expected to have a stable oxidation state of II and an unstable, possibly non-existent, IV state. Element 126 presents a variety of possibilities, including the chance that it might contain $5g$ electrons.

16.13 Relativistic effects

When the nuclear charge becomes large, the radial velocity of the inner electrons rises to become a significant fraction of the speed of light. This can be envisaged on a simple planetary model—as the positive charge on the nucleus rises, the negative electron has to move faster to remain in a particular orbit. Calculations show that, for elements around Hg with $Z = 80$, the average radial velocity of the $1s$ electron is about 60% of the speed of light. At such speeds, the special theory of relativity shows that the mass increases, by about 20%, and, as a result, the average radius of the orbital contracts by about 20%. The effect is most prominent for orbitals with electron density close to the nucleus, so the main effect is on s orbitals. There is a lesser effect on p orbitals with $m_1 = 0$ but not on the $+1$ or -1 values, causing a splitting of the p set into groups of 1 and 2 orbitals.

There are two principal consequences. First, the s orbitals of the heavier elements become more stable than otherwise expected, and therefore have higher ionization potentials for electron loss or more exothermic electron affinities for electron gain. The second consequence is that the more contracted s orbitals shield the outer orbitals of the d and f sets more effectively from the nuclear charge, so that these orbitals expand and their energies are less.

The chemical consequences are seen mainly in the heaviest elements. While the lanthanide contraction accounts for the similarity between the 4d and 5d series, it has been calculated that relativistic effects contribute about 20% to the contraction of Hf, so that the extremely close similarity between the earlier members of the two series depends in part on the relativistic effect. As we move along the 5d series, the resemblance to the 4d congener decreases (compare sections 15·6 to 15·10) and this is a combination of both relativistic effects. The reduced binding of the 5d electrons allows them to participate more fully, which is seen in the increasing number and stability of the higher oxidation states. Examples are Pt(IV) or Au(III) versus the much less stable Pd(IV) or Ag(III), and the existence of Pt(VI) or Au(V) with no Pd or Ag counterparts. Other examples are clear in Chapter 15.

A second effect is the stabilization of the $6s^2$ configuration. This can be clearly seen on comparing the properties of mercury with its neighbour, gold (and also to a lesser extent with thallium). Metallic mercury has anomalously low melting and boiling points, is monoatomic in the gas phase and the density of mercury ($13.53\,g\,cm^{-3}$) is markedly different from that of gold ($19.32\,g\,cm^{-3}$). Mercury thus appears to be behaving as if it has a rare-gas electronic configuration and this can be nicely accounted for by the stabilization of the $6s^2$ configuration by relativistic effects.

A similar effect, which has been encountered previously (section 15.9.2), is the striking tendency of gold to adopt the same $6s^2$ configuration in the Au^- ion. Thus, the intermetallic compound CsAu is a semiconductor rather than a metal and has a significant amount of ionic character, Cs^+Au^-. Consistent with this, gold has the highest electron affinity and electronegativity outside of the 'typical' electronegative elements, higher than sulfur and almost as high as iodine. Again, gold is showing a strong tendency to form a stable $6s^2$ configuratin. In addition, the surprisingly stable gas-phase Au_2 molecule has a stronger bond dissociation energy than I_2 ($221\ vs.\ 151\,kJ\,mol^{-1}$), and gold, in many respects, can be thought of as a pseudo-halide analogous to iodine, both elements being one electron short of an inert-gas electronic configuration. Another phenomenon in gold chemistry arising from relativistic effects is the strong tendency for gold to form $Au \cdots Au$ contacts in the solid state, similar to halogen $\cdots$ halogen interactions in organic compounds. The energy of this interaction has been estimated to be as strong as $30\,kJ\,mol^{-1}$, comparable with a hydrogen bond, and this is therefore a significant factor in determining the solid-state structures of gold complexes.

A further manifestation of relativistic effects in chemistry is the 'inert pair' effect in the heaviest Main Group elements Tl, Pb and Bi, where the most stable oxidation state is two less than the Group oxidation state (compare sections 17.1,

17.3 and the chemistries of these elements). This is again nicely accounted for by the relativistic stabilization of the $6s^2$ configuration.

The relativistic contraction does not change smoothly with Z, but increases markedly while the 5d shell is filling, with the maximum effect at Au. The effect diminishes to Bi and then changes only slowly through the 5f shell so that effects comparable to Au are seen only around Fm. It appears that the chemisry of the superheavy elements will eventually develop, though probably rather slowly, and the influence of relativistic effects on the chemistry of these elements should prove very interesting indeed. Recent relativistic calculations on the superheavy elements has suggested that these effects may be quite subtle, particularly with regard to p orbital occupation. One interesting pointer is the possible configuration of Lr where the last electron may be in the $7p$ level rather than the 6d one, as a consequence of the relativistic effect on the $m = 0$ level. It has also been recently suggested that the ground state electronic configuration for element 104 (rutherfordium) may be $[Rn]\ 5f^{14}\ 6d^1\ 7s^2\ 7p^1$, while on the other hand that of element 105 (hahnium) may be $[Rn]\ 5f^{14}\ 6d^3\ 7s^2$, by analogy with the configuration of tantalum. Around eka-lead (element 114) there exists the possibility of an 'inert quartet' effect.

The relativistic effect is not the only contribution to the unusual chemistry of the heavier elements, and other rationalizations of the inert pair effect have been proposed, but it is a substantial contribution. It is clear that calculated changes in the relative energies of the s and d levels are in accord with observation and account for many of the unusual features.

PROBLEMS

16.1 Discuss the following organometallic compounds in terms of the eighteen-electron rule:

$(\eta^6\text{-}C_6H_6)Cr(CO)_3$ $Mn_2(CO)_{10}$ $(\eta^5\text{-}C_5H_5)Fe(CO)_2Br$

$(\eta^5\text{-}C_5H_5)Ni(NO)$ $Fe_3(CO)_{12}$ $Mn(CO)_6^+$

16.2 Why is the average C–O stretching frequency in the infrared spectrum of $(Ph_3P)_3Cr(CO)_3$ lower in energy than that of $Cr(CO)_6$?

16.3 Explain the following infrared spectroscopic data:

Compound	C–O stretching frequency (cm^{-1})	M–C stretching frequency (cm^{-1})
$V(CO)_6^-$	1860	460
$Mn(CO)_6^+$	2090	416

17 The Elements of the 'p' Block

17.1 Introduction and general properties

Those elements which have their least tightly bound electron in a p orbital lie in the Main Groups of the Periodic Table headed by B, C, N, O, F, and He. Some aspects of the chemistry of these elements have been discussed in the earlier chapters. The hydrides of the p elements are included in Chapter 9, the structures are discussed in Chapter 4, solid state structures are treated in Chapter 5, while aspects of aqueous and non-aqueous chemistry, including properties of oxyacids, are covered in Chapter 6.

This Chapter aims to give a broad picture of Main Group chemistry, illustrated largely by the simpler compounds. A very limited selection from the huge range of topics of current interest is surveyed in the next Chapter.

Many of the applications of p block elements are listed under the individual elements. One general theme is polymers with inorganic backbones which include the silicones, the polyphosphazenes, and the sulfur nitrides. These are studied to find polymers with specific advantages over organic ones in properties such as electrical or thermal conductivity, or resistance to heat and oxidation. Another application which ranges across various groups is that of high performance ceramics which include B_4C, SiC, Si_3N_4, BN, AlN, Al_2O_3, MgO and mixed oxides with transition elements such as Al_2TiO_5 or $PbZrTiO_3$. In the important field of electronics, the semiconductors Ge and Si are now supplemented by the isoelectronic mixed species like GaAs, InP, ZnSe, and ternary analogues. The use of chemical vapour deposition, as a means of building up precisely oriented layers of specific composition and thickness by pyrolysis of volatile compounds of the contributing elements, has demanded the preparation and handling of very high purity hydrides like SiH_4 or GeH_4, and alkyls like Me_3As Me_3Ga or $InEt_3$. Solar cells involve deposit of Si, or oxides, by similar processes.

The maximum oxidation state shown by a p element (the 'Group oxidation state'), is equal to the total of the valency electrons, i.e. to the sum of the s and p electrons, and is the same as the Group Number in the Periodic Table. (or the Group Number less 10 in the 1 to 18 form of the Table). In addition to this oxidation state, p elements may show other oxidation states which differ from the Group state by steps of two. Clearly, the number of possible oxidation states increases towards the right of the Periodic Table. The most important oxidation states in the various Groups of p elements are shown in Table 17.1. Where oxidation states other than these occur, they usually arise either from multiple bonding, as in the nitrogen oxides, or from the existence of element-element bonds as in hydrazine, $H_2N - NH_2$, $(N = -II)$ or disilane, $H_3Si - SiH_3$, $(Si = III)$.

As fluorine is the most electronegative element, it can show only negative oxidation states and always exists in the $-I$ state. Oxygen is also always negative, except in its fluorides, and is found in the $-I$ state in peroxides (due to the $O-O$ link) and in the $-II$ state in its general chemistry. The other highly electronegative elements, nitrogen, sulfur, and the halogens also show stable negative oxidation states in which they form anions, hydrides, and organic derivatives.

In the boron, carbon, and nitrogen Groups, the Group oxidation state is the most stable state for the lighter elements

TABLE 17.1 Oxidation states among the p elements

Group headed by	B	C	N	O	F	He
Group oxidation state	III	IV	V	VI	VII	VIII
Other states	I	II	III	IV	V	VI
		$(-IV)*$	I	II	III	IV
			$-III$	$-II$	I	II
					$-I$	

*As the electronegativity of C lies between that of H and those of O, N, or halogen, carbon in CH_4 is $-IV$ while carbon in CF_4 is IV, and all intermediate cases occur. The other carbon Group elements are formally in the IV state in their hydrides, and show II and IV in their general chemistry.

in the Group while the state two less than the Group oxidation state is the most stable one for the heaviest element in each Group. The relative stabilities of these two states varies down the Group; thus, in the carbon Group, lead(II) is stable and lead(IV) is strongly oxidizing, tin(II) and (IV) are about equal in stability, germanium(II) is represented by a handful of compounds only, and germanium(IV) is the stable state, while silicon shows only the IV state.

In the oxygen and fluorine Groups the position is more complicated because of the wider ranges of oxidation states. Oxygen and the halogens are most stable in the negative states. The VI state is stable for sulfur and falls in stability for the heavier elements of the oxygen Group, but both the IV and the II states are relatively stable for the heavy elements. Among the halogens, chlorine and iodine both show the Group state of VII in oxyions; bromine(VII) was only recently found. Iodine is also fairly stable in the V and III states. Thus, the general trend for the lower oxidation states to be more stable for the heavier elements is shown by elements such as iodine or tellurium, but more than one state is involved and no simple rule may be given. Of the rare gas compounds, the difluorides are known to exist for radon and krypton as well as for xenon. The other oxidation states in Table 17.1 are shown by xenon.

The general pattern of oxidation state stabilities in the Main Group elements may be summarized: the common oxidation states vary in steps of two, with states lower than the Group state being the most stable for the heavier elements. This trend is opposite to that found among the transition elements.

This pattern of stabilities is reflected in the oxides, sulfides, and halides formed by the Main Group elements, as given in Tables 17.2a and b. A stable oxidation state will form the oxide, sulfide, and all four halides, while a state which is unstable and oxidizing, such as the Group states of the heavy elements, thallium, lead, and bismuth, will not form compounds of the readily oxidizable sulfide or heavy halide ions. Similarly, an oxidation state which is unstable and reducing, such as gallium(I), will form the sulfide and heavier halides, but not the oxide or fluoride.

A similar picture of the stabilities of the oxidation states, but this time in solution, is given by the oxidation state free energy diagrams of Figures 17.5, 17.18, 17.27, 17.43 and 17.55. These diagrams show how the Group oxidation states become unstable with respect to lower oxidation states for the heavier elements in each Group, and they also indicate the instability, with respect to disproportionation, of the intermediate oxidation states in Groups, such as the halogens, which show a number of states.

The effects shown in Tables 17.2 and 17.3 and in the oxida-

TABLE 17.2a Oxides and sulfides of p elements

Elements in oxidation state = number of valency electrons

B_2O_3	B_2S_3	CO_2	CS_2	N_2O_5		
Al_2O_3	Al_2S_3	SiO_2	SiS_2	P_4O_{10}	P_4S_{10}	SO_3
Ga_2O_3	Ga_2S_3	GeO_2	GeS_2	As_2O_5	As_2S_5	SeO_3
In_2O_3	In_2S_3	SnO_2	SnS_2	Sb_2O_5	Sb_2S_5	TeO_3
Tl_2O_3		PbO_2		$(Bi_2O_5?)$		$(PoO_3?)$

Elements in an oxidation state lower by two

		CO		N_2O_3		
				P_4O_6		SO_2
Ga_2O	Ga_2S	GeO	GeS	As_4O_6	As_4S_6	SeO_2
$In_2O?$		SnO	SnS	Sb_4O_6*	Sb_2S_3	TeO_2
Tl_2O	Tl_2S	PbO	PbS	Bi_2O_3	Bi_2S_3	PoO_2

Elements in an oxidation state lower by four

				N_2O		
						TeO
						PoO	PoS

Other compounds

$(BO)_x$	GaS	C_3O_2	$NO, NO_2, N_2O_4,$		
	Ga_4S_5		$(NO_3 \text{ or } N_2O_6)$		
	InS		N_4S_4		
	In_6S_7		(PO_2)	S_2O	
			P_4S_3, P_4S_5, P_4S_7		
			As_4S_3, As_4S_4		
			(SbO_2)		
		Pb_3O_4			

*Antimony trioxide also occurs as $(Sb_2O_3)_n$ chains.

TABLE 17.2b Halides of the elements

Fluorides of the Groups headed by				Chlorides of the Groups headed by				Bromides of the Groups headed by				Iodides of the Groups headed by			
B	C	N	O	B	C	N	O	B	C	N	O	B	C	N	O

elements in oxidation state = Number of valency electrons

MF_3	MF_4	MF_5	MF_6	MCl_3	MCl_4	MCl_5	MCl_6	MBr_3	MBr_4	MBr_5	MBr_6	MI_3	MI_4	MI_5	MI_6
B	C			B	C			B	C			B	C		
Al	Si	P	S	Al	Si	P		Al	Si	P		Al	Si		
Ga	Ge	As	Se	Ga	Ge			Ga	Ge			Ga	Ge		
In	Sn	Sb	Te	In	Sn	Sb		In	Sn			In	Sn		
Tl	Pb	Bi	Po?	(Tl)	(Pb)			(Tl)							

elements in oxidation state = Two less than the number of valency electrons

MF	MF_2	MF_3	MF_4	MCl	MCl_2	MCl_3	MCl_4	MBr	MBr_2	MBr_3	MBr_4	MI	MI_2	MI_3	MI_4
		N				(N)				(N)				(N)	
		P	S			P	(S)			P				P	
	Ge	As	Se	Ga?	Ge	As	Se	Ga?	Ge	As	Se	Ga?	Ge	As	
	Sn	Sb	Te	In	Sn	Sb	Te	In	Sn	Sb	Te	In	Sn	Sb	Te
Tl	Pb	Bi	Po	Tl	Pb	Bi	Po	Tl	Pb	Bi	Po	Tl	Pb	Bi	Po

Other compounds

Fluorides B	Fluorides C	Fluorides N	Fluorides O	Chlorides B	Chlorides C	Chlorides N	Chlorides O	Bromides B	Bromides C	Bromides N	Bromides O	Iodides B	Iodides C	Iodides N	Iodides O
B_2F_4		N_2F_2 N_2F_4 NF_2	OF_2 O_2F_2 O_3F_2 O_4F_2	B_2Cl_4 B_nCl_n ($n=4$, 8–12) $(B_9Cl_8)_2$											
	Si_nF_{2n+2} (up to $n=14$)		S_2F_2 S_2F_{10}		P_2Cl_4 Si_nCl_{2n+2} (up to $n=10$)		S_nCl_2 (up to n ca.100)		Si_2Br_6 Si_3Br_8 Si_4Br_{10}		S_2Br_2		Si_2I_6	P_2I_4	
Ga_2F_4			Se_2F_{10}	Ga_2Cl_4 Ge_2Cl_6			$SeCl_2$ Se_2Cl_2	Ga_2Br_4			$SeBr_2$ Se_2Br_2	Ga_2I_4			
InF_2?			Te_2F_{10}	In_2Cl_3 In_2Cl_4			$TeCl_2$	In_2Br_4			$TeBr_2$	In_2I_4		(Sb_2I_4)	
				Tl_2Cl_3 Tl_2Cl_4	$BiCl$*		$PoCl_2$	Tl_2Br_3 Tl_2Br_4			$PoBr_2$	$Tl^I(I_3)$			

*See section 17.6 for a discussion of this compound.

tion state free energies are also reflected in the ionization potentials (Table 2.8), the electron affinities (Table 2.9), and in the electronegativities (Table 2.14), of the p elements. It will be seen that the five elements in a p Group, although they have many properties in common, split into three sets when their detailed chemistry is examined. This division is:

(i) the lightest element
(ii) the three middle elements
(iii) the heaviest element

The lightest element shows the most marked differences from the rest of the Group, in properties which are discussed in detail in the next section. The heavier elements are discussed in the following section; the division between the heaviest element and the rest is not so marked as that between the first element and the Group, but it is quite distinctive and has been noted above in the stability of the lower oxidation states. It might also be noted that the middle element in the Group does not always fit between the second and fourth

TABLE 17.3 Some bond energies kJ mol^{-1}

(a) From diatomic molecules

H−H	436·0	H−F	566	F−F	158	F−Cl	257
O=O	497·3	H−Cl	431	Cl−Cl	242	F−Br	234
N≡N	945·6	H−Br	366	Br−Br	194	F−I	197
C≡O	1075	H−I	299	I−I	153	Cl−Br	222
NO	626·3					Cl−I	211
						Br−I	179

(b) E values from polyatomic molecules

C−C	348	N−N	160	O−O	146	C−H	416	C−F	485
Si−Si	297	P−P	215	S−S	265	Si−H	326	Si−F	582
Ge−Ge	260	As−As	134	Se−Se	172	Ge−H	289	C−Cl	327
Sn−Sn	240	Sb−Sb	126	Te−Te	138	Sn−H	251	Si−Cl	391
Si−C	301	Bi−Bi	105	O−F	190	N−H	391	Ge−Cl	342
Ge−C	270	N−F	272	O−Cl	205	P−H	322	Sn−Cl	320
Sn−C	226	N−Cl	193	S−F	326	As−H	247	Pb−Cl	244
Pb−C	130	P−F	490	S−Cl	250	O−H	467	C−Br	285
C−N	292	P−Cl	319	S−Br	212	S−H	374	Si−Br	310
C−P	264	P−Br	264	Se−F	285	Se−H	277	Ge−Br	276
C−O	336	P−I	184	Se−Cl	243	Te−H	238	Sn−Br	272
C−S	272	As−F	464	Te−F	335			C−I	213
		As−Cl	317					Si−I	234
		As−Br	243	O−Si	369			Ge−I	213
		As−I	178					Sn−I	272
		Sb−Cl	212	S−Si	227				
		Bi−Cl	278						

multiple bonds		C=C	682	C=N	640	C≡C	962	C≡O	1075
		N=N	450	C=O	732	N≡N	946	C≡N	937
		O=O	402	C=S	477				
				N=O	481				

(c) Some bond dissociation energies, D

O−ClO	243	O=CO	536	F$_3$P=O	544	H−CH$_3$	435
HO−Br	239	H$_2$C=CH$_2$	699	Cl$_3$P=O	511	H−NH$_2$	448
O−NO	306	HC≡CH	963	Br$_3$P=O	498	H−OH	498
O−NN	167	H−CN	540			H−CF$_3$	444
HO−OH	213	NC−CN	607			CH$_3$−F	452
F$_2$N−NF$_2$	83						

elements (compare electronegativities and chemical evidence such as the non-existence of AsCl$_5$), but these deviations are minor and there is a fairly regular trend of properties among the four heavy elements in each Group.

It is useful here to summarize some bond energies of the p elements and hydrogen. Two different types of bond energy are encountered. First, if the bond between each pair of atoms in most molecules is independent of the rest of the molecule, we might hope to compile a list of the energies of individual bonds which would allow us to predict the energy of formation of the molecule. For example, the energy of the process of forming SiH$_3$F from its atoms

$$Si + F + 3H \rightarrow SiH_3F$$

should be given by three times the Si−H energy plus the Si−F energy. Bond energies determined in this way are given the symbol E and are written $E(Si−H)$, etc. Experimentally, such energies are obtained from heats of formation and heats of atomization.

It is also possible to determine the energy required to break a particular bond, leaving two fragments in their ground state and this is called the bond dissociation energy and given the symbol D. Thus $D(HO−OH)$ is the energy required to break the O−O bond in hydrogen peroxide to give two OH radicals. In general, D and E values differ, and D values for successive steps also differ: thus $D(SiH_2F−H)$ is not the same as $E(Si−H)$, or $D(SiHF−H)$ or $D(SiF−H)$. We can also define the average value of the Si−H dissociation energy in SiH_3F, written $\bar{D}$, and this is usually fairly close in value to the corresponding E value. Clearly, for a diatomic molecule, D and E values are equal.

In Table 17.3 are listed bond energies for some diatomic molecules, which may also be used as E values for these bonds, together with a set of other E values chosen to reasonably reproduce known heats of formation. Also included are a few D values to emphasize the difference between the two. For prediction of heats of formation, the E values should be used, but the D values give a better indication of relative bond strengths. Note that values for first row elements forming single bonds, for example the X_2 energies for the halogens, are often unusually small compared with the heavier members. This is another effect distinguishing the lightest elements.

17.2 The first element in a *p* Group

In a *p* Group, the first element differs sharply from the remaining members. The valency shell configuration of the first row elements of the *p* Groups is $2s^2 2p^n$, and the orbital next in energy is the $3s$ level. This is separated from the $2p$ level by a considerable energy gap and is not used in bonding by these elements. The first row elements are accordingly restricted to a maximum coordination number of four (using the $2s$ and the three $2p$ orbitals). In contrast, the second element of a *p* Group, with the configuration $3s^2 3p^n$, has the $3d$ orbitals lying between the $3p$ level and the $4s$ level in energy (compare Figure 8.2). The second row elements make use of these *d* orbitals in bonding to a substantial extent. For this reason, and because of the larger size, coordination numbers greater than four occur. Compare, for example, the fluorine complexes of boron and aluminium where boron can form only BF_4^- while aluminium gives the AlF_6^{3-} ion. Similar behaviour is shown by the third and subsequent members of each *p* Group, which also have *d* orbitals available in the valence shell. A further effect reflects the fact that the first element is unique in the Group in having only the underlying $1s^2$ shell. As a result, the $2p$ orbitals are less diffuse relative to the $2s$ ones than is the case for the higher quantum shells. Thus, the contributions of the $2s$ and $2p$ orbitals to the bonding and lone pairs in a compound are more nearly matched than for the later members. Related to this is the size effect which leads to larger repulsion effects: for example, for an angle of $90°$, the separation of the H atoms in an EH_2 unit is about 14 pm if E is the first member of a Group, and around 25 pm is a second member. In general, see for example Table 4.3, bond angles at second or later members are narrower than at the first Group member. Other, more general, effects of size also contribute to the differences: as with the lightest *s* elements, Li and Be, the biggest relative change of size is between the first and second Group members.

A second, major, difference between the first and subsequent members of a *p* Group is the ability of the first row elements to form strong double bonds using only *p* orbitals (named $p_\pi - p_\pi$ bonds). One striking example is provided by carbon and silicon dioxides. Carbon dioxide is a volatile, monomeric molecule in which all the valency electrons are

used in forming σ and π bonds between carbon and oxygen (section 4.8). In silicon dioxide, π-bonding between the silicon and oxygen does not occur. Instead, each silicon forms a single bond to four oxygen atoms, and each oxygen links two silicon atoms, to form a giant molecule with a three-dimensional structure of single $Si - O$ bonds.

Other examples are provided by the oxides of other first row elements, given in Table 17.2a. In each case (compare, for example, the electronic structure of NO in Figure 3.19), the oxide of the first row element, is a small molecule with $p_\pi - p_\pi$ bonds—CO, CO_2, NO, NO_2, N_2O_5, etc. The elements of the second, and subsequent, rows in the Main Groups rarely form $p_\pi - p_\pi$ bonds, probably because their greater size leads to a much weaker π overlap between *p* orbitals. The situation is illustrated diagrammatically in Figure 17.1 which shows that the sideways extension of the $3p$ orbital is insufficient to give good π overlap in association with the longer σ bond formed by the larger second row atom.

A second factor in this contrast between the tendency of first and second row elements to form $p_\pi - p_\pi$ bonds is the strength of the homonuclear *single* bonds of many of the first row elements. For example, in Table 17.3 it may be seen that the bond energies of $F - F$, $O - O$ or $N - N$ are distinctly smaller than, respectively, $Cl - Cl$, $S - S$ or $P - P$. This reflects repulsions between the nonbonding electrons on these atoms when they are brought close enough together to form the single bond. (Although the doubly-bonded atoms are even closer, some of the electrons are now used in the π bond.) Thus, it seems that there are two factors to be balanced, first that the $p_\pi - p_\pi$ bond is stronger for the first row elements, and second that the sigma bond may be weaker. Thus these elements are found to form a sigma and pi bond between them, rather than using the second orbital and electron to form a second sigma bond to a third atom. For the heavier element the converse case holds, and it is more stable forming

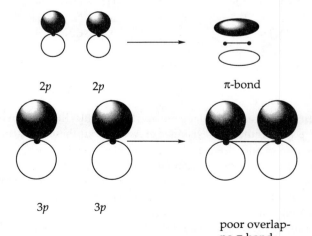

$2p$ $2p$ π-bond

$3p$ $3p$

poor overlap-
no π-bond

FIGURE 17.1 *The poor overlap of 3p and higher p orbitals in π-bonding*
As the second row elements are larger than those of the first row, the sigma bond formed between them is longer. Thus the sideways extension of the $3p$ orbital is not sufficiently large to give good π overlap. Consequently, a double bond is less favourable than two single bonds.

two different sigma bonds rather than one sigma and one pi bond between the same pair of atoms. Strong π bonding using only p orbitals is confined to the first element of a p Group and often leads to small, volatile molecules, as with the carbon and nitrogen oxides.

The ability to form $p_\pi - p_\pi$ bonds, and the weakness of simple sigma bonds, are effects of the small size of the first row elements. Other properties, such as ionization potential and electron affinity, which depend in part on size, also change sharply between the first and second elements of a p Group. This shows up very clearly in the electronegativity values in Table 2.14, where there is a major drop in value between the first and second elements in each p Group, and then a much slower fall in values down the rest of the Group.

The size effect also shows up in the bond strengths, as long as π bonding plays no part. The larger atoms usually form weaker sigma bonds, so that bond strengths fall on passing down a p Group, and the largest relative changes come between the first two elements of the Group. This point is illustrated by the stabilities of the hydrides, discussed in Chapter 9, and by a similar fall in the stability of element-carbon bonds in the organic analogues of the hydrides. These changes are illustrated by the values of the element-hydrogen and element-element bond energies shown in Figure 17.2.

The characteristic properties of the first element in each p Group may be summarized as:
(i) small size, with the greatest relative size change occurring between the first and second members of a Group
(ii) other properties which result from the size effect change in a similar way: for example, electronegativity and the tendency to form anions and negative oxidation states
(iii) the ability to form strong π bonds using the p orbitals is restricted to the first row elements, with the result that many simple compounds such as the oxides are small molecules instead of polymers
(iv) The first element has no low-lying d orbital and is limited to a maximum coordination number of four.

The effects sketched above are not independent of each

other, but they add up to a unique character for the first member of each p Group.

17.3 The remaining elements of the p Group

The last member of each p Group shows distinct differences from the other elements in the Group, though these differences are less pronounced than those which mark off the first member. The most obvious property of the last row of p elements is the stability of an oxidation state two less than the Group state in thallium, lead, and bismuth. The other two heavy elements, polonium and astatine, occur only as radioactive isotopes and their chemistry is less well known but, in polonium at least, the trend appears to continue and the IV, and also the II, states of polonium are more stable than the Group state of VI. This effect has been termed the inert pair effect as it corresponds to the valency which arises if two of the valency electrons are inactive. This phenomenon is due, at least in large part, to the relativistic effect (see section 16.13). The stability of the lower state, and the oxidizing properties of the Group state, in these heavy elements is clearly shown in Tables 17.2 and 17.3 and in the oxidation state free energy diagrams. The heavy elements show the Group oxidation state only in their oxides and fluorides, while sulfides and other halides of the elements exist only for the lower oxidation states.

The heaviest element in each p Group is also the most metallic and least electronegative member of its Group (as the larger size means that the outer electrons are less tightly held). Most of the heavy elements are metallic except polonium, which is metalloid, and astatine whose character is not clear. They all form basic oxides and appear in solution in cationic forms. The single bond strengths of these elements with hydrogen, organic groups, and halogens are lower than for the lighter elements in the p block and the very existence of some of the hydrides is dubious.

The chemistry of the middle three elements in a p Group follows a graded transition in properties from the small electronegative element at the head of the Group to the large, much more basic element at the foot. This is shown by the increasing stability of the oxidation state two less than the Group state, the increasing basicity of the oxides, and by the lower stability of bonds with hydrogen and organic groups as the Group is descended. This discontinuity between first and second element is most marked in the middle Groups of the p block. Boron and aluminium have each an empty p orbital in the valence shell in their trivalent compounds and thus have acceptor properties, forming four-coordinate species. The fact that Al enters into further bonding to form 5- or 6-coordinate species makes a difference in degree, but not in kind of behaviour so that boron and aluminium resemble each other reasonably closely. In sharp contrast, the ability of silicon, phosphorus, and sulfur to use the $3d$ orbitals, and the ability of carbon, nitrogen and oxygen to form p_π bonds combine to create marked differences in properties between the first and second members of these Groups. The ability to increase the coordination number above four has important effects on reactivity, and thus on the stability of compounds to air and

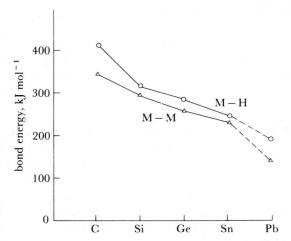

FIGURE 17.2 *The element-hydrogen and element-element bond energies in the carbon Group. Similar trends are formed for the other Main Groups.*

water. The most striking manifestation is in the carbon Group, where there is neither an empty low-lying orbital nor unshared electron pairs to provide a low-energy reaction intermediate. Among the hydrides and halides, compounds such as CH_4 and CCl_4 are stable in air and react only under vigorous conditions. In contrast, silicon hydrides inflame or explode in air and the silicon tetrahalides are hydrolysed violently on contact with water. Some part of this difference, particularly in the case of the hydrides, may be ascribed to the high $Si-O$ bond energy, but this cannot be the sole explanation. For example, the heats of the reaction:

$$MCl_4 + 2H_2O = MO_{2(aq)} + 4HCl$$

are exothermic and very similar in the two cases, $M = C$ and $M = Si$. A simple explanation of the difference arises if the mechanism of the reaction is examined. Carbon tetrachloride can react with water only if the strong $C-Cl$ bond is first broken, or considerably weakened, as there is no other way in which the water molecule can coordinate to the carbon. In contrast, silicon tetrachloride may be readily attacked by water if the $3d$ orbitals are used to expand the silicon coordination number above four, to give a reaction intermediate such as $Si(OH_2)Cl_4$ which then loses HCl. In support of this theory, silicon tetrachloride is known to form complexes, $SiCl_4.D$ or $SiCl_4.D_2$, with donor ligands, such as amines or pyridine (see Figure 17.3). Thus, the great difference in reactivity between carbon and silicon compounds may be ascribed, in considerable part, to the availability of a low-energy reaction path which makes use of the silicon d orbitals. Similar effects are probably present in the chemistry of the other p Groups, but, as boron has an empty p level, and

nitrogen, oxygen, and fluorine have lone pairs in their compounds, other means for providing low-energy reaction intermediates are present, so differences are less marked in these Groups.

At oxygen, and still more at fluorine, the tendency to enter the negative oxidation states as ions or covalent molecules becomes more important and re-introduces some resemblance in the general chemistry, especially between fluorine and chlorine. These trends may be linked with those in the s block elements so that, in the Main Groups as a whole, the first row element differs from the rest of its Group. This difference is least at the ends of the Periods in the lithium and fluorine Groups, and is most marked in the centre, in the carbon and nitrogen Groups.

Size effects in the chemistry of the heavier elements are much less distinctive than the changes observed at the head of the Group. There is a general tendency for the heavier elements to show higher coordination numbers but this is off-set to some extent by the weakening of element-ligand single bonds already noticed. In oxyacids, antimony, tellurium, and iodine give compounds in the Group oxidation state where the coordination to oxygen is six-fold in place of the four-coordinated oxyions of the lighter elements. In halogen compounds there is evidence for TeF_7^- species and IF_7 exists. It is not known whether this trend continues for polonium and astatine.

One further feature of the chemistry of the p Groups is the slight discontinuity in properties observed at the middle element in each Group (i.e. in the Period from gallium to bromine) which was mentioned above. These middle elements do not always have properties which interpolate between those of the second and fourth element. These effects are ascribed to the insertion of the first d shell immediately preceding this row of Main Group elements—the effect is reminiscent of the lanthanide contraction following the appearance of the first f shell and has even been termed the 'scandide contraction'. The effects are much less striking however than in the case of the lanthanide contraction. In the chemistry of the elements, the effect shows up in minor anomalies such as the low stability of $AsCl_5$ compared to phosphorus and antimony pentachlorides. Similarly, GeH_4 is found to be stable to dilute alkali while SiH_4 and SnH_4 are rapidly attacked. This is not a major effect, and most of the chemistry of the second, third, and fourth elements in any p Group follows a smooth trend, but the second order anomaly clearly exists.

These points may be summarized:

(i) The heaviest element in a p Group differs from the rest in the stability of oxidation states lower than the Group state, in its more metallic character with more basic oxides, and in its weak bonding to hydrogen and related ligands.

(ii) The properties of the middle three elements in a p Group form a transition to those of the heaviest element, with the lower oxidation states becoming more stable and the oxides more basic, etc.

(iii) There is a sharp discontinuity in properties with the first

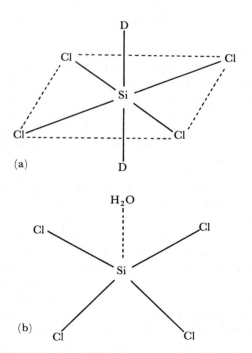

(a)

(b)

FIGURE 17.3 (a) *Structure of* $SiCl_4.2D$, (b) *possible structure for intermediate in hydrolysis of* $SiCl_4$

element, especially in the carbon, nitrogen, and oxygen Groups.

(iv) The heavier elements tend to show higher coordination numbers.

(v) Minor anomalies are observed in the chemistry of the elements of the middle row of the *p* block as compared with the properties of the rows above and below.

The extent of *d* orbital occupation appears, from modern calculations, to be less than implied by the classical hybridization ideas like sp^3d or sp^3d^2. Substantial energy contributions can arise for *d* populations of only a fraction of an electron. We shall use the more familiar terminology in this chapter, but see the discussion of this in Chapter 18, section 9. See also the comment on relativistic effects in section 16.13.

17.4 The boron Group $ns^2 np^1$

17.4.1 *The elements, general properties, and uses*
References to the properties of the boron Group elements given in earlier chapters include:

Ionization potentials   Table 2.8 and Figure 8.6
Electronegativities   Table 2.14
Hydrides   Chapter 9, especially electron-deficient hydrides, section 9.6

Table 17.4 lists some properties of the elements and Figures 17.4 and 17.5 give the variations of certain properties with Group position. It will be seen from the free energy diagram, Figure 17.5, that the III state is the most stable one, except for thallium.

Aluminium, the most common metallic element in the earth's crust, is extracted from the hydrated oxide, bauxite, by electrolysis of the oxide (after purification by alkaline treatment) dissolved in molten cryolite, sodium hexafluoro-aluminate. Boron is found in concentrated deposits as borax, $Na_2B_4O_5(OH)_4.8H_2O$, and similar tri-, tetra-, and penta-borates of Na and Ca. The element is formed by magnesium reduction of the oxide. The other three elements are found only in the form of minor components of various minerals, and the elements are produced by electrolytic reduction in aqueous solution. Gallium, indium, and thallium are rela-

TABLE 17.4 Properties of the elements of the boron group

Element	Symbol	Oxidation states	Common coordination numbers	Availability
Boron	B	III	3, 4	common
Aluminium	Al	III	3, 4, 6	common
Gallium	Ga	(I), III	3, 6	rare
Indium	In	I, III	3, 6	very rare
Thallium	Tl	I, (III)	3, 6	rare

All the elements have high boiling points (above 2000 °C), but gallium has an unusually low melting point at 29·8 °C which gives it the longest liquid range of any element.

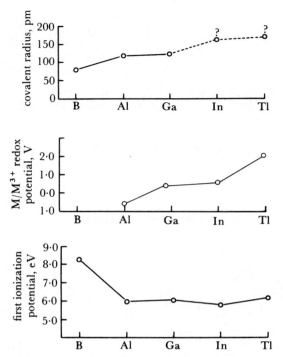

FIGURE 17.4 *Some properties of the boron Group elements*
The figure shows the covalent radii, the first ionization potentials, and the oxidation potentials as functions of Group position.

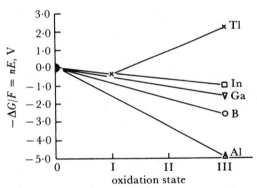

FIGURE 17.5 *The oxidation state free energy diagram for the boron Group elements*

tively soft and reactive metals which readily dissolve in acids. Aluminium is also a reactive metal but is usually found with a protective, coherent oxide layer which renders it inert to acids, although it is attacked by alkalies. Boron is non-metallic and the crystalline form is very hard, inert, and non-conducting. However, boron often shows considerable analogies with metals, specifically the transition metals, in its chemistry. This is particularly apparent when clusters containing these elements are considered (see sections 16.8 and 18.4). The amorphous form of boron, which is more common than the crystalline variety, is much more reactive. Figure 17.6 shows some typical reactions of the boron Group elements.

Boron reacts directly with most metals to give hard, inert, binary compounds of various formulae. These borides some-what resemble the interstitial carbides and nitrides. Table

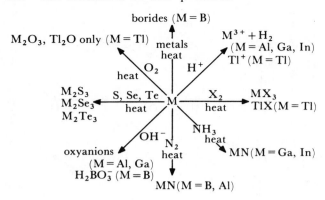

FIGURE 17.6 *Reactions of the elements of the boron Group*

17.5 lists some typical formulae, and some structures, which all involve chains, sheets, or clusters of boron atoms, are shown in Figure 17.7. The binary compound, boron nitride, BN, is interesting as it is isoelectronic with carbon and occurs in two structural modifications. One has a layer structure, like graphite (but is light in colour) and is soft and lubricating, while the other, formed under high pressure, has a very hard, stable, tetrahedral structure as in diamond.

In Li_3Ga_{14} there are Ga_{12} icosahedra linked into a three-dimensional network.

Boric oxide and borates find extensive application. Borax, and other borates find uses in water treatment, and in preserving timber from insects. Larger amounts of sodium or calcium borates, and of boric acid or oxide, are used in glass

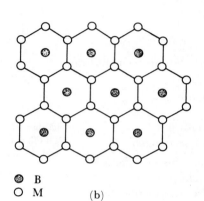

● B
○ M   (a)

● B
○ M   (b)

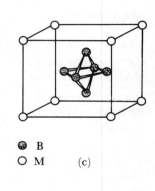

● B
○ M   (c)

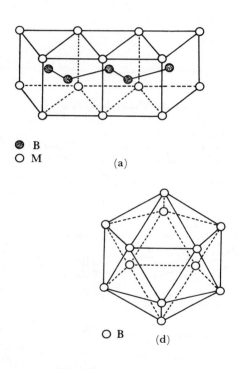

○ B   (d)

FIGURE 17.7 *Examples of the structures of boron units in borides*
(a) shows the configuration of the boron chain in MB compounds, (b) the sheet structure found in MB_2 and (c) the B_6 cluster in MB_6 compounds. The B_{12} icosahedron shown in (d) is found in boron itself, in some of the boron hydrides, and in BeB_{12} and AlB_{12}. The other MB_{12} species have a related structure. In the tetraborides, hexaborides, and **dodecaborides**, the boron clusters (such as the B_6 unit in Figure (c)) are themselves linked so that a bonded boron framework extends completely through the compound.

TABLE 17.5 Some typical borides

Formula	Boron atom structure	Boride examples
M_2B	single atoms	Be, Cr, Mn, Fe, Co, Ni, Mo, Ta, W
M_2M'	pairs	M = Al, Ti, Mo; M' = Cr, Fe, Ni
MB	single chain (Figure 17.7a)	V, Cr, Mn, Fe, Co, Ni, Nb, Ta, Mo, W
M_3B_4	double chains	V, Cr, Mn, Ni, Nb, Ta
MB_2	sheets (Figure 17.7b)	Be, Mg, Al, Sc, Ti, V, Cr, Mn, Y, Lu, U, Pu, Zr, Hf, Nb, Ta, Tc, Re, Ru, Os, Ag, Au
MB_4	sheets linking B_6 octahedra	Mg, Ca, Mn, Y, Ln, Th, U, Pu, Mo, W
MB_6	B_6 octahedra (which occupy Cl^- positions in a CsCl structure with the metals (Figure 17.7c))	Na, K, Be, Mg, Ca, Sr, Ba, Sc, Y, La, Th, Pu
MB_{12}	three-dimensional lattice consisting of linked B_{12} clusters and with metal atoms in the middle of each cluster	Be, Mg, Al, Sc, Y, Ln from Tb to Lu, U, Zr

Other more unusual species include CuB_{22}, $B_{13}M_2(M = P, As)$ and $Ru_{11}B_8$.

manufacture. Borosilicate glasses have a lower coefficient of thermal expansion than the more conventional ones, and are therefore more robust under heating. Sodium perborate, approximately $NaBO_3.4H_2O$ in composition, is widely used as a bleach. This material was once formulated as borate with H_2O_2 of crystallization, but it is now established as a true peroxyborate with $B-O-O$ links.

Aluminium oxide, alumina, and related complex oxides include a number of widely used species. Recall also that the clay minerals, the feldspars, zeolites and other important groups are aluminium-silicon-oxygen species (see section 18.6). Aluminium phosphates, which are isoelectronic with the silicates, have a similar range of structures and show promise in similar uses, for example in catalysis like the zeolites. The compact, inert alpha-alumina occurs as the minerals co-rundum (Figure 5.5b) and emery, both very hard, which find extensive uses in arbrasives and in refractory and ceramic materials. The lower temperature form, gamma-alumina, has a much more open structure and can be prepared in forms with very high surface areas which are widely used as catalysts and as supports for catalysts (compare dehydrosulfuriza-tion, for example, section 15.4). Such 'activated' alumina is also used for chromatography. Alumina, with zirconium dioxide, has been produced in the form of very fine, strong fibres which have important uses as reinforcing in lightweight materials. Thus, aluminium metal reinforced by alumina fibres can have up to five times the strength of the metal alone. As well as the clay minerals, which in turn give ceramics, two other important aluminium – metal ternary oxides are spinel (Figure 5.5d) and the species called β-alumina which actu-ally contains sodium ions and has the approximate formula $NaAl_{11}O_{17}$. This is used as the electrolyte in solid state electrical cells. The structure is blocks of spinel separated by layers of composition NaO. These sodium oxide layers are very open, with a number of equivalent Na^+ positions, so that sodium ions move readily, giving rise to the very low resistivity. Finally, we note that a major constituent of Portland cement is $Ca_9Al_6O_{18}$ which contains a ring formed by six AlO_4 tetrahedra linked by sharing corner oxygens (compare the similar 3-tetrahedron ring in silicates, section 18.6).

The elements form sulfides, selenides and tellurides, of similar formulae to the oxides, whose structures are based on four or six-coordination. Several have interesting electronic properties. Two unusual structures are those of $Ga_4S_{10}^{8-}$, which is the same as P_4O_{10} (Figure 17.32b) while $Ga_6Se_{14}^{10-}$ consists of six edge-sharing tetrahedra (Figure 17.8), and an extended form related to the many M_2X_6 examples (compare Figure 17.13). The other important compounds formed by the heavier elements of this Group are the 'III/V' binaries, especially GaAs and InP, which are very important semi-conductors. Structures are ZnS type where each element is tetrahedral, and related to Si with which they are isoelectronic.

17.4.2 *The (III) state*

The Group state of III is shown by all the elements, and is the

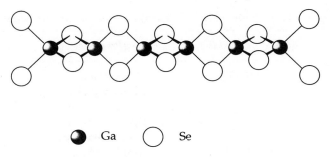

FIGURE 17.8 *The structure of the $Ga_6Se_{14}^{10-}$ ion*

most stable state for all except thallium. It is represented by a wide variety of compounds, of which the oxygen and halogen compounds are typical. There is no evidence for a free M^{3+} ion, either in the solid or in solution. A number of solids, especially fluorides and oxides, are high-melting and strongly bonded, but the bonds are intermediate between ionic and covalent and the stabilities of the solids are due to the forma-tion of giant molecules with uniform bonding. For example, aluminium chloride, bromide, and iodide are volatile, co-valent solids, while aluminium trifluoride is high-melting and a giant molecule. Similar effects are seen for the other tri-fluorides, except for BF_3, and for the oxides. In solution, extensive hydration and hydrolysis occur and ionic species (though often written as cations for convenience) are actually much more complex, e.g. $Al(OH)(H_2O)_5^{2+}$ has been shown to occur in 'Al^{3+}' solutions.

All the elements form the trioxides, usually as hydrated species by precipitation from solution or by hydrolysis of the trihalides. Chemical and structural properties are given in Table 17.6.

Oxides of the I state are treated later.

Hydration of the oxides gives a variety of hydrates and hydroxy-species. Boric oxide gives boric acid, $B(OH)_3$, on hydration which forms crystals in which the $B(OH)_3$ units are linked together by hydrogen bonding (Figure 9.18). When boric acid is heated, it dehydrates first to metaboric acid, HBO_2, and ultimately to boric oxide:

$$B(OH)_3 \underset{+H_2O}{\overset{-H_2O}{\rightleftharpoons}} HBO_2 \underset{+H_2O}{\overset{-H_2O}{\rightleftharpoons}} B_2O_3$$

Metaboric acid exists in three crystalline forms, one of which contains the cyclic unit shown in Figure 17.9. The structures of the other two are not known with certainty but appear to contain chains of BO_3 and BO_4 units. The cyclic anion is also found in sodium and potassium metaborates. A wide variety of other oxyanions of boron exists with very varied structural types. Not only are discrete ions, rings, chains, sheets, and three-dimensional structures found—as with the silicates—but boron occurs both in planar BO_3 units and in tetrahedral BO_4 units, and many borates contain OH groups. It is impossible to discuss all the borates, but Figure 17.10 gives a few representative borate structures.

Aluminium and gallium form hydrated oxides of two types —MO(OH) and $M(OH)_3$. These are precipitated from

TABLE 17.6 Oxides of the III state of the boron Group elements

Oxide	Properties	Structure
B_2O_3	Weakly acidic Many metal oxides give glasses with B_2O_3 as in the 'borax bead' test.	Glassy form—random array of planar BO_3 units with each O linking two B atoms. Crystalline form—BO_4 tetrahedra linked in chains.
Al_2O_3 and Ga_2O_3	Amphoteric	α-form—inactive, high-temperature form. Oxide ions ccp with metal ions distributed regularly in octahedral sites. γ-form — low-temperature form, more reactive. Metal ions arranged randomly over the octahedral and tetrahedral sites of a spinel structure.
In_2O_3 and Tl_2O_3	Weakly basic Tl_2O_3 gives O_2 and Tl_2O on heating to 100 °C	The structure has the metal ions in irregular six-coordination, and four-coordinated oxygens. The same structure is adopted by most oxides of the lanthanide elements, Ln_2O_3.

Other Oxides: $(BO)_x$ formed by heating $B + B_2O_3$ at 1050 °C. This probably contains both B − O − B and B − B links as it reacts with BCl_3 to give B_2Cl_4.

gem forms of alumina: ruby—Al_2O_3 + traces of Cr^{3+}

blue sapphire—Al_2O_3 + traces of Fe^{2+}, Fe^{3+} or Ti^{4+}

white sapphire—this is the gem form of alumina itself

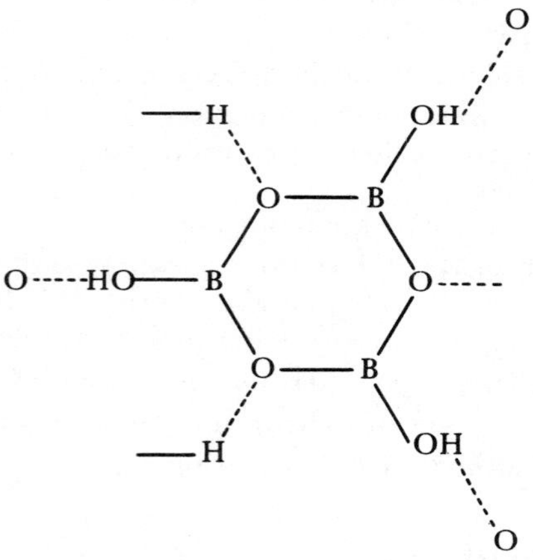

FIGURE 17.9 *The cyclic form of metaboric acid*

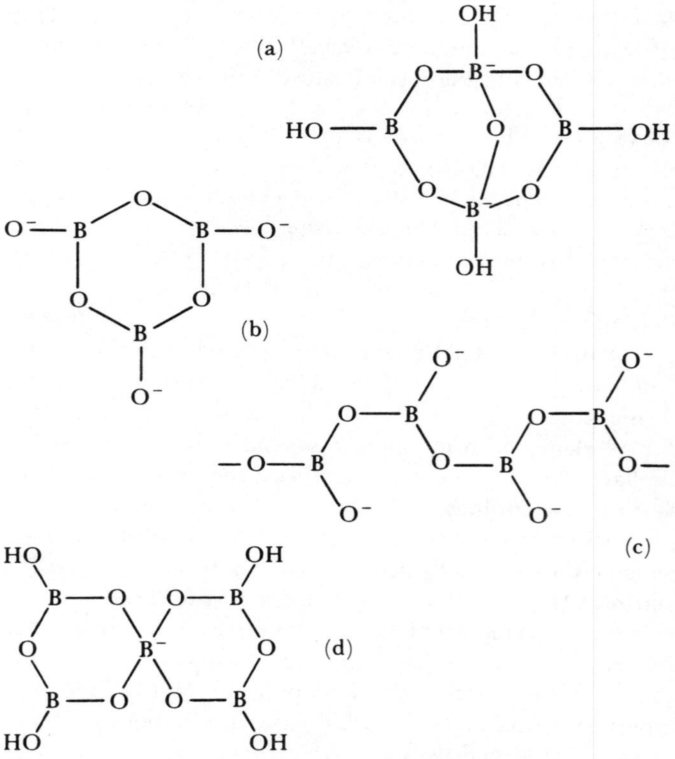

solution by, respectively, ammonia and carbon dioxide. Indium gives a hydrated oxide, $In(OH)_3$. In these compounds the metal is six-coordinated to oxygen.

Hydroxy species also occur in solution. Thus boric acid accepts an OH^- group in dilute solution and polymerizes in more concentrated solutions:

$$B(OH)_3 + 2H_2O = B(OH)_4^- + H_3O^+$$
$$3B(OH)_3 = B_3O_3(OH)_4^- + H_3O^+ + H_2O$$

The hydrates of the other elements of the Group behave similarly. For example:

$$M(H_2O)_6^{3+} \rightleftharpoons M(H_2O)_5(OH)^{2+} \rightleftharpoons \text{intermediate stages}$$
$$\rightleftharpoons M(OH)_6^{3-}$$

FIGURE 17.10 *Examples of borate structures:* (a) *in borax,* $Na_2B_4O_7 \cdot 10H_2O$, (b) *in metaborates,* $M_3B_3O_6$ *(cyclic anion),* (c) *in linear metaborates,* CaB_2O_4, (d) $B_5O_{10}H_4^-$

TABLE 17.7 Acceptor and structural properties of the trihalides of the boron Group elements

Halide	Structural use of empty p orbital	Halide complex
BF_3	Internal $p_\pi - p_\pi$ bonding: see note (1) and Figure 17.11	BF_4^-
BX_3	Possibly slight π bonding in BCl_3, otherwise none.	BX_4^-
AlF_3	Accepts lone pair from fluorine (as do two d orbitals) to give AlF_6 units in a highly polymerized solid. M.p. above 1000 °C.	AlF_6^{3-}, AlF_4^-
AlX_3 (Note 2)	Accept one lone pair from a halide to give Al_2X_6 dimer (Figure 17.12). m.p. 100–200 °C.	AlX_4^-
MF_3 (M = Ga, In, Tl)	As AlF_3 (m.p. about 1000 °C).	MCl_6^{3-}, MBr_6^{3-},
GaX_3, InX_3	As AlX_3 (m.p. 100–600 °C).	MCl_5^{2-}, MCl_4^-
TlX_3	As AlX_3 (see note 3)	MBr_4^- (M = Ga, In)
	In all cases above, X = Cl, Br and I.	$TlCl_5^{2-}$

Note (1) The evidence for internal π bonding in BF_3 derives from two sources. First, the $B-F$ bond length is shortened compared with that in the BF_4^- ion, 130 pm compared with 142 pm. Second, the order of acceptor strengths for the boron trihalides (forming $BX_3.D$) is $BBr_3 > BCl_3 > BF_3$. As ability to accept an electron-pair depends on the electron density at the boron, the strongly electronegative fluoride would be the strongest acceptor unless other effects intervene.

Note (2) For X = Cl, the dimer is found in the vapour but solid $AlCl_3$ exists as a slightly deformed $CrCl_3$ layer lattice structure (compare Table 5.3) with 6-coordination of aluminium.

Note (3) $TlBr_3$ decomposes to $TlBr$ and Br_2 at room temperature, and $TlCl_3$ loses chlorine similarly at 40 °C.

and a compound, $Ca_3[Al(OH)_6]_2$ has been isolated. Among species identified in solution, by a range of methods including ^{27}Al nmr, are $Al(H_2O)_6^{3+}$, $Al(H_2O)_5(OH)^{2+}$, $Al_2(OH)_2(H_2O)_8^{4+}$ and more highly polymerized entities such as $Al_8(OH)_{20}(H_2O)_n^{4+}$ and $Al_{13}O_4(OH)_{24}(H_2O)_{12}^{7+}$. The Al environments are probably all based on AlO_6 octahedra.

All the trihalides of all the boron Group elements exist and all correspond to the III state, except TlI_3 which is the tri-iodide, I_3^-, of Tl^+. The normal trihalides are planar molecules which have an empty p orbital in the valence shell. Most of the trihalides make use of this empty orbital, both in the structure of the trihalide, and in the formation of complexes of the form $MX_3.D$, where D is a lone pair donor. Table 17.7 lists these applications for the halides and for the halide complexes. Aluminium, and the heavier elements, also use their d orbitals to become six-coordinate. It will be seen that the formation of a $p_\pi - p_\pi$ bond in BF_3, and the use of d orbitals, especially in the fluorides, mirrors the discussion of these effects in section 17.2. An interesting example of a mixed dimer is provided by $NbAlCl_8$ which has octahedral $NbCl_6$ linked to tetrahedral $AlCl_4$ by sharing an edge (compare Figures 15.6b and 17.12).

Most of the trihalides react with water to give the hydrated oxides, but boron trifluoride gives 1:1 and 1:2 adducts, $BF_3.H_2O$ and $BF_3.2H_2O$, which are not ionized in the solid state. The 1:1 adduct has the expected donor structure, $F_3B.OH_2$, but the structure of the second is unknown. When these adducts are melted, they each ionize:

$$2BF_3.H_2O = (H_3O.BF_3)^+ + (HO.BF_3)^-$$
and $$BF_3.2H_2O = H_3O^+ + (HOBF_3)^-$$

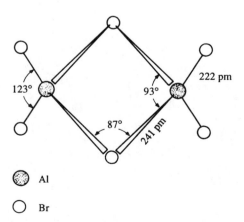

FIGURE 17.12 *The structure of the aluminium tribromide dimer*
The aluminium makes use of its empty p orbital to accept a lone pair from a bromine atom in a second $AlBr_3$ molecule, giving Al_2Br_6.

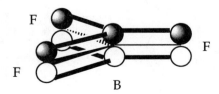

FIGURE 17.11 *Internal p_π-p_π bonding in boron trifluoride*
The p orbital on the boron which is not used in the sigma bonds, accepts electrons from the three corresponding fluorine orbitals to give internal pi-bonding in BF_3.

Among the complex halides, all the MX_4^- species are tetrahedral while the MX_6^{3-} ones are octahedral. The $[AlF_4]^-$ ion has been isolated by the reaction of $AlMe_3$ with HF and pyridine (py) to give $[pyH]^+[AlF_4]^-$. All the boron Group trihalides act as catalysts in the Friedel-Crafts reaction, where their function is to abstract a halide ion (giving MX_4^-) from the organic molecule, leaving a carbonium ion. As well as the MX_4^- and MX_6^{3-} ions shown in the Table, the isolation of MCl_5^{2-} ions has been reported for indium and thallium. A crystal study shows that, in its tetraethylammonium salt, $InCl_5^{2-}$ is a square pyramid, similar to the structures described in section 13.6. This is the first report of this structure for a main group element (not to be confused with species like IF_5 with a lone pair in addition). The form probably results from the way the ions pack with the large cation into the crystal. There are only a few polynuclear ions, including the binuclear Tl(III) complex $[Tl_2Cl_9]^{3-}$ with the structure shown in Figure 17.13. In $Sm(II)AlF_5$, there are two different types of aluminium ion. The structure of the $[Al_2F_{10}]^{4-}$ ions is that of two octahedra sharing an edge and the second type is the long chain ion formed by linking AlF_6 octahedra through *trans* apices.

The trihalides, and other trivalent MX_3 species, readily form tetrahedral complexes such as BH_4^-, $AlX_3.NR_3$, or $GaH_3.NMe_3$. A wide selection of 1:1 $BX_3.D$ complexes exist where D is a lone pair donor such as ammonia, amine, water or ether, phosphane, sulfide, etc., and X is halogen, hydrogen, or an organic group. The organic compounds $R_3B.D$ have been much studied to find the factors, such as electron attracting power and steric effects, which most influence Lewis acid-base behaviour. One anion of analytical importance is the tetraphenylboronate ion $B(C_6H_5)_4^-$, which forms insoluble salts with potassium and the heavier alkali metals and is used in their gravimetric determination.

One interesting application of $GaCl_3$, involving about 100 tons of a concentrated aqueous solution, is in the experiment (called GALLEX) to detect low energy solar neutrinos produced in the sun as a result of fusion reactions. In this experiment, ^{71}Ga nuclei are converted, in *extremely* low yield, by reaction with a neutrino and an electron into ^{71}Ge atoms, which are then separated and counted after conversion to germane, GeH_4. As an indication of the difficulties which had to be overcome, less than one ^{71}Ge atom was produced per day—consequently it was no mean technological feat to separate the germanium atoms from the $GaCl_3$!

Complexes of the elements other than boron include both tetrahedral types as above and also octahedral complexes, of which important examples are the β-diketone complexes shown in Figure 17.14 and the 8-hydroxyquinoline complex of Figure 17.15 which is used in the gravimetric determination of aluminium. These elements form hexahydrates, $M(H_2O)_6^{3+}$, which hydrolyse in solution. Also, hydrated salts with this cation and a variety of oxyanions are known. Aluminium also forms the well-known series of double salts, $MAl(SO_4)_2.12H_2O$, called alums. M is any univalent

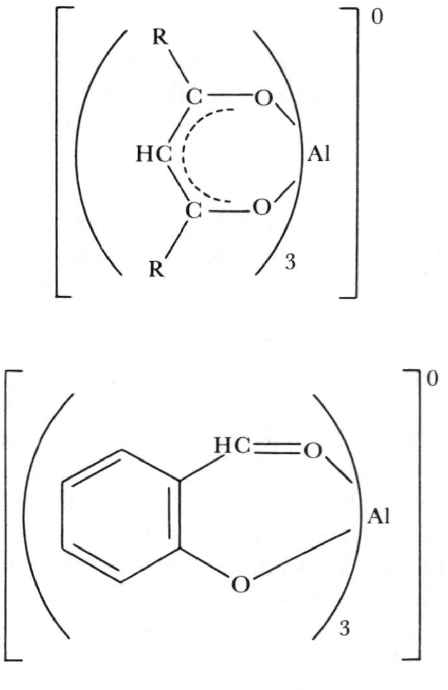

FIGURE 17.14 *β-diketone complexes of aluminium*

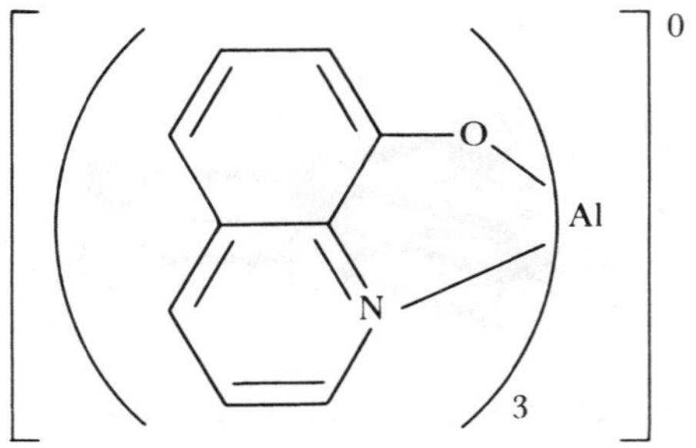

FIGURE 17.15 *The 8-hydroxyquinoline complex of aluminium*
This is commonly used to determine aluminium gravimetrically.

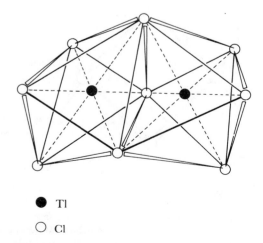

● Tl

○ Cl

FIGURE 17.13 *The structure of the ion, $Tl_2Cl_9^{3-}$*

cation and the aluminium may be replaced by a variety of trivalent ions such as Cr^{3+}, Fe^{3+}, Co^{3+}, Ga^{3+} or Ti^{3+}. The crystals contain $M(H_2O)_6^+$, $Al(H_2O)_6^{3+}$ — or other $M(H_2O)_6^{3+}$ ions—and sulfate ions.

17.4.3 Organic compounds and Ziegler–Natta catalysts

All the boron Group elements form organic compounds, R_3M, and also mixed types, R_2MX and RMX_2, with halogens and related groups. The boron compounds are monomers with planar BC_3 skeletons, but the later members of the Group give dimeric or polymeric compounds. The halides dimerize through halogen bridges similar to those in Al_2X_6. The purely organic compounds polymerize through electron deficient carbon bridges, as in $Al_2(CH_3)_6$, shown in Figure 9.13, similar to the bonding in diborane.

Organo-aluminium compounds are involved in a number of related, and commercially important, catalytic processes which arose from the work of Ziegler. The basic discovery was that $Al-H$ bonds add across the alkene double bond coupled with the fact that aluminium metal reacts under fairly easy conditions with hydrogen in the presence of aluminium alkyl. Thus, $Al-H$ bonds may be formed and converted into $Al-C$ in the overall process shown in equation (1).

$$Al + \tfrac{3}{2}H_2 + 3RCH=CH_2 \xrightarrow{AlR_3} Al(CH_2CH_2R)_3 \ldots (1)$$

The addition is to terminal double bonds. The $Al-C$ bond in turn will add across a terminal double bond, in a series of steps which leads to growth of the alkyl chain

$$R_2Al-R + R'CH=CH_2 \to R_2AlCH_2CH(R)R' \text{ etc.} \quad (2)$$

Finally, the process may be terminated by the reverse step to (1), yielding a long-chain α-alkene, or hydrolysis and oxidation is used to yield an alcohol

$$R_2Al-CH_2CHRR' \xrightarrow{O_2} R_2AlOCH_2CHRR' \xrightarrow{H_2O} R_2AlOH + RR'CHCH_2OH \ldots (3)$$

All the $Al-R$ links are eventually broken in step (3).

There are several applications of this process: firstly, using ethylene and high pressures at 160 °C, steps (1), (2) and (3) can be arranged to produce alcohols with chain length about C_{14} which are used in the production of bio-degradable detergents. Secondly, using ethylene at about 100 °C, steps (1), many repetitions of (2), terminated by the reverse of (1) produces polythenes with average chain-length about C_{200}. Thirdly, longer chain olefins may be dimerized, as in the formation of isoprene via the dimerization of propene

$$2CH_3CH=CH_2 \xrightarrow{(1) \text{ and } (2)} CH_3CH_2CH_2C(CH_3)=CH_2$$
$$\xrightarrow{heat} CH_4 + CH_2=CHC(CH_3)=CH_2$$

Isoprene polymerization itself takes place in the presence of catalysts including aluminium alkyls.

A further process, developed in part by Ziegler and partly by Natta, is an extension of the olefin polymerization. The process above, steps (1), (2) and the reverse of (1), gives a wide spectrum of chain lengths and disordered polymers which are soft and low-melting. In the Ziegler-Natta process, a transition metal halide, such as $TiCl_4$, is added to an aluminium trialkyl and the resulting reaction mixture is found to catalyse the polymerization of alkenes, giving a stereoregular product (one where, for example, all the sidechains lie the same way) and these regular polymers are much higher melting and more crystalline. For example, while ordinary polythene softens below 100 °C, polythene from the Ziegler-Natta process melts at 130–135 °C. Later improvements include 'second generation' catalysts involving Ti(III) or Cr, often on $MgCl_2$ supports, and also the use of homogeneous catalysts functioning in solution, involving organometallic Al and metal compounds. Such improvements aim at narrower ranges of molecular weights in the polymers, particular geometrical order, or greater ease of processing.

The exact nature of the catalytic process is still under study. A variety of transition metal halides may be used together with other active organometallic species in place of the AlR_3. For $TiCl_4$, it is established that the titanium is reduced to the III (or lower) state and one theory is that the catalysis takes place on the crystal surface of the reduced species. A chain growth process like (2) occurs but, as it is on a surface, the approach of the incoming olefin is oriented, giving a regular polymer. The aluminium alkyl acts as the reducing agent and also forms $Ti-R$ groups on the surface to provide growth sites. An alternative theory suggests some $Ti \ldots X \ldots Al$ bridge which provides the active site, where X may be an organic group or a halogen. Here again, the orientation of substituents is postulated to restrict the attacking alkene into a regular and repeatable orientation, giving a regular orientation of the product.

This is supported by the observation that optical isomers may be separated during the polymerization. The active site is regarded as an ordering matrix analogous to, though simpler than, the active sites on enzymes in biochemical catalyses.

17.4.4 The I oxidation state and mixed oxidation state compounds

The I oxidation state is most important in thallium chemistry where it is the most stable state. The few common thallium(III) compounds, such as the oxide and halides, are strongly oxidizing, and the potential Tl^{3+}/Tl^+ of 1·3 V in acid solution makes thallium(III) in solution as oxidizing as chlorate or MnO_2. Thallium(I) compounds are stable and show some resemblances to both lead(II) compounds and to those of alkali metals. The oxide, Tl_2O, and the hydroxide, TlOH, are strongly basic, like the alkali metal compounds, and absorb carbon dioxide from the atmosphere. The halides resemble lead halides in being more soluble in hot water than in cold and behave in analysis like the lead or silver compounds. Tl^I forms a number of stable salts which are generally isomorphous with the alkali metal ones; examples include the cyanide, perchlorate, carbonate, sulfate, and phosphates. TlF has a deformed sodium chloride structure

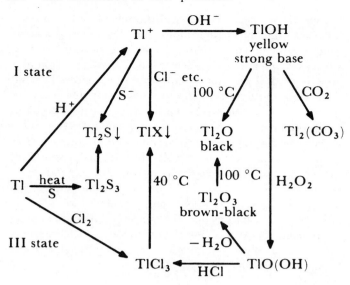

FIGURE 17.16 *Reactions of thallium compounds*

in the solid, while the other thallous halides crystallize with the cesium chloride structure. The chemistry of thallium is indicated in Figure 17.16 which shows the interrelation between the two oxidation states.

The I oxidation state becomes rarer and less stable as the Group is ascended. Gallium forms unstable monohalides and a sulfide. When Ga and HCl are heated to 1200 K, red GaCl forms as a gas and can be recovered by rapid cooling. It disproportionates to Ga and $GaCl_3$ above $0\,°C$. The oxide Ga_2O is formed by heating Ga_2O_3 with Ga, and disproportionates at higher temperatures. Indium(I) halides are more stable, but In_2O needs further study, while both gallium and indium form various sulfide phases as well as M_2S_3. The monohalides are interesting. The yellow form of InCl has the distorted NaCl structure like TlF. In its red modification, together with InBr, InI, and TlI, a 7:7 coordinated structure is found, where the coordination shell is 1:4:2—that is, like an octahedron but with two M atoms spanning the site of one apex. This structure may be seen as intermediate between the 6:6 NaCl one, and the 8:8 CsCl structure of TlCl, TlBr, and the second form of TlI.

Gallium, indium and also thallium, form halides of formula MX_2 which are mixed valency species containing the I state cation and the III state complex anion, $M^+[MX_4]^-$. A similar mixed valency formulation is found for In_2Cl_3, Tl_2Cl_3 and Tl_2Br_3 which have octahedral M(III) complex anions, $M_3^+[MX_6]^{3-}$. Related compounds with these anions such as $Ga^+[AlCl_4]^-$ or $Tl_3^+[InCl_6]^{3-}$, are known. In contrast, M_2Br_3, for M = Ga or In, contain the diatomic $[M-M]^{2+}$ cation, formally M(I), and the $M_2Br_6{}^{2-}$ anion mentioned below.

Crystal structures of these mixed halides have allowed the determination of the M^+ radii, which show less variation than the alkali metal ions. Thus Goldschmidt radii are $Ga^+ = 141\,pm$, $In^+ = 143\,pm$, and $Tl^+ = 149\,pm$, compare Table 2.12.

Gallium(II) is found in the $Ga_2X_6{}^{2-}$ species for X = Cl, Br, I, prepared electrolytically from gallium in strong acid. The structure appears to be $X_3Ga-GaX_3{}^{2-}$, like the isoelectronic Ge_2Cl_6. The formal $+II$ state then arises because of the metal-metal bond. The halides Ga_4X_6 are mixed oxidation state species containing the ion

$$(Ga^+)_2(Ga_2X_6)^{2-}.$$

Formal (II) states are shown in the ME species when M = Ga, In or Tl and E = S, Se, or Te. Two structures are common. The lighter combinations have the GaS structure where there are successive layers in the sequence $\cdots—S-Ga-Ga-S—S-Ga\cdots$ where the (II) states arises because of the Ga-Ga bonds, and the elements are four-coordinated. For InTe and the Tl species, the apparent II state arises from mixed (I) plus (III) as in $Tl(I)[Tl(III)S_2]$ where the TlS_2 formula represents an infinite chain of TlS_4 tetrahedra linked by shared edges.

An extremely interesting group of compounds involves benzene and methyl-substituted benzenes. It was reported in 1881 that the compounds of formula GaX_2 were unexpectedly soluble in benzene and could be isolated as benzene-containing solids. Studies a hundred years later isolated Ar_2Ga^+ complexes, where Ar = C_6H_6 or $C_6H_3Me_3$ and the Ga was equidistant from the six carbons (compare dibenzene chromium, section 16.3). A similar Tl(I) species is also found. When a more hindered benzene is used, at a lower ratio, the compound $ArGa^+$ was isolated as the $GaBr_4^-$ salt for Ar = C_6Me_6. The ring is almost planar and the Ga lies $255\,pm$ above its centre completing a hexagonal pyramid. The interaction is stronger than in the Ar_2Ga^+ ions (Ga to ring = $267\,pm$) and there is a weak interaction (distances 320 to $360\,pm$) with 5 Br atoms from $GaBr_4^-$ ions (the Ga(III)-Br distance is $232\,pm$). Analogous, isoelectronic Ar-Sn(II) complexes are also known.

Although Al_2O and AlO have been identified in the vapour phase above $1000\,°C$, aluminium chemistry at ordinary temperatures is almost entirely of the III state. The formal II state, arising from an Al-Al bond is found in species formed by

$$R_2AlX + 2K \rightarrow R_2Al-AlR_2 + 2KX$$

where R is a bulky organic group like $(CH_3)_2CH$. With smaller R groups, or on heating, R_3Al and aluminium metal are formed. $AlCl_2$ has also been reported.

The metastable compound AlCl containing aluminium in the I oxidation state, has recently been prepared in solid form by reaction of aluminium metal with HCl gas at 1200 K and low pressure, trapping out at low temperature (77 K) dark red, and thermally sensitive, AlCl. Upon warming, the expected disproportionation reaction occurs:

$$3AlCl \rightarrow AlCl_3 + 2Al$$

Triethylamine-stabilized aluminium(I) bromide $[AlBr.NEt_3]$ has been similarly obtained as a tetramer. An X-ray crystal structure study shows that the compound contains a square

Al_4 ring with Al–Al bonds. Each Al atom is also bonded to one terminal Br and one triethylamine ligand.

Boron is found in low formal oxidation states in the hydrides and in a variety of halides. The latter include the dihalides, B_2X_4, and lower-valent boron fluorides which are related to BF_3 by replacing F by a BF_2 group: $(F_2B)_nBF_{3-n}$ for $n = 0, 1, 2,$ and 3. These are prepared by treating B_2Cl_4 with SbF_3 or by strongly heating BF_3 with boron. $FB(BF_2)_2$ disproportionates at $-30\,°C$, to give BF_3 and B_8F_{12}, while the $n = 3$ member is not stable alone but does form adducts with $L = CO$ or PF_3, $(BF_2)_3BL$ analogous to H_3BCO or H_3BPF_3. It has been suggested that the rather unstable B_8F_{12} is the analogue of diborane with the structure

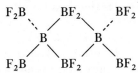

Like diborane, it gives monomer adducts

$$B_8F_{12} + CO \rightarrow (BF_2)_3BCO$$

(compare $B_2H_6 + CO \rightarrow H_3BCO$).

17.5 The carbon group, ns^2np^2

17.5.1 General properties of the elements, uses

References to the properties of the carbon Group elements include:

Ionization potentials	Table 2.8
Atomic properties and electron configuration	Table 2.5
Radii	Table 2.10
Electronegativities	Table 2.14
Redox potentials	Table 6.3
Hydrides	Chapter 9
Structures of elements	Sections 5.9, 19.3
Structures of silicates	Section 18.6

Table 17.8 summarizes some properties of the elements, and the variation with Group position of ionization potentials, radii and oxidation state free energy is indicated in Figures 17.17 and 18. The use of carbon in metal extraction is discussed in section 8.7.

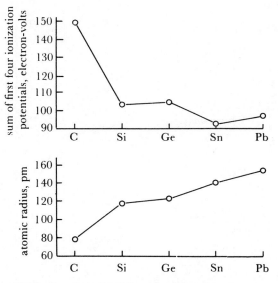

FIGURE 17.17 *Some properties of the carbon Group elements*
The Figure shows the variation, with Group position, of the atomic radii and the sum of the first four ionization potentials. The characteristic differences between first and second elements, and the similarity between second and third elements, are noticeable here.

All the elements are common except germanium, which occurs as a minor component in some ores, and also in trace amounts in some coals. Carbon, of course, occurs in all living things, and in deposits derived from them such as coals, oils, and tars. Hard coals like anthracite have high carbon contents. Heating coals to form coke removes hydrogen components leaving carbon containing a low percentage of metal compounds. Pure carbon is formed by pyrolysis of hydrocarbons and, as graphite, finds substantial industrial and electrical uses. While production of artificial diamonds is feasible, large pressures and temperatures are needed. More recently diamond films have been produced on metals by deposition from vapour-phase decomposition of methane, by plasma methods or by ion-beam deposition. Such interest in diamonds stems from their optical and semiconductor properties, and the ability to deposit films would be of great advantage where wear resistance is required.

Silicon, with oxygen, is the major component of the earth's crust and the vast majority of rocks, minerals, and their

TABLE 17.8 Properties of the elements of the carbon Group

Element	Symbol	Structures of elements	Oxidation states	Coordination numbers	Availability
Carbon	C	G, D	IV	4, 3, 2	common
Silicon	Si	D	IV	4, (6)	common
Germanium	Ge	D	(II), IV	4, 6	rare
Tin	Sn	D, M	II, IV	4, 6	common
Lead	Pb	M	II, (IV)	4, 6	common

G = graphite, D = diamond, and M = metallic forms

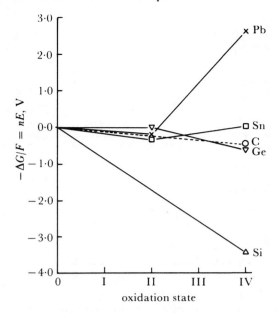

FIGURE 17.18 *The oxidation state free energy diagram for elements of the carbon Group*
It will be seen that Ge(II) is unstable, Sn(II) of nearly the same stability as Sn(IV), and Pb(IV) is very unstable relative to Pb(II).

breakdown products the sands and clays, are silicates. Tin occurs in concentrated deposits as the oxide *cassiterite*, SnO_2. Lead also occurs in concentrated form as the sulfide, *galena*, PbS. As both Sn and Pb are readily formed from their ores by heating with carbon in the form of wood fires, these two metals were among the earliest to be produced and used by man. Lead was particularly widely used by the Romans for water pipes ('plumbing' comes from Latin *Plumbum*, lead) while tin was the vital additive to copper (also an ancient metal) to form the much harder alloy, bronze.

Current uses still reflect the ancient patterns, with much of tin consumption being in alloys, and lead being phased out from water supply uses only in the second half of this century as awareness of its toxicity increased. About half of modern tin production is used in tin plate, where its inertness protects the underlying steel. Another use is in glass manufacture where the absolutely smooth surface required for the formation of sheet glass is provided by a bath of molten tin (making use of the inertness and non-toxicity of tin). A major outlet for both metals is in alloys—*solders* which are basically Sn/Pb, *type metals* which are Sn/Pb/Sb, *pewters* which are now mainly Sn/Sb but formerly contained some Pb, *bearing alloys* which are around 9 Sn to 1 Pb, and the tin-copper alloys *bronze* and *brass* (with Zn). All such alloys have a range of compositions optimized to different uses, and contain a number of other elements as minor components to refine the properties. The toxicity of lead is starting to limit some of its traditional uses, such as ceramic glazes. (It is thought that some of the less rational behaviour of the Roman emperors was due to lead poisoning from glazes used on wine jars.) It is widely used in batteries, in priming pigments, in sheathing for heavy-duty

cables, as well as in alloys. The structures of the elements are discussed in section 5.8 and illustrated in Figure 5.14.

First germanium, and then silicon, became extensively used in very pure forms in semi-conductor devices, which are at the basis of the whole electronic industry, including computer hardware. Germanium now accounts for only a few percent of the electronic use, and other materials, like the binary GaAs, make significant contributions, but the major use is of silicon. Since the detailed tailoring of a semiconductor demands the controlled addition of different elements at less than the parts per million level, silicon or germanium are first produced from purified oxide, via a cycle of MCl_4 distillation, and reduction to the element. They are then produced in very high degrees of purity (better than $1:10^9$) by zone refining. In this process, the element is formed into a rod which is heated near one end to produce a narrow molten zone. The heater is then moved slowly along the rod so that the molten zone travels from one end to the other. Impurities are more soluble in the molten metal than in the solid and thus concentrate in the liquid zone which carries them to the end of the rod.

Uses of compounds of the carbon Group elements include all the well-known industrial organic chemicals like the oil and plastics industries (see below for uses of organometallic compounds of the other elements of the Group). Silicon dioxide and silicates are the major components of glasses, ceramics, pottery and related products, while pure SiO_2 is important in finely-divided high surface area forms for adsorption uses in industry, medicine and the laboratory. Fused silica is a very high-melting, inert, glass with very low expansion coefficient. Silicon carbide, SiC, is the abrasive carborundum.

Tin dioxide is a component of heterogeneous catalysts, is used widely in ceramics and enamels, and—as a very thin film—in electroluminescent devices. Treatment of glass with $SnCl_4$ deposits a thin film of SnO_2 on the surface which adds markedly to the toughness. All the carbon Group elements are fairly unreactive, with reactivity greatest for tin and lead. They are attacked by halogens, alkalies, and acids. Silicon is attacked only by hydrofluoric acid, germanium by sulfuric and nitric acids, and tin and lead by a number of both oxidizing and non-oxidizing acids.

Carbon reacts, when heated, with many elements to give binary carbides. Numerous silicides also exist and these are similar to the borides in forming chains, rings, sheets, and three-dimensional structures. Table 17.9 summarizes the various carbide types and Figures 17.19 and 5.13 give some of the structures.

The carbon Group shows the same trend down the Group towards metallic properties as in the boron Group. The II state becomes more stable and the IV state less stable from carbon to lead. Carbon is a non-metal and occurs in the tetravalent state. Silicon is metalloidal, but nearer non-metal than metal, and forms compounds only in the IV state, apart from the occurrence of catenation. Germanium is a metalloid with a definite, though readily oxidizable, II state. Tin is a metal and its II and IV states are both reasonably stable and interconverted by moderately active re-

TABLE 17.9 Types of binary carbide

State of aggregation of the carbon atoms	Properties	Examples and structures
Single atoms (a) salt-like carbides (b) transition element carbides	Yield mainly CH_4 on hydrolysis (i) Conducting, hard, high melting, chemically inert (ii) Conducting, hard, high melting, but chemically active: give C, H_2, and mixed hydrocarbons on hydrolysis	Be_2C (antifluorite) Al_3C_4 MC, M = Ti, Zr, Hf, Ta, W, Mo (sodium chloride) W_2C, Mo_2C Compounds of the elements of the later transition Groups, e.g. M_3C where M = Fe, Mn, Ni
Linked carbon atoms (a) C_2 units 'acetylides'	(i) CaC_2 type—ionic, give only acetylene on hydrolysis (ii) ThC_2 type—apparently ionic, give a mixture of hydrocarbons on hydrolysis	MC_2, M = Ca, Sr, Ba: structure related to NaCl (Figure 5.13) Also Na_2C_2, K_2C_2, Cu_2C_2, Ag_2C_2 ThC_2 and MC_2 for M = lanthanide element. Structure (Figure 17.19) also related to sodium chloride
(b) C_3 chains	Gives propyne, $H_3C–C{\equiv}CH$, on hydrolysis	Li_4C_3, Mg_2C_3 and $Ca_3Cl_2C_3$ all contain C_3^{4-} ions. $C{=}C$ bond length = 134.4 pm, angle at C = 169–176°.
(c) C_n chains	C—C spacing in chain is similar to that in hydrocarbons	Cr_3C_2.—C—C—C— chains running through a metal lattice, compare FeB
Carbon sheets (lamellar structures derived from graphite) (a) Buckled sheets	Non-conducting, carbon atoms are four-coordinated	(i) 'graphite oxide' from the action of strong oxidizing agents on graphite. C:O ratio is 2:1 or larger and the compounds contain hydrogen. $C{=}O$, C—OH, and C—O—C groups have been identified (ii) 'graphite fluoride' from the reaction with F_2. White, idealized formula is $(CF)_n$, with n about 1·1. The C atoms within the sheets are bonded to one F, while the sheet edges are CF_2 units. These species have advantageous properties as high temperature lubricants.
(b) Planar sheets	Conducting π system is preserved	(i) Large alkali metal compounds—of K, Rb, or Cs: e.g. C_8K. The metal is ionized and the electron enters the π system, while the metal ions are held between the sheets (ii) Halogen compounds. X^- ions are held between the sheets and positive holes are left in the π system which increase the conductivity

agents. The Sn^{4+}/Sn^{2+} potential is $-0·15$ V and tin(II) in acid is well-known as a mild reducing agent. Lead is a metal with a stable II state. Lead(IV) is unstable and strongly oxidizing.

The elements of the carbon Group are particularly characterized by their tendency to *catenation*, i.e. to form chains with links between like atoms. Carbon, of course, has this property in an exceptional degree. In the hydrides, chains of up to ten atoms are established for silicon and germanium, as in $Si_{10}H_{22}$, and distannane, Sn_2H_6 is known. Silicon also forms long chain halides but germanium is limited to $GeCl_4$ and Ge_2Cl_6 as far as present studies go. However, when the chain is fully substituted by organic groups, as in $M_n(CH_3)_{2n+2}$, there is no apparent limit to n for M = Si, Ge, and Sn. In these compounds, as with silanes and germanes, the restriction is the experimental difficulty of

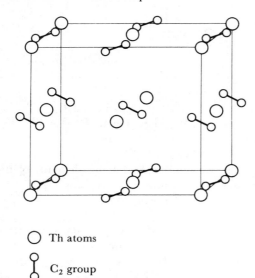

○ Th atoms

○—○ C$_2$ group

FIGURE 17.19 *Structure of thorium carbide, ThC$_2$*
As in CaC$_2$, the C$_2$ units occupy halide ion positions in a NaCl-type structure but differ in being oppositely aligned in successive layers (compare Figure 5.13).

handling high molecular-weight compounds. While the hydrides readily oxidize, and the halides hydrolyse rapidly in air, the organo-derivatives are moderately stable to attack by air. Lead compounds are more restricted, but Pb$_2$R$_6$ species are well-known. In addition, these elements form compounds (MR$_2$)$_n$ which are ring compounds. Rings with $n = 4$, 5 and 6 are well-known for M = Si, Ge and Sn, for R = Ph or related aryls, and some alkyl species are reported as well. For example, Me$_{10}$Si$_5$ has been made and converted to Me$_9$Si$_5$X (X = Cl, Br) which allows further reaction. A few three-membered rings occur but only with bulky ligands. Larger rings are established for Si and Sn. There is also definite evidence for the formation of branched chains. The compounds (Ph$_3$Ge)$_3$GeH and a number of mixed silylgermanes, (H$_3$Si)$_n$GeH$_{4-n}$ ($n = 2, 3, 4$), and methyl derivatives, (H$_3$Si)$_n$(CH$_3$)GeH$_{3-n}$ ($n = 1, 2, 3$), have been reported. Branched chain hydrides of M$_4$ and M$_5$ forms (M = Si, Ge) are also indicated by chromatographic experiments. For tetrasilanes and tetragermanes, the *n*- and *iso*- forms have been separated on the macro-scale, as have two of the three Ge$_5$H$_{12}$ isomers and related silicon-germanium species (see section 9.5). Recently, a number of compounds (Ph$_3$M)$_4$M' (M = Ge, Sn, Pb and M' = Sn, Pb) have been reported. The pentaplumbane of this *neo*-form, Pb(PbPh$_3$)$_4$, is the only other species with Pb–Pb bonds.

17.5.2 *The (IV) state*

The IV state is found for all the elements of the Group, and is stable for all but lead. Its properties are well illustrated by the oxygen and halogen compounds. The oxygen compounds are listed in Table 17.10. All the dioxides are prepared by direct reaction between the elements and oxygen. They are also precipitated in hydrated form (except CO$_2$ of course) by

addition of base to their solutions in acid. No true hydroxide, M(OH)$_4$, exists for any of the elements.

The very marked effect of $p_\pi - p_\pi$ bonding on the structures of the carbon, as compared with the silicon, compounds is obvious, as is the tendency towards a higher coordination number to oxygen for the heavier elements.

The oxyanions are also listed in Table 17.10 and comprise a wide array of compounds because of the strong tendency to form condensed polynuclear species based on EO$_4$ coordination for E = Si, Ge (compare silicates, section 18.6) or EO$_6$ units for the heavier elements. While the naturally-occurring silicates tend strongly towards ring, sheet, and three-dimensional structures, use of different counterions may give alternative structures. It is perhaps significant that the first short chain silicate (Si$_4$O$_{13}$)$^{10-}$ was recently isolated as the Ag$^+$ salt.

The formation of the heavier chalcogenides also reflects the relative stabilities of the II and IV states. While Si, Ge, and Sn, form such compounds as GeS$_2$ or SnSe$_2$, there are no such compounds of lead. Among the interesting complex species (again underlining the IV state stability) are Sn$_2$Te$_6$$^{4-}$ (two tetrahedra sharing an edge) and Si$_2$Se$_8$$^{2-}$ where the apparently anomalous oxidation state arises from a Se–Se bridge which links two SiSe$_3$ units, completing a tetrahedron at each Si.

All the tetrahalides, MX$_4$, are found except for lead (IV) which is too oxidizing to form the tetrabromide or iodide. All may be made from the elements, from the action of hydrogen halide on the oxide, or by halogen replacement. All the carbon tetrahalides, all the chlorides, bromides, and iodides, and also SiF$_4$ and GeF$_4$, are covalent, volatile molecules. The volatility and stability fall in a regular manner with increasing molecular weight of the tetrahalide. By contrast, SnF$_4$ and PbF$_4$, are involatile solids with melting or sublimation points at 705 °C and 600 °C respectively. They have polymeric structures based on MF$_6$ octahedra, with partially ionic bonding, as has aluminium trifluoride. Thus the tetrafluorides of the carbon Group parallel the trifluorides of the boron Group in changing from volatile to involatile and polymeric, but the change-over comes further down the Group. Bond lengths are given in Table 2.10b.

Carbon tetrafluoride (and all the fluorinated hydrocarbons) and carbon tetrachloride are very stable and unreactive, though CCl$_4$ will act as an oxidizing and chlorinating agent at higher temperatures. Carbon tetrabromide and iodide are stable under mild conditions, but act as halogenating agents on warming, and are also decomposed by light.

The silicon tetrahalides, except the fluoride, are hydrolysed rapidly to 'silicic acid' which is hydrated silicon dioxide. The heavier element tetrahalides also hydrolyse readily, but the hydrolysis is reversible and, for example, GeCl$_4$ can be distilled from a solution of germanium (IV) in strong hydrochloric acid.

A very wide range of 6-coordinate complexes MX$_4$.2D is formed by the tetrahalides, and 5-coordinate species such as R$_n$SnX$_{5-n}^-$ ($n = 1, 2, 3$) are also well-known. The latter have

TABLE 17.10 Oxygen compounds of carbon Group elements

Compound	Properties	Structure	Notes
Dioxides			
CO_2	Monomer, weak acid	Linear, $O=C=O$	$p_\pi - p_\pi$ bonding between first row elements giving π-bonded monomer
SiO_2	Involatile, weak acid	4:2 coordination with SiO_4 tetrahedra (see Figure 5.4b)	There is some weak $p_\pi - d_\pi$ bonding in some $Si-O-Si$ systems, though not in the oxide.
GeO_2	Amphoteric	Two forms: one with a 4:2 silica structure, and one with the rutile (6:3) structure	The Ge/O radius ratio is on the borderline between six- and four-coordination
SnO_2	Amphoteric	Rutile structure	
PbO_2	Inert to acids and bases	Rutile structure	Strongly oxidizing
Other Oxides (excluding monoxides)			
C_3O_2	Carbon suboxide, prepared by dehydrating malonic acid $$CH_2(COOH)_2 \xrightarrow[300\,°C]{P_2O_5} O=C=C=C=O$$	Linear, $C-C$ and $C-O$ distances are intermediate between those expected for single and for double bonds	The molecule contains extended π bonding of the same type as in CO_2.
Pb_3O_4	Red lead: oxidizing, contains both Pb(IV) and Pb(II)	Figure 17.20. The structure consists of $Pb^{IV}O_6$ octahedra linked in chains; the chains are joined by pyramidal $Pb^{II}O_3$ groups	
Oxyanions			
Carbonate	CO_3^{2-}	Planar with π bonding	Again C and O give p_π bonds
Silicates	Wide variety (section 18.6)	Formed from SiO_4 units	
Germanates	Variety of species	Contain both GeO_4 and GeO_6 units	Compare the two forms of germanium dioxide
Stannates } Plumbates }	e.g. $M(OH)_6^{2-}$	Contain octahedral MO_6 units	
Oxyhalides			
carbonyl halides			Rapidly hydrolysed
COF_2	b.p. $-83\,°C$	All are planar $\begin{smallmatrix}X\\ \\X\end{smallmatrix}\!\!\diagdown\!\!\diagup\, C=O$	Very poisonous: has been
$COCl_2$ (phosgene)	Stable		used as nonaqueous solvent
$COBr_2$	Fumes in air		

Other, mixed, oxyhalides such as COClBr are known

silicon oxyhalides
These are all single-bonded species containing $-Si-O-Si-O-$ chains (for example, $Cl_3Si-(OSiCl_2-)_nOSiCl_3$ with $n = 4, 3, 2, 1$ or 0) or rings (for example $(SiOX_2)_4$ where $X = Cl$ or Br)

a trigonal bipyramidal structure. Seven-coordination is found in $Me_2Sn(NCS)_2$(terpy) where the 3 nitrogens of the terpy (see Appendix B) and the two NCS groups form an almost regular pentagon with the methyls on the axis completing a pentagonal bipyramid. Eight-coordinate lead is found in $(C_6H_5)_2(CH_3COO)_3Pb^-$. The structure is a hexagonal bipyramid with the phenyl groups on the axis and the three bidentate acetates lying in the central plane.

17.5.3 *Hydride and organic derivatives*

Hydride-halides of the types MH_3X, MH_2X_2, and MHX_3 are also formed. Most representatives of these formulae (for $X = F, Cl, Br, I$) are found for silicon and germanium, but a few tin compounds, such as SnH_3Cl, are also known. Such compounds are key members of synthetic routes to organic and other derivatives, as in reactions such as:

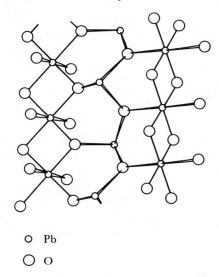

○ Pb

◯ O

FIGURE 17.20 *The structure of* Pb_3O_4
The structure contains $Pb^{IV}O_6$ octahedra linked together into chains by sharing edges. These chains are, in turn, linked by $Pb^{II}O_3$ pyramidal units which both link two of the Pb^{IV} chains and form a chain of $Pb^{II}O_3$ units.

$$SiH_3Br + RMgX \rightarrow RSiH_3 \ (R = \text{organic radical})$$

or $$GeH_3I + AgCN \rightarrow GeH_3CN + AgI$$

It has recently been discovered that the higher hydrides of silicon and germanium behave similarly, and all the compounds M_2H_5X, for M = Si or Ge, and X = F, Cl, Br, and I, have been prepared.

The elements from silicon to lead have an extensive organometallic chemistry and a wide variety of MR_4 and M_2R_6 compounds exist, with sigma metal-carbon bonds of considerable stability. Organotin and organolead compounds have been studied for their pharmaceutical and biocidal properties and organogermanium compounds may have similar properties. Butyl-tin compounds find widespread use in marine anti-fouling paints but concerns about the accumulation of organo-tin in the environment is leading to a search for alternatives. A number of organo-tin complexes show promising anti-tumour activity. Tetra-ethyl lead is manufactured on a large scale as an anti-knock agent for petrols, though it is being reduced in usage. All the tetra-alkyl and tetra-aryl compounds are stable, although stability falls from silicon to lead and the aryls are more stable than the alkyls. For example, tetraphenylsilicon, Ph_4Si, boils at 530 °C without decomposition, tetraphenyllead, Ph_4Pb, decomposes at 270 °C, while tetraethyllead, Et_4Pb, decomposes at 110 °C. A wide variety of organocompounds, with halogen, hydrogen, oxygen, or nitrogen linked to the metal, is also known and this class includes the *silicone polymers*. These are prepared by the hydrolysis of organosilicon halides:

$$R_2SiCl_2 \rightarrow \ -\overset{\displaystyle R}{\underset{\displaystyle R}{Si}}-O-\overset{\displaystyle R}{\underset{\displaystyle R}{Si}}-O-\overset{\displaystyle R}{\underset{\displaystyle R}{Si}}-O-$$

This long-chain polymer is linked by the very stable silicon – oxygen skeleton and the organic groups are also linked by strong bonds so the polymer has high thermal stability. The organic groups also confer water-repellent properties. The chain length is controlled by adding a proportion of R_3SiCl to the hydrolysing mixture to give chain-stopping $-OSiR_3$ groups, while the properties of the polymer may also be varied by introducing cross-links with $RSiCl_3$:

$$R_2SiCl_2 + R_3SiCl + RSiCl_3 \longrightarrow$$

The elements from silicon to lead use their d orbitals to form six-coordinated complexes which are octahedral. All four elements give stable MF_6^{2-} complexes with a wide variety of cations. The MCl_6^{2-} ion is formed for M = Ge, Sn, and Pb, and tin also gives $SnBr_6^{2-}$ and SnI_6^{2-}. In addition, a variety of MX_4L_2 complexes are formed by the tetrahalides with lone pair donors such as amines, ethers, or phosphanes. The chemistry of the tin compounds is particularly well-explored, and both *cis* and *trans* compounds are known. When stannic chloride was reacted with phosphorus pentachloride, Cl^- transfer took place (compare PCl_5 itself section 17.6.2) to yield $(PCl_4^+)_2SnCl_6^{2-}$, and the mixture also yielded $Sn_2Cl_{10}^{2-}$ (with a structure involving two octahedra sharing an edge) and $SnCl_5^-$ (trigonal bipyramid). Six-coordinate complexes may also be formed by the organic derivatives of Ge, Sn, and Pb, as long as there are enough electronegative substituents to give reasonable acceptor power. Thus we find $Me_2SnCl_4^{2-}$, and $Me_4Sn_2Cl_6^{4-}$, each with octahedral Sn and trans Me groups, and the latter with two bridging Cl.

There are a much more limited number of five-coordinated species including MF_5^- for M = Si, Ge, or Sn, and MCl_5^- for Ge and Sn. The chlorides are trigonal bipyramids, as expected, as are the fluorides in the presence of large cations. However, with smaller cations the fluorides form a *cis* fluorine-bridged polymeric structure.

The d orbitals are also used in internal π bonding, especially in silicon compounds. The classic case is trisilylamine, $(SiH_3)_3N$ Figure 17.22. This has a quite different structure from the carbon analogue, trimethylamine, $(CH_3)_3N$, shown in Figure 17.21. The pyramidal structure of trimethylamine is similar to that of NH_3 and reflects the steric effect of the unshared pair on the nitrogen. The NSi_3

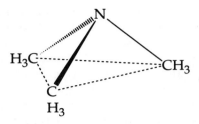

FIGURE 17.21 *The structure of trimethylamine,* $(CH_3)_3N$

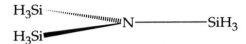

FIGURE 17.22 *The structure of trisilylamine,* $(SiH_3)_3N$

skeleton, by contrast, is flat and the nitrogen in trisilylamine shows no donor properties. This is due to the formation of a π bond involving the nitrogen p orbital and d orbitals on the silicon, as shown in Figure 17.23. The lone pair electrons donate into the empty silicon d orbitals and become delocalized over the NSi_3 group, and hence there is no donor property at the nitrogen. Structural evidence comes from the infra-red and Raman spectra and from the zero dipole moment. Although trisilylamine is a gas at room temperature, it has recently been possible to carry out an X-ray study of the solid at low temperatures and this has confirmed the planar Si_3N skeleton. It is noteworthy that $N(CF_3)_3$ is nearly planar $(CNC = 118°)$ and $N(C_2F_5)_3$ even closer $(CNC = 119.3°)$ showing the effect when strongly electron-withdrawing ligands remove much of the lone pair electron density.

Tetrasilylhydrazine, $(SiH_3)_2NN(SiH_3)_2$, also shows differences in symmetry compared with the methyl analogue, arising from similar π bonding. Another clear case is the isothiocyanate, MH_3NCS. Where $M = C$, the $C-N-C-S$ skeleton is bent at the nitrogen atom due to the steric effect of the nitrogen lone pair while, when $M = Si$, the $Si-N-C-S$ skeleton is linear and there is $d_\pi-p_\pi$ bonding between the Si and N atoms. Similar effects are observed in $M-O$ bonds: dimethyl ether, CH_3OCH_3, has a $C-O-C$ angle of $110°$ which is close to the tetrahedral value, while the bond angle in disilyl ether, SiH_3OSiH_3, is much greater—$140°$ to $150°$—indicating delocalization of the non-bonding pairs on the oxygen.

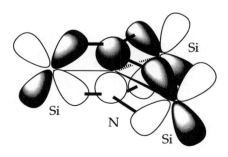

FIGURE 17.23 *Pi-bonding in trisilylamine*

This formation of π bonds using metal d orbitals and a p orbital on a first row element is most marked in the case of silicon, but there is some evidence that it occurs in germanium compounds as well. For example, the $Ge-F$ bond in GeH_3F is very short, which may indicate $d_\pi-p_\pi$ bonding but GeH_3NCO and GeH_3NCS have bent $Ge-N-C$ skeletons in contrast to their silicon analogues. Other evidence for π bonding by germanium is indirect and derived from acidities and reaction rates in substituted phenylgermanes.

17.5.4 *The (II) state*

The II state is the stable oxidation state for lead, and the IV state of lead, like thallium(III) in the last Group, is strongly oxidizing. Pb^{2+} ions exist in a number of salts, though hydrolysis occurs readily in solution:

$$Pb^{2+} + 2H_2O = PbOH^+ + H_3O^+$$

and a further equilibrium is found:

$$4Pb^{2+} + 12H_2O = Pb_4(OH)_6^{2+} + 6H_3O^+$$

A considerable variety of lead(II) salts is known, and these generally resemble the coresponding alkaline earth compounds in solubility, e.g., the carbonate and sulfate are very insoluble. The halides are less similar and $PbCl_2$ is, like $TlCl$, insoluble in cold water though more soluble in hot water.

Addition of alkali to lead(II) solutions gives a precipitate of the hydrated oxide, which dissolves in excess alkali to give plumbites. The hydrated oxide may be dehydrated to PbO, called litharge, which is yellow-brown in colour. The structure of PbO is an irregular one in which the lead is co-ordinated to four oxygen atoms at corners of a trigonal bipyramid, with the fifth position occupied by an unshared pair of electrons. Just as lead forms the mixed oxidation state oxide (see Table 17.10), a mixed oxidation state oxyanion species is formed in compounds like $KPbO_2$. In a related example, $KNa_7[PbO_4][PbO_3]$, a crystal structure shows an approximately tetrahedral $Pb(IV)O_4^{4-}$ ion and a pyramidal $Pb(II)O_3^{4-}$ ion. Another interesting structure is found in the $Pb_8O_4^{8+}$ cluster ion whose core is a cubane Pb_4O_4 unit with a further Pb bonded to each O corner, so that these 4 Pb atoms transcribe a tetrahedron around the cubane.

Lead forms all the heavier chalcogenides, PbE, as do Ge and Sn (E = S, Se, Te). Where known, the structures are layer lattices.

The II state of tin is mildly reducing but otherwise resembles lead(II). In solution, Sn^{2+} ions hydrolyse as in the first equation for Pb^{2+} above. Addition of alkali to stannous solutions precipitates $SnO.xH_2O$ (neither $Sn(OH)_2$ nor $Pb(OH)_2$ exist), and the hydrated oxide dissolves in excess alkali to give stannites. Dehydration gives SnO, which has a similar structure to PbO. One stannite (formally named oxostannate(II)) structure is known. $K_2Sn_2O_3$ is built of trigonal pyramidal SnO_3 units with $Sn-O-Sn$ bridges. Tin(II) gives all four dihalides and a number of oxysalts. In the vapour phase, $SnCl_2$ exists as monomeric molecules with the V-shape characteristic of species with two bonds and one lone pair. One molecule of water adds to this molecule to give the pyramidal hydrate, $SnCl_2 . H_2O$. SnF_2 has a tetrameric structure based on an Sn_4F_4 puckered 8-membered ring. The Sn−F−Sn bridge angles are 135°, the SnF distances average 215 pm, and each Sn also carries external F with a 207 pm bond length. The Sn atom is thus at the apex of a trigonal pyramid with FSnF angles of 84°. Much longer Sn−F bridges of about 290 pm link these 8-membered rings together.

Germanium gives a number of compounds in the II state; GeO, GeS, and all four dihalides are well established. These compounds all appear to have polymeric structures and are not too unstable, probably because attack on the polymeric molecules is relatively slow. The known structures are those of GeI_2, which has the CdI_2 layer structure, and of GeF_2 which has a long chain structure similar to that of SeO_2. The II state is readily oxidized to the IV state: thus the dihalides all react rapidly with halogen to give the tetrahalides, while the corresponding reaction of tin(II) compounds is slow. Yellow GeI_2 disproportionates to red GeI_4 and germanium on heating. The divalent compounds are involatile and insoluble, in keeping with a polymeric formulation. One complex ion of the II state is known, $GeCl_3^-$, in the well-known salt $CsGeCl_3$ and adducts $R_3P.GeI_2$ are also reported. The $GeBr_3^-$ ion has been characterized as the Rb^+ salt. The structure shows a pyramidal ion (Ge−Br = 253 pm, BrGeBr = 95.5°) which is more weakly linked to three Br of other ions (Ge···Br = 324 pm) giving distorted octahedral geometry at Ge. Germanium(II) may also be obtained in acid solution, in the absence of air, and addition of alkali precipitates the yellow hydrated oxide, $GeO.xH_2O$.

The divalent Si(II) halides, SiX_2(X = F, Cl, Br, and I), can be prepared as reactive species and have angular molecules. For example, $SiBr_2$ shows a Br−Si−Br angle of 103°, and Si−Br bond distances of 224 pm.

Tin and lead form more complexes in the II state than does Ge, though fewer than in the IV state. Halogen complexes, MX_3^- are pyramidal monomers for the heavier halides while polymeric forms are found for X = F. Thus SnF_3^- is an infinite chain of SnF_4 square pyramids with Sn at the apices and linked by sharing F atoms at *trans* corners. Similarly, MOX or $M(OH)X_2$ compounds are polymers, all with sterically active lone pairs. Cation complexes MX^+ are also polymers. Formally divalent organometallic compounds R_2M are mostly ring or long-chain species with M−M bonds, where M = Si, Ge, or Sn (see above). Rings with 4, 5 and 7 Si, Ge or Sn atoms are relatively more stable than their carbon analogues. However, simple molecular R_2M species, known as silylenes, germylenes, stannylenes and plumbylenes (for Si, Ge, Sn and Pb, respectively) are known to be highly reactive species which can be trapped in matrices and studied. These divalent organic derivatives, especially of Si, Ge or Sn, are also proven as reaction intermediates. A few examples have been more fully characterized. Large ligands are necessary to allow stable solids, as in the trifluoromethylphenyl compound $[(CF_3)_3C_6H_2]_2Sn$, where the CSnC bond

angle is 98°. The corresponding lead compound is also known and has CPbC = 94°. Similarly, carbenes such as CCl_2 are well-known in organic chemistry and provide a route for the synthesis of cyclopropanes, by the addition to alkenes. The use of large ligands allows stable compounds to be isolated and compounds like $Ge[CH(SiMe_3)]_2$ or $Sn[N(CMe_3)]_2$ are monomers. In a series of recent reports, a number of stabilized carbenes, silylenes and germylenes (shown in Figure 17.24) have been described which show extraordinary thermal stability, a property not expected for such reactive species. The silylene is stable for at least four months at 150 °C. This stability has opened up the debate on the electronic structure of such species, and whether or not they are true carbenes, silylenes or germylenes. It is likely that the lone pairs on the nitrogen atoms of such species are donating electron density to the electron-deficient carbon, silicon or germanium atom. It has even been suggested that these species are aromatic in nature, due to the possibility of having six π-electrons (two from the C = C double bond, and two from each of the nitrogens), similar to benzene.

$(C_5H_5)_2M$ (M = Ge, Sn, Pb) are monomers in the gas phase with ring−M−ring angles about 145° and the lone pairs pointing away from the rings. In the solid, they are polymeric. With the very bulky substituent in $[Ph_5C_5]_2Sn$ the rings are planar and parallel (compare Figure 16.7). A few mixed species, RMX, RM(OH) or $(RM)_2O$ are found. All are polymeric structures with sterically active lone pairs.

Although the monomers in the II state have a lone pair, and thus could act as donors (Lewis bases), such behaviour is much rarer than the acceptor modes above. One case is that of $(RNM)_4.2AlCl_3$ (for M = Ge or Sn and R is the bulky CMe_3 group) where the M_4N_4 skeleton is a cubane (with R on N) and two of the four M atoms form donor bonds to $AlCl_3$ groups.

Polymetal clusters such as Sn_5^{2-} are discussed in section 18.4.

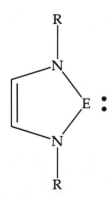

E = C, Si, Ge
R = -C(CH_3)_3 or adamantyl

FIGURE 17.24 *Structure of stabilized carbene, silylene and germylene species*

17.5.5 *Reaction mechanisms of silicon*

Work on inorganic reaction mechanisms is less developed than the corresponding area in organic chemistry, mainly because of the large variety of systems and of experimental difficulties. In particular, many non-organic reactions are extremely fast. Of all the *p* elements (for mechanisms at *d* elements, see section 13.9), silicon presents one of the most favourable cases for study, and we illustrate something of what is known about Main Group mechanisms by this outline of mechanisms at silicon.

Mechanisms are postulated (and remember that all reaction mechanisms are only hypotheses) on the basis of reaction kinetics, and study of silicon has the major advantage that kinetic work may be independently supported by evidence from optically-active compounds. The isolation and resolution of active silicon species has given a powerful tool which has been used, particularly by Sommer and his colleagues, to study mechanisms of substitution.

A number of optically active silicon species have been reported, one of the first being $Ph(\alpha\text{-}Nt)(Me)SiX$, where $\alpha\text{-}Nt = \alpha$-naphthyl. This was resolved using $X = (-)$ menthoxide (menthol being a naturally-occurring optically active species) by recrystallization from pentane at $-78\ °C$. We shall abbreviate the optically active species as R_3Si*X. This isolation was greatly aided by the presence of bulky aromatic groups which reduce the rate of reaction. Even so, R_3Si*X commonly reacts about a thousand times faster than similar carbon compounds.

It was first shown that stereospecific substitutions did occur by cycles of changes analogous to the Walden cycle, e.g.

$$(+)R_3Si*H + Cl_2 = (-)R_3Si*Cl \xrightarrow{\text{LiAlH}_4} (-)R_3Si*H$$

$[\alpha]_D$
values: $+34°$ $-6°$ $-34°$

Thus one of these steps must occur with inversion, and one with retention, and both must be highly stereospecific. Later work showed that the same relative configuration occurred in the following species R_3Si*X, shown with the rotations

X	H	Cl	OH	OMe	Br	F
$[\alpha]_D$	$+34°$	$-6°$	$+20°$	$+17°$	$-22°$	$+47°$.

Thus, the chlorination above is retention, while the reduction involves inversion of configuration.

These observations establish that stereospecific substitutions do take place. Extensive further work has led to the postulation of four main mechanisms at silicon. These are briefly outlined.

S_N2. This is similar to the mechanism at carbon, but is much faster. It is found for R_3Si*X in polar, but poorly ionizing, solvents and particularly when X is a halogen. The reaction takes place with inversion of configuration, and is postulated to proceed through a trigonal bipyramidal intermediate conformation, in which the organic groups are in the central plane

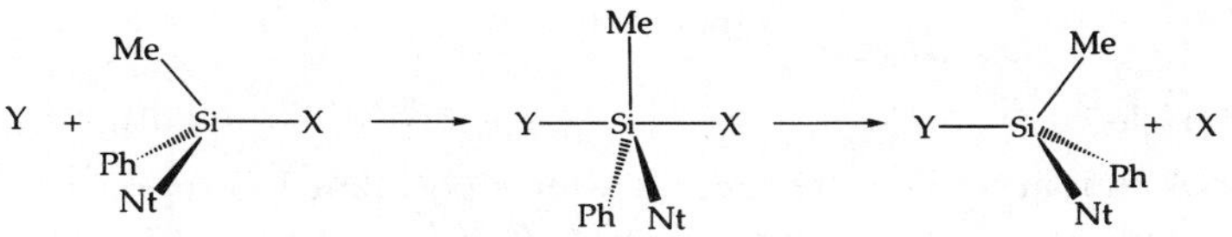

Typical examples are hydrolyses, or other replacements of Si–X by Si–OR, and the formation of the hydride above. The addition product between SiH_3Cl and dimethyl ether can be considered to be a model for the intermediate formed in hydrolysis reactions; the unreactive O–CH_3 groups prevent any further reaction in the case of the reactive O–H groups. The structure, Figure 17.25, has been found to be the expected trigonal bipyramid.

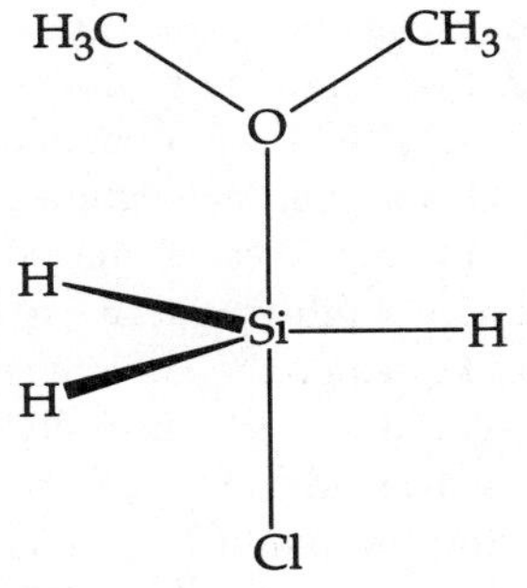

FIGURE 17.25 *The structure of the adduct* $SiH_3Cl[O(CH_3)_2]$ *formed from* SiH_3Cl *and dimethyl ether*

While such a process is assisted by using one of the silicon *d* orbitals to achieve five-coordination (which probably accounts for the speed of reaction) it does not necessarily follow that a stable intermediate forms. This could happen, or the effect of the *d* orbital may simply be to lower the activation energy compared with the carbon analogue.

S_Ni. When $X = OR$, and hydride or organometallic reagents are used in non-polar solvents, a slow reaction is found which proceeds with retention of configuration. This cannot be S_N2, as the intermediate would undergo fast loss of H or R (leading to racemization) rather than undergo cleavage of the very strong Si–O bond. It is therefore postulated that the reaction proceeds via a four-centre intermediate, and it is termed *internal nucleophilic substitution*. The intermediate may be represented

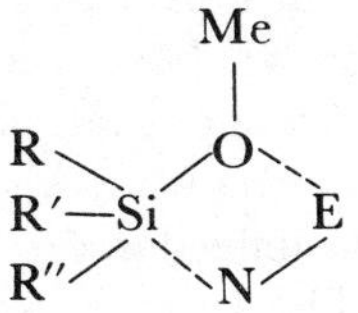

where E is the electrophilic and N the nucleophilic part of the reagent. Thus, for a Grignard reagent, $N = R$, and $E = MgX$; or for AlH_4^-, $N = H$ and $E = AlH_3$. The process may be understood as a nucleophilic attack assisted by the electrophilic coordination to oxygen which helps to overcome the strong Si–O binding energy. As an example,

$$R_3Si*OMe + LiAlH_4 \xrightarrow{\text{ether, 16 h}} R_3Si*H$$
$$[\alpha]_D = 16° \qquad\qquad\qquad [\alpha]_D = 30° \text{ (90\% retention)}$$

A similar four-centred mechanism is postulated for the very wide range of reactions called *hydrometallations* in which an M−H bond adds across a double bond. These are found for many metals, M, of which the most important are for M = B, Al (see last section), Si or Sn.

S_N1. This is less common than in carbon chemistry, and is found typically for halides in polar solvents of high dielectric constant. Thus, while R_3Si*Cl is recovered unchanged from solution in CCl_4 or an ether, when it is dissolved in acetonitrile or nitromethane (CH_3CN or CH_3NO_2, both with high dielectric constants) racemization takes place rapidly. This is postulated to proceed through a solvent-stabilized silyl cation, $RR'R''Si^+(solv)$. The silyl cation is analogous to the carbonium ion $RR'R''C^+$ known in organic chemistry.

In fact, the search for the tricoordinated and unsolvated silyl cation is an ongoing area of activity. Since the silyl cation $RR'R''Si^+$ is a highly reactive species, it tends to coordinate a 'ligand', either the solvent or the anion itself. Accordingly, the search for the free silyl cation has been paralleled by the search for the least coordinating anion. Classical non-coordinating anions, such as ClO_4^-, BF_4^-, PF_6^- and $CF_3SO_3^-$, for example, have all been found to coordinate to metals in all parts of the Periodic Table. The use of the even more poorly coordinating anions $B(C_6F_5)_4^-$ and carboranes, e.g. $CB_9H_{10}^-$, which have their negative charges spread over a large number of atoms, has allowed the isolation of compounds which contain an even weaker, but still persistent, interaction between the R_3Si^+ ion and the anion or the solvent. One example is the compound $[(toluene)SiEt_3]^+ [B(C_6F_5)_4]^-$ which contains an interaction between the toluene solvent molecule and the silicon atom. Other poorly coordinating anions, such as $[B(OTeF_5)_4]^-$ and $[Nb(OTeF_5)_6]^-$ which have their negative charge delocalized over a large number of electronegative fluorine atoms, also reduce the potential for these anions to act as 'ligands' towards Si^+ and other reactive species. The use of anions of this type in the isolation of silver carbonyls is described in section 15.9.1.

EO (expanded octet). One special mechanism is sometimes involved when X = F. For most reactions, fluorides behave as other halides and give the above mechanisms. However, the Si−F bond is much stronger than, for example, Si−Cl and this allows a further mechanism. An example is the reaction in which R_3Si*F is racemized in dry pentane solution by the addition of MeOH. This reaction has the following characteristics which exclude any of the three mechanisms outlined above:

(a) the rate is retarded in formic acid, a solvent of high dielectric constant, hence the reaction is not S_N1. Further, addition of HF retards the reaction so that the reaction does not proceed by loss of F^- as this would be stabilized as HF_2^-.

(b) A mixture of R_3Si*F and $R_3Si*OMe$ plus MeOH gives unchanged $R_3Si*OMe$ and racemic R_3SiF. Thus the racemization is not via R_3SiOMe or any species which could give rise to it, excluding S_N2 and S_Ni.

These features led to the postulate of an expanded-octet mechanism with a five- or six-coordinate intermediate formed by addition of OMe, and which subsequently loses OMe again:

As the intermediates are labile or inactive, racemization occurs. There is no breaking of the very strong Si−F bond. Formation of an expanded octet would be assisted by the presence of the fluorine substituent. Note that this mechanism must be rare or there could be no isolation of optically active silicon compounds at all.

These conclusions from optical studies may be supported by kinetic studies in favourable cases. Thus, the formation in a fast step of a relatively stable intermediate, followed by a slow dissociation to products

$$A + X-Y \underset{\leftarrow}{\xrightarrow{\text{fast, } k_1}} A-X-Y \underset{\leftarrow}{\xrightarrow{\text{slow, } k_2}} A-X+Y$$

would be characterized by a dependence on k_2 alone, and by the fact that the rate of consumption of A was not equal to the rate of appearance of Y. Thus, in S_Ni, EO, and some S_N2 reactions (if the intermediate was relatively long-lived) the above difference in rates would be detected.

Conversely, if the intermediate was unstable and immediately gave the product (i.e. if the second step above was very fast) the rate of appearance of Y would equal the rate of loss of A, and k_1 would be rate-determining. This reaction would thus be second order. Such kinetics would characterize the normal S_N2 reaction.

Finally, the S_N1 reaction is first order and the determining step is the dissociation into cation and anion.

While many mechanisms give rise to an intermediate kinetic picture (there may be a wide range of lifetimes for the A−X−Y intermediate, for example) if kinetic and optical studies agree, the postulated mechanism is quite strongly supported. As far as the silicon mechanisms outlined above are concerned, such kinetic studies as are reported do validate the proposed mechanisms.

17.6 The nitrogen Group, $ns^2 np^3$

17.6.1 *General properties*

References to the properties of the nitrogen Group elements which have occurred in the earlier part of the book include:

Ionization potentials	Table 2.8
Atomic properties and electron configurations	Table 2.5

Radii   Table 2.10. Table 2.11
Electronegativities   Table 2.14
Redox potentials   Table 6.3
Structures   Chapter 5

Table 17.11 lists some of the properties of the elements and the variation with Group position of important parameters is shown in Figure 17.26. The oxidation free energy diagram is shown in Figure 17.27.

Fixation of nitrogen is discussed under titanium (section 14.2), and nitrogen-complexes in sections 15.6 and 16.10.

Of these elements, nitrogen and bismuth are found in only one form while the others occur in a number of allotropic forms. Nitrogen exists only as the triply-bonded N_2 molecule, and bismuth forms a metallic layer structure shown in Figure 17.28. There are a number of allotropes of phosphorus. In white phosphorus, and also in the liquid and vapour states, the unit is the P_4 molecule where the four phosphorus atoms form a tetrahedron. If the vapour is heated above 800 °C, dissociation to P_2 units starts and rapid cooling of the vapour from 1000 °C gives an unstable brown form of phosphorus

TABLE 17.11 Properties of the nitrogen Group elements

Element	Symbol	Oxidation states	Co-ordination number	Availability
Nitrogen	N	− III, III, V	3, 4	common
Phosphorus	P	(− III), (I), III, V	3, 4, 5, 6	common
Arsenic	As	III, V	3, 4, (5), 6	common
Antimony	Sb	III, V	3, 4, (5), 6	common
Bismuth	Bi	III, (V)	3, 6	common

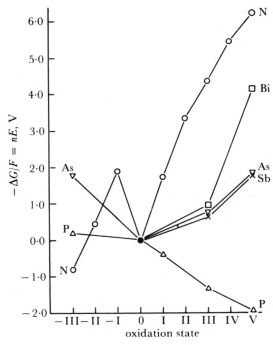

FIGURE 17.27 *Oxidation state free energy diagrams of elements of the nitrogen Group*

This is the most complex oxidation state diagram of all the Main Groups. The properties of nitrogen are the most individual, with the element and the − III states as the most stable. All the positive states between 0 and V tend to disproportionate in acid solution (though many form gaseous species in equilibrium with the species in solution) and the − I state is markedly unstable. The curves for P, As, Sb, and Bi form a family in which the − III state becomes increasingly unstable (values for Sb and Bi in this state are uncertain) and the V state becomes less stable with respect to the III state from P to Bi. All intermediate states of phosphorus tend to disproportionate to PH_3 plus phosphorus(V). The diagram also illustrates the very close similarity between As and Sb and the strongly oxidizing nature of bismuth(V).

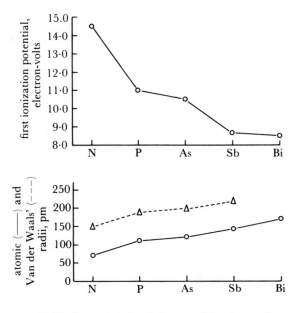

FIGURE 17.26 *Some properties of elements of the nitrogen Group*

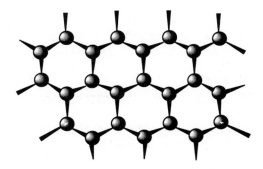

FIGURE 17.28 *The structure of bismuth*

which probably contains these P_2 units. When white phosphorus is heated for some time above 250 °C, the less reactive red form is produced. The structure of this form is not yet established and it exists as a number of modifications with various colours—violet, crimson, etc. These might be different structures or due to different crystal sizes, but they may also be due to the incorporation of part of the catalysts used in the transformation. When white phosphorus is heated under high pressure, or treated at a lower temperature with mercury as a catalyst, a dense black form results which has a layer structure like bismuth. A further vitreous form is also reported which results from heating and pressure. The structures of some of these allotropes are shown in Figure 17.29 and the interconversion of the allotropes in Figure 17.30. White

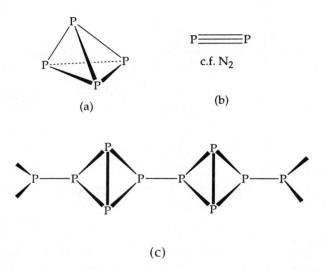

(a)

(b)

c.f. N_2

(c)

chains of linked, opened tetrahedra

FIGURE 17.29 *Structures of phosphorus allotropes:* (a) *white phosphorus,* (b) *brown phosphorus,* (c) *red phosphorus (postulated)*

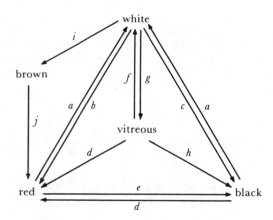

FIGURE 17.30 *The interconversion of the allotropes of phosphorus*
The reaction conditions are indicated by letters as follows:
(a) melting followed by quenching, or vacuum sublimation
(b) heating above 250 °C
(c) heating to 220 °C under pressure
(d) heating above 450 °C, or on prolonged standing at room temperature
(e) at 25 °C under high pressure
(f) vacuum sublimation
(g) heating above 250 °C under pressure
(h) heating to 400 °C under pressure
(i) rapid cooling of vapour with liquid nitrogen
(j) warming above liquid nitrogen temperature (-196 °C)
Many of these interconversions may also be brought about by catalysis, especially by mercury.

phosphorus is the most reactive of the common allotropes, red phosphorus is much less reactive, and black phosphorus is inert. Substantial further insight is provided by the recently established structures of the polyphosphorus hydrides and anions outlined in section 18.3.

Arsenic and antimony each occur in two forms. The most reactive is a yellow form which contains M_4 tetrahedral units and resembles white phosphorus. These yellow allotropes readily convert to the much less reactive metallic forms which have the same layer structure as bismuth.

Nitrogen gas is now produced on a substantial scale, largely as a byproduct of the isolation of oxygen from air for steelmaking. It is used to provide a cover for processes which are sensitive to air, and as liquid, it is in ever-increasing use as a coolant. One of the largest manufacturing applications using nitrogen is the Haber process for producing NH_3 from N_2 and H_2 under pressure over an iron catalyst. Recent work on catalysts such as ruthenium on carbon, allows the reaction to occur at atmospheric pressure and ordinary temperatures, foreshadowing a cheaper Haber process. The ammonia is used mainly as a fertilizer, as salts, and part is converted to nitrate by catalytic oxidation, again mainly for use as a fertilizer. Hydrazine, N_2H_4 is used in agricultural chemicals and herbicides and as an intermediate in pharmaceuticals.

Elementary phosphorus, made by carbon reduction of phosphate minerals, is extensively used in matches. The old formulation, based on poisonous white phosphorus, has now been entirely superseded by red phosphorus mixtures which also include sulfides of P and Sb. The oxidant is chlorate, and safety matches have separate oxidant in the match, with the phosphorus species on the striking surface. The major production of phosphorus compounds is that of phosphate fertilizer, mostly using the family of *apatite* ores, $Ca_5(PO_4)_3X$ (X = OH, Cl or F). These are of low solubility and superphosphate fertilizer, of higher solubility, is produced by treatment of the crude phosphate rock with sulfuric acid. Phosphoric acid is produced from the phosphate rock by treatment with sulfuric acid followed by purification by a solvent extraction process (see section 7.3). Phosphate, especially as a component of DNA and RNA, is essential to all life processes. The major component of teeth and the main bone mineral is hydroxyapatite, accompanied in the bone by amorphous calcium phosphate. In teeth F replacement, giving fluoroapatite, increases toughness and resistance to caries. Phosphates are widely used as food additives (such as phosphoric acid in cola drinks and sodium phosphates in processed foods), as flame retardants and in the prevention of scale and corrosion in water-cooling systems. Phosphorus sulfides are also manufactured on a large scale as starting materials for the synthesis of organophosphorus compounds which are widely used as pesticides, for example malathion. Many of the military nerve gases are organophosphorus compounds, e.g. sarin $CH_3P(O)(OPr^i)F$.

Traditional uses of arsenic compounds in insecticides and fungicides, e.g. as wood preservatives, are being phased out as less toxic alternatives become available. All three heavier elements find uses in alloys (see under tin in the last section).

As and Sb are becoming increasingly important in the III/V semiconductor materials mentioned under gallium above. Sb_2O_3 finds substantial use, along with various phosphorus compounds, as flame retardants in plastics.

Binary compounds of these elements are similar to those of previous Groups. Nitrides range from those of the active metals, which are definitely ionic with the N^{3-} ion, through the transition metal nitrides which resemble the carbides, to covalent nitrides like BN and S_4N_4. Heated Ba metal reacts with N_2 giving a product which hydrolyses to give 30% N_2H_4 and 70% NH_3. This suggests that the N_2^{4-} ion is formed as well as N^{3-}. The phosphides are similar. The interesting polyphosphide ion, P_3^{4-}, is isoelectronic with ClO_2 and contains an unpaired electron. The structure is V-shaped with an angle of 118° and a P–P length of 218.3 pm. The heavier elements form compounds with metals which become more alloy-like as one passes from phosphorus to bismuth. One important feature is the appearance of ionic nitrides, as compared with monatomic carbides which are not ionic. Ionic nitrides include the Li, Mg, Ca, Sr, Ba, and Th compounds. They are prepared by direct combination or by deammonation of the amides:

$$3Ba(NH_2)_2 \rightarrow Ba_3N_2 + 4NH_3$$

By contrast, the corresponding carbides either contain polyanions, like the acetylides, or are intermediate between ionic and giant covalent molecules.

Nitrogen, because of the high strength of the triple bond (heat of dissociation = 962 kJ mol^{-1}) is inert at low temperatures and its only reaction is with lithium to form the nitride. At higher temperatures it undergoes a number of important reactions including the combination with hydrogen to form ammonia (Haber process), with oxygen to give NO, with magnesium and other elements to give nitrides, and with calcium carbide to give cyanamide:

$$N_2 + CaC_2 = CaNCN + C$$

The other elements react directly with halogens, oxygen, and oxidizing acids.

This Group shows a richer chemistry than the boron and carbon Groups, as there are more than two stable oxidation states and there is a wider variety of shapes and coordination numbers. Nitrogen shows a stable $-III$ state in ammonia and its compounds, as well as in a wide variety of organo-nitrogen compounds. Phosphorus has an unstable $-III$ state in the hydride and also forms acids and salts, which contain direct P–H bonds, in the III and I states. In the normal states of V and III a variety of coordination numbers is found. The MX_3 compounds, where M = any element in the Group and X = H, halogen or pseudohalogen, or organic group, are pyramidal with a lone pair on M. The MX_5 compounds are trigonal bipyramids in the gas phase and adopt a variety of structures in the solid. It is interesting that while PPh_5 and $AsPh_5$ are trigonal bipyramids, $SbPh_5$ is a square pyramid both in the solid and in solution. This is a second (with $InCl_5^{2-}$) main group example of the square pyramid as the alternative five electron pair structure.

A number of MX_4^+ species are known, as well as $MX_3 \cdot A$ (where A = any acceptor molecule such as BR_3) and these are tetrahedral. A few $M^{III}X_5^{2-}$ complexes also exist and these are square pyramids with a lone pair in the sixth position. Finally, a variety of $M^VX_6^-$ and $M^VX_5 \cdot A$ compounds are found which are octahedral. Many of these shapes are repeated in compounds which include π bonding. A full set of examples is gathered in Table 17.12.

The nitrogen Groups shows a significant tendency to catenation, if less markedly than does the carbon Group. Polynuclear compounds are discussed in sections 18.3 and 4. Simpler chain compounds of nitrogen involve both single bonds, as in hydrazine and its derivatives, R_2N-NR_2, where R = H, F, or organic groups; and multiple bonds as in the diazenes RN = NR (R = F or organic groups, R = H is unstable unless coordinated—see section 16.10). Nitrogen is found in a chain of three atoms in hydrazoic acid, HN_3, and in the azides. These are prepared from amide and nitrous oxide:

$$2NaNH_2 + N_2O \rightarrow NaN_3 + NaOH + NH_3$$

The azide ion is linear, and isoelectronic with CO_2. The acid, called hydrazoic acid, is bent at the nitrogen bonded to the substituent, with the NNH angle equal to 114° and the bond distances indicating the presence of single and triple bonds

TABLE 17.12 Coordination numbers and stereochemistry in the nitrogen Group

Number of electron pairs	Number of π bonds	Number of non-bonding pairs	Shape	Examples
4	0	0	Tetrahedron	NH_4^+, MR_4^+, PX_4^+ (X = all halogens)
4	0	1	Pyramid	MH_3, MR_3, MX_3 (X = all halogens)
4	1	1	V	NO_2^-
5	0	0	Bipyramid	MF_5, PCl_5, PBr_5, PPh_5 ($SbPh_5$—see text)
5	1	0	Tetrahedron	MOX_3, MO_4^{3-}, HPO_3^{2-}, $H_2PO_2^-$
5	2	0	Plane	NO_3^-
6	0	0	Octahedron	MF_6^-, PCl_6^-, $SbPh_6^-$
6	0	1	Square pyramid	SbF_5^{2-}

(M = P, As, Sb and Bi, R = simple alkyl radical, Ph = phenyl)

(a)

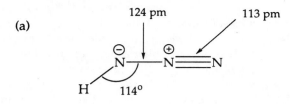

(b)

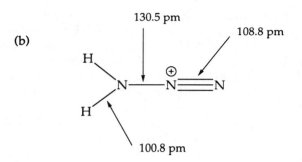

FIGURE 17.31 *Structures of* (a) *hydrazoic acid,* HN₃ *and* (b) *aminodiazonium cation,* $H_2N_3^+$
Only one resonance form is shown for each.

(Figure 17.31a). Organic azides, RN_3, have similar structures. In the N_3^- ion the two N—N distances are equal at 116 pm, showing delocalization of the π bonding. The aminodiazonium cation, $H_2N_3^+$ has been recently prepared and can be thought of as a protonated hydrazoic acid with both hydrogens residing on one nitrogen, and single and triple bonds between the nitrogens (Figure 17.31b).

Bond lengths of 121 pm for the central bond and 143 pm for the outer ones, and angles at the central N of 109°, are found in N_4H_4, formed from $R_2N = NH$, which is a sublimable solid containing a chain of 4N atoms, $H_2N–N = N–NH_2$. A few organic derivatives R_4N_4 are also reported, together with one or two longer chain organic compounds, but all are relatively unstable. The branched chain isomer of tetrazene is reported in

$$Cl_5W—N \quad \overset{\displaystyle N}{\underset{\displaystyle N}{|}} \quad N—WCl_5$$

where the planar tetrazene completes an octahedron around each W. Only this coordinated form is known, but this is the first example of a branched nitrogen chain.

M – M links for the heavier elements of the Group are found for all elements in the organic compounds $R_2M – MR_2$, although the Bi examples are unstable. The methyl-arsenic, *cacodyl*, $Me_2As—AsMe_2$, has been known since 1760. The corresponding hydrides are less stable and limited to P_2H_4 and As_2H_4, and all the halides P_2X_4 are also known. Longer chains, rings, nets, and open and closed clusters are

found for P, and to some extent for As, and are reviewed in section 18.3.

Trends within the Group are similar to those observed in the boron and carbon Groups. The acidic character of the oxides, and the stability of the V state, decrease in going from nitrogen to bismuth, so that bismuth has only a handful of compounds in the V state and these are unstable and strongly oxidizing. Antimony(V) and arsenic(V) are moderately oxidizing. Phosphorus(V) is very stable, while nitrogen is again oxidizing in the V state, reflecting the differences in formulae and coordination compared with phosphorus. The III state increases in stability as the V state becomes unstable: P(III) and As(III) are reducing, Sb(III) is mildly reducing and Bi(III) is stable. The III state also becomes increasingly stable in cationic forms for antimony and bismuth. Bi^{3+} probably exists in the salts of strong acids, such as the fluoride, and Sb^{3+} may be present in the sulfate. Both elements exist in solution, and in many salts, as the oxycation, MO^+.

17.6.2 *The(V) state*

All the oxides of the V state are known and their structures, where known, are shown in Figure 17.32. N_2O_5 is usually made by dehydrating nitric acid, but a cleaner synthesis results from the action of FNO_2 with excess $LiNO_3$. The solid is ionised with a planar nitrate ion and a linear nitronium cation, NO_2^+. Linear nitronium, or nearly so, is found with other stable anions: in $NO_2^+ ClO_4^-$, the ONO angle is 175°. In the gas phase, the N_2O_5 molecule exists with the structure of Figure 17.32. Higher oxides of nitrogen, NO_3 and N_2O_6, have been reported from the reaction of ozone on N_2O_5 but little is known of them.

(a)

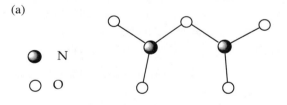

(b)

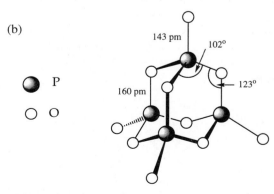

FIGURE 17.32 *Structures of pentoxides,* M_2O_5, *of the elements of the nitrogen Group:* (a) N_2O_5, (b) P_4O_{10}

Phosphorus burns in an excess of air to give the pentoxide which has the molecular formula, P_4O_{10}, and, in the vapour, the tetrahedral structure shown in Figure 17.32. This form is also found in the liquid and solid, but prolonged heating of either gives polymeric forms. Both 12- and 20-membered rings of PO_4 tetrahedra linked by corners have been identified. The environments of each phosphorus atom in P_4O_{10} and in the polymeric forms are similar. Each phosphorus atom is linked tetrahedrally to four oxygen atoms, three of which are shared with three other phosphorus atoms, while the fourth link is a $P = O$. The pentasulfide exists in the same P_4S_{10} form as the pentoxide (Figure 17.32b) and the mixed oxide-sulfides $P_4O_6S_4$ (with terminal $P = S$) and $P_4O_4S_6$ (with terminal $P = O$), completing a satisfying series of symmetric molecules. $P_4S_9N^-$ is also isoelectronic and isostructural with the N^- in an edge-bridging position.

The pentoxides, M_2O_5, of arsenic, antimony, and bismuth are made by oxidizing the element or the trioxide. Increasingly powerful oxidizing agents are needed from arsenic to bismuth, and Bi_2O_5 is not obtained in a pure stoichiometric form. As_2O_5 is a polymeric structure consisting of AsO_4 tetrahedra and AsO_6 octahedra sharing corners. The structures of Sb_2O_5 and Bi_2O_5 are not known but are probably based on MO_6 octahedra like their anions. All three lose oxygen readily on heating to give the trioxides.

Oxyacids and oxyanions of the V state are a very important class of compounds in the chemistry of this Group. The nitrogen, phosphorus, and arsenic compounds are included in Table 17.13. Antimony and bismuth do not form acids in the V state. The oxyanions may be made by reaction of the pentoxides in alkali or by oxidation of the trioxides in an alkaline medium. The bismuthates are strongly oxidizing, the best-known being the sodium salt, $NaBiO_3$, which is used to identify manganese in qualitative analysis by oxidizing it to permanganate. The antimonates are oxidizing, but more stable than the bismuthates, and are octahedral ions, $Sb(OH)_6^-$. This is in contrast to the tetrahedral coordination to oxygen shown in the phosphates and arsenates.

Both phosphorus and arsenic form acids in the V state, both the mononuclear acid, H_3MO_4, and polymeric acids. The polyphosphoric acids include chains, rings, and more complex structures, all formed from PO_4 tetrahedra sharing oxygen atoms. A wide variety of polyphosphate ions also exists, and structures of the first three members are indicated in Figure 17.36 (e), (f), and (g). The detailed structure of the fourth member, $P_4O_{13}^{6-}$, isolated as the $[Co(NH_3)_6]^{3+}$ salt, shows a significant alternation of $P - O$ bond lengths in the chain. Tripolyphosphates are used in detergents as they are excellent sequestering agents, though use has diminished as worries about eutrophication have grown (see section 20.3.1). These pyro-, meta- and other poly-phosphates are stable and hydrolyse only slowly to the ortho- acid, H_3PO_4. Poly-arsenates also exist in similar forms, but these are much less stable and readily revert to H_3AsO_4.

Nitrogen in the V state forms nitric acid, HNO_3, and the nitrate ion, NO_3^-. Here, the coordination number of only three to oxygen, and the planar structure, reflect the presence of $p_\pi - p_\pi$ bonding between the first row elements, nitrogen and oxygen. Although usually found as the free anion, nitrate does sometimes coordinate to metals. It occurs as a monodentate, $M - O - NO_2$, bidentate $M \diagup_{O}^{O} \diagdown NO$ or bridging, $M - O(NO) - O - M$, ligand. A second, unexpected, salt has recently been reported, NO_4^{3-}. A structural study of the sodium salt shows a tetrahedral structure with a NO distance of 139 pm, compared with 122 pm in NO_3^- (Figure 17.34) and 122 pm for $N - O$ and 141 pm for $N - OH$ in the isolated $HONO_2$ molecule. The $N - O$ bond order in NO_4^{3-} is thus near to one, suggesting a relatively weak interaction, and a description in terms of semi-polar bonds is probably the most appropriate, compare section 18.9.

The stable V state halides are limited to the pentafluorides, PF_5, AsF_5, SbF_5, and BiF_5, together with PCl_5, $SbCl_5$ and PBr_5. This reflects the decreasing stability of the V state from phosphorus to bismuth. $AsCl_5$ has only recently been identified as a product of the UV photolysis of $AsCl_3$ and Cl_2 at $-105\,^\circ C$. It decomposes at $-50\,^\circ C$. This behaviour is an example of the 'middle element anomaly' already discussed. All the structures so far determined show that the pentahalides are trigonal bipyramidal in the gas phase, but these structures alter in the solid, reflecting the instability of five-coordination in crystal lattices. PCl_5 ionizes in the solid to $PCl_4^+PCl_6^-$ while PBr_5 ionizes to $PBr_4^+Br^-$. The cations are tetrahedral and PCl_6^- is octahedral. In addition there is another metastable modification of solid PCl_5 which has been found by an X-ray diffraction study to be $(PCl_4^+)_2(PCl_6^-)(Cl^-)$, with a significant interaction between the PCl_4^+ and Cl^- ions. This form slowly reverts to the normal $PCl_4^+PCl_6^-$ form on standing at room temperature. SbF_5 also attains a more stable configuration in the solid, this time by becoming six-coordinated through sharing fluorine atoms between two antimony atoms in a chain structure. Except for BiF_5, all the pentafluorides readily accept F^- to form the stable octahedral anion, MF_6^-. PF_5, especially, is a strong Lewis acid and forms $PF_5 . D$ complexes with a wide variety of nitrogen and oxygen donors (D). Antimony also forms the dimeric ion $Sb_2F_{11}^-$, consisting of two SbF_6 octahedra sharing a corner.

The pentachlorides are similar but weaker acceptors. They do accept a further chloride ion, and $SbCl_6^-$ in particular is well established and stable: PCl_6^- is mentioned above. It has recently been shown that $AsCl_6^-$ may also be prepared if it is stabilized by a large cation, thus $Et_4N^+AsCl_6^-$ has been prepared. All these pentahalides may be made by direct combination or by halogenation of the trihalides, MX_3. The pentafluorides, and PCl_5, are relatively stable, but PBr_5 and $SbCl_5$ readily lose halogen at room temperature (to give the trihalides) and are strong halogenating agents, as is BiF_5 which is by far the most reactive pentafluoride. A number of mixed pentahalides, such as PF_3Cl_2, are also known. They are formed by treating the trihalide with a different halogen:

$$PF_3 + Cl_2 = PF_3Cl_2$$

These are similar to the pentahalides in many ways. For example, PF_3Cl_2 is covalent on formation but passes over into the ionic form $PCl_4^+ PF_6^-$ on standing. Another example is AsF_3Cl_2 which also appears to be ionized to $AsCl_4^+ AsF_6^-$, showing that the second ionic component of the unstable $AsCl_5$ exists. It has also been shown that $AsClF_4$ is quite stable but that, though $AsCl_4F$ exists at low temperatures, it readily loses Cl_2 to give $AsCl_4^+ AsCl_6^-$ and $AsCl_3$. $AsCl_3F_2$ is not accessible from $AsCl_3$, but results from the treatment of $AsCl_2F_3$ with $CaCl_2$. Like PCl_3F_2, the Cl are equatorial in the trigonal bipyramid. Overall, As(V) with 4 or 5 bonded Cl is unstable with respect to As(III) plus Cl_2, though it can be stabilized by suitable donors or in the presence of suitable counter-ions. The set of AsX_4^+ ions was completed by the discovery of AsI_4^+ as the $AlCl_4^-$ salt.

Although nitrogen cannot form a penthalide, as only four valency orbitals are present in the second level, it is interesting that the cation NF_4^+ exists showing that nitrogen(V) can bond to fluorine, in a species where only sigma interactions are possible. The oxychloro cation of nitrogen(V), $(NOCl_2)^+$, has been made as the $[AsF_6]^-$ or $[SbCl_6]^-$ salt. The structure is a flattened pyramid with $O=N-Cl = 119°$.

In their covalent forms, the mixed pentahalides PX_nY_{5-n}, such as PF_3Cl_2, have structures in which the most electronegative halogens occupy the two axial positions. Although PH_5 is unknown, mixed hydride-fluorides of P(V), are established e.g. PHF_4 and PH_2F_3. These form anions, HPF_5^- and $H_2PF_4^-$, related to PF_6^-.

Nitrogen, in the V state, forms the oxyfluoride NOF_3 from NF_3 and O_2 or from the elements. It readily transfers F^-, e.g. to form $(NOF_2)^+ (BF_4)^-$. The structure is pyramidal: see remarks above about the formulation of H_3NO_4. Phosphorus and arsenic form a range of oxyhalides in the V state. Three compounds of phosphorus are known, POX_3 where X = F, Cl, or Br, and for arsenic $AsOF_3$ and $AsOCl_3$. $POCl_3$ may be made from PCl_3 or from the pentachloride and pentoxide:

$$PCl_3 + \tfrac{1}{2}O_2 \rightarrow POCl_3$$
$$\text{or} \quad P_4O_{10} + 6PCl_5 \rightarrow 10POCl_3$$

The other phosphorus compounds are made from the oxychloride. $AsOF_3$ is made by the action of fluorine on a mixture of $AsCl_3$ and As_2O_3 while $AsCl_3 + O_3$ gives $AsOCl_3$ which decomposes at $-30°C$. All these are tetrahedral $X_3M = O$, with $p_\pi - d_\pi$ bonding between the O and M atoms (Figure 4.23). Phosphorus gives the corresponding sulfur and selenium compounds, PSX_3 and $PSeX_3$, again illustrating the marked stability of four-coordinated P(V).

Three pentasulfides are found in this Group. P_4S_{10} has the same structure as P_4O_{10}—and there is also a compound $P_4O_6S_4$ which again has the same structure, with the oxygen atoms bridging along the edges of the tetrahedron, and a sulfur atom attached directly to each phosphorus. The structures of As_2S_5 and Sb_2S_5 are unknown. All three sulfides may be formed by direct reaction between the

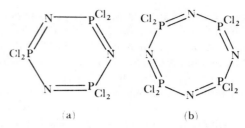

FIGURE 17.33 (a) *Trimeric and* (b) *tetrameric phosphonitrilic chlorides*

elements, and arsenic and antimony pentasulfides are also formed by the action of H_2S on As(V) or Sb(V) in solution.

One final important class of phosphorus (V) compounds is that of the phosphonitrilic halides. If PCl_5 is heated with ammonium chloride, compounds of the formula $(PNCl_2)_x$ result:

$$PCl_5 + NH_4Cl = (PNCl_2)_3 + (PNCl_2)_4 + (PNCl_2)_x + HCl$$

The corresponding bromides may be made in a similar reaction, and the chlorines may also be replaced by groups such as F, NCS, or CH_3 and other alkyl groups, either by substitution reactions, or by using the appropriate starting materials. When $x = 3$ or 4, the six- or eight-membered rings shown in Figure 17.33, are formed. Similar rings have been identified for x values up to 17 in the case of chlorides and fluorides and for $x = 6$ for the bromides. In addition, for large values of x, linear polymers are formed, of accurate formula $Cl(PNCl_2)_xPCl_4$. In the ring compounds, the trimer and pentamer are planar, while the tetramer and hexamer are puckered. The nature of the bonding in the rings, and also in the chain compounds, is not yet clearly determined but probably involves π bonding between nitrogen p orbitals and phosphorus d orbitals. In the trimeric chlorides, it has been suggested that this π bonding involves a strong interaction above and below the plane of the ring, as in benzene, and also a weaker interaction in the plane of the

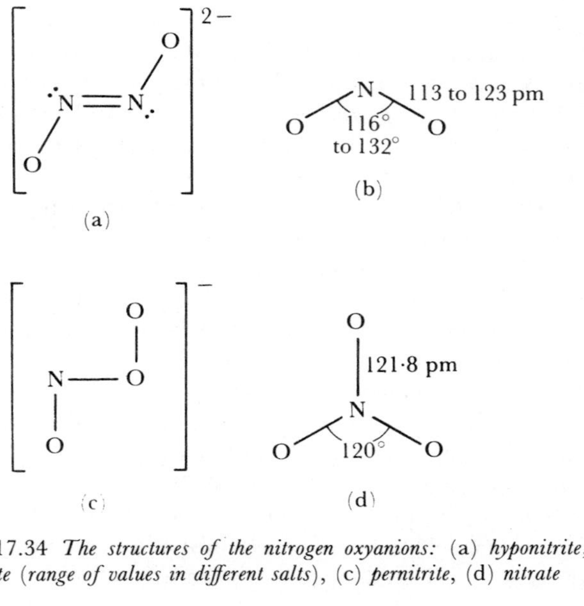

FIGURE 17.34 *The structures of the nitrogen oxyanions:* (a) *hyponitrite,* (b) *nitrite (range of values in different salts),* (c) *pernitrite,* (d) *nitrate*

ring. In the non-planar tetramer, this second type of π bonding can make a stronger contribution. In polymers, with OR groups such as OCH_2CF_3 replacing the halides, useful properties are found, especially resistance to oxidation and burning. These polyphosphazenes are finding application in special performance rubbers, gaskets and insulating materials. As these polymers are compatible with tissues, they have potential value in biomedical devices.

17.6.3 The (III) state
The III state is reducing for nitrogen, phosphorus, and arsenic, and the stable state for antimony and bismuth.

TABLE 17.13 Oxyacids and oxyanions of the nitrogen Group

Nitrogen

The nitrogen acids and anions all show nitrogen two- or three-coordinated to oxygen and all (except hyponitrous acid) have $p_\pi - p_\pi$ bonding between N and O.

$H_2N_2O_2$ hyponitrous	$N_2O_2^{2-}$ hyponitrite	Reduction of nitrite by sodium amalgam. Weak acid. Readily decomposes to N_2O.
HNO_2 nitrous	NO_2^- nitrite	Acidify nitrite solution. Free acid known only in gas phase. Weak acid. Aqueous solution decomposes reversibly, $3HNO_2 = HNO_3 + 2NO + H_2O$.
HOONO pernitrous	$(OONO)^-$ pernitrite	$H_2O_2 + HNO_2$. Free acid postulated as reaction intermediate.
HNO_3 nitric	NO_3^- nitrate	Oxidation of NH_3 from Haber process. Strong acid. Powerful oxidizing agent in concentrated solution.

The structures of these species are shown in Figure 17.34

Phosphorus

The phosphorus acids and anions all contain four-coordinate phosphorus. In the phosphorus (V) acids, all four bonds are to oxygen, while P—H and P—P bonds are present in the acids and ions of the I and III states. Various intermediate oxidation states are found in anions which have P—P bonds, often with P—H ones as well. The most striking case is $[P(O)(OH)]_6$. The crystal structure of the cesium salt shows a six-membered, puckered, P_6 ring.

H_3PO_2 hypophosphorous	$H_2PO_2^-$ hypophosphite	White P plus alkaline hydroxide. Monobasic acid, strongly reducing.
H_3PO_3 phosphorous	HPO_3^{2-} phosphite	Water plus P_2O_3 or PCl_3. Dibasic acid, reducing.
$H_4P_2O_5$ pyrophosphorous	$H_2P_2O_5^{2-}$ pyrophosphite	Heat phosphite: dibasic acid with P—O—P link. Reducing.
$H_4P_2O_6$ hypophosphoric	$P_2O_6^{4-}$ hypophosphate	Oxidation of red P, or of P_2I_4, in alkali, gives sodium salt, which gives the acid on treatment with H^+. Tetrabasic acid with a P—P link. Resistant to oxidation to phosphoric acid.
H_3PO_4 (ortho) phosphoric	PO_4^{3-} phosphate	P_2O_5 or PCl_5 plus water. Stable.
Also pyrophosphate $(O_3POPO_3)^{4-}$ polyphosphates $(O_3P[OPO_2]_nOPO_3)^{(4+n)-}$ metaphosphates $(PO_3)_m^{m-}$		Formed by heating orthophosphate. Linear polyphosphates with n up to 5 and cyclic metaphosphates with $m = 3$ to 10 have been individually identified (cf. Figure 7.2).

The structures of these phosphorus species are shown in Figure 17.36.

Arsenic

H_3AsO_3?, or $As_2O_3.xH_2O$ arsenious acids	$HAsO_3^{2-}$ and more complex forms arsenites	Formed from the trioxide or trihalides. The acid does not contain As—H and is weak, but the arsenites are well-established, in mononuclear and polynuclear forms. The arsenic(III) species are reducing and thermally unstable.
H_3AsO_4 arsenic acid	AsO_4^{3-} arsenate	$As + HNO_3 \rightarrow H_3AsO_4.\frac{1}{2}H_2O$. Tribasic acid and moderately oxidizing. Arsenates are often isomorphous with the corresponding phosphates.
Condensed arsenates		A number of these exist in the solid state but are less stable than polyphosphates and rapidly hydrolyse to AsO_4^{3-}.

As far as they are known, arsenic anions and acids have the same structures as the corresponding phosphates.

Antimony and bismuth give no free acids, though salts of the III and V states are found, and are discussed in the text. Coordination to oxygen is always six, not four as with phosphorus and arsenic.

Among the oxygen compounds, all the oxides, M_2O_3, and all the oxyanions are known, but the free acids of the III state are found only for nitrogen, phosphorus, and possibly arsenic. This points to an increase in basicity down the Group, as expected.

The oxide of nitrogen, N_2O_3, is found as a deep blue solid or liquid, melting at about $-100\,°C$. It is formed by mixing equimolar proportions of NO and NO_2, and reverts to these two components in the gas phase at room temperature. An X-ray structure determination, carried out at low temperature, showed the compound to be 'nitrosyl nitrite' with an N–N bond:

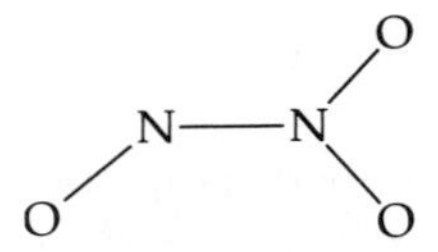

N_2O_3, or an equimolar mixture of NO and NO_2, gives nitrous acid when dissolved in water, and nitrites when dissolved in alkali.

Phosphorus(III) oxide results when phosphorus is burned in a deficiency of air. In the vapour phase, it has the formula, P_4O_6, and a structure derived from that of P_4O_{10} by removing the terminal oxygen atoms, see Figure 17.32. The phosphorus trioxide is acidic and reducing, and dissolves in water to give phosphorous acid. We note here that there is now a complete series of mixed P(III)/P(V) oxides, P_4O_x with $x = 7$ (Figure 17.35, compare Figure 17.32b), 8, and 9, where terminal oxygens are progressively added to P_4O_6 giving all intermediate structures between that and P_4O_{10}.

When arsenic is burned in air, the only product is the trioxide, which has the formula As_4O_6, and a similar structure to P_4O_6, in both the gas phase and in the solid. A second form also occurs in the solid, but this structure is not known. Arsenic trioxide is acidic.

Antimony and bismuth also burn in air to give only the trioxides. Antimony trioxide has the form, Sb_4O_6, both in the gas and in the solid, and there is a second solid form. This has a structure consisting of long double chains made up of $\dots -O-Sb-O-Sb-O-\dots$ single chains linked together through an oxygen atom on each antimony. Antimony tri-

oxide is amphoteric. Bi_2O_3 is yellow (all the other compounds are white) and exists in a number of solid forms. These are not known in detail, but some at least contain BiO_6 units in a distorted prism arrangement. Bismuth trioxide is basic only. Antimony and bismuth, in the III state, commonly exist in solution and in their salts as the MO^+ ion, as already noted.

The oxyanions and acids of phosphorus(III) and arsenic(III) are included in Table 17.13 and in Figure 17.36. All the structures known contain four-coordinated phosphorus or arsenic, and the III oxidation state results from the presence of a direct P–H or As–H bond (which, of course, does not ionize to give a proton). Although the stable form of phosphorous acid is the tetrahedral form shown in Figure 17.36, there is some evidence from exchange studies (compare section 2.3) for the transient existence of the pyramidal $P(OH)_3$ form. Organic derivatives of this form, $P(OR)_3$, are well-known. Phosphorous acid, and the phosphites, are reducing, and also disproportionate readily as the oxidation state free energy diagram, Figure 17.27, shows:

$$4H_3PO_3 \rightarrow 3H_3PO_4 + PH_3$$

Arsenites are also mildly reducing with a potential, in acid, of $0·56\,V$ with respect to arsenic(V) acid, so that arsenites are rather weaker reducing agents than iron(II) in acid solution.

(a)

(b)

(c)

(d)

(e)

(f)

(g)

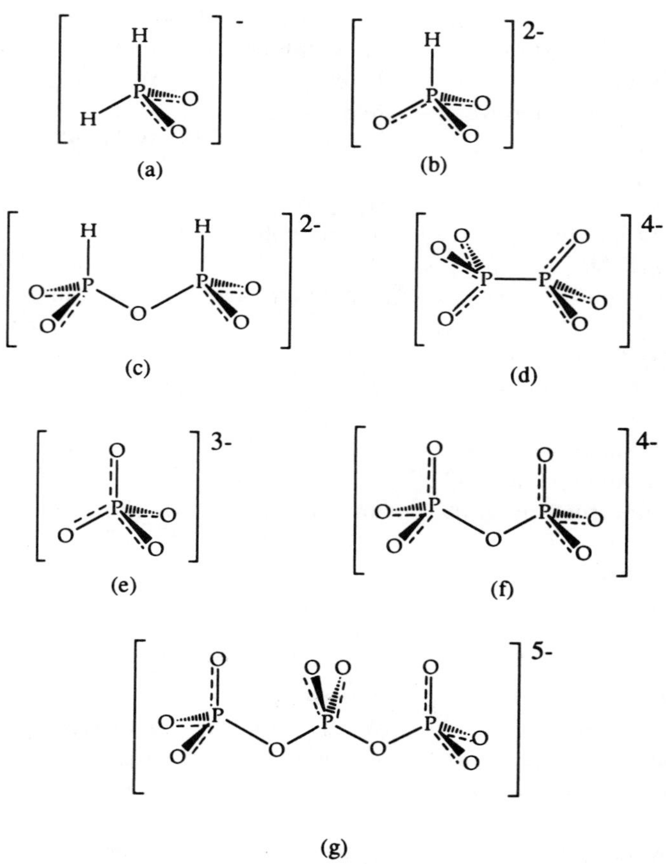

FIGURE 17.36   *The structures of phosphorus oxyanions:* (a) *hypophosphite,* (b) *phosphite,* (c) *pyrophosphite,* (d) *hypophosphate,* (e) *orthophosphate,* (f) *pyrophosphate,* (g) *tripolyphosphate*

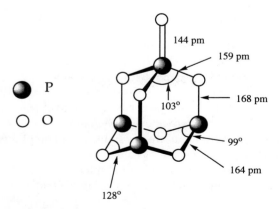

144 pm
159 pm
168 pm
103°
99°
164 pm
128°

P

O

FIGURE 17.35   *The structure of* P_4O_7

One interesting bismuth (III) compound is the complex oxycation $[Bi_6O_4(OH)_4]^{6+}$ whose structure contains six Bi atoms in an octahedron with 4O and 4OH groups triply bridging the triangular faces in a regular manner. This is formed by Bi_2O_3 in perchloric acid. Analogous Sn, V, and Ce cluster ions of this $M_6O_4(OH)_4$ formula are found.

Arsenic, antimony, and bismuth all form trisulfides, M_2S_3. The arsenic compound exists as As_4S_6, while the antimony and bismuth sulfides have polymeric chain structures. All are formed by the action of H_2S on solutions of the element in the III state.

The tale of the lower phosphorus sulfides is more complicated, as the main series have P–P bonds and may be seen as derived from the P_4 tetrahedron. The lowest S content is found in P_4S_3 (Figure 17.37a) where S atoms are inserted in three of the six edges of P_4. The mixed P/As analogue, PAs_3S_3 has the P at the unique apex position and an As_3 triangle in the base. Further edge insertion is found in the two isomers of P_4S_4, where the remaining two P – P bonds are either contiguous or at opposite edges of the tetrahedron. In P_4S_5, S is added to a terminal position (Figure 17.37b), while P_4S_7 sees edge insertion and terminal addition (Figure 17.37c).

The remaining well-established compound is P_4S_9 where the structure is like the oxide, with one terminal S removed from P_4S_{10} (Figure 17.32b). We note that, as S is isoelectronic with P^- or PH, the species P_7H_3 and P_7^{3-} (and also As_7^{3-}) have structures analogous to Figure 17.37a (compare section 18.2). Arsenic also forms a sulfide, As_4S_3, and a selenide, As_4Se_3, with the P_4S_3 structure. A third arsenic sulfide, As_4S_4 (called *realgar*), is also known and its structure is shown in Figure 17.37d. A phosphorus selenide of composition P_2Se_5 has been found to have the structure shown in Figure 17.37e. In marked contrast to phosphorus and arsenic, nitrogen sulfides are rare and a good example is given by the contrast in stabilities between the very stable N_2O and the transient sulfur analogue N_2S which has only been detected spectroscopically.

The five elements of the nitrogen Group all give all the trihalides, MX_3. The least stable are the nitrogen compounds of which only NF_3 is stable and occurs as the expected pyramidal molecule formed by the reaction of nitrogen with excess fluorine in the presence of copper. NF_3 has almost no donor power and has only a very low dipole moment as the strong N–F bond polarizations practically cancel out the effect of the lone pair. Nitrogen trichloride decomposes explosively. NI_3 is typically prepared as an unstable ammoniate such as $NI_3.6NH_3$, which is highly shock sensitive when freed from excess ammonia. The structure of one ammoniate, $NI_3.NH_3$, has been determined crystallographically. It consists of chains of NI_4 tetrahedra formed by sharing corners and the NH_3 molecules are bonded to the non-bridging iodines. NI_3 free from ammonia has been prepared only recently by the reaction of boron nitride with IF:

$$BN + 3IF \rightarrow NI_3 + BF_3$$

The free NI_3 is rather unstable but can be sublimed as a deep red solid which reacts with ammonia to form the ammoniate $NI_3.3NH_3$.

The other sixteen trihalides of the Group are all relatively stable molecules, with the expected trigonal pyramidal structure in the gas phase. The pyramidal structure is also found in many of the solids, but other solid state structures are also found, particularly among the tri-iodides which adopt layer lattices with the metal atoms octahedrally surrounded by six halogen atoms. PI_3 has three P – I distances of 246 pm (and IPI angles of 102°) and three of 367 pm (angles 60°) showing the P lone pair is still sterically important. In addition to the simple trihalides, MX_3, a wide variety of mixed halides, MX_2Y or MXYZ, are found.

All the trihalides are readily hydrolysed, giving the oxide, oxyanion, or—in the cases of antimony and bismuth—the oxycation, MO^+. They may act as donor molecules, by virtue of the lone pair, and PF_3 in particular has been widely studied. It is rather less reactive to water and more easily handled than the other trihalides. PF_3 complexes resemble the corresponding carbonyls, for example, $Ni(PF_3)_4$ is similar to $Ni(CO)_4$. The trihalides also show acceptor properties, especially the trifluorides and chlorides. Complex

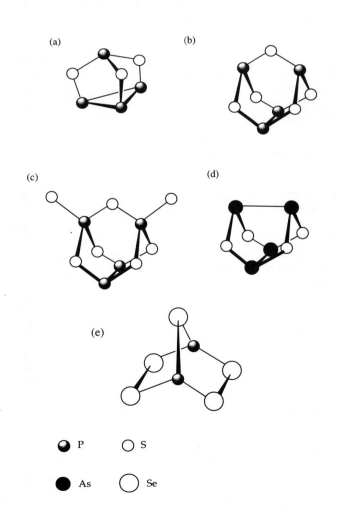

(a)   (b)

(c)   (d)

(e)

● P   ○ S

● As   ○ Se

FIGURE 17.37 *The structures of* (a) P_4S_3, (b) P_4S_5, (c) P_4S_7, (d) As_4S_4, (e) P_2Se_5.

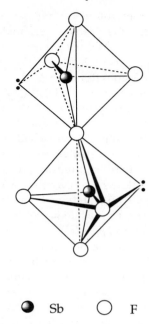

FIGURE 17.38 *The structure of the ion,* $Sb_2F_7^-$

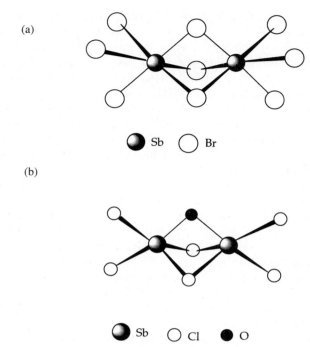

FIGURE 17.39 *The structures of* (a) *the* $M_2Br_9^{3-}$ *ion,* (b) *the* $[Sb_2OCl_6]^{2-}$ *ion*

ions, such as SbF_5^{2-}, are formed, and SbF_3 also gives the interesting dimeric ion, $Sb_2F_7^-$ (Figure 17.38). In mixed oxidation state fluorides, such as Sb_8F_{30}, antimony(III) cations are found as well as the $[Sb^VF_6]^-$ anion. This compound contains $[Sb_2F_5]^+$ and $[Sb_3F_7]^{2+}$ together with three anions. The cations are linked through single, nearly linear, F bridges, e.g. $[F_2Sb-F-SbF_2]$, and with angles at Sb around 80°. For the heavier halogens, X = Cl, Br or I, three types of complex ion are found: SbX_4^-, SbX_5^{2-} and SbX_6^{3-}. The first two have the AB_4L (Figure 4.4b) and AB_5L (Figure 4.5b) structures expected from the presence of the lone pair, but the SbX_6^{3-} species are regular octahedra. In this they parallel the seven-electron-pair MX_6^{2-} species formed by the heavy elements of the oxygen Group (compare section 17.7.4). In complex ions of formula $M_2X_9^{3-}$, a range of structures is found, from the binuclear (Figure 17.39a; two octahedra sharing a face) through the tetrameric structure of $Bi_4Cl_{18}^{6-}$ to the polymeric $[Sb_2Cl_9^{2-}]_x$ unit where each antimony carries three terminal Cl atoms, and is linked by three single Cl bridges to three different neighbours in a double-chain. Hydrolysis yields $[Sb_2OCl_6]^{2-}$ with the triply-bridged structure of Figure 17.39b. A similar structure is found for $[As_2SBr_6]^{2-}$ and for $[Sb_2SCl_6]^{2-}$. The trihalides are common reaction intermediates, and, for example, react with silver salts to give products such as $P(NCO)_3$, and with organometallic reagents to give a wide variety of organic derivatives, MR_3. Among such analogues we note species like $As(OTeF_5)_3$ and $Sb(CF_3)_3$ where the ligands behave very similarly to F and have a similar extensive chemistry.

17.6.4 *Other oxidation states*

A number of oxidation states other than V and III are found, especially among the oxides and oxyacids. Nitrogen forms the I, II, and IV oxides, N_2O, NO, and NO_2 or N_2O_4. Nitrous oxide, N_2O, is formed by heating ammonium nitrate solution or hydroxylamine and is fairly unreactive. It has a π-bonded linear structure, NNO. Nitric oxide has already been discussed from the structural point of view. Although it has an unpaired electron, it shows little tendency to dimerize, though some association to rather loose dimers,

$$\begin{matrix} N & \cdots\cdots & O \\ | & & | \\ O & \cdots\cdots & N \end{matrix},$$

occurs in the liquid and solid. It rapidly reacts with oxygen to form NO_2. This is also an odd electron molecule but does dimerize readily to dinitrogen tetroxide. The solid is entirely N_2O_4 and this dissociates slightly in the liquid and increasingly in the gas phase until, at 100 °C, the vapour contains 90 per cent NO_2. NO_2 is brown and paramagnetic and has an angular structure with ONO = 134°. N_2O_4 is colourless and diamagnetic with the symmetrical structure

$$\begin{matrix} O & & O \\ \diagdown & & \diagup \\ & N-N & \\ \diagup & & \diagdown \\ O & & O \end{matrix}$$

and a very long N−N bond of 175 pm (compare 145 pm for the N−N single bond length in a molecule like hydrazine).

The compound called *Angeli's salt*, $Na_2N_2O_3 \cdot H_2O$, is formally N(II). The anion has the structure

$$\begin{matrix} & & O \\ & & \diagup \\ N=N & & \\ \diagup & & \diagdown \\ O & & O \end{matrix}$$

Here the unique NO distance is 135 pm and the others are 131–2 pm.

The unusual nitrogen oxide N_4O has recently been prepared by reaction of NaN_3 with $NOCl$, and is formulated as nitrosyl azide with the following Lewis representation being proposed:

$$O = N - \overset{\ominus}{N} - \overset{\oplus}{N} = N$$

When P_2O_3 is heated above $210\,°C$ a third oxide, PO_2, is formed, along with red phosphorus. This compound has a vapour density corresponding to P_8O_{18} and it behaves chemically as if it contains both $P(V)$ and $P(III)$. It may also contain $P-P$ bonds as it reacts with iodine to give P_2I_4. Its structure is unknown.

The phosphorus oxide PO, which is isoelectronic to the nitrogen analogue NO, is believed to be the most abundant phosphorus-containing molecule in interstellar clouds. Several metal complexes containing PO as a bridging ligand are known.

Heating either Sb_4O_6 or Sb_2O_5 in air above $900\ °C$ gives an oxide of formula SbO_2. This consists of a network of fused SbO_6 octahedra containing both $Sb(III)$ and $Sb(V)$. A corresponding AsO_2 may exist.

There are also two oxyacids of low oxidation states in the Group. These are hyponitrous acid, $H_2N_2O_2$, with nitrogen (I), and hypophosphorous acid, H_3PO_2, with phosphorus (I). These are included in Table 17.13. Nitrogen forms its low oxidation state in hyponitrous acid by $p_\pi - p_\pi$ and $N-N$ bonding, while phosphorus in hypophosphorous acid is tetrahedral and the low oxidation state arises from two direct $P-H$ bonds. Intermediate phosphorus oxidation states in oxypolyphosphorus compounds resulting from $P-P$ and $P-H$ bonds are shown in Table 17.13 and Figure 17.36c, d.

Lower oxidation state nitrogen fluorides are made by direct combination using less fluorine than required for NF_3. N_2F_2 has a planar structure

$$\underset{F}{\overset{}{\diagdown}} N = N \overset{F}{\diagup}$$

which is most stable in this trans form, but which may also occur in the cis form

$$\overset{N=N}{\underset{F\qquad F}{\diagup\qquad\diagdown}}$$

. There is a π bond between the two N atoms which have each a lone pair. N_2F_4 is a gas with a skew structure similar to that of hydrazine;

$$\overset{F}{\underset{F}{\diagdown}} N - N \overset{F}{\underset{F}{\diagup}}$$

In the gas and liquid phases it undergoes reversible dissociation to NF_2:

$$N_2F_4 = 2NF_2$$

similar to that of N_2O_4. NF_2 is an angular molecule and contains an unpaired electron. For an odd-electron species, it has fairly high stability resembling NO, NO_2, and ClO_2 in this respect.

Phosphorus forms X_2P-PX_2 lower halides which probably have a skew structure as in N_2F_4.

The lower halides of bismuth present a much more complicated picture. When bismuth is dissolved in molten bismuth trichloride, an intensely coloured solution results from which may be isolated a compound with the accurate formula $Bi_{12}Cl_{14}$. This is a complicated structure with 48 Bi atoms and 56 Cl atoms in the unit cell. These are arranged as $4Bi_9^{5+}$, $8BiCl_5^{2-}$, and $2Bi_2Cl_8^{2-}$ units. The $BiCl_5^{2-}$ ion is a square pyramid with Bi(III) and resembles the SbF_5^{2-} ion mentioned earlier. In the structure these units are weakly linked to each other to form a chain. The $Bi_2Cl_8^{2-}$ unit contains Bi(III) and consists of two square pyramids sharing an edge of the base with their apices trans to each other. The Bi_9^{5+} unit has six Bi atoms at the corners of a somewhat distorted trigonal prism, and the other three Bi atoms above the rectangular faces. These units are shown in Figure 17.40. Thus the solid may be written $(Bi_9^{5+})_2(BiCl_5^{2-})_4Bi_2Cl_8^{2-}$. In $Bi_{10}Hf_3Cl_{18}$, the Bi^+ ion has been recognized along with Bi_9^{5+} and $HfCl_6^{2-}$ ions. Further work has produced other cluster compounds of Bi and Sb, and these are reviewed in section 18.4.

In BiI, an infinite chain structure is found, where Bi atoms are in two environments (Figure 17.41). In the A chain, Bi is bonded only to three other Bi atoms, and is formally Bi(0). In

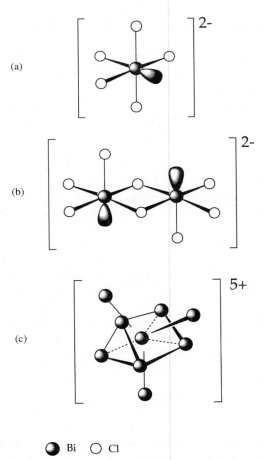

● Bi   ○ Cl

FIGURE 17.40   *The structures of* (a) $BiCl_5^{2-}$, (b) $Bi_2Cl_8^{2-}$, *and* (c) Bi_9^{5+}

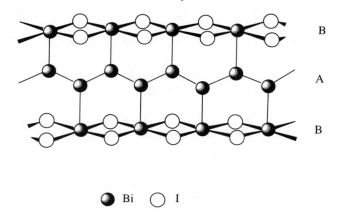

Bi ◯ I

FIGURE 17.41   *The structure of* BiI

the B chain, there are four Bi−I bonds, each shared with a second Bi, and thus with the formal oxidation state Bi(II). These Bi atoms form a fifth bond to a bismuth of the A chain.

In the lower bromides we find BiBr, isostructural with BiI, and also $Bi_{12}Br_{14}$, comparable with $Bi_{12}Cl_{14}$.

Negative oxidation states appear in the hydrides, MH_3, and in the organic compounds, NR_3. The other elements, except possibly arsenic (see Table 2.13), are of lower electronegativity than carbon in alkyl groups, and their organic compounds correspond to positive oxidation states. This distinction is not a useful one and is best regarded—as in the case of carbon chemistry in general—as an accidental result of the definitions. The organic compounds, MR_3, and organohydrides such as R_2PH, behave in a similar way to the hydrides but with the M−C bond stronger than M−H. An extensive organometallic chemistry of this Group exists which requires a textbook of its own for review. Here we note only some direct comparisons with the hydrides. As well as MR_3, analogous to MH_3, ions NR_4^+, PR_4^+, AsR_4^+, SbR_4^+ and BiR_4^+ exist which are tetrahedral and analogous to NH_4^+. The phosphorus analogue of NH_4^+ exists, for example in PH_4I which is prepared from PH_3 and HI. However, the phosphonium halides are relatively unstable and readily decompose to phosphane and hydrogen halide. They are much more covalent than the ammonium salts. The analogous AsH_4^+ and SbH_4^+ ions have also been recently prepared as their SbF_6^- salts by protonation of the parent hydride MH_3 using the superacid HF/SbF_5.

Although no pentavalent hydride, MH_5, exists, the pentaphenyl phosphoranes MPh_5 of P, As, Sb and Bi exist, as does PMe_5. PPh_5 and $AsPh_5$ are trigonal bipyramids in shape but $SbPh_5$ is a square pyramid. $SbPh_5$ reacts with PhLi to give the octahedral $SbPh_6^-$ ion. $BiMe_6^-$ is also known and has the expected octahedral geometry.

A number of mixed chloro-alkyl or chloro-aryl phosphoranes, R_nPX_{5-n} (R = alkyl, aryl; X = halide), are also known, comparable with PH_2F_3 and similar hydridehalides. As a result of recent studies the chemistry of these species has been found to be more complex than first thought. Species of this type can occur in a variety of forms and as a general illustration we take Ph_3PI_2. Reaction of

Ph_3P with di-iodine in ether gives the molecular four-coordinate compound $Ph_3P–I–I$ and neither ionic $[Ph_3P–I^+]I^-$ or five-coordinate covalent phosphorane $Ph_3P(I)(I)$ which were previously thought to be the stable forms of this compound. $Ph_3P–I–I$ can be thought of as a donor−acceptor compound between Ph_3P and I_2. Analogous compounds $Ph_3P–Br–Br$ and $Ph_3As–I–I$ can also be formed. There appears to be a delicate balance of factors determining whether a covalent or ionic form is produced since $PhPCl_4$ is a molecular compound whilst the corresponding methyl analogue exists as $[MePCl_3]^+Cl^-$. In addition, when $R_3P–I–I$ compounds are dissolved in chloroform they ionize to $[Ph_3P–I]^+I^-$.

Nitrogen is found in the −II state in hydrazine and its organic derivatives, R_4N_2, and in the −I state in hydroxylamine, NH_2OH.

17.7 The oxygen Group, $ns^2 np^4$

17.7.1 *General properties*

References to the properties of oxygen Group elements will be found in the following places:

Ionization potentials	Table 2.8
Atomic properties and electron configurations	Table 2.5
Radii	Table 2.10, 2.11, 2.12
Electronegativities	Table 2.14
Redox potentials	Table 6.3

Polysulfides and polyselenides are discussed in fuller detail in section 18.2 and metal-polysulfur compounds are covered in section 19.2. Metal dioxygen species are covered in section 16.9 with structural parameters in Table 16.5. Table 17.14 lists some of the properties of the elements and Figures 17.42 and 17.43 show the variations with Group position of a number of parameters and of the oxidation state free energies.

Oxygen occurs both as the O_2 molecule and as ozone, O_3. O_2 is paramagnetic (section 3.5) and has a dissociation energy of 489 kJ mol^{-1} It is pale blue in the liquid and solid states. Ozone is usually formed by the action of an electric discharge on O_2. Pure O_3 is deep blue as the liquid with m.p. = −250 °C and b.p. = −112 °C. It is diamagnetic and explodes readily as the decomposition to oxygen is

TABLE 17.14 Properties of the elements of the oxygen Group

Element	Symbol	Oxidation states	Coordination numbers	Availability
Oxygen	O	−II, (−I)	1, 2, (3), (4)	common
Sulfur	S	−II, (II), IV, VI	2, 4, 6	common
Selenium	Se	(−II), (II), IV, VI	2, 4, 6	common
Tellurium	Te	II, IV	6	common
Polonium	Po	II, IV		very rare

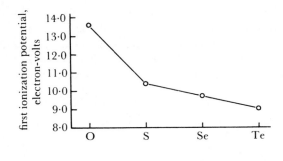

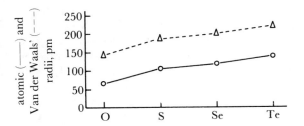

FIGURE 17.42 *Some properties of the oxygen Group elements*
The radii of the anions, X^{2-}, are almost identical with the corresponding Van der Waals' radii.

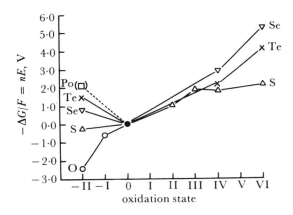

FIGURE 17.43 *Oxidation state free energies of the oxygen Group elements*
Oxygen shows only negative oxidation states. The $-$ II state becomes decreasingly stable from O to Po. The positive oxidation states show the drop in stability of the VI state after S and the tendency for intermediate states to be the more stable for Se and Te. Polonium values are not known.

exothermic and easily catalysed:

$$O_3 = 3/2\, O_2, \quad \Delta H = -142 \text{ kJ mol}^{-1}$$

Ozone has an angular structure with the OOO angle equal to 117°. The bonding in ozone has been described briefly previously in section 4.8 and Figure 4.30. Of the eighteen valency electrons, four are held in the sigma bonds and eight in lone pairs on the two terminal oxygens. Two are present as a lone pair on the centre oxygen, leaving four electrons and the three p orbitals perpendicular to the plane of the molecule to form a pi system. The three p orbitals combine

to form a bonding, a nonbinding and an antibonding three-centre π orbital and the four electrons occupy the first two. There is thus one bonding π orbital over the three O atoms giving, together with the σ bonds, a bond order of about one and a third. The bond length is 128 pm which agrees with this; O$-$O for a single bond in H_2O_2 is 149 pm while O$=$O in O_2 is 121 pm. Adding one more electron to form the O_3^- ion starts to populate the antibonding π^* orbital, and we find the bond length increases to 133 pm with an angle of 114°.

Ozone occurs in the upper atmosphere, where it is formed photochemically from O_2, and has the important property of absorbing harmful middle and far-UV radiation. The recent 'hole' in the ozone layer, caused by chlorofluorocarbons and related materials, is of significant current concern and this topic is discussed in greater detail in Chapter 20.

Ozone is a strong oxidizing agent, especially in acid solution where the potential is 2.07 V (Table 6.3). It is exceeded in oxidizing power only by fluorine, oxygen difluoride, and some radicals.

Since the S$-$S single bond is strong, sulfur chains form readily, and polysulfur species are one of the major classes of catenated compounds. The S$-$S bond is also labile, so that a particular chain compound, such as S_6Cl_2, readily redistributes into an equilibrium mixture of different chain lengths, leading to difficulties in characterizing such compounds. The element sulfur itself occurs as chains or rings (which are simply closed chains), and its structural complexities are now understood.

Sulfur shares with phosphorus the ability to form a wide variety of allotropic forms in all three phases. However, many of these varieties of sulfur turn out to be mixtures of long chains and rings. The main interrelations are shown in Figure 17.44.

Under normal conditions, the thermodynamically stable form of sulfur is the S_8 ring which has the crown structure shown in Figure 17.45. If we consider a short chain of S atoms (say S$-$S$-$S$-$S$-$S), then the central S forms two bonds and has two lone pairs. The SSS angle is thus expected

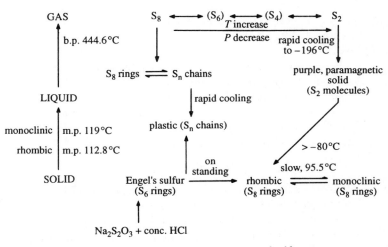

FIGURE 17.44 *The interconversion of the allotropes of sulfur*

FIGURE 17.45 *The structure of the* S_8 *or* Se_8 *ring*

to be around $105°$ and the chain is a zig-zag. Further, the arrangement will twist to move lone pairs as far apart as possible, so we find an optimum *dihedral angle* (angle between successive SSS planes), which minimizes lone pair repulsions, of around $85-100°$. Putting these preferred angles together, we find that a long chain of S atoms will tend to coil up and 'bite back' on itself. With an arrangement of 8 S atoms, the resulting ring allows an optimum choice of angles.

Several of the allotropes (more specifically polymorphs) of sulfur contain the S_8 ring and differ in the ways in which the rings are packed in the solid. When sulfur is heated to about $160°C$, there is a sudden large increase in viscosity and this is ascribed to the $S_8 \rightarrow S_\infty$ change from rings to long chains of S atoms. The S_8 ring is also found in the gas phase, together with smaller units.

A second, long-known, orange-red form of sulfur was first reported by Engel. It has S_6 rings, arranged in the chair form. This, and the many more recently discovered sulfur rings, is treated in section 18.2.

A further interesting modification of sulfur is the S_2 unit which occurs in the gas at high temperatures. On rapid cooling it condenses to a purple solid, which is paramagnetic like the isoelectronic O_2.

Selenium also forms a number of allotropes. Two different red forms containing Se_8 rings are found but the stable modification is the grey form. This contains infinite spiral chains of selenium atoms with a weak, metallic interaction between atoms in adjoining chains. The chain form of sulfur also contains spiral chains but has no metallic character.

Tellurium has only one form in the solid. This is silvery-white, semi-metallic, and isomorphous with grey selenium, but with rather more metallic interaction as the self-conductivity is higher. In the vapour, selenium and tellurium have a greater tendency than sulfur to exist as diatomic and monatomic species.

Polonium is a true metal with two allotropic forms, in both of which the coordination number is six. Polonium is found in small amounts in thorium and uranium minerals (where it was discovered by Marie Curie) as one of the decay products of the parent elements, Th or U. It can now be more readily made by bombarding bismuth in a reactor:

$$^{209}_{83}\text{Bi} + ^{1}_{0}\text{n} \rightarrow ^{210}_{83}\text{Bi} \rightarrow ^{210}_{84}\text{Po} + ^{0}_{-1}\text{e}$$

All polonium isotopes have relatively short half-lives and this isotope, polonium 210, is the longest-lived with $t_{\frac{1}{2}} = 138$ days. The isotopes all have high activity and the handling problems are severe so that polonium chemistry is not fully explored.

The trends observed in the nitrogen Group appear in a more marked degree in the chemistry of the oxygen Group. The $-II$ state is well-established, not only for oxygen, but also for sulfur and even selenium, and it often occurs as the M^{2-} ion in the compounds of these elements. Apart from the Group state of VI, both the IV and the II states are observed in the Group, the IV state becoming the most stable one for tellurium, and the IV or II state the most stable one for polonium. Apart from the fluorine compounds, oxygen appears only in the $-II$ state, except for the $-I$ state in the peroxides. A variety of shapes is again observed. The VI state is usually octahedral or tetrahedral while the presence of the non-bonding pair in compounds of the IV state gives distorted tetrahedral shapes in the halides, MX_4, and pyramidal or V shapes in oxyhalides or oxides. Coordination numbers higher than six are uncommon but TeF_6 does add F^- to give TeF_7^-.

Oxygen has little in common with the other elements in the Group apart from formal resemblances in the $-II$ states. Polonium shows signs of a distinctive chemistry, similar to that of lead or bismuth, but the difficulties of studying the element mean that there are many gaps in its known chemistry. Tellurium resembles antimony in showing a strong tendency to be six-coordinate to oxygen, while sulfur and selenium show four-coordination. Sulfur is stable in the VI state, while selenium(VI) is mildly oxidizing.

The tendency to catenation shown by the elements is continued in the compounds, especially those of sulfur. Oxygen forms O–O links in the peroxides and in O_2F_2, and three- and four-membered chains may be present in the unstable fluorides, O_3F_2 and O_4F_2. Sulfur forms many chain compounds, particularly the dichlorosulfanes, S_xCl_2, where x may be as high as 100, and the polythionates, $(O_3SS_xSO_3)^{2-}$ where the compounds with $x = 1$ to 12 have been isolated. Chain-forming tendencies are slight for selenium and absent in tellurium chemistry.

It is convenient to discuss oxygen chemistry separately from that of the other elements.

17.7.2 *Oxygen*

Oxygen is prepared on a large scale by fractionation of liquid air, and had its first large-scale use in the Bessemer steel-making process where it is used to reduce the carbon content of iron. It has become more widely used in metallurgy to assist the combustion of heavy fuel oils, allowing replacement of coke. It is also used directly in various large scale organic syntheses such as the formation of ethylene or propylene oxides from the alkenes. Smaller-scale uses, often via peroxides, are in bleaching, biological and medical work, and sewage treatment.

Oxygen combines with all elements except the lighter rare gases, and most of the oxides are listed in Tables 13.3 and 17.2. Their properties have been discussed already. The change in acidity from the s element oxides to the oxides of the light p elements will already be familiar to the reader. Oxygen shares with the other first row elements the ability to form $p_\pi - p_\pi$ bonds to itself and to the other first row elements, and it is sufficiently reactive and electronegative to form $p_\pi - d_\pi$ bonds with the heavier elements.

Dioxygen coordinates to a variety of transition metals in species which have been variously formulated as containing neutral O_2, superoxide O_2^-, or peroxide O_2^{2-}. These are discussed in section 16.9.

As fluorine is more electronegative than oxygen, compounds of the two are oxygen fluorides and are discussed here. The halogen oxides of Cl, Br, and I are to be found in section 17.8. Four oxygen fluorides are well established— OF_2, O_2F_2, O_3F_2 and O_4F_2—and two more have been reported relatively recently—O_5F_2 and O_6F_2. In all of these compounds oxygen is in a formally positive oxidation state and these are the only compounds where this occurs. OF_2 is formed by the action of fluorine on dilute sodium hydroxide solution, while the other five result from the reaction of an electric discharge on an O_2/F_2 mixture. An increasing proportion of O_2 and decreasing temperature are required to make the highest members of the series. Thus O_5F_2 and O_6F_2 were prepared in a discharge at $-213\,°C$ using a 5/1 mixture of O_2 and F_2. The use of higher proportions of oxygen does not give O_7 or higher species at this temperature but it might be possible to make these by working at a still lower temperature. All six compounds are very volatile with boiling points well below room temperature.

OF_2 is the most stable of the six oxygen fluorides. It does not react with H_2, CH_4, or CO on mixing, although such mixtures react explosively on sparking. Mixtures of OF_2 and halogens explode at room temperature. OF_2 reacts slowly with water:

$$OF_2 + H_2O \rightarrow O_2 + 2HF$$

and is readily hydrolysed by base:

$$OF_2 + 2OH^- \rightarrow O_2 + 2F^- + H_2O$$

The structure of OF_2 is V-shaped, like H_2O, with the FOF angle $= 103.2°$ and the bond length, $O-F = 141.8$ pm.

The other five oxygen fluorides are much less stable. O_2F_2 decomposes at its boiling point of $-57\,°C$, O_3F_2 at $-157\,°C$, O_4F_2 at $-170\,°C$, and O_5F_2 and O_6F_2 are stable only to $-200\,°C$. Indeed, O_6F_2 can explode if it is warmed rapidly up to $-185\,°C$. All are red or red-brown in colour. Electron spin resonance studies on O_2F_2 and O_3F_2 have shown that these contain the OF radical to the extent of 0.1 per cent for O_2F_2 and 5 per cent for O_3F_2. It is likely that the higher members also contain free radicals, and these may account for the colour. O_2F_2 has a skew structure, $\begin{smallmatrix} & & F \\ & \diagup \\ O-O \\ \diagup \\ F \end{smallmatrix}$, similar to that of hydrogen peroxide and the others may be chains, FO_nF.

When O_2F_2 is reacted at low temperatures with molecules which will accept F^-, such as BF_3 or PF_5, oxygenyl compounds result:

$$O_2F_2 + BF_3 \rightarrow (O_2)BF_4 + \tfrac{1}{2}F_2 \text{ (at } -126\,°C)$$

Oxygenyl tetrafluoroborate decomposes at room tempera-

ture to BF_3, O_2, and fluorine. The oxygenyl group may be replaced by the nitronium ion, NO_2^+:

$$O_2BF_4 + NO_2 \rightarrow (NO_2)BF_4 + O_2$$

The oxygenyl ion may also be formed directly from gaseous oxygen by reaction with the strongly oxidizing platinum hexafluoride molecule:

$$O_2 + PtF_6 \rightarrow (O_2)^+(PtF_6)^-$$

Other $O_2^+(MF_6)^-$ species have M = P, As, Sb, Bi, Nb, Ru, Rh, Pd or Au (note the unusual Au(V) state). $O_2^+(M_2F_{11})^-$ for M = Sb, Bi, Nb and Ta are also known. Recently $O_2^+(GeF_5)^-$ was reported. In all these species the $O-O$ stretching lies in the range 1825–65 cm^{-1}, reflecting a stronger bond than in O_2.

Oxygen forms two compounds with hydrogen, water and hydrogen peroxide, H_2O_2. Water, and its solvent properties, is discussed in Chapter 6 and hydrogen-bonding in O–H systems is included in section 9.7 and Figures 9.1, 9.2.

Hydrogen peroxide may be prepared by acidifying an ionic peroxide solution (section 10.2) or, on a large scale, by electrolytic oxidation of a sulfate system:

$$2HSO_4^- = S_2O_8^{2-} + 2H^+ + 2e^-$$
$$S_2O_8^{2-} + 2H_3O^+ = 2H_2SO_4 + H_2O_2.$$

(The intermediate is called persulfate.) However, most hydrogen peroxide today is manufactured by the anthraquinone autoxidation process. In this a functionalized anthraquinone dissolved in an organic solvent is first reduced (using hydrogen and a catalyst, typically palladium) to give the corresponding hydroquinone which is then oxidized using air, giving hydrogen peroxide and regenerating the anthraquinone, shown in Figure 17.46. The hydrogen peroxide is extracted with water in a counter-current process and may be concentrated by fractionation.

Pure H_2O_2 is a pale blue liquid, m.p. $-0.89\,°C$, b.p. $150.2\,°C$, with a high dielectric constant and is similar to water in its properties as an ionizing solvent. It is, however, a strong oxidizing agent and readily decomposes in the presence of catalytic amounts of heavy metal ions:

$$H_2O_2 \rightarrow H_2O + \tfrac{1}{2}O_2$$

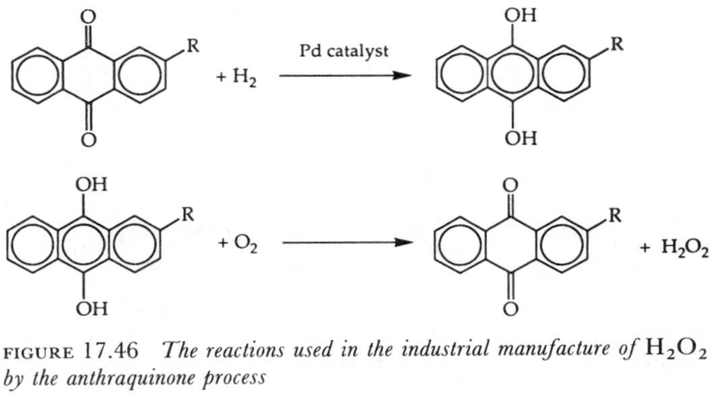

FIGURE 17.46 *The reactions used in the industrial manufacture of* H_2O_2 *by the anthraquinone process*

Accordingly, stabilizers, such as EDTA, which are good complexing agents for these metal ions, need to be added to prevent catalytic decomposition. Hydrogen peroxide has the skew structure shown in Figure 9.7b.

The ionic peroxides, and the peroxy compounds of the transition elements, have been discussed in previous Chapters (10 and 16). There are also a number of covalent peroxy species, acids or oxyanions, which may be regarded as derived from the normal oxygen compounds by replacing $-O-$ by $-O-O-$; just as $H-O-H$ is related to $H-O-O-H$. The best known examples are permonosulfuric acid (Caro's acid) H_2SO_5— which occurs as an intermediate in the persulfate oxidation above—perdisulfuric acid (persulfuric) $H_2S_2O_8$, perphosphoric acid, H_3PO_5, and perdicarbonic acid, $H_2C_2O_6$. The structures of the sulfuric acids have been definitely established and are related to the oxygen compounds as shown in Figure 17.47. An X-ray structure determination has recently been carried out on explosive crystals of Caro's acid, confirming the expected tetrahedral geometry about sulfur and the presence of an $S-O-O-H$ linkage. The molecules are linked by hydrogen bonding involving both the OH and the OOH groups— $S=O \cdots HOS$ and $S=O \cdots HOOS$. The O–O bond length is 146·4 pm, compared with 145·8 pm in H_2O_2. The other acids

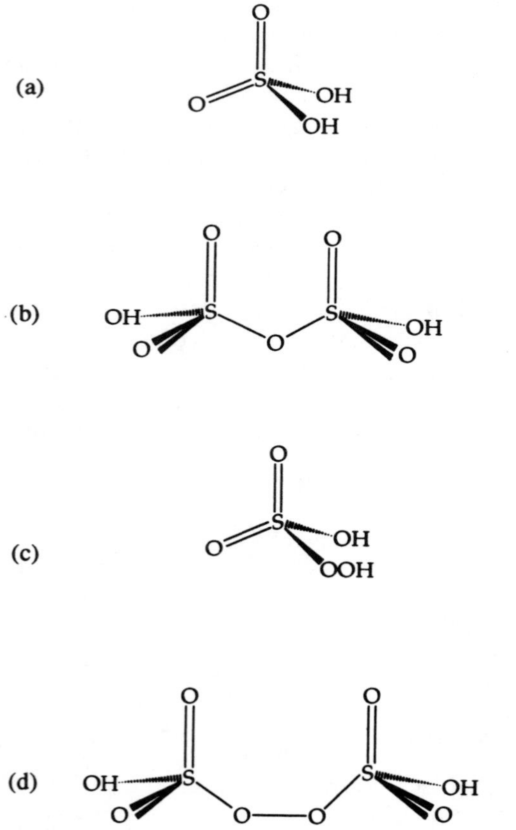

are probably $(HO)_2(HOO)P=O$ and $HO_2C-O-O-CO_2H$, and the latter has been structurally characterized by an X-ray study as its cyclohexyl ester, $CyO_2C-O-O-CO_2Cy$. The structure of the peroxydiphosphate ion, $[P_2O_8]^{4-}$, is similar to that of peroxydisulfate but with the two PO_3 units lying *trans* across the planar P–O–O–P link. The overall structure is C_{2h}. The O–O length is 149 pm.

A number of other compounds are commonly termed per-acids but most are only simple acids with hydrogen peroxide of crystallization. For example, the so-called percarbonic acid salts, such as $Li_2CO_4.H_2O$, are more probably carbonates, $Li_2CO_3.H_2O_2$. However, structural information is not available to finally determine these cases.

The percarbonate group is known to exist in transition metal complexes such as:

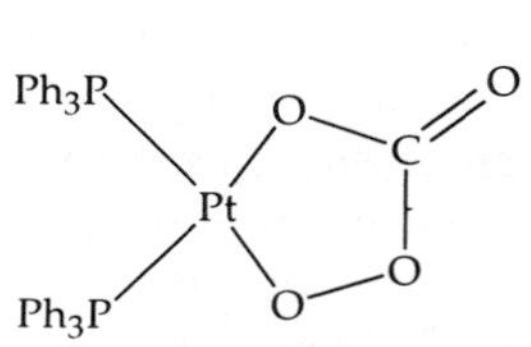

Sodium perborate has been established as a true peroxy compound with B–O–O linkages, see section 17.4.1.

Finally, it must be noted that the terms, peracid (better, peroxoacid) and peroxide, are properly applied only to compounds which contain the $-O-O-$ group, which may be regarded as derived from hydrogen peroxide. In the older literature, and in much technical literature, some higher oxides such as MnO_2 are termed peroxides. This usage is incorrect on the modern convention of nomenclature.

17.7.3 *The other elements: the (VI) state*

Sulfur occurs as the free element, especially in volcanic areas around hot springs, and was one of the elements known in ancient times. Its main current source is from the desulfurization of oils and natural gas, mining in volcanic deposits, as a byproduct of extracting sulfide ores, and by the Frasch process where it is recovered from underground deposits by hot water under pressure. Its major use is in sulfuric acid manufacture, with minor consumption in many industries, especially rubber manufacture. Sulfuric acid is made by burning S to SO_2, followed by catalytic oxidation to SO_3 and solution in sulfuric acid and dilution. More than half the sulfuric acid is used in fertilizer manufacture, with other uses spread over more than a hundred industries. Selenium and tellurium are byproducts of copper extraction. Selenium is used in glass, as a decolorizer and to produce red and yellow colours. It is used in photoconductors and electronic devices, and is the photoreceptor which is basic to xerographic photocopying. Tellurium finds uses in metallurgy and alloys.

The trioxides, MO_3, of the VI state are formed by sulfur, selenium, tellurium, and polonium, and all these elements (except possibly polonium) form oxyanions. The acids are included in Table 17.15. Sulfur and selenium are four-coordinated to oxygen in the acid (H_2MO_4) and in the anion (MO_4^{2-}), while tellurium forms the dibasic, six-coordinated

FIGURE 17.47 *The structures of the sulfuric and per-sulfuric acids* (a) Sulfuric acid, (b) pyrosulfuric acid, (c) peroxymonosulfuric acid, (d) peroxydi-sulfuric acid.

acid, $Te(OH)_6$, and two series of six-coordinated tellurates, $TeO(OH)_5^-$ and $TeO_2(OH)_4^{2-}$. In addition, the compound $Rb_6(Te_2O_9)$ has been shown to contain equal numbers of tetrahedral $[TeO_4]^{2-}$ and trigonal bipyramidal $[TeO_5]^{4-}$ ions. Sulfur, selenium, and tellurium all burn in air to form the dioxide, MO_2, and the trioxides are made by oxidizing these. Sulfuric acid, H_2SO_4, is formed by dissolving SO_3 in water, but selenic acid, H_2SeO_4, is more readily made by oxidizing the selenium(IV) acid, H_2SeO_3, made by dissolving SeO_2. Sulfuric acid and the sulfates are stable, while selenic acid, the selenates, telluric acid, and the tellurates are all oxidizing agents, although they are typically slow in reacting.

Only sulfur gives a condensed acid in the VI state, and it forms only the binuclear species, pyrosulfuric acid $H_2S_2O_7$. Condensed anions include $S_3O_{10}^{2-}$ and $S_5O_{16}^{2-}$ as well as $M_2O_7^{2-}$ (M = S, Se), but tellurium gives no polyanions at all. As condensation is also restricted in the IV state to binuclear compounds of sulfur and selenium, it will be seen that the tendency to condense is much slighter in this Group than it was in the nitrogen Group.

Oxidizing power of the VI state increases down the Group, and polonium(VI) is strongly oxidizing, so that the existence of the oxide and oxyanions is in some doubt.

The other main representatives of the VI state are the hexafluorides, MF_6. These compounds are all relatively stable but with reactivity increasing from sulfur to tellurium. SF_6 is extremely stable and inert, both thermally and chemically, and is used as an inert dielectric medium in a number of electrical devices. However, in the presence of oxygen an electrical discharge can generate species such as $F_5S-O-SF_5$ and the peroxy analogue $F_5S-O-O-SF_5$. SeF_6 is more reactive and TeF_6 is completely hydrolysed after 24 hours contact with water. The hexafluorides are more stable, and much less reactive as fluorinating agents, than the tetrafluorides of these elements. All the hexafluorides result from the direct reaction of the elements, and dimeric molecules S_2F_{10} or Se_2F_{10} are also found (but not Te_2F_{10}). These compounds are rather more reactive than the hexafluorides, but still reasonably stable. The coordination is octahedral, F_5M-MF_5. They hydrolyse readily, with fission of the M−M bond, and S_2F_{10} reacts with chlorine to give the mixed hexahalide, SF_5Cl. Addition of ClF to SeF_4 gives SeF_5Cl and tellurium forms similar species, TeF_5X for X = Cl, OF or OCl. Pseudohalide compounds such as SF_5CN are also known. Tracer studies reveal the existence of a volatile polonium fluoride, probably PoF_6.

TeF_6 adds F^- to form a seven-coordinated anion and also reacts with other Lewis bases, such as amines, to give eight-coordinated adducts like $(Me_3N)_2TeF_6$.

The VI state is also found in sulfur and selenium oxyhalides, and in halosulfonic acids. Sulfur gives sulfuryl halides, SO_2X_2, with fluorine and chlorine, and also mixed compounds SO_2FCl and SO_2FBr. Only one selenium analogue, SeO_2F_2, is found. If one OH group in sulfuric acid (Figure 17.47a) is replaced by F, fluorosulfonic acid results. The anion has been characterized and shows the

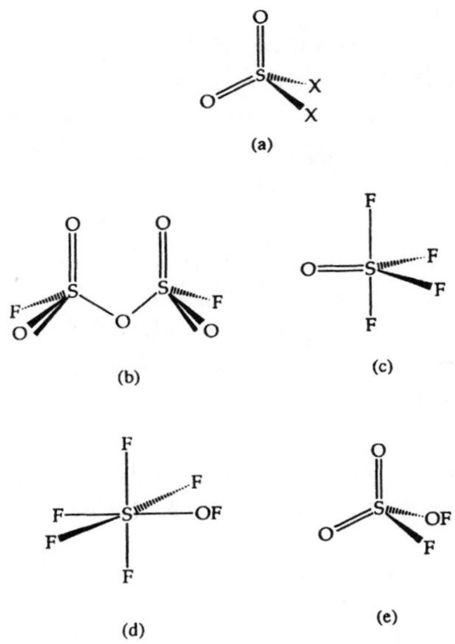

FIGURE 17.48   *Oxyhalides of sulfur;* (a) SO_2X_2, (b) $S_2O_5F_2$ *(or* Cl_2*)*, (c) SOF_4, (d) SOF_6, (e) SO_3F_2

expected tetrahedral configuration around S with the S−F bond distance of 165 pm and the other three S−O distances equal to 143 pm: the SOF bond angle is 108°. Sulfur also gives a number of more complex oxyhalides, some of which are shown in Figure 17.48. The oxyhalides are formed from halogen and dioxide, or from the halosulfonates. SO_2F_2 is inert chemically but the other compounds are much more reactive and hydrolyse readily.

Tellurium, in particular, forms a number of oxyfluorides of complex formulae such as $Te_5O_4F_{22}$ or $Te_7O_6F_{30}$. Structural studies show that many of these are derived from TeF_6 by successive replacement of F by the $OTeF_5$ group. There is evidence for $F_4Te(OTeF_5)_2$, $F_2Te(OTeF_5)_4$, $FTe(OTeF_5)_5$ and $Te(OTeF_5)_6$ (e.g. Figure 17.49a). The structures are

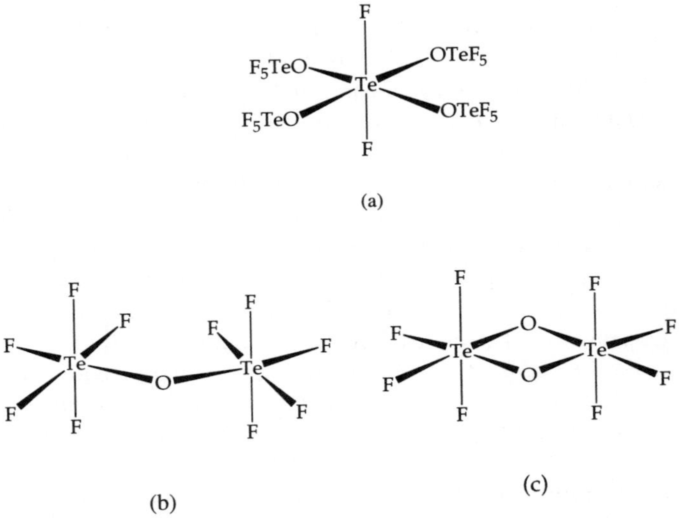

FIGURE 17.49   (a) $F_2Te(OTeF_5)_4$, (b) $(F_5Te)_2O$, (c) $(F_4Te)_2(O)_2$.

octahedral, and *cis* and *trans* isomers of both the F_4TeO_2 and F_2TeO_4 species are known. Related species are $[TeOF_4]_2$ which is a pair of octahedra linked by two oxygen bridges, and $(TeF_5)_2O$ (Figure 17.49b, c). The selenium analogue is also found. (These structures contrast with the transition metal analogues, such as WOF_4, which are linked by fluorine bridges.)

The $OTeF_5$, and $OSeF_5$, groups are found to bond to other atoms. Thus we find $V(OTeF_5)_6$ with an octahedral VO_6 configuration. Similar species include $I(OSeF_5)_5$, $B(OTeF_5)_3$ and $Xe(OTeF_5)_n$ for $n = 2$, 4 and 6. There is a range of angles at O from XeOTe (125°), TeOTe (139°) to VOTe (170°) which suggests an increasing degree of $M-O-Te$ π bonding. The simple species F_5MOMF_5 is found for M = S, Se and Te, each with similar MOM angles of 142–5°, and in each case the fluorines are eclipsed, again arguing for a $p\pi - d\pi$ interaction of the oxygen lone pairs.

As this field develops, it is becoming apparent that the $OSeF_5$ and $OTeF_5$ groups are behaving very like F, as ligands of high electronegativity. Many species occur in which stepwise replacement of F is found, as above and also in oxyfluorides like $OXe(OTeF_5)_4$ (cf. $OXeF_4$), as ionic species e.g. $Na^+(OSeF_5)^-$, and halo-species as $ClW(OTeF_5)_5$.

Three halosulfonic acids are known, FSO_3H, $ClSO_3H$, and $BrSO_3H$. The structures are tetrahedral

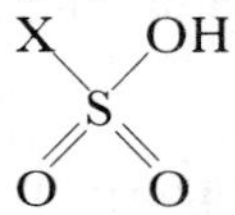

$$\begin{matrix} X & & OH \\ & S & \\ O & & O \end{matrix}$$

and are formed from sulfuric acid by replacing an OH group by a halogen atom. They are strong monobasic acids but only fluosulfonic acid is stable and forms stable salts, fluosulfonates FSO_3^-, which are similar in structure and solubilities to the perchlorates.

17.7.4 *The (IV) state*

The IV state is represented by oxides, MO_2, halides, MX_4, oxyhalides, MOX_2, acids, and anions, for all four elements. Sulfur(IV) is reducing, selenium(IV) is mildly reducing (going to Se(VI)) and also weakly oxidizing (going to the element). tellurium(IV) is the most stable state of tellurium, while polonium(IV) is weakly oxidizing (going to Po(II)). In addition, a number of complexes are known.

The oxides are given in Table 17.2a and the acids and anions in Table 17.15. The dioxides result from direct reaction of the elements. SO_2 and SeO_2 dissolve in water to give the acids, but TeO_2 and PoO_2 are insoluble and the parent acids do not exist. Tellurium and selenium oxyanions result from the solution of the oxides in bases. Salts of one condensed acid exist: these are the pyrosulfites, $S_2O_5^{2-}$, which have the unsymmetrical structure with a S–S bond, $^-O_3SSO_2^-$.

Figure 17.50 gives the structures of some of the oxides and oxyions. The IV state oxides and oxyanions have an unshared electron pair (in monomeric structures) and thus have unsymmetrical structures. The stable form of the dioxide shows an interesting transition from monomeric covalent molecule, SO_2, through polymeric covalent, SeO_2, to ionic forms for TeO_2 and PoO_2. Sulfur, selenium, and

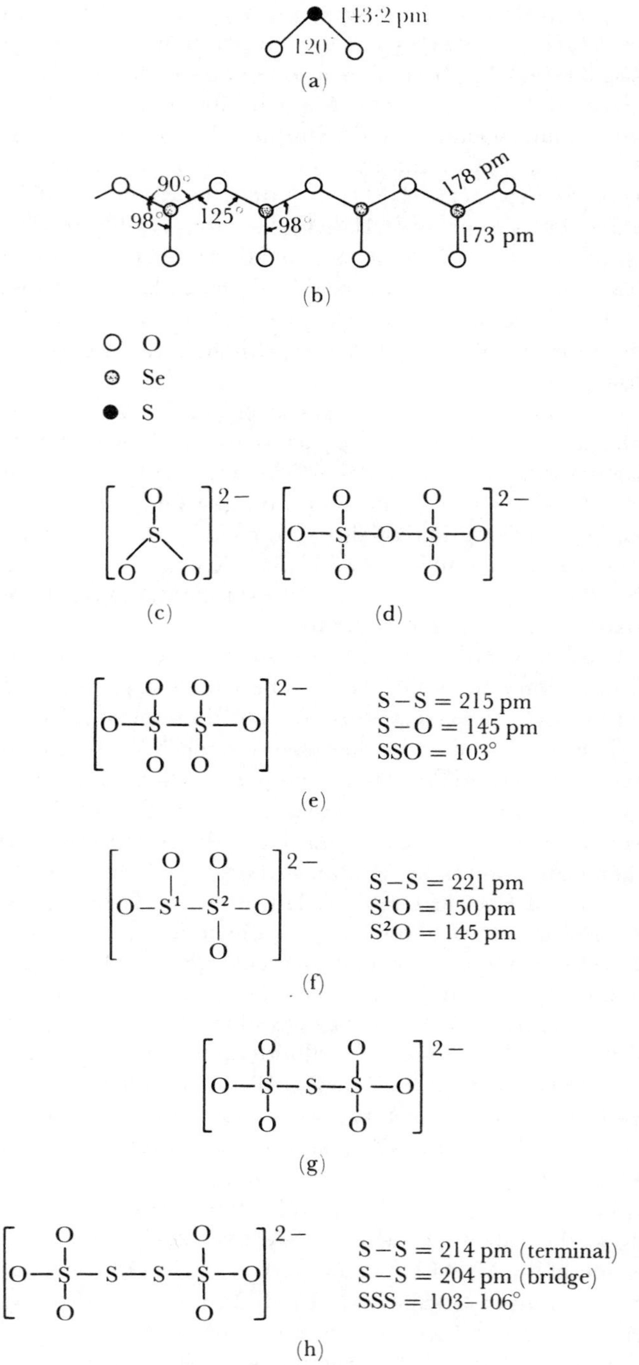

FIGURE 17.50 *Oxygen compounds of sulfur and selenium:* (a) SO_2, (b) SeO_2, (c) S(*or* Se)O_3^{2-}, (d) $S_2O_7^{2-}$, (e) $S_2O_6^{2-}$, (f) $S_2O_5^{2-}$, (g) $S_3O_6^{2-}$, (h) $S_4O_6^{2-}$ (*and similar* $S_nO_6^{2-}$)

S–S = 215 pm
S–O = 145 pm
SSO = 103°

S–S = 221 pm
S^1O = 150 pm
S^2O = 145 pm

S–S = 214 pm (terminal)
S–S = 204 pm (bridge)
SSS = 103–106°

tellurium dioxides are acidic and dissolve in bases. PoO_2 appears to be more basic than acidic and dissolves in acids as well as forming polonites with strong bases.

One compound intermediate between the IV and VI oxides is reported. This is Se_2O_5, formed by controlled heating of SeO_3. This compound is conducting in the fused state and it has been suggested that it is $SeO^{2+}SeO_4^{2-}$, a salt of Se(IV) and Se(VI).

The known tetrahalides are listed in Table 17.2b. The missing ones are SBr_4, SI_4 and SeI_4 while SCl_4 and $SeBr_4$ are also unstable, SCl_4 decomposing at -31 °C. The tellurium tetrahalides are markedly more stable, even TeI_4 being stable up to 100 °C. $TeCl_4$, $TeBr_4$ and $SeCl_4$ are all stable up to 200 °C and the tellurium compounds to 400 °C. All the tetrafluorides are known and these are rather more stable than the other tetrahalides to thermal decomposition. They are much more reactive than the hexafluorides and act as strong, though selective, fluorinating agents. All four polonium tetrahalides are known and resemble the tellurium analogues, though they seem to be rather less stable. The structures of most tetrahalides are known. SF_4, SeF_4, $TeBr_4$ and $TeCl_4$, are all found as the distorted tetrahedron derived from the trigonal bipyramid with one equatorial position occupied by an unshared pair of electrons (see section 4.2). Tellurium tetrafluoride forms a polymeric structure in which square pyramids are linked into chains by sharing corners, $TeF_3(F_{2/2})$. TeI_4 is a tetramer consisting of four TeI_6 octahedra sharing edges. Thus the lone pair is still sterically active. SCl_4 and $SeCl_4$ exist only in the solid and vaporize as $MCl_2 + Cl_2$— the sulfur compound at -30 °C and the selenium one at 196 °C. Raman spectra suggest the solid is $MCl_3^+ Cl^-$ and this is supported by crystallographic evidence for EBr_3^+ and EI_3^+ ($E = S$, Se) in addition to more complex cations like $Se_2I_4^{2+}$.

The selenium and tellurium, and probably the polonium, tetrahalides readily add one or two halide ions to give complex anions such as SeF_5^- and MX_6^{2-} ($M = Se$, Te, Po: $X = $ any halogen). These hexahalo (IV) complex ions are interesting from the structural point of view as they contain seven electron pairs. Structural determinations on a number of compounds show that the MX_6^{2-} group is octahedral, but the group may be regular or distorted. The electron pair repulsion theory, discussed in Chapter 4, would imply that the non-bonded pair occupied a spatial position and should lead to a distorted octahedron. It is possible, however, that with large, heavy, central atoms, the non-bonded pair might be accommodated in an inner orbital rather than in a particular spatial direction. Structural studies so far reported, indicate that two groups of structures are found. In the hexafluorides of the IV state, MF_6^{2-}, as in the isoelectronic IF_6^- and XeF_6, the structure is distorted octahedral, indicating a sterically active lone pair. However, in the heavier halides like MCl_6^{2-}, the structure is regular octahedral. It is proposed here that the dominant factor is repulsion between halogen atoms which is minimized in the regular structure. In this form, the lone pair would have to occupy an s orbital, presumably.

The structures of pentafluoro anions, $[EF_5]^-$ for $E = S$, Se, Te and various cations, are all square pyramids with the lone pair to complete the octahedron. The TeF_5^- units link into chains by weak $Te-F \cdots Te$ bonding while the Se analogue shows weakly-bonded tetramers. As in TeF_4, the E atoms lie below the base of the square pyramid, that is, the base F atoms are bent up away from the lone pair position. It will be recalled from section 13.6, that the opposite distortion occurs when there are five ligands and no lone pair. Selenium forms an interesting pair of dinuclear complexes in the IV state. $Se_2Cl_{10}^{2-}$ has two octahedra sharing an edge, while $Se_2Cl_9^-$ has two octahedra sharing a face (alternatively seen as two Se atoms linked respectively by 2 or 3 bridging Cl).

Oxyhalides of the IV state are formed only by sulfur and selenium and have the formula MOX_2. Thionyl halides, SOF_2, $SOCl_2$, $SOBr_2$, and $SOFCl$ are known while the selenyl fluoride, chloride, and bromide exist. Thionyl chloride is made from SO_2 and phosphorus pentachloride:

$$SO_2 + PCl_5 \rightarrow SOCl_2 + POCl_3$$

The other thionyl halides are derived from the chloride. Selenyl chloride is obtained from SeO_2 and $SeCl_4$. These oxyhalides are stable near room temperature but decompose on heating to a mixture of dioxide, halogen, and lower halides. SOF_2 is relatively stable to water, but all the other compounds hydrolyse violently. These compounds have an unshared electron pair on the central element and are pyramidal in structure (Figure 17.51). In $SeOF_2$, $FSeF = 92 \cdot 6°$ and $OSeF = 105 \cdot 5°$. The oxyhalides act as weak donor molecules through the lone pairs on the oxygen atoms, and also as acceptors using d orbitals on the sulphur or selenium. $SeOX_3^-$ complex ions are known for $X = F$, Cl, and Br. With large cations they exist as isolated square pyramidal structures. In $SeOF_2$, $FSeF = $ tendency to polymerise for $X = F < Cl < Br$. In $SeOCl_2.2py$ (py = pyridine) the structure is a square pyramid with a sterically active lone pair. The di-oxygen complex fluorides, $MO_2F_2^{2-}$ and MO_2F^- are found for both $M = Se$ and Te, together with $TeOF_4^{2-}$.

17.7.5 The (II) state: the ($-II$) state

If poly-sulfur compounds are excluded, the II state is represented by a more limited range of compounds than the IV or VI states, but it is more fully developed than the I state of the nitrogen Group. Tellurium and polonium form readily oxidized monoxides, MO, and a number of dihalides, MX_2, exist. No difluorides occur (apart from a possible SF_2) but the general pattern of stability of the other dihalides resembles that of the tetrahalides, stability decreasing from tellurium to sulfur and from chloride to iodide. $TeCl_2$, $TeBr_2$, $PoCl_2$, and $PoBr_2$ are stable; $SeCl_2$ and $SeBr_2$ decompose in the vapour while SCl_2, the only sulfur dihalide, decomposes at 60 °C. No di-iodide is known although there is a lower iodide

FIGURE 17.51   *The structure of thionyl halides,* SOX_2

TABLE 17.15 Oxyacids and oxyanions of sulfur, selenium, and tellurium

---

SULFUR

Sulfur shows coordination numbers up to four, the most common being tetrahedral; S—S bonds are common among the lower acids.

H_2SO_4 sulfuric acid	HSO_4^- and SO_4^{2-} sulfate	Stable, strong, dibasic acid: formed from SO_3 and water. Structure, Figure 17.47.
$H_2S_2O_7$ pyrosulfuric acid	$HS_2O_7^-$ and $S_2O_7^{2-}$ pyrosulfate	Strong, dibasic acid: formed from SO_3 and H_2SO_4 (anions by heating HSO_4^-), loses SO_3 on heating. Sulfonating agent. Structure, Figure 17.47. Also anions $S_3O_{10}^{2-}$ and $S_5O_{16}^{2-}$.
H_2SO_3 sulfurous acid	HSO_3^- and SO_3^{2-} sulfites	Existence of free acid doubtful. $SO_2 + H_2O$ gives a solution containing the anions but this loses SO_2 on dehydration. Reducing and weak, dibasic acid. Structure, pyramidal SO_3^{2-} ion; lone pair on S.
	$S_2O_5^{2-}$ pyrosulfite	No free acid. Formed by heating HSO_3^- or by $SO_2 + SO_3^{2-}$.

polythionic acids

$H_2S_2O_6$ dithionic acid	$S_2O_6^{2-}$ dithionates	Acid stable only in dilute solution, anions stable. Formed by oxidation of sulfites and stable to further oxidation, or to reduction. Strong, dibasic acid. Structure, Figure 17.50.
	$S_nO_6^{2-}$ $(n = 3$ to $6)$ polythionates	Free acids cannot be isolated; anions formed by reaction of SO_2 and H_2S or arsenite. Unstable and readily lose sulfur, reducing Structures contain chains of S atoms, Figure 17.50.
$H_2S_2O_4$ dithionous acid (or hyposulfurous)	$S_2O_4^{2-}$ dithionite	Acid prepared by zinc reduction of sulfurous acid solution, and salts (called also hyposulfites or hydrosulfites) prepared by zinc reduction of sulfites. Unstable in acid solution, but salts are stable in solid or alkaline media, powerful reducing agents. Decompose to sulfite and thiosulfate. Structure, Fig. 17.52, contains S—S link.
	$S_2O_3^{2-}$ thiosulfate	Prepared in alkaline media by action of S with sulfites. Perfectly stable in absence of acid, but gives sulfur in acid media. Mild reducing agent, as in action with I_2 which gives tetrathionate, $S_4O_6^{2-}$. Structure, Figure 17.50, derived from sulfate.
	SO_2^{2-} sulfoxylate	Best known as the cobalt salt from $CoS_2O_4 + NH_3 = CoSO_2 + (NH_4)_2SO_3$. The zinc salt may also exist. Unstable and reducing, structure probably V-shaped.

The peroxy acids H_2SO_5 and $H_2S_2O_8$, corresponding to sulfuric acid and pyrosulfuric acid, also exist. Structures are shown in Figure 17.47.

SELENIUM

Selenium commonly shows a coordination number of four: a smaller number of selenium acids than sulfur acids is found as the Se — Se bond is weaker.

H_2SeO_4 selenic acid	$HSeO_4^-$ and SeO_4^{2-} selenates	Formed by oxidation of selenites. Strong dibasic acid, oxidizing. Similar structures to sulfur compounds.
	$Se_2O_7^{2-}$ pyroselenate	No acid, formed by heating selenates.
H_2SeO_3 selenous acid	$HSeO_3^-$ and SeO_3^{2-} selenite	Selenium dioxide solution. Similar to S species, but less reducing and more oxidizing. Structure contains pyramidal SeO_3^{2-} ion.

Chain anions of selenium have not been found, but selenium (and tellurium) may form part of the polythionate chain; as in $SeS_4O_6^{2-}$ and $TeS_4O_6^{2-}$ where the Se or Te atoms occupy the central position in the chain.

TELLURIUM

Tellurium, like the preceding elements in this Period, is six-coordinate to oxygen.

H_6TeO_6 telluric acid	$TeO(OH)_5^-$ and $TeO_2(OH)_4^{2-}$ tellurates	Prepared by strong oxidation of Te or TeO_2. Structure is octahedral $Te(OH)_6$, and only two of the protons are sufficiently acidic to be ionized, and then, only weakly. The acid and salts are strong oxidizing agents.
	tellurites	TeO_2 is insoluble and no acid of the IV state is formed. Tellurites, and polytellurites, are formed by fusing TeO_2 with metal oxides.

of tellurium, $(TeI)_n$. SCl_2 is unstable with respect to dissociation to $S_2Cl_2 + Cl_2$.

The lower iodide of sulfur has been shown to be S_2I_2: there is no evidence for SI_2.

After considerable confusion, the properties of the fluorides of low oxidation state sulfur are now fairly clear. SF_2, S_2F_4, and two isomers of S_2F_2 exist. SF_2 readily disproportionates into SF_4 and S_8 (probably driven by the high $S-S$ bond energy). It is made from SCl_2 plus KF or HgF_2 under careful conditions: this system also readily yields S_2F_2. SF_2 is a V-shaped molecule with bond angle $98°$. It forms the unsymmetrical dimer F_3S-SF.

S_2F_2 exists in the *trans* chain and in the branched form

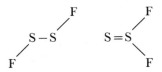

In the FSSF form, SS is 189 pm, similar to the distances in S_2 and shorter than HSSH or ClSSCl. In the SSF_2 form, SS is reduced further to 186 pm, the FSS angle is $108°$ and the FSF angle is $93°$, arguing some increased electron density in SS.

These species are stable at low temperatures in an inert environment. The FSSF form readily converts to the more stable SSF_2, which can be heated above $200°C$ before decomposition. However, decomposition is rapid in presence of species like HF.

The complex, $SeBr_2L$, where L is tetramethylthiourea, is an example of the rare T-coordination. The BrSeBr angle is $175°$ and the S atom of the ligand completes the T. A number of planar halide anions, such as $SeBr_4^{2-}$ and $Se_2Br_6^{2-}$, are known as are mixed oxidation state Se(II)–Se(IV) anions, such as $Se_2Br_8^{2-}$ and $Se_3Br_{10}^{2-}$.

The –II state is found in the hydrides, H_2M, and in the anions, M^{2-}. With the more active metals, sulfur, selenium, and tellurium all form compounds which are largely ionic, although they have increasing metallic, and alloy-like, properties as the electronegativity of the metal increases. The tendency to form anions increases from the nitrogen Group (where the evidence for M^{3-} ions, apart from N^{3-}, is limited), through the oxygen Group, (where the M^{2-} ion is more widely found), to the halogens where the X^- ion is the most stable form for all the elements. This trend in anionic behaviour reflects the increasing electronegativity of the elements, and the greater ease of forming singly-charged species over doubly- or triply-charged ones.

17.7.6 *Compounds with an* $S-S$ *or* $Se-Se$ *bond*

Many compounds with S–S bonds are known, together with a smaller range of Se–Se species and an even smaller range of Te–Te species. In this section we cover those compounds with one such bond, together with all the oxygen polysulfur compounds. The remaining catenated species with three or more bonded S or Se are discussed in section 18.2.

Disulfides, S_2^{2-}, are formed by reacting sulfides with S and addition of acids yields H_2S_2. This has a skew structure

similar to H_2O_2 but with angles at S of $92°$. Reacting Cl_2 or Br_2 with molten sulfur gives S_2Cl_2 or S_2Br_2 which are more stable than the corresponding SX_2 and also have the skew $X-S-S-X$ structure with SSX angles about $103°$. Selenium analogues, Se_2X_2 (X = F, Cl, Br), have similar structures. Photolysis of the FE=EF species gives the isomeric form $F_2E=E$ (E=O, S or Se). H_2Te_2 has recently been found to be stable in the gas phase.

Among the oxides, only S_2O contains a $S-S$ bond. This compound was, for a long time, reported as SO but the most recent work has established the existence of S_2O and suggests that SO is a mixture of S_2O and SO_2. S_2O is formed by the action of an electric discharge on sulfur dioxide and it is unstable at room temperature. The structure SSO is proposed.

A variety of oxyacids and oxyanions with $S-S$ bonds exists, among which are the polythionic acids, $H_2(O_3S-S_n-SO_3)$ where n varies from 0 to 12, and a miscellaneous group of compounds including thiosulfate, dithionite and pyrosulfite. All these compounds are included in Table 17.15. In sulfur-metabolizing organisms, it would be expected that species representing partial oxidation of sulfur would occur on the path to sulfite and sulfate. It is intriguing to find that chromatographic investigations show the presence of all the polythionates from $n = 1$ to 20 in cultures of *Thiabacillus ferrooxidans*.

In the polythionic acids, there is a marked difference in stability between dithionic acid, $H_2S_2O_6$, and the higher members. As dithionic acid contains no sulfur atom which is bonded only to sulfur, it is much more stable than the other polythionic acids which do contain such sulfurs, see Table 17.15. Dithionates result from the oxidation of sulfites:

$$2SO_3^{2-} + MnO_2 + 4H_3O^+ \rightarrow Mn^{2+} + S_2O_6^{2-} + 6H_2O$$

and the parent acid may be recovered on acidification. Dithionic acid and the dithionates are moderately stable, and the acid is a strong acid. The structure, O_3S-SO_3, has an approximately tetrahedral arrangement at each sulfur, with π bonding between sulfur and the oxygen atoms.

The reaction of H_2S and SO_2 gives a mixture of the polythionates from $S_3O_6^{2-}$ to $S_{14}O_6^{2-}$, while specific preparations for each member also exist, as in the preparation of tetrathionate in volumetric analysis by oxidation of thiosulfate by iodine:

$$2S_2O_3^{2-} + I_2 \rightarrow S_4O_6^{2-} + 2I^-$$

These compounds have very unstable parent acids, which readily decompose to sulfur and sulfur dioxide, but the anions are somewhat more stable. The structures are all established as $O_2S-S_n-SO_3$ with sulfur chains which resemble those in sulfur polyanions. As with the other examples of sulfur chain compounds, each polythionic acid readily disproportionates to an equilibrium mixture of all the others.

Thiosulfate, $S_2O_3^{2-}$ is formed by the reaction of sulfite with sulfur. The free acid is unstable but the alkali metal salts are well-known in photography, volumetric analysis, in

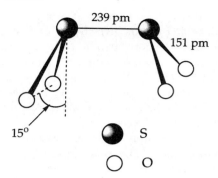

FIGURE 17.52   *The structure of the dithionite ion,* $S_2O_4^{2-}$

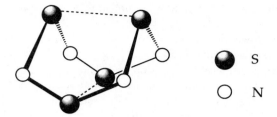

FIGURE 17.53   *The structure of tetrasulfur tetranitride*

other applications such as paper making and as an antidote for cyanide poisoning, since thiosulfate can convert cyanide to non-toxic products *via* the reaction:

$$CN^- + S_2O_3^{2-} \rightarrow SCN^- + SO_3^{2-}$$

The thiosulfate ion has been found to have the expected tetrahedral geometry, analogous to sulfate, but with one oxygen replaced by a sulfur atom, and with sulfur–sulfur and sulfur–oxygen bond distances of 201 and 147 pm respectively. Exchange studies with radioactive sulfur have also demonstrated the presence of two types of sulfur atom. The structure of the ion with one added H has been determined by Raman spectroscopy supported by calculations. In $[HS_2O_3]^-$ the H is attached to the S, giving the linkage $[H-S-SO_3]^-$ showing that the formal S=S bond in $[S_2O_3]^{2-}$ is easily attacked.

Dithionite ions, $S_2O_4^{2-}$, result from the reduction of sulfite with zinc dust. The free acid is unknown, and the salts are used in alkaline solution as reducing agents. Dithionite (also called hyposulfite or hydrosulfite) decomposes readily:

$$2S_2O_4^{2-} + H_2O \rightarrow S_2O_3^{2-} + 2HSO_3^-$$

and the solutions are also oxidized readily in air. The dithionite ion has the unusual structure shown in Figure 17.52. The oxygen atoms are in the eclipsed configuration and the S–S bond is very long. The S–O bond lengths show that π bonding exists between the sulfur and oxygen atoms, and there is also a lone pair on each sulfur atom. The S atoms must thus make use of *d* orbitals and it is proposed that the unusual configuration and the long S–S distance arise from the use of a *d* orbital.

17.7.7 Sulfur-nitrogen ring compounds

Sulfur and nitrogen form a number of ring compounds, of which the best-known is tetrasulfur tetranitride, S_4N_4. This is formed by reacting the chlorosulfane of approximate composition SCl_3 with ammonia. It is a yellow-orange solid which is not oxidized in air, although it can be detonated by shock. The structure is shown in Figure 17.53. The four nitrogen atoms lie in a plane, and the four atoms of sulfur form a flattened tetrahedron, interpenetrating the N_4 square. Alternatively, the structure may be regarded as an S_8 ring with every second sulfur replaced by a nitrogen.

S–S distances are fairly short, 259 pm, showing that there is some weak bonding across the puckered ring between opposite sulfur atoms. The selenium analogue, Se_4N_4, has a similar structure with Se–N = 178 pm and the long Se$\cdots$Se distance = 274 pm. In the $S_4N_4^{2+}$ cation, the loss of electrons removes the cross-ring bonds. The cation is found in two forms, planar and boat-shaped, where all the S–N bond lengths are equal and the charge is delocalized. Similar planar $S_4N_3^+$ and $S_5N_5^+$ cations with equal bonds are known.

A number of other ring compounds exist, including $S_4N_5^-$ where the additional N bridges the two S atoms at the top of the molecule, $F_4S_4N_4$, where the substituent F atoms are on the sulfur atoms of the S_4N_4 ring, and $S_4N_4H_4$ with the H atoms on the nitrogens. A different eight-membered ring is found in S_7NH, again with H bonded to the nitrogen. This hydrogen is fairly acidic and can be replaced by a number of metals. Different ring sizes are also found, as in $S_3N_3Cl_3$ and S_4N_3Cl. In S_4N_2 the structure is a 'half-chair' where there is a planar S–N–S–N–S section with the last S linking the two ends any lying above the plane.

Related to these species is polymeric $(SN)_x$ which has metallic conducting properties, and is a semi-conductor at 0·26 K. This compound oxidizes too readily for large-scale use, but it has directed interest towards other 'covalent metals', and it is possible that polymers including S–F or S–O units to protect the SN backbone may be more resistant to oxygen.

17.8 The fluorine Group, *ns²np⁵* (the halogens)

17.8.1 *General properties*

Reference to properties of the halogens are included in the following Tables and Figures:

Ionization potentials	Table 2.8
Electron affinities	Table 2.9
Atomic properties and electron configurations	Table 2.5
Radii	Table 2.10, 2.11, 2.12
Electronegativities	Table 2.14
Redox potentials	Table 6.3
Uses as nonaqueous solvents	Chapter 6

Table 17.16 lists some of the properties of the elements, Figure 17.54 shows the variations with Group position of a number of important parameters, and Figure 17.55 gives the oxidation state free energies.

TABLE 17.16 Properties of the elements of the fluorine Group

Element	Symbol	Oxidation states	Coordination numbers	Availability
Fluorine	F	$-I$	$1, (2)$	common
Chlorine	Cl	$-I, I, III, V, VII$	$1, 2, 3, 4$	common
Bromine	Br	$-I, I, III, V, VII$	$1, 2, 3, 5$	common
Iodine	I	$-I, I, III, V, VII$	$1, 2, 3, 4, 5, 6, 7$	common
Astatine	At	$-I, I, III?, V$		very rare

Astatine exists only as radioactive isotopes, all of which are very short-lived. Work has been done using either ^{211}At ($t_{\frac{1}{2}} = 7\cdot2$ h) or ^{210}At ($t_{\frac{1}{2}} = 8\cdot3$ h) and the very high activity necessitates working in $10^{-14}M$ solutions, and following reactions by coprecipitation with iodine compounds. The chemistry of astatine is therefore little known. Preparation is by the action of α-particles on bismuth, for example:

$$^{209}_{83}Bi + {}^{4}_{2}He \rightarrow {}^{211}_{85}At + 2{}^{1}_{0}n$$

Coprecipitation with iodine compounds indicates the existence of the oxidation states shown in Table 17.16. The negative state is probably the At^- ion and the positive states, the oxyions, AtO^-, AtO_2^-, and AtO_3^-. Astatine appears to differ from iodine in not giving a VII state, in which it parallels the behaviour of the other heavy elements in its Period, and in not forming a cation, At^+. Two interhalogens are known for astatine, AtI and AtBr, and there is evidence for polyhalide ions including astatine.

It is possible to synthesize organic astatine compounds, $RAtO_2$, R_2AtCl and $RAtCl_2$, in the III and V states. The compounds in the commonest state, RAt, correspond to the well-known organic halides in the $-I$ state. A very wide range of organic groups R have been explored, including biologically important molecules, such as steroids. These have the potential for specific radiotherapy by taking At to the site of a tumour. A more detailed account of the applications of inorganic compounds in medicine is given in Chapter 20.

The other elements of the Group are all well-known and occur mainly as salts containing the halide anion. The free elements are too reactive to exist in nature. Free fluorine, F_2, is outstandingly reactive and can be handled only in extremely dry glass systems (otherwise it reacts with glass to give SiF_4), in Teflon, or in apparatus made of metals which form a protective layer of fluoride on the surface, such as copper, nickel, or steel. Provided these conditions are rigorously adhered to, fluorine can be handled quite readily in the laboratory.

F_2 is made by the electrolysis of KF in anhydrous HF. Because of its extreme reactivity, it is almost always used *in situ*, either to make directly products such as UF_6 used in atomic fuel manufacture or volatile element fluorides such as WF_6 used to deposit metal coatings, or to produce more controllable fluorine compounds such as ClF_3 or BrF_5 which in turn are used to form fluorides. Of fluorine compounds, the

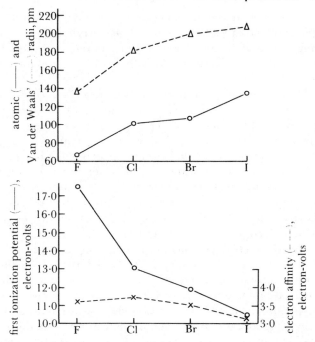

FIGURE 17.54 *Some properties of the halogens*
The radii of the anions, X^-, are almost identical with the corresponding van der Waals' radii. Notice also, that the electron affinities do not follow a regular trend from fluorine to iodine.

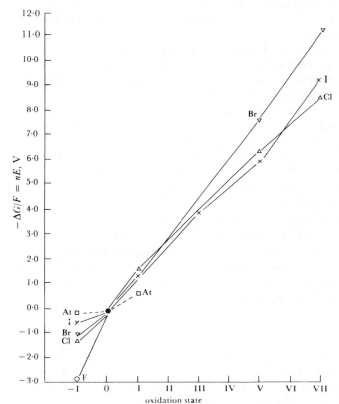

FIGURE 17.55 *Oxidation state free energies of the halogens*
This Group, like the earlier ones, shows a negative state which decreases in stability from the first to the last member of the Group. The positive states are all fairly similar: bromine has the least stable V and VII states while I^V is more stable relative to I^{VII} than is Cl^V relative to Cl^{VII}.

widest use is of the very inert fluorocarbons used in all sorts of applications from high-performance lubricants to non-stick frying pans. Simpler members are used as refrigeration liquids and propellants, mostly mixed fluorochloromethanes or ethanes.

Chlorine is made by electrolysis of brines and isolated either as Cl_2, as aqueous base solutions containing OCl^- initially and converted to chlorate or perchlorate, or converted with hydrogen to HCl. Principal uses of the element or the oxychloro compounds are in bleaching (wood pulp for paper, textiles) or sterilization (water supply, swimming pools, some stages of sewage treatment, domestic bleaches). Organochlorine compounds in industry are often produced by the action of Cl_2 over metal chloride catalysts, especially Fe or Cu. These products, such as vinyl chloride, $CH_2 = CHCl$, are themselves used largely in polymer formation giving the plastics of commerce (e.g. polyvinyl chloride, PVC). Cl_2 is used to make Br_2 from bromides, found in brines or from seawater in general. Bromine is mainly used to prepare organobromine compounds which are used as insecticides, as fire retardants for fibres, as an anti-knock agent in petrol (ethylene dibromide), and in some applications paralleling chlorine in water treatment and sterilization. Iodine is recovered from iodide salts and brines by chlorine oxidation, often with an intermediate stage of iodide concentration. It finds a wide range of uses, mainly as organoiodine compounds, in pharmaceuticals, photography, pigments, sterilization, dyestuffs, and rubber manufacture.

Fluorine is the most electronegative element and can therefore exist only in the $-I$ state. The $-I$ state is also the most common and stable state for the other elements, although positive states up to the Group state of VII occur. The positive states are largely found in oxyions and in compounds with other halogens, but iodine does appear to exist in certain systems as the coordinated I^+ cation.

The reactivity of the elements decreases from fluorine to iodine. Fluorine is the most reactive of all the elements and forms compounds with all elements except helium, neon, and argon. In all cases, except with the other rare gases, oxygen, and nitrogen, fluorine compounds result from direct, uncatalysed reactions between the elements: the exceptions react in the presence of metal catalysts such as nickel or copper. Chlorine and bromine also combine directly, though less vigorously, with most elements while iodine is less reactive and does not react with some elements such as sulfur.

17.8.2 *The positive oxidation states*

The positive oxidation states occur in the compounds of chlorine, bromine, and iodine with oxygen and fluorine, and of the heavier halogens with the lighter ones. The oxides of the elements are shown in Table 17.17.

Stability of the oxides is greatest for iodine, then chlorine, with bromine oxides the least stable. The higher oxides are rather more stable than the lower ones. Typical preparations are:

$$2Cl_2 + 2HgO \rightarrow Cl_2O + HgCl_2.HgO \text{ (also for } Br_2O)$$

TABLE 17.17 The oxides of the halogens

Average oxidation state	Chlorine	Bromine	Iodine
I	Cl_2O (b. 2°)	Br_2O (d. $-16°$)	
II	$Cl-ClO_2$		
III		$Br-O-BrO_2$ (d. $-40°$)°	
IV	ClO_2 (b. 11°)	$Br-O-BrO_3$	I_2O_4 (d. 130°) I_4O_9 (d. $> 100°$)
V		Br_2O_5 (d. $-20°$)	I_2O_5 (d. $> 300°$)
VI	Cl_2O_6 (b. 203°)	BrO_3 or Br_3O_8 (d. 20°)	
VII	Cl_2O_7 (b. 80°)	Br_2O_7(?)	

(b = boils d = decomposes, temperatures in °C)

$$2KClO_3 + 2H_2C_2O_4 = 2ClO_2 + 2CO_2 + K_2C_2O_4 + 2H_2O$$

(also with sulfuric acid: the method using oxalic acid gives the very explosive ClO_2 safely diluted with carbon dioxide).

ClO_3, BrO_3, and I_4O_9 are prepared by the action of ozone while Cl_2O_7 and I_2O_5 are the result of dehydrating the corresponding acids.

The most stable of all the oxides is iodine pentoxide, which is obtained as a stable white solid by heating HIO_3 at 200 °C. It dissolves in water to re-form iodic acid, and is a strong oxidizing agent which finds one important use in the estimation of carbon monoxide:

$$I_2O_5 + 5CO = I_2 + 5CO_2$$

The iodine is estimated in the usual way. The other iodine oxides are less stable and decompose to iodine and I_2O_5 on heating. I_2O_4 contains chains of IO units which are cross-linked by IO_3 units. The structure of I_4O_9 is less certain but it has been formulated as $I(IO_3)_3$. I_2O_7 has been reported as an orange polymeric solid formed by dehydrating HIO_4 with oleum.

The most stable chlorine oxide is the heptoxide which is the anhydride of perchloric acid, $HClO_4$. It is a strong oxidizing agent and dissolves readily in water to give perchloric acid. It has the structure $O_3Cl-O-ClO_3$ with tetrahedral chlorine. Cl_2O_6 is a red oil which melts at 4 °C. It is dimeric in carbon tetrachloride solution but the pure liquid may contain some of the monomer, ClO_3. In the solid, it is ionic, $ClO_2^+ClO_4^-$. The ClO_2^+ ion is known in a few other compounds with large anions such as $ClO_2^+ GeF_5^-$. It readily decomposes to ClO_2 and oxygen and reacts explosively with organic materials. Like the other chlorine oxides, it can be detonated by shock. In this, it is more sensitive than the heptoxide but more stable than Cl_2O and ClO_2. Chlorine dioxide is a yellow gas with an odd number of electrons. It detonates readily but is gradually being used on a large scale in industrial processes, for example as a more environmentally

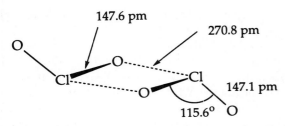

147.6 pm

270.8 pm

147.1 pm

115.6°

FIGURE 17.56 *The structure of chlorine dioxide,* ClO_2, *in the solid state* The ClO_2 molecules dimerize *via* a weak interaction.

friendly bleaching agent. In these processes it is made *in situ* and always kept well diluted with an inert gas. Notwithstanding the extremely hazardous nature of ClO_2, its molecular structure has been determined in the solid and gaseous states. In the gas phase ClO_2 is an angular molecule with an O–Cl–O angle of 118°. In the solid state, however, ClO_2 is dimerized, Figure 17.56, though the interactions are relatively weak. ClO_2 dissolves freely in water to give initially the hydrate, $ClO_2.8H_2O$. On exposure to visible or ultraviolet radiation this gives HCl and $HClO_4$ solution.

Dichlorine monoxide, Cl_2O, is an orange gas which dissolves in water to give a solution containing hypochlorous acid, HOCl. It is a symmetrical angular molecule with a Cl–O–Cl angle of 110°. It is a powerful oxidizing agent and highly explosive—indeed, most manipulations of Cl_2O are carried out with a substantial wall between the experimenter and the compound.

The mixed oxidation state chlorine oxide, chloryl chloride, $Cl–ClO_2$ has also been recently described. This compound decomposes to ClO_2 and Cl_2 in the gas phase. In solid matrices, $Cl–ClO_2$ can be isomerized to Cl–O–Cl–O and Cl–O–O–Cl.

The bromine oxides are the least stable of all the halogen oxides and all decompose below room temperature, though there is some indication that they may be less explosive than the chlorine analogues. Br_2O is dark brown, Br_2O_3 and Br_2O_5 are orange and colourless respectively. It is not clear whether the next oxide has the formula BrO_3 or Br_3O_8. This is also colourless and it decomposes in vacuum with the evolution of Br_2O, leaving a white solid which could possibly be Br_2O_7. Reaction of ozone with bromine gives Br_2O_3,

formulated as bromine bromate, or, under longer reaction times, dibromine pentoxide, Br_2O_5, having the structures

One of the possible compounds having the composition BrO_2 has been recently shown to be bromine perbromate, $Br–O–BrO_3$. The bromine oxides dissolve in water or alkali to give mixtures of the oxyanions.

The oxyacids of the halogens are shown in Table 17.18 Most are obtainable only in solution, although salts of nearly all can be isolated. HOF has also recently been established.

Hypochlorous acid, HOCl, occurs to an appreciable extent, 30 per cent, in solutions of chlorine in water, but only traces of HOBr and no HOI are found in bromine or iodine solutions. All three halogens give hypohalite on solution in alkali:

$$X_2 + 2OH^- = XO^- + X^- + H_2O \dots\dots\dots\dots (1)$$

$$K = \frac{[X^-][XO^-]}{[X_2]} = 7 \times 10^{15} \text{ for Cl}, 2 \times 10^8 \text{ for Br, 30 for I}$$

but the hypohalites readily disproportionate to halide and halate:

$$3XO^- = 2X^- + XO_3^- \dots\dots\dots\dots\dots (2)$$

with equilibrium constants of 10^{27} for Cl, 10^{15} for Br, and 10^{20} for I. Thus the actual products depend on the rates of these two competing reactions. For the case of chlorine, the formation of hypochlorite by reaction (1) is rapid, while the disproportionation by reaction (2) is slow at room temperatures so that the main products of dissolution of chlorine in alkali are chloride and hypochlorite. For bromine, reactions (1) and (2) are both fast at room temperature so that the products are bromide, hypobromite, and bromate, the proportion of bromate being reduced if the reactions occur at 0 °C. In the case of iodine, reaction (2) is very fast and iodine dissolves in alkali at all temperatures to give iodide and iodate quantitatively:

$$3I_2 + 6OH^- = IO_3^- + 5I^- + 3H_2O$$

The only halous acid definitely established is chlorous acid,

TABLE 17.18 Oxyacids of the halogens

Type	Name		Stability
HOX	hypohalous	X = F, Cl, Br, I	F > Cl > Br > I. All are unstable and are known only in solution.
HOXO	halous	X = Cl	$HBrO_2$ possibly exists also.
$HOXO_2$	halic	X = Cl, Br, I	Cl < Br < I. Chloric and bromic in solution only but iodic acid can be isolated as a solid.
$HOXO_3$	perhalic	X = Cl, Br, I	Free perchloric, perbromic, and periodic acids occur.

Also $(HO)_5IO$ and $H_4I_2O_9$ forms of periodic acid

$HClO_2$, though salts containing the bromite ion BrO_2^- are now well defined. An X-ray structure determination on sodium bromite trihydrate confirms the presence of an angular BrO_2^- ion with an O–Br–O angle of 105·3°. $HClO_2$ does not occur in any of the disproportionation reactions above and is formed by acidification of chlorites. The latter are themselves formed by reaction of ClO_2 with bases:

$$2ClO_2 + 2OH^- = ClO_2^- + ClO_3^- + H_2O$$

Chlorites are relatively stable in alkaline solution and are used as bleaches. In acid, chlorous acid rapidly disproportionates to chloride, chlorate, and chlorine dioxide.

For the halic acids, stability is greatest for iodine. Salts of all three acids are well-known and stable, with a pyramidal structure. The halic acids are stronger acids than the lower ones and are weaker oxidizing agents.

All perhalates exist and perchloric acid and perchlorates are well-known. $HClO_4$ is the only oxychlorine acid which can be prepared in the free state. It, and perchlorates, although strong oxidizing agents, are the least strongly oxidizing of all the oxychlorine compounds. The perchlorates of many elements exist. The ion is tetrahedral and has the important property of being weakly coordinated by cations. It is thus very useful in the preparation of complexes, as metal ions may be introduced to a reaction as the perchlorates, with the assurance that the perchlorate group will probably remain uncoordinated—contrast this with the behaviour of ions such as halide, carbonate, or nitrite which are often found as ligands in the complexes, e.g.:

$$Co^{2+} + NH_3 + Cl^- \xrightarrow{\text{oxidize}} Co(NH_3)_6^{3+} + Co(NH_3)_5Cl^{2+} +, \text{etc.}$$

$$\text{but } Co^{2+} + NH_3 + ClO_4^- \xrightarrow{\text{oxidize}} Co(NH_3)_6^{3+} \text{ only}$$

After eluding attempts to make it for many years, perbromic acid and its salts were synthesized in 1968. Perbromate, BrO_4^- was prepared electrolytically, alternatively by oxidation with XeF_2, but the most convenient synthesis was found to be oxidation of bromate in alkali by molecular fluorine. Acidification yields perbromic acid which is a strong monobasic acid, stable in solutions up to about 6 M. $KBrO_4$ contains tetrahedral BrO_4^- ions, and this species is predominant in solution with no evidence of a second form as found in periodates. The electrode potentials have been assessed for the reaction

$$XO_4^- + 2H^+ + 2e^- = XO_3^- + H_2O$$

as 1·23 V for X = Cl, 1·76 V for X = Br, and 1·64 V for X = I. Thus perbromate is a somewhat stronger oxidant than perchlorate or periodate, but its oxidizing reactions are sluggish. Thus, the oxidizing power is not the reason for the difficulties found in the synthesis of perbromate. It seems that the preparation from Br(V) requires the surmounting of an activation barrier, and any process proceeding by one-electron additions might have failed because of the instability of the intermediate species.

Thus, although the long-standing anomaly about perbromate no longer exists, it is clear that bromine(VII) is less stable in compounds with oxygen than either chlorine(VII) or iodine(VII). To this extent, bromine still reflects the middle element anomaly.

Periodic acid and periodates occur, like perrhenates, in four-, five- and six-coordinated forms. Oxidation of iodine in sodium hydroxide solution gives the periodate, $Na_2H_3IO_6$, and the three silver salts, $AgIO_4$, Ag_5IO_6, and Ag_3IO_5 may be precipitated from solutions of this sodium salt under various conditions. Deliquescent white crystals of the acid H_5IO_6 may be obtained from the silver salt and this loses water in two stages:

$$(HO)_5IO \xrightarrow{80\ °C} H_4I_2O_9 \xrightarrow{100\ °C} (HO)IO_3$$

Salts, such as $K_4I_2O_9$, of the binuclear acid may be obtained. The $I_2O_9^{2-}$ ion has the $O_3IO_3IO_3$ structure, of two IO_6 octahedra sharing one triangular face. In the periodates, the IO_6 group is octahedral and the IO_4 group is tetrahedral, as expected. In the related ion $[(HO)_2I_2O_8]^{4-}$, there are two IO_6 octahedra sharing an edge with distance $I-OH = 190\ pm$, $I-O$ (terminal) $= 181\ pm$, and $I-O$ (bridge) $= 202\ pm$. In K_3IO_5, the IO_5^{3-} ion has a square pyramidal structure. The iodine oxyanions thus continue the trend, already observed in the earlier members of this Period, to become six-coordinated to oxygen.

It should be noted that halates, and oxyhalides generally, are potentially explosive in contact with oxidizable materials.

Positive oxidation states for halogens are also found in the interhalogen compounds, where the lighter halogen is in the −I state and the heavier one in a positive state. A similar situation pertains for the mixed polyhalide ions. Table 17.19 lists these compounds.

All the interhalogens of type AB are known, although IF and BrF are very unstable. All the AF_3 and AF_5 types also occur, although, again, some are unstable. Only IF_7 is found in the VII oxidation state, and the only other higher interhalogen is iodine trichloride. All are made by direct combination of the elements under suitable conditions, apart from IF_7 which results from the fluorination of IF_5. All the compounds are liquids or volatile solids except ClF, which boils at −100 °C. Most boiling points fall between 0 °C and 100 °C.

The halogen fluorides are all very reactive and act as strong fluorinating agents. Reactivity is highest for chlorine trifluoride, which fluorinates as strongly as elemental fluorine. Reactivity falls from the chlorine to the bromine and iodine fluorides, and also falls off as the number of fluorine atoms in the molecule increases. BrF_3 and IF_5 are particularly useful as fluorinating agents for the production of fluorides of elements in intermediate oxidation states. These two interhalogens, along with the two iodine chlorides ICl and ICl_3, undergo self-ionization and are useful as solvent systems, as discussed in Chapter 6. The anions of these systems, BrF_4^-,

TABLE 17.19 Interhalogens and polyhalide ions

AB	AB_3	AB_5	AB_7	A_3^-		AB_4^-	AB_6^-	AB_8^-	A_n^-	A_n^{2-}
ClF	ClF_3	(ClF_5)	IF_7	Br_3^-	ICl_2^-	ClF_4^-	ClF_6^-	IF_8^-	I_5^-	I_8^{2-}
BrF	BrF_3	BrF_5		I_3^-	$ClBr_2^-$	BrF_4^-	BrF_6^-		I_7^-	
$(IF)_n$	$(IF_3)_n$	IF_5		ClF_2^-	IBr_2^-	IF_4^-	IF_6^-		I_9^-	
BrCl	I_2Cl_6			BrF_2^-	$IBrF^-$	ICl_4^-			$I_2Cl_3^-$	
ICl				$BrCl_2^-$	$IBrCl^-$	ICl_3F^-			$I_2Cl_2Br^-$	
IBr					BrI_2^-	(IF_6^{3-})				

IF_6^-, ICl_2^-, and ICl_4^-, respectively, are among the polyhalide ions in the Table. The cations, BrF_2^+, IF_4^+, I^+, and ICl_2^+, are less familiar but may be isolated by adding halide ion acceptors to the interhalogen, for example:

$$BrF_3 + SbF_5 = (BrF_2)^+(SbF_6)^-$$

Although these compounds are formulated as ions, structural studies show that there may be an interaction between the cation and anion. For example, ICl_2SbCl_6, (from ICl_3 and $SbCl_5$) consists of $SbCl_6$ octahedra and angular ICl_2 units, but with a weak coordination of two of the chlorines in an octahedron to two different iodines to give a chain structure, Figure 17.57. This structure seems to be quite general, $BrF_2^+SbF_6^-$ adopting a very similar form to that shown in Figure 17.57. In both cases, the AB_2 unit may be described as the cation, AB_2^+, forming two further weak bridges to the anion, SbX_6^-, or as covalent with a very distorted square planar AB_4 unit. The two A–B distances are sufficiently different to make the former description the more acceptable. For FCl_2^+ the vibrational spectrum and calculation points to the asymmetric V-shape structure with the linkage Cl–Cl–F.

The other interhalogens of the AB type have properties and reactivities which are roughly the average of the properties of the constituent halogens.

The structures of the interhalogens have been largely covered in Chapter 3. ClF_3 and BrF_3 are T-shaped, BrF_5, and probably the other pentafluorides, are square pyramids of fluorine atoms with the bromine atom just below the plane of the four fluorines of the base. IF_7 is approximately a

A_2^+	A_3^+	A_5^+	AB_2^+	AB_4^+	AB_6^+
Br_2^+	Cl_3^+	Br_5^+	ClF_2^+	ClF_4^+	ClF_6^+
I_2^+	Br_3^+	I_5^+	BrF_2^+	BrF_4^+	BrF_6^+
also	I_3^+	(also	FCl_2^+	IF_4^+	IF_6^+
I_4^{2+}		I_7^+)	$BrCl_2^+$		
			ICl_2^+		
			IBr_2^+		

pentagonal bipyramid. The remaining compound is iodine trichloride which has the dimeric structure shown in Figure 17.58.

A variety of polyhalide ions is known. These are all prepared by the general method of adding halogen or interhalogen to a solution of the halide. The examples with three halogen atoms are linear with the heaviest atom in the middle in the mixed types (although it is not known whether $ClBr_2^-$ and BrI_2^- are symmetrical or obey this rule and have structures I–I–Br or Br–Br–Cl). The three-halogen ion, $ClBrI^-$ forms a disordered crystal so its structure is not definitely known. The data are best fitted by the arrangement Br–I–Cl$^-$ with I–Cl longer than Br–I (291 pm and 251 pm respectively). These anions are AB_2L_3 structures, while the three-atom cations like ClF_2^+ have an electron pair less and are V-shaped. Established atom sequences include F–Cl–F$^+$ and Cl–Cl–F$^+$. The AB_4 compounds are probably all planar ions, with two lone pairs in the remaining octahedral positions. The structures of ICl_4^- and

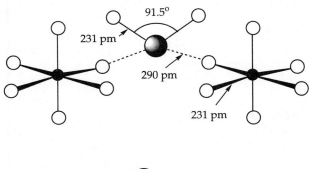

91.5°
231 pm
290 pm
231 pm

● Sb ◐ I ○ Cl

FIGURE 17.57 *The structure of* ICl_2SbCl_6

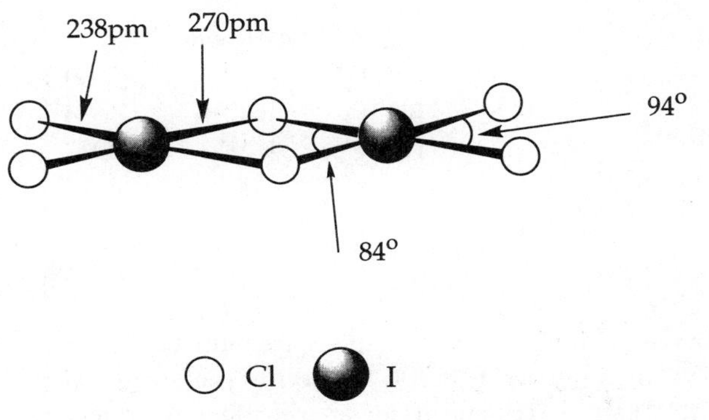

238pm 270pm 94° 84°

○ Cl ◐ I

FIGURE 17.58 *The structure of iodine trichloride,* I_2Cl_6

BrF_4^- have been determined and they are planar. The structure of one of the AB_4^+ cations, IF_4^+, with one pair of electrons fewer, has also been determined. This has the same form as SF_4, which is isoelectronic with it, and is a trigonal bipyramid with a lone pair in an equatorial position. IF_6^+ and IF_6^-, which again differ by one unshared electron pair, present an interesting structural pair. The cation, which has only the six-bond pairs around the iodine, is a regular octahedron as expected. The anion, with a lone pair in addition, is an example of AB_6L, and resembles some other fluorides with this same number of electrons in forming a distorted octahedron. That is, the lone pair is sterically active, as in XeF_6 or SeF_6^{2-}. In contrast, however, a recent X-ray structure determination on $Cs^+BrF_6^-$ shows that the BrF_6^- ion in this species is an almost perfect regular octahedron. In the analogous $[Me_4N]^+[ClF_6]^-$ the vibrational spectrum is very similar to that of BrF_6^- so the Cl species also has a regular octahedral geometry. In these cases the lone pair on the central atom is sterically inactive and presumably lies in an s orbital which is poorly screened from the nucleus.

The structure of IF_8^- has recently been determined and found to have a square antiprismatic geometry which is overall very similar to the structure of the isoelectronic species XeF_8^{2-} (see section 17.9.4).

The other type of polyhalide is limited to iodine and the structures of I_5^-, I_7^-, I_8^{2-}, and I_9^- are all irregular chains. These compounds are only stable in presence of large cations such as cesium or ammonium and alkylammonium ions. The structures are shown in Figure 17.59. The shorter I–I distances are similar to those in iodine but the longer ones correspond to only weak interactions, and all the polyiodides may be regarded as composed of I^-, I_2, and I_3^- groups weakly bonded together. None of the polyiodides

survive in solution and they go to iodide, iodine, and triiodide. The recently discovered mixed species $I_2Cl_3^-$ and $I_2Cl_2Br^-$ have bent structures like I_5^-.

There are also a number of compounds which contain both halogen-oxygen and halogen-halogen bonds. These are listed in Table 17.20, along with their ions. Syntheses include fluorination of the corresponding anion

$$\text{e.g.} \quad ClO_4^- + HSO_3F \rightarrow ClO_3F$$

reaction of the interhalogen with OF_2

$$ClF_5 + OF_2 \rightarrow ClOF_3$$

or oxidations as

$$ClO_2F + ClO_2 \rightarrow ClO_3F$$

Cations are formed by reactions with fluoride ion acceptors such as MF_5 or BF_3 while addition of F^- gives anions. Most of the species are strong fluorinating and oxidizing agents though perchloryl fluoride, ClO_3F, is relatively unreactive. The species are relatively stable thermally except for ClOF, the bromine compounds usually being the least stable.

The structures of selected oxyfluoro-halogens are shown in Figure 17.60 and are those predicted by VSEPR considerations. For IO_2F_3 the structure is the trimer shown in Figure 17.61. Polymerization also occurs in place of cation formation when MF_5 species are added to IO_2F_3.

The structure of IOF_6^- is of particular interest since it is a seven-electron pair species and one which therefore could adopt one (or more) of a number of possible structures close in energy—see section 4.2.7 for a discussion on the VSEPR of such species. An X-ray crystallographic study has shown IOF_6^- to have a pentagonal bipyramidal structure with the oxygen atom in an apical position.

TABLE 17.20 Halogen oxyfluorides and their ions

Oxidation state								
VII		*oxypentafluoride*		*dioxytrifluoride*			*perhalylfluoride*	
		$ClOF_5$		$ClO_2F_2^+$ ClO_2F_3 $ClO_2F_4^-$			ClO_3F	(a)
				BrO_2F_3			BrO_3F	
		IOF_5 IOF_6^-	(b)	$[IO_2F_3]_3$ $IO_2F_4^-$			$(IO_3F)_n$	$(n=4?)$
V		*oxytrifluoride*					*halylfluoride*	
	$ClOF_2^+$	$ClOF_3$	$ClOF_4^-$			ClO_2^+	ClO_2F	$ClO_2F_2^-$
	$BrOF_2^+$	$BrOF_3$	$BrOF_4^-$			BrO_2^+	BrO_2F	$BrO_2F_2^-$
	IOF_2^+	IOF_3	IOF_4^-			$(IO_2^+)_n$	IO_2F	$IO_2F_2^-$
III				*oxyfluoride*				
				[ClOF]				
				(unstable)				

The commonly-used trivial names are indicated.

  (a) the isomer $(FO)ClO_2$, chloryl hypofluorite exists but is less stable.

  (b) $IO_2F_3 . MF_5$ etc. are oxygen-bridged polymers.

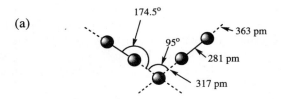

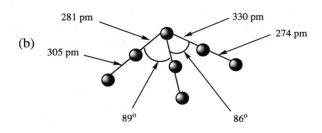

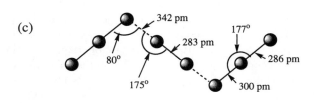

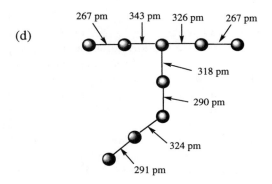

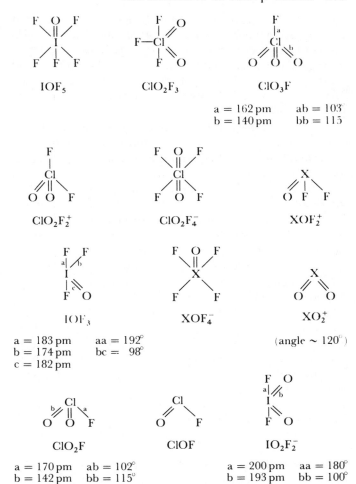

IOF₅ ClO₂F₃ ClO₃F

a = 162 pm   ab = 103°
b = 140 pm   bb = 115

ClO₂F₂⁺ ClO₂F₄⁻ XOF₂⁺

IOF₃ XOF₄⁻ XO₂⁺

a = 183 pm   aa = 192°
b = 174 pm   bc = 98°
c = 182 pm

(angle ∼ 120°)

ClO₂F ClOF IO₂F₂⁻

a = 170 pm   ab = 102°   a = 200 pm   aa = 180°
b = 142 pm   bb = 115°   b = 193 pm   bb = 100°

FIGURE 17.60 *Structures of oxyfluoro-halogens*
Exact determinations are those for which values of parameters are given. The remainder are compatible with nmr and vibrational spectra.

FIGURE 17.59 *Structures of polyiodide ions:* (a) I₅⁻, (b) I₇⁻, (c) I₈²⁻, (d) I₉⁻

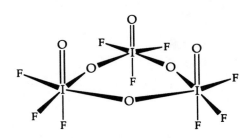

FIGURE 17.61 IO₂F₃
The suggested structure is a *cis* bridged trimer.

Other electronegative groups may be bonded to halogens in a similar way to F and O. Thus treatment of BrF_5 with sodium nitrate yields $BrOF_4^-$, but if $LiNO_3$ is used, $BrOF_3$ is evolved and bromine(I) nitrate remains, presumably because the small lithium ion is insufficient to stabilize the large $BrOF_4^-$ anion. The structure is covalent, $OBrONO_2$, and the Cl analogue is the same. Action of O_3 yields O_2BrONO_2, a bromine(V) compound. It is probable that other oxyion products, especially with Cl or Br in positive oxidation states, are similar covalent species.

Although most of the halogen compounds in positive oxidation states are best described as covalent, there is limited evidence available to support an ionic formulation, especially for iodine(III) compounds such as the acetate,

$I(OCOCH_3)_3$, and phosphate, IPO_4, and the fluosulfonate, $I(SO_3F)_3$. If a saturated solution of iodine triacetate in acetic anhydride is electrolysed, iodine is found at the cathode and, when a silver cathode is used, AgI is formed and current is used as required by the equation

$$I^{3+} + 3e^- + Ag = AgI.$$

Although no other structural evidence is available, these observations indicate that the compounds could be formulated as containing the I^{3+} cation, though such an ion would be expected to interact strongly with the anion and the solids are not to be regarded as simple salts. It is also possible to formulate the oxides I_4O_9 and I_2O_4 as $I^{3+}(IO_3^-)_3$ and $(IO)^+(IO_3^-)$. A related polymeric iodine–oxygen cation $(I_3O_6)_n^{n+}$ has recently been described which has the structure shown in Figure 17.62.

A large number of well-characterized compounds are found which contain the I^+ ion stabilized by coordination. For example, the pyridine complexes $(Ipy_2)^+X^-$ are known for a wide variety of anions, and other lone pair donors coordinate similarly. In all these complexes the iodine appears at the cathode on electrolysis. Chlorine and bromine form analogous, coordinated X^+ species. In these compounds, the arrangement of the halogen and ligands is linear, $L\!-\!X\!-\!L^+$.

One system which was thought for several years to contain the I^+ cation has recently been reformulated. This is the blue solution formed by iodine dissolved in oleum and other strong acids (compare section 6.9). The blue species is also formed by the oxidation of I_2 in IF_5. It was shown, mainly by work in HSO_3F, that the blue species is not I^+ but I_2^+. Magnetic, conductivity and freezing point depression measurements all fit better for the I_2^+ species. Addition of further iodine gives rise to the I_3^+ and I_5^+ ions. Both the I_2^+ and the I_5^+ ions tend to disproportionate to the I_3^+ species. Detailed study of the freezing-point depression shows the equilibrium

$$2I_2^+ = I_4^{2+}$$

which lies well to the right at low temperatures. This I_4^{2+} ion can also be synthesized directly in superacid systems.

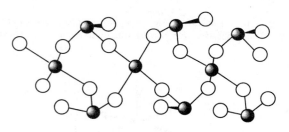

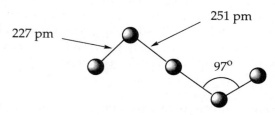

● I   ○ O

FIGURE 17.62 *The structure of the polymeric cation* $(I_3O_6)_n^{n+}$
Two types of iodines are observed: square-planar iodine(III) and pyramidal iodine(V) bridged by oxygen atoms, forming a chain-like structure.

227 pm   251 pm   97°

FIGURE 17.63 *The structure of the* Br_5^+ *ion*

Work with superacid systems has similarly demonstrated the existence of Br_2^+ and Br_3^+. The Br_5^+ ion has also been recently described. Its structure has been found to be a zigzag chain of bromines, as shown in Figure 17.63. The most base-sensitive ions Cl_2^+ or ClF^+ cannot be formed, even in these media, contrary to initial reports. The Cl_3^+ cation does exist and has been isolated as $(Cl_3^+)(AsF_6^-)$.

It is interesting that a structural study of Br_2^+ gives $Br\!-\!Br = 213$ pm, compared with 227 pm in neutral Br_2. This shortening is expected as the electron which is removed in forming the ion comes from an antibonding orbital.

17.8.3 *The* $-I$ *oxidation state*

In this state, the chemistry of the halogens is well-known and many compounds have already been discussed in the earlier sections. Ionic compounds with X^- are formed with the s elements, except beryllium, and in the II and III oxidation states of the transition elements. The change from ionic to covalent character comes further to the left in the Periodic Table for the heavier halogens than for fluorine, as expected from the electronegativities. The region of change is marked by the occurrence of polymeric structures such as AlF_3, which is a giant molecule with $Al\!-\!F$ bonds which can be described as intermediate between ionic and covalent.

In the covalent halides, the main differences between the compounds of the different halogens may be ascribed to the differences in size and reactivity of fluorine compared to the heavier halides. This is often shown in the formation of hexavalent fluorides and tetravalent halides of the rest of the Group, or, similarly, by the formation of six-coordinated instead of four-coordinated complexes. Examples include the formation of SF_6 but only SCl_4, or of CoF_6^{3-} but only $CoCl_4^-$.

In this oxidation state, the halogens show their strongest resemblance and fluorine fits into place as the most reactive of all. This high reactivity derives in part from the relative weakness of the $F\!-\!F$ bond in fluorine (similar effects are found for the $O\!-\!O$ and $N\!-\!N$ single bonds in hydrogen peroxide and hydrazine). The heat of dissociation of F_2 is only 129.3 kJ mol^{-1}, compared with 237.8 kJ mol^{-1} for Cl_2, 188.9 kJ mol^{-1} for Br_2 and 147.9 kJ mol^{-1} for I_2. A value extrapolated from those of the heavier halogens would be about twice the observed F_2 value. The decrease is considerable when it is recalled that most element-halogen bond strengths are in the order $F > Cl > Br > I$.

17.8.4 *Pseudohalogens or halogenoids*

A number of univalent radicals are found which resemble the halogens in many of their properties, and the name pseudohalogen has been given to these. For example, consider the cyanide ion, CN^-. This resembles the halides in the following respects:
(i) it occurs as $(CN)_2$—cyanogen—and forms an HX acid, HCN
(ii) it forms insoluble salts with Ag^+, Hg^+ and Pb^{2+}
(iii) it also gives complex ions of similar formulae to the halogens, e.g. $Co(CN)_6^{3-}$ or $Hg(CN)_4^{2-}$

(iv) it forms covalent compounds and ionic compounds with similar ranges of elements as the halogens
(v) it gives 'interhalogen' compounds such as ClCN or ICN
The analogy should not be pressed too far. Thus, the CN^- ion has different donor and acceptor properties from the halogens, so its transition metal complexes differ in stability and reactions.

Other radicals with similar properties include cyanate, OCN, thiocyanate SCN, selenocyanate SeCN, and azidocarbondisulfide, $SCSN_3$: all these form R_2 molecules. In addition, the ions azide, N_3^-, and tellurocyanide, $TeCN^-$, act as pseudohalides although no molecule, R_2, is formed. Despite their explosive nature, the halogen azides, XN_3, have found application in the synthesis of organic azides and related nitrogen compounds. Bromine azide has a zigzag molecular structure, Br–N–N–N, with Br–N = 190 pm, the central N–N = 123 pm and the outer N–N = 113 pm. The BrNN angle is 110° and the NNN one = 171°—all parameters very similar to those of BrNCO (Br–N = 186 pm, BrNC = 118°, NCO = 172°). In contrast, the solid-state structure of IN_3 consists of I–N–I–N– chains with each azide bridging two iodines through a single terminal nitrogen. Other iodine–azide compounds are the $I(N_3)_2^+$ cation (analogous to ICl_2^+) and $I_2N_3^+$, which has been proposed to have an I–I–N–N–N chain structure.

17.9 The helium Group

The elements of the helium Group are termed the rare gases, or the inert or noble (implying unreactive) gases. None of these terms is now particularly appropriate. The elements are rare only by comparison with the very abundant components of the atmosphere, oxygen and nitrogen. In terms of absolute composition of the crust and atmosphere, the lighter elements of this Group are common. Neither are the gases inert, as has recently been shown. Probably the term noble gases is least unsatisfactory but no general agreement has been reached as yet. The IUPAC recommended name is rare gas and this will be used here.

The rare gases occur as minor components of the atmosphere, ranging in abundance from argon (0.9 per cent by volume) to xenon (9 parts per million). Helium also occurs in natural hydrocarbon gases in some oilfields and is found

occluded in some rocks. In both cases, this helium probably arises from α-particles emitted during radioactive decay. The heaviest member, radon, is radioactive and is found in uranium and thorium minerals, where it is produced in the course of the decay of the heavy elements. There is increasing public concern over the harmful effects of radon gas, which is found naturally, accumulating in mines and the basements of houses, etc. The other elements are usually produced by fractional distillation of liquid air. The main properties are given in Table 17.21. The low boiling points and heats of vaporization reflect the very low interatomic forces between these monatomic elements: the rise in these values with atomic weight shows the increasing polarizability of the larger electronic clouds.

The main isotope of helium is helium-4, and if this is cooled below 2.178 K surprising properties appear. In this form, called helium-II, the viscosity is too low to be detected, the liquid becomes superfluid, and it appears to flow in thin films without friction and is able to flow uphill from one vessel to another. No full theoretical explanation of these phenomena is yet available.

Until 1962, all attempts to form compounds of the rare gases had failed. Transient species, such as HHe, had been observed in electric discharges but these had very short lifetimes. The rare gases were also found in solids, such as $3C_6H_4(OH)_2.0.74Kr$, but these are not true compounds but *clathrates*. A clathrate is formed when a compound crystalizes in a rather open 'cage' lattice which can trap suitably sized atoms or molecules within them. An example is provided by para-quinol (p-$C_6H_4(OH)_2$; p-dihydroxybenzene) which, when crystallized under a high pressure of rare gas, forms an open, hydrogen-bonded cage structure which holds the rare gas atoms in compounds like the krypton one above, or like $3C_6H_4(OH)_2.0.88Xe$. When the quinol is dissolved or melted, the rare gas escapes. That the clathrates are not true compounds is shown by the large variety of atoms and molecules which may enter the cages. Not only are quinol clathrates formed by krypton and xenon, but also by O_2, NO, methanol, and many others. The only requirement is that the clathrated species should be small enough to fit the cages and not so small that it can diffuse out: thus helium and neon are too small to form clathrates with p-quinol. Other

TABLE 17.21 Properties of the rare gases

Element	Symbol	B.p. (K)	Heat of vaporization (kJ mol⁻¹)	Ionization potential (kJ mol⁻¹)	Uses
Helium	He	4.18	0.092	2371	Refrigerant at low temperatures: airships.
Neon	Ne	27.1	1.84	2080	Lighting.
Argon	Ar	87.3	6.27	1520	Inert atmosphere for chemical and technical applications.
Krypton	Kr	120.3	9.66	1359	
Xenon	Xe	166.1	13.68	1170	
Radon	Rn	208.2	17.99	1037	Radiotherapy.

compounds give clathrates with the rare gases. In particular, the reported hydrates of the rare gases are clathrates of these elements in ice, which crystallizes in an unusually open cage form. Although all clathrates do not involve hydrogen-bonded species, for example the benzene clathrate, $Ni(CN)_2$. $NH_3.C_6H_6$, clathrate formation by hydrogen-bonded molecules is common, as open structures are more readily formed.

17.9.1 Xenon compounds

All other attempts to form rare gas compounds, including many studies of possible donor action to yield compounds such as $Xe \rightarrow BF_3$, failed until 1962. Then Bartlett reported that xenon reacted with PtF_6 to form a compound which he formulated as $Xe^+(PtF_6)^-$. He was led to try this reaction after his discovery of $O_2^+(PtF_6)^-$—see section 15.8—by the consideration that the ionization potential of xenon was close to that $(914\,kJ\,mol^{-1})$ of the O_2 molecule, so that if PtF_6 could oxidize O_2 to O_2^+, there was the chance of its oxidizing Xe to Xe^+. Further exploration of this field was extremely rapid. A fuller investigation of the reaction led to the discovery of XeF_4 in the second half of 1962. Interest was then concentrated on simple fluorides, oxides, oxyfluorides and species present in aqueous solution. The compounds of these classes are listed in Table 17.22. The existence of $XeCl_2$ and $XeCl_4$ has been indicated in Mössbauer experiments using ICl_2^- or ICl_4^- as sources. These compounds decompose below room temperature, but $XeCl_2$ was found at $20\,K$ in the products formed by photolysis or by passing Xe and Cl_2 through a microwave discharge. Spectroscopic studies suggest that $XeCl_2$ is linear, though more recent studies have questioned the very nature of this molecule. It has been suggested that $XeCl_2$ is in fact a van der Waals molecule either linear, of the type $Xe \cdots Cl-Cl$ (analogous to the known van der Waals molecules $HeCl_2$, $NeCl_2$ and $ArCl_2$), or a T-shaped molecule where the Xe atom bridges the $Cl-Cl$ bond.

Liquid xenon is finding increasing use as an 'inert' solvent in which oxidative addition reactions (see sections 13.9 and 15.7) of highly reactive metal complexes with the $C-H$ bonds of alkanes (usually thought to be fairly inert themselves!) can be investigated. As with the earlier discussion on non-coordinating anions (section 17.5.5), a completely non-coordinating solvent is also somewhat of a 'Holy Grail' in chemistry. This is demonstrated by the detection, at low temperatures, of short-lived donor complexes of the type $X-M(CO)_5$, where X is either Kr or Xe, and M is either Cr or W. The strength of the $W-Xe$ bond in $Xe-W(CO)_5$ has been determined to be around $35\,kJ\,mol^{-1}$ and the complex has a lifetime of about 1.5 minutes at $170\,K$ in liquid Xe.

17.9.2 Preparation and properties of simple compounds

Preparations of the fluorides are all by direct reaction under different conditions. Thus a $1:4$ mixture of xenon and fluorine passed through a nickel tube at $400\,°C$ gives XeF_2, a $1:5$ ratio heated for an hour at 13 atmospheres in a nickel can, at $400\,°C$, gives XeF_4, while heating xenon in excess fluorine at 200 atmospheres pressure gives XeF_6. Since these original routes were discovered, new routes have been added. In fact, exposing a mixture of xenon and fluorine to sunlight provides a route to XeF_2 which crystallizes on the walls of the reaction vessel. In a recent report the reaction of Xe and F_2 over a hot filament gives goods yields and purity of XeF_6, and XeF_4 can be correspondingly prepared from O_2F_2 and Xe. In the presence of a fluoride ion acceptor such as BF_3, AgF_2 is also a strong enough oxidant to convert xenon gas to XeF_2.

Most other compounds result on hydrolysis of the fluorides:

$$XeF_6 + H_2O \rightarrow XeOF_4 + 2HF$$

with an excess of water,

$$XeF_6 + 3H_2O \rightarrow XeO_3 + 6HF$$
$$\text{or} \quad XeF_4 + H_2O \rightarrow Xe + O_2 + XeO_3 + HF,$$

here, about half the $Xe(IV)$ disproportionates to Xe and $Xe(VI)$ while the other half oxidizes water to oxygen, form-

TABLE 17.22 Simple rare gas compounds

Oxidation State	Fluorides	Oxides	Oxyfluorides	Acids and Salts
II	KrF_2 XeF_2 RnF_2			
IV	XeF_4		$(XeOF_2)$	
VI	XeF_6	XeO_3	$XeOF_4$ XeO_2F_2	$HXeO_4^-$ $Xe(OH)_6$? Ba_3XeO_6 (and other salts)
VIII		XeO_4	XeO_3F_2 (XeO_2F_4)	$HXeO_6^{3-}$ Ba_2XeO_6 (and similar salts)

Compounds in brackets are unstable at room temperature.
$XeCl_2$ exists at low temperatures. $XeCl_4$ and $XeBr_2$ exist transiently.

ing xenon. Xenon trioxide is hydrolysed in water, probably according to the equilibrium

$$XeO_3 \rightleftharpoons Xe(OH)_6$$

Interaction of XeO_3 and $XeOF_4$ gives XeO_2.

The xenon(IV) oxyfluoride, $XeOF_2$, is formed in a low temperature matrix by reacting Xe with OF_2, or by the low temperature hydrolysis of XeF_4. At about $-20\,^{\circ}C$, it disproportionates

$$XeOF_2 \rightarrow XeO_2F_2 + XeF_2$$

XeO_2F_2 itself gives $XeF_2 + O_2$ on standing at room temperature.

Solution of xenon trioxide, or hydrolysis of the hexafluoride in acid, gives xenates such as Ba_3XeO_6. In neutral or alkaline solution, xenates rapidly disproportionate to perxenates, xenon(VIII). Thus is observed an overall reaction such as

$$2XeF_6 + 4Na^+ + 16OH^-$$
$$\rightarrow Na_4XeO_6 + Xe + O_2 + 12F^- + 8H_2O$$

The addition of acid to a perxenate yields the tetroxide

$$Ba_2XeO_6 + 2H_2SO_4 \rightarrow XeO_4 + 2BaSO_4 + 2H_2O$$

The reaction mixture must be kept cold as explosive decomposition of XeO_4 readily occurs. At room temperature, decomposition to $Xe + O_2$ occurs readily.

Reaction of XeF_6 with solid Na_4XeO_6 yields the xenon(VIII) oxyfluoride, XeO_3F_2, together with much $XeOF_4$. XeO_3F_2 is volatile and was characterized by its mass spectrum. $XeOF_4$ itself shows some interesting reactions. With CsF, it forms $XeOF_5^-$ which has a distorted octahedral structure (compare XeF_6 below). With an excess of $XeOF_4$ a second product forms, $[(XeOF_4)_3F]^-$ where the central F is bonded to the three Xe atoms in a shallow pyramid (angle $116 \cdot 5^{\circ}$) which a very weak bond of 262 pm which contrasts with 190 pm for the $Xe-F$ bond within the $XeOF_4$ units. An improved procedure for the synthesis of $XeOF_4$ from XeF_6 employs readily prepared POF_3 as the oxidising agent. This avoids the generation of explosive XeO_3 and the byproduct PF_5 is readily removed from the $XeOF_4$ product because of its higher volatility.

The fluorides are all strong oxidizing and fluorinating agents, and most reactions give free xenon and oxidized products:

$$XeF_4 + 4KI \rightarrow Xe + 2I_2 + 4KF$$
$$XeF_2 + H_2O \rightarrow Xe + \tfrac{1}{2}O_2 + 2HF$$
$$XeF_4 + Pt \rightarrow Xe + PtF_4$$
$$XeF_4 + 2SF_4 \rightarrow Xe + 2SF_6$$

The three xenon fluorides are all formed in exothermic reactions (heats of formation of the gases are about 85, 230, and 335 kJ mol^{-1} for XeF_2, XeF_4, and XeF_6 respectively). The trioxide is endothermic by 402 kJ mol^{-1}—largely due to the high dissociation energy of O_2. The bond energies of the $Xe-F$ bonds in the fluorides range from 120 to 134 kJ mol^{-1} from the difluoride to the hexafluoride, a difference which is not far outside the experimental errors. The $Xe-O$ energy is about 85 kJ mol^{-1}.

The fluorides are all white volatile solids with the volatility increasing from the difluoride to the hexafluoride. The trioxide is also white but non-volatile, while the tetroxide, XeO_4, is a yellow, volatile solid which is unstable at room temperature. The oxyfluorides are also white, volatile solids.

While all the early compounds of Xe showed bonds only to F or O, an increasing number of $Xe-N$ and $Xe-C$ bonded species are now known, although these invariably require electronegative groups, often fluorines, on the ligand. As an illustration of a $Xe-N$ bonded species we note the XeF_2 derivative, $FXeN(SO_2F)_2$, which has a linear $F-Xe-N$ unit and a pyramidal N bonded to 1 Xe and 2 $S(O)_2F$ units. A number of nitrile–krypton or – xenon compounds of the type $RCN-NgF^+$ ($Ng = Kr$, Xe) have recently been prepared. In these, the ligand again generally contains an electron-withdrawing fluorinated R group, such as CF_3 or C_2F_5. The unsubstituted krypton ion $HCN-KrF^+$ has been reported and, in addition, a theoretical study has suggested that the argon analogue, $HCN-ArF^+$, should be stable and experimentally accessible. These compounds have the linear structures predicted by VSEPR.

In recent years there has been an upsurge of interest in the synthesis of compounds containing xenon–carbon bonds. In these, the organic group again generally has a high level of fluorine substitution. Some of the compounds of this type which have been prepared are shown in Figure 17.64. A number of pentafluorophenyl compounds have been described, including $MeCN-Xe-C_6F_5^+$, Figure 17.64a. These have the expected approximately linear geometry at Xe with a $Xe-C$ bond length of 209 pm. Studies using fluorophenyl groups with lower numbers of fluorine atoms (replacing them by hydrogens) results in a decrease in stability of the $Xe-C$ bonds. Other interesting, new organic derivatives of Xe include the alkenyl and alkynyl derivatives shown in Figure 17.64b and c. It is also interesting to note that a pentafluorophenyl xenon derivative is the first example to contain a xenon–carboxylate group, in $C_6F_5-Xe-OC(O)C_6F_5$, Figure 17.64d.

17.9.3 Structures

The structures of most of the xenon compounds are those predicted by the simple electron pair considerations outlined in Chapter 4, and correspond to the structures of isoelectronic iodine compounds. Some structures are shown in Figure 17.65. XeO_2F_2 has a structure like $IO_2F_2^-$ (Figure 4.6) with a nearly linear $F-Xe-F$ arrangement and an OXeO angle of 106°. XeF is 190 pm and XeO equals 171 pm. XeO_4 is probably a regular tetrahedron, XeO_6^{4-} is octahedral, while the spectrum of $XeOF_2$ shows it is a further example of the T-shape expected for AB_3L_2 species:

(a)

209 pm
268 pm
174.5°

(b)

(c)

R————C≡≡C————Xe⊕

R = C₂H₅, t-butyl, SiMe₃

(d)

237 pm

FIGURE 17.64 *The structures of some compounds containing xenon–carbon bonds*

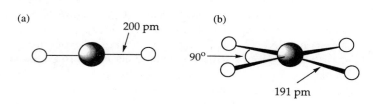

(a)

200 pm

(b)

90°
191 pm

(c)

176 pm
103°

(d)

92°
173 pm
190 pm

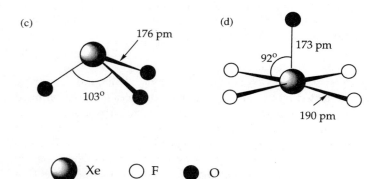

Xe   ○ F   ● O

FIGURE 17.65 *The structures of xenon compounds:* (a) XeF_2, (b) XeF_4, (c) XeO_3, (d) $XeOF_4$
These structures are the same as those of the iodine analogues ICl_2^-, IF_4^- and IO_3^-.

The most interesting structural problem is that presented by XeF_6. A monomer would have the AB_6L structure and would not be expected to be a regular octahedron. In the gas phase, a distorted structure is clearly indicated and three different molecular shapes may be present in equilibrium.

However, XeF_6 appears to exist in a polymeric form in the solid and liquid phases. In solution, nmr studies indicate a tetramer, $(XeF_6)_4$, with equal interaction at room temperature between all the F atoms and the four Xe atoms. This can happen only if the fluorines are rapidly exchanging.

The structure of the solid is very complex with the unit cell containing 144 XeF_6 units. These are present as 24 tetramers and 8 hexamers. The configuration of both aggregates is most simply described as XeF_5^+ and F^-. The XeF_5^+ units are square pyramids and are linked together by bridges … F_5Xe^+ … F^- … XeF_5^+ … In the tetramers, the F^- bridges pairs of XeF_5^+ units to form a puckered ring, with unsymmetric bridges of 223 and 260 pm $Xe—F^-$ distances and $Xe(F^-)Xe$ angles of 121°. In the hexamer, each F^- bridges three XeF_5^+ units lying symmetrically with $Xe—F^-$ equal to 256 pm and an angle of 119°. (Compare these distances with the bonded XeF values given in Figure 17.67d.) Such bridging would, of course, allow the ready fluorine exchange observed in solution. We note that XeF_5^+ is an AB_5L species and the square pyramid is the shape expected for sterically active lone pair.

17.9.4 *Reactions with fluorides*

Xenon fluorides and oxyfluorides react with the fluorides of many elements to give a variety of species. For example, in BrF_5 as solvent

$$XeF_2 + SbF_5 \rightarrow nXeF_2 \cdot mSbF_5$$

$$(n:m = 1:1, 1:2, 2:1, 1:1\cdot5, 1:6)$$

By direct reaction, or using SbF_5 as solvent, the species $XeF_4 \cdot SbF_5$, $XeF_4 \cdot 2SbF_5$, $XeOF_4 \cdot SbF_5$, $XeOF_4 \cdot 2SbF_5$ and $XeO_2F_2 \cdot 2SbF_5$ may be formed. While SbF_5 forms the widest range of species, many other MF_3, MF_4, MF_5 and MOF_4 species react similarly, especially with XeF_2. Under strong fluorinating conditions, $XeF_6 \cdot AuF_5$ and $2XeF_6 \cdot AuF_5$ are formed (containing the unusual Au(V) oxidation state).

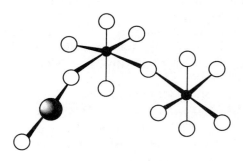

Xe   ○ F   ● Sb

FIGURE 17.66 *Structure of the adduct* $XeF_2 \cdot 2SbF_5$
The structure is intermediate between that expected for a full covalent Xe—F—Sb bridge and that corresponding to $XeF^+Sb_2F_{11}^-$. Mean non-bridging Sb—F bond length = 183 pm.

The crystal structures of a number of these species have been determined, and they are best described as containing the fluoroxenon cation formed by transferring one F^-. The cation is then weakly bonded to the anion through a fluorine bridge. The 2:1 adducts contain more complex cations with $Xe-F-Xe$ bridges while 1:2 adducts contain dimeric anions such as $Sb_2F_{11}^-$ whose structure is two octahedra linked by a shared F (Figure 17.66). Further, weaker interactions occur in some crystals, so that the species contain the cations listed in Table 17.23. The structures of some examples are included in Figure 17.67. The vibrational spectra are compatible with T-shaped XeF_3^+, pyramidal XeO_2F^+, the AB_4L type with equatorial oxygen for $XeOF_3^+$, and square pyramidal $XeO_2F_3^-$. The crystal structure shows XeO_3F^- is a polymer of XeO_3 units (similar to xenon trioxide) bridged through F^- (see Figure 17.67e). The more complex xenon(VI) ion, $Xe_2F_{11}^+$ consists of two square pyramids joined through one of the base corners and this is linked to the AuF_6^- ion in $(Xe_2F_{11})^+(AuF_6)^-$ by two more shared fluorines giving an Xe_2Au triangle with an F in each edge. The $Au-F\ldots Xe$ links are very unsymmetrical. A further example of the tendency of Xe(VI) to form complex structures is provided by $XeF_5^+AsF_6^-$. The XeF_5^+ ion is linked to the AsF_6^- one in a very unsymmetric bridge $Xe\ldots F-As$ $(Xe\ldots F = 265$ pm, $F-As = 173$ pm) and this $XeF_5^+AsF_6^-$ unit forms a weak dimer by two pairs of $Xe(F)_2As$ bridges with long $Xe-F$ distances of 270 and 281 pm.

All these weak $Xe\ldots F$ interactions should be compared with the van der Waals non-bonding contact distance of 350 pm.

It has been possible to isolate the high-coordination complex fluorides as NF_4^+ salts. Crystal structures show that XeF_8^{2-} is a regular square antiprism (compare Figure 15.5b)

while XeF_7^- has a similar structure with one position empty. Thus the seven-coordinate species shows a sterically active lone pair while the eight-coordinate one does not, presumably as a size effect.

A crystal structure has recently been carried out on the XeF_5^- ion, and it was found to have a pentagonal planar structure with the two lone pairs adopting *trans*-axial positions.

One of the few compounds to have a bond to Xe from an element other than O or F is formed in the reaction

$$O_2BF_4 + Xe \xrightarrow{-100\,°C} BXeF_3 \xrightarrow{-30\,°C} Xe + BF_3$$

A planar structure based on XeF_2 with a BF_2 unit replacing one F is indicated

$$F-Xe-B\begin{smallmatrix}\diagup F\\[2pt]\diagdown F\end{smallmatrix}$$

In reactions with strong acids, HOY, stepwise substitution of F in XeF_2 occurs: e.g.

$$XeF_2 + HOY \rightarrow FXeOY \rightarrow Xe(OY)_2$$

This is known for $Y = TeF_5, SeF_5, SO_2F, SO_2CF_3, SO_2CH_3, ClO_2$ and CF_3CO and possibly for NO_2. Some analogous Xe(IV) and Xe(VI) species also occur. A related synthesis uses the anhydride $(YO)_2O$ as in

$$XeF_2 + P_2O_3F_4 \rightarrow FXeOPOF_2 \rightarrow Xe(OPOF_2)_2$$

The structure of FXeOY is not dissimilar to the $FXe^+MF_6^-$ species, as shown in Figure 17.67a and b. Similarly, $Xe(OY)_2$ has a linear bridge as in $F_5SeO-Xe-OSeF_5$.

TABLE 17.23 Some rare gas ions

Oxidation State					
II	KrF^+	$Kr_2F_3^+$			
	XeF^+	$Xe_2F_3^+$			
	RnF^+	$Rn_2F_3^+(?)$			
IV	XeF_3^+	$XeOF_3^-$			
	XeF_5^-				
VI	XeF_5^+	$Xe_2F_{11}^+$	$XeOF_3^+$	KrO_2F^+	XeO_3F^-
	XeF_7^-		$XeOF_5^-$	XeO_2F^+	XeO_3Cl^-
	XeF_8^{2-}			$XeO_2F_3^-$	

Notes
1. See text for structures: note especially that there are weak F bonds between cations and fluoro-anions.
2. Counter cations are usually large M^+ e.g. Rb^+, Cs^+, NO^+.
3. Counter anions include MF_6^- (M = P?, As, Sb, V?, Nb, Ta, Ru, Os, Ir, Pt and Au)
   MF_4^- (M = B, Al)
   MF_6^{2-} (M = Ge, Sn, Ta, Hf, Pd)
   $M_2F_{11}^-$ (M = As, Sb, Nb, Ta, Ir, Pt)
   MOF_4^- (M = W).

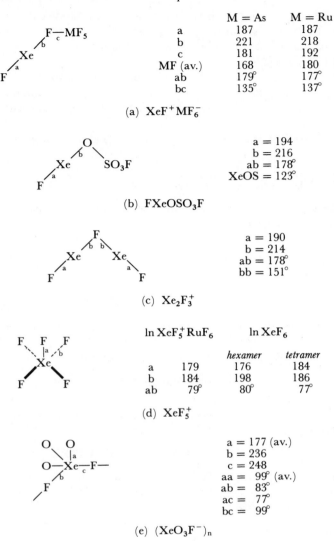

	M = As	M = Ru
a	187	187
b	221	218
c	181	192
MF (av.)	168	180
ab	179°	177°
bc	135°	137°

(a) $XeF^+MF_6^-$

| a = 194 |
| b = 216 |
| ab = 178° |
| XeOS = 123° |

(b) $FXeOSO_3F$

| a = 190 |
| b = 214 |
| ab = 178° |
| bb = 151° |

(c) $Xe_2F_3^+$

	ln $XeF_5^+RuF_6^-$	ln XeF_6	
		hexamer	*tetramer*
a	179	176	184
b	184	198	186
ab	79°	80°	77°

(d) XeF_5^+

| a = 177 (av.) |
| b = 236 |
| c = 248 |
| aa = 99° (av.) |
| ab = 83° |
| ac = 77° |
| bc = 99° |

(e) $(XeO_3F^-)_n$

FIGURE 17.67 *Structures of some xenon-fluorine species* (All bond lengths are in pm).

In the IV state analogous compounds are also becoming established. $Xe(OTeF_5)_4$ is square planar with $Xe-O = 203$ pm. The VI state is represented by $O_2Xe(OTeF_5)_2$ and $OXe(OTeF_5)_4$. The first shows the expected 5-electron pair structure with equatorial oxygens with distinctly weak double bond repulsion ($Xe=O = 173$ pm and an $O=Xe=O$ angle of 106°). The axial single bond, $Xe-OTeF_5$, distance is 202 pm.

Reaction of xenon gas with $XeF^+Sb_2F_{11}^-$ in SbF_5 solvent gives rise to a green colouration which has been assigned to the dixenon cation, Xe_2^+, on the basis of spectroscopic measurements.

17.9.5 *Krypton and radon*

The only simple compound of krypton is KrF_2 (earlier reported as KrF_4). This is thermodynamically unstable and the $Kr-F$ bond energy of about 50 kJ mol^{-1} is the lowest known element-fluorine value. KrF_2 should therefore find use as a very reactive fluorinating agent. The compound is linear,

with $Kr-F$ equal 189 pm, and it decomposes at room temperature at the rate of 10% an hour. Its chemistry is little-explored but the 1:1, 1:2 and 2:1 adducts with MF_5 (M = Sb, Nb and Ta) are reported. These are formulated as the KrF^+ or $Kr_2F_3^+$ salts of MF_6^- or $M_2F_{11}^-$, as with the xenon analogues. These species are rather more stable than KrF_2. Evidence for compounds containing Kr–O bonds comes from $KrO_2F^+Sb_2F_{11}^-$, made like the Xe analogue, and $Kr(OTeF_5)_2$, prepared by the reaction of KrF_2 with $B(OTeF_5)_3$.

KrF_2, often in conjunction with xenon fluorides, finds application as a powerful fluorinating agent. Thus treatment of Ln(III) compounds yields the Ln(IV) species LnF_4, LnF_7^{3-}, and $LnOF_2$, not only for the readily oxidized Ce, but also for Ln = Pr, Nd, Tb and Dy (compare section 11.4).

Extrapolating from Kr through Xe would indicate that radon should form a range of compounds. However, all the isotopes are radioactive, and not much work is reported— RnF_2 is established and the Rn(II) ions, RnF^+ and $Rn_2F_3^+$ are likely.

It is suggested from the ionization potentials (Table 17.21) that argon, neon or helium are unlikely to form compounds in positive oxidation states corresponding to the xenon and krypton compounds above. Any compounds formed by the smaller rare gases are unlikely to follow such an energy-demanding route and fluorides or oxides are likely to be of only transient existence. A possible alternative mode of reaction is provided by the earlier suggestion that the rare gases would act as electron pair donors to strong Lewis acids. In this context it is interesting that a number of XHe^+ species (X = a range of common electron-deficient radicals) have been identified in interstellar clouds. (Recall that He is the second most abundant element in the Universe.) The search for the next derivative is almost certain to be focused on argon chemistry and a recent theoretical calculation has suggested that salts of the type ArF^+ might possibly be stable enough to be isolated.

A calculation reported in 1987, using He as the most difficult case for compound formation, suggested that a number of species might be sufficiently stable to allow isolation, including $(HeCCHe)^+$ and other similar acetylene derivatives. Also quite stable could be the species HeBeO which could be formed by implanting BeO molecules (found in the vapour above solid BeO) into a helium matrix at low temperatures. Argon, krypton and xenon analogues have been recently reported.

It is finally worth noting that the dimeric species He_2 has been identified spectroscopically in the gas phase. The forces holding together the two He atoms are very weak van der Waals forces and, due to the fact that each He atom only posseses two electrons, these interactions are exceedingly weak. The reader will recall (section 3.3) that the molecular orbital description of the bonding in He_2 produces a zero bond order—the weak van der Waals forces 'holding together' the He_2 dimer are present between all types of atoms but are exceedingly weak by comparison to chemical bonds.

17.10 Bonding in Main Group compounds: the use of d orbitals

The bonding in xenon compounds has been the subject of some controversy which raises the wider question of the degree to which the valence shell d orbitals participate in bonding in any compound where more than four electron pairs surround a Main Group atom. The case of XeF_2 is the simplest to discuss. The two ligands occupy axial positions and three lone pairs occupy equatorial positions in a trigonal bipyramid. To accommodate these five electron pairs requires five orbitals which are formed from the s, the three p, and a d orbital on the central atom. But it has been objected that, in the case of xenon, the energy gap between the p and d orbitals is very large, equal to about 960 kJ mol^{-1}, and it is unlikely that the bond energies are sufficient to compensate for the energy required to make use of a d orbital in such a scheme. Instead, three-centre sigma bonds between xenon and fluorine are proposed. The molecular axis is taken as the z-axis, as in Figure 17.68, and the relevant orbitals are the xenon p_z orbital, which contains two electrons, and the p_z orbital on each fluorine which hold one electron each (the other six valency electrons on xenon and fluorine fill the s, p_x, and p_y orbitals on each atom, all non-bonding). The three p_z orbitals can be combined to give three sigma orbitals, centred on the three atoms, one of which is bonding, one non-bonding, and one anti-bonding (Figure 17.68). The four electrons fill the bonding and non-bonding orbitals to give an overall bonding effect. The position is similar to that in the three-centre $B-H-B$ bond in the boranes, except that there are four electrons instead of two to be accommodated (compare section 9.5). The cases of XeF_4 and XeF_6 can be explained similarly using xenon p_x (or p_x and p_y for XeF_6) orbitals as well as p_z in the two cases.

Similar descriptions apply to the interhalogen compounds and ions such as ICl_2^-, ICl_4^-, BrF_5, and so on. All these species can be described either in terms of full electron pair

bonds, plus non-bonding pairs, by using one or two of the d orbitals, or the use of d orbitals may be avoided by using polycentred molecular orbitals at the price of reducing the bond order. It has been reported that calculations using s and p orbitals only are very successful in reproducing the bond angles and bond lengths found in interhalogen compounds.

The theory using three-centred orbitals formed by the p orbitals would predict that XeF_6, and the isoelectronic IF_6^-, should be regular octahedra while the electron pair repulsion theory suggests that these species with seven pairs around the central atoms should be distorted. It is now established that XeF_6, IF_6^- and TeF_6^{2-} are distorted octahedra, while the isoelectronic MX_6^{2-} species (where M = Se, Te, Po and X = Cl, Br, I but *not* F) are undistorted. This again shows the delicate balance which must exist between the various energies making up these two different structures.

Recent theoretical work on the general problem of d orbital contributions to bonding in Main Group compounds has made considerable progress. This is mainly because use of large computers has allowed work to be carried out with fewer initial assumptions and approximations.

(1) In compounds with second row and heavier elements bonded to highly electronegative elements like oxygen and fluorine, there is evidence of substantial d orbital participation in the bonding orbitals. For example, in PF_3, the population of the phosphorus orbitals is calculated to be:

sigma bonding

$3s$ 1·51 electrons $3p$ 1·05 electrons $3d$ 0·13 electrons

pi bonding $3p$ 0·84 electrons $3d$ 0·62 electrons

Thus, the use of the d orbital gives a small stabilization to the sigma bonds, but makes a significant contribution to π bonds.

Notice that this contribution is made in a molecule with only four electron pairs on the phosphorus, that is where there are enough s and p orbitals to hold all the valency electrons. In higher oxidation state fluorides and oxides, such as PF_5, OPF_3, SF_6 or XeF_6, there is distinct d orbital involvement in the bonding. This would seem to be supported experimentally by the existence of IF_7 and non-octahedral MF_6L species.

The test of the calculation is its agreement with experimentally-determined ionization or promotion energies (showing the differences between orbital energies) and the agreement between calculated and experimental dipole moments (which reflect the distribution and spatial density of the electron clouds).

(2) In compounds where the bonded atoms are of lower electronegativity, like C or H, only a very low d orbital population, of the order of a few percent, is found for the central atom. This has little effect on the calculated energies, but does greatly improve the agreement between calculated and observed dipole moments. That is, the introduction of d orbitals into the calculations has, as its main benefit, a better description of the dispersion of the electrons in space.

For further discussion of bonding in compounds of the Main Group elements, see section 18.9.

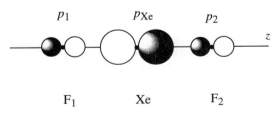

p_1 p_{Xe} p_2

                                                z

F$_1$ Xe   F$_2$

FIGURE 17.68 *The formation of three-centre bonds in* XeF_2: *the constituent atomic orbitals*

If we define the positive direction of the p orbital as for the Xe one drawn, the three-centre combinations give the following orbitals
bonding $\psi = -p_1 + cp_{Xe} - p_2$
non-bonding $\psi = -p_1 + p_2$
antibonding $\psi = -p_1 - c'p_{Xe} + p_2$
where the constants c and c' are of similar size and the expressions are to be regarded as normalized. The bonding orbital and the non-bonding orbital both contain two electrons. As only one pair of bonding electrons exists between three atoms, the $Xe-F$ bond order is only one-half.
XeF_4 may be described similarly using p orbitals in both the x and y directions.

PROBLEMS

The systematic chemistry given in this chapter is best assimilated by working through it in as many ways as possible. Compare behaviour within the same Period, within the same Group, with other species of the same oxidation state or of the same valence electron configuration. See the remarks on the transition element problems and also correlate particularly with Chapters 3 and 4.

1 Illustrations of the topics which could be reviewed are:

(a) each oxidation state (properties, stability)
(b) each coordination number (electron pair counting, stability, extent of distribution)
(c) polymeric species
(d) element − element bonding
(e) $p\pi − p\pi$ bonding
(f) d orbital participation in main Group chemistry
(g) distinguishing behaviour of first element in a Group.

2 The first dissociation constants K_1 of some oxyacids are listed below. Compare with their structures (see also Table 6.2). Find out the values for the other oxyacids of the main Group elements and discuss any anomalies.

K_1 (mol l^{-1})	HNO_2	H_3PO_4	$H_4P_2O_7$	H_3AsO_3	H_3AsO_4
	10^{-3}	7×10^{-3}	10^{-1}	6×10^{-10}	5×10^{-3}

3 Compare the oxidation state free energy diagram for Cl with that of Mn. Discuss the similarities and differences and correlate with the chemistry. Extend this comparison to the other halogens and to rhenium. Carry out a similar analysis for other Groups.

4 Plot the first ionization potential against the atomic radius for the p elements across each of the periods. Comment on trends and anomalies. If the sum of the potentials involving all the valence electrons is plotted similarly, do the same trends emerge?

5 Write an essay on Main Group species with ring structures, including the structures of the elements as well as compounds. Compare ring compounds with the corresponding chain species.

6 Survey the structures found for the chlorides of the p elements. Compare and contrast with those of the same formula found for (a) s elements, (b) d elements, (c) f elements.

7 Recently, $NaPO_3$ has been isolated at low temperatures as a monomeric species. Discuss its likely structure. What compound or compounds would you expect it to form on warming to room temperature?

8 The structural parameters of some sulfuryl halides (compare Figure 17.48a) and thionyl halides (Figure 17.51) are tabulated below. Discuss the variations in bond lengths and angles in terms of VSEPR theory and the relative electron-withdrawing effects of the halogens.

	S=O (pm)	OSO(°)	S−F (pm)	XSX(°)	S−Cl(Br) (pm)
SO_2F_2	138.6	125	151.4	99	
SO_2FCl	140.7	123	153.8	99	196.4
$SOCl_2$	141.8	122		77	198.0
$SOCl_2$	143.9			96	206.8
$SOBr_2$	145			99	(225)

9 $Se_2O_2F_8$ is prepared via $SeOF_4$. Predict possible structures for the monomer and dimer. Compare $Se_2O_2F_8$ with other oxyhalides and discuss whether $Se−O−Se$ or $Se−F−Se$ bridging is the more likely. How do your predictions match with the following observed parameters:

For $Se_2F_{10}O$, $Se−O = 178$ pm, $SeOSe = 98°$,
Se . . . . . Se $= 267$ pm
For $Se_2F_8O_2$, $Se−O = 170$ pm, $SeOSe = 142°$,
Se . . . . . Se $= 321$ pm

10 What structure would you expect for the $Si_4O_{13}^{10-}$ ion, recently isolated as the silver salt? Compare this with the species of overall formula $Si_4O_{12}^{8-}$, $Si_4O_{11}^{6-}$ and $Si_4O_{10}^{4-}$ (compare section 18.6). What formulae would you find for each of the phosphorus analogues?

11 A bulky cation, I_8^{2-}, has an almost planar structure

$$I \overset{a}{-} I \overset{b}{-} I \diagdown_c$$

a = 283 pm ab = 174°
b = 304 pm bc = 132°
c = 339 pm cd = 168°
d = 277 pm

(a) Discuss this in terms of the idea that polyiodides can be seen as associations of I_2 and I_3^- units.
(b) Compare with the structure of Cs_2I_8 shown in Figure 17.59.

12 In the gas phase, ClF_3O shows the following parameters:

lengths (pm) Cl−O = 141, Cl−F(1) = 160, Cl−F(2) = 171
angles (°) OClF(1) = 109, OClF(2) = 95, F(1)ClF(2) = 88 and F(2)ClF(2) = 171.

Discuss these values in terms of the structure predicted by the VSEPR approach. Compare the values with the related compounds shown in Figure 17.60.

13 The ClO_2^+ ion is bent, with an angle of 119° and Cl−O lengths of 141 pm. In ClO_2 the bond is longer, 148 pm, but the angle is similar, 118°. Discuss these observations in terms of the expected bonding.

14 N_2O_5, PCl_5, and Cl_2O_6 share the property of being covalent in the gas phase, but forming an ionic solid. Find other examples, from both the d and p blocks. Discuss the various reasons accounting for this behaviour.

15 Discuss the molecular parameters listed below:

	$M-F_{eq}$	$M-F_{ax}$	Angle (°)
ClF_3	160 pm	170 pm	87·5
BrF_3	172 pm	181 pm	86·2
XeF_3^+	184 pm	191 pm	81

16 One compound containing the XeF_3^+ ion has the SbF_6^- counterion. Discuss the likely preparation. As the $Sb-F$ distance is 191 pm and there is one of the anion F atoms at 241 pm from Xe, discuss secondary interactions in this compound and compare with related compounds.

17 Discuss the parameters found for SeF_4 and $SeOF_2$:

$Se-F(a) = 177·1$ pm, $F-Se-F$ angle(a) $= 100·6°$ for SeF_4
$Se-F(b) = 168·2$ pm, $F-Se-F$ angle(b) $= 169·2°$

$Se-O = 157·6$ pm, $O-Se-F$ angle $= 105°$ for $SeOF_2$
$Se-F = 172·95$ pm, $F-Se-F$ angle $= 92°$

18 Selected Topics in Main Group Chemistry and Bonding

In this chapter, some aspects of the chemistry of the elements of the p block raised in earlier chapters (particularly Chapter 4, 8, and 17) are discussed in a little more depth. Some leading references are cited to allow further reading. (see Appendix A).

As for Chapter 16, only a limited and arbitrary selection of topics is possible. There are many other areas of current intense interest and development, and these are indicated in the general reading list given in Appendix A.

18.1 The formation of bonds between like Main Group atoms

The chemistry of carbon is so enormously rich, extensive, and diverse that it is easy to think that the property of forming bonds between like atoms, called *catenation*, is unique to carbon. Other long-chain species have been known for a long time, for example the plastic form of sulfur is S_∞, but because of the difficulties of handling and identifying such materials, they have often been dismissed as obscure amorphous deposits. More refined, modern experimental approaches have greatly changed this picture. Crystal structures may be determined much more readily, so that solid compounds are more readily identified. For noncrystalline compounds, spectroscopic methods are now much more powerful and give excellent structural information.

In Chapter 17 we have already seen examples of species with groups of like atoms bonded together. The forms of the elements themselves cover the main types. Small finite molecules are represented by the halogens, X_2, while S and Se give *chains* or *rings* (Figure 17.45) formed by two-coordinate atoms. Chains are also found in some of the borides and carbides. Graphite (Figure 5.14c) and bismuth (Figure 17.28) are examples of layer structures, while full three-dimensional clusters include boron and the borides of Figure 17.7, white phosphorus, P_4 (Figure 17.29), the recently discovered form of molecular carbon, C_{60} (Figure 5.14d), and in infinite extension, diamond (Figure 5.14a).

Chapter 17 was generally limited to simple cases.

Molecules with a single E–E bond are found for the majority of elements while chains E_nR_{2n+2} or rings E_nR_{2n} (R = H, organic group) are represented by the Group 14 compounds E = Si, Ge or Sn.

From Table 17.3, we see that most single-bond energies are quite high, and similar for the majority of Main Group elements within a factor of two (Single bonds N – N, O – O, or F – F are unusually weak.) While there is a general tendency for bond strength to fall with mass, the changes are not abrupt. It is no surprise to find that E – E links form for many of the Main Group elements. The resulting compounds range from diatomic molecules for the halogens, to rings, chains, networks, and open or closed clusters for the elements of the remaining p Groups. The following sections give illustrations.

18.2 Polysulfur and polyselenium rings and chains

The survey of sulfur allotropes in section 17.7 included the classical S_6 and S_8 rings and the long chains of plastic sulfur, together with the analogous Se_8 and Se_∞. Discussion also included compounds with single S – S or Se – Se bonds, and some chain oxygen species like the polythionates $[(O_3S) – S_n – (SO_3)]^{2-}$, for n up to about 20. These species were selected rather arbitrarily to represent the phenomenon of polysulfur and polyselenium chemistry. In this section, we develop this theme further.

If sulfur is compared with its neighbours, oxygen and chlorine, we can see that it has a much greater potential to form extended arrays of – S – S– S –links. Chlorine, with one electron more, has a fully satisfied electron configuration in diatomic Cl – Cl, and the halogens only form longer chains to a limited extent, with weaker interactions, and mainly involving iodine (compare Tables 17.19 and 20, and Figure 17.59). Oxygen differs from S in its ability to form a stable double bond in O=O, and species with O_n chains readily break up with formation of O=O. Thus polyoxygen species are limited to $n = 3$ and 4 (section 17.7.1). In contrast, sulfur has one electron less than Cl, and any S=S is

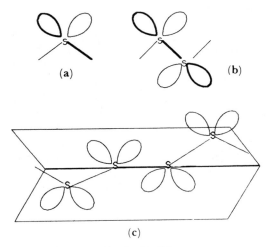

(a)   **(b)**

(c)

FIGURE 18.1 *Component units of polysulfides.*
A divalent S with two lone pairs (a) will tend to form a skew arrangement with its neighbour (b). For a chain, there will be an optimum dihedral angle between successive SSS planes (c).

unstable with respect to $-S-S-$, encouraging chain formation. The S_2 molecule does occur in the gas at high temperatures and gives a purple, paramagnetic solid on rapid cooling, but it reverts to S_8 when allowed to warm to room temperature.

Sulfur chains are built from divalent S atoms with two lone pairs, and will extend in a preferred conformation which minimizes lone-pair/lone-pair interaction as indicated in Figure 18.1. The long-known S_8 ring attains the optimum array of angles, and the S_6 ring of Engel's sulfur is more strained and readily converts (section 17.7.1). Clearly, if the ring opens out into a chain, there are no constraints but the chain ends are reactive so that only very long chains survive, as in plastic sulfur where the chains probably coil into a helix.

From this, it is reasonable to find that the further polysulfur species fall into three groups (Table 18.1):

(a) Rings, with those larger than 8-membered more stable
(b) Anions S_n^{2-} and cations S_n^{x+} where $x = 1$ or 2
(c) $XS-S_n-SX$ where monovalent X groups take up the terminal valencies. An example is the polythionates with $X = SO_2(OH)^-$.

18.2.1 *Rings*
Development of a range of new ring compounds depended on the discovery of syntheses giving specific rings, rather than an intractable mixture. One such reaction is between polysulfur hydrides and polysulfur chlorides, e.g.

$$H_2S_8 + S_4Cl_2 \rightarrow S_{12} + 2HCl \dots\dots\dots (1)$$

This procedure gives a convenient route to S_6 and also yields the new rings S_{12}, S_{18} and S_{20}. A modification involves the cleavage of an S_5 unit from the stable, six-membered MS_5 rings formed by a number of transition metals. This route leads to rings with odd numbers of S atoms

$$(C_5H_5)_2TiS_5 + S_2Cl_2 \rightarrow S_7 + (C_5H_5)_2TiCl_2 \dots (2)$$

Use of S_4Cl_2 or S_6Cl_2 in (2) gives S_9 and S_{11} respectively. An attempt to make a ring containing the SO_2 unit by using SO_2Cl_2 in (2) yielded instead S_{10}.

Further, more convenient, routes have been found for some of these rings. Knowing the properties of the isolated species, it has been possible to identify them in classical melts and mixtures. Thus S_{12} is soluble in CS_2, but 150 times less so than S_8, and can thus be separated. All the polysulfur rings are relatively unstable to heat and light, with S_{12}, S_{18} and S_{20} the most stable after S_8. S_6 is very sensitive and a saturated solution of S_6 gives S_{12} after a short exposure to light. The odd-membered rings are also all very reactive with S_7 polymerizing as low as $45\,°C$.

18.2.2 *Ions*
When sulfur is dissolved in strong acids (see section 6.9), ring cations are formed. Typical are the long-known coloured solutions of sulfur in oleum, which are red, yellow or blue according to concentration. In such systems, S_4^{2+}, S_8^{2+} and S_{16}^{2+} have been identified. In addition, radical ions S_n^+ give paramagnetic properties. Of these S_4^+ and S_8^+ (formed respectively from S_8^{2+} and S_{16}^{2+}) are reasonably well-established.

Table 18.1 summarizes some properties of the polysulfur species, and Figure 18.2 indicates some of the rings.

The structure of the ion S_8^{2+} has been determined (Figure 18.3); the bond lengths and angles are very similar to S_8, but the shape changes from the open crown to a half-closed form and the 283 pm $S-S$ distance across the ring corresponds to a weak transannular bond. If we compare S_8 and S_8^{2+} we see that the two missing electrons are balanced by the extra $S-S$ link. This process continues in S_4N_4 with only 44 valency electrons compared with 48 in S_8. As the structure of Figure 17.53 shows there are now two trans-annular $S-S$ interactions and the structure is closed up.

It is convenient here to mention S_8O, formed by the reaction

$$H_2S_x + SOCl_2 \rightarrow S_8O + 2HCl$$

The species is stable in the dark at $-20°$, but soon forms SO_2 and sulfur on warming. The structure is that of the S_8 ring with the SO bond angled towards the crown. The S_nO series has been extended to include $n = 6, 7, 9$ and 10—all with similar preparations, stabilities and structures to S_8. There is also one disubstituted species, S_7O_2, formed as dark orange crystals by further oxidation of S_7O with CF_3COOH. It decomposes at $60\,°C$, evolving SO_2, and calculations show the most likely structure to be a chair with the two O atoms 1,3 or 1,4. S_7I^+ has a similar structure to S_7O.

If two electrons are added to a polysulfur chain, to give an anion S_n^{2-}, the chain ends are no longer radicals, and such anions are expected to be stable chains. (Alternatively, we can see that adding two electrons opens a ring by breaking one bond.) Relatively few anions of sulfur are known, though S_3^{2-}, which gives rise to the blue and green colours of ultramarines, is now fully characterized. They are generally made by boiling S^{2-} solutions with sulfur and are stabilized by large cations, as in Cs_2S_6.

TABLE 18.1 *Properties of polysulfur species*

	Colour	*m.pt.* (°C)	*S—S* (*pm*)	*SSS* (°)	*Dihedral* (°)	
S_2	purple		189·0			Stable in gas at 800°C, para-magnetic. $S_3 - S_5$ also in gas
S_6	orange-red	50d	205·7	102·2	74·5	unstable to light
S_7	yellow	39	partial structure reported			unstable to heat and light, polymerizes 45°C
S_8	yellow	119	206·0	108·0	98·3	stable, several crystal modifications
S_9	deep yellow	c. 50d				
S_{10}	yellow		203·3	103	75	two different S environments in crystal
			207·8	110	79	
S_{11}						no details available
S_{12}	yellow	148	205·3	106·6	86·1	most stable after S_8
S_{18}	lemon	128	205·9	106·3	84·4	stable in absence of light
S_{20}	pale yellow	124	204·7	106·5	83·0	stable solid, unstable in solution
S chains			206·6	106	85·3	mixtures of different chain lengths —averaging up to 10^5 S atoms per chain in melt
S_4^{2+}	pale yellow		198			square
S_8^{2+}	deep blue		204·8	108	90	Figure 18.2b
S_{19}^{2+}	red		187 to 221			
S_3^{2-}	blue			103		chain
S_5^{2-}*		terminal	205	109	76	chain
		central	207	106	69	

d = decomposes. *also S_4^{2-} and S_6^{2-} (chains)

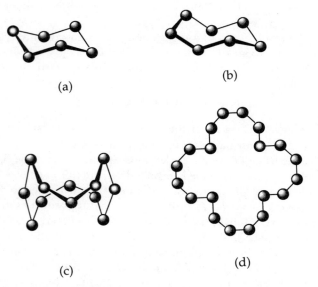

FIGURE 18.2 *Some sulfur rings*
(a) S_6, (b) S_7, (c) S_{12} and (d) S_{20}.

The longest known chain is found in the related Te_6^{2-} which has a central Te — Te distance of 276 pm, and rather longer outer bonds of 292 and 298 pm. Reduction of this forms further weak links (298 and 316 pm) between chains giving an overall cubane-like $[Te_{12}^{6-}]_\infty$.

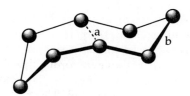

FIGURE 18.3 *The structure of the S_8^{2+} ion*
For S_8, the S–S distance is 204 pm, for S_8^{2+}, a = 283 pm and b = 204 pm. (For the Se analogues, the distances are 234 pm in the molecule and 284 and 232 pm in the cation).

18.2.3 $X - S_n - X$ species

Compounds with X = halogen are well known, and the disulfur members ($n = 2$) are included in section 17.7.6. Longer chains are found in the chloro- and bromo-polysulfanes, S_nCl_2 and S_nBr_2. In the case of the chlorosulfanes, there is evidence that n may be as high as 100. However, individual members of the series have been isolated only up to $n = 8$ for both the chloro- and bromo-sulfanes. These compounds are all formed by dissolving sulfur in S_2Cl_2. The latter results from the reaction of Cl_2 on molten sulfur. It has a skew Cl–S–S–Cl structure, similar to H_2O_2, with an SSCl angle of 103°. The higher chlorosulfanes probably have chain structures similar to those of the anions. Individual compounds may be isolated by careful distillation or by chromatography. Mild conditions are necessary, as individual members readily disproportionate into mixtures of various chain lengths.

The hydrides, $X = H$, called *sulfanes*, are formed either from the polysulfides or from the chlorosulfanes:

$$S_n^{2-} + 2HCl \rightarrow H_2S_n + 2Cl^-$$

$$S_nCl_2 + 2H_2S \rightarrow H_2S_{n+2} + 2HCl$$

Compounds up to $n = 8$ have been isolated and the existence of higher members has been demonstrated by chromatographic studies. All are yellow and range from gaseous H_2S to increasingly viscous liquids as the chain length increases. The sulfanes may be interconverted by heating—indeed any one member readily converts to an equilibrium mixture of the others—and all ultimately revert to sulfur and H_2S, although slowly. The structures are not known but are probably chains like the polysulfide ions. The sulfanes plus chlorosulfanes give syntheses of specific rings as in equations (1) and (3), which gave the first clean synthesis of Engel's sulfur:

$$H_2S_2 + S_4Cl_2 \rightarrow S_6 + 2HCl. \qquad (3)$$

A third important group has $X = $ a transition metal plus ligands. Both chain compounds and rings where the two ends of the S_n chain bond to the same M, are well known and these compounds are reviewed in section 18.7. Equation (2) illustrates their use to synthesize particular rings.

18.2.4 *Polyselenium and polytellurium rings and chains*

As in sulfur, recent studies using chromatography for separation, and spectroscopy or crystallography for characterization, have established the existence of a number of selenium rings of different sizes. In the vapour of molten selenium, the main components are Se_5 and Se_6, whereas Se_7 and Se_8 are minor. Se_5, however, cannot be isolated as it converts very rapidly to the larger rings (S_5 behaves similarly). In solution, there is an equilibrium, $2Se_7 \rightleftharpoons Se_6 + Se_8$, which can be demonstrated by high pressure liquid chromatography. Separation of Se_6 and Se_7 is possible but difficult, as they readily convert to stable Se_8. Perhaps the easiest route will be by synthesis, as in

$$(C_5H_5)_2TiSe_5 + Se_2Cl_2 \rightarrow Se_7 + (C_5H_5)_2TiCl_2 \qquad (4)$$

which has been established. A crystal structure determination shows Se_6 has a chair form like S_6 (compare Figure 18.2a), and Se_7 has probably the structure of Figure 18.2b. Se_8 is stable and known in several crystal modifications which all contain the crown ring (compare Figure 17.45) packed in different ways. There is evidence for the existence of Se_9 and Se_{10} rings at very low pressures over subliming selenium, and rings such as Se_{16} have been proposed as intermediates in the conversion of Se_8 to the grey form Se_∞. None of these larger ring molecules has been isolated, though they might be more accessible via direct syntheses analogous to equations (1) or (4). However, selenium differs from sulfur in the higher stability of Se_∞ relative to the rings, and the large-ring chemistry of selenium may remain limited. The established bond lengths and angles are listed in Table 18.2.

TABLE 18.2 Some parameters of polyselenium species

	Se–Se (pm)	SeSeSe (°)	Dihedral (°)
Se_6	236	101	76
Se_8	234	106	101
Se_∞	237	103	101

The very low dihedral angle means greatly increased interaction between neighbouring lone pairs in Se_6, and accounts for its ready transformation into Se_8. Note the somewhat smaller Se_3 and larger dihedral angles of the stable Se_∞ and Se_8 forms compared with their sulfur analogues.

Selenium, and also tellurium, forms cations in strong acid media and also in melts (compare section 6.9). Yellow Se_4^{2+}, green Se_8^{2+} and red Se_{10}^{2+} have been isolated and their structures are known. There is also an indication of Se_n^{2+} ($n = 2, 12$ and 16) in melts. Te_n^{2+} species are known for $n = 2, 4$ and 6. The M_4^{2+} species are planar:

$$\begin{array}{cc} M{-}M & Se{-}Se = 234\,pm \\ {\mid}\quad{\mid} & \\ M{-}M & Te{-}Te = 266\,pm \end{array}$$

Se_8^{2+} has the same structure as S_8^{2+} (see Figure 18.3). In contrast, Te_8^{2+} has the structure of two boats sharing an edge, or alternatively described as an 8-ring which is nearer in structure to the crown form of S_8, and with a weak transannular link. The ring Te–Te distances are in the range 270–278 pm and the transannular Te---Te distance is 299 pm. These units are further weakly associated (Te⋯Te = 342 pm) into chains of rings.

The Se_{10}^{2+} ion is interesting. It forms a structure where a Se_6 ring in a boat-like form is joined 'prow to stern' by a chain of four Se atoms, shown schematically in Figure 18.4. A range of distances are seen implying bonding interactions of quite varying strengths. As with Se_∞, we see the evidence for weak additional interactions, (compare the Van der Waals' distance of 380–400 pm) a step along the path to a delocalized cluster.

In liquid ammonia, selenium forms polyanions Se_n^{2-}. The species with $n = 3, 4$ and 6 have been isolated while Se_2^{2-} exists only in equilibrium with Se^{2-} and Se_3^{2-}. The structure of

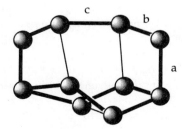

FIGURE 18.4 *The structure of* Se_{10}^{2+}
The bonds from the three-coordinate Se atoms a are longer (241–246 pm) than their neighbours b (223–227 pm) with an alternation along the chain making $c = 236$ pm. The lighter links are shorter than non-bonded distance at 330–350 pm.

TABLE 18.3 Some parameters of polyselenium anions in salts with large counterions

	Se–Se (inner) pm	Se–Se (outer) pm	SeSeSe (°)	Torsion (°)
Se_5^{2-}	230	235	107–109	98
Se_6^{2-}	230	234–236	109–113	82–98
Se_7^{2-}	230	233–235	109–113	82–98

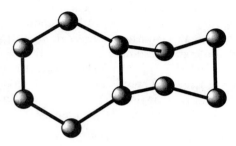

FIGURE 18.5 *The structure of the* Se_{10}^{2-} *ion*
The structure can be viewed as equivalent to the carbon analogue decalin, with two fused cyclohexane-type chain rings, but with each selenium having two extra valence electrons. The geometry about the two central Se atoms is derived from a trigonal bipyramid, with the two lone pairs on each Se in equatorial positions.

Se_6^{2-}, isolated as the black $(Ph_4P)^+$ salt, is a kinked chain with Se–Se distances of 229 pm (terminal) and 231 pm (inner). Angles at Se are in the range 105–109°. Slightly different values are found when the counter-ion is the larger Rb^+ or Cs^+ stabilized by crown ethers or cryptands (Table 18.3).

As an example of a structurally more complex anion, Se_{10}^{2-} has been isolated in the presence of a bulky cation and has the bicyclic structure shown in Figure 18.5 (contrast the cation, Figure 18.4).

There also exist mixed-element rings between S, Se and Te such as Se_nS_{8-n} ($n = 1, 2, 3$), Se_2S_{10}, and Se_7TeCl_2. Mixed cations include $Te_2Se_8^{2+}$, which has the same structure as Se_{10}^{2+} (Figure 18.4) with the two Te atoms in the three-coordinate positions. $Te_2Se_2^{2+}$ is square with alternate atoms. There are two mixed species which have new structures: $Te_3S_3^{2+}$ (Figure 18.6b) and $Te_2Se_4^{2+}$ (Figure 18.6a) have a boat structure with one transannular bond. These can be related to S_6 in the chair form, where removal of the two electrons in the cations requires a further bond. Again, this structure could be regarded in another way as resulting from the P_4S_3 type by removing one S atom (compare Figure 17.37a).

Tellurium forms the only 4 + ion so far found, Te_6^{4+}, which has the structure of Figure 18.6c. Here, instead of the removal of two electrons adding one more bond to the 18.6a structure, a more symmetric structure is adopted, but the bonds linking the two end triangles are distinctly long and weak.

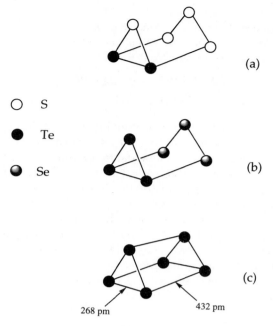

○ S
● Te
◐ Se

268 pm   432 pm

FIGURE 18.6 *Structures of* (a) $Te_2Se_4^{2+}$ (b) $Te_3S_3^{2+}$ (c) Te_6^{4+} Compared with S_6 (Figure 18.2a), these show the effect of removing electrons and compensating by added bonds.

A simple mixed element anion, TeS_3^{2-}, is a pyramid with Te–S = 233 pm and STeS = 106°. The more complex TeS_{10}^{2-} ion is an example of a simple linked-ring species (compare next section). It has two chair-form TeS_5 rings linked by the common Te with Te–S = 270 pm, S–S averaging 205 pm, the STeS angle is 89°, and angles at S in the range 104–109°.

18.3 Nets and linked rings

When more than two valencies are available, structures may grow in two or three dimensions. For cases where both 2- and 3-valent atoms are available, we find linked rings (more 2- than 3-valencies) or open cages or networks (more 3- than 2-valencies). Such behaviour is found in phosphorus and related compounds reviewed here.

18.3.1 *Polyphosphorus compounds and related species*

In section 18.2, compounds containing chains of S or Se atoms have been described. If we move one place to the left in the Periodic Table, we see that a suitable structural unit is formed by taking an electron from one of the lone pairs of S, leaving P with one lone pair and three valence electrons (Figure 18.7a). This unit has greater flexibility, and poly-phosphorus compounds show more variety of structure than polysulfur species. The simplest closed unit using the building element of Figure 18.7a is the P_4 tetrahedron found in white phosphorus (Figure 17.29a), and other ways of combining trivalent P units are the linked opened tetrahedra of Figure 17.29c, or the layer structure of black phosphorus thought to be similar to Figure 17.28. An interesting range of mixed tetrahedra incorporating such P units is found for metal derivatives, as in $PCo_3(CO)_9$.

Further development of polyphosphorus species requires

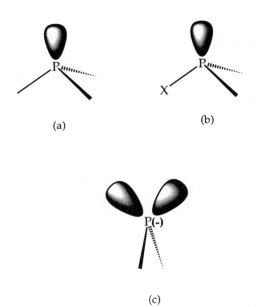

(a)   (b)

(c)

FIGURE 18.7 *Some building blocks for polyphosphorus compounds.* (a) a single P atom which may build in three dimensions, (b) a X–P unit which may form chains, (c) the P^- unit isoelectronic with S

some saturation of the trivalency. A P–X unit (Figure 18.7b) has two remaining valencies as does a P^- unit (Figure 18.7c) which is isoelectronic with S (Figure 18.1a). For X = halogen, examples are largely limited to the P_2X_4 species discussed in section 17.6.4, where the last two X atoms terminate the chain.

However, when X = H, giving *polyphosphanes*, a large group of compounds has now built up. Phosphanes of formula $H_2P-(PH)_n-PH_2$ are hydrocarbon analogues, with the two extra hydrogens allowing chain termination. Straight chains up to $n = 3$ are established while branched chains become preferred for higher n values. Thus a second pentaphosphane, $(H_2P)_2P(PH)PH_2$ has been identified, and branched chain isomers of compounds with up to 9P atoms are known. In a second series, $(PH)_n$ for $n = 3$ to 10, rings are formed. In addition to the arrangement of the phosphorus skeleton, geometrical isomers are possible depending on the arrangement of successive lone pairs. A simple example is P_2H_4 which can give *cis*, *trans*, or *skew* forms.

The hydrides are synthesized by the long-known hydrolysis of Ca_3P_2, which gives a mixture of hydrides which can be interconverted by gentle heating (compare section 9.5), for example

$$2P_2H_4 \rightarrow P_3H_5 + PH_3$$

Reaction of a hydride with $LiPH_2$ or an alkyl-lithium gives the Li salt which is more stable to further reaction or disproportionation. Thus in monoglyme at $-20\,°C$

$$9P_2H_4 + 3LiPH_2 \rightarrow Li_3P_7 + 14PH_3$$

The addition of an acid converts the Li salt to P_7H_3. These preparations are now well understood, so that they may be directed to specific materials.

This is a very striking example of the development of a new

chemistry out of a field which was seen to be very difficult. The hydrides are very air-sensitive, readily interconvert, the higher molecular weight species are amorphous, and no solvents were readily available. Because of these difficulties, the initial steps were very slow, with only PH_3, P_2H_4, and P_4 well established. The very richness of the chemistry made interpretation and characterization more difficult.

With much more powerful techniques, including mass spectroscopy and particularly modern ^{31}P nmr carried out at low temperatures to avoid interconversion, the first few additional compounds were characterized. Then it was possible to probe further and identify new species when they occurred in the presence of known ones. Synthetic strategies were further improved and could be monitored, allowing characterization of more compounds, until the whole field began to develop rapidly.

To give a flavour of the chemistry, a few examples are selected. The known hydrides are listed in Table 18.3.

P_4H_6. The two geometrical isomers of the straight chain have been identified, together with the branched chain isomer (Figure 18.8).

In Figure 18.8(a) and (b), all substituents on the two central P atoms are different, and thus they are chiral centres—(a) has d and l forms while (b) is *meso*. Different preparation routes give different proportions of (a), (b), and (c), allowing their identification in mixtures. Replacing

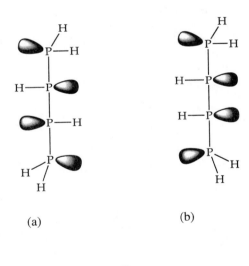

(a)   (b)

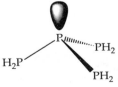

(c)

FIGURE 18.8 *The isomers of* P_4H_6
(a) and (b) show the geometrical isomers of n-P_4H_6 and (c) is the branched chain isomer

H by bulky substituents favours the less-hindered isomer. Thus, $P_4(t\text{-}Bu)_4H_2$ has alternating *tert*-butyl groups in an (a) structure. Branching is preferred as chain length increases. Thus the P_6H_8 samples characterized so far are $H_2P-P(PH_2)-PH-PH-PH_2$, $H_2P-PH-P(PH_2)-PH-PH_2$, and $H_2P-P(PH_2)-P(PH_2)-PH_2$ but not $H_2P-PH-PH-PH-PH-PH_2$.

Simple rings, P_nH_n, and planar rings P_n^{x-}. The smallest ring, P_3H_3, has been identified, but it very easily rearranges, forming P_5H_5. Its organic derivatives, P_3R_3, are stable and readily synthesized. The lone pair on the P atoms may act as donor, as in $(RP)_3Cr(CO)_5$ where the phosphorus ring occupies one position in the octahedron around Cr. In $(RP)_3[Cr(CO)_5]_2$, two different P atoms donate to the two Cr atoms and so the ring links the two octahedra.

A large variety of mixed-element three-membered rings is known of the general type of Figure 18.9a, where E may be CR_2, $C=CR_2$ or related species, SiR_2, GeR_2, SnR_2, BNR_2, NR, AsR, SbR, S or Se. For E = Si, Ge, or Sn, two rings participate in the *spiro* unit of Figure 18.9b.

Larger rings are non-planar, and the five-membered one is relatively stable. Above $n = 5$, there is a strong tendency to form rings with side-chains: thus P_6H_6 is as shown in Figure 18.10a. The four-membered ring size is not favoured, and is found only when bulky substituents are present, as in Figure 18.10b. Such stabilized rings are also found in the sequence of molecules $P_nBu^t_{n-2}$ for $n = 8, 9$ and 10 which consist of two P_4 rings slightly bent across a diagonal. For $n = 8$ the link is directly between an apex P atom of each ring and for $n = 9$ and 10 there are, respectively, a PBu^t and a PBu^t-PBu^t bridging unit. Larger rings are also found with

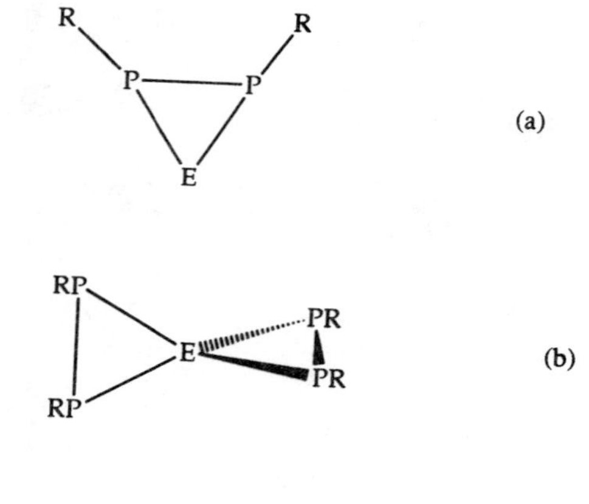

(a)   (b)

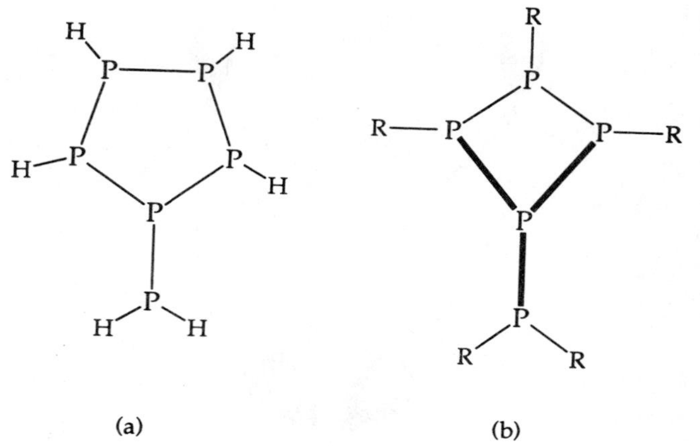

FIGURE 18.10 *Rings with sidechains*
(a) The stable form of P_6H_6, (b) a four-membered ring stabilized by $C(CH_3)_3$ groups in $(RP)_5$.

hetero-elements. To quote only one example, we find the mixed-ring *spiro* skeleton of Figure 18.9c in $GeP_6(CMe_3)_6$.

The above rings correspond to the saturated hydrocarbons like cyclopentane. It was surprising, initially, to find a different formula type in P_5^-. This was formed in the reaction of white phosphorus with alkali metals in diglyme, giving for example $[Na(\text{diglyme})_x]^+P_5^-$, and also with $LiPH_2$ in THF giving $[Li(THF)_x]^+P_5^-$. Its properties suggest that P_5^- is a planar ion with delocalized charge, isoelectronic with cyclopentadienide, $C_5H_5^-$ (compare section 16.4). That is, the P_5^- ion is analogous to aromatic hydrocarbon ions. Strong support for this view came from the isolation of the novel structure

$$(C_5Me_5)Cr(P_5)Cr(C_5Me_5)$$

where the two (C_5Me_5) rings and the P_5 ring are all planar and parallel, making a triple-decker sandwich exactly analogous to that of Figure 16.7c. The mixed ring, P_4CH^-, has been characterized and is also planar with delocalized charge.

A second member of the aromatic series is found in $(C_5Me_5)Mo(P_6)Mo(C_5Me_5)$, the analogous triple-decker molecule which contains the planar P_6 ring analogous to benzene. This compound was prepared by reacting white phosphorus, P_4, with a molybdenum precursor. Thus the P_6 ring formed spontaneously in the course of the reaction, arguing a significant stability in this environment.

Networks or open clusters. Perhaps the most intriguing polyphosphorus compounds are those where the number of H atoms is less than the number of P atoms. These are built of frameworks involving both trivalent and divalent units, Figure 18.7. Out of the wide range of compounds, we may select one in particular as examplar. The P_7^{3-} cluster is the starting point. This is isoelectronic with P_4S_3 (Figure 17.37a) and has the same structure with P^- replacing S (Figure 18.11).

Then more complex structures are built up by adding P_x units at the P^- positions, converting some of the divalent bridge P atoms to trivalent ones.

(a)

(b)

(c)

FIGURE 18.9 *Some mixed-element polyphosphorus rings.*
(a) The smallest ring, $(RP)_2E$, known for many E groups (see text) (b) a *spiro* structure formed by E = Si or Ge (c) a *spiro* structure as part of a larger ring in $Ge(PR)_6$. R = *tert*-butyl or similar bulky group.

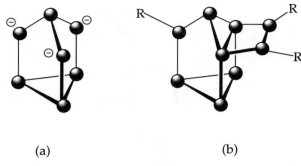

(a)   (b)

FIGURE 18.11   *The structure of* (a) P_7^{3-} *and* (b) R_3P_9.

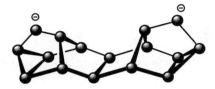

FIGURE 18.12   *The* P_{16}^{2-} *ion*

FIGURE 18.14   *The* P_{21}^{3-} *ion*

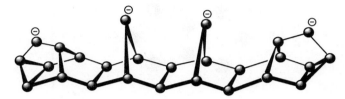

FIGURE 18.15   *The extended polyphosphorus ion,* P_{26}^{4-}

The simplest example is found in the R_3P_9 molecules where a P–P unit bonds to two of the bridging positions (Figure 18.10b). With a bulky R group, this compound can be crystallized and the structure determined. It is likely that P_9H_3 and P_9^{3-} have the same framework.

Larger molecules may be built up if units are linked together. In the P_{16}^{2-} structure, two P_7 units are linked using a common P–P unit bonded to two of the bridging P^- positions in each P_7 (Figure 18.12). The third bridge position in each P_7 retains its negative charge. Thus the whole P_7 unit acts as a divalent building block, essentially bonding through a P–P edge involving two of the P^- positions.

By opening up the P_7 skeleton of Figure 18.11a—by adding two electrons—we find the structural unit of the anion $P_7H_4^-$ (Figure 18.13a) which results from breaking one of the bonds in the base P_3 triangle. This open P_7 unit then links with a closed one in the ion $P_{14}H_2^{2-}$ (Figure 18.13b).

This is built up by removing two H atoms from one edge of Figure 18.13a, and using these two positions to join to two bridge positions of Figure 18.11a.

As final examples, we find the elegant structures of P_{21}^{3-} (Figure 18.14) and P_{26}^{4-} (Figure 18.15), which have two end units of the Figure 18.11a form joined by one and two, respectively, units of the Figure 18.13a form. As Table 18.4 shows, there is a further wide variety of polyphosphorus compounds. As far as current experimental evidence goes, the main structural theme is one of relatively open clusters, linked together by network units, as in the above examples. Structures will generally become more closed as the H/P ratio decreases.

18.3.2   *Polyarsenic compounds and other analogues*

Included in Table 18.4 are the polyarsenic compounds of similar formula to the polyphosphanes. Structural studies are

TABLE 18.4   Polyphosphorus and polyarsenic compounds

Formula	$E = P; R = H$	$E = As;$ $R = C(CH_3)_3$
	(or the corresponding anions of P or As)	
E_nR_{n+2}	$n = 1$ to 9	
E_nR_n	$n = 3$ to 10	
E_nR_{n-2}	$n = 4$ to 12	$n = 4$ to 9
E_nR_{n-4}	$n = 5$ to 13	$n = 6$ to 13
E_nR_{n-6}	$n = 7$ to 15	$n = 8$ to 16
E_nR_{n-8}	$n = 10$ to 17	$n = 10$ to 14, 20
E_nR_{n-10}	$n = 12$ to 20	$n = 12, 13$
E_nR_{n-12}	$n = 13$ to 20	$n = 14, 16$
E_nR_{n-14}	$n = 15$ to 21	$n = 16$
E_nR_{n-16}	$n = 17, 19, 20$ to 22	
E_nR_{n-18}	$n = 19$ to 22	

The ion P_{26}^{4-} is the first case of the E_nR_{n-22} family.

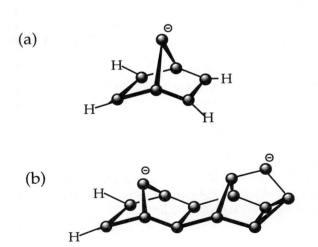

(a)

H   H
H   H

(b)

H
H

FIGURE 18.13   *More open polyphosphorus framework ions:* (a) $P_7H_4^-$, (b) $P_{14}H_2^{2-}$

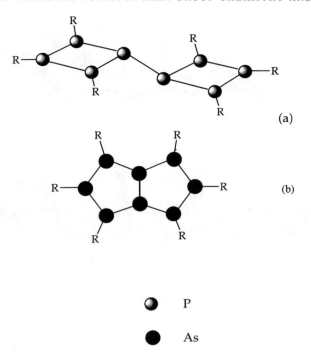

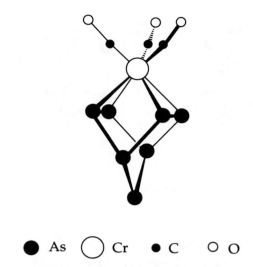

As ●   Cr ◯   C ●   O ◯

FIGURE 18.17   *The chromium–arsenic cluster* $[As_7Cr(CO)_3]^{3-}$

P ◗

As ●

FIGURE 18.16 *Contrasting structures of polyphosphorus and polyarsenic compounds* (a) P_8R_8 (b) As_8R_8 (R = CMe_3)

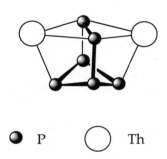

P ◗   Th ◯

FIGURE 18.18   *The thorium–phosphorus core of the cluster* $Cp''_2Th_2P_6$. Each thorium atom also bears two Cp'' ligands (not shown) where $Cp'' = \eta^5$-1,3-di-*tert*-butylcyclopentadienyl.

less developed—in particular, there is no useful arsenic nmr nucleus equivalent to ^{31}P, so this powerful technique is not available. While it is likely that some structure types will be common to P and As, it is already clear that there are differences. Thus, both elements form E_8R_6 with R = CMe_3, but with different structures, as shown in Figure 18.16.

In general, the As–H bond is weaker than the P–H one, so that there are fewer hydrogen-rich arsenic compounds and more use of organic derivatives is to be expected in developing the field. Rings are established for $(AsR)_x$ in the cases R = Me, CMe_3, or Ph, $x = 3, 4, 5$ or 6. The structure of $(PhAs)_6$ shows a chair formation for the As_6 ring. The $(PhSb)_6$ ring is similar, and 4- or 5-membered antimony rings are also found.

18.3.3 *Mixed systems*

We have already indicated the close relation between the polyphosphorus anions and mixed P–S compounds. There seems to be no reason why an extensive chemistry of polyphosphorus–sulfur should not exist, and this should also extend to As and Se, together with mixed species of three (compare PAs_3S_3) or all four of these elements.

Simple mixed hydrides of phosphorus and Si or Ge, such as GeH_3PH_2 or $(SiH_3)_2PH$, have long been known, and a range of mixed Si–P skeletons is emerging, particularly those where a divalent R_2Si unit replaces P^-. Thus we find $P_4(SiMe_2)_3$ where the Me_2Si units lie in the bridging positions of Figure 18.11. In $P_4(SiMe_2)_6$ the $SiMe_2$ units bridge all six edges of the phosphorus tetrahedron in a structure analogous to P_4O_6 (Figure 17.35). In such structures, together with many compounds where the $SiMe_3$ group is terminally bonded to P (acting as a bulky substituent), an

extensive chemistry of mixed phosphorus–silicon compounds is growing rapidly. Again, there seems no reason why we should not find similar compounds involving As and Ge.

As an extension of this we note that metals can also be incorporated in such cluster systems, and the polyphosphorus, -arsenic or other element cluster can be viewed, in one sense, as a ligand coordinating to the metal. Examples of recently prepared species are shown in Figures 18.17 and 18.18 for an arsenic and a phosphorus cluster containing a chromium and two thorium atoms, respectively. Related polysulfur and -selenium species also act as very good ligands towards metal centres and we develop this theme further in section 19.2.

18.4 Cluster compounds of the *p* block elements

When valencies of 3 or more are available, the atoms form three-dimensional units, open or closed clusters. Clusters may be described in terms of localized bonds along the edges, or of delocalized bonds. For closed clusters, as in this section, the latter description is more valid.

18.4.1 *General*

Closed clusters have been seen for some borides (Figure 17.7 and Table 17.5), for the E_4 tetrahedral form of P, As, and Sb

(Figure 17.29a), and for the Bi_9^{5+} ion which is one component of $Bi_{12}Cl_{14}$ (Figure 17.40c). In these species, the structure is closed, often a regular solid with equilateral triangular faces. Such regular structures are also found for the boron hydride ions of formula $(BH)_n^{2-}$ (compare section 9.6), while all the other boron hydrides, and related species like the carboranes, are fragments of such regular structures formed by removing one or more apices. Other electron deficient compounds behave similarly: for example, methyl-lithium crystallises as $(LiMe)_4$ with a Li_4 tetrahedron bridged on each face by a methyl group, while a chair structure is found in $(EtLi)_6$.

In addition to these classes of compound, there is a further, steadily growing, range of cluster compounds formed by p block elements. The structures are regular solids, or nearly so, and we shall look at two different types; (a) the sub-halides of boron, and (b) the 'naked metal' clusters of the later members of the carbon, nitrogen, and oxygen Groups.

It is convenient first to review one general approach to clusters, which involves the rationalization of structures in terms of the electron count. While there is significant theoretical support for the approach, it is best seen as a model and rationalization of observed structures. It involves counting the electrons in a cluster, allocating certain of them to bonds to external substituents, and assigning the rest to hold the skeleton together. The expected skeletal structure is derived from the number of cluster atoms and the number of these electrons. The scheme was first developed for boron hydrides, but has now been extended to all types of transition element and Main Group clusters.

18.4.2 Skeletal electron pairs

The *skeletal electron pair* rationalization postulates that $(N + 1)$ such pairs are required to give a stable regular cluster with N vertices. The regular clusters are those with all the faces equilateral triangles. The structures expected for some of the simpler cases of $(N + 1)$ skeletal electron pairs and N atoms are listed in Table 18.5 which is a fuller version of Table 9.6. The skeletal electron pair (SEP) method is also commonly talked of as applying Wade's Rules, after one of the early developers of the approach.

The rules for counting skeletal pairs are simplest for the s and p elements. Let us start with the $(BH)_n^{2-}$ ions mentioned in section 9.6. For these, we note that each $B - H$ bond points outwards from the cluster, and has all the properties of a normal 2-electron bond. Thus we allocate one H per B, together with 2 electrons and an outward-pointing orbital on B, to form these external $B - H$ links. The remaining electrons, and the remaining three orbitals per boron, remain to bond the cluster. For N boron atoms, $N + 1$ of these orbitals are stable and bonding, so the cluster forms if there are $N + 1$ skeletal electron pairs to fill them.

If there are fewer than N atoms, the structure is formed by removing an appropriate number of vertices from the regular figure. Thus, for 7 electron pairs, $N = 6$, and the structures are (i) an octahedron if there are 6 atoms in the cluster (ii) a square pyramid if there are only 5 atoms, and (iii) a square or a butterfly (two triangles sharing an edge) if there are only 4 atoms. Likewise, if there are more than N atoms the extra ones fit over one or more faces of the regular solid in a *capping* position (placed in a regular way so that the new faces are all regular triangles).

To describe these forms, a general terminology is used. The regular figure is called *closo* (simply indicating closed), a structure lacking one apex is called *nido* (from Greek indicating nest-shaped), one lacking two apices is *arachno* (Greek spider, or web-like), one lacking three apices is *hypho* (Greek, net). Extra atoms give *mono-*, *bi-*, *tri-*, etc. *capped* structures. Table 18.5 shows examples.

The skeletal electron pair counting rules for different types of cluster are summarized as follows.

Boron hydrides, carboranes, and similar species. Regard the hydride as $(BH)_nH_x$ so that any hydrogens in excess of 1 per B contribute to the cluster. Such extra hydrogens are found bonded to the basic B_n cluster, often face- or edge-bridging, or bonded terminally to B atoms at open faces. Any charges are added or subtracted from the total count, and for mixed element clusters such as the carboranes, each element contributes its valency electrons, and each is assumed to bear one terminal hydrogen. Thus, $C_2B_3H_7$ has $(2 \times 4) + (3 \times 3) + 7 = 24$ electrons, of which 4 are needed for 2CH and 6 for 3BH terminal groups. This leaves 7 skeletal pairs and the structure should be a square pyramid. Similar treatments apply to other additional elements. Note that the electron counting process gives no prediction of where heteroatoms are placed, nor does it define where any additional hydrogens are bonded. However, it should be emphasized that the skeletal electron pair theory is very successful in rationalizing literally hundreds of boron clusters.

Transition metal clusters. An account of the application of SEP or Wade's Rules to metal carbonyl clusters is given in section 16.8 with a number of examples. Some additional considerations are summarized here. A guide to the structures is the realization that a group such as $M(CO)_3$ is similar to BH in the orbitals available for bonding a cluster. BH has a centrally pointing orbital (for example the sp hybrid opposite to that forming the B–H bond) together with two

TABLE 18.5 Shapes and skeletal electron pairs

No. of skeletal electron pairs $N + 1$	Shape and no. of atoms in the cluster			
	closo N	nido $N - 1$	arachno $N - 2$	capped $N + 1$
5	4 tetrahedron	3 triangle		5 trigonal bipyramid
6	5 trigonal bipyramid	4 tetrahedron	3 triangle	6 bicapped tetrahedron
7	6 octahedron	5 square pyramid	4 square	7 capped octahedron
8	7 pentagonal bipyramid	6 pentagonal pyramid	5 irregular pyramid	

p orbitals at right angles to this, tangential to the cluster. The $M(CO)_3$ unit has a centrally pointing orbital (say d_{z^2}) together with two tangential ones (say d_{xz} and d_{yz}) of similar energies to the HB ones. Such a situation is called *isolobal*, and isolobal groups often behave in similar ways, and substitute one for another. Note that there do not need to be 3 CO groups present on each metal; those of the 12 electrons not needed for M–CO bonding, remain non-bonding: 12 electrons per M are subtracted whatever the structure. Thus, $Rh_6(CO)_{16}$ has $(6 \times 9) + (16 \times 2) = 86$ electrons, less 12 per Rh leaves 7 skeletal pairs, and the structure is octahedral. In the actual structure, there are two terminal CO per Rh and the other four CO face bridge every second face.

Other ligands contribute electrons appropriately: H or halogens provide 1, other donors like PF_3 give 2, while NO is usually a 3-electron source. Encapsulated atoms, such as C in $Ru_6(CO)_{17}C$, contribute all their valency electrons. Thus $Fe_5(CO)_{15}C$ counts $(5 \times 8) + (15 \times 2) + 4 = 74$ less (12×5) gives 7 skeletal electron pairs, and the predicted structure is a square pyramid, as found.

The skeletal electron counting approach becomes less useful for transition metal clusters of more than 8 or 9 atoms, and has limitations even with smaller clusters. However, it is generally successful with 3 to 8 atom species, and it is also very valuable for dealing with the metalloboranes which may be regarded as mixed species between the boranes and the metal clusters.

Other cases. The application of electron counting to Main Group clusters is based on the postulate that there is an outward pointing lone pair on each cluster atom (equivalent to the two electrons of the BH unit in the boranes) which does not contribute to the bonding. There are also mixed cases, such as metal clusters incorporating Main Group atoms, for which the approach has moderate success.

Overall, skeletal electron pair counting has had a remarkable success over a very wide range of compounds. It has played an important role, not only in ordering many structures and highlighting anomalies for further study, but also in emphasizing the underlying similarities between classes of compound which seem, at first sight, to be quite different.

18.4.3 Boron subhalides

As well as B_2X_4 and the lower fluorides, discussed in section 17.4.4, boron forms $(BCl)_n$ for $n = 4, 8, 9, 10, 11,$ and 12, $(BBr)_n$ for $n = 7$ to 10, and $(BI)_n$ for $n = 8$ and 9, together with a few related compounds like mixed chloride-bromides.

The structures of three of the compounds have been determined. Each boron bears one terminal chlorine atom: B_4Cl_4 has a regular B_4 tetrahedron; in B_8Cl_8, the boron atoms form a dodecahedron, while in B_9Cl_9 the nine boron atoms lie at the apices of a tricapped trigonal prism. All these are regular arrangements for these numbers of atoms, and it is assumed that the remaining compounds are also regular. We note that the boron subhalides are all short of one electron pair for a closed cluster but the extent of electron donation

from the halides is unknown.

The chlorides were prepared from B_2Cl_4 (itself prepared by a radio-frequency discharge on BCl_3) by heating to $1000\,°C$, which gives a mixture of all n species. One specific synthesis is of B_9Cl_9 by treating $B_9H_9^{2-}$ with $SOCl_2$. B_2Br_4 is prepared similarly, and this disproportionates on standing at room temperature to give the tribromide and a mixture of the sub-bromides.

Particular members of the group can also be made from the chloride by halogen exchange with Al_2Br_6. B_8I_8 and B_9I_9 were formed when B_2I_4 was melted at $100\,°C$. Characterization depends heavily on mass spectra, helped by the fact that B and all the halogens are polyisotopic, so their combinations give characteristic intensity patterns.

Only a limited study of reactions has so far been made. Treatment of B_9Br_9 with $SnMe_4$ gives a number of methyl derivatives $B_9Br_{9-x}Me_x$ for $x = 1$ to 6, with apparently the same B_9 skeleton.

One link between the halides and the carboranes is the compound $(BCl)_{10}(CH)_2$, which is icosahedral, like the isoelectronic $B_{12}H_{12}^{2-}$. These compounds, taken with the hydrides and binary borides, show boron has a strong tendency to form clusters, whose exploration will undoubtedly continue to attract interest.

18.4.4 Naked metal cluster ions

The first observations of compounds which are now known to be polynuclear anions came a century ago when it was first observed that lead and other heavier p block elements gave a sequence of colours when they were treated with solutions of sodium in liquid ammonia. Later work up to the 1930s showed that such solutions give the poly-sulfur and poly-selenium chain anions, and suggested the other elements gave more condensed species.

For example, germanium, tin and lead give polyanions in liquid ammonia. The tin or lead dihalides react with sodium in ammonia to give, first, the insoluble Na_4M species, and these add extra metal ions to give intensely-coloured solutions of Na_4Sn_9 or Na_4Pb_9. This latter compound was formulated as an ionic species with Pb_9^{4-} ions, on the basis of conductance and electrolytic experiments. For example, electrolysis was found to deposit 2.25 lead equivalents for each faraday passed. The compounds are heavily ammoniated and decompose when the ammonia is removed. GeI_2 behaved similarly. When ethylenediamine (en) was added to alloys of the p elements with alkali metals, products could be isolated, such as $Na_4(en)_5Ge_9$, since the solvating interaction was strong enough to avoid loss of the en. It is probable that, if the cation is in close contact with the large polynuclear anion, reverse electron transfer occurs and a metallic alloy, such as Na_4Ge_9 forms once the solvent NH_3 is removed, and this was avoided in the en complex cation. Crystal structure determinations on such complexes in the 1970s, at last gave some certainty to the field.

By this time, the cryptates and crown ethers (Figure 10.5) had been evolved specifically to form stable complexes with cations such as the alkali metals. By using cryptate-stabilized

(a)

(b)

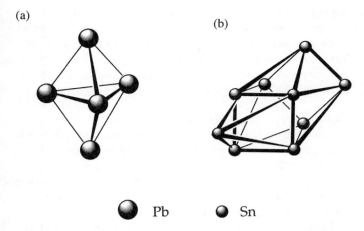

Pb   Sn

FIGURE 18.19 *The structure of (Na crypt)$^+$-stabilized (a)* Pb_5^{2-} *and* (b) Sn_9^{4-}
(a) is a trigonal bipyramid with equatorial Pb–Pb = 300 pm and axial-equatorial Pb–Pb = 323 pm. (b) is a square antiprism capped on one square face with Sn–Sn distances of 293 to 302 pm, except that the edges of the capped face are expanded to about 325 pm.

TABLE 18.6 Some clusters

Examples	Total electrons	Cluster electron pairs	Shape
Ge_4^{2-}, Sn_4^{2-}	18	5	Elongated tetrahedron
P_4, As_4, Sb_4, $Sn_2Bi_2^{2-}$ $Pb_2Sb_2^{2-}$	20	6	Tetrahedron
$Tl_2Te_2^{2-}$	20	6	Butterfly
As_4^{2-}, Sb_4^{2-}, Bi_4^{2-}, Se_4^{2+}, Te_4^{2+}	22	7	Square
Sn_5^{2-}, Pb_5^{2-}, Bi_5^{3+}	22	6	Trigonal bipyramid
Bi_8^{2+}	38	11	Square antiprism
Ge_9^{2-}, $TlSn_8^{3-}$	38	10	Tricapped trigonal prism
Sn_9^{3-}	39	$10\frac{1}{2}$	Tricapped trigonal prism
Bi_9^{5+}	40	11	Tricapped trigonal prism
Ge_9^{4-}, Sn_9^{4-}, Pb_9^{4-}, $Sn_xGe_{9-x}^{4-}$, $Sn_xPb_{9-x}^{4-}$, Sn_8Sb^{3-}, Sn_8Tl^{5-}	40	11	Capped square antiprism
$TlSn_9^{3-}$	42	11	Bicapped square antiprism

alkali metal ions (see section 10.5), polyanions have been isolated as solids. Crystal structures are known for Na(crypt)$^+$ salts of red Pb_5^{2-} (Figure 18.19a), orange Sn_5^{2-}, and dark red Sn_9^{4-} (Figure 18.19b). The green Pb_9^{4-} species, strongly indicated by conductivity in solution, has not yet been isolated: perhaps this requires use of an even larger cation.

One particularly striking product resulted from the reaction of an alloy of composition KGe, which gave $K(crypt)_6(Ge_9)^{2-}(Ge_9)^{4-}.2.5en$, where the difference of two electrons between the two germanium clusters gives different structures. A second unexpected species is the odd-electron Sn_9^{3-} cluster. Table 18.6 lists established clusters.

Most of the structures have been determined crystallographically, or by comparison of spectroscopic properties with those of established compounds. The mixed Sn/Pb and Sn/Ge nine-atom clusters were characterized by ^{119}Sn and ^{205}Pb nmr and all ten species occur in the mixture. Interestingly, the Pb_9^{4-} species is seen here. It is noteworthy that some ions with high charge/metal ratio indicated in the solution experiments, such as M_3^{3-} and M_5^{3-} for M = As, Sb, or Bi, have also not been isolated as solids. These ions may need an even larger cation than the crypt species for stability, as may the preparation of clusters with more than the present upper limit to cluster size of about 10 atoms.

An alternative synthesis derives from the early observation that Bi dissolved in $BiCl_3$ gives the cluster Bi_9^{5+} (Figure 17.40c) as part of a mixture of products (see section 17.6.4). Other fused salts provide similar media, and $AlCl_3$ melts have been generally used. Addition of NaCl increases the basicity of such melts and this type of non-aqueous system has proved to be a source of cluster ions, as well as other unusual species (compare section 18.9). Using $NaAlCl_4$ melts, two more bismuth clusters, Bi_5^{3+} and Bi_8^{2+} have been prepared. These ions are probably trigonal bipyramidal and

square anti-prismatic structures respectively. Reinvestigation of bismuth species in liquid ammonia, stabilized by large solvated alkali metal cations (compare section 10.5) has led to the isolation of the cluster anion, Bi_4^{2-}, isoelectronic with the known Te_4^{2+} and Se_4^{2+}, with a square of metal atoms.

While the structures of the clusters are clearly related to the number of electrons, there are obviously finely-balanced factors as shown by the two different structures of the 40 electron species. If we assume that there is one outward pointing lone pair on each atom, the remaining electrons, listed as *cluster electron pairs* in Table 18.6 may be used in the skeletal electron pair approach. The 18-electron, M_4^{2-} compounds have 5 skeletal pairs and we predict a *closo* figure, a tetrahedron. Similarly, the 22-electron M_5 (Figures 18.19a and 18.20a), the 38-electron M_9 (Figure 18.20b), and the 42-electron $TlSn_9^{3-}$ clusters are *closo* figures as expected from the 6, 10, or 11 skeletal pairs respectively.

The 20-electron M_4 species have 6 skeletal electron pairs, so a 5-atom cluster would be a trigonal bipyramid, but there are only four M atoms so the structure is a trigonal bipyramid less one vertex—which is a tetrahedron. Likewise, all the 40-electron M_9 species should give the *nido* 10-vertex figure, corresponding to the observed capped square antiprism (Figure 18.19b and 18.21). Thus the Bi_5^{5+} structure is the anomalous one, and we note that its trigonal prism is much more elongated than those found for the 38 electron *closo* structures. The odd-electron Sn_9^{3-} adopts the 10-pair rather than the 11-pair configuration.

The square 22-electron M_4 species correspond to an *arachno* octahedron. Alternatively, we can describe this as the result of adding two electron pairs to the 18-electron tetrahedron, breaking two bonds and leaving a square. The square antiprism Bi_8^{2+} is likewise the *arachno* 10-vertex structure.

The closed clusters may be correlated with more open species, as indicated above for the M_4 case. Thus the various E_6 species found for S, Se, and Te, may be seen as resulting from the successive opening of an octahedron, by adding electrons: six pairs removes six of the twelve edges, leaving a chair (Figure 18.6), for example.

It is also possible to make connections between the clusters and the extended structures of solids. In many of the mixed phases, called *Zintl* phases, formed between electropositive and less electropositive elements, cluster units may be identified as building blocks of the three-dimensional arrays. Tetrahedra are common, but the other isolated clusters are rare. Instead, more extended linkages occur giving rise to puckered layers or extended nets.

18.4.5 *Group 14 prismanes and related structures*

An interesting development in organic chemistry was the synthesis of closed carbon frameworks such as tetrahedrane (C_4H_4) or cubane (C_8H_8) with C–C bonds forming each edge of a regular solid. Naturally, making analogous structures has been taken up as a challenge by chemists working with Si, Ge or Sn.

Theoretical studies indicated that a range of closed $(EH)_n$ structures should be stable and that the preferred shape was the *n*-prismane. Prismanes have two regular plane faces in eclipsed positions joined by apex–apex bonds forming a series of square or rectangular faces. We have already met the trigonal (or triangular) prism (3-prismane) and a cube is a special case of a square prism (4-prismane).

The first syntheses of Si, Ge and Sn prismanes were reported almost simultaneously in 1988–9. The key was to use a large bulky substitutent, R, such as *tert*-butyl $(CH_3)_3C-$, $(Me_3Si)_3C-$ or a polysubstituted phenyl. The common syntheses involved removing halogen from REX_3 by using an active metal species like an organolithium. Such reductive dehalogenation of $RGeCl_3$ gave the trigonal prism $[RGe]_6$ (Figure 18.22a), for $R = CH(SiMe_3)_2$, and the cubane $[RGe]_8$ (Figure 18.22b) when R was CEt_2Me. The alternative product shown in Figure 18.22c was also isolated from a similar dehalogenation of $Bu^tGeBr_2GeBr_2Bu^t$. This may be seen as an intermediate *en route* to the cubane, with the two square faces twisted apart. The Si analogues were prepared by similar routes.

In an alternative synthesis thermolysis of the cyclotristannane, $[(2,6-Et_2C_6H_3)_2Sn]_3$, at 200 °C gave not only the

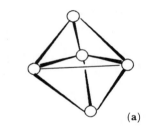

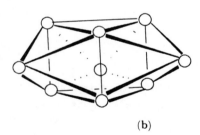

(a)

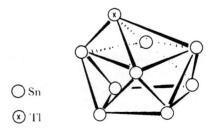

(b)

FIGURE 18.20 *Structures of the tin clusters* (a) Sn_5^{2-} *and* (b) Sn_9^{3-} The 38-electron Ge_9^{2-} ion is of similar structure to (b)

FIGURE 18.21 *The capped square antiprismatic structure of* $TlSn_8^{5-}$, *and* Sn_9^{4-}
Compare Figure 18.19b.

$\bigcirc$ Sn
$\otimes$ Tl

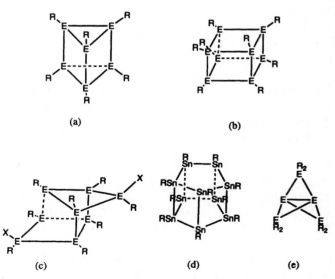

(a)   (b)

(c)   (d)   (e)

FIGURE 18.22 *Some cluster structures for germanium and tin*

TABLE 18.7 Parameters of representative Group 14 prismanes.

Figure 18.22	ER	E–E (pm) prism face	E–E (pm) square face	Angles (°)
(a)	GeCH(SiMe$_3$)$_2$	258.0	252.1	60° and 90°
(b)	GeC$_6$H$_4$Et$_2$	247.8–250.3	same	88.9–91.1°
(b)	SnC$_6$H$_4$Et$_2$	285.4	same	88–92°
(d)	SnC$_6$H$_4$Et$_2$	285.6	285.6	108°, 90°

cubane (Figure 18.22b) but also the pentagonal prismane (Figure 18.22d) and the propellane (Figure 18.22e). This can be seen as three triangles sharing a common edge. Typical parameters found for representative n-prismane clusters are listed in Table 18.7. Note that all the E–E bonds are somewhat longer than indicated by the values of Table 2.10a which are derived from simple E–E species like H$_3$Ge–GeH$_3$. They are comparable with the lengthened bonds found in simple R$_6$E$_2$ compounds with bulky ligands.

The structures are essentially ideal ones. In the tin propellane (Figure 18.22e) the outer bonds are normal for Sn–Sn (284–287 pm) while the central bond is very long (Sn–Sn = 336.7 pm). Further structures found for silicon include the striking tetrahedrane Si$_4$R$_4$, by making R the 'supersilyl' highly bulky SiBu$^t_3$ ligand, and the mixed-element cubane (Bu^{t}SiP)$_4$, the analogue of Figure 18.22b with alternating Si and P in a distorted cage which can be described as a large P$_4$ tetrahedron interpenetrating a smaller Si$_4$ one.

18.5 Polynuclear ions and the acid strength of preparation media

In Chapter 17 and in sections 18.2, 3, and 4, we have seen a wide range of examples of polynuclear ions. Almost all the preparations of these ions involve non-aqueous media such as liquid ammonia (section 6.7), superacids (section 6.9), or AlCl$_3$ and other melts. Polynuclear cations are very sensitive to base attack and require to be prepared in media of high acidity; conversely, polynuclear anions need strongly basic media. The media also act as vehicles for strong reducing agents (alkali metals in liquid ammonia) or oxidizing agents (F$_2$ in liquid HF or S$_2$O$_6$F$_2$ in HSO$_3$F), but the acidity appears to be the dominant factor.

Sufficient work has been done to allow these factors to be drawn together in an interesting picture which underlines some of the similarities between various non-aqueous solvents.

As an example, we can consider the iodine cations I$_2^+$, I$_3^+$ and I$_5^+$ (section 17.8). We use the Hammett acidity function, H_0, as a measure of the relative acidity in the superacid media (the larger the negative value of H_0, the higher the acidity). First, it may be noted that I$_x^+$ may be prepared in HSO$_3$F by oxidation of I$_2$ with S$_2$O$_6$F$_2$ in an appropriate stoichiometry: thus 3:1 gives I$_3^+$ and 5:1 gives I$_5^+$. However, the species that is recovered depends also on the acidity. The full equation for the formation of I$_2^+$ is

$$S_2O_6F_2 + 2I_2 \rightarrow 2I_2^+ + 2SO_3F^- \quad \cdots\cdots\cdots (1)$$

but the I$_2^+$ slowly transforms to I$_3^+$ plus I(SO$_3$F)$_3$. Now, SO$_3$F$^-$ is the base in HSO$_3$F according to the self-dissociation (compare Chapter 6):

$$2HSO_3F \rightleftharpoons H_2SO_3F^+ + SO_3F^- \cdots\cdots\cdots (2)$$

Thus the SO$_3$F$^-$ formed in equation (1) reduces the acidity of the medium. In fact, the H_0 value for HSO$_3$F is −15·1, while the SO$_3$F$^-$ will reduce H_0 to about −14·1. If, on the other hand, SbF$_5$ is added this removes SO$_3$F$^-$ (see equation A2 in section 6.9), and I$_2^+$ is very stable (H_0 −16 to −18 depending on SbF$_5$ concentration). Therefore, the higher the ratio of positive charge to iodine atoms in the ion (the higher the formal oxidation state) the more susceptible the ion to base attack and the higher the required acidity, so I$_3^+$ is stable in (relatively!) low acidity with charge ratio 1:3, whereas I$_2^+$ (ratio 1:2) needs higher acidity.

In the early reports of polyiodine cations, the medium was 100% H$_2$SO$_4$ ($H_0 = -11·9$), and I$_3^+$ and I$_5^+$ were stable while I$_2^+$ readily converted to I$_3^+$. In 60% oleum (i.e. 60% added SO$_3$ forming H$_2$S$_2$O$_7$), with $H_0 = -14·8$, I$_2^+$ was stable. Conversely, in H$_2$SO$_4$/HSO$_4^-$ ($H_0 = -11·9$), only I$_3^+$ was stable. Thus, while oxidant is needed to convert I(0) in I$_2$ to the fractional positive oxidation state in the I$_x^+$ ions, the ion recovered depends on the acidity.

Similar results are obtained using HF as a solvent ($H_0 = -15·1$) with acidity varied by adding the base F$^-$ or GeF$_4$ (which removes F$^-$ by forming GeF$_6^{2-}$). Since HF has a very small self-dissociation, the acidity is extremely sensitive to [F$^-$]. Thus reaction of F$_2$ in HF with I$_2$ gives I$_3^+$ as the F$^-$ formed reduces H_0 to around −10·5. When GeF$_4$ was added (H_0 about −5), the product was I$_2^+$.

The result of increasing basicity is the disproportionation of the cations into a cation of lower charge/atoms ratio plus a covalent IX$_3$ species:

$$8I_2^+ + 3X^- \rightarrow 5I_3^+ + IX_3$$
$$7I_3^+ + 3X^- \rightarrow 4I_5^+ + IX_3$$
$$3I_5^+ + 3X^- \rightarrow 7I_2 + IX_3 \quad (X = F^-, SO_3F^- \text{ or } HSO_4^-).$$

A different form of acidic medium is provided by molten AlCl$_3$, and the acidity may be adjusted by adding varying proportions of KCl (providing the base Cl$^-$ or AlCl$_4^-$). Thus the iodine species formed in the acid melt 2AlCl$_3$ + 1KCl is I$_2^+$, while I$_3^+$ is stable in the neutral melt 1AlCl$_3$ + 1KCl (and this melt also gave I$_5^+$AlCl$_4^-$ at the appropriate stoichiometry).

Finally, a non-basic medium like SO$_2$ or AsF$_3$ allows the formation of the higher polyiodides:

$$I_2 + AsF_5 \rightarrow I_3^+AsF_6^- + AsF_3 \text{ in } SO_2$$
$$I_2 + SbF_5 \rightarrow I_5^+SbF_6^- + SbF_3 \text{ in } AsF_3$$
$$I_2 + 3SbF_5 \rightarrow I_2^+Sb_2F_{11}^- + SbF_3 \text{ in } SO_2$$

because none of the species SO$_2$, AsF$_3$ or SbF$_3$ are sufficiently basic to attack the I$_3^+$ or I$_5^+$. The last equation underlines a further contribution—that of specific and usually large counterions in stabilizing a particular cation. This contribution is significant in many cases but is not as dominant as the acidity.

These ideas may be further reinforced if we look at the smaller halogens. Since the charge density per atom is higher in, say, Br_3^+ than in I_3^+, the acidity required to stabilize the bromine species will be higher than for the iodine analogue, while chlorine species will be even more demanding. It is found that Br_3^+ is stable in the highly acidic $HSO_3F - SO_3 - SbF_5$ system ($H_0 = -19$), but is only marginally stable at $H_0 = -13.8$ where I_3^+ is stable. Br_2^+ is observed at $H_0 = -19$ but is not stable. No polychlorine cation survives in solution in even the most acidic media, though Cl_3^+ has been isolated as a solid. Cl_2^+ is not known.

The similar situation for the chalcogen polycations in sulfuric acid systems is summarized in Table 18.8. The H_0 values range from -10 in 95% H_2SO_4 to -14.1 in an oleum with 40% SO_3.

All these observations on polyatomic cations reinforce the general discussion in Chapter 6 on the similarities between different solvent systems. For example we see the exact parallel between the oxidant F_2 in liquid HF giving the solvent base F^- and the oxidant $(SO_3F)_2$ in HSO_3F giving the base SO_3F^-. We see the parallel between these protonic acid solvents and the non-protonic $AlCl_3$ melt.

Finally we note the difference between such 'participating' solvents and SO_2, which is functioning as a non-interacting medium. It should be noted, of course, that acidity is not the only factor in these reactions. Clearly the intrinsic properties of the reagents are significant—especially where there are small differences in charge density, as in the formation of I_3^+ versus I_5^+, or in the contrasting formation of Se_{10}^{2+} and S_{19}^{2+} for low charge density. For the isolation of solids, the presence of a counterion of appropriate size may be vital.

The considerations above apply also to simpler cations which are unstable in less acidic media, such as water. Thus U^{3+} may be prepared by reacting U with HF, whose acidity is enhanced by adding BF_3, whereas a more basic system containing F^- gives UF_4. U^{3+} was also prepared in an acid 2:1 $AlCl_3/KCl$ melt and disproportionated to U plus U(IV) in a 1:1 melt. Similarly, Ti^{2+} (as $Ti[HF]_6^{2+}$) is formed from HF/SbF_5 and $TiCl_2$ in $AlCl_3$. The formation of $Sm^{2+}GeF_6^{2-}$ by reduction of Sm^{3+} in HF with added GeF_4 reflects the effects both of acidity and of a suitable precipitating anion.

All these examples indicate how the control of acidity (supplemented by choice of medium, oxidant, and counterions) may lead to the directed synthesis of low-oxidation-state, polynuclear cations. An exactly similar approach in highly basic media would apply to polynuclear anions.

TABLE 18.8 Values of H_0 where X_n^{2+} cations are stable.

Formula	X_6^{4+}	X_4^{2+}	X_8^{2+}	X_n^{2+}
Oxidation State	0.67	0.5	0.25	< 0.25
X = Te	−13	−11		
X = Se		−11.9	−11.9	Se_{10}^{2+} at −10
X = S		−14.1	−13.2	S_{19}^{2+} at −12.5

18.6 Silicates, aluminosilicates and related materials

Having discussed a number of ring and chain species formed by the p-block elements, we now turn our attention to another very important class of species formed by these elements—the silicates. The Si–O–Si linkage forms very readily by elimination of water between two Si(OH) groups and a very wide range of silicates is found. In almost all cases the coordination number of silicon with respect to oxygen is four, and thus the vast majority of silicates are built up from SiO_4 tetrahedra. The only exceptions occur in materials formed under conditions of extreme pressure and temperature—such as in meteorite impacts—where the SiO_2 mineral known as *stishovite*, having silicon in six-coordination, can be found. Our discussion of the silicates will begin with a survey of the various silicate species formed and will then move on to more complex materials having network structures, such as the aluminosilicates, including the very important *zeolite* class of materials, and finally to the related aluminophosphates. One recurring theme in the structural chemistry of silicates is that highly complex structures can be built up from the basic SiO_4 tetrahedron sharing zero, one, two, three or four vertices with other SiO_4 (or MO_4) tetrahedra.

18.6.1 Simple silicate anions: SiO_4^{4-} and $Si_2O_7^{6-}$

The simplest type of silicate contains the isolated SiO_4^{4-} ion shown in Figure 18.23a. A very common mode of representation of the SiO_4 unit is as a tetrahedral SiO_4 framework unit which has the oxygen atoms at the vertices, as shown in Figure 18.23a, and this representation will be used extensively to simplify the structures of the more complex silicates discussed later. An example of a material containing the SiO_4^{4-} ion is Mg_2SiO_4 where the magnesium ions are surrounded octahedrally by oxygen atoms. The magnesium ions may be readily replaced by other divalent cations of about 80 pm in diameter, such as Fe^{2+} and Mn^{2+}. As a result, natural materials of this type have a wide range of compositions. The mineral olivine is a naturally occurring example of such a magnesium silicate in which about one in ten of the Mg^{2+} ions is replaced by a Fe^{2+} ion. *Garnets* also contain isolated SiO_4^{4-} ions.

A less simple silicate which contains discrete silicate ions is the rather rare scandium silicate $Sc_2Si_2O_7$, known as *thortveitite*, containing the unusual dinuclear ion $Si_2O_7^{6-}$. This ion is formed by two SiO_4 tetrahedra sharing a common oxygen atom, as shown in Figure 18.23b. The scandium ions in this mineral have octahedral coordination.

At this point it is convenient to mention the calculation of charge on a silicate ion—this can be simply done since each terminal oxygen contributes one negative charge while each shared oxygen contributes zero charge. The charge on the Si_2O_7 species is thus confirmed as $6-$ since the six terminal oxygens each contribute $1-$ while the single shared oxygen does not contribute to the charge.

18.6.2 Rings and chains: $[SiO_3]_n^{2-}$

If each SiO_4 tetrahedron shares an oxygen with each of two neighbouring tetrahedra, then ring or chain silicate species

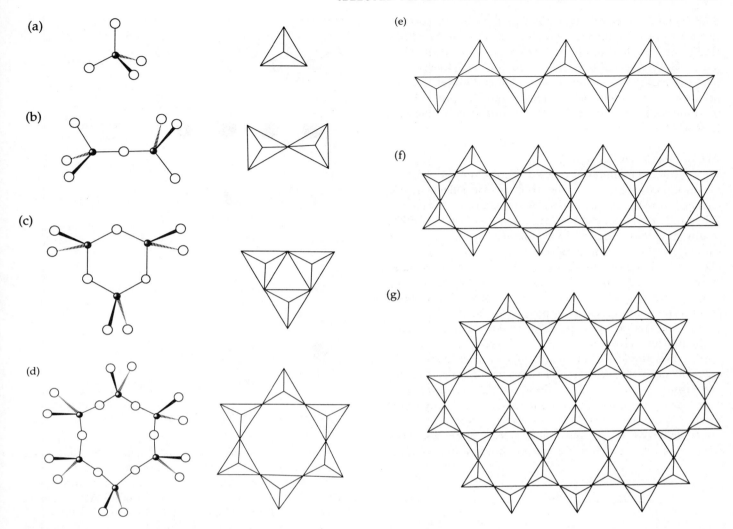

FIGURE 18.23 *Representations of silicate anions*
(a) SiO_4^{4-}, (b) $Si_2O_7^{6-}$, (c) $Si_3O_9^{6-}$, (d) $Si_6O_{18}^{12-}$, (e) an SiO_3^{2-} chain, (f) an $Si_4O_{11}^{6-}$ double chain, and (g) an $Si_4O_{10}^{4-}$ sheet.

can result. Two different ring sizes are known. The smaller ring contains three SiO_4 tetrahedra linked together as a six-membered ring $Si_3O_9^{6-}$ (Figure 18.23c) which occurs in, for example, *benitoite*, $BaTiSi_3O_9$, while the larger ring has six linked tetrahedra in the ion $Si_6O_{18}^{12-}$ (Figure 18.23d) found in the mineral *beryl*, $Be_3Al_2Si_6O_{18}$. In the green variety of this mineral, known as *emerald*, the green Cr^{3+} ions substitute some of the Al^{3+} ions. In the structure of these minerals the rings are arranged in sheets with their planes parallel and the metal ions lie between the sheets, binding the parallel rings together by electrostatic forces. The Ba, Ti and Al atoms are coordinated to six O atoms while the smaller Be atom is four-coordinated to O.

The chain structure shown in Figure 18.23e is found in a very wide range of minerals and these are known collectively as the *pyroxene* group. Examples are $MgSiO_3$ and $CaMg(SiO_3)_2$, known as the mineral *diopside*. In these compounds the silicate chains lie parallel to each other and the cations lie between the chains, binding them together by electrostatic forces. The magnesium ions are six-

coordinated while the larger calcium ions are eight-coordinated.

Two $[SiO_3]_n^{2-}$ silicate chains can also be linked together to form *double chains* (Figure 18.23f) which correspond to the formula $Si_4O_{11}^{6-}$. Materials containing these double-chain silicates are also relatively common and are represented by the class of minerals called *amphiboles* which includes most *asbestos* minerals. One example is $Ca_2Mg_5(Si_4O_{11})_2(OH)_2$ in which the magnesium ions are coordinated to six oxygens, three from the silicate and three from the hydroxyl groups. The presence of hydroxyl groups in the structure is a characteristic feature of the amphibole series of minerals. *Tremolite* is the mineral to which the name asbestos was first given although the term is now applied more widely. As with all minerals, the above composition is an idealized one and other ions of a similar size may be present. In particular, some of the magnesium may be replaced by iron and when the iron content rises to about 2 per cent the mineral is termed *actinolite*.

Minerals having chain silicate ions, particularly the

double-chain silicates, typically adopt fibrous forms—this is readily explained since the bonds between the silicate chains are weaker than the very strong covalent Si–O bonds within the chains themselves and so the mineral tends to cleave parallel to the long silicate chains. This relationship between the structure of the silicate and the properties of the minerals is particularly well demonstrated by the next class of silicate materials.

18.6.3 Sheet structures: $Si_4O_{10}^{4-}$

We continue our survey through the structures adopted by silicate materials by considering structures formed when SiO_4 tetrahedra share three vertices with adjacent tetrahedra. When this occurs the sheet silicate structure of Figure 18.23g results, having the empirical formula $Si_4O_{10}^{4-}$. Again, a wide variety of minerals containing sheet silicates are known and, like the amphiboles, these usually contain hydroxyl groups, as in the well-known mineral *talc*, $Mg_3(OH)_2Si_4O_{10}$ (Figure 18.24). The weak forces between the silicate sheets result in the material cleaving parallel to the sheets, similar to graphite (section 5.8) and MoS_2 (section 15.4.3), resulting in the lubricating properties of these layered materials.

We have already mentioned that it is very common in naturally occurring silicate minerals to find that one element can substitute for another of the same size, resulting in a wide range of mineral compositions. A very important substitution, which occurs extensively in nature and which can be achieved readily in synthetic silicate materials, is the substitution of aluminium for silicon. Aluminium, which is

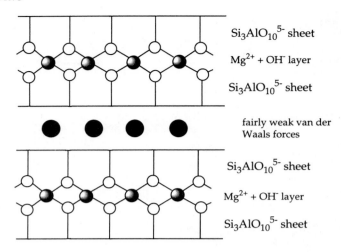

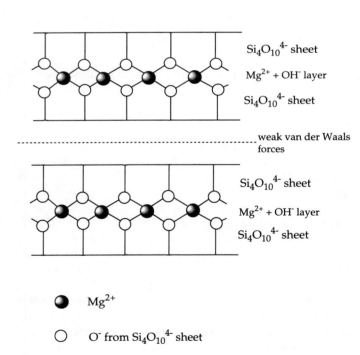

FIGURE 18.24 *Diagrammatic representation of the layered structure of talc,* $Mg_3(OH)_2Si_4O_{10}$

FIGURE 18.25 *Diagrammatic representation of the layered structure of phlogopite mica,* $KMg_3(OH)_2Si_3AlO_{10}$

the same size as silicon, readily substitutes but since aluminium is only a $3+$ ion, an additional cation is required for charge balance. Thus an Si^{4+} in the silicate framework can be substituted by the combination of an Al^{3+} with an additional (non-framework) K^+ ion. Substitution of this type happens in the *mica* group of minerals, such as in *phlogopite* $KMg_3(OH)_2Si_3AlO_{10}$ (Figure 18.25). In this sheet-silicate mineral, one in four silicon atoms of the talc structure is replaced by an aluminium atom plus a potassium atom. In micas the additional cations reside between the silicate sheets and help to bind them together by electrostatic forces. Though these binding forces are still much weaker than the bonds within the sheets themselves, micas are harder than talc but still preferentially cleave parallel to the sheets.

In addition to the micas, the main minerals in the group of sheet-silicate structures are the clay minerals such as the *kaolins*, used in ceramics, *vermiculite*, used as a soil conditioner, and *bentonite* which is used as a binder and adsorbent.

18.6.4 Three-dimensional structures

If SiO_4 tetrahedra share all four oxygen atoms with adjacent tetrahedra, the inifinite three-dimensional network structure of silica, SiO_2, results (see Figure 5.3c). Since all of the oxygen atoms are shared, and there are no terminal Si–O$^-$ groups, there is no net charge and the strongly bonded Si–O covalent network results in the crystalline forms of SiO_2 (e.g. *quartz* or *cristobalite*) which are very hard materials.

As for the sheet silicates, it is again possible to replace some of the silicon atoms by aluminium atoms, with additional cations to balance the charge, and when this occurs three-dimensional framework aluminosilicates result. These structures are represented by the important *feldspar* group of minerals which are the most abundant of all the rock-forming minerals. Examples of feldspars include *orthoclase*, $KAlSi_3O_8$ (often found as large pink crystals in granites), and the continuous solid–solution series formed between the end-members of *albite* ($NaAlSi_3O_8$) and the aluminium-rich *anorthite* ($CaAl_2Si_2O_8$).

18.6.5 *Zeolites*

A second series of framework aluminosilicates, of significant importance to the chemist, are the *zeolites*. The name zeolite comes from the Greek word for 'boiling stone'—naturally occurring zeolites are heavily hydrated minerals which, when heated, release their water giving the impression of 'boiling'! While many naturally occurring zeolites are known, these materials can also be synthesised industrially and in the laboratory, and since they have a range of industrial applications, we will discuss them in some detail.

Zeolites are typified by having aluminosilicate frameworks with very open structures and large, but highly regular, channels and cavities present in them. A common structural unit in a number of zeolites is the sodalite cage found in the zeolite mineral of the same name. The sodalite cage is depicted as a framework structure in Figure 18.26 in which the vertices represent the positions of the aluminium or silicon atoms. Conceptually, this building block can be viewed as an octahedron which has had each vertex 'sliced off', yielding a polyhedron with both square and hexagonal faces. A number of different zeolites can then be built up from the sodalite cage building blocks. When sodalite cages are joined *via* their hexagonal faces, the mineral *faujasite*, $Na_2[Al_2Si_5O_{14}]$. $10H_2O$, results whereas when they join in a cubic pattern *via* the square faces, as shown in Figure 18.27, the synthetic

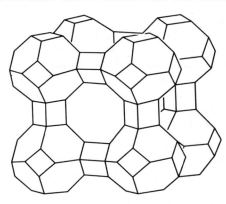

FIGURE 18.27   *The structure of zeolite A showing how the sodalite units join via linking of their square faces*
The vertices mark the position of the Si or Al atoms.

zeolite A is formed which has the composition $Na_{12}[Al_{12}Si_{12}O_{48}]$. $27H_2O$. This arrangement of sodalite cages results in a very large central cavity, called the α-cage, with an opening diameter of 420 pm. These large cages ordinarily contain the water molecules of hydration, together with the cations (e.g. Na^+). An important characteristic of zeolites is that this water can be readily removed upon heating *without destroying the open framework structure*. The resulting dehydrated zeolites then have a very strong affinity for water and one of the most important applications of these materials is as drying agents.

The synthesis of zeolites in the laboratory can be a relatively straightforward procedure in which an aqueous aluminate/silicate mixture, formed from SiO_2, Al_2O_3 and alkali, (e.g. NaOH), is heated, often under pressure, to form the zeolite. It is believed that the open structure forms around a hydrated cation and one of the common techniques for obtaining zeolites with different cage sizes is to use a large organic cation, such as the *tetra*-butyl ammonium cation, as a 'template' around which the new zeolite structure forms. The material is then heated to a high temperature in air or oxygen to 'burn off' the organic template, leaving behind the desired zeolite. By using different *tetra*-alkyl ammonium cations, different sized cavities can be formed in the zeolite. In a similar fashion it has been possible to prepare zeolites containing no aluminium whatsoever. These materials, consisting entirely of SiO_2 but with very open zeolite structures, are called *silicalite* zeolites.

The cations present inside the cages and channels of a zeolite can also be readily exchanged for other cations. One well-known application of this occurs in commercial water softeners in which the hard-water causing Mg^{2+} and Ca^{2+} ions are exchanged with the Na^+ ions of the zeolite, thereby softening the water. The recent concerns over the use of polyphosphates (another class of inorganic chain species) as water softeners has led to their replacement by zeolites in many countries. However, recent research might appear to suggest that the overall environmental impact of zeolite manufacture and use is similar to that of phosphates: the environmental influences of phosphates and zeolites are discussed in greater detail in Chapter 20.

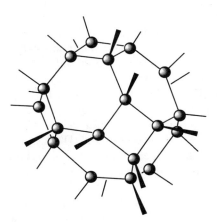

FIGURE 18.26   *The structure of the sodalite cage, the building block of a number of zeolites.*
The structure is based on a truncated octahedron. The shaded circles mark the positions of the silicon or aluminium atoms, hence each edge of the polyhedron is bridged by an oxygen atom, forming the usual M—O—M (M = Si or Al) linkages. In addition, each silicon or aluminium atom bears one terminal oxygen

The replacement of the alkali metal ions of a zeolite by appropriate transition metals, such as rhodium, chromium or vanadium, is of interest for the development of new types of catalysts. It is in catalysis that zeolites are perhaps of the greatest interest to the chemist since the presence of large cages and channels within the structure offers the opportunity to use these as a kind of 'molecular-sized reaction vessel'. Since zeolites are crystalline materials, they have very regularly sized cavities and this allows the design of very selective catalysts. The ability of zeolites to incorporate molecules of one type into their internal cavities, while excluding others, has led to the commonly used term 'molecular sieves' being coined for them.

The synthetic zeolite ZSM-5 is an excellent catalyst widely used in a number of industrial processes. A good example is the manufacture of synthetic gasoline from methanol. Since methanol is readily formed from any carbon source through the catalysed conversion of synthesis gas (CO plus H_2), this route provides a source of gasoline from non-petroleum feedstuffs and is therefore an increasingly important chemical technology, given the decline in world oil reserves. This process is carried out on a large scale in New Zealand and produces about one third of the country's gasoline requirements. A good account of the chemistry involved is given in the article by Hutchings in the reading list (Appendix A). As a result of the regular structure of the zeolite, the process is highly selective in favour of straight-chain hydrocarbons, such as octane. Branched-chain species, which do not fit into the zeolite channels, are selected against—the net result is a high octane gasoline. The acid form of zeolites is used in this type of reaction (protons are the catalytically active species involved in hydrocarbon synthesis, cracking and rearrangement reactions). Protons are produced by modifying the ammonium salt of the zeolite by ion-exchange, replacing the Na^+ ion with NH_4^+, followed by strong heating to drive off NH_3:

$$Na^+[zeolite] + NH_4^+ \rightarrow NH_4^+[zeolite] + Na^+$$
$$NH_4^+[zeolite] \rightarrow H^+[zeolite] + NH_3$$

These solid acid catalysts are of great importance as replacements for existing acid catalysts, such as concentrated sulfuric acid, used industrially in a number of processes. The zeolites have advantages in that the acidity is contained within the molecular pores of the solid, resulting in simple handling of the catalyst. In addition, the zeolite can be very easily filtered from the reaction vessel for recycling.

The regular cavities in zeolites also lead to their use in gas separation processes. Other novel uses of zeolites are certain to be developed in the not too distant future and the search for new zeolites which have different types of framework structures is an important part of the study of the chemistry of these interesting materials.

This overview of silicate structures only gives an introduction to the rich variety of structures which are possible for these materials. The large number of structures arises from the different ways in which the framework SiO_4 tetrahedra can link up and from the possibility of replacing some of the silicon atoms by others of similar size, especially aluminium. We note, however, that such linked polyoxo species are not restricted to the silicates. Polyphosphates also form important ring and chain species—chain phosphates are used as builders in detergents and are found in ATP, the energy source of life. The reader is referred to sections 7.2 and 17.6 for polyphosphates, and to Chapter 20 for a discussion of the environmental chemistry of these species. Similar polymeric species are found for the borates (section 17.4) and a very extensive series of synthetic polyoxometallate ions are known, almost beginning to rival the silicates in number and complexity. A brief summary is presented in sections 14.3 and 15.4. In the final part of this discussion we turn our attention briefly to a number of interesting hybrid materials—metal phosphates and metal phosphonates.

18.6.6 *Metal phosphates and metal phosphonates*

Metal phosphates are a very well-known class of materials which occur naturally. However, there has been a recent resurgence of interest in the structural chemistry of these (and other) materials since novel types of structures have been found. One of the simplest, yet structurally most interesting, materials is aluminium phosphate, $AlPO_4$. Since two silicons are isoelectronic to one aluminium plus one phosphorus, $AlPO_4$ is isoelectronic with $2\ SiO_2$ and has similar structural properties. As with silicates in the previous discussions, an imbalance between the number of Al and P atoms in the framework will create charged frameworks, in a similar manner to the formation of aluminosilicates. Aluminophosphates having structures similar to those of aluminosilicate zeolites are known and it is anticipated that such phosphate materials will lead to the development of new, improved and more selective catalysts. Alternative framework atoms can be introduced, such as gallium which produces the material $GaPO_4$, known as *cloverite*, named as a result of the very large clover-shaped cavities present in the structure.

The ability to synthesise highly regular solids with interesting structures has led to the investigation of a wide range of metal phosphate materials. A particularly interesting recent example is the vanadium phosphate $[(CH_3)_2NH_2]K_4[V_{10}O_{10}(H_2O)_2(OH)_4(PO_4)_7] \cdot 4H_2O$, formed in high yield from a 'one-pot' mixture of simple reactants. This compound contains a novel chiral double-helix structure—a kind of inorganic DNA!

When one of the oxygen atoms of the phosphate ion PO_4^{3-} is replaced by an organic group (e.g. a methyl or a phenyl), then a phosphonate anion, RPO_3^{2-}, results. Whereas aluminophosphates have three-dimensional structures, similar to network silicates such as SiO_2 or zeolites, the structures of metal phosphonates resembles those of sheet silicates (clays and micas) (see Figure 18.28). Such layered structures, like clays, are of interest for their catalytic and ion-exchange properties. Clay minerals have been found to swell by the intercalation between their layers of various small molecules such as alcohols. Metal phosphonates also display the same

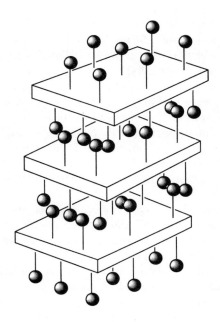

●   Phosphonate alkyl group

FIGURE 18.28 *Schematic diagram of a layered metal phosphanate*
The 'slabs' represent the metal and oxygen part of the structure with the organic substituents bonded to the phosphorus atoms depicted by shaded circles.

behaviours but have a very distinct advantage in that the organic substituent on phosphorus may be readily varied, providing the potential for the preparation of a very large range of derivatives with fine-tuning of different chemical properties. Layered materials are not restricted to phosphonates, however, and acid phosphates, in which one P–O group is protonated, such as in the zirconium compound $Zr(HPO_4)_2$, also have layered structures and are of interest as solid acid catalysts.

It can be seen from the above discussion that novel materials, having interesting structures and useful applications, can be formed from very simple inorganic precursors. It is these possibilities, together with the large number of inorganic elements available for study, which has set about a renaissance in the chemistry of materials.

18.7 Multiple bonds involving heavier Main Group elements

The discussion in section 17.2 about the restriction of multiple bonds to first row elements using $2p$ orbitals is generally valid for species which are stable at room temperature, and with ordinary ligands. There has been evidence both for unsaturated species and low-valent intermediates (such as Me_2Si or $H_2Si = SiH_2$) as the reactive species in reactions such as pyrolysis, and such compounds are realistically postulated as reaction intermediates. The isolation of stable compounds with double bonds has also been accomplished, but requires special conditions.

An example is the reaction sequence

$$R_2ECl_2 + LiAr \rightarrow (R_2E)_3 \xrightarrow{\text{photolysis}} R_2E \!=\! ER_2$$

Here, E is Si, Ge or Sn, and R is a sterically hindered ligand like Me_3C-, $(Me_3Si)_{3-n}CH_n$- (for $n = 0$ or 1 usually), or a polymethylbenzene group with at least one, and usually two, methyl groups *ortho* to the E – Ar bond. With smaller R groups, the cyclic trimer or larger rings, $(ER_2)_n$, form the stable products. In such systems, it is also possible to produce the divalent monomer, R_2E, which may be in equilibrium with double-bonded $R_2E \!=\! ER_2$. The product formed depends both on R and E, and also on the reductant and the activating energy.

The properties of the products are shown in Table 18.9.

The structures show some interesting features. The E–E distance is less than the sum of single bond radii, but the shortening is very variable, ranging from about 2% to values similar to that for a $C \!=\! C$ double bond. The $R_2E \!=\! ER_2$ species also differ from C analogues in being increasingly non-planar as E gets heavier. The angle given in Table 18.9 is that between the R_2E plane and the all-planar position. These structures imply that the R_2E unit retains some lone-pair character in the dimer.

Similar features are found for the heavier elements of the nitrogen Group. The chemistry of $RP=C$ species is particularly well explored and $P \equiv C$ compounds are reported. Mixed species both within and between Groups are established, such as $RSb=AsR$ or $R_2Ge=PR$. Those involving light elements C, N or O, such as $R_2C=SiR'_2$, are currently being explored with particular vigour. Stable compounds result when R and R′ are large groups which hinder polymerization or other further reaction. Further recent examples include a germa- or stanna-imine, with E=N, and a Te=C bond in $F_2C=Te$. As expected, the lack of bulky shielding groups results in very rapid dimerization of the latter to a four-membered Te–C–Te–C ring above $-196\,^\circ C$. The B=B system has been established and there is even evidence for a triple bond between N and Si in C_6H_5NSi formed transiently in the gas phase.

TABLE 18.9 Parameters for multiply-bonded species

E	E – E (pm)	Single E – E bond length (pm)	Non-planar angle (°)
Si	213–216	232	0–18
Ge	221–235	244	15–32
Sn	276	281	41
P	200–203	221	
As	222–225	243	
P = As	212	232	

Note. For E = Si, Ge, Sn, values refer to $R_2E \!=\! ER_2$: for E = P, As values are for $RE \!=\! ER$. Range of values are for a variety of hindered ligands R.

Finally, we note that Main Group–Transition Element multiple bonds are also seen. Not only do we have the long-established metal carbenes, $L_xM{=}CR_2$, and carbynes, L_xMCR, but similar heavy-element compounds are known. One example of a growing class is $[(CO)_5Mn]_2Ge$, which contains a linear heavy atom skeleton which is formulated as $Mn = Ge = Mn$ on the basis of this unexpected geometry, and also on the bond shortening.

Calculations suggest that, while rings involving only single bonds, e.g. $(R_2E)_n$ or $(RE)_n$, are the most stable, both the double-bonded species and the monomers, R_2E or RE, are not substantially less stable, so that steric effects may be enough to tip the balance.

These compounds generally show reactions similar to those of classical unsaturated species, particularly addition reactions across the double bond by $H - X$ or $R - X$.

18.8 Commentary on VSEPR

As Chapter 17 shows, the Valence Shell Electron Pair Repulsion approach is very successful in predicting and rationalizing the shapes of molecules, despite its very simple premises. As with all successful models, it has been subject to further development and to criticism, and its potential and limits have become well-defined. To illustrate its successful application, we note that rare gas chemistry, and almost every halogen oxyfluoro compound listed in Table 17.21 and Figure 17.60, are post-VSEPR and the structures were found to be as predicted. In these, and many other cases, the prediction has been extremely good, *within the limits of the theory*. For example, VSEPR, does not predict the preferred linkage isomer or the degree of polymerization (for example the formation of tetrameric XeF_6 in the solid) but it will predict the structure corresponding to each specific formulation.

Many improvements to VSEPR have been suggested, but most would only marginally extend its scope at the price of the essential simplicity of the approach. It is becoming clear that it is probably best to settle for a straightforward version of VSEPR, and leave the finer details of structures to be properly treated at a more sophisticated level.

Indeed, in order to preserve the enormous value of the VSEPR approach as a rule-of-thumb prediction, it might even be appropriate to reduce the weight placed on the rationalizing ideas and reformulate the guideline: *The shape of a polyatomic species is the one which would result if the shape is governed by the repulsion of electron-rich regions of valency electrons: viz., the bond pairs, the lone- pairs, and multiple bond, as set out in Tables 4.1, 4.2 and 4.4.*

This is not simply a quibble; it means that VSEPR may be preserved even if quite different explanations appear from deeper treatments of the shapes. For example, it is clear that simple ideas of steric hindrance would also account for the shapes of Table 4.1, as would a model of repulsion (or attraction!) by the electron clouds of the ligands. In the reading list of Appendix A are references to a number of such alternative approaches which the reader should assess.

It is useful to examine three areas in a little more detail.

18.8.1 *Experimental electron densities*

A major premise of VSEPR is that lone pairs are sterically active, that is that they occupy a definite direction in space and are not spherically symmetric. As valency electron distribution is not measurable by standard structural techniques, the presence of sterically active lone pairs has had to be deduced less directly. For example, the existence of a dipole rules out a planar structure for NH_3 or a linear structure for GeF_2. The most common evidence is the atom positions from diffraction experiments. Where these match VSEPR predictions, the postulated lone pair position is supported, though not proved.

The recent development of methods of determining the density distribution of valence shell electrons (see section 19.1) has allowed the location of lone pair electron densities. The experiments are difficult, so the number of examples is small, but the results are basically in accord with VSEPR predictions. One case which is particularly striking, because the difficulties are enhanced with heavy atoms, is a study of Me_2TeCl_2. VSEPR makes this an $AB_2B_2'L$ case where the more electronegative Cl atoms should be in the apex positions and the two Me groups and the lone pair in the equatorial position. Figure 18.29 shows schematic electron density difference maps (see section 19.1 for explanation) in the $TeMe_2$ plane and in the plane through ClTeCl bisecting the CTeC angle. The position of the lone pair is exactly as expected.

Such studies thus amount to direct observation of sterically active lone pairs, completing the indirect evidence from atom positions.

18.8.2 *Limiting cases*

(1) VSEPR is at the limits of its usefulness (i) for d^n configurations which are better treated as in Chapter 13, and (ii) for more than six electron pairs. In such species, the energy difference between alternative possible configurations is small, and the predictive power of the theory fades out. For example, although IF_7 is a pentagonal bipyramid, as expected, the very small angle of $72°$ in the equatorial plane is relieved by distortion. While this is expected from the basic VSEPR rationalization, the difference between a distorted pentagonal bipyramid and, say, a distorted capped octahedron, is small and almost a matter of semantics only.

In the particular case of AB_6L compounds, where VSEPR predicts that the structure will be distorted from a regular octahedron by the presence of the lone pair, both distorted and regular structures are seen (section 17.7.4). In structures determined so far, distorted octahedra are usually seen when B is a highly electronegative ligand such as F. However, the balance is clearly a very fine one since $[BrF_6]^-$ is a regular octahedron while $[IF_6]^-$ is distorted. When less electronegative ligands like Cl, Br or I are present, the structures reported so far are regular octahedra. It is likely that, as the use of very large cations is explored, more dichotomies will be found. A number of explanations, including a simple steric one, have been put forward. From a simple approach to VSEPR, it is probably better to note that this case marks one limit of applicability of the system.

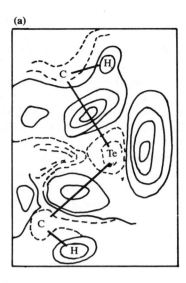

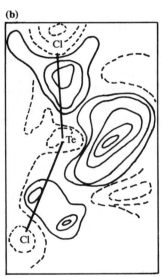

FIGURE 18.29 *Representation of the electron density difference distribution in Me_2TeCl_2 (a) in the equatorial plane (b) in the axial plane.*

(2) Molecules like $(SiH_3)_3N$ (Figure 17.22) and related compounds with bonds between Si and N or O, violate the predictions. The AB_3L nitrogen is planar not pyramidal, and the AB_2L_2 oxygen is linear, or at least has a much wider angle than the VSEPR prediction of 'less than tetrahedral'. These cases were explained by additional pi-bonding (Figure 17.23) and the function of the VSEPR prediction is essentially to highlight a compound class where an extra effect occurs.

(3) Some anomalous structures involve fewer than six electron pairs, as in the symmetrical sandwich structure found for $(Ph_5C_5)_2Sn$, where the lone pair on tin should lead to a bent structure. In such cases the explanation is a simple steric one. This is part of a more general series of observations that very bulky groups, like *tert*-butyl or $(Me_3Si)_3C$-, force unusual stereochemistries. Similarly, rigid ligands with a specific coordination site may force unusual coordinations, such as the square planar Si or Ge in their porphyrin

complexes. The overriding effects of such constraints apply, not only to VSEPR, but to other structural approaches.

18.8.3 AB_2 dihalides

Steric effects cannot account for some anomalous AB_2 dihalide structures. Normal VSEPR predictions apply to relatively covalent species: the AB_2 dihalides of Zn, Cd, and Hg are linear, and the AB_2L dihalides of the carbon Group are bent (with angle sequences such as $105°$, $101°$, and $97°$ for CF_2, SiF_2, and GeF_2 respectively, paralleling Table 4.3 changes). However, work on the triatomic MX_2 species found in the gas phase above heated alkaline earth element dihalides shows the following pattern of angles ($°$):

	Be	Mg	Ca	Sr	Ba
MF_2	linear	linear	140	108	100
MCl_2	linear	linear	linear	120	ca. 100
MBr_2	linear	linear	linear	linear	bent
MI_2	linear	linear	linear	linear	bent

These results arise from three different types of experiment: (i) detection of polar molecules by electric field deflection and mass spectrometry in a high-temperature sample; (ii) electron diffraction of gas samples; (iii) infrared study of isolated MX_2 species deposited from a high-temperature beam into a krypton matrix at 20 K.

For all these dihalides, VSEPR predicts a linear structure, as does a simple electrostatic model $X^- M^{2+} X^-$. Explanations proposed have usually involved participation of d orbitals, which involve less excitation energy for the larger atoms. From the general aspect of VSEPR we simply have to accept these bent species as anomalous, and it may be that other transient, high-temperature, ion multiplets will also show unexpected structures (through the dipole method shows linear $Li^+ O^{2-} Li^+$). It may be noted that the alternative approach to molecular geometry through Valence Bond theory proposed by Smith (see reference, Appendix A) does account for these MX_2 cases, in addition to covering the VSEPR-accurate structures.

18.8.4 *The problem of s^2 configurations*

In section 4.5, we noted that molecules with bond angles close to $90°$ could be discussed in terms of bonds involving the p orbitals, leaving non-bonding electrons in the s orbital. Further, the photoelectron spectrum of water suggests that the most stable molecular orbital is very close in energy to the unperturbed oxygen $2s$ orbital (Figures 4.20 and 21). These examples represent the general case where VSEPR states that there is at least one sterically active lone pair, and bonding theory places an essentially non-bonding electron pair in an s orbital which is spherically symmetric. Allied to this, we might add the complaint that while there is evidence of one sterically active lone pair, there is no real evidence for two.

Such a problem arises because ideas at different levels of sophistication are being mixed. VSEPR is essentially an empirical model supported by a relatively low-level rationalization. For example, VSEPR arrives at the AB_2L_2 electron count for H_2O and deduces an angle less than tetrahedral. While the relation with AB_4, e.g. for CH_4, is taken to imply some type of sp^3 hybrid, that explanation is not intrinsic to VSEPR. We can take VSEPR at a basic level of sophistication as predicting an angle less than $109\frac{1}{2}°$ for AB_2L_2 without requiring any further interpretation along the lines of 'and the two non-bonding pairs are in sp^3 hybrids'. The shape prediction is at one level while bonding theory discussion is at another, and the difficulty arises because these two are mixed in a way that is not logically required (even if almost irresistible for the chemist!)

A response at the level of bonding theory is to note that the s orbital will only be completely independent if this is imposed by other constraints (for example, by symmetry). Thus, in the water molecule, the $(s + s)$ combination of the H orbitals has the same symmetry, in the $\mathbf{C_{2v}}$ point group of the molecule (see Appendix C), as the oxygen $2s$ orbital. Thus, some interaction will occur, even if the energy difference means this will be very small.

In general, there will not be an absolutely unperturbed s^2 configuration, and only minor contribution is needed to remove the spherical symmetry. This would allow a reasonable description of PH_3, for example, as containing a lone pair with at least some excess electron density concentrated in the direction away from the three $P-H$ bonds. For two lone pairs, it is not necessary that they be in equivalent orbitals, and it is difficult to see how this could be distinguished by experiment.

It should be emphasized that all the exceptions and anomalies discussed in this section amount to only a tiny fraction of the number of structures where VSEPR predictions are fulfilled.

From a broader viewpoint, it is best to keep the two processes, (i) prediction/rationalization of shape by VSEPR, and (ii) description of the bonding by VB or MO or other approach, quite separate, remembering that the strength of VSEPR is in its simple and qualitative approach. Other accounts of molecular shape are available, and it may be that future developments—based perhaps on greatly increased computing power—will supersede VSEPR. At present, it is valuable at its own level, and with its now well-defined limits.

18.9 Bonding in compounds of the heavier Main Group elements

In Chapter 17, and in the earlier discussion in Chapter 4, the general picture was developed that the heavier Main Group elements differed from the second row, B, C, N, O, and F, by (i) not forming pi orbitals from their p orbitals, and (ii) using their valence shell, but higher-energy, d orbitals to allow expansion of the octet. Although double bonds between the heavier elements are found (section 18.7), the pi contribution to the bond strength is only about half the sigma contribution (contrast C, N, and O where the pi contribution is equal or larger than the sigma). While there is significant shortening, compared with the single bond, in molecules like $R_2Si = SiR_2$ or $RP = PR$ this will reflect not only p pi contributions but also contributions from d orbitals and from reductions in repulsion energy. Thus, it is still sound to regard

pi bonding by p orbitals as characteristic of the light elements only.

The degree of involvement of the Main Group element d orbitals has been made clearer by substantial theoretical investigation over the last two or three decades. Take a representative example, PF_2H_3, where the F atoms are axial and the H atoms equatorial in a trigonal bipyramid. Regard the planar PH_3 unit as bonded by the s, p_x and p_y orbitals on P, leaving the p_z orbital pointing towards the two F atoms. If three-centre orbitals are formed with the p_z orbitals on each F and on P (compare Figure 17.68) then PF_2H_3 would be constructed without using d orbitals. However, adding some contribution from the d_{z^2} orbital on P to the non-bonding $(-p_1 + p_2)$ combination of Figure 17.68 creates a $(-p_1 + d_{z^2} + p_2)$ orbital which is now bonding in character (and an out-of-phase antibonding equivalent). As the originally non-bonded $(-p_1 + p_2)$ orbital was occupied, the effect of adding d contribution is to stabilise these electrons and increase the total bonding. This does not require the d contribution to be as high as the p one. The calculation arrives at a contribution of about 25% by the use of d orbitals to the overall energy, with a d orbital population of around a quarter of an electron.

A further case where d orbitals have commonly been invoked is where it seems that double bonds to O are needed, as in $F_3P = O$; compare Figures 4.23 and 4.24. A calculation on H_3PO suggests that including a d orbital contribution shortens the bond length by about 15 pm compared with a single $P-O$ bond, but with a d orbital population of only about 0.5 electron. As there are two equivalent d orbitals available, a more appropriate description is a partial triple bond. There is no suggestion of full d orbital participation, and the description as a semipolar bond, $H_3P^+ - O^-$ is preferred, with the d orbital involved in partial back-donation to reduce the charge separation. The semi-polar description is the only one appropriate for H_3NO.

Thus, in all Main Group oxygen compounds which are formally written with $E = O$ bonds, the better description is probably that of semi-polar bonds modified by back-donation. This does involve some use of d orbitals, but does not amount to full d participation. For NO_4^{3-} and NOF_3, the semipolar description is the most useful one, and the bond order for $N-O$ is unity.

In summary, Main Group compounds which cannot be accounted for by up to four two-centre two-electron bonds, are probably best described in terms of multicentred bonds involving p orbitals, and in terms of semi-polar bonds where appropriate, rather than by the older descriptions (e.g. sp^3d^2 hybrids) implying d orbital populations of several electrons. Addition of some d character, of up to one electron, substantially improves the energy, bond length and other parameters. None of these approaches seems, at present, to be particularly useful for species with coordination numbers greater than six, or with AB_6L compounds. Overall, the limited contribution of d orbital occupation reflects the penalty in activation energy required to populate them.

In a further reorientation of thinking about Main Group bonding, the emphasis on substantial hybridization between s

and p orbitals is seen to be more appropriate for the $2s$ and $2p$ orbitals than for those of higher quantum number. As the $2s$ and $2p$ orbitals are less extended, and also localized in roughly the same region of space, hybridization is more effective. At the same time, the shorter distances mean that repulsions become more important, and the ability to remove lone pair density from bonding directions by hybridization is significant. For the more diffuse $3p$ orbitals, hybridization with the relatively compact $3s$ orbital is less effective, and the larger size means that repulsions are less important: similarly for higher quantum numbers. Thus the low bond angles found for compounds like PH_3 or H_2Se are the 'normal' ones, and the larger angles found for NH_3 or H_2O are the exceptional ones.

Finally, we note that relativistic effects (section 16.3) increasingly separate the s and p energy levels as the quantum number increases, stabilizing $6s$ especially. Thus the pattern of the behaviour of the elements of a Main Group (see section 17.3) re-emerges, although with different emphases in the theoretical explanations

(i) The first Group member is unique, as $2s - 2p$ hybridization is significant, repulsion effects are important because of the small size, and participation of higher level orbitals in bonding is negligible.

(ii) The remaining elements show the reverse behaviour. Participation of the d orbitals with the same principal quantum number is not as high as implied by (iii) The heaviest member of the Group has distinct behaviour, in significant part because relativistic stabilization of the $6s$ orbital reduces its role in bonding.

19 Three General Topics

In this chapter, we review three topics which tie in discussions in different parts of the preceding text. The first is the study of the distribution of the valency electrons which is basic to the whole idea of chemical bonds. In addition, it underlies the experimental ionic radii (section 2.15.2), and has given evidence for sterically active lone pairs (section 18.8.1). Current work involves both experimental and theoretical contributions.

The second topic is chosen to reflect the strong current interest in experimental fields which span transition metal and Main Group chemistries. As we have discussed transition metal–sulfur species involving single S atoms (e.g. sections 14.6.3 and 16.8.3) and also polysulfur rings (section 18.2), we have chosen to link these by surveying rings containing transition metals and sulfur, together with some discussion on related selenium and tellurium species.

The third topic is designed to illustrate the very recent advances which have been made in 'materials chemistry' and to demonstrate that materials such as carbon, long thought only to exist in the diamond and graphite allotropes (see section 5.9), can atually form a whole range of polyhedral structural forms. This 'molecular carbon', as will be seen, can form organometallic complexes with transition metals (see Chapter 16, sections 16.2 to 16.5). The existence of other nominally layered materials, such as MoS_2 (section 15.4) and boron nitride which also adopt novel polyhedral structures, suggests that this area of chemistry will become increasingly important.

19.1 Electron density determinations

The improvement in X-ray and neutron diffraction methods since about 1970 has allowed the determination of the distribution of the electron density of bonds, lone pairs and other features. This has been paralleled by improvements in the power and sophistication of molecular orbital calculations. Each approach, through experiment, or through calculation, is hedged by difficulties, but methods have been refined to the point where there is good agreement in key cases

and the results may be used with some confidence.

For the experimental approach we note (section 7.4) that X-rays are scattered by electrons, and the electron density may be calculated from the scattering pattern. The major contribution comes from the inner core electrons, which are highly concentrated, while the valency electrons give only a residual effect which has to be separated from the heavy core scattering and other contributions like those arising from thermal motion. The normal X-ray structure determination refines the atom positions to the centres of electron density, and therefore produces positions which are slightly biased by the distribution of the bond and lone pair electrons. However, accurate atom positions may be measured by neutron scattering, which is a nuclear process, not an electronic one. Alternatively, X-ray scattering at large angles depends only on the central core, and this can be used to determine atom positions.

The majority of X-ray experiments, while fully adequate for structure determination, are not sufficiently refined for electron density determination. Careful work, at low temperatures, and obtaining good high angle data (which is intrinsically less accurate), or carried out in parallel with neutron diffraction, has allowed electron density determination of a good number of molecules.

In calculations, the valency electron density is the small difference between the total electron density in the molecule and that arising from the *promolecule*, which is made up of free atoms with spherically symmetric electron density placed at the positions of the nuclei. Although computing power has increased enormously, an accurate electron density calculation requires a very good degree of approximation in the calculation (indicated by the number and type of *basis functions* used to approximate the true wave function. As the computing time goes up as the fourth power of the number of basis functions, there are clearly limits.) One important criterion for good electron density calculations (which need not handicap total energy calculations, for example) is that the chosen function should be an equally good approximation

FIGURE 19.1 *The idealized electron density distribution plot for a single bond.* Dark shading, positive density, light shading, negative density.

in all regions of the molecule to avoid introducing artefacts.

However, it is now well understood how to get the best results for both experiment and calculation, and each may be used as a test of the other. The best function to evaluate is the *density difference* which is found by subtracting, either the free atom densities from the observed one to get the *deformation density*, or else the inner shell density to get the *valence density*. A number of ways of creating such density difference maps are used (see the references in Appendix A). Such difference maps accumulate all the errors in both the total density and the free atom or core densities, so only major features are significant.

When an electron density difference map is plotted for an X–Y bond, we find a result which may be idealized as in Figure 19.1. A bond takes electron density away from the atoms to concentrate it in the bonding region and the plot is the difference in electron density between the free atoms and the molecule. Thus, a zone of negative electron density appears close to the atom positions and a peak of electron density appears in the bond region. Minor density differences are seen on the remote sides of the atoms. Contours are usually plotted at intervals of electron density of 0.05 electrons per cubic ångstrom—perhaps 0.01 for more refined results—and experimental errors are usually in the range -0.03–$0.05\,\mathrm{e\,\AA^{-3}}$ in general regions, but are much higher very close to the nucleus.

Figure 19.2 shows the electron density difference through a $\mathrm{Na^+CN^-}$ pair in a crystal of $\mathrm{NaCN.2H_2O}$, measured at 150 K. We see the very high charge density in the CN triple bond, the negative contours around N and C showing where electrons have been removed, and electron density on the remote sides of C and N corresponding to lone pairs. The lone

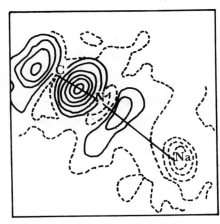

FIGURE 19.2 *Electron density difference map for a Na–N–C vector in* $\mathrm{NaCN.2H_2O}$.

pair on C is less tightly bound, and we know the cyanide ion always donates through the C when it forms complexes. Finally, we note the entirely negative, and almost spherical, distribution of electron density around the sodium ion, corresponding to the loss of the electron in forming the ion. Note that the peak electron density in the triple bond is about $0.5\,\mathrm{e\,\AA^{-3}}$.

As a second example, Figure 19.3 shows the electron density difference around Co, and along one Co–N bond in the octahedral d^6 complex ion $\mathrm{Co(NO_2)_6^{3-}}$. The plane shown passes through two opposite N atoms and bisects the equatorial plane so the remaining four $\mathrm{NO_2}$ groups are above and below the figure. The two O atoms on the N shown are also out of plane. The figure shows the region close to the N where there is extra electron density, corresponding to lone pair donation, with shift of electron density from the remote side of the N. The most interesting feature is that electron density has increased along two axes at 45° to the Co–N bond directions, and has been removed from the Co–N direction in the region

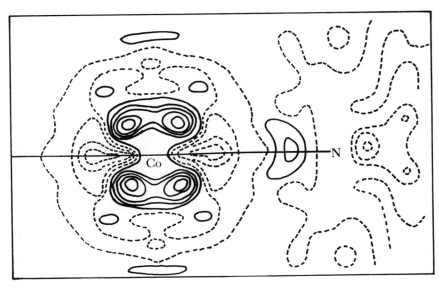

FIGURE 19.3 *A cross-section of the electron density difference in the* $\mathrm{Co(NO_2)_6^{3-}}$ *ion in a plane bisecting opposite edges of the* $\mathrm{CoN_6}$ *octahedron*

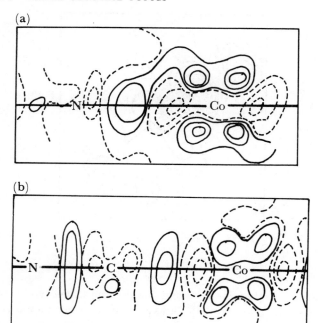

FIGURE 19.4 *The electron density differences in* (a) *the cation and* (b) *the anion of* $[Co(NH_3)_6]^{3+}[Co(CN)_6]^{3-}$

near the Co, exactly as expected for electrons moving from the e_g orbitals into the t_{2g} orbitals. When the plot in three dimensions is considered, there are eight zones, one in each quadrant (of which Figure 19.3 shows a cross-section of four) where there is an increase of electron density as expected for the filled t_{2g}^6 configuration of Co(III).

A very similar situation was seen, though less clearly, in an earlier study on $[Co(NH_3)_6][Co(CN)_6]$ where both the Co(III) complex ions showed a loss of charge density near Co along the bond directions and a gain in regions at 45° to these, though the maxima were only about 0.3 e$\mathring{A}^{-3}$. It is interesting to find that the changes in electron density were approximate ellipsoids which were much more elongated at right angles to the bond for Co–CN than for Co–NH$_3$, matching the pi back-donation postulated for such ligands. (Figure 19.4).

19.2 Metal–polychalcogenide compounds

19.2.1 *Metal polysulfur compounds*
Just as sulfur itself forms rings reflecting its preferred bond and dihedral angles, so do we find rings containing polysulfur units bonded to metal atoms. As both the metals and the sulfur chain are reasonably flexible in their steric demands, a wide variety of compounds is found.

In compounds containing an S–S unit, a range of coordination modes is found. Three-membered rings, describable also as sideways bonded S–S, are common and are analogous to the peroxides. Other coordination modes include two metal atoms linked *trans* across an S–S unit or in a double-sideways mode where each M bonds to each S.

Rings of varying sizes are formed. There are only a few examples of a four-membered ring, illustrated by

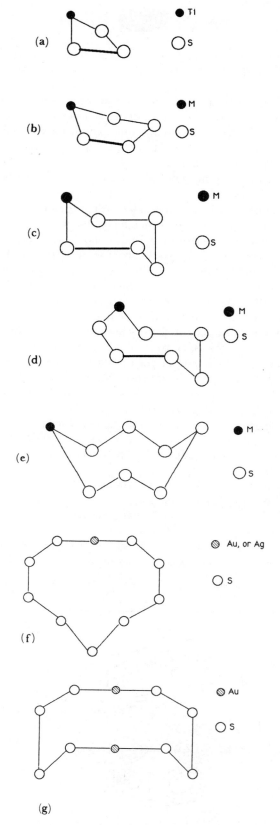

FIGURE 19.5 *Ring shapes in compounds containing* MS$_x$ *rings* (a) *four-membered ring in* $(Me_5C_5)_2TiS_3$ (b) *five-membered rings in half-chair form* (c) *the chair form of the six-membered rings* (d) *seven-membered ring* (e) *eight-membered ring* (f) *the nine-membered ring of* AuS$_9^-$ (g) Au$_2$S$_8^{2-}$

$(Me_5C_5)_2TiS_3$ where the ring is bent with the middle S atom 49° above the STiS plane (Figure 19.5a). Five-membered rings are more common, and examples include $M(S_4)_2^{2-}$ for M = Ni, Pd, Zn or Hg where the central MS_4 configuration is tetrahedral, and $(C_5H_5)_2MS_4$ for M = Mo or W. Rings are not confined to transition metals, and $Sn(S_4)_3^{2-}$ has three SnS_4 five-membered rings. These MS_4 rings are in the half-chair form of Figure 19.5b.

The best-known of all metal-sulfur ring compounds is $(C_5H_5)_2TiS_5$ which is prepared from $(C_5H_5)_2TiCl_2$ by reaction with Li_2S_2 and sulfur. It is used to form other sulfur species, as in the syntheses of some of the parent polysulfur rings (compare section 17.7). The Zr, Hf, and V analogues of the Ti compounds are known, and other examples of the MS_5 ring include $(NH_3)_2Cr(S_5)_2$ and $Pt(S_5)_3^{2-}$. These rings have the chair structure (Figure 19.5c), which is also found for the parent S_6 (compare Figure 17.45).

Larger rings are found including the 7-membered MS_6 in $[M(S_6)_2]^{2-}$ for M = Zn, Cd, Hg, 8-membered in $(R_3P)_3MS_7$ for M = Ru or Os, and also in the Ti_2S_6 ring of $[(C_5H_5)Ti]_2(S_3)_2$. The MS_6 ring has the extended chair form of S_7 and $Ti(S_3)Ti(S_3)$ rings have the crown form of S_8 (Figures 19.5d and e, compare Figures 17.45 and 17.44 respectively).

In the above rings, the metal atoms remain in tetrahedral or octahedral coordination. The preference of Au and Ag for linear 2-coordination, is accommodated by the 10-membered rings of MS_9^-, and $Au_2S_8^{2-}$, shown in Figures 19.5f and 19.5g. The large ring size allows approximately linear sections through the M atoms.

Much more complex polysulfides are known. A relatively open structure is found in $M_2S_{20}^{4-}$ (M = Cu or Ag) in which there are two MS_6 rings linked by two S_4 units joined *trans* in an overall kinked chain. Similarly, $Bi_2S_{34}^{4-}$ has two Bi atoms linked by a 6-membered chain and each bearing two S_7 units in 8-membered BiS_7 rings, all crowns. Most polymetal-polysulfur species tend to assume more condensed forms. A good example is $Re_4S_4(S_3)_6^{4-}$ where the central Re_4S_4 core has an S–S–S chain linking each possible pair of Re atoms. Another interesting rhenium example displaying a diverse range of metal-sulfur bonding types is the complex $[Re_2S_{16}]^{2-}$, shown in Figure 19.6. The complex contains two Re atoms linked by an Re–Re bond and bridged by two S^{2-} ions as well as by two S_3^{2-} groups. In addition, each Re atom is part of an ReS_4 ring system and overall displays a distorted octahedral geometry. Clearly, the ultimate end of progressive condensation is the formation of the metal sulfide, and it has been suggested that the polysulfur metal compounds are involved in the mobilization of metals in the geochemical formation of metal sulfide minerals.

The very wide range of M–S species includes the binary sulfides, often forming layer lattices, the thioanions which are sulfur equivalent of oxyanions, the M–S–M species of the types discussed particularly in sections 14.6, 15.4 and 15.10, and the species containing S–S links discussed above, together with many cases of mixtures of the above, as in thioanions joined by polysulfur bridges.

19.2.2 *Metal polyselenide and -telluride complexes*

Related polyselenide and polytelluride complexes showing many of the structural types of the sulfur analogues have also been described. This research area continues to be an actively studied one and one which continues to turn up an extensive range of diverse compounds. However, the chemistry of the higher chalcogenides often differs significantly from the sulfur chemistry. Multinuclear nmr spectroscopy is a very useful tool for the study of the selenium and tellurium complexes since both selenium (^{77}Se, nuclear spin $\frac{1}{2}$, 7.6% natural abundance) and tellurium (^{125}Te, nuclear spin $\frac{1}{2}$, 7% natural abundance) are readily accessible nuclei, and this technique can be used to provide information on solution-state structures of these complexes.

Examples of metal–polyselenide complexes include $[M(Se_4)_2]^{2-}$ (M = e.g. Ni, Pd, Mn, Zn, Cd and Hg), $[Pt(Se_4)_3]^{2-}$ and $[Pt(Se_4)_2]^{2-}$ which all contain five-membered MSe_4 rings analogous to the sulfur complexes (Figure 19.5(b)). Metal–polytelluride complexes appear to generate the most 'unconventional' bonding patterns, such as in the complexes $[AgTe_7]^{3-}$ and $[HgTe_7]^{2-}$, shown in Figure 19.7, where the metal shows a trigonal planar coordination geometry. The polytelluride ligand is an unusual $[\eta^3-Te_7]^{4-}$ ligand with the metals assigned to their usual Ag(I) and Hg(II) oxidation states in which trigonal planar coordination has a well-defined precedent.

Somewhat similar, in many regards, to the polychalcogenide–metal complexes are metal complexes of mixed Se/S/N compounds. The compound S_4N_4 has been known for many years (it was first reported in 1835) and is a well-known explosive (see section 17.7.7). However, by coordination to

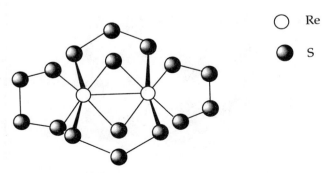

FIGURE 19.6 *The structure of the $[Re_2S_{16}]^{2-}$ ion*

○ Re

● S

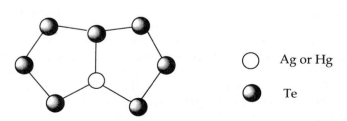

○ Ag or Hg

● Te

FIGURE 19.7 *The structures of the ions $[AgTe_7]^{3-}$ and $[HgTe_7]^{2-}$*

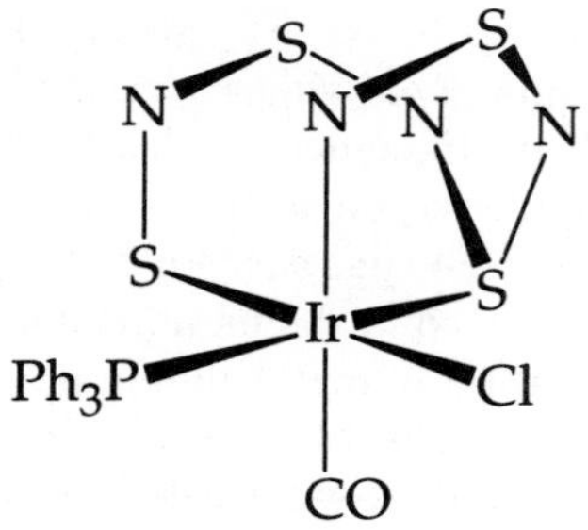

FIGURE 19.8 *The structure of the sulfur–nitride iridium complex* $IrCl(CO)(PPh_3)(S_4N_4)$

metal centres the S_4N_4 (and related Se_4N_4) species are stabilized. The first derivative prepared was the iridium complex shown in Figure 19.8. Since then quite a diverse range of metal-stabilized unusual chalcogen-nitrides have since been synthesised, and the interested reader is referred to the additional reading in Appendix A.

19.3 Fullerenes, nanotubes and carbon 'onions'—new forms of elemental carbon

19.3.1 *Fullerenes and their metal derivatives*

It was believed for many years that there were only two well-defined allotropes of the element carbon—the three-dimensional diamond and the planar-sheet structure of graphite. However, the chemistry of carbon has recently taken on a new dimension with the discovery of molecular allotropes such as C_{60} (see section 5.9 and Figure 5.15).

The discovery of C_{60} was a rather serendipitous one. The investigators were looking at carbon-clustering experiments in the gas phase aimed at simulating the chemistry of molecules observed in cold, dark clouds in interstellar space. (The chemistry of outer space is a fascinating subject in itself and a whole range of molecules, such as the polyalkynyl cyanides, HC_xN where x is 5,7 or 9, have been identified. A very brief overview is given in section 8.7.) The C_{60}^+ molecular ion was regularly observed as a strong peak in the mass spectra and this led to a closed-shell spheroidal structure being proposed for the C_{60} molecule, as illustrated in Figure 5.15d. (The name Buckminsterfullerene was coined for this new molecule after R. Buckminster-Fuller, the architect who first designed geodesic domes.) Not until 1990 was it found that carbon soots could contain relatively large amounts of fullerenes, predominantly C_{60} but also a significant amount of C_{70} and larger fullerenes. The fullerenes can be recovered from the soot simply by extracting it with an aromatic hydrocarbon and chromatography can be used to separate C_{60} from C_{70}.

The availability of macroscopic amounts of C_{60} turned it, almost overnight, from what was essentially a curiosity to a material with an enormous number of potential applications. The synthetic procedure for making fullerenes is deceptively simple and essentially involves striking an arc between two graphite electrodes in a low-pressure helium atmosphere. In this regard it is highly surprising that the fullerenes were not discovered much earlier. The procedure is also highly suitable for use in the undergraduate chemistry laboratory and reference to this is made in Appendix A.

Since their discovery fullerenes have also turned up in some rather extraordinary places! C_{60} and C_{70} have been detected in *fulgurite*, a glassy rock formed where lightning strikes the ground—presumably the intense conditions of the lightning strike provide sufficient energy for the fullerenes to form. Fullerenes have also very recently been detected in an impact crater on a spacecraft which had been in orbit for almost six years. The actual mechanism of formation of these carbon polyhedra is currently the subject of intensive research.

The structures of the fullerenes are now well established as there have been many crystallographic determinations of the structures, both of the parent fullerenes and their derivatives. The structure of the C_{70} molecule, given in Figure 19.9, is similar to that of C_{60} except that it is elongated by insertion of additional hexagons. Again it contains the twelve pentagons necessary to form a closed polyhedron. It was originally envisaged that C_{60} was a kind of 'three-dimensional' benzene, or graphite, with fully aromatic properties. The amount of aromatic character is still under scientific debate, though it appears from [3]He nmr data of helium atoms trapped within the fullerene cage that C_{70} has the greater amount of aromatic character, based on the diamagnetic shifts of the [3]He resonance. In fact the C_{60} molecule contains 'shorter' and 'longer' C–C bonds rather than having all bonds equal which would be expected if the molecule were fully conjugated. This influences the chemistry of these molecules as described later.

Not surprisingly, chemists, physicists and materials scientists all around the world rapidly began investigating the properties of the fullerenes almost as if a completely new element had been discovered. Given the pace of developments in this field, we can give only a brief mention of this chemistry here and without doubt this will be thoroughly out-of-date by the time this book reaches print. We do not even attempt to apologise for this since it clearly indicates the vigorous

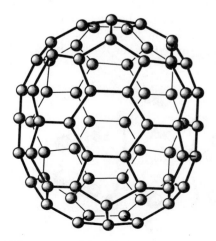

FIGURE 19.9 *The structure of the fullerene* C_{70}

activity which is occurring in this area of chemistry today. A selection of review articles from the recent literature is included in Appendix A and almost any journal in the current literature, particularly *Science* and *Nature*, will have the latest findings.

Of particular interest are fluorinated derivatives, such as the perfluorinated $C_{60}F_{60}$. Due to the excellent lubricating properties both of graphite and fluorinated polymers, such as Teflon, it was envisaged that $C_{60}F_{60}$ would be a 'molecular lubricant', though whether or not such applications arise remains to be seen.

The coordination chemistry of the fullerenes was one of the earliest areas studied and it also serves a very useful purpose in structurally characterizing these molecules. The problem lies with the high symmetry of the fullerene molecules which in turn produces rotational disorder of the molecules in the crystal lattice. Making a metal derivative disrupts the symmetry, thereby reducing the probability of disorder.

The first metal derivatives prepared were osmylated complexes formed by addition of OsO_4 and pyridine to the C_{60} framework. The complex formed is shown schematically in Figure 19.10. Very high selectivity is typically observed, with the osmium adding to a 6:6 ring junction where the highest degree of carbon–carbon double bond character exists. (C_{60} contains both 6:6 and 6:5 ring junctins but no 5:5 ring junctions—a soccerball provides an extremely useful model!) This chemistry has also been used to form the first example of an optically active element since the fullerene C_{82} is chiral (it contains either a left-handed or a right-handed helical structure). By using OsO_4 with a chiral pyridine ligand, a kinetic resolution of the different enantiomers is achieved. One enantiomer reacts quicker with one form of the base, giving diasteroisomers which can then be separated by physical methods. The osmium can then be removed by reduction with $SnCl_2$, giving the resolved enantiomers of C_{82}.

A range of organometallic derivatives of fullerenes has also been synthesized and in all of these the general properties of the complexes point towards the fullerenes behaving more like electron-deficient alkenes than electron-rich benzene-like ligands. For example, the electron-rich zero-valent platinum–ethylene complex $Pt(\eta^2\text{-}C_2H_4)(PPh_3)_2$ (see section 16.4 for a discussion on metal–alkene complexes) reacts with C_{60} forming the $\eta^2\text{-}C_{60}$ complex $Pt(\eta^2\text{-}C_{60})(PPh_3)_2$ in which again the platinum bonds to the more reactive 6:6 ring junction. Quite a wide range of complexes with other metals and fullerenes has been reported. Perhaps the most compelling evidence for the electron-deficient character of C_{60} comes from ruthenium derivatives. The pentamethylcyclopentadienyl complex $[Cp^*Ru(NCCH_3)_3]^+$ contains labile methyl cyanide ligands which are readily substituted by benzene and most other arenes, to form η^6-arene complexes:

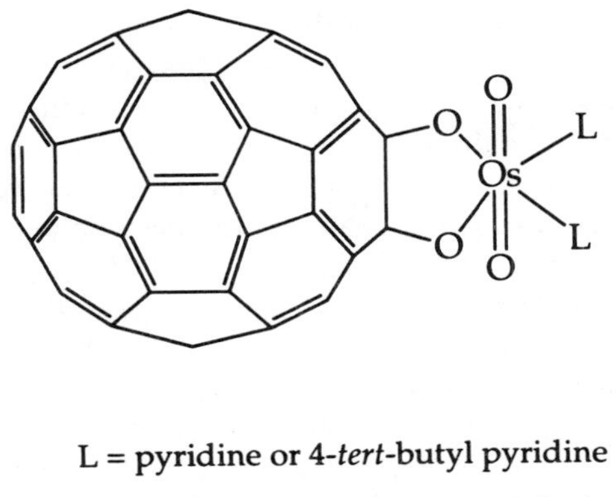

$$[Cp^*Ru(NCCH_3)_3]^+ + \text{arene} \rightarrow [Cp^*Ru(\eta^6\text{-arene})]^+ \\ + 3CH_3CN$$

Compare this hybrid sandwich complex with both ferrocene, Cp_2Fe, and dibenzenechromium, $(\eta^6\text{-}C_6H_6)_2Cr$ (section 16.4). However, when C_{60} is reacted in the same way only one of the three methyl cyanide ligands is displaced from each ruthenium and three rutheniums add to each C_{60}:

$$\text{excess } [Cp^*Ru(NCCH_3)_3]^+ + C_{60} \rightarrow \\ [Cp^*Ru(NCCH_3)_2]_3C_{60}^{3+} + 3CH_3CN$$

This is more reminiscent of the reaction of $[Cp^*Ru(NCCH_3)_3]^+$ with electron-deficient alkenes which also only substitute a single methyl cyanide ligand from the ruthenium.

In addition to these 'traditional' complexes of fullerenes, another completely different class of metal derivative has been found, though these are much less well-defined at the present time. It has been found that the vaporization of a composite rod composed of graphite and a metal oxide, for example La_2O_3, under similar conditions as for the preparation of fullerenes, leads to a mixture of fullerenes together with species in which the lanthanum (or other metal atom) is trapped within the fullerene cage. These derivatives have been termed *endohedral* complexes (in order to form a distinction with the organometallic and osmylated derivatives which are *exohedral*). The terminology $La@C_{60}$ has been coined to describe the endohedral lanthanum complex of C_{60}. The properties of these endohedral complexes appear to be somewhat different from those of the parent fullerenes, presumably because the metal causes a significant perturbation in the electronic structure of the fullerene. Larger fullerenes also allow the encapsulation of two or more metal atoms, for example in $Y_2@C_{82}$.

Another class of metal derivatives of the fullerenes are the fulleride salts formed by reaction of the fullerenes, e.g. C_{60}, with alkali metals. The many vacant molecular orbitals of the C_{60} molecule allow it to accept electrons (from the alkali metals) forming compounds containing the fulleride anions, e.g. M_3C_{60} and M_6C_{60}. The former of these is of current interest since it has been found to superconduct (section 16.1) at low temperatures. The solid-state structure of M_3C_{60} is also worthy of comment. C_{60} itself, as might be expected of a highly symmetrical, pseudo-spherical molecule, packs

L = **pyridine** or 4-*tert*-**butyl pyridine**

FIGURE 19.10 *The structure of the osmylated* C_{60} *derivative formed on reaction of* C_{60} *with* OsO_4 *and pyridine/t-butyl pyridine*

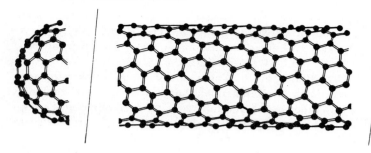

FIGURE 19.11 *Structure of a single carbon nanotube ('Buckytube')*
The structure can be considered as a graphite sheet folded to form a cylinder with fullerene-like hemispherical caps at both ends to eliminate all 'dangling' bonds. Carbon nanotubes prepared in the laboratory typically consist of many concentric, or 'nested', nanotubes. Similar structures are observed for other layered materials, e.g. WS_2

together quite efficiently and the compound crystallizes in a face-centred cubic lattice (section 5.6). It will be recalled that for every atom in a close-packed lattice there are two tetrahedral holes and one octahedral hole per lattice unit, in this case a C_{60} molecule. The structure of M_3C_{60} can therefore be derived based on a close-packed C_{60} array with all of the octahedral and tetrahedral holes filled by metal ions.

19.3.2 *Carbon nanotubes and giant fullerenes*

In addition to the simple fullerenes C_{60}, C_{70} and their larger analogues described above, several other completely new forms of elemental carbon have been discovered in the last couple of years. Variations of the procedure for forming fullerenes have led to carbon nanotubes, commonly known as buckytubes. These can be thought of as being formed from a number of sheets of graphite folded round on themselves to form 'nested' cylinders the ends of which are closed with hemispherical fullerene-like caps, as shown in Figure 19.11. Yet again, 12 pentagons provide the curvature at the fullerene-like ends of the tubes. A number of novel applications can be envisaged—on heating in air in the presence of lead, the caps are oxidized away, opening the tubes which then act as 'nanopipettes' and fill with the molten lead. Such materials chemistry has enormous potential for the fabrication of nanowires which could be used in electronic devices and the like.

Another form of carbon, related to the fullerenes, is the giant nested closed-shell fullerene structures which have been termed 'carbon onions'. It has been found possible to encapsulate moisture-sensitive materials, such as LaC_2, inside these giant structures, thereby protecting them from atmospheric moisture. In this regard these materials resemble the endohedral fullerene complexes described earlier.

19.3.3 *Polyhedral structures formed by other materials: transition metal chalcogenides and metallacarbohedranes*

The recent discovery that elemental carbon forms molecular species naturally led to the investigation of other materials. The question was asked: 'If graphite, which nominally adopts a layered structure, can be converted into polyhedral forms such as fullerenes and nanotubes, can the same be done for other layered materials?' Molybdenum and tungsten dichalcogenides $MX_2 (X = S$ or $Se)$ are materials which normally adopt layered structures. These are typified by having strong bonding interactions in two dimensions (in the layer) and weaker interactions in the third direction perpendicular to the layers, as typified by the layered silicates talc and mica (section 18.6). The structure of MoS_2 consists of close-packed sulfide layers with the Mo atoms lying in trigonal prismatic holes between these layers. These slabs then stack by weak forces, as mentioned in section 15.4. This results in the material finding application as a lubricant. Recent research has found that the MX_2 compounds also form nanotube and nested polyhedral structures akin to those formed by carbon. A recent report has described the formation of similar structures by boron nitride, another layered material isoelectronic and isostructural with graphite.

Polyhedral cluster species containing a metal and carbon have also recently been prepared by the laser vaporization of a metal in a hydrocarbon atmosphere. This has been accomplished for a range of metals including Ti, Zr, V, Cr, Fe and Cu and the resulting clusters have been termed metallacarbohedranes or 'met-cars'. The most stable of these is the M_8C_{12} cluster. There is general agreement that these M_8C_{12} species form closed-shell polyhedra but the actual structures have been the subject of much recent research and a number of structures have been suggested for then, including a distorted pentagonal dodecahedron containing twelve pentagonal faces each with two metal and three carbon atoms. Other species which have been detected include the $M_{14}C_{13}$ cluster which has been proposed to have a face-centred cubic structure, essentially a $3 \times 3 \times 3 (= 27)$ atom fragment of the cubic sodium chloride lattice (Figure 5.1a). It is noteworthy that this solid-state structure is adopted by a wide number of binary metal carbides (sections 5.1 and 5.6). Preliminary studies indicate that the M_8C_{12} met-cars display a coordination chemistry and species of the type $M_8C_{12}(H_2O)_n$ $(n = 1–8)$ have been detected in which each metal atom can coordinate a donor ligand, in this case water.

Compounds such as these are changing the way that chemists think of materials, and this research is opening up a whole new and exciting area of chemistry. New discoveries and applications of these materials are certain to follow.

20 Biological, Medicinal and Environmental Inorganic Chemistry

In this chapter we have selected three topics of current interest regarding 'natural' systems where inorganic chemistry plays a significant part. One of these topics, biological inorganic chemistry or bio-inorganic chemistry, is effectively a sub-discipline on its own. The fact that nature utilizes inorganic chemistry to a large extent and with very great effect suggests that we can learn a great deal by studying the way in which nature solves a particular problem. After all, nature has had millions of years to solve problems—our own efforts, by comparison, are often dwarfed into insignificance. Often, nature's solution to a problem involves the use of a metal ion of some sort with various ligands bonded to it. This chapter covers in some detail the inorganic chemistry found and utilized in biological systems.

Having given an overview of inorganic biochemistry, we then move on to cases where nature's solutions break down and drugs are necessary for the maintainance of well-being. Increasingly, drugs made of inorganic compounds are finding application in the treatment of diseases and we have selected several topics, which include elements from various parts of the Periodic Table, to illustrate the general use of inorganic medicinal compounds both for diagnostic and therapeutic purposes.

In the third and final part of this chapter we turn our attention environmental inorganic chemistry. This is a field which is currently of great importance and concern, and consequently we feel that an inorganic chemistry textbook should at least present some discussion on this area.

20.1 Biological inorganic chemistry

20.1.1 *Introduction*

In the preceeding chapters we have surveyed the chemistry of the elements, starting from relatively simple compounds, such as oxides and halides, and have seen that the Periodic Table can be used to largely predict the properties of the elements and their compounds. Natural systems also use a very diverse range of elements. Table 20.1 shows the elements in the Periodic Table which are utilized by biological systems. Many, such as potassium, phosphorus and sulfur, are ubiquitous whilst elements such as selenium, molybdenum, copper and many others are essential to life in trace amounts. The role of many of these elements is well understood though for others educated speculation is necessary. In the first section we describe several aspects of biological inorganic chemistry, using the Periodic Table as a framework for this discussion. The use of model complexes, to mimic biological systems, is a very important experimental method and we refer to this at several points in the discussion.

20.1.2 *The s elements in biochemistry*

The ions Na^+, K^+, Mg^{2+} and Ca^{2+} are of great importance in biochemistry. Potassium is an essential plant nutrient and the major production of potassium compounds is for use in fertilizers. In animals Na^+ and K^+ ions are mobile throughout the body and participate in many cell functions, such as nerve impulse transmissions, which depend on the ratio of Na^+ to K^+ and the concentration gradient across the cell membrane. The ions are thought to traverse the cell membrane by means of channels whose surface contains donor groups in an array similar to the multidentate donors of Figure 10.5. The 6-oxygen donors (like the crown ether, Figure 10.5a) have a cavity size which matches the size of K^+ better than Na^+ while a similar 5-oxygen donor favours Na^+ over K^+ so the channels through the cell membrane may select for K^+ or Na^+ under different cell conditions. This allows the build-up of concentration gradients between the inside and outside of the cell. It also results in the creation of charge differences since the total number of cations in the cell changes and anions transer less readily or not at all. Thus, a nerve cell starts with a higher concentration of K^+ inside the cell than outside and this concentration gradient is maintained by a corresponding negative membrane potential of $-70\,mV$. On

TABLE 20.1 The Periodic Table showing the naturally occurring elements required by living systems

								Group number									
1	2	3	4	5	6	7	8	9	10	11	12	13	14	15	16	17	18
H																	*He*
Li	Be											**B**	*C*	*N*	*O*	**F**	Ne
Na	*Mg*											Al	**Si**	*P*	*S*	*Cl*	Ar
K	*Ca*	Sc	Ti	**V**	**Cr**	**Mn**	**Fe**	**Co**	**Ni**	**Cu**	**Zn**	Ga	Ge	*A̲s̲*	**Se**	*B̲r̲*	Kr
Rb	Sr	Y	Zr	Nb	**Mo**	Tc	Ru	Rh	Pd	Ag	Cd	In	*S̲n̲*	Sb	Te	**I**	Xe
Cs	Ba	Ln	Hf	Ta	W	Re	Os	Ir	Pt	Au	Hg	Tl	Pb	Bi	Po	At	Rn
Fr	Ra	Ac	Th	Pa	U												

XX Bulk biological elements
YY Trace elements believed to be essential for plants or animals
Z̲Z̲ Possible essential trace elements

stimulation the membrane becomes more permeable to Na^+ ions which flow from the higher external concentration into the lower internal one faster than the K^+ can now move out, causing a brief period when the cell loses its large internal negative charge. Thus charge may be transferred down a line of cells as each stimulates the next. The concentrations of Na^+ and K^+ are restored by chemical action *via* the hydrolysis of adenosine triphosphate (ATP) by the Na^+/K^+ *ATP-ase* enzyme.

Similar processes occur for Ca^{2+} and Mg^{2+} ions. For these doubly-charged ions major functions include the triggering of sets of biochemical transformations, probably by coordination to donor atoms and changing the configuration of macromolecules or by bringing the reacting groups closer together. Thus Ca^{2+} is accumulated in exchange for Na^+ in heart mitochondria and contraction of the heart muscle is triggered by the release of this Ca^{2+}. Proposed mechanisms for PO_4^{3-} transfer have involved Mg^{2+} ions, whose coordination changes from one phosphate group to another, in molecules such as ATP and its congeners. This reaction is critical to the pathway by which the energy obtained from the oxidation of food compounds is stored and utilized. In many mechanisms Ca^{2+} and Mg^{2+} have opposing effects just as in the Na^+/K^+ changes in nerves.

Ca^{2+} and Mg^{2+} also have more localized functions. For example, calcium is the main cation accompanying phosphate in bones and teeth. Nature uses inorganic materials as structural components to great effect, often as composite materials with proteins, etc., and we can learn much about the design of new synthetic materials by studying natural materials. Calcium-based biominerals predominate over the other Group 2 metals because of the low solubility of many calcium salts (such as carbonates, sulfates and phosphates) together with the relatively high calcium levels in extracellular fluids. By comparison, magnesium salts are typically more soluble than their calcium analogues and no simple magnesium biominerals have yet been found. Living things also utilize many other solid-state inorganic compounds for structural and other functional means.

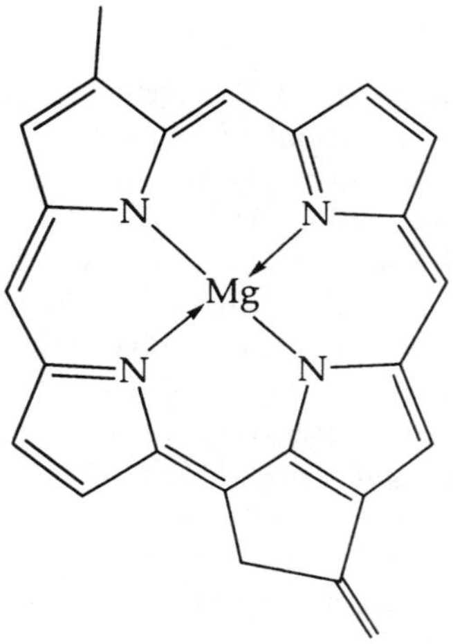

FIGURE 20.1 *The structure round the magnesium ion in chlorophyll a*

Magnesium is found in chlorophylls, the pigments which are responsible for nearly all the conversion of CO_2 and H_2O into organic molecules using the energy of sunlight, i.e. photosynthesis. This is of great importance when the overall carbon balance of the planet is considered. Plants act as sinks for the CO_2 added to the atmosphere as a result of anthropogenic activity (see section 20.3.3). The central coordination of magnesium in chlorophyll is to four nitrogen atoms in a macrocyclic *porphyrin* ring, as shown in Figure 20.1. This environment is very similar to that found for iron in haemoglobin (see section 20.1.4.2) or for cobalt in vitamin B_{12} coenzyme (see Figure 20.7). Chlorophyll absorbs radiation at both the red and at the blue to near-ultraviolet ends of the spectrum. Radiation between these wavelengths is not absorbed thus giving the green colour which is characteristic of chlorophyll-containing plants.

20.1.3 The p elements in biochemistry

The *p* elements have a diverse range of roles in biological systems ranging from carbon, nitrogen, oxygen, sulfur and hydrogen, the building blocks of organic molecules, *via* phosphates and sulfates as structural components, to trace elements such as selenium and silicon.

Phosphate plays an extremely important role in the energy changes of cell processes and also in buffering acidity changes. The classical interconversion between adenosine triphosphate (ATP) and the diphosphate (ADP) plus phosphate was one of the earliest biochemical processes to be worked out. The role of chloride, sulfate, phosphate and carbonate/bicarbonate in controlling the ion balance and flow of protons is ubiquitous in living systems. These processes are fully described in biochemistry texts. Here we focus on two examples of p element roles of current interest.

The role of silicon in biology is a very interesting one. Amorphous silica is used as a structural compoment of the exoskeletons of radiolaria and diatoms, and has also been found in various grasses. Silicon may have an important role to play—it has been found that rats and chicks when deprived of silicon in their diet show reduced weight gains and it has been suggested that the major role of silicon could be in the minimization of aluminium toxicity. Aluminium is a toxic element due largely to its high charge/size ratio which results in it forming much more stable complexes than Ca^{2+} or Mg^{2+} with hard oxygen donor groups (such as phosphates). Aluminium has been linked with Alzheimer's disease in humans, though the exact causes have not yet been fully determined. Human uptake of aluminium can come from various sources including water treatment (floculation) chemicals and in particular, aluminium saucepans (especially when used for cooking acidic foodstuffs such as rhubarb—the combination of high acidity and good complexing agents for aluminium, such as citrates, results in relatively large amounts of soluble, and thus available, aluminium). In the presence of silicon, hydroxyaluminosilicate species may form thereby reducing the bioavailability of the toxic aluminium. It is possible that silicon starvation in laboratory animals increases the availability of aluminium, with its consequent toxic effects.

The small molecule nitric oxide, NO, and its role in biological systems has recently attracted a substantial amount of interest. It may appear as a surprise to the reader that NO is involved in biochemistry at all—it is a highly reactive, toxic gas. Nevertheless, nature uses this molecule in a variety of elegant ways, including neurotransmission, blood clotting and blood pressure control, and it also plays a role in the immune system. Since NO is a small diatomic molecule, which is uncharged, its transport is very different to the channels or other specific transport systems used for other species, such as K^+ or Na^+, and diffusion is its primary mode of movement. During muscle relaxation (which is one of the keys to understanding high blood pressure) NO is produced by endothelial cells on the insides of arteries. The NO then diffuses into the muscle cells where it binds to the iron atom which is at the active site of the enzyme responsible for controlling muscle relaxation, forming an Fe–NO complex (see section 16.2.5). In fact, now that the role of NO in blood pressure control has come to light, it explains

the longstanding success of using glyceryl trinitrate as a drug for the treatment of angina (narrowing of the heart arteries). In the body glyceryl trinitrate is converted into NO which then causes dilation of the arteries. Another drug used for vasodilation (widening of the arteries) in surgery is the NO-coordination complex sodium nitroprusside, $Na_2[Fe(CN)_5NO]$ (see Table 20.3). What glyceryl trinitrate and sodium nitroprusside have in common is the potential to generate NO in the body.

In a completely different role, NO is involved in the body's immune system, for example to kill any alien bacteria which have invaded the body. Certain cells in the defence system, called *macrophages*, kill bacteria by injecting NO into them. Once the NO has had its toxic effect it is oxidized to NO_2^- and NO_3^- and excreted from the body. It appears that an enzyme, called *NO-synthase*, is responsible for the production of NO in the body by the oxidation of the amino acid arginine (Figure 20.2).

NO also acts as a messenger molecule in the brain—electrical stimulation of certain cells at synapses results in the production of NO and it has been suggested that it may play a role in the memory function of the brain. Research reported in 1993 has also found that NO may have antiviral properties.

It is clear that these recent discoveries indicate that a small reactive molecule has very important biochemical roles to play and is a further example of the elegance of nature in utilizing certain inorganic compounds. It is certainly possible that other new biochemical roles for this simple molecule will be determined in the future. The interested reader is referred to Appendix A for a number of recent articles on this topic which reflect on the current interest.

20.1.4 *The transition metals in biochemistry*

When the roles of the various transition elements in biology are studied it is clear that the range of functions is even more diverse than with the p or s elements. This can be readily

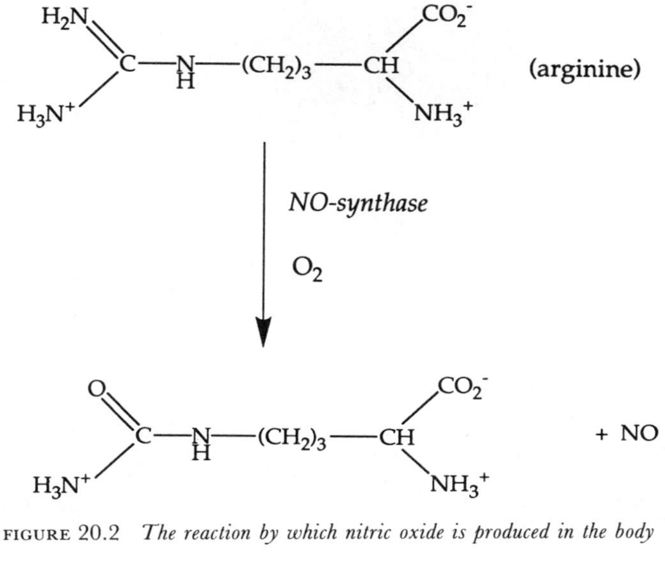

FIGURE 20.2 *The reaction by which nitric oxide is produced in the body*

understood in terms of the general chemistry of these elements where typically a range of coordination geometries and oxidation states are accessible for each element allowing many more 'options' in a biological system. A good illustration comes from enzymes—many of the most active of these contain a metal, or metals, at the active site. The presence of the metal allows small molecules (such as CO_2, H_2O, CH_4, etc.) to be selectively coordinated, and subsequently reacted, and good parallels come from a study of coordination and organometallic complexes where many of the fundamental chemical steps have been studied in detail. While we have not attempted to give a comprehensive survey, selected examples of transition elements in biological systems have been chosen to illustrate some of the chemistry. A particular emphasis is placed on iron, since it is the most abundant metallic element in the human body and plays a crucial role in the biochemistry of all living organisms.

20.1.4.1 Structural materials and iron storage.

As with the s and p Groups of elements, the transition metals, in particular iron but also manganese, are found in bioinorganic minerals. Several types of *magnetotactic* bacteria contain solid-state, magnetic, mixed-valence oxide or sulfide crystals, which are used as a means of navigation in the ambient magnetic field of the Earth. Most bacteria contain *magnetite*, Fe_3O_4, whereas some, growing in sulfur-rich regions such as geothermal areas, contain the mineral *greigite* Fe_3S_4. The single crystals, as shown in Figure 20.3, are aligned in chains. Other bacteria are able to mineralize a fairly diverse range of inorganic materials such as cadmium, zinc, copper and lead sulfides.

Related to the deposition of single-phase mineral particles within organisms is the iron storage protein *ferritin* which

contains solid iron oxide as nanometre-sized particles. Ferritin has been colourfully called 'biological rust' and the composition of the active site approximates $[FeO(OH)]_8$ $[FeO(OPO_3H_2)]$. The body needs to store and mobilize substantial amounts of iron in a form which prevents the precipitation of iron hydroxide species. Various iron carboxylate complexes have been synthesised, with different degrees of complexity, some of which are models for ferritin. An example is the complex $Fe_{11}O_6(OH)_6(O_2CPh)_{15}$ in which the iron atoms are triply bridged by O or OH and are also linked by the carboxylate groups Fe–O–CPh–O–Fe in a cubane-like structure which lacks some of the edges.

20.1.4.2 Haemoglobin and related compounds.

Perhaps the most famous complex of iron in a biological system is the complex *haem* which exists in haemoglobin. The central porphyrin ring system is shown in Figure 20.4. Side chains are attached to the porphyrin skeleton and an imidazole ring on one of

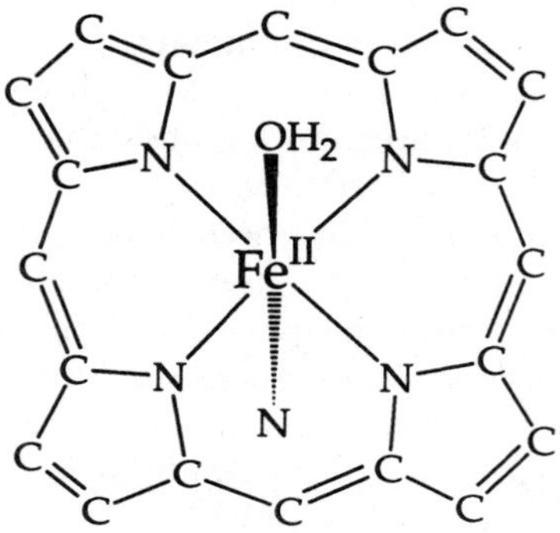

FIGURE 20.4 *The environment of the iron atom in haemoglobin*
The four N atoms of the porphyrin ring are coplanar with the iron. The fifth position on the iron is occupied by a nitrogen atom from a long side-chain on one of the rings, leaving the sixth site in the octahedron around the iron to hold an oxygen molecule, a water molecule or some other group. Side-chains and links to the rest of the protein are attached to the outer carbon atoms.

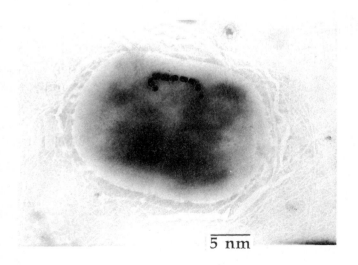

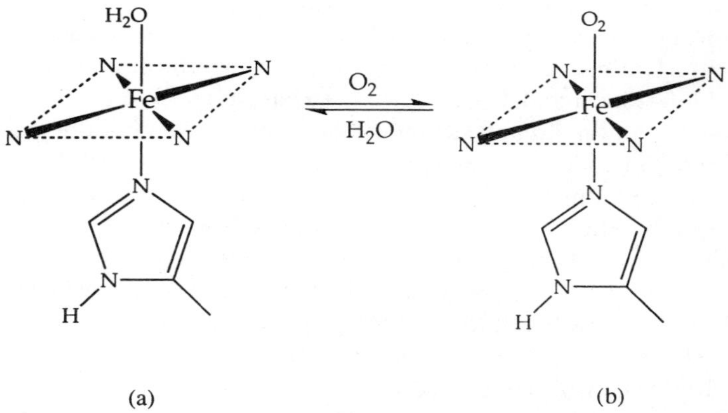

FIGURE 20.3 *A scanning electron microscopic image (see chapter 7) of a coccus-type bacterium which contains a chain of nine dark magnetite (Fe_3O_4) single crystals*
The bacterium utilizes these microscopic 'bar-magnets' for orientation in the Earth's magnetic field.
Taken from S. Mann, *Journal of the Chemical Society, Dalton Transactions*, 1993, 1.

FIGURE 20.5 *The mode of oxygen carriage by haemoglobin:* (a) *with a water molecule which is reversibly replaced by an oxygen molecule* (b)

these is coordinated to a fifth position on the iron atom. A water molecule occupies the sixth position, as shown in Figure 20.5a. This water molecule may be replaced by an O_2 molecule and this process is reversible (Figure 20.5b), providing the mechanism for the transport of oxygen by red blood cells to various parts of the body (compare section 16.7). The oxygen site on the iron may also be occupied by CN^-, CO or PF_3 and the coordination in these cases is strong and irreversible. This is one reason for the poisonous nature of these substances (although cyanide, in particular, influences other reactions in the body as well). Iron also occurs in myoglobin which is used to store oxygen in muscles.

It has been disputed for a long time whether the oxygen was carried in these respiratory proteins in an 'end-on' or 'side-on' configuration. Both coordination modes are well-known in coordination complexes of dioxygen, as described in section 16.10. Haemoglobin and myoglobin are large molecules whose structures have not been determined in the detail required to see the O_2 coordination. When simpler molecules (containing the basic porphyrin unit of Figure 20.4) are used as model complexes, the difficulty has been to obtain a species which reacts *reversibly* with O_2 as do the natural proteins. One approach is to protect the oxygen site on the Fe by using the so-called 'picket-fence' porphyrins. In these, substituents are placed on the C bridges between the C_4N rings of the porphyrin which lie above the FeN_4 ring, as indicated schematically in Figure 20.6. It was found that such molecules did oxygenate reversibly and that crystalline dioxygen complexes could be isolated. Although the crystals were marginal for X-ray work, an end-on

unit with an angle at O of about 130° was suggested. Later, and even more demanding, structural work showed an angle at O of 156° in haemoglobin itself. The O–O stretching frequencies of the picket-fence model compounds were in the

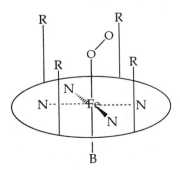

FIGURE 20.6 *'Picket-fence' porphyrin (diagrammatic)*
R = NHC(O)CMe₃, B = base, e.g. THF. Compare with Figure 20.4.

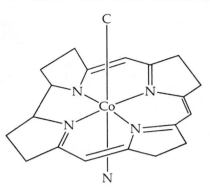

FIGURE 20.7 *The geometry about the cobalt in vitamin B_{12}*
Note that the cobalt atom is bonded to a CH_2 group of an adenosine group and is thus an example of a naturally occurring organometallic complex. Compare with the structures of chlorophyll (Figure 20.1) and haemoglobin (Figure 20.4).

range 1140–1165 cm⁻¹, similar to those for myoglobin and the cobalt analogue of haemoglobin. This has shown that the end-on, or superoxo (see section 16.10), description is the best for all the oxygenated respiratory proteins and that model compounds can be put to good use in understanding the structures of the inorganic components of highly complex biological molecules.

Similar, though not identical, to the role of iron in haemoglobin is the natural cobalt(III) complex, vitamin B_{12}. The cobalt is situated in the middle of a porphyrin-type structure (Figure 20.7) and coordinated by the four ring nitrogens and a fifth nitrogen from a side-chain group. The sixth site, completing the octahedron, is the active site and a number of derivatives are known with different groups occupying this site including CN^-, hydrogen and organometallic compounds containing a direct σ cobalt–aliphatic carbon bond.

Iron is also found in various cytochrome pigments. In these a chain of electron-transfer occurs which links dehydrogenation of alcohols or fatty acids with the conversion of O_2 to H_2O *via* a series of oxidation–reduction steps involving Fe(II)/Fe(III) (or Cu(I)/Cu(II)) conversions in various cytochromes.

20.1.4.3 Ferredoxins: biological electron-transfer agents. A further important group of iron-containing natural products are the *ferredoxins* which contain Fe and S atoms and are used in many organisms as electron-transfer reagents. Two basic structural units (Figure 20.8) have been observed, (a) Fe_2S_2 and (b) Fe_4S_4. These structures are linked to the protein skeleton by further Fe–S bonds to the sulfur-containing amino acid cysteine, $HSCH_2CH(NH_2)CO_2H$. For example, in one isolated ferredoxin (from *Peptococcus aerogenes*) there are two well-separated Fe_4S_4 units each linked to four positions, again well-separated, by the cysteine bonds. It is possible that the electron-transfer functions *via* the formation of Fe–Fe bonds across the face diagonals of the (b) units. Many iron–sulfur compounds have been synthesised as inorganic models for such systems.

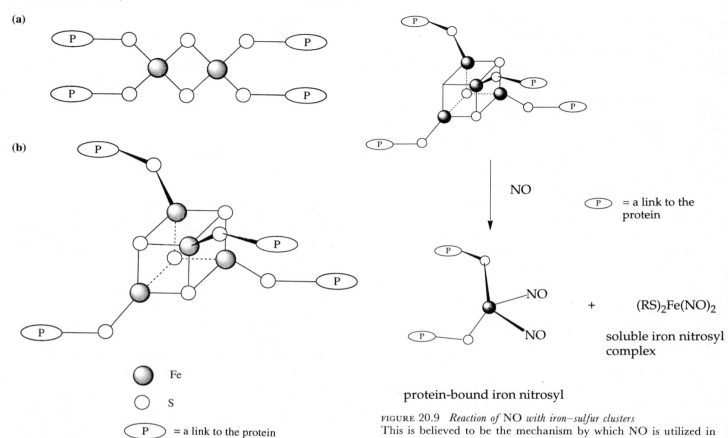

(a)

(b)

● Fe

○ S

⟨ P ⟩ = a link to the protein

FIGURE 20.8 *Two iron–sulfur binding units found in ferredoxins, nitrogenase and other iron proteins*

NO

⟨ P ⟩ = a link to the protein

+ (RS)$_2$Fe(NO)$_2$

soluble iron nitrosyl complex

protein-bound iron nitrosyl

FIGURE 20.9 *Reaction of NO with iron–sulfur clusters*
This is believed to be the mechanism by which NO is utilized in biological systems to destroy invading bacteria.

In the preceeding section the increasingly important role of NO in biochemical processes was discussed and it has been found that NO can damage cells by attacking and breaking up the Fe–S clusters present in important proteins, as shown schematically in Figure 20.9.

20.1.4.4 Enzymes—nature's catalysts. A wide range of enzymes are known to occur in biological systems, and these accomplish a large number of chemical transformation steps rapidly and with high reactant and product specificity. Many enzymes find industrial and household applications, such as lipase enzymes in detergents and glucose isomerase for the conversion of glucose to fructose in the food industry. Enzymes which occur in thermophilic, or 'heat-loving', bacteria found in geothermal regions, are also of practical importance since they function most efficiently at high temperatures. In marked contrast most other enzymes are denatured at high temperatures. Because of their specificity for chemical transformations, chemists are becoming increasingly aware of the utility of thermophilic bacteria in synthetic procedures, such as in making enantiomerically pure materials.

Many enzymes contain a metal, or a number of metals, at the 'active site' where the chemical transformation is accomplished. Table 20.2 shows the metal atoms present in

TABLE 20.2 The functions of some metal-containing enzymes

Metal ion	Small molecule reactant	Examples
Co (in B$_{12}$ cofactor)	Glycols, ribose	Rearrangements, reduction
Zn	CO$_2$, H$_2$O	Carbonic anhydrase
Zn, Mg	phosphate esters	Alkaline phosphatase
Fe, Mn	RNA	Acid phosphatase
Mo, Fe	N$_2$	Nitrogenase
Ni, Fe	CH$_4$, H$_2$	Methanogenesis, hydrogenase
Fe	O$_2$	Cytochrome P-450
Fe, Se	H$_2$O$_2$/Cl$^-$, Br$^-$, I$^-$	Catalase, peroxidase
Ni	H$_2$O/urea	Urease

a number of different enzymes, together with their functions. The determination of the structure of the metal-containing active site of an enzyme poses significant challenges since it represents only a small part of the protein structure as a whole. Nevertheless, the active sites of a number of enzymes have been determined. The study of the properties (electrochemical, spectroscopic) of model transition-metal coordination complexes again has a significant role to play in these studies. Learning about the structures of enzyme active sites may lead to the design of better synthetic catalysts and good examples are the ongoing searches to try and find practical catalysts for nitrogen fixation (see sections 16.9 and 20.1.4.5) and for artificial photosynthesis. If this can be accomplished, we will be readily able to convert the abundant nitrogen in the atmosphere into chemical feedstocks (the Haber process, section 17.6.1, can already do this but only at high temperatures and pressures using a catalyst) and harness the sun's energy directly into stored chemical energy.

As a recent example of an enzyme where the structure of the active site has been largely determined by a series of elegant experiments, we have chosen to describe a hydroxylase enzyme which catalyses the conversion of methane to methanol in methanotropic bacteria. These bacteria derive their carbon and energy solely from methane *via* the reaction:

$$CH_4 + O_2 + H^+ + NADH \rightarrow CH_3OH + H_2O + NAD^+$$

The study of the biochemical processes occurring in such bacteria is prompted by several factors, including the implication of methane as one of the greenhouse gases in global warming (see section 20.3.3), together with the possibility of using such bacteria to convert readily available natural gas into chemical feedstocks.

While spectroscopic techniques, together with biochemical methods, have given a substantial amount of information on the structure of the active site of the hydroxylase enzymes, X-ray crystallography (section 7.4) is the most powerful technique, giving direct three-dimensional structural informa-

tion. However, obtaining crystals of such large biomolecules can be painstaking work—more than 1000 crystallization trials were attempted before a suitable crystal of the hydroxylase enzyme was obtained! Even with this, since the molecule is so large, the resolution of the structure is only around 2 Å.

There are two identical active sites in the enzyme which are dinuclear, hydroxide-bridged units, the structure of which is shown in Figure 20.10. The two iron(III) atoms are bridged by a hydroxide ligand, by a semibridging carboxylate ligand from an amino acid residue (Glu144) of the protein, and by another ligand, suggested to be an acetate, which has come from the ammonium acetate buffer used in the solution from which the enzyme crystals were grown. It has been proposed that another ligand (other than acetate) bridging the two iron atoms is present in the native enzyme. One possibility is the related formate ion, HCO_2^-, since this is produced by metabolic oxidation of methanol by the bacteria.

20.1.4.5 Nitrogenase enzymes. Molybdenum-containing enzymes are responsible for nitrogen fixation in which dinitrogen in the atmosphere is converted to ammonia. Nature thus accomplishes (at about 15 °C and 0.8 atmospheres pressure of N_2) what we have been able to accomplish only at high temperatures and pressures using a catalyst. Such nitrogen-assimilating enzymes, the *nitrogenases*, occur in root nodules of clovers and other legumes. Given the increased demand for fertilizers, together with the increased cost of the energy required for high energy industrial syntheses of ammonia, there is underway a major research effort to understand, reproduce or model the N_2 fixation process. A discussion on dinitrogen complexes of the transition metals, with emphasis on the chemistry relevant to nitrogen fixation, is given in section 10.9.

Much lead-up work has established that there are two major proteins in the nitrogenases. One (a) is the nitrogen-binding protein (with a molecular weight around 230 000). This was found by spectroscopic and other methods to contain 2 Mo atoms, some 28 S atoms and 24–32 Fe atoms, and it is probable that the structural units are Fe_4S_4 or Fe_3MoS_4 cubanes. The second (b) is a smaller protein containing Fe and no Mo which carries electrons and contains Fe_4S_4 clusters.

In a major triumph of X-ray crystallography the structure of three FeMo nitrogenases was achieved at 27 pm resolution, the best so far. This, coupled with much other data, establishes that the Fe protein (b) (M_r about 60 000) contains two identical protein sub-units bridged by an Fe_4S_4 cubane unit (compare Figure 20.8). This functions as a one-electron donor for the FeMo protein, which is the major site for the transformation of N_2, by adding overall 8 H^+ and 8e to give NH_3 plus H_2.

The FeMo protein (a) is a tetramer containing two FeMo cofactor units and two units termed P-clusters. The two cofactors are 700 pm apart and so do not interact directly, contrary to earlier ideas. Each P-cluster is separated from

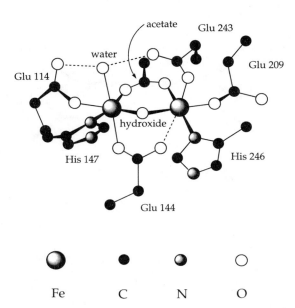

FIGURE 20.10 *The structure of the dinuclear iron centre in the active site of a methane hydroxylase enzyme*
The coordinated amino acid residues are labelled by His = histidine, Glu = glutamine.

the nearest cofactor by only 190 pm and is probably involved in the fixation steps. These metallic centres are surrounded by the protein chain which provides the S, N and other coordinating atoms found in the structure.

The metal centres of the P-clusters consist of two Fe_4S_4 cubanes placed side by side and linked by two S bridges between Fe atoms of the cubane. All these S atoms are part of cysteine components of the proteins which wrap around these centres.

The FeMo cofactor has the structure shown in Figure 20.12b which compares interestingly with the model compound of Figure 20.12a which has been prepared earlier (see below). The cofactor consists of two cubane fragments, Fe_4S_3 and Fe_3MoS_3, which are each missing one S corner. The two fragments are bridged by linking the three Fe atoms thus exposed—two by S and the third by a group which may be O or NH. The Mo is six-coordinate: to the three S of the cubane, to an N from a histidine and by two O from a homocitrate residue.

The role of the Mo is thought to be critical because uptake of N_2 seems to depend on its presence in the large majority of systems studied. The determination of the crystal structure is a major advance and will guide further studies. However, discoveries of vanadium-containing nitrogenases, and even of all-Fe species, show there is a long way to go before full understanding is achieved.

The exact mode of action between N_2 and nitrogenase is still undefined. The strategy was developed of synthesizing transition-metal clusters and exploring their chemistry as models for the natural process. This has so far been the main contribution of inorganic chemistry to this side of the nitrogen fixation problem. This chemistry of iron–sulfur clusters is described in section 20.1.4.3 and for Mo species in section 15.4. In some ways tungsten is more manageable than molybdenum and its chemistry has been studied as a parallel, creating a major area of the chemistry of these elements. This work has led to many intriguing and interesting compounds but not yet to a detailed scheme for nitrogen fixation.

To illustrate this field we start with an experiment where MoS_2 was treated with CN^- in aqueous solution. One theory of the pre-life evolution of important biomolecules emphasizes the role of cyanide in its polymeric forms, so such species may have been significant in the evolution of Mo-containing enzymes. Four products were isolated where there were increasing numbers of Mo and S atoms bound together and each Mo was made up to six-coordination by cyanide groups. The skeletons were as shown in Figure 20.11. These occurred in the ions (a) $[Mo_2S(CN)_{12}]^{6-}$, (b) $[Mo_2S_2(CN)_8]^{n-}$ for $n = 6$ or 8, (c) $[Mo_3S_4(CN)_9]^{5-}$ and (d) $[Mo_4S_4(CN)_{12}]^{4-}$.

The skeleton in (d) is a cubane cluster, a common structural unit in transition metal–sulfur cluster chemistry. The other three skeletons are formally fragments of the cubane structure: (b) is a face and (c) is formed by removing one metal corner. Similar structures are widespread both in synthetic compounds and in biological molecules. The units of Figure 20.11 may be linked together in a wide variety of ways, including double cubanes sharing a face in $[M_6S_6X_{18}]^{6-}$ species and many more open structures. One illustration is given in Figure 20.12a which parallels the nitrogenases in containing both Fe, Mo and S atoms in the cluster unit. The two cubanes are linked by three bridging S atoms through the Mo corners: note that the Mo atoms are 6-coordinate while the Fe atoms are tetrahedral.

To give an illustration of the experiments directed towards directly mimicking the natural fixation process we note also the following approach. Because of the similarity to

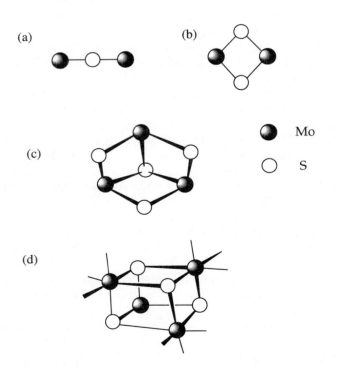

(a)

(b)

(c)

● Mo

○ S

(d)

FIGURE 20.11 *Molybdenum–sulfur skeletal units*

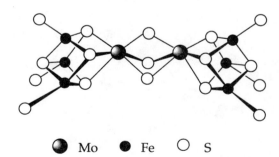

◐ Mo ● Fe ○ S

FIGURE 20.12a *The heavy atom arrangement in the ion* $[Fe_6Mo_2S_8$-$(SEt)_9]^{3-}$

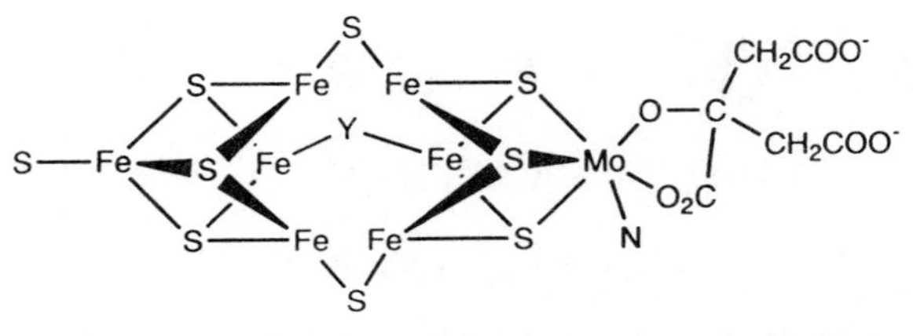

FIGURE 20.12b *Structure of the* FeMo *cofactor (Y may be O or NH)*

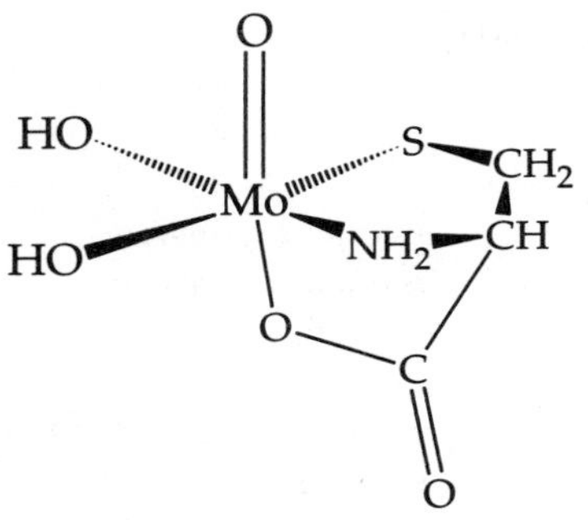

FIGURE 20.13 *A molybdenum–cysteine model complex for nitrogenase studies*

ferredoxins (section 20.1.4.3) and because no N_2 uptake occurs in the absence of molybdenum, a model system was studied consisting of $[(cysteine-S)Fe]_4S_4$ as the electron transfer agent and the molybdenum-cysteine complex shown in Figure 20.13 to react with N_2. In this system conversion of N_2 to NH_3 at about 1% of the nitrogenase rate was achieved and the postulated pathway is *via* diimine and hydrazine.

20.1.4.6 Other transition metal-containing systems. Over 80 zinc-containing enzymes have been discovered and among the more important ones are *carbonic anhydrase* and *carboxypeptidase A*. In these enzymes the principal role of the zinc is to act as a Lewis acid. For example, in carboxypeptidase A, an enzyme which catalyses the hydrolysis of the terminal peptide bond at the carboxyl end of a protein, the zinc coordinates to the carbonyl oxygen. This polarizes it more strongly, promoting nucleophilic attack at the carbon by the oxygen atom of a Zn–OH group. A similar nucleophilic Zn–OH group undergoes attack at the C=O bond of carbonic anhydrase, thereby greatly speeding up the normally slow hydration of carbon dioxide:

$$CO_{2(aq)} + H_2O = H_2CO_3$$

Copper is one of the more important elements in biological systems being the third most abundant metal ion in the body after iron and zinc. It is particularly concerned in the uptake of inorganic sulfur into organic molecules and copper(I)/copper(II) changes are involved in redox transfer systems in cytochromes in a similar way to iron(II)/(III) ones. Intensely blue copper(II)-containing proteins are typical and are involved in oxidation steps whereas copper(I) is found in haemocyanins.

Vanadium has also been found to be a biologically essential element and perhaps one of the best known occurrences is in the bromo- and iodo-peroxidase enzymes found in seaweeds. Several species of tunicates, better known as 'sea-squirts', contain relatively high concentrations of vanadium in their blood. The vanadium has been shown to exist as V(III) though small amounts of V(IV) as VO^{2+} are also detectable. Current research is directed at the establishing the detailed role of vanadium in such organisms.

Nickel occurs in the enzyme urease which catalyses the hydrolysis of urea:

$$H_2NC(O)NH_2 + H_2O \rightarrow CO_2 + 2NH_3$$

While urease was the first enzyme ever to be crystallized, the X-ray structure has only recently been carried out. This showed the active site to contain two Ni centres in accord with previous evidence obtained using other techniques.

Manganese is also extensively utilized by biological systems. As an example we quote the Mn–O–OH cluster species implicated in photosynthesis. Discussion of model complexes synthesised in relation to these biological systems is given in section 14.5 to which the reader is referred.

Many other elements are found in trace amounts in enzymes and other biological molecules, and while we cannot cover them here the interested reader is referred to the various reviews given in Appendix A.

It is finally worth noting that the biological utilization of transition metals is believed to have varied with time. The very primitive forms of life on Earth lived in a high energy, reducing environment with a planetary atmosphere of N_2, H_2S and CH_4 prevalent. It is believed that the metals which were largely employed by organisms living under these conditions employed sulfide-based metal clusters at the active sites with the metals ranging from iron, cobalt and nickel to molybdenum and vanadium, all in their lower oxidation states. At this stage metals such as copper, which forms highly insoluble sulfides, were generally not available to biological systems in any significant amounts. On the advent of photosynthesis, large-scale introduction of O_2 into the atmosphere occurred, changing it irrevocably (this can be thought of as the biggest single act of 'air pollution' the planet has observed). As a result, H_2S became oxidized to SO_4^{2-} and the sulfide-based chemistry of cobalt and nickel became less important. Oxygen-containing structural units such as the dinuclear Fe–O–Fe (present in the methane hydroxylase enzyme discussed previously in this section) began to take on an increased role in biological systems. It is believed that the inorganic biochemistry of cobalt and nickel is becoming generally of lesser importance. With the exceptions of hydrogenase enzymes and urease, nickel-based biochemistry is largely restricted to the *archaebacteria*, very primitive forms of life occurring in sulfur-rich geothermal regions. These bacteria are believed to be largely unchanged since the earliest forms of life existed. Interestingly, archaebacteria use very little copper in their metabolism. The introduction of an oxidizing planetary atmosphere mobilized copper into the environment and allowed organisms to develop copper biochemistry which is believed to have come relatively recently in evolutionary history.

20.2 Medicinal inorganic chemistry

20.2.1 Overview

The treatment of human (and animal) ailments is one of the major objectives for many researchers and medical practitioners around the world. Although the majority of

TABLE 20.3 Selected examples of inorganic medicines and imaging agents in use today

Element	Trade name	Compound	Use
Li	Camcolit	Li_2CO_3	manic depression
Mg	Magnesia	MgO	antacid, laxative
Al	Alludrox	$Al(OH)_3$	antacid
Ca	Settlers	$CaCO_3$	antacid
Fe	Nipride	$Na_2[Fe(CN)_5NO]$	hypertensive, vasodilator
Zn	Calamine	ZnO	skin ointment
Se	Selsun	SeS_2	antiseborrhoeic (shampoos)
Sr	Sensodyne	$Sr(acetate)_2$	toothpastes
Zr	Deodorants	Zr(IV) glycinate	antiperspirant
Tc	Ceretec	^{99m}Tc propyleneamine oxime	diagnostic imaging
Ag	Flamazine	Ag sulfadiazine	antibacterial
Ba	Baridol	$BaSO_4$	X-ray contrast
Gd	Magnevist	$[Gd(DTPA)](meglumine)_2$	MRI contrast
Pt	Cisplatin	cis-$[PtCl_2(NH_3)_2]$	anticancer
Au	Myocrisin	$Na_2Au(I)$ thiomalate	antiarthritic
Bi	De-Nol	Bi(III) citrate	antacid, antiulcer

new drugs result from research on organic compounds, inorganic derivatives are finding an increased role in this field. The large number of inorganic elements available, together with the chemical complexity of inorganic compounds when compared to organic ones, clearly indicates that a great many more combinations of elements are possible with inorganic systems. Of course, only a small number of these are going to be synthesized and of these only a miniscule fraction will have any useful medicinal action. Nevertheless, the search for new drugs among inorganic compounds is a search which is likely to have very fruitful outcomes. Many new drugs are based on metals which are not utilized by natural systems.

While there are a number of inorganic compounds used in medicine (summarized in Table 20.3), some of which will be highlighted in this chapter, the use of inorganic compounds has had a long but somewhat checkered history. In the past many 'treatments' using toxic inorganic compounds, such as mercury salts and even radioactive radium 'tonics', invariably did the patient more harm than good! An organoarsenic compound, arsphenamine, is considered to be the first of the modern chemotherapeutic agents, introduced in 1910 for the treatment of syphilis.

While we do not have space for more than a brief overview, we have chosen a selection of topics to illustrate some of the more important uses of inorganic compounds in medicine. Very broadly speaking these 'drugs' fall into two categories. Firstly, diagnostic agents can be used to investigate ailments without attempting to effect any treatment. As examples of these we discuss the use of radioactive technetium compounds and non-radioactive, but strongly paramagnetic, gadolinium compounds as imaging agents. The second category of drugs are used for therapeutic purposes and the examples chosen

include the use of platinum and boron compounds in cancer chemotherapy, and gold drugs in the treatment of rheumatoid arthritis. The use of lithium salts in the treatment of hypertension and chelating agents for heavy metal toxicity represent some other examples of therapeutic inorganic-based drugs.

20.2.2 Inorganic diagnostic (imaging) agents in medicine

Perhaps the most well-known 'imaging' agent is a simple inorganic salt, barium sulfate ($BaSO_4$), which is widely used as a 'barium meal' for imaging stomach ulcers and the like. Although barium, like other heavy metal ions, is highly toxic the extreme insolubility of $BaSO_4$ gives it a very low toxicity and it is the opaqueness of this material to X-rays which imparts its usefulness in this application.

In general, however, most imaging agents are based on a gamma-emitting nucleide which localizes in a specific organ of the body after being injected. The structure of the organ under examination can then be obtained by looking at the radiation emitted by the radioisotope using a scintillation camera. A number of requirements must be met by any radioactive imaging agent. These are:

(i) low toxicity
(ii) a highly specific biodistribution so that the organ in question can be clearly imaged—this is clearly important to minimize the dose of imaging agent required
(iii) low radiation dose—this means that a pure gamma emitter is required. At first sight it may appear that the injection of radioactive material into the body is a very dangerous procedure but in fact modern imaging agents expose the patient to little more radiation than a routine X-ray

(iv) short isotope half-life so that the patient's long-term exposure to radiation is minimized, though if the half-life is too short this makes transport of the imaging agent rather difficult.

20.2.3 *Technetium imaging agents*

Technetium provides an excellent, and the most widely used, example of a modern imaging agent. Various technetium complexes, containing different ligands, are commercially available to image different organs of the body. In other words, the coordination sphere of the technetium can be modified to alter its biodistribution. An overview of the chemistry of technetium is given in section 15.5.

The medically useful isotope of technetium is the metastable ^{99m}Tc. This is manufactured at the point of use by means of a parent–daughter nucleidic pair in which the relatively long-lived parent ^{99}Mo spontaneously decays to the ^{99m}Tc daughter which is then recovered for imaging. The ^{99}Mo is made in a nuclear reactor by thermal neutron fission of ^{235}U:

$$^{235}\text{U} + \text{n} \rightarrow {}^{99}\text{Zr} \xrightarrow[\beta^-]{30\,\text{s}} {}^{99}\text{Nb} \xrightarrow[\beta^-]{3\,\text{min}} {}^{99}\text{Mo}$$

Alternatively, thermal neutron activation of a MoO_3 target also gives ^{99}Mo. The ^{99}Mo is then adsorbed as $^{99}MoO_4^{2-}$ (molybdate) onto an alumina column to form a ^{99m}Tc 'generator' which is transported to the point of use. The ^{99}Mo decays on the column with a 66-hour half-life, generating ^{99m}Tc:

$$^{99}\text{Mo} \rightarrow {}^{99m}\text{Tc} + \beta^-$$

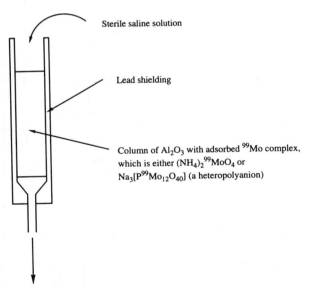

Sterile saline solution

Lead shielding

Column of Al_2O_3 with adsorbed ^{99}Mo complex, which is either $(NH_4)_2{}^{99}MoO_4$ or $Na_3[P^{99}Mo_{12}O_{40}]$ (a heteropolyanion)

Sterile Na$^+[^{99m}$TcO$_4^-]$ solution
eluted from column

FIGURE 20.14 *Schematic diagram of a generator for producing ^{99m}TcO$_4^-$ solution for use as an imaging agent.*
The ^{99}Mo complex is strongly adsorbed to the column but the daughter ^{99m}TcO$_4^-$ only weakly adsorbs and is readily eluted from the column using saline solution.

The technetium is produced in the generator in the form of ^{99m}TcO$_4^-$ which, on account of its much lower affinity for the alumina column (compared to MoO_4^{2-}), is readily eluted with sterile saline solution. A schematic diagram of a Mo–Tc generator is given in Figure 20.14. The ^{99m}Tc decays (half-life of 6 hours) to 'stable' ^{99}Tc (half-life 2×10^5 years) emitting a 140 keV gamma ray which can be easily imaged. The useful lifetime of the technetium generator is therefore limited to about two half-lives of the parent, i.e. about 130 hours.

The ^{99m}TcO$_4^-$ solution so produced can either be used directly, to image the thyroid gland, the brain, the kidneys or salivary glands, or it can be converted into a different technetium complex which will image different parts of the body. This is accomplished by adding the ^{99m}TcO$_4^-$ solution to a pre-made sterile 'kit' containing the appropriate ligand and any other reagents necessary (such as a reducing agent, to reduce the technetium from oxidation state VII to a lower oxidation state, such as III or IV). It is this versatility, together with the excellent characteristics of the ^{99m}Tc isotope, which has made technetium the most widely used radio-imaging agent available. Some other examples of technetium imaging agents and their applications include the cationic Tc(III) complexes of the type $TcL_2X_2^+$ (where L is a diphosphane or a diarsane—see Appendix B) which image the heart, and technetium phosphonate complexes which are used to image bone tumours.

In addition to technetium imaging agents other radionucleides are also used for specific applications. Examples include thallium-201, gallium-67, iodine-123 and astatine-211.

20.2.4 *Magnetic resonance imaging*

Another method of imaging bodily organs uses the paramagnetism of certain metal ions, in particular Fe^{3+}, Mn^{2+} and Gd^{3+}, in a technique known as magnetic resonance imaging (MRI). The basic principles of magnetic resonance involving protons (and other nmr active nuclei) have already been introduced in section 7.8. The imaging technique is based on the fact that the hydrogen atoms of water molecules in different tissues will 'relax' at different rates—the magnetic resonance experiment excites hydrogen atoms into a higher energy state from which they relax back to their original state. Introduction of paramagnetic metal ions into the desired tissues causes the water molecules to relax much more quickly, thereby allowing the desired tissue to be imaged more readily.

Some of the most widely used imaging agents utilize gadolinium complexes (see Chapter 11 for the chemistry of the lanthanide elements). Since simple gadolinium compounds are quite toxic it is necessary to complex the Gd in a stable form which will not break down under physiological conditions but which still has a site at which tissue water molecules can coordinate in order to impart the relaxation effect in magnetic resonance imaging. One example is the diethylenetriamine penta-acetate complex (Figure 20.15) which is highly stable in water, of low toxicity and approved

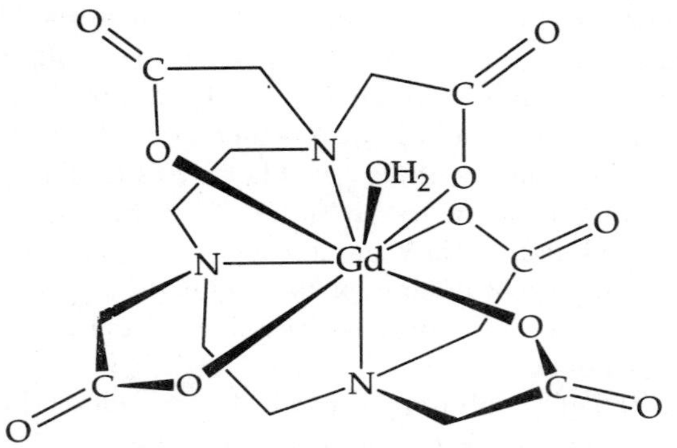

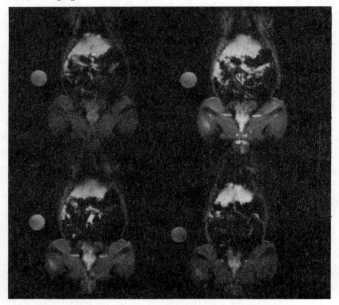

FIGURE 20.15   *The paramagnetic gadolinium(III) complex* [Gd(H₂O)(DTPA)]²⁻ *(DTPA = diethylenetriamine penta-acetate) used in magnetic resonance imaging*

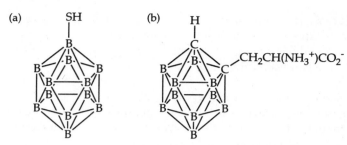

FIGURE 20.17   *Examples of* (a) *a polyhedral thiol-containing borane, and* (b) *a carborane used in boron neutron capture therapy for the treatment of cancer*

FIGURE 20.16   *An* MRI *image of a rat taken at (left to right, top to bottom) 0, 3, 100 and 450 minutes after injecting a gadolinium imaging agent* The presence of an implanted tumour in the rat's thigh can be clearly seen.

for use in MRI. An illustration of the type of visual image which can be obtained using the MRI technique enhanced by a gadolinium imaging agent is depicted in Figure 20.16 which shows an MRI scan of a rat which has had a tumour implanted into its thigh. It is clear that this non-invasive detection of abnormalities such as tumours has become a relatively straightforward process for medical personnel.

The discovery that transition metal (and Main Group) compounds have biological activity and can be used as drugs has opened up a new field of research. In the next section we will look at a couple of specific examples—the use of boron and platinum compounds in the treatment of cancer and the use of gold drugs for arthritis.

20.2.5   *Boron neutron capture therapy (BNCT)*
An example of the use of an inorganic drug in cancer treatment is boron neutron capture therapy (BNCT) in

which a boron compound, with a specific biodistribution, becomes localized in a tumour. Irradiation of the tumour with neutrons, which are readily absorbed by the boron atoms, results in the emission of high energy alpha particles within the tumour, destroying it. This treatment causes only minimal damage to the surrounding healthy bodily tissues since the penetrative power of alpha particles is quite low.

The method relies on the capture of neutrons by boron-10 nuclei which make up about 20% of naturally occurring boron. The role of the chemist in the development of improved versions of this treatment lies in the discovery of new boron compounds which localize specifically in the tumour, again to minimise damage to healthy tissues. Among the early candidates were simple inorganic boron compounds, such as borates (section 17.4.2), though recent clinical trials have turned to the polyhedral borane clusters, such as those shown in Figure 20.17. Such closed shell, or *closo*, boranes are relatively stable and have high boron contents. Further developments in the chemistry of these species may lead to new agents for selective BNCT. The reader is referred to section 9.6 for an introduction to boron-hydride clusters.

20.2.6   *Platinum-based drugs in cancer chemotherapy*
The most widely used drug in the Western world (plus Japan) for the treatment of cancer is a simple inorganic coordination complex, *cis*-[PtCl₂(NH₃)₂], commonly known as *cisplatin*. This drug is particularly effective in the treatment of solid tumours, especially of the genito-urinary tract. Like a number of other important scientific discoveries, the observation that cisplatin kills tumour cells was discovered by accident. In fact the *trans* isomer of PtCl₂(NH₃)₂ is inactive against tumour cells and it is thought that the mode of action of cisplatin is the substitution at the N(7) site of guanosine (abbreviated G) in DNA in place of the Cl ligands. The chloride ligands of cisplatin are replaced slowly by water molecules and the resulting aquo complex *cis*-[Pt(OH₂)₂(NH₃)₂]²⁺ is more reactive than cisplatin. It is believed that this aquo complex forms inside the tumour cells. The role of the (NH₃)₂Pt unit is thus to bridge across two neighbouring G residues of DNA thereby interrupting the replication of the DNA of the tumour cell. It turns out that in cisplatin the Pt–Cl bonds are of just about the right strength since studies on related platinum complexes have shown that if the platinum–ligand bonds are too strong, then

binding of the platinum to the N(7) of guanosine does not occur.

Despite the effectiveness of cisplatin in treating certain types of tumours, it has substantial drawbacks in that it is also rather toxic towards healthy cells as well as tumour cells and this produces the rather severe side effects often associated with cancer chemotherapy. As a result of this there has been a fairly intensive search for other platinum complexes showing improved anticancer activity but with reduced side effects to the patient. In particular, variations on the amine and the anionic X groups (in place of chloride) have been intensively studied, resulting in a number of second-generation anticancer drugs reaching clinical use. Figure 20.18 shows the structure of cisplatin (for comparison) together with some other platinum anticancer drugs in clinical use. In these compounds it is apparent that the amine needs to have at least one hydrogen atom, suggesting that hydrogen bonding plays an important role, and the most promising substitutes for ammonia seem to be diamines, such as cyclohexane-1,2-diamine $1,2\text{-}C_6H_{10}(NH_2)_2$, while the best replacements for Cl may be dicarboxylates, such as cyclobutane-1,1-dicarboxylate in *carboplatin*.

While we have concentrated here solely on platinum-based anticancer drugs, anticancer activity is not restricted to platinum. A wide range of other metal complexes have been found to exhibit good anticancer activity and a number have reached the clinical trials stage. Examples include the

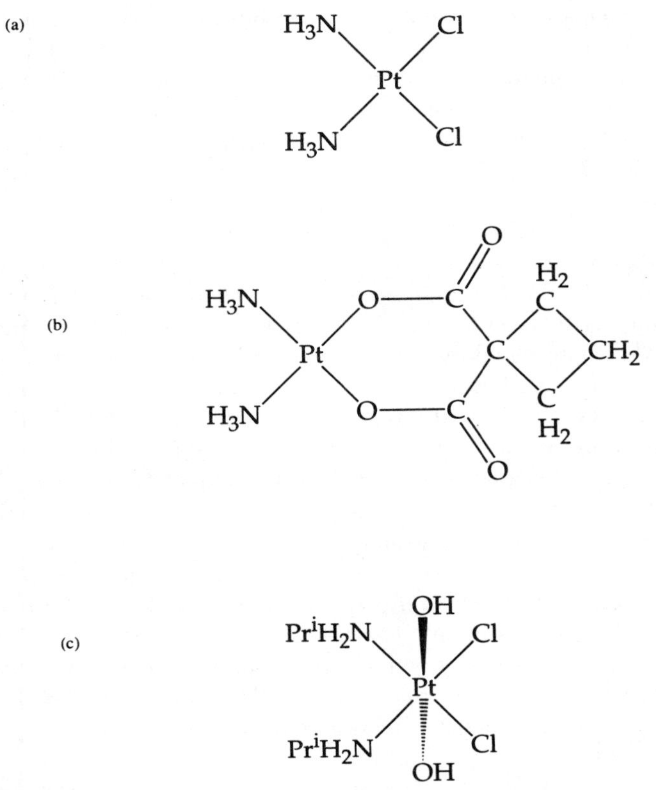

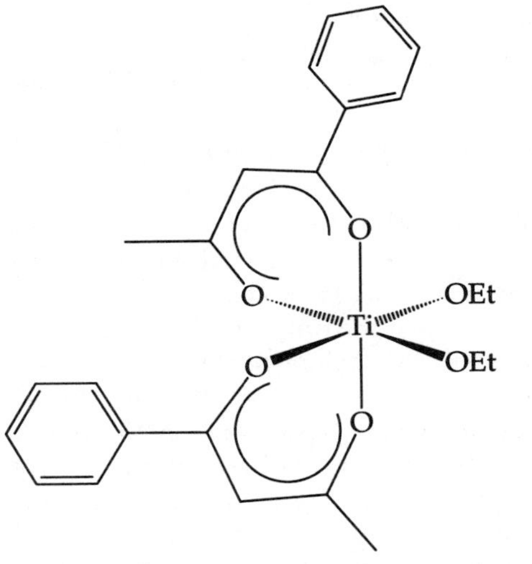

FIGURE 20.19 *The titanium anticancer drug Budotitane*

titanium benzoylacetonate complex *budotitane* (Figure 20.19) and the cyclopentadienyl titanium complex $(\eta^5\text{-}C_5H_5)_2TiCl_2$.

20.2.7 *Therapeutic gold drugs*

Gold has been used in medicine for quite some time and the Chinese employed it in medicine as far back as 2500 BC. A number of gold drugs are particularly effective in the treatment of arthritis though their mechanism of action is not well understood. Examples of older drugs of this type include gold(I) thiomalate and gold(I) thioglucose. These materials have been found to contain linearly two-coordinated gold(I), the most common coordination geometry for gold(I), though they are polymeric in the solid state. However, it is only with the advent of powerful new instrumental techniques (Chapter 7) that the structures of some of these drugs have been established. It is believed that on introduction into the body the polymeric structures react with thiol groups of proteins and the gold is subsequently distributed around the body.

A more recent example of an orally administered anti-arthritic gold drug is Auranofin (Figure 20.20), which contains a triethylphosphane ligand together with an acetylated glucose thiolate ligand. Both the phosphane and the thiolate are 'soft' ligands and show a strong affinity for the soft gold(I) centre, making this durg much more stable than its predecessors. In the body the acetate groups of the thioglucose are hydrolysed off and ultimtely the phosphane is displaced from the gold and excreted as triethylphosphane oxide. The gold becomes bound to the thiol groups of blood proteins. There is currently much interest in determining the

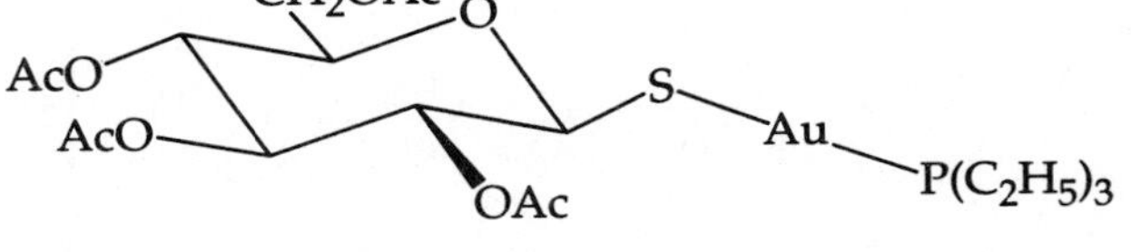

FIGURE 20.18 *The structures of some platinum anticancer drugs in clinical use:* (a) *cisplatin* (b) *carboplatin* (c) *iproplatin*

FIGURE 20.20 *The antiarthritic gold drug Auranofin*

mechanism of action of gold in the treatment of arthritis since this will hopefully lead to improved drugs being developed.

20.2.8 *Chelation therapy for metal poisoning*

For the final part of this section we view the topic of medicinal inorganic chemistry from a slightly different angle in which a drug has to be administered to negate the effects of poisoning by a metal. In this case the aim of the drug is such that the metal and the drug will combine to form a complex which is then much more readily excreted from the body than is the free metal ion. Hence, the principles of coordination chemistry (see Chapter 13) and ligand design (Appendix B) are of great importance in designing drugs to treat poisoning by specific metals. For the 'soft' polarizable metals, such as thallium, mercury and lead, etc., sulfur or other soft donor atom ligands are required whereas for the hard, oxophilic metals, such as the lanthanides and actinides, oxygen (and nitrogen) donors are necessary in order to form stable complexes. By forming chelate complexes (section 13.7) optimal stability for a given metal can be attained, hence the term *chelation therapy*.

As a specific example, we will look at the development of drugs for the treatment of plutonium poisoning. Plutonium (section 12.6) is the most toxic element in the Periodic Table—the lethal dose lies at the *microgram* level. Relatively few people are likely to come into contact with plutonium but the extreme toxicity of this element makes it essential that effective treatments are available.

Plutonium readily migrates and concentrates in bone and it is essential that as much of the plutonium is removed before this occurs. A good complexing agent for plutonium is diethylenetriamine penta-acetate (DTPA). This ligand, which is related to EDTA, another excellent complexing agent (see Appendix B), forms stable plutonium complexes

analogous to the DTPA-gadolinium complex shown in Figure 20.15 and used in magnetic resonance imaging. DTPA is potentially an octadentate ligand which can satisfy the coordination number and donor ligand requirements of the oxophilic plutonium ion. The effectiveness of DTPA is shown by the fact that up to 90% of ingested plutonium is excreted within a week when DTPA is administered to the patient within 30 minutes of poisoning occurring. However, given the extreme toxicity of the element, it is naturally desirable to develop much more selective ligands for plutonium chelation therapy.

Two examples of ligands developed for use in chelation therapy in cases of plutonium are shown in Figure 20.21. These 1,2-dihydroxybenzene, or *catechol*, ligands can bond strongly to a plutonium ion and, like DTPA, are octadentate. The stability constant for the formation of the Pu^{4+} complex of the flexible ligand in Figure 20.21b has been reported as 10^{52}, making it the most effective, yet non-toxic, ligand for plutonium chelation therapy yet discovered.

20.3 Environmental inorganic chemistry

In the 1990s there is an increased awareness of the impact of human activities on our environment. Phenomena such as 'the ozone hole', 'the greenhouse effect', 'acid rain', 'eutrophication' (of lakes and rivers) and many others are terms which are in the consciousness of the general public. There is much inorganic chemistry behind these phenomena. Therefore we feel that it is fitting that an inorganic chemistry textbook should reflect on the inorganic chemistry of these current environmental issues. It is our aim in this section to give a general overview of a selected number of relevant environmental topics in which inorganic chemistry is involved to a significant extent. Readers who wish to study the topics in greater depth are encouraged to consult the specialized readings given in Appendix A.

20.3.1 *Phosphates in detergents: eutrophication*

Phosphates have been utilized as a 'builder' in detergent formulations since around the 1930s. Their role is to complex any hard-water forming ions present (specifically Ca^{2+} and Mg^{2+}) and hold them in a complexed form in solution. This prevents the formation of a solid 'scum' on the water and enables the surfactant to work much more efficiently, preventing soil from being redeposited on the item being washed. Because of their ability to hold Ca^{2+} and Mg^{2+} ions in solution, phosphates have also been used for the prevention of scale formation in industrial cooling water systems. The most widely used phosphate builder is sodium tripolyphosphate, $Na_5P_3O_{10}$, containing a P–O–P–O–P chain (see section 17.6.2 and Figure 17.36). The performance-cost of sodium tripolyphosphate is excellent, however a serious questionmark lurks over the use of phosphates in detergents as a result of the phenomenon of *eutrophication*.

Since phosphate is an essential nutrient for plant growth, and is often the limiting reagent, the supply of phosphate to the environment from detergents (and other sources,

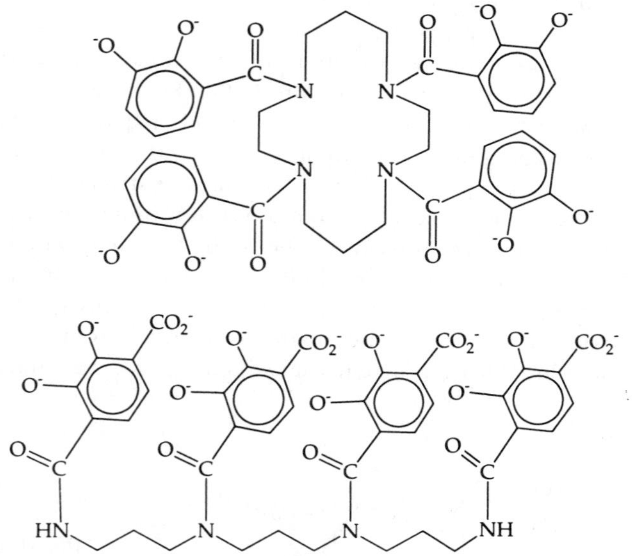

FIGURE 20.21 *Two highly selective ligands used in plutonium chelation therapy*

especially the heavy use of phosphate fertilizers in many areas) may encourage excessive plant growth. This may occur particularly in situations where there is a comparatively low throughput of water, such as lakes and rivers. The large amount of plant biomass produced subsequently decays, recycling carbon dioxide plus the nutrient elements nitrogen, phosphorus and potassium to continue the cycle. Eutrophication has been around for millions of years—it is responsible for the vast peat and coal deposits present on the Earth—yet phosphate has been critiqued for accelerating the problem. As a consequence of this, alternative builders were sought in the 1980s to replace the phosphates used in detergents.

One of the earliest replacements for phosphate was nitrilotris(acetate), $N(CH_2CO_2)_3^{3-}$, known as NTA. However, studies have suggested that this compound is carcinogenic and it has consequently been discontinued as a builder. The discovery that zeolites were efficient ion-exchangers indicated that these materials would be a good replacement for phosphates and they have become the single largest replacement builder in detergents where the phosphates are prohibited. A discussion on the chemistry of zeolites is given in section 18.6. The introduction of zeolite A was initially viewed to be a complete solution to the phosphate eutrophication problem.

More recent studies, however, have suggested that zeolites themselves may have environmental problems associated with them. When the performance cost of zeolites is compared to that of phosphates, zeolites are not quite as good. In addition, since zeolites are solid materials, they are used as a suspension (unlike tripolyphosphate which is completely soluble in water). Other chemicals need to be added as a 'co-builder' in order to assist in the dispersion of the zeolite particles, to control the pH and to prevent dirt redeposition. Long-chain organic polymers bearing large numbers of carboxylic acid groups are one such class of compound employed for this purpose. Over the last few years thick brown algal mats have developed off the Adriatic coast and research has suggested that these are caused by the use of zeolites and polycarboxylic acids in detergents. On the other hand, studies have suggested that phosphates are causing destruction of coral reefs off the northeast coast of Australia. The evidence seems to suggest that both types of builder have some significant detrimental effect on the environment.

A recent study has suggested that the *overall* environmental impact of phosphate versus zeolite builders should be considered in order to obtain a fair comparison. The mining, manufacturing, use and disposal operations all clearly have their own environmental impact which must be included in the equation when considering the relative merits of the two detergent builders. When this is done, it has been suggested that zeolites and phosphates are similar in overall environmental impact. This is due, in part, to the nature of bauxite mining, the source of aluminium for the zeolites, which produces large quantities of alkaline wastes, termed 'red muds'. In addition, most bauxite is mined in Australia

so significant amounts of energy must be employed to transport it to the point of use. On the other hand, the mining of phosphates produces large quantities of gypsum ($CaSO_4$) which is discharged into the ocean, smothering the ocean floor and releasing toxic heavy metals originally present in the phosphate rock.

Given that the overall environmental impact of zeolites seems to be comparable to phosphates, it would appear that there is little to choose between them. However, technology is available to recycle phosphates from sewage and convert them to a form which can be re-used in detergents. This is achieved very simply by the addition of calcium ions to the effluent stream and recovery of the insoluble calcium phosphate produced. Pilot plants demonstrating this technology are currently in operation in the Netherlands. However, at present, there is no known technology to recover zeolites from sewage.

It is clear that the phosphate debate is still going on and is likely to do so for some time to come. Ideally, one would like to dispense with both zeolites and phosphates as builders. However, it is clear from the above case histories that any replacement must be thoroughly vetted before introduction since detrimental effects may not be discovered until it is too late.

20.3.2 *Chlorofluorocarbons: the ozone hole*

Chlorofluorocarbons, otherwise known as CFCs, are 'inert' compounds with a number of applications: as propellants of aerosol cans, blowing agents for polyurethane foams and fluids in refrigeration units. Examples of such materials include CCl_3F (CFC-11), CCl_2F_2 (CFC-12) and C_2ClF_5 (CFC-115). When they were introduced they were thought to be extremely non-toxic and inert, and as a result found large-scale use. Worldwide production of CFCs has reached levels of around half a million metric tons per year. It was first thought that due to the stability of these compounds large quantities could be released into the atmosphere with no effect. However, in 1974 it was suggested that CFCs are responsible for the destruction of stratospheric ozone, O_3 (see section 17.7.1). Ozone in the upper atmosphere is highly beneficial to life on earth since it filters out hard (i.e. short wavelength) ultraviolet radiation. Such radiation is responsible for causing cancer and other ailments including eye cataracts. (At this point it is worth noting that ozone in the *lower* atmosphere is highly undesirable—ozone is a toxic, powerful oxidant which causes rubber and other polymers to perish and causes major health problems.)

In the upper atmosphere high energy radiation causes cleavage of the carbon–chlorine bonds of CFCs, generating chlorine radicals:

$$\text{e.g. } CF_2Cl_2 \xrightarrow{h\nu} Cl\cdot + CF_2Cl\cdot$$

These chlorine radicals then react with ozone producing dioxygen and ClO radicals which then react with a number of species present in the upper atmosphere, such as oxygen atoms (produced by the photolysis of ozone: $O_3 \rightarrow O_2 + O\cdot$)

or with nitric oxide, NO.

$$Cl\cdot + O_3 \rightarrow O_2 + ClO\cdot$$

The net effect is to set up a series of chain reactions the overall result of which is the destruction of the beneficial ozone in the upper atmosphere. Examples of reactions occurring are:

$$ClO\cdot + O\cdot \rightarrow Cl\cdot + O_2$$

then

$$Cl\cdot + O_3 \rightarrow ClO\cdot + O_2$$

the net effect of which is the ozone-destroying reaction

$$O_3 + O\cdot \rightarrow 2O_2$$

and the regeneration of $Cl\cdot$ to repeat the cycle. Similarly, ozone can react with NO, again with the net destruction of ozone:

$$O_3 + NO \rightarrow NO_2 + O_2$$

Perhaps the most well-known incidence of ozone depletion is the Antarctic ozone hole which is characterized by up to a 50% depletion in polar stratospheric ozone in the late Antarctic winter and early spring. Under normal conditions the reaction of atmospheric NO_2 with ClO radicals serves to limit the amount of Cl atoms available for ozone destruction:

$$ClO\cdot + NO_2 \rightarrow ClONO_2 \text{ (chlorine nitrate)}$$

However, with temperatures below $-70\,^\circ C$ in the cold polar stratospheric clouds, NO_2 is frozen out as compounds such as nitric acid hydrate, $HNO_3.3H_2O$. In addition, $ClONO_2$ can generate, in the ice crystals, various ozone-destroying chlorine species, as shown in the following reactions:

$$ClONO_2 + H_2O \rightarrow HOCl + HNO_3$$

$$ClONO_2 + HCl \rightarrow Cl_2 + HNO_3 \overset{h\nu}{\rightarrow} 2Cl\cdot$$

$$HOCl \overset{h\nu}{\rightarrow} HO\cdot + Cl\cdot$$

Bromine has also been implicated in the destruction of atmospheric ozone. *Halons* are compounds related to CFCs but which contain bromine atoms. Examples include $CBrF_3$ (Halon-1301) and $C_2Br_2F_4$ (Halon-2402). Such compounds have been used in fire extinguishers since they generate halogen radicals which are very efficient at preventing combustion. However, the major source of bromine in the atmosphere remains uncertain since large quantities of methyl bromide (CH_3Br) are released by natural sources.

The very real potential for the massive destruction of the Earth's protective ozone layer has led to measures to eliminate the production of CFCs. As a result, the inception of the Montreal Protocol (on substances that deplete the ozone layer) in 1987 calls for a number of measures including a cutback in CFC production to 50% of 1986 levels by 1998 and a complete ban on the production of CFCs in the USA by the end of 1995. What is rather surprising, yet highly

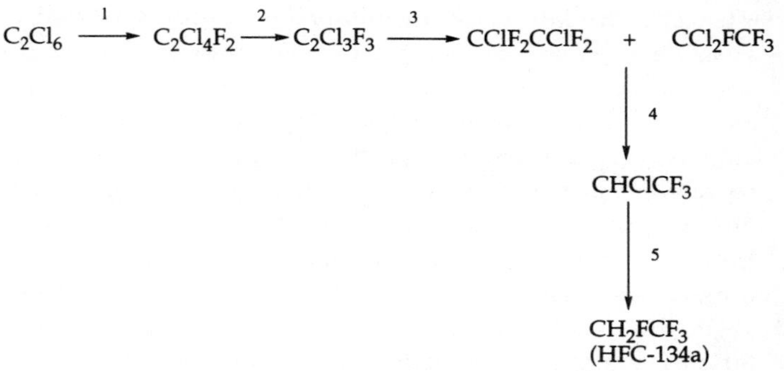

FIGURE 20.22 *Scheme for the synthesis of the CFC replacement* CH_2FCF_3 *(HFC-134a)*
Steps 1–3 use a Cr_2O_3 catalyst while steps 4 and 5, which involve hydrogenolysis, use a supported palladium catalyst.

pleasing, is that countries are voluntarily speeding up their compliance to the Montreal Protocol and the various amendments introduced since its inception.

The Montreal Protocol has led to a massive search for new replacements for CFCs which do not damage the ozone layer. Among the more promising candidates are materials which contain hydrogen atoms in addition to fluorine (and sometimes chlorine). The C–H bonds are much more readily broken than are C–Cl or C–F bonds and so these compounds can degrade much more rapidly in the atmosphere, particularly by reaction with OH radicals, typically causing only 1 to 10% of the ozone destruction of CFCs. A good example is HFC-134a (CH_2FCF_3) where the HFC stands for hydrofluorocarbon. This HFC is now being manufactured for use as a replacement of CFC-12 in refrigeration systems.

One of the major roles of inorganic chemistry in the search for new CFC replacements has been in the development of catalytic processes for their manufacture. The catalytic chemistry used in one of the possible routes for the manufacture of HFC-134a is shown in Figure 20.22. Other studies are concerned with the gas phase radical chemistry of CFCs and HCFCs, the range of experiments which can be conducted and the information that can be ascertained from them (see Appendix A). The introduction of the Montreal Protocol is hopefully the start of many similar agreements for the minimization of the causes of other current environmental problems. The ultimate replacement compounds will have no C–Cl bonds (e.g. $C_2F_4H_2$) but these are less accessible and more expensive at present.

20.3.3 *The greenhouse effect*

The greenhouse effect is an environmental problem of particular concern. Solar radiation (visible light) is absorbed by the Earth which reaches thermal equilibrium by re-radiation in the infrared region. In the greenhouse effect this infrared radiation is absorbed by the so-called 'greenhouse gases' of the atmosphere and lower energy radiation re-emitted towards the Earth. As might be interpreted from the name, it is the same principle which accounts for the warming action of glass greenhouses and the effect on the Earth is similar. The principal greenhouse gases are water and carbon dioxide. The latter is causing the greatest concern due to the increase of CO_2 in the atmosphere from the almost exponential increase in the use of carbon-based fuels.

Measurements over the last 30 years have indicated that the amount of CO_2 in the atmosphere is increasing at the rate of about 1 part per million (ppm) per year. It has been predicted that CO_2 levels will double by the middle of the 21st century and this will produce, as a result of the greenhouse effect, an overall increase in the surface temperature of the Earth predicted to be between 1·5 and 4·5 °C. Though this may not sound significant, it will cause melting of substantial portions of the polar ice caps and expansion of the oceans, causing massive destruction of coastal areas close to or below sea level. As a good comparison, a temperature rise of this magnitude is what has occurred since the last ice age. What adds to the problem is that one of the 'sinks' for carbon dioxide are forests which, through the process of photosynthesis, fix vast amounts of CO_2 as organic carbon. The current world policy of deforestation can only add to the problem.

Evidence for the increase in atmospheric CO_2 levels comes from the analysis of air trapped in ice cores from the polar regions (in addition to direct measurements). These experiments suggest that the composition of the air at the time of the last ice age contained CO_2 levels 25% less than pre-industrial levels (it is generally accepted that the first significant human input of CO_2 into the atmosphere started at the time of the industrial revolution).

In addition to water and carbon dioxide, other trace constituents in the atmosphere also contribute to global warming. Among these are nitrous oxide, N_2O (see section 17.6.4), the chlorofluorocarbons discussed earlier in this chapter and methane. Along with CO_2, human activities are also increasing the CH_4 content of the atmosphere. This is due to factors such as natural gas spillages, evolution of gas from landfill sites and methane emitted from the digestive tracts of cattle, among others. Methanotropic bacteria, discussed in section 20.4.4, utilize methane as their source of carbon and energy, and may find use in the bioremediation of land, such as in landfill sites where substantial quantities of methane can be produced.

Various factors also operate to decrease the CO_2 content of the atmosphere. One of these the photosynthetic conversion of CO_2 to organic carbon, has already been mentioned. Another vast sink of carbon are the oceans—these contain about sixty times the amount of CO_2 found in the atmosphere. However, due to relatively poor mixing of the atmosphere with the oceans, transfer of CO_2 is relatively slow and while the ultimate end for atmospheric CO_2 is likely to be carbonate sediments (such as $CaCO_3$) deposited in the oceans, it is as yet unclear what the effects of increased CO_2 input into the atmosphere are likely to be. Much effort is being spent in computer-modelling of the greenhouse effect. However, the interrelationships between a large number of parameters are difficult to model and predictions are likely to be approximate at the very best.

It is noteworthy that the next, and final, environmental aspect which we will consider, the phenomenon of acid rain, may have a counteractive influence on the greenhouse effect due to the formation of light-reflecting sulfuric acid clouds in the atmosphere. This would clearly be a completely unacceptable solution and emissions of CO_2 must surely be severely curbed in the near future. Perhaps the forthcoming decline in carbon fossil fuel resources may effect a change to a 'hydrogen economy' (using hydrogen as a fuel in relatively efficient and pollution-free electrochemical devices such as fuel cells) rather than the current 'carbon economy' with which the world operates. Such an enforced change would have a beneficial effect on CO_2 emissions.

Finally, it is worth casting our eyes a little further afield for a look at one of our solar system neighbours—Venus. It is generally considered that Venus is a case history of a 'runaway' greenhouse effect. As a result of greenhouse emissions the surface of the planet is believed to have heated up, driving off any liquid volatiles into the atmosphere to act further as greenhouse gases. The surface of Venus is a thoroughly inhospitable place—temperatures of several hundred degrees, the rocks glowing a dull red colour with sulfuric acid clouds lurking overhead. Though Venus is closer to the Sun than Earth, the temperature far exceeds that expected as a result of a runaway greenhouse effect.

20.3.4 Acid rain

One of the consequences of industrial activity is that large quantities of sulfur dioxide and nitrogen oxides are introduced into the atmosphere. The principal source of SO_2 comes from the combustion of fossil fuels which contain various sulfur compounds. Similarly, nitrogen oxides come from emissions from various industrial processes as well as automobile exhaust emissions. In the atmosphere these gases are oxidized and converted to sulfuric and nitric acids:

$$SO_2 + \tfrac{1}{2}O_2 + H_2O \rightarrow H_2SO_4$$

$$2NO_2 + \tfrac{1}{2}O_2 + H_2O \rightarrow 2HNO_3$$

These acids (when combined with other acids such as hydrochloric acid, also added to the atmosphere by emissions from industrial processes) render precipitation acidic—*acid rain*.

Acid rain is an international problem. Emissions resulting from one country's activities may be swept by weather systems to deposit acid rain up to several thousand kilometers away. Perhaps the classical example of this is the occurrence of acid rain in Scandinavia, originating from emissions in the densely industrialized regions of Western Europe.

There are many damaging effects of acid rain including destruction of forests which may occur directly (as a result of the low pH) or indirectly (since decreased pH increases leaching of Al^{3+} from soils and rocks) with concomitant detrimental effects on plant health. Acid rain also produces an increased incidence of human respiratory ailments, acidification of lakes and general corrosion of limestone and steel items.

Various moves are underway to curb acid rain. Naturally,

tackling the major source of SO_2 and NO_2 emissions is the logical way to proceed. There is an increased awareness of the need to use relatively sulfur-free coal and oil in power stations and to fit scrubbing systems to minimize SO_2 emissions. For automobile emissions catalytic converters provide significant reductions in emissions of pollutants. Catalytic converters typically use a platinum metal-based catalyst (see section 15.8) and serve a dual role. First, they ensure complete oxidation of hydrocarbons and carbon monoxide, and second, they aim to reduce any nitrogen oxides (to dinitrogen) which have been produced in the combustion process. The development of new, improved catalytic systems, using cheaper platinum group metals such as palladium, should decrease the cost of such converters and make them more widely available. This will also have benefits on other more localized environmental problems, such as photochemical smog, to which hydrocarbons in the atmosphere are a major contributor. Again, movement from a carbon-based economy to cleaner fuel sources (such as those based on elemental hydrogen) should provide future solutions, especially when coupled to low-pollution energy sources (such as solar energy or, ultimately, hydrogen fusion power).

20.3.5 *Other inorganic environmental problems*

In concluding this discussion on environmental chemistry we will look, rather superficially, at some of the major environmental problems which concern inorganic chemistry. There are many others. Indeed, the impact of any human activity on the global environment needs to be much more closely regulated. Lead from petrol is one such example. For many years lead has been added to petrols as an antiknocking agent in the form of an organic derivative, tetraethyllead. Naturally, the lead enters the environment and studies have found cases of impaired learning in children living near busy roads. In the late 1980s and the 1990s, a strong 'green' movement is campaigning for the total elimination of lead from petrols. While such a stance must be applauded, we also need to consider carefully the replacements for lead. One such replacement antiknock compound is the cyclopentadienyl manganese compound $(\eta^5\text{-}C_5H_4CH_3)Mn(CO)_3$ (see section 16.4 for a discussion on cyclopentadienyl complexes). The environmental impact of manganese emissions needs to be very carefully studied. It has also been suggested that the elimination of lead from petrols is likely to cause an increase in the concentration of carcinogenic polycyclic aromatic hydrocarbons in the environment.

Heavy metals in the environment must be closely monitored. Biological systems are able to accumulate heavy metals, especially in fatty tissues, and with humans residing at the top of the food chain this can result in significant human uptakes of these toxic metals. Perhaps the classical example comes from mercury poisoning in Japan where mercury-laden fish were consumed resulting in a large number of deaths from the so-called 'Minamata' disease. The increased consumption of fish such as tuna, which accumulate mercury and other heavy metals, is also increasing the human uptake of this element. At the time of writing, there is a debate on the relative contribution to human mercury uptake from the mercury present in amalgam dental fillings. It is worth noting that mercury in inorganic forms has a much lower toxicity when compared with organometallic derivatives, such as CH_3Hg^+, which can be produced by biomethylation of Hg^{2+} by certain microorganisms.

Finally, our planet, as a result of continued population growth and industrialization, has an ever-increasing demand for energy. The continued headlong consumption of declining fossil fuel reserves is not a solution, as we have seen from the earlier discussions on the greenhouse effect and acid rain. Perhaps in the future power from hydrogen nuclear fusion will provide us with as much energy as we need—we have already mentioned the possible advantages of a hydrogen-based energy economy. The advent of readily available fusion power is likely to lie in the next century. Other energy sources—where the technology is generally already available—include solar and wind power. In the interim, howeve, energy sources must be found. Nuclear power, using nuclear *fission*, is perhaps the most widespread, particularly in regions which are not blessed with geothermal or hydroelectric energy sources. This energy source, when operated safely, is a relatively clean source of energy compared to energy generated by the combustion of a high sulfur content coal. Nevertheless, incidents such as those at Three Mile Island, Windscale and, more recently, Chernobyl, serve to generate a significant distrust of this energy source. In addition, factors such as the continued problem of the long-term disposal of radioactive waste, plus the ever-growing threat of nuclear terrorism using stolen uranium or plutonium, must also be considered in the overall equation.

It is easy to despair when faced with such problems, or to yearn for a return to a simpler age. However, although earlier ages may have been simpler, they were not golden! Descriptions of pollution in most mediaeval cities make modern problems seem marginal by comparison. The 'idyllic' rural life of early farmers, or even more so of early hunter-gatherers, was characterized by life expectancies of less than 30 years and high mortality of infants and in childbirth.

Problems appear to pile up in part because we currently have the knowledge to see that there is indeed a problem. Conflict and uncertainty arise because we are attempting to understand very complex systems. Information is becoming firmer and the people of Earth are moving slowly towards solutions.

The first problem is to understand and develop the science then the technology and engineering in order to determine and provide adequate and acceptable solutions. The second problem is to develop the social and political understanding, and then the social concensus, to choose good solutions. While everyone can contribute to the second problem, the challenge to scientists, engineers and technologists is to contribute to the solution of both.

Appendix A
Further Reading

Note to the Reader

You may wish to consult other sources for a number of reasons, and it helps to be clear about your aims. Nothing is more confusing and frustrating than turning up an explanation in great depth or a review in great detail when only minor clarification is sought.

(1)   Explanation of points not understood in text or lectures is best sought, in the first place, from other sources written at a similar level. So often, a different form of words or different approach will clarify a problem. It is surprising how often the process of consulting source A, then B, then C, then A again solves some puzzling point.

If you suspect there is a lack of basic information, then lower-level sources are useful. If often pays to revise a little in school textbooks.

(2)   A search for supplementary information, explanations in greater depth, or a desire to know where a topic leads are best satisfied by first reading one or more of the advanced texts, Honours or specialist, (section 1). Follow this by consulting the appropriate parts of the multi-volume works listed. These will give the desired material or refer in turn to specialist sources.

(3)   To find much fuller information on a limited topic (such as 'chromium chemistry' or 'metal-metal bonds' or 'inorganic biochemistry') the specialist references or reviews should be consulted first. A recent review will normally yield more than enough information for an essay topic, for example. Suitable sources are given in section (2), while many individual articles are listed in section (3).

(4)   For complete information on a topic (not usually required from an undergraduate), specialized reviews or multivolume references (section 1) will provide a terminus. A convenient source for updating these is provided by the appropriate *Specialist Periodical Reports* and *Annual Reports* published by the Chemical Society (section 2). For even fuller or more recent information the next step is a search through *Chemical Abstracts* or the *Citation Index* to bring the survey up to date. Any science library will provide such sources and guidance on their use. Completely up-to-date information involves scanning the latest few issues of the major journals. A range of computer-search services are available.

(5)   Note also the various detailed compilations of data, including those listed in section (3).

(6)   For more detailed guides to the chemical literature see *Use of the Chemical Literature*, R T BOTTLE, Butterworths, 3rd Edition, 1979, or *Guide to Basic Information Sources in Chemistry*, A. ANTONY, Wiley, 1980; *How to Find Chemical Information: A Guide for Practicing Chemists, Teachers and Students*, R. E. MAIZELL, Wiley, 2nd Edition, 1987.

Electronic access to chemical information *via* the World Wide Web is gaining in popularity and developments in this area are proceeding with a rapid pace. A good introduction to chemical information on the internet is provided in the September 1995 issue of *Chemistry in Britain*.

(7)   As an aid to writing, *The Chemist's English*, R. SCHOENFELD, VCH, 3rd Edition, 1989, may be consulted with profit and entertainment.

(1) Books

Textbooks at Honours level

Advanced Inorganic Chemistry, F. A. COTTON AND G. WILKINSON, Interscience, 5th Edition, 1988.

This is the most widely-used Honours textbook in inorganic chemistry. It set a new style some thirty years ago, and many other books follow the same tradition. More recently, books with alternative approaches and emphases have appeared. Among titles worth consulting are:

Inorganic Chemistry, D. F. SHRIVER, P. W. ATKINS AND C H. LANGFORD, Oxford university Press, 2nd Edition, 1994.

Inorganic Chemistry, Principles of Structure and Reactivity, J. E. HUHEEY, Harper Collins College Publishers, 4th Edition, 1993.

Inorganic Chemistry, A. G. SHARPE, Longman, 3rd Edition, 1992.

Concise Inorganic Chemistry, J. D. LEE, Chapmen & Hall, 5th Edition, 1997.

Principles of Inorganic Chemistry, W. L. JOLLY, McGraw-Hill, International (and revised) Edition, 1985.

Chemistry of the Elements, N. N. GREENWOOD AND A. EARNSHAW, Pergamon Press, 1984.

Inorganic Chemistry, A Unified Approach, W. W. PORTERFIELD, Addison-Wesley, 1984.

The student wishing to undertake a fuller study of Inorganic Chemistry, or of a particular field, should consult several of the above texts. These will give further references which may be followed up in detail.

A good library will also contain a number of older textbooks which should not be neglected. The experimental facts will be unaltered, and it is often easier to grasp a theory by following its evolution from an earlier form. It is very important to check the date of publication, and to be aware that most of modern chemistry has evolved very rapidly over the past thirty years.

Other texts

There are a number of useful general titles which are less wide-ranging than the Honours texts. We also note sources for a number of areas which are not covered in detail in this text.

A series of chemistry primers have been recently published by Oxford University Press. These cover a range of topics and provide an excellent, concise introduction. The following titles have been published in the year indicated:

Organometallics 1: Complexes with Transition Metal–Carbon σ bonds, M. BOCHMANN (1994).

Organometallics 2: Complexes with Transition Metal–Carbon π bonds, M. BOCHMANN (1994).

Biocoordination Chemistry, D. E. FENTON (1995).

Cluster Molecules of the p-Block Elements, C. E. HOUSECROFT (1994).

Essentials of Inorganic Chemistry 1. D. M. P. MINGOS (1995).

Periodicity and the p-Block Elements, N. C. NORMAN (1994).

Inorganic Materials Chemistry, M. T. WELLER (1995).

Chemical Bonding, M. J. WINTER (1994).

d-Block Chemistry, M. J. WINTER (1994).

See also:

Basic Inorganic Chemistry, F. A. COTTON, G. WILKINSON, AND P. L. GAUS, John Wiley, 2nd Edition, 1987.

Main Group Chemistry, A. G. MASSEY, Ellis Horwood, 1990.

Inorganic Substances—A Prelude to the Study of Descriptive Inorganic Chemistry, D. W. SMITH, Cambridge University Press, 1990.

The Periodic Table of the Elements, R. J. PUDDEPHATT AND P. K. MONAGHAN, Oxford University Press, 2nd Edition, 1986.

Inorganic Energetics, W. E. DASENT, Cambridge University Press, 2nd Edition, 1982

Some Thermodynamic Aspects of Inorganic Chemistry, D. A. JOHNSON, Cambridge University Press, 2nd Edition, 1982

Though now somewhat dated, a classical text which focuses on significant topics is still well worth consulting.
Modern Aspects of Inorganic Chemistry, H. J. EMELEUS AND J. S. ANDERSON, 4th edition with A. G. SHARPE, Routledge and Kegan Paul, 4th Edition, 1973.

The Elements on Earth; Inorganic Chemistry in the Environment, P. A. COX, Oxford University Press, 1995.

Inorganic Geochemistry, P. HENDERSON, Pergamon Press, 1982.

Inorganic Chemistry and the Earth, J. E. FERGUSSON, Pergamon, 1982.

A short volume which overviews chemical resources, extraction, uses, and environmental effects.

Safety in the chemical laboratory:
Hazards in the Chemical Laboratory, ed. L. BRETHERICK, Royal Society of Chemistry, 4th Edition, 1986

Handbook of Reactive Chemical Hazards, L. BRETHERICK, Butterworths, 2nd Edition, 1979

General data:
CRC Handbook of Chemistry and Physics, ed. R. WEAST, CRC Press. Reissued annually (68th edition was 1987–88) with intermittent revision: check the actual date of the specific data which you use, as this may be much older than the date of the edition consulted.

Practical inorganic chemistry

See *Inorganic Experiments*, J. D. WOOLLINS, VCH, 1995.

Guide to Practical Inorganic and Organo-metallic Chemistry, R. J. ERRINGTON, Blackie, 1995.

Technical and industrial:

To expand on the details given in the text, a good starting point for any technical query is

Kirk-Othmer Concise Encyclopedia of Chemical Technology, eds. M. GRAYSON AND D. ECKROTH. John Wiley, 1985. There, is also the 26-volume full work, *Kirk-Othmer Encyclopedia of Chemical Technology*, 3rd Edition, Wiley, 1984. See also

McGraw Hill Dictionary of Scientific and Technical Terms, 5th Edition, 1994.

Ullmann's Encyclopedia of Industrial Chemistry, ed. W. GERHARTZ, VCH, 1985–1993.

Industrial Inorganic Chemistry, W. BÜCHNER, R. SCHLIEBS, G. WINTER AND K. H. BÜCHEL, translated by R. TERRELL, VCH, 1989. (An excellent, well-written overview of the manufacture and uses of inorganic materials, including ceramics, glasses, etc.)

Industrial Chemistry, E. STOCCHI, Ellis Horwood, 1990.

An Introduction to Industrial Chemistry, ed. C. A. HEATON, Blackie, 2nd Edition, 1991. (Of interest to those with a leaning towards chemical engineering, economics, etc.)

Catalytic Chemistry, B. C. GATES, Wiley, 1992.

Industrial Applications of Homogeneous Catalysis, eds A. MORTREUX AND F. PETIT, Catalysis by Metal Complexes Series, D. Reidel Publishing Company, 1988. (See also other volumes in this series.)

For those interested in household applications of chemicals, *Chemistry in the Marketplace*, B. SELINGER, Harcourt Brace Jovanovich Publishers, 4th Edition, 1989, provides an excellent overview.

Another, often useful, source is the latest edition of any of the major encyclopedias.

For current information it is worth scanning recent issues of *Chemical & Engineering News* or *Chemtech*. A useful guide to current world production of inorganic chemicals is given by the annual 'Facts and Figures for the Chemical Industry' published in *Chemical & Engineering News*, usually in June or July (e.g. 4th July issue in 1994; 28th June, 1993 and 29th June, 1992).

For more detail, see:

The Modern Inorganic Chemicals Industry, Royal Society of Chemistry Special Publication 31, 1977 (reprinted 1986)

Speciality Inorganic Chemicals, Royal Society of Chemistry Special Publication 40, 1981

Fine Chemicals for the Electronic Industry, ed. P. BAMFIELD, Royal Society of Chemistry, Special Publication 60, 1986.

Syn-fuels from Boiling Stones, G. J. HUTCHINGS, *Chemistry in Britain*, 1987, 762 (use of zeolites as catalysts).

Recent Achievements, Trends and Prospects in Homogeneous Catalysis, F. J. WALLER, *J. Molecular Catalysis*, 1986 Review Issue, 43–61: see also the preceding article on Heterogeneous Catalysis by Metals.

Ziegler-Natta Catalysis, H. SINN AND W. KAMISKY, *Advances in Organometallic Chemistry* **18**, 1980, 99–143

Trends and Opportunities for Organometallic Chemistry in Industry, G. W. PARSHALL, *Organometallics* **6**, 1987, 687–692.

Cluster Chemistry Broadens Scope of Organometallic Research: Commercial Breakthroughs in Catalysis, Ceramics, Electronic Materials, J. HAGGIN, *Chemical and Engineering News*, Oct. 5th 1987, 31–44.

The Importance of Chemistry in the Development of High Performance Ceramics, F. ALDINGER AND H-J. KALZ, *Angewandte Chemie, International Edition in English* **26**, 1987, 371–81: The Development of Bioglass Ceramics for Medical Applications, W. VOGEL AND W. HOLAND, *Angewandte Chemie, International Edition in English* **26**, 1987, 527.

Organometallic chemistry:

Several parts of organometallic chemistry are converted briefly in the text, and the survey of organolanthanide chemistry in section 16.5 gives a more detailed impression of one, small but finite, part of the field. The literature of organometallic chemistry matches that of inorganic chemistry in volume. A basic, wide-ranging account is

Organometallic Compounds, G. E. COATES, Chapman & Hall: all editions are worth consulting. 4th Edition, Volume 1—The Main Group Elements, with B. J. AYLETT AND K. WADE, Part 1, 1980; Part 2, 1979: Volume 2, The Transition Elements, with M. L. H. GREEN AND D. M. P. MINGOS, 1982.

Comprehensive Organometallic Chemistry, eds. G. WILKINSON, F. G. A. STONE, AND E. W. ABEL, Pergamon Press, 1982. (This is a multi-volume survey with many co-authors.)

Comprehensive Organometallic Chemistry II (Elsevier, 1995) is a fourteen volume update of the organometallic literature from 1982–1994.

Trends and Opportunities for Organometallic Chemistry in Industry, G. W. PARSHALL, *Organometallics*, **6** 1987, 687–692.

The Close Ties Between Organometallic Chemistry, Surface Science, and the Solid State, R. HOFFMANN, S. D. HOFFMANN, S. D. WIJEYESEKEVA AND S.-S. SUNG, *Pure and Applied Chemistry*, **58**, 1986, 481–94.

Organometallic Chemistry, Coordination Chemistry, Main Group Chemistry, Where are the Frontiers at Present? J. G. RIESS, *J. Organometallic Chemistry* **281**, 1985, 1–14.

Finally, it is worth looking at the following title, for aesthetic reasons as well as for its wide range of simply-presented ideas.
Marvels of the Molecule, L. SALEM, illustrated by C. RATTRAY, VCH Publishers, 1987. A splendidly illustrated sweep from atoms to solids and giant biological molecules.

Comprehensive surveys and multi-volume works

While a full and detailed literature search will not normally be asked of an undergraduate student, the techniques for undertaking such a search are an important part of the chemist's armoury. The multi-volume works listed below present a starting-point which should then be brought up to date using *Chemical Abstracts* and the current literature.

An excellent, and up-to-date, entry point to the inorganic literature is the recent (1994) eight-volume series *Encyclopedia of Inorganic Chemistry*, Editor in Chief R. B. KING, John Wiley. This provides a comprehensive coverage of a wide range of inorganic chemistry topics and provides a substantial number of references to the primary journal literature.

Comprehensive Inorganic Chemistry, eds. J. C. BAILOR, H. J. EMELEUS, R. S. NYHOLM AND A. F. TROTMAN-DICKENSON, Pergamon Press, 1974, consists of five volumes made up of essay reviews covering all the elements systematically, together with a number of broad topics such as organo-transition metal chemistry. The literature is scanned to the date of the manuscript (note that some terminate as early as 1969) with adequate keys to the older literature.

M.T.P. International Review of Science: Inorganic Chemistry, Series 2, General Ed., H. J. EMELEUS (10 volumes edited by M. F. Lappert, D. B. Sowerby. V. Gutmann, B. J. Aylett, D. W. A. Sharp, M. Mays, K. W. Bagnall, A. G. Maddock, M. L. Tobe and L. E. J. Roberts respectively), Butterworths 1975, contains more concentrated reviews.
The above two are to some extent complementary. Unfortunately, both are rather out-of-date and initially-promised regular revisions have not yet appeared. A substantial proportion of inorganic chemistry is covered in

Comprehensive Coordination Chemistry, eds. G. WILKINSON, R. D. GILLARD AND J. A. McCLEVERTY, Pergamon Press, 1987, (A major multi-volume work.) There is also much inorganic

chemistry in *Comprehensive Organometallic Chemistry*, cited earlier.

A further multi-volume work with emphasis on preparations is starting to appear:
Inorganic Reactions and Methods, eds. J. J. ZUCKERMAN, VCH, 1986 onwards. Volumes 1–13, Formation of Bonds: Volumes 14–18, Selected Themes.

The multi-volume series *Inorganic Syntheses* (Wiley, 1939–) provides detailed, checked preparations of selected common inorganic compounds.

The problems of producing a fully comprehensive treatise are immense, due to the very rapid modern expansion of inorganic chemistry. A major effort to produce an up-to-date comprehensive treatise stems from *Handbuch der anorganische Chemie*, L. GMELIN, 1924 onwards.
First published as a multi-volume work, the Gmelin Institute in now pursuing an extensive programme of up-dating by supplementary volumes, for example, *Tellurium compounds*, 8th Edition, main series, Supplement Volume B, 1977. This handbook now runs to literally hundreds of volumes, is expensive, not held by every library, and is mainly in German. However, for a comprehensive search it is an excellent starting point if the topic is covered by a recent volume, so it is well worth consulting your library catalogue for it.

For information on specific compounds the *Dictionary of Inorganic Compounds* (Executive Editor J. E. MACINTYRE, Chapman & Hall, 1992) and the *Dictionary of Organometallic Compounds* (Chapman & Hall, 1984, with four supplements) can be very useful. The corresponding *Dictionary of Organic Compounds* can be useful when searching for information on ligands etc.

(2) Reviews and Journals

More specific information may be obtained from reviews, and ultimately from papers in the journals. There are now many review series published which survey fairly specific areas of chemistry, such as organometallic chemistry, fluorine chemistry, transition metal compounds, spectroscopy, and so on. Series which cover all of inorganic chemistry are

Advances in Inorganic Chemistry and Radiochemistry, ed. H. J. EMELEUS AND A. G. SHARPE, Academic Press, approximately annual, 1959 onwards: title changed to *Advances in Inorganic Chemistry* from Volume 31, 1987.

Progress in Inorganic Chemistry, ed. F. A. COTTON, Interscience, 1959 onwards
Both these series give accumulated contents lists in the latest volume, and every tenth volume in each series has a full accumulated index.

Coordination Chemistry Reviews, ed. A. B. P. LEVER, Elsevier, 1966

onwards. Covers both Main Group and Transition Metal coordination compounds, with some emphasis on methods.

Also important as initial sources of fuller information are *Quarterly Reviews of the Chemical Society*, 1949–1971, titled *Chemical Society Reviews*, 1972 onwards. The American Chemical Society publishes *Accounts of Chemical Research and Chemtech*, and general reviews also appear in the German Chemical Society's *Angewandte Chemie* (see the International Edition, which is in English).

These set out to publish articles on any part of chemistry, aimed at the general chemical reader. The student will find a number of useful articles, especially in the earlier volumes.
Chemistry in Britain, Royal Society of Chemistry.
Chemical and Engineering News, American Chemical Society.
These are the regular news-journals of these societies, which often include survey articles, particularly on topics of current interest, and on industrial activities. Similar informal journals are produced by a number of other Societies, e.g. *Chemistry in Australia*, *Chemistry in New Zealand*, *Canadian Chemical News*. Other sources are *Education in Chemistry* (Royal Society of Chemistry) and *Journal of Chemical Education* (American Chemical Society)

The ultimate source of chemical information is the journal article where working chemists publish the results of their investigations, after their papers have been vetted by referees. A journal reader is assumed to be knowledgeable in the field, and modern publication is under pressure to be concise. Students should consult journal articles only after surveying the information available in textbooks and reviews. Often, the most useful parts of a paper for the student are the introduction, setting the background and outlining earlier work, and the discussion-conclusion section. Many journals publish inorganic chemistry. The two most important English language titles are *Inorganic Chemistry* (American Chemical Society) and *Journal of the Chemical Society, Dalton Transactions* (Royal Society of Chemistry).

(3) Bibliographies for Particular Sections of the Text

In addition to the appropriate sections of the advanced texts listed above, more information about particular themes may be gained from the following books and articles. This list is not intended to be exhaustive, but is designed to give a lead in to published material in each area. More specialized and advanced references may be traced through the given titles.

SI UNITS AND CHAPTER 1

For a fuller account of units and nomenclature see:

1979 manual of symbols and terminology for physicochemical quantities and units, *Pure and Applied Chemistry* **51**, 1979, 1–41.

A Dictionary of Scientific Units, H. G. JERRARD AND D. B. MCNEILL, Chapman & Hall, 1986.

UPAC Nomenclature of Inorganic Chemistry, Recommendations 1990, ed. G. J. LEIGH, Blackwell Scientific, 1990.

For an interesting account see:

Orgin of the Names of Chemical Elements, V. RINGNES, *Journal of Chemical Education* **66**, 1989, 731–738. For naming of the superheavy elements see references for Chapter 16.

For examples of the range of inorganic chemistry, we quote:

Silicon and Silicones, About Stone-age Tools, Antique Pottery, Modern Ceramics, Computers, Space Materials, and How They All Got That Way, E. G. ROCHOW, Springer-Verlag, 1987.

Pinpointing the Past, M. COWELL, *Chemistry in Britain* **28**, 1992, 892–896.

For an interesting discussion on clay minerals see *Clay Minerals and the Origin of Life*, A. G. CAIRNS-SMITH AND H. HARTMAN, Cambridge University Press, 1986.

ATOMIC PROPERTIES (Chapter 2)

Energy Levels in Atoms and Molecules, W. G. RICHARDS AND P. R. SCOTT, Oxford Chemistry Primer, Oxford University Press, 1994. See also: *Atomic Spectra*, T. P. SOFTLEY (1994) in the same series.

Principles of Atomic Orbitals, N. N. GREENWOOD, Royal Society of Chemistry Monographs for Teachers No. 8, 3rd Edition, 1980. See also the references in Chapter 3.

Relative atomic masses (atomic weights): *Pure and Applied Chemistry* Revised values appear every even year in this IUPAC publication.

Table Talk, D. H. ROUVRAY (coordinator), *Chemistry in Britain*, 1994, 371–386. A historical account of the development of the Periodic Table, on its 125th anniversary.

Ionisation Energies Revisited, N. C. PYPER AND M. BERRY, *Education in Chemistry*, September 1990, 135–137.

The Periodicity of Electron Affinity, R. T. MEYERS, *Journal of Chemical Education* **67**, 1990, 307–308.

Revised Effective Ionic Radii in Halides and Chalcogenides, R. D. SHANNON, *Acta Crystallographica* **A32**, 1976, 751–767: O. JOHNSON, *Inorganic Chemistry* **12**, 1973, 780.

Soft-sphere Ionic Radii for Alkali and Halogenide Ions, L. PAULING, *J. Chem. Soc., Dalton Transactions*, 1980, 645 and references therein.

Absolute Electronegativity and Hardness, R. G. PEARSON, *Chemistry in Britain*, May 1991, 444–447.

Principles of Electronegativity, R. T. SANDERSON, *Journal of Chemical Education* **65**, 1988, 112–118.

Electronegativities of Elements in Valence States, Y. ZHANG, *Inorganic Chemistry* **21**, 1982, 3886–9; Applications to Strengths of Lewis Acids, 3889–93.

The 'Inert-Pair' Effect on Electronegativity, R. T. SANDERSON, *Inorganic Chemistry* **25**, 1986, 1856–8; also *J. Amer. Chem. Soc.* **105**, 1983, 2259–61 and references therein.

An Electronegativity Scale Based upon Geometry Changes on Ionisation, P H. BLUSTIN AND W. T. RAYNES, *J. Chem. Soc., Dalton Transactions*, 1981, 1237.

The Covalent Potential: A Simple and Useful Measure of the Valence-State Electronegativity for Correlating Molecular Energetics, Y.-R. LUO AND S. W. BENSON, *Accounts of Chemical Research* **25**, 1992, 375–381.

Bent's Rule: Energetics, Electronegativity, and the Structures of Nonmetal Fluorides, J. E. HUHEEY *Inorganic Chemistry* **20**, 1981, 4033–4035.

MOLECULAR SHAPES AND BONDING (Chapters 3 and 4)

The Shape and Structure of Molecules, C. A. COULSON revised by R. MCWEENY, 2nd Edition, Clarendon, 1982.

Valence, C. A. COULSON, Oxford University Press, 3rd Edition, 1979, by R. MCWEENY.

This is a very clear account of atomic and molecular structure, ranging over inorganic and organic molecules. The general student will find it useful if he or she is willing to 'read round' the more detailed sections.

See also the following:

An Introduction to Molecular Orbitals, Y. JEAN, F. VOLATRON AND J. K. BURDETT, Oxford University Press, 1993.

Bonding and Structure: Structural Principles in Inorganic and Organic Chemistry, N. W. ALCOCK, Ellis Horwood, 1990.

Chemical Bonding Theory, B. WEBSTER, Blackwell, 1990.

The Chemical Bond, J. N. MURRELL, S. F. A. KETTLE, AND J. M. TEDDER, Wiley, 2nd Edition, 1985.

Symmetry and Structure, S. F. A. KETTLE, Wiley, 1985.

Valency, M. F. O'DWYER, J. E. KENT AND R. D. BROWN, Springer, 2nd Edition, 1978, reprinted 1986.

The Nature of the Chemical Bond, L. PAULING, Cornell University Press (Oxford University Press in UK), 3rd Edition, 1960.

This is Pauling's classical text on chemical bonding, which should be dipped into by every chemist. See Also *The Chemical Bond*, Oxford University Press, 1967, in which Pauling gives a shortened and updated survey.

Lewis Structures, Formal Charge, and Oxidation Numbers. A More user-Friendly Approach, J. E. PACKER AND S. D. WOODGATE, *Journal of Chemical Education* **68**, 1991, 456–458. A good summary account of the rules for working out Lewis structures of molecules.

The Simplest Molecule, I. R. MCNAB, *Chemistry in Britain*, 1992, 538–542. A discussion of the bonding in H_2^+.

Describing Electron Distribution in the Hydrogen molecule. A New Approach. C. J. WILLIS, *Journal of Chemical Education* **68**, 1991 743–747.

The Relative Energies of Molecular Orbitals for Second-Row Homonuclear Diatomic Molecules. The Effect of $s–p$ Mixing, A. HAIM, *Journal of Chemical Education* **68**, 1991, 737–738.

The Three Forms of Molecular Oxygen, M. LAING, *Journal of Chemical Education* **66**, 1989, 453–454.

The Significance of the Bond Angle in Sulfur Dioxide, G. H. PURSER, *Journal of Chemical Education* **66**, 1989, 710–713.

Bonding Considerations of the Nitrate Anion, G. R. WILLEY, *Education in Chemistry*, 1989, 78–82.

The electron pair repulsion theory was introduced by N. V. Sidgwick and H. M. Powell, *Proc. Royal Soc.* **176A** (1940), 153, and developed extensively by R. J. Gillespie and R. S. Nyholm, as in *Quarterly Reviews*, **11** (1957), 261, and more recently in the following articles:

Shaping up with EAN and VSEPR, M. LAING, *Education in Chemistry*, July 1995, 102–105.

The VSEPR Model Revisited, *Chemical Society Reviews*, 1992, 59–69, and Multiple Bonds and the VSEPR Model, *Journal of Chemical Education* **69** (1992), 116–121.

Precise data on molecular structure, in the gaseous and condensed states, is to be found in The Chemical Society's Special Publications 11 and 18, *Interatomic Distances and Supplement*.

SOLIDS (Chapter 5)

Inorganic Structural Chemistry, U. MÜLLER, Wiley, 1993.

Structural Inorganic Chemistry, A. F. WELLS, Oxford University Press, 5th Edition, 1984.

Molecular and Crystal Structure Models, A. WALTON, Ellis Horwood, 1978.

Perovskites—Chemical Chameleons, A. RELLER AND T. WILLIAMS, *Chemistry in Britain*, 1989, 1227–1230.

Lattice energies—a detailed treatment of basic work is given by T. C. WADDINGTON, *Advances in Inorganic Chemistry and Radiochemistry* **1**, 1959, 158–221. For more recent discussions see:

Lattice Enthalpies of Ionic Halides, Hydrides, Oxides, and Sulfides, J. B. HOLBROOK, R. SABRY-GRANT, B. C. SMITH, AND T. V. TANDEL, *Journal of Chemical Education* **67**, 1990, 304–306; The Calculation of Lattice Energy: Some Problems and Some Solutions, H. D. B. JENKINS, *Revue de Chimie Minérale* **16**, 1979, 134–150.

A Tetrahedron of Bonding, M. LAING, *Education in Chemistry*, 1993, 160–163.

SOLVENTS (Chapter 6)

Redox Mechanisms in Inorganic Chemistry, A. G. LAPPIN, Ellis Horwood, 1993.

Ions in Solution. Basic Principles of Chemical Interactions, J. BURGESS, Ellis Horwood, 1988.

Recent Advances in the Concept of Hard and Soft Acids and Bases, R. G. PEARSON, *J. Chemical Education* **64**, 1987, 561–567. Makes this longstanding theory more quantitative.

An Acidity Scale for Binary Oxides, D. W. SMITH, *J. Chemical Education* **64**, 1987, 480–481.

The Useage of the Terms 'Equivalent' and 'Normal', H. M. N. H. IRVING (T. S. WEST), *Pure and Applied Chemistry* **50**, 1978, 325.

See *Modern Aspects of Inorganic Chemistry*, H. J. EMELEUS AND A. G. SHARPE, listed above, for a general survey of non-aqueous solvents.

Nonaqueous Solution Chemistry, O. POPOVYCH AND R. P. T. TOMKINS, Wiley, 1981.

Inorganic Chemistry in Liquid Ammonia, D. NICHOLLS, Elsevier, 1979.

EXPERIMENTAL METHODS (Chapter 7)

Structural Methods in Inorganic Chemistry, E. A. V. EBSWORTH, D. W. H. RANKIN, AND S. CRADOCK, Blackwell, 2nd Edition, 1991.

A brief summary of solvent extraction processes can be found in the following:

Liquid-Liquid Extraction: Metals, P. J. BAILES, C. HANSON, AND M. A. HUGHES, *Chemical Engineering*, 1976, 86–94.

For more recent updates see the various sections on solvent extraction and hydrometallurgy in the *Kirk-Othmer Encyclopedia of Chemical Technology*.

Molecular Structure: Its Study by Crystal Diffraction, J. C. SPEAKMAN, Royal Society of Chemistry, Monograph for Teachers **30**, 1977.

Handbook of X-ray and Ultraviolet Photoelectron Spectra, ed. D. BRIGGS, Heyden, 1978.

The Partnership of Gas-Phase Core and Valence Photoelectron Spectroscopy, W. L. JOLLY, Accounts of Chemical Research **16**, 1983, 370–376.

NMR spectroscopy:

For basic, readable introductions, see

NMR in Chemistry, W. KEMP, Macmillan, 1986.

Nuclear Magnetic Resonance, P. J. HORE, Oxford Chemistry Primers, Oxford University Press, 1995.

For more advanced students:

Modern NMR Spectroscopy. A Guide for Chemists, J. K. M. SANDERS AND B. K. HUNTER, Oxford University Press, 2nd Edition, 1993. See also the accompanying *Workbook of Chemical Problems*, J. K. M. SANDERS, E. C. CONSTABLE, B. K. HUNTER, AND C. M. PEARCE, Oxford University Press, 2nd Edition, 1993.

NMR and Chemistry. An Introduction to Modern NMR Spectroscopy, J. W. AKITT, Chapman & Hall, 3rd Edition, 1992.

State of the Art for Solids, R. K. HARRIS, *Chemistry in Britain*, 1993, 601–604. A good introduction to the technique and applications of solid-state nmr spectroscopy.

Scanning Tunneling Microscopy, C. M. LIEBER, *Chemical & Engineering News Special Report*, April 18, 1994, 28–43. See also:

Applications of Scanning Tunneling Microscopy to Inorganic Chemistry, X. L. WU AND C. M. LIEBER, *Progress in Inorganic Chemistry* **39**, 1991, 431–510.

Scanning Tunneling Microscopy and Atomic Force Microscopy in Organic Chemistry, J. FROMMER, *Angewandte Chemie, International Edition* **31**, 1992, 1298–1328.

Probing Surfaces, J. LECKENBY, *Chemistry in Britain*, March 1995, 212.

GENERAL PROPERTIES OF THE ELEMENTS
(Chapter 8)

Nucleogenesis see

Supernova 1987A—New Evidence for Nucleosynthesis, C. H. ATWOOD, *Chemistry in Britain*, May 1990, 423–426.

Stellar Alchemy—The Origin of the Chemical Elements, E. B. NORMAN, *Journal of Chemical Education* **71**, 1994, 813–820.

See also *Inorganic Geochemistry*, P. HENDERSON, Pergamon Press, 1982, for an overview on nucleosynthesis.

Nobel Lecture, W. A. FOWLER, *Science* **226**, 1984, 922–935.

For interstellar chemistry see

Semistable Molecules in the Laboratory and in Space, H. W. KROTO, *Chemical Society Reviews* **11**, 1982, 435–491.

Chemistry of the Solar System, H. E. SUESS, Wiley, 1987.

General

Thermochemistry of Inorganic Fluorine Compounds, A. A. WOOLF, *Advances in Inorganic Chemistry and Radiochemistry* **24**, 1981, 1–56.

Conditions for Stability of Oxidation States Derived from Photo-electron Spectra and Inductive Quantum Chemistry, C. K. JØRGENSEN, *Z. anorganische. allg. Chemie* **540/541**, 1986, 91–105.

A very broad sweep through the whole of the Periodic Table arranging oxidation states by their Kossel numbers (Kossel number is Z minus the ionic charge) and comparing with the ionization energy from photoelectron spectra. Article is in English.

Oxidation Potentials, W. M. LATIMER, Prentice-Hall, 2nd Edition, 1952. This is a complete survey of oxidation potential data and still the standard source (though more recent values are available for many of the more difficult determinations).

A Graphical Method of Representing the Free Energies of Oxidation–Reduction Systems, E. A. V. EBSWORTH, *Education in Chemistry* **1**, 1964, 123.

For properties and extraction of elements

see appropriate sections in the *Kirk-Othmer Encyclopedia of Chemical Technology*.

See also:

The Extraction of Metals from Ores Using Bacteria, D. K. EWART AND M. N. HUGHES, *Advances in Inorganic Chemistry* **36**, 1991, 103–135.

Principles of the Extraction of Metals, D. J. G. IVES, R. I. C. Monograph.

HYDROGEN (Chapter 9)

Liquid Water—The Story Unfolds, M. C. R. SYMONS, *Chemistry in Britain*, 1989, 491–494. An account of current thinking concerning the structure of liquid water.

Active MgH_2–Mg Systems for Reversible Chemical Energy Storage, B. BOGDANOVI'C, A. RITTER, AND B. SPLIETHOFF, *Angewandte Chemie, International Edition* **29**, 1990, 223–234.

Taking Stock: The Astonishing Development of Boron Hydride Cluster Chemistry (Ludwig Mond Lecture), N. N. GREENWOOD, *Chemical Society Reviews* **21**, 1992, 49–57.

The Hydrides of Aluminium, Gallium, Indium, and Thallium: A Re-evaluation, A. J. DOWNS AND C. R. PULHAM, *Chemical Society Reviews* 1994, 175–184.

The Hunting of the Gallium Hydrides, A. J. DOWNS AND C. R. PULHAM, *Advances in Inorganic Chemistry* **41**, 1994, 171.

Recent Developments in the Chemistry of Alane (AlH_3) and Gallane (GaH_3), C. L. RASTON, *Journal of Organometallic Chemistry* **475**, 1994, 15–24.

Rings, Clusters and Polymers of Main Group and Transition Elements, H. W. ROESKY, Elsevier, 1989. See also: *Boranes and Metalloboranes*, C. E. HOUSECROFT, Ellis Horwood, 1990.

Hydrido Complexes of the Transition Metals, D. S. MOORE AND S. D. ROBINSON, *Chemical Society Reviews* **12**, 1983, 415–452.

A New Stage in the Development of Transition Metal Alumohydrides, B. M. BULYCHEV, *Polyhedron* **9**, 1990, 387–408.

Very Strong Hydrogen Bonds, J. EMSLEY, *Chemical Society Reviews* **9**, 1980, 91–124.

's' ELEMENTS (Chapter 10)

Alkali and Alkaline Earth Metal Cryptates, D. PARKER, *Advances in Inorganic Chemistry and Radiochemistry* **27**, 1983, 1–26.

Electrides, Negatively Charged Metal ions and Related Phenomena, J. L. DYE, *Progress in Inorganic Chemistry* **32**, 1984, 327–441.

First Electride Crystal Structure, S. B. DAWES, D. L. WARD, R. H. HUANG, AND J. L. DYE, *J. Amer. Chem. Soc.* **108**, 1986, 3534–3535. (in the compound $Cs(18\text{-crown-}6)_2^+ \cdot e^-$).

To be or not to Be—The Story of Beryllium Toxicity, D. N. SKILLETER, *Chemistry in Britain*, 1990, 26–30. Toxicology, chemistry and applications of Be and its compounds.

Beryllium Coordination Chemistry, C. Y. WONG AND J. D. WOOLLINS, *Coordination Chemistry Reviews* **130**, 1994, 243–273.

The Great Radium Scandal, R. M. MACKLIS, *Scientific American*, August 1993, 78–83. An interesting account of the use of a radioactive radium-laced medicine in the 1920s.

Structures of Organo Alkali Metal Complexes and Related Compounds, E. WEISS, *Angewandte Chemie, International Edition* **32**, 1993, 1501–1523.

Strontium—a neglected element, J. W. NICHOLSON AND L. R. PIERCE, *Education in Chemistry*, May 1995, 74–76.

'f' ELEMENTS (Chapters 11 and 12)

Lanthanides and Actinides, S. COTTON, Macmillan, 1991.

The Discovery of the Rare Earth Element, C. H. EVANS, *Chemistry in Britain*, 1989, 880–882.

Application of Lanthanide Reagents in Organic Synthesis, G. A. MOLANDER, *Chemical Reviews* **92**, 1992, 29–68.

Reduced Halides of the Rare Earth Elements, G. MEYER, *Chemical Reviews* **88**, 1988, 93–107.

The Chemistry of the Actinide Elements, eds J. J. KATZ, G. T. SEABORG AND L. R. MORSS, 2nd Edition, 2 volumes, 1986.

Preparation and Purification of Actinide Metals, J. C. SPIRLET, J. R. PETERSON, AND L. B. ASPREY, *Advances in Inorganic Chemistry* **31**, 1987, 1–41.

Plutonium—The Element of Surprise, G. R. CHOPPIN AND B. E. STOUT, *Chemistry in Britain*, 1991, 1126–1129.

The Most Useful Actinide Isotope: Americium-241, J. D. NAVRATIL, W. W. SCHULTZ, AND G. T. SEABORG, *Journal of Chemical Education* **67**, 1990, 15–16.

The Chemistry of Berkelium, J. P. PETERSON AND D. E. HOBART, *Advances in Inorganic Chemistry and Radiochemistry* **28**, 1984, 29–64.

Actinide Alkoxide Chemistry, W. G. VAN DER SLUYS AND A. P. SATTELBERGER, *Chemical Reviews* **90**, 1990, 1027–1040.

TRANSITION ELEMENTS (Chapters 13, 14 and 15)

Synthesis of Complex Metal Oxides by Novel Routes, C. N. R. RAO AND J. GOPALAKRISHANAN, *Accounts of Chemical Research* **20**, 1987, 228–235.

Solid State Structures of the Binary Fluorides of the Transition Metals, A. J. EDWARDS, *Advances in Inorganic Chemistry and Radiochemistry* **27**, 1983, 83–112.

Preparations and Reactions of Oxide Fluorides of the Transition Metals, the Lanthanides, and the Actinides, J. H. HOLLOWAY AND D. LAYCOCK, *Advances in Inorganic Chemistry and Radiochemistry* **28**, 1984, 73–100.

The Chemistry and Spectroscopy of Mixed-Valence Compounds, R. J. H. CLARK, *Chemical Society Reviews* **13**, 1984, 219–244.

Mixed Valence Compounds of d^5–d^6 Metal Centres, C. CREUTZ, *Progress in Inorganic Chemistry* **30**, 1983, 1–73.

Vanadium Peroxide Complexes, A. BUTLER, M. J. CLAGUE, AND G. E. MEISTER, *Chemical Reviews* **94**, 1994, 625–638.

Polyoxometalate Chemistry: An Old Field with New Dimensions in Several Disciplines, M. T. POPE AND A. MÜLLER, *Angewandte Chemie, International Edition* **30**, 1991, 34–48.

Peroxo and Superoxo Complexes of Chromium, Molybdenum, and Tungsten, M. H. DICKMAN AND M. T. POPE, *Chemical Reviews* **94**, 1994, 569–584.

The Active Sites in Manganese-Containing Metalloproteins and Inorganic Model Complexes, K. WIEGHARDT, *Angewandte Chemie, International Edition* **28**, 1989, 1153–1172.

The Coordination Chemistry of Technetium, J. BALDAS, *Advances in Inorganic Chemistry* **41**, 1994, 1.

Ferric Iodide as a Nonexistent Compound, K. B. YOON AND J. K. KOCHI, *Inorganic Chemistry* **29**, 1990, 869–874.

Nickel: An Element with Wide Application in Industrial Homogeneous Catalysis, W. KEIM, *Angewandte Chemie, International Edition* **29**, 1990, 235–244.

High Oxidation State Organometallic Chemistry, A Challenge—the Example of Rhenium, W. A. HERRMANN, *Angewandte Chemie, International Edition* **27**, 1988, 1297–1313.

Homogeneous Catalysis with Compounds of Rhodium and Iridium, R. S. DICKSON, D. Reidel, 1985.

Chlorotris(triphenylphosphine)rhodium(I): Its Chemical and Catalytic Reactions, F. H. JARDINE, *Progress in Inorganic Chemistry* **28**, 1981, 64–202.

One-Dimensional Inorganic Platinum-Chain Electrical Conductors, J. M. WILLIAMS, *Advances in Inorganic Chemistry and Radiochemistry* **26**, 1983, 235–268.

Coordination Chemistry of Halocarbons, R. J. KULAWIEC AND R. H. CRABTREE, *Coordination Chemistry Reviews* **99**, 1990, 89–115.

Homoleptic Nobel Metal Carbonyl Cations, L. WEBER, *Angewandte Chemie, International Edition* **33**, 1994, 1077–1078.

Compounds of Gold in Unusual Oxidation States, H. SCHMIDBAUR AND K. C. DASH, *Advances in Inorganic Chemistry and Radiochemistry* **25**, 1982, 239–262.

Homo- and Heteronuclear Cluster Compounds of Gold, K. P. HALL AND D. M. P. MINGOS, *Progress in Inorganic Chemistry* **32**, 1984, 237–325.

Fluorides of Copper, Silver, Gold, and Palladium, B. G. MÜLLER, *Angewandte Chemie, International Edition* **26**, 1987, 1081–1097.

The Crystal Engineering of Non-Molecular Metal Compounds with Anionic Chalcogenide Ligands E^{2-} and RE^-, I. G. DANCE in *Perspectives in Inorganic Chemistry*, eds. A. F. WILLIAMS, C. FLORIANI, AND A. E. MERBACH, VCH, 1992, 165–181.

As an interesting example of a new class of ruthenium coordination complex based on a dendritic structure, see *Chemical & Engineering News*, February 1, 1993, 28–30.

TRANSITION METAL TOPICS (Chapter 16)

Superconductors

Few issues of the journals from 1987 onwards lack a reference to warm superconductors! *Nature* **329**, 1987, 763 and *Chemistry in Britain*, 1987, 962–966 give summaries to that time. For recent review articles see the following:

Chemistry in Britain, September 1994. This issue is devoted to superconductors, with particular emphasis on the cuprate systems.

Structure, Composition and Properties of High-Temperature Cuprate Superconductors, J. K. BURDETT, *Perspectives in Coordination Chemistry*, VCH, 1992, 293–319.

The New Superconductors, *Chemical & Engineering News Special Report*, December 21, 1992, 24–41.

Superconductors Beyond 1–2–3, R. J. CAVA, *Scientific American*, August 1990, 24–31.

Structural Chemistry and the Local Charge picture of Copper Oxide Superconductors, R. J. CAVA, *Science* **247**, 1990, 656–662.

Structural Aspects of High-Temperature Cuprate Superconductors, C. N. R. RAO AND B. RAVEAU, *Accounts of Chemical Research* **22**, 1989, 106–113.

Chemical Aspects of Solution Routes to Perovskite-Phase Mixed-Metal Oxides from Metal-Organic Precursors, C. D.

CHANDLER, C. ROGER, AND M. J. HAMPDEN-SMITH, *Chemical Reviews* **93**, 1993, 1205–1241.

Transition metal carbonyl compounds

100 years of Metal Carbonyls, E. W. ABEL, *Education in Chemistry*, March 1992, 46–49. A very readable historical summary of the developments in this field.

Highly Reduced Metal Carbonyl Anions: Synthesis, Characterisation, and Chemical Properties, J. E. ELLIS, *Advances in Organometallic Chemistry* **31**, 1990, 1–51.

Organometallic compounds

The Organometallic Chemistry of the Transition Metals, R. H. CRABTREE, 2nd Edition, Wiley, 1993.

Organometallics, C. ELSCHENBROICH AND A. SALZER, VCH, 1989.

Organometallic Chemistry. An Overview., J. S. THAYER, VCH, 1987.

An Introduction to Organometallic Chemistry, A. W. PARKINS AND R. C. POLLER, Macmillan, 1986.

Carbene Complexes in Organic Synthesis, K. H. DÖTZ, *Angewandte Chemie, International Edition* **23**, 1984, 587–608.

Metal–Carbon and Metal–Metal Bonds as Ligands in Transition-Metal Chemistry: the Isolobal Connection., F. G. A. STONE, *Angewandte Chemie, International Edition* **23**, 1984, 89–99.

Cyclopentadienyl compounds

Metallocenes Come of Age, *Chemistry in Britain*, February 1994, 87–88.

Bulky or Supracyclopentadienyl Derivatives in Organometallic Chemistry, C. JANIAK, AND H. SCHUMANN, *Advances in Organometallic Chemistry* **33**, 1991, 291–393.

Organometallic chemistry of the lanthanides and actinides

Organometallic Chemistry of the lanthanides, C. J. SCHAVERIEN, *Advances in Organometallic Chemistry* **36**, 1994, 283–362.

Zero Oxidation State Compounds of Scandium Yttrium, and the Lanthanides, F. G. A. CLOKE, *Chemical Society Reviews* **22**, 1993, 17–24.

Monocyclopentadienyl Halide Complexes of the *d*- and *f*-Block Elements, R. POLI, *Chemical Reviews* **91**, 1991, 509–551.

See also W. J. EVANS, *Polyhedron* **6**, 1987, 803–835; *Advances in Organometallic Chemistry* **24**, 1985, 131–173.

Multiple metal–metal bonds

Metal–Metal Bonds of Order Four, J. L. TEMPLETON,

Progress in Inorganic Chemistry **26**, 1979, 211–300.

Multiple Bonds Between Metal Atoms. F. A. COTTON AND R. A. WALTON, Wiley 1982.

Metal clusters
Introduction to Cluster Chemistry, D. M. P. MINGOS AND D. J. WALES, Inorganic and Organometallic Chemistry Series, Prentice Hall, 1990.

The Chemistry of Metal Cluster Complexes, D. F. SHRIVER, H. D. KAESZ, AND R. D. ADAMS, VCH, 1990.

Large Clusters and Colloids. Metals in the Embryonic State, G. SCHMID, *Chemical Reviews* **92**, 1992, 1709–1727.

Metal Clusters Revisited, J. LEWIS, *Chemistry in Britain*, 1988, 795–800.

Metal Clusters in Catalysis, eds. B. C. GATES, L. GUCZI AND H. KNÖZINGER, Elsevier, 1986.

High Nuclearity Carbonyl Clusters: Their Synthesis and Reactivity, M. D. VARGAS AND J. N. NICHOLLS, *Advances in Inorganic Chemistry and Radiochemistry* **30**, 1986, 123–222.

For a good example of a giant cluster see: A New Copper Selenide Cluster with PPh_3 Ligands: $[Cu_{146}Se_{73}(PPh_3)_{30}]$, H. KRAUTSCHEID, D. FENSKE, G. BAUM, AND M. SEMMELMANN, *Angewandte Chemie, International Edition* **32**, 1993, 1303–1305.

Bonding in Molecular Clusters and Their Relationship to Bulk Metals, D. M. P. MINGOS, *Chemical Society Reviews* **15**, 1986, 31–61.

How Chemistry and Physics Meet in the Solid State, R. HOFFMANN, *Angewandte Chemie, International Edition* **26**, 1987, 846–878 Bond theory and band theory—how they are reconciled.

Metal-dioxygen complexes
For comprehensive reviews on metal–dioxygen complexes and closely related topics, see the May 1994 issue of *Chemical Reviews* which is entirely devoted to this topic.

The Structure and Reactivity of Dioxygen Complexes of the Transition Metals, M. H. GUBELMANN AND A. F. WILLIAMS, *Structure and Bonding* **55**, 1983, 1–65.

Metal-Nitrogen Complexes
The Discovery of $[Ru(NH_3)_5N_2]^{2+}$. A Case of Serendipity and the Scientific Method, C. V. SENOFF, *Journal of Chemical Education* **67**, 1990, 368–370.

The Chemistry of Nitrogen Fixation and Models for the Reactions of Nitrogenase, R. A. HENDERSON, G. J. LEIGH, AND C. J. PICKETT, *Advances in Inorganic Chemistry and Radiochemistry*

27, 1983, 198–292.

Metal-hydrogen complexes
Coordination Chemistry of Dihydrogen, D. M. HEINEKEY AND W. J. OLDHAM JR, *Chemical Reviews* **93**, 1993, 913–926.

Dihydrogen Complexes: Some Structural and Chemical Studies, R. H. CRABTREE, *Accounts of Chemical Research* **23**, 1990, 95–101.

Molecular Hydrogen Complexes: Coordination of a σ Bond to Transition Metals, G. J. KUBAS, *Accounts of Chemical Research* **21**, 1988, 120–128.

Superheavy elements

Names and Symbols of Transfermium Elements, IUPAC Recommendations 1994, *Pure and Applied Chemistry* **66**, 1994, 2420–2421.

Naming Names, J. WILLIAMS, *Chemistry in Britain*, September 1995, 692–694.

Searching for the Transactinides, S. A. COTTON, *Education in Chemisty*, May 1995, 67–70.

110, 111 . . . . and counting, D. C. HOFFMAN, *Nature* **373**, 1995, 471–472.

Transuranium Elements: Past, Present and Future, G. T. SEABORG, *Accounts of Chemical Research* **28**, 1995, 257–264.

The Heaviest Elements, D. C. HOFFMAN, *Chemical & Engineering News*, Special Report, 2 May 1994, 24–34. An excellent account of the methods by which the superheavy elements are synthesised and their chemistry studied.

Creating Superheavy Elements, P. ARMBRUSTER AND G. MUNZENBERG, *Scientific American*, May 1989, 36–42.

The Elements Beyond Uranium, G. T. SEABORG AND W. T. LOVELAND, John Wiley, 1990.

Relativity
Relativistic Effects in Structural Chemistry, P. PYYKKÖ, *Chemical Reviews* **88**, 1988, 563–594.

Relativistic Effects on Periodic Trends, P. PYYKKÖ, in *The Effects of Relativity in Atoms, Molecules and the Solid-State*, eds. S. WILSON, I. P. GRANT, AND B. L. GYORFFY, Plenum, 1990.

Why is Mercury Liquid? Or, Why Do Relativistic Effects Not Get Into Chemistry Textbooks?, L. J. NORRBY, *Journal of Chemical Education* **68**, 1991, 110–113.

p ELEMENTS (Chapter 17)

Main Group Chemistry, A. G. MASSEY, Ellis Horwood, 1990.

The Resurgence of Main Group Element Chemistry, N. N. GREENWOOD, *Journal of the Chemical Society, Dalton Transactions*, 1991, 565–573. An excellent overview of recent developments in this field.

Preparations and Reactions of Inorganic Main-Group Oxide Fluorides, J. H. HOLLOWAY AND D. LAYCOCK, *Advances in Inorganic Chemistry and Radiochemistry* **27**, 1983, 157–179.

The Metallic Face of Boron, T. P. FEHLNER, *Advances in Inorganic Chemistry* **35**, 1990, 199–233.

Chemistry of Aluminium, Gallium, Indium, and Thallium, ed. A. J. DOWNS, Blackie, 1993.

Subvalent Aluminium Halides as Examples of High-Temperature Species in Inorganic Syntheses, J. J. SCHNEIDER, *Angewandte Chemie, International Edition* **33**, 1994, 1830–1832.

Reactions of Group 13 Alkyls with Dioxygen and Elemental Chalcogens: From Carelessness to Chemistry, A. R. BARRON, *Chemical Society Reviews* **22**, 1993, 93–99.

The Chemistry of GALLEX—Measurement of Solar Neutrinos with a Radiochemical Gallium Detector, E. HENRICH AND K. H. EBERT, *Angewandte Chemie, International Edition* **31**, 1992, 1283–1297.

The Race for Glittering Prizes, *Chemistry in Britain* **28**, 1992, 686–687. Fabrication of diamond films.

Silicon Nitride—From Powder Synthesis to Ceramic Materials, H. LANGE, G. WÖTTING, AND G. WINTER, *Angewandte Chemie, International Edition* **30**, 1991, 1579–1597.

Where are the Lone-Pair Electrons in Subvalent Fourth-Group Compounds? S.-W. NG AND J. J. ZUCKERMAN, *Advances in Inorganic Chemistry and Radiochemistry* **29**, 1985, 297–326.

PECVD—a Technique for New Technologies, P. M. MARTINEAU AND P. B. DAVIES, *Chemistry in Britain* **25**, 1989, 1018–1022. Discusses deposition of silicon films from silane.

The Industrial Uses of Tin Chemicals, S. J. BLUNDEN, P. A. CUSACK AND R. HILL, Royal Society of Chemistry, 1985.

The Nitrogen Fluorides and Some Related Compounds, H. J. EMELÉUS, J. M. SHREEVE, AND R. D. VERMA, *Advances in Inorganic Chemistry* **33**, 1989, 139–196.

Solid-State Chemistry with Nonmetal Nitrides, W. SCHNICK, *Angewandte Chemie, International Edition* **32**, 1993, 806–818.

The Stereochemistry of Sb(III) Halides and Some Related Compounds, J. F. SAWYER AND R. J. GILLESPIE, *Progress in Inorganic Chemistry* **34**, 1986, 65–113. Notes VSEPR relationships including lone pair effects.

The Structures of the Group 15 Element(III) Halides and Halogenoanions, G. A. FISHER AND N. C. NORMAN, *Advances in Inorganic Chemistry* **41**, 1994, 233–271.

The True Allotrops of Sulphur, G. RAYNER-CANHAM AND J. KETTLE, *Education in Chemistry* 1991, 49–51.

There is no such thing as H_2SO_3, M. LAING, *Education in Chemistry* 1993, 140.

Thiosulfate. An Interesting Sulfur Oxoanion That Is Iseful in Both Medicine and Industry—But Is Implicated in Corrosion, S. W. DHAWALE, *Journal of Chemical Education* **70**, 1993, 12–14.

Developments in Chalcogen–Halide Chemistry, B. KREBS AND F.-P. AHLERS, *Advances in Inorganic Chemistry* **35**, 1990, 235–317.

The Modern Sulphuric Acid process (of manufacture), A. PHILLIPS, *Chemistry in Britain* **13**, 1977, 471; see also 459.

Fluorine: the First Hundred years (1886–1986) ed. R. E. BANKS, D. W. A. SHARP AND J. C. TATLOW, Elsevier, 1987.

Astatine: organonuclear Chemistry and Biomedical Applications, I. BROWN, *Advances in Inorganic Chemistry* **31**, 1987, 43–88.

The Chemistry of Iodine Azide, K. DEHNICKE, *Angewandte Chemie, International Edition* **18**, 1979, 507–514. see also: The Chemistry of the Halogen Azides, *Advances in Inorganic Chemistry and Radiochemistry* **26**, 1983, 201–234 (includes characterization and uses in organic and inorganic synthesis). See also: Covalent Inorganic Azides, I. C. TORNIEPORTH-OETTING AND T. M. KLAPÖTKE, *Angewandte Chemie, International Edition* **34**, 1995, 511–520.

A Noble Cause, G. M. R. GRANT, *Chemistry in Britain* 1994, 388–390.

One or Several Pioneers? The Discovery of Noble-Gas Compounds, P. LASZLO AND G. J. SCHROBILGEN, *Agnewandte Chemie, International Edition* **27**, 1988, 479–489.

Radon: Not So Noble, J. D. LEE AND T. E. EDMONDS, *Education in Chemistry*, 1991, 152–154. See also: The Menace Under the Floorboards, A. F. GARDNER, R. S. GILLETT AND P. S. PHILLIPS, *Chemistry in Britain*, 1992, 344–348.

Chemical and Physical Properties of Some Xenon Compounds, J. L. HUSTON, *Inorganic Chemistry* **21**, 1982, 685–688, A useful review and systematization in terms of acid-base chemistry.

MAIN GROUP TOPICS (Chapter 18)

Chains, rings, nets, and clusters
Cluster Molecules of the p-Block Elements, C. E. HOUSECROFT,

Oxford University Press, 1994.

Cluster Chemistry, G. GONZÁLEZ-MORAGA, Springer-Verlag, 1993.

The Chemistry of Inorganic Homo- and Heterocycles, I. HAIDUC AND D. B. SOWERBY, Academic Press, 1987 (deals with Main Group species).

Homocyclic Selenium Molecules and Related Cations, R. STEUDEL AND E.-M. STRAUSS, *Advances in Inorganic Chemistry and Radiochemistry* **28**, 1984, 135–167; see also **18**, 1976, for sulfur.

Synthesis and Reactions of Phosphorus-rich Silylphosphines, G. FRITZ, *Advances in Inorganic Chemistry* **31**, 1987, 171–243.

Carbosilanes, G. FRITZ, *Angewandte Chemie, International Edition* **26**, 1987, 1111–1132.

Chemistry of the Polyhedral Boron Halides and the Diboron Tetrahalides, J. A. MORRISON, *Chemical Reviews* **91**, 1991, 35–48.

Early Carboranes and Their Structural legacy, R. E. WILLIAMS, *Advances in Organometallic Chemistry* **36**, 1994, 1–55.

Open-Chain Polyphosphorus Hydrides (Phosphanes), M. BAUDLER AND K. GLINKA, *Chemical Reviews* **94**, 1994, 1273–1297.

Monocyclic and Polycyclic Phosphanes, M. BAUDLER AND K. GLINKA, *Chemical Reviews* **93**, 1993, 1623–1667. A comprehensive review on the structural units found in polyphosphorus species.

Bridging Chasms with Polyphosphides, H.-G. VON SCHNERING AND W. HÖNLE, *Chemical Reviews* **88**, 1988, 243–273.

Clusters of Phosphorus: A Theoretical Investigation, M. HÄSER, U. SCHNEIDER AND R. AHLRICHS, *J. Amer. Chem. Soc.* **114**, 1992, 9551–9559.

Recent Aspects of the Structure and Reactivity of Cyclophosphazenes, V. CHANDRASEKHAR AND K. R. J. THOMAS, *Structure and Bonding* **81**, Springer-Verlag, 1993, 41–113.

Polyatomic Zintl Anions of the Post-transition Elements, J. D. CORBETT, *Chemical Reviews* **85**, 1985, 383–397.

The Polyborane, Carborane, Carbocation Continuum: Architectural Patterns, R. E. WILLIAMS, *Chemical Reviews* **92**, 1992, 177–207. Other reviews on boranes and carboranes are to be found in the same issue.

For Skeletal Electron Pairs see K. WADE, *Advances in Inorganic Chemistry and Radiochemistry* **18**, 1976, 1–66.

Solvents

Stabilisation of Unusual Cationic Species in Protonic Superacids and Acid Melts, T. A. O'DONNELL, *Chemical Society Reviews* **16**, 1987, 1–43.

Superacids, G. A. OLAH, G. K. S. PRAKASH, AND J. SOMMER, Wiley, 1985.

Double bonds

Strained-Ring and Double-Bond Systems Consisting of the Group 14 Elements Si, Ge, and Sn, T. TSUMURAYA, S. A. BATCHELLER, AND S. MASAMUNE, *Angewandte Chemie, International Edition* **30**, 1991, 902–939.

Unsaturated Molecules Containing Main Group Metals, M. VIETH, *Angewandte Chemie, International Edition* **26**, 1987, 1–14.

Organoelement Compounds with Al–Al, Ga–Ga, and In–In Bonds, W. UHL, *Angewandte Chemie, International Edition* **32**, 1993, 1386–1397.

The Chemistry of the Silicon–Silicon Double Bond, R. WEST, *Angewandte Chemie, International Edition* **26**, 1987, 1201–1211.

Transition Metal Complexes of Silylenes, Silenes, Disilenes, and Related Species, P. D. LICKISS, *Chemical Society Reviews* **21**, 1992, 271–279.

Multiply Bonded Germanium Species. Recent Developments, J. BARRAU, J. ESCUDIÉ, AND J. SATGÉ, *Chemical Reviews* **90**, 1990, 283–319.

The Chemistry of Diphosphenes and Their Heavy Congeners: Synthesis, Structure, and Reactivity, L. WEBER, *Chemical Reviews* **92**, 1992, 1839–1906.

The Syntheses, Properties, and Reactivities of Stable Compounds Featuring Double Bonds Between Heavier Group 14 and 15 Elements, A. H. COWLEY AND N. C. NORMAN, *Progress in Inorganic Chemistry* **34**, 1986, 1–63.

Double Bonds Between Phosphorus and Carbon, R. APPEL AND F. KNOLL, *Advances in Inorganic Chemistry* **33**, 1989, 259–361.

Phosphoalkynes: New Building Blocks in Synthetic Chemistry, M. REGITZ, *Chemical Reviews* **90**, 1990, 191–213.

Synthesis of Di-, Tri-, and Polyphosphane and Phosphene Transition-Metal Complexes, A.-M. CAMINADE, J.-P. MAJORAL, AND R. MATHIEU, *Chemical Reviews* **91**, 1991, 575–612.

Unconventional Multiple Bonds: Coordination Compounds of Unstable Vth Main Group Ligands, G. HUTTNER, *Pure and Applied Chemistry* **58**, 1986, 585–596.

VSEPR

Experimental observation of the Tellurium(IV) Bonding and Lone-pair Electron Density in Dimethyltellurium Dichloride by X-ray Diffraction Techniques, R. F. ZIOLO AND J. M. TROUP, *J. Amer. Chem. Soc.* **105**, 1983, 229–234.

A New Electrostatic Model of Molecular Shapes, J. L. BILLS AND S. P. STEED, *Inorganic Chemistry* **22**, 1983, 2401–2405.

The Valence Bond Interpretation of Molecular Geometry, D. W. SMITH, *J. Chemical Education* **57**, 1980, 106–109.

Directional Character, Strength, and Nature of the Hydrogen Bond in Gas-phase Dimers, A. C. LEGON AND D. J. MILLER, *Accounts of Chemical Research* **20**, 1987, 39–46. 'dimer' = Mol···HX: H-bond is found in the direction expected for the lone pair position. This gives a different type of evidence for sterically-active lone pairs.

A Theoretical Study of the Linear Versus Bent Geometry for Several MX_2 Molecules: MgF_2, CaH_2, CaF_2, CeO_2 and $YbCl_2$, R. L. DEKOCK, M. A. PETERSON, L. K. TIMMER, E. J. BAERENDS, AND P. VERNOOIJS, *Polyhedron* **9**, 1990, 1919–1934.

Bonding

Chemical Bonding in Higher Main Group Elements, W. KUTZELNIGG, *Angewandte Chemie, International Edition* **23**, 1984, 272–295.

The Chemistry of Hypervalent Compounds, J. I. MUSHER, *Angewandte Chemie, International Edition* **8**, 1969, 54–68. A hypervalent compound is essentially one where the octet is exceeded and the approach in this reference is *via* multicentred bonds using *p*-orbitals.

Studies of Silicon-Phosphorus Bonding, K. J. DYKEMA, T. N. TRUONG, AND M. S. GORDON, *J. Amer. Chem. Soc.* **107**, 1985, 4535–4541 An example of a characteristic *ab initio* calculation with interesting diagrams of calculated bond densities.

Silicates

For a more detailed account of silicate structures and for a more geochemical perspective see *Introduction to Mineral Sciences*, A. PUTNIS, Cambridge University Press, 1992.

Structural Chemistry of Silicates, F. LIEBAU, Springer-Verlag, 1985.

Curiosity, Chance, Paradox and Perspective in the Chemistry of Materials, J. M. THOMAS, *Journal of the Chemical Society, Dalton Transactions*, 1991, 555–563. An account of recent developments in materials chemistry.

Synthetic Zeolites, G. T. KERR, *Scientific American*, July 1989, 82–87.

Solid Acid Catalysts, J. M. THOMAS, *Scientific American*, April 1992, 82–88 (use of clays, zeolites and polyoxometallate anions in catalysis).

Zeolite Molecular Sieves, B. M. LOWE, *Education in Chemistry* 1992, 15–18.

Catalysis in Intracrystalline Space, G. HUTCHINGS, *Chemistry in Britain*, 1992, 1006–1009. See also 991–994 of the same issue.

Zeolitic and Layered Materials, S. L. SUIB, *Chemical Reviews* **93**, 1993, 803–826.

Syn-fuels from Boiling Stones, G. J. HUTCHINGS, *Chemistry in Britain*, 1987, 762 (use of zeolites in gasoline manufacture).

Layered Metal Phosphates and Phosphonates: From Crystals to Monolayers, G. CAO, H.-G. HONG AND T. E. MALLOUK, *Accounts of Chemical Research* **25**, 1992, 420–427.

An Inorganic Double Helix: Hydrothermal Synthesis, Structure, and Magnetism of Chiral $[(CH_3)_2NH_2]K_4$ $[V_{10}O_{10}(H_2O)_2(OH)_4(PO_4)_7].4H_2O$, V. SOGHOMONIAN, Q. CHEN, R. C. HAUSHALTER, J. ZUBIETA, AND C. J. O'CONNOR, *Science* **259**, 1993, 1596–1599.

THREE GENERAL TOPICS (Chapter 19)

Electron densities

Electron Density Distributions in Inorganic Compounds, K. TORIUMI AND Y. SAITO, *Advances in Inorganic Chemistry and Radiochemistry* **27**, 1983, 28–79.

Experimental Electron Densities and Chemical Bonding, P. COPPENS, *Angewandte Chemie, International Edition* **16**, 1977, 32–40.

Metal-sulfur and -selenium rings

Polysulphide Complexes of Metals, A. MULLER AND E. DIEMANN, *Advances in Inorganic Chemistry* **31**, 1987, 89–122.

Transition Metal Polysulphides: Coordination Compounds with Purely Inorganic Chelate Ligands, M. DRAGANJAC AND T. B. RAUCHFUSS, *Angewandte Chemie, International Edition* **24**, 1985, 742–757.

New Developments in the Coordination Chemistry of Inorganic Selenide and Telluride Ligands, L. C. ROOF AND J. W. KOLIS, *Chemical Reviews* **93**, 1993, 1037–1080. A comprehensive review illustrating the diversity in the coordination chemistry of these systems.

Caged Explosives: Metal-Stabilised Chalcogen Nitrides, P. F. KELLY, A. M. Z. SLAWIN, D. J. WILLIAMS, AND J. D. WOOLLINS, *Chemical Society Reviews* **21**, 1992, 245–252.

Fullerenes and related species

The Fullerenes. new Horizons for the Chemistry, Physics and Astrophysics of Carbon, eds. H. W. KROTO AND R. M. WALTON, Royal Society, 1993.

Simple Generation of C$_{60}$ (Buckminsterfullerene), D. W. IACOE, W. T. POTTER, AND D. TEETERS, *Journal of Chemical Education* **69**, 1992, 663.

C$_{60}$: Buckminsterfullerene, The Celestial Sphere that Fell to Earth, H. W. KROTO, *Angewandte Chemie, International Edition*, **31**, 1992, 111–246 A very readable and personal account of the discovery of the fullerenes.

Accounts of Chemical Research **25**, 1992, 97–175 A special issue on buckminsterfullerenes including sections on metal derivatives and fullerides.

The Chemistry of Fullerenes, R. TAYLOR AND R. M. WALTON, *Nature*, **363**, 1993, 685–693.

Organometallic Derivatives of Fullerenes, J. R. BOWSER, *Advances in Organometallic Chemistry* **36**, 1994, 57–94.

Atoms in Carbon Cages: the Structure and Properties of Endohedral Fullerenes, D. S. BETHUNE, R. D. JOHNSON, J. R. SALEM, M. S. DE VRIES, AND C. S. YANNONI, *Nature* **366**, 1993, 123–128.

Growth and Dissociation of Metal-Carbon Nanocrystals Probed, *Chemical & Engineering News*, October 25, 1993, 29–30. See also May 23, 1994, 35–36.

BIOLOGICAL, MEDICINAL AND ENVIRONMENTAL INORGANIC CHEMISTRY (Chapter 20)

Inorganic Biochemistry

Bioinorganic Chemistry: Inorganic Elements in the Chemistry of Life, W. KAIM AND B. SCHWEDERSKI, Wiley, 1995.

Principles of Inorganic Biochemistry, S. J. LIPPARD AND J. M. BERG, University Science Books, 1994.

Inorganic Biochemistry. An Introduction. J. A. COWAN, VCH, 1993.

The Biological Chemistry of the Elements. The Inorganic Chemistry of Life, J. J. R. FRAUSTO DA SILVA AND R. J. P. WILLIAMS, Clarendon Press, 1991.

Metal Ions in Biology, ed. T. G. SPIRO, Wiley, 1987.

Inorganic Chemistry of Biological Processes, M. N. HUGHES, Wiley, 3rd Edition, 1987.

Missing Information in Bio-Inorganic Chemistry, R. J. P. WILLIAMS, *Coordination Chemistry Reviews* **79**, 1987, 175–193.

Bio-inorganic Chemistry: Its Conceptual Evolution, R. J. P. WILLIAMS, *Coordination Chemistry Reviews* **100**, 1990, 573.

The Chemical Elements of Life, R. J. P. WILLIAMS, *Journal of the Chemical Society, Dalton Transactions* 1991, 539–546.

Biomineralization: the Hard part of Bioinorganic Chemistry! S. MANN, *Journal of the Chemical Society, Dalton Transactions*, 1993, 1–9. See also Solid-State Bioinorganic Chemistry: Mechanisms and Models of Biomineralization, S. MANN AND C. C. PERRY, *Advances in Inorganic Chemistry* **36**, 1991, 137–200.

The Role of Silicon in Biology, J. D. BIRCHALL, *Chemistry in Britain*, 1990, 141–144.

For nitric oxide in biochemistry see:

Nitric Oxide—Small but Powerful, A. R. BUTLER, *Education in Chemistry*, 1993, 120.

The Surprising Life of Nitric Oxide, P. L. FELDMAN, O. W. GRIFFITH AND D. J. STUEHR, *Chemical & Engineering News*, Special Report, December 1993, 26–38.

NO—Its Role in the Control of Blood Pressure, A. R. BUTLER, *Chemistry in Britain*, 1990, 419–421.

Biological Roles of Nitric Oxide, S. H. SNYDER AND D. S. BREDT, *Scientific American*, May 1992, 28–35.

The Physiological Role of Nitric Oxide, A. R. BUTLER AND D. L. H. WILLIAMS, *Chemical Society Reviews*, 1993, 233–241.

Chemistry Relevant to the Biological Effects of Nitric Oxide and Metallonitrosyls, M. J. CLARKE AND J. B. GAUL, *Structure and Bonding* **81**, Springer-Verlag, 1993, 147–181.
For iron-sulfur proteins, see volume 38 of *Advances in Inorganic Chemistry*, published in 1992. The entire issue is devoted to this topic.

Les Séismes Moléculaires de L'Hémoglobine, J.-L. MARTIN AND J.-C. LAMBY, *La Recherche* **24**, 1993, 572–575.

Determining the Structure of a Hydroxylase Enzyme That Catalyses the Conversion of Methane to Methanol in Methanotropic bacteria, A. C. ROSENZWEIG AND S. J. LIPPARD, *Accounts of Chemical Research* **27**, 1994, 229–236.

For model complexes see also: Oxo- and Hydroxo-Bridged Diiron Complexes: A Chemical Perspective on a Biological unit. D. M. KURTZ, Jr., *Chemical Reviews* **90**, 1990, 585–606.

Bioinorganic Chemistry of Vanadium, D. REHDER, *Angewandte Chemie, International Edition* **30**, 1991, 148–167. See also Vanadium: A Biologically Relevant Element, R. WIEVER AND K. KUSTIN, *Advances in Inorganic Chemistry* **35**, 1990, 81–115.

Some Aspects of the Bioinorganic Chemistry of Zinc, R. H. PRINCE, *Advances in Inorganic Chemistry and Radiochemistry* **22**, 1979, 240–302.

Molybdenum Enzymes, ed. T. SPIRO, Wiley, 1985. See also The Mo-, V-, and Fe-Based Nitrogenase Systems of Azotobacter, R. E. EADY, *Advances in Inorganic Chemistry* **36**, 1991, 77–102.

Metals in the Nitrogenases, R. R. EADY AND G. J. LEIGH, *Journal of the Chemical Society, Dalton Transactions*, 1994, 2739–2747.

Medicinal Inorganic Chemistry
Inorganic Chemistry and Drug Design, P. J. SADLER, *Advances in Inorganic Chemistry* **36**, 1991, 1–48. See also 49–75 for a review on the use of lithium compounds in medicine.

Targeting Metal Complexes, D. PARKER, *Chemistry in Britain*, 1994, 818–822.

Medical Diagnostic Imaging with Complexes of ^{99m}Tc, M. J. CLARKE AND L. PODBIELSKI, *Coordination Chemistry Reviews* **78**, 1987, 253–330.

Nuclear Medicine and Positron Emission Tomography: an Overview, T. J. MCCARTHY, S. W. SCHWARZ, AND M. J. WELCH, *Journal of Chemical Education* **71**, 1994, 830–836.

Technetium Chemistry and Technetium Radiopharmaceuticals, E. DEUTSCH, K. LIBSON, S. JURASSIN, AND, L. F. LINDOY, *Progress in Inorganic Chemistry* **31**, 1984, 75–139. See also Coordination Compounds in Nuclear Medicine, S. JURISSON, D. BERNING, W. JIA AND D. MA, *Chemical Reviews* **93**, 1993, 1137–1156; Technetium Radiopharmaceuticals–Fundamentals, Synthesis, Structure, and Development, K. SCHWOCHAU, *Angewandte Chemie, International Edition* **33**, 1994, 2258–2267.

Inorganic Chemistry and Medicine, P. J. SADLER, *Education in Chemistry*, May 1992, 80–83 An excellent concise summary of this topic.

Metal Compounds in Therapy and Diagnosis, M. J. ABRAMS AND B. A. MURRER, *Science* **261**, 1993, 725–730.

Metal Compounds in Cancer Chemotherapy, I. HAIDUC AND C. SILVESTRU, *Coordination Chemistry Reviews* **99**, 1990, 253–296.

Non-Platinum Group Metal Antitumor Agents: History, Current Status, and Perspectives, P. KÖPF-MAIER AND H. KÖPF-MAIER, *Chemical Reviews* **87**, 1987, 1137–1152.

Boron Neutron Capture Therapy, J. H. MORRIS, *Chemistry in Britain*, 1991, 331–334.

Boron Neutron Capture Therapy for Cancer, R. F. BARTH, A. H. SOLOWAY AND R. G. FAIRCHILD, *Scientific American*, October 1990, 100–107.

For a detailed review see:

The Role of Chemistry in the Development of Boron Neutron Capture Therapy of Cancer, M. F. HAWTHORNE, *Angewandte Chemie, International Edition* **32**, 1993, 950–984.

The Chemistry of the Gold Drugs Used in the Treatment of Rheumatoid Arthritis, D. H. BROWN AND W. E. SMITH, *Chemical Society Reviews* **9**, 1980, 217–240.

MRI Primer, W. OLDENDORF AND W. OLDENDORF, JR., Raven Press, 1991. Theory, methods and applications of magnetic resonance imaging.

The Chemistry of Chelating Agents in Medical Sciences, R. A. BULMAN, *Structure and Bonding* **67**, Springer-Verlag, 1987.

Environmental inorganic chemistry
Environmental Chemistry, S. E. MANAHAN, Lewis Publishers, 5th Edition, 1991.

For discussion on the relative environmental impact of phosphates versus zeolites, see *The Phosphate Report* published by Landbank Environmental Research and Consulting, 1994. A summary is given in *New Scientist*, 5th February 1994, 10. See also *Science* **263**, 1994, 1086 and *New Scientist*, 17 April 1993, 7.

Ozone Depletion: 20 Years After The Alarm, *Chemical & Engineering News*, August 15, 1994, 8–13. A good historical account of the discovery of ozone depletion, and subsequent measures taken.

Polar Stratospheric Clouds and Ozone Depletion, O. B. TOON AND R. P. TURCO, *Scientific American*, June 1991, 40–47.

Towards a Laboratory Strategy for the Study of Heterogeneous Catalysis in Stratospheric Ozone Depletion, M. R. S. MCCOUSTRA AND A. B. HORN, *Chemical Society Reviews*, 1994, 195–204.

Ozone Depletion's Recurring Surprises Challenge Atmospheric Scientists, *Chemical & Engineering News*, 24 May 1993, 8–18.

New Routes to Alternative Halocarbons, G. WEBB AND J. WINFIELD, *Chemistry in Britain*, 1992, 996–997.

Environmental issues—CFC alternatives, R. TWEDDLE, *Education in Chemistry*, January 1995, 17–19.

Looming Ban on Production of CFCs, Halons Spurs Switch to Substitutes, *Chemical & Engineering News*, 15 November 1993, 12–18.

Atmospheric Lifetime, Its Application and Its Determination: CFC-Substitutes as a Case Study, A. R. RAVISHANKARA AND E. R. LOVEJOY, *J. Chem. Soc., Faraday Transactions* **90**, 1994, 2159–2169.

Chemistry of the Atmosphere, M. J. MCEWAN AND L. E. PHILLIPS, Edward Arnold, 1975.

Ocean Carbon Cycle, J. L. SARMIENTO, *Chemical & Engineering News*, Special Report, May 1993, 30–43.

Aluminium: A Neurotoxic Product of Acid Rain, R. B. MARTIN, *Accounts of Chemical Research* **27**, 1994, 204–210.

GROUP THEORY (Appendix C)

The best introductions are to be found in the following two texts:

Group Theory for Chemists, G. DAVIDSON, Macmillan, 1991. A well-written text suitable for undergraduate courses in group theory.

Chemical Applications of Group Theory, F. A. COTTON, Interscience, 2nd Edition, 1971.

Appendix B
Some Common Polydentante Ligands

Ligand	Formula	Mode of Coordination
BIDENTATE LIGANDS		
ethylenediamine (en)	$H_2NCH_2CH_2NH_2$	Through both N atoms giving a five-membered ring (Figure 13.13a)
dicarboxylic acids (and S analogues)	$(CH_2)_n(COOH)_2$ ($n = 0, 1, 2$ etc)	A proton is lost from each COOH group and the two O^- atoms coordinate giving a $(5 + n)$ ring (compare Figure 13.17 for $n = 0$)
acetylacetone (acac)	$CH_3C(OH) = CHCOCH_3$ (in enol form)	Enol proton lost and coordination is through both O atoms (Figures 11.2, 12.5, 14.13)
(and other β-diketones: also as a monodentate ligand in keto form)		
8-quinolinol (oxine or 8-hydroxyquinoline)		Proton lost and coordinates through O and N giving five-membered ring (Figure 13.13d)
biuret	$H_2NCONHCONH_2$	One NH_2 proton lost and coordinates through two outer N atoms to give a six-membered ring
dimethylglyoxime (DMG)	$CH_3C(= NOH)C(= NOH)CH_3$	One proton lost and coordinates through both N atoms giving five-membered ring. Further six-membered rings formed by hydrogen bonding between NOH and ON (Figure 14.31)
salicylaldehyde (salic)		Hydroxyl proton lost and bonds through both O atoms giving six-membered ring (Figure 10.4)
(Similarly salicylaldimines: $C_6H_4(OH)(CH = NR)$ where $R = H$, alkyl or OH. Here bonding is through O and N giving six-membered rings. See Figure 13.13e)		
2:2'-dipyridyl (bipy)		Through both N atoms giving five-membered ring (Figure 14.7)
(Also numerous substituted dipyridyls and analogues)		

Ligand	Formula	Mode of Coordination
BIDENTATE LIGANDS		
1:10-phenanthroline (phenan)		As above
dithiols	$HS(CH_2)_nSH$; $HSCH_2(SH)CH_2R$; $o\text{-}(C_6H_4)(SH)_2$ 1,2 unsaturated thiols $RC=CR$, and dithioketones $R-C-C-R$ with HS SH and S S substituents	Loss of one proton and coordination through two S atoms generally giving four-, five-, or six-membered rings
diphosphanes (diphos) and diarsanes (diars)	$R_2MCH_2CH_2MR_2$ or benzene ring with MR_2, MR_2 $M = P$ or As	Through two P or two As (or P and As) giving five-membered rings
dimethoxyethane (DME)* (glyme)	$MeOCH_2CH_2OMe$	Through two O atoms giving a five-membered ring
TRIDENTATE LIGANDS		
diethylenetriamine (dien)	$H_2NCH_2CH_2NHCH_2CH_2NH_2$	Through three N atoms to the same M ion giving two five-membered rings (Figure 13.13c)
2:6-bis(α pyridyl) pyridine (terpyridine or terpy)		To one metal through three N atoms giving two five-membered rings
oxydiacetic acid (oda)	$O=C-CH_2-O-CH_2-C=O$ with OH and OH substituents	Loss of two protons from OH and coordination through these two O and the central one to give two five-membered rings
triarsanes (triars) Similar molecules with three As, three P or three (As + P) atoms	e.g. $Me_2As(CH_2)_3As(Me)(CH_2)_3AsMe_2$	To one metal atom through three As atoms giving two six-membered rings
QUADRIDENTATE LIGANDS		
porphyrins and phthalocyanins	See Figure 14.21	Through four N atoms forming square-planar coordination to M
bis(acetylacetone)ethylenediamine (acacen) (A Schiff's base)	$MeC=CHC=NCH_2CH_2N=CCH=CMe$ with HO Me Me OH substituents	Two protons lost (from OH) and bonds through two O and two N atoms to an M ion forming two six-membered rings and one five-membered ring (compare Figure 13.14 for a related ligand)
triethylenetetramine	$H_2NCH_2CH_2NHCH_2CH_2NHCH_2CH_2NH_2$	Through four N atoms forming four five-membered rings around a metal atom
triaminotriethylamine (tren) Also similar tetrarsanes and tetraphosphanes where N is replaced by As or P	$N(CH_2CH_2NR_2)_3$ where $R = H$ or alkyl	Through four N atoms as above

Ligand	Formula	Mode of Coordination
tris(o-diphenylarsanophenyl)arsane (QAS) Also the phosphane where the central As is replaced by P		Through four As atoms giving four five-membered rings. The geometric requirements of the ligand give it a pyramidal coordination

QUINQUEDENTATE LIGANDS

Ligand	Formula	Mode of Coordination
tetraethylenepentamine (tetren)	$H_2NCH_2CH_2NHCH_2CH_2NHCH_2CH_2NHCH_2CH_2NH_2$	To one metal through all five N atoms
The Schiff's Base	$CH = NCH_2CH_2NHCH_2CH_2N = CH$ $\vert$ $\vert$ OH HO	Loses two protons and bonds through three N and two O atoms. (Compare Figure 13.14 for the corresponding tetradentate ligand)

SEXADENTATE LIGANDS

Ligand	Formula	Mode of Coordination
ethylenediaminetetraacetic acid (EDTA)	$(HO_2CCH_2)_2NCH_2CH_2N(CH_2CO_2H)_2$	Loses four H^+ and bonds through four O and two N atoms to give six five-membered rings. (Figure 10.5)
pentaethylenehexamine	$H_2NCH_2CH_2(NHCH_2CH_2)_4CH_2CH_2NH_2$	To one metal through all six N atoms
Schiff's Bases e.g.	$CH = NCH_2CH_2SCH_2CH_2SCH_2CH_2N = CH$ $\vert$ $\vert$ OH HO	Loses two protons and bonds to one metal through two O, two S and two N atoms giving five rings

Also the corresponding compounds with NH in place of S, or with pyridine in place of the $-C_6H_4-OH$ ring. (Compare Figure 13.14 for a tetradentate analogue)

Ligand	Formula	Mode of Coordination
An Octadentate Ligand? diethylenetriaminepentaacetic acid (Compare EDTA)		Bonding unknown but it is potentially eight coordinating through 5O and 3N atoms and it gives a 1:1 zirconium compound

*Analogous ligands $MeO(CH_2CH_2O)_nMe$ (called diglyme, DIME; triglyme, etc.) bond through 3, 4, 5 etc. O giving five-membered rings.

Appendix C
Molecular Symmetry and
Point Groups

The symmetry of a molecule or ion is a property of fundamental importance in more advanced study of its properties. For example, the choice of atomic orbitals which may be combined into molecular sigma or pi orbitals is restricted by symmetry considerations. Thus, the basic reason why the combinations indicated in Figure 3.12 are not allowed is that the two orbitals in each pair belong to different symmetry classes in a diatomic molecule. Similarly, the number and type of transitions expected in the electronic or vibrational spectra of a molecule depend fundamentally on the molecular symmetry. For example, in section 7.5, the predictions of the number of fundamentals in the infrared and Raman spectra which are indicated there are derived solely from the consideration of the molecular symmetry.

While it is not our purpose, in an introductory text, to discuss these topics it is useful for the student to be able to determine the symmetry of a molecule at an early stage. This may be done quite simply, as explained below, and can conveniently be tackled in conjunction with the determination of molecular shapes discussed in Chapter 4. Familiarity with the nomenclature of symmetry, and the ability to determine the formal symmetry of a molecule, are the necessary first steps to a fuller understanding of bonding and spectroscopy.

There are two steps in the process: first we define symmetry elements and symmetry operations and then the sum of the symmetry elements is used to determine the point group to which the molecule belongs.

Symmetry elements and symmetry operations

A symmetry *operation* is some transformation of the molecule, such as a rotation or a reflection, which leaves the molecule in a configuration in space which is indistinguishable from its initial configuration. A symmetry *element* is that point, line, or plane in the molecule about which the symmetry operation takes place. The number of symmetry elements and operations which apply to single, real molecules is quite small and these are detailed in Table C.1. (When dealing with a crystal, with extended repeated units, extra symmetry operations become possible which transform one molecule into the neighbouring position which, in an infinitely extended lattice, also gives rise to an indistinguishable configuration. We shall not discuss such elements here.)

These operations will be more readily understood from a few examples. It is very much easier to follow the description using a molecular model and it would be well worth the reader's while to make models and carry out the symmetry operations described.

Consider first the ammonia molecule which is pyramidal in shape, Figure C.1. There is a three-fold axis passing

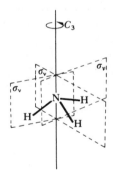

FIGURE C.1 *Symmetry elements of* NH$_3$
Ammonia belongs to the **C$_{3v}$** point group. The C$_3$ axis passes through N and the midpoint of the triangle defined by the three H atoms. There are three vertical planes, σ_v, each containing C$_3$ and one N—H bond and bisecting the opposite HNH angle.

through the nitrogen atom and perpendicular to the plane containing the three hydrogens. Rotation about this axis by $360°/3 = 120°$ leaves the N unchanged and moves each H into the position of the next. The resulting configuration is indistinguishable (though different) from the original one. The molecule may also be rotated twice about this axis by $120°$ to produce another indistinguishable configuration, but if the molecule is rotated three times successively by $120°$ it is returned to the original configuration. This symmetry element is labelled C_3, and the operations are distinguished

TABLE C.1. Symmetry elements and symmetry operations

Element	Symbol	Operation
identity	E	Leaves each particle in its original position
n-fold axis (proper axis)	C_n	Rotation about the axis by $360°/n$, or by some multiple of this. Only $n = 1, 2, 3, 4, 5, 6$ and ∞ need be considered for real molecules.
plane	σ	Reflection in the plane
centre	i	Inversion through the centre
n-fold alternating axis (improper axis)	S_n	Rotation by $360°/n$ (or by a multiple) followed by reflection in a plane perpendicular to the axis.

as C_3, C_3^2, and C_3^3. As the last produces the original configuration, we may write

$$C_3^3 = E$$

Similarly, $C_3^4 = C_3$, $C_3^5 = C_3^2$ and so on.

In the NH_3 molecule, there are also three planes of symmetry. Each one contains the N atom and one H, and bisects the angle between the other two H atoms. Reflection in this plane leaves the N and contained H unaltered, and exchanges the other two H atoms. Clearly, if this reflection operation is repeated, the original configuration is restored. That is, $\sigma^2 = E$.

The symmetry elements of the ammonia molecule are thus E, C_3, and the three planes. It is the convention to align the molecule so that the axis of symmetry is vertical. Then a symmetry plane which contains this axis is a *vertical plane*, symbol σ_v, while a plane perpendicular to the axis is a *horizontal plane* with the symbol σ_h. Thus the symmetry planes in ammonia are vertical planes.

If more than one symmetry axis is present, the one of highest order (highest value of n) is termed the *principal axis*,

and this is the one which is placed vertically. There are a few cases where a molecule has improper axes but no proper axis, and then the alternating axis of highest order is chosen as the principal axis.

As a further example, consider BF_3 which is planar, Figure C.2. This molecule has the C_3 axis through the B atom, and the three vertical planes, each containing one BF bond, which correspond to the symmetry elements found in NH_3. In addition, there are three C_2 axes, one along each B—F bond. A rotation of $360°/2$ about such an axis leaves the contained F and B unchanged, and interchanges the other two F atoms. Furthermore, the plane of the molecule is a symmetry plane as reflection in it leaves all the atoms unchanged. By the definitions above, the principal axis is the C_3 axis, as this is the axis of highest order. Thus the plane of the molecule is a horizontal plane, σ_h, while the planes through the BF bonds are vertical planes. Further, as there is a C_3 axis and a horizontal plane of symmetry, there is necessarily an S_3 axis coincident with the C_3 one. Thus, the symmetry elements of BF_3 are E, C_3, $3C_2$, σ_h, $3\sigma_v$, and S_3.

A centre of symmetry (or an inversion centre) is a point in a molecule such that if an atom is moved from its position, through the centre of inversion for an equal distance in the same direction, it lands in the position of an identical atom. For example, the centre of an octahedron or of a *trans*-MA_2B_4 species is a centre of symmetry. There is also a centre at the mid-point of the C—C bond in ethane in the staggered configuration. On the other hand, there is no inversion centre in a tetrahedron, or in ethane in the eclipsed configuration.

While an alternating axis is necessarily present when both a proper axis and a horizontal plane are present, as in the case of the S_3 axis in BF_3, the alternating axis is an independent symmetry element and it may be present when neither the proper axis nor the plane exist as symmetry elements. An example is provided by ethane in the staggered configuration. If the C—C bond is taken as an axis, rotation by $60°$ followed by reflection in a plane perpendicular to the C—C bond and containing its mid-point will exchange the two carbons and the hydrogens. Thus there is an S_6 axis along the C—C bond while the highest order proper axis is only C_3 along this bond. There is no horizontal plane of symmetry in staggered ethane. It may be noted that a centre of inversion, i, is equivalent to an S_2 axis.

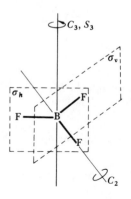

FIGURE C.2 *Symmetry elements of* BF_3

Boron trifluoride belongs to the point group $\mathbf{D_{3h}}$. This has the elements, C_3 (through B perpendicular to the plane of the molecule) and three σ_v (containing one B   F bond and bisecting the opposite angle) similar to those in $\mathbf{C_{3v}}$ (Figure C.1) together with a horizontal plane of symmetry, σ_h (which is the plane of the molecule), an S_3 axis coincident with C_3, and three C_2 axes, one along each B—F bond. Only one of the three vertical planes and one of the two-fold axes is shown.

Point Groups

The complete list of symmetry operations which may be performed on a molecule serves to define the *point group* to which the molecule belongs. Conversely, knowing the point group is equivalent to knowing all the symmetry operations. Thus the fact that ammonia may be subjected to the operations E, C_3, C_3^2 and three different σ_v reflections determines that NH_3 belongs to the $\mathbf{C_{3v}}$ point group.

The point groups are named by symbols, which are related to those for the symmetry elements, and are usually distinguished in print by the use of bold type. The point groups which span molecules are the following:

(a) $\mathbf{C_n}$ to which belong molecules with only a proper axis of symmetry C_n.

Note that $\mathbf{C_1}$ contains only $E = C_1$, and is the point group in which molecules with no symmetry at all are placed.

(b) $\mathbf{C_{nv}}$ to which belong molecules with a C_n axis and n vertical planes only (note that if there is one plane, there must be n as the n-fold rotation about the axis transforms one plane into the next).

(c) $\mathbf{C_{nh}}$ to which belong molecules with only a C_n axis, a horizontal plane, σ_h, and the resulting S_n axis. Note that $\mathbf{C_{1h}} \equiv \mathbf{C_{1v}}$ is more commonly labelled $\mathbf{C_s}$.

(d) $\mathbf{D_n}$ is the point group to which belong molecules containing only one C_n axis together with n C_2 axes (note that again, if there is one C_2 axis there must be n of them).

(e) $\mathbf{D_{nd}}$ covers molecules with one C_n and n C_2 axes, together with n vertical planes, and the S_{2n} axis which these imply. Note that the subscript is $\mathbf{d}$ (for dihedral) and not $\mathbf{v}$.

(f) $\mathbf{D_{nh}}$ to which belong molecules with C_n and n C_2 axes, together with a horizontal plane. These elements imply the presence of an S_n axis and n vertical planes.

In chemical structures, axes of higher order than six are extremely rare, thus the point groups specified under (a) to (f) have $\mathbf{n} = 1, 2, 3, 4, 5$ and 6, apart from linear molecules where $\mathbf{n}$ is infinity.

(g) $\mathbf{S_n}$. A few cases exist where molecules have no planes of symmetry and the highest order axis is S_n, together with the $C_{n/2}$ which this implies. Only even values of n are found. $\mathbf{S_2}$ is the group with one S_2 axis, which is equivalent to i, and this group is usually labelled $\mathbf{C_i}$.

(h) $\mathbf{T_d}$, $\mathbf{O_h}$, $\mathbf{I_h}$ are respectively the labels for groups with full tetrahedral, octahedral, or icosahedral symmetry. These have

TABLE C.2. Some common point groups

Point group	Diagnostic elements	Other elements	Examples
$\mathbf{C_1}$	E only		SiHClBrI
$\mathbf{C_s}$	E, σ only		SiH$_2$ClBr
$\mathbf{C_i}$	E, i only		*trans*-HClBrSiSiBrClH
$\mathbf{C_2}$	E, C_2 only		H$_2$O$_2$ (non-planar)
$\mathbf{C_{2v}}$	$E, C_2, 2\sigma_v$		H$_2$O, SiH$_2$Cl$_2$
$\mathbf{C_{3v}}$	$E, C_3, 3\sigma_v$		NH$_3$, SiHCl$_3$
$\mathbf{C_{4v}}$	$E, C_4, 4\sigma_v$	$C_2 = C_4^2$	BrF$_5$, SF$_5$Cl
$\mathbf{C_{2h}}$	E, C_2, σ_h	i	*trans*-C$_6$H$_2$Cl$_2$Br$_2$
$\mathbf{C_{3h}}$	E, C_3, σ_h		B(OH)$_3$ in form
$\mathbf{D_2}$	$E, C_2, 2C_2$		
$\mathbf{D_{2d}}$	$E, C_2, 2C_2, 2\sigma_v$	S_4	H$_2$C=C=CH$_2$
$\mathbf{D_{3d}}$	$E, C_3, 3C_2, 3\sigma_v$	i, S_6	C$_2$H$_6$, Si$_2$Cl$_6$ (staggered)
$\mathbf{D_{4d}}$	$E, C_4, 4C_2, 4\sigma_v$	$C_2 = C_4^2, S_8$	S$_8$ (puckered ring)
$\mathbf{D_{5d}}$	$E, C_5, 5C_2, 5\sigma_v$	i, S_{10}	(C$_5$H$_5$)$_2$Fe (staggered)
$\mathbf{D_{2h}}$	$E, C_2, 2C_2, \sigma_h$	$i, 2\sigma_v$	B$_2$Cl$_4$, *trans*-A$_2$B$_2$C$_2$M
$\mathbf{D_{3h}}$	$E, C_3, 3C_2, \sigma_h$	$S_3, 3\sigma_v$	BF$_3$, PF$_5$
$\mathbf{D_{4h}}$	$E, C_4, 4C_2, \sigma_h$	$i, S_4, C_2, 4\sigma_v$	PtCl$_4^{2-}$, *trans*-A$_2$B$_4$M
$\mathbf{D_{5h}}$	$E, C_5, 5C_2, \sigma_h$	$S_5, 5\sigma_v$	C$_5$H$_5$, (C$_5$H$_5$)$_2$Ru (eclipsed)
$\mathbf{D_{6h}}$	$E, C_6, 6C_2, \sigma_h$	$i, S_6, S_3, C_3, C_2, 6\sigma_v$	C$_6$H$_6$
$\mathbf{T_d}$	$E, 4C_3, 3C_2, 3S_4$, and $6\sigma_v$		SiH$_4$, GeCl$_4$, TiCl$_4$
$\mathbf{O_h}$	$E, 3C_4, 4C_3, 6C_2, i, 3S_4, 4S_6, 3\sigma_h, 6\sigma_v$		SF$_6$, ML$_6$, PF$_6^-$
$\mathbf{I_h}$	$E, 6C_5, 10C_3, 15C_2, i, 10S_6, 6S_{10}, 15\sigma$		B$_{12}$H$_{12}^{2-}$, B$_{12}$ (Figure 17.7d)

For the $\mathbf{C_{3h}}$ row, B(OH)$_3$ structure:

```
          H
          |
          O         O — H
           \       /
            \     /
              B
              |
              O
              |
              H
```

large numbers of symmetry elements, listed in Table C.2, and it is impossible not to recognize them. There are a few related groups, such as **T** which has the axes but not the planes of a full tetrahedron, but these are very rarely encountered.

In the discussion which follows, the groups **S₄**, **S₆**, **S₈** from (g) and **T**, **Tₕ** and **O** which are related to those in (h) are omitted as they are seldom represented by real molecules in their most stable configurations.

Although the more symmetric groups are defined by quite large numbers of symmetry operations, the fact that the presence of the element automatically implies the presence of all the operations of that element (including those giving rise to related elements such as $C_4^2 = C_2$) means that the point group may be decided by considering the symmetry elements. Furthermore, as the presence of certain elements implies the presence of others (as the C_3 and σ_h in BF_3 implied the S_3), the point group to which a molecule belongs may be *diagnosed* by looking for a small number of elements *in a particular order*. If the diagnostic elements are present, then the other symmetry elements, and all the corresponding symmetry operations, are necessarily present.

It then becomes a simple matter to determine the point group to which a molecule belongs by looking for symmetry elements in a fixed order. This order is:

(1) Find the principal axis.
(2) Look for n C_2 axes perpendicular to the principal axis.
(3) Look for horizontal planes.
(4) Look for vertical planes.
(5) If none of the above are present, check for a centre of inversion.

It is important that the search is always carried out in this order or higher symmetry groups may be missed because too much weight is put on minor elements. Notice, for example, that **Dₙₕ** groups have vertical as well as horizontal planes of symmetry so that, if vertical planes are looked for before the horizontal plane is excluded, **Dₙₕ** might be mistakenly assigned as **Dₙd**.

The process of diagnosing the point group may then be set out as follows:

(A) A molecule of high symmetry?	Assign as **Tₐ** (tetrahedron), **Oₕ** (octahedron) or **Iₕ** (icosahedron) after checking that all the requisite symmetry elements are present from Table C.2.
(B) Find the axis of highest order, C_n	If present, go to (C). If absent assign as **Cₛ** (plane only), **Cᵢ** (centre of inversion, $i \equiv S_2$, only) or **C₁** (no element).
(C) Look for n C_2 axes perpendicular to the principal axis	If present, go to (F): if none, go to (D).
(D) Look for horizontal plane	If present, assign as **Cₙₕ**, if absent, go to (E).

(E) If no horizontal plane, look for n vertical planes. — If present, assign as **Cₙᵥ**: if no planes then assign as **Cₙ**.

(F) If C_n and n C_2 axes are present, look for
  (i) a horizontal plane — Assign as **Dₙₕ**.
  (ii) if no horizontal plane, look for n vertical planes — Assign as **Dₙd**.
  (iii) If no planes at all — Assign as **Dₙ**.

A network form of this search is shown in Figure C.3, and

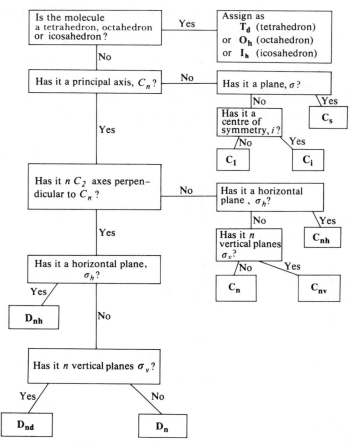

FIGURE C.3 *Diagnostic network for point groups*

Table C.2 lists all the symmetry elements of the chemically important point groups. It will be seen that while the simpler groups contain only the diagnostic elements, the more symmetric groups have a considerable number of consequent elements.

A linear molecule fits naturally into the above scheme once it is realized that the molecular axis is an infinity-fold one, because any rotation of the molecule about this axis—however small—produces an indistinguishable configuration. Thus the principal axis is C_∞ and a molecule like N_2 or CO_2 has an infinite number of C_2 axes, perpendicular to the principal axis, passing through the centre of the molecule (mid-point of $N-N$ or through the C in $O-C-O$). Thus the point group is of the D type and, as there is a symmetry plane, σ_h through the midpoint and perpendicular to the molecule axis, these molecules belong to the $\mathbf{D}_{\infty h}$ group. Similar analysis shows that non-symmetric linear molecules like NO or N_2O (arranged $N-N-O$) are members of the $\mathbf{C}_{\infty v}$ group.

For further use of the point group, it is necessary to refer to its *character table* which is effectively a summary of all the symmetry properties implied by the point group. For use of the character tables, the reader should refer to one of the more advanced treatments indicated in the references in Appendix A. However, for convenience of reference, we append the characters tables of those point groups which occur most commonly, Table C.3.

From the list, it may be seen that the character table is divided into sections. The symbol for the point group comes first, and then there is listed along the top row all the symmetry operations of the group. The main body of the table shows all the possible ways in which a function might transform under the operations of the group. Here, if the operation transforms the function into itself we find a 1, if the function is reversed we find -1, while if the function is mixed with others we find some other symbol. For example, think of the p orbitals on N in ammonia (that is, the unperturbed atomic orbitals before any combination or hybridization is carried out). Ammonia belongs to the point group $\mathbf{C_{3v}}$ and the p_z orbital will be taken to coincide with the C_3 axis. It can be seen that if the operation C_3, rotation by $120°$ about the z axis,

is carried out the p_z orbital is unaltered. Similarly the operations E or reflection in any of the σ_v planes leaves the p_z orbital unaltered. Thus we would write for p_z

$$\begin{array}{ccc} E & C_3 & 3\sigma_v \\ 1 & 1 & 1 \end{array}$$

corresponding to the row labelled A_1. On the other hand, rotation of the p_x orbital about the C_3 axis (i.e. the z axis) by $120°$ would bring it to a position which must be expressed as a combination of the p_x and p_y orbitals, $-p_x \cos 60 + p_y \sin 60$. Similarly, C_3 operating on p_y mixes it with p_x. Thus the p_x and p_y orbitals are degenerate in the $\mathbf{C_{3v}}$ point group and show the characters indicated in the table opposite E in the left-hand column. In general, the left-hand column in the character table gives symbols which summarize the symmetry properties of the different possible classes of functions in that point group. The exact nomenclature need not concern us, but note that non-degenerate classes are labelled A or B, doubly degenerate ones E and triply degenerate ones T (or F in the older nomenclature). The final two columns give functions which transform as the various classes. In the fourth column are vectors in the x, y or z directions (such as a p orbital or a translation along one axis) and rotations around each axis. In the last column are the second order products, such as the d orbitals.

TABLE C.3. Character tables for common point groups

$\mathbf{C_1}$	E		
A	1	all functions	

$\mathbf{C_s}$	E	σ_h		
A'	1	1	T_x, T_y, R_z	x^2, y^2, z^2, xy
A''	1	-1	T_z, R_x, R_y	yz, xz

$\mathbf{C_i}$	E	i		
A_g	1	1	R_x, R_y, R_z	$x^2, y^2, z^2, xy, xz, yz$
A_u	1	-1	T_x, T_y, T_z	

$\mathbf{C_2}$	E	C_2		
A	1	1	T_z, R_z	x^2, y^2, z^2, xy
B	1	-1	T_x, T_y, R_x, R_y	yz, xz

$\mathbf{C_{2v}}$	E	C_2	$\sigma_v(xz)$	$\sigma_v(yz)$		
A_1	1	1	1	1	T_z	x^2, y^2, z^2
A_2	1	1	-1	-1	R_z	xy
B_1	1	-1	1	-1	T_x, R_y	xz
B_2	1	-1	-1	1	T_y, R_x	yz

$\mathbf{C_{3v}}$	E	$2C_3$	$3\sigma_v$		
A_1	1	1	1	T_z	x^2+y^2, z^2
A_2	1	1	-1	R_z	
E	2	-1	0	$(T_x, T_y) (R_x, R_y)$	$(x^2-y^2, xy) (xz, yz)$

$\mathbf{C_{4v}}$	E	$2C_4$	C_2	$2\sigma_v$	$2\sigma_d$		
A_1	1	1	1	1	1	T_z	x^2+y^2, z^2
A_2	1	1	1	-1	-1	R_z	
B_1	1	-1	1	1	-1		x^2-y^2
B_2	1	-1	1	-1	1		xy
E	2	0	-2	0	0	$(T_x, T_y) (R_x, R_y)$	(xz, yz)

$\mathbf{C_{2h}}$	E	C_2	i	σ_h		
A_g	1	1	1	1	R_z	x^2, y^2, z^2, xy
B_g	1	-1	1	-1	R_x, R_y	xz, yz
A_u	1	1	-1	-1	z	
B_u	1	-1	-1	1	x, y	

$\mathbf{D_{2h}}$	E	$C_2(z)$	$C_2(y)$	$C_2(x)$	i	$\sigma(xy)$	$\sigma(xz)$	$\sigma(yz)$		
A_g	1	1	1	1	1	1	1	1		x^2, y^2, z^2
B_{1g}	1	1	-1	-1	1	1	-1	-1	R_z	xy
B_{2g}	1	-1	1	-1	1	-1	1	-1	R_y	xz
B_{3g}	1	-1	-1	1	1	-1	-1	1	R_x	yz
A_u	1	1	1	1	-1	-1	-1	-1		
B_{1u}	1	1	-1	-1	-1	-1	1	1	T_z	
B_{2u}	1	-1	1	-1	-1	1	-1	1	T_y	
B_{3u}	1	-1	-1	1	-1	1	1	-1	T_x	

$\mathbf{D_{3h}}$	E	$2C_3$	$3C_2$	σ_h	$2S_3$	$3\sigma_v$		
A_1'	1	1	1	1	1	1		x^2+y^2, z^2
A_2'	1	1	-1	1	1	-1	R_z	
E'	2	-1	0	2	-1	0	(T_x, T_y)	(x^2-y^2, xy)
A_1''	1	1	1	-1	-1	-1		
A_2''	1	1	-1	-1	-1	1	T_z	
E''	2	-1	0	-2	1	0	(R_x, R_y)	(xz, yz)

D_{4h}	E	$2C_4$	C_2	$2C_2'$	$2C_2''$	i	$2S_4$	σ_h	$2\sigma_v$	$2\sigma_d$		
A_{1g}	1	1	1	1	1	1	1	1	1	1		$x^2+y^2,\ z^2$
A_{2g}	1	1	1	-1	-1	1	1	1	-1	-1	R_z	
B_{1g}	1	-1	1	1	-1	1	-1	1	1	-1		x^2-y^2
B_{2g}	1	-1	1	-1	1	1	-1	1	-1	1		xy
E_g	2	0	-2	0	0	2	0	-2	0	0	(R_x, R_y)	(xz, yz)
A_{1u}	1	1	1	1	1	-1	-1	-1	-1	-1		
A_{2u}	1	1	1	-1	-1	-1	-1	-1	1	1	T_z	
B_{1u}	1	-1	1	1	-1	-1	1	-1	-1	1		
B_{2u}	1	-1	1	-1	1	-1	1	-1	1	-1		
E_u	2	0	-2	0	0	-2	0	2	0	0	(T_x, T_y)	

D_{5h}	E	$2C_5$	$2C_5^2$	$5C_2$	σ_h	$2S_5$	$2S_5^3$	$5\sigma_r$		
A_1'	1	1	1	1	1	1	1	1		$x^2+y^2,\ z^2$
A_2'	1	1	1	-1	1	1	1	-1	R_z	
E_1'	2	$2\cos 72°$	$2\cos 144°$	0	2	$2\cos 72°$	$2\cos 144°$	0	(T_x, T_y)	
E_2'	2	$2\cos 144°$	$2\cos 72°$	0	2	$2\cos 144°$	$2\cos 72°$	0		$(x^2-y^2,\ xy)$
A_1''	1	1	1	1	-1	-1	-1	-1		
A_2''	1	1	1	-1	-1	-1	-1	1	T_z	
E_1''	2	$2\cos 72°$	$2\cos 144°$	0	-2	$-2\cos 72°$	$-2\cos 144°$	0	(R_x, R_y)	(xz, yz)
E_2''	2	$2\cos 144°$	$2\cos 72°$	0	-2	$-2\cos 144°$	$-2\cos 72°$	0		

D_{6h}	E	$2C_6$	$2C_3$	C_2	$3C_2'$	$3C_2''$	i	$2S_3$	$2S_6$	σ_h	$3\sigma_d$	$3\sigma_v$		
A_{1g}	1	1	1	1	1	1	1	1	1	1	1	1		$x^2+y^2,\ z^2$
A_{2g}	1	1	1	1	-1	-1	1	1	1	1	-1	-1	R_z	
B_{1g}	1	-1	1	-1	1	-1	1	-1	1	-1	1	-1		
B_{2g}	1	-1	1	-1	-1	1	1	-1	1	-1	-1	1		
E_{1g}	2	1	-1	-2	0	0	2	1	-1	-2	0	0	(R_x, R_y)	(xz, yz)
E_{2g}	2	-1	-1	2	0	0	2	-1	-1	2	0	0		$(x^2-y^2,\ xy)$
A_{1u}	1	1	1	1	1	1	-1	-1	-1	-1	-1	-1		
A_{2u}	1	1	1	1	-1	-1	-1	-1	-1	-1	1	1	T_z	
B_{1u}	1	-1	1	-1	1	-1	-1	1	-1	1	-1	1		
B_{2u}	1	-1	1	-1	-1	1	-1	1	-1	1	1	-1		
E_{1u}	2	1	-1	-2	0	0	-2	-1	1	2	0	0	(T_x, T_y)	
E_{2u}	2	-1	-1	2	0	0	-2	1	1	-2	0	0		

D_{2d}	E	$2S_4$	C_2	$2C_2'$	$2\sigma_d$		
A_1	1	1	1	1	1		$x^2+y^2,\ z^2$
A_2	1	1	1	-1	-1	R_z	
B_1	1	-1	1	1	-1		x^2-y^2
B_2	1	-1	1	-1	1	z	xy
E	2	0	-2	0	0	(x, y) (R_x, R_y)	(xz, yz)

$\mathbf{D_{3d}}$	E	$2C_3$	$3C_2$	i	$2S_6$	$3\sigma_d$		
A_{1g}	1	1	1	1	1	1		x^2+y^2, z^2
A_{2g}	1	1	-1	1	1	-1	R_z	
E_g	2	-1	0	2	-1	0	(R_x, R_y)	(x^2-y^2, xy) (xz, yz)
A_{1u}	1	1	1	-1	-1	-1		
A_{2u}	1	1	-1	-1	-1	1	z	
E_u	2	-1	0	-2	1	0	(x, y)	

$\mathbf{T_d}$	E	$8C_3$	$3C_2$	$6S_4$	$6\sigma_d$		
A_1	1	1	1	1	1		$x^2 + y^2 + z^2$
A_2	1	1	1	-1	-1		
E	2	-1	2	0	0		$(2z^2 - x^2 - y^2,$ $x^2 - y^2)$
T_1	3	0	-1	1	-1	(R_x, R_y, R_z)	
T_2	3	0	-1	-1	1	(T_x, T_y, T_z)	(xy, xz, yz)

$\mathbf{O_h}$	E	$8C_3$	$6C_2$	$6C_4$	$3C_2 (= C_4^2)$	i	$6S_4$	$8S_6$	$3\sigma_h$	$6\sigma_d$		
A_{1g}	1	1	1	1	1	1	1	1	1	1		$x^2 + y^2 + z^2$
A_{2g}	1	1	-1	-1	1	1	-1	1	1	-1		
E_g	2	-1	0	0	2	2	0	-1	2	0		$(2z^2-x^2-y^2,$ $x^2 - y^2)$
T_{1g}	3	0	-1	1	-1	3	1	0	-1	-1	$(R_x, R_y,$ $R_z)$	
T_{2g}	3	0	1	-1	-1	3	-1	0	-1	1		(xz, yz, xy)
A_{1u}	1	1	1	1	1	-1	-1	-1	-1	-1		
A_{2u}	1	1	-1	-1	1	-1	1	-1	-1	1		
E_u	2	-1	0	0	2	-2	0	1	-2	0		
T_{1u}	3	0	-1	1	-1	-3	-1	0	1	1	$(T_x, T_y,$ $T_z)$	
T_{2u}	3	0	1	-1	-1	-3	1	0	1	-1		

Index

461

Relative Atomic Masses (based on $C^{12} = 12 \cdot 000$)

Element	Mass	Element	Mass	Element	Mass	Element	Mass
Actinium*	227·03	Erbium	167·26	Mercury	200·59	Samarium	150·36
Aluminium	26·982	Europium	151·97	Molybdenum	95·94	Scandium	44·956
Americium*	241·06	Fermium*	257·10	Neodymium	144·24	Selenium	78·96
Antimony	121·76	Fluorine	18·998	Neon	20·180	Silicon	28·086
Argon	39·948	Francium*	223·02	Neptunium*	237·05	Silver	107·87
Arsenic	74·922	Gadolinium	157·25	Nickel	58·69	Sodium	22·990
Astatine*	209·99	Gallium	69·723	Niobium	92·906	Strontium	87·62
Barium	137·33	Germanium	72·61	Nitrogen	14·007	Sulfur	32·066
Berkelium*	249·08	Gold	196·97	Nobelium*	259·10	Tantalum	180·95
Beryllium	9·0122	Hafnium	178·49	Osmium	190·2	Technetium*	98·906
Bismuth	208·98	Helium	4·0026	Oxygen	15·999	Tellurium	127·60
Boron	10·811	Holmium	164·93	Palladium	106·42	Terbium	158·93
Bromine	79·904	Hydrogen	1·0079	Phosphorus	30·974	Thallium	204·38
Cadmium	112·41	Indium	114·82	Platinum	195·08	Thorium*	232·04
Calcium	40·078	Iodine	126·90	Plutonium*	239·05	Thulium	168·93
Californium*	252·08	Iridium	192·22	Polonium*	209·98	Tin	118·71
Carbon	12·011	Iron	55·845	Potassium	39·098	Titanium	47·867
Cerium	140·12	Krypton	83·80	Praseodymium	140·91	Tungsten	183·85
Cesium	132·91	Lanthanum	138·91	Promethium*	146·92		
Chlorine	35·453	Lawrencium*	262	Protactinium*	231·04	Uranium*	238·03
Chromium	51·996	Lead	207·2	Radium*	226·03	Vanadium	50·942
Cobalt	58·933	Lithium	6·941	Radon*	222·02	Xenon	131·29
Copper	63·546	Lutetium	174·97	Rhenium	186·21	Ytterbium	173·04
Curium*	244·06	Magnesium	24·305	Rhodium	102·91	Yttrium	88·906
Dysprosium	162·50	Manganese	54·938	Rubidium	85·468	Zinc	65·39
Einsteinium*	252·08	Mendelevium*	258·10	Ruthenium	101·07	Zirconium	91·224

Note: The above values are rounded off to values sufficiently accurate for normal calculations: for accurate values based on the 1995 revision, see Table 2.5. Transfermium elements, see Table 16.8

*Element with no stable isotope.

Periodic Table of the Elements

$d^n s^x$ (n = 1 to 10; x = 0, 1 or 2)

s^1	s^2											$s^2 p^1$	$s^2 p^2$	$s^2 p^3$	$s^2 p^4$	$s^2 p^5$	$s^2 p^6$
1s						1 H		2 He									10 Ne
2s 2p	4 Be											5 B	6 C	7 N	8 O	9 F	10 Ne
3 Li	12 Mg											13 Al	14 Si	15 P	16 S	17 Cl	18 Ar
3s 3p 11 Na	20 Ca	21 Sc	22 Ti	23 V	24 Cr	25 Mn	26 Fe	27 Co	28 Ni	29 Cu	30 Zn	31 Ga	32 Ge	33 As	34 Se	35 Br	36 Kr
4s 3d 4p 19 K	38 Sr	39 Y	40 Zr	41 Nb	42 Mo	43 Tc	44 Ru	45 Rh	46 Pd	47 Ag	48 Cd	49 In	50 Sn	51 Sb	52 Te	53 I	54 Xe
5s 4d 5p 37 Rb	56 Ba	57* La	72 Hf	73 Ta	74 W	75 Re	76 Os	77 Ir	78 Pt	79 Au	80 Hg	81 Tl	82 Pb	83 Bi	84 Po	85 At	86 Rn
6s (4f) 5d 6p 55 Cs	88 Ra	89** Ac	104 Db	105 Jl	106 Rf	107 Bh	108 Hn	109 Mt	110 Uun	111 Uuu	112 Uub						
7s (5f) 6d 87 Fr																	

$f^p d^n s^2$ (p = 1 to 14, n = 0 or 1 (2 for Th))

*Lanthanide series 4f	58 Ce	59 Pr	60 Nd	61 Pm	62 Sm	63 Eu	64 Gd	65 Tb	66 Dy	67 Ho	68 Er	69 Tm	70 Yb	71 Lu
**Actinide series 5f	90 Th	91 Pa	92 U	93 Np	94 Pu	95 Am	96 Cm	97 Bk	98 Cf	99 Es	100 Fm	101 Md	102	103 Lr

Contents

IV
NEW DIRECTIONS

Foreword

Carole Klein

THIS COLLECTION OF writings by American Jewish women represents the work of a group of people who, though diverse, are linked to each other by a common heritage. No matter what they write about—the search for identity, the need to rebel, ambition versus sublimation—their perspective is touched by the fact that they are women who were born into the Jewish faith.

A good deal has been written about how being a woman affects one's creative expression. But when a woman is also Jewish, the shape and content of her expression are even more molded to particular experience. Who our grandparents are, how we are viewed by the dominating culture, what pressures we feel to assimilate into that culture, become part of our identity and follow us to the typewriter or writing table.

Reading this book one gets, finally, beyond the mythology that has defined the American Jewish woman. And it is high

Carole Klein is the author of *Aline,* the biography of stage designer Aline Bernstein. Born and brought up in New York City, Klein attended Goddard College, where she now teaches. Her other books are *The Single Parent Experience* and *The Myth of The Happy Child.*

time. As women of all religious persuasions demand new images of themselves, Jewish women too must refuse to accept
the easy generalizations that have been made about their lives.
We are not the stereotypic, self-indulged daughters or grasping
wives or salacious seductresses; we are not the smothering,
obscenely over-protective mothers who imprison sons into
cages of guilt and heterosexual conflict. These are the distorted
images of male writers.

In truth, as these stories prove, women are thinking, caring,
sensitive people with a high degree of consciousness about
themselves and their world. They struggle to make themselves
and the world better, to live in that world with joy, pride, and
integrity. These are wonderful stories, and as we read them, we
experience the full range of human emotion, from giddy delight
to tragic terror.

Perhaps just as important as learning through this anthology
what Jewish women are writing about is learning that they are
writing at all. As Ms. Mazow suggests, one would not realize this
by looking at anthologies compiled by men. Masculine editors
have successfully created a self-fulfilling prophecy: they have
drawn a trivial portrait of the Jewish woman's experience and
then not taken seriously the often-powerful renditions of her
experience that have come from her pen.

It is not easy for any woman in this society to be a writer. The
cultural demands and expectations of the female role run counter to the freedom to create. It seems especially difficult for the
Jewish woman writer, however. A Jewish daughter grows up
with strong pulls toward family and domestic life, and internal-
izes values that say she should put herself and her own needs
after her family's. In such an atmosphere, going into "a room
of one's own" to write is no easy matter. It is a constant and
wearying challenge to find a comfortable, guilt-free balance
between personal accomplishment and family life.

The more I reflect on it, the more I am convinced that Jewish-
American artists who are also women have the most confusing
socialization. Ambivalence trails them from childhood; they are
encouraged every bit as much as sons to do well in school, and
then, when they reach marriageable age, are pressured (however lovingly), to spend energies on less intellectual pursuits. It

is only recently, with the spreading feminist consciousness, that Jewish women artists have been able to measure the effects of that ambivalence on their creative well-being. Young Jewish women today are attempting to fashion new, less dissonant images of themselves, as some of the more recent writings in this anthology illustrate.

These rich pages merit the attention of men as well as women, of people of all religious faiths and cultural backgrounds. For by seeing what Jewish women are really like, and what Jewish women writers are really saying, we all move deeper into the circle of understanding that makes every life fuller. Certainly these stories are enriched by thousands of years of Jewish experience; yet they are far more universal than parochial in content.

Out of her particular history comes a new, strong voice of the Jewish-American woman writer: an occasion for pride and excitement to anyone who reads, writes, or reflects on the evolving human condition.

Acknowledgments

I WISH TO thank those who participated in the nine months of planning for Houston's 1976 lecture series on the Jewish woman. Our discussions helped me formulate the ideas behind this book, presented first as a lecture entitled "The Jewish Mother and the Jewish-American Princess: Where Are the Stereotypes of Yesteryear?" Given in a different form before the American Studies Association of Texas, it laid the groundwork for the introduction to this anthology.

The women of *Lilith* magazine introduced me to the network of American Jewish women writing today. June Arnold, Jack Mazow, Rosemary S. Minard, Miriam Kass, and Yaal Silberberg read early versions of the manuscript. Denise Weinberg, Rena Reisman, Mary Ross Taylor, Effie Feld, Beverly Freedenthal, and Robbie Odom Moses provided many valuable suggestions, as well as their friendship.

Most of all, I want to express gratitude to the women who wrote the works this anthology contains. They had the courage to tell their part of the story. And they have enriched the experience of us all.

J. W. M.

Introduction

Julia Wolf Mazow

ANY DISCUSSION OF Jewish women in America must acknowledge a history and a tradition which go back more than 5000
years. For generations, being a Jew has meant living as an
outsider on the periphery of the larger society, often as a scape-
goat. Therefore, Jews also understand the meaning of discrimi-
nation and persecution. It is not surprising, then, that many
Jews worked hard for the Civil Rights movement in this country.
And it seems appropriate, as others have noted, that so many
of the theoreticians and leaders of the recent women's movement should be Jewish women.[1]

Jewish tradition is rich in ambiguous attitudes and ambiva-
lent feelings towards women. On the one hand, the Talmud tells
us that a woman is "lightheaded" and that she is seductive to
men; on the other hand, the picture that emerges in Proverbs
31 is of great competence and energy. No matter how repres-
sive a given historical period may have been towards both Jews
and women, there have always been individuals who managed

---

[1]See, for example, Ann Wolfe's "The Jewish Woman," *Dialogue on Diversity*
(New York: Institute on Pluralism and Group Identity, 1976).

to overcome the restrictions of their times.[2]

Following the emigration of Jews to America from Western Europe in the 1840s, and from Eastern Europe in the 1880s and well into the 1900s, many changes occurred in the lives of women and men. The man who had been a respected scholar in the *shtetl*, for example, might have become a shoemaker without a place in the world, like Schmendrik in Anzia Yezierska's "My Own People." Of course, not all men in the *shtetl* were scholars, and not all were without work in this country. And it was just as common for women to work alongside their husbands in the marketplace here as it was for them to contribute to the family upkeep in Europe.

Having bought the middle-class American dream of success, many immigrant Jews became economically well-off. For some, success was accompanied by assimilation into the larger society. Scholarship ceased to be a cultural ideal for most Jews; achievement in business and the professions replaced it—for the men. But their mothers, wives, daughters, and sisters, whose worlds once included both home and workplace, were relegated to the sphere of home and family. One sign of the man's success was that the family did not need the wife's financial contribution. Previously, perpetuating Jewish tradition in the home had been one of her many duties; now, it became one of her few options.[3]

I began this project with a search for short fiction and autobiographical writings by American Jewish women from the 1930s to the present, because I wanted to find out who these women were and what had happened to them.[4] What were the concerns of the post-immigrant generations? What kinds of stories did they tell about themselves? Were they really the spoiled daughters of castrating mothers, as they were sometimes depicted in books written by American Jewish men?

---

[2] See Sondra Henry and Emily Taitz, *Written Out of History: A Hidden Legacy of Jewish Women Revealed Through Their Writings and Letters* (New York: Bloch Publishing Company, Inc., 1978).

[3] *The Jewish Woman in America* by Charlotte Baum, Paula Hyman, and Sonya Michel (New York: The Dial Press, 1976) contains a full discussion.

[4] For a thorough listing, see Aviva Cantor's annotated *Bibliography on the Jewish Woman* (Fresh Meadows, New York: Biblio Press, 1979).

My first discovery was the relative absence of Jewish women authors from well-known anthologies. Azriel Eisenberg's *The Golden Land: A Literary Portrait of American Jewry, 1654 to the Present*, contains excerpts from eighty-eight works, only eight of them by women. Irving Malin and Irwin Stark's 1964 *Breakthrough: A Treasury of Contemporary American Jewish Literature* contains thirty-one pieces, two of them by women, while Harold Ribalow's 1968 *Autobiographies of American Jews 1880–1920* includes selections by six women out of a total of twenty-five. *The Rise of American Jewish Literature*, edited by Charles Angoff and Meyer Levin (1970), lists twenty-two entries in its table of contents; none are by women. Theodore Gross's *The Literature of American Jews* (1973) includes the work of forty-one writers, six of them women. Of those six, only Grace Paley and Muriel Rukeyser published in this century. Finally, Irving Howe's anthology *Jewish American Stories* (1977) lists twenty-six works; four of them are by women.

What does this cursory survey mean? To some it may indicate that Jewish women are not writing. But the sampling also suggests that some male anthologizers are oblivious to the work of writers who are women.

When we look at some of the better-known female creations of American Jewish male writers we may find a clue. What can we say about these women, the Sophie Portnoys, the Marjorie Morningstars?

Generally, they are competent enablers. And they are also destroyers—a familiar stereotype on whose shoulders, we are sometimes told, the survival of the Jewish people rests.

But for the reader concerned with reality rather than appearance, the negative stereotype of the Jewish American Princess who grows up to be a Jewish mother is limiting, divisive, and simplistic. It tends to stress inevitability, to deny individual natures their due.

When we examine works that American Jewish women have written about themselves and their relationships, a different picture emerges. Even stereotypes appear in a different light. For when we encounter a mother, deeply concerned with the lives of her children and her home, we see much more: a woman who feels, one who is introspective. The stereotype ceases to be

rigid. Because stereotypes, by definition, do not allow for individual differences, once they are filled in and humanized they
no longer exist. Another pattern evolves when the woman tells
her story from her own point of view.

Instead of the conventional negative stereotypic qualities, the
women in this collection exhibit vast amounts of energy which
take diverse shapes and occur in works as dissimilar as Andrea
Dworkin's "First Love," Nancy Datan's "Making Jews," and
Gertrude Friedberg's "Where Moth and Rust." For Datan, the
process of "Making Jews" is a conscious, ongoing choice that
requires the hard work of constant, rigorous self-examination.
Dworkin writes of the intense drive to know and tell and do all
that is possible, remembering always that, "There is no place
on earth, no day or night, no hour or minute, when one is not
a Jew or a woman." Friedberg's Mrs. Fortrey, on the other hand,
uses her energy to forestall the disaster which would occur if
she were to stop the obsessive performance of household tasks.
Once she redirects her energy, she uncovers a new idea of what
is possible for her.

In these pieces there is no single vision. In some there is no
overtly Jewish content; in others, of course, there is. As Cheike
says in "My Mother Was a Light Housekeeper," "Anarchy and
variety are our beginning characteristics." There are no simplistic stereotypes. Rather, there are diverse women telling
their stories.

The divisions of the book reflect the themes of the stories as
they recur in the lives of many women, past and present. Section One, "Familiar Connections," deals with tradition, with
families, and the need for roots. The families portrayed here
range from Emma Goldman's to the newly created unit in
"What Must I Say to You?" Rosen's story reveals an unexpected
communion between the woman and her baby's nurse, one that
is closer than other, more "familiar" relationships—for familiar
also refers to the traditional aspects of our lives, those that we
have become accustomed to. Sophie Sapinsky of Yezierska's
"My Own People" comes to see just how strong her familiar
connections are. The mother of five in Tillie Olsen's "I Stand
Here Ironing" attempts to understand her experience with her
first-born child, while the mother in Narell's "Papa's Tea" is a

source of consternation to her small daughter, because she is *not* the stereotypic Jewish mother.

A sense of alienation and sometimes rebellion pervades the stories of "Seeking." Mrs. Fortrey of "Where Moth and Rust" ceases to perform her household duties, doing what she wants to do rather than what she should do. Some of the selections deal with personal loss. In *Riverfinger Women,* Inez, who has rejected many of her family's values, reveals her sadness that she has not yet found the voice to speak to her parents.

These pieces are about the need to re-establish a connection with the past. In "The Long-Distance Runner," Faith literally runs to her past and back to the present. Aviva Cantor's "The Phantom Child" shows a similar movement: as she weighs the question of abortion for herself, she explores her relationship with her mother and re-examines Judaism for its present meaning. And A of Andrea Dworkin's "First Love" writes a letter describing a hard-won self-knowledge, a synthesis of all that has happened to her.

The search for meaning is also an attempt to overcome aloneness, as it is for Sarah in "The Woman Who Lost Her Names." She marries, bears children, travels through America, and moves to Israel. But despite her gains, the loss of her names implies the loss of the power of naming, of language itself.

The selections listed under "Finding" tell of the acceptance that may evolve through time and understanding, as in Nancy Datan's "Making Jews." Sometimes it comes with recognition, as it does in "My Mother's Story" and "My Mother Was a Light Housekeeper," where it also signals a renewed relationship, based upon mutual knowledge of a shared past. Where conflicts once existed between husband and wife ("Anniversary"), young and old ("The Four Leaf Clover Story"), there is now some sense of affirmation. In this group of stories, to connect with the past becomes one way to find meaning in the present and hope for the future.

The two works that point toward "New Directions" are set in worlds different from those which most people encounter today. "X" is a fable of egalitarian child-rearing, far from the experience of all of us. The excerpt from *A Weave of Women* depicts a group of women in the Old City of Jerusalem striving to create their lives anew. The first is set in another time; the second, in another place.

For those with dreams of challenging the patriarchal bias of American culture, the inference is inescapable upon reading these two selections: Our dedication to change is not yet strong enough. We may have the imagination, but most of us lack the courage to risk and to make our visions reality.

Some selections could just as easily have been placed under another heading. For example, "The Fraychie Story" says as much about reconciliation, about finding, as it does about "familiar connections." And the life of Emma Goldman is certainly the life of a rebel. But the part of her autobiography reprinted here also shows the strength of her relationship with her family. The four divisions are thus suggestive rather than definitive.

This collection is only a beginning. It is not meant to be exhaustive, a task requiring many volumes. In the writings at hand, however, a picture of multiplicity emerges. The women portrayed are various; the thoughts and situations they have recorded are diverse. It is true that their Jewishness exists in greater or lesser degrees. It is also true that in some instances the language used to describe their experiences is somewhat rough-hewn.

But the American Jewish woman who is defining herself with language, as these writers are, is, in a sense, still at the beginning of a tradition. While she may once have lost the power of naming, she is about to regain it.

Meanwhile, those of us who wish can continue to search for each woman's story. We must learn what that story was and is. For some of us, the way to educate ourselves will be to write some fragment of it.

I
FAMILIAR
CONNECTIONS

My Own People

Anzia Yezierska

WITH THE SUITCASE containing all her worldly possessions under her arm, Sophie Sapinsky elbowed her way through the noisy ghetto crowds. Pushcart peddlers and pullers-in shouted and gesticulated. Women with market-baskets pushed and shoved one another, eyes straining with the one thought— how to get the food a penny cheaper. With the same strained intentness, Sophie scanned each tenement, searching for a room cheap enough for her dwindling means.

In a dingy basement window a crooked sign, in straggling, penciled letters, caught Sophie's eye: "Room to let, a bargain, cheap."

The exuberant phrasing was quite in keeping with the extravagant dilapidation of the surroundings. "This is the very

In 1907, *Anzia Yezierska* and her family emigrated from Sukovoly, Russia, to New York. There she worked in a factory, in a Delancey Street sweatshop, and in a restaurant. A naturalized U.S. citizen, *Yezierska* began writing stories of ghetto life in 1918. *The Open Cage: An Anzia Yezierska Collection*, edited by Alice K. Harris, was published in 1979 by Persea Books. "My Own People" is reprinted from *Hungry Hearts* by permission of copyright owner, Louise Levitas Henriksen.

place," thought Sophie. "There could n't be nothing cheaper in all New York."

At the foot of the basement steps she knocked.

"Come in!" a voice answered.

As she opened the door she saw an old man bending over a pot of potatoes on a shoemaker's bench. A group of children in all degrees of rags surrounded him, greedily snatching at the potatoes he handed out.

Sophie paused for an instant, but her absorption in her own problem was too great to halt the question: "Is there a room to let?"

"Hanneh Breineh, in the back, has a room." The old man was so preoccupied filling the hungry hands that he did not even look up.

Sophie groped her way to the rear hall. A gaunt-faced woman answered her inquiry with loquacious enthusiasm. "A grand room for the money. I'll let it down to you only for three dollars a month. In the whole block is no bigger bargain. I should live so."

As she talked, the woman led her through the dark hall into an airshaft room. A narrow window looked out into the bottom of a chimney-like pit, where lay the accumulated refuse from a score of crowded kitchens.

"Oi weh!" gasped Sophie, throwing open the sash. "No air and no light. Outside shines the sun and here it's so dark."

"It ain't so dark. It's only a little shady. Let me only turn up the gas for you and you'll quick see everything like with sunshine."

The claw-fingered flame revealed a rusty, iron cot, an inverted potato barrel that served for a table, and two soapboxes for chairs.

Sophie felt of the cot. It sagged and flopped under her touch. "The bed has only three feet!" she exclaimed in dismay.

"You can't have Rockefeller's palace for three dollars a month," defended Hanneh Breineh, as she shoved one of the boxes under the legless corner of the cot. "If the bed ain't so steady, so you got good neighbors. Upstairs lives Shprintzeh Gittle, the herring-woman. You can buy by her the biggest bargains in fish, a few days older.... What she got left over from the Sabbath, she sells to the neighbors cheap.... In the front lives Shmendrik, the shoemaker. I'll tell you the truth, he ain't no real shoemaker. He never yet made a pair of whole shoes in his life. He's a learner from the old country—a tzadik, a saint; but every time he sees in the street a child with torn feet, he calls them in and patches them up. His own eating, the last bite from his mouth, he divides up with them."

"Three dollars," deliberated Sophie, scarcely hearing Hanneh Breineh's chatter. "I will never find anything cheaper. It has a door to lock and I can shut this woman out. . . . I'll take it," she said, handing her the money.

Hanneh Breineh kissed the greasy bills gloatingly. "I'll treat you like a mother! You'll have it good by me like in your own home."

"Thanks—but I got no time to shmoos. I got to be alone to get my work done."

The rebuff could not penetrate Hanneh Breineh's joy over the sudden possession of three dollars.

"Long years on you! May we be to good luck to one another!" was Hanneh Breineh's blessing as she closed the door.

Alone in her room—*her* room, securely hers—yet with the flash of triumph, a stab of bitterness. All that was hers—so wretched and so ugly! Had her eager spirit, eager to give and give, no claim to a bit of beauty—a shred of comfort?

Perhaps her family was right in condemning her rashness. Was it worth while to give up the peace of home, the security of a regular job—suffer hunger, loneliness, and want—for what? For something she knew in her heart was beyond her reach. Would her writing ever amount to enough to vindicate the uprooting of her past? Would she ever become articulate enough to express beautifully what she saw and felt? What had she, after all, but a stifling, sweatshop experience, a meager, night-school education, and this wild, blind hunger to release the dumbness that choked her?

Sophie spread her papers on the cot beside her. Resting her elbows on the potato barrel, she clutched her pencil with tense fingers. In the notebook before her were a hundred beginnings, essays, abstractions, outbursts of chaotic moods. She glanced through the titles: "Believe in Yourself," "The Quest of the Ideal."

Meaningless tracings on the paper, her words seemed to her

now—a restless spirit pawing at the air. The intensity of experience, the surge of emotion that had been hers when she wrote—where were they? The words had failed to catch the lifebeat—had failed to register the passion she had poured into them.

Perhaps she was not a writer, after all. Had the years and years of night-study been in vain? Choked with discouragement, the cry broke from her, "O—God—God help me! I feel—I see, but it all dies in me—dumb!"

Tedious days passed into weeks. Again Sophie sat staring into her notebook. "There's nothing here that's alive. Not a word yet says what's in me . . .

"But it *is* in me!" With clenched fist she smote her bosom. "It must be in me! I believe in it! I got to get it out—even if it tears my flesh in pieces—even if it kills me! . . .

"But these words—these flat, dead words . . .

"Whether I can write or can't write—I can't stop writing. I can't rest. I can't breathe. There's no peace, no running away for me on earth except in the struggle to give out what's in me. The beat from my heart—the blood from my veins—must flow out into my words."

She returned to her unfinished essay, "Believe in Yourself." Her mind groping—clutching at the misty incoherence that clouded her thoughts—she wrote on.

"These sentences are yet only wood—lead; but I can't help it—I'll push on—on—I'll not eat—I'll not sleep—I'll not move from this spot till I get it to say on the paper what I got in my heart!"

Slowly the dead words seemed to begin to breathe. Her eyes brightened. Her cheeks flushed. Her very pencil trembled with the eager onrush of words.

Then a sharp rap sounded on her door. With a gesture of irritation Sophie put down her pencil and looked into the burning, sunken eyes of her neighbor, Hanneh Breineh.

"I got yourself a glass of tea, good friend. It ain't much I got to give away, but it's warm even if it's nothing."

Sophie scowled. "You mustn't bother yourself with me. I'm so busy—thanks."

"Don't thank me yet so quick. I got no sugar." Hanneh Brei-neh edged herself into the room confidingly. "At home, in Poland, I not only had sugar for tea—but even jelly—a jelly that would lift you up to heaven. I thought in America everything would be so plenty, I could drink the tea out from my sugar-bowl. But ach! Not in Poland did my children starve like in America!"

Hanneh Breineh, in a friendly manner, settled herself on the sound end of the bed, and began her jeremiad.

"Yosef, my man, ain't no bread-giver. Already he got consumption the second year. One week he works and nine weeks he lays sick."

In despair Sophie gathered her papers, wondering how to get the woman out of her room. She glanced through the page she had written, but Hanneh Breineh, unconscious of her indifference, went right on.

"How many times it is tearing the heart out from my body—should I take Yosef's milk to give to the baby, or the baby's milk to give to Yosef? If he was dead the pensions they give to widows would help feed my children. Now I got only the charities to help me. A black year on them! They should only have to feed their own children on what they give me."

Resolved not to listen to the intruder, Sophie debated within herself: "Should I call my essay 'Believe in Yourself,' or would n't it be stronger to say, 'Trust Yourself'? But if I say, 'Trust Yourself,' would n't they think that I got the words from Emerson?"

Hanneh Breineh's voice went on, but it sounded to Sophie like a faint buzzing from afar. "Gotteniu! How much did it cost me my life to go and swear myself that my little Fannie—only skin and bones—that she is already fourteen! How it chokes me the tears every morning when I got to wake her and push her out to the shop when her eyes are yet shutting themselves with sleep!"

Sophie glanced at her wrist-watch as it ticked away the precious minutes. She must get rid of the woman! Had she not left her own sister, sacrificed all comfort, all association, for solitude and its golden possibilities? For the first time in her life she had the chance to be by herself and think. And now, the thoughts which a moment ago had seemed like a flock of fluttering birds had come so close—and this woman with her sordid wailing had scattered them.

"I'm a savage, a beast, but I got to ask her to get out—this very minute," resolved Sophie. But before she could summon the courage to do what she wanted to do, there was a timid knock at the door, and the wizened little Fannie, her face streaked with tears, stumbled in.

"The inspector said it's a lie. I ain't yet fourteen," she whimpered.

Hanneh Breineh paled. "Woe is me! Sent back from the shop? God from the world—is there no end to my troubles? Why did n't you hide yourself when you saw the inspector come?"

"I was running to hide myself under the table, but she caught me and she said she'll take me to the Children's Society and arrest me and my mother for sending me to work too soon."

"Arrest me?" shrieked Hanneh Breineh, beating her breast. "Let them only come and arrest me! I'll show America who I am! Let them only begin themselves with me! ... Black is for my eyes ... the groceryman will not give us another bread till we pay him the bill!"

"The inspector said ..." The child's brow puckered in an effort to recall the words.

"What did the inspector said? Gotteniu!" Hanneh Breineh wrung her hands in passionate entreaty. "Listen only once to my prayer! Send on the inspector only a quick death! I only wish her to have her own house with twenty-four rooms and each of the twenty-four rooms should be twenty-four beds and the chills and the fever should throw her from one bed to another!"

"Hanneh Breineh, still yourself a little," entreated Sophie.

"How can I still myself without Fannie's wages? Bitter is me! Why do I have to live so long?"

"The inspector said ..."

"What did the inspector said? A thunder should strike the inspector! Ain't I as good a mother as other mothers? Would n't I better send my children to school? But who'll give us to eat? And who'll pay us the rent?"

Hanneh Breineh wiped her red-lidded eyes with the corner of her apron.

"The president from America should only come to my bitter heart. Let him go fighting himself with the pushcarts how to get the eating a penny cheaper. Let him try to feed his children on the money the charities give me and we'd see if he would n't better send his littlest ones to the shop better than to let them starve before his eyes. Woe is me! What for did I come to America? What's my life—nothing but one terrible, never-stopping fight with the grocer and the butcher and the landlord . . ."

Suddenly Sophie's resentment for her lost morning was forgotten. The crying waste of Hanneh Breineh's life lay open before her eyes like pictures in a book. She saw her own life in Hanneh Breineh's life. Her efforts to write were like Hanneh Breineh's efforts to feed her children. Behind her life and Hanneh Breineh's life she saw the massed ghosts of thousands upon thousands beating—beating out their hearts against rock barriers.

"The inspector said . . ." Fannie timidly attempted again to explain.

"The inspector!" shrieked Hanneh Breineh, as she seized hold of Fannie in a rage. "Hellfire should burn the inspector! Tell me again about the inspector and I'll choke the life out from you—"

Sophie sprang forward to protect the child from the mother. "She's only trying to tell you something."

"Why should she yet throw salt on my wounds? If there was enough bread in the house would I need an inspector to tell me to send her to school? If America is so interested in poor people's children, then why don't they give them to eat till they should go to work? What learning can come into a child's head when the stomach is empty?"

A clutter of feet down the creaking cellar steps, a scuffle of broken shoes and a chorus of shrill voices, as the younger children rushed in from school.

"Mamma—what's to eat?"

"It smells potatoes!"

"Pfui! The pot is empty! It smells over from Cohen's."

"Jake grabbed all the bread!"

"Mamma—he kicked the piece out from my hands!"

"Mamma—it's so empty in my stomach! Ain't there nothing?"

"Gluttons—wolves—thieves!" Hanneh Breineh shrieked. "I should only live to bury you all in one day!"

The children, regardless of Hanneh Breineh's invectives, swarmed around her like hungry bees, tearing at her apron, her skirt. Their voices rose in increased clamor, topped only by their mother's imprecations. "Gotteniu! Tear me away from these leeches on my neck! Send on them only a quick death! . . . Only a minute's peace before I die!"

"Hanneh Breineh—children! What's the matter?" Shmendrik stood at the door. The sweet quiet of the old man stilled the raucous voices as the coming of evening stills the noises of the day.

"There's no end to my troubles! Hear them hollering for bread, and the grocer stopped to give till the bill is paid. Woe is me! Fannie sent home by the inspector and not a crumb in the house!"

"I got something." The old man put his hands over the heads of the children in silent benediction. "All come in by me. I got sent me a box of cake."

"Cake!" The children cried, catching at the kind hands and snuggling about the shabby coat.

"Yes. Cake and nuts and raisins and even a bottle of wine."

The children leaped and danced around him in their wild burst of joy.

"Cake and wine—a box—to you? Have the charities gone crazy?" Hanneh Breineh's eyes sparkled with light and laughter.

"No—no," Shmendrik explained hastily. "Not from the charities—from a friend—for the holidays."

Shmendrik nodded invitingly to Sophie, who was standing in the door of her room. "The roomerkeh will also give a taste with us our party?"

"Sure will she!" Hanneh Breineh took Sophie by the arm. "Who'll say no in this black life to cake and wine?"

Young throats burst into shrill cries: "Cake and wine—wine and cake—raisins and nuts—nuts and raisins!" The words rose in a triumphant chorus. The children leaped and danced in time to their chant, almost carrying the old man bodily into his room in the wildness of their joy.

The contagion of this sudden hilarity erased from Sophie's mind the last thought of work and she found herself seated with the others on the cobbler's bench.

From under his cot the old man drew forth a wooden box. Lifting the cover he held up before wondering eyes a large frosted cake embedded in raisins and nuts.

Amid the shouts of glee Shmendrik now waved aloft a large bottle of grape-juice.

The children could contain themselves no longer and dashed forward.

"Shah—shah! Wait only!" He gently halted their onrush and waved them back to their seats.

"The glasses for the wine!" Hanneh Breineh rushed about hither and thither in happy confusion. From the sink, the shelf, the windowsill, she gathered cracked glasses, cups without handles—anything that would hold even a few drops of the yellow wine.

Sacrificial solemnity filled the basement as the children breathlessly watched Shmendrik cut the precious cake. Mouths —even eyes—watered with the intensity of their emotion.

With almost religious fervor Hanneh Breineh poured the grape-juice into the glasses held in the trembling hands of the children. So overwhelming was the occasion that none dared to taste till the ritual was completed. The suspense was agonizing as one and all waited for Shmendrik's signal.

"Hanneh Breineh—you drink from my Sabbath wine-glass!"

Hanneh Breineh clinked glasses with Schmendrik. "Long years on us all!" Then she turned to Sophie, clinked glasses once more. "May you yet marry yourself from our basement to a millionaire!" Then she lifted the glass to her lips.

The spell was broken. With a yell of triumph the children gobbled the cake in huge mouthfuls and sucked the golden liquid. All the traditions of wealth and joy that ever sparkled from the bubbles of champagne smiled at Hanneh Breineh from her glass of California grape-juice.

"Ach!" she sighed. "How good it is to forget your troubles,

and only those that's got troubles have the chance to forget them."

She sipped the grape-juice leisurely, thrilled into ecstasy with each lingering drop. "How it laughs yet in me, the life, the minute I turn my head from my worries!"

With growing wonder in her eyes, Sophie watched Hanneh Breineh. This ragged wreck of a woman—how passionately she clung to every atom of life! Hungrily, she burned through the depths of every experience. How she flared against wrongs—and how every tiny spark of pleasure blazed into joy!

Within a half-hour this woman had touched the whole range of human emotions, from bitterest agony to dancing joy. The terrible despair at the onrush of her starving children when she cried out, "O that I should only bury you all in one day!" And now the leaping light of the words: "How it laughs yet in me, the life, the minute I turn my head from my worries."

"Ach, if I could only write like Hanneh Breineh talks!" thought Sophie. "Her words dance with a thousand colors. Like a rainbow it flows from her lips." Sentences from her own essays marched before her, stiff and wooden. How clumsy, how unreal, were her most labored phrases compared to Hanneh Breineh's spontaneity. Fascinated, she listened to Hanneh Breineh, drinking her words as a thirst-perishing man drinks water. Every bubbling phrase filled her with a drunken rapture to create.

"Up till now I was only trying to write from my head. It was n't real—it was n't life. Hanneh Breineh is real. Hanneh Breineh is life."

"Ach! What do the rich people got but dried-up dollars? Pfui on them and their money!" Hanneh Breineh held up her glass to be refilled. "Let me only win a fortune on the lotteree and move myself in my own bought house. Let me only have my first hundred dollars in the bank and I'll lift up my head like a person and tell the charities to eat their own cornmeal. I'll get myself an automobile like the kind rich ladies and ride up to their houses on Fifth Avenue and feed them only once on the eating they like so good for me and my children."

With a smile of benediction Shmendrik refilled the glasses and cut for each of his guests another slice of cake. Then came the handful of nuts and raisins.

As the children were scurrying about for hammers and iron lasts with which to crack their nuts, the basement door creaked. Unannounced, a woman entered—the "friendly visitor" of the charities. Her look of awful amazement swept the group of merrymakers.

"Mr. Shmendrik!—Hanneh Breineh!" Indignation seethed in her voice. "What's this? A feast—a birthday?"

Gasps—bewildered glances—a struggle for utterance!

"I came to make my monthly visit—evidently I'm not needed."

Shmendrik faced the accusing eyes of the "friendly visitor." "Holiday eating . . ."

"Oh—I'm glad you're so prosperous."

Before any one had gained presence of mind enough to explain things, the door had clanked. The "friendly visitor" had vanished.

"Pfui!" Hanneh Breineh snatched up her glass and drained its contents. "What will she do now? Will we get no more dry bread from the charities because once we ate cake?"

"What for did she come?" asked Sophie.

"To see that we don't over-eat ourselves!" returned Hanneh Breineh. "She's a 'friendly visitor'! She learns us how to cook cornmeal. By pictures and lectures she shows us how the poor people should live without meat, without milk, without butter, and without eggs. Always it's on the end of my tongue to ask her, 'You learned us to do without so much, why can't you yet learn us how to eat without eating?' "

The children seized the last crumbs of cake that Shmendrik handed them and rushed for the street.

"What a killing look was on her face," said Sophie. "Couldn't she be a little glad for your gladness?"

"Charity ladies—gladness?" The joy of the grape-wine still rippled in Hanneh Breineh's laughter. "For poor people is only cornmeal. Ten cents a day—to feed my children!"

Still in her rollicking mood Hanneh Breineh picked up the baby and tossed it like a Bacchante. "Could you be happy a lot with ten cents in your stomach? Ten cents—half a can of condensed milk—then fill yourself the rest with water! . . . Maybe yet feed you with all water and save the ten-cent pieces to buy you a carriage like the Fifth Avenue babies! . . ."

The soft sound of a limousine purred through the area grating and two well-fed figures in sealskin coats, led by the "friendly visitor," appeared at the door.

"Mr. Bernstein, you can see for yourself." The "friendly visitor" pointed to the table.

The merry group shrank back. It was as if a gust of icy wind had swept all the joy and laughter from the basement.

"You are charged with intent to deceive and obtain assistance by dishonest means," said Mr. Bernstein.

"Dishonest?" Shmendrik paled.

Sophie's throat strained with passionate protest, but no words came to her release.

"A friend—a friend"—stammered Shmendrik—"sent me the holiday eating."

The superintendent of the Social Betterment Society faced him accusingly. "You told us that you had no friends when you applied to us for assistance."

"My friend—he knew me in my better time." Shmendrik flushed painfully. "I was once a scholar—respected. I wanted by this one friend to hold myself like I was."

Mr. Bernstein had taken from the bookshelf a number of letters, glanced through them rapidly and handed them one by one to the deferential superintendent.

Shmendrik clutched at his heart in an agony of humiliation. Suddenly his bent body straightened. His eyes dilated. "My letters—my life—you dare?"

"Of course we dare!" The superintendent returned Shmendrik's livid gaze, made bold by the confidence that what he was doing was the only scientific method of administering philanthropy. "These dollars, so generously given, must go to those most worthy. . . . I find in these letters references to gifts of fruit and other luxuries you did not report at our office."

"He never kept nothing for himself!" Hanneh Breineh broke in defensively. "He gave it all for the children."

Ignoring the interruption Mr. Bernstein turned to the "friendly visitor." "I'm glad you brought my attention to this case. It's but one of the many impositions on our charity . . . Come . . ."

"Kossacks! Pogromschiks!" Sophie's rage broke at last.

"You call yourselves Americans? You dare call yourselves Jews? You bosses of the poor! This man Shmendrik, whose house you broke into, whom you made to shame like a beggar—he is the one Jew from whom the Jews can be proud! He gives all he is—all he has—as God gives. *He is* charity.

"But you—you are the greed—the shame of the Jews! *All-right-niks*—fat bellies in fur coats! What do you give from yourselves? You may eat and bust eating! Nothing you give till you've stuffed yourselves so full that your hearts are dead!"

The door closed in her face. Her wrath fell on indifferent backs as the visitors mounted the steps to the street.

Shmendrik groped blindly for the Bible. In a low, quavering voice, he began the chant of the oppressed—the wail of the downtrodden. "I am afraid, and a trembling taketh hold of my flesh. Wherefore do the wicked live, become old, yea, mighty in power?"

Hanneh Breineh and the children drew close around the old man. They were weeping—unconscious of their weeping—deep-buried memories roused by the music, the age-old music of the Hebrew race.

Through the grating Sophie saw the limousine pass. The chant flowed on: "Their houses are safe from fear; neither is the rod of God upon them."

Silently Sophie stole back to her room. She flung herself on the cot, pressed her fingers to her burning eyeballs. For a long time she lay rigid, clenched—listening to the drumming of her heart like the sea against rock barriers. Presently the barriers burst. Something in her began pouring itself out. She felt for her pencil—paper—and began to write. Whether she reached out to God or man she knew not, but she wrote on and on all through that night.

The gray light entering her grated window told her that beyond was dawn. Sophie looked up: "Ach! At last it writes itself in me!" she whispered triumphantly. "It's not me—it's their cries—my own people—crying in me! Hanneh Breineh, Shmendrik, they will not be stilled in me, till all America stops to listen."

from Living My Life

Emma Goldman

I HAD WORKED IN factories before, in St. Petersburg. In the winter of 1882, when Mother, my two little brothers, and I came from Konigsberg to join Father in the Russian capital, we found that he had lost his position. He had been manager of his cousin's drygoods store; but, shortly before our arrival, the business failed. The loss of his job was a tragedy to our family, as Father had not managed to save anything. The only bread-winner then was Helena. Mother was forced to turn to her brothers for a loan. The three hundred roubles they advanced were invested in a grocery store. The business yielded little at first, and it became necessary for me to find employment.

Knitted shawls were then much in vogue, and a neighbour told my mother where I might find work to do at home. By keeping at the task many hours a day, sometimes late into the night, I contrived to earn twelve roubles a month.

At the age of seventeen, *Emma Goldman* came to the U.S. from Russia. Married briefly, she worked in sweatshops and factories and was a leader in the radical-anarchist circles of her time, lecturing and writing on diverse topics. She was deported to Russia after World War I and was never allowed to return to America. She died in 1940. Chapter 2 of *Living My Life*, Vol. II, by Emma Goldman is reprinted with permission of Dover Publications.

The shawls I knitted for a livelihood were by no means masterpieces, but somehow they passed. I hated the work, and my eyes gave way under the strain of constant application. Father's cousin who had failed in the dry-goods business now owned a glove factory. He offered to teach me the trade and give me work.

The factory was far from our place. One had to get up at five in the morning to be at work at seven. The rooms were stuffy, unventilated, and dark. Oil lamps gave the light; the sun never penetrated the work-room.

There were six hundred of us, of all ages, working on costly and beautiful gloves day in, day out, for very small pay. But we were allowed sufficient time for our noon meal and twice a day for tea. We could talk and sing while at work; we were not driven or harassed. That was in St. Petersburg, in 1882.

Now I was in America, in the Flower City of the State of New York, in a model factory, as I was told. Certainly, Garson's clothing-works were a vast improvement on the glove factory on the Vassilevsky Ostrov. The rooms were large, bright, and airy. One had elbow space. There were none of those ill-smelling odours that used to nauseate me in our cousin's shop. Yet the work here was harder, and the day, with only half an hour for lunch, seemed endless. The iron discipline forbade free movement (one could not even go to the toilet without permission), and the constant surveillance of the foreman weighed like stone on my heart. The end of each day found me sapped, with just enough energy to drag myself to my sister's home and crawl into bed. This continued with deadly monotony week after week.

The amazing thing to me was that no one else in the factory seemed to be so affected as I, no one but my neighbour, frail little Tanya. She was delicate and pale, frequently complained of headaches, and often broke into tears when the task of handling heavy ulsters proved too much for her. One morning, as I looked up from my work, I discovered her all huddled in a heap. She had fallen in a faint. I called to the foreman to help me carry her to the dressing-room, but the deafening noise of the machines drowned my voice. Several girls near by heard me and began to shout. They ceased working and rushed over to

Tanya. The sudden stopping of the machines attracted the foreman's attention and he came over to us. Without even asking
the reason for the commotion, he shouted: "Back to your machines! What do you mean stopping work now? Do you want to
be fired? Get back at once!" When he spied the crumpled body
of Tanya, he yelled: "What the hell is the matter with her?" "She
has fainted," I replied, trying hard to control my voice. "Fainted, nothing," he sneered, "she's only shamming."

"You are a liar and a brute!" I cried, no longer able to keep
back my indignation.

I bent over Tanya, loosened her waist, and squeezed the juice
of an orange I had in my lunch basket into her half-opened
mouth. Her face was white, a cold sweat on her forehead. She
looked so ill that even the foreman realized she had not been
shamming. He excused her for the day. "I will go with Tanya,"
I said; "you can deduct from my pay for the time." "You can go
to hell, you wildcat!" he flung after me.

We went to a coffee place. I myself felt empty and faint, but
all we had between us was seventy-five cents. We decided to
spend forty on food, and use the rest for a street-car ride to the
park. There, in the fresh air, amid the flowers and trees, we
forgot our dreaded tasks. The day that had begun in trouble
ended restfully and in peace.

The next morning the enervating routine started all over
again, continuing for weeks and months, broken only by the
new arrival in our family, a baby girl. The child became the one
interest in my dull existence. Often, when the atmosphere in
Garson's factory threatened to overcome me, the thought of the
lovely mite at home revived my spirit. The evenings were no
longer dreary and meaningless. But, while little Stella brought
joy into our household, she added to the material anxiety of my
sister and my brother-in-law.

Lena never by word or deed made me feel that the dollar and
fifty cents I was giving her for my board (the car fare amounted
to sixty cents a week, the remaining forty cents being my pin-
money) did not cover my keep. But I had overheard my brother-
in-law grumbling over the growing expense of the house. I felt
he was right. I did not want my sister worried, she was nursing
her child. I decided to apply for a rise. I knew it was no use
talking to the foreman and therefore I asked to see Mr. Garson.

I was ushered into a luxurious office. American Beauties were on the table. Often I had admired them in the flower shops, and once, unable to withstand the temptation, I had gone in to ask the price. They were one dollar and a half apiece—more than half of my week's earnings. The lovely vase in Mr. Garson's office held a great many of them.

I was not asked to sit down. For a moment I forgot my mission. The beautiful room, the roses, the aroma of the bluish smoke from Mr. Garson's cigar, fascinated me. I was recalled to reality by my employer's question: "Well, what can I do for you?"

I had come to ask for a rise, I told him. The two dollars and a half I was getting did not pay my board, let alone anything else, such as an occasional book or a theatre ticket for twenty-five cents. Mr. Garson replied that for a factory girl I had rather extravagant tastes, that all his "hands" were well satisfied, that they seemed to be getting along all right—that I, too, would have to manage or find work elsewhere. "If I raise your wages, I'll have to raise the others' as well and I can't afford that," he said. I decided to leave Garson's employ.

A few days later I secured a job at Rubinstein's factory at four dollars a week. It was a small shop, not far from where I lived. The house stood in a garden, and only a dozen men and women were employed in the place. The Garson discipline and drive were missing.

Next to my machine worked an attractive young man whose name was Jacob Kershner. He lived near Lena's home, and we would often walk from work together. Before long he began calling for me in the morning. We used to converse in Russian, my English still being very halting. His Russian was like music to me; it was the first real Russian, outside of Helena's, that I had had an opportunity to hear in Rochester since my arrival.

Kershner had come to America in 1881 from Odessa, where he had finished the *Gymnasium.* Having no trade, he became an "operator" on cloaks. He used to spend most of his leisure, he told me, reading or going to dances. He had no friends, because he found his co-workers in Rochester interested only in money-making, their ideal being to start a shop of their own. He had heard of our arrival, Helena's and mine—had even seen me on the street several times—but he did not know how to get acquainted. Now he would no longer feel lonely, he said brightly; we could visit places together and he would lend me his books to read. My own loneliness no longer was so poignant.

I told my sisters of my new acquaintance, and Lena asked me to invite him the next Sunday. When Kershner came, she was favourably impressed; but Helena took a violent dislike to him from the first. She said nothing about it for a long time, but I could sense it.

One day Kershner invited me to a dance. It was my first since I came to America. The very anticipation was exciting, bringing back memories of my first ball in St. Petersburg.

I was fifteen then. Helena had been invited to the fashionable German Club by her employer, who gave her two tickets, so she could bring me with her. Some time previously my sister had presented me with a lovely blue velvet for my first long dress; but before it could be made up, our peasant servant walked off with the material. My grief over its loss made me quite ill for several days. If only I had a dress, I thought, Father might consent to my attending the ball. "I'll get you material for a dress," Helena consoled me, "but I'm afraid Father will refuse." "Then I will defy him!" I declared.

She bought another piece of blue stuff, not so beautiful as my velvet, but I no longer minded. I was too happy over the prospect of my first ball, of the bliss of dancing in public. Somehow Helena succeeded in getting Father's consent, but at the last moment he changed his mind. I had been guilty of some infraction during the day, and he categorically declared that I would have to stay home. Thereupon Helena said she also would not go. But I was determined to defy my father, no matter what the consequences.

With bated breath I waited for my parents to retire for the night. Then I dressed and woke Helena. I told her she must come with me or I would run away from home. "We can be back before Father wakes up," I urged. Dear Helena—she was always so timid! She had infinite capacity for suffering, for endurance, but she could not fight. On this occasion she was carried away by my desperate decision. She dressed and we quietly slipped out of the house.

At the German Club everything was bright and gay. We found

Helena's employer, whose name was Kadison, and some of his
young friends. I was asked for every dance, and I danced in
frantic excitement and abandon. It was getting late and many
people were already leaving when Kadison invited me for another dance. Helena insisted that I was too exhausted, but I
would not have it so. "I will dance!" I declared; "I will dance
myself to death!" My flesh felt hot, my heart beat violently as my
cavalier swung me round the ball-room, holding me tightly. To
dance to death—what more glorious end!

It was towards five in the morning when we arrived home.
Our people were still asleep. I awoke late in the day, pretending
a sick headache, and secretly I gloried in my triumph of having
outwitted our old man.

The memory of that experience still vivid in my mind, I accompanied Jacob Kershner to the party, full of anticipation. My
disappointment was bitter: there were no beautiful ball-room,
no lovely women, no dashing young men, no gaiety. The music
was shrill, the dancers clumsy. Jacob danced not badly, but he
lacked spirit and fire. "Four years at the machine have taken the
strength out of me," he said; "I get tired so easily."

I had known Jacob Kershner about four months when he
asked me to marry him. I admitted I liked him, but I did not want
to marry so young. We still knew so little of each other. He said
he'd wait as long as I pleased, but there was already a great deal
of talk about our being out together so much. "Why should we
not get engaged?" he pleaded. Finally I consented. Helena's
antagonism to Jacob had become almost an obsession; she
fairly hated him. But I was lonely—I needed companionship.
Ultimately I won over my sister. Her great love for me could
never refuse me anything or stand out against my wishes.

The late fall of 1886 brought the rest of our family to Rochester—Father, Mother, my brothers, Herman and Yegor. Conditions in St. Petersburg had become intolerable for the Jews, and
the grocery business did not yield enough for the ever-growing
bribery Father had to practice in order to be allowed to exist.
America became the only solution.

Together with Helena I had prepared a home for our parents,
and on their arrival we went to live with them. Our earnings
soon proved inadequate to meet the household expenses. Jacob

Kershner offered to board with us, which would be of some help, and before long he moved in.

The house was small, consisting of a living-room, a kitchen, and two bedrooms. One of them was used by my parents, the other by Helena, myself, and our little brother. Kershner and Herman slept in the living-room. The close proximity of Jacob and the lack of privacy kept me in constant irritation. I suffered from sleepless nights, waking dreams and great fatigue at work. Life was becoming unbearable, and Jacob stressed the need of a home of our own.

On nearer acquaintance I had grown to understand that we were too different. His interest in books, which had first attracted me to him, had waned. He had fallen into the ways of his shopmates, playing cards and attending dull dances. I, on the contrary, was filled with striving and aspirations. In spirit I was still in Russia, in my beloved St. Petersburg, living in the world of the books I had read, the operas I had heard, the circle of the students I had known. I hated Rochester even more than before. But Kershner was the only human being I had met since my arrival. He filled a void in my life, and I was strongly attracted to him. In February 1887 we were married in Rochester by a rabbi, according to Jewish rites, which were then considered sufficient by the law of the country.

My feverish excitement of that day, my suspense and ardent anticipation gave way at night to a feeling of utter bewilderment. Jacob lay trembling near me; he was impotent.

My first erotic sensations I remember had come to me when I was about six. My parents lived in Popelan then, where we children had no home in any real sense. Father kept an inn, which was constantly filled with peasants, drunk and quarrelling, and government officials. Mother was busy superintending the servants in our large, chaotic house. My sisters, Lena and Helena, fourteen and twelve, were burdened with work. I was left to myself most of the day. Among the stable help there was a young peasant, Petrushka, who served as shepherd, looking after our cows and sheep. Often he would take me with him to the meadows, and I would listen to the sweet tones of his flute. In the evening he would carry me back home on his shoulders, I sitting astride. He would play horse—run as fast as his legs could carry him, then suddenly throw me up in the air, catch me in his arms, and press me to him. It used to give me a peculiar sensation, fill me with exultation, followed by blissful release.

I became inseparable from Petrushka. I grew so fond of him that I began stealing cake and fruit from Mother's pantry for him. To be with Petrushka out in the fields, to listen to his music, to ride on his shoulders, became the obsession of my waking and sleeping hours. One day Father had an altercation with Petrushka, and the boy was sent away. The loss of him was one of the greatest tragedies of my child-life. For weeks afterwards I kept on dreaming of Petrushka, the meadows, the music, and reliving the joy and ecstasy of our play. One morning I felt myself torn out of sleep. Mother was bending over me, tightly holding my right hand. In an angry voice she cried: "If ever I find your hand again like that, I'll whip you, you naughty child!"

The approach of puberty gave me my first consciousness of the effect of men on me. I was eleven then. Early one summer day I woke up in great agony. My head, spine, and legs ached as if they were being pulled asunder. I called for Mother. She drew back my bedcovers, and suddenly I felt a stinging pain in my face. She had struck me. I let out a shriek, fastening on Mother terrified eyes. "This is necessary for a girl," she said, "when she becomes a woman, as a protection against disgrace." She tried to take me in her arms, but I pushed her back. I was writhing in pain and I was too outraged for her to touch me. "I am going to die," I howled, "I want the *Feldscher* (assistant doctor)." The *Feldscher* was sent for. He was a young man, a new-comer in our village. He examined me and gave me something to put me to sleep. Thenceforth my dreams were of the *Feldscher*.

When I was fifteen, I was employed in a corset factory in the Hermitage Arcade in St. Petersburg. After working hours, on leaving the shop together with the other girls, we would be waylaid by young Russian officers and civilians. Most of the girls had their sweethearts; only a Jewish girl chum of mine and I refused to be taken to the *konditorskaya* (pastry shop) or to the park.

Next to the Hermitage was a hotel we had to pass. One of the clerks, a handsome fellow of about twenty, singled me out for his attentions. At first I scorned him, but gradually he began to exert a fascination on me. His perseverance slowly undermined my pride and I accepted his courtship. We used to meet in some quiet spot or in an out-of-the-way pastry shop. I had to invent all sorts of stories to explain to my father why I returned late from work or stayed out after nine o'clock. One day he spied me in the Summer Garden in the company of other girls and some boy students. When I returned home, he threw me violently against the shelves in our grocery store, which sent the jars of Mother's wonderful *varenya* flying to the floor. He pounded me with his fists, shouting that he would not tolerate a loose daughter. The experience made my home more unbearable, the need of escape more compelling.

For several months my admirer and I met clandestinely. One day he asked me whether I should not like to go through the hotel to see the luxurious rooms. I had never been in a hotel before—the joy and gaiety I fancied behind the gorgeous windows used to fascinate me as I would pass the place on my way from work.

The boy led me through a side entrance, along a thickly carpeted corridor, into a large room. It was brightly illumined and beautifully furnished. A table near the sofa held flowers and a tea-tray. We sat down. The young man poured out a golden-coloured liquid and asked me to clink glasses to our friendship. I put the wine to my lips. Suddenly I found myself in his arms, my waist torn open—his passionate kisses covered my face, neck, and breasts. Not until after the violent contact of our bodies and the excruciating pain he caused me did I come to my senses. I screamed, savagely beating against the man's chest with my fists. Suddenly I heard Helena's voice in the hall. "She must be here—she must be here!" I became speechless. The man, too, was terrorized. His grip relaxed, and we listened in breathless silence. After what seemed to me hours, Helena's voice receded. The man got up. I rose mechanically, mechanically buttoned my waist and brushed back my hair.

Strange, I felt no shame—only a great shock at the discovery that the contact between man and woman could be so brutal and so painful. I walked out in a daze, bruised in every nerve.

When I reached home I found Helena fearfully wrought up. She had been uneasy about me, aware of my meeting with the boy. She had made it her business to find out where he worked, and when I failed to return, she had gone to the hotel in search of me. The shame I did not feel in the arms of the man now overwhelmed me. I could not muster up courage to tell Helena of my experience.

After that I always felt between two fires in the presence of men. Their lure remained strong, but it was always mingled with violent revulsion. I could not bear to have them touch me.

These pictures passed through my mind vividly as I lay alongside my husband on our wedding night. He had fallen fast asleep.

The weeks went on. There was no change. I urged Jacob to consult a doctor. At first he refused, pleading diffidence, but finally he went. He was told it would take considerable time to "build up his manhood." My own passion had subsided. The material anxiety of making ends meet excluded everything else. I had stopped work; it was considered disgraceful for a married woman to go to the shop. Jacob was earning fifteen dollars a week. He had developed a passion for cards, which swallowed up a considerable part of our income. He grew jealous, suspecting everyone. Life became insupportable. I was saved from utter despair by my interest in the Haymarket events.

After the death of the Chicago anarchists I insisted on a separation from Kershner. He fought long against it, but finally consented to a divorce. It was given to us by the same rabbi who had performed our marriage ceremony. Then I left for New Haven, Connecticut, to work in a corset-factory.

During my efforts to free myself from Kershner the only one who stood by me was my sister Helena. She had been strenuously opposed to the marriage in the first place, but now she offered not a single reproach. On the contrary, she gave me help and comfort. She pleaded with my parents and with Lena in behalf of my decision to get a divorce. As always, her devotion knew no bounds.

In New Haven I met a group of young Russians, students mainly, now working at various trades. Most of them were socialists and anarchists. They often organized meetings, generally inviting speakers from New York, one of whom was A. Solotaroff. Life was interesting and colourful, but gradually the strain of the work became too much for my depleted vitality. Finally I had to return to Rochester.

I went to Helena. She lived with her husband and child over their little printing shop, which also served as an office for their steamship agency. But both occupations did not bring in enough to keep them from dire poverty. Helena had married Jacob Hochstein, a man ten years her senior. He was a great Hebrew scholar, an authority on the English and Russian classics, and a very rare personality. His integrity and independent character made him a poor competitor in the sordid business life. When anyone brought him a printing order worth two dollars, Jacob Hochstein devoted as much time to it as if he were getting fifty. If a customer showed a tendency to bargain over prices, he would send him away. He could not bear the implication that he might overcharge. His income was insufficient for the needs of the family, and the one to worry and fret most about it was my poor Helena. She was pregnant with her second child and yet had to drudge from morning till night to make ends meet, with never a word of complaint. But, then, she had been that way all her life, suffering silently, always resigned.

Helena's marriage had not sprung from a passionate love. It was the union of two mature people who longed for comradeship, for a quiet life. Whatever there had been of passion in my sister had burned out when she was twenty-four. At the age of sixteen, while we were living in Popelan, she had fallen in love with a young Lithuanian, a beautiful soul. But he was a *goi* (gentile) and Helena knew that marriage between them was impossible. After a great struggle and many tears Helena broke off the affair with young Susha. Years later, while on our way to America, we stopped in Kovno, our native town. Helena had arranged for Susha to meet her there. She could not bear to go away so far without saying good-bye to him. They met and parted as good friends—the fire of their youth was in ashes.

On my return from New Haven Helena received me, as always, with tenderness and with the assurance that her home was also mine. It was good to be near my darling again, with little Stella and my young brother Yegor. But it did not take me long to discover the pinched condition in Helena's home. I went back to the shop.

Living in the Jewish district, it was impossible to avoid those one did not wish to see. I ran into Kershner almost immediately after my arrival. Day after day he would seek me out. He began to plead with me to go back to him—all would be different. One day he threatened suicide—actually pulled out a bottle of poison. Insistently he pressed me for a final answer.

I was not naive enough to think that a renewed life with Kershner would prove more satisfactory or lasting than at first. Besides, I had definitely decided to go to New York, to equip myself for the work I had vowed to take up after the death of my Chicago comrades. But Kershner's threat frightened me: I could not be responsible for his death. I remarried him. My parents rejoiced and so did Lena and her husband, but Helena was sick with grief.

Without Kershner's knowledge I took up a course in dressmaking, in order to have a trade that would free me from the shop. During three long months I wrestled with my husband to let me go my way. I tried to make him see the futility of living a patched life, but he remained obdurate. Late one night, after bitter recriminations, I left Jacob Kershner and my home, this time definitely.

I was immediately ostracized by the whole Jewish population of Rochester. I could not pass on the street without being held up to scorn. My parents forbade me their house, and again it was only Helena who stood by me. Out of her meagre income she even paid my fare to New York.

So I left Rochester, where I had known so much pain, hard work, and loneliness, but the joy of my departure was marred by separation from Helena, from Stella, and the little brother I loved so well.

The break of the new day in the Minkin flat still found me awake. The door upon the old had now closed for ever. The new was calling, and I eagerly stretched out my hands towards it. I fell into a deep, peaceful sleep.

I was awakened by Anna Minkin's voice announcing the arrival of Alexander Berkman. It was late afternoon.

I Stand Here Ironing

Tillie Olsen

I STAND HERE IRONING, and what you asked me moves tormented back and forth with the iron.

"I wish you would manage the time to come in and talk with
me about your daughter. I'm sure you can help me understand
her. She's a youngster who needs help and whom I'm deeply
interested in helping."

"Who needs help." ... Even if I came, what good would it do?
You think because I am her mother I have a key, or that in some
way you could use me as a key? She has lived for nineteen years.

Born in Omaha, Nebraska, educated in public libraries, and a San Franciscan
most of her life, *Tillie Olsen* has received fellowships and grants from the Ford
Foundation, the National Endowment for the Arts, the Radcliffe Institute, and
the Guggenheim Memorial Foundation. She is the recipient of the Literary
Award of the American Academy and National Institute of Arts and Letters and
of an honorary Doctor of Arts and Letters degree from the University of
Nebraska. Her published works include *Yonnondio, Silences,* and *Tell Me a Riddle,*
a collection of her stories. The title selection in the latter won first prize in the
1961 O. Henry Awards. Olsen's works have been anthologized more than 45
times in various publications, among them *The Best American Short Stories* 1957,
1961, 1971 and *Fifty Best American Stories,* 1915–1965. "I Stand Here Ironing" is
excerpted from *Tell Me a Riddle* by Tillie Olsen, copyright © 1956 by Tillie
Olsen, and reprinted by permission of Delacorte Press/Seymour Lawrence.

There is all that life that has happened outside of me, beyond me.

And when is there time to remember, to sift, to weigh, to estimate, to total? I will start and there will be an interruption and I will have to gather it all together again. Or I will become engulfed with all I did or did not do, with what should have been and what cannot be helped.

She was a beautiful baby. The first and only one of our five that was beautiful at birth. You do not guess how new and uneasy her tenancy in her now-loveliness. You did not know her all those years she was thought homely, or see her poring over her baby pictures, making me tell her over and over how beautiful she had been—and would be, I would tell her—and was now, to the seeing eye. But the seeing eyes were few or nonexistent. Including mine.

I nursed her. They feel that's important nowadays. I nursed all the children, but with her, with all the fierce rigidity of first motherhood, I did like the books then said. Though her cries battered me to trembling and my breasts ached with swollenness, I waited till the clock decreed.

Why do I put that first? I do not even know if it matters, or if it explains anything.

She was a beautiful baby. She blew shining bubbles of sound. She loved motion, loved light, loved color and music and textures. She would lie on the floor in her blue overalls patting the surface so hard in ecstasy her hands and feet would blur. She was a miracle to me, but when she was eight months old I had to leave her daytimes with the woman downstairs to whom she was no miracle at all, for I worked or looked for work and for Emily's father, who "could no longer endure" (he wrote in his good-bye note) "sharing want with us."

I was nineteen. It was the pre-relief, pre-WPA world of the depression. I would start running as soon as I got off the street-car, running up the stairs, the place smelling sour, and awake or asleep to startle awake, when she saw me she would break into a clogged weeping that could not be comforted, a weeping I can hear yet.

After a while I found a job hashing at night so I could be with her days, and it was better. But it came to where I had to bring her to his family and leave her.

It took a long time to raise the money for her fare back. Then she got chicken pox and I had to wait longer. When she finally came, I hardly knew her, walking quick and nervous like her father, looking like her father, thin, and dressed in a shoddy red that yellowed her skin and glared at the pockmarks. All the baby loveliness gone.

She was two. Old enough for nursery school they said, and I did not know then what I know now—the fatigue of the long day, and the lacerations of group life in the kinds of nurseries that are only parking places for children.

Except that it would have made no difference if I had known. It was the only place there was. It was the only way we could be together, the only way I could hold a job.

And even without knowing, I knew. I knew the teacher that was evil because all these years it has curdled into my memory, the little boy hunched in the corner, her rasp, "why aren't you outside, because Alvin hits you? that's no reason, go out, scaredy." I knew Emily hated it even if she did not clutch and implore "don't go Mommy" like the other children, mornings.

She always had a reason why we should stay home. Momma, you look sick. Momma, I feel sick. Momma, the teachers aren't there today, they're sick. Momma, we can't go, there was a fire there last night. Momma, it's a holiday today, no school, they told me.

But never a direct protest, never rebellion. I think of our others in their three-, four-year-oldness—the explosions, the tempers, the denunciations, the demands—and I feel suddenly ill. I put the iron down. What in me demanded that goodness in her? And what was the cost, the cost to her of such goodness?

The old man living in the back once said in his gentle way: "You should smile at Emily more when you look at her." What *was* in my face when I looked at her? I loved her. There were all the acts of love.

It was only with the others I remembered what he said, and it was the face of joy, and not of care or tightness or worry I turned to them—too late for Emily. She does not smile easily, let alone almost always as her brothers and sisters do. Her face is closed and sombre, but when she wants, how fluid. You must have seen it in her pantomimes, you spoke of her rare gift for comedy on the stage that rouses a laughter out of the audience so dear they applaud and applaud and do not want to let her go.

Where does it come from, that comedy? There was none of it in her when she came back to me that second time, after I had had to send her away again. She had a new daddy now to learn to love, and I think perhaps it was a better time.

Except when we left her alone nights, telling ourselves she was old enough.

"Can't you go some other time, Mommy, like tomorrow?" she would ask. "Will it be just a little while you'll be gone? Do you promise?"

The time we came back, the front door open, the clock on the floor in the hall. She rigid awake. "It wasn't just a little while. I didn't cry. Three times I called you, just three times, and then I ran downstairs to open the door so you could come faster. The clock talked loud. I threw it away, it scared me what it talked."

She said the clock talked loud again that night I went to the hospital to have Susan. She was delirious with the fever that comes before red measles, but she was fully conscious all the week I was gone and the week after we were home when she could not come near the new baby or me.

She did not get well. She stayed skeleton thin, not wanting to eat, and night after night she had nightmares. She would call for me, and I would rouse from exhaustion to sleepily call back: "You're all right, darling, go to sleep, it's just a dream," and if she still called, in a sterner voice, "now go to sleep, Emily, there's nothing to hurt you." Twice, only twice, when I had to get up for Susan anyhow, I went in to sit with her.

Now when it is too late (as if she would let me hold and comfort her like I do the others) I get up and go to her at once at her moan or restless stirring. "Are you awake, Emily? Can I get you something?" And the answer is always the same: "No, I'm all right, go back to sleep, Mother."

They persuaded me at the clinic to send her away to a convalescent home in the country where "she can have the kind of food and care you can't manage for her, and you'll be free to concentrate on the new baby." They still send children to that place. I see pictures on the society page of sleek young women planning affairs to raise money for it, or dancing at the affairs, or decorating Easter eggs or filling Christmas stockings for the children.

They never have a picture of the children so I do not know if the girls still wear those gigantic red bows and the ravaged looks on the every other Sunday when parents can come to visit "unless otherwise notified"—as we were notified the first six weeks.

Oh it is a handsome place, green lawns and tall trees and fluted flower beds. High up on the balconies of each cottage the children stand, the girls in their red bows and white dresses, the boys in white suits and giant red ties. The parents stand below shrieking up to be heard and the children shriek down to be heard, and between them the invisible wall "Not to Be Contaminated by Parental Germs or Physical Affection."

There was a tiny girl who always stood hand in hand with Emily. Her parents never came. One visit she was gone. "They moved her to Rose Cottage," Emily shouted in explanation. "They don't like you to love anybody here."

She wrote once a week, the labored writing of a seven-year-old. "I am fine. How is the baby. If I write my leter nicly I will have a star. Love." There never was a star. We wrote every other day, letters she could never hold or keep but only hear read—once. "We simply do not have room for children to keep any personal possessions," they patiently explained when we pieced one Sunday's shrieking together to plead how much it would mean to Emily, who loved so to keep things, to be allowed to keep her letters and cards.

Each visit she looked frailer. "She isn't eating," they told us.

(They had runny eggs for breakfast or mush with lumps, Emily said later, I'd hold it in my mouth and not swallow. Nothing ever tasted good, just when they had chicken.)

It took us eight months to get her released home, and only the fact that she gained back so little of her seven lost pounds convinced the social worker.

I used to try to hold and love her after she came back, but her body would stay stiff, and after a while she'd push away. She ate little. Food sickened her, and I think much of life too. Oh she had physical lightness and brightness, twinkling by on skates, bouncing like a ball up and down up and down over the jump rope, skimming over the hill; but these were momentary.

She fretted about her appearance, thin and dark and foreign-looking at a time when every little girl was supposed to look or thought she should look a chubby blonde replica of Shirley Temple. The doorbell sometimes rang for her, but no one seemed to come and play in the house or be a best friend. Maybe because we moved so much.

There was a boy she loved painfully through two school semesters. Months later she told me how she had taken pennies from my purse to buy him candy. "Licorice was his favorite and I brought him some every day, but he still liked Jennifer better'n me. Why, Mommy?" The kind of question for which there is no answer.

School was a worry to her. She was not glib or quick in a world where glibness and quickness were easily confused with ability to learn. To her overworked and exasperated teachers she was an overconscientious "slow learner" who kept trying to catch up and was absent entirely too often.

I let her be absent, though sometimes the illness was imaginary. How different from my now-strictness about attendance with the others. I wasn't working. We had a new baby, I was home anyhow. Sometimes, after Susan grew old enough, I would keep her home from school, too, to have them all together.

Mostly Emily had asthma, and her breathing, harsh and labored, would fill the house with a curiously tranquil sound. I would bring the two old dresser mirrors and her boxes of collections to her bed. She would select beads and single earrings, bottle tops and shells, dried flowers and pebbles, old postcards and scraps, all sorts of oddments; then she and Susan would play Kingdom, setting up landscapes and furniture, peopling them with action.

Those were the only times of peaceful companionship between her and Susan. I have edged away from it, that poisonous feeling between them, that terrible balancing of hurts and needs I had to do between the two, and did so badly, those earlier years.

Oh there are conflicts between the others too, each one human, needing, demanding, hurting, taking—but only between Emily and Susan, no, Emily toward Susan that corroding resentment. It seems so obvious on the surface, yet it is not obvious. Susan, the second child, Susan, golden- and curly-haired and chubby, quick and articulate and assured, everything in appearance and manner Emily was not; Susan, not able to resist Emily's precious things, losing or sometimes clumsily breaking them; Susan telling jokes and riddles to company for applause while Emily sat silent (to say to me later: that was *my* riddle, Mother, I told it to Susan); Susan, who for all the five years' difference in age was just a year behind Emily in developing physically.

I am glad for that slow physical development that widened the difference between her and her contemporaries, though she suffered over it. She was too vulnerable for that terrible world of youthful competition, or preening and parading, of constant measuring of yourself against every other, of envy, "If I had that copper hair," "If I had that skin. . . ." She tormented herself enough about not looking like the others, there was enough of the unsureness, the having to be conscious of words before you speak, the constant caring—what are they thinking of me? without having it all magnified by the merciless physical drives.

Ronnie is calling. He is wet and I change him. It is rare there is such a cry now. That time of motherhood is almost behind me when the ear is not one's own but must always be racked and listening for the child cry, the child call. We sit for a while and I hold him, looking out over the city spread in charcoal with its soft aisles of light. "*Shoogily*," he breathes and curls closer. I carry him back to bed, asleep. *Shoogily.* A funny word, a family word, inherited from Emily, invented by her to say: *comfort.*

In this and other ways she leaves her seal, I say aloud. And startle at my saying it. What do I mean? What did I start to gather together, to try and make coherent? I was at the terrible, growing years. War years. I do not remember them well. I was working, there were four smaller ones now, there was not time for her. She had to help be a mother, and housekeeper, and shopper. She had to set her seal. Mornings of crisis and near hysteria trying to get lunches packed, hair combed, coats and shoes found, everyone to school or Child Care on time, the baby ready for transportation. And always the paper scribbled on by a smaller one, the book looked at by Susan then mislaid, the homework not done. Running out to that huge school where she was one, she was lost, she was a drop; suffering over the unpreparedness, stammering and unsure in her classes.

There was so little time left at night after the kids were bedded down. She would struggle over books, always eating (it was in those years she developed her enormous appetite that is legendary in our family) and I would be ironing, or preparing food for the next day, or writing V-mail to Bill, or tending the baby. Sometimes, to make me laugh, or out of her despair, she would imitate happenings or types at school.

I think I said once: "Why don't you do something like this in the school amateur show?" One morning she phoned me at work, hardly understandable through the weeping: "Mother, I did it. I won, I won; they gave me first prize; they clapped and clapped and wouldn't let me go."

Now suddenly she was Somebody, and as imprisoned in her difference as she had been in anonymity.

She began to be asked to perform at other high schools, even in colleges, then at city and statewide affairs. The first one we went to, I only recognized her that first moment when thin, shy, she almost drowned herself into the curtains. Then: Was this Emily? The control, the command, the convulsing and deadly clowning, the spell, then the roaring, stamping audience, unwilling to let this rare and precious laughter out of their lives.

Afterwards: You ought to do something about her with a gift like that—but without money or knowing how, what does one do? We have left it all to her, and the gift has as often eddied inside, clogged and clotted, as been used and growing.

She is coming. She runs up the stairs two at a time with her light graceful step, and I know she is happy tonight. Whatever it was that occasioned your call did not happen today.

"Aren't you ever going to finish the ironing, Mother? Whistler painted his mother in a rocker. I'd have to paint mine standing over an ironing board." This is one of her communicative nights and she tells me everything and nothing as she fixes herself a plate of food out of the icebox.

She is so lovely. Why did you want me to come in at all? Why were you concerned? She will find her way.

She starts up the stairs to bed. "Don't get me up with the rest in the morning." "But I thought you were having midterms." "Oh, those," she comes back in, kisses me, and says quite lightly, "in a couple of years when we'll all be atom-dead they won't matter a bit."

She has said it before. She *believes* it. But because I have been dredging the past, and all that compounds a human being is so heavy and meaningful in me, I cannot endure it tonight.

I will never total it all. I will never come in to say: She was a child seldom smiled at. Her father left me before she was a year old. I had to work her first six years when there was work, or I sent her home and to his relatives. There were years she had care she hated. She was dark and thin and foreign-looking in a world where the prestige went to blondeness and curly hair and dimples, she was slow where glibness was prized. She was a child of anxious, not proud, love. We were poor and could not afford for her the soil of easy growth. I was a young mother, I was a distracted mother. There were the other children pushing up, demanding. Her younger sister seemed all that she was not. There were years she did not want me to touch her. She kept too much in herself, her life was such she had to keep too much in herself. My wisdom came too late. She has much to her and probably little will come of it. She is a child of her age, of depression, of war, of fear.

Let her be. So all that is in her will not bloom—but in how many does it? There is still enough left to live by. Only help her to know—help make it so there is cause for her to know—that she is more than this dress on the ironing board, helpless before the iron.

1953–1955

The Fraychie Story

Harriet Rosenstein

THERE WERE, IN the beginning, seven children, each rising out of my great-grandmother's darkness every twelve or thirteen months like little full moons, following, even in birth, the quirky Jewish calendar. The sons came first: Mendel, Mischa, Isaac, and Schmuel. And then the pale and round-faced daughters: Sarah, Fraychie, and Sosiesther. My great-grandmother conceived and bore them, I am told, with bemused passivity, as tolerant as the moon must be of her own swellings and thinnings and equally unconscious. Four babies. Seven. A miscarriage or two. The number became finally a matter of indifference. For, of the Jews in Vischay, Lithuania, they were the least poor; my great-grandfather owned the *schvitzbod* (sweatbath), which usually kept the family in kopeks and spared the children the terrible winters. After the sons stoked the low

Harriet Rosenstein lives in Cambridge, Massachusetts, and teaches fiction writing at Tufts University. Her fiction and literary criticism have appeared in many publications, among them *The New Review* (London), *The New York Times Book Review*, *Ms.* magazine, and *Tri-Quarterly*. "The Fraychie Story," copyright © 1974 by Harriet Rosenstein, is reprinted from *Ms.* magazine (March 1974) by permission of the author.

brick bathhouse ovens with firewood for the night, they would stretch out on them to sleep, contented from December through March like no other boys in the village. Once the girls were old enough not to topple, they too were accommodated on the warm brick ledge. They ate well enough, always had a bit of braid or ribbon to trim their coats, and knew that when the peddler passed through town they could count on a whistle or a top from his wagon. All year round plants hung in pots inside the windows and in summer vines grew round the house. It could have gone on indefinitely—summers, winters, whistles, babies—except for two other facts of life: Cossacks and conscription. The Cossacks you've heard about, galloping giants with fine uniforms and swinging sabers. They would fill themselves full of vodka, then plunge through the villages behind the main road, one after the other, never stopping, whacking off Jewish heads as if they were turnips. In less than a minute they had made a clean sweep and vanished. A pastime, a sport. I would not tell you about this without reason; the Cossacks killed the first and last sons of my great-grandparents. Mendel and Schmuel died on the same day, Mendel trying to protect his little brother, whose howling must have enraged the men on horseback.

You know that scientists ignore what is random; only if something happens often enough to be measured will they acknowledge its reality by giving it a name. So it was with the Cossacks. Pretty soon their raids became systematic; they didn't even need the vodka to get them going; their casualties were no longer haphazard. In other words, these massacres were admitted into the world of known phenomena and thus they got a name: pogroms. And my great-grandparents' family suffered still further. Now, like the rest of the villagers, they heard rumors, advance warnings, and attempted to flee the streets and the house and the *schvitzbod* before the sabers came. On one occasion my great-grandmother, murmuring passages from Lamentations as she pulled her apron over her head and her cloak around her daughters, ran to the frozen riverbank to hide. They covered themselves, all four, first with her cloak, then with icy twigs and rushes, and burrowed into the ground, facedown. There they lay from late morning until moonrise, too scared and chilled even to whisper prayers. When it was safe, my great-grandmother, blue and almost paralyzed from the cold, managed to stand up, pulling away the cape and the reeds that shrouded her daughters' backs. All three seemed dead. She massaged and slapped and pounded them in turn, trying to beat the life back into her girls. Sarah and Fraychie survived. Sosiesther's body was purple, rigid as the glassy rushes clinging to her hair. They carried her in silence back to the town.

It was then they decided to come to America. In a year, my great-grandfather figured, he could save enough for steerage fare for his wife and four children. But, hard as it is to believe, the remaining sons were lost. Mischa and Isaac had hair on their faces and the voices of men. Elsewhere these facts might have been neutral; here they meant conscription. Friends of Mischa and Isaac, awaiting the Czar's officer who would carry them off forever, had gone out to the woods with axes and chopped off their own toes to save themselves from induction. Now the sons of my great-grandparents had four choices: to mutilate themselves or to give themselves to the Czar (and between them there was very little difference), to try to escape the country immediately, or to wait out the months till the whole family went, counting on sheer good luck to keep the officer from their door in the meantime. Luck they knew better than to trust. So my great-grandfather bribed a neighboring Gentile farmer to hide Mischa and Isaac under his hayrick and transport them to a border from which they would eventually make their way to America. Their parting was full of tears and promises. My great-grandfather improvised a prayer: "With the help of the Almighty, through Whom all things are possible, may we never again be parted after this parting. May we meet to live in safety in the new country in time to celebrate the next Purim." Then they went round the room, one by one, repeating my great-grandfather's words. Mischa and Isaac embraced their little round-faced sisters; their father, weeping, embraced the sons now taller than himself; but their mother could only sob without looking up—mourning her boys while they still stood in her kitchen. Finally, with purses tucked inside their sashes, Mischa and Isaac walked away from their parents and sisters in the middle of the August night. The farmer did his job. But Mischa and Isaac were never heard from.

When they were allowed to leave Ellis Island with the Anglicized name the official there had given them, the family went immediately to a Jewish agency. Before even asking for temporary shelter, they asked for their sons. Had Mischa and Isaac from Vischay in Lithuania, tall boys with thick black hair and gray-green eyes, seventeen and sixteen years old, come or written here in the past five months to say where they were and what name they now had? The lady at the desk spoke good Yiddish. The boys, she said, had not yet come; perhaps they were registered with a bureau in another part of the city or perhaps they were living in a different town. She would investigate. It would take time; there were many immigrants looking for relatives; surely, though, Mischa and Isaac would be found. And for seven years, from the time that my great-grandparents settled into a two-room flat, through the time that my great-grandfather, always a modest entrepreneur, opened a tiny dry-goods shop whose best customers were black and for whom, in his hoarse, cracked English, he would sing little package-wrapping songs—"Sam Brown he went to town," "Bill Jones he boughten cotton" —the family searched for Mischa and Isaac. They put ads in the *Forward;* every Jewish service bureau in the country and most of the Orthodox synagogues in the East knew to watch for the young men. At the end of the seven years, against the wishes of her husband, my great-grandmother lit *yahrzeit* candles (memorial candles for the dead) for her two lost boys. She, whose relation to maternity had been so passive, now took command of mortality. There had been enough waiting, she told him.

This was the story that Sarah, my grandmother, told me as I grew up. She told me it in summer as we spooned cold spinach borscht from tall glasses and in winter as we transferred her gold-brown *challahs* from the oven to a long mahogany sideboard in the dining room. There, lined up on the special sabbath cloth she had herself embroidered, they would cool. And we would talk. About the *shtetl* (Eastern European peasant village), about her brothers and her parents and her husband, all dearly loved, all dead. What she never talked about was her sister Fraychie. And I never dared to ask. Even my mother, her mother's favorite, could say very little. She knew only that there had been some rivalry or antagonism between the sisters when they were twenty or so. That they had argued bitterly and vowed never to do so much as nod at each other again. Their parents, by then well enough established, able to afford a house with two maple trees in the back and a toilet on the second floor, were helpless before their daughters. They pleaded; they scolded; they invoked the memories of the children who were dead and the mystery of those who were lost. But it was useless. The way they stared through each other, the sisters might as well have been windows.

My grandmother met a man named Sammy, an immigrant with a good heart who worked long hours selling wholesale produce at the market, who courted her with fresh peaches and my great-grandmother with luxuriant bouquets of beets. Even Sammy was enlisted in my great-grandparents' campaign to open the mouths and eyes of their daughters. He brought them green grapes, nectarines, melons, the sweetest of sweets. He might as well have brought lemons. When the couple became engaged, it seemed certain that the sisters would make up their differences, that love would soften Sarah and that she would ask Fraychie to be her bridesmaid. Perhaps Sarah tried. Perhaps Fraychie was envious. All my mother could tell me—and that she had pried from her grandparents when she was small—was that on the day Sarah and Sammy were to be married, Fraychie packed a handgrip with some underwear and a shift or two and she left. My great-grandparents went out of their heads with mortification and grief but the ceremony still took place. They lived long enough to receive a few cards from Fraychie, always postmarked from somewhere in the state, and to receive their first grandchildren, my uncle Willy and my mother, Annie. Sarah, I am told, refused even to utter her sister's name because she had cast such a blight on the family and, in particular, on her wedding day.

Yet she was not otherwise bitter. She had loved her husband with joy and pride and, like her own mother, had produced seven children; they had the good fortune to live. She was devout. She never entered her bedroom without first kissing her fingers and pressing them to the mezuzah (a parchment, bearing twenty-two lines from the Bible, rolled in a metal or wooden case) on the doorpost. Then she would station herself over the heat register to undress. (I remember her always standing on some heat register or other. Her circulation had gone bad, perhaps because of the frostbite she had suffered as a girl. Whatever the reason, she was always cold, she never sat—drawn to warm air with the fidelity of a migrating bird. Whether it flowed out of the oven or up from the register, there she would alight; so when she was not baking, she was leaning against a wall over an air duct. And, like the birds, she left tokens of habitation. Her resting places were marked for all time by the grease spots that her behind eventually pressed on walls above every register in her house.) So—my grandmother Sarah, braced by her grease spot, soothed from the bottom up by so much heat, ended the day by uttering prayers from the instant she started unlacing her corset until she had pulled the covers up to her chin.

I used to watch this performance in amazement. The dress and the nightgown traveled a parallel course, the dress stopping at her knees, the nightgown at her neck. Her head covered in layers of flannel, her fingers seemed, nonetheless, to have eyes of their own. They would unlace the corset—it seemed always to be the same corset, although she must have had many of the same pale orange color, the same stays made of whalebone, the same terrible dimensions. This corset stretched from the invisible underside of my grandmother's breasts to the equally invisible point where her hips met her thighs. Her flesh was invisible because beneath the corset she wore one of her husband's cotton undershirts, the kind with the skinny shoulder straps, cut in deep ovals front and back, designed for sweaty summers. But she wore them always, her great breasts swaddled in cotton, hoisted over the top of the corset, and then left to plunge disastrously downward. In such a fashion my grandmother retained her waistline and her vanity and kept her naked body forever a wonder and a secret to me. The corset undone, it slipped down at the exact rate as her dress below and her nightgown above—a perfect orchestration. Once this operation was completed, she managed, because her flowered flannel nightgowns were invariably enormous, to get the undershirt off while keeping the nightgown on. Throughout it all, she would pray like a woman possessed. Since I never learned Hebrew, I had no idea what she was talking about, and in my first years assumed that her sounds and the ritual of uncorseting had some mystical connection. Without this strange language, perhaps, there would be a failure: the dress too soon, the gown too late, the breasts that symbolized for me all the mysteries of the universe triumphantly, enormously revealed. But the prayers always worked. She would lie beside me in the cool double bed, praying still, and I would feel her flannel backside, and wonder.

I slept with her on those evenings that my parents deposited me in her house because her bed was half-empty. Her husband had died when I was only three. He had had great success with his fruits and vegetables and, my grandmother assured me, was very like her own father. He too would chant little ryhmes to his customers—a trick he had picked up to improve business and continued because it pleased him so. To his favorite black lady he would sing, "Esther Green, you are a queen," and his poetry made him loved, gave him honor. No one left his living room without a bag of apples and, when times were difficult, a bag of potatoes, too. As generous with his love as with his produce, he literally sang my grandmother's praises, even unto her pale orange corsets. If business was slow, he wrote courtly stanzas in black marking crayon on the backs of brown-paper sacks; her eyes dark as raisins, her skin white as milk, her face round as a melon, her mouth red and plump as a plum. It was grocer's verse. The grammar was bad. But my grandmother kept the brown bags. Of his children he was no less adoring: his three daughters were princesses and his four sons, every one a prince. He paid them tribute in rhyme, in kisses, in embraces and, one winter, in a bounty worthy of Romanoffs. At that time the youngest, Nomi, was three, and Willy, two years past Bar Mitzvah. The snows had been as dense, my grandmother said, as those she had endured in the old country, and the winds savage as a Cossack's blade. So my grandfather left his place early one afternoon and went to visit a friend, a wholesaler in furs. He came home with seven identical coats, thick-piled and white like Siberian bears, with matching muffs for the girls and flat-topped, steepsided hats for the boys. "Let the winter find us now!" he had laughed. That story I knew as a duet. My grandmother and my mother loved to tell it, my grandmother describing the opening of the boxes, my mother the trying on of the coats, my grandmother the hurtling of the children out of the house and into the twilight like oversize snowballs, and then, in unison, my grandfather's delighted cry.

The mourning for him, they told me, had been tremendous. More than a thousand people came to his coffin and to the home of his family, not because he was rich or powerful, but because he treasured life, because he made up little poems. People like Esther Green traveled from the other end of the city simply to let his wife know that they had cherished him.

My grandmother stayed on in the house for another ten years, no public figure, but a presence still. Small, pale, round-faced, even in old age remarkably beautiful. Her hair was pure white, pulled into a smooth circle at the base of her neck. Just leaning against the wall, despite the secret chill her bones held, she seemed a heat source in herself: her quick, dark eyes, her arms and fingers moving emphatically, then subtly, then quizzically to illustrate her stories or her meditations, her nostrils always flaring as if emotions had scents that she alone could recognize. Standing in her living room, surrounded by her children and her children's children, all of us slack in our easy chairs, she was the center, the life from which all this life had sprung. She knew more Torah than her sons, more prayers than her daughters, more Talmud than her rabbi. She knew from childhood the worst of loss and grief and terror and therefore did not fear them, would not capitulate to them. She knew that anger and love were next-door neighbors and with equal passion gave both to others. Women brought woes and confusions to her as they would to a Solomon and they left her light and easy, like the leaves after a rain. What she believed she believed absolutely; what she doubted she doubted silently, consigning her uncertainties perhaps to the cold places that hid inside her bones. When she laughed it was with such force that her undershirt and corset combined could not stop her breasts from laughing too.

At her death I wept for days. It was the springtime. She was beginning to set out flowers and hanging plants on her front porch, preparing to leave her place by the wall and to move outside where the warmth was real. Once *shivah* (seven-day mourning period) was over, her daughters sorted her possessions. They found my grandfather's crayoned love poems and found, too, poems that their mother had written to him. Over these they wept with surprise: they had lost more through her death than they had ever known they possessed. They found old photographs, old documents, newspaper advertisements for Mischa and Isaac. They placed in rows her corsets, her nightgowns, her aprons, her dresses—each daughter taking something as a remembrance. And they found a handful of yellow postcards going back almost fifty years. The cards were addressed to their grandparents and were signed by Fraychie, their mother's sister. It was decided first by my mother and my aunts Dora and Nomi, and then agreed upon by my uncles Willy, Norman, David, and Daniel, that Fraychie should be found. That, for the sake of rightness and reconciliation, she should be informed of her sister's death. None of them knew whether Fraychie had married and thus what name she went by. She could be anywhere. She could be dead. With nothing more than these yellow cards to go on, they set about investigating. God knows how they did it, but what they finally came up with was this: there was a woman with the strange name of Fraychie who lived on a small farm in the north of the state where she kept chickens and where she sold eggs. In all likelihood, that Fraychie was their mother's sister.

By now it was the middle of May; the trees had leapt overnight from that sweet sad yellow-green of budding into the strong deep green of leaves who know their rights. And we assembled, Sarah's seven children, and their husbands and wives and children, crowding my grandmother's porch the way the leaves were crowding the trees. My uncle Willy with his family was to lead the caravan and in our cars, one after another, we were to travel to that place where Fraychie lived. Among my parents, my sister, and me there was more silence than talk, more staring than looking. For three hours we drove, going higher, climbing hills separating impossible pastures where the cows tilted and the boulders showed, passing villages that lasted for a minute then disappeared, and then, one by one, the cars slowed and turned and stopped. My oldest uncle walked into a general store and came out nodding. And everyone in the seven cars nodded back. Again he started out and again we followed him. Past the hamlet, down a dirt road that cut in half a forest, the first forest I had ever seen. Through these woods, where suddenly it was chill and damp and musky and dark, we slowly moved, finally halted. There were thirty of us, going on tiptoe over the dead leaves till we left the forest and the quiet it had imposed on us, and found the sun once more. Before us was a tiny house, whitewashed, slate-roofed, with vines that were just beginning to color. Nobody had planned what to do when we arrived; we simply stood there with the sun on us and the noises of chickens coming from someplace unseen.

Then from that place, behind the house, walked a woman carrying a crate of eggs. The woman could not have been Fraychie because she was my grandmother. The same face, the same eyes, the same hair, even the same flaring nostrils. And when she smiled, it was my grandmother smiling, inviting me and my sister in for borscht and tales of the old country. She was no more surprised at this congregation than my grandmother was to see us arrive for Pesach or Purim. And this woman—Fraychie, Sarah, whoever she was—opened her mouth and out of it came my grandmother's voice and, more mysterious still, my grandmother's words. She looked at us and she said, "You are Willy, yes? And you must be Annie. And you, with such a grin on you, you are Daniel." And this woman went on, identifying each of her nieces and nephews, and then, when I had stopped breathing, each of us, their children. She asked us into her house and we went. Into her small living room whose walls were covered with photographs of bearded men and beardless boys and corseted women and skinny girls and these were the photographs of my grandmother's stories. The vivid lines her hands had drawn in telling them were fixed and mounted here. The woman brought us tall glasses of tea. She smiled. The conversation was sweet and delicate. She knew us and all about us. And there was no question of knowing her. We had known her all of our lives. I knew the prayers she said at night and the way she got into her nightgown. I knew the way she absorbed and erased the suffering of anyone who brought it to her. I knew the way she kneaded dough and grated beets till her fingers bled. I knew her pallor and the beauty of her white hair when she let it down to brush before bed. As we sat on the floor and in chairs in that little room while she stood there talking to us, all my grief at my grandmother's death left me. My senses said what was as obvious as the sun. Here was Sarah, going by the name of Fraychie, merely transported, tending chickens, cupping eggs in her palms, living in a whitewashed house at the edge of a forest.

For two hours we sat in her house. She explained that she had been told of her sister's death, as she had of the births of her nieces and nephews and of their children by an old friend of a still older relative. There was nothing new to say. There was nothing old to question. There had never been a breach and there was now no reason to heal it. As quietly, as sweetly, as we had begun, we rose to end our talk. We filed out of her house and into the cars smiling, touching her, kissing her, feeling her lips on our cheeks, her fingers on our hair, watching her grow smaller as we backed down the dirt road, still beautiful, still circled by sun, still holding one arm up whether in greeting or good-bye I could not have said, until the woods and the road were gone and we faced the known direction home.

Papa's Tea

Irena Narell

PAPA LIKED HIS tea strong and piping hot. He would come into the kitchen, his dignified, fastidious figure in utter contrast to the disorder that reigned all over the room. With a helpless glance at the large kitchen table on which, as usual, lay an infinite variety of objects of non-culinary character, he would ask:

"Well, where is Mama? I'd like some tea."

Where was Mama? Where was the wind, he could just as well have asked. She had flown off an hour, maybe two hours ago, no one knew precisely when. Where was she off to? The bakery maybe or the dry goods store? She was going to look at that piece of blue wool for my sister Bessie's new coat, so maybe that's where she went . . .

"Max, where did Mama go?"

Irena Narell is the author of *Joshua: Fighter for Bar Kochba*, winner of the 1979 National Jewish Book Award. The Polish-born historian, novelist, translator, and lecturer came to the U.S. in 1939. She is now working on a study of San Francisco's early Jewish settlers and speaks frequently on California Jewish history. "Papa's Tea," copyright © 1969 by Irena Narell, was first published by Brook Press of New Jersey.

"I don't know," came a disgruntled reply from my brother. "Don't bother me."

"Oh, for a glass of tea," sighed my father. "One glass, what am I saying? I could drink a dozen glasses."

What was the matter with Mama? She never could remember the time Papa was due home. Even if she did, she was sure to find some errand that couldn't wait and forget all about him.

Not that there was anything personal she had against Papa. Oh, no. Didn't she often forget when we were supposed to come home from school for a quick, half-hour lunch, and didn't we end up crying in the street dozens of times until some neighbor would feel sorry for us and let us in? And when we would berate her afterwards, she would say:

"I know, I know. I only went shopping for a short while. So why couldn't you wait five minutes?"

That was Mama. You couldn't change her and you couldn't make her understand certain simple facts of life, I thought, as I went into the kitchen, determined to take care of Papa. One just had to accept Mama as she was. And she was wonderful in so many ways! When she sang near the open window, it was as if all the sound and glory of a spring morning in the country had found its way into our Bronx apartment. The neighbors would hang their heads out of the windows to listen and comment:

"Mrs. Siegel, how you can sing! Like a regular opera star."

And Mama was always smiling. Sometimes there really wasn't anything to be cheerful about, but Mama invariably found some little thing to make her happy. Maybe it was a concert in the park, or a ticket to the opera, or the neighbor's fat new baby, or the pretty dress she had just made for me with her beautiful, clear stitches. I often wondered how a person so untidy in her household could sew so perfectly and be so painstakingly neat in this work. And she never, never hit us.

Her mind was always on beautiful things. And how she loved pretty clothes! Next to music, I guess she loved clothes just about the best. When the tailor got through making her a handsome new suit, Mama would go out and buy a twenty dollar silk blouse to wear with it.

We all thought she looked simply stunning. She was a little thing with a lovely head of brown hair, an elfish face, and a good figure, despite her lack of height. She carried her clothes well. But what happened to these clothes when she took them off!

Papa was fond of saying:

"Mama needs nice clothes so she can keep them under the kitchen sink."

And he wasn't far from wrong. Her things were everywhere except in the closet where they should have been. Once she got it into her head that she wanted a mink stole. Papa only said:

"Mama wants a mink stole so she can keep it underneath the washtub." And that was the end of that.

In the kitchen, I put my arms around Papa's neck.

"How is my Lily today?" he said, and smiled in that special way he reserved only for me. Even though there were four of us children, Papa's preference for me was acknowledged by the whole family and resented most by my brother Max.

"Papa's darling!" he would jeer at me when he got good and mad.

"Go in the living room, Papa," I said. "I'll make you some tea."

"Thank you, Lily dear, that would be very nice," Papa answered and he walked out.

The kitchen table was a real mess! There was the fabric for my skirt, and the pieces for Bessie's new hat, and two pots from last night's supper, and seven dirty spoons and forks, and the cookie jar, and the pot of jam from breakfast, and last Sunday's paper! The disarray was not exactly out of the ordinary. Only the items changed from time to time. Even when Mama happened to be home the table was so piled up with bundles that there was barely enough room for a plate or two.

I took off the two pots and put them in the sink and began moving the other things, one by one. For once Papa would have a completely clear table to drink his tea!

I found a pretty plate with rose clusters to put under the glass, and a white napkin. Where was there a clean spoon? I got the kettle going and prepared the tea. I took Mama's bundle of fabrics to her bedroom, the "dump" room. This was where we hastily dumped everything that lay all week long on chairs and couches and the floor, when a visitor was announced. Why did the house always have to be so disorderly? It would look exactly the same when we came back from school as when we had left it in the morning.

"But Mama, I'm ashamed to even bring a friend home," I would complain.

"If you don't like it, stay home and do it yourself. I got too many other things to do."

"But you know I have to go to school, Mama."

"So, is that my fault?"

What could you say?

The kettle was whistling. I ran back to the kitchen. What was that Papa had said?

"I could drink a dozen glasses." That was it!

He shall have a dozen glasses then! My heart was bursting with love for Papa who was good and kind. I knew how determined he was that we, his children, should have a better life than he himself had led, and how hard he labored to make it come true.

Papa's father was an innkeeper in a Russian town. His first wife, Papa's mother, came from a wealthy family and stayed with him just long enough to bear him two sons. Then she divorced him and married the town mayor. My grandfather promptly married a widow with two little girls. One of those girls was Mama! Grandfather was of sturdy stock and proceeded to outlive this wife of his, and then married twice more, each time outliving his much younger wives. Finally there were seventeen children in the house, and Papa, as the oldest, bore most of the responsibility for them. Mama was a very determined little girl even then. As Papa told it:

"When we were children, Mama decided that we would get married some day. As far as I was concerned, I could have stayed single all of my life!"

Mama had a good voice and wanted to go on the stage but instead was apprenticed to a dressmaker. Still, she managed to sneak out every now and then and sing in cabarets, until her stepfather would find out and drag her home virtually by the hair. But did this stop her? No. At the very next opportunity she was out, like a spark of flame, dancing, singing, enjoying herself. Life beckoned to Mama with irresistible force.

Papa felt responsible for her even then. For her, and the whole brood of children, until he was drafted into the czar's army. This was a calamity and Mama was in despair. She saved money from her earnings as a seamstress and it was arranged, over Papa's protests, to have him smuggled over the border the first time he came home on leave. Mama took care of getting him a passport and a change of clothes, bribing whomever was necessary, but Papa resisted to the very last. It wasn't legal, he said. It was not the proper way of doing things and Papa was a very proper person. Well, they got him over the border somehow and then onto a boat bound for England where an uncle lived.

There Papa who wanted to be an artist became a house painter so that he could earn a living. As soon as he had mastered the trade he was shipped off to America with a letter of recommendation to a "lantzman" on the Lower East Side of Manhattan. He worked and saved and sent for Mama. She too moved into a "lantzman's" railroad flat and went to work, sewing in a factory.

They didn't get married right away. How could they afford it, with so many brothers and sisters back home to send for? First, there was Aunt Esther. As soon as she came she fell in love with a cousin and the foursome worked to bring more relatives to America. Next came my Aunt Minnie. But that was after Mama and Papa got married and moved uptown to a five room apartment. Aunt Esther and Uncle Abe also moved in with them, and when Minnie came, naturally, she moved in as well.

Papa said to Minnie:

"Now it's your turn to go to work and send for your sisters."

But my Aunt Minnie wasn't that altruistic. Besides, she hated to work.

"I should work for them?" she said. "To hell with that!" Aunt Minnie didn't mince words.

"But Minnie, how can you behave like that?" Papa said. "Didn't we send for you?"

"So what!" was her answer. "I don't feel like it."

"Well, if that's the case, then maybe we'll let your sisters stay back in the old country," said Papa.

Aunt Minnie finally went to work for about a year but hated

it so much that as soon as Uncle Sam proposed to her, she married him, although she was only sixteen and he over twenty-five. From that moment on, she wiped her hands of her family in Russia. It took years for Papa to save enough money to try to bring more of them over, what with his own growing family to support.

By that time it was very difficult to locate them. They were scattered all over what was now the Soviet Union and had families and roots of their own. Besides, the government made it impossible for them to leave the country. Papa never saw them again.

I was rummaging in Mama's cupboard, trying to find twelve clean glasses. Then I tiptoed to the living room to steal a look at what Papa was doing. I didn't want him to discover my surprise a moment too soon!

Papa was sitting with his dictionary and a pad, writing down his daily quota of words. Papa read a lot, mostly Jewish books and Jewish newspapers. But he also made it a point to study Webster's dictionary, increasing his vocabulary by at least three words each day. He would write the definitions painstakingly on a little pad and never forget them.

"Why aren't you a businessman like Uncle Hymie, Papa?" I remember asking him once. Uncle Hymie lived in Brooklyn and had a fancy car.

"You really are so clever, Papa. You could make a lot of money, I'm sure!"

"Because, Lily darling, I'm just not made to be a businessman. A businessman, he has to be a little bit crooked. Don't let anybody tell you it isn't so. You can't make money in business being a hundred per cent honest. But your uncle Hymie, he is made to be a businessman. Even when we were children together in Russia, I knew he would be a businessman one day."

"Why, Papa?"

"Well, I'll tell you. On Saturdays when our father would go to shule, your uncle Hymie would set up a little business of his own. There was a window at the inn that went out onto a side street. And under this window was a high, long table. Your uncle Hymie would have a show right under that window, on top of that table, and he would charge money for it. A kopeck apiece

to all the children in our town, and they could see the show from
the outside. And business was good! Sometimes he would make
a whole ruble on a Saturday!"

"But how, Papa?"

"Well, he would just pull down his pants and show his behind
to the other children for a kopeck, that's how!"

"No, Papa!" I collapsed, convulsed with laughter.

"So, you see, Lily, that's how I knew he would be a businessman," Papa concluded triumphantly.

Here were the glasses at last. I counted an even dozen. Some
of them didn't look any too clean, so I took them down from the
shelf and reached under the sink for soap. I pulled out Mama's
skirt instead, the one with the front pleat. She had only had it
a month! Oh, well, the soap must be around somewhere. There
was Bessie's slipper that she'd been looking for all week! Here
was the soap, finally. I washed the glasses with care, making
them shine for Papa. Where could Mama be all this time? I bet
she went to the movies. There was a new musical at the Prospect, that's where she must be! Mama and her movies!

I smiled at the memory of all the movies she and we had seen
together. We, the children, I mean. You see, Papa works very
hard and during the week he likes to be in bed early. Nine
o'clock and he is snoring away. But not Mama. Night does
things to her. Her eyes begin to shine, her little feet begin to
dance with impatience. She must be off and going somewhere.
Sometimes I think that only with the advent of night does Mama
really begin to realize fully her fierce joy of living. And it is not
until the early hours of the morning that she begins to lose her
nocturnal sparkle and makes ready to surrender herself to
death-simulating sleep.

Well, the four of us children have to be put to bed. That is,
four of us since a year and half ago, when Claire was born. Until
then there was just Max, Bessie, and I. Papa needs his rest and
Mama must go out to see a movie. Since we children make a lot
of noise all alone by ourselves, Mama worked out a perfect
compromise. She'd take us along!

As soon as Papa was asleep Mama would dress us all, including Claire, the baby, and off we would be, to the movies. We

would settle ourselves as comfortably as we could, for we were good for the night. Mama had to see the picture several times over and often we would not go home until the movie house shut down for the night. Mama watched the movie and we slept until it was time to go home. She even got these nightly excursions down to a science. We would be wearing our pajamas underneath our clothes so that we could jump right into bed when we got home. Mama would open the door as quietly as she could, and would warn us as we staggered into bed:

"Shh! Don't wake Papa!"

It was nearly three years before Papa caught onto the fact that his whole family was missing, night after night. Claire at that point had been going with us for over a year. It seems that he had an attack of indigestion and got up at about 11 P.M., looked for Mama, then took a peek into our rooms. Nobody home. Well!

When we tiptoed in that night who should be sitting in the kitchen but Papa! What a rumpus there was! I have never seen Papa so mad. He took a plate and broke it right before our eyes! We all hid but we could hear his roaring.

"The children have to go to school! What do you mean by keeping them out so late? And a baby, too!"

He went on like this, fuming and shouting for quite a long time. Mama just sat there.

Finally she said, calm as calm could be:

"You're finished?"

"Yes, I'm finished!" Papa roared.

"What do you expect me to do? I'm not going to stay home night after night. If you want, I'll leave them home."

And that's just what she did. From then on our nightly escapades were over. All went well until one night Claire had a fit. She demanded Mama's presence and Papa couldn't calm her down. Finally, exasperated and worn to a frazzle, he gave her a good spanking. The next day, sheepishly, he said to Mama:

"From now on, you take *her* along." So Claire was now the only one of us children qualified to be a movie critic.

I had the gleaming glasses neatly arranged on the table. An even dozen! I brought over the sugar bowl, a little lemon. That's how Papa liked it. The tea smelled delicious. I poured it carefully, and didn't spill a drop.

"Papa, your tea is ready!" I called.

As he came into the kitchen, I pointed proudly.

"A dozen glasses, just like you wanted."

Papa looked surprised.

"But Lily, darling, I didn't really mean ..."

He reached for his eyeglasses and his handkerchief, and wiped them. Then he took me around and began to laugh. It was a happy kind of laughter, and I was relieved, because for a moment I thought he was going to cry. I guess he was pleased with his tea. But instead of drinking it, Papa kept laughing and laughing. I don't remember when I had heard Papa laugh like that before.

My, it was good to hear Papa laugh!

What Must I Say to You?

Norma Rosen

WHEN I OPEN the door for Mrs. Cooper at two in the after-noon, three days a week, that is the one time her voice fails us both. She smiles over my left shoulder and hurries out the words "Just fine," to get past me. She is looking for the baby, either in the bassinet in the living room or in the crib in the baby's room. When she finds her, she can talk more easily to me—through the baby. But at the doorway again, in the early evening, taking leave, Mrs. Cooper speaks up in her rightful voice, strong and slow: "I am saying good night." It seems to me that the "I am saying" form, once removed from herself, frees her of her shyness. As if she had already left and were standing in the hall, away from strangers, and were sending back the message "I am saying good night."

Maybe. I know little about Mrs. Cooper, and so read much into her ways. Despite the differences between us, each of us

Norma Rosen's work has appeared in *The New Yorker, Commentary, Ms.,* and other periodicals. She received a Radcliffe Institute fellowship and a New York State Creative Artist Public Service grant. Rosen is the author of *Joy to Levine, Green,* and *Touching Evil.* "What Must I Say to You?" copyright © 1963 by Norma Rosen, originally appeared in *The New Yorker* magazine, October 26, 1963.

seems to read the other the same—tender creature, prone to suffer. Mrs. Cooper says to me, many times a day, "That is all right, that is all right," in a soothing tone. I say to her, "That's such a help, thank you, such a help." What can I guess, except what reflects myself, about someone so different from me?

Mrs. Cooper is from Jamaica. She is round-faced and round-figured. She is my age, thirty, and about my height, five-five. But because she is twice my girth (not fat; if there is any unfavorable comparison to be drawn, it may as well be that I am, by her standard, meager) and because she has four children to my one, she seems older. She is very black; I am—as I remember the campus doctor at the women's college I attended saying—"surprisingly fair." Though, of course, not Anglo-Saxon. If you are not Anglo-Saxon, being fair counts only up to a point. I learned that at the women's college. I remember a conversation with a girl at college who had an ambiguous name—Green or Black or Brown. She said in the long run life was simpler if your name was Finkelstein. And I said it was better to be dark and done with it.

Mrs. Cooper has been coming to us, with her serious black bulk and her beautiful voice, for some months now, so that I can get on with my work, which is free-lance editorial. The name is lighthearted enough, but the lance is heavy and keeps me pinned to my desk. Mrs. Cooper's work, in her hands, seems delightful. Though she comes to relieve me of that same work, it is a little like watching Tom Sawyer paint a fence—so attractive one would gladly pay an apple to be allowed to lend a hand. Even the slippery bath, the howls as my daughter's sparse hairs are shampooed, become amusing mites on the giant surface of Mrs. Cooper's calm. They raise Mrs. Cooper's laugh. "Ooh, my! You can certainly sing!"

I sneak from my desk several times an afternoon to watch the work and to hear Mrs. Cooper speak. Her speech, with its trotty Jamaican rhythm, brings every syllable to life and pays exquisite attention to the final sounds of words. When she telephones home to instruct the oldest of her children in the care of the youngest, it is true that her syntax relaxes. I hear "Give she supper and put she to bed." Or "When I'm coming home I am going to wash the children them hair." But the tone of her voice is the same as when she speaks to me. It is warm, melodious. Always the diction is glorious—ready, with only a bit of memorizing, for Shakespeare. Or, if one could connect a woman's voice with the Old Testament, for that.

"God is not a God of confusion." Mrs. Cooper says that to me one day while the baby naps and she washes baby clothes in the double tub in the kitchen. I have come in to get an apple from the refrigerator. She refuses any fruit, and I stand and eat and watch the best work in the world: rhythmic rubbing-a-dubbing in a sudsy tub. With sturdy arms.

She says it again. "God is not a God of confusion, that is what my husband cousin say." A pause. "And that is what I see."

She washes; I suspend my apple.

"It is very noisy in these churches you have here." She has been in this country for three years—her husband came before, and later sent for her and the children, mildly surprising her mother, who had other daughters and daughters' kids similarly left but not reclaimed—and still she is bothered by noisy churches. Her family in Jamaica is Baptist. But when she goes to the Baptist church in Harlem, she is offended by the stamping and handclapping, by the shouted confessions and the tearful salvations. "They say wherever you go you are at home in your church. But we would never do that way at home."

She lifts her arms from the tub and pushes the suds down over her wrists and hands. "But I will find a church." The purity of her diction gives the words great strength. The tone and timbre would be fitting if she had said, "I will build a church."

Again she plunges her arms in suds. "Do you ever go," she asks me, "to that church? To that Baptist church?"

Now is the time for me to tell her that my husband and I are Jewish—and so, it occurs to me suddenly and absurdly, is our three-month-old daughter, Susan.

It is coming to Christmas. I have already mentioned to my husband that Mrs. Cooper, who has said how her children look forward to the tree, will wonder at our not having one for our child. "I don't feel like making any announcements," I tell my husband, "but I suppose I should. She'll wonder."

"You don't owe her an explanation." My husband doesn't know how close, on winter afternoons, a woman is drawn to another woman who works in her house. It would surprise him to hear that I have already mentioned to Mrs. Cooper certain intimate details of my life, and that she has revealed to me a heartache about her husband.

"But I think I'll tell her," I say. "Not even a spray of balsam. I'd rather have her think us Godless than heartless."

My husband suggests, "Tell her about Chanukah"—which with us is humor, because he knows I wouldn't know what to tell.

Mrs. Cooper stands before my tub in the lighted kitchen. I lean in the doorway, watching her. The kitchen window is black. Outside, it is a freezing four o'clock. Inside, time is suspended, as always when the baby sleeps. I smell the hot, soaped flannel, wrung out and heaped on the drainboard, waiting to be rinsed in three pure waters. "We don't attend church," I say. "We go—at least, my husband goes—to a synagogue. My husband and I are Jewish, Mrs. Cooper."

Mrs. Cooper looks into the tub. After a moment, she says, "That is all right." She fishes below a cream of suds, pulls up a garment, and unrolls a mitten sleeve. She wrings it and rubs it and plunges it down to soak. Loving work, as she performs it—mother's work. As I watch, my body seems to pass into her body.

I am glad that my reluctance to speak of synagogues at all has led me to speak while Mrs. Cooper is working. That is the right way. We never, I realize while she scrubs, still seeming to be listening, talk face to face. She is always looking somewhere else—at the washing or the baby's toy she is going to pick up. Being a shy person, I have drilled myself to stare people in the eyes when I speak. But Mrs. Cooper convinces me this is wrong. The face-to-face stare is for selling something, or for saying, "Look here, I don't like you and I never have liked you," or for answering, "Oh, no, Madam, we never accept for refund after eight days."

The time Mrs. Cooper told me her husband had stopped going to church altogether, she was holding Susan, and she uttered those exquisite and grieved tones—"He will not go with me, or alone, or at all any more"—straight into the baby's face, not mine.

Mrs. Cooper now pulls the stopper from the tub and the suds

choke down. While she is waiting, she casts a sidelong look at me, which I sense rather than see, as I am examining my apple core. She likes to see the expression on my face after I have spoken, though not while I speak. She looks back at the sucking tub.

When Mrs. Cooper comes again on Friday, she tells me, as she measures formula into bottles, "My husband says we do not believe Christmas is Christ's birthday."

I, of course, do not look at her, except to snatch a glance out of the corner of my eye, while I fold diapers unnecessarily. Her expression is calm and bland, high round cheekbones shining, slightly slanted eyes narrowed to the measuring. "He was born, we believe, sometime in April." After a bit, she adds, "We believe there is one God for everyone."

Though my husband has told me over and over again that this is what Jews say, Mrs. Cooper's words move me as though I have never heard them before. I murmur something about my work, and escape to my desk and my lance again.

Mrs. Cooper has quoted her husband to me several times. I am curious about him, as I am sure she is about my husband. She and my husband have at least met once or twice in the doorway, but I have only seen a snapshot of her husband: a stocky man with a mustache, who is as black as she, with no smiles for photographers. Mrs. Cooper has added, in the winter afternoons, certain details important to my picture.

Her husband plays cricket on Staten Island on Sundays and goes on vacations in the summer without her or the children, sometimes with the cricketers. But to balance that, he brings her shrimps and rice when he returns at 1:00 A.M. from cricket-club meetings on Friday nights. His opinion of the bus strike in the city was that wages should go up but it was unfair to make bus riders suffer. About Elizabeth Taylor he thought it was all just nonsense; she was not even what he called pretty—more like skinny and ugly.

In most other respects, it seems to me, he is taking on the coloration of a zestful America-adopter. There are two kinds of immigrants, I observe. One kind loves everything about America, is happy to throw off the ways of the old country, and thereafter looks back largely with contempt. The other kind dislikes, compares, regrets, awakens to *Welt-* and *Ichschmerz* and feels the new life mainly as a loss of the old. Often, the two marry each other.

Mr. Cooper, though he still plays cricket, now enjoys baseball, the fights on television, his factory job and union card, and the bustle and opportunity of New York. I mention this last with no irony. Mr. Cooper's job opportunities here are infinitely better than in Jamaica, where there aren't employers even to turn him down. He goes to school two nights a week for technical training. He became a citizen three years ago, destroying his wife's hopes of returning to Jamaica in their young years. But she dreams of going back when they are old. She would have servants there, she told me. "Because there aren't enough jobs, servants are cheap." Her husband, in her dream, would have a job, and so they would also have a car. And a quiet, gossipy life. She likes to move slowly, and this, as she herself points out, is very nice for my baby.

Christmas Week comes, and we give Mrs. Cooper presents for her children. And since Christmas Day falls on the last of her regular three days a week, we pay her for her holiday at the end of the second day. "Merry Christmas, Mrs. Cooper," I say. "Have a happy holiday."

Mrs. Cooper looks with interest at the baby in my arms, whom she had a moment before handed over to me. Suddenly she laughs and ducks her knees. Her fingers fly with unaccustomed haste to her cheek and she asks, "What must I say to you?"

"You can wish me the same," I say. "We have a holiday. My husband gets the day off, too."

I am glad that Mrs. Cooper has not grown reticent, since her embarrassment at Christmas, in speaking to me of holidays. Soon she is telling me how her children are looking forward to Easter. The oldest girl is preparing already for her part in a church play.

I fuss with the can of Enfamil, helping Mrs. Cooper this way when what I want is to help her another way. "Will your husband come to the play?" I ask casually.

"I am not sure," she says. After a while, "We haven't told him yet." Another little while. "Because it seems also he is against these plays." Then, with just enough of a pause to send those tones to my heart, she says, "I think he will not come."

Because the Judaeo-Christian tradition will have its little joke, Passover Week sometimes coincides with Easter Week, overlaying it like a reproach. It does the year Mrs. Cooper is with us. First, Good Friday, then in a few days is the first day of Passover.

"This year," my husband says, "because of Susie, to celebrate her first year with us, I want us to put a mezuzah outside our door before Passover."

"I'm not in favor." I manage to say it quietly.

"You don't understand enough about it," my husband says.

"I understand that much."

"Do you know what a mezuzah is? Do you know what's in it?" Taking my silence as an admission of ignorance, my husband produces a Bible. "Deuteronomy," he says. He reads:

"Hear, O Israel: The Lord our God, the Lord is one Lord:
And thou shalt love the Lord thy God with all thine heart, and with
   all thy soul, and with all thy might.
And these words, which I command thee this day, shall be in thine
   heart:
And thou shalt teach them diligently unto thy children, and shalt
   talk of them when thou sittest in thine house, and when thou
   walkest by the way, and when thou liest down, and when thou
   risest up. . . ."

All this and more is written on a parchment that is rolled up tight and fitted into the metal or wooden mezuzah, which is no more than two inches high and less than half an inch across and is mounted on a base for fastening to the doorframe. My husband finishes his reading.

"And thou shalt write them upon the door-posts of thine house,
   and upon thy gates:
That your days may be multiplied, and the days of your children, in
   the land which the Lord sware unto your fathers to give them, as
   the days of heaven upon the earth."

The words might move me if I allowed them to, but I will not allow them to.

My husband closes the Bible and asks, "What did your family observe? What was Passover like?"

"My grandfather sat on a pillow, and I was the youngest, so I found the matzos and he gave me money."

"No questions? No answers?"

"Just one. I would ask my grandfather, 'Where is my prize?' And he would laugh and give me money."

"Is that all?" my husband asks.

"That was a very nice ceremony in itself," I say. "And I remember it with pleasure, and my grandfather with love!"

"But besides the food, besides the children's game. Didn't your grandparents observe anything?"

"I don't remember."

"You sat at their table for eighteen years!"

"Well, my grandmother lit Friday-night candles, and that was something I think she did all her life. But she did it by herself, in the breakfast room."

"Didn't they go to a synagogue?"

"My grandmother did. My grandfather did, too, but then I remember he stopped. He'd be home on holidays, not at the services."

"Your parents didn't tell you anything?"

"My parents were the next generation," I say. "And I'm the generation after that. We evolved," I say—and luckily that is also humor between my husband and me.

But my husband rubs his head. It's different now, and not so funny, because this year we have Susan.

My husband was born in Europe, of an Orthodox family. He is neither Orthodox nor Reform. He is his own council of rabbis, selecting as he goes. He has plenty to say about the influence of America on Jewishness, Orthodox or not. "The European Jew," my husband says, "didn't necessarily feel that if he rose in the social or economic scale he had to stop observing his Jewishness. There were even a number of wealthy and prominent German Jews who were strictly observant."

"I'm sure that helped them a lot!" This is as close as I come to speaking of the unspeakable. Somewhere in the monstrous testimony I have read about concentration camps and killings are buried the small, intense lives of my husband's family. But why is it I am more bitter than my husband about his own experiences? And why should my bitterness cut the wrong way?

It is the word "German" that does it to me. My soul knots in hate. "German!" Even the softening, pathetic sound of "Jew" that follows it now doesn't help. All words fail. If I could grasp words, I would come on words that would jump so to life they would jump into my heart and kill me. All I can do is make a fantasy. Somewhere in New York I will meet a smiling German. In his pocket smile the best export accounts in the city—he is from the land of scissors and knives and ground glass. Because I am surprisingly fair, he will be oh, so surprised when I strike at him with all my might. "For the children! For the children!" My words come out shrieks. He protests it was his duty and, besides, he didn't know. I am all leaking, dissolving. How can a mist break stone? Once we exchange words it is hopeless; the words of the eyewitness consume everything, as in a fire:

> "The children were covered with sores. . . . They screamed and wept all night in the empty rooms where they had been put. . . . Then the police would go up and the children, screaming with terror, would be carried kicking and struggling to the courtyard."

How is it my husband doesn't know that after this there can be no mezuzahs?

"It's too painful to quarrel," my husband says. He puts his hands on my shoulders, his forehead against mine. "This is something I want very much. And you feel for me. I know you feel for me in this."

"Yes, I do, of course I do." I use Mrs. Cooper's trick, and even at that close range twist my head elsewhere. "Only that particular symbol—"

"No, with you it's all the symbols." My husband drops his hands from my shoulders. "You don't know enough about them to discard them."

I don't have the right to judge them—that is what I feel he means. Since I was not even scorched by the flames of their futility. As he was, and came out cursing less than I.

"But besides everything else"—I take hasty shelter in practicalness—"a mezuzah is ugly. I remember that ugly tin thing nailed to the door of my grandmother's room. If I spend three weeks picking out a light fixture for my foyer, why should I have something so ugly on my door?"

Then, as my husband answers, I see that this shabby attack has fixed my defeat, because he is immediately reasonable. "Now, that's something else. I won't argue aesthetics with you. The outer covering is of no importance. I'll find something attractive."

The next night my husband brings home a mezuzah made in the East. It is a narrow green rectangle, twice the normal size, inlaid with mosaic and outlined in brass. It does not look Jewish to me at all. It looks foreign—a strange bit of green enamel and brass.

"I don't like it," I say. "I'm sorry."

"But it's only the idea you don't like?" My husband smiles teasingly. "In looks, you at least relent?"

"It doesn't look bad," I admit.

"Well, that is the first step." I am happy to see the mezuzah disappear in his dresser drawer before we go in to our dinner.

When Mrs. Cooper comes next day, she asks, "What have you on your door?"

I step out to look, and at first have the impression that a praying mantis has somehow hatched out of season high on our doorway. Then I recognize it. "Oh, that's . . ." I say. "That's . . ." I find I cannot explain a mezuzah to someone who has never heard of one.

While Mrs. Cooper changes her clothes, I touch the mezuzah to see if it will fall off. But my husband has glued it firmly to the metal doorframe.

My husband's office works a three-quarter day on Good Friday. I ask Mrs. Cooper if she would like time off, but she says no, her husband will be home ahead of her to look after things. I have the impression she would rather be here.

My husband comes home early, bestowing strangeness on the rhythm of the house in lieu of celebration. I kiss him and put away his hat. "Well, that was a nasty thing to do." I say it lazily and with a smirk. The lazy tone is to show that I am not really involved, and the smirk that I intend to swallow it down like bad medicine. He will have his way, but I will have my say—that's all I mean. My say will be humorous, with just a little cut to it, as is proper between husband and wife. He will cut back a little, with a grin, and after Mrs. Cooper goes we will have our peace-

ful dinner. The conversation will meander, never actually prick-
ing sore points, but winding words about them, making pads
and cushions, so that should they ever bleed, there, already
softly wrapped around them, will be the bandages our words
wove. Weave enough of these bandages and nothing will ever
smash, I say. I always prepare in advance a last line, too, so that
I will know where to stop. "When mezuzahs last in the doorway
bloomed," I will say tonight. And then I expect us both to laugh.

But where has he been all day? The same office, the same
thirty-minute subway ride to and from each way, the same lunch
with the same cronies. . . . But he has traveled somewhere else
in his head. "Doesn't anything mean anything to you?" he says,
and walks by me to the bedroom.

I follow with a bandage, but it slips from my hand. "I know
a lot of women who would have taken that right down!" It is
something of a shout, to my surprise.

He says nothing.

"I left it up. All I wanted was my say."

He says nothing.

"I live here, too. That's my door also."

He says nothing.

"And I don't like it!"

I hear a loud smashing of glass. It brings both our heads up.
My husband is the first to understand. "Mrs. Cooper broke a
bottle." He puts his arms around me and says, "Let's not quarrel about a doorway. Let's not quarrel at all, but especially not
about the entrance to our home."

I lower my face into his tie. What's a mezuzah? Let's have ten,
I think, so long as nothing will smash.

Later, I reproach myself. I am in the living room, straighten-
ing piles of magazines, avoiding both kitchen and bedroom. A
woman, I think, is the one creature who builds satisfaction of
the pleasure she gets from giving in. What might the world be
if women would continue the dialogue? But no, they must give
in and be satisfied. Nevertheless, I don't intend to take back
what I've given in on and thereby give up what I've gained.

I am aware of Mrs. Cooper, boiling formula in the kitchen,
and of my baby, registering in sleep her parents' first quarrel
since her birth. "What must I say to you?" I think of saying to

my daughter—Mrs. Cooper's words come naturally to my mind.

I go to the kitchen doorway and look at Mrs. Cooper. Her face indicates deaf and dumb. She is finishing the bottles.

When Mrs. Cooper is dressed and ready to leave, she looks into the living room. "I am saying good night."

"I hope you and your family will have a happy Easter," I say, smiling for her.

I know in advance that Mrs. Cooper will ask, "What must I say to you?"

This time she asks it soberly, and this time my husband, who has heard, comes in to tell Mrs. Cooper the story of Passover. As always in the traditional version, there is little mention of Moses, the Jews having set down from the beginning not the tragedy of one but their intuition for the tragedy of many.

When my husband leaves us, Mrs. Cooper takes four wrapped candies from the candy bowl on the desk, holds them up to be sure I see her taking them, and puts them in her purse. "I do hope everything will be all right," she says.

"Oh, yes," I say, looking at the magazines. "It was such a help today. I got so much work done. Thank you."

I hear that she is motionless.

"I will not be like this all the days of my life." It is a cry from the heart, stunningly articulated. I lift my head from the magazines, and this time I do stare. Not be like what? A Jamaican without a servant? A wife who never vacations? An exile? A baby nurse? A woman who gives in? What Mrs. Cooper might not want to be flashes up in a lightning jumble. "I am going to find a church," she says, and strains her face away from mine.

I think of all the descriptions of God I have ever heard—that He is jealous, loving, vengeful, waiting, teaching, forgetful, permissive, broken-hearted, dead, asleep.

Mrs. Cooper and I wish each other a pleasant weekend.

II
SEEKING

Where Moth and Rust

Gertrude Friedberg

1

IN RECENT YEARS there have been many precautions one has had to take against disaster. The disaster that menaced Mrs. Fortrey was not the obvious one that frightened all others. It was the destruction and ruin that would engulf her home were she to fail in her performance of the many small tasks of protection and repair that custom, direction leaflets, and her vigilant sister prescribed.

No sooner was a new acquisition, whether a furnishing, an ornament, or merely a child, brought proudly into the home than it began to succumb to a thousand effects of use, age, and chemistry, which might be held in unwilling abeyance only by the indicated rituals. So seductively did the written prescriptions belittle the amount of time needed for each particular care that Mrs. Fortrey for a long time failed to feel deprived as she

A New Yorker and a graduate of Barnard College, *Gertrude Friedberg* is the author of *The Revolving Boy*, a science fiction novel. Her short stories have appeared in *The Atlantic, Harper's, Esquire*, and elsewhere. "Where Moth and Rust," copyright © 1957 by The Atlantic Monthly Company, Boston, Massachusetts, is reprinted with permission.

raced in a treadmill of small palliative measures which produced nothing, improved nothing, but merely retarded the processes of inevitable decay or returned things to a state that was surely not as good as new.

So every Monday morning Mrs. Fortrey started out bravely with a pitcher of water and a can of 3-in-1 oil to perform eight measures of prophylaxis in one grand round from kitchen to front room. The electric fan, the vacuum cleaner, and the sewing machine needed a paltry libation of only three drops of oil (after every ten hours of use) to keep their motors from burning out or worse. It took but a few minutes more to oil the wall can opener, the knife sharpener, the children's roller skates, the flute keys, and the drapery pulley tracks.

On the return trip from front room to kitchen Mrs. Fortrey poured a little water into the philodendron plants, moistened the rubber plant and the ivy, put a bit in the parakeet's cup, poured some into the radiator pan to keep the air from drying the furniture, and put a little in a pan under the broiler to keep the stove from smoking up the white kitchen. Then she usually stopped to empty the pencil sharpener. She once told her sister that she oiled all the way from the back of the apartment to the front and watered from the front to the back.

Mrs. Fortrey knew of four kinds of rot and how to avoid them. The chrome fixtures, if not shined with a special cream, would succumb to green rot; the mahogany tables, if not rubbed with polish, would yield to dry rot; clothes left too long in a hamper would get wet rot; dead batteries left in a flashlight produced chemical rot. And a leaflet that the children brought home from school cautioned Mrs. Fortrey that if the frayed electric cords were not taped up, her family would be electrocuted.

Every month or so a bit of Vaseline had to be spread on the washing machine spindle to keep the agitator from sticking, and a bit of black stuff had to be rubbed on Mr. Fortrey's electric razor to sharpen it while the electricity hummed and Mrs. Fortrey counted to sixty in a loud voice as bidden. She rubbed ashes on the glass rings left on table tops, leather conditioner on the desk top, and, almost with a sense of wanton mischief, soap on the edges of all the doors.

When the carpets were brand-new, Mrs. Fortrey's sister made

the family take their shoes off as soon as they entered the house. This zeal did not survive one week, for as small unavoidable spots gathered despite their care, their hearts hardened and their feet stepped more boldly. At last only Mrs. Fortrey frowned at the dark patches before each door. To avoid increasing dirt and wear on the most traveled paths of the carpet, she took to sidling close against the walls and leaping from a point about two feet before each threshold to a point two feet within the room. She took great pleasure in the thought that in this way the carpet was spared perhaps hundreds of extra scuffs. Once Mr. Fortrey, coming back for a moment for an umbrella, saw her sidling and leaping about the apartment. He thought it best to say nothing, but quietly let himself out again.

Finally she contented herself with brushing a glamorizing powder into the dark patches twice a month. This usually reminded her to sprinkle talcum powder on the slide rule where it customarily slid, and fuller's earth on the food spots on the dining-room chairs.

2

Mrs. Fortrey's sister always had many useful suggestions for the preservation of things. It was she who pointed out that all the ash trays, lamps, and bric-a-brac which stood on the fine tables would leave dreadful scratches each time they were moved if Mrs. Fortrey did not line their bases with some gentle stuff. Looking for scraps of felt for this purpose, Mrs. Fortrey was happy to find in the children's room a great pile of round felt pieces of exactly the right conformation. They were camp awards which, accumulated for six years or more of earnest athletic endeavor, presented at last a grave problem of storage. It was therefore doubly satisfying to Mrs. Fortrey when each morning, for many mornings (work divided into time in this way gave her the illusion of being somehow diminished), she pasted to the base of a lamp or vase a bright yellow patch. Only the truly curious would ever lift them to read "Camp Katchewan —For Proficiency in Intermediate Archery."

Shoes had to be reheeled and resoled. Hems fell and needed to be raised. Trouser cuffs drooped and were retacked. Elbows wore out and were patched, and everybody in the family dropped buttons and stood still while Mrs. Fortrey sewed them on again. She sewed the little ribbon tape back on the umbrella, the fringe back on the chaise longue, and the hanger tape back on Mr. Fortrey's overcoat, and restitched three brief cases with a handy little leather stitcher.

It seemed quite clear from the start that a bright new upholstered chair ought not to be sat on. When Mr. Fortrey, who sometimes forgot the rules, plunked down in the beige chair, she ran at him with a lace doily to put under his head.

"Darling, nobody uses such things any more," her sister said. "You have your upholsterer make false covers for the backs and arms out of the leftover pieces of material."

Soon every chair, and even the couch, had extra pieces of matching material over the original coverings. They were merely tacked in place and were not too noticeable. Mr. Fortrey never got the knack of sitting properly in the living room. He managed to sit on each piece in turn and always disarranged the protective disguise pieces, so that after an evening the carpet would be littered with pages of newspaper and pieces of backs or arms.

When summer came, all the true coverings and false coverings had to be covered with covers which were neither true nor false but just slip. You slipped the chairs and cushions into them much as you slipped a heavy woman into a girdle, and zipped. They were very attractive and her sister thought it a pity that they should be ruined so quickly by the dust which rained in through the windows all day. "I'd just put a plastic covering over the very tops," she said, and Mrs. Fortrey added, "And maybe just where your hands rest."

Every night a plastic cover had to be put over the parakeet cage, and every winter a plastic cover had to be put over the air conditioner, and every summer plastic covers were put on all the lamp shades. Mrs. Fortrey once considered making a plastic cover for Mr. Fortrey, who habitually spilled soup on his tie and sloshed his cuffs around in the gravy, but she ended by rubbing him down instead with Renuzit.

The stapling machine and the ball-point pens had to be reloaded, the filters replaced in the air conditioner, a fresh throwaway bag inserted into the vacuum cleaner, a fresh ribbon put into the typewriter, and fresh fuel oil dripped into the cigarette lighters.

Every summer the drapes had to be taken down, cleaned, folded away, and in the fall put up again. This large familiar effort each year stung the wound of passing time.

"Why couldn't I just let them hang?" Mrs. Fortrey asked her sister.

"You'll have to throw them out in the fall," said her sister. "It's not only the dust. It's the sun too."

She took them down.

Her coat needed periodic reglazing, the rubbertile foyer needed periodic rewaxing, her graying hair needed periodic retinting, the springs of the couch needed retying, the mirrors needed resilvering, the silver pieces needed reglazing to protect their resilvering, the bathroom wallpaper needed repasting, the kitchen needed repainting, the library steps needed revarnishing, the dictionary needed rebinding, and one of the children, changing schools, needed extensive readjusting.

Moths were a constant threat. Mr. Fortrey had once told her very firmly that it unnerved him if during an evening in which they had been quietly reading, leaning with only half their weight on the false backs of their living-room chairs and dangling their hands unsupported three inches above the chair arms, she would suddenly leap up with a loud cry and clap her hands together directly above his head.

"It's too noticeable," he complained.

After that Mrs. Fortrey tried to be more discreet in her pursuit, but she continued to feel that to be mothproof was to achieve virtue. She sprayed every woolen suit, coat, and dress, put them in clothes bags, hung moth gas containers in all the closets, and put the children's mittens in the little cellophane bags which she saved from bunches of carrots.

She used Easy-off on the stove, Soil-off on the walls, Dust-off —full of static electricity—on the coffee table, Slipit on the window tracks, Weldit on the loose tile in the kitchen wall, Kaukit on the spot under the hot-water pipe where the plaster had come off, and Exit under the sinks. She used Jubilee on the enamels, Pride on the kitchen linoleum, Cheer in the clothes washer, Joy in the sink, and had death in her heart.

One winter the handle came off the door of the refrigerator, the drawers stuck, a piano key broke, a radio knob cracked, the screws fell out of the stove, the ironing board collapsed, the plates chipped, the magnetic can holder let the cans drop, the hair drier smoked, the wire recorder raveled a whole spool of wire, all the tubes on the horizontal circuit of the television set burned out, and one of Mrs. Fortrey's own tubes was a bit inflamed.

When she got back from the hospital, she found that the agitator in the washing machine was wobbly, there were moth holes in the suit that Mr. Fortrey had been saving until he got thinner, and the children had spots. There were cracks in the window shades, feathers were coming out of the pillows, the eight-day clock went only three and a half days, there was a steady hum on the record player, and the ceiling drier had fallen.

It was not long after this that Mrs. Fortrey stopped. She told Mr. Fortrey while he was having breakfast.

"I'm not going to do anything any more," she said. "It's all too much trouble for me."

"Name me one thing," said Mr. Fortrey, "that's too much for you to do."

Mrs. Fortrey thought a moment. "I don't want to reload the stapling machine," she said.

Reloading the stapling machine was really a small thing, but Mrs. Fortrey didn't want to mention all the other things to Mr. Fortrey while he was having his breakfast. Actually she stopped everything that very day.

At first nobody noticed, and Mrs. Fortrey didn't say any more. She started to play the piano, learn Spanish, and practice tumbling tricks. As she walked through the apartment she stepped boldly on the middle of the carpet and wiped her feet on the dark spot at the threshold.

The calendar leaves weren't torn off, and since nobody knew when October was over, the library books were overdue. As if despoiled by an autumn trapped unexpectedly indoors with all its colors ready to flow, her furnishings started to turn color. The silver turned brown. The chrome fixtures turned green. The copper casserole turned pink. The blue chair turned gray. The white bedroom drapes turned yellow.

One day Mr. Fortrey pushed back his dining chair, took off his plastic cuffs, and told Mrs. Fortrey that he was leaving her. "I refuse to be seen in public with a woman in an unglazed mink coat," he said firmly. He would have left that day, but the zipper on his valise got stuck midway and he had to wait two weeks to have the zipper teeth realigned by a woman whose business it was to realign zipper teeth. When he left, Mrs. Fortrey turned off the air conditioners.

The parakeet was the next to leave. His cage had no carpet to darken at the threshold, but the gravel paper which served him instead had not been changed for three weeks and was piled high with empty seed shells. He found an open window and headed toward Park and 56th Street, where he felt he might find a tidier home.

The children put up with her for a few months longer, but relations broke down at last when they found that Mrs. Fortrey had given up using the pliers on the rings of their loose-leaf notebooks to ply them into tight embrace, and had neglected to Scotch-tape the edges of the music books, which were now beginning to rain bits of dried yellow paper all over the piano, the carpet, and their hands as they practiced. This was too much and both children went to join their father, taking the Erector set with them.

Mrs. Fortrey moved into one room and lived, on the small allowance she received, not at all as she should. She found out how umbrellas started, did lots of geometry, and learned Speedwriting, which she wrote slowly, embellishing each capital letter with extraordinary flourishes and tiny clever pictures. She set up a basketball net over her front door, looked up how to pronounce *disheveled,* and practiced bad posture.

She ate only foods that came ready-baked in heat-and-throw-away containers. Rather than wash each item of her no-iron dip-and-hang clothing separately, she bathed with her clothes on and sat on the radiator, reading, while she dried. Late at night when the city was quiet she could hear the moths feeding in the closet. From time to time she opened the closet door and threw them a sweater.

When Mrs. Fortrey died, it was after a period of six months during which nobody had seen or spoken to her. Two men broke down her door and took a good two hours to cart away rubbish piled ceiling high before they reached her bed.

She had died of what is still called a natural cause. A small valve in her heart, diseased from a childhood illness, had become rigid, its sides had fused, and the closure had provoked a gradual deterioration of her circulatory system and finally death. A surgeon, with the gentlest pressure of his index finger, might have split the fused valve open and restored her to a state almost as good as new. It was the sort of small mechanical repair that Mrs. Fortrey had once performed on a stenosed perfume atomizer in the days of her greatest zeal.

The Long-Distance Runner

Grace Paley

ONE DAY, BEFORE or after forty-two, I became a long-distance runner. Though I was stout and in many ways inadequate to this desire, I wanted to go far and fast, not as fast as bicycles and trains, not as far as Taipei, Hingwen, places like that, islands of the slant-eyed cunt, as sailors in bus stations say when speaking of travel, but round and round the county from the sea side to the bridges, along the old neighborhood streets a couple of times, before old age and urban renewal ended them and me.

I tried the country first, Connecticut, which being wooded is always full of buds in spring. All creation is secret, isn't that true? So I trained in the wide-zoned suburban hills where I wasn't known. I ran all spring in and out of dogwood bloom, then laurel.

Grace Paley is the author of two short-story collections: *The Little Disturbances of Man* and *Enormous Changes at the Last Minute*. A recipient of grants from the Guggenheim Foundation and the National Council on the Arts, and an award from the National Institute of Arts and Letters, she teaches at Sarah Lawrence College. "The Long-Distance Runner," from *Enormous Changes at the Last Minute* by Grace Paley, copyright © 1974 by Grace Paley, is reprinted with the permission of Farrar, Straus and Giroux, Inc.

People sometimes stopped and asked me why I ran, a lady in silk shorts halfway down over her fat thighs. In training, I replied and rested only to answer if closely questioned. I wore a white sleeveless undershirt as well, with excellent support, not to attract the attention of old men and prudish children.

The summer came, my legs seemed strong. I kissed the kids goodbye. They were quite old by then. It was near the time for parting anyway. I told Mrs. Raftery to look in now and then and give them some of that rotten Celtic supper she makes.

I told them they could take off any time they wanted to. Go lead your private life, I said. Only leave me out of it.

A word to the wise . . . said Richard.

You're depressed Faith, Mrs. Raftery said. Your boy friend Jack, the one you think's so hotsy-totsy, hasn't called and you're as gloomy as a tick on Sunday.

Cut the folkshit with me, Raftery, I muttered. Her eyes filled with tears because that's who she is: folkshit from bunion to topknot. That's how she got liked by me, loved, invented and endured.

When I walked out the door they were all reclining before the television set, Richard, Tonto and Mrs. Raftery, gazing at the news. Which proved with moving pictures that there *had* been a voyage to the moon and Africa and South America hid in a furious whorl of clouds.

I said, Goodbye. They said, Yeah, O.K., sure.

If that's how it is, forget it, I hollered and took the Independent subway to Brighton Beach.

At Brighton Beach I stopped at the Salty Breezes Locker Room to change my clothes. Twenty-five years ago my father invested $500 in its future. In fact he still clears about $3.50 a year, which goes directly (by law) to the Children of Judea to cover their deficit.

No one paid too much attention when I started to run, easy and light on my feet. I ran on the boardwalk first, past my mother's leafleting station—between a soft-ice-cream stand and a degenerated dune. There she had been assigned by her comrades to halt the tides of cruel American enterprise with simple socialist sense.

I wanted to stop and admire the long beach. I wanted to stop in order to think admiringly about New York. There aren't many rotting cities so tan and sandy and speckled with citizens at their salty edges. But I had already spent a lot of life lying down or standing and staring. I had decided to run.

After about a mile and a half I left the boardwalk and began to trot into the old neighborhood. I was running well. My breath was long and deep. I was thinking pridefully about my form.

Suddenly I was surrounded by about three hundred blacks.

Who you?

Who that?

Look at her! Just look! When you seen a fatter ass?

Poor thing. She ain't right. Leave her, you boys, you bad boys.

I used to live here, I said.

Oh yes, they said, in the white old days. That time too bad to last.

But we loved it here. We never went to Flatbush Avenue or Times Square. We loved our block.

Tough black titty.

I like your speech, I said. Metaphor and all.

Right on. We get that from talking.

Yes my people also had a way of speech. And don't forget the Irish. The gift of gab.

Who they? said a small boy.

Cops.

Nowadays, I suggested, there's more than Irish on the police force.

You right, said two ladies. More, more, much much more. They's French Chinamen Russkies Congoleans. Oh missee, you too right.

I lived in that house, I said. That apartment house. All my life. Till I got married.

Now that *is* nice. Live in one place. My mother live that way in South Carolina. One place. Her daddy farmed. She said. They ate. No matter winter war bad times. Roosevelt. Something! Ain't that wonderful! And it weren't cold! Big trees!

That apartment. I looked up and pointed. There. The third floor.

They all looked up. So what! You blubrous devil! said a dark young man. He wore horn-rimmed glasses and had that intelligent look that City College boys used to have when I was eighteen and first looked at them.

He seemed to lead them in contempt and anger, even the littlest ones who moved toward me with dramatic stealth singing. Devil, Oh Devil. I don't think the little kids had bad feeling because they poked a finger into me, then laughed.

Still I thought it might be wise to keep my head. So I jumped right in with some facts. I said, How many flowers' names do you know? Wild flowers, I mean. My people only knew two. That's what they say now anyway. Rich or poor, they only had two flowers' names. Rose and violet.

Daisy, said one boy immediately.

Weed, said another. That *is* a flower, I thought. But everyone else got the joke.

Saxifrage, lupine, said a lady. Viper's bugloss, said a small Girl Scout in medium green with a dark green sash. She held up a *Handbook of Wild Flowers.*

How many you know, fat mama? a boy asked warmly. He wasn't against my being a mother or fat. I turned all my attention to him.

Oh sonny, I said, I'm way ahead of my people. I know in yellows alone: common cinquefoil, trout lily, yellow adder's-tongue, swamp buttercup and common buttercup, golden sorrel, yellow or hop clover, devil's-paintbrush, evening primrose, black-eyed Susan, golden aster, also the yellow pickerelweed growing down by the water if not in the water, and dandelions of course. I've seen all these myself. Seen them.

You could see China from the boardwalk, a boy said. When it's nice.

I know more flowers than countries. Mostly young people these days have traveled in many countries.

Not me. I ain't been nowhere.

Not me either, said about seventeen boys.

I'm not allowed, said a little girl. There's drunken junkies.

But *I! I!* cried out a tall black youth, very handsome and well dressed. I am an African. My father came from the high stolen plains. *I* have been everywhere. I was in Moscow six months, learning machinery. I was in France, learning French. I was in

Italy, observing the peculiar Renaissance and the people's sweetness. I was in England, where I studied the common law and the urban blight. I was at the Conference of Dark Youth in Cuba to understand our passion. I am now here. Here am I to become an engineer and return to my people, around the Cape of Good Hope in a Norwegian sailing vessel. In this way I will learn the fine old art of sailing in case the engines of the new society of my old inland country should fail.

We had an extraordinary amount of silence after that. Then one old lady in a black dress and high white lace collar said to another old lady dressed exactly the same way, Glad tidings when someone got brains in the head not fish juice. Amen, said a few.

Whyn't you go up to Mrs. Luddy living in your house, you lady, huh? The Girl Scout asked this.

Why she just groove to see you, said some sarcastic snickerer.

She got palpitations. Her man, he give it to her.

That ain't all, he a natural gift giver.

I'll take you, said the Girl Scout. My name is Cynthia. I'm in Troop 355, Brooklyn.

I'm not dressed, I said, looking at my lumpy knees.

You shouldn't wear no undershirt like that without no runnin number or no team writ on it. It look like a undershirt.

Cynthia! Don't take her up there, said an important boy. Her head strange. Don't you take her. Hear?

Lawrence, she said softly, you tell me once more what to do I'll wrap you round that lamppost.

Git! She said, powerfully addressing *me*.

In this way I was led into the hallway of the whole house of my childhood.

The first door I saw was still marked in flaky gold, 1A. That's where the janitor lived, I said. He was a Negro.

How come like that? Cynthia made an astonished face. How come the janitor was a black man?

Oh Cynthia, I said. Then I turned to the opposite door, first floor front, 1B. I remembered. Now, here, this was Mrs. Gore-ditsky, very very fat lady. All her children died at birth. Born,

then one, two, three. Dead. Five children, then Mr. Goreditsky
said, I'm bad luck on you Tessie and he went away. He sent $15
a week for seven years. Then no one heard.

I know her, poor thing, said Cynthia. The city come for her
summer before last. The way they knew, it smelled. They
wropped her up in a canvas. They couldn't get through the front
door. It scraped off a piece of her. My uncle Ronald had to help
them, but he got disgusted.

Only two years ago. She was still here! Wasn't she scared?

So we all, said Cynthia. White ain't everything.

Who lived up here, she asked, 2B? Right now, my best friend
Nancy Rosalind lives here. She got two brothers, and her sister
married and got a baby. She very light-skinned. Not her
mother. We got all colors amongst us.

Your best friend? That's funny. Because it was *my* best friend.
Right in that apartment. Joanna Rosen.

What become of her? Cynthia asked. She got a running shirt
too?

Come on Cynthia, if you really want to know, I'll tell you. She
married this man, Marvin Steirs.

Who's he?

I recollected his achievements. Well, he's the president of a
big corporation, JoMar Plastics. This corporation owns a steel
company, a radio station, a new Xerox-type machine that lets
you do twenty-five different pages at once. This corporation has
a foundation, The JoMar Fund for Research in Conservation.
Capitalism is like that, I added, in order to be politically useful.

How come you know? You go over their house a lot?

No. I happened to read all about them on the financial page,
just last week. It made me think: a different life. That's all.

Different spokes for different folks, said Cynthia.

I sat down on the cool marble steps and remembered Joanna's cousin Ziggie. He was older than we were. He wrote a poem
which told us we were lovely flowers and our legs were petals,
which nature would force open no matter how many times we
said no.

Then I had several other interior thoughts that I couldn't
share with a child, the kind that give your face a blank or melancholy look.

Now you're not interested, said Cynthia. Now you're not gonna say a thing. Who lived here, 2A? Who? Two men lives here now. Women coming and women going. My mother says, Danger sign: Stay away, my darling, stay away.

I don't remember, Cynthia. I really don't.

You got to. What'd you come for, anyways?

Then I tried. 2A. 2A. Was it the twins? I felt a strong obligation as though remembering was in charge of the *existence* of the past. This is not so.

Cynthia, I said, I don't want to go any further. I don't even want to remember.

Come on, she said, tugging at my shorts, don't you want to see Mrs. Luddy, the one lives in your old house? That be fun, no?

No. No, I don't want to see Mrs. Luddy.

Now you shouldn't pay no attention to those boys downstairs. She will like you. I mean, she is kind. She don't like most white people, but she might like you.

No Cynthia, it's not that, but I don't want to see my father and mother's house now.

I didn't know what to say. I said, Because my mother's dead. This was a lie, because my mother lives in her own room with my father in the Children of Judea. With her hand over her socialist heart, she reads the paper every morning after breakfast. Then she says sadly to my father, Every day the same. Dying . . . dying, dying from killing.

My mother's dead Cynthia. I can't go in there.

Oh . . . oh, the poor thing, she said, looking into my eyes. Oh, if my mother died, I don't know what I'd do. Even if I was old as you. I could kill myself. Tears filled her eyes and started down her cheeks. If my mother died, what would I do? She is my protector, she won't let the pushers get me. She hold me tight. She gonna hide me in the cedar box if my Uncle Rudford comes try to get me back. She *can't* die, my mother.

Cynthia—honey—she won't die. She's young. I put my arm out to comfort her. You could come live with me, I said. I got two boys, they're nearly grown up. I missed it, not having a girl.

What? What you mean now, live with you and boys. She pulled away and ran for the stairs. Stay way from me, honky lady. I know them white boys. They just gonna try and jostle my black womanhood. My mother told me about that, keep you white honky devil boys to your devil self, you just leave me be you old bitch you. Somebody help me, she started to scream, you hear. Somebody help. She gonna take me away.

She flattened herself to the wall, trembling. I was too frightened by her fear of me to say, honey, I wouldn't hurt you, it's me. I heard her helpers, the voices of large boys crying. We coming, we coming, hold your head up, we coming. I ran past her fear to the stairs and up them two at a time. I came to my old own door. I knocked like the landlord, loud and terrible.

Mama not home, a child's voice said. No, no, I said. It's me! a lady! Someone's chasing me, let me in. Mama not home, I ain't allowed to open up for nobody.

It's me! I cried out in terror. Mama! Mama! let me in!

The door opened. A slim woman whose age I couldn't invent looked at me. She said, Get in and shut that door tight. She took a hard pinching hold on my upper arm. Then she bolted the door herself. Them hustlers after you. They make me pink. Hide this white lady now, Donald. Stick her under your bed, you got a high bed.

Oh that's O.K. I'm fine now, I said. I felt safe and at home.

You in my house, she said. You do as I say. For two cents, I throw you out.

I squatted under a small kid's pissy mattress. Then I heard the knock. It was tentative and respectful. My mama don't allow me to open. Donald! someone called. Donald!

Oh no, he said. Can't do it. She gonna wear me out. You know her. She already tore up my ass this morning once. Ain't *gonna* open up.

I lived there for about three weeks with Mrs. Luddy and Donald and three little baby girls nearly the same age. I told her a joke about Irish twins. Ain't Irish, she said.

Nearly every morning the babies woke us at about 6:45. We gave them a bottle and went back to sleep till 8:00. I made coffee and she changed diapers. Then it really stank for a while. At this time I usually said, Well listen, thanks really, but I've got to go I guess. I guess I'm going. She'd usually say, Well, guess again.

I guess you ain't. Or if she was feeling disgusted she'd say, Go on now! Get! You wanna go, I guess by now I have snorted enough white lady stink to choke a horse. Go on!

I'd get to the door and then I'd hear voices. I'm ashamed to say I'd become fearful. Despite my wide geographical love of mankind, I would be attacked by local fears.

There was a sentimental truth that lay beside all that going and not going. It *was* my house where I'd lived long ago my family life. There was a tile on the bathroom floor that I myself had broken, dropping a hammer on the toe of my brother Charles as he stood dreamily shaving, his prick halfway up his undershorts. Astonishment and knowledge first seized me right there. The kitchen was the same. The table was the enameled table common to our class, easy to clean, with wooden undercorners for indigent and old cockroaches that couldn't make it to the kitchen sink. (However, it was not the same table, because I have inherited that one, chips and all.)

The living room was something like ours, only we had less plastic. There may have been less plastic in the world at that time. Also, my mother had set beautiful cushions everywhere, on beds and chairs. It was the way she expressed herself, artistically, to embroider at night or take strips of flowered cotton and sew them across ordinary white or blue muslin in the most delicate designs, the way women have always used materials that live and die in hunks and tatters to say: This is my place.

Mrs. Luddy said, Uh huh!

Of course, I said, men don't have that outlet. That's how come they run around so much.

Till they drunk enough to lay down, she said.

Yes, I said, on a large scale you can see it in the world. First they make something, then they murder it. Then they write a book about how interesting it is.

You got something there, she said. Sometimes she said, Girl, you don't know *nothing*.

We often sat at the window looking out and down. Little tufts of breeze grew on that windowsill. The blazing afternoon was around the corner and up the block.

You say men, she said. Is that men? she asked. What you call—a Man?

Four flights below us, leaning on the stoop, were about a dozen people and around them devastation. Just a minute, I said. I had seen devastation on my way, running, gotten some of the pebbles of it in my running shoe and the dust of it in my eyes. I had thought with the indignant courtesy of a citizen, This is a disgrace to the City of New York which I love and am running through.

But now, from the commanding heights of home, I saw it clearly. The tenement in which Jack my old and present friend had come to gloomy manhood had been destroyed, first by fire, then by demolition (which is a swinging ball of steel that cracks bedrooms and kitchens). Because of this work, we could see several blocks wide and a block and a half long. Crazy Eddy's house still stood, famous 1510 gutted, with black window frames, no glass, open laths. The stubbornness of the supporting beams! Some persons or families still lived on the lowest floors. In the lots between, a couple of old sofas lay on their fat faces, their springs sticking up into the air. Just as in wartime a half-dozen ailanthus trees had already found their first quarter inch of earth and begun a living attack on the dead yards. At night, I knew animals roamed the place, squalling and howling, furious New York dogs and street cats and mighty rats. You would think you were in Bear Mountain Park, the terror of venturing forth.

Someone ought to clean that up, I said.

Mrs. Luddy said, Who you got in mind? Mrs. Kennedy?—

Donald made a stern face. He said, That just what I gonna do when I get big. Gonna get the Sanitary Man in and show it to him. You see that, you big guinea you, you clean it up right now! Then he stamped his feet and fierced his eyes.

Mrs. Luddy said, Come here, you little nigger. She kissed the top of his head and gave him a whack on the backside all at one time.

Well, said Donald, encouraged, look out there now you all! Go on I say, look! Though we had already seen, to please him we looked. So the stoop men and boys lounged, leaned, hopped about, stood on one leg, then another, took their socks off, and scratched their toes, talked, sat on their haunches, heads down, dozing.

Donald said, Look at them. They ain't got self-respect. They got Afros *on* their heads, but they don't know they black *in* their heads.

I thought he ought to learn to be more sympathetic. I said, There are reasons that people are that way.

Yes, ma'am, said Donald.

Anyway, how come you never go down and play with the other kids, how come you're up here so much?

My mama don't like me do that. Some of them is bad. Bad. I might become a dope addict. I got to stay clear.

You just a dope, that's a fact, said Mrs. Luddy.

He ought to be with kids his age more, I think.

He see them in school, miss. Don't trouble your head about it if you don't mind.

Actually, Mrs. Luddy didn't go down into the street either. Donald did all the shopping. She let the welfare investigator in, the meterman came into the kitchen to read the meter. I saw him from the back room, where I hid. She did pick up her check. She cashed it. She returned to wash the babies, change their diapers, wash clothes, iron, feed people, and then in free half hours she sat by that window. She was waiting.

I believed she was watching and waiting for a particular man. I wanted to discuss this with her, talk lovingly like sisters. But before I could freely say, Forget about that son of a bitch, he's a pig, I did have to offer a few solid facts about myself, my kids, about fathers, husbands, passers-by, evening companions, and the life of my father and mother in this room by this exact afternoon window.

I told her for instance, that in my worst times I had given myself one extremely simple physical pleasure. This was cream cheese for breakfast. In fact, I insisted on it, sometimes depriving the children of very important articles and foods.

Girl, you don't know nothing, she said.

Then for a little while she talked gently as one does to a person who is innocent and insane and incorruptible because of stupidity. She had had two such special pleasures for hard times she said. The first, men, but they turned rotten, white women had ruined the best, give them the idea their dicks made of solid gold. The second pleasure she had tried was wine. She said, I do like wine. You *has* to have something just for yourself by yourself. Then she said, But you can't raise a decent boy when you liquor-dazed every night.

White or black, I said, returning to men, they did think they were bringing a rare gift, whereas it was just sex, which is common like bread, though essential.

Oh, you can do without, she said. There's folks does without.

I told her Donald deserved the best. I loved him. If he had flaws, I hardly noticed them. It's one of my beliefs that children do not have flaws, even the worst do not.

Donald was brilliant—like my boys except that he had an easier disposition. For this reason I decided, almost the second moment of my residence in that household, to bring him up to reading level at once. I told him we would work with books and newspapers. He went immediately to his neighborhood library and brought some hard books to amuse me. *Black Folktales* by Julius Lester and *The Pushcart War*, which is about another neighborhood but relevant.

Donald always agreed with me when we talked about reading and writing. In fact, when I mentioned poetry, he told me he knew all about it, that David Henderson, a known black poet, had visited his second-grade class. So Donald was, as it turned out, well ahead of my nosy tongue. He was usually very busy shopping. He also had to spend a lot of time making faces to force the little serious baby girls into laughter. But if the subject came up, he could take *the* poem right out of the air into which language and event had just gone.

An example: That morning, his mother had said, Whew, I just got too much piss and diapers and wash. I wanna just sit down by that window and rest myself. He wrote a poem:

> Just got too much pissy diapers
> and wash and wash
> just wanna sit down by that window
> and look out
> > ain't nothing there.

Donald, I said, you are plain brilliant. I'm never going to forget you. For God's sakes don't you forget me.

You fool with him too much, said Mrs. Luddy. He already

don't even remember his grandma, you never gonna meet someone like her, a curse never come past her lips.

I do remember, Mama, I remember. She lying in bed, right there. A man standing in the door. She say, Esdras, I put a curse on you head. You worsen tomorrow. How come she said like that?

Gomorrah, I believe Gomorrah, she said. She know the Bible inside out.

Did she live with you?

No. No, she visiting. She come up to see us all, her children, how we doing. She come up to see sights. Then she lay down and died. She was old.

I remained quiet because of the death of mothers. Mrs. Luddy looked at me thoughtfully, then she said:

My mama had stories to tell, she raised me on. *Her* mama was a little thing, no sense. Stand in the door of the cabin all day, sucking her thumb. It was slave times. One day a young field boy came storming along. He knock on the door of the first cabin hollering, Sister, come out, it's freedom. She come out. She say, Yeah? When? He say, Now! It's freedom now! Then he knock at the next door and say, Sister! It's freedom! Now! From one cabin he run to the next cabin, crying out, Sister, it's freedom now!

Oh I remember that story, said Donald. Freedom now! Freedom now! He jumped up and down.

You don't remember nothing boy. Go on, get Eloise, she want to get into the good times.

Eloise was two but undersized. We got her like that, said Donald. Mrs. Luddy let me buy her ice cream and green vegetables. She was waiting for kale and chard, but it was too early. The kale liked cold. You not about to be here November, she said. No, no. I turned away, lonesomeness touching me and sang our Eloise song:

> Eloise loves the bees
> the bees they buzz
> like Eloise does.

Then Eloise crawled all over the splintery floor, buzzing wildly.

Oh you crazy baby, said Donald, buzz buzz buzz.

Mrs. Luddy sat down by the window.

You all make a lot of noise, she said sadly. You just right on noisy.

The next morning Mrs. Luddy woke me up.

Time to go, she said.

What?

Home.

What? I said.

Well, don't you think your little spoiled boys crying for you? Where's Mama? They standing in the window. Time to go lady. This ain't Free Vacation Farm. Time we was by ourself a little.

Oh Ma, said Donald, she ain't a lot of trouble. Go on, get Eloise, she hollering. And button up your lip.

She didn't offer me coffee. She looked at me strictly all the time. I tried to look strictly back, but I failed because I loved the sight of her.

Donald was teary, but I didn't dare turn my face to him, until the parting minute at the door. Even then, I kissed the top of his head a little too forcefully and said, Well, I'll see you.

On the front stoop there were about half a dozen mid-morning family people and kids arguing about who had dumped garbage out of which window. They were very disgusted with one another.

Two young men in handsome dashikis stood in counsel and agreement at the street corner. They divided a comment. How come white womens got rotten teeth? And look so old? A young woman waiting at the light said, Hush . . .

I walked past them and didn't begin my run till the road opened up somewhere along Ocean Parkway. I was a little stiff because my way of life had used only small movements, an occasional stretch to put a knife or teapot out of reach of the babies. I ran about ten, fifteen blocks. Then my second wind came, which is classical, famous among runners, it's the beginning of flying.

In the three weeks I'd been off the street, jogging had become popular. It seemed that I was only one person doing her thing, which happened like most American eccentric acts to be the most "in" thing I could have done. In fact, two young men ran alongside of me for nearly a mile. They ran silently beside me and turned off at Avenue H. A gentleman with a mustache, running poorly in the opposite direction, waved. He called out, Hi, senora.

Near home I ran through our park, where I had aired my children on weekends and late-summer afternoons. I stopped at the northeast playground, where I met a dozen young mothers intelligently handling their little ones. In order to prepare them, meaning no harm, I said, In fifteen years, you girls will be like me, wrong in everything.

At home it was Saturday morning. Jack had returned looking as grim as ever, but he'd brought cash and a vacuum cleaner. While the coffee perked, he showed Richard how to use it. They were playing tick tack toe on the dusty wall.

Richard said, Well! Look who's here! Hi!

Any news? I asked.

Letter from Daddy, he said. From the lake and water country in Chile. He says it's like Minnesota.

He's never been to Minnesota, I said. Where's Anthony?

Here I am, said Tonto, appearing. But I'm leaving.

Oh yes, I said. Of course. Every Saturday he hurries through breakfast or misses it. He goes to visit his friends in institutions. These are well-known places like Bellevue, Hillside, Rockland State, Central Islip, Manhattan. These visits take him all day and sometimes half the night.

I found some chocolate-chip cookies in the pantry. Take them, Tonto, I said. I remember nearly all his friends as little boys and girls always hopping, skipping, jumping and cookie-eating. He was annoyed. He said, No! Chocolate cookies is what the commissaries are full of. How about money?

Jack dropped the vacuum cleaner. He said, No! They have parents for that.

I said, Here, five dollars for cigarettes, one dollar each.

Cigarettes! said Jack. Goddamnit! Black lungs and death! Cancer! Emphysema! He stomped out of the kitchen, breathing. He took the bike from the back room and started for Central Park, which has been closed to cars but opened to bicycle riders. When he'd been gone about ten minutes, Anthony said, It's really open only on Sundays.

Why didn't you say so? Why can't you be decent to him? I asked. It's important to me.

Oh Faith, he said, patting me on the head because he'd grown so tall, all that air. It's good for his lungs. And his muscles! He'll be back soon.

You should ride too, I said. You don't want to get mushy in your legs. You should go swimming once a week.

I'm too busy, he said. I have to see my friends.

Then Richard, who had been vacuuming under his bed, came into the kitchen. You still here, Tonto?

Going going gone, said Anthony, don't bat your eye.

Now listen, Richard said, here's a note. It's for Judy, if you get as far as Rockland. Don't forget it. Don't open it. Don't read it. I know he'll read it.

Anthony smiled and slammed the door.

Did I lose weight? I asked. Yes, said Richard. You look O.K. You never look too bad. But where were you? I got sick of Raftery's boiled potatoes. Where were you, Faith?

Well! I said. Well! I stayed a few weeks in my old apartment, where Grandpa and Grandma and me and Hope and Charlie lived, when we were little. I took you there long ago. Not so far from the ocean where Grandma made us very healthy with sun and air.

What are you talking about? said Richard. Cut the baby talk.

Anthony came home earlier than expected that evening because some people were in shock therapy and someone else had run away. He listened to me for a while. Then he said, I don't know what she's talking about either.

Neither did Jack, despite the understanding often produced by love after absence. He said, Tell me again. He was in a good mood. He said, You can even tell it to me twice.

I repeated the story. They all said, What?

Because it isn't usually so simple. Have you known it to happen much nowadays? A woman inside the steamy energy of middle age runs and runs. She finds the houses and streets where her childhood happened. She lives in them. She learns as though she was still a child what in the world is coming next.

from Riverfinger Women

Elana Nachman

DISOWN THEM AS she may try, Mr. & Mrs. Bramanoi and their youngest child, Albert (re-christened Al Bear Riverfingers by Inez, ten years his senior), write a few words to their only daughter as she travels. They love her, they try to love her, they try anyway to give her a sense that she has a family background.

August, 1963

Dear Inez,

    Albert is now going to type a few words to you all by himself.

    sally runs tothe car.Dick play ball.

    Thank uyou very much for therecord. Itis a nice record.I love

    ey you very much.

    Jane fell down.Spot bit Puff.

                  Albert

Elana Nachman has traveled extensively in the Northwest, and lived there for a time. In addition to her novel, *Riverfinger Women,* she is the author of a short story collection, *They Will Know Me by My Teeth.* This excerpt from *Riverfinger Women,* copyright © 1974 by Elana Nachman, is reprinted by permission of Daughters Publishing Co., Inc.

May, 1964

Dearest daughter,

It was so good speaking to you on the phone. Things here follow their usual course. Mommy is usually in pain but bears up well. And I go on working. Now that I have given up psychiatry, I am lost. Instead of weird dreams, I have nightmares. No one ever knows the private hells that exist in the minds of others or the loneliness. When I had my kidney stone all I could think about besides the pain was you and how awful it would have been if this happened when I was alone. I started to worry about your being alone. I am concerned that man because of his frailty and lack of physical ability to cope with his environment is a gregarious animal. He *needs* people to help him live. We all live needing other people. Enough philosophy. How's kicks? Can't wait to see you. All my love,

Daddy

June, 1966

Dear Inez,

I love you very much. School is very good and my report card is best and I got all A's and B's, too. I'm doing very good in school but school is out now and we have a summer vacation. School is out, play begun.

Mike just had his birthday party and he's 11 years old and he had a picture of two bats and a ball on his cake. I'm doing very good in my swimming. I went to the beach yesterday and Mike got a fishing rod for his birthday from his dad and I gave him Flubber but that was nothing much. I think that's all.

Love,
Albert

June, 1967

Dear Inez,

I only wish I could express myself as well as you do, could put thoughts and feelings on paper in the magic, meaningful ways you do. I'm slowly getting back to my typewriter (your father insists I should write more) but still the mechanical instrument fights back. But I shall try to tell you what reactions your letter brought, what latent feeling and emotions within me.

First of all, your letter about you and Abby clarified, in a way, what I was mulling over in my heart and mind these past months and I tried to express to you during your visit here spring vacation: all of us here love you very much—if a child does well she should be happy and of course mother and father will be at ease. But, as your mother, perhaps greatly aided by five years of my own psychiatry and gaining more insight thereby, I came to realize that I have always loved you (fat, thin, bright, stupid, responsive, withdrawn, etc.) and I don't even know exactly if it's because you are my daughter, though I know it may be trite to repeat the old cliches of "mother's love is a birthright" and "even a beetle is beautiful in the eyes of its mother." I say these things to you now for several reasons: perhaps I myself have been too "hung up" to express them before to you and you have been out of the family environment for so long that you cannot really know that you would have actually felt this love constantly had you lived with all of us, even in the midst of fights, yellings, scenes, worries and problems.

Your relationship with Abby, as you described it so well, comes as no great shock or surprise to me or would to those who have known you best through these years. I know that you have been kind of wavering back and forth about homosexuality for a long time (forgive my nonscientific language) and that was evident often from other graspings for relationships you had; that whatever you feel you must do to maintain your happiness and dispel the fears that go bump in the night can only bring happiness to me, too . . . I have long since learned that all is not black or white and mine is not to "approve" or "disapprove" but just try to understand and accept—this I do, in all good spirit and love (should add that, in my role of mother and omnivorous reader, I shall continue to caution against anything but brief, limited, experimental activities involving "mind-blowing" drugs and the like, including traveling around without using extreme caution and sense, which I think you have developed better than most—not that my cautions will prevent anything but it makes me feel better that I haven't been entirely remiss in "parental advice").

As for Abby herself, she did immediately strike me as a warm, sensitive, yearning person, the kind I instinctively like and want to be with, and I'm very glad now that you gave us the opportunity to meet her at your graduation. I wonder, not actually knowing them personally, if her parents are not aware somehow of your situation, if only subconsciously, and of course, most parents do not want to recognize any "divergencies" unless directly confronted. At any rate, I hope, for both of you, that your relationship is as warm and good as you make it sound.

You asked me to accept your life and you for what it is and you are. This you can count on always. We are here to be counted on and to be of help and to love you.

But also, there will be times when neither I nor your old friends will be available to you in Portland. I have mulled this over and feel it is fair enough to ask, in return for my understanding (not for my love—that is not on the "bargain" table), you contact, at the *very first* opportunity when you get to Portland, a "good" psychiatrist, and try to make a start at regular appointments with him (or her)—believe me, Inez dear, this suggestion is not made because I have any notions that you are "sick" or that your relationship with Abby is "evil" or "unhealthy"—I just feel that there will be times, certainly, when you will need absolutely someone with whom you can talk and work things out.

I have never known many women—or men, for that matter—who were homosexuals, perhaps there were some who never "let it out," but it will be difficult, I imagine, for you to adjust to college life with this extra facet, and I want to make sure you can get whatever help you might need.

Daddy had not been feeling well for a couple of weeks and he desperately needs some rest. Frankly, Inez, I don't really know how *he* was hit emotionally by your letter because he doesn't talk much about these things (I used to think it was just him who was so reticent, but I am beginning to feel most men don't have the same facility for talking about their emotions as women do; that's nature, I guess). But you are his love—that you know without my rhetoric.

My arm is dropping off! Write (or rite, as Albert says)! All my love,

Mother

January, 1968

Dearest Inez,

I realize that it has been too long since my last letter (I can plead work, of which there is always so much I don't know what to do) or my own father's death, but these things are my worries —and you are my joy. Possibly that's why I have not written in so long—the Puritan in me keeps me from those I most enjoy.

But you worry me sometimes too. I never answered/responded to your summer letter about you and Abby partially because I thought Mommy did such a good job and partially because I did not then, nor now, know what it is I should say, or want to say.

In some sense there is a parental feeling of failure—I know that is a social hang-up but it exists. The only homosexuals I ever knew were when I was in college, they were all men and seemed to me caught up in a trap of self-centered, egotistical (and brief) relationships to each other and the world, characterized chiefly by their vanity. I certainly hope that you will not become like that and I realize that these were only what they call the visible homosexuals. As I have always told you, I only want that you should be happy; if this is how you want to live your life, I accept it so long as it does not hurt you or anyone else and does not hold back or freeze your capacity as a productive person.

More than that I don't know what to say. I hope you are enjoying school and not working as hard as I have to, to keep you there. Seriously, I know the competition is rough, but I'm sure you're doing well. You're my daughter, after all. I miss you all the time and wish you were here. You'll always be the girl of my dreams.

Love,
Daddy

Mom, I'm Glad You're Not a Scorpio

A mother is a strange object in my mind, one who scrapes out the lining of my intestines with a furcovered fork, whom I would please, whom I would be patient for, through the time when the

question is asked and considered: When will you admit to knowing something about me? But there was never anything—a
space you tried to cover with the cards of your boredom (boredom?).

mother

I have no comfort to give you, only shame. Mother, so many
impossible gestures built into the blood because I can say to
you, "I love you, I am trying to" and you can tell me that a
mother always loves its child; but the lioness in the wood and
the bear by the stream full of their appetites love nothing but
themselves, with a tedious and lumbering instinct, and not with
the knowledge of what it is to die and what precautions must be
taken in the human state to keep the children of knowledge
from going crazy, wandering the earth in small bands waking
in the night for fear of noises and afraid of the water and the
sky. It takes more than a blind rat gnawing at the trap door to
tap the hidden spring of what people can do for each other.

mother

This is a letter from me to you which I didn't start out at. The
Investigating Committee has my name now because I joined the
L.A. G.L.F. not to mention Gay Women's Liberation in San
Francisco and other things I can't tell you about. They are afraid
because we are no longer separating out our private lives from
our public lives and they use always the tools of the liberals to
lobotomize the revolutionaries. To be a lesbian is to be implic-
itly revolutionary and I am just beginning to find out what that
means and so the slick men question and where they can they
jail or hospitalize. I'm not making this up, mother, I've seen it.

Mother I am not sure where I am. I think I am in love again
but sometimes it slips from me while I am out walking after
sundown and I look at the stars and feel as if no one will ever
look up and see me. The distances are too great, the buildings
are too long. There is a certain span where one hopes and after
that time of being able to hope one must give in to being
disappointed. One cannot reach in voids every day, day after
day, when day after day the silver fish that flashes in the bright
waters, which our fingers touch, is gone. She does not hear and
I walk sometimes, over my body a kind of ache after which the
time of hope is gone. The pain, once one has shrugged finally

and said "I hear nothing either," remains a kind of bitterness, a tearing hunger, final impoverishment, and that is what comes to me sometimes though I am still hoping as yet and not really preparing for rejection. But I am afraid and if I am going to be hurt again a year since Abby's going I want it to be simple, clean anesthetic

mother

what could I teach you but how to look gently beyond your own sorrow and fear into how human that is, and how to move past our terrible isolation. My dreams are of you by the sea and large buildings and elevators and the impossibility of it all, about misunderstandings and betrayals and fleeing and not being able to find the right apartment and lost in the alarm clock lost in the work.

mother

these things granted: that you will never speak to me and will never listen to my voice or the stories I mean to tell you and the understanding and the shield, the refuge, sanctuary, will never be given me, nor clemency—you will not know what I mean because you will answer "well, you have a way with words" and gradually the letters will drift back to being about how well you did in the last tennis tournament and my letters will be about nothing. With all of that it is enough you are a Sagittarius and not Scorpio because the women with their black hair, their fresh-bread-smelling breasts and their spider-lady eyes are almost always those (intertwined with the gentle Tauruses who come to rescue me when it becomes too hard). With the Scorpio women it is always life or death (but mother understand I have been hurt so by women because I am an intense asking person. I am always asking and if they had been men instead it would have been much much worse. What I am saying is do not harbor a reserve, a pity, because what a shame it is that she can't be making these motions with the opposite sex. If it had been Scorpio men, mother, I would be dead). And if you had been, had had that hooked jaw behind the careless breath, the accidental brushing of hips at dinner and the denial always of your intentions I would have been mad long past. It is well you are a Sagittarius and tried to do your best and although nothing came of it at least it is not rape itself having to be faced each

time you march across my memory, but only the sense of loss,
a certain kind of long moon sadness, a fading Japanese print of
it, with the sun going down and its red beams like searchlights
over the highways
          mother open this letter
          I cannot bring myself to write it

Naomi Riverfingers turned to Inez in the Gay Liberation
Office and hugged her, having read this letter. Naomi had only
come to see how Inez was, to talk about a project Inez was
planning, in Tangier. Naomi was not altogether comfortable in
the Gay Liberation office; it was just to see Inez that she had
come. They would sit and talk together, Naomi and Inez, about
writing and revolutions, how the world is changing. Naomi
liked the idea of strong women coming together to make a
women's revolution. She liked it as an idea she might want to
put into a dance. Naomi quickly saw that there were not many
women in the Gay Liberation Front office. Inez was slowly coming to the same conclusion.

Naomi was seventeen, she was a dancer, and she lived with
men. She had violet eyes and long black hair. They heard things
about a women's movement, but for Naomi it would mean
reconsidering the men. She wasn't ready. To Inez it meant
calling Naomi her sister, which wasn't true. Naomi was a friend
who had a strong taste for adventure. For Naomi knowing Inez
was an experiment.

"I'm sorry. I can't," Naomi said after she read the letter.
Naomi was not a Scorpio, but she knew she would do as well;
she did not mean to get entangled in the part that was not an
adventure, in the part that is about: will we be working for a
good common life? She hugged Inez, and walked out. They
would meet next week.

Inez folded the letter carefully into her notebook. Inez
watched from the window, Naomi and her hair walking down
the street.

The Phantom Child

Aviva Cantor

SEVEN YEARS AGO I became pregnant and had to decide whether or not to have an abortion.

The circumstances were these: I was thirty-one, not overage, and physically healthy. I'd been married five years. I was not poverty-stricken; financially, we could probably afford a child if we cut down on some of the material goodies.

I knew I was pregnant even before the second test verified it, knew it by a kind of sharp sucking-in feeling in the belly. I was only a week late, but I Knew. Still, I was not prepared for the shock I felt upon getting the official confirmation. I couldn't breathe. I could barely move. I was frozen with terror.

Yes or no? Day and night, I cried interminably. The only

Aviva Cantor, the daughter of Russian Jewish immigrants, graduated from Barnard College, having also attended the Hebrew University. Since receiving her master's degree in journalism from Columbia, she has served on the staff of the *London Jewish Chronicle, Israel Horizons,* and *Hadassah* magazine. A Socialist Zionist, she worked for the Jewish radical movement and edited the *Jewish Liberation Journal.* She is a founding editor of *Lilith,* the Jewish feminist magazine. She lives in Manhattan with her husband and two cats.

previous anxiety attack of such magnitude and scope I'd experienced was when my mother was dying piece by piece. Except that then, I escaped into work, into sleep, into eating, and into an intense romance (despite my guilt). This time, those escape routes were closed. My husband withdrew into a self-protective shell, sicklied o'er with the pale cast of pseudo-feminist rhetoric: "It's your body, you decide," he said. My terror was compounded by his withdrawal.

Yes or no? Back and forth like a tortured laboratory animal in an impossible maze: there are two exits, but both give the animal an electric shock.

Yes or no? Could I handle being responsible for another human being? And yet, could I actually destroy this life? And for such ... frivolous reasons? Could I be so selfish, so petty? When one of my cats gave birth to a kitten who couldn't walk without falling over, I could not bring myself to "put him to sleep."

Could I kill my own *child*? Excise it from my body like a cancer?

There is no doubt in my mind that part of me wanted a child.

For the preceding three years I had been involved body and soul in work for the Jewish radical movement. I'd had the feeling that all my previous experience and education were but a *hachshara*, a preparation for this great moment in history of which I was a part. At last, I thought, at last I found my Destiny. But now the movement was collapsing ... my Destiny had eluded me, it was a phantom. With the movement gone, the thought teased: "Well, if it happens, it happens."

Had I fallen, then, into the typical woman's trap of becoming pregnant because I perceived other paths to be blocked? Was it because everything was collapsing around me that I sought "refuge" in (creating a new) family? For I had no family: my parents were dead, my aunts and uncles far away in Canada and Russia, my siblings nonexistent. My husband was not "family": a couple is not a family.

Yes or no? As the thoughts swirl in my head, the fetus digs

its claws deeper into my uterus. I try not to think of this thing growing in me: growing means like a growth. Growing uncontrollably. I have no control over it. It will grow and grow and take over. Take over my body and my life.

What, oh what would my mother have said had she been alive—and thank God she was dead. This would have killed her—and me, too, had she known. But while my mother had been buried in 1966, she was alive inside my head, exhorting, begging, pleading, threatening, inducing guilt: nagging. How can you do this? You'll always regret it, always. You remember your Aunt Dora, how poor they were when they got married, it was the Depression? Well, she thought she was pregnant and took hot baths and her period came and she never got pregnant again. Some women only have one chance. (Remembering this, I immerse myself in super-hot baths and emerge faint and breathless but still pregnant. I also contemplate throwing myself down the stairs as Ava Gardner did in *The Snows of Kilimanjaro* but am too cowardly.) And didn't Simone de Beauvoir once say that the one thing she regretted was never having a child?

Not to speak of my mother herself, who lost the baby she was carrying three years after I was born. And who was still weeping over it when I was nine and she told me how babies are born (an appropriate confluence of information-sharing, of course). And whose passionate desire was to have another baby because she saw herself first and foremost as a Mother: who went to numerous doctors who stuck painful objects up her womb to Find Out Why she couldn't get pregnant again, but who never discovered The Secret. My mother who considered the childless to be victims of a cruel fate from which she was but one step removed. I was not to speak of children, *any* children, in front of Aunt Dora, it was too painful. Do you know, she almost went *crazy* when she found out she couldn't get pregnant?

Yes, women were always going Crazy when they couldn't get pregnant and here I was going crazy when I did. Our Fourmothers (foremothers) were constantly tearing their hair out when they had a prolonged nonintentional breather between children. (Of course, since the birth of daughters was never recorded, they may have been popping out unwanted girl-children, we'll never know.) And remember that scene in *The Wall* when Symka tries to tell Berson "It's impossible" for her to have children but he's busy discussing philosophy with a friend and tosses off a nonchalant "That's too bad, dear ... come and discuss Kant with us" and she flings herself into the bedroom to sob (while the men continue discussing philosophy).

At last I understand the look on my mother's face in the picture album I'd made at the age of nine in imitation of the one on *I Remember Mama* and as a talisman to transform us into a "happy family," American-style. Before she'd had the miscarriage, her face is all trusting smiles, love. Afterward, a look of hard determination, of steely stubbornness. A Never Again look. She blamed my father for the miscarriage and resolved never to trust him again. From now on, *she* would make the important decisions in the family, she would be Boss. And so it was.

How I hated her for it! How I wanted us to be a "happy family," i.e., where the man was boss. How I longed for Daddy, a gentle and peaceful man (whom I saw then as weak), to put his foot down when she began nagging. And yet ... maybe she was as she was because she knew that men cannot be counted on, that the basic unit in society has always been the mother and her children, that men are not dependable, men collapse, men withdraw, men abandon.

My mother had learned about abandonment a long, long time before this. She was born in Dubno (then Russian Poland) several years before World War I, the second of four girls. When she was only six, her father Lazar died of gangrene; her mother Esther of typhoid. Superimposed on this primal abandonment was a secondary one: the rest of the family, the uncles and aunts and grandparents, did not take in the four little girls, they left them to fend for themselves all alone, to dig for potatoes and to steal apples in the middle of the night. Why they abandoned the children is a family mystery. (Many, many years later, when one of the aunts survived the Holocaust and came to North America, she begged forgiveness of my mother, but had no explanation for the family's behavior.) And finally, the third abandonment. A committee of well-meaning Jews in Montreal decided to bring war orphans to Canada. They were farmed out to foster homes. The "people I lived with" (as my mother called them) treated her like a servant, forcing her to leave high school to take care of the children and the store. Another abandonment. Finally, she and her two younger sisters ran away.

The oldest sister, Gittel, had remained in Dubno because she was already in love with Shmuel, whom she later married. There's a picture in the album of their two sweet little boys who wrote my mother from Dubno how they were going to "make something of themselves." Fertilizer the whole family became. Shmuel, his arms torn off during a deportation. Gittel shot trying to reach him. And the little boys, nobody knows what happened to them, I don't even know their names. . . .

When I was twelve, my mother's first cousin came to New York to seek other survivors. He sat in the kitchen at night telling my parents horror stories by the hour. He and his mother had survived because a peasant hid them in a pit with other Jews for two years. You couldn't stand in the pit, only lean. Little did they know that I was listening. After that I graduated to my own extensive Holocaust research.

I remember how in *The Wall* Rutka and Mordechai decide to have a baby and how that baby makes it impossible for her (but not him) to be in the Resistance. (What if, God forbid, there's another Holocaust? If I have a child I'll have to worry about her getting killed or tortured, having to hide her, give her away. Without a child, the worst that can happen is that I'd be tortured.) Someone asks Rutka why did they deliberately have the baby in such brutal times and she tells them how Chaim, a worker in Vilna, had told her, "Rutka, you'll never know what it means to be a Jew until you're the parent of one."

For the preceding five years, I had written and lectured passionately about "Jewish values" and how we had to rediscover them and preserve them and develop them. Now I sought to find one Jewish reason to have this child when the idea so terrified me and I could not.

I thought of great scholars, thinkers, leaders. What if Ber Borochov's mother had had an abortion? Or Moshe's mother? Or Emma Goldman's? What if this baby is destined to do great deeds, to be a light unto the world?

What if my baby is the Messiah?

(I don't have enough to feel guilty about, I also have to lacerate myself for denying Redemption to the world.)

My friend Fanya, who is well along in years, listens to the whole story. Like all my women friends, she is warm and supportive. She tells me this is not a world to bring a child into.

But wasn't it always thus? Didn't people in every period of history think their times the worst? My mother told me how she'd convinced my father that it was time to become parents. My father, having married at forty-two after an enjoyable bachelorhood, hadn't given children much thought; moreover, he was underemployed. "Let's wait for the right time," said Joseph. "There'll never *be* a right time," from Naomi. So I was conceived and brought into the world, loved and welcomed. The pictures of my mother in the album have a "hey, I actually did it" look on her thin girlish face. My father looks pleasantly surprised and vastly relieved: it wasn't so bad, after all.

Fanya doesn't feel guilty about her nine illegal abortions, she tells me. And yet, had she also aborted her only child, a daughter, who would be taking care of her now? If I abort this child, I'll be alone in my old age. I'll be a shopping-bag lady on the streets of New York, a prey of muggers. I'll be helpless, a prisoner in a Bergman nursing home. I'll be dying of cancer in great pain with no one to even bring me exit pills.

My mother, though, did have a daughter and that was as good as having none at all. My mother died of cancer in great pain, and I didn't help her. I hadn't known what to do, I was afraid of the doctors and nurses, afraid to pester them again and again for more morphine lest they vindictively cut it off entirely. Why didn't I scream at the doctor? find another one? ten other doctors? fifty other treatments? Why did I abandon her as she'd been abandoned so many times before?

So I blew my last chance to reconcile, to atone. To make up for having been an insensitive and inconsiderate daughter. For having tormented her with words of rage when we fought, telling her how worthless her life was because she didn't have more children and how glad I was that she'd had a miscarriage and lost the baby because she would have made my sister suffer as she made me suffer. Every word a poisoned dart. But somehow, this revenge had no sweetness, no relief. She would cry. And then I would feel my guts turning over with pity and guilt, and I would cry. And we'd swear never to fight again . . . until the next time.

There in the hospital, she lies dying and I can't talk to her. We are playing a Game. I don't tell her she's dying to "protect" her; she doesn't tell me she knows she's dying to "protect" me. The doctors have told me this is the best way. One night I dream we are shopping and she buys me a coat because, she says, she'll never be able to buy me anything ever again. I wake up sobbing: I realize she knows. . . . But I do nothing. I am paralyzed. I cannot accept that she is dying.

If I'd had a good relationship with my mother, wouldn't I want to be a mother myself? Am I afraid of repeating with *my* child the relationship I had with my mother? Do I want to have an abortion to spite my mother (even though she will never know)? Or do I want to have the baby to placate her? to diminish my guilt (it can never be removed) by doing something that she considered the Right Thing? To do something *for* her? To do this thing that I don't want to do as punishment for killing her so that I can stop punishing myself in other ways? Do I see having a child as a punishment? ("May your children treat you as you treat me," was one of her favorite curses.) As penance?

Although I can find no Jewish reason to have the child, I find plenty of Jewish reasons for having an abortion. Consulting my trusty resource *Birth Control and Abortion in Jewish Law*, I confirm that abortion is permitted when the mother's physical *or mental* health is in danger. Then it's a matter of self-defense and the mother's life comes first. Didn't the rabbis permit abortions during the Holocaust because a pregnant mother's life was automatically in danger? And further, doesn't the Talmud say that if two people are in a desert and one has a bottle of water, and if sharing it with his companion would save neither, he (the bottle owner) has the right to drink it all himself, because his life is worth just as much as his friend's. Isn't my life worth saving just as much as the baby's? I wouldn't go so far as saying more, but as much. . . . See, all those years of studying Talmud and learning *pilpul* have not been in vain.

After a month of agonizing, it comes to this: it's either me or the child. And I have a right to live.

I have to sacrifice the child in self-defense.

February 4, 1972. A bright crisp morning eight days before my thirty-second birthday (makes it easy to remember, as if I needed that *zets* every year). My husband drives me to the doctor's office. Getting out of the car, I have a brief flash of running away screaming, "No, I won't do it!" like April in *The Best of Everything* (the movie not the book).

But I didn't run away.

The operation takes all of twenty minutes, only about five of which are painful. I'm willing to endure the pain: I deserve it. During these endless minutes, I see myself in Block X in Auschwitz, being sterilized by Dr. Mengele or a cohort, without anesthesia or other medication. And Dr. Brewda (the brave woman doctor there: not for naught have I done my Holocaust research) holds my hand and says, "Be brave, my child, just a little longer." I hold the nurse's hand very tight and she actually squeezes back. I love her like a mother, a real mother, who is tender and loving, not nagging, pushing, yelling, rejecting.

It's all over.

I resolve not to ask or think about what they do with the fetuses after the "procedure" as they call it, otherwise I'll weep forever.

As I walk out, I feel I've made the first adult decision of my entire life.

P. S.: Five years later. I go to a NOW-New York abortion rap. A woman there talks of the illegal abortion she had thirty years ago. "You never forget," she said. "Every year you think of its birthday, how old it would be now.

"It's a phantom child."

First Love

Andrea Dworkin

Yet if I care to care
force loving into being, then I pry open

all memory's charnal house of sores
—Robin Morgan, "Credo"

E,

It is so hard to write you. Why am I doing it this way, not intending ever to send this letter, still with one eye to publication, a grand concept for a book in some sense, and still with one eye, that poets conscience, to a future which becomes increasingly impossible to imagine. It seems the only way I can bear the passion behind the language, the memory, the desire, the only way not to be burnt up by what I feel. You come over me in waves of memory, especially when I sleep, and I wake up in sweat and trembling, not knowing where I am, not remembering the years that separate us.

So often I wanted to write, dear E, now I am this person, I look this way (you wouldnt like it), I do this, I feel this, lists, details, it was warm or cold on that day when that happened and then my life changed in this way and that—but I cant, I never could, and I cant now. In writing this letter, not to be sent, perhaps I can find the signs that will tell you who I have become.

Dearest E, I loved you. Now that love is memory, sometimes haunting, sometimes buried, forgotten, as if dead. I see yr face, yes, I know, as it was, I remember you as I remember the sun, always, burned in my brain; somehow you are part of me, mixed up in me, for all the days of my life. I left you when you were life to me, when to be physically separated from you was sheer and consuming pain, as if a limb had been cut off, amputated. Leaving you was the hardest, and perhaps the bravest, thing I have ever done.

Dearest E, I want to describe in some way the *drive to become* that impelled me to go to you and to go from you, that has driven me from person to person, place to place, bed to bed, street to street, and which somehow coheres, finds cogency and true expression, when I say, I want to write, or I want to be a writer, or I am a writer. I want to tell you that this *drive to become* is why I left you and why I never returned as I had promised.

I was 19 when I knew you. I wanted to be a writer. I didnt want to go mad or suffer or die. I was 19. I wasnt afraid of anything, or, as I sometimes thought, I was equally afraid of everything so that nothing held a special terror and no action that interested me was too dangerous. I wanted to do everything that I could imagine doing, everything I had ever read about, anything any poet or hero had ever done. I loved Rimbaud. I loved Plato and through him Socrates. I loved Sappho. I loved Dostoevsky, and sweet Shelley, and Homer. I loved cold Valery, and warm D.H. Lawrence, and tortured Kafka, and raging tender Ginsberg.

I didnt have questions in words in my mind. I had instead

these surging impulses that welled up and were spent. I had a hunger to know and to tell and to do everything that could be done. I had an absolute faith in my own will to survive.

What I didnt want to do was to say, look Im this height, and I went to school here and there, and then that year I did this and that, and then I knew so and so, and then the next one was so and so, and then this situation occurred, and then that one, and the room was red and blue and three by four, and then I was that old and went there and did that and then that and then, naturally, that.

I wanted instead to write books that were fire and ice, wind sweeping the earth. I wanted to write books that, once experienced, could not be forgotten, books that would be cherished as we cherish the most exquisite light we have ever seen. I had contempt for anything less than this perfect book that I could imagine. This book that lived in my imagination was small and perfect and I wanted it to live in person after person, forever. Even in the darkest of human times, it would live. Even in the life of one person who would sustain it and be sustained by it, it would live. I wanted to write a book that would be read even by one person, but always. For the rest of human time some one person would always know that book, and think it beautiful and fine and true, and then it would be like any tree that grows, or any grain of sand. It would *be,* and once it was it would never not be.

In my secret longings there was another desire as well, not opposite but different, not the same but as strong. There would be a new social order in which people could live in a new way. There would be this new way of living which I could, on the edges of my mind and in the core of my being, imagine and taste. People would be free, and they would live decent lives, and those lives would not be without pain, but they would be without certain kinds of pain. They would be lives untouched by prisons and killings and hunger and bombs. I imagined that there could be a world without institutionalized murder and systematic cruelty.

I imagined that I could write a book that would make such a world possible.

So my idea of my book that I would write sometimes took another turn. It had less to do with the one person who would always, no matter how dark the times, somewhere be reading it, and it had more to do with here and now, change, transformation, revolution. I had some idea of standing, as one among many, my book as my contribution, at one point in history and changing its course and flow. I thought, imagine a book that could have stopped the Nazis, imagine a life strong and honest enough to enable one to make such a book. I began to think of writing as a powerful way of changing the human condition instead of as a beautiful way of lamenting it or as an enriching or moving way of describing it.

I had wanted to make Art, which was, I had been led to believe, some impeccable product, inhuman in its process, made by madmen, inhuman in its final form, removed from life, without flaw, perfect, crystal, monumental, pain turned beautiful, sweat turned cold and stopped in time, suffering turned noble and stopped in time.

But I also wanted to write a book that could be smelled and felt, that was total human process, the raw edges left as raw as any life, real, with a resolution that took one to a new beginning, not separate from my life or the lives of the multitudes who were living when I was living. I wanted to write a book that would mean something to people, not to dead people past or future, but to living people, something that would not only sustain them but change them, not only enhance the world in the sense of ornament, but transform, redefine, reinvent it.

When I knew you I was 19. I did not know many things. How could I? I wanted to make Art, and I had a passion for life, and I wanted to act in the world so that it would be changed, and I knew that those things nourished one another but I did not know how. I did not know that they could be the same, that for me they must be the same, for they all had to live in this one body as one or they could not live at all.

The teachers I had had did not know or tell the truth. They did not care about how artists lived in the world. They seemed to find the lives of artists shoddy and cheap, even as they found works of art marble and pure. They never talked about art as if it had anything at all to do with life. They thought that the texts were there to be analyzed, or memorized, one after another. They thought that art was better than life, better than the artists who made the art and lived their lives. They had no notion of process, how one made something out of the raw impulses of the imagination, how one ransacked or chiseled experience in search of meaning, how one cried out or mourned or raged in images, in language, in ideas. So they taught that ideas were fixed, dead, sacred or profane, right or wrong, to be studied but not created, to be learned but not lived. They did not seem to know that the whole of human literature is a conversation through time, each voice speaking to the whole of human living.

And I did not understand so much. I did not understand, for instance, that people really die. I did not understand that death is irrevocable. I did not understand the grief of those who remember the dead. I did not understand that the horrors of history, those textbook cases of genocide, rape, and slaughter, would happen in my lifetime to people I knew. And so I did not understand that the earth is real, and that what happens on it happens to real people just like me. I did not understand that as I grew older my life would continue with me. I thought instead that each event in my life was discrete, each person of that moment only. I did not understand that the people I knew I would always know, one way or another, for the rest of my life. I did not know that one never stops knowing anything, that time continues to pass relentlessly, though without any particular vengeance, taking each of us with it. I did not understand then that there is no choice, that one always writes for the living, that there is no other way to create the future or to redeem the past. I also did not know that each human life is precious, brief, an agony, filled with pain and struggle, sorrow and loss.

I love books the way I love nature. I can imagine now that someday there will be no nature, at least not as we knew it together on Crete, no mysterious ocean, no luminous sky, no stark and unsettled mountains. I can imagine now that a time will come, that it is almost upon us, when no one will love books, that there will be no people who need them the way some of us need them now—like food and air, sunshine and warmth. It is no accident, I think, that books and nature (as we know it) may disappear simultaneously from human experience. There is no mind-body split.

I never think of you without remembering the ocean. It is an emblem for me of that time in my life, of the depth and tumult of my feelings, of how my life broke out of my skin and beyond itself into an unknown, primal realm. The ocean does not signify anything whimsical, cheap, romantic, or self-indulgent. It signifies the true mysteries, not the mystifying ones. It signifies the light years between galaxies, as well as ones tie to everything on earth. It signifies ones tie to the enormity of being, to the mystery of this universe—stars, moon, sun, black holes, rings around Saturn. It makes one aware that this universe is a tapestry of the most awesome magnificence. It does compel awe.

It has always been to me, the ocean, overwhelming, monstrous, deep, dark, green and black, so foreign that it requires respect, silence, humility. It is boundless and deep, no human sense of time can circumscribe it, it rumbles with cavernous sounds, it is filled with grotesque forms, luminous colors, shapes that defy imagination. All of the life in it is menacing, compelling, exquisite, with nothing consoling.

I love books too in the same way. They are the human ocean, life before and through and beyond this self, footsteps on the sand in the largest desert, the wind blows, the tracks are sometimes obscured, covered over, hidden, waves of human experience in which one drowns, which carry one, against which one struggles with every life force, forced sometimes under, struggling for any breath, the weight of that water bearing down mercilessly on one, or floating, effortless, calm, at the precise point between earth and sky. They are the human ocean of our time, the quest of people through time to know, ask, feel, survive, to survive beyond the limits of an awful, or insignificant, or invisible, or painful or ordinary life, beyond the limits of this mortal body, sick, needful, the vessel of so much suffering and despair. They are the meaning of life as fully as we can render it. They are the human ocean of everything that has been experienced, thought, felt, wondered, suffered, recognized, realized, imagined, affirmed—messages sent through time from one finite human who asked questions, attempted answers, described, felt, needed, wanted, endured, resisted, to another who is different yet the same.

A book is at once connected to eternity and to one persons mortal flesh. It is whatever this flow is that connects us one to the other throughout human time, but it is also the fruit of one persons specific moment. It is the present, just as the ocean, whatever it was before, whatever it will be later, exists for the one who sees it when she sees it. Think of it, each book is what it is for one person to be alive, in her particular present, what it is, anguish, joy, fear, duration, process, hope. Each person asks the question of her time and place. Each persons life inhabits and informs every word written. Sitting somewhere, ancient Greece or Manhattan 1974, hoping that the words will come and make the feeling in the body bearable, fill the need, make the day or night endurable, that one will be able to give shape to the chaos of feeling, needing, not knowing. The world takes form when one writes, for the writer. The world becomes knowable, its meaning revealed and affirmed. Struggling with the present, with death, with pain, with love, articulating the present, imagining it as it is and as it might be, asking every question but also taking time itself and giving it shape, substance, weight: revealing it to those who share it.

Ive been reading Kafka, his letter to his father, his diaries, his letters to his woman friend Felice. Discovering the person behind those monuments of consciousness, discovering the tortured man who subsumed the person. Discovering the fragile, vulnerable, terrified being behind those monuments of ravaged and ravaging male consciousness. What is it about genius that it can inhabit the body of a tubercular, frightened, insignificant German Jew and that he can then force the world into a new shape so profound, so recognizable, its vision so deeply rooted in the nature of things as they are, so tangled in the gut and psyche of life as we experience it, that one says, I dont know where this story ends and life begins, I dont know the difference anymore between this story and life, I dont know at what point I became part of this story and forgot that it is print on a page, I dont know how these words were ever put together this way, or how these images were formed, but I know that this writing embodies the world as profoundly as a male could embody it in words.

To me, the real mystery is, what made him a writer, how did he become a writer, what in his life determined it, how was it even possible. He was a writer, how can I say it, the way that a fish is a fish. Not fragmented, a bit here, a bit there, sometimes choosing this, sometimes doing that, not with other ways of being, e.g. sometimes we walk and sometimes we sit and sometimes we run. He was a writer as a fish is a fish, always, all the time, knowing nothing else, without any other possibility. Imagine being a writer like that. (In a footnote I read: "Kafka was survived by three sisters. All three sisters, including Kafka's favorite, Ottla, and the larger part of their families, were killed by the Nazis.")

The first book I remember reading was *Squanto and the Indians*. The Pilgrims, an austere religious group, came to Amerika from England where they were persecuted for their religious beliefs. The voyage was long and hard and many died and many more were gravely ill. In the new land life was no easier. Winters were freezing and hard, the soil was barren and nothing they planted grew. They suffered terribly, starving and dying. Then an Indian named Squanto befriended them. He showed them how to plant corn and how to live off the land. He helped them to plant their crops. They reaped a good harvest which Amerikans commemorate as Thanksgiving. Then they slaughtered Squanto and his tribe.

On the one hand, the genius, the kindness, the fragile, single human being who can, through an act of being, a simple act of simple giving—writing, teaching, planting—do so much more than endure, who can transform, who can make life both possible and meaningful. Then, always, on the other hand, vicious slaughter, insane, impossible, relentless slaughter.

Squanto, Kafka, the Nazis, those first English interlopers, the tanks entering Athens, my friends, fragile human beings every one, rounded up like cattle, herded into jails, there tortured, there their bodies broken, terror, violation, killings and ravagings on a grand scale, always the grand scale, mad ambition, hundreds or thousands or millions, victims, tanks, rifle butts, machine guns, searches, seizures, arrests, terror, death. I am always asking, will it never end. I am always vowing, we will end it.

I remember one letter you wrote me after the colonels took over. You said that yr life was bitter, that the earth had turned to poison. You said, what do you know about any of this? And, after all, what did I know? I didnt know then most of what I have had to learn—slow, dimwitted, dull, fighting always the romantic self-indulgences into which I was born. I didnt know then that I wont be spared anything. I didnt know then that none of us will be spared anything. Anyway, there is really no way to describe white Amerikan ignorance (and it is not only middle-class, it is *Amerikan,* an ignorance democratically distributed). Who would believe that this ignorance is real as villages burn and people die? And there is really no way to talk about white Amerikan innocence, except to say that some of us have lost it. Except to say that years later I learned that I was a woman, and so learned most of everything.

I came to Crete. I was 19. I was running from Amerika. I was dislocated, wounded, confused.

I had spent four days in jail, yes, only four days, New York Citys Womens House of Detention, a brutal, dirty, archaic jail. I had been arrested for demonstrating at the United States Mission to the United Nations. Adlai Stevenson, then the conscience of liberal Amerika, walked by us into the building as the police dragged us away. Inside the jail I was given a brutal internal examination by two male doctors. As a result I bled for 15 days after that, terrified, afraid to go to a doctor, afraid of doctors, afraid to tell my parents, afraid to ask anyone for help. At that time I was living with two men, and they had what Ive since learned to appreciate as a typical male reaction to Blood Down There, a kind of *his*terical stiffening of every muscle, a stony indifference, a strained withdrawing of mind and body. But at the time I thought that they, two particular persons, were horrified by me, one particular person, who was bleeding,

bleeding, bleeding. At that time, I also had another lover, an older black man named Arthur. I liked him a lot, and so on the phone he said, where you been, and I said, the House of D, and they did that and that and that to me, and he said, white girl, thats what they do. I felt his contempt for me, and also knew more than I could stand to know about his real life, and so I never saw him again. Wherever I turned trying to say what had happened to me, I met that same contempt, or silence, or indifference—but of course, I always turned to men. When finally, choked and enraged, filled with fury and confusion, I did turn to two women (I barely knew them), they knew viscerally, absolutely, what had happened to me, they knew what it had been. But then, in those years, I didnt turn to women very often or understand that men could not dare to know.

I felt alone, enraged, furious, violated, hurt, and so afraid. I did not know how to contain or to understand what had happened to me. I didnt know how to contain or to understand what I had seen happening in that jail to other women. There was no language to describe it.

There are themes in ones life, themes which resonate. One theme in my life, an important piece of who I am, the Nazi slaughter, resonated then. What had happened to me, the blood, my fear, the brutality, conjured up the Nazi doctors who had tortured flesh of my flesh and blood of my blood, and an aunt who had survived to tell me, retching in terror and memory. The doctors in that jail when they were abusing me—my aunts Nazi violation resonated then; in the nightmares I had after—it resonated then. It was what it was, the violation of one woman by two particular men, but it also conjured up that near history of my living flesh and so it had a resonance beyond itself—a sound, an echo, through 6,000,000 bodies.

I didnt know then about the 9,000,000 witches burned alive, or the billions of women raped, abused, bloodied, and abandoned all over this planet. I didnt know then. I felt it, womans fury, but I couldnt name it, or call it out, and so I anguished, isolated, confused, unable to name, the very power of speech, and so of knowing, taken away from me. I didnt know then that we women were a sisterhood united in blood and toil on this earth, each one speechless, experiencing the unspeakable, robbed of the power of naming and so of speaking and so of

knowing. I knew then only about the several hundred women in that one jail, each speechless, each experiencing the unspeakable, robbed of the power of naming and so of speaking and so of knowing. This is happening to *us*, I remember was the phrase that turned over and over again in my mind those days in jail, this is happening to *us*.

A Jew, a woman, my ties to the dead, my commitment to the living. There is no place on earth, no day or night, no hour or minute, when one is not a Jew or a woman. There is no time or place on this earth that does not resonate through 6,000,000 bodies tortured and gassed, through 9,000,000 bodies tortured and burned.

In all the years Ive been involved with leftist politics, in Greece, Holland, England, and Amerika, you were the only man who ever told me this story: "I was a member of the young communists, an illegal group in Greece since the Civil War; there was a woman comrade, and we had all done actions with her, and slept with her, and then she had a political difference with the others and they, who had been her lovers, refused to speak to her or to associate with her, she was ostracized and cursed; and I quit because I thought, *if these are the people who are making the revolution and if this is the way they act, then I dont want to live in the kind of world they would make.*"

I didnt understand this story until many years later. When I knew you, I was a committed leftist. I had seen many women used then abandoned, I had been used then abandoned myself; still, I could not make sense of what I had seen or of my own experience, I did not make sense of it for several more years. The story you told me stayed with me, embarrassed me somewhat because I didnt entirely understand what you had done or why you had done it. Still, I liked you for it.

I learned from you, from my landlady, that ancient ruined woman, that the Nazis had come to Crete because (as reasons go) they wanted a small airstrip on the island. In the course of their occupation, they did carnage, annihilated whole villages—sometimes they killed all of the men and boys and left the women and girls (when they finally did leave, after rape) to mourn, to go hungry, to survive mutilated as if each husband

and boy child killed were a limb that had been severed and she, the woman, was left with the stumps bleeding.

I did not mistake where I was or what had happened there. Each day there was an echo, almost hissing in the air, the Nazi slaughter. Each day that slaughter was sounded in the bodies of the old women, dressed in black, mourning still, remembering still, faces older than this old earth, faces weighted down with the years of loss and murder. And before that, the Cretans were murdered by the Turks, 400 years of occupation and tyranny. And the Cretans murdered the Jews—each year over the centuries pogroms on Easter avenged the death of that other Jew, Jesus. And of course the women belonged to the occupied or to the occupiers, the living or the dead. The women were murdered and the women were raped and the women were left to mourn their dead. It was the human family, bound together in a web of murder and pain, and each member of that family had murders mark on her.

Living on Crete brought me to a new sensitivity, acute and intolerable. I felt the resonances of those dead, all of them, and the lives of those living, all of them, in my own body, and I came to know who I was—that self tied to the past which was ever present in a way that was not melancholy or romantic. In Amerika, each person is new, like hemp before the rope is made. On Crete the rope was used, bloodstained, it smelled of everything that had ever touched it.

And so we, you and I, in ways so different, each were suffused with Crete. You loved the land, the mountains and what they held, the sea and what it brought and took away. Amerikans for the most part dont know what that means. The land moved you, you knew its story, and you were bound to it. I was Amerikan, Jew, female, who knew nothing at first of the land and what it held—I grew up first in a city, cement, telephone poles, and then in a suburb, boxlike houses, small plotted lawns, an occasional tree. But yr land and its people entered into me and in me I began to discover the memory, passion, and experience of all the peoples of whom I was a part. In that way we touched each other, and in that way we were brother and sister.

But now comes the harder part, how we were lovers. Who was

I then, I barely remember her, that woman. She doesnt live in me very much anymore.

I was in Greece (Athens, Piraeus, Crete). I was 19. I wrote. I saw, for the first time, the mountains, the light, that luminous Greek light, the ocean which from the shore was filled with bright strips of color. I had many lovers, all men.

I was a person who always had her legs open, whose breast was always warm and accommodating, who derived great pleasure from passion with tenderness, without tenderness, with brutality, with violence, with anything any man had to offer.

I was a person who always had her legs open, who lived entirely from minute to minute, from man to man. I was a person who did not know that there was real malice in the world, or that people were driven—to cruelty, to vengeance, to rage. I had no notion at all of the damage that people sustain and how that damage drives them to do harm to others.

I was a person who was very much a woman, who had internalized certain ways of being and of feeling, ways given to her through books, movies, the full force of media and culture— and through the real demands of real men.

I was a person who was very much a woman, accommodating, adoring of mens bodies, needful, needing above all to be fucked, to be penetrated, loving that moment more than any other.

I was a person who was very much a woman, who loved men, who loved to be fucked, who gloried in cock, who called every sexual act, tender, violent, brutal, the same name, "lovemaking."

I thought, how can I even explain it now, that life was what Miller and Mailer and Lawrence had said it was. I believed them. I thought that they were creative and brilliant truth tellers. I thought that the world was as they said it was, that to be a hero, one must be as their heroes were. I wanted to be a hero-writer, outside the bounds of stifling convention. I thought that I was becoming as they were by doing that which they admired and advocated. I did not know, or feel, or realize what was being done to me by those who were as they were. I did not experience myself or my body as my own.

I did not feel what was being done to me until, many years

later, I read Kate Milletts *Sexual Politics*. Something in me moved then, shifted, changed forever. Suddenly I discovered something inside me, to feel what I had felt somewhere but had had no name for, no place for. I began to feel what was being done to me, to experience it, to recognize it, to find the right names for it. I began to know that there was nothing good or romantic or noble in the myths I was living out; that, in fact, the effect of these myths was to deprive me of my bodily integrity, to cripple me creatively, to take me from myself. I began to change in a way so fundamental that there was no longer any place for me in the world—I was no longer a woman as I had been a woman before. I experienced this change as an agony. There was no place for me anywhere in the world. I began to feel anger, rage, bitterness, despair, fury, absolute fury, as I began to know that they, those writers and their kind, had taken cruelty and rape and named it for me, "life," "sex," "lovemaking," "freedom," and I hated them for it, and I hate them for it still.

There was a particular part of *Sexual Politics* that began this change in me, a small moment in a vast book. Millett described Henry Millers depictions of sex acts in a voice I had never heard before. She said, simply-it seems now, *look at this, this is what he does and then this is what he calls it.* Then I saw it—the cruelty of it—as what it was, no matter what others, the whole world, called it. No one who has ever had this experience denies the revolutionary power of language or the absolute importance of naming, or the violation which inheres in being robbed of speech even as one experiences the unspeakable.

E, you see, this is what is so hard to describe to those who have not experienced it: that as a woman, ones body is *colonialized*, ones flesh is actually taken from one, named and owned by others, all experience *their* experience, all value *their* value. The process for a woman of becoming whole, herself, cannot even be described as reclaiming ones flesh (ones land), ones personality (ones land), ones own integrity (ones land), because one has been deprived of both core and vessel for too long, over too many generations and centuries. One can say that the French colonialized Algeria, and conjure up a vision of a free Algeria, because one has a memory that the French did not always own Algeria. But Algerian women have no memory of a time when they were not owned by Algerian men. Algerian women, and all women, have been robbed of any memory of freedom. Our bondage is so ancient, so absolute, it is every inch of the past that we can know. So we cannot reclaim, because no memory of freedom animates us. We must invent, reinvent, create, imagine the scenarios of our own freedom against the will of the world. At the same time we must build the physical and psychic communities that will nourish and sustain us. For in reality, as the Three Marias of Portugal have written, "there is no bread for us at the table of man," that is, unless we are first willing to prepare and serve the meal. And, of course, the men own the bread and the table and the women who serve and the beds we must sleep in at night.

I am saying that my body was colonialized, owned by others, imperialists who robbed it of its richest resources—possessed, taken, conquered, all the words those male writers use to describe ecstatic sexuality. And I am saying that I was that slave woman, that caricature of a human being, that servant whose core and vessel belonged to those who had conquered it. I was that slave woman who accepted the conquerors naming of my experience and called it, their dreadful brutality, their possessing and taking, "lovemaking," "ecstasy," "freedom." That was the woman you knew.

I tremble when I know that you loved her, and only her. I am afraid, cynical, bitter, when I want to believe that you were also better than that, as I was in some not yet living part of myself; when I think, over these 10 long years, this is a man who could know me now, who could love me now, whom I could know and love. In some part of me—a part I do not dare trust or respect— I believe, but am also afraid to believe, and also do not believe, that in you there lives one who is not committed to oppressing women. I dare allow myself, sometimes, to imagine (or is it remember) that we did touch each other in those hidden parts.

I arrived in Athens on my 19th birthday. I was very lonely. There had been riots in Athens, Papandreou Senior had been ousted from the government by King Constantine; the people rioted in protest. I met an officer in the Greek Army, we drank ouzo on a mountain top, we looked down on the thousands in the streets, then we went to some crummy hotel and he fucked me. It was a horrible moment afterwards, when I looked at him and *saw* him and said, you really hate women you know. I saw the muscles in his arm tighten, and the impulse to strike animate his body, and his insane vanity, and then the decision that it wasnt important after all. I had never known that, that there were men who hated women, and yet at that moment I knew that I had just been fucked by one. It was the gift of my 19th birthday. I never forgot that man or that moment.

There seemed to be only two possibilities. To be a housewife like my mother, limited, boring, irritating. How we have been robbed of our mothers, I knew only the narrowness of her life, nothing of its depths. She was kept from me, cloistered, covered from my gaze by impenetrable layers of cultural lies. She was kept from me. We were set against each other, every mother Clytemnestra, every daughter Electra. I did not want to be her. I wanted to be Miller, or Mailer, or Rimbaud, my Rimbaud, a hero, nothing of the world closed to me, an initiator, an inventor, a creator. No one said the truth: phallic initiator, phallic inventor, phallic creator. Those were the false, vicious choices.

There were so many, and each was the one I was with. One after another, over and over.

I had been on Crete maybe three months when I first saw you. Glorious, golden moment. I was drinking vermouth at an outdoor café. The day was dark and drizzly. You stepped out of a doorway, looked around, stepped back in out of sight. You were so beautiful, so incredibly beautiful, radiating light, yr eyes so huge and deep and dark. I dont remember how we began to talk or when we first made love, but it really did happen that way, I saw you and the earth stood still, everything in me opened up and reached out to you. Later I understood that you were too beautiful, that yr physical beauty interfered with yr life, stood between you and it, that it created an almost unbridgeable distance between you and others, even as it drew them to you.

I was happy. I loved you. I was consumed by my love for you.

It was as if I breathed you instead of the air. Sometimes I felt a peace so great that I thought it would lift me off the earth. I felt in you and through you and because of you. Later, when you were so much a part of me that I didnt know where you ended and I began, I would still sometimes step back and marvel at yr physical beauty. Sometimes I would think that my life would be complete if I would always be able to look at you.

I dont know what you felt. I never questioned it or thought about it. What was admitted of no other possibility. What was had no words, no language. I remember that a time came when we no longer made love all day and all night, but only twice each day, once in the night and once in the morning, and I asked a woman I knew if she thought you still loved me.

I was ecstatic with you. What are the words? I loved you, I breathed you. What does that mean? What does it mean that two people, a man and a woman, who share no common language, come together and for almost a year share every day in an erotic ecstasy, die in each other, are born in each other, rise and fall and intertwine and cry out, breathe in and through each other, are nourished and sustained by mutual touch, are one in the way that the sun is one, when the coming together of those two people embodies every possible feeling, sound, silence?

And towards the end, before I left, when we began to fight, to have those monstrous wordless fights composed of a passion as large as the love we were—what was that? What does it mean that two people, a man and a woman, who require each other for the sake of life itself, like water or food or air, who do not share a common language, who speak only pidgin bits of French, English, Greek, but know each other completely, understand whole sentences and speeches composed in three languages at a time, begin to tear and rend each others insides—using gestures, fragments, emblems, signs. What does it mean when these two people, a man and a woman, have a fight, a monstrous fight, that lasts all night, through every fury and silence (but he will not leave her, he will not go from her house), a fight that begins when she tries to kill him, literally to tear the life out of him with her bare hands because he dares to touch her (and she would die without that touch), and their pain is so great, so physically unbearable, that still they have only each other, because only they in all the world share that pain and grief? What is that?

I swear I dont know, all these years later I still dont know. When I left you I thought that the pain would kill me, literally, physically. I felt a physical pain so acute, all through my body, in every part of it, for well over a year I felt this pain, it kept me awake, it filled my sleep, nothing around me was as real as the pain inside me, and still, ten years later, sometimes I wake up from a dream that has forced me to feel it again.

I have always wanted to know why I left you. I have wanted to know what in me was stronger than my love for you—what nameless drive, in me but not claimed by me as part of me, moved me to decide to leave you, to make the arrangements necessary to leave you, to walk to the boat, to get on the boat, to stay on the boat even as you called to me from the shore.

I remember that you hated it that I was a writer. It was all right as long as it meant that I had been at home all day, nothing more. But when a small collection of my poetry was privately printed by some friends, on the day I held that book in my hands, you hated me. You were jealous as you never would have been of another lover. (I remember that one night I woke up to find you rifling through my papers, searching fiercely, not able to read English—searching for what?—searching, I think, for the strength that did not breathe in you and because of you.)

I dont know exactly when or why yr anger took explicit sexual forms. You began fucking me in the ass, brutally, brutally. I began to have rectal bleeding. I told you, I implored you. You ignored my screams of pain, my whispers begging you to stop. You said, a woman who loves a man stands the pain. I was a woman who loved a man; I submitted, screamed, cried out, submitted. To refuse was, I thought, to lose you, and any pain was smaller than that pain, or even the contemplation of that pain. I wondered even then, how can he take such pleasure when I am in such pain. My pain increased, and so did yr pleasure.

Once you stopped speaking to me (had I resisted in some way?).

When finally (was it a day or two?) you came to me I waited for an explanation. Instead you touched me, wanting to fuck me, as if no explanation were necessary, as if I was yrs to take, no matter what. Had I been strong enough, I would have killed you with my bare hands. As it was, you were weak in yr surprise, and I hurt yr neck badly. I was glad (Im still glad). We fought the whole night long, with long stretches of awful silence and a desperate despair. In the course of that night you told me that we would marry. It was towards morning, and after you had raped me as is the way with men who are locked in a hatred which is bitter and without mercy, you said, thats all thats left, to get married, isnt that what people do, isnt this the way that married people feel. Bored and dead and utterly bound to each other. Miserable and sick and without freedom or hope. Yr body moving above me during that rape, my body absolutely still in resistance, my eyes wide open staring at you in resistance, and you said, now Ill fuck you the way I fuck a whore, now youll know the difference, how I loved you before and how I hate you now. I said, numb and dead and dying, no, I wont marry you, I cant stand this, its worse than anything. You said, we cant be apart, youll see, it wont be so bad. I remember that then you lay between my legs, both of us on our backs, and we didnt move until dawn. Then you left.

The next day I took my razor blades to a woman friend and I said, keep these, I dont want to be silly but I think that at any minute I wont be able to stand it anymore, to stand this excruciating pain, to take one more second of this being alive without him, and I will be happy to be dead before the next second comes, but I dont want to be dead, and I need help not to be. She knew that it was the truth and my friends didnt leave me alone for one minute after that. I was in despair. I had no hope. Time was anguish. I learned how many seconds there were in a day.

I left Crete a few weeks later. Somehow we endured. Somehow we survived that agony. Somehow only we had suffered it and all the others were outside of it. Somehow we became tender with each other again. Somehow we made love again, with such great sadness and softness that it was new. It was as if either of us might break into a million pieces at any minute, as if there was nothing to save or to hide or to redeem either, as if the only parts of us still living were as fragile as dust in the wind.

And it was very important, I think, that our last week together was spent celibate. You had, after that terrible night, gone to Athens and there gotten the clap from some young man, and me from you, and so our last week together we didnt make love. We went to Athens in yr fathers truck full of tomatoes to take the tomatoes to the market. I cried the whole time, hysterically, doubled up on the car seat, from market to market, howling, wailing, screaming like a banshee, the tears never stopped. You were very kind, tender, and so we began to laugh together again, and on the day I left we were closer than we had perhaps ever been.

If you loved him, why did you leave him? My friends asked me that often, and it was strange that I had left, that any woman would leave a man she loved the way I loved you. I answered in many ways. Sometimes I said that I had become sick. It was true. I had gonorrhea, and my ass had been torn apart. I had an operation on my rectum and as I lay in the hospital wracked with pain, I received letters from you which were completely indifferent to my physical condition. You did not want to know. A woman who loves a man accepts the pain. I did accept it, but not gracefully. Sometimes I told people that I had left because we had begun to fight. That too was true, though when I left I knew that I could stay, that you would not leave me, that we could even marry, if I wanted.

The decision to leave was not rational. It was made, in fact, long before the worst happened. It was a feeling, an impulse, that inhabited my body like a fever. Once I felt it I knew that I would leave no matter what. I describe it to you now as the *drive to become* that lives in the part of me that did not breathe in you, that is a writer, and that even my identity as a woman could not entirely silence. It is that part of me that enraged you even as it enthralled you, the part that could not be subsumed by seduction or anal assault or any sort of domination. It is that part that could not even be conquered, or quieted, by tenderness. It is the part of me that was, even then, most alive, and that no man, not even you who were for me the air I breathed, could ever take from me.

If you had truly loved him, you never would have left him. Some have said that to me, but I say no, I loved you, and I left you. I had a drive to become, to live, to imagine, to create, and it could not be contained in what took place between us.

I wanted to come back. I expected to come back. I planned to come back. I started to come back. But I never did return to you.

Two years after I had left, as I had promised, I started on my way back to you. I went to Amsterdam. I wrote you, Im coming back. I received a letter from you that said, my life is bitter, you dont know whats happening here, Amerikans are stupid and you are an Amerikan, tanks and death are everywhere, my friends are being imprisoned and tortured and killed, come if you can bear it, I cant promise you anything. You said that you yourself knew only bitterness, and, indeed, yr letter was bitterness.

I had exactly enough money for fare one way, nothing more. I had wanted you to say, come, come now, I need you now, now in this time of desperate trouble I need you.

I did not return. A few months later I married.

I was married for three years. During those years, I dreamed of you. I would wake up in a cold sweat, desperate just to hear the sound of yr voice. I never understood why I had not gone to you.

A year after my marriage ended, talking with a friend, I understood why I had not gone to you. Whatever the false (male-determined) values that still infuse my judgment of myself—e.g. I betrayed you, abandoned you, deserted you, had no right not to return to you given yr desperate situation—I discovered, in my failure to return, the dimensions of my own cowardice. I had been so afraid, E, so afraid of the reality of what had happened/ was happening to you. The real guns. The real police. The real torture. The real dying. I had stayed in Amsterdam to pursue a life of "radical" pleasure—smoking dope, fucking, the romance of radical ideas without the reality of dangerous opposition. And I realized too that I had not been able to accept the letter you had written me—"I am only bitter"—no image of romantic love was there to propel me toward you, toward self-sacrifice, toward bravery.

I wrote to you then, after my marriage ended, saying, I am living alone, writing a book, and in November I would like to come see you if you are still willing to see me. Miraculously, you wrote back, saying where you would be, warm, saying to come.

But as I worked on my book and struggled with this new clarity, I saw that in Greece I could do nothing, and that my struggle was in Amerika. I saw that I had to come back here, to Amerika, to hone my book into an instrument of revolution. I had to confront the real danger here—not give myself in service to the romance of Amerikas male "radicals," but instead to confront the hatred of women, male power over women, from which, I believe, all other illegitimate power is derived. Here, knowing the language, I could take responsibility. Elsewhere, I would still be running, still hiding. I saw that this assumption of responsibility must be at the center of my life. I saw that I could not be any mans woman, not even yours; that I myself must act in the world directly, develop and use all my strength in the pursuit of my vision, a vision no male could have birthed. I knew that I had discovered my true faith.

I wrote you again, saying, E, I am returning to Amerika, when I finally do come to Crete will you see me? No answer from you. I ask our Greek friends in Paris for news of you, but there is none.

Now, more time has passed. I dont think that I will ever come back to you or see you again. Sometimes I wish that were not so. But I have one choice to make in life, to make and to keep making—will I seek freedom, or will I dress myself in chains? I am on a journey long forbidden to women. I want the freedom to become. I want that freedom more than I want any other thing life has to offer. I no longer believe that yr freedom is more important than mine, that yr pleasure or pain is more important than mine. I no longer believe that the torture of a man in prison is worse than the torture of a woman in bed.

I began this letter in desire; I end in anger. I dream that love without tyranny is possible.

A.

The Woman Who Lost Her Names

Nessa Rapoport

SHE WAS NAMED after her grandmother, Sarah, a name no one else had then because it was considered old-fashioned. Eight days after the naming her father's brother died, and they gave her a middle name. The brother was Yosef—Joseph—so her mother went down to City Hall, Bureau of Births and Deaths, and Josephine was typed in the space after Sarah. "A name with class," her Aunt Rosie said. Sarah hated it.

When she got to school the kindergarten teacher sent home a note. The family read it together, sitting around the kitchen table. "Dear Mrs. Levi, we have decided to call the child Sally for the purpose of school as it will help integrate her and make the adjustment easier."

"What's to adjust?" the brother next to her asked.

"Shah," her father said.

Her father was a gentle man, remote, inaccessible. The books that covered the tables and chairs in his small apartment were the most constant factor in his daily life, and the incongruity of

Nessa Rapoport is a writer and editor living in New York. She is currently working on a novel. "The Woman Who Lost Her Names," copyright © 1979 by Nessa Rapoport, originally appeared in *Lilith* magazine and is reprinted by permission. All rights reserved.

raising seven Orthodox children in the enlightened secularism of the Upper West Side never penetrated his absorption in Torah. Sarah grew up next to the families of Columbia intellectuals who were already far enough from Europe to want to teach their children civilization. The girls in her class had radios, then TVs, then nose jobs and contact lenses. They grew more graceful in their affluence, and she grew a foot taller than all of them, early. There were many blond girls in her class each year, and she'd stare at their fair delicate arms whose hair was almost invisible. "Sally, how does your garden grow?" the boys would pass her in the hall, staring at her breasts, the thick dark hair covering her arms to her wrists, the wild hair that sprung from her head independent of her. She'd look down at herself, her bigness, ungainly, and think "peasant, you peasant" to herself and the grandmother who'd bequeathed her these outsized limbs. No one would fall in love with her.

Her mother was fierce, intense, passionately arguing, worrying about people, disdainful. "She married for money," "he could have been a scholar," indicting these neighbors who were changing their names, selling their birthright. "Sarele," she'd suddenly gather her in her arms late at night when her brothers were sleeping, "remember who you are and you'll have yourself. No matter what else you lose—" She never finished the sentence. Sarah would look into her mother's face, full of shadows, ghosts, and touch the cheek that was softer than anything. "You're a big girl, Sarah," her mother shook herself free, always, "go to bed." Her mother would sit at the kitchen table alone, head in her hands, thinking. Once long past midnight Sarah saw her that way, shaking her head between her clenched fingers, and tiptoed in to say, "Mama, I understand." Her mother looked up, uncomprehending.

When she was seventeen and had given up hope she suddenly bloomed. Her hair calmed down, and a kind of beauty emerged from under her skin. The boys in her youth movement started to talk to her after meetings, inviting her places. First she said no, then she believed them and went gladly to rallies, campfires, lectures to raise money for the new state of Israel. She dreamed of Israel often, dancing the folk dances in the orange groves of her imagination, fighting malaria, drowning in jasmine. None of the boys touched her heart.

At school boys were thinking of college, and girls were thinking of boys. Graduation came on a hot day in June, and her parents watched her get a special award in poetry, poems she had written that no one but the teacher would ever read. "Poetry, Sarele," her father was pleased. "My dreamer," her mother whispered. "We have a surprise for you. From Israel," the word was still strange to her mother's tongue.

Yakov Halevi was her cousin, a first cousin from Jerusalem she'd never met. He got up to greet her from the couch in her parents' living room, the room reserved for company, and she watched his thin energized frame spring forward. He was meant to be dressed in black like the rest of her cousins whose pictures she'd seen in her mother's hand, sidecurls swaying in an overseas wind. But his hair was short, startling, red, and the hair of his chest showed in his open-neck shirt. He spoke seriously, with a heavy accent, and she loved to watch the words form in his mouth before he released them. Yakov was a poet, only twenty and known already for his fervent lines. Great things were expected of him, and he carried their weight on his narrow Hebrew shoulders. Her Bible knowledge wasn't enough, and she struggled with the new language to read his book, tracing the letters of the title page, alarmed: Jerusalem Fruit, by Yakov Peniel.

"Who's Peniel?" she asked him. "Why did you change your name?"

"I didn't change it, I lost it," he laughed. "When the editor wrote to me accepting my poems he had to ask my name, for I hadn't sent it. 'Hagidah na shmekha,' he wrote, what Yakov our forefather asked when he wrestled the angel. 'Why is it that you ask my name?' I wrote back, as the angel answered, and he published me under the name Yakov gave that place—Peniel."

"But your letters come to you in that name. How did it happen?"

He shrugged. "People wanted to meet the bright young poet, Peniel. Then I was asked to talk, introduced that way. On the street they would say 'That's Peniel,' and so it came to be."

"But you're the tenth generation of Jerusalem Halevis. You can't give it up, it's your name."

"Just a name," he smiled. "The soul underneath is the same, in better and worse."

She loved that humility, and the heart of Eretz Yisrael she heard pounding in his chest when he held her. He loved her and loved her America. "It's not mine," she'd insist, "I don't belong here." But he stood in the middle of New York looking up. "So big," he would cry in his foreign tongue. "So big."

He wanted to cross the country sea to sea, to marvel at mountains and chasms. She had waited so long to go home, to Israel, she could wait a little longer. Then he was her home, she became him, she loved every bone of his self, every line. The words of his mouth were her thoughts, what he touched she found worthy. It thrilled her, their sameness, and she'd wake in the mornings eager for the coming confirmations. They would say the same things at one time, and reach for each other, marveling. She wasn't alone anymore, she had found a companion. When she tried to explain about the Upper West Side and the girls in her class he would say, "Every one is alone. Man is alone before God, that's our state." Hearing him say it bound her even more. She wanted to breathe his breath, use his language, and searched through his poems, word after word, for her hidden presence.

"My muse," he sorted her hair. "Sarele," saying it the old Jewish way as he'd heard his mother, also a Sarah, being called.

They knew they would marry, she floated for months on that knowledge, walking down Broadway to the rhythm of Solomon's Songs. "Sarele," Yakov drew her to him one night. "We must talk of the name."

"It's OK. I don't mind Peniel. It's better in a way than Halevi, which is almost my name. No one would know I was married."

"It's the other name—Sarah. My mother's name. A man cannot marry a woman with his mother's name."

She turned white. "A man cannot marry?"

He noticed her face. "Oh no no, he can marry, but she, she must change the name."

She said in relief, "But what name? Sarah was my name."

"Do you have a middle name?"

She scowled. "Josephine."

"A *goyische* name, Josephine. So what do we do?"

She thought for a while. "I don't know. Josephine is Josie, but that's no good."

"Jozzi," it was clumsy, "Jozzi. There is no Jozzi in Hebrew." She had no suggestion.

"Wait," he told her. "Jozzi is Joseph, Yosef, is that right?" She nodded.

"Well then it's Yosefah. Yosefah," he tried it on. The Hebrew sounds spun in the air. "Yosefah," he turned her around and around till the trees flew in front of her, dizzy.

They married in spring and all summer they traveled as he fell in love with America. He loved New York City, the place where they'd started, he fingered the wheat of Midwestern fields, and stood on a rock high over the ocean as if he'd discovered the water. In the evenings and as she woke up he was scribbling. Poems, letters, stories poured from his hands. She sat amazed as the papers grew and multiplied in hotel rooms, in the trunk of the car. The strange Hebrew letters leaped from the pages, keeping their secrets against her straining will. "Are you writing of me?" she wanted to know. He smiled, "They are all of you." He told her he sometimes took phrases she said and transplanted them into his work. She was grateful and mystified, peering through the foreign marks for herself, not finding resemblance.

Yearning for Israel they moved to New York. He studied small Talmudic matters on which great things depended. She had a son, then another. The boys laughed and cried with her all day, alone in the house, surrounded by papers, and books of ancient cracked binding. When he finished his doctorate they would live in Israel, and she counted the days as they lengthened to months, alone with her children and the fierceness of her desire. She was America to him, aspiring to be free, and he envied her readiness to leave such abundance behind. "It's your home," he tried to soften her, frightened by her single burning.

"It has nothing to do with me," she'd deny. He trembled in the face of seduction—the grandness of America's gestures, hundreds of plains crossed by rivers whose opposite shores were too far to see.

"We must leave," he said. "We must go."

She stopped her daydream of years and started to pack. The boat left in winter, and the grey piers of New York, city of her birth but not her death, she was sure, were left behind her, unmourned. He stood watching the gap of water widen, then turned to her and was thankful.

There were cypress and palm trees, traveling in, and a perfume heavier than air. Jerusalem approached them at twilight, her gold roofs and domes aching for heaven. She recognized the city as a lover, missing past time, a shock of remembrance that stirred her body like a child. The boys were crying, tired, afraid, and she sheltered them under the sleeves of her coat, "We are home."

Jerusalem was designed for the world to come more than for now, and she washed, cleaned, shopped, scrubbed over and over again, as the dust blew in the summer and the winter wind seeped through the cracks. The boys got sick, and well, and she was sick in the morning, pregnant with the next child. Yakov laughed as he smoothed her hair. "A girl, a *maidele* next."

From America letters came. A brother was hired, her mother was sick, Aunt Rosie was worried, family troubles, her mother was sick, Papa retired, a nephew converted, her mother was going, dying, gone. There was no money for planes, and the pregnancy was hard, so half a world away she guarded her pain, talking to her mother in dreams about the coming daughter. Her mother, sitting now at a kitchen table that was not of the earth, holding her head in her hands. Night after night she lay on her back, her stomach a dome in her arms.

When the child was born she could hardly know, groaning in a voice unknown to herself, stuffed between Arabs and old men in this not-American hospital. Outside the war in Sinai was sending soldiers into her ward, and sending her into the hall and then home, almost before she could stand. She bore the child alone, Yakov at the front, and when she looked down at her daughter, resting on her breast, she was full and at peace with this breathing body of her secret prayers, in love with the child, flesh of her flesh, bone of her bone, not a stranger. Yakov came home, exhausted, off for three days to see her and rejoice in his daughter. When he finally was there she stared from her bed unbelieving, two ones loved so dearly, both whole in limb. It was wondrous to her, and she ran her fingers up and down those tiny arms and legs hundreds of times. And Yakov, unmutilated, only tired, spoke to her saying, "We must give the child a name."

"But I know the name," and she did, waiting for him to come home from the war, reading the Book of Psalms. "She is Ayelet Hashachar—the dawn star."

Yakov smiled over her, indulgent, "This is not a name."

"This is the name," she said firmly. "Ayelet, Ayelet Hashachar, it's beautiful."

"Yes," he said gravely. "It is beautiful. But it's not the child's name. Yosefele," the smile, "your mother."

"My mother would love the name. She would love it," remembering her mother alone at the table dreaming her dreams long past midnight.

"Your mother was Dinsche," he told her, "and the child must be Dina. It is the Jewish way."

She looked at him, trying to find in herself some agreement, even a small accord and she'd bend to his will. But there was nothing.

"Yakov," she pleaded, "my mother won't care. I represent her, I know." Her mother holding her head in her hands. "It cannot be Dina. It can't be." Her voice was rising, new sounds that surprised her. "There's blood on that name."

She rose from the bed, "Look," and held out the Bible, shaking, to him. "Read," she tore through the page in her haste. " 'And when Schechem the son of Hamor the Hivite, the prince of the land, saw her, he seized her and lay with her and humbled her.' "

He stood before her in silence. "Rape," she said. "You want a daughter named for a rape."

"It is out of respect," he said. "For your mother. I don't understand what you want. A *goyische* name like Diana?" he asked. "If it's better we'll call her Diana."

"Ayelet Hashachar," she whispered in mourning, swaying like a rebbe in prayer.

"So what is it?" he asked.

"I don't care what you do," came the words in that voice, the one she had heard from herself giving birth. "Do what you want," turning her face to the wall.

He stroked her hair straight back from her forehead until finally she was asleep.

When the day of the naming arrived she was numb, jabbing the pins of her headcovering into her hair. She walked with her sons and her husband to the synagogue, and left them to climb the steps to the balcony for women. Below her the men were lifting the Torah, opening and closing it, dressing, undressing it, reading the day's portion. The people in the synagogue were singing quite loud, and some of the women sang too. The women around her moved their lips to the words. She stood still. She stood in her place, the place where the mothers always sat for their children. She closed her mouth, her lips pressed together, one on top of the other, and waited to hear her daughter's name.

My Mother's Story

Roberta Kalechofsky

"TELL US A story," the children said.

My mother turned at the door with her night-time look of weariness. I, being older by nine years than the other two, understood her longing to be finished with the day.

"Jim," she said to me, "you read to the children tonight."

"No, no," they said, "*you* tell us a story." They spoke like imperious rulers and my mother took her hand off the door-knob and sat down on the edge of their bed.

"Well," she said, twisting her wedding band as she always did when she was driven to think something out against her will, "what shall I tell you about? I am quite empty tonight." She laughed about her confession to the small children, yet pleasure came into her eyes as Bobby, who was eight, curled his boyhood into her lap and Lucille, seven, draped her arms around her

Roberta Kalechofsky's work has appeared in *Epoch, Western Humanities Review, The Jewish Advocate,* and other publications, including *The Best American Short Stories 1972;* in the 1975 volume "My Mother's Story" was listed as a distinctive short story. A teacher and lecturer, Kalechofsky lives in Marblehead, Massachusetts, where she founded Micah Publications in 1975. "My Mother's Story," copyright © 1974 by Ball State University Forum, is reprinted with permission.

neck, and my mother was held captive by what she longed to escape and what she hoped to rescue.

"Tell us a story about when you were a child," Bobby said.

"Oh, my," my mother laughed, "and who told you I was ever a child?" She looked at me sitting in the rocking chair in the corner with my science manual in my lap. "This was Jim's night to tell you about space ships and rockets and how you can fly away from this world. Wouldn't you rather hear about that?"

"Tell us a story," Lucille whispered sternly, knowing that this banter was wasting time because Mother rarely refused them anything, least of all a story.

"Very well," she said. "A story. About my own childhood." I dimmed the light over my shoulder and smiled at her with the understanding that she enjoyed this more than she would admit. She smiled back at me vacantly, frowning, reaching into her mind for something to tell them. She stretched out her hands over Bobby's head and looked at them in that complex way I knew better than the children, as if she were struggling to repossess something she had lost, struggling and losing. I sat under my dim light and watched her, not sure of the sympathy I felt for her, because I knew she wanted sympathy and understanding from me. She looked back at me, resenting my determination to be a child and keep her in her place. Then her eyes and her hands fluttered down to Bobby, little boy rightfully in his mother's lap, and she found her voice.

"Well, you know, Mommy used to take piano lessons a long time ago. That's why she knows so many songs to play to you. At that time I lived in a city far north of here and it used to be cold almost all the year round. Somehow I remember always going for my lessons in the snow, although I am sure there must have been months when there was no snow. But it seems to me," she smiled, "that I was always walking in the snow and that it was miles and miles to my teacher's house. It was an old-fashioned stone house with a stone wall and green awnings, but I never saw the awnings unrolled because there was never enough sunshine for that.

"The house was old, older than my piano teacher. There were eight stone steps to the front door, and some of the steps were cracked. There was a little garden in front of the house and a pine tree in the middle of it. When I sat at the piano in front of the bay window I used to watch the snow shape the branches of the tree, until my teacher would rap on the piano top with her pencil and call me back to attention. Not kindly, either. She had no patience with dreaming or with mistakes. I'm not even sure she liked me."

She took her hands off Bobby's head and clasped them under her chin. "No, I'm sure she didn't like me," she said in a voice as if an old question had suddenly been answered. "But she gave me such careful attention that it sometimes confused me. I paid for an hour's lesson three times a week and she often gave me two hours and sometimes more, while I played and played and watched the snow grow shaggy on the pine tree. She gave me more than I wanted." My mother stretched her hands out and laughed. "More than I've known what to do with. She gave and I took, and I'm not sure why I took so much." She looked at me sitting in the dim light. "Children so often have such a sense of obligation, so often feel obliged to stay where they are put, forgetting they have legs under them." She looked down at her wedding band again. "I was most unhappy in her house, but I took all she had to give."

Bobby twisted his head in her lap and assumed a thoughtful expression. "Was she a witch?" he whispered.

My mother looked at him blankly for a moment. Then catching the light in his eyes, she said, "Quite. Yes, she must have been a witch and she cast a spell on me. She took the tongue out of my head so that I could not tell her when my time was up and that I'd rather play in the snow, and she took my legs away so that once I sat down on the piano bench I could not get up again. All that she left me was my hands that played on and on like the red slippers that danced by themselves."

"What did she look like?" Lucille said, apparently pleased that my mother was getting down to fundamentals.

But my mother was confused by the question. I saw her struggling between her memory and my sister's imagination. "Well, like a witch, I suppose," she said. "Certainly not at all like a piano teacher. She had a small, plump, bustling body that seemed as impatient as I was to be doing something else. She had short, thin legs. She had tiny feet that led me from the door to the piano bench with a quick, sharp step as if they were always saying, quickly, quickly, before we think of something else. She took my coat, hung it away, and usually said, 'Very well, begin.' Sometimes she held her hand out the door and said, 'Snowing? Again? Well, well.' Certainly she had a sharp, witchlike nose and she kept rhythm with it. Her nose was as sharp and as perfect as the metronome that clicked on the piano top, tick-tock, tick-tock, for two hours or more until I forgot the sound of the snow and the sound of the fire in the fireplace. Tick-tock, tick-tock, cheerlessly and perfectly.

"Her ears were very large. She sat on a chair next to my bench with one ear turned to me, and I could see that her ear was most witchlike. It was as large as a tunnel and all the melody was running into it like a river. She listened with her eyes too, for she never closed them and dreamed like people sometimes do when they have to listen to music for a long time. She watched the air with her small, brown eyes as if each note was before her.

"It might interest you to know," my mother said, "that as sharp and as small and as witchlike as this teacher was she had a husband who was large and slow and who sat in a wheelchair and dreamed by the fire."

"I never heard of a witch who was married," Bobby said with suspicion.

My mother saw that she had taken a wrong turn. "He was a witch too, and maybe they weren't married. I think they weren't. They lived together because they had bewitched one another. He lived in his wheelchair by the side of the fire and she lived on a little stool by the side of the piano. That was all I ever knew of them at that time, and even after he died I did not come to know much more, except for one fact."

She pulled Lucille's arms around her neck. "I tell you this because music teachers have a way of lingering in one's mind. It is a special kind of friendship, having almost nothing to do with anything else in life. All we had together was the music and the hour, and the beat of the metronome and the snow outside. She was the only teacher in town who gave music lessons, and she gave lessons to all the young children because at that time parents had a strange hope about their children learning to play the piano; but I was the only one to whom she gave more than I needed.

"I went there for twelve years, three times a week. The years rolled by, carrying me through Chopin and Beethoven and Mozart against my will. And yet." My mother looked about the room, straining at a thought. Then her face lit with a pleasurable concession, and her fingers played on the air with a soundless passion. They bent with tension as if the piano board were right beneath her fingertips. "And yet surely with all my will," she said, "because I never refused to go there. Chopin and Beethoven and Mozart were in my fingers, whether I wanted them there or not." She smiled at Bobby, "But suddenly one grows up and asks: where is all the rest of the life?" Her smile lapsed and she looked away from my brother. "I asked and I was answered. The morning after he died I heard my mother talking to a friend on the telephone. 'Died, has he?' she said. 'Well, I don't know whether to say it's a shame or a blessing. Too late to be a blessing, I suppose.'

"I asked my mother why she said such an awful thing. She hung the telephone back in place and looked at me with some doubts as to what she should say or how she should go about it. She took my hands and looked at them in a way that she would sometimes look at my face, absorbed with a vision. 'Tain't nothing,' she said, 'except that she had such promise. But everbody has promise sometime or other.'

"But I had caught hold of a tale and I would not let go. I pursued my mother with questions. My mother dropped my hands and turned sternly back to her baking. 'Tain't nothing else but that,' she said.

"My teacher did not stop giving lessons and the next week when I went to her my heart was beating very fast. I did not know how she would look now that he had died. I imagined that people changed fearfully when something like that happened. I did not know what I could say to her. The snow was banked on either side of me as high as my hips and I ran down the narrow path wanting not to go there, wanting not to face a house where someone had died, not knowing what I could say to her. I felt a terrible sorrow for her in spite of her sharp, loveless impatience with me, and the steady, larger patience that hovered behind her and me like a threat.

This day she did not put out her hand to feel for the snow. Her step was slower. She drifted towards the piano. The wheelchair was empty. I carefully did not notice it. But my heart was beating fast. Yet it was not the death in the house I was afraid of."

My mother stretched out her hands again. Bobby's eyes remained steadfast, faithful to the story. "I did not know then what it was I was afraid of and I did not think about it very carefully."

"Was it the she-witch?" Lucille asked.

My mother nodded her head with gratitude for the clue. "Yes, that was it. Strange how I did not know it all these years. But that's what it must have been," she shook her head gravely, "because as I began to play, a very unnatural thing happened. I sat down at the piano and did my lesson. In spite of the death and my beating heart there was nothing else to do. The hour ticked by with the metronome. I did not watch the snow shapes on the pine tree because I was too old for that. The second hour ticked its way up and the shadows began to fall. Suddenly, sitting on her stool in the crook of the piano, the witch began to disappear. First her eyes grew smaller. They became as small as yours or mine. Then her nose shrunk and became soft. Her eyes clouded over and took to dreaming; and her hands, that had always lain studiously still in her lap, began to move in the air with rhythm as if she were playing the piece with me. I finished the second hour with determination to end it right there. She looked at me with surprise, disapproval, and disappointment.

" 'You must go on,' she said.

" 'I can't,' I said, 'not today.'

"She looked at me for the first time with a granule of sad and apparent patience. 'Oh, but you mustn't mind death, or life,' she said.

"But I had grown up and discovered my will. I turned to her and said the only sympathetic words I could think of. 'I would like to hear you play.'

"She looked at me for a long time, for a very long time. I do believe it was a look fetched from all the untiring patience she had had with me for twelve years, all the unloved, wasted patience."

My mother stopped there as if the story was finished, but the children knew better. She looked at Bobby and saw him waiting.

She felt Lucille's nudging arms. She looked at me and I moved restlessly in my chair.

"And did she play?" Bobby asked.

My mother rolled him off her lap and laid him down in his bed. She stood up and straightened her dress. "Now, Lucille," she said, "you lie down too."

"But did she play?" Lucille asked, getting under her covers.

My mother turned distractedly at the door. Her hand fluttered to the doorknob, grasping it with determination. "Oh, yes," she said, "she played too well." She signaled to me to come out and let the children go to sleep. I put out the light and went unwillingly because I did not wish to meet her resigned expression in the dark hallway.

Making Jews

Nancy Datan

W HY ARE YOU going to fast when you're an atheist?"
    "Why do we go to synagogue for Kol Nidre?"
"I asked you first."

"I answered you," says my firstborn daughter, and a private
chill takes me, the same chill which comes to the pit of my
stomach at Pesach when my son asks, each year the curiosity
greater, "Why is this night different from all other nights?" *I
started this* says the voice of the private chill, and indeed I am
convinced that this voice speaks to every parent of a Jew, for
they are made, not born.

Choice is more obvious with my own, who came into the
world with fair hair, snub noses, an Anglo-Saxon family name,
and a paternal lineage which traces back to the Mayflower by

Nancy Datan says that "Making Jews" is a fair indication of her personal history.
A resident of Israel for ten years, she is a married mother of three children and
is Associate Professor of Psychology at West Virginia University. Her most
recent works include *Transitions of Aging* and *Forbidden Fruits and Sorrow*, a collection of papers on myth, folklore, and fairy tales. "Making Jews" appeared in
Moment magazine and is reprinted by permission of Nancy Datan and Raines
& Raines. Copyright © 1976 by Continuity, Inc.

way of the Daughters of the American Revolution. Against these odds I have only the stiff neck of the Jews and the relative ignorance of my upbringing, including decisive acts of dropping out from—if my daughters will pardon the expression—Sunday School. I brought to my impermanent intermarriage the memories of festivals, wine and white tablecloths, an uncle who would never miss Yom Kippur services and conducted quiet games of gin rummy at the back of what I later learned was not really a temple, but only called a temple by ignorant Jews. "We're all ignorant Jews," a friend told me once, but: "*L'hav-dil*," said someone else years later, a friend who gave himself to history on Yom Kippur in 1973 and left me the memory of his commandment to differentiate between the sacred and the profane. My own ignorance extended to embrace both, but:

"There is scattered about in this land a peculiar people, whose ways are not the king's ways, and therefore it does not profit the king to suffer them," echoes the voice of Haman over the centuries; at that potent convergence of forces which marks a first pregnancy I discover that my father-in-law registered as a pacifist in 1941, perhaps the very year in which my father's cousins were registered for death, and whatever else this pregnancy will yield, it is going to be a stiff-necked Jew, and so it was.

Now, more than a dozen years later, Merav Datan—the first of those names chosen and given in Chicago, in anticipation of the forthcoming trip to Israel, and the second, the heretic's name, taken by me a decade later and taken by my children to my surprise—remarks one day: "I'd have more friends if I were Christian."

"Why?"

"That's just the way kids are here," in a small city in West Virginia.

"Okay, baby, it's the easiest thing in the world to fix, we'll just march you down to the local church and give them your pedigree and register you for the D.A.R.," which met often, according to the paper, and to the amusement of the children, when they noticed it.

"Well, obviously I don't want those kinds of friends." It is not altogether obvious, but the pact with history is sealed within some months at her bat mitzvah, and both of us are shocked for different reasons by the fact that I had none myself.

Of course I had *mishpoche,* and she did not. Indeed, the seasons of the life cycle are marked at the festivals: the early years, when the seder takes forever until the aunts bring out the food; the middle years, when suddenly the service seems too brief; and then adulthood, the aunts are far away and the uncles are dead and I am making seder myself.

"*You* make seder?" asks a colleague who knows of me only that my daughters cook for the family: that, among other things, is what makes this night different from all other nights. My secondborn daughter is near to her bat mitzvah, but I consider my son, eight, who calls *matzot* "Joosh bread," still ignorant of the answers to the four questions. At least I am going to teach him how to ask.

"Do we believe in God?" he asks.

"It's not a corporate decision"—and therefore different from all other decisions made in this family, ranging from country of residence to the color of the new toilet seat: "Don't forget that Gidon never lifts it," I am told. "Pick something dark."

"Do *you* believe in God?" persists my son.

"No."

"Then I don't too."

"That isn't the way to decide. It's a very personal issue, and you have to give it years of thought before you can even come to the beginning of a conclusion."

"Well, I'm *eight.*"

"Well, try to give it another four or five years at least," and the chill touches me once more: his older sister, whose countdown without food or water has already begun, asked me irritably when she was just his age:

"Why do we have to go to synagogue for Kol Nidre?"

"Because we're Jews."

"We never went to synagogue in Israel." That is certainly true: I would never set foot under a roof which featured a walled gallery for women, and it was years before I learned that a reform movement, with a synagogue, existed in Jerusalem; but prayer of any sort made me uneasy. In Jerusalem it made no difference: I had my own variety of religious experience, one not catalogued by William James, when I read the account by the Roman historian Tacitus of the fall of Jerusalem in 70 A.D.: "As I am about to relate the last days of a famous city, it seems appropriate to throw some light upon its origin," sitting in my own front yard, the bright Jerusalem sun shining on the ersatz constructions of the Jewish Agency, built to accommodate the ingathered Jews of the twentieth century.

The record of the battle was terrible and brief: ". . . (Titus) pitched his camp . . . before the walls of Jerusalem, and displayed his legions in order of battle. The Jews formed their line close under their walls . . . I have heard that the total number of the besieged, of every age and both sexes, amounted to six hundred thousand. All who were able to bore arms, and a number, more than proportionate to the population, had the courage to do so. Men and women showed equal resolution, and life seemed more terrible than death, if they were to be forced to leave their country. Such was this city and nation; and Titus Caesar, seeing the position . . . determined to proceed by earthworks and covered approaches . . . There was a cessation from fighting, till all the inventions, used in ancient warfare or devised by modern ingenuity for the reduction of cities, were constructed." Thus Tacitus concluded, and turned his attention to the Roman campaigns in Germany.

I closed the book as a neighbor paused, on her way home, and I said, "You know what? If you read about the battles of Jerusalem in 70 C.E. and in 1948, it really does sound as if the Jews are one people."

She is Orthodox, and smiles: "I don't need a Roman to tell me that."

The apperceptive mass of middle age hangs heavier on the Jew at the High Holidays: that cumulative lens of personal experience through which the world is seen, upon which the years' memories are inscribed, clouds over, distorts. My earliest Yom Kippur, a Yom Kippur of innocence when parents brought me and decisions were years away, is a happy recollection of a recitation of sins which seemed to whet my appetite, for somewhere through the rabbi's talk (the discovery that this might as well be called a sermon, and that the Orthodox did not have such things, lay ahead of me) I turned to the carnation pinned to my shoulder, and consumed it inconspicuously, petal by petal. "I ate my Yom Kippur corsage"—a memory worthy of Alexander Portnoy: but for me there was no aftertaste of guilt, indeed, I learnt to eat carnations, and couldn't be trusted around floral arrangements for years.

No doubt I would do the same today, but in every other respect Yom Kippur has been transformed. The small community of Jews in West Virginia has had to serve for *mishpoche,* and though it serves us well, I am a stranger and a sojourner here, an expatriate on two continents now, destined to be homesick no matter what my home address. My own years are scattered all over the world, while in Morgantown other children's grandparents pinch my children's cheeks. Two years ago on Kol Nidre night the rabbi spoke of the sirens which sounded and would sound forever on Yom Kippur, and my secondborn daughter sobbed in my arms with a grief she would never have known if I had not set with such ferocity on making a Jew of her: *I started this,* and, *what have I done?*

But it's done now. Yes, my son, who must be just the age I was when I discovered carnations, asks with my firstborn daughter's voice, Do I have to go to synagogue on Yom Kippur? Yes, I tell him. Will there be food afterward? asks this child, flesh of my flesh, appetites like mine. Yes, after sunset there will be food. I tell him, and the pact is sealed: women wiser than myself have prepared the way to a child's heart and soul.

And so much for the internal power politics of the High Holidays. On this campus one is as likely as not to see the Homecoming football game scheduled for Yom Kippur, and I am the only Jew in my department who never comes in to the office over the holidays. Born among the most ignorant of all the ignorant Jews, and still quite passably ignorant, the abundance of my Jewish heritage is in that stiff neck. My course calendars were ready for the start of the first academic year and so was I when, early in September, I was invited to speak on a panel which dealt with religion and the social sciences.

"Sure." I had just returned from the meetings of the International Society for the Study of Behavioral Development: it seemed unnatural to speak without a microphone in front of me, colleagues beside me, and a sea of faces before me.

"Don't you want the date?" for I was about to conclude the conversation.

"Oh. Yes."

"September 27."

"Wait a minute. That's Rosh Hashanah. I've canceled classes, and I don't think I can speak on a panel, even when I think a Jew should be there."

"Well, I do understand that, of course, but I hope you will think it over before you refuse. If it will make any difference at all to you, there *is* going to be another Jew on the panel"— whom I met shortly afterward, who told me with surprise, "But I'm a Unitarian!" and I replied, "Haven't you learned that in small towns there are no Jewish Unitarians?"

"I'm in a peculiar position," I told the organizer—"We're *all* in peculiar positions," my friend might have said—"because I don't go to synagogue but I don't hold classes: I can see the contribution I might make, but I don't feel right about participating."

"Perhaps you could think it over for a day."

In the interval, I inquired of the director of developmental psychology whether the department had any vested interest in representing itself on the panel; he assured me that the decision was entirely mine.

"I don't teach classes on Rosh Hashanah . . ."

"Well, Rosh Hashanah is a designated Day of Special Concern. You aren't supposed to give tests or lectures which can't be made up by reading, but you aren't supposed to cancel class . . ."

"*I* don't teach on Rosh Hashanah!"

"Well, if it's against your religion, of course . . ."

"It's not my *religion,* it's my fucking *holiday!*" and I had my answer at last, and called the organizer: "I can't participate on the panel, and furthermore, I don't think a lecture restricted by default to *goyim* and nonobservant Jews can be considered the Distinguished Religious Lecture of the Year," a belligerent assertion which generated an enduring friendship on this contradictory campus in this contradictory city, where the fires of the crosses of the Ku Klux Klan could be seen burning on the hills just forty years ago; where a gentile shoemaker sent in a donation to Israel in the dark days of the Yom Kippur War which lay just ahead; where from time time bomb calls threaten synagogue services; where, for the eight days of Pesach, I let the crumbs of my peanut butter and matzot sandwich work an inverse baptism as they sprinkle over the faculty dining room, in violation of the injunction to bring no food into a place where food is served. But if it comes to a question of who is to violate whom, my answer is quick. Years ago I told a friend, who had learned with me to worship Socrates in his adolescence and whose interdisciplinary program now looked threatened, "If the hemlock comes, fling it at the bearer. It's better to do an evil than to suffer one." Not more than a year ago I discovered myself to be obeying another of the six hundred thirteen commandments: "If a man comes to kill you, strike first."

So, slowly, over the years, do we all discover ourselves to be keepers of the commandments, I suspect. "In Judaism we each have to discover our own road," the rabbi told me when I sought his counsel on the question of the panel convened on Rosh Hashanah. No! said the voice of the preceding decade, *Halachah* tells us. But it doesn't tell *me:* indeed, I am among those who argue that Jews make the law, law does not make us Jews.

If so—and it is so for many of us—I am recalled once more to my peculiar position, our peculiar position, that of this peculiar people. "How can you possibly rule your children's lives?" a nonJewish colleague asked me recently, and I told him, "God made man in His own image, didn't He?" My colleague, no more religious than I am, is taken aback: but it isn't blasphemy which shakes me but rather the foreknowledge of alternating currents of heresy and devotion, as it is said, for all our generations.

The Four Leaf Clover Story

Beverly Schneider

WE WERE STILL four blocks from the cemetery, waiting for the left turn arrow onto North and South Road, when Aunt Bertha began apprising us of the local IT victims: Mr. Gumpfer, from the dry cleaners, had IT; Label's son was taking radiation treatments (you know what that means); and a TV soap opera star had IT (I don't know whether she had IT in real life or only on stage; Aunt Bertha was unclear). Three years earlier, my mother had died from cancer and it was since then that Aunt Bertha had refused to let the dreaded word fall from her lips, as though the name itself evoked the disease's destruction. Protected in this manner, she was free to talk about IT as frequently as possible. My other aunts were not quite ready to play the IT game. They were reading the laundry label of Aunt Lottie's half-priced raincoat. I was confident that Aunt Bertha would be but momentarily thwarted. Having sheltered my own vulnerabilities within a cynical, condescending anticipation of an-

Beverly Schneider is coordinator of Kol Nashim, the Orthodox women's davening group of New York's Upper West Side. A former teacher and social worker, she lived in Israel for three years; she now works as a financial assistant on Wall Street.

other Kaplan sisters' play of fools, I listened, half-humored, to Aunt Bertha's horror stories as I drove the car down North and South Road, past the house in which I had lived with my family, and pulled into the B'nai Amoona cemetery where my mother was buried.

This visit had started as inauspiciously as the other five visits we had made together these past three years. Exactly one week earlier, Aunt Bertha's daily phone calls had begun. "Becky, I'm going with you. With Aunt Sissy and Aunt Lottie." She was always afraid of being excluded. "We'll go to Ida's grave, I mean Mama's, I mean your mother's grave, first. Then, you'll drive us to the other cemetery, Cheser Shamos, uh, no, Chaim, no. You know where it is. To see Grandma and Grandpa. You remember them." It was futile to again remind her that her parents had died long before I was born.

I always called my father at the nursing home before I left on these visits. By tacit agreement, we never spoke of my mother's death, which had coincided with his first stroke, but I liked to hear his voice anyway. The phone was busy all morning so I called my sister in Chicago instead. She was bored by these visits, but I felt obliged to tell her in spite of her condescension. "Why don't you just drop off the three Fates at the cemetery and wait in the car while they have a good cry?" she snickered. We chatted briefly about the baby and her new kitchen wallpaper, but she was in a hurry to go bowling. "You don't have anything in particular you want, do you? I've got to be running. I'm late as it is." I hung up, hoping that somewhere between Aunt Bertha's fixations and my sister's denial, there was some room for me.

Nothing seemed to alter these visits. They unfolded like some inexorable ritual upon which I exercised no control. So, the production continued as we parked the car a short distance from my mother's gravesite. "Did you bring enough Kleenex, Becky, should we roll up the windows, it'll get hot, it may rain, are you bringing a purse, Bertha, we won't have to roll up the windows." Eventually, they removed themselves clumsily from the car, with three purses, two umbrellas, and Kleenex for an army.

Aunt Bertha jabbered incessantly.

"Look, isn't it nice how they keep this cemetery," she said, remarking on the black man mowing the grass in the distance. "Not like that awful place where Mama is. Wait till we get to Ida's grave. I'll bet it's clean and neat."

"Quiet, Bertha. You shouldn't say that about Mama's grave. There weren't any weeds on it last year." Aunt Sissy could not tolerate ugliness. It was not respectful and threatened her vanity.

So, this year's theme is Cemetery Upkeep. I sneered silently and began to muse about my grandparents' cemetery with its trees spreading shade and leaves over crowded tombstones and its shabby old men with beards who hovered about with charity boxes, insinuating eyes, and mumbled Yiddish prayers. There, death was really Death and any form of life, even weeds, was counted as victory. I didn't want some hired hand mowing my produce when I died. Manicured cemeteries. Humpf. My poor aunts were certainly of the Depression generation.

"Oh, it was horrible! What do you mean you don't remember? Would you want to lie in a place like that? Forever?" Aunt Bertha baited her.

"When I die, I want to be buried here. Near Ida," Aunt Sissy responded.

"Do you hear that, Lottie? She doesn't want to be buried near Mama and Papa."

"Bertha!" Aunt Sissy was angry but felt obliged to offer an explanation; "I knew Ida longer."

"Oh," Aunt Bertha was stopped by the logic. So was I.

Aunt Bertha had difficulty keeping up with us. Her legs were shorter and she had to read the complete inscription on almost every gravestone. "*Jennie Greenspan,*" she whimpered. "Remember her? *Died 1952.* Lottie, what was her boyfriend's name? You know, the one with the limp ... *Sadie Schwartz.* Wasn't that a shame! She's buried near Izzy Yawitz, from the South Broadway dairy. How they used to fight!"

Curiosity overcame Aunt Sissy, who dropped back to join her older sister. They continued their banter, arguing now over biographic details. Aunt Lottie, as usual, walked close behind me. I tried to believe that she identified my silence with quiet dignity, but more often I feared she was simply guarding over me, watching, worrying about my secrecy. I was relieved when she too would stop to glance at the names and dates on the tombstones. Periodically, when she would release a quiet sigh, I would peer at her from out of the corners of my eyes.

This year, the completion of another row of graves and the healthy thickness of ivy that now covered my mother's gravesite gave the illusion of change. It was the same grave, the double tombstone, as yet half-inscribed: *IDA SCHREIBER*, no middle name, *1908-1963.* The four of us formed an automatic semicircle around it. My aunts quietened abruptly and appeared to lose themselves in thought. This was the very moment I dreaded. I never knew what to do, never was in the right frame of mind, never had the right feelings. I resorted to an old exercise, closed my eyes, imagined the white bones, the lost flesh that was now my mother, trying to remember, to feel something, anything. I pictured her skeleton deep beneath the green ivy. "This was my mother. This was my mother," I repeated silently over and over again. But without conviction. I could not bridge the gap between my mother and those bones. I grew irritated with my aunts whose shadows seemed to have returned to haunt me. They took my feelings, they felt more, loved more, missed more. They cried. I didn't. They cried in descending order. Aunt Bertha with a red nose and real tears. Aunt Sissy with pink eyes and sniffling. Aunt Lottie teary-eyed. They always cried in that order. Maybe, if they weren't here, maybe I would dare explode, dare cry. But they were here, watching me, waiting. Thank God my sister wasn't here too. She would mock everything. Even me. Once again I tried to imagine my mother, but even the white bones had been snatched away by these images of my aunts' tears and my sister's sarcasm.

Irritated and dry-eyed, I ended the silence with a signal that I was about to read the *Kaddish*, the traditional Jewish prayer of mourning. I never told them that this recitation with four women was not proper. They would have condemned the religion, would have insisted that the rightful way for a Jewish woman to mourn is to recite the *Kaddish* from her Jewish heart at the cemetery. I did not want to persuade them otherwise for I too liked to hear the incantation of that prayer. Relieved to have a prepared script before me, I read in a rolling, rumbling monotone the practised syllables of generations, until the last cadence was maintained and the silence held for them, a sufficient time to allow for respect. They all stared, motionless, a bit teary. Three sisters, with slightly stooped shoulders, scarves carefully covering their heads, arms forming a "V" at their pocketbooks. I always repented of my anger when I saw them like this.

After a moment or two I bent down, tore up some grass, and solemnly spread it over the ivy-covered grave according to the custom in my family. Moving slowly, significantly, my aunts followed suit. I loved this form, although I never learned its source. It was sweet and simple. Final. Another visit was completed. Thank God, I made it without . . .

"A four leaf clover!" Aunt Sissy exclaimed, startling our solemnity. We stared in bewilderment as she separated a four leaf clover from the tuft of grass she had uprooted and displayed it to us.

"What're you going to do with that?" Aunt Bertha demanded.

Aunt Sissy spread the grass over my mother's grave with care and delicacy—she loved ceremony; and then, with as noble a gesture as she could manage, she bestowed the single clover upon me.

"Here, Becky. This is yours. From your mother. Now you can have everything you ever wanted."

"Thank you, Aunt Sissy. Thank you." I started to put the four leaf clover into my prayerbook with borrowed pomp while Aunt Bertha, so touched by this gesture, began hunting for a clover of her own discovery, "so I can give good luck to Becky, too."

Trying to understand, I stared blankly at her fragile gift. Is this all it takes to stop the writhing and suffering: this delicate, green clover I cradled in my palm? It seemed so easy, and yet such a betrayal. I did not know why my mother died, only that I needed something more to explain her death than this petite symmetry, mute to eternity.

Aunt Lottie stood by me with a puzzled look as though she were trying to understand a naughty game her sisters played. When I turned to her, she urged Aunt Bertha to get up and reminded Aunt Sissy that we still had to stop at Rabbi Halpern's grave before we could leave the cemetery.

I guided them to his grave at the very front of the cemetery. We always stopped there. Nothing spoiled our routine, although Aunt Bertha availed herself of various clover patches surrounding his large, simple tombstone and interrupted our meditation more than once with premature victory cries of discoveries. Aunt Sissy and Aunt Lottie remained solemn, until one of them finally sighed, "Remember how Ida used to say, 'If only Rabbi Halpern were alive now?' " Yes, my mother believed in miracles, too—not in four leaf clover miracles, she was too practical for such flimsy promises—but in Rabbi Halpern. If only he had been alive, her cancer would have gone away.

After corralling Aunt Bertha, we returned to the car and began the short trip to my grandparents' cemetery, a trip which was always a relief to me. Somewhere between the two cemeteries, Aunt Bertha, as usual, started calling her sisters by the Yiddish diminutives used when they were children. The names were soft and enchanting, "Liba" and "Lehya." From the moment I turned left from Olive Street Road into the Chesed Shel Emeth Cemetery, the sisters would argue regarding the location of the grave. It was lane no. 113b or no. 117b, across path no. 7 or no. 3. The argument continued even as we walked. I never minded. Here were the noble dead of a generation I never knew. I imagined myself immersed in that immigrant culture. Many of the graves had faded oval pictures of the dead permanently attached to the tombstones. I was intrigued by the faces. Aunt Bertha's cemetery geography fascinated me. And even though we walked for awhile, we were never lost. She seemed to know everyone on the block.

I forget which aunt was right—they probably each claimed to be—but we finally found the gravesite. "Oh, I knew it all the time. Right across from Minnie Klein. They used to walk to Soulard Market together."

Once again I waited for an appropriate length of time for silent meditation. I looked at my aunts. They did not seem so old. I tried to make them even younger, tried to imagine their bereavement when their mother died. Did they remember her kiss, her face, her voice after all these years? What she wore when she died? Did they try to communicate with her now? "Mama, I've been good, Mama. Honest, Mama. I still learn things the hard way, Mama; I'm learning only now what you meant to me. I never really knew you, Mama." Did they think that she understood her children better now that she was dead? My mother used to make these visits with them. I could not imagine her having had these thoughts and then coming home to us as she did to iron and fry liver. The thought almost made me cry.

Aunt Sissy signaled me to begin the performance of the *Kaddish*. Slow, mournful, illegitimate, I tried to read as correctly as possible, substituting the European pronunciation familiar to them for the modern Israeli one I commonly used. I felt the moment transformed. The sun withdrew respectfully behind a dark cloud, erasing all shadows. An orchestra of birds and spring breezes played an accompaniment in the trees above. My voice chronicled the death of generations.

I was sure I heard my aunts whimpering and slowed my reading to prolong the soulful mood. It seemed the birds honored the changed tempo. The leaves lay still. Only my breathing, somehow, was not completely right; the short gasps incorrectly punctuated the rhythm.

"Oh, I see one. Over there." Aunt Bertha stood next to me with her head bent downwards, eyes studying the weeds. I pretended to ignore her but raised my voice and read in a deliberate style, hoping this gentle reprimand would silence her. It did. Soon after, though, I noticed Aunt Sissy, on my left side, buckling slightly. I cast a reproving glance at her during an awkward breath stop and she straightened guiltily, but momentarily. Soon my peripheral vision became a seesaw. Right side up respectfully; left side down irresistibly until the two sisters succumbed completely.

I continued to read falteringly. The birds and breezes seemed to have deserted me. Only Aunt Lottie stood by me, a faithful attendant to my *Kaddish*. Eventually, when her knees began to bend and then straighten alternately, I knew that in a moment I would remain alone. I read on, compulsively, to silence my fears. Aunt Lottie, don't leave me. Don't leave me without my shams. Don't leave me unprotected from that wild wind that weaves through these stones, from the vultures that eye my flesh.

"Lehya, look. Here's one. Just look." Her eldest sister bid her join the search. Aunt Lottie smiled at me. Her myopic eyes beseeched forgiveness from me. They begged permission for her indulgence. And then she giggled and stooped to join her sisters.

I managed still to read. Quietly. For spite. I fulfilled the family obligation, alone, deserted by my mother's sisters, self-right-eous before the crowds that mingled through the cemetery. And while I read, I thought of the story I could tell of the Kaplan sisters' play of fools. I planned to thank them for providing this grande finale after I finished the *Kaddish,* but I was haunted by Aunt Lottie's smile and could not take my mockery seriously.

So, I read to the last amen and then having finished I turned to look at them.

They did not realize I had completed the *Kaddish* or that it was time to tear up some grass and, in a gesture, end this visit. They were somewhere else. Three women in their fifties crawling over the unkempt grass, as oblivious as they were forty years ago when they made clover rings in Carondelet Park, innocent of the story I would tell. Happily, happily, they understood nothing and trusted everything.

With some coaxing, I gathered them up and brought them safely home. When I said goodbye to Aunt Bertha at her apart-ment, she promised she would find me a four leaf clover; and, if not, well, she would try to find something else that would be just as good. "So, call me. Don't keep away. I've got a telephone, too, you know. Instead of calling Aunt Sissy next time, you can call me."

That was the last time we visited the cemeteries together. Shortly thereafter my father died and I moved to Boston. My trips to St. Louis did not coincide with Mother's Day or fall in the month before Rosh Hashanah when my family customarily made these visits. I kept the four leaf clover that Aunt Sissy had given me in my wallet, grey and rigid, until six years later when I laid it solemnly on a solitary plot, several rows behind my parents' gravesite. It was a freshly sodded, manicured grave. Everything was just as she would have wanted it. Aunt Bertha and Aunt Lottie must have brought the pot of plastic flowers that rested at the head of her grave. They never put flowers on my mother's grave, but I didn't mind; Aunt Sissy was the lady, beautiful and refined, and she loved pretty things. I remembered Aunt Lottie's letters to me describing Aunt Sissy's long battle with cancer and I tried to imagine the angelic expression she was said to have worn when she died.

When I left, instead of tearing some grass, I placed the flattened clover ceremonially on the center of her grave because the irony comforted me and because Aunt Sissy would have appreciated the sentiment. Otherwise, I did not know what else to do.

Anniversary

Elaine Marcus Starkman

WE'VE GROWN so alike standing before our bedroom mirror this morning. Today, our fifteenth wedding anniversary. Even our reflections look alike. Two full faces, steady hazel eyes hidden beneath silver-rimmed glasses, sensitive lips, he, balding, I with the color of my hair gone drab.

Wordless he shaves, black stubble from his electric razor falling into our double-bowl vanity, onto our white carpet. Hum of his razor and my toothbrush droning under harsh fluorescent light.

He pulls out the plug, stuffs razor into our toothpaste-smudged drawer, that drawer like all our dresser drawers. Cluttered with indispensable litter that ties us together. Inside: notes, change, mismatched socks, graying underwear. Outside: surgical journals, poems, math tests, tennis shoes.

Elaine Marcus Starkman is the author of *Coming Together*, a collection of poems. She teaches Jewish literature and journal writing at the College of Marin, J.F.K. University, and St. Mary's. Her works have appeared in *Hadassah, Present Tense, Jewish Frontier, Contemporary California Poets, Tunnel Road, Studies in American Jewish Literature, Blue Unicorn,* and other publications. She lives in Walnut Creek, California, with her husband and four children.

He throws his pajamas into the old straw laundry basket and dresses, clothes unstylish and conservative as the day we met. Dark suit, narrow lapels. Reluctantly pulls on blue-flowered shirt I gave him for his birthday.

Banging of bathroom door, burnt toast, bitter coffee, frantic search for jackets and lunch money fade into morning routine. House silent. Only whirring of washer and dryer. When did I start them?

An ordinary morning. Hair unclipped, terrycloth robe stained. Yet different. Fifteenth anniversary, and I find his face among towels. How he rushed in late last night, high beneficent forehead pink with chill, glasses steamy.

"Sorry, got stuck again. Just a standard procedure. Nothing interesting. Maybe we could go to a movie tomorrow. The one by that woman film director is playing in Berkeley."

"*Swept Away?* I thought you said it was too violent."

"It's the old taming of the shrew updated, isn't it?"

"Yes. You won't like it."

"Probably not, but you said it was important to see. Ingmar Bergman's got a new film out. Maybe I'll come home early. We can go to either one."

What sudden extravagance, when has he ever cared about films? Has he remembered our anniversary this year? The year of my migraines, his weight gain, our son's running away, our daughter's school failure? The summer we fought on the Santa Cruz beach in front of a crowd of strangers, I screaming, "Doctor, doctor, doctor," he slapping me across the cheek, voice seething, "Shut up; I'm sick of you. From now on I'll do what *I* want." We drove home from his day off, the children shocked into silence. Those endless hot months, no classes, no poems to sustain me, only family responsibility, I hated what I'd once loved. Maury, a doctor before he was born. He *had* to be. And I his wife.

Throw wet towels into the dryer. What would I do in my spare time without laundry. Golf, shop, drive to the city, join the hospital auxiliary?

Brush my hair and dress. Our wedding portrait. How voguish I look. Blonde teased pageboy. Squinting through my contact lenses. A tiny pearl crown on my head. My mother sighing with

relief: *A doctor she's marrying, thank God; she's nearly twenty-three.*
Except for his baldness, he hasn't changed a bit. Same soft eyes,
same earnest smile, same uncompromising values, same rabbinical look. In those days I'd look away modestly and say, "I
love you." Even when he cut short our three-day honeymoon to
return to his residency. How understanding I was.

How quickly we fell into routine. The long weekends of his
final year, endless hours in the emergency room, his exhaustion, my posing with a huge belly in the St. Louis snow. After
lovemaking he'd stand awkwardly in the hall in his white gown
almost afraid to say those words: "I have to go now, Nina. I'll
be back tomorrow night."

"All right," I lied.

"Maybe you should phone your mother so you won't be lonely."

"No, no, everything's fine. I'd rather be alone."

"Then why are you crying? You knew it would be this way."

"It's okay. Go on." Listen to footsteps resound in the dark
hallway of our apartment building.

Five years and we never quarreled. *Nina, my friend needs a loan.
But Maury, we haven't gone anywhere for so long. I've got to lend him
some money. All right, Maury. Time to visit my family. Can't we skip this
Friday night? You don't have to say anything. Just sit with them. Listen,
let's invite them here next week. Maury, you know I can't cook anything
your mother does. Make something simple; it doesn't matter.*

How grateful I was. His family called me lucky. What could I
do to please such a giving man? Only conceive and that I
managed despite our obstetrician-friend's prognosis: *her pelvis
is small.* But they came anyway, three births in five years, the
miracle of our children's lives escaping us, distracting us from
our own goals. They ran through the town house shrieking,
crying, tumbling down the stairs, biting each other, demanding
love and attention. He loved them; they bewildered him. "Why
are they so wild?" he scratched his head.

"Because they're spoiled. Didn't your mother spoil you?"

"She didn't have time. She always worked in the grocery to
save money for our schooling."

"Why don't you humor them? Why are you so strict?"

"Because I love them, because I want them to know right from wrong."

How problematic they were—Joey repeating first grade, Amy moody and withdrawn, Rachel with her temper tantrums. Every year we were unprepared for their growing pains. And our own as well. Every year he said, *the medical profession's got to change. No, Maury, you've got to adapt to it. What you want is impossible.* Six times we moved chasing impossibilities. All the time our parents hovering in the background: *Tell him medicine is a business; his income depends on that.*

Our final move to the west coast nearly destroyed us. Here away from friends and family among capricious San Francisco temperaments, Rachel crawls out of diapers and into kindergarten. "I'm sick of your ideals. Sick of being called 'doctor's wife.' I'm going after mine now," I rant.

"What are you going to do?"

"I don't know, but I'll find out."

"Do what you want. Have I ever stopped you from doing what you want?"

"Your very presence stops me. Your—"

"That's your fault, not mine. I could never please you anyway. Just don't ask *me* to change; just leave me alone."

I don't ask. I sit in the kitchen of our new California-style house late at night writing poems. For the first time in years. Those poems, that passion, born of my guts, my joy alone, not his, *mine.* Published in an obscure feminist university journal. I'm a feminist, liberated. Suburban mother of three, I howl with delight. Dance with the children, drive to the top of Mt. Diablo, shout from its peaks: *Mamma's gone mad.*

With that madness 1970 breaks, and I'm transformed. Nina the Iconoclast. Chipping away at his pedestal. He's rigid, unathletic, boring, *too moral.* I'll leave. Go to live in a women's commune in Berkeley, fly to Israel and build the desert, make up for lost time. Drown myself in Bergman, look for the young Chicano I met in the poetry stacks. But I watch him say *Kaddish* for the death of his patient, listen to the sound of his Danish rocker and favorite Oistrakh, look into his deep hazel eyes that refuse to see the senselessness of our lives, and bound off to the laundry room like a hurt puppy. There in that familiar maze of underwear, old texts and paperbacks, I announce, "I'm starting school again."

"Good idea, but don't become too compulsive about it."

How well he knows my compulsiveness. School's not so easy now. My forehead tingles, my stomach cramps. Rachel asks why I'm not room-mother. "Because I have to study," I snap, a dragon at dinner hour. Course after course. Women and Madness, La Raza Studies, Seminar in Faulkner, Tai Chi, the Yiddish Novel.

The children's diapers are replaced by braces, lessons, tutors, car pools, conferences, puberty. They alone grow. We merely age, I studying at night toward some unknown pursuit, Maury reading journals and Middle East news reports.

"Ten o'clock. I'm on call. I've got to get some rest."

"Go ahead. I'm studying."

"Why do you play this game every night?"

"What game?"

"All right, study. Study until you beg me to touch you. I'm fed up. You and your games."

"I'm not playing games. I just have no desire."

"Then why don't you leave? It's fashionable for women to leave men these days. I'll give you anything you want. Just go so I can have some peace."

"Don't be ridiculous; you think I'd leave you and the children?"

"Why not? You probably have a lover anyway."

"Are you mad? Are you *absolutely* mad?"

"Then why do you act like this?"

"Like what?"

"Your lack of desire, your inability to manage the kids. Why don't you go for help?"

"Because I'm trying to change? Because I'm trying to do something for myself? You're the one who should go. You've lost touch with reality."

"Go in the kitchen and study. You don't respect anything I believe in. You want something I can't give you. I'm just a small man."

The yellow wallpaper rages in silence. All these years—wasted. He with his kindness has made me afraid of the world, he with his selflessness has made me nothing, and I've let it happen. What am I next to him?

The phone rings. He dresses and leaves. I close my text and creep into our bedroom, lie by the telephone on his side of the bed. Browse in a new novel. What reckless heroines, what cads the men are. Why can't we be like that?

He's gone an hour now. The rage dissolves into a whimper. The wind bangs at the patio windows, dogs bark, children moan, the faucet drips. What if Maury dies? Would I mourn him for years? Would I ever take a lover?

Phone again. A woman I met at school.

"One partner in fourteen years? This is the Age of Aquarius."

"For him it isn't."

"And for you?"

"Me? I don't know."

"You're still hung up because he's a doctor."

"It's not that. It's the kind of person he is."

"You mean *you are*. You'll change, you'll see. By the end of the year you'll change."

"Not in *that* way. You know I still have some traditional values."

"You and he are just alike, killing yourselves with an outmoded moral code. You have to move with the times."

"I know, I know."

"No you don't. Otherwise you'd forget your fantasies or admit you still care for him."

"I can't."

At 2:00 A.M. he comes in, face sallow, tired, blood on his undershorts. "A terrible case, Nina, a young man. Motorcycle accident. He'll be dead by morning. Let me hold you. I can't do my work without you."

A winter moon floods our bedroom. Succumb but don't surrender. Don't share his act of love. Separate mind, nourish it with illusions of young men in my classes. He falls asleep immediately, not waking to wash. Listen to his snoring. Anesthetic on his fingers, spittle on his lip. How could I ever leave? He'd work himself to death. I'm the small one, not he. If only

I'd let go of my anger, if only I'd stop blaming him for my failures. Sleep, morning obligation, make lunch, the children—

We wake, shut off from one another, two strangers. Wordless, he shaves, runs off to the hospital. Phones me at ten. A sudden joy in his voice. "Nina, he's going to *live.* There *must* be a God, do you hear? When I told that to the men, they laughed. Nina, say something."

I can't. He waits for my silence to end. "Listen, I know this has been a hard month. One of the men told me about a place in Mexico. Maybe we could drive down with the kids when they're on spring vacation."

"I don't think they'll want to go. Last time we took them with us it was a disaster. You hollered at Joey the whole time, and Amy's complaints drove me crazy."

"We could leave them with Mrs. Larsen."

"Rachel hates Mrs. Larsen."

"How about Mrs. Bertoli?"

"Too expensive."

"Well, I just wanted to check before I take extra call this month. This malpractice insurance will either kill me or make me retire."

"Take the call, Maury; we'll stay at home."

"You're sure?"

"I'm sure; you're not ready to retire. Maury—I'm glad about your patient."

"Don't be glad for me, be glad for him."

He didn't even mention our anniversary; he didn't remember. Every year he forgets. Yes, we'll stay home this time as we have all the other times. And yet there's something different this year as I sort the towels. His key in the door, breakfast dishes unwashed. He's home so soon flipping on the news.

"It's too early for lunch."

"I know." Hands me a single rose, kisses my neck. "Happy Anniversary." Looks so helpless I want to reach out but don't. My eyes wet. "Come on, I'll make you lunch. What time do you have to be back?"

"I don't; the anesthesiologist is on strike. No work."

He changes into his gardening pants. When did he let his sideburns grow? They're nearly gray. How he loves watering the garden. Always has been more nurturing than I. Awkward eating without the children. What to say to each other—just two of us alone. A new phase we're entering this fifteenth year.

"The salad's great. You know, we should become vegetarians," he suddenly laughs.

"Can you see Joey existing without his sloppy joes and hot dogs?"

"Give him five more years, he will. He get off okay this morning? He was mumbling about some science project."

"Fine, he's even turning into a decent kid. Decent but different from us."

"He's growing up in such different times; it's hard on him. Good to be home. After seeing that patient this morning I promised myself not to worry about anything. Are you angry at me?"

"Why do you ask?"

"You seemed upset this morning."

"I was angry at myself."

"What for?"

"Because I've let myself become too much like you."

"Why do you say that? We have our separate friends, interests—"

"Even Bob Stoller said we're getting to look alike."

He grins. "Only our noses. Did you know that Stoller and his wife broke up last week?"

"Nothing surprises me these days. How come you didn't tell me?"

"What's the use of telling bad news? Nina, let's sell the house; let's go live in the country. I'm sick and tired of everything. I'm going to look at a place this weekend."

"Are you kidding?"

"I'm not. Let's finish lunch and talk about it in bed."

"But Rachel comes home at two, and I've got to finish a poem."

"Forget the poem. If you work too hard, you'll become a poet. Then what would I do?"

"Maury, you're a chauvinist!"

"No, I'm not, just jealous."

"*You're* jealous of *me*?"

"You've more time to think, to be yourself."

"I suppose I do."

"More time to do what you want."

"Oh, come on now."

"Well, I can't do what I want either. People like us have too many responsibilities. We live limited lives."

He stands there in his underwear, his pale skin covered with black down. The veins in his neck pulsate like the days in our old St. Louis apartment when we were first married. March twenty-first, 1960. First day of spring. *I still care, I still care.* I didn't know that I did, but I do.

My Mother Was a Light Housekeeper

Thyme S. Seagull

Hello, Berkeley, your acid sparkled streets glistening through the smog, heartland in the production of rhetoric, land of free boxes, neighborhood warning systems, block dances, food conspiracies, and my relatives. Aunt Riba and Uncle Roy's bright one-story stucco, catty-corner to a decaying mansion—where some of their children live communally.

Rosh Hashonah. I wonder why Jewish New Year begins in the Fall instead of the Spring. The Spring would feel more natural. I reach their doorbell just as mom and dad drive up in their camper. Everyone on time for the holiday dinner. Me, mom, and dad have never been here before because Aunt, Uncle, and Sarah, the Old Mother, have just moved to Berkeley to be near their children. Everyone very keyed up. Riba, Roy, and Sarah all answer the door. Excited talking, sharing information about

Thyme S. Seagull lives in Eugene, Oregon, where she does lesbian feminist radio programming. She is the author of "Rima Comes to Emerald City" and "Get It Right the First Time," unpublished sequels to "My Mother Was a Light Housekeeper." "My Mother Was a Light Housekeeper," copyright © 1978 by Thyme S. Seagull, was first published in *Sinister Wisdom* 6 (Summer 1978).

missing members of the clan. Showing family photos. I drift to the rear and notice Sarah, the Old Mother, standing at the doorway to the back porch.

"We had a nice house in D.C.," Sarah says to me in familiar yet surprising Yiddish accents, a sound I have not heard for years. She is very short and stooped, wrinkled, not older but ancient. "It had a lot of room, big backyard, lots of flowers and trees." I can see into the backyard here: tiny.

"Yes, but they left that house and moved here?"

"They wanted to be near the children."

"Your great-grandchildren."

"Yes, them especially, Lillakah and Lil'Umbillakah."

I look wonderingly at the Old Mother, a fantastic survivor. A Ukrainian. She is so short it is easy to overlook her wrinkled face. But I look at her carefully. I had overlooked my own grandmother.

"You knew my Bubie?" (I ask after my grandmother.)

"Yes," Sarah whispers. "A very progressive woman, very progressive. I knew her in Elsinore, when I lived in Los Angeles." The rest of the family is floating towards us. I see mom's attention about to catch onto our conversation. I tighten up and respond quickly, "Yes! I was there too."

"Elsinore is a Yiddishe shtetl," Sarah comments.

"Yes, I know," I whisper quickly, as mom interjects. "Yiddishe shtetl! She doesn't know what you mean, grandma, these kids don't know what it means—" Ha ha ha ha, they laugh about how these kids don't understand.

"I know what it means," I say, but she and Sarah are talking away in Yiddish. (Oy Gottenyu, she's auf tsurus, A Shaynim Donk in Pupik.) It is true I don't understand much.

"It means Jewish ghetto," mom stops long enough to tell me.

"I *know.*"

"Well, that's better than the rest of these kids."

"You only spoke Yiddish to each other, not to us." You and dad spoke in your thick private flavors. We were outside of it; we spoke English in harsh unhappy tones, coping with public school in rural America . . . "But I *was* in Elsinore and I'm *not* a total idiot."

Mom blue-eyes me, smiling. Is it ok to come out with it like that? Did I hurt her? A record in Hebrew I never heard before is playing on the stereo. "Anachnu Ve'Atem" is the refrain, a powerful chant from Israel.

"What's that, mom?"

"A song about Us and Them."

I don't know how to ask mom if I have hurt her. It is very beautiful music so it catches us for awhile, then she wanders off to join a conversation with my aunt, who is setting the table. I turn back to the Old Mother. "I know Elsinore is the West Coast shtetl. I could easily see that. The only ones in the streets and the stores are elderly Jews."

"Yes, and they have a community center," says Sarah.

"Yes, I know. Bubie and I went there. To a concert. A woman was singing in an evening gown. Bubie started singing too."

"Yes, she loved to sing. Such a good voice too."

"The people who were running the show didn't think she should sing because it wasn't her concert. A man came over and told her to stop singing. She got really pissed off. We had to leave. She was mad and didn't want to stay anymore. As we walked out, she muttered resentful curses at the people who ran the community center, all the way out, real loud, everyone heard and turned around as we left."

"She was an individual, very stubborn, very progressive."

The dinner table is covered with white paper, centered with candelabra and dotted with red wine bottles. For a moment I recall the ancient ceremonies, chanting in Hebrew for three hours around the table before we could eat, the ancient ceremonies during which I stared at the matzoh crumbs and the red wine stains on the white paper. While the grandfather chanted in a language I didn't learn, and all the relatives jabbered away in yet another language I did not learn. What does anything mean? I long ago tired of asking, or maybe I never did ask. I am a foreigner in the ancient culture, the languages, the traditions. I am light years away. Or has it only been a generation? In me, the tradition of five thousand years dies. I am the daughter who cannot carry on the family name, practice the ceremonies. I am the daughter who searches for a new tribe. Who searches for primitive solidarity in a new culture of women. What was coherent for them has never been coherent for me.

"Maybe you'd want to write a book review for *Freiheit?*" asks Uncle Roy. (*Freiheit* is a socialist journal which he edits, modeled after the old *Freiheit.*)

"No," I say, flatly and glumly. "I've done that already. Book reviews are fine, but first I want a different space."

"She doesn't want to do journalism anymore, Roy." Aunt Riba. I look at her in surprised gratitude.

"What do you want to write? For whom? To be published in a book? Magazine, where?" Uncle Roy demands.

"A book maybe."

"What, a novel, fiction, what?"

"Something which emanates from the dream consciousness. *Freedom begins in the realm of dreams,*" I quote.

"Freedom is only in the realm of dreams," says Sarah, in Yiddish. Mom translates to the rest of us, the younger generation.

A silence follows. Then Uncle Roy continues dominating the dinner conversation with social, political issues. Mom, dad, and Aunt Riba follow along avidly while the rest of us red-diaper babies, in various stages of dress and stress, delicately slurp and gobble the split pea soup, chicken, and mushrooms.

"Do you think the working class is becoming more progressive?" asks Uncle Roy of my father.

Dad clears his throat. "Well, I can tell you, from traveling around the country a lot and living in trailer parks these last several years, I can tell you they are still very reactionary," says dad, in his slow laryngitic voice.

"Or maybe it's the union leaders who are," adds mom.

Why is it that I am always surrounded by people speculating on the working class—is it capable of transforming itself in the U.S.? can there be a revolution? I am still in the heartland of theory, with my relatives. I haven't seen mom and dad in a year; we've all been traveling around. I want to talk about experiences.

"Who have you met lately? Who travels in these trailers?"

"We met some travelers in Montana who used to live down the road from us," says mom.

"Really, who?"

"Oh, we never knew them when we lived there. They were on their way to moving to Grants Pass, Oregon. . . . *We* could never move to Grants Pass," mom ends sharply.

"Why not?" I ask, still not getting it.

"Because we're Jews." (Get it, get it?)

"I can't just move anywhere I want either," I say. "I mean lesbians can't." Mom looks at me, somewhat pained.

"Isn't that where you spent the summer, Cheike, around there? At a festival?" Dad.

"What kind of festival?" asks Uncle Roy, perking up.

"A women's spiritual gathering."

"What's that?" he persists.

"Ummm, uh, uhmmm," I look around the table. Everyone is staring down at their chicken, listening intently. What to say? "It was women getting together on land, like a campout, to center ourselves as a group. We're developing our own culture." Unfortunately, I get into it and go on. "But we didn't own the land, and the man who owned the land insisted on being there."

"You've gone too far," says dad. "Why do you have to have all-women gatherings and always say, 'Too bad there was a man there'? It's like colored people when they got money at first they went out and bought Cadillacs but now some of 'em drive VW's."

I can't follow the logic of dad's remark so I turn to my aunt and mom, who are sitting together, seeming to agree with dad. "You have to experience working with all women and being with all women. You've been married for decades and decades, how would you know?"

"Oh, I have experience," answers mom, with a touch of hauteur.

"The women at B'nai B'rith used to get together every week," adds Aunt Riba.

"It's not the same thing. You went home to husbands afterwards. Living with men and visiting or going to meetings is different than living with all women . . ."

At this point my uncle gets up, grandiosely announcing that he will wash the dishes.

"I'm going to help him," says mom, with laughing emphasis.

"How long can a group of women stay together?" asks dad. "It's not like you have kids to keep you together."

I move my mouth to protest. Some lesbians have kids. Why do you need kids to stay together? What does stay-together mean? How important is it? But how can I ask them? *They* don't know.

"Women living just with women, ay meshugge," continues dad. "It's absurd." Almost everyone laughs, thinking of the absurdity of it. Not maliciously, though, it is just ludicrous to them. Nothing I say can make it real; I don't have the answers, the experiences, five thousand years of ritual at my fingertips, five thousand years of red wine at my lips, to make them speak the words from beyond the heartland of rhetoric. I try to imagine my tribe, we who will shout *KinsWoman!* to each other in a tribal tongue. I imagine us in a dance to the light, our own festival of light, into ourselves and into each other, dancing around candles on the Solstice. I imagine saying to the children, "We have danced this dance and chanted this chant for five thousand years. It celebrates the eternal return of the light."

These are the words, but I do not have them.

Walking along the streets of the colonial town where I went to high school, I am with a very open being. A being who has just come into the country from Nepal or Ceylon. But I sense she is the same one from high school, the mysterious Sing, who appeared those years to us in stately silence, elegant, impeccable. Sing, forever the profile of the kiss not taken. I am showing her Main Street.

Then the scene shifts to the farm: Nuovev Jasmine. She has a room under our roof, like a nest in the eaves. We go up to my little bedroom overlooking the pond, right under her nest. As we lie in bed together, we can see the sun rising over the trees around the pond. We are touching, I am touching her face, so tender and clear. She is open to me.

I ask her questions. I take off the last of my clothes and then she does too, smiling, watching me. The softness of her caress across my breast is like silky seaweed on underwater skin. Everywhere I am opened, sensual.

Suddenly, the sense of another. Vashti is sitting hidden in the blankets at the foot of the bed, staring at the sunrise. My younger sister. I feel around the blankets. "So there you are. Why don't you open the downstairs drapes?" (Same view.)

"That's OK, I'm going outside," she says, and goes. I lie back. We're

spending the whole day in bed together, timelessly. "I don't know what time it is because I don't have any clocks around," I explain, although it is not necessary. Then the dream changes, turns round. I'm waking up.

Cheike! Cheike! "Are you ready to face reality?" asks my mother, with her own brand of gentle sarcasm. The fragile film of the dream is dissolving. Oh, what was it? The feeling of spending the whole day in bed with a woman, a very large woman who holds me, with long, thick red-gold hair. And cascading matzoh boxes at the very end.

Daylight streams into the trailer. Mom turns on the nine o'clock news. Patty Hearst capture news. Reality. I am not lying with my lover. I have no lover. I am with mom and dad in their trailer. California sunshine is bouncing off all the cement and metal in this trailer court.

"Can I turn off the radio?" I leap up, turning off the radio.

"Don't you want to hear what's really happening in the world?"

"I can't do anything to help Patty Hearst. I want to write in my journal."

Mom is washing the dishes, standing firmly in sneakers and anklets. She glances over at the marbled cardboard notebook I open, like the ones from grade school. Back on the farm, it hadn't occurred to me to write in front of anyone, and I had never seen anyone else write. It was a deep dark secret, like what happened behind closed doors in bathrooms and bedrooms.

"Do you ever read it, or do you just write in it?" mom asks.

"No read, just write, ha ha ha." No, I don't show my writing. What if I opened my books to mom and dad, showed my whole self, dark green and silent childhood? But my flippant manner has closed the subject.

"Yes, I reread it sometimes. I learn from it." Whoever wrote in a journal on the farm? All I ever saw was a locked wooden box of mom and dad's love letters. And my older sister's locked diary, in which she vowed never to tell a future husband she had lost her chastity. Nuovev Jasmine, a paradisiacal Jewish chicken farm in the forties and fifties, when adventuring into the interior was called "too subjective." Politics and economics were going to save the era. And politics was what got you in trouble and made people hate you. The childhood faces of the Rosen-berg boys haunted my childhood from the back cover of *Death-House Letters.* Which sat on their dresser for years, like a permanent fixture. I recall that bedroom, which looked out over the pond. The double bed I had to cover so precisely—but never lie on. The night table next to it on mom's side, upon which sat *The Well of Loneliness* (in paper) and a lamp, also for years.

"Why don't we ever share what it felt like to live at Nuovev Jasmine?"

Mom looks at me warily. I used to be so critical of her. She finishes drying her hands on a dish towel. "What do you want to talk about?"

I feel shaky, on thin ice, but ready to skate—"I hear you calling my name. Like a shattering reverberation over fifty acres of corn stubbles, rattling through the woods. 'Cheikk-kkeeeeeeee. . . .'

"It is winter, the corn stubbles are frozen, my feet crunch in the snow. I want to explore how the eastern woods look in the snow, the density, the designs, the feeling of the snow transforming an already secret place. It was not a place where anyone accompanied me. I could see dad driving the pick-up on a far western cornfield, heading toward the garbage pit. You were calling from the porch, you wanted me to help with the laundry . . ." Mom's tan face is frowning.

Is it ok to talk like this? Dad is outside the trailer, I can see him setting up a vise-grip on the back of the camper, his new workroom. A condensation of the garage and feedroom at Nuovev Jasmine. I can see his white hair curling up the back of his neck as he bends forward in concentration. He is making earrings out of nickels. Mom gets up to sweep the floor, commenting that what she does and what dad does is what is most comfortable for each. There is no problem. As she moves around, the trailer shakes, the floor shakes, and the ceiling shakes. Reminding me of dawn. (At dawn the rocking trailer woke mom up. "What's going on?" she asked in a startled urgent voice. From my berth I stopped wiggling and didn't tell her what was going on. Just laughed that I had had a farm dream again.)

"So what? I was doing the laundry . . . dad was dumping the garbage, you were—"

"What? What was I?"

Mom's eyes light up to answer, "I didn't want you to do the laundry, Cheike. Actually, I wanted you to do the eggs!"

"Ah yes, you wanted me to do the eggs. Remember when the bouncing trailer woke us this morning? I was having a dream of packing eggs!" (I still won't say I was masturbating.) "I was bouncing in my dream." At this point dad comes in and leans against the door frame, listening to us. "Only I'm not packing eggs for Nuovev Jasmine, I'm packing eggs for *Amazonia*. It's clearly printed in blue on the side of each thirty-dozen box. The eggs are all packed strangely, all different sizes—jumbos and peewees together, checked and cracked, flats and fillers not filled to the proper amounts . . ."

We all laugh. Yea, the egg room, grading and packing all those years, a fragile egg myself, thin and unexposed, not like those hardy egos developing in the outside world. Eggs falling off the machine and smashing on the cement—eaten by wild chicken coop cats, darting in at the sound—eggs crunched together on the way from the washer to the grader, squeezed and falling, eggs outside crunched by tractor tires, eggshells in the mud. Huge white eggs and little inverts, soft-shelled eggs, malleable in fingers, floppy, lacking—calcium? Lacking the right support . . . The Large go in the large, the Mediums go in the medium, the Pullets go in the pullet . . . Experiences rolling down the blue runways of the grader table, and you pack them away fast, so they don't pile up, and ship them in caseloads to the Cities . . .

But the soft ones, maybe they didn't smash as hard and fast as the firm ones. Eggshells in Amazonia, which exists in our northwoods minds like an Atlantis on land. Like the new continent of Mu. The trick is to reach Amazonia without these old, almost fatal wounds. Might this trailer understand Amazonia— understand it as the total space, separate culture, the earth and sky. And eggs, mom and dad understood: eggs, seeds, and ponds.

"I want to ask you something, Cheike." Mom, in a serious tone. Oh, no! What is it? "Why did you change your name?"

Reflex: defense. "Because these Anglos can't pronounce Cheike properly."

"It doesn't make sense to me. People will mispronounce your new name too."

"It's a tribal name. I belong to a different tribe now. But I have many names. To you I am still Cheike."

"To me, you are still Baby." She laughs. I don't mind her laugh, but I bring her up to date. "I'm not a baby anymore though, you know, I'm a lesbian woman . . ."

"Why do you always have to bring up being a lesbian?"

"I hardly ever do—that you hear."

"Like last night."

"Because it fit in. You always think you're different and can't fit in anywhere in small town America, and that's true, but there are lots of reasons why people don't fit in and can't move anywhere they like. It goes double for me. There are very few places I can live. You wonder why I don't settle down, and that's why. I have to find a women's community I can live in . . ." I look up at dad, who is shaking his head. "What, no?"

"Tell me, Cheike," his old hoarse voice, "do you have to be a lesbian to be a feminist?"

Oh no, that question again. I can pretend I don't know. "I don't know. I did—have to become a lesbian before I could be anything but a feminist literary critic. But that was because my ideas were years ahead of my behavior. Some of my ideas."

"Don't you think women can oppress other women?" pursues dad. "I know some very domineering women."

"Yea, sure. The difference is in a group awareness. Now we try to avoid the roles that reinforce oppression."

"That's you and your friends. There are domineering women."

"I'm talking about what lesbians all over the country are trying to do these days . . . Let's go to the beach!" I say in sudden inspiration, my fingers almost burning against the hot silver metal outside the open door.

It is our last day together. Mom and dad are heading for Arizona in their trailer and I to Oregon in my VW bug. Everyone agrees to the trip to the beach. We do not rush.

"Tell me, what are you going to do from here? Are you going to settle down?" mom continues.

"You sound like Bubie."

"We're worried about you."

"I see my life as a story: at first I was very threatened. I escaped by running into the pond and going underwater. I held out for a long time, long enough, although once when I did pop up I got shot at. I went under again—long enough to be in a later age, a different time and place. I come up. It feels quiet now, safe now. I go to a house where I feel attracted; a woman is there I want to see. To leave, we have to climb out the second story window down an escape, because of her husband, who wants to continue to possess her. But I want to leave, and she follows me. We go out on a road, which is a ledge, a narrow ledge of ourselves. To the right is an enormous drop; to our left, a sheer cliff wall. We are following the path to Amazonia."

"Where's Amazonia?"

"It's an erotic hillside in Brazil. It's a scenario."

"A scenery? A Brazil scenery?"

"Huh. It's women, all kinds of women, egg women, crying young egg girls growing up in so many shapes and sizes, checked and cracked, who grew up in so many conditions but together at last. Not sorted out into hierarchies, you might say . . . or I might say, anyway . . . anarchy and variety are our beginning characteristics."

Silence. Then, "Your generation has had a much better opportunity to associate with each other from different social, religious, and economic backgrounds." Wow! She understands. "Then there's psychological backgrounds and astrological backgrounds . . . !" I have to add.

"Let's go."

On the way we stop to pick up Daro, a young woman who wants a ride with me to Oregon. May as well start early. Although I don't know Daro at this point, I know that with her along I won't be outnumbered. I associate her as one on my path.

Daro smiles calmly and swings in the front seat of my bug. Mom and dad are in back. Her vibrant brown eyes beam familiarity.

"Why do women wear beards these days?" asks dad. Oh no, he has noticed Daro's strong stray chin hairs. Daro looks at me suddenly and then turns around to the back seat. "Why don't you?"

Mom comes in here, rendering a textbook lecture on hor-

monal balances in each sex. She remembers a dark-haired girl-
friend from high school who had had chin hair. I glance over
at Daro's shoulder, covered with glistening reddish-dark hair.
"Was that the one who didn't want you to get married, mom?"
I pipe up. "Yes, that was the one."

"It's defiance," says dad. Dad still believes in some of the
hierarchies of family and society.

"Defiance! It takes courage to be ourselves in a society with
its overpowering demands for a certain look!"

Mom agrees. Dad goes on, "You'd be safer on the streets if
your appearance conformed. That's why the Jews got away from
wearing payess [side curls], yarmulkahs, and black robes. Why
do you think? I see some women walking around in ripped, dirty
jeans. People aren't tolerant. They'll attack you if you look different."

I can tell dad is recalling his boyhood in eastern Poland. He
knows what it is like to be hated because of how you look. He
is trying to warn us compassionately, but I am pretty sure he is
offending Daro, who stares down at her velvet jeans patches.

"How you look is not reason for men to attack women," says
mom.

"Yes, but it makes you stand out, that's when people attack
you, when they think you hate them."

"We're into love, not hate, dad."

"The world isn't ready for love."

I stare out along the gorgeous coastline. What is the world
ready for, then?

"Let's pull over," I say, and I pick a beach, a pull-in near an
artichoke field. Beyond that, there is a promontory with a lighthouse on the end.

"Can you pick any beach?" "Yes," I assure them, "that is what
I always do." They have not been to a Pacific beach before. Daro
bounds out over the dunes as soon as we stop. Mom and dad
settle faces down on a blanket. Dad goes to sleep, and mom
reads a *Country Women* magazine I gave her. She unzips the back
of her sundress. Only the skin around her neck is freckled and
tan. Dad's whole back is dark-red bronze, like their faces. The
contrast with their white hair strikes me. Mom's blue eyes sparkle out to me. Together with them at last on a Pacific beach, we

are a flock of water-sharers. No one else is on the beach, just us refugees from the East—and Daro. Here the ocean is not blocked into swimming areas like the Middle Atlantic, no boardwalks, no saltwater taffy stands, no screaming rides. Just dunes, wildflowers, cliff walls, spun sponge rocks, phosphorescent caves. A freshwater stream—flowing through a crater of sand nearby. Around us—trails of seagull feathers and transparent sacs of jelly.

Daro reappears, examining the purple sac inside one, a glob of jelly fish, watching the bubbles blow within it. She hands me a feather and sinks down on the blanket, resting her hand on my knee. Mom on my other side reaches her hand around me to touch Daro. Her hand doesn't quite reach so she wiggles around and moves up an inch so she can reach. Daro's self-absorption is like soft tissue wrapped around a sharp object. Mom wants to make up for the car conversation. Dad decides to tell a joke. "Once there was a kid who dared her father to eat half a worm: 'I'll eat half if you eat half.' The father takes up the dare and eats half the worm. He gives the other half to the kid, but the kid says, 'You ate the wrong half!' " Daro likes this story and laughs, bowing her head over my lap, laughing.

"I didn't mean anything before about how you look, honey, I was just talking." Dad's laryngitic kindly voice. He has a cold. Daro looks at him and relaxes. He is a grandfather, and tired, smiling inside his thinning curly mop of white hair, his red skin criss-crossed with lines.

"Mom and dad used to be farmers, Daro, now they are gypsies, like us." I want her to know them, to feel safe and accepted with them, as I do, because I feel how suspicious she is of parents. Later, I discover she is more like them than I will ever be. Those three are of the earth, and I am of the air. For years I resisted mom and dad, not just their parental intonations, but as representatives of the material plane itself. The maintenance of a real farm. They had long ago freely admitted to not being philosopher-poets—my one requirement for the people I live with—but bread-and-butter folks, the people who keep the home front repaired. I was a little Rimbaud monster hatched from the cosmic egg, seeking visions. Later, I learned that air needs the protection of earth, and the earth needs the air to breathe. We unite in the sight of the water and the fire. It would be neat to stay and have a camp fire on the beach.

"I wish I hadn't felt so separated from you two for so long. We could have shared a lot more."

"You wouldn't unlock your door," says mom.

"You didn't ask the right questions."

"We didn't realize you felt separated. Parents assume they can tell what's going on, they don't ask questions," says dad.

"Your father didn't converse with you, draw you out, did he, dad?" Dad exhales sharply, like *Are you kidding?* His father was a rabbi from the Old Country, not attentive to the psychology of childhood.

"You know," says mom, "my mother always asked a lot of questions and I resented it. I kept my distance from you. I didn't want to intrude like that."

"Oh, I'm a bird, I love like a bird," I say. What is the use of going over the past, the years I spent behind a closed door, a little alcove where I wrote things I never admitted to and gazed at the pond from my attic window. "I was a bird, trapped in an attic." I laugh. And then lapse into broody silence. Daro smiles brightly sparkling brown eyes at me and reaches around, tentatively touching my shoulder. I reach around mom and knead her shoulders. Smooth the little knots. Imagine she is doing the same to me. Dad goes back to sleep. I concentrate on breathing in and out. Mom looks over the *Country Women* magazine I gave her, an issue on "older women." After awhile she says, "What about older men? I don't want to leave my boyfriend." And puts her arm around dad's sleeping form. I don't know; I don't know. My eyes feel tired, either from the sun or from uncried tears. *When* am I going to be with whomever I should be with? And *when* am I going to be wherever I should be? And they, who have each other only, where can they settle down, their new Jerusalem? Not in the same direction as Amazonia. We are cultural refugees together for only a few weeks each year, on our separate roads. Nuovev Jasmine is gone, the pond a swamp again. I haven't been back, but I saw that in a dream. Never again can we skinny dip, boat, or skate on it. The pond is shrunk and dirty, receded from the pier, ukky. The man who bought it turned it into a sewage disposal for his trailer park.

I jump up, facing the vast expanse before us. I want to swim; I rip off my clothes and run down into the tide. Raising my arms high and jumping, a tremendous wave crests beneath me. There is only the Adventure left, only the quest for a new native land. Look at them on the shore, here for only one moment do we inhabit the same reality, here at land's end, mom, dad, Daro and me in the same shot, the same frame. Click!

Over the dunes a man in street dress appears. He says, "This isn't a public beach, it's my beach." So we have to leave here. Dad wants to go anyway, he's not feeling that well. I slip into my soft gym pants, but the waist drawstring is lost down its slot. I feel tired, I want to flop down, not leave. I feel like a baby, wanting to cry in frustration because I can't get the string to come through the slot. To scream and cry like a baby.

"Mom . . .?"

"You do it with a safety pin. Here." She stands in front of me, patiently threading the knot through the opening. I stand silently watching, like a pacified toddler. I can see the lighthouse behind her, at the edge of the promontory. Victorious, finished, she stands straight and smiles. Sardonically, she says, "Your mother is a light housekeeper . . ."

Virgo woman, steadiness and continuation in seas of change, *I salute you.*

IV
NEW
DIRECTIONS

X: A Fabulous Child's Story

Lois Gould

ONCE UPON A time, a baby named X was born. This baby was named X so that nobody could tell whether it was a boy or a girl. Its parents could tell, of course, but they couldn't tell anybody else. They couldn't even tell Baby X, at first.

You see, it was all part of a very important Secret Scientific Xperiment, known officially as Project Baby X. The smartest scientists had set up this Xperiment at a cost of Xactly 23 billion dollars and 72 cents, which might seem like a lot for just one baby, even a very important Xperimental baby. But when you remember the prices of things like strained carrots and stuffed bunnies, and popcorn for the movies and booster shots for camp, let alone 28 shiny quarters from the tooth fairy, you begin to see how it adds up.

Also, long before Baby X was born, all those scientists had to

Lois Gould is a former police reporter who served as editor of the *Long Island Star Journal* and as executive editor of the *Ladies' Home Journal.* Her four published novels are *Such Good Friends, Necessary Objects, Final Analysis,* and *A Sea Change;* a collection of her essays, *Not Responsible for Personal Articles,* appeared in 1978. "X: A Fabulous Child's Story," copyright © 1978 by Lois Gould, is reprinted by permission of Daughters Publishing Co., Inc.

be paid to work out the details of the Xperiment, and to write the *Official Instruction Manual* for Baby X's parents and, most important of all, to find the right set of parents to bring up Baby X. These parents had to be selected very carefully. Thousands of volunteers had to take thousands of tests and answer thousands of tricky questions. Almost everybody failed because, it turned out, almost everybody really wanted either a baby boy or a baby girl, and not Baby X at all. Also, almost everybody was afraid that a Baby X would be a lot more trouble than a boy or a girl. (They were probably right, the scientists admitted, but Baby X needed parents who wouldn't *mind* the Xtra trouble.)

There were families with grandparents named Milton and Agatha, who didn't see why the baby couldn't be named Milton or Agatha instead of X, even if it *was* an X. There were families with aunts who insisted on knitting tiny dresses and uncles who insisted on sending tiny baseball mitts. Worst of all, there were families that already had other children who couldn't be trusted to keep the secret. Certainly not if they knew the secret was worth 23 billion dollars and 72 cents—and all you had to do was take one little peek at Baby X in the bathtub to know if it was a boy or a girl.

But, finally, the scientists found the Joneses, who really wanted to raise an X more than any other kind of baby—no matter how much trouble it would be. Ms. and Mr. Jones had to promise they would take equal turns caring for X, and feeding it, and singing it lullabies. And they had to promise never to hire any baby-sitters. The government scientists knew perfectly well that a baby-sitter would probably peek at X in the bathtub, too.

The day the Joneses brought their baby home, lots of friends and relatives came over to see it. None of them knew about the secret Xperiment, though. So the first thing they asked was what kind of baby X was. When the Joneses smiled and said, "It's an X!" nobody knew what to say. They couldn't say, "Look at her cute little dimples!" And they couldn't say, "Look at his husky little biceps!" And they couldn't even say just plain "kitchy-coo." In fact, they all thought the Joneses were playing some kind of rude joke.

But, of course, the Joneses were not joking. "It's an X" was absolutely all they would say. And that made the friends and

relatives very angry. The relatives all felt embarrassed about
having an X in the family. "People will think there's something
wrong with it!" some of them whispered. "There *is* something
wrong with it!" others whispered back.

"Nonsense!" the Joneses told them all cheerfully. "What
could possibly be wrong with this perfectly adorable X?"

Nobody could answer that, except Baby X, who had just finished its bottle. Baby X's answer was a loud, satisfied burp.

Clearly, nothing at all was wrong. Nevertheless, none of the
relatives felt comfortable about buying a present for a Baby X.
The cousins who sent the baby a tiny football helmet would not
come and visit any more. And the neighbors who sent a pink-
flowered romper suit pulled their shades down when the
Joneses passed their house.

The *Official Instruction Manual* had warned the new parents
that this would happen, so they didn't fret about it. Besides, they
were too busy with Baby X and the hundreds of different Xer-
cises for treating it properly.

Ms. and Mr. Jones had to be Xtra careful about how they
played with little X. They knew that if they kept bouncing it up
in the air and saying how *strong* and *active* it was, they'd be
treating it more like a boy than an X. But if all they did was
cuddle it and kiss it and tell it how *sweet* and *dainty* it was, they'd
be treating it more like a girl than an X.

On page 1,654 of the *Official Instruction Manual*, the scientists
prescribed: "plenty of bouncing and plenty of cuddling, *both*. X
ought to be strong and sweet and active. Forget about *dainty*
altogether."

Meanwhile, the Joneses were worrying about other problems.
Toys, for instance. And clothes. On his first shopping trip, Mr.
Jones told the store clerk, "I need some clothes and toys for my
new baby." The clerk smiled and said, "Well, now, is it a boy or
a girl?" "It's an X," Mr. Jones said, smiling back. But the clerk
got all red in the face and said huffily, "In *that* case, I'm afraid
I can't help you, sir." So Mr. Jones wandered helplessly up and
down the aisles trying to find what X needed. But everything in
the store was piled up in sections marked "Boys" or "Girls."
There were "Boys' Pajamas" and "Girls' Underwear" and
"Boys' Fire Engines" and "Girls' Housekeeping Sets." Mr.

Jones went home without buying anything for X. That night he and Ms. Jones consulted page 2,326 of the *Official Instruction Manual.* "Buy plenty of everything!" it said firmly.

So they bought plenty of sturdy blue pajamas in the Boys' Department and cheerful flowered underwear in the Girls' Department. And they bought all kinds of toys. A boy doll that made pee-pee and cried, "Pa-pa." And a girl doll that talked in three languages and said, "I am the Pres-i-dent of Gen-er-al Mo-tors." They also bought a storybook about a brave princess who rescued a handsome prince from his ivory tower, and another one about a sister and brother who grew up to be a baseball star and a ballet star, and you had to guess which was which.

The head scientists of Project Baby X checked all their purchases and told them to keep up the good work. They also reminded the Joneses to see page 4,629 of the *Manual,* where it said, "Never make Baby X feel *embarrassed* or *ashamed* about what it wants to play with. And if X gets dirty climbing rocks, never say 'Nice little Xes don't get dirty climbing rocks.'"

Likewise, it said, "If X falls down and cries, never say 'Brave little Xes don't cry.' Because, of course, nice little Xes *do* get dirty, and brave little Xes *do* cry. No matter how dirty X gets, or how hard it cries, don't worry. It's all part of the Xperiment."

Whenever the Joneses pushed Baby X's stroller in the park, smiling strangers would come over and coo: "Is that a boy or a girl?" The Joneses would smile back and say, "It's an X." The strangers would stop smiling then, and often snarl something nasty—as if the Joneses had snarled at *them.*

By the time X grew big enough to play with other children, the Joneses' troubles had grown bigger, too. Once a little girl grabbed X's shovel in the sandbox, and zonked X on the head with it. "Now, now, Tracy," the little girl's mother began to scold, "little girls mustn't hit little—" and she turned to ask X, "Are you a little boy or a little girl, dear?"

Mr. Jones, who was sitting near the sandbox, held his breath and crossed his fingers.

X smiled politely at the lady, even though X's head had never been zonked so hard in its life. "I'm a little X," X replied.

"You're a *what*?" the lady exclaimed angrily. "You're a little b-r-a-t, you mean!"

"But little girls mustn't hit little Xes, either!" said X, retrieving the shovel with another polite smile. "What good does hitting do, anyway?"

X's father, who was still holding his breath, finally let it out, uncrossed his fingers, and grinned back at X.

And at their next secret Project Baby X meeting, the scientists grinned, too. Baby X was doing fine.

But then it was time for X to start school. The Joneses were really worried about this, because school was even more full of rules for boys and girls, and there were no rules for Xes. The teacher would tell boys to form one line, and girls to form another line. There would be boys' games and girls' games, and boys' secrets and girls' secrets. The school library would have a list of recommended books for girls, and a different list of recommended books for boys. There would even be a bathroom marked BOYS and another one marked GIRLS. Pretty soon boys and girls would hardly talk to each other. What would happen to poor little X?

The Joneses spent weeks consulting their *Instruction Manual* (there were 249½ pages of advice under "First Day of School"), and attending urgent special conferences with the smart scientists of Project Baby X.

The scientists had to make sure that X's mother had taught X how to throw and catch a ball properly, and that X's father had been sure to teach X what to serve at a doll's tea party. X had to know how to shoot marbles and how to jump rope and, most of all, what to say when the Other Children asked whether X was a Boy or a Girl.

Finally, X was ready. The Joneses helped X button on a nice new pair of red-and-white checked overalls, and sharpened six pencils for X's nice new pencilbox, and marked X's name clearly on all the books in its nice new bookbag. X brushed its teeth and combed its hair, which just about covered its ears, and remembered to put a napkin in its lunchbox.

The Joneses had asked X's teacher if the class could line up alphabetically, instead of forming separate lines for boys and girls. And they had asked if X could use the principal's bathroom, because it wasn't marked anything except BATHROOM. X's teacher promised to take care of all those problems. But nobody could help X with the biggest problem of all—Other Children.

Nobody in X's class had ever known an X before. What would they think? How would X make friends?

You couldn't tell what X was by studying its clothes—overalls don't even button right-to-left, like girls' clothes, or left-to-right, like boys' clothes. And you couldn't guess whether X had a girl's short haircut or a boy's long haircut. And it was very hard to tell by the games X liked to play. Either X played ball very well for a girl, or else X played house very well for a boy.

Some of the children tried to find out by asking X tricky questions, like "Who's your favorite sports star?" That was easy. X had two favorite sports stars: a girl jockey named Robyn Smith and a boy archery champion named Robin Hood. Then they asked, "What's your favorite TV program?" And that was even easier. X's favorite TV program was "Lassie," which stars a girl dog played by a boy dog.

When X said that its favorite toy was a doll, everyone decided that X must be a girl. But then X said that the doll was really a robot, and that X had computerized it, and that it was programmed to bake fudge brownies and then clean up the kitchen. After X told them that, the other children gave up guessing what X was. All they knew was they'd sure like to see X's doll.

After school, X wanted to play with the Other Children. "How about shooting some baskets in the gym?" X asked the girls. But all they did was make faces and giggle behind X's back.

"How about weaving some baskets in the arts and crafts room?" X asked the boys. But they all made faces and giggled behind X's back, too.

That night, Ms. and Mr. Jones asked X how things had gone at school. X told them sadly that the lessons were okay, but otherwise school was a terrible place for an X. It seemed as if Other Children would never want an X for a friend.

Once more, the Joneses reached for the *Instruction Manual.* Under "Other Children," they found the following message: "What did you Xpect? *Other Children* have to obey all the silly

boy-girl rules, because their parents taught them to. Lucky X—
you don't have to stick to the rules at all! All you have to do is
be yourself. P.S. We're not saying it'll be easy."

X liked being itself. But X cried a lot that night, partly because
it felt afraid. So X's father held X tight, and cuddled it, and
couldn't help crying a little, too. And X's mother cheered them
both up by reading an Xciting story about an enchanted prince
called Sleeping Handsome, who woke up when Princess Charm-
ing kissed him.

The next morning, they all felt much better, and little X went
back to school with a brave smile and a clean pair of red-and-
white checked overalls.

There was a seven-letter-word spelling bee in class that day.
And a seven-lap boys' relay race in the gym. And a seven-layer-
cake baking contest in the girls' kitchen corner. X won the
spelling bee. X also won the relay race. And X almost won the
baking contest, except it forgot to light the oven. Which only
proves that nobody's perfect.

One of the Other Children noticed something else, too. He
said: "Winning or losing doesn't seem to count to X. X seems
to have fun being good at boys' skills *and* girls' skills."

"Come to think of it," said another one of the Other Children, "maybe X is having twice as much fun as we are!"

So after school that day, the girl who beat X at the baking
contest gave X a big slice of her prizewinning cake. And the boy
X beat in the relay race asked X to race him home.

From then on, some really funny things began to happen.
Susie, who sat next to X in class, suddenly refused to wear pink
dresses to school any more. She insisted on wearing red-and-
white checked overalls—just like X's. Overalls, she told her
parents, were much better for climbing monkey bars.

Then Jim, the class football nut, started wheeling his little
sister's doll carriage around the football field. He'd put on his
entire football uniform, except for the helmet. Then he'd put
the helmet *in* the carriage, lovingly tucked under an old set of
shoulderpads. Then he'd start jogging around the field, pushing the carriage and singing "Rockabye Baby" to his football
helmet. He told his family that X did the same thing, so it must
be okay. After all, X was now the team's star quarterback.

Susie's parents were horrified by her behavior, and Jim's parents were worried sick about his. But the worst came when the twins, Joe and Peggy, decided to share everything with each other. Peggy used Joe's hockey skates, and his microscope, and took half his newspaper route. Joe used Peggy's needlepoint kit, and her cookbooks, and took two of her three baby-sitting jobs. Peggy started running the lawn mower, and Joe started running the vacuum cleaner.

Their parents weren't one bit pleased with Peggy's wonderful biology experiments, or with Joe's terrific needlepoint pillows. They didn't care that Peggy mowed the lawn better, and that Joe vacuumed the carpet better. In fact, they were furious. It's all that little X's fault, they agreed. Just because X doesn't know what it is, or what it's supposed to be, it wants to get everybody *else* mixed up, too!

Peggy and Joe were forbidden to play with X any more. So was Susie, and then Jim, and then *all* the Other Children. But it was too late; the Other Children stayed mixed up and happy and free, and refused to go back to the way they'd been before X.

Finally, Joe and Peggy's parents decided to call an emergency meeting of the school's Parent's Association, to discuss "The X Problem." They sent a report to the principal stating that X was a "disruptive influence." They demanded immediate action. The Joneses, they said, should be *forced* to tell whether X was a boy or a girl. And then X should be *forced* to behave like whichever it was. If the Joneses refused to tell, the Parents' Association said, then X must take an Xamination. The school psychiatrist must Xamine it physically and mentally, and issue a full report. If X's test showed it was a boy, it would have to obey all the boys' rules. If it proved to be a girl, X would have to obey all the girls' rules.

And if X turned out to be some kind of mixed-up misfit, then X should be Xpelled from the school. Immediately!

The principal was very upset. Disruptive influence? Mixed-up misfit? But X was an Xcellent student. All the teachers said it was a delight to have X in their classes. X was president of the student council. X had won first prize in the talent show, and second prize in the art show, and honorable mention in the science fair, and six athletic events on field day, including the potato race.

Nevertheless, insisted the Parents' Associaton, X is a Problem Child. X is the Biggest Problem Child we have ever seen!

So the principal reluctantly notified X's parents that numerous complaints about X's behavior had come to the school's attention. And that after the psychiatrist's Xamination, the school would decide what to do about X.

The Joneses reported this at once to the scientists, who referred them to page 85,759 of the *Instruction Manual.* "Sooner or later," it said, "X will have to be Xamined by a psychiatrist. This may be the only way any of us will know for sure whether X is mixed up—or whether everyone else is."

The night before X was to be Xamined, the Joneses tried not to let X see how worried they were. "What if—?" Mr. Jones would say. And Ms. Jones would reply, "No use worrying." Then a few minutes later, Ms. Jones would say, "What if—?" and Mr. Jones would reply, "No use worrying."

X just smiled at them both, and hugged them hard and didn't say much of anything. X was thinking, What if—? And then X thought: No use worrying.

At Xactly 9 o'clock the next day, X reported to the school psychiatrist's office. The principal, along with a committee from the Parents' Associaton, X's teacher, X's classmates, and Ms. and Mr. Jones, waited in the hall outside. Nobody knew the details of the tests X was to be given, but everybody knew they'd be *very* hard, and that they'd reveal Xactly what everyone wanted to know about X, but were afraid to ask.

It was terribly quiet in the hall. Almost spooky. Once in a while, they would hear a strange noise inside the room. There were buzzes. And a beep or two. And several bells. An occasional light would flash under the door. The Joneses thought it was a white light, but the principal thought it was blue. Two or three children swore it was either yellow or green. And the Parents' Committee missed it completely.

Through it all, you could hear the psychiatrist's low voice, asking hundreds of questions, and X's higher voice, answering hundreds of answers.

The whole thing took so long that everyone knew it must be

the most complete Xamination anyone had ever had to take. Poor X, the Joneses thought. Serves X right, the Parents' Committee thought. I wouldn't like to be in X's overalls right now, the children thought.

At last, the door opened. Everyone crowded around to hear the results. X didn't look any different; in fact, X was smiling. But the psychiatrist looked terrible. He looked as if he was crying! "What happened?" everyone began shouting. Had X done something disgraceful? "I wouldn't be a bit surprised!" muttered Peggy and Joe's parents. "Did X flunk the *whole* test?" cried Susie's parents. "Or just the most important part?" yelled Jim's parents.

"Oh, dear," sighed Mr. Jones.

"Oh, dear," sighed Ms. Jones.

"*Sssh,*" ssshed the principal. "The psychiatrist is trying to speak."

Wiping his eyes and clearing his throat, the psychiatrist began, in a hoarse whisper. "In my opinion," he whispered—you could tell he must be very upset—"in my opinion, young X here—"

"Yes? Yes?" shouted a parent impatiently.

"*Sssh!*" ssshed the principal.

"Young *Sssh* here, I mean young X," said the doctor, frowning, "is just about—"

"Just about *what?* Let's have it!" shouted another parent.

"... just about the *least* mixed-up child I've ever Xamined!" said the psychiatrist.

"Yay for X!" yelled one of the children. And then the others began yelling, too. Clapping and cheering and jumping up and down.

"SSSH!" SSShed the principal, but nobody did.

The Parents' Committee was angry and bewildered. How *could* X have passed the whole Xamination? Didn't X have an *identity* problem? Wasn't X mixed up at *all?* Wasn't X *any* kind of a misfit? How could it *not* be, when it didn't even *know* what it was? And why was the psychiatrist crying?

Actually, he had stopped crying and was smiling politely through his tears. "Don't you see?" he said. "I'm crying because it's wonderful! X has absolutely no identity problem! X isn't one bit mixed up! As for being a misfit—ridiculous! X knows perfectly well what it is! Don't you, X?" The doctor winked. X winked back.

"But what *is* X?" shrieked Peggy and Joe's parents. "*We* still want to know what it is!"

"Ah, yes," said the doctor, winking again. "Well, don't worry. You'll all know one of these days. And you won't need me to tell you."

"What? What does he mean?" some of the parents grumbled suspiciously.

Susie and Peggy and Joe all answered at once. "He means that by the time X's sex matters, it won't be a secret any more!"

With that, the doctor began to push through the crowd toward X's parents. "How do you do," he said, somewhat stiffly. And then he reached out to hug them both. "If I ever have an X of my own," he whispered, "I sure hope you'll lend me your instruction manual."

Needless to say, the Joneses were very happy. The Project Baby X scientists were rather pleased, too. So were Susie, Jim, Peggy, Joe, and all the Other Children. The Parents' Association wasn't, but they had promised to accept the psychiatrist's report, and not make any more trouble. They even invited Ms. and Mr. Jones to become honorary members, which they did.

Later that day, all X's friends put on their red-and-white checked overalls and went over to see X. They found X in the back yard, playing with a very tiny baby that none of them had ever seen before. The baby was wearing very tiny red-and-white checked overalls.

"How do you like our new baby?" X asked the Other Children proudly.

"It's got cute dimples," said Jim.

"It's got husky biceps, too," said Susie.

"What kind of baby is it?" asked Joe and Peggy.

X frowned at them. "Can't you tell?" Then X broke into a big, mischievous grin. *"It's a Y!"*

from A Weave of Women

E. M. Broner

THE BIRTH

THEY EMBRACE AND face Jordan. They are turned golden in the evening light, like the stone. There are several of them. The weeping one will not turn to salt. In fact, she drinks her tears. She is a full-bellied sabra and her fruit is sucking away inside.

The women breathe with Simha. Heavy labor has not started yet. They sit in the doorway of their stone house in the Old City. They move inside and breathe with Simha into the stones.

Simha groans. They echo. Simha loses her water. The Dead Sea. The River Jordan. The Nile. They wait for the baby to float on a basket from Simha's reeds.

They have notified two midwives. It is a mistake. The doctors have refused home delivery. A nurse warned them against it.

E. M. Broner commutes regularly to a loft in New York from Detroit, where she is Associate Professor and Writer-in-Residence at Wayne State University. Co-editor of *A Lost Tradition: A History of Mothers and Daughters in Literature,* she is at work on her fifth novel. This excerpt from *A Weave of Women* by E. M. Broner, copyright © 1978 by E. M. Broner, is reprinted by permission of Holt, Rinehart and Winston, Publishers.

Deedee leans against the stones and tells of a friend who bore a son on Christmas Eve under the Christmas tree.

"Was it planned?"

"No. It was an accident."

"Which?"

"Both the pregnancy and the birth."

"What happened to the baby?"

"He was put into the Dirty Baby Ward at the hospital because he was covered with pine needles."

The midwives arrive. One is a Jews for Jesus girl, dressed modestly in kerchief, long skirt, covered arms. She knows her Old and New Testaments but not her Biology. The other is a Swede who is into health. She worked as a volunteer at a border kibbutz and volunteered herself to a kibbutz member. She arrives nursing her new baby. The Swede is intent on smearing ointment onto her sore nipples. The other midwife is reading Leviticus in variant texts.

Simha exclaims. She is in pain. The women are joyous. Pain quickens the moment to birth. Pain quickens the excitation.

They see the cervix. It is purple, a reddish purple.

The midwives look disturbed.

"Hot water?" asks the social worker.

They don't think so.

Simha is sitting on a stool. The women put a soft blanket between Simha's knees.

One midwife tries to hear the baby's heartbeat. She cannot. The Swede listens and hears the regular beat. The midwives do not know how many fingers Simha is dilated. Simha shouts at them from her stool. They leave.

The social worker goes to a house with a phone and, from there, she telephones accomplished women. She phones Martin Buber's granddaughter, Golda Meir's cousin, Ben Gurion's niece. She phones women in the parliament, the law courts, at the university.

Gradually the experts gather. They surround Simha, crouched on her stool. They have books of instructions in several languages.

Antoinette is in the room, a Shakespearean from London. Joan is there, a playwright, a Britisher, but from Manchester. A scientist arrives, originally from Germany. Dahlia is there, a singer from Beer Sheva. Tova has been there all along, the curly-haired actress from New York. Hepzibah jitneys down from Haifa wearing her padded scarf against the wrath of the Father-Lord. Mickey arrives in the midst of her divorce. Gloria, the redhead, has been there for the fun. She came over from California for the fun. Another social worker arrives from Tel Aviv, serious Polish Vered.

Simha has left her stool for the mattress on the floor. Her labor has stopped and she dozes. Why is the baby so bad? It will do nothing on time or within reason. It will cry and the mother will never know why. Simha, matted hair, parched lips, is having nightmares.

The scientist says this is not her field of expertise but nothing seems wrong, the head down, labor proceeding.

Hepzibah barely refrains from covering Simha's full, long hair with a kerchief. Simha is not married and, therefore, does not need a head covering, but perhaps the rules are different when it comes to unwed mothers.

Deedee is both intense and amused. The Jews are something else! She is Irish and prefers her own, and then the Greeks and then the Jews. But Israel is warm and she has women friends here who will neither let her starve nor weep.

They are up all night on this night of vigil. The women brew tea. The redhead, Gloria, knows a special way to make Turkish coffee that she learned from an Arab cab driver. He added cardamom seed. She talks on about the cab ride. All of her tales wander, often with cab drivers, jitney drivers, young students from the university who cannot resist her and follow her across the wadi near campus.

The women drink the thick coffee. They are not sleepy. It becomes chilly and they put naphtha into the heater. They open the door a moment. The bad odor escapes, the blue flames light and the early morning air is warmed.

The Shakespearean has a crisp accent and a bustling manner. "Did someone call the father?" she asks.

Gloria from California did.

At dawn the father runs into the little stone house. He has hitchhiked from his kibbutz. Gloria's message was delivered to the dining room and shouted from table to table. Ah! He is on his shift in the cow shed. His brother takes over and his mother gives him a basket of freshly baked bread from the kibbutz oven, also her own jam and a plant from her garden.

The kibbutz, accepting paternity, sends wishes of easy birth and warm blankets to receive the son.

A daughter is born.

By the time of the birth the father does not care what comes from Simha's cave, what small, furtive animal that is gnawing at her and making her scream. When the animal slides out, it has soft hair on its head and soft fur on its body. It is a daughter-puppy.

The women prepare the mother.

Mickey, in the midst of her divorce and of her hatred, combs Simha's hair. She braids it. She brushes it out again and puts it up. She lets it hang down Simha's back.

Antoinette gathers flowers from the valley of the Old City.

"There's rosemary, that's for remembrance. . . . And there is pansies, that's for thought. There's fennel for you, and columbines . . . daisy . . . violets . . ."

"That's morbid," says Joan. "Ophelia of all things!"

Antoinette does not deign to reply. What does a Manchester person know of repartee, of the deft thrust? She wraps herself in London imperial dignity.

Gloria stares at the father of the child. He fell asleep wearily. He interests her, the natural man who eats hungrily without good or bad manners, sleeps easily, loves shyly. Gloria looks at his kibbutz stomach stretching the buttons of his cotton shirt. His hand is curled. She touches it and it jerks.

Hepzibah takes out her Daily Prayer Book. Terry, the social worker, readies a manifesto. Gerda, the scientist born in Germany, looks at this new marvel of science.

Dahlia from Beer Sheva goes out to exercise her face and chest. The total person sings, not just the vocal cords. She returns red-cheeked and lively-eyed. She sings to Simha of happiness, of bodies, of seas of the moon, of inner caverns that hold tropical fish, of waters that flood and the fish swims clear. The fish has no gills and now must sing instead of gulping.

Tova, the actress, strokes and smoothes the baby daughter.

No one will spank her, croons Tova. Life is not a slap on the behind.

The women stroke the animal, pet it behind its ears, pet its scaly legs, its stretching turtle neck, its lumpy head.

Tova brings the stage property—an Indian bedspread with monkeys and fruit-bearing trees. Simha is sat up on the chattering monkey and the lush tree. A garland is twined for her hair.

Gloria tries again to awaken the kibbutznik. He opens his eyes, looks into Gloria's goyishe pale blue and smiles. He has forgotten Simha, who was his happiness. He has forgotten the new daughter. He looks at Gloria's incandescent hair, her straight nose, her American long legs.

The women watch him silently. The kibbutznik feels a crick in his neck. He is a father. He must speak up at kibbutz meetings, in public, and ask for her maintenance. He is the lover of Simha, and she is a head taller than he. She has wide hips, pendulous breasts, large brown eyes, thick hair. He sighs. Never will he feel that canopy of red above him, those blue eyes sunlit upon his arrival. He belongs to furry animals and hairy, big women.

They surround the mattress and chant:

"Welcome to the new mother. Welcome to the new daughter."

Simha is happy but sloppy. She grabs at her breasts, which are beginning to fill and hurt. She cannot close her legs. The womb is sore. But she looks young again. That old woman who went into childbirth is rejuvenated.

Their home is a busy one and across the narrow street is a busier one. That is the Home for Wayward Girls, actually the Home for Jewish Wayward Girls. They are wayward because they ran away. Or because they stand along the roadside near the universities, secular or religious, crying out in the holy tongue, "Come and fuck me!"

Committees have met on these girls. Terry has sat on such a committee as Director of the Home. It is the committee decision that the girls are prostitutes because they are psychologically disturbed.

"No," says Terry, "economically disturbed."

She brings out charts on inflation, lack of education, illiteracy.

"What is left for these girls to do?" asks Terry.

"No one tells them to," says the committee.

But they, the girls, know it's what they have to do. And, afterward, their mothers scratch at their eyes. Their fathers whip them. The daughters have failed the religion and the socialist state.

The girls leave both religion and socialism. They wear pointy bras, beehive hairdos or they straighten and blow-dry Arabic-Jewish-African hair. They wear high-heeled chunky shoes and miniskirts when no one is mini anymore, not the Israeli policewomen or the women in the army, who have lengthened skirts to mid-knee.

Rina arrives for the ceremony, Rina from development town, court and prison. Rina wears a long skirt. She wears eyeliner around her almond eyes. She covers her pointy breasts with a shawl. Her parents threw her out because she was lazy in the house. Her lovers threw her out because she was fearful in bed.

Shula is invited, Shula the Westerner, the Pole. She is blonde and buxom. She spits on her parents. She does not honor their days in the land. She spits on her teachers. She does not follow the command to honor the teacher above all. She spits on the police. She spits on all uniforms: traffic, civil guard, nurses, beauty parlor.

"Who taught you to spit, Shula?"

"My grandmother. She spat on me when she saw me and said, 'An ugliness. May the Evil Eye leave her on this earth. May she only be seen as ugly and undesirable so that we never lose her.' And so I am—ugly and undesirable."

The ceremony begins.

Simha wears a wraparound skirt to cover either taut or flabby belly. The kibbutznik is present, bearing another basket from his kibbutz, sensible girls' clothing of strong kibbutz material.

He stares at both of the women, his daughter and his love. And he feels jealous. His daughter fits on the mother's hip, on her breast. There is no need for a father as there was no need for a husband, and he has traveled by bus a winding route for four hours to arrive in Jerusalem.

"Find a girl in the kibbutz," advises the kibbutz psychologist. "Forget her. She takes advantage of you."

Ah, she did! She did! His first love. She took advantage of him and deflowered him.

Gloria is there wearing a low-cut dress. She has found sun during this chilly weather, and her freckles reach into her cleavage.

The kibbutznik stares into the basket he is carrying.

Tova from New York is there. She has an Arab lover. He is too shy to attend, but he sends a song.

Mother, I would leap from the mountain for you.
Mother, I will never forget you.

Shh. It is Simha's prayer.

"I come into your house, O Mother God. You inclined your ear toward me, and I will whisper into it all the days of my life.

"The cords of life and death encompassed me. From the hollow of the grave, from the cave of the mouth of birth I called. I knew happiness and anguish. You delivered my soul from death into birth, my eyes from tears, my feet from falling. I shall walk before You in the land of the living."

The baby is poking at her own eyes. She grimaces. A thin cry.

Hepzibah says, "Let her become a comfort to Simha in her old age."

Hepzibah puts down her prayer book and kisses Simha.

"She may need it," murmurs Hepzibah.

Terry says, "Let there be peace. Grant this daughter of her people peace."

Ah, they do not know what will befall or that there will be no room for the daughter or for her people.

The kibbutznik says: "May she serve however she wishes, in the chicken house, the barn, the rose garden, the cotton field, the kitchen, the guesthouse. May she be a member of the community of people."

Simha knows why she chose him to be the father.

CEREMONIES ON
THE THRONE OF MIRIAM

It is during the second orange harvest that Simha's baby has her hymenotomy. The air smells of orange blossoms. The clementines are sweet, the grapefruits not tart. Cucumbers are light green and long. Radishes are large and mild. Nothing growing in this season is bitter.

Simha has not taken her baby from her breast these eight days. If she does for an hour or two, she worries about what else she should be doing.

The wayward girls, Rina and Shula, wander in, washing their faces, arms and hands carefully before they sponge Simha or handle the baby.

Mickey attends them between appointments at the rabbinical house. She burps the baby and says, *Momzera*, little illegitimate girl.

Vered, the Polish social worker from Tel Aviv, comes with warnings. She tells Terry about the statistics dealing with bastards. There is no longer the old, benign kibbutz attitude about marriage or the lack of it. This is a conservative time. The baby will suffer.

Vered's clients these days are the Russians, the dwindling number of them who still want to settle in The Land. Vered wants the Israelis to provide a good example. Simha, lying indolently, her robe always open, her breasts leaking, does not provide such an example.

Nor later will Vered.

The women gather for the ceremony.

They giggle nervously.

"What should it be called?"

"Hymenectomy."

"No. That sounds like an appendectomy."

"It sounds like you're getting rid of your Uncle Hymen."

They wait for Gerda. They enclose the house, covering the windows with cloth, placing lace on the chosen chair. Tova has brought lace that she uses in the theater, old, torn, delicate lace she found in the suk in Old Jaffa. Tova drapes it. Simha sits in lace, the baby on her lap. Deedee, the oldest of many sisters and brothers, has made a christening dress for this baby.

"Christening?"

Deedee sighs. Jews and etymology. Jews and sensitivity. Here in the land of her God, also, they are still fighting any mention of His name. "B.C." they call "Before the Common Era." "Christendom" they don't mention at all. December 25 falls into whatever Jewish month it falls into—Tevet or Shevet the twenty-fifth. No one is named Christine or Christopher. No one buys Christian Brothers wine. They tease her about New Year's Eve. It's either Christ's circumcision or they call it Sylvester, and no one gives a party but a few lonesome Americans. The flora is not called Christmas fern, Christmas rose, Christmas berry. There is no Christ's thorn that she can find. Christmas trees they call a German invention. . . .

"It's such a beautiful dress," says Simha. "We're proud in it."

Deedee relaxes and Gloria tenses. Too much attention on Deedee.

Gerda enters.

"Name the baby."

They tell her the name of the baby.

"Hava, mother of everything living."

Hava translates into Eve, they explain to Gloria, a recent convert. Gloria tells them about her conversion and about the young Reform rabbi laying her on the floor of his study.

"Is this part of the conversion?" she asked him.

The girls don't want to hear Gloria.

"Name the mother," says Gerda.

Simha is the mother.

Simha says, "Here am I acting upon the command that is not yet written that the daughter of eight days shall be pierced."

Simha rises and hands the baby to the godmother, her roommate Terry. Simha walks with difficulty to her bed. Terry sits in the draped chair.

"This is the Throne of Miriam the Prophetess," says Gerda. "May she be remembered for greatness."

Gerda lifts Hava from Terry's arms.

"This is a daughter of Eve."

"What do you mean?"

"We trace our descent from Eve."

Rina and Shula are whispering. They object to the baby being called "a daughter of Eve."

"They use that name for us," says Shula. "It is another name for whore."

"But the son of Adam means human being," says Tova.

"So it is," say the girls.

Terry says her part.

"This baby is descended from Eve. This baby is descended from Sara. Sara's name is princess, the origin of 'Israel.' From her will come tribes and ceremonies."

Tova lifts Hava.

"This child Hava, may she one day be great."

Hepzibah cannot bring herself to say unfamiliar prayers. They are difficult, as well, for Antoinette, a traditionalist.

The women of the room bless the fruit of the vine. They bless the fruit from Simha's womb.

Simha begins to cry from her bed. Terry's arms are shaking. The wayward girls dry Simha's tears and avert their eyes from the ceremony.

Gerda hesitates.

"Do you want it?"

"Yes," says Simha sobbing.

"Yes," says Terry trembling.

"No, no," say Rina and Shula. "She should be a virgin at her wedding."

"I stop here," says Gerda and refuses to continue.

"Explain it to us, Gerda," says Dahlia from Beer Sheva. "No one ever explains to us so how can we choose?"

Gerda lectures easily at the university, over the phone, at the homes of friends.

She diagrams. Two views. A lovely mouth. It is surrounded by lips, the labia majora and the labia minora.

"*Menorah?*" ask the wayward girls. "Lamp?"

It is like an opening, parted lips, with a kind of jagged tooth, with the jagged bits of membrane.

Gerda makes another diagram. The tiny hymen.

"Just a poke with a pick, with something sharp," says Gerda, "that will pierce it. The hymen is so tiny in a baby."

A third diagram. The entrance, the tissue that blocks it, the vaginal wall, the cervix.

"It is usually broken," says Gerda, "by the time young girls see doctors, either by accident or through intent."

"Do it, Gerda," says Simha.

"I did not invent this ceremony," says Gerda. "You did. I'm the scientist. You're the mystic. I will not proceed alone without agreement."

"What is it called, Gerda?"

"It has no name. It has not been done before."

"Name it."

"Hymenotomy."

"Sounds like lobotomy," says Gloria.

"Why hymenotomy?"

"*Tomy*," says Gerda, "a cutting of a tissue."

Antoinette becomes alert.

"Actually," says Antoinette in her crisp London accent, "it is from the root *tem*, to cut."

"What are you using?" ask the women of Gerda.

"Any sharp cutting instrument will do," says Gerda. "I have one here."

"Will it bleed?"

"Not really."

"Will it hurt?"

"Scarcely at all."

"And," says Antoinette to Hepzibah, who is also trying not to participate, "*tome* is a cutting from a larger roll of paper, one book in a work of several volumes."

Gerda says, "There is the thin membrane that separates the outside from the inside of the vagina. It is a vestigial structure."

"Cut it," says Terry, against all vestigial structures, social, physical.

Antoinette's voice rises. "*Dichotomy, anatomy.*"

Gerda is swift. The hymeneal membrane is pierced. She has poked easily through the hymeneal ring.

Hava is startled and wails. Simha gives Hava her nipple, first dabbing wine on it.

There is a loud sigh from the group.

Tova says, "Now you are one of us."

Terry, the godmother, says, "May all orifices be opened."

Dahlia says, "May she not be delivered intact to her bridegroom or judged by her hymen but by the energies of her life."

The baby drinks and lets go of the nipple to cry. Hepzibah

frowns. The wayward girls are shocked. There is so much to shock wayward girls, much they encounter that others do not—physical violence, rape, but this devirginizing of an eight-day-old is the most shocking of all.

"Who will love her?" they whisper to each other, "once they find out?"

Deedee awakens. She fills a large silver cup with sacramental wine.

"May she never suffer again from piercing," says Deedee, "of the body or of the heart."

The lace at the window stirs. It is old Shlomo Sassoon. They shoo him away.

There is a knock at the door. It is the father. He is too early. Would he mind waiting? Gloria brings him a cup of her Turkish coffee. The bitter brew will cool him, will rest him from his hard journey.

"Did you ask his permission?" asks Vered from Tel Aviv.

Simha says, "I told him I wanted to do it, but I did not ask his permission."

"Did he want to do it?" asks Vered.

"No," says Simha.

"Then you did it without his permission."

"I do not have to ask his permission."

"It's being done to his baby," says Vered.

"It's being done to a girl child," says Simha, "and he knows nothing of that."

"Of what?" ask the wayward girls.

"Of the ways that women must open themselves."

Shula sulks. She opens herself. She is a wishbone, sometimes bending her legs around small, bony men, sometimes around large, fat men. One day she will crack in half with no wishes ever granted her.

Dahlia sings in her throaty voice: "I will open my lips with song and my bones declare, Blessed is she that cometh in the name of the Shehena, the womanly God."

They seat themselves on chairs, pillows, an Arab rug and around Simha's bed.

Dahlia sings of women's chairs: the birthing stool, the throne of Miriam, the chair of the longing woman at the window, of the cooking woman at the table, of chariots of war and chariots of angels ascending.

The song ends shrilly. The women worry. Does little Hava so soon have to think of war?

"We are always at war," says Dahlia, "in our land and in our lives."

Deedee is bored. These Jewish friends never cease from whining and dining. The Irish talk, lovely, lively chatter, but this endless talmudic tract, spiritual gossip, this daily poop of prayer. Who is she, Deedee, to such commentary and ceremony?

Tova says, "It is not only Simha's baby who has been pierced. We should each tell of the piercing."

You first, Tova.

"I was pierced," says Tova, "by the angry glance of my enemy."

"Arab lover!" says Mickey. "But what's the difference? Your friends become your enemies too. I stayed a virgin and I will always regret it—that I was pierced by the bastard, my husband. I saved it for the man who threw it away."

"Threw what away?" asks Rina.

"Threw me away, stupid," says Mickey. "What are *they* doing here?" she asks about the girls from across the way. "Isn't it their bedtime?"

The girls huddle together.

"He had me," Mickey says, "and told me, 'What do I need you for? You're not a virgin anymore, you're common and any man can have you.'"

"He was not a son of Adam," says Hepzibah, reaching for Mickey's shaking hands. "He wasn't human."

Deedee relaxes with her friends. They take a long time getting there but, eventually, they do.

"My boss pierced me," says Deedee. "Small, fat boss. I, his secretary, towered over him. Only when I was at my desk and he standing next to me each morning was he taller than I. On my nineteenth birthday he took me for drinks to the bar in our building. Sitting next to him in the booth I could see down on the top of his head where his pink scalp shone through the thinning hair. He bought me a cocktail, one, two, a few. He helped me to his car to take me home. We ended up at the park. Not dark yet. Dusk. People walking around us. What did he care? He parked near the fountain. The water was turned off because it was getting cold and the pipes would freeze. Not his. He had his heater on. I was dozing. 'Want to see who's bigger,' he said, 'you or me?' I bled on his car seat, which he didn't appreciate. He took me home. 'It's my period, ma,' I said and ran upstairs. My ma followed me. When she saw me sobbing, she knew it wasn't and slapped me down into the bed. 'You're of no use to anybody, including yourself,' said my ma. In those days I, only nineteen, was supporting my ma and all the kids, for my dad had left us years before."

Deedee's head is on Tova's shoulder. Terry holds Deedee's hand.

"Hava," says Deedee, "let your piercing be among friends. Let it be ceremonious and correct. Let it be supervised. Let it be done openly, not in anger, not in cars. Ah, Hava," weeps Deedee, "how I envy you."

Gloria says, "The guy who pierced my ears pierced my hymen. On his jewelry shop window was a sign, FREE. EARS PIERCED FREE. I went in. 'I've pierced women all over the world,' said the jeweler. 'Is it free?' I asked. 'You doubt it?' said the jeweler. 'I could be arrested for false advertising.' I said, 'I'm afraid. I'm afraid of pain.' He said, 'I'll be gentle. You won't feel a thing. Only choose an earring and I'll pierce you with it.' 'What are those numbers on the earrings?' 'Nothing. My bookkeeping,' he says. I chose pearls with the number forty. He smiled, seated me in the Piercing Chair and pierced me with the pearls. 'That's forty dollars,' he said, 'for the earrings.' Of course I didn't have it. 'Ask your mother.' 'I don't have one.' 'Ask your father.' 'He'd beat you first and me afterward.' He said, 'Only one way I'll get my money's worth.'"

Gerda, the scientist, says, "I pierced myself. 'Why make a fuss?' I asked myself. 'If I ever marry do I need blood on a sheet? Do the neighbors have to be called in? I went through my Ph.D. to be judged this way?' I had no boyfriend, no one really interested in me, but I could not wait. One must control one's destiny. I am my own bridegroom."

Antoinette has not married. She is shy about not marrying

and about having a hymen and about the stories the women have told.

"It is there," says Antoinette softly, "like tonsils or appendix. It never really gave me any trouble. It's there, a part of me, like a scab or a scar, a lock or protection. Let it be there. I like politeness, someone knocking, announcing. All doors and windows need not be open."

Antoinette sits self-consciously. She double-crosses her legs.

Hepzibah does not speak. She refastens her scarf more securely. Hepzibah always smiles. She refers to her husband and to her male sons as her "soldiers," and to her daughter as "one day the mother of soldiers." Yet she is with these women and full of love for them. But she cannot speak of what is under the scarf, the blouse, the skirt, the stockings, between the legs.

Vered says, "I was lucky. A fifty-two-year-old man did it, skillfully, carefully. No young man could have been so kind to me."

They listen.

"And even now," says Vered, who is twenty-nine, "I can be with no man younger than fifty."

Some have not spoken yet—the wayward girls, Dahlia, Joan from Manchester, Terry or Simha.

It is getting dark.

"Ach! Where's my daughter's father?" asks Simha.

Gloria opens the door. The kibbutznik has gone. The Turkish coffee cup is on its side. There is a thick residue of the coffee on the step.

"I'm such a shit!" says Simha.

Abruptly she smiles. "I always was. I lived in an isolated part of town—Jerusalem—where no houses had been built. We lived together, Jews and Arabs. I played with everything in The Land, stones in the wadi, the insides of caves, snakes, insects. I played with the Arab shepherds and their goats. I lay with them in the grass and we had each other, watching the goats."

Hepzibah sighs. Why does she come here after all? This is irrelevant to her. She rises and leaves silently. The women do not always notice when she goes but long for her afterward. Her hand in theirs is dry and steady.

"And you, Dahlia?" asks Simha.

"I pierced myself with a high note," says Dahlia.

That is good enough. The women applaud.

Terry said, "I lived in New York. I worked there but hated it. I had a scholarship offer to one of the schools in the state system to work on my M.S.W. and I accepted. But to leave the city a virgin, to be pierced in Schenectady or Binghamton. What would the choices be? A midget or someone with stumps or a Paul Bunyan who would cripple me? The week before the school semester began I quit my job and spent the day looking for my piercer. I found him easily, a fellow from the lunch counter where I ate, the fastest waiter, smooth, everything in one motion, athletic, younger than I. I waited until his shift was over and asked him to accompany me home. I said I was afraid to walk alone in the dark. But he was street smart. He never held my hand or looked at me; he just accompanied me up the stairs to my room and closed the door himself. I was right—everything in one motion, elegant, smooth. I never regretted it and did not eat at the counter again."

Joan, the playwright, said, "I was pierced on the Mediterranean. I was starting a new life and had to start with a new body. He was a tourist from some northern country, one of those that invaded England. I let him invade me on deck, under the bright moon."

The wayward girls are shocked. They would not dream of speaking about such a private matter.

The women cut cheese, eat it with fresh white bread and slices of mild radish. They cut sweet cucumbers, tomatoes and green peppers into small pieces for a kibbutz salad. They drink wine and say the blessing afterward.

"As it shall be written, may the mother be gladdened with the fruit of her body and the father rejoice in his offspring. May they live to rear her, may her friends learn to help her, may the law accept her and wisdom guide her."

Gerda says, "Terry, all the guests present and the mother Simha have observed this hymenotomy. Let us rejoice that we have performed this deed of piety."

The women make a tent around the baby.

Terry says, "Let us send forth the tidings that a daughter of the blood has joined us."

Simha is weary. She and the baby have both bled into the chair, Simha through her stitches, the baby through its tiny wound. Tova's lace is stained, torn, lying awry. The ceremony has ended.

Rina and Shula decide not to tell the girls of Wayward what they have witnessed. It might get Terry into more trouble. The young girls return to their house wondering and silently crying.

No one welcomed them at birth. Rina's mother said:

"I had hoped for a daughter that looked like an angel, and you, dark one, came along. I had hoped for a son, and a daughter was sent. I had hoped for a good baby, a crier lay. I had hoped for an eater, a fusser was in my arms. Why did God punish me? I had hoped for a circumcision, instead there was blood on her diaper. I screamed for the nurse. She laughed. 'Well, you have a girl for sure,' she said."

The women are dozing. Gloria stays over another night. She has some data for Deedee.

"In Japan," says Gloria, "you can have an operation and have a hymen sewed in. They stick in goatskin for one hundred thirty-seven dollars and no one ever knows. When it tears, you bleed."

"Gross," says Deedee.

They sleep and dream.

They are all virgins again.